Of Related Interest
from the Addison Wesley Longman
Series in the Life Sciences

General Biology

C. Barnard, F. Gilbert, and P. McGregor
*Asking Questions in Biology: Design,
Analysis, and Presentation in Practical
Work* (1993)

N.A. Campbell, J. B. Reece, and
L. G. Mitchell
Biology, Fifth Edition (1999)

N. A. Campbell, L. G. Mitchell, and
J. B. Reece
Biology: Concepts and Connections,
Second Edition (1997)

J. Dickey
Laboratory Investigations for Biology
(1995)

R. J. Ferl and R. A. Wallace
Biology: The Realm of Life,
Third Edition (1996)

J. Hagen, D. Allchen, and F. Singer
Doing Biology (1996)

A. Jones, R. Reed, and J. Weyers
Practical Skills in Biology,
Second Edition (1998)

A. Lawson and B. D. Smith
Studying for Biology (1995)

R. Liebaert
Interactive Study Partner CD ROM (1999)

M. C. Mix, P. Farber, and K. I. King
Biology: The Network of Life,
Second Edition (1996)

J. G. Morgan and M. E. B. Carter
*Investigating Biology: A Laboratory Manual
for Biology,* Third Edition (1999)

J. Pechenik
A Short Guide to Writing Biology,
Third Edition (1997)

G. L. Sackheim
*An Introduction to Chemistry for
Biology Students,* Sixth Edition (1999)

R. M. Thornton
The Chemistry of Life CD ROM (1998)

R. A. Wallace
Biology: The World of Life,
Seventh Edition (1997)

R. A. Wallace, G. P. Sanders, and R. J. Ferl
Biology: The Science of Life, Fourth Edition
(1996)

Biochemistry

R. F. Boyer
Modern Experimental Biochemistry,
Second Edition (1993)

D. J. Holme and H. Peck
Analytical Biochemistry,
Third Edition (1998)

C. K. Mathews and K. E. van Holde
Biochemistry, Second Edition (1996)

Cell Biology

W. M. Becker, J. B. Reece, and M. F. Poenie
The World of the Cell, Third Edition
(1996)

J. T. Hancock
Cell Signalling (1997)

L. J. Kleinsmith and V. M. Kish
Principles of Cell and Molecular Biology,
Second Edition (1995)

Genetics

J. P. Chinnici and D. J. Matthes
Genetics: Practice Problems and Solutions
(1999)

D. S. Falconer and T. F. C. Mackay
Introduction to Quantitative Genetics,
Fourth Edition (1995)

R. P. Nickerson
*Genetics: A Guide to Basic Concepts and
Problem Solving* (1990)

A. Radford, D. J. Cove, and S. Baumberg
A Primer of Genetics (1995)

P. J. Russell
Fundamentals of Genetics (1994)

P. J. Russell
Genetics, Fifth Edition (1998)

P. Sudbery
Human Molecular Genetics (1998)

Molecular Biology

M. V. Bloom, G. A. Freyer, and
D. A. Micklos *Laboratory DNA Science*
(1996)

R. J. King
Cancer Biology (1996)

R. Reed, D. Holmes, J. Weyers, and
A. Jones *Practical Skills in Biomolecular
Sciences* (1998)

J. D. Watson, N. H. Hopkins, J. W. Roberts, J.
A. Steitz, and A. M. Weiner
Molecular Biology of the Gene,
Fourth Edition (1987)

GENETICS
ANALYSIS AND PRINCIPLES

ROBERT J. BROOKER
UNIVERSITY OF MINNESOTA, TWIN CITIES

An imprint of Addison Wesley Longman, Inc.

Menlo Park, California • Reading, Massachusetts • New York • Harlow, England
Don Mills, Ontario • Sydney • Mexico City • Madrid • Amsterdam

Publisher: Jim Green
Sponsoring Editor: Nina Horne
Senior Developmental Editor: Suzanne Olivier
Editorial Assistant: Erika Buck
Managing Editor: Laura Kenney
Production/Composition/Photo Research: Electronic Publishing Services Inc., NYC
Principal Artists: Electronic Publishing Services Inc., NYC; Precision Graphics
Market Development Manager: David Horwitz
Marketing Manager: Gay Meixel
Index: Shane-Armstrong Information Systems
Permissions Editor: Mary Shields
Cover Design: Yvo Riezebos
Cover Photograph: Drosophila melanogaster, © Dennis Kunkel

Library of Congress Cataloging-in-Publication Data

Brooker, Robert J.
    Genetics: analysis and principles / Robert J. Brooker.
    p. cm.
    Includes bibliographical references and index.
    ISBN 0-8053-9175-4
    1. Genetics.   I. Title
  QH430.B766 1999
  576.5 — dc21   98-34247
                          CIP

ISBN 0-8053-9175-4

1 2 3 4 5 6 7 8 9 10—RNV—02 01 00 99 98

Benjamin/Cummings, an imprint of Addison Wesley Longman, Inc.
2725 Sand Hill Road
Menlo Park, CA 94025

DEDICATION

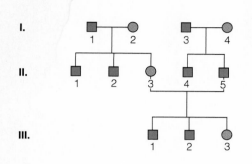

It seems wonderfully appropriate to dedicate this genetics textbook to my family (see pedigree at left). Most importantly, this book is dedicated to my wife, with love, and to my three children. My family provides me a with a daily reminder of the continuity of life and its true meaning.

I-1, Stanley Verrill; I-2, Genevieve Verrill; I-3, Charles Brooker; I-4, Katherine Brooker
II-1, Harland Verrill; II-2, Andrew Verrill; II-3, Deborah (Verrill) Brooker; II-4, David Brooker; II-5, Robert Brooker
III-1, Daniel Brooker; III-2, Nathan Brooker; III-3, Sarah Brooker

ABOUT THE AUTHOR

Robert J. Brooker (Ph.D., Yale University) is a Professor in the Department of Genetics, Cell Biology, and Development at the University of Minnesota, Twin Cities. He received his B.A. in Biology at Wittenberg University in 1978. Working with Dr. Caroline W. Slayman, he obtained a Ph.D. in Human Genetics at Yale University. In 1982, he began a four year post-doctoral fellowship in the laboratory of Dr. Thomas H. Wilson at Harvard University. At Harvard, he studied the lactose permease, the product of the *lacY* gene of the *lac* operon. He continues that work at the University of Minnesota, where he has an active research lab composed of undergraduates, Ph.D. students, post-doctoral fellows, and laboratory technicians (see photo). In addition to the lactose permease, Dr. Brooker's laboratory also investigates the structure, function, and expression of amino acid transporters found in various bacterial species. He regularly publishes research papers in scientific journals such as *Protein Science, Journal of Bacteriology,* and *The Journal of Biological*

Chemistry. He attends many scientific meetings, is a member of the editorial board of *The Journal of Biological Chemistry,* and is a member of the Metabolic Biochemistry Panel of the National Science Foundation.

At the University of Minnesota, Dr. Brooker teaches undergraduate courses in biology, genetics, human genetics, and cell biology. He has received the Outstanding Teacher of the Year Award from the Biological Sciences Student Association at the University of Minnesota. Through the Department of Independent Study at the University of Minnesota, he has adapted several of his undergraduate courses so that students can complete them through correspondence or by Interactive Television (ITV). In 1996, Dr. Brooker was appointed the Director of Graduate Studies for the Ph.D. program in Molecular, Cellular, Developmental Biology, and Genetics at the University of Minnesota, Twin Cities. He also teaches students at the graduate level and greatly enjoys the more informal method of discussing research papers in class.

Front row: Aileen Green, Rob Brooker, and Jason Patzlaff.
Second row: Paul Judd, Shane Cain, Jerry Johnson, Sara Brokaw, and Doug Stemke.

Rob Brooker with his undergraduate genetics professor, Dr. Elizabeth Powelson.

My desire to write this genetics textbook actually stemmed from a cell biology course that I introduced at the University of Minnesota many years ago. Our department decided to offer an intermediate level cell biology class to bridge the gap between our basic undergraduate cell biology and a more advanced graduate course, and I was asked to create it. I developed and taught this course for many years. It was a particularly fun experience, but I must admit that it was very scary the first time I taught it. Let me explain.

I developed an intermediate cell biology course which combined an undergraduate level of explanation but incorporated experiments from actual research papers. For each class session, the students were expected to read one or two research papers and understand them, and then we were supposed to discuss them in class. So as a young, naive, yet enthusiastic assistant professor, this seemed like a good plan. During the first few weeks of this course, however, I would walk into the classroom each day and ask the class, "So tell me about the first experiment in the paper you were supposed to read for today." I was expecting half the class to raise their hands and want to rush to the blackboard and explain the experiment. But instead, I basically got the "deer in the headlights" response. Obviously something was wrong.

I knew that the problem wasn't the students in the course, because many of our brightest and hardest working students were enrolled. And I (desperately) hoped that I wasn't the problem. So it had to be something else. Over time, the students and I uncovered the problem. Their previous coursework was not training them to analyze experiments according to the scientific method. They had learned many "facts" and bits and pieces of experiments, but they hadn't been exposed, in a rigorous way, to the scientific process. We needed to fix that. The good news was that the intermediate cell biology class turned out to be really fun for the students (and for me), because it was the first time that many of them had taken a college course that allowed them to unleash their curiosity and utilize their critical thinking skills.

For the intermediate cell biology course to be a success, the class and I had to develop a strategy to coherently discuss the experiments within research papers. Looking back on it, the solution seems obvious. We needed to follow the scientific method as a way to dissect the parts of each experiment. We ended up doing that. For each experiment, I would ask the class a series of questions.

1. **Why was this experiment done?** The students' answers would have two parts. First, they would need to consider the background work that led to the experiment. And second, the students would have to explicitly state the hypothesis that the researchers were trying to test.
2. **How was this experiment done?** As a class, we would generate a flow diagram that described the experimental methods.
3. **What are the results?** We would look at the raw data and quantitatively analyze it.
4. **What do the results mean?** We would discuss our interpretations of the data. Sometimes, our interpretations were quite different from the interpretations of the researchers who had actually done the experiments. The occasional conflict between our interpretations and those of the researchers provided the most meaningful moments in the classroom. It caused many students to realize that the scientific principles they had learned about in previous courses are the result of scientists' interpretations of their own data. And maybe, they aren't always right!

I have to say that the intermediate cell biology course has been my most inspiring teaching experience. What I mean is that the students were inspiring to me. (I hope that I was inspiring to them as well). On the positive side, it was impressive to see how the students improved their abilities to analyze experiments as the course progressed. On the negative side, however, it left me with a deeply disappointing sense that most college-level science courses do not provide students with an adequate exposure to the scientific process. To some extent, I felt that textbooks were to blame. In an effort to include too much information, science textbooks had deleted the essence of science: the scientific method. At that point, I decided to write a textbook that would incorporate the scientific method into each chapter. Since my primary background is in genetics rather than cell biology, it was logical for me to write a genetics textbook, although I strongly feel that all scientific disciplines would benefit from this approach.

As you will see when you use this book, each chapter (beginning with Chapter 2) has one or two experiments which are incorporated into the chapter and which are analyzed according to the scientific method. These experiments are not "boxed off" from the rest of the chapter for you to read in your spare time. Rather, they are within the chapters and cannot be skipped. As you are reading the experiments, you will simultaneously explore the scientific method and the genetic principles that are derived from the method. For students, I hope this book will help you to see the fundamental connection between scientific analysis and principles. For both students and instructors, I expect that this strategy will make genetics much more fun to explore.

UNIFYING THEME

When I first began to write this book, my acquisitions editor said that it needed to have a unifying theme. I must confess that, at the time, I wasn't aware that any textbooks are actually written with a particular theme in mind. Nevertheless, she insisted that "all good books have a theme," so this caused me to consider an appropriate theme for my genetics text. It seemed to me that the theme should capture the essence of genetics. Eventually, I stumbled upon the obvious. The essence of genetics is the relationship between genes and traits. More explicitly, the theme of this book is to examine how the molecular expression of genes leads to the outcome of traits in individuals and in populations. As such, this theme encompasses the three traditional subdivisions of genetics: transmission, molecular, and population genetics. With regard to transmission genetics, we need to understand the relationships between the patterns of gene transmission, and how those patterns ultimately influence the outcome of traits in offspring. At the molecular level, we also need to appreciate how the biochemical expression of genes determines the phenotypic characteristics of cells and of complete organisms. Finally, at the populational level, geneticists want to understand the relationship between genetic variation and the prevalence of genes within populations.

As a way to remind the student of this theme, each chapter has one or more figures in which the figure legends emphasize the theme. The theme legends are shown in blue type. Each theme legend connects the material within the chapter to the broad theme of the text. A unifying theme throughout this book provides a framework on which to relate many different genetic concepts.

CHAPTER ORGANIZATION

Each chapter is organized into a limited number of sections, usually two to four. Each section begins with an overview of the material that will be covered in that section. The subsections that follow are given subtitles in the form of complete sentences that emphasize the key points of the material. Important terms are shown in **bold-faced type,** and these are included in the glossary. Other less commonly used terms are shown in *italic,* as are the names of specific genes. Each chapter ends with a Conceptual Summary which emphasizes the principles and concepts that were covered in the chapter and also reminds the student of the terms that they have learned. In addition, there is an Experimental Summary which discusses some of the key experiments that led to the formulation of the genetic principles.

TEXT ORGANIZATION

The text is organized into six parts. Following an overview of genetics in Part I, Part II "Patterns of Inheritance" (Chapters 2–8) covers the traditional field of transmission genetics. It emphasizes patterns of genetic transmission and how those patterns affect the outcome of traits, both in eukaryotes and prokaryotes. We consider how genetic crosses and gene transfer can be analyzed as a way to map genes. In addition, this section emphasizes the relationships between chromosome transmission, structure, and number in the outcome of traits. Part III "Structure and Replication of the Genetic Material" (Chapters 9–11) begins an in-depth discussion of molecular genetics. This section is primarily devoted to nucleic acid structure rather than function. It also considers how the genetic material is copied. The emphasis on molecular genetics continues in Part IV "Molecular Properties of Genes" (Chapters 12–18). In these chapters, however, the emphasis is primarily on gene function. Much of the section pertains to the relationships between DNA sequences within genes, and how they are expressed at the molecular level. The last two sections of the text include Part V "Genetics and Technology" (Chapters 19–22), which focuses on the many techniques, tools, and strategies that researchers use to investigate genetic questions, and Part VI "Genetic Analysis of Individuals and Populations" (Chapters 23–27).

In our surveys of genetics instructors, we found that about two-thirds liked to cover transmission genetics first and then molecular genetics, while the other one-third preferred the opposite approach. In this text, we have presented transmission genetics first (Chapters 2-8) and molecular genetics later (Chapters 9-18). Nevertheless, as we have constructed this book, we have been mindful of those instructors (which includes the author of this text, incidentally) who choose to discuss molecular genetics first. The book is written in such a way that it is perfectly fine to begin with Chapters 9-18 and then follow them with Chapters 2-8. We have provided adequate explanations and term definitions so that either strategy will work. Also, the unifying theme within the book provides a framework that can be followed with either approach.

EXPERIMENTS AND ILLUSTRATIONS

A special strength of this text is the quality of illustrations, particularly those that involve experimentation and analysis. There are hundreds of illustrations that emphasize experimentation in this text. Not only do these figures describe techniques, but they also often contain raw data. In this way, students can appreciate the relationship between experimentation and quantitative analysis. Each figure, whether it be a conceptual or experimental one, has been carefully designed to follow closely with the textual material. Great effort has been made to define all the labels in each figure, either in the figure legend or in the text.

This textbook also has a completely novel way to illustrate the one or two experiments in each chapter that are rigorously examined according to the scientific method. For these experiments, there are carefully worded descriptions that explain each step in the procedure. Next to these descriptions, you will find parallel illustrations that depict how the experiment was actually conducted (i.e., at the Experimental Level), and what is happening at the Conceptual Level. With this layout, the student can correlate experimental procedures with their aim of elucidating scientific principles. These unique illustrations are then followed by the data of the experiment, as it was reported in the original research paper. This visual integration of techniques, concepts, and data provides the student with a direct way to view the scientific method.

PROBLEM SETS

The problems sets continue to bolster the relationship between concepts and experimentation. The conceptual questions are aimed at testing a student's ability to understand the basic genetic principles. The student is given many questions with a wide range of difficulty. Some of these questions require critical thinking skills, and some require the student to be able to write coherent essay questions. Thus, the conceptual questions should improve a student's cognitive and writing skills. The experimental questions are aimed at testing the student's ability to analyze data, design experiments, and/or appreciate the relevance of experimental techniques. At the end of the problem sets, there are also student discussion/collaboration questions that can be given to groups of students. Some of these are meant to foster discussion of practical problems or to consider broad concepts within a chapter. Other questions may require a substantial amount of computational activities which can be worked on as a group.

SUPPLEMENTS

To enhance the teaching and learning of genetics, several supplements have been developed to help instructors and students. An **Instructor's Presentation CD-ROM** is available that includes nine animations on key genetic concepts and [all] full color figures from the text. The set of **transparency acetates** includes 102 key conceptual figures, including several of the integrated experiments from the text, in full color. A **transparency sampler** of selected key figures is available to allow instructors to easily preview the transparency acetates.

With this edition we also offer student subscriptions to The Biology Place™, a web-based learning environment created and maintained by Peregrine Publishers, Inc. The Biology Place™ offers students interactive learning activities, articles from *Scientific American,* links to other high-quality science related web sites, comprehensive quizzes, and constantly updated news about reasearch developments. These subscriptions can be packaged with the text at a reduced price for the student.

The solutions to half of the problem sets in this text are printed in the back of the book. The other half of the solutions are available on-line at the **Benjamin/Cummings** web site: http://www.awl.com/bc. *Genetics: Practice Problems and Solutions,* by Joseph Chinnici of Virginia Commonwealth University and David Matthes of San Jose State University, provides over 400 additional problems for students.

ACKNOWLEDGMENTS

It is perhaps unfair that this textbook has a single author since I truly feel that the writing of this book has been an enormous collaborative effort. Each chapter went through three rigorous revisions and was reviewed by many scientists and editors. The collective contributions of all these people are reflected in the final outcome. In particular, I wish to acknowledge Cathy Pusateri, my acquisitions editor, who initially contacted me about writing a genetics textbook. When I discussed my vision of including complete experiments within each chapter of a textbook, her response was overwhelmingly enthusiastic. My first interactions with Cathy were instrumental in galvanizing my resolve to complete this textbook. In addition, I'm extremely grateful to my development editor, Suzanne Olivier, who has worked most closely with me on the creation of this book. Her imprint is found on nearly every page. Not only did she help me develop my own skills at writing and organization, but she also had a substantial amount of creative input into the pedagogy of this book. I think she even enjoyed learning genetics, which made our phone conversations particularly fun.

I would also like to thank Nina Horne, my sponsoring editor, for her skillful guidance of this project, and Laura Kenney, for her help during the production phase of the book. I am also very grateful to several people at Electronic Publishing Services Inc., who worked very hard to produce a well crafted book. In particular, Patty O'Connell did an excellent job of coordinating the art, text, and photographs, and working with me on the necessary changes. In addition, many thanks go to Andrew Schwartz and Michael Gutch, my copy editors, who did a superb job of improving the text, and Francis Hogan, who worked tirelessly to find many photographs which are found in this book.

I'm also deeply indebted to my wife, Deborah, not only for her support but also for her scientific and artistic input. Her tenacious ability to proofread chapters and examine art was a constant source of improvement. My kids, Dan, Nate, and Sarah, also helped me by asking me two questions on a weekly basis: 1. What chapter are you working on? 2. When is your book going to be done? The second question became progressively more annoying as the years wore on, but it probably provided a subconscious source of motivation.

I also want to thank the many scientists who have reviewed chapters of this book. Their critical input was an important factor that shaped its final content and organization. I am truly grateful for their time and effort.

ROB BROOKER

REVIEWERS

Acton, Gwen, Harvard University
Altschuler, Marsha, Williams College
Angus, Robert, University of Alabama, Birmingham
Asleson, Catherine M., University of Minnesota
Backer, James S., St. Vincent's College
Belote, John, Syracuse University
Benner, Michael S., Rider University
Bird, Margaret A., Marshall University
Bloom, Kerry, University of North Carolina, Chapel Hill
Boussy, Ian, Loyola University of Chicago
Cain, Shane M., University of Minnesota, Twin Cities
Cassill, Aaron, University of Texas, San Antonio
Chase, Bruce A., University of Nebraska, Omaha
Chinnici, Joseph P., Virginia Commonwealth University
Conboy, John, Lawrence Berkeley Laboratory
Courtright, James B., Marquette University
Curran, James F., Wake Forest University
Davisson, Vincent J., Purdue University
Egelman, Edward H., Universtiy of Minnesota Medical School
Farish, Guy E., Adams State College
Feldman, Jerry, University of California, Santa Cruz
Fox, Thomas D., Cornell University
Franco, Peter, Harvard Medical School
Fromson, David R., California State University, Fullerton
Ganetzky, Barry, University of Wisconsin
Girton, Jack, Iowa State University
Granholm, Nels, South Dakota State University
Green, Aileen L., University of Minnesota, Twin Cities
Hudack, George, Indiana University, Bloomington
Johnson, Blaine, University of Nebraska
Karcher, Susan J., Purdue University
Klein, Anita, University of New Hampshire
Lamberson, Bill, University of Missouri, Columbia
Largaespada, David, University of Minnesota
Mathies, Margaret, Claremont Colleges
McClung, C. Robertson, Dartmouth College
Niles, Mary Jane, University of San Francisco
Osterman, John C., University of Nebraska, Lincoln
Piers, Kevin, Red Deer College
Prestridge, Dan, University of Minnesota
Ranum, Laura, University of Minnesota Medical School
Shaw, Ruth, University of Minnesota
Rolfes, Ronda J., Georgetown University
Rougvie, Ann, University of Minnesota
Sanders, Mark, University of California, Davis
Shotwell, Mark, Slippery Rock University
Steglich, Carolyn, Slippery Rock University
Stone, W.H., Trinity University
Strecker, Teresa R., Pomona College
Sturtevant, Mark A., Northern Arizona University
Tansey, Teresa R., Georgetown University
Voytas, Daniel F., Iowa State University

Genetics: Analysis and Principles

Robert J. Brooker
University of Minnesota, Twin Cities

Robert J. Brooker emphasizes the five-step scientific method to reinforce for students the connection between genetics concepts and the experiments conducted by geneticists. Thirty-six text-integrated experiments reveal to students the "how" and "why" behind the most important research that led to our current knowledge of the science. This approach, along with a focus on the link between genes and traits, the fundamental principle of genetics, helps students develop their own scientific reasoning skills.

IDENTIFICATION OF DNA AS THE GENETIC MATERIAL   235

Hershey and Chase provided evidence that the genetic material injected into the bacterial cytoplasm is T2 phage DNA

EXPERIMENT 9A

A second experimental approach indicating that DNA is the genetic material came from the studies of A. D. Hershey and Martha Chase at Cold Spring Harbor. Their research centered on the study of a virus known as T2. This virus infects *Escherichia coli* bacterial cells and is therefore known as a **bacteriophage** or simply a **phage**. As shown in Figure 9-4, the external structure of the T2 phage, known as the *phage coat*, contains a *capsid*, *sheath*, *baseplate*, and *tail fibers*. Biochemically, the phage coat is composed entirely of protein, which includes several different polypeptides. DNA is found inside the T2 capsid. From a molecular point of view, this virus is rather simple, since it is only composed of two types of macromolecules: DNA and proteins.

Although the viral genetic material contains the blueprint to make new viruses, a virus itself cannot synthesize new viruses. Instead, a virus must introduce its genetic material into the cytoplasm of a living cell. In the case of T2, this first involves the attachment of its tail to the bacterial cell wall and the subsequent injection of its genetic material into the cytoplasm of the cell (Figure 9-5). The phage coat remains attached on the outside of the bacterium and does not enter the cell. After the entry of the viral genetic material, the bacterial cytoplasm provides all the synthetic machinery necessary to make viral proteins and DNA. The viral proteins and DNA assemble to make new viruses, which are subsequently released from the cell by **lysis** (i.e., cell breakage).

 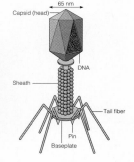

65 nm

Capsid (head)

DNA

Sheath

Tail fiber

Pin

Baseplate

225 nm

FIGURE **9-4**
Structure of the T2 bacteriophage. The T2 bacteriophage is composed of a phage coat and genetic material inside the phage head. The phage coat is divided into regions called the capsid, sheath, baseplate, and tail fibers. These components are composed of proteins. The genetic material is composed of DNA.

GENES → TRAITS: The genetic material of a bacteriophage contains many genes, which provide the blueprint for making new viruses. When the bacteriophage injects its genetic material into a bacterium, these genes are activated and direct the host cell to make new bacteriophages as described in Figure 9-5.

To verify that DNA is the genetic material, Hershey and Chase devised a method to separate the phage coat that is attached to the outside of the bacterium from the genetic material that is injected into the cytoplasm. They were aware of microscopy experiments by T. F. Anderson showing that the T2 phage attaches itself to the outside of a bacterium by its tail. Hershey and Chase reasoned that this is a fairly precarious attachment that could be disrupted by subjecting the bacteria to high shear forces such as those produced in a blender. As described in this experiment, their method was to expose bacteria to T2 phage, allowing sufficient time for the viruses to attach to bacteria and inject their genetic material. They then sheared the phage coats from the surface of the bacteria by a blender treatment. In this way, the phage's genetic material, which had been injected into the cytoplasm of the bacterial cell, could be separated from the rest of the phage coat, which was sheared away.

Before discussing this experiment, it should be mentioned that radioisotopes were used to distinguish between DNA and proteins. Hershey and Chase obtained T2 phages that were radiolabeled with either ^{35}S (a radioisotope of sulfur) or ^{32}P (a radioisotope of phosphorus). Sulfur atoms are found in proteins but not in DNA, whereas phosphorus atoms are found in DNA but not in viral proteins. Therefore, ^{35}S and ^{32}P were used in this experiment to specifically label proteins and DNA, respectively.

THE HYPOTHESIS

This experiment tests the hypothesis that bacteriophage T2 injects DNA rather than protein into the bacterial cytoplasm, where it functions as the viral genetic material.

Step 1: Background Observations

Each experiment begins with a description of the information that led researchers to study an experimental problem. Detailed information about the researchers and the experimental challenges they faced helps students to understand actual research.

Link Between Genes and Traits

The important relationship between genes and traits is emphasized both in text and in illustrations. This constant theme reinforces the relationship between abstract concepts and concrete physical expressions.

Step 2: Hypothesis

The student is given a statement describing the possible explanation for the observed phenomenon that will be tested. *The Hypothesis* section reinforces the scientific method and allows students to experience the process for themselves.

TESTING THE HYPOTHESIS

Starting materials: The starting materials are *E. coli* cells and two preparations of T2 phage. One preparation is labeled with ^{35}S to label the phage proteins; the other preparation is labeled with ^{32}P to label the phage DNA.

Experimental Level   Conceptual Level

1. Grow bacterial cells. Divide into two flasks.

Solution of
E. coli cells

^{35}S-labeled protein capsid

2. Into one flask, add ^{35}S-labeled phage; in the second flask, add ^{32}P-labeled phage.

^{35}S-labeled
T2 phage

^{32}P-labeled
T2 phage

^{32}P-labeled DNA

3. Allow infection to occur.

4. Spin solutions in blenders for different lengths of time to shear the empty phages off the bacterial cells.

Bacterial cell   After blending

Viral
genetic material

Sheared empty phage

Solution of *E. coli*
infected with
^{35}S-labeled phage

Solution of *E. coli*
infected with
^{32}P-labeled phage

Bacterial
cell

Viral
genetic
material

Sheared empty
phage

5. Centrifuge at 10,000 rpm.

6. Note: The heavy bacterial cells sediment to the pellet, while the lighter phages remain in the supernatant. (See appendix for an explanation of centrifugation.)

Sheared labeled phages

Supernatant
with ^{35}S-labeled
empty phage

Supernatant
with
unlabeled
empty phage

Pellet with
unlabeled
infected
E. coli cells

Pellet with
^{32}P-labeled
DNA in
infected
E. coli cells

7. Count the amount of radioisotope in the supernatant with a scintillation counter (see the appendix). Compare it with the starting amount.

THE DATA

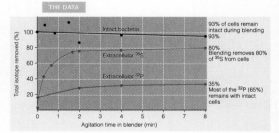

93% of cells remain intact during blending
93%

Intact bacteria

80%
Blending removes 80% of ^{35}S from cells

Extracellular ^{35}S

Extracellular ^{32}P

35%
Most of the ^{32}P (65%) remains with intact cells

INTERPRETING THE DATA

Following infection, most of the phage protein was sheared from the bacterial cells and ended up in the supernatant. This indicated that the empty phages contain primarily protein. Furthermore, less than 40% of the DNA was found in the supernatant following shearing. Therefore, most of the DNA was located within the bacterial cells in the pellet. These results are consistent with the idea that the DNA is injected into the bacterial cytoplasm during infection. This is the expected result if DNA is the genetic material.

By themselves, the results described in Experiment 9A were not conclusive evidence that DNA is the genetic material. For example, you may have noticed that less that 100% of the viral protein was found in the supernatant. Therefore, some of the phage protein could have been introduced into the bacterial cells (and could function as the genetic material). Nevertheless, the results of Hershey and Chase leaned toward the conclusion that DNA is the genetic material rather than protein. Overall, their studies of the T2 phage, published in 1952, were quite influential in promoting the belief that DNA is the genetic material.

RNA functions as the genetic material in some viruses

We now know that bacteria, protozoa, fungi, algae, plants, and animals all use DNA as their genetic material. As mentioned, viruses also have their own genetic material. Hershey and Chase concluded from their experiments that this genetic material is DNA. In the case of T2 bacteriophage, that is the correct conclusion. However, many viruses use RNA, rather than DNA, as their genetic material. In 1956, A. Gierer and G. Schramm at the Max Planck Institute in Germany isolated RNA from the tobacco mosaic virus (TMV), which infects plant cells. When this purified RNA was applied to plant tissue, the plants developed the same types of lesions that occurred when they were exposed to intact TMV viruses. Gierer and Schramm correctly concluded that the viral genome of tobacco mosaic virus is composed of RNA. Since that time, many other viruses have been found to contain RNA as their genetic material. Table 9-1 compares the genetic compositions of several different types of viruses.

NUCLEIC ACID STRUCTURE

Geneticists, biochemists, and biophysicists have been interested in the molecular structure of nucleic acids for many decades. Both DNA and RNA are large macro-

Step 3: Testing the Hypothesis

This section illustrates the experimental process, including the actual steps followed by scientists to test their hypothesis. Science comes alive for students with this detailed look at experimentation.

At each step in the experiment, experimental illustrations show what happened in the laboratory. Students gain insight into the physical nature of experimentation.

To connect concepts to research, experimental illustrations are paired with representational illustrations to show what occurs at the conceptual level. Here, the author directly relates underlying scientific principles to experimental procedure.

Step 4: The Data

Actual data from the original research paper helps students understand how actual research results are reported. Each experiment's results are discussed in the context of the larger genetic principle to help students understand the implications and importance of the research.

Step 5: Interpreting the Data

This discussion, which examines whether the experimental data supported or disproved the hypothesis, gives students a feel for the art of scientific interpretation. Through this analysis, students can experience the similarities between classic experiments and their own work.

CONCEPTUAL SUMMARY

The molecular structure of nucleic acids underlies their function. *Nucleotides*, which are composed of a sugar, phosphate, and nitrogenous base, form the repeating structural unit of nucleic acids. The *primary structure* is a *strand* that contains a linear sequence of nucleotides. The formation of *secondary structure* occurs because *complementary* regions of DNA (and RNA) can form hydrogen bonds between adenine and thymine (or uracil), and between guanine and cytosine. *Base stacking* also stabilizes a double-stranded structure. The most common form of DNA is a right-handed helix; the *backbone* in the *double helix* is composed of *sugar–phosphate linkages*, with the bases projecting inward from the backbone and hydrogen bonding with each other. The two strands are *antiparallel*. In DNA, different helical conformations have been identified, including A-DNA, B-DNA, and Z-DNA. B-DNA is the predominant form found in living cells; but short regions of Z-DNA may play an important functional role in gene transcription. Within chromosomes, DNA is folded into a *tertiary conformation* with the aid of proteins. The structure of chromosomes will be discussed in the next chapter. It is also common for short segments within RNA to form double helical structures such as stem-loops, bulges, loops, and junctions. RNA secondary structures can also play many important functional roles. The final tertiary structure of RNA is dictated by several factors including double helical regions, *base stacking*, hydrogen bonding between bases and backbone, and interactions with other molecules.

EXPERIMENTAL SUMMARY

Two different experimental approaches were used to show that DNA is the genetic material. Avery, MacLeod, and McCarty took advantage of Griffith's observations regarding transformation in pneumococci. They purified DNA from type IIIS strain

From the Conceptual to the Experimental

The relationship between what scientists do in the laboratory (*the experimental level*) and the processes that occur during the experiment (*the conceptual level*) is further explored with the conceptual and experimental summaries and the end-of-chapter conceptual and experimental problems.

PROBLEM SETS

SOLVED PROBLEMS

1. A naturally occurring DNA has the following sequence:

5′–A–A–G–G–A–A–A–A–G–G–G–A–G–G–A–G–A–G–3′
3′–T–T–C–C–T–T–T–T–C–C–C–T–C–C–T–C–T–C–5′

What sequence of DNA molecule could form triplex DNA with this double helix?

Answer: 5′–T–T–C–C–T–T–T–T–C–C–C–T–C–C–T–C–T–C–3′

2. As we will discuss in future chapters, the formation of stem-loop structures within RNA can be crucial in many functional ways. For example, the formation of a stem-loop at the beginning of mRNA can influence the rate at which that mRNA is translated into protein. A hypothetical sequence at the beginning of an mRNA molecule is

5′–AUUUGCCCUAGCAAACGUAGCAAACG. . .rest of the coding sequence

Using two out of the three underlined sequences, draw two possible models for potential stem-loop structures at the 5′ end of this mRNA.

Answer:

CONCEPTUAL QUESTIONS

1. What is the meaning of the term genetic material?

2. After the DNA from type IIIS bacteria is exposed to type IIR bacteria, list all of the steps that you think must occur for the bacteria to start making a type IIIS capsule.

3. What are the building blocks of a nucleotide? With regard to the 5′ and 3′ positions on a sugar molecule, how are nucleotides linked together to form a strand of DNA?

4. Draw the structure of guanine, guanosine, and deoxyguanosine triphosphate.

5. Draw the structure of a phosphodieste

6. Describe how bases interact with each

12. List the structural differences between DNA and RNA.

13. Draw the structure of deoxyribose, and number the carbon atoms. Describe the numbering of the carbon atoms in deoxyribose with regard to the directionality of a DNA strand. In a DNA double helix, what does the term antiparallel mean?

14. Write out a sequence of an RNA molecule that could form a stem-loop with 24 nucleotides in the stem and 16 nucleotides in the loop.

15. Compare the structural features of a double-stranded RNA stem

---

20. Some viruses contain single- or double-stranded RNA as their genetic material. If a virus contains the following amounts of nucleotides, would you conclude that its genetic material is single-stranded or double-stranded: A = 15%, U = 29%, G = 28%, and C = 28%?

21. Let's suppose that you have recently identified an organism that was scraped off of an asteroid that hit the earth. (Fortunately, no one was injured.) When you analyze this organism, you discover that its DNA is a triple helix, composed of six different nucleotides: A, T, G, C, X, and Y. You measure the chemical composition of the bases and find the following amounts of these six bases: A = 24%, T = 23%, G = 11%, C = 12%, X = 21%, Y = 9%. What rules would you propose govern triplex DNA formation in this organism? Note: There is more than one possibility.

22. Upon further analysis of the DNA described in problem 21, you discover that the triplex DNA in this alien organism is composed of a double helix, with the third helix wound within the major groove (just like the DNA in Figure 9-18). How would you propose that this DNA is able to replicate itself? In your answer, be specific about the

base pairing rules within the double helix, and which part of the triplex DNA would be replicated first.

23. A DNA-binding protein recognizes the following double-stranded sequence:

5′–GCCCGGGC–3′
3′–CGGGCCCG–5′

This type of double-stranded structure could also occur within the stem region of an RNA molecule. Discuss the structural differences between RNA and DNA that might prevent this DNA-binding protein from recognizing a double-stranded RNA molecule.

24. Within a protein, certain amino acids are positively charged (e.g., lysine and arginine), some are negatively charged (e.g., glutamate and aspartate), some are polar but uncharged, and some are nonpolar. If you knew that a DNA-binding protein was recognizing the DNA backbone rather than base sequences, which amino acids in the protein would be good candidates for interacting with the DNA?

EXPERIMENTAL QUESTIONS

1. In the experiment described in Figure 9-3, list several possible reasons that only a small percentage of the type IIR bacteria were converted to type IIIS.

2. Another interesting trait that some bacteria exhibit is resistance to killing by antibiotics. For example, certain strains of bacteria are resistant to tetracycline, whereas other strains are sensitive. Describe an experiment that you would carry out to demonstrate that tetracycline resistance is an inherited trait encoded by the DNA of the resistant strain.

3. In Experiment 9A, give possible explanations why less than 30% of the DNA is in the supernatant.

4. Plot the results of Experiment 9A if the radioactivity in the pellet, rather than in the supernatant, had been measured.

5. In Experiment 9A, why were ^{32}P and ^{35}S chosen as radioisotopes to label the ph

7. It is possible to specifically label DNA or RNA by providing bacteria with radiolabeled thymine or uracil, respectively. With this type of tool, design an experiment to show whether a newly identified bacteriophage contains DNA or RNA as its genetic material. Describe your expected results depending on whether the genetic material is DNA or RNA.

8. The type of model building that was used by Pauling, Watson, and Crick involved the use of small ball-and-stick units. Now we can do model building on a computer screen. Even though you may not be familiar with this approach, discuss some potential advantages computers might provide in molecular model building.

9. In Chargaff's experiment (Experiment 9B), what is the purpose of paper chromatography?

10. Would Chargaff's experiments have been convincing if they had been done on only one species? Discuss.

Problem Sets

Crafted to aid students in developing a wide range of skills, the problems develop students' cognitive, writing, analytical, computational, and collaborative abilities.

CONTENTS

Part I
Introduction

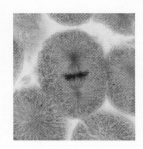

CHAPTER 1

OVERVIEW OF GENETICS

An appreciation for the concept of heredity can be traced far back in human history. Hippocrates, a famous Greek physician, was the first person to provide an explanation for hereditary traits (*ca.* 400 B.C.). He suggested that "seeds" are produced by all parts of the body and are then collected and transmitted to the offspring at the time of conception. Furthermore, he hypothesized that these "seeds" cause certain traits of the offspring to resemble those of the parents. This theory, known as **pangenesis**, was the first attempt to explain the transmission of hereditary traits from generation to generation.

For the next 2000 years, the ideas of Hippocrates were accepted by some and rejected by many. After the invention of the microscope in the late 17th century, some people observed sperm and thought they could see a tiny creature inside, which they termed a homunculus (little man). This homunculus was hypothesized to be a miniature human waiting to develop within the womb of its mother. These spermists, therefore, suggested that only the father was responsible for creating future generations and that any resemblance between mother and offspring was due to influences "within the womb." During the same time, an opposite school of thought also developed. According to the ovists, it was the egg that was solely responsible for human characteristics. The only role of the sperm was to stimulate the egg to begin on its path of development. Eventually, the ideas of the ovists and spermists were refuted by agricultural breeders who showed convincingly that certain traits were determined by both the male and female parents.

During the late 19th century, ideas concerning the forces of heredity were dramatically changed by two scientists, Gregor Mendel and Charles Darwin. Their theories were to greatly influence the ways that scientists viewed genetics. As will be

THE RELATIONSHIP
BETWEEN GENES
AND TRAITS

FIELDS OF GENETICS

discussed in Chapter 2, Gregor Mendel's work with pea plants provided insight into the patterns of inheritance from parents to offspring.

Charles Darwin's contributions were related to his understanding of species in their native environments. On his famous voyage of the *Beagle*, which lasted from 1832 to 1836, he carefully studied many different species in their natural habitats. In his book, *The Origin of Species*, Darwin proposed that some natural variation occurs in the traits of species and these varied traits are passed from parent to offspring. Most species produce many more offspring than can survive and reproduce, creating, Darwin pointed out, a "*struggle for existence*" that results in the "*survival of the fittest*." Over the course of many generations, those individuals who possess the most favorable traits will come to dominate the composition of the population. This, Darwin proposed, leads to the evolution of new species.

Of course, much has happened since the time of Mendel and Darwin. At the beginning of the 20th century, around the time that Mendel's work was rediscovered, many scientists began to investigate the cellular and biochemical basis for genetics. In 1903, Walter Sutton and Theodore Boveri independently proposed the chromosomal theory of inheritance. This theory identified the chromosomes as the carriers of the genetic material. However, it took almost another 40 years before scientists, including Oswald Avery, Colin MacLeod, Maclyn McCarty, A. D. Hershey, and Martha Chase, showed that the genetic material is actually deoxyribonucleic acid—DNA.

In the 1950s, James Watson and Francis Crick, working with their colleagues Rosalind Franklin and Maurice Wilkins, elucidated the structure of DNA. The structural features of the DNA double helix provided profound insights into its role as the genetic material. During the 1960s and 1970s, researchers developed many laboratory techniques, such as DNA cloning and sequencing, to probe the structure and function of DNA at the molecular level. Since that time, the numbers of scientists working in the areas of molecular biology and molecular genetics has expanded greatly. These researchers have found that the genetic blueprint is extraordinarily complex, and the details they continue to uncover are endlessly interesting. During the past 50 years, we have learned a great deal about the genetic material at the molecular level. An exciting challenge of the future will be to further unravel its intricate complexity.

THE RELATIONSHIP BETWEEN GENES AND TRAITS

Genetics is the branch of biology that deals with heredity and variation. It is the unifying discipline in biology, because genetics allows us to explain how life can exist at all levels of complexity, ranging from the molecular to the populational level. Genetic variation is the root of the natural diversity that we observe among members of the same species and among different species.

To a large extent, genetics is the study of genes. A **gene** is defined as a unit of heredity. We often describe genes according to the way that they affect the **traits** or characteristics of an organism. In humans, for example, we speak of genes that govern eye color, hair texture, and the propensity to develop diseases such as hemophilia (a blood-clotting disorder) and cancer. The ongoing theme of this textbook will be the relationship between genes and traits. As an organism grows and develops, its collection of genes provides a blueprint to determine its characteristics. We shall also see that an organism's interactions with its environment play a key role.

In this introductory chapter, we will examine the general features of life, beginning with the molecular level and ending with populations of organisms. As will become apparent, genetics is the common thread that explains the existence of life on our planet.

Living cells are composed of biochemicals

To understand the relationship between genes and traits, we need to begin with an examination of the composition of living organisms. Every cell is constructed from intricately organized chemical substances. Small organic molecules such as glucose and amino acids are produced from the linkage of atoms via chemical bonds. The chemical activity of organic molecules is essential for cell vitality in several ways. For example, the breakage of chemical bonds during the degradation of small molecules provides energy to drive cellular processes. A second important function of organic molecules is that they serve as the building blocks for the synthesis of macromolecules. The four types of biological macromolecules are **nucleic acids** (i.e., DNA and RNA), **proteins**, **carbohydrates**, and **lipids**.

The formation of cellular structures relies on the interactions of macromolecules. For example, cell membranes are formed from the association of lipids into a bilayer. This bilayer also includes certain proteins and carbohydrates. Another example is the eukaryotic chromosome, which is composed of DNA along with a variety of proteins. As shown in Figure 1-1, DNA is composed of nucleotides, which are linked together within chromosomal DNA. The chromosomes are contained within a membrane-bound organelle called the nucleus. As a general theme, the formation of large cellular structures arises from interactions among different macromolecules. These cellular structures, in turn, are organized to make a complete living cell.

Each cell contains many different proteins that play an active role in cellular structure and function

To a great extent, the characteristics of a cell depend on the types of proteins that it makes. As we will learn throughout this text, proteins can perform various functions. Some proteins help determine the shape and structure of a given cell. For example, the protein known as tubulin can assemble into large structures known as microtubules, which provide the cell with internal structure and organization. Other proteins are inserted into cell membranes and aid in the transport of ions and small molecules across the membrane. Another interesting category of proteins are those that function as biological motors. An example is the protein known as myosin, which is involved in the contractile properties of muscle cells. Within multicellular organisms, certain proteins also function in cell-to-cell recognition and signaling. For example, hormones such as insulin are secreted by endocrine cells and bind to the insulin receptor protein found within the plasma membrane of target cells.

A particularly important category of proteins are **enzymes**, which accelerate chemical reactions within the cell. Some enzymes play a role in the breakdown of molecules or macromolecules into smaller units. These are known as *catabolic enzymes* and are important in generating cellular energy. Alternatively, *anabolic enzymes* function in the synthesis of molecules and macromolecules. Throughout the cell, the synthesis of molecules and macromolecules relies on enzymes and accessory proteins. Thus, the construction of a cell greatly depends on its proteins involved in anabolism, since these are required to synthesize all cellular macromolecules.

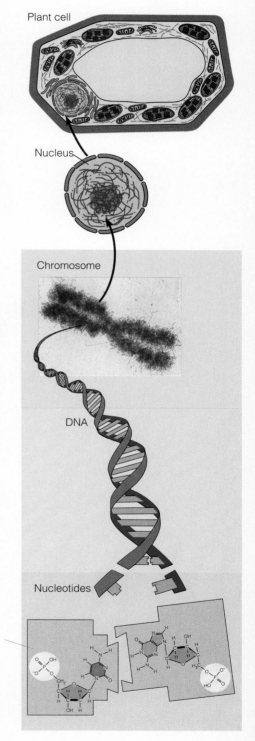

FIGURE **1-1**

Chemical composition of living cells. Cellular structures are constructed from smaller building blocks. In this example, DNA is formed from the linkage of nucleotides to produce a very long macromolecule. The DNA associates with proteins to form a chromosome. The chromosomes are located within a membrane-bound organelle called the cell nucleus. Many different types of organelles are found within a complete cell.

Molecular biologists have come to realize that the functions of proteins underlie the cellular characteristics of every organism. At the molecular level, proteins can be viewed as the "active participants" in the enterprise of life.

DNA stores the information for protein synthesis

The genetic material is composed of a substance called **deoxyribonucleic acid**, abbreviated **DNA**. The DNA stores the information needed for the synthesis of all cellular proteins. In other words, the main function of the genetic blueprint is to code for the production of cellular proteins in the correct cell, at the proper time, and in suitable amounts. This is an extremely complicated task, since living cells make thousands of different proteins. Genetic analyses have shown that a typical bacterium can make a few thousand different proteins, and estimates among eukaryotes range from tens of thousands to more than one hundred thousand.

DNA's ability to store information is based on its molecular structure. In Chapter 9, we will examine this molecular structure in greater detail. DNA is composed of a linear sequence of **nucleotides**. Each nucleotide contains one nitrogen-containing base, either adenine (A), thymine (T), guanine (G), or cytosine (C). The sequence of these bases along a DNA molecule stores information much as the letters of the alphabet represent words. For example, the "meaning" of the sequence of bases A–T–G–G–G–C–C–T–T–A–G–C differs from that of the sequence T–T–T–A–A–G–C–T–T–G–C–C. To understand the meaning of any sequence, we must know the **genetic code**, which is read in groups of three bases. The specific details of the genetic code will be discussed in Chapter 12. In the code, a three-base sequence specifies one particular **amino acid** among the 20 possible choices. In this way, the sequence of nucleotides within DNA can store the information to specify the sequence of amino acids within a protein:

DNA Sequence	*Amino Acid Sequence*
A–T–G – G–G–C – C–T–T – A–G–C	methionine–glycine–leucine–serine
T–T–T – A–A–G – C–T–T – G–C–C	phenylalanine–lysine–leucine–alanine

In living cells, the DNA is found within large structures known as **chromosomes**. Figure 1-2 is a photograph of the 46 chromosomes contained in a typical human cell. The DNA of an average human chromosome is an extraordinarily long, linear, double-stranded structure, that contains well over a hundred million nucleotides. Along the immense length of a chromosome, the genetic information is parceled into shorter segments known as genes. Most genes code for the sequence of amino acids within a protein. An average-sized human chromosome is expected to contain a few thousand different genes.

The information within the DNA is accessed during the process of gene expression

To synthesize its proteins, a cell must be able to access the information that is stored within its DNA. This process of using a gene sequence to synthesize a cellular protein is referred to as **gene expression**. At the molecular level, the information within genes is accessed in a stepwise process. In the first step, known as **transcription**, the DNA sequence within a gene is copied into a nucleotide sequence of **ribonucleic acid (RNA)**. The information within most genes codes for the synthesis of a particular protein. For the synthesis to occur, the sequence of nucleotides transcribed in the RNA must be **translated** (using the genetic code) into the amino acid sequence of a protein (Figure 1-3).

The expression of a gene thus results in the production of a protein with a specific structure and function. The unique relationship between a gene's sequence and

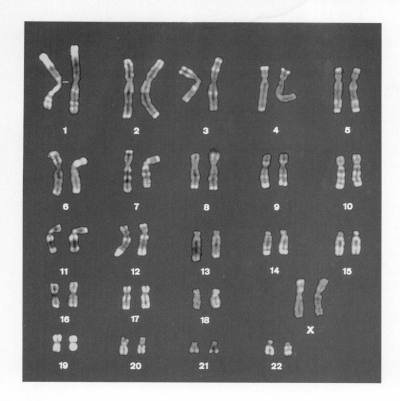

FIGURE **1-2**

A micrograph of the 46 human chromosomes.

a protein's structure is of paramount importance, because the distinctive structure of each protein determines its function within a living cell or organism. Mediated by the process of gene expression, therefore, the sequence of nucleotides in DNA stores the information required for synthesizing proteins with specific structures and functions.

The molecular expression of genes within cells leads to an organism's outwardly visible traits

A **trait** is any characteristic that an organism displays. In genetics, we often focus our attention on **morphological traits** that affect the appearance of an organism. The color

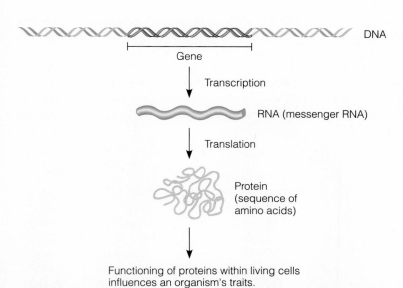

FIGURE **1-3**

Gene expression. The expression of a gene is a two-step process. During transcription, one of the DNA strands is used as a template to make an RNA strand. During translation, the RNA strand is used to specify the sequence of amino acids within a protein. This protein then functions within the cell, thereby influencing an organism's traits.

of a flower or the height of a pea plant are morphological traits. Geneticists often study these types of traits, because they are easy to evaluate. For example, an experimenter can simply look at a plant and tell if it has red or white flowers. However, not all traits are morphological. **Physiological traits** affect the ability of an organism to function. For example, the rate at which a person metabolizes glucose is a physiological trait. The ability of plants to wilt in response to dehydration is another example. Like morphological traits, physiological traits are governed by the expression of genes.

A difficult, yet very exciting aspect of genetics is that our observations and theories span four levels of biological organization: molecular, cellular, organismal, and populational. This can make it difficult to appreciate the relationship between genes and traits. To understand this connection, we need to relate the following phenomena:

1. Genes are expressed at the **molecular level**. In other words, gene transcription and translation lead to the production of a particular protein. This is a molecular process.
2. Proteins function at the **cellular level**. The function of a protein within a cell will affect the structure and workings of that cell.
3. An organism's traits are determined by the characteristics of its cells. We do not have microscopic vision, yet when we view morphological traits, we are really observing the properties of an individual's cells. For example, a red flower has its color because the flower cells make a red pigment. The trait of red flower color is an observation at the **organismal level**. Yet it is rooted in the molecular characteristics of the organism's cells.
4. The occurrence of a trait within a species is an observation at the **populational level**. Along with learning how a trait occurs, we also want to understand why a trait becomes prevalent in a particular species. In many cases, we discover that a trait predominates within a population because it promotes the survival or reproduction of the members of the population.

As a schematic example to illustrate the four levels of genetics, Figure 1-4 shows the trait of pigmentation in butterflies. Some are light colored, and others are very dark. Now let's consider how we can explain this trait at the molecular, cellular, organismal, and populational levels.

At the molecular level, we need to understand the nature of the gene or genes that govern this trait. As shown in Figure 1-4a, a gene, which we will call the pigmentation gene, is responsible for the amount of pigment that is produced. The pigmentation gene can exist in two different forms, called **alleles**. In this example, the two alleles are called the light and dark alleles. Each of these alleles encodes a protein that functions as a pigment-synthesizing enzyme. However, the DNA sequences of the two alleles differ slightly from each other. This difference in the DNA sequence leads to a variation in the structure and function of the respective pigmentation enzymes.

At the cellular level (Figure 1-4b), the functional differences between the pigmentation enzymes affect the amount of pigment that is produced. The dark allele shown on the left encodes a protein that functions very well. Therefore, when this gene is expressed in the cells of the wings, a large amount of dark pigment is made. By comparison, the gene of the light allele encodes an enzyme that functions poorly. Therefore, when the wing cells express the light allele, little pigment is made.

At the organismal level (Figure 1-4c), the amount of pigment in the wing cells governs the color of the wings. If the pigment cells produce high amounts of pigment, the wings are dark colored; if the pigment cells produce little pigment, the wings are light.

Finally, at the populational level (Figure 1-4d), population geneticists would like to know why a species of butterfly would contain some members with dark wings and other members with light wings. One explanation could involve predation. The butterflies with dark wings might avoid being eaten by birds if they live within the dim light of a forest. The dark wings would help to camouflage the but-

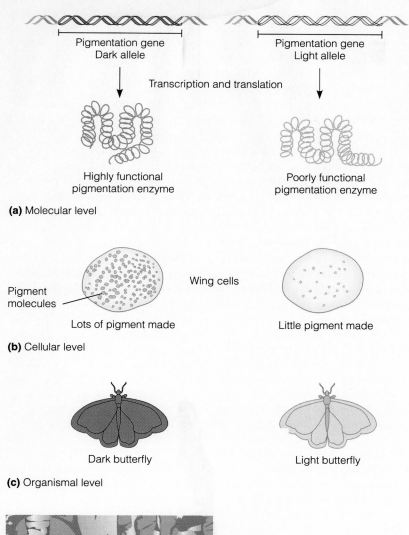

Pigmentation gene
Dark allele

Pigmentation gene
Light allele

Transcription and translation

Highly functional
pigmentation enzyme

Poorly functional
pigmentation enzyme

(a) Molecular level

Pigment
molecules

Wing cells

Lots of pigment made

Little pigment made

(b) Cellular level

Dark butterfly

Light butterfly

(c) Organismal level

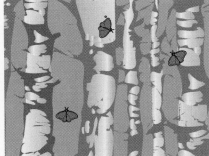

Dark butterflies are usually
in forested regions.

Light butterflies are usually
in unforested regions.

(d) Populational level

FIGURE **1-4**

The relationship between genes and traits at the (a) molecular, (b) cellular, (c) organismal, and (d) populational levels.

terfly if it was perched on a dark surface such as a tree trunk. In contrast, the lightly colored wings would be an advantage if the butterfly inhabited a brightly lit meadow. Under these conditions, a bird may be less likely to notice a light-colored butterfly that is perched on a sunlit surface. A population geneticist might study this species of butterfly and find that the dark-colored members usually live in forested areas, the light-colored members in unforested regions.

Inherited differences in traits are due to genetic variation

In Figure 1-4, we considered how gene expression can lead to variation in a trait of the organism (e.g., dark- versus light-colored butterflies). Variation in traits among members of the same species is very common. For example, some people have brown hair, while others have blond hair; some petunias have white flowers, while others have purple flowers. These are examples of **genetic variation**. This term describes the differences in inherited traits among individuals within a population.

In large populations that occupy a wide geographic range, genetic variation can be quite striking. In fact, morphological differences have often led geneticists to misidentify two members of the same species as belonging to separate species. As an example, Figure 1-5 compares four garter snakes that are members of same species, *Thamnophis ordinoides*. They display dramatic differences in their markings. Such contrasting forms within a single species are termed **morphs**. It is easy to imagine how someone might erroneously conclude that these four snakes were not members of the same species.

Changes in the molecular structure of DNA underlie the genetic variation that we see among individuals. Throughout this text, we will routinely examine how variation in the genetic material results in changes in the outcome of traits. At the molecular level, genetic variation can be attributed to different types of modifications. These are as follows:

1. Small differences can occur within gene sequences. These are called **gene mutations**. This type of variation, which produces two or more alleles of the same gene, was previously described in Figure 1-4. In many cases, gene mutations will alter the function of the protein that the gene specifies.
2. Major alterations can also occur in the structure of a chromosome. A large segment of chromosome can be lost or reattached to another chromosome. As will be discussed in Chapter 8, these types of changes can greatly affect the characteristics of an organism.
3. There can be variations in the total number of chromosomes. In some cases, there may be one too many or one too few chromosomes. In other cases, an offspring may inherit an extra set of chromosomes.

FIGURE **1-5**

Four garter snakes showing different morphs within a single species.

(a)   (b)

FIGURE **1-6**

Examples of chromosome variation. (a) A person with Down syndrome competing in the Special Olympics. This person has 47 chromosomes rather than the normal number of 46. **(b)** A wheat plant. Modern wheat is derived from the contributions of three related species with two sets of chromosomes each, producing an organism with six sets of chromosomes.

Variations within the sequences of genes are a common source of genetic variation among members of the same species. In humans, familiar examples of variation involve genes for eye color, hair texture, and skin pigmentation. Chromosome variation (i.e., changes in chromosome structure and/or number) is also found and may lead to substantial changes in the characteristics of an individual. In humans and other animals, variations in the structure and/or number of chromosomes are usually detrimental. Many human genetic disorders are the result of chromosomal alterations. The most common example is Down syndrome, which is due to the presence of an extra chromosome (Figure 1-6a). By comparison, chromosome variation in plants is common and often can lead to strains of plants with superior characteristics, such as increased resistance to disease. This observation has been frequently exploited by plant breeders. The cultivated variety of wheat, for example, has many more chromosomes than the wild species (Figure 1-6b).

Traits are governed by genes and by the environment

In our discussion thus far, we have considered the role that genes play in the outcome of traits. There is another important factor—the environment. A variety of factors in an organism's environment profoundly affect its morphological and physiological features. For example, a person's diet will greatly influence many traits such as height, weight, and even intelligence. Likewise, the amount of sunlight that a plant receives will affect its growth rate and the color of its flowers.

External influences may dictate the way genetic variation is manifested in an individual. An interesting example of this phenomenon is the human genetic disease

phenylketonuria (PKU). Humans possess a gene that encodes an enzyme known as *phenylalanine hydroxylase*. Most people have two normal copies of this gene, which encodes a normal phenylalanine hydroxylase protein. People with one or two normal copies of the gene can eat foods containing the amino acid phenylalanine and metabolize it correctly.

A rare variation in the sequence of the phenylalanine hydroxylase gene results in a nonfunctional version of this protein. Individuals with two copies of this rare defective allele cannot metabolize phenylalanine. When given a standard diet containing phenylalanine, these individuals are unable to break down this compound and as it accumulates within their bodies it becomes highly toxic. Under these environmental conditions, PKU individuals manifest a variety of detrimental traits including mental retardation, underdeveloped teeth, and foul-smelling urine. In contrast, when PKU individuals are identified at birth and raised on a restricted diet that is free of phenylalanine, they develop normally (Figure 1-7). This is a dramatic example of how the environment and an individual's genes can interact to influence the traits of the organism.

During sexual reproduction, genes are passed from parent to offspring

Now that we have considered how genes and the environment govern the outcome of traits, we can turn to the issue of inheritance. A centrally important matter in genetics is the manner in which traits are passed from parents to offspring. The foundation for our understanding of inheritance came from the studies of Gregor Mendel. His work revealed that genetic determinants, which we now call genes, are passed from parent to offspring in an unaltered form. As will be discussed in Chapter 2, we can now predict the outcome of genetic crosses based on Mendel's laws of inheritance.

The inheritance patterns of Mendel can be explained by the existence of chromosomes and their behavior during cell division. This topic will be examined in Chapter 3. As in Mendel's pea plants, it is common for sexually reproducing species to be **diploid**. This means they contain two copies of each chromosome. The two copies are called homologues of each other. Since genes are located on the chromosomes, diploid organisms have two copies of most genes. In humans, for example, there are 46 chromosomes, which are found in 23 homologous pairs (Figure 1-8a). With the exception of the sex chromosomes (namely, X and Y), each homologous pair contains the same kinds of genes. For example, both copies of

FIGURE **1-7**

Environmental influences on the outcome of PKU within a single family. All three children pictured here have inherited the alleles that cause PKU. The child in the middle was raised on a phenylalanine-free diet and developed normally. The other two children were born before the benefits of a phenylalanine-free diet were known and were raised on diets that contained phenylalanine. Therefore, they manifest a variety of symptoms, including mental retardation. (Photo from the March of Dimes Birth Defects Foundation.)

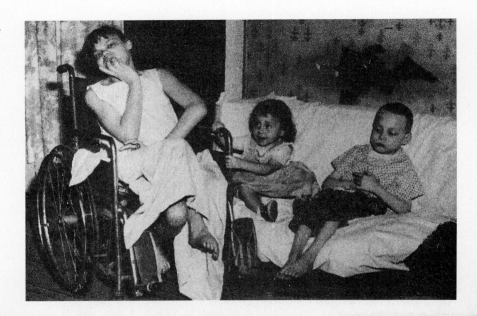

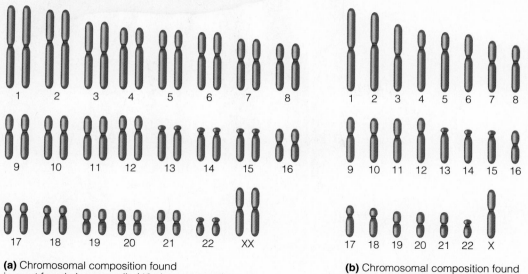

(a) Chromosomal composition found in most female human cells (46 chromosomes)

(b) Chromosomal composition found in a human gamete (23 chromosomes)

FIGURE **1-8**

The complement of human chromosomes in body cells and gametes. (a) A schematic drawing of the 46 chromosomes of a human. With the exception of the sex chromosomes, these are always found in homologous pairs. **(b)** The chromosomal composition of a gamete, which contains only 23 chromosomes, one from each pair.

human chromosome 12 carry the gene that encodes phenylalanine hydroxylase, which we discussed previously. Therefore, an individual has two copies of this gene; these two copies may or may not be identical alleles.

Most cells of the human body contain 46 chromosomes. The exceptions are the gametes (i.e, sperm and egg cells), which contain half that number (Figure 1-8b). The gametes are needed for sexual reproduction. The union of gametes during fertilization restores the diploid number of chromosomes. The primary advantage of sexual reproduction is that it enhances genetic variation. For example, a tall person with blue eyes and a short person with brown eyes may have short offspring with blue eyes or tall offspring with brown eyes. Therefore, sexual reproduction can result in new combinations of two or more traits that differ from those of either parent.

The genetic composition of a species evolves over the course of many generations

As we have just seen, sexual reproduction has the potential to enhance genetic variation. This can be an advantage for a population of individuals as they struggle to survive and compete within their natural environment. The term **biological evolution** refers to the phenomenon that the genetic makeup of a population can change over the course of many generations.

As suggested by Darwin, the members of a species are in competition with each other for essential resources. Random genetic changes (i.e., mutations) occasionally occur within an individual's genes, and sometimes these changes lead to a modification of traits that promotes survival. For example, over the course of many generations, random gene mutations have lengthened the neck of the giraffe, enabling it to feed on the leaves that are high in the trees. When a mutation creates a new allele that is beneficial, the allele may become prevalent within future generations because the individuals carrying the allele are more likely to survive and/or reproduce and pass the beneficial allele to their offspring. This process is known as **natural selection**. In this way, a species becomes better adapted to its environment. In addition, as we will learn in Chapter 27, neutral mutations, which have little or no impact on survival, can also accumulate in natural populations.

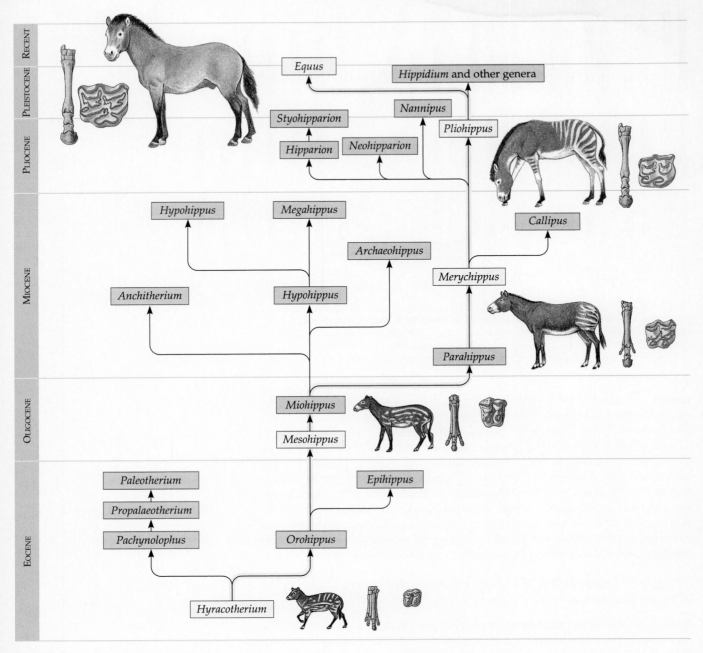

FIGURE **1-9**

The evolutionary changes that led to the modern horse. Three important morphological changes that occurred were larger size, fewer toes, and a shift toward a jaw structure suited for grazing.

Over a long period of time, the accumulation of many genetic changes leads to rather striking modifications in a species' characteristics. As an example, Figure 1-9 depicts the evolutionary changes that led to the development of the modern day horse. A variety of morphological changes are apparent, including an increase in size, fewer toes, and modified jaw structure.

FIELDS OF GENETICS

Genetics is a very broad discipline encompassing molecular, cellular, organismal, and population biology. Many scientists who are interested in genetics have been trained in supporting disciplines such as biochemistry, biophysics, cell biology, mathematics, microbiology, population biology, ecology, agriculture, and medicine. The study of genetics has been traditionally divided into three areas: transmission, molecular, and population genetics. In this section, we will examine the general questions that scientists in these areas are attempting to answer.

Transmission genetics explores the inheritance patterns of traits as they are passed from parents to offspring

Transmission genetics is the oldest field of genetics. A scientist working in this field explores the relationship between the transmission of genes from parent to offspring and the outcome of the offspring's traits. For example, how can two brown-eyed parents produce a blue-eyed child? Or why do tall parents tend to produce tall children, but not always? Our modern understanding of transmission genetics began with the studies of Gregor Mendel. His work provided the conceptual framework for transmission genetics. In particular, he originated the idea that genetic determinants, which we now call genes, are passed unaltered from parent to offspring via gametes (sperm and egg cells). We will examine his work in Chapter 2. Since these pioneering studies of the 1860s, our knowledge of genetic transmission has greatly increased. Many patterns of genetic transmission are more complex than the simple Mendelian patterns that are described in Chapter 2. These additional complexities of transmission genetics are examined in Chapters 3 through 8 of your text.

Experimentally, the fundamental approach of a transmission geneticist is the **genetic cross**. A genetic cross is a mating between two individuals and analysis of their offspring in an attempt to understand how traits are passed from parent to offspring. In the case of experimental organisms, the experimenter will choose two parents with particular traits and then categorize the offspring according to the traits they possess. In many cases, this analysis is quantitative in nature. For example, an experimenter may cross two tall pea plants and obtain 1000 offspring that fall into two categories: 750 tall and 250 dwarf. As we will see in Chapter 2, the ratio of tall and dwarf offspring provides important information concerning the inheritance pattern of this trait.

Throughout Chapters 2 to 8, we will learn how transmission genetics seeks to answer many fundamental questions concerning the passage of traits from parents to offspring. Some of these questions are as follows:

What are the common patterns of inheritance for genes?   Chapters 2–4

When two or more genes are located on the same chromosome, how does this affect the pattern of inheritance?   Chapters 5–6

Are there unusual patterns of inheritance that cannot be explained by the simple transmission of genes located on chromosomes in the cell nucleus?   Chapter 7

How do variations in chromosome structure or chromosome number affect the outcome of traits?   Chapter 8

Molecular genetics seeks a biochemical understanding of the hereditary material

As the name of the field implies, the goal of molecular genetics is to understand how the genetic material works at the molecular level. In other words, molecular geneticists want to understand the molecular features of DNA, and how these molecular features underlie the expression of genes. Experimentally, molecular geneticists often study organisms such as the bacterium *Escherichia coli*, *Saccharomyces cerevisiae* (baker's yeast), *Drosophila*, or mice and mammalian cell lines. The experiments are usually conducted within the confines of a laboratory. The efforts of a molecular geneticist frequently progress to a detailed analysis of DNA, RNA, and/or protein using a variety of techniques that will be described throughout Sections III, IV, and V of this text.

Studies within molecular genetics interface with other disciplines such as biochemistry, biophysics, and cell biology. In addition, advances within molecular genetics have shed considerable light in the areas of transmission and population genetics. In Chapters 19 through 22, we will see that our quest to understand molecular genetics has spawned a variety of modern molecular technologies (Chapters 19 and 20) and computer-based approaches to understand genetics (Chapter 22). It is particularly exciting that discoveries within molecular genetics have had widespread applications in agriculture, medicine, and biotechnology (Chapter 21).

Some general questions within the field of molecular genetics are the following:

What are the molecular structures of DNA and RNA?   **Chapter 9**

What is the composition and conformation of chromosomes?   **Chapter 10**

How is the genetic material copied?   **Chapter 11**

How are genes expressed at the molecular level?   **Chapters 12–14**

How is gene expression regulated so that it occurs at the appropriate time?
Chapters 15–16

What is the molecular nature of mutations? How are mutations repaired?
Chapter 17

How does the genetic material become rearranged at the molecular level?
Chapter 18

What is the underlying relationship between genes and genetic diseases?
Chapter 23

How do genes govern the development of multicellular organisms?   **Chapter 24**

Population genetics is concerned with the prevalence of genes within a population and how those genes change as a species evolves

The foundations of population genetics arose during the first few decades of the 20th century as a way to correlate the work of Mendel and Darwin. Mendel's work and that of many succeeding geneticists gave insight into the nature of genes and how they are transmitted from parents to offspring. The work of Darwin provided a natural explanation for the various types of characteristics observed among the members of a species. To relate these two phenomena, population geneticists have developed mathematical theories to explain the occurrence of genes within populations of individuals. The work of population geneticists helps us understand how the forces of nature have produced and favored the existence of individuals that carry particular genes.

Population geneticists are particularly interested in genetic variation and how that variation is related to an organism's environment. In this field, the prevalence of alleles within a population is of central importance. Some general questions in population genetics are the following:

What are the contributions of genetics and environment in the outcome of a trait?
Chapter 25

How are quantitative traits, such as size and weight, influenced by genetics and the environment?   **Chapter 25**

Why are two or more different alleles of a gene maintained in a population?
Chapter 26

What factors alter the prevalence of alleles within a population?   **Chapter 26**

What factors have the most impact on the process of evolution?   **Chapter 27**

How does evolution occur at the molecular level?   **Chapter 27**

Genetics is an experimental science

Science is a way of knowing about our natural world. The science of genetics allows us to understand how the expression of our genes produces the traits that we possess. The **scientific method** is the basis for conducting science. It is a standard process that scientists follow so that they may reach verifiable conclusions about the world in which they live. Although scientists arrive at their theories in different ways, the scientific method provides a way to validate (or invalidate) a particular hypothesis.

In traditional science textbooks, the emphasis often lies on the product of science. Namely, many textbooks are aimed primarily at teaching the student about the observations that scientists have made and the theories that they have proposed to explain these observations. Along the way, the student is provided with many bits and pieces of experimental techniques and data. Likewise, this textbook also provides you with many observations and theories. In addition, however, it attempts to go one step further. Each of the following chapters contains one or two experiments that have been "dissected" into five individual components:

1. Some *background* information will be provided so that you may appreciate what previous observations were known prior to conducting the experiment.
2. We will then examine the *hypothesis* that the scientist (or scientists) was trying to test. In other words, what scientific question was the researcher trying to answer?
3. We will follow the experimental steps that the scientist took to *test the hypothesis*. The steps necessary to carry out the experiment will be listed in the order they were conducted. In other words, it will be shown how the experiment was done.
4. The raw *data* for each experiment will then be presented.
5. And finally, *an interpretation* of the data will be offered.

The rationale behind this approach is that it will enable you to see the experimental process from beginning to end. Hopefully, you will find this a more interesting and rewarding way to learn about genetics. As you read through the chapters, the experiments will help you to see the relationship between science and scientific theories.

As a student of genetics, this textbook gives you the opportunity to involve your mind in the experimental process. As you are reading an experiment, you may find yourself thinking about different approaches and alternative hypotheses. You do not necessarily have to accept the interpretations that are given in this text. Even though the experiments in this text are typically classic experiments that seem to have withstood the test of time, it is still the case that any interpretation is a personal decision based on an individual's analysis of the data. Different people can view the same data and arrive at very different conclusions. As you progress through the experiments in this text, you will enjoy genetics far more if you try to develop your own skills at formulating hypotheses, designing experiments, and interpreting data. Also, some of the questions in the problems sets are aimed at refining these skills.

Finally, it is worthwhile to point out that science is a very social discipline. As you develop your skills at scrutinizing experiments, it is fun to discuss your ideas with other people, including fellow students and faculty members. Importantly, you do not need to "know all the answers" before you enter into a scientific discussion. Instead, it is more rewarding to view science as an ongoing and never-ending argument.

Part II
Patterns of Inheritance

CHAPTER 2

MENDELIAN INHERITANCE

MENDEL'S LAWS OF INHERITANCE

PROBABILITY AND STATISTICS

As we discussed in Chapter 1, many interesting and erroneous notions concerning the underlying principles of inheritance had been proposed as far back as 400 B.C. The first systematic studies of genetic crosses were carried out by Joseph Kölreuter from 1761–1766. In crosses between different strains of tobacco plants, he found that the offspring were usually intermediate in appearance between the two parents. This led Kölreuter to conclude that both parents make equal genetic contributions to their offspring. Furthermore, his observations were consistent with the prevailing **blending theory of inheritance**. According to this view, the "seeds" that dictate hereditary traits can blend together from generation to generation. The blended traits would then be passed on to the next generation. Another erroneous idea that was prevalent during this time period was the theory of **pangenesis**. It suggested that hereditary traits could be modified depending on the lifestyle of the individual (see Chapter 1). Pangenesis predicted, for example, that a person who practiced a particular skill (e.g., archery) would produce offspring that would be better at that skill. Overall, the popular view before the 1860s was that hereditary traits were rather malleable and could change and blend over the course of one or two generations. However, the pioneering work of Gregor Mendel would prove instrumental in refuting this notion.

In this chapter, we will examine some simple inheritance patterns that are found in pea plants, humans, mice, and fruit flies. We begin our inquiry into genetics here, because the inheritance of traits is the most fundamental concept in the study of heredity. Mendel's insights into the pattern of inheritance in pea plants revealed some simple rules that govern this process. In Chapters 3 through 8, we will explore more complex patterns of inheritance and also consider the role that the chromosomes play as the carriers of the genetic material.

   In the second part of this chapter, we will become familiar with some general concepts in probability and statistics. This is useful in two ways. First, probability calculations allow us to predict the outcomes of the genetic crosses described in this chapter, as well as the outcomes of more complicated crosses described in later chapters. In addition, we will learn how to use statistics to test the validity of genetic hypotheses that attempt to explain the inheritance patterns of traits.

MENDEL'S LAWS OF INHERITANCE

Gregor Johann Mendel, born in 1822, is now remembered as the father of genetics (Figure 2-1). He grew up on a small farm in Heinzendorf in northern Moravia, which was then a part of Austria and is now a part of the Czech Republic. As a young boy, he worked with his father grafting trees to improve the family orchard. Undoubtedly, his success at grafting taught him that precision and attention to detail are important elements of success. These qualities would later be important in his experiments as an adult scientist. Instead of farming, however, Mendel was accepted into the Augustinian monastery of St. Thomas, completed his studies for the priesthood, and was ordained in 1847. Soon after becoming a priest, Mendel worked for a short time as a substitute teacher. To continue that role, he needed to obtain a teaching license from the government. Surprisingly, he failed the licensing exam due to poor answers in the areas of physics and natural history. Therefore, Mendel then enrolled at the University of Vienna to expand his knowledge in these two areas. Mendel's training in physics and mathematics taught him to perceive the world as an orderly place, governed by natural laws. In his studies, Mendel learned that these natural laws could be stated as simple mathematical relationships. Mendel realized that a supreme achievement for a scientist is to deduce a mathematical expression that explains an important natural phenomenon.

   In 1853, Mendel returned to the monastery of St. Thomas, where he continued to teach physics and natural history and to conduct independent research studies. It was there, in 1856, that Gregor Mendel began his historic studies on pea plants. For eight years, he grew and crossed thousands of pea plants on a small $115' \times 23'$ plot. He kept meticulously accurate records that included quantitative data concerning the outcome of his crosses. He published his work, entitled "Experiments on Plant Hybrids," in 1866. This paper was largely ignored by scientists at that time, possibly because of its title or because it was published in a rather obscure journal (*The Proceedings of the Brünn Society of Natural History*). Nevertheless, his work allowed him to propose the natural laws that now provide a framework for our understanding of genetics.

   Prior to his death in 1884, Mendel reflected, "My scientific work has brought me a great deal of satisfaction and I am convinced that it will be appreciated before long by the whole world." Sixteen years later, in 1900, the work of Mendel was independently rediscovered by three biologists with an interest in plant genetics: Hugh de Vries of Holland, Carl Correns of Germany, and Erich von Tschermak of Austria. Within a few years, the impact of Mendel's studies was felt around the world. In this chapter, we will describe Mendel's experiments and consider their significance in the field of genetics.

FIGURE **2-1**

Gregor Johann Mendel, the father of genetics.

Mendel chose pea plants as his experimental organism

Mendel's study of genetics grew out of his interest in ornamental flowers. Prior to his work with pea plants, many plant breeders had conducted experiments aimed at obtaining flowers with new varieties of colors. When two plants of the same species but with different characteristics are mated (or *crossed*) to each other, this is called a **hybridization** experiment and the offspring are referred to as **hybrids**. For example, a hybridization experiment could involve a cross between a purple-flowered

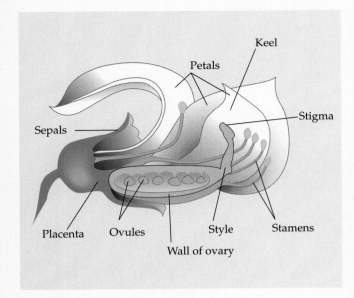

(a) Structure of a pea flower

(b) A flowering pea plant

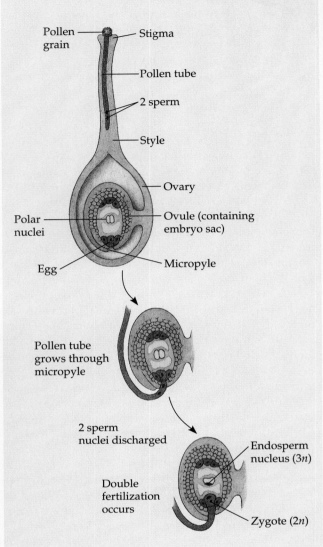

(c) Pollination and fertilization in angiosperms

FIGURE **2-2**

Flower structure and pollination in pea plants. (a) The pea flower can produce both pollen and egg cells. The pollen grains are produced within the anthers, and the egg cells are produced within the ovary. A modified petal, called a keel, encloses the anthers and ovaries. **(b)** Photograph of a flowering pea plant. **(c)** A pollen grain containing the male gamete must first land on the stigma. After this occurs, the pollen sends out a long tube, through which two sperm nuclei travel toward the ovules in order to reach the egg cells. The fusion between a sperm nucleus and an egg cell results in fertilization to create a zygote. A second sperm nucleus fuses with two polar nuclei to create endosperm. The endosperm provides storage material for the developing embryo.

plant and a white-flowered plant. From his knowledge of many hybridization studies, Mendel was particularly intrigued by the regularity with which hybrid characteristics appeared in the offspring. His intellectual foundation in physics and the natural sciences led him to consider that this regularity might be rooted in natural laws that could be expressed mathematically. To uncover these laws, he realized that he would need to carry out quantitative experiments in which the numbers of offspring carrying certain traits were carefully recorded and analyzed.

Mendel chose the garden pea, *Pisum sativum*, to investigate the natural laws that govern plant hybrids. The morphological features of this plant are shown in Figure 2-2a and b. Several properties of this species were particularly advantageous for studying plant hybridization. First, the species was available in several varieties that had decisively different physical characteristics. As we will discuss, many strains of the garden pea were available that varied in the appearance of their seeds, pods, flowers, and stems.

A second important issue is the ease of making crosses. In flowering plants, reproduction occurs by a pollination event (Figure 2-2c). The male gamete is contained within a *pollen grain*, and the female gamete is an *egg cell*. The pollen grains are produced within the *stamens*, while the eggs are produced within the *ovules*. For fertilization to occur, a pollen grain lands on the *stigma* and a *sperm nucleus* enters the stigma and migrates toward an egg cell. Fertilization occurs when a sperm nucleus fuses with an egg cell. This process will be described in greater detail in Chapter 3.

In some experiments, Mendel wanted to carry out **self-fertilization**. This means that the pollen and egg are derived from the same plant. In peas, the structure of the flowers greatly favors self-fertilization. The stamens (which produce pollen) and the ovules (which produce eggs) are covered by a modified petal known as the *keel*. Because of this covering, pea plants naturally reproduce by self-fertilization. In fact, pollination occurs even before the flower opens.

In other experiments, however, Mendel wanted to make crosses between different plants. Pea plants contain relatively large flowers that are easy to manipulate, making it possible to make crosses between two particular plants and study their outcomes. This process, known as **cross-fertilization**, requires that the pollen from one plant be placed on the stigma of another plant. This procedure is shown in Figure 2-3. Mendel was able to pry open premature flowers and remove the stamens before they produced pollen. Therefore, these flowers could not self-fertilize. He would then obtain pollen from another plant by gently touching its mature stamens with a paint brush. Mendel then applied this pollen to the stigma of the flower that already had its stamens removed. In this way, Mendel was able to cross-fertilize his pea plants and thereby obtain any type of hybrid he wanted.

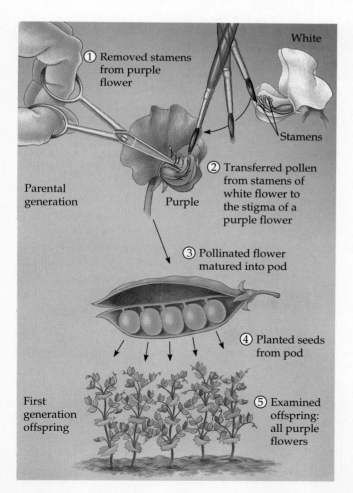

FIGURE **2-3**

How Mendel cross-fertilized two different pea plants. This illustration depicts a cross between a plant with white flowers and one with purple flowers. The offspring obtained from this cross are the result of pollination of the purple flower using pollen from a white flower.

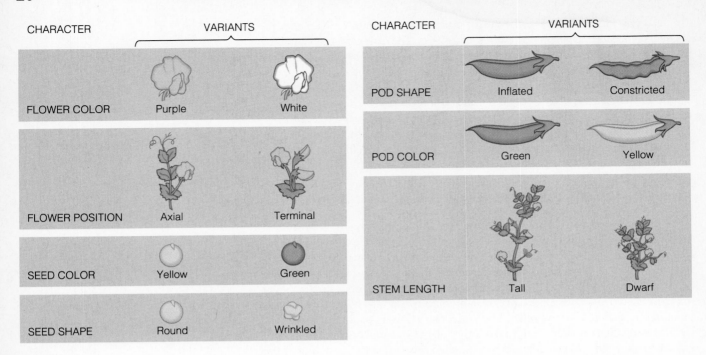

FIGURE **2-4**

An illustration of the seven traits that Mendel studied. Each trait was found as two variants that were decisively different from each other.

Mendel studied seven traits that bred true

When he initiated his studies, Mendel contacted several seed dealers and obtained a total of 34 varieties of peas that were considered to be distinct. These plants were different with regard to many morphological characteristics. The characteristics of an organism are called **characters** or **traits**. Over the course of two years, Mendel tested the strains to determine if their characteristics *bred true*. This means that a trait did not vary in appearance from generation to generation. For example, if the seeds from a pea plant were yellow, the next generation would also produce yellow seeds. Likewise, if these offspring were allowed to self-fertilize, all of their offspring would also produce yellow seeds, and so on. A variety that continues to produce the same characteristic after several generations of self-fertilization is called a **true-breeding line**.

Mendel next concentrated his efforts on the analysis of characteristics that were clearly distinguishable between different true-breeding lines. Figure 2-4 illustrates the seven characteristics that Mendel eventually chose to follow in his breeding experiments. All seven of Mendel's traits were found in two **variants**. For example, one trait he followed was height. He had variants for this trait, tall and dwarf plants, which he crossed; he then followed the outcomes of these crosses. A cross in which an experimenter is only observing one trait is called a **monohybrid** cross. This type of cross produces single-trait hybrids.

EXPERIMENT 2A

Around 1860, Mendel followed the outcome of monohybrid crosses in pea plants for two generations

As we have mentioned, Mendel became interested in the inheritance of traits that were found in pea plants. Mendel did not, prior to conducting his experiments, already have a hypothesis to explain the formation of hybrids. Instead, his experiments were designed to determine the quantitative relationships from which the laws could be discovered. This is known as an **empirical approach**. Laws that are deduced from an empirical approach are known as **empirical laws**.

Before discussing the details of this experiment, let's consider some terminology that pertains to the series of steps in Mendel's crosses. The true-breeding plants that he obtained were considered the parental generation, or **P generation**. When the true-breeding parents were crossed to each other, this is called a P cross, and the offspring comprise the first filial or **F₁ generation**. When members of the F₁ generation were crossed, this produced the **F₂ generation**. And so on. This terminology will be used throughout this text.

THE HYPOTHESIS

A careful, quantitative analysis of plant hybrids may uncover mathematical relationships that provide the basis for deducing a natural law.

TESTING THE HYPOTHESIS

Starting material: Mendel began his experiments with true-breeding pea plants that varied with regard to only one of seven different traits (see Figure 2-4).

Experimental Level   **Conceptual Level**

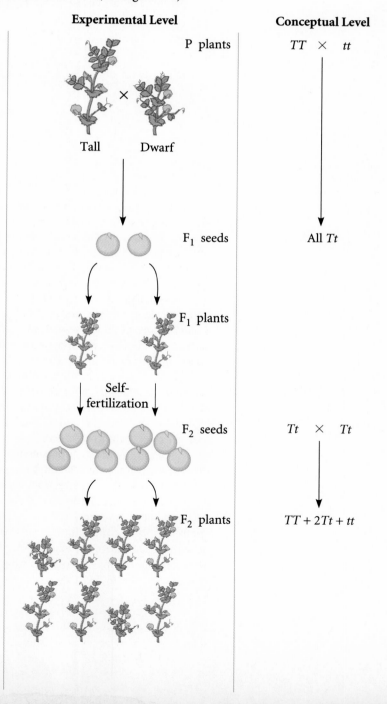

1. For each of seven traits, Mendel cross-fertilized two different true-breeding lines. Keep in mind that each cross involves two plants that differ only with regard to one trait. The illustration at the right is showing one cross between a tall and dwarf plant. This is called a P (parental) cross.

Note: The P cross produces seeds that are part of the F₁ generation.

2. Collect many seeds. The following spring, plant the seeds and allow the plants to grow. These are the plants of the F₁ generation.

3. Allow the F₁ generation plants to self-fertilize. This cross produces seeds that are part of the F₂ generation.

4. Collect the seeds and plant them the following spring to obtain the F₂ generation plants.

5. Analyze the characteristics found in each generation.

P plants $TT \times tt$

Tall   Dwarf

F₁ seeds   All Tt

F₁ plants

Self-fertilization

F₂ seeds $Tt \times Tt$

F₂ plants $TT + 2Tt + tt$

THE DATA

P Cross	F_1 Generation	F_2 Generation	Ratio
Round × wrinkled seeds	All round	5,474 round, 1,850 wrinkled	2.96 : 1
Yellow × green seeds	All yellow	6,022 yellow, 2,001 green	3.01 : 1
Purple × white flowers	All purple	705 purple, 224 white	3.15 : 1
Smooth × constricted pods	All smooth	882 smooth, 299 constricted	2.95 : 1
Green × yellow pods	All green	428 green, 152 yellow	2.82 : 1
Axial × terminal flowers	All axial	651 axial, 207 terminal	3.14 : 1
Tall × dwarf stem	All tall	787 tall, 277 dwarf	2.84 : 1
Total	All dominant	14,949 dom., 5,010 rec.	2.98 : 1

INTERPRETING THE DATA

Mendel's data argued strongly against a blending mechanism of heredity. In all seven cases, the F_1 generation displayed characteristics distinctly like one of the two parents rather than traits intermediate in character. Using genetic terms that Mendel originated and are still used today, we would say that one trait is always **dominant** over another trait. For example, the trait of green pods is dominant to that of yellow pods. The term **recessive** is used to describe a trait that is masked by the presence of a dominant trait. Yellow pods and dwarf stems are examples of recessive traits.

When a true-breeding plant containing a dominant trait was crossed to a true-breeding plant with a recessive trait, the dominant trait was always observed in the F_1 generation. In the F_2 generation, some offspring displayed the dominant characteristic while a lower proportion showed the recessive trait. However, none of the offspring exhibited intermediate traits. Since the recessive trait turned up in the second generation, Mendel also concluded that the genetic determinants of traits are passed along in an unaltered way from generation to generation. In contrast to a blending theory, his data were consistent with a **particulate theory of inheritance**, in which traits are inherited as discrete units that remain unchanged as they are passed from parent to offspring.

A second important interpretation of Mendel's data is related to the proportions of offspring. When Mendel compared the numbers of dominant and recessive offspring in the F_2 generation, he noticed a recurring pattern. Although there was some experimental variation, he always observed approximately a 3:1 ratio between the dominant trait and the recessive trait. Mendel was the first scientist to apply this type of quantitative analysis in a biological experiment. As described next, this quantitative approach allowed Mendel to conclude that genetic determinants segregate from each other during gamete formation.

The 3:1 phenotypic ratio that Mendel observed is consistent with the segregation of genes, now known as Mendel's law of segregation

In the 1860s, Mendel's research was aimed at understanding the laws that govern the inheritance of traits. At that time, scientists did not understand the molecular composition of the genetic material or its mode of transmission during gamete formation and fertilization. We now know that the genetic material is composed of deoxyribonucleic acid (DNA), a component of chromosomes. Each chromosome contains hundreds or thousand of shorter segments that function as **genes**. A gene is defined as a *unit of heredity* that may influence the outcome of an organism's traits. The term was originally coined by the Danish botanist Wilhelm Johannsen in 1909. Each of the seven traits that Mendel studied is controlled by a particular gene.

As we will discuss in Chapter 3, many eukaryotic species, such as pea plants and humans, have their genetic material organized into pairs of chromosomes. For this reason, there are two copies of most eukaryotic genes. These copies may be the same, or they may differ. The term **allele** refers to different versions of the same gene. With this modern knowledge, the results of Experiment 2A are consistent with the idea that each parent transmits only one copy of each gene (i.e., one allele) to each offspring. This is **Mendel's law of segregation**: *the two copies of a gene* **segregate** *from each other during transmission from parent to offspring*. During gamete (i.e., sperm or egg cell) formation, the two copies of a gene segregate from each other so that only one copy is found in each gamete. At fertilization, two gametes combine randomly, producing different allelic combinations.

Let's use Mendel's cross of tall and dwarf pea plants as an example. The letters T and t are used to represent the alleles of the gene that determine plant height; by convention, the uppercase letter represents the dominant allele (T for tall height, in this case), and the recessive allele is represented by the same letter in lowercase. For the P cross, both parents are true-breeding plants; therefore, we know they have identical copies of the plant height gene, a condition called being **homozygous**. In the P cross, the tall plant is homozygous for the tall allele T, while the dwarf plant is homozygous for the dwarf allele t. The term **genotype** refers to the genetic composition of an individual; TT and tt are the genotypes of the P generation in this experiment. The term **phenotype** refers to the outward appearance of an organism. In the P generation, half of the plants are phenotypically tall and half are dwarf (see Figure 2-5).

In contrast, the F_1 generation is **heterozygous**, with the genotype Tt, because every individual carries one copy of the tall allele and one copy of the dwarf allele. A heterozygous individual carries different alleles of a gene. Although these plants are heterozygous, their phenotypes are all tall, because they have a copy of the dominant tall allele.

The law of segregation predicts that the phenotypes of the F_2 generation will be tall and dwarf in a ratio of 3:1 (see Figure 2-5). Since the parents of the F_2 generation are heterozygous, each parent can transmit either a T allele or a t allele to their offspring, but not both, because each gamete carries only one of the two alleles. Therefore, TT, Tt, and tt are the possible genotypes of the F_2 generation (note that the genotype Tt is the same as tT). By randomly combining these alleles, the genotypes are produced in a 1:2:1 ratio. Because TT and Tt both produce tall phenotypes, a 3:1 phenotypic ratio is observed in the F_2 generation.

PHENOTYPE

GENOTYPE

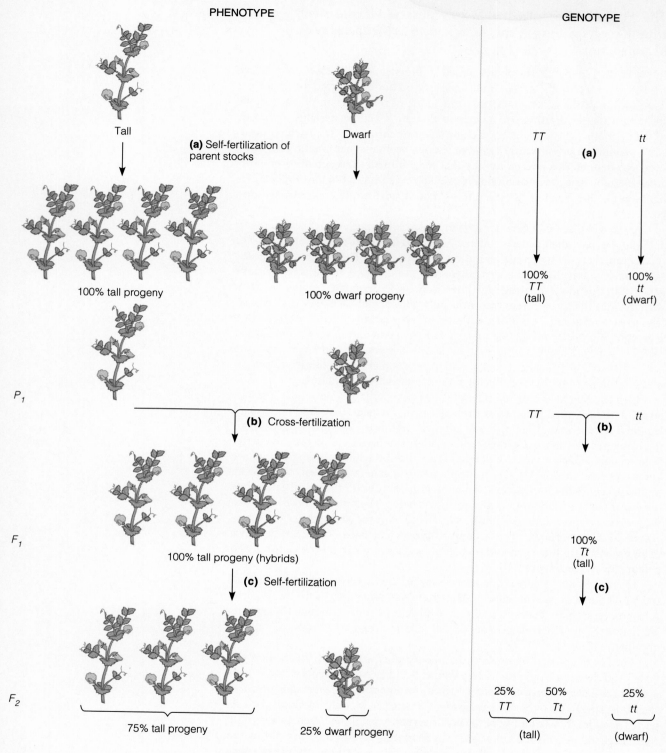

FIGURE **2-5**

Mendel's law of segregation. This illustration shows a cross between a tall and a dwarf plant and the subsequent segregation of the tall (*T*) and dwarf (*t*) alleles in the F$_1$ and F$_2$ generations.

A Punnett square can be used to predict the outcome of crosses

An easy way to predict the outcome of simple genetic crosses is to use a **Punnett square**, a method originally proposed by Reginald Punnett in the early 1900s. To construct a Punnett square, you must know the genotypes of the parents. With this information, the Punnett square enables you to predict the types of offspring the parents will produce and in what proportions. This section provides a step-by-step description of the Punnett square approach using the F_1 self-fertilization of heterozygous tall plants as an example.

Step 1 *Write down the genotypes of both parents.* We will cross a heterozygous tall plant with another heterozygous tall plant. The plant providing the pollen is considered the male parent and the plant providing the eggs, the female parent.

Male parent: *Tt*

Female parent: *Tt*

Step 2 *Write down the possible gametes that each parent can make.* Remember that the law of segregation tells us that a gamete can contain only one copy of each gene.

Male gametes: *T* or *t*

Female gametes: *T* or *t*

Step 3 *First, you must create an empty Punnett square.* The number of columns equals the number of male gametes, and the number of rows equals the number of female gametes. In our example, there are two rows and two columns. Place the male gametes across the top of the Punnett square and the female gametes along the side.

Step 4 *Fill in the possible genotypes of the offspring by combining the gametes in the empty boxes.*

Step 5 *Determine the relative proportions of genotypes and phenotypes of the offspring.* The genotypes are obtained directly from the Punnett square. They are contained within the boxes you have filled in. In this example, the genotypes are *TT*, *Tt*, and *tt* in a 1:2:1 ratio. To determine the phenotypes, you must know the dominant/recessive relationships between the alleles. For plant height, we know that *T* (tall) is dominant to *t* (dwarf). The genotypes *TT* and *Tt* are tall, whereas the genotype *tt* is dwarf. Therefore, our Punnett square shows us that the ratio of phenotypes is 3:1, or 3 tall plants to 1 dwarf plant.

Around 1860, Mendel also followed the outcome of dihybrid crosses in pea plants

In Experiment 2A, the analysis of monohybrid crosses allowed Mendel to understand how the variants for a single trait (in that case, tall versus dwarf plants) are segregated during future generations. From a modern point of view, it also allowed us to understand how the alleles for a particular gene are segregated into gametes. Mendel realized, however, that additional laws of inheritance might be uncovered if he conducted more complicated experiments.

In this experiment, Mendel conducted crosses in which he simultaneously investigated the pattern of inheritance for two different traits (in this case, seed shape and seed color). Therefore, he carried out **dihybrid crosses**, in which he followed the inheritance of two different traits within the same individual. One of the traits was seed texture, found in round or wrinkled variants. The second trait was seed color, which existed as yellow and green variants. In this dihybrid cross, he followed the inheritance pattern for both traits simultaneously. As we will see, he crossed a true-breeding plant with round, yellow seeds with one that had wrinkled, green seeds and then analyzed the outcome of future generations for *both* traits.

Before we discuss Mendel's results, let's consider different possible patterns of inheritance for two traits. One possibility is that the genetic determinants for two different traits are *linked* to each other and are inherited as a single unit. A second possibility is that they are not linked and can assort themselves *independently* into gametes. These two possibilities are shown here. (Keep in mind that the results of Experiment 2A have already shown us that a gamete carries only one allele for each gene.) Using the symbols R = round, r = wrinkled, Y = yellow, and y = green:

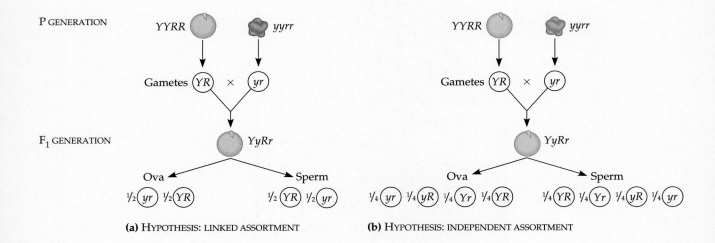

(a) HYPOTHESIS: LINKED ASSORTMENT   **(b)** HYPOTHESIS: INDEPENDENT ASSORTMENT

We can now turn to Mendel's experiment, which decisively supported one of these two models.

The goal of this experiment is to understand the natural laws that govern the inheritance patterns of two different genes relative to each other.

TESTING THE HYPOTHESIS

Starting material: In this experiment Mendel began with two types of true-breeding pea plants that were different with regard to two traits. One plant had round, yellow seeds (*RRYY*); the other plant had wrinkled, green seeds (*rryy*).

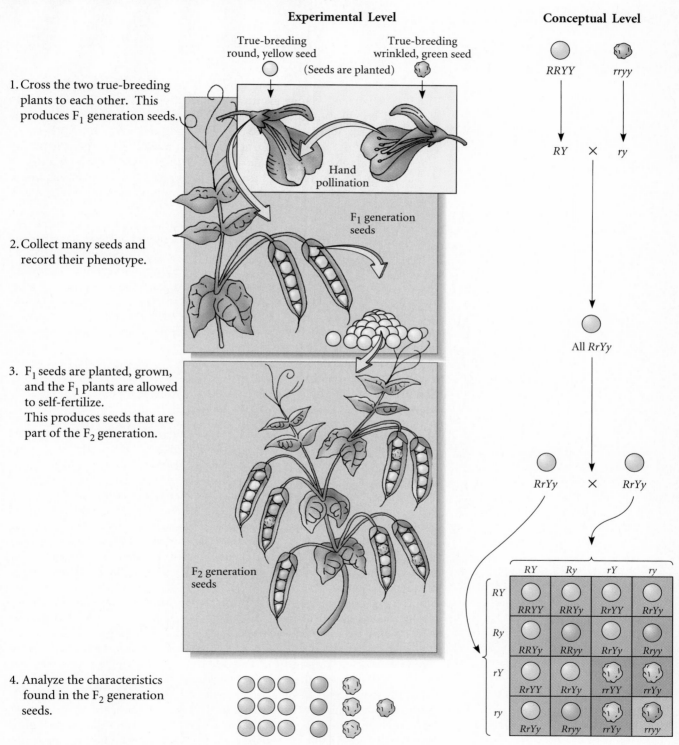

Experimental Level

True-breeding round, yellow seed

True-breeding wrinkled, green seed

(Seeds are planted)

1. Cross the two true-breeding plants to each other. This produces F₁ generation seeds.

Hand pollination

F₁ generation seeds

2. Collect many seeds and record their phenotype.

3. F₁ seeds are planted, grown, and the F₁ plants are allowed to self-fertilize.
This produces seeds that are part of the F₂ generation.

F₂ generation seeds

4. Analyze the characteristics found in the F₂ generation seeds.

Conceptual Level

RRYY   *rryy*

RY × *ry*

All *RrYy*

RrYy × *RrYy*

	RY	Ry	rY	ry
RY	RRYY	RRYy	RrYY	RrYy
Ry	RRYy	RRyy	RrYy	Rryy
rY	RrYY	RrYy	rrYY	rrYy
ry	RrYy	Rryy	rrYy	rryy

P Cross	F₁ Generation	F₂ Generation
Round, yellow seeds	All round, yellow	315 round, yellow seeds
× wrinkled, green seeds		101 wrinkled, yellow seeds
		108 round, green seeds
		32 wrinkled, green seeds

The P cross is true-breeding plants with round, yellow seeds (*RRYY*) mated to plants with wrinkled, green seeds (*rryy*). As expected, this produces an F₁ generation of *RrYy* seeds that display a phenotype of round and yellow. It is the F₂ generation that supports the independent assortment model and refutes the linkage model. In the F₂ generation, there are seeds that are round and green, and also seeds that are wrinkled and yellow. These two categories of F₂ seeds are called **nonparentals**, because these combinations of traits were not found in the true-breeding plants of the parental generation.

The occurrence of nonparental offspring contradicts the linkage model. According to the linkage model, the *R* and *Y* variants should be linked together and so should the *r* and *y* variants. If this were the case, the F₁ self-fertilization could only produce gametes that are *RY* or *ry*. These would combine to produce *RRYY* (round, yellow), *RrYy* (round, yellow), or *rryy* (wrinkled, green) in a 1:2:1 ratio. There would not be any nonparental seeds produced. However, Mendel did not obtain this result. Instead, he observed a phenotypic ratio of 9:3:3:1 in the F₂ generation. As described next, this result supported the idea that different traits can assort from each other during gamete formation.

The law of independent assortment was deduced from Mendel's experiments with dihybrid crosses

Mendel's results from many dihybrid crosses supported the idea that different traits assort themselves independently during reproduction. Using the modern notion of genes, the **law of independent assortment** states that *two different genes will randomly assort their alleles during gamete formation*. Using the example given in Experiment 2B, the round and wrinkled alleles will be assorted into gametes independently of the yellow and green alleles. Therefore, a heterozygous *RrYy* parent can produce four different gametes—*RY*, *Ry*, *rY*, and *ry*—in equal proportions. In an F₁ self-fertilization experiment, any two gametes can combine randomly during fertilization. This allows for 4^2, or 16, possible offspring, although some offspring will be genetically identical to each other. These 16 possible combinations are shown in Figure 2-6.

The 16 genotypes shown in Figure 2-6 result in seeds with the following phenotypes: 9 round, yellow; 3 wrinkled, yellow; 3 round, green; and 1 wrinkled, green. This 9:3:3:1 ratio is the expected outcome of a dihybrid cross from parents that are heterozygous for both genes. Mendel was clever enough to realize that the data for his dihybrid crosses were close to a 9:3:3:1 ratio. In Experiment 2B, for example, his F₁ cross produced F₂ seeds with the following characteristics: 315 round, yellow seeds; 101 wrinkled, yellow seeds; 108 round, green seeds; and 32 wrinkled, green seeds. If we divide each of these numbers by 32 (the number of plants with wrinkled, green seeds), the phenotypic ratio of the F₂ generation is 9.8 : 3.2 : 3.4 : 1.0. Within experimental error, Mendel's data approximated the predicted 9:3:3:1 ratio for the F₂ generation.

identify the protein that is defective or missing in the white petals but functionally active in the purple ones. The identification and characterization of this protein would provide a molecular explanation for this phenotypic characteristic.

In Chapter 5, we will also see that the identification of two alleles for a particular gene makes it possible to determine the gene's chromosomal location by genetic mapping techniques. In Chapter 19, we will learn how researchers can make many copies of a particular gene by cloning techniques. This allows a geneticist to better understand the relationship between the molecular characteristics of a gene and its effect on the phenotype of the organism.

Pedigree analysis can be used to follow the Mendelian inheritance of traits in humans

Before we end our discussion of simple Mendelian traits, it is interesting to consider how we can analyze inheritance patterns among humans. In Mendel's experiments, he selectively made crosses and then analyzed a large number of offspring. When studying human traits, however, it is not ethical to control parental crosses. Instead, we must rely on the information that is contained within family trees. This type of approach, known as a **pedigree analysis**, is aimed at determining the type of inheritance pattern that a gene will follow. Although this method may be less definitive than the results described in Experiments 2A and 2B, a pedigree analysis can often provide important clues concerning the pattern of inheritance of traits within human families.

Before discussing the applications of pedigree analyses, we need to understand the symbols and organization of a pedigree (Figure 2-7). The oldest generation is at the top of the pedigree, and the most recent generation is at the bottom. Vertical lines connect each succeeding generation. A man (square) and woman (circle)

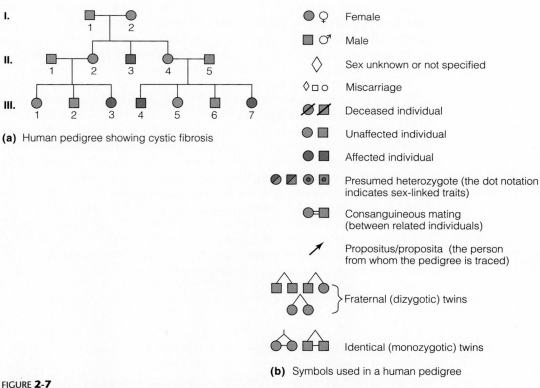

(a) Human pedigree showing cystic fibrosis

⚫ ♀	Female
⬛ ♂	Male
◇	Sex unknown or not specified
◇ □ o	Miscarriage
⊘ ⧄	Deceased individual
⚫ ⬛	Unaffected individual
⚫ ⬛	Affected individual
⊘ ⧄ ⊙ ⊡	Presumed heterozygote (the dot notation indicates sex-linked traits)
⚫—⬛	Consanguineous mating (between related individuals)
↗	Propositus/proposita (the person from whom the pedigree is traced)
	Fraternal (dizygotic) twins
	Identical (monozygotic) twins

(b) Symbols used in a human pedigree

FIGURE **2-7**

Pedigree analysis. (a) A family pedigree in which some of the members are affected with cystic fibrosis. Individuals I-1, I-2, II-1, II-2, II-4, and II-5 could also be depicted as presumed heterozygotes because they produce affected offspring. **(b)** The symbols used in a pedigree analysis.

who produce one or more offspring are directly connected by a horizontal line (see Figure 2-7). A vertical line connects parents with their offspring. If parents produce two or more offspring, the group of siblings (brothers and sisters) is denoted by two or more individuals projecting from the same horizontal line.

When a pedigree involves the transmission of a human trait or disease, affected individuals are depicted by filled symbols (in this case, red) that distinguish them from unaffected individuals (shown in purple). Each generation is given a Roman numeral designation, and individuals within the same generation are numbered from left to right. Some examples of the genetic relationships in Figure 2-7 are described here:

Individuals I-1 and I-2 are the grandparents of individual III-2

Individuals III-1, III-2, and III-3 are brothers and sisters

Individuals II-3 and III-3 are affected by a genetic disease

Pedigree analysis is commonly used to determine the inheritance pattern of human genetic diseases. Human geneticists are routinely interested in knowing whether a genetic disease is inherited as a recessive or dominant trait. One way to discern the dominant/recessive relationship between two alleles is by a pedigree analysis. Mutations in genes that cause inherited disease often exist as a normal allele versus the mutant allele that causes disease symptoms. If the disease follows a simple Mendelian pattern of inheritance and is caused by a recessive allele, an individual must inherit two copies of the abnormal allele to exhibit the disease. Therefore, a recessive pattern of inheritance makes two important predictions. First, two heterozygous normal individuals will, on average, have 1/4 of their offspring affected. Second, all the offspring of two affected individuals will be affected. Alternatively, a dominant trait predicts that affected individuals will have inherited the gene from at least one affected parent.

The pedigree in Figure 2-7a concerns a human genetic disease known as cystic fibrosis (CF). Among Caucasians, approximately 3% of the population are heterozygous carriers of this recessive allele. In homozygotes, the disease symptoms include abnormalities of the pancreas, intestine, sweat glands, and lungs. These abnormalities are caused by an imbalance of ions across the cell membrane. In the lungs, this leads to a buildup of thick dry mucous. Respiratory problems may lead to early death, although modern treatments have greatly increased the life span of CF patients. In the late 1980s, the gene for CF was identified. It encodes a protein called the cystic fibrosis transmembrane conductance regulator (CFTR). This protein regulates the ion balance across the cell membrane in tissues such as the pancreas, intestine, sweat glands, and lungs. The CF allele for this gene contains a mutation that alters the encoded CFTR protein. The altered CFTR protein functions incorrectly and thereby causes the ionic imbalance. As seen in the pedigree, the pattern of affected and unaffected individuals is consistent with a recessive mode of inheritance. There are several cases in which two unaffected individuals produce affected offspring. Although not shown in this pedigree, a recessive mode of inheritance is also characterized by the observation that two affected individuals will produce 100% affected offspring. However, for human genetic diseases that affect survival and/or fertility, there may never be cases where two affected individuals produce offspring.

PROBABILITY AND STATISTICS

A powerful application of Mendel's work is that the laws of inheritance can be used to predict the outcomes of genetic crosses. This is useful in many ways. In agriculture, for example, plant and animal breeders are concerned with the types of offspring that their crosses will produce. This information is used to produce commercially important crops and livestock. In addition, people are often interested in predicting the types of children they may have. This is particularly important to individuals who may carry alleles that cause inherited diseases. Of course, it is not possible to see into the future and definitively predict what will happen. Nevertheless, genetic counselors may help couples to predict the likelihood of having an affected child. This probability is one factor that may influence a couple's decision whether to have children.

In this section, we will see, based on Mendel's laws, how probability calculations are used in genetic problems to predict the outcome of crosses. To compute probability, we will use three mathematical operations known as the sum rule, the product rule, and the binomial expansion formula. These methods allow us to determine the probability that a cross between two individuals will produce a particular outcome. To apply these operations, we must know the genotypes of the parents and the laws that govern their pattern of inheritance. In several examples we will determine the probability of having certain types of offspring.

Probability calculations can also be used in hypothesis testing. In many situations, a researcher would like to discern the genotypes and patterns of inheritance for traits that are not yet understood. A traditional approach to this problem is to conduct crosses and then analyze their outcomes. The proportions of offspring may provide important clues that allow the experimenter to propose a hypothesis explaining the transmission of the trait from parent to offspring. Statistical methods, such as the chi square test, can then be used to evaluate how well a genetic hypothesis fits the observed data from crosses. We will end this chapter by considering a problem that uses the chi square test as a way to evaluate the validity of a genetic hypothesis.

Probability is the likelihood that an event will occur

The chance that an event will occur in the future is called the event's **probability**. For example, if you flip a coin, the probability is 0.50, or 50%, that the head side will be showing when the coin lands. Probability depends on the number of possible outcomes. In this case, there are two possible outcomes (heads and tails), which are equally likely. This allows us to predict that there is a 50% chance that a coin flip will produce heads. The general formula for probability is

$$\text{Probability} = \frac{\text{Number of times an event occurs}}{\text{Total number of events}}$$

$$P_{\text{heads}} = 1 \text{ heads}/(1 \text{ heads} + 1 \text{ tails}) = 1/2 = 50\%$$

In genetic problems, we are often interested in the probability that a particular type of offspring will be produced. For example, when two heterozygous tall pea plants (Tt) are crossed, the phenotypic ratio of the offspring is 3 tall : 1 dwarf. This information can be used to calculate the probability for either type of offspring:

$$\text{Probability} = \frac{\text{Number of individuals with a given phenotype}}{\text{Total number of individuals}}$$

$$P_{\text{tall}} = 3 \text{ tall}/(3 \text{ tall} + 1 \text{ dwarf}) = 3/4 = 0.75 = 75\%$$

$$P_{\text{dwarf}} = 1 \text{ dwarf}/(3 \text{ tall} + 1 \text{ dwarf}) = 1/4 = 0.25 = 25\%$$

The probability of obtaining a tall plant is 75%, a dwarf plant 25%. When we add together the probabilities of all the possible outcomes (tall and dwarf), we should get a sum of 100% (here, 75% + 25% = 100%).

A probability calculation allows us to predict the likelihood that an event will occur in the future. The accuracy of this prediction, however, depends to a great extent on the size of the sample. For example, if we toss a coin 6 times, our probability prediction would suggest that 50% of the time we should get heads (i.e., 3 heads and 3 tails). In this small sample size, however, we would not be too surprised if we came up with 4 heads and 2 tails (66.7% heads, 16.7% different from the prediction). Each time we toss a coin, there is a random chance that it will be heads or tails. The deviation between the observed and expected outcomes is due to **random sampling error**. In a small sample, the error between the predicted percentage of heads and the actual percentage observed may be quite large (in this case, it was 50% vs. 66.7%). By comparison, if we flipped a coin 1000 times, the percentage of heads would be fairly close to the predicted 50% value. In a larger sample, we expect the random sampling error to be much smaller.

The sum rule can be used to predict the occurrence of mutually exclusive events

Now that we have an understanding of probability, we can see how mathematical operations with probability values can be used to predict the outcome of genetic crosses. Our first genetic problem will involve the use of the **sum rule**, which states that *the probability that one of two or more mutually exclusive events will occur is equal to the sum of the individual probabilities of the events.*

As an example, let us consider a cross between two mice that are both heterozygous for genes affecting the ears and tail. One gene can be found in an allele designated *de*[1], which is a recessive allele that causes droopy ears; the normal allele is *De*. An allele of a second gene causes a crinkly tail. This crinkly tail allele (*ct*) is recessive to the normal allele (*Ct*). If a cross is made between two heterozygous mice (*Dede/Ctct*), the predicted ratio of offspring is 9 with normal ears and normal tails, 3 with normal ears and crinkly tails, 3 with droopy ears and normal tails, and 1 with droopy ears and a crinkly tail. These four phenotypes are mutually exclusive. For example, a mouse with droopy ears and a normal tail cannot have normal ears or a crinkly tail.

The sum rule allows us to determine the probability that we will obtain any one of two or more different types of offspring. For example, in a cross between two heterozygotes (*Dede/Ctct* × *Dede/Ctct*), we can ask the following question: "What is the probability that an offspring will have normal ears and a normal tail *or* have droopy ears and a crinkly tail?" In other words, if we closed our eyes and picked an offspring out a litter from this cross, what are chances that we would be holding a mouse that has normal ears and a normal tail *or* a mouse with droopy ears and a crinkly tail? In this case, the investigator wants to predict whether one of two mutually exclusive events will occur. A strategy for solving such genetic problems using the sum rule is described here.

The Cross *Dede/Ctct* × *Dede/Ctct*.

The Question What is the probability that an offspring will have normal ears and a normal tail, or have droopy ears and a crinkly tail?

Step 1 *Calculate the individual probabilities of each phenotype.* This can be accomplished using a Punnett square.

The probability of normal ears and a normal tail is $9/(9 + 3 + 3 + 1) = 9/16$
The probability of droopy ears and a crinkly tail is $1/(9 + 3 + 3 + 1) = 1/16$

Step 2 *Add together the individual probabilities.*

$$9/16 + 1/16 = 10/16$$

This means that 10/16 is the probability that an offspring will have either normal ears and a normal tail *or* droopy ears and a crinkly tail. We can convert 10/16 to 0.625, which means that 62.5% of the offspring are predicted to have normal ears and a normal tail, or droopy ears and a crinkly tail.

---

[1]*de* is a two-letter abbreviation for the allele that causes *droopy* ears.

The product rule can be used to predict the probability of independent events that occur in a particular order

We can use probability to make predictions about the order of independent outcomes from a genetic cross. For example, let's consider a rare, recessive human trait known as congenital indifference to pain (known as congenital analgesia). Persons with this trait can distinguish between sharp and dull, and hot and cold, but extremes of sensation are not perceived as being painful. The first case of congenital analgesia, described in 1932, was a man who made his living entertaining the public as a "human pincushion."

For a phenotypically normal couple, each being heterozygous for the recessive allele causing congenital analgesia, we can ask the question, "What is the probability that their first three offspring will have congenital analgesia?" To answer this question, the **product rule** is used. According to this rule, *the probability that two or more independent events will occur in a particular order is equal to the product of their individual probabilities*. A strategy for solving this type of problem is shown here.

The Cross $Pp \times Pp$ (where P is the normal allele and p is the recessive congenital analgesia allele)

The Question What is the probability that the first three offspring will have congenital analgesia?

Step 1 *Calculate the individual probability of this phenotype.* As described previously, this is accomplished using a Punnett square.

The probability of an affected offspring is 1/4.

Step 2 *Multiply the individual probabilities.* In this case, we are asking about the first *three* offspring, and so we multiply 1/4 three times:

$$1/4 \times 1/4 \times 1/4 = 1/64 = 0.016$$

Thus the probability that the first three offspring will have this trait is 0.016. In other words, we predict that 1.6% of the time the first three offspring of a couple, each heterozygous for the recessive allele, will all have congenital analgesia.

In the problem described here, we have used the product rule to determine the probability that the first three offspring will all have the same phenotype (congenital analgesia). We can also apply the rule to predict the probability of a sequence of events that involves combinations of different offspring. For example, consider the question, "What is the probability that the first offspring will be normal, the second offspring will have congenital analgesia, and the third offspring will be normal?" Again, to solve this problem, begin by calculating the individual probability of each phenotype:

$$\text{normal} = 3/4$$
$$\text{congenital analgesia} = 1/4$$

The probability that these three phenotypes will occur in this specified order is

$$3/4 \times 1/4 \times 3/4 = 9/64 = 0.14$$

In words, this sequence of events is expected to occur only 14% of the time.

The binomial expansion can be used to predict the probability of an unordered combination of events

A third predictive problem in genetics is to determine the probability that a certain proportion of offspring will be produced with certain characteristics; here they can be produced in an unspecified order. For example, let us consider a cross between two heterozygous tall (Tt) pea plants. A geneticist might want to know, "What is the probability that two out of five plants will be dwarf?"

In this case, we are not concerned with the order in which the offspring are produced. Instead, we are only concerned with the final numbers of dwarf and tall offspring. Say we obtain five seeds from this cross and plant them in five separate pots, which we label 1 through 5. One desired outcome would be the following: pot 1, tall plant; pot 2, dwarf plant; pot 3, tall plant; pot 4, dwarf plant; and pot 5, tall plant. Another desired outcome could be pot 1, dwarf plant; pot 2, dwarf plant; pot 3, tall plant; pot 4, tall plant; and pot 5, tall plant. Either of these two situations would satisfy the desire to obtain two dwarf offspring and three tall offspring out of a total of five. In fact, there are many other possible ways to obtain two dwarf and three tall offspring.

To solve this type of question, the binomial expansion equation can be used:

$$P = \frac{n!}{x!\,(n-x)!}\,p^{x}q^{n-x}$$

where

P = the probability that the unordered number of events will occur

n = total number of events

x = number of events in one category (e.g., dwarf plants)

p = individual probability of x

q = individual probability of other category (e.g., tall plants)

Note: In this case, $p + q = 1$. (The symbol ! denotes a *factorial*. n! is the product of all integers from n down to 1. For example, $4! = 4 \times 3 \times 2 \times 1 = 24$. An exception is 0!, which equals 1.)

The use of the binomial expansion equation is illustrated next.

The Cross *Tt × Tt*

The Question What is the probability that two out of five offspring will be dwarf plants?

Step 1 *Calculate the individual probabilities of the dwarf and tall phenotypes.* If we constructed a Punnett square, we would find that the probability of dwarf plants is 1/4 and the probability of tall plants is 3/4:

$$p = 1/4$$
$$q = 3/4$$

Step 2 *Determine the number of events in category x (in this case, dwarf plants) versus the total number of events.* In this example, the number of events in category x is two dwarf plants among a total number of five plants:

$$x = 2$$
$$n = 5$$

Step 3 *Substitute the values for p, q, x, and n in the binomial expansion equation:*

$$P = \frac{n!}{x!\,(n-x)!}\,p^{x}q^{n-x}$$

$$P = \frac{5!}{2!\,(5-2)!}\,(1/4)^{2}(3/4)^{5-2}$$

$$P = \frac{5 \times 4 \times 3 \times 2 \times 1}{(2 \times 1)(3 \times 2 \times 1)}\,(1/16)(27/64)$$

$$P = 0.26 = 26\%$$

Thus the probability is 0.26 that two out of five offspring will be dwarf plants. In other words, 26% of the time we expect a $Tt \times Tt$ cross yielding five offspring to contain two dwarf plants and three tall plants.

The chi square test can be used to test the validity of a genetic hypothesis

We will now look at a different issue in genetic problems, namely **hypothesis testing**. Our goal here is to determine if the data from genetic crosses are consistent with a particular pattern of inheritance. For example, a geneticist may study the inheritance of body color and wing shape in fruit flies over the course of two generations. The following question may be asked about the F_2 generation: "Do the observed numbers of offspring agree with the predicted numbers based on Mendel's laws of segregation and independent assortment?" As we will see in Chapters 3 through 6, not all traits follow a simple Mendelian pattern of inheritance. We will see that some genes do not segregate and independently assort themselves the same way that Mendel's seven traits did in pea plants.

To distinguish between inheritance patterns that obey Mendel's laws and those that do not, a conventional strategy is to make crosses and then quantitatively analyze the offspring. Based on the observed outcome, an experimenter may make a tentative hypothesis. For example, it may seem that the data are obeying Mendel's laws. Hypothesis testing provides an objective, statistical method to evaluate whether or not the observed data really agree with the hypothesis. In other words, we use statistical methods to determine whether the data that have been gathered from crosses are consistent with predictions based on quantitative laws of inheritance.

The rationale behind a statistical approach is to evaluate the **goodness of fit** between the observed data and the data that are predicted from a hypothesis. If the observed and predicted data are very similar, we can conclude that the hypothesis is consistent with observed outcome. In this case, it is reasonable to accept the hypothesis. Alternatively, statistical methods may reveal that there is a poor fit between hypothesis and data. If this occurs, the hypothesis is rejected. Hopefully, the experimenter can subsequently propose an alternative hypothesis that has a reasonable fit with the data.

One commonly used statistical method to determine goodness of fit is the **chi square test** (often written χ^2). We can use the chi square test to analyze population data in which the members of the population fall into different categories. This is the kind of data we have when we evaluate the outcome of genetic crosses, because these usually produce a population of offspring that differ with regard to phenotypes. The general formula for the chi square test is

$$\chi^2 = \sum \frac{(O - E)^2}{E}$$

where

O = observed data in each category
E = expected data in each category based on the experimenter's hypothesis
Σ means to sum this calculation for each category. For example, if the population data fell into two categores, the chi square calculation would be:

$$\chi^2 = \frac{(O_1 - E_1)^2}{E_1} + \frac{(O_2 - E_2)^2}{E_2}$$

We can use the chi square test to determine if a genetic hypothesis is consistent with the observed outcome of a genetic cross. The strategy described here provides a step-by-step outline for applying the chi square testing method. In this problem, the experimenter wants to determine if a dihybrid cross is obeying the laws of Mendel. The experimental organism is *Drosophila melanogaster* (the common fruit-fly), and the two traits affect wing shape and body color. Normal wing shape and curved wing shape are designated by c^+ and c, respectively; normal (gray) body color and ebony body color are designated by e^+ and e, respectively.[2]

The Cross A true-breeding fly with straight wings and a gray body ($c^+c^+\,e^+e^+$) is crossed to a true-breeding fly with curved wings and an ebony body (*cc ee*). The flies of the F_1 generation are then allowed to mate with each other to produce an F_2 generation.

The Outcome

     F_1 generation: all offspring have straight wings and gray bodies

     F_2 generation: 193 straight wings, gray bodies

                       69 straight wings, ebony bodies

                       64 curved wings, gray bodies

                       26 curved wings, ebony bodies

           Total: 352

Step 1 *Propose a hypothesis that allows us to calculate the expected values based on Mendel's laws.* The F_1 generation suggests that the trait of straight wings is dominant to curved wings and gray body coloration is dominant to ebony. Looking at the F_2 generation, it appears that offspring are following a 9:3:3:1 ratio. If so, this is consistent with an independent assortment of the two traits.

Based on these observations, the hypothesis is:

Straight (c^+) is dominant to curved (c), and gray (e^+) is dominant to ebony (e).
The two traits assort independently from generation to generation.

Step 2 *Based on the hypothesis, calculate the expected values of each of the four phenotypes.* We first need to calculate the individual probabilities of each of the four phenotypes. According to our hypothesis, there should be a 9:3:3:1 ratio in the F_2 generation. Therefore, the expected probabilities are

                9/16 = straight wings, gray bodies

                3/16 = straight wings, ebony bodies

                3/16 = curved wings, gray bodies

                1/16 = curved wings, ebony bodies

The observed F_2 generation contained a total of 352 individuals. Our next step is to calculate the expected numbers of each type of offspring when the total equals 352. This can be accomplished by multiplying each individual probability by 352:

     $9/16 \times 352 = 198$ (expected number with straight wings, gray bodies)

     $3/16 \times 352 = 66$ (expected number with straight wings, ebony bodies)

     $3/16 \times 352 = 66$ (expected number with curved wings, gray bodies)

     $1/16 \times 352 = 22$ (expected number with curved wings, ebony bodies)

---

[2]In certain species, such as *Drosophila melanogaster*, the convention is to designate the wild-type allele with a plus sign. Recessive mutant alleles are designated with lowercase letters, and dominant mutant alleles with capital letters.

Step 3 *Apply the chi square formula, using the data for the observed values and the expected values that have been calculated in step 2.* In this case, there are four categories within the population, and thus the sum has four terms:

$$\chi^2 = \frac{(O_1 - E_1)^2}{E_1} + \frac{(O_2 - E_2)^2}{E_2} + \frac{(O_3 - E_3)^2}{E_3} + \frac{(O_4 - E_4)^2}{E_4}$$

$$= \frac{(193 - 198)^2}{198} + \frac{(69 - 66)^2}{66} + \frac{(64 - 66)^2}{66} + \frac{(26 - 22)^2}{22}$$

$$= 0.13 + 0.14 + 0.06 + 0.73 = 1.06$$

Step 4 *Interpret the calculated chi square value. This is done using a chi square table.*

Before interpreting the chi square value we have obtained, we must understand how to use Table 2-1. The probabilities, called **P values**, listed in the chi square table allow us to determine the likelihood that a chi square value is due to random chance alone. For example, let's consider the values listed in row 1. (The meaning of the rows will be explained presently.) Chi square values that are equal to or greater than 0.00393 are expected to occur 95% of the time when a hypothesis is correct. A low chi square value indicates a high probability that the observed deviations could be due to random chance alone. By comparison, chi square values that are equal to or greater than 3.841 are only expected to occur less than 5% of the time due to random sampling error. If a high chi square value is obtained, an experimenter becomes suspicious that the high deviations have occurred because the hypothesis is incorrect. It is a common convention to reject a hypothesis if the chi square value results in a probability that is less than 0.05 (i.e., less than 5%).

In our problem involving flies with straight or curved wings and gray or ebony bodies, we have calculated a chi square value of 1.06. Before we can determine the probability that this deviation would have occurred as a matter of random chance, we must first determine the degrees of freedom (df) in this experiment. The **degrees of freedom** is a measure of the number of categories that are independent of each other. It is typically $n - 1$, where n equals the total number of categories. In the problem above, $n = 4$ (the categories are the phenotypes: straight wings and

TABLE 2–1 Chi Square Values and Probability

Degrees of Freedom	$P = 0.99$	0.95	0.80	.050	0.20	0.05	0.01
1	0.000157	0.00393	0.0642	0.455	1.642	3.841	6.635
2	0.020	0.103	0.446	1.386	3.219	5.991	9.210
3	0.115	0.352	1.005	2.366	4.642	7.815	11.345
4	0.297	0.711	1.649	3.357	5.989	9.488	13.277
5	0.554	1.145	2.343	4.351	7.289	11.070	15.086
6	0.872	1.635	3.070	5.348	8.558	12.592	16.812
7	1.239	2.167	3.822	6.346	9.803	14.067	18.475
8	1.646	2.733	4.594	7.344	11.030	15.507	20.090
9	2.088	3.325	5.380	8.343	12.242	16.919	21.666
10	2.558	3.940	6.179	9.342	13.442	18.307	23.209
15	5.229	7.261	10.307	14.339	19.311	24.996	30.578
20	8.260	10.851	14.578	19.337	25.038	31.410	37.566
25	11.524	14.611	18.940	24.337	30.675	37.652	44.314
30	14.953	18.493	23.364	29.336	36.250	43.773	50.892

From Fisher, R. A., and Yates, F. (1943) *Statistical Tables for Biological, Agricultural, and Medical Research.* Oliver and Boyd, London.

gray body; straight wings and ebony body; curved wings and gray body; and curved wings and ebony body); thus, the degrees of freedom equals 3. We now have sufficient information to interpret our chi square value of 1.06. With df = 3, the chi square value of 1.06 is slightly greater than 1.005 ($P = 0.80$). Values equal to or greater than 1.005 are expected to occur 80% of time based on random chance alone. In this experiment, it is quite probable that deviations between the observed and expected values can be explained by random sampling error. To reject the hypothesis, the chi square would have to be greater than 7.815. Since it was actually far less than this value, we are inclined to accept that the hypothesis is correct.

We must keep in mind that the chi square test does not *prove* that a hypothesis is correct. It is a statistical method for evaluating whether or not the data and hypothesis have a good fit.

CONCEPTUAL SUMMARY

The work of Gregor Mendel has led to great insight into the transmission of traits from parent to offspring. First, his work showed us that the determinants of traits are discrete units that are passed along unaltered from generation to generation. This *particulate mechanism of hereditary* refuted previous theories such as the blending theory of inheritance and pangenesis. In addition, Mendel deduced two fundamental laws that pertain to the transmission of many traits in eukaryotic organisms. The *law of segregation* tells us that two variants for a single trait will segregate during their passage to offspring. This means that the two alleles for a particular gene will segregate from each other during gamete formation, so that a gamete will only contain one copy of an allele. Mendel's *law of independent assortment* says that two different genes assort independently of each other during gamete formation.

An understanding of genetic inheritance allows us to make predictions about future offspring and also to test the validity of genetic hypotheses. Probability is the likelihood that an event will occur in the future. By knowing the genotypes of parents, we can determine the probability of obtaining an offspring with a particular trait. Furthermore, mathematical operations can be used to predict combinations of offspring that may, or may not, occur in a specified order. The *sum rule* is used in genetic problems that involve mutually exclusive events; the *product rule* is used when independent events occur in a specified order; and the *binomial expansion formula* is used when unordered events are involved. In other genetic problems, the goal is to deduce the pattern of inheritance from parents to offspring. This requires an analysis of genetic crosses in order to propose a genetic hypothesis that is consistent with the observed data. A chi square test can be used to evaluate whether or not the hypothesis is likely to be correct. In this approach, the amount of deviation between the observed and expected values is used to determine the goodness of fit between the data and hypothesis. If the deviation is so large that it would be expected to occur less than 5% of the time as a matter of random sampling error, then it is appropriate to reject the hypothesis. If, on the other hand, the chi square value is low, the hypothesis is accepted.

Chapter 2 has begun our foundation in the inheritance pattern of simple traits. The two laws that we examined, segregation and independent assortment, provide us with insight into the transmission of traits from parent to offspring. While they form a cornerstone for our understanding of genetics, we will see in later chapters that not all genes obey these laws. In Chapter 3, we will take a close look at the composition and behavior of chromosomes, particularly during meiosis. As we will see, the segregation and assortment of chromosomes provides a cellular explanation of Mendel's laws.

In this chapter, we have examined the research studies of Gregor Mendel. By making monohybrid crosses and examining the traits of offspring for two generations, Mendel found that the determinants of traits do not blend; rather, they are discrete units that are passed along unaltered from generation to generation. He also quantitatively analyzed the outcome of his crosses and showed that there was a 3:1 phenotypic ratio for the F_2 generation. This observation allowed him to deduce the law of segregation.

In dihybrid crosses, Mendel followed the inheritance pattern of two genes relative to each other. His data supported the idea that two different genes assort independently of each other during gamete formation, producing a 9:3:3:1 phenotypic ratio in the F_2 generation. The law of independent assortment held true for all seven traits that Mendel studied. However, as we will see in Chapter 5, some genes follow a linked pattern of inheritance.

PROBLEM SETS

SOLVED PROBLEMS

1. A heterozygous pea plant that is tall with yellow seeds, *TtYy*, is allowed to self-fertilize. What is the probability that an offspring will be either tall with yellow seeds, tall with green seeds, or dwarf with yellow seeds?

Answer: This problem involves three mutually exclusive events, and so we use the sum rule to solve it.

First, we must calculate the individual probabilities for the three phenotypes. The outcome of the cross can be determined using a Punnett square.

	TY	*Ty*	*tY*	*ty*
TY	*TTYY*	*TTYy*	*TtYY*	*TtYy*
Ty	*TTYy*	*TTyy*	*TtYy*	*Ttyy*
tY	*TtYY*	*TtYy*	*ttYY*	*ttYy*
ty	*TtYy*	*Ttyy*	*ttYy*	*ttyy*

$P_{\text{Tall with yellow seeds}} = 9/(9 + 3 + 3 + 1) = 9/16$

$P_{\text{Tall with green seeds}} = 3/(9 + 3 + 3 + 1) = 3/16$

$P_{\text{Dwarf with yellow seeds}} = 3/(9 + 3 + 3 + 1) = 3/16$

Sum rule: $9/16 + 3/16 + 3/16 = 15/16 = 0.94 = 94\%$

We expect to get one of these three phenotypes 15/16 or 94% of the time.

2. As described in this chapter, a human disease known as cystic fibrosis is inherited as a recessive trait. A normal couple's first child has the disease. What is the probability that their next two children will not have the disease?

Answer: A phenotypically normal couple has already produced an affected child. To be affected, the child must be homozygous for the disease allele and, thus, has inherited one copy from each parent. Therefore, since the parents are unaffected with the disease, we know that both of them must be heterozygous carriers for this recessive disease. With this information, we can calculate the probability that they will produce an unaffected offspring. Using a Punnett square, this couple should produce a ratio of 3 unaffected to 1 affected offspring:

	N	*n*
N	*NN*	*Nn*
n	*Nn*	*nn*

N = normal allele
n = cystic fibrosis allele

The probability of a single unaffected offspring is

$$P_{\text{Unaffected}} = 3/(3 + 1) = 3/4$$

To obtain the probability of getting two unaffected offspring in a row (i.e., in a specified order), we must apply the product rule:

$$3/4 \times 3/4 = 9/16 = 0.56 = 56\%$$

There is a 56% chance that their next two children will be unaffected.

3. A pea plant that is heterozygous for three genes (*TtRrYy*), where *T* = tall, *t* = dwarf, *R* = round seeds, *r* = wrinkled seeds, *Y* = yellow seeds, and *y* = green seeds is self-fertilized. What are the predicted phenotypes of the offspring?

Answer: One could solve this problem by constructing a large Punnett square and filling in the boxes. However, in this case, there are 8 possible male gametes and 8 possible female gametes: *TRY, TRy, TrY, tRY, trY, Try, tRy,* and *try*. It would become rather tiresome to construct and fill in this Punnett square, which would contain 64 boxes. As an alternative, we can consider each gene separately and then algebraically combine them by multiplying together the expected phenotypic outcomes for each gene. In the cross *TtRrYy* × *TtRrYy*, the following Punnett squares can be made for each gene:

	T	*t*
T	TT	Tt
t	Tt	tt

3 tall : 1 dwarf

	R	*r*
R	RR	Rr
r	Rr	rr

3 round : 1 wrinkled

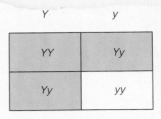

	Y	*y*
Y	YY	Yy
Y	Yy	yy

3 yellow : 1 green

To determine the phenotypic outcome of this trihybrid cross, we can simply multiply these three combinations together:

(3 tall + 1 dwarf)(3 round + 1 wrinkled)(3 yellow + 1 green)

This multiplication operation can be done in a stepwise manner. First, multiply (3 tall + 1 dwarf) by (3 round + 1 wrinkled):

(3 tall + 1 dwarf)(3 round + 1 wrinkled)= 9 tall, round + 3 dwarf, round + 3 tall, wrinkled + 1 dwarf, wrinkled

Next, multiply this product by (3 yellow + 1 green):

(9 tall, round, + 3 dwarf, round + 3 tall, wrinkled + 1 dwarf, wrinkled)(3 yellow + 1 green) = 27 tall, round, yellow + 9 tall, round, green + 9 dwarf, round, yellow + 3 dwarf, round, green + 9 tall, wrinkled, yellow + 3 tall, wrinkled, green + 3 dwarf, wrinkled, yellow + 1 dwarf, wrinkled, green

Even though the multiplication steps are also somewhat tedious, this approach is significantly easier than making a Punnett square with 64 boxes, filling them in, deducing each phenotype, and then adding them up!

CONCEPTUAL QUESTIONS

1. Why did Mendel's work refute the blending theory of inheritance?

2. What is the difference between cross-hybridization and self-fertilization?

3. Describe the difference between genotype and phenotype. Give three examples. Is it possible for two individuals to have the same phenotype but different genotypes? Is it possible for two organisms to have the same genotypes and different phenotypes?

4. With regard to alleles, what is a true-breeding organism?

5. How can you determine whether an organism is heterozygous or homozygous for a dominant trait?

6. In your own words, describe what Mendel's law of segregation means. Do not use the word segregation in your answer.

7. Based on genes in pea plants that we have considered in this chapter, which statement(s) is not correct?

    A. The gene for tall plants is an allele of the gene for dwarf plants.

    B. The gene for tall plants is an allele of the gene for purple flowers.

    C. The alleles for tall plants and purple flowers are dominant.

8. In a cross between a heterozygous tall pea plant and a dwarf plant, predict the ratios of the offspring's genotypes and phenotypes.

9. Do you know the genotype of an individual with a recessive trait and/or a dominant trait? Explain your answer.

10. A cross is made between a pea plant that has constricted pods (a recessive trait) and is heterozygous for seed color (yellow is dominant to green) and a plant that is heterozygous for both pod texture and seed color. Construct a Punnett square that depicts this cross. What are the predicted outcomes of genotypes and phenotypes of the offspring?

11. A dwarf pea plant that is heterozygous with regard to seed color (yellow is dominant to green) is allowed to self-fertilize. What are the predicted outcomes of genotypes and phenotypes of the offspring?

12. A tall pea plant with axial flowers is crossed to a dwarf plant with terminal flowers. Tall plants and axial flowers are dominant traits.

The following offspring were obtained: 27 tall, axial flowers; 23 tall, terminal flowers; 28 dwarf, axial flowers; and 25 dwarf, terminal flowers. What are the genotypes of the parents?

13. Describe the significance of nonparentals with regard to the law of independent assortment. In other words, explain how the appearance of nonparentals refutes a linkage hypothesis.

14. For the following pedigrees, describe what you think is the most likely inheritance pattern (dominant versus recessive). Explain your reasoning. Red symbols indicate affected individuals.

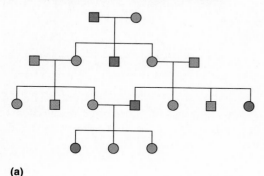

(a)

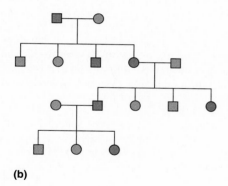

(b)

15. In humans, the allele for brown eye color (B) is dominant to blue eye color (b). If two heterozygous parents produce children, what are the following probabilities?

A. The first two children have blue eyes.
B. A total of four children, two with blue eyes and the other two with brown eyes.
C. The first child has blue eyes, and the next two have brown eyes.

16. Albinism is a recessive human trait. If a normal couple produces an albino child, what is the probability that their next child will be albino?

17. A true-breeding tall plant is crossed to a dwarf plant. Tallness is a dominant trait. The F_1 offspring are allowed to self-fertilize. What are the following probabilities for the F_2 generation:

A. The first plant is dwarf.
B. The first plant is dwarf or tall.
C. The first three plants are tall.
D. For any seven plants, three are tall and four are dwarf.
E. The first plant is tall, and then among the next four, two are tall and the other two are dwarf.

18. A true-breeding pea plant with round and green seeds is crossed to a true-breeding plant with wrinkled and yellow seeds. Round and yellow seeds are the dominant traits. The F_1 plants are allowed to self-fertilize. What are the following probabilities for the F_2 generation?

A. An F_2 plant with wrinkled, yellow seeds.
B. Three out of three F_2 plants with round, yellow seeds.
C. Five F_2 plants in the following order: two have round, yellow seeds, one has round, green seeds, and two have wrinkled, green seeds.
D. An F_2 plant will not have round, yellow seeds.

19. A true-breeding tall pea plant is crossed to a true-breeding dwarf plant. What is the probability that an F_1 offspring will be true breeding? What is the probability that an offspring will be a true-breeding tall plant?

20. What are the expected phenotypic ratios from the following cross: $TtRryyAa \times TtrrYYAa$, where T = tall, t = dwarf, R = round, r = wrinkled, Y = yellow, y = green, A = axial, a = terminal; T, R, Y, and A are dominant alleles. Note: See solved problem 3 for help in answering this problem.

21. A cross is made between two strains of plants that are agriculturally important. One strain is disease resistant but herbicide sensitive; the other strain is disease sensitive but herbicide resistant. A plant breeder crosses the two plants and then allows the F_1 generation to self-fertilize. The following results are obtained:

F_1 generation:	all offspring are disease sensitive and herbicide resistant
F_2 generation:	157 disease sensitive, herbicide resistant
	57 disease sensitive, herbicide sensitive
	54 disease resistant, herbicide resistant
	20 disease resistant, herbicide sensitive
Total:	288

Formulate a hypothesis that you think is consistent with the observed data. Test the goodness of fit between the data and your hypothesis using a chi square test. Explain what the chi square results mean.

22. A cross is made between a plant that has blue flowers and purple seeds to a plant with white flowers and green seeds. The following results were obtained:

F_1 generation:	all offspring have blue flowers with purple seeds
F_2 generation:	103 blue flowers, purple seeds
	49 blue flowers, green seeds
	44 yellow flowers, purple seeds
	104 white flowers, green seeds
Total:	300

Start with the hypothesis that blue flowers and purple seeds are dominant traits and that the two genes are assorting independently. Calculate a chi square value. What does this value mean with regard to your hypothesis? If you decide to reject your hypothesis, which aspect of the hypothesis do you think is incorrect (i.e., blue flowers and purple seeds are dominant traits, or the idea that the two genes are assorting independently)?

EXPERIMENTAL QUESTIONS

1. Describe three advantages of using pea plants as an experimental organism.

2. Explain the technical differences between a cross-hybridization experiment and a self-fertilization experiment.

3. How long did it take Mendel to complete Experiment 2A?

4. For all seven traits described in Experiment 2A, Mendel allowed the F_2 plants to self-fertilize. He found that when F_2 plants with recessive traits were crossed to each other, they always bred true. However, when F_2 plants with dominant traits were crossed, some bred true while others did not. A summary of Mendel's results is shown in the table on the right.

When considering the data in this table, keep in mind that it describes the characteristics of the F_2 generation parents that had displayed a dominant phenotype. These data were deduced by analyzing the outcome of the F_3 generation. Based on Mendel's laws, explain the 1:2 ratio obtained in these data.

The Ratio of True-Breeding and Non-true-Breeding Parents of the F_2 Generation

F_2 Parents	True Breeding	Non-true Breeding	Ratio
Round	193	372	1 : 1.93
Yellow	166	353	1 : 2.13
Gray	36	64	1 : 1.78
Smooth	29	71	1 : 2.45
Green	40	60	1 : 1.5
Axial	33	67	1 : 2.03
Tall	28	72	1 : 2.57
Total:	525	1059	1 : 2.02

5. From the point of view of crosses and data collection, what are the experimental differences between a monohybrid and a dihybrid cross?

QUESTIONS FOR STUDENT DISCUSSION/COLLABORATION

1. Consider the following cross in pea plants: $TtRryyAa \times TtrrYyAa$, where T = tall, t = dwarf, R = round, r = wrinkled, Y = yellow, y = green, A = axial, a = terminal. What is the expected phenotypic outcome of this cross? Have one group of students solve this problem by making one big Punnett square, and have another group solve it by making four single-gene Punnett squares and using multiplication. Time each other to see who gets done first.

2. A cross is made between two pea plants, $TtAa$ and $Ttaa$, where T = tall, t = dwarf, A = axial, and a = terminal. What is the probability that the first three offspring will be tall with axial flowers or dwarf with terminal flowers, and the fourth offspring will be tall with axial flowers? Discuss what operation(s) (sum rule, product rule, and/or binomial expansion) you used to solve the problem, and in what order they were used.

3. Consider the following tetrahybrid cross, $TtRryyAa \times TtRRYyaa$, where T = tall, t = dwarf, R = round, r = wrinkled, Y = yellow, y = green, A = axial, a = terminal. What is the probability that the first three plants will have axial flowers? What is the easiest way to solve this problem?

CHAPTER 3

REPRODUCTION AND CHROMOSOME TRANSMISSION

In Chapter 2, we considered some patterns of inheritance that explain the passage of traits from parent to offspring. This chapter will survey reproduction at the cellular level, and we will pay close attention to the inheritance of chromosomes. It's rather exciting that an examination of chromosomes at the microscopic level provides us with insights to understand the inheritance patterns of individuals' traits. To appreciate this relationship, we will first consider how cells distribute their chromosomes during the process of cell division. We will see that in bacteria and certain unicellular eukaryotes, simple cell division provides a way to reproduce asexually. Then we will explore a form of cell division that occurs during gamete formation. By closely examining the process of gamete formation, we will see how the transmission of chromosomes accounts for the inheritance patterns of traits that were observed by Mendel.

GENERAL FEATURES OF CHROMOSOMES

CELLULAR DIVISION

SEXUAL REPRODUCTION

THE CHROMOSOME THEORY OF INHERITANCE

GENERAL FEATURES OF CHROMOSOMES

The **chromosomes** are the structures within living cells that contain the genetic material. In other words, genes are physically located within the chromosomes. Biochemically, chromosomes contain a very long segment of DNA, which is the genetic material, and proteins, which are bound to the DNA and provide it with an organized structure. In this chapter, we will study the relationship between the patterns of gene transmission that we considered in Chapter 2 and the cellular mechanics of chromosome transmission. In particular, we will examine, in a general way, how chromosomes are copied and sorted into newly made cells. Later chapters, particularly Chapters 8, 10, and 11, will examine the structural features of chromosomes in greater detail.

Before we begin a description of chromosome transmission, we need to consider the distinctive cellular differences between bacteria and eukaryotic species. Bacteria are referred to as **prokaryotes**, from the Greek meaning *prenucleus*, because their chromosomes are not contained within a separate nucleus of the cell. Prokaryotes usually contain a single type of circular chromosome in the cytoplasm of the cell. In contrast, **eukaryotes**, from the Greek meaning *true nucleus*, are more complex organisms whose cells contain nuclei that are bounded by cell membranes. Some simple eukaryotic species are single-celled protists and yeast; more complex multicellular species include fungi, plants, and animals. As we will examine next, eukaryotic species contain genetic material that comes in sets of linear chromosomes, which are found within the nucleus of the cell.

Eukaryotic chromosomes are examined cytologically to yield a karyotype

We have gained our understanding of inheritance, to a significant extent, by observing the behavior of chromosomes under the microscope. **Cytogenetics** is the field of genetics that involves the microscopic examination of chromosomes. The most basic observation that a cytogeneticist can make is to examine the chromosomal composition of a particular cell. For eukaryotic species, this is usually accomplished by observing the chromosomes as they are found in actively dividing cells. As discussed later in this chapter, chromosomes become more condensed (i.e., thicker) when a cell is preparing to divide. This makes it much easier to view them with a light microscope.

Figure 3-1 shows the general procedure for viewing chromosomes in a eukaryotic cell. In this example, the cells were obtained from a sample of human blood; more specifically, the chromosomes within lymphocytes (a type of white blood cell) were examined. Blood cells are a type of **somatic cell**. This term refers to any cell of the body that is not a gamete. The gametes (i.e., sperm and egg cells) are types of **germ cells**. As we will see, the chromosomal compositions of somatic cells and germ cells are fundamentally different.

After the blood cells have been removed from the body, they are treated with drugs that stimulate them to divide. In step 1 of Figure 3-1, these actively dividing cells are centrifuged to concentrate them. In step 2, the concentrated preparation of lymphocytes is mixed with a hypotonic solution that makes the cells swell. This swelling causes the chromosomes to spread out within the cell and thereby makes it easier to see each individual chromosome. In step 3, the cells have been treated with a fixative, which chemically "freezes" the cells so that the chromosomes will no longer move around. The cells are then treated with a chemical dye that binds to the chromosomes and stains them. This greatly enhances the visualization of the chromosomes. The cells are then placed on a slide in step 4 and viewed with a light microscope.

In a cytogenetics laboratory, the microscopes are equipped with a camera, which can photograph the chromosomes. In recent years, advances in technology have allowed cytogeneticists to scan microscopic images into a computer. On a computer screen, the chromosomes can then be organized in a standard way, usually from largest to smallest. As shown in step 5, the human chromosomes have been lined up, and a number is given to designate each type of chromosome. An exception are the sex chromosomes, which are designated with the letters X and Y. We will consider the sex chromosomes later in this chapter. A photographic representation of all the chromosomes within a cell, as in Figure 3-1, is called a **karyotype**. A karyotype reveals how many chromosomes are found within an actively dividing somatic cell.

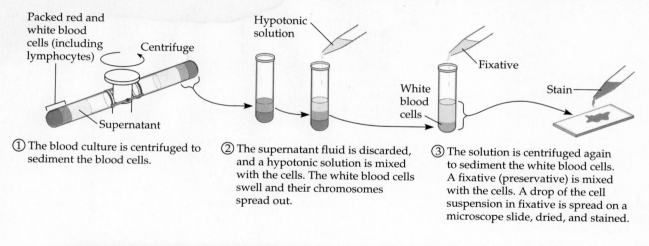

Packed red and white blood cells (including lymphocytes)

Centrifuge

Supernatant

Hypotonic solution

Fixative

White blood cells

Stain

① The blood culture is centrifuged to sediment the blood cells.

② The supernatant fluid is discarded, and a hypotonic solution is mixed with the cells. The white blood cells swell and their chromosomes spread out.

③ The solution is centrifuged again to sediment the white blood cells. A fixative (preservative) is mixed with the cells. A drop of the cell suspension in fixative is spread on a microscope slide, dried, and stained.

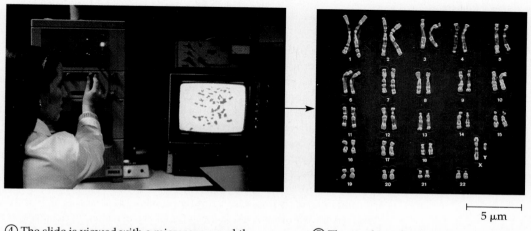

5 μm

④ The slide is viewed with a microscope, and the chromosomes are photographed. The photograph is entered into a computer, and the chromosomes are electronically rearranged into pairs according to size and shape.

⑤ The resulting display is the karyotype.

FIGURE **3-1**

The procedure for making a karyotype.

Eukaryotic chromosomes are inherited in sets

As we have just seen, cytogeneticists can determine the chromosomal composition of cells via microscopy. As seen in the karyotype of Figure 3-1 and examined in Figure 3-2, many eukaryotes are **diploid**, which means that each type of chromosome is a member of a pair. Therefore, diploid cells contain two sets of chromosomes. In Figure 3-1, the human karyotype contains two sets of chromosomes with 23 chromosomes per set (46 total). Other diploid species, however, can have different numbers of chromosomes. For example, the dog has 78 chromosomes (39 per set), the fruit fly has 8 chromosomes (4 per set), and the tomato has 24 chromosomes (12 per set).

When a species is diploid, the members of a pair of chromosomes are referred to as **homologues**. For a diploid organism, each type of chromosome is found in a homologous pair. As shown in Figure 3-1, for example, there are two copies of chromosome 1, two copies of chromosome 2, and so forth. Within each pair, the chromosome on the left is a homologue to the one on the right (and vice versa). Homologous chromosomes are very similar to each other. Note that each of the two

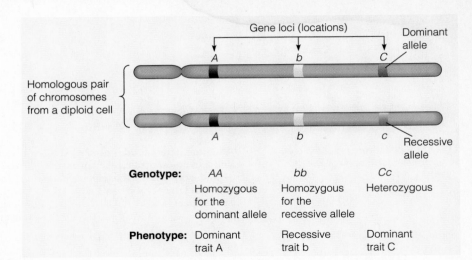

FIGURE **3-2**

A comparison of homologous chromosomes. Each pair of homologous chromosomes carries the same types of genes, but as shown here the alleles may be different.

chromosomes in a homologous pair are approximately the same size (except for the sex chromosomes). Furthermore, homologous chromosomes contain a similar composition of genetic material. If a particular gene (designated with a letter) is found on one copy of a chromosome, it is also found on the other homologue.

However, as illustrated in Figure 3-2, homologous chromosomes are not usually identical to each other. Even though two homologues carry copies of the same genes, the two copies of a gene may be different alleles. In the figure, the two copies of gene *A* and gene *B* are identical on the two homologues. Gene *A* is homozygous for the dominant allele; gene *B* is homozygous for the recessive allele. The alleles of gene *C*, however, are not identical on the two homologues. A dominant allele is found on one chromosome, a recessive allele on the other. The individual with these chromosomes is heterozygous for this gene. The physical *location* of a gene is called its **locus** (plural = **loci**). As seen in Figure 3-2, for example, the locus of gene *C* is toward one end of this chromosome while the locus of gene *B* is more in the middle.

CELLULAR DIVISION

Now that we have an appreciation for the chromosomal composition of living cells, we can examine how chromosomes are copied and transmitted when cells divide. One purpose of cell division is for **asexual reproduction**. This is the way that some unicellular organisms produce new individuals. In this process, a preexisting cell divides to produce two new cells. By convention, the original cell is usually called the *mother cell*, and the new cells are the two *daughter cells*. When species are unicellular, the mother cell is judged to be one individual and the two daughter cells are two new separate individual organisms. Asexual reproduction is how bacterial cells proliferate. In addition, certain unicellular eukaryotes, such as the amoeba and baker's yeast (*Saccharomyces cerevisiae*), can reproduce asexually.

A second important reason for cell division is to produce multicellularity. Multicellular species such as fungi, plants, and animals are derived from a single cell (e.g., a fertilized egg) that has undergone repeated cellular divisions. Humans, for example, begin as a single fertilized egg; repeated cellular divisions produce an adult with several trillion cells. As you might imagine, it is critical to have the precise transmission of chromosomes during every cell division, so that all the cells of the body receive the correct amount of genetic material.

In this section, we will consider how the process of cell division requires the duplication, organization, and sorting of the chromosomes. In bacteria, which have a single circular chromosome, the sorting process is less complicated. Prior to cell division, bacteria duplicate their circular chromosome; they then distribute a copy into each of the two daughter cells. This process, known as binary fission, will be described next. By comparison, eukaryotes have multiple numbers of chromosomes that occur as sets. This added complexity requires a more complicated sorting process to ensure that each newly made cell will receive the correct number and types of chromosomes. As described later in this section, a mechanism known as mitosis entails the organization and sorting of eukaryotic chromosomes during cell division.

Prokaryotes reproduce asexually by binary fission

Bacterial species are unicellular, although individual bacteria may associate with each other to form pairs, chains, or even clumps. Bacteria contain a cell wall, a cell membrane, and cytoplasm. The cell wall provides a rigid protective surface for the bacterium, and the cell membrane is a permeable barrier that regulates the passage of small molecules into and out of the cell. The genetic and metabolic activities of the cell occur within the cytoplasm. Unlike eukaryotes, which have their chromosomes in a separate nucleus, the circular chromosomes of bacteria are in direct contact with the cytoplasm. In Chapter 10, we will consider the molecular structure of bacterial chromosomes in greater detail.

The capacity of bacteria to divide is really quite astounding. Some species, such as *Escherichia coli* (a common bacterium of the intestine), can divide every 20 to 30 minutes. Prior to cell division, bacterial cells copy, or **replicate**, their chromosomal DNA (a topic that will be discussed in Chapter 11). This produces two identical copies of the genetic material as shown at the top of Figure 3-3. Following DNA replication, bacterial cells divide into two daughter cells by a process known as **binary fission**. During this event, the two daughter cells become divided by the formation of a *septum*. As seen in the figure, each cell receives a copy of the chromosomal genetic material. Except when rare mutations occur, the daughter cells are usually genetically identical, because they contain exact copies of the genetic material from the mother cell.

Binary fission is an asexual form of reproduction, since it does not involve genetic contributions from two different gametes. On occasion, bacteria can exchange small pieces of genetic material with each other. We will consider some interesting mechanisms of genetic exchange in Chapter 6.

The transmission of chromosomes during the division of eukaryotic cells requires a sorting process known as mitosis

The goal of eukaryotic cell division is to produce two daughter cells that have the same number and types of chromosomes as the original mother cell. This requires a replication and sorting process that is more complicated than simple binary fission. Eukaryotic cells that are destined to divide progress through a series of stages known as the **cell cycle** (see Figure 3-4). The phases are G for growth, S for synthesis (of the genetic material), and M for mitosis. There are two G phases, G_1 and G_2. Most of a cell's lifetime and metabolic activity is spent during the G_1, S, and G_2 phases, which are collectively known as **interphase**.

During the G_1 phase, a cell prepares to divide. Depending on the cell type and the conditions that a cell encounters, it may accumulate molecular changes that will cause it to progress through the rest of the cell cycle. When this occurs, cell biologists say that a cell has reached a *restriction point* and is committed on a pathway to cell division. Once past the restriction point, the cell will then advance to the S phase, during which the chromosomes are replicated. After replication, the two copies of chromosome are called **sister chromatids**; the two sister chromatids are attached to each other at a structure known as the **centromere** (see

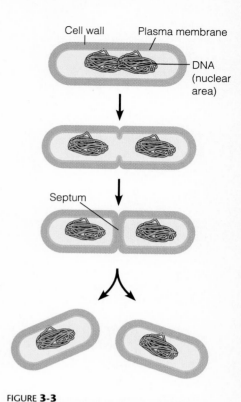

FIGURE **3-3**

Binary fission produces two bacterial cells. Prior to division, the chromosome replicates to produce two identical copies. These two copies segregate from each other, with one copy going to each daughter cell.

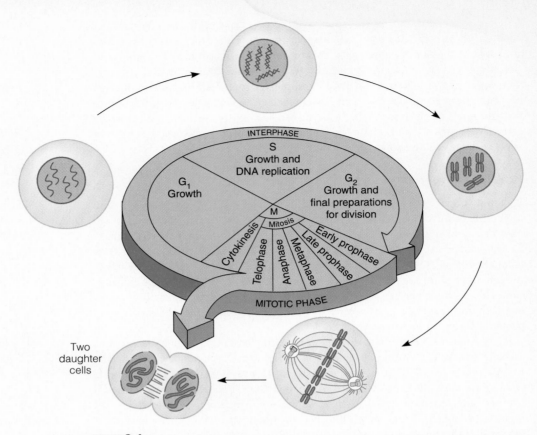

FIGURE **3-4**

The eukaryotic cell cycle. Cells that are destined to divide progress through a series of stages, denoted G_1, S, G_2, and M phases (mitosis). This diagram shows the progression of a cell through mitosis to produce two daughter cells. The original diploid cell had two pairs of chromosomes, for a total of four individual chromosomes. During S phase, these have replicated to yield eight sister chromatids. After mitosis is complete, there are two daughter cells each containing four chromosomes.

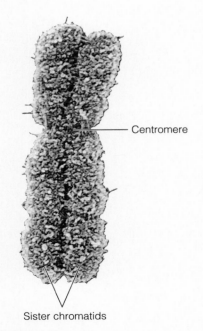

Centromere

Sister chromatids

FIGURE **3-5**

A photomicrograph of a metaphase chromosome.

Figure 3-5). The functional role of the centromere will be described later. When the S phase is completed, a cell actually has twice as many sister chromatids compared to the number of chromosomes in the G_1 phase. For example, a human cell in the G_1 phase has 46 separate chromosomes whereas in G_2 it would have 92 chromatids or 46 pairs of sister chromatids.

During the G_2 phase, the cell accumulates the materials that are necessary for nuclear and cell division. It then progresses into the M phase of the cell cycle when **mitosis** occurs. The primary purpose of mitosis is to distribute the replicated chromosomes, dividing one cell nucleus into two nuclei, so that each daughter cell will receive the same complement of chromosomes. For example, a human cell in the G_2 phase has 92 sister chromatids, which are found in 46 pairs. It is during mitosis that these pairs of chromatids are separated and sorted so that each daughter cell will receive 46 chromosomes. Mitosis is the name given to this sorting process.

Mitosis was first observed microscopically during the 1870s. The most careful and influential studies were carried out by the German biologist Walter Flemming, who coined the term mitosis (from the Greek *mitos*, meaning *thread*). He studied the large, transparent epithelial cells of salamander larvae and noticed that chromosomes are constructed of two "threads" that are double in appearance along their length. These double threads are divided and move apart, one going to each of the two daughter nuclei. By this mechanism, Flemming pointed out that the two daughter cells receive an identical group of threads, of quantity comparable to the number of threads in the parent cell.

In Figure 3-6, the process of mitosis is shown for a diploid animal cell. In this simplified diagram, the original mother cell contained four chromosomes; it was diploid ($2n$) and contained two chromosomes per set ($n = 2$). One set is shown in blue, and the homologous set is shown in red. Mitosis is subdivided into phases known as prophase, metaphase, anaphase, and telophase. At the start of mitosis, in **prophase**, the chromosomes have already replicated to produce eight sister chromatids. The organization and distribution of chromosomes during mitosis is brought about by the construction of a *spindle apparatus*. This apparatus is organized from two structures called the *centrosomes*. Each centrosome organizes the construction of protein fibers called *microtubules* (MTs). As mitosis progresses, the

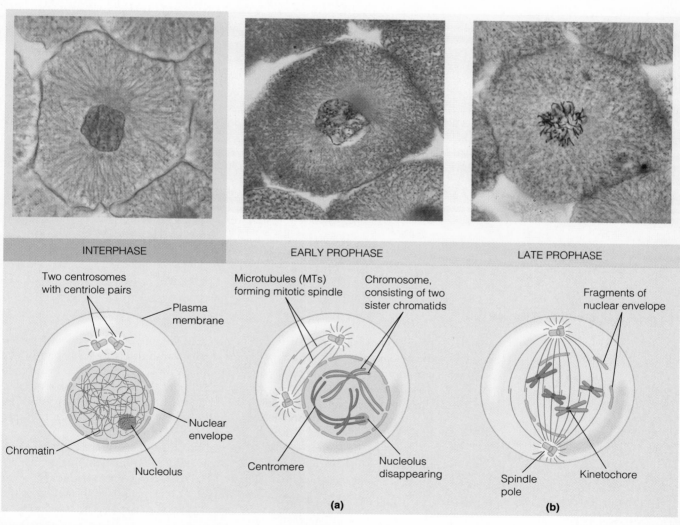

FIGURE **3-6**

The process of mitosis in an animal cell. The top panels illustrate the cells of a fish embryo progressing through mitosis. The bottom panels are schematic drawings that emphasize the sorting and separation of the chromosomes. In this case, the original diploid cell had four chromosomes (two in each set). At the start of mitosis, these have already replicated into eight sister chromatids. The ultimate result is two daughter cells each containing four chromosomes.

(continued)

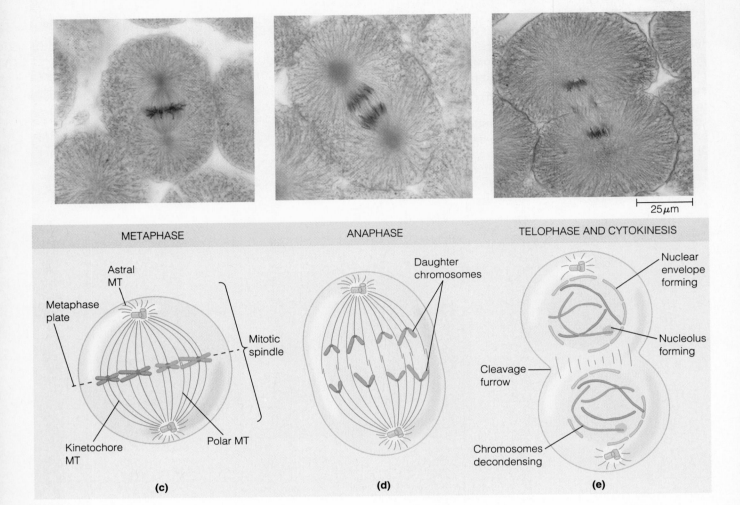

25μm

| METAPHASE | ANAPHASE | TELOPHASE AND CYTOKINESIS |

Astral MT

Metaphase plate

Mitotic spindle

Kinetochore MT

Polar MT

(c)

Daughter chromosomes

(d)

Nuclear envelope forming

Nucleolus forming

Cleavage furrow

Chromosomes decondensing

(e)

FIGURE **3-6**

The process of mitosis in an animal cell.
(continued)

centrosomes will move apart and demarcate two poles, one within each of the future daughter cells. The function of the microtubules is to guide half of the chromosomes to each of these poles. As prophase proceeds, the nuclear membrane dissociates into small vesicles (see Figure 3-6b). At the same time, the replicated chromosomes condense into more compact structures that are readily visible by light microscopy. The condensation is necessary to organize the chromosomes during division.

After condensation is finished, the sorting process begins. The compact chromosomes (which are attached to each other as sister chromatids) line up along a plane called the *metaphase plate* (Figure 3-6c). As its name suggests, this alignment occurs during late prophase and **metaphase**. At metaphase, the formation of the spindle apparatus is complete. Each pair of sister chromatids is attached to both poles by a type of spindle fiber known as a *kinetochore microtubule*. Kinetochore microtubules emanate from each pole and are connected to a structure on the centromere known as the *kinetochore*. Figure 3-7 illustrates the connection between the kinetochore microtubules and the kinetochore. During metaphase of mitosis, the pairs of sister chromatids have become organized into a single row along the metaphase plate. When this organization process is complete, the chromosomes can then be equally distributed into two daughter cells.

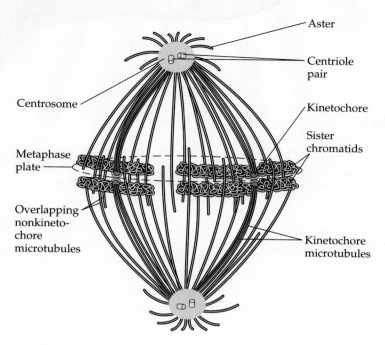

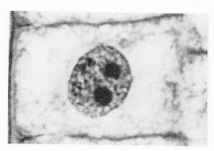

(a) Early prophase

(b) Late prophase

FIGURE **3-7**

Attachment of the kinetochore microtubules to a replicated chromosome during mitosis. The kinetochore is a collection of proteins that are attached to the centromere. The kinetochore microtubules are attached to the kinetochore.

(c) Metaphase

The next step in the sorting process occurs during **anaphase** (Figure 3-6d). At this stage, the connection holding each pair of sister chromatids together is broken. After this occurs, each separate chromosome is only linked to one of the two poles. As anaphase proceeds, the chromosomes move toward the one pole to which they are attached. This involves a shortening of the kinetochore microtubules. In addition, the two poles themselves tend to move farther away from each other due to a lengthening of other types of spindle fibers known as polar microtubules.

During **telophase**, the chromosomes have reached their respective poles and decondense. The nuclear membrane now re-forms to produce two separate nuclei. In Figure 3-6e, this has produced two nuclei that contain four chromosomes each. In most cases, mitosis is quickly followed by **cytokinesis**, in which the two nuclei are segregated into separate daughter cells. In animal cells, cytokinesis involves the formation of a *cleavage furrow*, which constricts to separate the cells (Figure 3-6e). In plants, the two daughter cells are separated by the formation of a cell plate (Figure 3-8).

Mitosis and cytokinesis ultimately produce two daughter cells having the same number of chromosomes as the mother cell had during the G_1 phase of its cell cycle. Barring rare mutations, the two daughter cells are genetically identical to each other and to the mother cell from which they were derived. Thus, the critical consequence of this sorting process is to ensure genetic consistency from one cell to the next. The development of multicellularity relies on the repeated process of mitosis and cytokinesis. For diploid organisms that are multicellular, most of the somatic cells are diploid and genetically identical to each other. As described next, the gametes, which are germ cells, are distinctly different.

(d) Anaphase

FIGURE **3-8**

Mitosis in a plant cell. These photomicrographs show mitosis in cells of an onion root. Cytokinesis occurs via the formation of a cell plate between the two daughter cells.

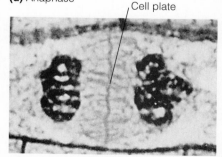

Cell plate

(e) Telophase

25 µm

SEXUAL REPRODUCTION

In the previous section, we considered how a cell can divide to produce two new cells with identical complements of genetic material. Now we will turn our attention to sexual reproduction, the most common way for eukaryotic organisms to produce progeny. During sexual reproduction, parents make gametes that fuse with each other in the process of **fertilization** to begin the life of a new organism. Gametes are highly specialized cells designed to locate each other and provide nutrients to the developing embryo.

Some simple eukaryotic species are **isogamous**, which means that the gametes they produce are morphologically similar. Examples of isogamous species include many species of fungi and algae. Most eukaryotic species, however, are **heterogamous**—they produce two morphologically different types of gametes. Male gametes, **sperm cells**, are relatively small and usually travel far distances to reach the female gamete. The mobility of the male gamete is an important cellular characteristic making it likely to come in close proximity to the female gamete. The sperm of animals often contain flagella that enable them to swim. In plants, the sperm cells are contained within pollen grains. Pollen is a small mobile structure that can be carried by the wind or on the feet or hairs of insects. By comparison, the female gamete, known as the **egg cell** or **ovum**, is usually very large and nonmotile. In animal species, the egg stores a large amount of nutrients that will be available to the growing embryo.

Gametes are **haploid**, which means that they contain half the number of chromosomes as somatic cells. For a species that is diploid, a haploid gamete contains a single set of chromosomes. In this case, the haploid gametes are $1n$ while the somatic cells are $2n$. For example, a diploid human cell contains 46 chromosomes, but a gamete (sperm or egg cell) contains only 23 chromosomes, one from each of the 23 pairs.

In many species, including humans, haploid gametes are descended from germ cells that are originally diploid. To produce haploid gametes, the chromosomes must be correctly sorted and distributed in a way that reduces the chromosome number to half its original value. In the case of human gametes, for example, each one must receive half of the total number of chromosomes, but not just any 23 chromosomes will do. A gamete must receive one chromosome from each of the 23 pairs. The sorting process that results in the production of haploid gametes from a diploid cell is called **meiosis**. In this section, we will examine the cellular events of gamete development and how the stages of meiosis lead to the formation of gametes with a haploid complement of chromosomes.

Meiosis produces gametes that are haploid

In 1883, the Belgian biologist Edouard van Beneden was the first to observe that gamete formation produces cells that contain half the number of chromosomes. In many ways, this process bears striking similarities to mitosis. Like mitosis, meiosis begins after a cell has progressed through the G_1, S, and G_2 phases of the cell cycle. However, after G_2, meiosis involves two successive divisions rather than one (as in mitosis). In this section, we will emphasize the differences in meiosis that reduce the amount of genetic material.

The stages of meiosis are described in Figures 3-9 and 3-10. These simplified diagrams depict a diploid cell ($2n$) that contains a total of four chromosomes (as in our look at mitosis). Prior to meiosis, the chromosomes are replicated in S phase to produce pairs of sister chromatids. This single replication event is then followed by two sequential cell divisions. These two divisions are called *meiosis I* and *meiosis II*. Like mitosis, each of these are subdivided into prophase, metaphase, anaphase, and telophase. Figure 3-9 emphasizes some of the important events that occur during prophase I, which is further subdivided into periods known as leptotene, zygotene, pachytene, diplotene, and diakinesis.

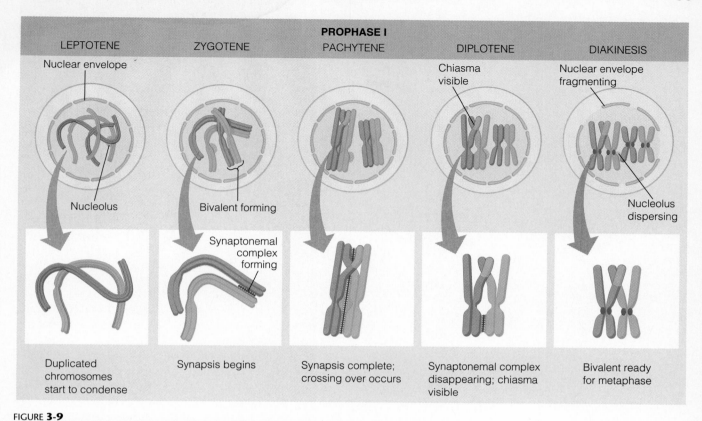

FIGURE **3-9**

The events that occur during prophase of meiosis I.

During *leptotene*, the chromosomes begin to condense and become visible with a light microscope. At first, they appear as threadlike structures, and the individual sister chromatids are not readily discernable. In Figure 3-9 (leptotene and zygotene stages), you must look closely to notice that each structure is actually a pair of sister chromatids.

Unlike mitosis, the *zygotene* stage of prophase I involves a recognition process known as **synapsis**. The homologous chromosomes recognize each other and then align themselves along their entire lengths. This recognition process is described at the molecular level in Chapter 18. A group of cellular proteins known as the *synaptonemal complex* facilitates the synapsing of homologous chromosomes. The associated chromosomes are known as *bivalents*. Each bivalent contains two pairs of sister chromatids, or a total of four sister chromatids.

In *pachytene*, when synapsis is complete, an event known as **crossing over** commonly occurs. Crossing over involves a physical exchange of chromosome pieces. In Figure 3-9, crossing over is occurring at a single site between two of the (larger) homologous chromosomes. This site of crossing over is called a **chiasma** (plural, chiasmata), because it physically resembles the Greek letter chi, χ. We will consider the genetic consequences of crossing over in Chapter 5 and the molecular process of crossing over in Chapter 18.

During the *diplotene* stage, the synaptonemal complex begins to disappear. Microscopically, it becomes easier to see that a bivalent is actually composed of four sister chromatids. This association between four sister chromatids is also called a **tetrad** (from the prefix *tetra-*, meaning *four*). In the last stage of prophase I, *diakinesis*, the synaptonemal complex completely disappears.

Figure 3-9 was meant to emphasize the pairing and crossing over that occurs during prophase I. In Figure 3-10, we will turn our attention to the interactions between

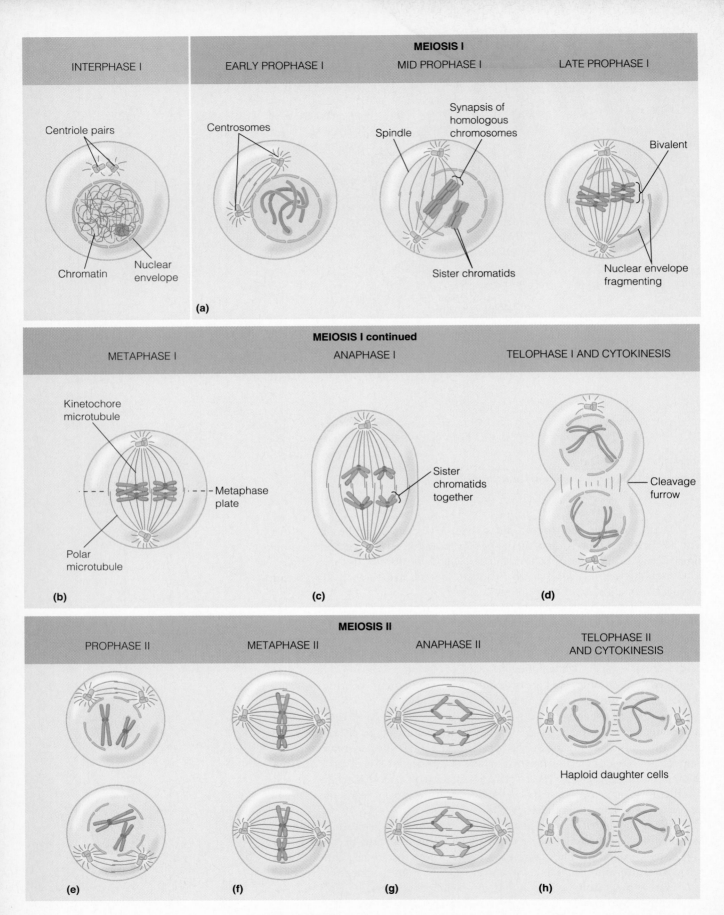

FIGURE **3-10**

The stages of meiosis in an animal cell. See text for details.

the chromosomes and spindle apparatus, and how these interactions promote chromosome sorting. In early prophase I, the spindle apparatus is complete, and the chromosomes are attached via kinetochore microtubules (Figure 3-10a).

At metaphase I (Figure 3-10b), the tetrads are organized along a plane called the metaphase plate. This pattern of alignment is different from that observed during mitosis (refer back to Figure 3-6). Before we consider the rest of meiosis I, a particularly critical feature for you to appreciate is how the tetrads are aligned along the metaphase plate. During metaphase I, the alignment of chromosomes is in a double row rather than a single row (as in mitosis). The arrangement of chromatids within this double row is random with regard to the (blue and red) homologues. In Figure 3-10, one of the blue homologues is above the metaphase plate and the other is below, while one of the red homologues is below the metaphase plate and other is above. In an organism that is producing many gametes, meiosis in other cells could produce a different arrangement of homologues (e.g., two blues above and none below, or none above and two below). Since most eukaryotic species have several chromosomes per set, there are many possible ways that the homologues can randomly align themselves along the metaphase plate.

In addition to the random arrangement of homologues within a double row, a second distinctive feature of metaphase I is the attachment of kinetochore microtubules to a pair of sister chromatids (see Figure 3-11). One pair of sister chromatids is linked to one of the poles, and the homologous pair of sister chromatids is linked to the opposite pole. This arrangement is different from the kinetochore attachment sites during mitosis (compare Figures 3-7 and 3-11).

During anaphase I, the tetrads separate from each other. However, the sister chromatids do not separate. Instead, each joined pair of sister chromatids migrates to one pole, and the homologous pairs of sister chromatids move to the opposite pole.

During telophase I, the sister chromatids have reached their respective poles, and decondensation occurs. The nuclear membrane now re-forms to produce two separate nuclei. In this example, each nucleus contains two pairs of sister chromatids that are still joined to each other. Meiosis I is followed by cytokinesis and then meiosis II.

The cellular events that occur during meiosis II are identical to those that occur during mitosis. However, the starting point is different. For a diploid organism with four chromosomes, mitosis begins with eight sister chromatids that are joined as four pairs (refer back to Figure 3-6). By comparison, the two cells that begin meiosis II each have four sister chromatids that are joined as two pairs. Otherwise, the steps that occur during prophase, metaphase, anaphase, and telophase of meiosis II are analogous to a mitotic division.

If we compare the outcome of meiosis (Figure 3-10) with that of mitosis (Figure 3-6), the results are quite different. In these simplified examples, mitosis produced two diploid daughter cells with four chromosomes each, whereas

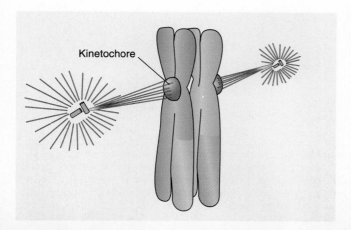

Kinetochore

FIGURE **3-11**

Attachment of the kinetochore microtubules to replicated chromosomes during meiosis. The kinetochore microtubules are attached to one replicated chromosome in a tetrad, but not to both.

meiosis produced four haploid daughter cells with two chromosomes each. In other words, meiosis has halved the number of chromosomes per cell. With regard to alleles, the results of mitosis and meiosis are also different. The daughter cells produced by mitosis are genetically identical. However, the haploid cells produced by meiosis are not genetically identical to each other, because they only contain one homologous chromosome from each pair. Later in this chapter (Figures 3-14 and 3-15), we will consider how the gametes may differ in the alleles that they carry on their homologous chromosomes.

In mammals, spermatogenesis produces four haploid sperm cells and oogenesis produces a single haploid egg cell

In male animals, **spermatogenesis**, the production of sperm, occurs within glands known as the *testes*. In the testes are found spermatogonial cells that divide by mitosis to produce two cells. One of these will remain a spermatogonial cell, while the other cell becomes a *primary spermatocyte*. As shown in Figure 3-12a, the spermatocyte progresses through meiosis I and meiosis II to produce four haploid cells that are known as **spermatids**. These cells then mature into sperm cells. The structure of a sperm cell includes the long flagellum and a head. The head of the sperm contains little more than a haploid nucleus and an organelle at its tip, known as an *acrosome*. The acrosome contains digestive enzymes that are released when a sperm meets an egg cell. These enzymes enable the sperm to penetrate the outer protective layers of

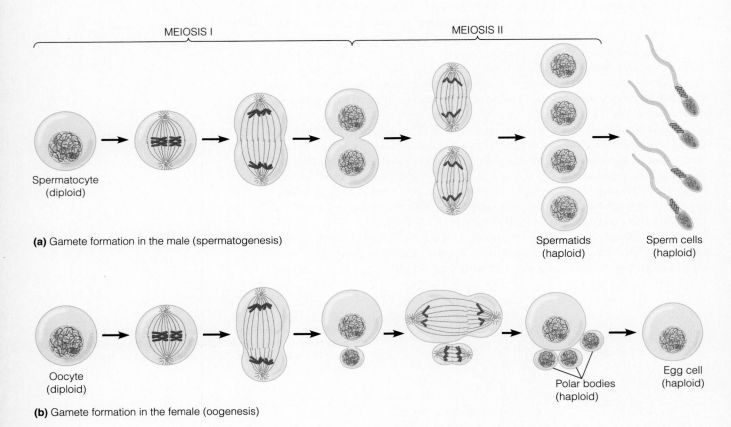

MEIOSIS I   MEIOSIS II

Spermatocyte
(diploid)

(a) Gamete formation in the male (spermatogenesis)

Spermatids
(haploid)

Sperm cells
(haploid)

Oocyte
(diploid)

(b) Gamete formation in the female (oogenesis)

Polar bodies
(haploid)

Egg cell
(haploid)

FIGURE **3-12**

Gametogenesis in animals. (a) Spermatogenesis. A diploid spermatocyte undergoes meiosis to produce four haploid (*n*) spermatids. These differentiate during spermatogenesis to become mature sperm. **(b)** Oogenesis. A diploid oocyte undergoes meiosis to produce one haploid egg cell and three polar bodies. This maximizes the amount of cytoplasm that the egg receives. The polar bodies degenerate.

the egg and gain entry into the egg cell's cytoplasm. In mature animal species without a mating season, sperm production is a continuous process. A mature human male, for example, produces several hundred million sperm each day.

In female animals, **oogenesis**, the production of egg cells, occurs within specialized cells of the *ovary* known as *oogonia* (see Figure 3-12b). Quite early in the development of the ovary, the oogonia initiate meiosis to produce *primary oocytes*. For example, in humans, approximately one million primary oocytes per ovary are produced before birth. These primary oocytes are arrested (i.e., enter a dormant phase) at prophase I of meiosis, remaining at this stage until the female animal becomes sexually mature. Beginning at this stage, primary oocytes are periodically activated to progress through the remaining stages of oocyte development. In humans, approximately one primary oocyte per month is activated to undergo this process.

During oocyte maturation in many animal species, meiosis produces only one cell (rather than four) that is destined to become an egg (Figure 3-12b). With each cycle, one or more primary oocytes within the ovary progress through the first meiotic division. This division is very asymmetric and produces a *secondary oocyte* and a much smaller cell, known as a *polar body*. Most of the cytoplasm is retained by the secondary oocyte and very little by the polar body. The secondary oocyte then begins meiosis II. While this process is under way, the secondary oocyte is released from the ovary (an event called *ovulation*) and travels down the Fallopian tubes toward the uterus. During this journey, if a sperm cell penetrates the secondary oocyte, it is stimulated to complete meiosis II to produce a haploid egg and a second polar body. The haploid egg and sperm nuclei then unite to create the diploid nucleus of a new individual.

Plant species alternate between haploid (gametophyte) and diploid (sporophyte) generations

Most species of animals are diploid, and their haploid gametes are considered to be a specialized type of cell. By comparison, the life cycles of plant species alternate between haploid and diploid generations. The haploid generation is called the **gametophyte**, whereas the diploid generation is called the **sporophyte**. In more complex plants, the organism that we think of as "a plant" is the sporophyte, while the gametophyte is very inconspicuous. In fact, the gametophytes of many plant species are microscopic haploid spores produced by meiosis within the much larger sporophyte. In simpler plants, a haploid spore can then produce a large multicellular gametophyte by mitosis. In more complex plants, however, spores develop into gametophytes that only contain a few cells. Certain cells within the haploid gametophytes then become specialized as gametes.

Figure 3-13 provides an overview of gametophyte development and **gametogenesis** in more complex plants. This diagram depicts a flower from an *angiosperm* (a true seed-producing plant). Meiosis occurs within two different structures of the sporophyte: the *anthers* and the *ovaries*. In the anther, diploid cells called *microsporocytes* undergo meiosis to produce four haploid microspores. These separate into individual microspores. Each microspore undergoes mitosis to produce a two-celled structure containing one tube cell and one generative cell (both of which are haploid). This structure differentiates into a *pollen grain*. Later, the generative cell will undergo mitosis to produce two haploid sperm cells. In most plant species, this mitosis only occurs if the pollen grain germinates (refer back to Figure 2-2c). A germinated pollen grain containing one tube cell and two sperm cells is considered a mature male gametophyte.

By comparison, female gametophytes are produced within ovules that are found inside the plant ovaries. A cell in the ovule, known as the *megasporocyte*, undergoes meiosis to produce four haploid *megaspores*. In most cases, three of the four megaspores degenerate. The remaining haploid megaspore then undergoes three successive mitotic divisions accompanied by asymmetric cytokinesis to produce

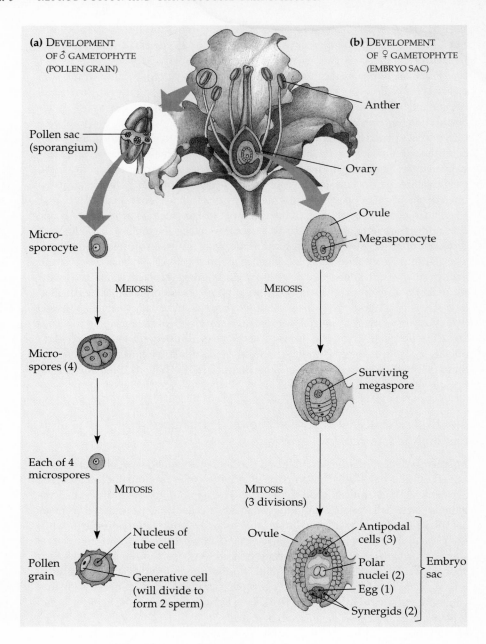

(a) DEVELOPMENT OF ♂ GAMETOPHYTE (POLLEN GRAIN)

(b) DEVELOPMENT OF ♀ GAMETOPHYTE (EMBRYO SAC)

Anther

Pollen sac (sporangium)

Ovary

Ovule

Micro-sporocyte

Megasporocyte

MEIOSIS

MEIOSIS

Micro-spores (4)

Surviving megaspore

Each of 4 microspores

MITOSIS

MITOSIS (3 divisions)

Nucleus of tube cell

Ovule

Antipodal cells (3)

Pollen grain

Generative cell (will divide to form 2 sperm)

Polar nuclei (2)

Egg (1)

Synergids (2)

Embryo sac

FIGURE **3-13**

The production of (a) male and (b) female gametes by the gametophyte of angiosperms.

seven individual cells. This seven-celled structure, also known as the *embryo sac*, is considered the mature female gametophyte.

For fertilization to occur, specialized cells within the male and female gametophytes must meet. The steps of plant fertilization were described in Chapter 2 (see pp. 18–19). To begin this process, a pollen grain lands on a stigma (see Figure 2-2c). This stimulates the tube cell to sprout a tube that grows through the style and ovary, eventually making contact with an ovule. As this is occurring, the generative cell divides to produce two haploid sperm cells. The sperm cell nuclei migrate through the pollen tube and into an ovule. One of the sperm nuclei enters the central cell that contains the two polar nuclei. This creates a cell that is *triploid* (**3n**). This cell will divide mitotically to produce endosperm, which acts as a food-storing tissue. The other sperm nucleus enters the egg cell. The egg and sperm nuclei fuse to create a diploid cell, which becomes a plant embryo. Therefore, plant fertilization is actually a double fertilization. It is intended to ensure that the endosperm (which utilizes a large amount of plant resources) will only develop when an egg cell has been fertilized. After fertilization is complete, the ovule develops into a seed, and the surrounding ovary develops into the fruit, which encloses one or more seeds.

THE CHROMOSOME THEORY OF INHERITANCE

Thus far, we have considered how chromosomes are transmitted during cell division and gamete formation. In this section, we will examine how chromosomal transmission is related to the pattern of inheritance of an individual's characteristics. This relationship, known as the chromosome theory of inheritance, was proposed in the early 1900s. It was a major breakthrough in our understanding of genetics, because it established the framework for understanding how chromosomes carry and transmit the genetic determinants that govern the outcome of traits.

In the late 19th and early 20th century, this theory dramatically unfolded as a result of three lines of scientific inquiry (Table 3-1). As discussed in Chapter 2, one avenue was Mendel's breeding studies, in which he analyzed the transmission of traits from parent to offspring. A second line of inquiry involved the material basis for heredity. In the 1880s, a German biologist, August Weismann and a Swiss botanist, Carl Nägeli, championed the idea that there is a substance within living cells that is responsible for the transmission of traits from parents to offspring. Nägeli also suggested that both parents contribute equal amounts of this substance to their offspring. At that time, the idea that heredity has a chemical and physical basis was a truly remarkable concept, since these scientists had little idea of its composition. The proposal of a genetic material by Nägeli and Weismann caused many scientists to ask, "Where is the genetic material located within living cells?" In 1884–5, several scientists, including Oscar Hertwig, Eduard Strasburger, and Walter Flemming, conducted studies suggesting that the chromosomes are the carriers of the genetic material. We now know that the DNA within the chromosomes is the genetic material.

Finally, the third line of evidence for the chromosomal theory of inheritance involved the microscopic examination of the processes of fertilization, mitosis, and meiosis. In the latter half of the 18th century, advances in microscopy enabled biologists to view and carefully study living cells. During this period, it became increasingly clear that the characteristics of organisms are rooted in the continuity of cells during the life of an organism and from one generation to the next. Some pivotal studies occurred during the 1870s and 1880s (see Table 3-1). Around 1900, when the work of Mendel was rediscovered, several scientists noted striking parallels between the segregation and assortment of traits noted by Mendel and the behavior of chromosomes during meiosis. Among them were Theodore Boveri, a German biologist, and Walter Sutton at Columbia University. In 1902–3, they independently proposed the chromosome theory of inheritance, which was a milestone in our understanding of genetics. The tenets of this theory are described next.

The chromosome theory of inheritance relates the behavior of chromosomes with the Mendelian inheritance of traits

According to the **chromosome theory of inheritance**, the inheritance patterns of traits can be explained by the transmission patterns of chromosomes during gametogenesis and fertilization. This theory is based on a few fundamental principles. First, Sutton and Boveri believed—correctly—that the chromosomes contain the genetic material. In addition, they understood Mendel's principles of inheritance (see Chapter 2). Mendel's ideas concerning the segregation and independent assortment of traits were of central importance. Sutton and Boveri realized that these principles of inheritance corresponded to the behavior of chromosomes during meiosis. Therefore, the chromosome theory of inheritance provides a way

TABLE 3–1	Chronology for the Development and Proof of the Chromosome Theory of Inheritance
1866	Gregor Mendel: analyzed the transmission of traits from parents to offspring and showed that it follows a pattern of segregation and independent assortment.
1876/7	Oscar Hertwig and Hermann Fol: observed that the nucleus of the sperm enters the egg during animal cell fertilization.
1877	Eduard Strasburger: observed that the sperm nucleus of plants (and no detectable cytoplasm) enters the egg during plant fertilization.
1878	Walter Flemming: described mitosis in careful detail.
1883	Carl Nägeli and August Weismann: proposed the existence of a genetic material, which Nägeli called idioplasm and Weismann called germ plasm.
1883	Wilhelm Roux: proposed that the most important event of mitosis is the equal partitioning of "nuclear qualities" to the daughter cells.
1883	Edouard van Beneden: showed that gametes contain half the number of chromosomes and that fertilization restores the normal diploid number.
1884/5	Hertwig, Strasburger, and Weismann: proposed that chromosomes are carriers of the genetic material.
1889	Theodore Boveri: showed that enucleated sea urchin eggs that are fertilized by sperm from a different species develop into larva having characteristics that coincide with the sperm's species.
1900	Hugh de Vries, Carl Correns, and Erich von Tschermak: rediscovered Mendel's work.
1901	Thomas Montgomery: observed that maternal and paternal chromosomes pair with each other during meiosis.
1901	C. E. McClung: showed that sex determination in insects is related to differences in chromosome composition.
1902	Boveri: showed that when sea urchin eggs were fertilized by two sperm, the abnormal development of the embryo was related to an abnormal number of chromosomes.
1902/3	Boveri and Sutton: independently proposed tenets of the chromosome theory of inheritance. Some historians primarily credit this theory to Sutton.
1903	Walter Sutton: showed that even though the chromosomes seem to disappear during interphase, they do not actually disintegrate. Instead, he argued, chromosomes must retain their continuity and individuality from one cell division to the next.
1910	Thomas Hunt Morgan: showed that a genetic trait (i.e., white-eyed phenotype in Drosophila) was linked to a particular chromosome.
1913	E. Eleanor Carothers: demonstrated that homologous pairs of chromosomes show independent assortment.
1916	Calvin Bridges: studied nondisjunction as a way to prove the chromosome theory of inheritance.

For a description of these experiments, the student is encouraged to read: Voeller, B. R. (1968) *The Chromosome Theory of Inheritance. Classic Papers in Development and Heredity.* Appleton-Century-Crofts, New York.

to explain how the cellular transmission of chromosomes passes genetic determinants (i.e., genes) from parent to offspring. According to this view:

1. Chromosomes contain the genetic material that is transmitted from parent to offspring (and from cell to cell).
2. Chromosomes are replicated and passed along generation after generation from parent to offspring. They are also passed along from cell to cell during the multicellular development of an organism. Each type of chromosome retains its individuality during cell division and gamete formation.
3. The nuclei of most eukaryotic cells contain chromosomes that are found in homologous pairs (i.e., they are diploid). One member of each pair is inherited from the mother, the other from the father. At meiosis, one of the

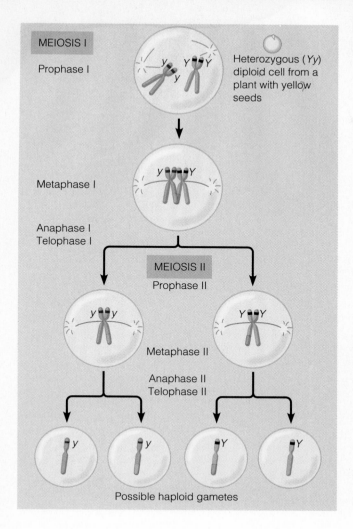

FIGURE **3-14**

Mendel's law of segregation can be explained by the segregation of homologues during meiosis. The two copies for a gene are contained on homologous chromosomes. In this example using pea seed color, the two alleles are *Y* (yellow) and *y* (green). During meiosis, the homologous chromosomes segregate from each other, leading to segregation of the two alleles into separate gametes.

GENES ⟶ TRAITS: The gene for seed color exists in two alleles, *Y* (yellow) and *y* (green). During gametogenesis, the homologous chromosomes that carry these alleles will segregate from each other, so that each gamete will receive the *Y* or *y* allele, but not both. If two gametes unite during fertilization, the alleles that they carry will affect the traits of the resulting offspring.

two members of each pair segregates into one daughter nucleus, and the other segregates into a different daughter nucleus. Therefore, gametes contain one set of chromosomes (i.e., they are haploid).

4. During gamete formation, different types of chromosomes segregate independently of each other.
5. Each parent contributes one set of chromosomes to its offspring. The maternal and paternal sets of homologous chromosomes are functionally equivalent; each set carries a full complement of genetic determinants (genes).

The chromosome theory of inheritance allows us to see the relationship between Mendel's laws and chromosomal transmission. As shown in Figure 3-14, Mendel's law of segregation of traits can be explained by the homologous pairing and segregation of chromosomes during meiosis. This figure depicts the behavior of a pair of homologous chromosomes from a pea plant that carry a gene for seed color. One of the chromosomes carries a dominant allele that confers yellow seed color, while the homologous chromosome carries a recessive allele that confers green color. As discussed in Chapter 2, the heterozygous individual would only pass one of these alleles to each offspring. In other words, a gamete may contain the yellow allele or the green allele but not both. This is because the chromosomes segregate from each other, so that a gamete will only contain one copy of each type of chromosome.

The law of independent assortment can also be explained by the behavior of chromosomes. Figure 3-15 considers the segregation of two types of chromosomes, each carrying a different gene. One pair of chromosomes carries the gene for seed

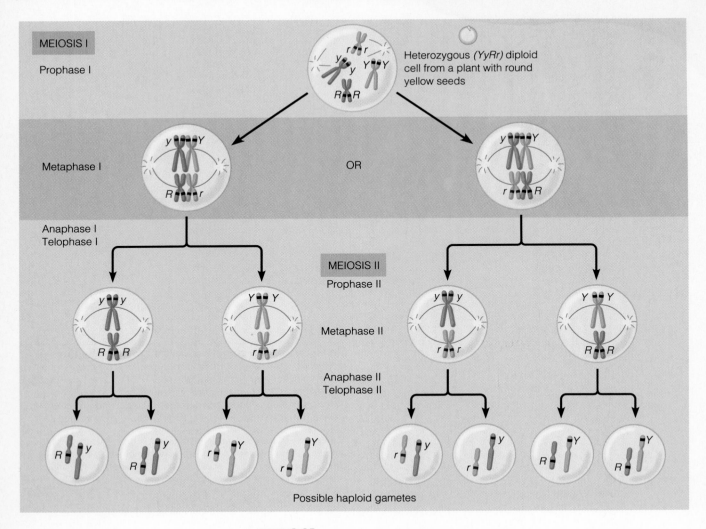

FIGURE **3-15**

Mendel's law of independent assortment can be explained by the random alignment of tetrads during prophase of meiosis I. This figure shows the assortment of two genes located on two different chromosomes, using pea seed color and shape as an example (*YyRr*). During metaphase of meiosis I, different possible arrangements of the homologues within tetrads can lead to different combinations of the alleles in the resulting gametes. For example, on the left, the *R* allele has segregated with the *y* allele; on the right, the *R* allele has segregated with the *Y* allele.

GENES ⟶ TRAITS: Most species have several different chromosomes that carry many different genes. In this example, the gene for seed color exists in two alleles, *Y* (yellow) and *y* (green), and the gene for seed shape is found as *R* (round) and *r* (wrinkled) alleles. The two genes are found on different chromosomes. During gametogenesis, the homologous chromosomes that carry these alleles will segregate from each other. In addition, the chromosomes carrying the *Y* or *y* alleles will independently assort from the chromosomes carrying the *R* or *r* alleles. As shown here, this provides a reassortment of alleles during gamete formation, perhaps creating combinations of alleles that are different from the parental combinations. If two gametes unite during fertilization, the alleles that they carry will affect the traits of the resulting offspring.

color: the yellow (*Y*) allele is on one copy, and the green (*y*) allele is on the homologous copy. The other pair of (smaller) chromosomes carries the gene for seed texture: one copy has the round (*R*) allele, and the homologous copy carries the wrinkled (*r*) allele. During metaphase I of meiosis, the different types of chromosomes randomly align themselves along the metaphase plate. As shown in Figure 3-15, this can occur in more than one way. On the left, the *R* allele has segregated with the *y* allele, while the *r* allele has segregated with the *Y* allele. On the right, the opposite situation has occurred. Therefore, the random alignment of chromosome types during meiosis I can lead to an independent assortment of genes that are found on different chromosomes. As we will see in Chapter 5, this law is violated if two different types of genes are located close to one another on the same chromosome.

Gender differences correlate with the presence of sex chromosomes

According to the chromosome theory of inheritance, the chromosomes carry the genes that determine an organism's traits. Some early evidence supporting this theory involved a consideration of gender. Many species are divided into male and female genders. In 1901, C. E. McClung, who studied fruit flies, was the first to suggest that male and female genders are due to the inheritance of particular chromosomes. Since McClung's initial observations, we now know that a pair of chromosomes, called the **sex chromosomes**, determines gender in many different species. Some examples are described in Figure 3-16.

In the X–Y system of gender determination, which operates in mammals, the male contains one X-chromosome and one Y-chromosome, while the female contains two X-chromosomes. In this case, the male is called the **heterogametic sex**, because he produces two types of sperm: one that carries only the X-chromosome, and another type that carries the Y. In contrast, the female is the **homogametic sex**, because all of her eggs carry an X-chromosome. This principle is illustrated in Figure 3-16a (also refer back to the karyotype in Figure 3-1). The 46 chromosomes carried by humans are made up of one pair of sex chromosomes and 22 pairs of **autosomes**,

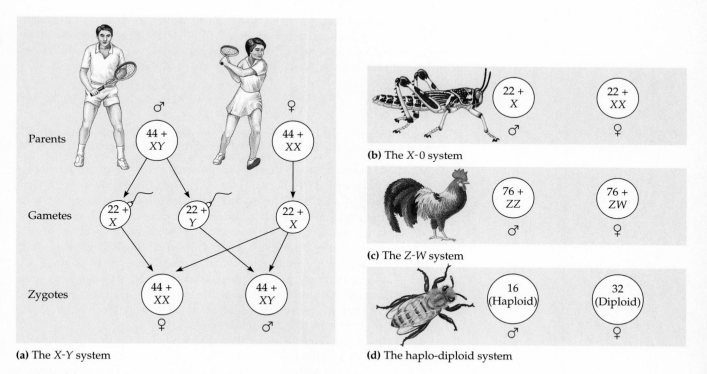

(a) The X-Y system

(b) The X-0 system

(c) The Z-W system

(d) The haplo-diploid system

FIGURE **3-16**

Different mechanisms of gender determination in animals. See text for a description.

GENES ⟶ TRAITS: The genes that are found on the sex chromosomes play a key role in the development of gender (male vs. female). For example, in mammals, genes on the Y-chromosome initiate male development. In the X–0 system, the balance between the genes on the X-chromosome and the genes on the autosomes plays a key role in governing the pathway of development toward male or female.

chromosomes that are not sex chromosomes. In the human male, each of the four sperm produced during gametogenesis contains 23 chromosomes; two sperm contain an X-chromosome, the other two a Y-chromosome. The gender of the offspring is determined by whether the sperm fertilizing an egg carries an X- or a Y-chromosome. As will be described in Chapter 8, it is the presence of the Y-chromosome that causes maleness in mammals. This is known from the analysis of rare individuals who carry chromosomal abnormalities. For example, mistakes that occasionally occur during meiosis may produce an individual who carries two X-chromosomes and one Y-chromosome. Such an individual develops into a male.

Other mechanisms of gender determination include the X–O, Z–W, and haplo-diploid systems. The *X–O system* of gender determination operates in many insects (Figure 3-16b). In some insect species, the male only has one sex chromosome (the X), and is designated XO, while the female has a pair (two Xs). In other insect species, such as *Drosophila melanogaster*, the male is XY. For both types of insect species (i.e., XO or XY males, and XX females), the ratio between X chromosomes and the number of autosomal sets determines gender. If a fly is diploid for the autosomes (2*n*), and has one X chromosome, the ratio is 1/2 or 0.5. This fly will become a male even if it does not receive a Y chromosome. In insects, in contrast to mammals, the Y-chromosome in the X–0 system does not determine maleness. If a fly is diploid and receives two X-chromosomes, the ratio is 2/2 or 1.0, and the fly becomes a female.

For the *Z–W system*, which determines gender in birds and some fish, the male is ZZ and the female is ZW. (The letters Z and W are used to distinguish these types of sex chromosomes from those found in the X–Y pattern of sex determination of other species.) As shown in Figure 3-16c, the male is the homogametic sex and the female is heterogametic.

Another interesting mechanism of sex determination, known as the *haplo-diploid system*, is found in bees (see Figure 3-16d). The male bee, also known as the drone, is produced from unfertilized haploid eggs. Female bees, both worker bees and queen bees, are produced from fertilized eggs and therefore are diploid.

EXPERIMENT 3A

Morgan's experiments in the early 1900s showed a relationship between a genetic trait and the inheritance of a sex chromosome in *Drosophila*

A particularly influential study that also supported the chromosome theory of inheritance was carried out by Thomas Hunt Morgan at Columbia University. Morgan was trained as an embryologist, and much of his early research involved descriptive and experimental work in that field. He was particularly interested in the ways that organisms change. He wrote, "The most distinctive problem of zoological work is the change in form that animals undergo, both in the course of their development from the egg (embryology) and in their development in time (evolution)." Throughout his life, he usually had dozens of different experiments going on simultaneously, many of them unrelated to each other. He jokingly said that there are three kinds of experiments—those that are foolish, those that are damn foolish, and those that are worse than that!

In 1908, he engaged one of his graduate students to rear the fruit fly *Drosophila melanogaster* in the dark, hoping to produce flies whose eyes would atrophy from disuse and disappear in future generations. Even after many consecutive generations, however, the flies appeared to have no noticeable changes in spite of repeated attempts at inducing mutations by treatments with agents such as X-rays and radium.

After two years, they finally obtained an interesting result when a true-breeding line of *Drosophila* produced a male fruit fly with white eyes rather than the normal red eyes. Since this had been a pure line of flies, this white-eyed male must have arisen from a new **mutation** (i.e., a permanent change in the genetic material) that created a white-eyed allele. Morgan is said to have carried this fly home with him in a jar, put it by his bedside at night while he slept, and then taken it back to the laboratory during the day.

Much like Mendel, Morgan studied the inheritance of this white-eyed trait by making crosses and quantitatively analyzing their outcome. In this experiment, he performed a **test cross,** which is a cross between a recessive individual and an individual whose genotype the experimenter wishes to test. w^+ designates the red eye (wild-type) allele, w designates the white eye allele. As described next, Morgan's results supported the idea that the inheritance of traits (in this case, eye color) parallels the inheritance of the X chromosome. The location of the w or w^+ allele within the X chromosome is designated X^w and X^{w+}, respectively.

THE HYPOTHESIS

Morgan did not have a preconceived hypothesis how this white-eyed trait would be passed along from parent to offspring. His general hypothesis was that rearing flies in the dark would result in the loss of traits that would be useless for flies that could not see. Nevertheless, once he obtained his white-eyed fly, he expected to gain insight into the inheritance pattern of this new trait by making the appropriate crosses.

TESTING THE HYPOTHESIS

Starting material: A true-breeding line of red-eyed fruit flies plus one white-eyed male fly that was discovered in the culture.

	Experimental Level	**Conceptual Level**
1. Cross the white-eyed male to red-eyed females.		$X^wY \times X^{w+}X^{w+}$
2. Record the results of the F_1 generation. This involves noting the eye color and gender of several thousand flies.		$X^{w+}Y$ male offspring and $X^{w+}X^w$ female offspring
3. Cross F_1 offspring with each other to obtain F_2 offspring. Also record the eye color and gender of the F_2 offspring.		$X^{w+}Y \times X^{w+}X^w$ $1\ X^{w+}Y : 1\ X^wY : 1\ X^{w+}X^{w+} : 1\ X^{w+}X^w$ 1 red-eyed male : 1 white-eyed male : 2 red-eyed females
4. In a separate experiment, perform a test cross between a white-eyed male and a red-eyed female from the F_1 generation. Record the results.		$X^wY \times X^{w+}X^w$ $1\ X^{w+}Y : 1\ X^wY : 1\ X^{w+}X^w : 1\ X^wX^w$ 1 red-eyed male : 1 white-eyed male : 1 red-eyed female : 1 white-eyed female

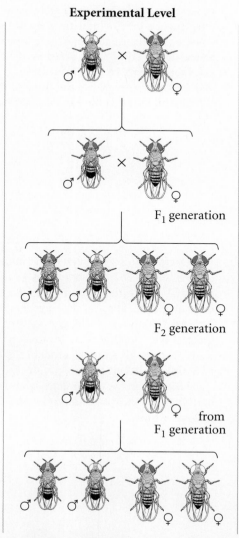

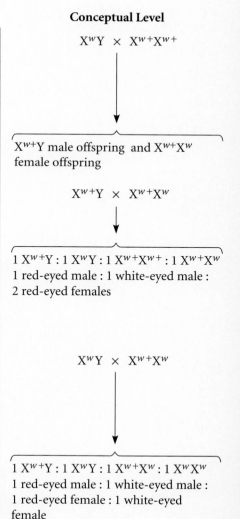

THE DATA

Cross	Results	
Original white-eyed male to red-eyed females	F₁ generation:	all red-eyed flies*
F₁ male to F₁ female	F₂ generation:	2459 red-eyed females
		1011 red-eyed males
		0 white-eyed females
		782 white-eyed males
White-eyed male to F₁ female	Test cross:	129 red-eyed females
		132 red-eyed males
		88 white-eyed females
		86 white-eyed males

*See question 1 for student discussion.

INTERPRETING THE DATA

Let's begin with an interpretation that is based on our modern knowledge of this trait. The F₁ generation contained all red-eyed flies. Therefore, the red-eyed allele is dominant to the white-eyed allele. The key observation suggesting that the eye color gene is located on the X-chromosome is found in the F₂ generation. In this generation, produced from a cross between an F₁ male and F₁ female, the ratios of offspring are greatly influenced by the gender of the F₂ offspring. There were no white-eyed females, although there were many red-eyed females and approximately an equal ratio (1011 : 782) of red-eyed males to white-eyed males. These results suggest that the pattern of transmission between male and female parents and male and female offspring is not the same. As shown in the Punnett square here, it is consistent with idea that the eye color alleles are located on the X-chromosome:

F₁ male is $X^{w+}Y$
F₁ female is $X^{w+}X^w$

Overall, the experimental ratio in the F₂ generation of red eyes to white eyes is (2459 + 1011) : 782, which equals 4.4 : 1. This ratio is not too far from the predicted ratio of 3 : 1 shown in the last Punnett square. More dramatically, however, the Punnett square predicts that the F₂ generation will not have any white-eyed females. This prediction was strikingly confirmed experimentally. These results indicate that the eye color alleles are located on the X-chromosome. Genes that are physically located within the structure of the X-chromosome are called **X-linked genes** or **X-linked alleles.**

Likewise, the test cross data are also consistent with an X-linked pattern of inheritance. As shown in the following Punnett square, the test cross predicts a 1 : 1 : 1 : 1 ratio:

Test cross: male is X^wY
F₁ female is $X^{w+}X^w$

The observed data are 129 : 132 : 88 : 86, which is 1.5 : 1.5 : 1 : 1. The lower than expected numbers of white-eyed males and females can be explained by a lower survival rate for white-eyed flies.

In his own interpretation, Morgan did not explicitly state that the gene for eye color was located on the X-chromosome. Instead, he concluded, "R (red eye color) and X (a sex factor that is present in two copies in the female) are combined and have never existed apart." Using modern terminology, we would interpret Morgan's comments as meaning that the gene for eye color is on the X-chromosome. Morgan was given the Nobel prize in 1933, the first geneticist to receive this award.

In 1914, Calvin Bridges, a graduate student in the laboratory of Morgan, also examined the transmission of X-linked traits. Bridges conducted hundreds of crosses involving several different types of X-linked alleles. In his crosses, he occasionally obtained offspring that had unexpected phenotypes and abnormalities in sex chromosome composition. In all cases, the parallel between the cytological presence of sex chromosome abnormalities and the occurrence of unexpected traits confirmed the idea that the sex chromosomes carry X-linked genes. Altogether, the work of Morgan and Bridges provided an impressive body of evidence confirming the idea that traits following an X-linked pattern of inheritance are governed by genes that are physically located on the X-chromosome. Bridges wrote, "There can be no doubt that the complete parallelism between the unique behavior of chromosomes and the behavior of sex-linked genes and sex in this case means that the sex-linked genes are located in and borne by the X-chromosomes"[1]

The inheritance of X-linked genes is not the same in reciprocal crosses

As we have seen, the traits of an organism may be governed by genes that are located on a sex chromosome. We have also learned that many species have males and females that differ in their sex chromosome composition. In these species, the transmission pattern of certain traits may depend on the gender of the parent. Experimentally, a **reciprocal cross** can be conducted to discern if a trait is carried on a sex chromosome or on an autosome; X-linked traits do not behave identically in reciprocal crosses. To explain what a reciprocal cross is, let's first consider a cross between a homozygous white-eyed female fly and a red-eyed male. As shown here, the outcome is all males with white eyes and all females with red eyes:

Male $X^{w+}Y$, is crossed to a female, $X^{w}X^{w}$

	Male gametes X^{w+}	Y
Female gametes X^{w}	$X^{w+}X^{w}$ red ♀	$X^{w}Y$ white ♂
X^{w}	$X^{w+}X^{w}$ red ♀	$X^{w}Y$ white ♂

In the reciprocal cross, a homozygous red-eyed female is crossed to a white-eyed male. The result is that 100% of the offspring have red eyes:

Male $X^{w}Y$, is crossed to a female, $X^{w+}X^{w+}$

	Male gametes X^{w}	Y
Female gametes X^{w+}	$X^{w+}X^{w}$ red ♀	$X^{w+}Y$ red ♂
X^{w+}	$X^{w+}X^{w}$ red ♀	$X^{w+}Y$ red ♂

---

[1]From Bridges, C. B. (1914) "Direct Proof Through Non-disjunction that the Sex-Linked Genes of Drosophila are Borne by the X-Chromosome." Science 40, 107–109.

When comparing the two Punnett squares, the outcome of the reciprocal cross did not yield same results. This is expected of X-linked genes, because the male only transmits the gene to his female offspring while the female transmits an X-chromosome to both male and female offspring. Since the male parent does not transmit the X-chromosome to his sons, he does not contribute to the outcome of their X-linked traits. This explains why X-linked traits do not behave equally in reciprocal crosses. Experimentally, the observation that reciprocal crosses do not yield the same results is an important clue that a trait may be X-linked.

When setting up a Punnett square involving X-linked traits, we must combine the alleles with their location on the X-chromosome. Also, keep in mind that the male makes two types of gametes, one that carries the X-chromosome and one that carries the Y. The Punnett square must also include the Y-chromosome even though this chromosome does not carry any X-linked genes. This is necessary to determine the relationships between the genotypes and genders of the offspring.

Genes located on human sex chromosomes can be transmitted in an X-linked, a Y-linked, or a pseudoautosomal pattern

Our previous discussion of the sex chromosomes has concentrated on X-linked genes that are located on the X-chromosome but not on the Y-chromosome. The term **hemizygous** is used to describe the single copy of an X-linked gene in the male. A male mammal is said to be hemizygous for his X-linked genes. In Chapter 23, we will consider how recessive X-linked traits are more likely to occur in males than in females due to hemizygosity.

The general term **sex linkage** is used to describe genes that are found on one of the two types of sex chromosomes but not on both. Hundreds of X-linked genes have been identified in humans and other mammals (examples will be given in Chapter 23). By comparison, there are relatively few **Y-linked genes** that are only located on the Y-chromosome. An example of a Y-linked gene is the *SRY* gene. Its expression is necessary for proper male development. Since the X-chromosome contains many more genes, the term sex linkage is sometimes used synonymously with X-linkage. However, in its strictest sense, sex linkage may refer to Y-linked genes as well.

Besides sex-linked genes, the X- and Y-chromosomes also contain short regions of homology where the X- and Y-chromosomes carry the same genes. As shown in Figure 3-17, the human sex chromosomes contain homologous regions at one end of the X- and Y-chromosomes. These regions promote the necessary pairing of the X- and Y-chromosomes that occurs during meiosis I of spermatogenesis. There are only a few genes that are located in this homologous region. One example is a gene called *MIC2*, which is involved in antibody production. The *MIC2* gene is found on both the X- and Y-chromosomes. It follows a pattern of inheritance that is called **pseudoautosomal inheritance**. The term pseudoautosomal refers to the idea that the inheritance pattern of the *MIC2* gene is the same as the inheritance pattern of a gene located on an autosome even though the *MIC2* is actually located on the sex chromosomes. As in autosomal inheritance, males contain two copies of pseudoautosomally inherited genes, and they can transmit the genes to both daughters and sons.

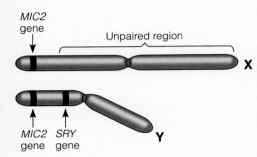

FIGURE **3-17**

A comparison of the homologous and nonhomologous regions of the X- and Y-chromosomes in humans. A few pseudoautosomal linked genes, such as *MIC2*, are found on both the X- and Y-chromosomes in the region of homology between the two chromosomes. Many X-linked genes are found in the unpaired region of the X-chromosome. A few Y-linked genes, such as *SRY*, are found on the Y-chromosome in the unpaired region.

CONCEPTUAL SUMMARY

In this chapter, we have considered the process of reproduction at the cellular level. During *binary fission* in bacteria and *mitosis* in eukaryotic cells, a mother cell divides to produce two daughter cells that are genetically identical to each other. In bacteria, or prokaryotes, which contain a single circular chromosome, binary fission provides a way to reproduce asexually. Eukaryotic species, which contain many linear chromosomes, must sort their chromosomes during cell division in a process known as mitosis. This ensures that the chromosomes will be correctly distributed to the daughter cells. In simple eukaryotes, mitotic divisions can produce new unicellular offspring. Otherwise, the primary reason that eukaryotic species undergo mitotic cellular divisions is to produce organisms that are multicellular. Most eukaryotes develop from a single fertilized egg cell that undergoes repeated mitotic and cellular divisions to produce an organism containing many cells. Most of the cells of the body are genetically identical to each other.

Sexual reproduction requires the formation of *gametes* (i.e., sperm cells or egg cells) that contain half the genetic material. For *diploid* organisms (those with two sets of chromosomes, or 2*n*), the process of *meiosis* results in gametes that are *haploid* (containing one set, or 1*n*). Meiosis requires two successive divisions. Meiosis I involves a pairing between homologous chromosomes that does not occur during mitosis. This pairing is necessary to ensure that the resultant gametes will contain one complete set of chromosomes. Meiosis II is very similar to a mitotic division, except that the cells entering meiosis II begin with half the normal number of chromosomes as compared with cells entering mitosis.

In eukaryotic species, meiosis occurs in conjunction with the cellular specialization of gametes. During *spermatogenesis*, meiosis coincides with the production of four motile sperm cells. By comparison, *oogenesis* is aimed at producing a single haploid egg that is rich in nutrients. Fertilization in animals involves the union between a single sperm and egg cell. In plants, sexual reproduction involves an alternation between diploid organisms (*sporophytes*) and haploid organisms (*gametophytes*). However, the gametophytes of more complex plants only contain a few cells, which are primarily involved in producing gametes. Plant reproduction is a double fertilization event in which one sperm fertilizes the egg and a second sperm fertilizes the *endosperm*. The proliferation of endosperm provides nutrients to the plant embryo.

A cellular understanding of mitosis, meiosis, and fertilization allows us to appreciate how the transmission of traits is due to the passage of chromosomes. Genes, which govern the outcome of traits, are physically located within the structure of chromosomes. The *chromosome theory of inheritance*, proposed by Boveri and Sutton, describes how the transmission of chromosomes accounts for Mendel's laws of segregation and independent assortment.

A central method that has been used to understand chromosomal transmission is light microscopy. Cytogeneticists can view the chromosomes in actively dividing cells under a microscope. This has allowed researchers to study the behavior of chromosomes during mitosis and meiosis. In the late 19th and early 20th century, microscopic examinations of chromosomes yielded observations that were consistent with the idea that the chromosomes are the carriers of genes. This idea, along with Mendel's experiments in pea plants, led to the formulation of the chromosome theory of inheritance. Many subsequent experiments have confirmed the validity of this theory. In particular, investigations involving the inheritance of the X-chromosome were instrumental in establishing that the inheritance of traits paralleled the inheritance of the X-chromosome. Morgan was the first scientist to show that inheritance of a recessive X-linked trait was consistent with the respective gene's location on the X-chromosome. Since that time, the chromosomal locations of many genes have been firmly established.

PROBLEM SETS

SOLVED PROBLEMS

1. A diploid cell has eight chromosomes, four per set. In the following diagram, what phase of mitosis, meiosis I, or meiosis II is this cell in?

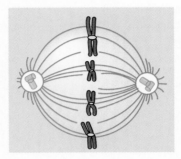

Answer: The cell is in metaphase of meiosis II. You can tell because the chromosomes are lined up along the metaphase plate, and it only has four pairs of sister chromatids. If it were mitosis, the cell would have eight pairs of sister chromatids.

2. An unaffected woman (i.e., without disease symptoms) who is heterozygous for the X-linked trait Duchenne's muscular dystrophy has children with a normal man. What are the probabilities of the following combinations of offspring?

    A. An unaffected son
    B. An unaffected son or daughter
    C. A family of three children, all of whom are affected

Answer: The first thing we must do is construct a Punnett square to determine the outcome of the cross. N represents the normal allele, n the recessive allele causing Duchenne's muscular dystrophy.

	X^N	Y
X^N	$X^N X^N$	$X^N Y$
X^n	$X^N X^n$	$X^n Y$

Phenotypic ratio is

2 normal daughters:
1 normal son:
1 affected son

    A. There are four possible children, one of whom is an unaffected son. Therefore, the probability of an unaffected son is 1/4.
    B. Use the sum rule: 1/4 + 1/2 = 3/4.
    C. Use the product rule: $1/4 \times 1/4 \times 1/4 = 1/64$.

CONCEPTUAL QUESTIONS

1. The process of binary fission begins with a single mother cell and ends up with two daughter cells. Would you expect the mother and daughter cells to be genetically identical? Explain why or why not.

2. What is a homologue? With regard to genes and alleles, how are homologues similar to and different from each other?

3. What are sister chromatids? Are they genetically similar or identical? Explain.

4. With regard to the chromosomes, which phase of mitosis is the organization phase and which is the separation phase?

5. A species is diploid containing three chromosomes per set. Draw what the chromosomes would look like in the G_1 and G_2 phases of the cell cycle.

6. Discuss what you think would happen if a sister chromatid was not attached to a kinetochore microtubule.

7. For the following events, specify whether they occur during mitosis, meiosis I, and/or meiosis II.

    A. Separation of sister chromatids
    B. Pairing of homologous chromosomes
    C. Alignment of chromosomes along the metaphase plate
    D. Attachment of the pairs of sister chromatids to both poles

8. Describe the key events during meiosis that result in a 50% reduction in the amount of genetic material per cell.

9. A cell is diploid and contains three chromosomes per set. Draw the arrangement of chromosomes during metaphase of mitosis and metaphase I and II of meiosis. In your drawing, make one set black and the other white.

10. The arrangement of homologues during metaphase I is a random process. In your own words, explain what this means.

11. A eukaryotic cell is diploid containing ten chromosomes (five in each set). Describe the results of mitosis and meiosis. In each process, how many daughter cells would be produced and how many chromosomes would each one contain?

12. If a diploid cell contains six chromosomes (i.e., three per set), how many possible random arrangements of homologues could occur during metaphase I?

13. A cell has four pairs of chromosomes. Assuming that crossing over does not occur, what is the probability that a gamete will contain all of the paternal chromosomes? If n equals the number of chromosomes in a set, which of the following expressions can be used to calculate the probability that a gamete will receive all of the paternal chromosomes: $(1/2)^n$, $(1/2)^{n-1}$, or $n^{1/2}$?

14. With regard to question 13, how would the phenomenon of crossing over affect the results? In other words, would there be a higher or lower probability of a gamete inheriting only paternal chromosomes? Explain your answer.

15. In corn, there are ten chromosomes per set and the species is diploid. If you performed a karyotype, what is the total number of chromosomes that you would expect to see in the following types of cells?

   A. A leaf cell
   B. The sperm nucleus of a pollen grain
   C. An endosperm cell
   D. A root cell

16. Let's suppose that a gene affecting pigmentation is found on the X-chromosome (in mammals or insects) or the Z-chromosome (in birds), but not on the Y- or W-chromosome. It is found on an autosome in bees. This gene is found in two alleles, D (dark), which is dominant to d (light). What would be the results of crosses between a true-breeding dark female and true-breeding light male, and the reciprocal crosses involving a true-breeding light female and true-breeding dark male in the following species?

   A. Birds
   B. *Drosophila*
   C. Bees
   D. Humans

17. Describe the cellular differences between male and female gametes.

18. At puberty, the testes contain a finite number of cells and produce an enormous number of sperm cells during the life span of a male. Explain why testes do not run out of spermatogonial cells.

19. Describe the timing of meiosis I and II during mammalian oogenesis.

20. Three genes (A, B, and C) are found on three different chromosomes. For the following diploid genotypes, describe all of the possible gamete combinations and their predicted ratios.

   A. *AaBbCc*
   B. *AABbCC*
   C. *AaBBCc*
   D. *Aabbcc*

21. In male mammals, X-linked genes are hemizygous. Propose a genetic term that would describe what Y-linked genes are in females.

22. Propose the most likely mode of inheritance for the following pedigrees. Affected individuals are shown in red.

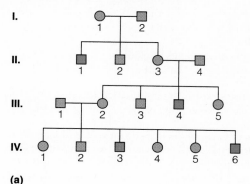

(a)

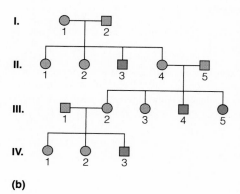

(b)

23. A human disease known as vitamin D resistant rickets is inherited as an X-linked dominant trait. If a male with the disease produces children with a female who does not have the disease, what is the expected proportion of their offspring?

24. Hemophilia is a recessive X-linked trait in humans. If a heterozygous woman has children with a normal man, what are the odds of the following combinations of children?

   A. An affected son
   B. Four unaffected offspring in a row
   C. An unaffected daughter or son
   D. Two out of five offspring that are affected

25. A form of color blindness is inherited as a recessive, X-linked trait. A couple has a colorblind daughter and a normal son. What are the genotypes and phenotypes of the parents? What is the probability that their next two children will be colorblind?

EXPERIMENTAL QUESTIONS

1. When studying living cells in a laboratory, researchers sometimes use drugs as a way to make cells remain at a particular stage of the cell cycle. For example, aphidicolin inhibits DNA synthesis in eukaryotic cells and causes them to remain in the G_1 phase because they cannot replicate their DNA. In what phase of the cell cycle—G_1, S, G_2, prophase, metaphase, anaphase, or telophase—would you expect somatic cells to stay, if the following types of drug were added?

A. A drug that inhibits microtubule formation
B. A drug that allows microtubules to form, but prevents them from shortening
C. A drug that inhibits cytokinesis
D. A drug that prevents chromosomal condensation

2. In Morgan's experiments, which result do you think is the most convincing piece of evidence pointing to X-linkage? Explain your answer.

3. In his original studies of Experiment 3A, Morgan first suggested that the original white-eyed male had two copies of the white-eyed allele. In this problem, let's assume that he meant the fly was X^wY^w instead of X^wY. Are his data consistent with this hypothesis? What crosses would need to be made to rule out the possibility that the Y-chromosome carries a copy of the eye color gene?

4. How would you set up crosses to determine if a gene was Y-linked versus X-linked?

5. Occasionally during meiosis, a mistake can happen whereby a gamete may receive zero or two sex chromosomes rather than one. In 1914, Calvin Bridges at Columbia University made a cross between white-eyed female flies and red-eyed males. As you would expect, most of the offspring were red-eyed females and white-eyed males. On rare occasions, however, he found a white-eyed female or a red-eyed male. These rare flies were not due to new mutations, but instead were due to mistakes during meiosis in the parent flies. Consider the mechanism of gender determination in fruit flies and propose how this could happen. In your answer, describe the sex chromosome composition of these rare flies. Do these rare flies support the chromosomal theory of inheritance?

6. Let's suppose that you have karyotyped a female fruit fly with red eyes and found that it has three X chromosomes instead of the normal two. Although you do not know its parents, you do know that this fly came from a mixed culture of flies in which some had red eyes, some had white eyes, and some had eosin eyes. (See Experiment 4A for a description of eosin eye color.) With regard to the eye color alleles, what are the possible genotypes of this female fly? You cross this female with a white-eyed male and count the number of offspring. You may assume that this unusual female makes half of its gametes with one X-chromosome and half of its gametes with two X-chromosomes. The following results are obtained:

	Females*	Males
Red eyes	50	11
White eyes	0	0
Eosin	20	0
Light eosin	21	20

*Note: A female offspring can be XXX, XX, or XXY.

Explain the 3:1 ratio between female and male offspring. What was the genotype of the original mother that had red eyes and three X-chromosomes? Construct a Punnett square that is consistent with these data.

QUESTIONS FOR STUDENT DISCUSSION/COLLABORATION

1. In Experiment 3A, Morgan obtained a white-eyed male fly in a population containing many red-eyed flies that he thought were true breeding. As mentioned in the experiment, he crossed this fly with several red-eyed sisters, and all the offspring had red eyes. But actually this is not quite true. Morgan observed 1237 red-eyed flies and 3 white-eyed males. Provide two or more explanations for why he obtained 3 white-eyed males in the F_1 generation.

2. A diploid eukaryotic cell has ten chromosomes (five per set). As a group, take turns having one student draw the cell as it would look during a phase of mitosis, meiosis I, or meiosis II; then have the other students guess which phase it is.

3. Discuss the tenets of the chromosome theory of inheritance. Which tenets were deduced by using light microscopy, and which were deduced from crosses? What modern techniques could be used to support the chromosome theory of inheritance?

CHAPTER 4

EXTENSIONS OF
MENDELIAN INHERITANCE

We use the term **Mendelian inheritance** to describe the transmission of most eukaryotic genes located on the chromosomes found within the cell nucleus. As we have seen in Chapters 2 and 3, there is a correlation between the behavior of chromosomes during meiosis and the transmission of traits. When the chromosomes are passed from parent to offspring during sexual reproduction, we expect that the pattern of Mendelian inheritance will obey two laws: the law of segregation and the law of independent assortment.

Until now, we have mainly considered traits, in pea plants and in fruit flies, that are affected by a single gene that is found in two different alleles. In these cases, one allele was dominant over the other. This type of inheritance is sometimes called *simple* Mendelian inheritance, because the observed ratios in the offspring readily obey Mendel's laws. For example, when two different true-breeding pea plants are crossed (e.g., tall and dwarf) and the F_1 generation is allowed to self-fertilize, the F_2 generation shows a 3:1 phenotypic ratio of tall and dwarf offspring. Similarly, in Chapter 3, we examined the implications of the idea that some genes are located on the sex chromosomes. Many of the experiments of Morgan and Bridges involved X-linked traits. The pattern of transmission of these traits also depended on the pattern of transmission of the X-chromosome, yielding differing but predictable ratios among male and female offspring.

In this chapter, we will extend our understanding of Mendelian inheritance in two ways. First, we will examine the molecular expression of genes to gain insight into the dominant and recessive relationships between two (or more) alleles. In addition, we will explore several traits that do not produce the ratios expected from

INHERITANCE PATTERNS
OF SINGLE GENES

GENE INTERACTIONS

simple Mendelian inheritance. As we will discover, there are many reasons why the transmission of traits may not produce simple Mendelian ratios even though the relevant genes are obeying Mendel's laws. This does not mean that Mendel was wrong. Rather, the inheritance patterns of many traits are more complex and interesting than he had realized. Later, in Chapters 5 and 7, we will examine eukaryotic inheritance patterns that actually violate the laws of segregation and independent assortment.

INHERITANCE PATTERNS OF SINGLE GENES

We begin this chapter with further exploration of the outcome of traits that are controlled by a single type of gene. We will first consider how the molecular expression of genes is involved in the production of an organism's traits. This examination will help us understand the inheritance patterns that Mendel observed, as well as more complex patterns that have been discovered since the early 1900s. Table 4-1 describes the general features of several types of inheritance patterns that have been observed by geneticists during the 20th century. As we will see, many of these patterns produce ratios of offspring that deviate from the expected ratios of simple Mendelian inheritance. These variant patterns occur because there are many different ways that two alleles may govern the outcome of a trait. In this section, these patterns of inheritance will be described in greater detail.

TABLE 4–1 Different Types of Mendelian Inheritance Patterns

Type	Description
Simple Mendelian inheritance	This term is commonly applied to the inheritance of alleles that follow a strict, dominant/recessive relationship. In previous chapters, we have considered genes that exist in two alleles. In this chapter, we will see that some genes can be found in three or more alleles.
X-linked inheritance	Described in Chapter 3. It involves the inheritance of genes that are located on the X-chromosome. In mammals and fruit flies, males are hemizygous for X-linked genes while females have two copies.
Lethal alleles	An allele in an essential gene that has the potential of causing the death of an organism.
Incomplete dominance	Occurs when the heterozygote has a phenotype that is intermediate between either corresponding homozygote. For example, a plant produced from a cross between red-flowered and white-flowered parents will have pink flowers.
Codominance	Occurs when the heterozygote expresses both alleles simultaneously. For example, in blood typing, an individual carrying the A and B alleles will have an AB blood type.
Overdominance	Occurs when the heterozygote has traits that are more beneficial than either homozygote.
Incomplete penetrance	Occurs when a dominant allele is not expressed even though it is present in an individual.
Sex-influenced inheritance	Refers to the impact of gender on the phenotype of the individual. Some alleles are recessive in one gender and dominant in the opposite gender.
Sex-limited inheritance	Refers to traits that only occur in only one of the two genders.

Recessive alleles often cause a reduction in the amount or function of the encoded protein

For any given gene, geneticists usually refer to the most prevalent allele in a population as the **wild-type allele**. In most instances, a wild-type allele encodes a protein that is made in the proper amount and functions normally. Compared with the wild-type allele, alleles that have been altered by mutation are called **mutant alleles**. Among Mendel's seven traits discussed in Chapter 2, the wild-type alleles are purple flowers, axial flowers, yellow seeds, round seeds, smooth pods, green pods, and tall plants. The mutant alleles are white flowers, terminal flowers, green seeds, wrinkled seeds, constricted pods, yellow pods, and dwarf plants. You may have already noticed that the seven wild-type alleles are dominant over the seven mutant alleles. Likewise, red eyes and normal wings are examples of wild-type alleles in *Drosophila*, and white eyes and miniature wings are recessive mutant alleles. As we will discuss in Chapter 17, it is far more common for a mutant allele to be defective in its ability to express a functional protein. In other words, mutations that produce mutant alleles are likely to decrease or eliminate the synthesis or functional activity of a protein. Such mutations are often inherited in a recessive fashion.

The idea that recessive alleles usually cause a substantial decrease in the expression of a functional protein is supported by analyses of many human genetic diseases. Keep in mind that a genetic disease is caused by a mutant allele. Table 4-2 lists several examples of human genetic diseases in which the recessive allele has been shown to prevent the proper production of a specific cellular protein in its active form. In many cases, molecular techniques have enabled researchers to clone these

TABLE 4–2 Examples of Recessive Human Diseases

Disease	Protein that is Produced* by the Normal Gene	Description
Phenylketonuria	Phenylalanine hydroxylase	Inability to metabolize the amino acid phenylalanine. The disease can be controlled by a phenyl-alanine-free diet. If the diet is not followed early in life, it can lead to symptoms that include severe mental retardation and physical degeneration.
Albinism	Tyrosinase	Lack of pigmentation in the skin, eyes, and hair.
Tay-Sachs disease	Hexosaminidase A	Defect in lipid metabolism. Leads to paralysis, blindness, and early death.
Sandhoff disease	Hexosaminidase B	Muscle weakness in infancy, early blindness, and progressive mental and motor deterioration.
Cystic fibrosis	Chloride-transporter	Inability to regulate ion balance across epithelial cells. Leads to overproduction of lung mucus and chronic lung infections.
Lesch-Nyhan syndrome	Hypoxanthine-guanine phosphoribosyl transferase	Inability to metabolize purines, which are bases found in DNA and RNA. Leads to mental retardation, poor motor skills, and kidney failure.

*Individuals who exhibit the disease are homozygous (or hemizygous) for a recessive allele that results in a defect in the amount or function of the normal protein.

genes and determine the differences between the normal and recessive alleles. They have found that the recessive allele usually contains a mutation that is expected to prevent the synthesis of a fully functional protein.

To understand why many defective mutant alleles are inherited recessively, we need to take a quantitative look at protein function. With the exception of sex-linked traits, diploid individuals have two copies for every gene. In a simple dominant/recessive relationship, the recessive allele does not affect the phenotype of the heterozygote. In other words, a single copy of the dominant (normal) allele is sufficient to mask the effects of the recessive allele. If the recessive allele prevents the expression of a functional protein, how do we explain the wild-type phenotype of the heterozygote? As described in Figure 4-1, the usual explanation is that 50% of the functional protein encoded by the normal gene is adequate to provide a normal phenotype. In this example, the *PP* homozygote and *Pp* heterozygote each make sufficient functional protein to yield purple flowers. This means that the normal homozygous individual is making much more of the wild-type protein than it really needs. Therefore, if the normal amount is reduced to 50%, as in the heterozygote, the individual still has plenty of this protein to accomplish whatever cellular function it performs. The phenomenon that "50% of the normal protein is enough" is fairly common among many genes. However, later in this chapter, we will see that this is not always the case.

Mutations that cause a loss of function in an essential gene result in a lethal phenotype

When the absence of a specific protein results in a lethal phenotype, the gene that encodes the protein is considered an **essential gene** for survival. An allele that causes the death of an organism is called a **lethal allele**. Several studies have attempted to estimate the total number of genes in experimental organisms, such as yeast and *Drosophila*, whose loss of function due to mutation will result in lethal alleles. There are thousands of such genes (exact results vary). For example, in 1986, Mark Goebl and Thomas Petes at the University of Chicago examined the percentage of essential genes in the simple eukaryote *Saccharomyces cerevisiae* (baker's yeast). They estimated that this yeast has approximately 5500 different genes, of which 1200 are essential for survival.

By comparison, **nonessential genes** are not absolutely required for survival, although they are likely to be beneficial to the organism. A loss of function mutation in a nonessential gene will not cause death. On rare occasions, however, a nonessential gene may acquire a mutation that causes the gene product to be abnormally expressed in a way that leads to a lethal phenotype. For example, a mutation that causes the overexpression of a nonessential gene may interfere with normal cell

---

FIGURE **4-1**

A comparison of protein levels among homozygous and heterozygous genotypes *PP*, *Pp*, and *pp*.

GENES ⟶ TRAITS: In a simple dominant/recessive relationship, 50% of the normal protein is sufficient to produce the normal phenotype, in this case purple flowers. A complete lack of the normal protein results in white flowers.

Normal allele: *P* (purple)
Recessive (defective) allele: *p* (white)

Genotype			
	PP	*Pp*	*pp*
Amount of functional protein P:	100%	50%	0%
Phenotype			
	Purple	Purple	White
Simple dominant/ recessive relationship:			

function and thereby cause lethality. Therefore, not all lethal mutations are in essential genes, although the great majority are.

Many lethal alleles prevent cell division, and thereby kill an organism at a very early stage. Others, however, may only exert their effects later in life, or under certain environmental conditions. For example, a human genetic disease known as Huntington disease is characterized by a progressive degeneration of the nervous system, dementia, and early death. The age when these symptoms appear, or the **age of onset**, is usually between 30 and 50. Therefore, this lethal allele acts relatively late in life. Other lethal alleles may only kill an organism when certain environmental conditions prevail. These are called **conditional lethal alleles**. They have been extensively studied in experimental organisms. For example, some conditional lethals will only kill an organism in a particular temperature range. These alleles, called *temperature-sensitive* (ts) *lethals*, have been observed in many organisms such as *Drosophila*. A *ts* allele may kill a developing larva at a high temperature (30°C), but it will survive if grown at a lower temperature (22°C). Finally, it is surprising that certain lethal alleles act only in some individuals. These are called **semilethal alleles**. Of course, any particular individual cannot be semidead. However, within a population, a semilethal allele will kill some individuals but not all of them. The reasons for semilethality are not always understood, but environmental conditions and the actions of other genes within the organism may help to prevent the detrimental effects of certain semilethal alleles.

Since genetic lethality decreases the number of surviving individuals, the prevalence of lethal alleles in a viable species must be kept very low. There are several ways that this is accomplished. First, as will be discussed in Chapter 17, the rate of spontaneous mutations is exceedingly small, and so the formation of new lethal alleles is relatively rare. Also, when lethality occurs before sexual maturity, the premature death of the organism decreases the frequency of the lethal allele in future generations by preventing its transmission. In other words, natural selection readily selects against the accumulation of lethal alleles in a natural population. Finally, a large proportion of lethal alleles are recessive. Since the occurrence of lethal alleles is relatively rare, it becomes very unlikely that an individual will inherit two copies of the same lethal allele. However, as we will discuss later in this text (Chapter 26), matings between genetically related individuals (known as inbreeding) can have the detrimental outcome of increasing the proportion of homozygous lethal traits.

Incomplete dominance occurs when two alleles produce an intermediate phenotype

In some cases, a heterozygote that carries two different alleles may exhibit a phenotype that is intermediate between the corresponding homozygous individuals. This phenomenon is known as **incomplete dominance**. In flowering plants such as the four-o'clock (*Mirabilis jalapa*), this is observed for alleles that affect flower color. This was first discovered by the German botanist Carl Correns in 1905. Figure 4-2 describes Correns's experiment in which a homozygous red-flowered four-o'clock plant was crossed to a homozygous white-flowered plant. The wild-type allele for red flower color is designated C^R, and the mutant allele for white color is C^W. As shown here, the offspring had pink flowers. If these F_1 offspring were allowed to self-fertilize, the F_2 generation showed a ratio of 1/4 red-flowered plants, 1/2 pink-flowered plants, and 1/4 white-flowered plants. The pink plants in the F_2 generation have the heterozygous genotype and the intermediate phenotype. As noted in the Punnett square in Figure 4-2, the F_2 generation displays a 1:2:1 ratio, which is different from the 3:1 ratio observed for simple Mendelian inheritance.

In Figure 4-2, an incompletely dominant relationship results in a heterozygote with an intermediate phenotype. At the molecular level, the mutant allele

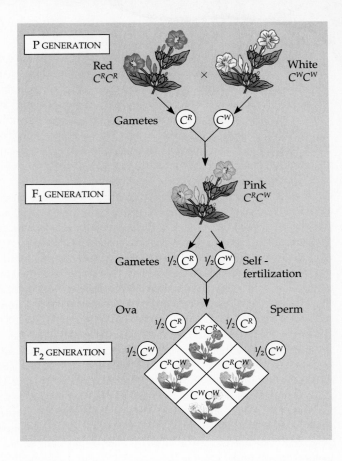

FIGURE **4-2**

Incomplete dominance in the four-o'clock plant.

GENES $\longrightarrow$ TRAITS: When two different homozygotes (C^RC^R and C^WC^W) are crossed, the resulting heterozygote, C^RC^W, has an intermediate phenotype of pink flowers. In this case, 50% of the protein encoded by the C^R gene is not sufficient to produce a red phenotype.

that causes a white phenotype is expected to result in a reduced amount of a functional protein required for pigmentation. The heterozygotes contain 50% of the normal protein, but this is not sufficient to produce the same phenotype as the homozygote that makes twice as much of this protein. In this example, 50% of the normal protein cannot accomplish the same level of pigment synthesis that 100% of the protein can.

Finally, our opinion of whether a trait is dominant or incompletely dominant may depend on how closely we examine the trait in the individual. The more closely we look, the more likely we are to discover that the heterozygote is not quite the same as the wild-type homozygote. For example, Gregor Mendel studied the characteristic of pea seed shape and visually concluded that the *RR* and *Rr* genotypes produced round seeds and the *rr* genotype produced wrinkled seeds.[1] The peculiar morphology of the wrinkled seed is caused by a great decrease in the amount of starch deposition in the seed due to a defective *r* allele. Since the time of Mendel's work, other scientists have dissected round and wrinkled seeds and examined their contents under the microscope. They have found that round seeds from heterozygotes actually contain an intermediate amount of starch grains compared with seeds from the corresponding homozygotes (Figure 4-3). Within the seed, an intermediate amount of the functional protein is not enough to produce as many starch grains as in the homozygote carrying the *R* allele. Nevertheless, at the level of our naked eyes, heterozygotes produce seeds that appear to be round. With regard to phenotypes, the *R* allele is dominant to the *r* allele at the level of visual examination, but the *R* and *r* alleles are incompletely dominant at the level of starch biosynthesis.

---

[1] The *RR*, *Rr*, and *rr* designations correspond to those described in Mendel's original work. We now know that the endosperm within the seed is triploid. Therefore, the genotypes of the endosperm within the homozygous seeds are *RRR* or *rrr*; the heterozygous seeds are *RRr* or *Rrr*.

Normal allele: *R* (round)
Recessive (defective) allele: *r* (wrinkled)

Genotype

	RR	*Rr*	*rr*
Amount of functional (starch-producing) protein	100%	50%	0%

Phenotype

	Round	Round	Wrinkled
With naked eye			
With microscope			

FIGURE **4-3**

A comparison of phenotype at the macroscopic and microscopic levels.[2]

GENES ⟶ TRAITS: This illustration shows the effects of the heterozygote having only 50% of the functional protein needed for starch production. This seed appears to be as round as those of the homozygote carrying the *R* allele, but when examined microscopically, it has produced only half the amount of starch.

Some genes exist as three or more different alleles

Within a population of organisms, some genes are found in three or more alleles. In other words, a gene can exist in **multiple alleles** that are different from each other. An interesting example is coat color in rabbits. Figure 4-4 illustrates the relationship between genotype and phenotype for a combination of four different alleles, which are designated *C* (full coat color), c^{ch} (chinchilla pattern of coat color), c^h (himalayan pattern of coat color), and *c* (albino). Any particular rabbit can only inherit two copies of these alleles. A rabbit's phenotype, therefore, depends on the dominant/recessive relationships among combinations of alleles. These are as follows:

C is dominant to c^{ch}, c^h, and *c*

c^{ch} is recessive to *C*, but dominant to c^h and *c*

c^h is recessive to *C* and c^{ch}, but dominant to *c*

c is recessive to *C*, c^{ch}, and c^h

These alleles are also interesting at the molecular level. The wild-type allele, *C*, provides full coat color. The albino allele *c*, which prevents pigmentation, is a defective allele that cannot produce a protein necessary for pigment synthesis. The chinchilla allele is only a partial defect in pigmentation. The himalayan allele is particularly interesting, because it results in pigmentation in certain parts of the body. We can understand how this works by considering gene expression at the molecular level.

The himalayan pattern of coat color is an example of a *temperature-sensitive conditional allele*. The mutation in this gene has caused a slight change in the structure of the encoded protein, so that it only works enzymatically at low temperature. Because of this property, the enzyme only functions in cooler regions of the body, primarily the ends of the extremities, the tail, and the tips of the nose and ears. As shown in Figure 4-5 and Figure 4-4c, similar types of temperature-sensitive alleles have been found in several species of domestic animals.

---

[2] See footnote on previous page.

(a) Full coat color *CC*, Cc^{ch}, Cc^h, or *Cc*

(b) Chinchilla coat color, $c^{ch}c^{ch}$, $c^{ch}c^h$, or $c^{ch}c$

(c) Himalayan coat color c^hc^h or c^hc

(d) Albino coat color, *cc*

FIGURE **4-4**

The relationship between genotype and phenotype in rabbit coat color.

(a) Siamese cat

(b) Himalayan rabbit

FIGURE **4-5**

Photographs of different species showing the expression of a temperature-sensitive conditional allele that produces a Siamese or himalayan pattern of coat color.

GENES ⟶ TRAITS: The allele shown here encodes a pigment-producing protein that only functions at lower temperatures. For this reason, the dark fur is only produced in the cooler parts of the animal, including the tips of the nose, ears, and feet.

Alleles of the ABO blood group can be dominant, recessive, or codominant

The ABO group of antigens that determine blood type in humans provides another example of multiple alleles. As shown in Table 4-3, the red blood cells of humans contain structures on their surface known as *surface antigens*. Antigens are substances that are recognized by *antibodies* produced by the immune system. There are three different possible types of surface antigens, known as A, B, and O. The synthesis of these surface antigens is controlled by three alleles designated I^A, I^B, and i, respectively. The i allele is recessive to both I^A and I^B. A person who is ii homozygous will have type O blood. A homozygous $I^A I^A$ or heterozygous $I^A i$ individual will have type A blood. The red blood cells for this individual will contain the surface antigen known as A. Similarly, a homozygous $I^B I^B$ or heterozygous $I^B i$ individual will produce the surface antigen B.

TABLE **4–3 The ABO Blood Group**

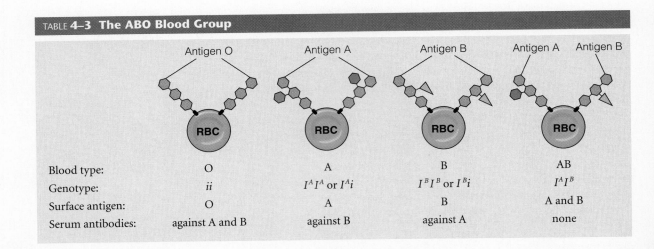

Blood type:	O	A	B	AB
Genotype:	ii	$I^A I^A$ or $I^A i$	$I^B I^B$ or $I^B i$	$I^A I^B$
Surface antigen:	O	A	B	A and B
Serum antibodies:	against A and B	against B	against A	none

As suggested by the drawing in Table 4-3, surface antigens A and B have significantly different molecular structures. A person who is $I^A I^B$ will have the blood type AB, and express both surface antigens A and B. The phenomenon in which two alleles are both expressed in the heterozygous individual is called **codominance**. In this case, the I^A and I^B alleles are codominant to each other.

Blood typing is essential for safe blood transfusions. When a person receives a blood transfusion, as for medical reasons, the donor's blood must be an appropriate match with the recipient's blood. Such a match is critical due to the occurrence of antibodies within the blood of the recipient. Antibodies bind very specifically to certain antigens. A person with type O blood naturally produces antibodies that recognize the A antigen and those that recognize the B antigen. If a person with type O blood is given type A, type B, or type AB blood, the antibodies in the recipient will react with the donated blood cells and cause them to agglutinate (clump together). This is a life-threatening situation, since it will cause the blood vessels to clog. Other incompatible combinations include a type A person receiving type B or type AB blood and a type B person receiving type A or type AB blood.

Morgan and Bridges found that the eosin eye color allele in fruit flies shows a gene dosage effect

As we have seen, dominant and recessive relationships often depend on the amount of protein necessary to produce a phenotype. In a simple dominant/recessive relationship, 50% of the functional protein allows the heterozygote to have the dominant phenotype. By comparison, incomplete dominance occurs when 50% will not produce the dominant phenotype. Here we will explore a very early experiment that supports the idea that the amount of functional protein can play an important role in influencing the phenotype of the individual.

As described in Chapter 3, Thomas Hunt Morgan and Calvin Bridges studied the pattern of inheritance of genes found in *Drosophila*. Much of their work centered on the study of X-linked genes. As was described in Experiment 3A, Morgan found an X-linked allele that caused a white-eyed phenotype. During the course of their work, Morgan and Bridges and their colleagues at Columbia University observed many thousands of fruit flies. As a matter of chance, they found many other interesting phenotypic changes in fruit flies. For example, in addition to red (normal) eye color and the recessive white color, they found another eye mutation called *eosin* (see Figure 4-6). This allele had the curious phenotypic effect of making the intensity of the eye color different in males and females. In female flies homozygous for this allele, the eye color is a yellowish pink; in males, the eyes are pinkish yellow, more translucent, and less intense in color.

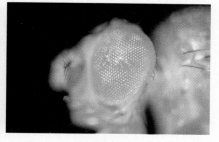

(a) Red eyes in a female fly   **(b)** Eosin eyes in a female fly

FIGURE **4-6**

Eosin eye color in *Drosophila*.

Morgan and Bridges proposed that this observed variation might be caused by the difference in the number of X-chromosomes between the female and male. They suggested that the eyes of the female contain more color because it is diploid with regard to the "eosin color producer" allele, whereas the paler eyes of the male are due to it having a single copy of this allele. This is an example of a **gene dosage effect**. In this case, two copies (and, thus, doses) of the allele provide more color than one copy of the allele.

To explore this phenomenon further, Morgan and Bridges conducted experiments to look at gene dosage effects among these three eye color alleles. In this experiment, we will consider the outcome of their crosses involving combinations of the red, white, and eosin alleles. These alleles are designated X^{w+}, X^w, and X^{w-e}, respectively.

THE HYPOTHESIS

The phenotypic effects of the eosin eye color allele are related to the number of copies of the allele.

TESTING THE HYPOTHESIS

Starting Material: True breeding strains of flies that have either the red-eye (normal) allele, white-eye allele, or eosin-eye allele.

	Experimental Level	**Conceptual Level**
1. Make the following crosses:		
A. White-eyed females × red-eyed males		A. $X^wX^w \times X^{w+}Y$
B. Red-eyed females × eosin-eyed males		B. $X^{w+}X^{w+} \times X^{w-e}Y$
C. Eosin-eyed females × white-eyed males		C. $X^{w-e}X^{w-e} \times X^wY$

2. Observe the outcome of the three types of crosses. Compare the eye colors of the true-breeding parents with their offspring.

Ratio of offspring

A. 1 red-eyed female ($X^{w+}X^w$):
   1 white-eyed male (X^wY)

B. 1 red-eyed female ($X^{w+}X^{w-e}$):
   1 red-eyed male ($X^{w+}Y$)

C. 1 (light) eosin-eyed female ($X^{w-e}X^w$): 1 eosin-eyed male ($X^{w-e}Y$)

THE DATA

Cross	Outcome*
White-eyed females (X^wX^w) × red-eyed males ($X^{w+}Y$)	225 red-eyed females ($X^{w+}X^w$) : 208 white-eyed males (X^wY)
Red-eyed females ($X^{w+}X^{w+}$) × eosin-eyed males ($X^{w-e}Y$)	679 red-eyed females ($X^{w+}X^{w-e}$) : 747 red-eyed males ($X^{w+}Y$)
Eosin-eyed females ($X^{w-e}X^{w-e}$) × white-eyed males (X^wY)	694 (light) eosin-eyed females ($X^{w-e}X^w$) : 579 eosin-eyed males ($X^{w-e}Y$)

Genotype	Phenotype	
Red-eyed females ($X^{w+}X^{w+}$)	Red eyes	No difference was observed in the redness of the eyes among these four groups.
Red-eyed females ($X^{w+}X^{w-e}$)	Red eyes	
Red-eyed females ($X^{w+}X^w$)	Red eyes	
Red-eyed males ($X^{w+}Y$)	Red eyes	
Eosin-eyed females ($X^{w-e}X^{w-e}$)	Eosin eyes	The heterozygous females had similar eye color to the hemizygous males.
Eosin-eyed females ($X^{w-e}X^w$)	Light eosin eyes	
Eosin-eyed males ($X^{w-e}Y$)	Light eosin eyes	
White-eyed females (X^wX^w)	White eyes	
White-eyed males (X^wY)	White eyes	

*From Morgan, T. H., and Bridges, C. B. (1913) Dilution Effects and Bicolorism in Certain Eye Colors of Drosophila. *J. Experimental Zool.* 15, 429–466.

INTERPRETING THE DATA

When Morgan and Bridges crossed females that were homozygous for an eye color allele to males that were hemizygous for a different allele, they always observed a 1:1 ratio among the offspring. This is consistent with the idea that these genes for eye color are X-linked and allelic to each other. Let's first consider flies that carried an X^{w+} allele. All these flies exhibited a red eye color. Furthermore, the eye color was bright red whether the fly had one or two copies of the red allele. The $X^{w+}X^{w+}$ females had red eyes that were indistinguishable from eyes of flies that had only one copy of the X^{w+} allele (namely, $X^{w+}X^{w-e}$ females, $X^{w+}X^w$ females, and $X^{w+}Y$ males). In this case, the red allele appears to be dominant to the eosin and white alleles, and no gene dosage effect occurs.

In contrast, the eosin phenotype is affected by the number of copies of the X^{w-e} allele. Homozygous females containing two copies of the X^{w-e} allele had eosin eyes, while flies with only one copy of this allele (namely, $X^{w-e}X^w$ females and $X^{w-e}Y$ males) had light eosin eyes. In heterozygous $X^{w-e}X^w$ females, Morgan and Bridges proposed that the white allele (which does not produce color) was having a "dilution effect" on the phenotype. Another way of considering this phenomenon is that the X^{w-e} allele shows a gene dosage effect. An increase in the dosage of this gene from one copy to two copies yields an increase in the degree of the eosin color in the eyes of the flies. This result supports the idea that the amount of a gene product can influence the outcome of a trait.

Overdominance occurs when heterozygotes have superior traits

Let's now turn our attention to another way that two alleles may interact to affect the phenotype of the individual. For certain genes, heterozygotes may display

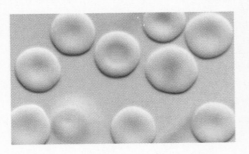

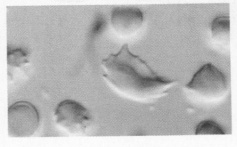

(a) Red blood cells containing normal hemoglobin, *HbᴬHbᴬ*

⊢————⊣ 10 μm

(b) Red blood cells containing sickle-cell hemoglobin, *HbˢHbˢ*

⊢————⊣ 10 μm

FIGURE **4-7**

A comparison of (a) normal red blood cells and (b) those from a person with sickle cell anemia. The cells in (b) exhibit an abnormal sickle cell morphology.

characteristics that are more beneficial to their survival and reproduction than either corresponding homozygote. For example, a heterozygote may be larger, disease resistant, or better able to withstand harsh environmental conditions. The phenomenon in which a heterozygote is more vigorous than both of the corresponding homozygotes is called **overdominance** or **heterozygote advantage**.

A well-documented example involves a human allele that causes *sickle cell anemia* in homozygous individuals. This disease is an autosomal recessive disorder in which the affected individual produces an abnormal form of the protein *hemoglobin*. This protein carries oxygen within red blood cells. Normal individuals carry the *Hbᴬ* allele and make hemoglobin A. Individuals affected with this disease are homozygous for the *Hbˢ* allele and only produce hemoglobin S. This causes their red blood cells to deform into a sickle shape under conditions of low oxygen concentration (Figure 4-7). The sickling phenomenon causes the life span of these cells to be greatly shortened, to only a few weeks compared with a normal span of four months. In addition, abnormal sickled cells become clogged in the capillaries throughout the body, leading to localized areas of oxygen depletion, tissue damage, and painful crises. For these reasons, the homozygote *HbˢHbˢ* individual is less likely to survive than an individual producing hemoglobin A.

In spite of the harmful consequences to homozygotes, the sickle cell allele has been found at a fairly high frequency among African populations that are exposed to *Plasmodium vivax*, *P. falciparum*, *P. ovale*, or *P. malariae*, the causative agents of malaria. This protozoan spends part of its life cycle within the *Anopheles* mosquito and another part within the red blood cells of humans who have been bitten by an infected mosquito. *Plasmodium sp.* do not survive as well within the red blood cells of persons who contain both hemoglobin S and hemoglobin A. This provides people who are heterozygous, *HbᴬHbˢ*, with better resistance to malaria compared with normal *HbᴬHbᴬ* homozygotes. Therefore, even though the homozygous *HbˢHbˢ* condition is detrimental, the greater survival of the heterozygote has selected for the presence of the *Hbˢ* allele within populations where malaria is prevalent. In Chapter 26, we will consider the role that natural selection plays in maintaining alleles that are beneficial to the heterozygote but harmful to the homozygote.

At the molecular level, overdominance is usually due to two alleles that produce polypeptides with slightly different amino acid sequences. One way to explain how two different polypeptides may work better than either polypeptide alone is related to the subunit composition of proteins. In some cases, proteins function as a complex of multiple subunits, each composed of a separate polypeptide. When a protein is composed of two subunits, it is known as a dimer. When both subunits are encoded by the same gene, the protein is called a *homodimer*. Figure 4-8 considers a situation where a gene exists in two alleles that encode polypeptides designated

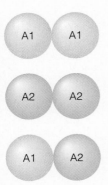

FIGURE **4-8**

Subunit composition of homodimers when a gene exists in two alleles. In this example, a gene exists in two alleles designated A1 and A2, which encode polypeptides also designated A1 and A2. The homozygotes that are A1A1 or A2A2 will make homodimers that are A1A1 and A2A2, respectively. The A1A2 heterozygote can make A1A1 and A2A2 and can also make A1A2 homodimers.

A1 and A2. Homozygous individuals can only produce A1A1 or A2A2 homodimers, whereas a heterozygote can also produce an A1A2 homodimer. (Thus, heterozygotes produce three forms of the homodimer, homozygotes only one.) For some proteins, A1A2 homodimers may have better functional activity because they are more stable or able to function under a wider range of conditions. The greater activity of the homodimer protein may be one underlying reason that a heterozygote has a phenotype that has characteristics superior to either homozygote.

A second molecular explanation of overdominance is that the protein encoded by each allele functions under different conditions. For example, suppose that a gene encodes a metabolic enzyme that can be found in two forms (corresponding to the alleles), one that functions better at a lower temperature and the other that functions optimally at a higher temperature. The heterozygote, which makes a mixture of both enzymes, may be at an advantage under a wider temperature range than either of the corresponding homozygotes.

Before ending this section, we will compare overdominance with a related phenomenon. Among plant and animal breeders, a common mating strategy is to begin with two different highly inbred strains and cross them together to produce hybrids. When the hybrids display traits that are superior to both corresponding parental strains, the outcome is known as **heterosis** or **hybrid vigor**. Within the field of agriculture, heterosis has been particularly valuable in improving quantitative traits such as size, weight, growth rate, and disease resistance. It was first described by George Shull in 1908 in regard to crosses of different strains of corn. Heterosis is different from overdominance, because the hybrid may be heterozygous for many genes, not just a single gene. Nevertheless, some of the beneficial effects of heterosis may be caused by the occurrence of overdominance in one or more heterozygous genes. However, as we will see in Chapter 25, heterosis can also result from the masking of deleterious recessive alleles that tend to accumulate in highly inbred domesticated strains.

Incomplete penetrance occurs when some dominant traits skip a generation

Dominant alleles are expected to influence the outcome of a trait when they are present in heterozygotes. Occasionally, however, this may not occur. Figure 4-9 illustrates a human pedigree for a dominant trait known as *polydactyly*. This trait causes the affected individual to have additional fingers and/or toes (Figure 4-10). A single copy of this gene is sufficient to cause this condition. Sometimes, however, individuals carry the dominant allele but do not exhibit the trait. In Figure 4-9, individual III-2 has inherited the polydactyly allele from his mother and passed the trait on to a daughter and a son. However, individual III-2 does not actually exhibit the trait himself, even though he is a heterozygote. This phenomenon is called **incomplete penetrance**. This term indicates that a dominant allele does not always "penetrate" into the phenotype of the individual. The measure of penetrance is described at the populational level. For example, if 60% of the heterozygotes carrying a dominant allele exhibit the trait, we would say that this trait is 60% penetrant. In any particular individual, of course, the trait is either present or it is not.

A second issue in evaluating the outcome of traits is the degree to which the trait is expressed. In the case of polydactyly, the number of extra digits can vary. For example, one individual may only have an extra toe on one foot, whereas a second individual may have extra digits on both the hands and feet. This variation is known as the **expressivity** of the trait. Using genetic terminology, a person with several extra digits would have high expressivity of this trait, while a person with a single extra digit would have a low expressivity. The extreme case, when a heterozygote does not express the trait at all, is incomplete penetrance.

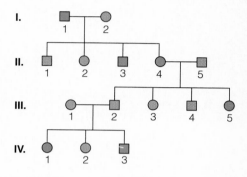

FIGURE **4-9**

A family pedigree for polydactyly, a dominant trait showing incomplete penetrance. Affected individuals are shown in red. Notice that an offspring may inherit the trait from a parent who is heterozygous and phenotypically normal.

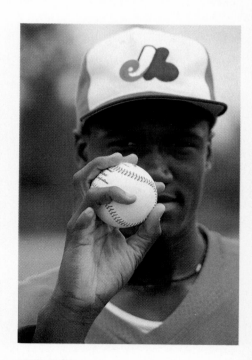

FIGURE **4-10**

Antonio Alfonseca, a baseball player with polydactyly.

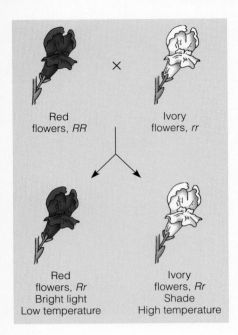

FIGURE **4-11**
Variation in the expression of a trait due to environmental effects.

GENES → TRAITS: In this example, snapdragon flower color in the heterozygote, *Rr*, is affected by the amount of light and temperature in the plant's environment. Presumably, the protein encoded by this gene is sensitive to changes in temperature and light intensity.

The expression of a trait can be influenced by the environment and by modifier genes

Throughout this text, our study of genetics often leads us to focus on the roles of particular genes in the outcome of certain traits. In addition to genetics, environmental conditions may have a great impact on the phenotype of the individual. For example, the temperature and degree of sunlight have been shown to affect the flower color of certain snapdragons (Figure 4-11). When a true-breeding red-flowered plant is crossed to a true-breeding ivory-flowered plant, the outcome of the F_1 generation depends on the lighting and temperature. When grown in bright light at relatively low temperatures, the heterozygotes develop red flowers, but under shady conditions and warmer temperatures, the flowers are ivory.

A dramatic example of the relationship between environment and phenotype can be seen in the human genetic disease known as phenylketonuria (PKU). This autosomal recessive disease is caused by a defect in a gene that encodes the enzyme phenylalanine hydroxylase. Homozygous individuals with this defective allele are unable to metabolize the amino acid phenylalanine. When given a normal diet containing phenylalanine, PKU individuals manifest a variety of detrimental traits including mental retardation, underdeveloped teeth, and foul-smelling urine. In contrast, when PKU individuals are diagnosed early and given a restricted diet that is free of phenylalanine, they develop normally (see Figure 1-7). Since the 1960s, testing methods have been developed that can determine if an individual is lacking the phenylalanine hydroxylase enzyme. These tests permit the identification of infants that have PKU; their diets can then be modified before the harmful effects of phenylalanine ingestion have occurred. As a result of government legislation, greater than 90% of the infants born in the United States are now tested for PKU. This test prevents a great deal of human suffering. Furthermore, it is cost-effective. In the United States, the annual cost of the PKU testing is estimated to be a few million dollars, whereas the cost of treating severely retarded individuals with the disease would be hundreds of millions of dollars.

The outcome of certain traits can be influenced by the gender of the individual

The inheritance pattern for certain traits is governed by the gender of the individual. The term **sex-influenced inheritance** refers to the phenomenon in which an allele is dominant in one gender but recessive in the opposite gender.

In humans, pattern baldness provides an example of a sex-influenced trait. As shown in Figure 4-12, it is characterized by a balding pattern on the male's head in which hair loss occurs on the front and top but not on the sides. The gene that causes pattern baldness is inherited as an autosomal trait. (A common misperception is that this gene is X-linked.) When a male is heterozygous for the baldness allele, he will become bald:

Genotype	Phenotype	
	Females	*Males*
BB	bald	bald
Bb	nonbald	bald
bb	nonbald	nonbald

In contrast, a heterozygous female will not be bald. Women who are homozygous for the baldness allele will develop the trait, but in women it is usually characterized by a significant thinning of the hair that occurs relatively late in life. The sex-influenced nature of pattern baldness appears to be related to the levels of the male sex hormones. In females, a rare tumor of the adrenal gland can cause the se-

(a) John Adams

(b) John Quincy Adams

(c) Charles Francis Adams

(d) Henry Adams

FIGURE **4-12**

Pattern baldness in the Adams's family line.

cretion of large amounts of male sex hormones. If this occurs in a heterozygous *Bb* female, she will become bald. When the tumor is removed surgically, her hair will return to its normal condition.

The autosomal nature of pattern baldness has been revealed by the analysis of many human pedigrees. An example is shown in Figure 4-13. A bald male may inherit the bald allele from either parent, and thus a striking observation is that bald fathers can pass this trait on their sons. This could not occur if the trait was X-linked, since fathers transmit Y-chromosomes to their sons. The analysis of many human pedigrees have shown that bald fathers, on average, have at least 50% bald

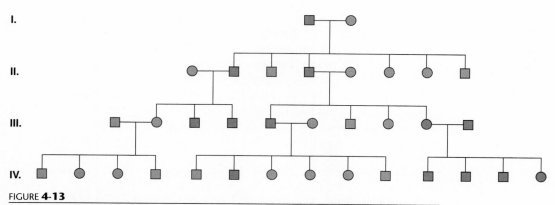

FIGURE **4-13**

A pedigree of human pattern baldness, an example of the sex-influenced expression of an autosomal gene. Bald individuals are shown in red.

(a) Cock-feathered male, *hh*

(b) Hen-feathered female, *HH, Hh,* or *hh*

FIGURE **4-14**

Differences in the feathering pattern between male and female chickens, an example of sex-limited inheritance. Most breeds of domesticated chickens are *hh*.

sons. They are expected to produce an even higher percentage of bald male offspring if they are homozygous for the bald allele and/or the mother also carries one or two copies of the bald allele. For example, a heterozygous bald male and heterozygous (nonbald) female will produce 75% bald sons, while a homozygous bald male or homozygous bald female will produce all bald sons.

An extreme example of gender influence on an organism's phenotype is provided by **sex-limited traits**, which are those that occur in only one of the two genders. In humans, for example, breast development is a trait that is normally limited to females, whereas beard growth is limited to males. Among many types of birds, the male of the species has more ornate plumage than the female. As shown in Figure 4-14, roosters normally have a larger comb and wattles, and longer neck, tail, and sickle feathers than do hens. These are sex-limited features that may be found in roosters but never in normal hens. However, some varieties of chickens have males that are hen-feathered. Hen-feathering is controlled by a dominant allele that is expressed both in males and females, whereas cock-feathering is controlled by a recessive allele that is only expressed in males:

Genotype	Phenotype	
	Females	*Males*
hh	hen-feathered	cock-feathered
Hh	hen-feathered	hen-feathered
HH	hen-feathered	hen-feathered

Like pattern baldness in humans, the pattern of hen-feathering depends on the production of sex hormones. If a newly hatched female with an *hh* genotype has her single ovary removed surgically, she will develop cock-feathering and look indistinguishable from a male.

Sex-influenced and sex-limited inheritance are often confused with the different phenomenon of sex linkage. As discussed in Chapter 3, some genes are located on the sex chromosomes, such as the X-chromosome in mammals. Since males have only one copy of the X-chromosome and females have two, this results in unusual patterns of transmission and expression. In the examples of sex-influenced and sex-limited traits described in this chapter, however, the genes are not located on the sex chromosomes. Baldness in humans and feather plumage in chickens are cases where the genes are located on autosomes. Nevertheless, the expression of these genes is affected by whether the individual is male or female.

GENE INTERACTIONS

In the preceding section, we considered the effects of a single gene on the outcome of a trait. Most traits, however, can be affected by the contributions of two or more genes. Morphological features such as height, weight, growth rate, and pigmentation are all affected by the expression of many different genes. We often discuss the effects of a single gene on the outcome of a single trait as a way to simplify the genetic analysis. Even so, the trait may not be completely determined by a single gene. For example, Mendel studied one gene that affected the height of pea plants (tall versus dwarf alleles). Actually, there are many other genes in pea plants that can also affect height, but Mendel did not happen to study them. In general, most traits of an organism are determined by complex contributions of many different genes that collectively influence the individual's phenotype. When two or more different genes influence the outcome of a single trait, this is known as a **gene interaction**.

In this section, we introduce the diverse ways that two genes may interact to affect a single trait. We will discuss three different examples, all involving two genes that exist in two alleles. To compare these examples, we will focus our attention on the outcome of a cross between two individuals that are heterozygous for both genes. In a general way, we can illustrate this cross as

$AaBb \times AaBb$ where A is dominant to a, and B is dominant to b

If these two genes govern two different traits, Mendel's laws of segregation and independent assortment then predict a 9:3:3:1 ratio among the offspring. However, in this section, the two genes will affect the same trait. As we will see, there are many different ways that the alleles of two genes may interact to affect a trait. We will appreciate their interactions by considering how they affect the 9:3:3:1 ratio.

A cross involving a two-gene interaction can produce a 9:3:3:1 ratio in offspring when four distinct phenotypes are produced

The first case of two different genes interacting to affect a single trait was discovered by William Bateson and Reginald Punnett in 1906, while they were investigating the inheritance of comb morphology in chickens. Several common varieties of fowl possess combs with different morphologies as illustrated in Figure 4-15a. In their studies, Bateson and Punnett crossed a Wyandotte breed with a rose comb to a Brahma having a pea comb. All the F$_1$ offspring had a walnut comb.

When these F$_1$ offspring were mated to each other, the F$_2$ generation consisted of chickens with four types of combs in the following phenotypic ratio: 9 walnut : 3 rose : 3 pea : 1 single. As we have seen in Chapter 2, a 9:3:3:1 ratio is obtained when the F$_1$ generation is heterozygous for two different genes and these genes assort independently. However, an important difference here is that we have four distinct categories of a single trait. Based on the 9:3:3:1 ratio, Bateson and Punnett reasoned that a single trait (comb morphology) was determined by two different genes:

> R (rose comb) is dominant to r
>
> P (pea comb) is dominant to p
>
> R and P are codominant (walnut comb)
>
> $rrpp$ produces single comb

As shown in the Punnett square of Figure 4-15b, each of the genes can exist in two alleles, and the two genes show independent assortment.

A cross involving a two-gene interaction can produce a 9:7 ratio in offspring when both genes are epistatic to each other

Bateson and Punnett also discovered an unexpected gene interaction when studying crosses involving the sweet pea, *Lathyrus odoratus*. The wild sweet pea normally has purple flowers. However, they obtained several true-breeding varieties with white flowers. Not surprisingly, when they crossed a true-breeding purple-flowered plant to a true-breeding white-flowered plant, the F$_1$ generation was all purple-flowered plants and the F$_2$ generation (produced by self-hybridization of the F$_1$ generation) contained purple- and white-flowered plants in a 3:1 ratio. The surprising result came in an experiment where they crossed two different varieties of white-flowered plants (see Figure 4-16). All the F$_1$ generation plants had purple flowers! When these were allowed to self-hybridize, the F$_2$ generation contained purple and white flowers in a ratio of 9 purple : 7 white. From this result, Bateson and Punnett deduced that two different genes were involved:

> C (one purple-color producing) allele is dominant to c (white)
>
> P (another purple-color producing) allele is dominant to p (white)
>
> cc or pp masks P or C alleles, producing white color

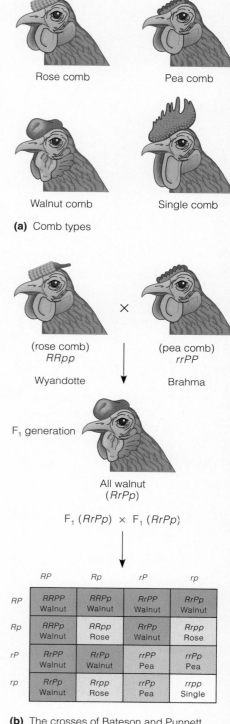

(a) Comb types

(b) The crosses of Bateson and Punnett

FIGURE **4-15**

Inheritance of comb morphology in chickens. This trait is influenced by two different genes, which can each exist in two alleles. **(a)** Four phenotypic outcomes are possible. **(b)** The crosses Bateson and Punnett performed to examine the interaction of the two genes.

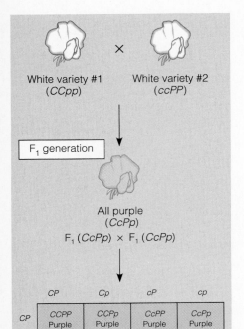

FIGURE **4-16**

A cross between two different white varieties of the sweet pea.

GENES → TRAITS: The color of the sweet pea flower is controlled by two genes, which are epistatic to each other. Each gene is necessary for the production of an enzyme required for pigment synthesis. The recessive allele of either gene prevents the synthesis of a functional enzyme. If an individual is homozygous recessive for either of the two genes, the purple pigment cannot be synthesized. This results in a white phenotype.

If a plant was homozygous for either recessive white allele, it would develop a white flower even though the other gene may contain a purple-producing allele. The term **epistasis** is used to describe the situation where one gene can mask the phenotypic effects of a different gene. In this case, homozygosity for the white allele at one gene will mask the purple-producing allele of another gene.

Epistatic interactions often arise because two (or more) different proteins participate in an enzymatic pathway leading to the formation of a single product. As an example, let's consider the formation of a purple pigment in the sweet pea:

$$\text{Colorless precursor} \xrightarrow{\text{Enzyme C}} \text{Colorless intermediate} \xrightarrow{\text{Enzyme P}} \text{Purple pigment}$$

In this simplified example, a colorless precursor molecule must be acted on by two different enzymes to produce the purple pigment. Gene C encodes a functional protein called enzyme C, which converts the colorless precursor into a colorless intermediate. The recessive allele (c) results in a lack of production of this enzyme in the homozygote. Gene P encodes a functional enzyme P, which converts the colorless intermediate into the purple pigment. Like the c allele, the recessive p allele results in an inability to produce enzyme P in the homozygote. If an individual is homozygous for either recessive allele (cc or pp), it will not make any functional enzyme C or enzyme P. When one of these enzymes is missing, it is impossible to make the purple pigment. Therefore, the flowers remain white.

EXPERIMENT 4B

Bridges observed an 8:4:3:1 ratio in offspring from fruit fly crosses when he found that the cream-eye allele can modify only the phenotypic expression of the eosin-eye allele, not the red or white alleles

A very early example of two different genes interacting to affect the phenotype of a single trait was also discovered by Calvin Bridges, in 1919. Within true-breeding cultures of flies with eosin eyes (see Experiment 4A), he occasionally found a fly that had a noticeably different eye color. For example, he identified a rare fly with cream-colored eyes. He reasoned that this new eye color could be the result of two different events. One possibility is that the cream-colored phenotype could be the result of a new mutation that changed the eosin allele into a cream allele. A second possibility is that a different gene may have incurred a mutation that modified the expression of the eosin allele. This second possibility is an example of a gene interaction. To distinguish between these two possibilities, he carried out the crosses described here. Keep in mind that Bridges already knew that the eosin allele is X-linked. However, he did not know whether the cream allele was in the same gene as the eosin allele, in another gene located somewhere else on the X-chromosome, or on an autosome. As we will see, the cream allele was found to be in a different gene located on an autosome. In this second gene, C represents the normal allele (which does not modify the eosin phenotype) and c^a is the cream allele that modifies the eosin phenotype.

THE HYPOTHESIS

Cream-colored eyes in fruit flies are due to the effect of a second gene that modifies the expression of the eosin allele.

TESTING THE HYPOTHESIS

Starting material: From a culture of flies with eosin eyes, Bridges obtained a fly with cream-colored eyes. The allele was called cream a. This fly was used to produce a true-breeding culture of flies with cream-colored eyes.

Experimental Level

Conceptual Level

1. Cross males with cream-colored eyes to wild-type females.

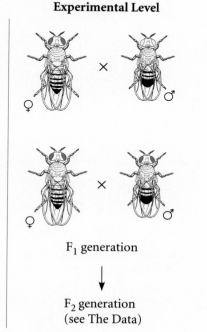

$$CCX^{w+}X^{w+} \times c^a c^a X^{w-e}Y$$

2. Observe the F_1 offspring and then allow the offspring to mate with each other.

F_1 generation

$Cc^a X^{w+}X^{w-e}$ and $Cc^a X^{w+}Y$
Red-eyed females and males

3. Observe and record the eye color and gender of the F_2 generation.

F_2 generation
(see The Data)

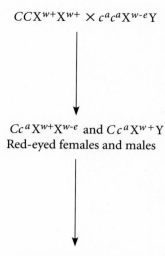

See Punnett square under Interpreting The Data

THE DATA

Cross	Outcome*
P cross:	
Cream-eyed male × wild-type female	F_1: all red eyes
F1 cross:	
F_1 brother × F_1 sister	F_2: 104 females with red eyes
	47 males with red eyes
	44 males with eosin eyes
	14 males with cream eyes

*From Bridges, C. B. (1919) Specific Modifiers of Eosin Eye Color in *Drosophila melanogaster.* *J. Experimental Zool.* 28, 337–384.

INTERPRETING THE DATA

The F_2 generation indicates that the cream allele is not in the same gene as the eosin allele. Bridges started this experiment with male flies that had cream-colored eyes. If the cream allele was in the same gene as the eosin allele, these male flies could have only the cream allele, not the eosin allele. Likewise, the parental female flies were wild-type flies that did not contain the eosin allele. Nevertheless, the F_2 generation produced 44 male flies with eosin eye color. These results are consistent with the idea that the male flies of the parental generation possessed both the eosin and cream alleles, implying that these two alleles must be in different genes. To explain the data, Bridges proposed that the cream allele is a "specific modifier" of eosin eye color.

As outlined here, the *cream a* allele is a recessive gene found on an autosome. The parental cross is expected to produce all red-eyed F_1 flies in which the males are $Cc^a X^{w+}Y$ and the females are $Cc^a X^{w+}X^{w-e}$. When these F_1 offspring are allowed to mate with each other, the following outcome should occur:

	CX^{w+}	CY	$c^a X^{w+}$	$c^a Y$
CX^{w+}	$CCX^{w+}X^{w+}$	$CCX^{w+}Y$	$Cc^a X^{w+}X^{w+}$	$Cc^a X^{w+}Y$
CX^{w-e}	$CCX^{w+}X^{w-e}$	$CCX^{w-e}Y$	$Cc^a X^{w+}X^{w-e}$	$Cc^a X^{w-e}Y$
$c^a X^{w+}$	$Cc^a X^{w+}X^{w+}$	$Cc^a X^{w+}Y$	$c^a c^a X^{w+}X^{w+}$	$c^a c^a X^{w+}Y$
$c^a X^{w-e}$	$Cc^a X^{w+}X^{w-e}$	$Cc^a X^{w-e}Y$	$c^a c^a X^{w+}X^{w-e}$	$c^a c^a X^{w-e}Y$

Outcome:

$1\ CCX^{w+}X^{w+} : 1\ CCX^{w+}X^{w-e} : 2\ Cc^a X^{w+}X^{w+} : 2\ Cc^a X^{w+}X^{w-e} : 1\ c^a c^a X^{w+}X^{w+} :$
$1\ c^a c^a X^{w+}X^{w-e}$ = **8 red-eyed females**

$1\ CCX^{w+}Y : 2\ Cc^a X^{w+}Y : 1\ c^a c^a X^{w+}Y$ = **4 red-eyed males**

$1\ CCX^{w-e}Y : 2\ Cc^a X^{w-e}Y$ = **3 eosin-eyed males**

$1\ c^a c^a X^{w-e}Y$ = **1 cream-eyed male**

This phenotypic outcome proposes that the specific modifier allele, c^a, can modify the phenotype of the eosin allele but not the red-eyed allele. The eosin allele can only be modified when the c^a allele is homozygous. The predicted 8:4:3:1 ratio agrees reasonably well with Bridges's data shown in the results. Bridges concluded:

> Specific modifications are clear and simple cases of "multiple genes." Each is the result of the coaction of a specific modifying gene (cream a), which by itself produces little or no visible effect, and of a particular gene (eosin) that is necessary as a "base" or "differentiator."

CONCEPTUAL SUMMARY

In this chapter, we have discussed many examples that extend our understanding of Mendelian inheritance. For single genes, the relationship between dominance and recessiveness has been broadened to include situations that deviate from the simple dominant/recessive relationship observed by Mendel. For example, many genes have *multiple alleles*. This creates a more complex relationship where one allele can be dominant to a second allele but recessive to a third allele. Alternatively, two alleles in the heterozygote may follow a pattern that is different from clear-cut dominance. In *codominance*, two alleles in the heterozygote both contribute to the phenotype. *Incomplete dominance* produces a heterozygote with an intermediate phenotype, whereas *overdominance* results in a heterozygote with a superior phenotype. Certain alleles can also exhibit a *gene dosage effect*, in which the degree of the phenotype is correlated with the number of copies of the allele. In addition, the outcome of a trait may exhibit a range of phenotypes. The *expressivity* of a trait may be very high, very low, or in the case of *incomplete penetrance*, even zero. Both environmental and genetic factors contribute to the expressivity of inherited traits.

We have also explored cases in which two (or more) different genes affect the outcome of a single trait. This is known as a *gene interaction*. Even though geneticists tend to study single genes that have a great impact on a single trait, it is more realistic to expect that traits are due to complex contributions from many genes. As in the case of chicken combs, two alleles may interact in several different ways to alter an organism's morphology. In addition, alleles in one gene may act to mask the phenotypic effects of the alleles at a second gene. This phenomenon is known as *epistasis*. For example, two genes affect flower color in the sweet pea; homozygosity for a white-flower allele at one gene masks the phenotypic expression of a purple-flower allele at a second gene.

Many particularly gratifying and exciting advances in genetics have involved explaining Mendelian inheritance at the molecular level. As described in Parts III and IV of this text, there continues to be a great amount of effort toward understanding molecular genetics. Recessive alleles often are the result of a defective gene that is unable to produce a functionally active protein. In many cases, the normal (dominant) gene can compensate for this defect and still produce a normal phenotype. However, in an incomplete dominant relationship, 50% of the normal protein is not sufficient to yield the normal phenotype. Overdominance can sometimes be explained by protein subunit associations or by enzymes that can function under different conditions. Codominant alleles may alter the function of cellular proteins in specific ways. Also, we have seen that conditional alleles may alter protein structure so that the protein will only function under certain circumstances.

To understand how inheritance may deviate from a simple Mendelian pattern, either a genetic or molecular approach can be taken. In a genetic approach, a researcher makes crosses and then analyzes the phenotypes of the offspring. In this chapter, we have seen that the experimenter must weigh many factors when analyzing the outcome of a cross. These include the dominant/recessive relationships of alleles (namely, dominant, recessive, codominant, incompletely dominant, or overdominant), the presence of multiple alleles in a population, the relationship between the gender of the offspring and their phenotype, and the influence of the environment. In addition, complicating factors such as incomplete penetrance and gene interactions must be considered. Based on these factors, a researcher can construct a Punnett square to see if a pattern of inheritance fits the available data obtained from crosses.

Another level of understanding in Mendelian inheritance is to appreciate how the molecular expression of genes underlies their impact on the phenotype of an organism. Using tools that are described later in this text (particularly in Chapter 19), researchers can quantitatively compare the level of gene expression between wild-type and mutant alleles. This helps us to understand how the amount of gene expression is correlated with the phenotype of the organism. Also, when gene interactions occur, an investigator may want to understand, at the molecular level, how the gene products participate in a common molecular pathway such as an enzymatic pathway.

PROBLEM SETS

SOLVED PROBLEMS

1. Coat color in rodents is determined by a gene interaction between two genes. If a true-breeding black rat is crossed to a true-breeding albino rat, the result is a rat with agouti (brownish/dark gray) coat color. If two agouti animals of the F_1 generation are crossed to each other, they produce agouti, black, and albino animals in a 9:3:4 ratio. Explain the pattern of inheritance for this trait.

Answer: Since the parental generation is true-breeding, we know that the F_1 offspring are heterozygous for two genes. One strategy for solving this problem is to first consider how a 9:3:4 ratio deviates from the 9:3:3:1 ratio we have already encountered in (two-gene) independent assortment problems. A 9:3:3:1 ratio has four categories of phenotypes. If we consider that the last two categories are combined via a gene interaction, this would yield a 9:3:4 ratio. With this idea in mind, we can then proceed to fill in a Punnett square in order to deduce the pattern of inheritance.

The F_1 offspring are heterozygous for two genes. Let's call them A (for agouti) and C (for colored):

$$CcAa \times CcAa$$

	CA	Ca	cA	ca
CA	CCAA Agouti	CCAa Agouti	CcAA Agouti	CcAa Agouti
Ca	CCAa Agouti	CCaa Black	CcAa Agouti	Ccaa Black
cA	CcAA Agouti	CcAa Agouti	ccAA Albino	ccAa Albino
ca	CcAa Agouti	Ccaa Black	ccAa Albino	ccaa Albino

In this case, C is dominant to c, and A is dominant to a. If an animal has at least one copy of both dominant alleles, it will have the agouti coat color. If an animal has a dominant C allele but is aa homozygous, it will develop a black coat. The other four cases of albino animals all are cc homozygous. This occurs even when an animal carries the dominant A allele. Therefore, c is epistatic to A. The converse, however, is not true. The a allele is not epistatic to C, since $Ccaa$ or $CCaa$ animals are black.

2. In Ayrshire cattle, the spotting pattern of the animals can be either red and white or mahogany and white. The mahogany and white pattern is caused by the allele M. The red and white phenotype is controlled by the allele m. When crossed to each other, the following results are obtained:

Genotype	Phenotype	
	Females	*Males*
MM	mahogany and white	mahogany and white
Mm	red and white	mahogany and white
mm	red and white	red and white

Explain the pattern of inheritance.

Answer: The inheritance pattern for this trait is sex-influenced inheritance. The M allele is dominant in males but recessive in females, while the m allele is dominant in females but recessive in males.

3. In 1905, George Shull, a botanist at Princeton University, conducted a genetic study of a weed known as shepherd's purse, a member of the mustard family. One trait he followed was the shape of the seed capsule, which can be triangular or a small ovate shape. When he first crossed a true-breeding plant with triangular capsules to a plant having ovate capsules, the F_1 generation all had triangular capsules. The surprising result came in the F_2 generation. Shull observed a 15:1 ratio of plants having triangular capsules to ovate capsules. Explain this ratio based on a two-gene interaction.

Answer: This result can be explained by a two-gene interaction in which each gene is found as a dominant or recessive allele. The presence of one dominant allele in either gene results in a triangular capsular. The results of Shull's crosses are outlined in the following:

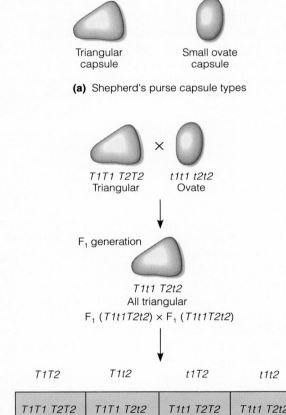

(a) Shepherd's purse capsule types

(b) The crosses of Shull

This early experiment of George Shull demonstrates an important phenomenon, confirmed in subsequent genetic studies. Many eukaryotic genes appear to be *redundant*. In other words, there are two (or more) genes that can play analogous roles in the life of the organism. In this example, either the *T1* or *T2* allele, which are alleles of different genes, are sufficient to cause triangular capsules. If the expression of one of the two genes is abolished by a recessive mutation, a dominant allele of the other gene can still produce the triangular phenotype. At the molecular level, *gene redundancy* can sometimes be explained by the formation of extra copies of a gene during evolution. This event, known as **gene duplication**, is described in Chapter 8.

4. Two pink-flowered four-o'clocks are crossed to each other. What are the following probabilities?

A. A red-flowered plant
B. The first three plants examined will be white
C. A plant will either be white or pink

Answer: The first thing we need to do is construct a Punnett square to determine the individual probabilities for each type of offspring. Since flower color is incompletely dominant, the cross is $Rr \times Rr$ and the expected phenotypic ratio is 1 red : 2 pink : 1 white. In other words, 1/4 are expected to be red, 1/2 pink, and 1/4 white.

A. The probability of a red-flowered plant is 1/4, which equals 25%.
B. Use the product rule: $1/4 \times 1/4 \times 1/4 = 1/64 = 1.6\%$.
C. Use the sum rule: $1/4 + 1/2 = 3/4 = 75\%$.

CONCEPTUAL QUESTIONS

1. Describe the differences among dominance, incomplete dominance, codominance, and overdominance.

2. What is a gene dosage effect? What types of alleles would you expect to exhibit a gene dosage effect?

3. Discuss the differences among sex-influenced, sex-limited, and sex-linked inheritance. What are the roles of sex hormones? Describe examples.

4. The term expressivity suggests that a trait can display a range of phenotypes. Using examples in this chapter, describe the factors that alter the expressivity of a trait.

5. What is meant by gene interaction? How can gene interaction be explained at the molecular level?

6. Let's suppose a recessive allele encodes a completely defective protein. If the normal allele is dominant, what does that tell us about the amount of the normal protein that is sufficient to cause the phenotype? What if the allele is incompletely dominant?

7. What is a common explanation for a recessive allele at the level of protein function?

8. A seed dealer wants to sell three different packs of four-o'clock seeds that will produce either red, white, or pink flowers. Explain how this should be done.

9. At the molecular level, how would you explain a gene dosage effect?

10. Which blood phenotypes (A, B, AB, and/or O) provide an unambiguous genotype?

11. A woman with type B blood has a child with type O blood. What are the possible genotypes and blood types of the father?

12. A type A woman was the daughter of a type O father and type A mother. If she has children with a type AB man, what are the following probabilities?

A. A type AB child
B. A type O child

C. The first three children with type AB blood
D. A family containing two children with type B blood and one child with type AB

13. A rabbit with chinchilla fur is mated to a himalayan. All their F_1 offspring have chinchilla fur. When the F_1 offspring are allowed to mate with each other, they sometimes produce albino F_2 offspring. What are the genotypes of the original parents and the F_1 offspring?

14. In Shorthorn cattle, fur color is controlled by a single gene that can exist as a red allele (R) or white allele (r). The heterozygotes (Rr) have a color called roan that looks less red than the RR homozygotes. However, when examined carefully, the roan phenotype in cattle is actually due to a mixture of completely red hairs and completely white hairs. Should this trait be called incompletely dominant, codominant, or something else? Explain your reasoning.

15. With regard to genes and the functions of the proteins they encode, suggest a molecular explanation for solved problem 1.

16. In chickens, the Leghorn variety has white feathers due to a dominant allele. In Silkies, white feathers are due to a recessive allele in a second (different) gene. If a true-breeding white Leghorn is crossed to a white Silkie, what is the expected phenotype of the F_1 generation? If members of the F_1 generation are mated to each other, what it the expected phenotypic outcome of the F_2 generation? (Assume that nonwhite birds will be brown.)

17. A woman who is not bald and whose mother is bald has children with a bald man whose father is not bald. What are their probabilities of having the following types of families?

A. Their first child will not become bald.
B. Their first child will be a male that will not become bald.
C. Their first three children will be normal females.

18. In chickens, some varieties have feathered shanks (legs) while others do not. In a cross between a Black Langhans (feathered shanks) and Buff Rocks (unfeathered shanks), the shanks of the F_1 generation are all feathered. When the F_1 generation is crossed, the F_2 generation contains chickens with feathered shanks to unfeathered shanks in a ratio of 15:1. Suggest an explanation for this outcome.

19. In sheep, the formation of horns is a sex-influenced trait; the allele that results in horns is dominant in males and recessive in females. Females must be homozygous for the horned allele to have horns. A horned ram is crossed to a polled (unhorned) ewe, and the first offspring they produce is a horned ewe. What are the genotypes of the parents?

20. In rabbits, the color of body fat is controlled by a single gene with two alleles, designated Y and y. The outcome of this trait is affected by the diet of the rabbit. When raised on a standard vegetarian diet, the dominant Y allele confers white body fat and the y allele confers yellow body fat. However, when raised on a xanthophyll-free diet, the homozygote yy animal has white body fat. If a heterozygous animal is crossed to a rabbit with yellow body fat, what are the proportions of offspring with white and yellow body fat when raised on a vegetarian diet? How do the proportions change if the offspring are raised on a xanthophyll-free diet?

21. In the clover butterfly, males are always yellow but females can be yellow or white. White is a dominant allele. Two yellow butterflies are crossed to yield an F_1 generation consisting of 50% yellow males, 25% yellow females, and 25% white females. Describe how this trait is inherited and the genotypes of the parents.

22. Duroc Jersey pigs are typically red, but a sandy variation is also seen. When two different varieties of true-breeding sandy pigs are crossed to each other, they produce F_1 offspring that are red. When these F_1 offspring are crossed to each other, they produce red, sandy, and white pigs in a 9:6:1 ratio. Explain this pattern of inheritance.

23. A true-breeding male fly with eosin eyes is crossed to a white-eyed female that is heterozygous for the normal (C) and cream alleles (c^a). What are the expected proportions of their offspring?

24. As discussed in this chapter, comb morphology in chickens is governed by a gene interaction. Two walnut comb chickens are crossed to each other. They produce only walnut comb and rose comb offspring, in a ratio of 3:1. What are the genotypes of the parents?

25. A farmer has two varieties of tomatoes. Variety A produces 1 pound tomatoes, and variety B produces 1.5 pound tomatoes. When these are crossed, the F_1 generation produces 2 pound tomatoes. When the F_1 generation is self-hybridized, the F_2 offspring produce 1, 2, and 1.5 pound tomatoes in a ratio of 1:2:1. Explain this result.

26. In certain species of summer squash, fruit color is determined by two interacting genes. A dominant allele, W, determines white color, while a recessive allele allows the fruit to be colored. In a homozygous ww individual, a second gene determines fruit color: G (green) is dominant to g (yellow). A white squash and a yellow squash are crossed, and the F_1 generation yields approximately 50% white fruit and 50% green fruit. What were the genotypes of the parents?

27. Starting with two heterozygous fowl that are hen-feathered, explain how you would obtain a true-breeding line that always produced cock-feathered birds.

28. In the pedigree shown here for a trait determined by a single gene (affected individuals are shown in red), state whether it would be possible for the trait to be inherited in each of the following ways:

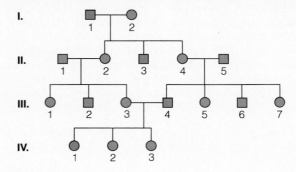

A. Recessive
B. X-linked recessive
C. Dominant, complete penetrance
D. Sex-influenced, dominant in males
E. Sex-limited, recessive in females
F. Dominant, incomplete penetrance

29. The pedigree shown here also concerns a trait determined by a single gene (affected individuals are shown in red). Which of the following patterns of inheritance are possible, and which one do you think is the most likely?

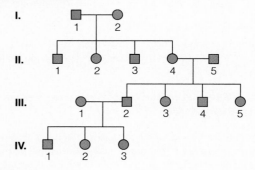

A. Recessive
B. X-linked recessive
C. Dominant
D. Sex-influenced, recessive in males
E. Sex-limited

30. Let's suppose that you have pedigree data from thousands of different families. For a particular genetic disease, how would you decide whether the disease is inherited as a recessive trait as opposed to one that is incompletely penetrant?

EXPERIMENTAL QUESTIONS

1. In Experiment 4A, explain why the heterozygous X^wX^{w-e} female fly had the same eye color as the $X^{w-e}Y$ male fly.

2. In *Drosophila*, many other eye colors have been found to be allelic to white. In addition to eosin, there are ivory, pearl, coral, cherry, apricot, and blood, to name a few. Based on the intensity of eye color alone, which alleles might you expect to have a gene dosage effect? Describe the types of observations and experiments you would carry out to determine if a gene dosage effect occurs.

3. A group of blue Andalusian fowls are crossed to each other and produce the following numbers of offspring: 117 black, 237 blue, 113 splashed white. Propose a hypothesis that explains this outcome, and then carry out a chi square analysis to see if your hypothesis is correct. What future experiments would you carry out to confirm this hypothesis?

4. For the data of Experiment 4B, conduct a chi square analysis to see if the experimental data has a good fit with the hypothesis.

5. In a species of plant, two genes control flower color. The red allele (R) is dominant to the white allele (r); the color-producing allele (C) is dominant to the non-color-producing allele (c). You suspect that either an rr homozygote or a cc homozygote will produce white flowers. In other words, rr is epistatic to C, and cc is epistatic to R. To test your hypothesis, you allow heterozygous plants ($RrCc$) to self-fertilize and count the offspring. You obtain the following results: 201 plants with red flowers and 144 with white flowers. Conduct a chi square analysis to see if your observed data are consistent with your hypothesis.

QUESTIONS FOR STUDENT DISCUSSION/COLLABORATION

1. Let's suppose a gene exists as a functional wild-type allele and a nonfunctional mutant allele. At the organismal level, the wild-type allele is dominant. In a heterozygote, discuss whether dominance occurs at the cellular and molecular levels. Discuss examples where the issue of dominance depends on the level of examination.

2. A true-breeding rooster with a rose comb, feathered shanks, and cock-feathering is crossed to a hen that is true-breeding for pea comb and unfeathered shanks but is heterozygous for hen-feathering. If you assume that the four genes involved can assort independently, what is the expected outcome of the F_1 generation?

3. In oats, the color of the chaff is determined by a two-gene interaction. When a true-breeding black plant is crossed to a true-breeding white plant, the F_1 generation is composed of all black plants. When the F_1 offspring are crossed to each other, the ratio produced is 12 black : 3 gray : 1 white. First, construct a Punnett square that accounts for this pattern of inheritance. Which genotypes produce the gray phenotype? Second, at the level of protein function, how would you explain this type of inheritance?

CHAPTER 5

LINKAGE AND GENETIC MAPPING IN EUKARYOTES

LINKAGE AND CROSSING OVER

GENETIC MAPPING IN DIPLOID EUKARYOTES

GENETIC MAPPING IN HAPLOID EUKARYOTES

In Chapter 2, we were introduced to the law of independent assortment. According to Mendel, we expect that two different genes will segregate and independently assort themselves during gamete formation. After Mendel's work was rediscovered at the turn of the 20th century, chromosomes were identified as the cellular structures that carry genes. The chromosome theory of inheritance explained how the transmission of chromosomes is responsible for the passage of genes from parent to offspring.

When geneticists first realized that chromosomes contain the genetic material, they began to suspect that a conflict may sometimes occur between Mendel's law of independent assortment of genes and the behavior of chromosomes during meiosis. In particular, geneticists assumed that each species of organism must contain thousands of different genes, yet cytological studies revealed that most species have at most a few dozen chromosomes. Therefore, it seemed likely (and turned out to be true) that each chromosome would carry many hundreds or even thousands of different genes. In 1911, Thomas Hunt Morgan conducted experiments showing that the transmission of genes located on the same chromosome violates the law of independent assortment.

In this chapter, we will consider the pattern of inheritance that occurs when different genes are situated on the same chromosome. We will also explore how the data from genetic crosses are used to construct genetic maps that describe the order of genes along a chromosome. A variety of newer strategies for gene mapping will also be described in Chapter 20. This chapter ends with a discussion of the eukaryotic microorganisms collectively known as fungi. This group includes yeast, such as *Saccharomyces cerevisiae* (baker's yeast), and molds, such as *Neurospora crassa* (red bread mold) and *Aspergillus nidulans* (green bread mold). We will consider the unique features of sexual reproduction that occur in certain fungal species. These features have provided the basis for distinctive genetic mapping approaches.

LINKAGE AND CROSSING OVER

In eukaryotic species, each linear chromosome contains a very long piece of DNA. As we have mentioned, a chromosome contains many individual functional units—called genes—that influence an organism's traits. A typical chromosome is expected to contain many hundreds or perhaps a few thousand different genes. The term **linkage** refers to the phenomenon that two or more genes can be located on the same chromosome. The genes are physically linked to each other, because each eukaryotic chromosome contains a single, continuous, linear piece of DNA.

As an example of linkage, let's consider a human karyotype that contains 46 chromosomes (shown back in Figure 3-1). All of the genes on each chromosome are linked to each other. Chromosomes thus are sometimes called **linkage groups**, because a chromosome contains a group of genes that are linked together. In a particular species, there are as many linkage groups as there are types of chromosomes. For example, humans have 46 chromosomes, which are composed of 22 types of autosomes that come in pairs and one pair of sex chromosomes. There are two types of sex chromosomes, the X and Y. Therefore, humans have 22 autosomal linkage groups, an X-chromosome linkage group, and a Y-chromosome linkage group.

Geneticists are often interested in the transmission of two or more traits in a genetic cross. When a geneticist follows two traits in a cross, this is called a **dihybrid cross**; when three traits are followed, it is a **trihybrid cross**; and so forth. The outcomes of dihybrid and trihybrid crosses depend on whether or not the genes are linked to each other along the same chromosome. In this section, we will examine how linkage affects the transmission patterns of two or more traits.

Crossing over may occur during the tetrad stage of meiosis; when it does, it produces recombinant chromosomes and ultimately recombinant phenotypes

Even though the alleles for different genes may be linked along the same chromosome, the linkage can be altered during the process of gamete formation. In diploid eukaryotic species, homologous chromosomes can exchange pieces with each other by crossing over. This event occurs frequently during prophase I of meiosis. As discussed in Chapter 3, the replicated chromosomes, known as sister chromatids, associate with the homologous sister chromatids to form a structure known as a *bivalent* or a *tetrad* (refer back to Figure 3-9). A tetrad is composed of two pairs of sister chromatids. In prophase I, a sister chromatid of one pair commonly will cross over with a sister chromatid from the homologous pair.

Figure 5-1 considers the formation of gametes when two genes are linked on the same chromosome. One of the parental chromosomes carries the *A* and *B* alleles, while the homologue carries the *a* and *b* alleles. In Figure 5-1a, no crossing over has occurred. Therefore, the gametes contain the same combination of alleles as the original chromosomes. In this case, two gametes carry the *A* and *B* alleles, and the other two gametes carry the recessive *a* and *b* alleles. The arrangement of linked alleles has not been altered.

In contrast, Figure 5-1b illustrates what can happen when crossing over occurs. Two of the gametes contain combinations of alleles (namely, *A* and *b*, *a* and *B*) that differ from those in the original chromosomes. In these two gametes, the grouping of linked alleles has been changed. An event such as this, leading to a new combination of alleles, is known as **genetic recombination**. The gametes carrying the *A*

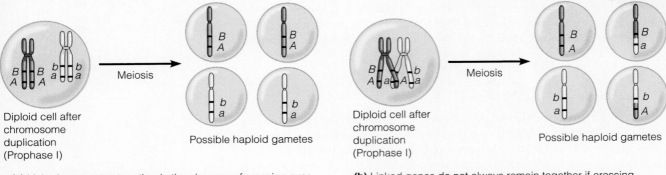

Diploid cell after chromosome duplication (Prophase I)

Meiosis

Possible haploid gametes

Diploid cell after chromosome duplication (Prophase I)

Meiosis

Possible haploid gametes

(a) Linked genes stay together in the absence of crossing over

(b) Linked genes do not always remain together if crossing over occurs

FIGURE **5-1**

Consequences of crossing over during meiosis. (a) In the absence of crossing over, the *A* and *B* alleles and the *a* and *b* alleles are maintained in the same arrangement as in the parental chromosomes. **(b)** Crossing over has occurred in the region between the two genes, creating two nonparental gametes with a new combination of alleles.

and *b*, or the *a* and *B*, alleles are called **nonparental** or **recombinant** gametes. Likewise, if these gametes participate in fertilization, the resulting offspring are called nonparental or recombinant offspring. These offspring can display combinations of traits that are different from those of either parent. In contrast, offspring that have inherited the same combination of alleles as their parents are known as **parental** or **nonrecombinant** offspring.

In this chapter, we will consider how crossing over affects the pattern of inheritance for genes linked on the same chromosome. In Chapter 18, we will consider the molecular events that allow crossing over to occur.

Bateson and Punnett discovered two traits that did not assort independently

The earliest study indicating that some traits may not assort independently was carried out by William Bateson and Reginald Punnett in 1905. As mentioned in Chapter 4, they were interested in the pattern of inheritance of genes in several organisms, including the chicken and the sweet pea. According to Mendel's law of independent assortment, a dihybrid cross between two individuals, heterozygous for two genes, should yield a 9:3:3:1 phenotypic ratio among the offspring. However, a surprising result occurred when Bateson and Punnett conducted a cross in the sweet pea involving two different traits, flower color and pollen shape (Figure 5-2).

As seen here, they began by crossing a true-breeding strain with purple flowers (*PP*) and long pollen (*LL*) to a strain with red flowers (*pp*) and round pollen (*ll*). This yielded an F_1 generation of plants that all had purple flowers and long pollen (*PpLl*). The unexpected result came from the F_2 generation. Even though there were four different phenotypic categories among the F_2 generation, the observed numbers of offspring did not conform to a 9:3:3:1 ratio. Bateson and Punnett found that the F_2 generation had a much greater proportion of the two parental types: purple flowers with long pollen, and red flowers with round pollen. Therefore, they suggested that the transmission of these two traits from the parental generation to the F_2 generation was somehow coupled and not easily assorted in an independent manner. However, Bateson and Punnett did not realize that this coupling was due to the linkage of the flower color gene and the pollen shape gene on the same chromosome.

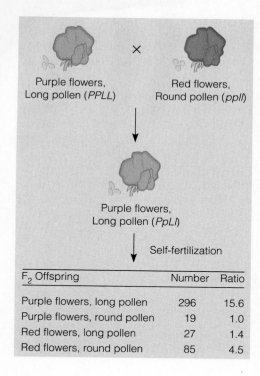

Purple flowers,
Long pollen (*PPLL*)

×

Red flowers,
Round pollen (*ppll*)

Purple flowers,
Long pollen (*PpLl*)

Self-fertilization

F₂ Offspring	Number	Ratio
Purple flowers, long pollen	296	15.6
Purple flowers, round pollen	19	1.0
Red flowers, long pollen	27	1.4
Red flowers, round pollen	85	4.5

FIGURE **5-2**

An experiment of Bateson and Punnett with sweet peas showing that independent assortment does not always occur.

GENES ⟶ TRAITS: Two genes that govern flower color and pollen shape are found on the same chromosome. Therefore, the offspring tend to inherit the parental combinations of alleles (*PL* or *pl*). Due to occasional crossing over, a lower percentage of offspring inherit nonparental combinations of alleles (*Pl* or *pL*).

Morgan provided evidence for the linkage of several X-linked genes and proposed that crossing over between X-chromosomes can occur

The first direct evidence that different genes can be physically located on the same chromosome came from the studies of Thomas Hunt Morgan. In Chapters 3 and 4, we considered some of Morgan's studies involving X-linked traits. These earlier studies provided the groundwork to demonstrate that the genes governing X-linked traits are physically linked on the X-chromosome.

Morgan investigated the inheritance pattern of many different traits that had been shown to follow an X-linked pattern of inheritance. Figure 5-3 illustrates an experiment involving three traits that Morgan studied. His parental crosses were normal male fruit flies to females that had yellow bodies (*yy*), white eyes (*ww*), and miniature wings (*mm*). The wild-type alleles for these three genes are designated *y⁺* (gray body), *w⁺* (red eyes), and *m⁺* (normal wings). As expected, the phenotypes of the F₁ generation were wild-type females and males with yellow bodies, white eyes, and miniature wings. The linkage of these genes was revealed when the F₁ flies were mated to each other and the F₂ generation examined.

Instead of equal proportions of the eight possible phenotypes, Morgan observed a much higher proportion of the parental combinations of traits. There were 758 flies with gray bodies, red eyes, and normal wings, and 700 flies with yellow bodies, white eyes, and miniature wings. The former combination (gray body, red eyes, and normal wings) was found in the males of the parental generation, the latter combination (yellow body, white eyes, and miniature wings) in the females of the parental generation. Morgan's proposed explanation for this higher proportion of parental combinations was that all three genes are located on the X-chromosome and, therefore, tend to be transmitted together as a unit.

However, to fully account for the data shown in Figure 5-3, Morgan needed to interpret two other key observations. First, he needed to explain why a significant proportion of the F₂ generation had nonparental combinations of alleles. Along with the two parental phenotypes, there were six other phenotypic combinations

FIGURE **5-3**

A trihybrid cross of Morgan's involving three X-linked traits in _Drosophila_.

GENES ⟶ TRAITS: Three genes that govern body color, eye color, and wing size are all found on the X-chromosome. Therefore, the offspring tend to inherit the parental combinations of alleles ($y^+ w^+ m^+$ or ywm). Figure 5-4 explains how single and double crossovers can create nonparental combinations of alleles.

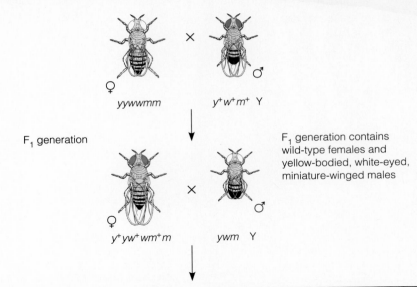

$yywwmm$ ♀ $y^+ w^+ m^+$ Y ♂

F₁ generation

F₁ generation contains wild-type females and yellow-bodied, white-eyed, miniature-winged males

$y^+ y w^+ w m^+ m$ ♀ ywm Y ♂

F₂ generation	Females	Males	Total*
Gray body, red eyes, normal wings	439	319	758
Gray body, red eyes, miniature wings	208	193	401
Gray body, white eyes, normal wings	1	0	1
Gray body, white eyes, miniature wings	5	11	16
Yellow body, red eyes, normal wings	7	5	12
Yellow body, red eyes, miniature wings	0	0	0
Yellow body, white eyes, normal wings	178	139	317
Yellow body, white eyes, miniature wings	365	335	700

that were not found in the parental generation. Second, he needed to explain why there was a quantitative difference between nonparental combinations involving body color and eye color versus eye color and wing length. This quantitative difference is revealed by reorganizing the data from Morgan's cross by pairs of genes:

	Total	
Gray body, red eyes	1159	
Yellow body, white eyes	1017	
Gray body, white eyes	17	} Nonparental
Yellow body, red eyes	12	offspring
	2205	

	Total	
Red eyes, normal wings	770	
White eyes, miniature wings	716	
Red eyes, miniature wings	401	} Nonparental
White eyes, normal wings	318	offspring
	2205	

We see that there were substantial differences between the numbers of nonparental offspring when pairs of genes were considered separately. It was fairly common for nonparental combinations to occur when just eye color and wing shape were examined (401 + 318 nonparental offspring). In sharp contrast, it was rare to obtain nonparental combinations when looking at body color and eye color (17 + 12 nonparental offspring).

To explain these data, Morgan considered the previous studies of the French cytologist, F. A. Janssens. In 1909, Janssens proposed that crossing over involves a physical exchange between homologous chromosomes that occurs prior to the microscopic visualization of chiasmata (refer back to Figure 3-9). Morgan shrewdly realized that crossing over between homologous X-chromosomes was consistent with his data. Overall, he made three important hypotheses to explain his results:

1. The genes for body color, eye color, and wing length are all located on the same chromosome (the X-chromosome). Therefore, it is most likely for all three traits to be inherited together.
2. Due to crossing over, the homologous X-chromosomes (in the female) can exchange pieces of chromosomes and create new (nonparental) combinations of alleles.
3. The likelihood of crossing over depends on the distance between two genes. If two genes are far apart from each other, it is more likely that crossing over will occur between them.

With these ideas in mind, Figure 5-4 illustrates the possible events that occurred in the F_1 female flies of Morgan's experiment. One of the X-chromosomes contained all three dominant alleles, the other all three recessive alleles. During oogenesis in the F_1 female flies, crossing over may or may not have occurred in this region of the X-chromosome. If no crossing over occurred, the parental phenotypes were produced in the F_2 offspring. Alternatively, a crossover sometimes occurred between the eye-color gene and the wing-length gene to create nonparental offspring (namely, gray body, red eyes, and miniature wings; or yellow body, white eyes, and normal wings). According to Morgan's proposal, this is a fairly likely event, because these two genes are far apart from each other on the X-chromosome. Because of the long distance, it was fairly likely for a crossover to occur in this region. In contrast, he proposed that the body color and eye color genes are very close together, which makes crossing over between them an unlikely event. Nevertheless, it occasionally occurred, yielding offspring with gray bodies, white eyes, and miniature wings, or with yellow bodies, red eyes, and normal wings. Finally, it was also possible for two homologous chromosomes to cross over twice. This double crossing over is expected to be a very unlikely event. Among the 2205 offspring Morgan examined, he only found 1 fly (gray body, white eyes, normal wings) that could be explained by this phenomenon.

A chi square analysis can be used to distinguish between linkage and independent assortment

Now that we have an appreciation for linkage and the production of recombinant offspring, let's consider how an experimenter can objectively decide whether two genes are linked or assort independently. In Chapter 2, chi square analysis was introduced to evaluate the goodness of fit between a genetic hypothesis and observed experimental data. This method is frequently used to determine if the outcome of a dihybrid cross is consistent with linkage or independent assortment.

To conduct a chi square analysis, we must first propose a hypothesis. In a dihybrid cross, the standard hypothesis is that the two genes are *not* linked. This hypothesis is chosen even if the observed data suggest linkage, because an independent assortment hypothesis allows us to calculate the expected number of offspring based on the genotypes of the parents and the law of independent assortment. In contrast, for two linked genes that have not been previously mapped, we cannot calculate the expected number of offspring from a genetic cross, because we do not know how likely it is for a crossover to occur between the two genes. Without expected numbers of recombinant and parental offspring, we cannot use a chi square test. Therefore, we begin with the hypothesis that the genes are not linked; then, we determine whether or not our data fit this hypothesis. If the chi square value is low and we cannot reject our hypothesis, we infer that the genes assort independently. On the other hand, if the chi square value is so high that our hypothesis is rejected, we will conclude that a linkage hypothesis is correct.

SEX CHROMOSOMES IN:

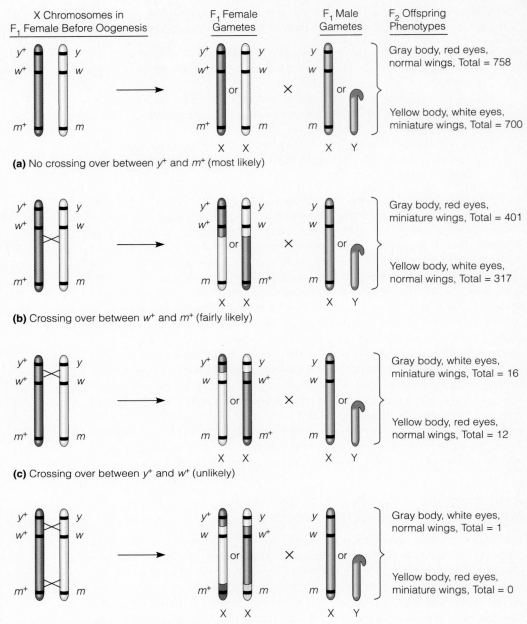

(a) No crossing over between y^+ and m^+ (most likely)

(b) Crossing over between w^+ and m^+ (fairly likely)

(c) Crossing over between y^+ and w^+ (unlikely)

(d) Double crossing over (very unlikely)

FIGURE **5-4**

The likelihood of crossing over provides an explanation of Morgan's trihybrid cross. In the absence of crossing over, the parental chromosomes carry the $y^+w^+m^+$ or ywm alleles. A crossover between the eye color and wing-length genes produces recombinant chromosomes that carry the y^+w^+m or ywm^+ alleles. Since the eye color and wing-length genes are relatively far apart, such recombinant offspring are not uncommon. A crossover between the body color and eye color genes produces recombinant chromosomes that contain the y^+wm or yw^+m^+ alleles. Since these two genes are very close together, these two types of recombinant offspring are fairly uncommon. Finally, a double crossover could produce chromosomes carrying the y^+wm^+ or yw^+m alleles. This is a very rare event.

In this experiment, crossing over occurred between homologous X-chromosomes in the female during oogenesis. In *Drosophila*, crossing over does not occur during spermatogenesis. This is a peculiar characteristic of *Drosophila*. In many other animal species, homologous recombination commonly occurs during spermatogenesis. Also note that this figure only shows a portion of the X-chromosome. A map of the entire X-chromosome is shown in Figure 5-8.

As an example, let's consider the data shown on page 104 concerning body color and eye color. This cross produced the following offspring: 1159 gray body, red eyes; 1017 yellow body, white eyes; 17 gray body, white eyes; and 12 yellow body, red eyes. However, when a heterozygous female ($y^+y\ w^+w$) is crossed to a hemizygous male ($yw\ Y$), an independent assortment hypothesis predicts the following outcome:

F_1 male gametes

		X^{yw}	Y	
	$X^{y^+w^+}$	$X^{yw}X^{y^+w^+}$	$X^{y^+w^+}Y$	= Gray body, red eyes
F_1 female gametes	X^{y^+w}	$X^{yw}X^{y^+w}$	$X^{y^+w}Y$	= Gray body, white eyes
	X^{yw^+}	$X^{yw}X^{yw^+}$	$X^{yw^+}Y$	= Yellow body, red eyes
	X^{yw}	$X^{yw}X^{yw}$	$X^{yw}Y$	= Yellow body, white eyes

The independent assortment hypothesis predicts a 1:1:1:1 ratio among the four phenotypes. The observed data mentioned above obviously seem to conflict with this hypothesis. Nevertheless, we stick to the strategy just discussed. First, we propose that the two genes are not linked, and then we use a chi square analysis to see if the data fit this hypothesis. If the data do not fit, we will reject the idea that the genes assort independently and conclude that the genes are linked.

An example of a chi square approach to determine linkage is shown here:

Step 1. *Propose a hypothesis.* Even though the observed data appear inconsistent with this hypothesis, we propose that the two genes for eye color and body color are X-linked but somehow are able to obey Mendel's law of independent assortment. This hypothesis will allow us to calculate expected values. We actually anticipate that the chi square analysis will allow us to reject the independent assortment hypothesis in favor of a linkage hypothesis.

Step 2. *Based on the hypothesis, calculate the expected values of each of the four phenotypes.* Each phenotype has an equal probability of occurring (see the Punnett square given previously). Therefore, the probability of each genotype is 1/4. The observed F_2 generation contained a total of 2205 individuals. Our next step is to calculate the expected numbers of offspring with each phenotype when the total equals 2205; 1/4 of the offspring should be each of the four phenotypes:

$$1/4 \times 2205 = 551 \qquad \text{(expected number of each phenotype)}$$

Step 3. *Apply the chi square formula, using the data for the observed values (O) and the expected values (E) that have been calculated in step 2.* In this case, there are four categories within the population:

$$\chi^2 = \frac{(O_1 - E_1)^2}{E_1} + \frac{(O_2 - E_2)^2}{E_2} + \frac{(O_3 - E_3)^2}{E_3} + \frac{(O_4 - E_4)^2}{E_4}$$

$$= \frac{(1159 - 551)^2}{551} + \frac{(17 - 551)^2}{551} + \frac{(12 - 551)^2}{551} + \frac{(1017 - 551)^2}{551}$$

$$= 670.9 + 517.5 + 527.3 + 394.1$$

$$= 2109.8$$

Step 4. *Interpret the calculated chi square value.* This is done with a chi square table, as discussed in Chapter 2. Since there are four experimental categories ($n = 4$), the degrees of freedom is $n - 1 = 3$.

The calculated chi square value is enormous! Thus, the deviation between observed and expected values is very large. According to Table 2-1, such a large deviation is expected to occur by chance alone less than 1% of time. Therefore, we reject the hypothesis that the two genes assort independently. In other words, we conclude that the genes are linked.

Creighton and McClintock correlated crossing over that produced new combinations of alleles with the exchange of homologous chromosomes

As we have seen, Morgan's studies were consistent with the hypothesis that crossing over occurs between homologous chromosomes to produce new combinations of alleles. In the experiment described here, which was published twenty years after the work of Morgan, Harriet Creighton and Barbara McClintock used an interesting strategy involving parallel observations. First, they made crosses involving two linked genes to produce parental and recombinant offspring. Second, they used a microscope to view the structures of the chromosomes in the parents and in the offspring. Because the parental chromosomes had some unusual structural features, they could microscopically distinguish the two homologous chromosomes within a pair. As we will see, this enabled them to correlate the occurrence of recombinant offspring with microscopically observable exchanges in segments of homologous chromosomes.

While working in the botany department at Cornell University, Creighton and McClintock focused much of their attention on the pattern of inheritance of traits in corn. In previous cytological examinations of corn chromosomes, some strains were found to have an unusual chromosome (#9) with a darkly staining knob at one end. McClintock also identified an abnormal version of this chromosome that had an extra piece of a different chromosome (#8) at the other end (called an **interchange** or a **translocation**). As shown in Figure 5-5, this unusual version of chromosome 9 had changes at both ends that could be distinguished under the microscope.

Creighton and McClintock insightfully realized that this abnormal chromosome could be used to demonstrate that two homologous chromosomes physically exchange segments as a result of crossing over. They knew that a gene was located near the knobbed end of chromosome 9 that provided color to corn kernels. It existed in two alleles, the dominant allele *C* (colored) and the recessive allele *c* (colorless). Toward the other end of the chromosome was located a second gene that affected the texture of the kernel endosperm. The dominant allele *Wx* caused starchy endosperm, while the recessive *wx* allele caused waxy endosperm. Creighton and McClintock reasoned that a crossover involving a normal chromosome 9 and a knobbed/translocated chromosome 9 would produce a chromosome that had either a knob or a translocation but not both. This chromosome would be distinctly different from either of the parental chromosomes (Figure 5-6).

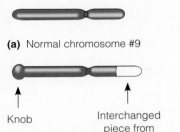

(a) Normal chromosome #9

Knob   Interchanged piece from chromosome #8

(b) Abnormal chromosome #9

FIGURE **5-5**

Normal and abnormal chromosome 9 in corn used by Creighton and McClintock. A normal chromosome 9 **(a)** is compared with an abnormal chromosome 9 **(b)** that contains a knob at one end and a translocation (i.e., an interchange) at the opposite end.

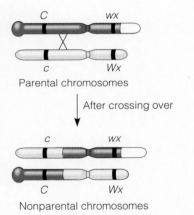

Parental chromosomes

After crossing over

Nonparental chromosomes

FIGURE **5-6**

Crossing over between normal and abnormal chromosome 9s in corn. A crossover produces a chromosome that only contains a knob at one end, and another chromosome that only contains a translocation at the other end.

THE HYPOTHESIS

Offspring with nonparental phenotypes are the product of a crossover. This crossover should create nonparental chromosomes via an exchange of chromosomal segments between homologous chromosomes.

Starting materials: Two different strains of corn. One strain (referred to as parent A) has an abnormal #9 (knobbed/translocation) with a dominant *C* allele and a recessive *wx* allele. It also contains a cytologically normal copy of chromosome #9 that carries the recessive *c* allele, and the dominant *Wx* allele. Its genotype is *Cc Wxwx*. The other strain (referred to as parent B) has two normal versions of chromosome #9. The genotype of this strain is *cc Wxwx*.

Experimental Level

Conceptual Level

1. Cross the two strains described. The tassel is the pollen-bearing structure and the silk (equivalent to the stigma and style) is connected to the ovary. After fertilization, the ovary will develop into an ear of corn.

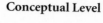
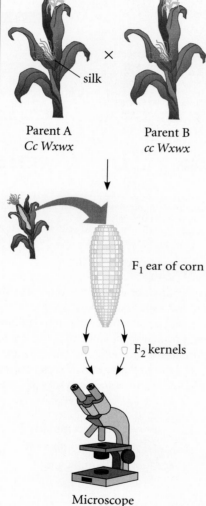
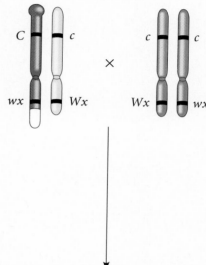

Parent A
Cc Wxwx

Parent B
cc Wxwx

2. Observe the kernels from this cross.

F₁ ear of corn

F₂ kernels

Each kernel is a separate seed that has inherited a set of chromosomes from each parent

3. Microscopically examine chromosome #9 in the kernels.

Microscope

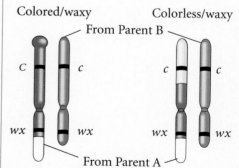

Colored/waxy

Colorless/waxy

From Parent B

C   *c*   *c*   *c*

wx   *wx*   *wx*   *wx*

From Parent A

This illustrates only two possible outcomes in the F₂ kernels. As shown in the data, there are six possible outcomes.

Phenotype of F₁ kernel	Number of kernels analyzed	Cytological appearance of a chromosome in F₁ offspring*		Did a crossover occur during gamete formation in Parent A?
Colored/waxy	3	Knobbed / translocation C wx	Normal c wx	No
Colorless/starchy	11	Knobless / normal c Wx	Normal c or Wx c wx	No
	4	Knobless / translocation c wx	Normal c Wx	Yes
Colorless/waxy	2	Knobless / translocation c wx	Normal c wx	Yes
Colored/starchy	5	Knobbed / normal C Wx	Normal c Wx	Yes

*In this table, the chromosome on the left was inherited from parent A and the chromosome on the right was inherited from parent B.
Data adapted from: Creighton, H. B., and McClintock, B. (1931). A Correlation of Cytological and Genetical Crossing-Over in *Zea mays*, *Proc. Natl. Acad. Sci. USA 17*, 492–497.

INTERPRETING THE DATA

In interpreting these results, we must note that Creighton and McClintock did not carry out this experiment like a standard cross, because neither of the parents were homozygous recessive for both genes. This adds some ambiguity in the relationship between the phenotypic categories and genetic recombination. In this experiment, we are interested in whether or not crossing over has occurred in parent A, which is heterozygous for both genes. This parent can produce four types of gametes, while parent B can only produce two types of gametes:

Parent A		*Parent B*	
C	wx (nonrecombinant)	c	Wx
c	Wx (nonrecombinant)	c	wx
C	Wx (recombinant)		
c	wx (recombinant)		

The following types of offspring can be produced:

	c Wx	c wx	
C wx	*CcWxwx* colored/starchy	*Ccwxwx* colored/waxy	Nonrecombinant
c Wx	*ccWxWx* colorless/starchy	*ccWxwx* colorless/starchy	Nonrecombinant
C Wx	*CcWxWx* colored/starchy	*CcWxwx* colored/starchy	Recombinant
c wx	*ccWxwx* colorless/starchy	*ccwxwx* colorless/waxy	Recombinant

Two of the phenotypic categories are ambiguous: colored/starchy (*Cc Wxwx*) and colorless/starchy (*cc WxWx*). These phenotypes can be produced whether or not recombination occurs in parent A. Therefore, let's begin by considering the two unambiguous phenotypic categories: colored/waxy (*Cc wxwx*) and colorless/waxy (*cc wxwx*). The colored/waxy phenotype can occur only if recombination did not occur in parent A and if parent A passed the knobbed/translocated chromosome to its offspring. As shown in the data table, three kernels were obtained with this phenotype, and all of them contained the knobbed/translocated chromosome. By comparison, the colorless/waxy phenotype can only be obtained if genetic recombination did occur in parent A and this parent passed a chromosome 9 that had a translocation but was knobless. Two kernels were obtained with this phenotype, and both of them contained the expected chromosome that had a translocation but was knobless.

Taken together, these results show a perfect correlation between genetic recombination of alleles and the cytological presence of a chromosome displaying a genetic exchange of chromosomal pieces in parent A. The other two phenotypic categories are ambiguous, because either a colorless/starchy or a colored/starchy phenotype can be produced in the presence or the absence of genetic recombination in parent A. Nevertheless, the results agree with the hypothesis that genetic recombination is correlated with an exchange of chromosome pieces.

Overall, the observations described in this experiment are consistent with the idea that a crossover occurred, in the region between the *C* and *wx* genes, involving an exchange of segments between two homologous chromosomes. As stated by the authors, "Pairing chromosomes, heteromorphic in two regions, have been shown to exchange parts at the same time they exchange genes assigned to these regions." These results support the view that genetic recombination involves a physical exchange between homologous chromosomes. This microscopic evidence helped to convince geneticists that recombinant offspring arise from the physical exchange of segments of homologous chromosomes.

Crossing over occasionally occurs during mitosis

In multicellular organisms, the union of egg and sperm is followed by many cellular divisions, which occur in conjunction with mitotic divisions of the cell nuclei. As discussed in Chapter 3, mitosis normally does not involve the homologous pairing of chromosomes to form a tetrad. Therefore, crossing over during mitosis is expected to be much less likely than during meiosis. Nevertheless, it does occur on rare occasions. When it happens, mitotic crossing over may produce a pair of recombinant chromosomes that have a new combination of alleles. This is known as **mitotic recombination**. If it occurs during an early stage of embryonic development, the daughter cells containing the recombinant chromosomes will continue to divide many times to produce a patch of tissue in the adult. This may result in a portion of tissue with characteristics different from those of the rest of the organism.

In 1936, while working at the University of Rochester, Curt Stern proposed that unusual patches on the bodies of certain *Drosophila* strains were due to mitotic recombination. As shown in Figure 5-7 (on the next page), he was working with strains carrying X-linked alleles affecting body color and bristle morphology. The recessive *y* allele confers yellow body color, and the recessive *sn* allele confers shorter body bristles that look singed. The corresponding wild-type alleles confer gray body color (y^+) and normal bristles (sn^+). Females that are $y^+y\ sn^+sn$ are expected to have gray body color and normal bristles. This was generally the case. However, when Stern carefully observed the bodies of these female flies under a low-power microscope, he occasionally noticed places in which two adjacent regions were different from the rest of the body. This is called a *twin spot*. He concluded that twin spotting was too frequent to be explained by the random positioning of two independent single spots that happened to occur close together. Instead, Stern proposed

FIGURE **5-7**

Mitotic recombination in *Drosophila* produces twin spotting.

GENES ⟶ TRAITS: In this illustration, the genotype of the larva was *y⁺y sn⁺sn*. A mitotic crossover occurred in a single larval cell. After mitotic crossing over, the separation of the chromatids occurred, so that one larval cell became *y⁺y⁺ snsn* and the adjacent sister cell became *yy sn⁺sn⁺*. The larval cells then continued to divide by normal mitosis to produce an adult fly. The cells in the adult that were derived from the *y⁺y⁺ snsn* larval cell produced a spot on the adult body that was gray and had singed bristles. The cells derived from the *yy sn⁺sn⁺* larval cell produced an adjacent spot on the body that was yellow and had normal bristles. In this case, mitotic recombination produced an unusual trait, known as twin spots. The characteristics of the twin spots differ from the surrounding tissue, which is gray with normal bristles.

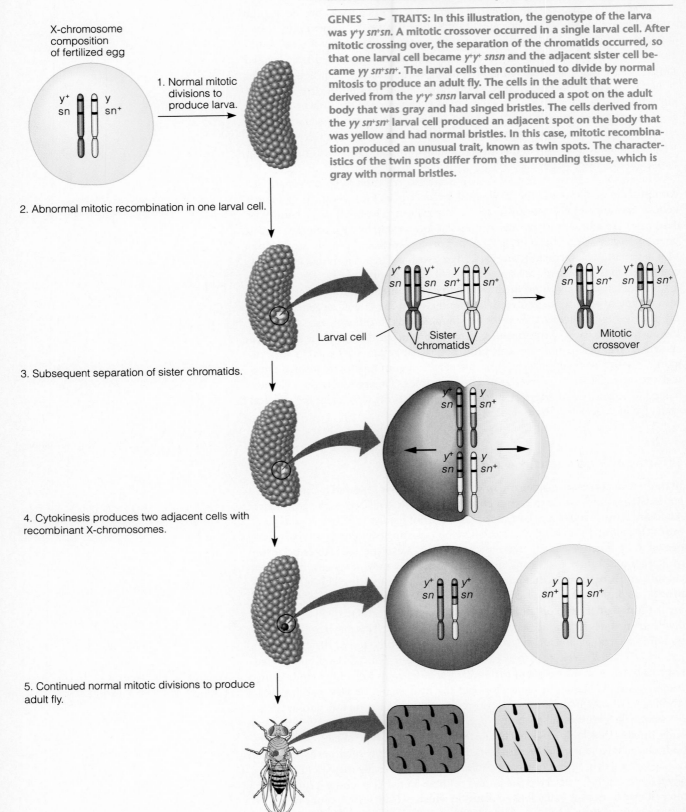

X-chromosome composition of fertilized egg

1. Normal mitotic divisions to produce larva.

2. Abnormal mitotic recombination in one larval cell.

Larval cell   Sister chromatids   Mitotic crossover

3. Subsequent separation of sister chromatids.

4. Cytokinesis produces two adjacent cells with recombinant X-chromosomes.

5. Continued normal mitotic divisions to produce adult fly.

that twin spots were due to a single mitotic recombination within one cell during embryonic development (Figure 5-7).

As shown here, the X-chromosomes of the female fly are $y^+ sn$ and $y sn^+$. Rarely, though, a crossover can occur during mitosis to produce two adjacent daughter cells that are $y^+y^+ snsn$ and $yy sn^+sn^+$. As embryonic development proceeds, the cell on the left will continue to divide to produce many cells, eventually producing a patch on the body that has gray color with singed bristles. The daughter cell next to it will produce a patch of yellow body color with normal bristles. These two adjacent patches (a twin spot) will be surrounded by cells that are $y^+y sn^+sn$ and, thus, have gray color and normal bristles. These infrequent twin spots provide evidence that mitotic recombination occasionally occurs.

GENETIC MAPPING IN DIPLOID EUKARYOTES

The purpose of **genetic mapping** (also known as gene mapping or chromosome mapping) is to determine the linear order of genes that are linked to each other along the same chromosome. Figure 5-8 illustrates a simplified genetic map of *Drosophila melanogaster* depicting the locations (i.e., loci) of many different genes along the individual chromosomes. As shown here, each gene has its own unique locus at a particular site within a chromosome. For example, the gene designated *vg*, which affects wing length, is located on chromosome 2. The

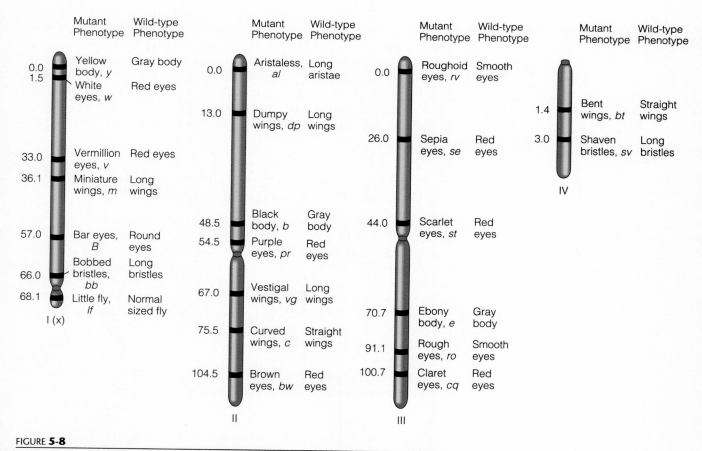

FIGURE **5-8**

A simplified genetic map of *Drosophila melanogaster*. This simplified map illustrates a few of the many hundreds of genes that have been identified in this organism.

gene designated *b*, which affects body color, is found a moderate distance away on the same chromosome.

Even though it is an enormous amount of work, the construction of a genetic map is useful in many ways. First, it allows geneticists to understand the overall complexity and genetic organization of a particular species. The genetic map of a species portrays the underlying basis for the inherited traits that an organism displays. In some cases, the known locus of a gene within a genetic map can help molecular geneticists to clone that gene and thereby obtain greater information about its molecular features. In addition, genetic maps are useful from a evolutionary point of view. A comparison of the genetic maps among different species can improve our understanding of the evolutionary relationships among these species.

Along with these scientific uses, genetic maps have many practical benefits. For example, many human genes that play a role in human disease have been genetically mapped. This information can be used to diagnose and perhaps someday to treat inherited human diseases. It can also help genetic counselors predict the likelihood that a couple will produce children with certain inherited diseases. In addition, genetic maps are gaining increasing importance in agriculture. A genetic map can provide plant and animal breeders with helpful information for improving agriculturally important strains through selective breeding programs.

In this section, we will begin by discussing conventional mapping techniques. These methods analyze crosses involving individuals heterozygous for two or more genes. The frequency of nonparental offspring provides a way to deduce the linear order of genes along a chromosome. As in Figure 5-8, this linear arrangement of genes is shown in a chart known as a **genetic linkage map**. This approach is particularly useful for analyzing organisms that are easily crossed and produce a large number of offspring in a short period of time. It has successfully mapped the genes of several plant species (with an annual generation time) and certain species of animals, such as *Drosophila*. For many organisms, however, conventional mapping approaches are difficult due to long generation times or the inability to carry out crosses (e.g., humans). Fortunately, many alternative methods of gene mapping have been developed in the past few decades. As will be described in Chapter 20, cytological and molecular approaches also can be used to map genes.

The frequency of recombination between two genes can be correlated with their map distance along a chromosome

Genetic mapping allows us to estimate the relative distances between linked genes, based on the likelihood that a crossover will occur between them. If two genes are very close together on the same chromosome, a crossover is unlikely to begin in the region between them. However, if two genes are very far apart, a crossover is more likely to be initiated in this region and thereby recombine the alleles of the two genes. Experimentally, the basis for genetic mapping is that the percentage of recombinant offspring is correlated with the distance between two genes. If two genes are far apart, many recombinant offspring will be produced. However, if two genes are close together, very few recombinant offspring will be observed.

To interpret a genetic mapping experiment, the experimenter must know if the characteristics of an offspring are due to crossing over during gamete formation in a parent. This is accomplished by conducting a **test cross**. Most test crosses are between an individual who is heterozygous for two or more genes and an individual who is recessive and homozygous for these same genes. The goal of the test cross is to determine if recombination has occurred during gamete formation in the heterozygous parent. New combinations of alleles cannot occur in the other parent, who is homozygous for these genes.

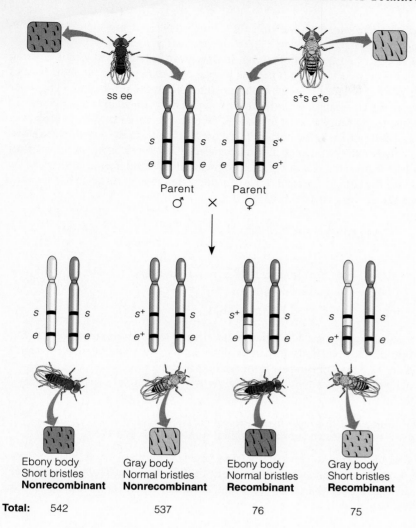

FIGURE **5-9**

An example of a test cross. The cross involves one *Drosophila* parent that is recessive for short bristles (*ss*) and ebony body (*ee*), and one parent heterozygous for these recessive traits (*s⁺s e⁺e*). (Note: *Drosophila* geneticists normally designate the short allele as *ss*, and a homozygous fly with short bristles as *ssss*. In the text, the allele causing short bristles is designated with a single *s* to avoid confusion between the allele designation and the genotype of the fly.)

Figure 5-9 illustrates how a test cross provides an experimental strategy to distinguish between recombinant and nonrecombinant offspring. This cross concerns two linked genes affecting bristle length and body color in fruit flies. The recessive alleles are *s* (short bristles) and *e* (ebony body), and the dominant (wild-type) alleles are *s⁺* (normal bristles) and *e⁺* (gray body). One parent displays both recessive traits. Therefore, we know that this parent is homozygous for the recessive alleles of the two genes (i.e., *ss ee*). The other parent is heterozygous for the linked genes affecting bristle length and body color. This parent was produced from a cross involving a true-breeding wild-type fly and a true-breeding fly with short bristles and an ebony body. Therefore, in this heterozygous parent, we know that the *s* and *e* alleles are linked on one chromosome and the corresponding *s⁺* and *e⁺* alleles are linked on the homologous chromosome.

Now let's take a look at the four possible types of offspring these parents can produce. The offspring's phenotypes are normal bristles/gray body, short bristles/ebony body, short bristles/gray body, and normal bristles/ebony body. All four types of offspring have inherited a chromosome carrying the *s* and *e* alleles from their homozygous parent (shown on the right in each pair). Focus your attention on the other chromosome. The offspring with short bristles and ebony bodies have also inherited a second chromosome carrying the *s* and *e* alleles from their other parent. This chromosome is not the product of a crossover in the heterozygous parent. The offspring with normal bristles and gray bodies have inherited a chromosome carrying the s^+ and e^+ alleles from the heterozygous parent. Again, this chromosome is not the product of a crossover.

The other two types of offspring, however, can only be produced if crossing over has occurred. Those with normal bristles and ebony bodies or short bristles and gray bodies have inherited a chromosome that is the product of a crossover during gamete formation in the heterozygous parent. As noted in Figure 5-9, the recombinant offspring are fewer in number than are the nonrecombinant offspring.

The data shown at the bottom of Figure 5-9 can be used to estimate the distance between the two genes. The map distance is defined as the number of recombinant offspring divided by the total number of offspring, multiplied by 100. With the data of Figure 5-9, we can calculate the map distance between the s and e alleles using this formula:

$$\text{Map distance} = \frac{\text{Number of recombinant offspring}}{\text{Total number of offspring}} \times 100$$

$$= \frac{76 + 75}{542 + 537 + 76 + 75} \times 100$$

$$= 12.3 \text{ map units}$$

The units of distance are called **map units** (mu) or sometimes **centimorgans** (cM) in honor of Thomas Hunt Morgan. In this example, we would say that the s and e alleles are 12.3 map units apart from each other along the same chromosome.

Alfred Sturtevant used the frequency of crossing over between two genes to produce the first genetic map in 1911

The first individual to construct a (very small) genetic map was Alfred Sturtevant, an undergraduate who spent time in the laboratory of Thomas Hunt Morgan. In 1965, more than fifty years after he constructed the first genetic map, Sturtevant wrote:

> In the latter part of 1911, in conversation with Morgan . . . I suddenly realized that the variations in the strength of linkage, already attributed by Morgan to differences in the spatial separation of the genes, offered the possibility of determining sequences [of different genes] in the linear dimension of a chromosome. I went home and spent most of the night (to the neglect of my undergraduate homework) in producing the first chromosome map, which included the sex-linked genes, y, w, v, m, and r, in the order and approximately the relative spacing that they still appear on the standard maps.

In the experiment described here, Sturtevant considered the outcome of crosses he conducted involving six different mutant alleles that altered the phenotype of normal flies. All of these alleles were known to be recessive and X-linked. They are y (yellow body color), w (white eye color), w-e (eosin eye color), v (vermillion eye color), m (miniature wings), and r (rudimentary wings). The w and w-e alleles are alleles of the same gene. In contrast, the v allele (vermillion eye color) is an allele of a different gene that also affects eye color. The two alleles that affect wing length, m and r, are also in different genes. Therefore, Sturtevant studied the inheritance of six recessive alleles, but since w and w-e are alleles of the same gene, his genetic map only contained five genes. The corresponding wild-type alleles are y^+ (gray body), w^+ (red eyes), v^+ (red eyes), m^+ (normal wings), and r^+ (normal wings).

THE HYPOTHESIS

When genes are located on the same chromosome, the distance between the genes can be estimated from the proportion of recombinant offspring. This provides a way to map the order of genes along a chromosome.

TESTING THE HYPOTHESIS

Starting materials: Sturtevant began with several different strains of *Drosophila* that contained the six alleles already described.

Experimental Level

Conceptual Level

1. Cross a female that is heterozygous for two different genes to a male that is hemizygous recessive for the same two genes. In this example, cross a female that is $X^{y^+w^+}X^{yw}$ to a male that is $X^{yw}Y$.

    This strategy was employed for many dihybrid combinations of the six alleles already described.

2. Observe the outcome of the crosses.

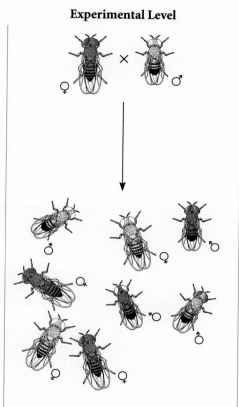

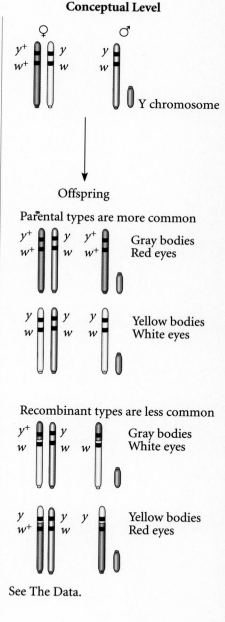

Offspring

Parental types are more common

Gray bodies
Red eyes

Yellow bodies
White eyes

Recombinant types are less common

Gray bodies
White eyes

Yellow bodies
Red eyes

3. Calculate the percentage of offspring that are the result of crossing over (# of nonparental/total).

See The Data.

THE DATA

Alleles Concerned*	Number Recombinant/ Total Number	Percent Recombinant Offspring
y and *w/w-e*	214/21,736	1.0
y and *v*	1,464/4,551	32.2
y and *r*	115/324	35.5
y and *m*	260/693	37.5
w/w-e and *v*	471/1,584	29.7
w/w-e and *r*	2,062/6,116	33.7
w/w-e and *m*	406/898	45.2
v and *r*	17/573	3.0
v and *m*	109/405	26.9

*Data from Sturtevant, A. H. (1914) The Linear Arrangement of Six Sex-Linked Factors in *Drosophila*, as Shown by Their Mode of Association. J. Exper. Zool. 14, 43–59. The allele designations have been changed to conform to modern allele designations and those used in this text. Since *w* and *w-e* are alleles of the same gene, the data from the crosses involving white and eosin alleles have been combined.

INTERPRETING THE DATA

Let's begin by contrasting the results between particular pairs of genes. In some dihybrid crosses, the percentage of nonparental offspring was rather low. For example, dihybrid crosses involving the *y* allele and the *w* or *w-e* allele yielded 1% recombinant offspring. This result suggests that these two genes are very close together. By comparison, other dihybrid crosses showed a higher percentage of nonparental offspring. Crosses involving the *v* and *m* alleles produced 26.9% recombinant offspring. These two genes are expected to be farther apart.

To construct his map, Sturtevant began with the assumption that the map distances would be more accurate between genes that are closely linked. Therefore, his map is based on the distance between *y* and *w* (1.0), *w* and *v* (29.7), *v* and *r* (3.0), and *v* and *m* (26.9). He also considered other features of the data to deduce the order of the genes. For example, the percentage of crossovers between *w* and *r* was 33.7. The percentage of crossovers between *w* and *v* was only 29.7, suggesting that *v* is between *w* and *r*, but closer to *r*. The proximity of *v* and *r* is confirmed by the low percentage of crossovers between *v* and *r* (3.0). Sturtevant collectively considered all these data and proposed the genetic map shown here:

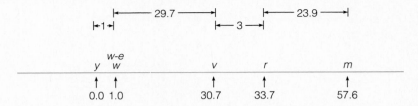

In this genetic map, Sturtevant began at the *y* allele and mapped the genes from left to right. For example, the *y* and *v* alleles are 30.7 map units apart, and the *v* and *r* alleles are 3.0 mu apart. This study by Sturtevant was a major breakthrough, since it showed how to map the locations of genes along chromosomes by making the appropriate crosses.

If you look carefully at Sturtevant's data, you will notice that there were two observations that do not agree very well with his genetic map. The percentage of

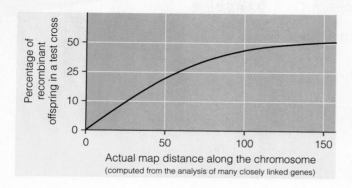

FIGURE **5-10**

Relationship between the percentage of recombinant offspring in a test cross and the actual map distance between genes. The *y*-axis depicts the percentage of recombinant offspring that would be observed in a dihybrid test cross. The actual map distance shown on the *x*-axis is calculated by analyzing the percentages of recombinant offspring from a series of many dihybrid crosses involving closely linked genes.

recombinant offspring for the *y* and *m* dihybrid cross was 37.5 (but the map distance is 57.6), and the crossover percentage between *w* and *m* was 45.2 (but the map distance is 56.6). As the percentage of recombinant offspring approaches a value of 50%, this value becomes a progressively more inaccurate measure of map distance (Figure 5-10). When the distance between two genes is large, the likelihood of multiple crossovers in the region between them causes the observed number of recombinant offspring to underestimate this distance. In addition, multiple crossovers set a quantitative limit on the relationship between map distance and the percentage of recombinant offspring. Even though two different genes can be on the same chromosome and more than 50 map units apart, a test cross is only expected to yield a maximum of 50% recombinant offspring.

Trihybrid crosses can be used to determine the order of and distance between linked genes

Until now, we have been considering the construction of genetic maps using dihybrid test crosses to compute map distance. The data from trihybrid crosses can also yield information about map distance and gene order. In a trihybrid cross, the experimenter crosses two individuals that differ in three traits. The following experiment follows a common strategy for using trihybrid crosses to map genes. In this experiment, the parental generation consists of fruit flies that differ in body color, eye color, and wing shape. We must begin with true-breeding lines so that we know which alleles are initially linked to each other on the same chromosome. In this example, all the dominant alleles are linked on the same chromosome.

Step 1. *Cross two true-breeding strains that differ with regard to three alleles.* In this example, we will cross a fly that has a black body (*bb*), purple eyes (*prpr*), and vestigial wings (*vgvg*) to a homozygous wild-type fly (*b⁺b⁺ pr⁺pr⁺ vg⁺vg⁺*):

Parental flies

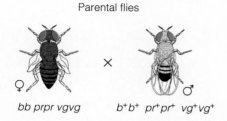

bb prpr vgvg   ×   b⁺b⁺ pr⁺pr⁺ vg⁺vg⁺

The goal in this step is to obtain F₁ individuals that are heterozygous for all three alleles. In the F₁ heterozygotes, all the dominant alleles are located on one chromosome, all the recessive alleles on the other homologous chromosome.

Step 2. *Perform a test cross by mating F_1 female heterozygotes to male flies that are homozygous recessive for all three alleles (bb prpr vgvg):*

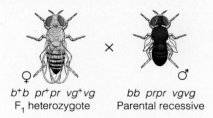

$b^+b\ pr^+pr\ vg^+vg$
F_1 heterozygote

$bb\ prpr\ vgvg$
Parental recessive

During gametogenesis in the heterozygous female F_1 flies, crossing over may occur to produce new combinations of the three alleles.

Step 3. *Collect data for the F_2 generation.* There are eight possible phenotypic combinations:

Phenotype	Number of Observed Offspring
Gray body, red eyes, normal wings	411
Black body, purple eyes, vestigial wings	412
Gray body, purple eyes, vestigial wings	30
Black body, red eyes, normal wings	28
Gray body, red eyes, vestigial wings	61
Black body, purple eyes, normal wings	60
Gray body, purple eyes, normal wings	2
Black body, red eyes, vestigial wings	1

Analysis of the F_2 generation flies will allow us to map these three genes. Since the three genes exist as two alleles each, there are $2^3 = 8$ possible combinations of offspring. If these alleles assorted independently, all eight combinations would occur in equal proportions. However, we see that the proportions of the eight phenotypes are far from equal.

The genotypes of the parental generation correspond to the phenotypes gray body, red eyes, and normal wings and black body, purple eyes, and vestigial wings. In crosses involving linked genes, the parental phenotypes occur most frequently in the offspring. The remaining six phenotypes are due to crossing over.

Two of the phenotypes (namely, gray body, purple eyes, and normal wings; and black body, red eyes, and vestigial wings) arise from a double crossover between two combinations of genes. The double crossover is always expected to be the least frequent category of offspring. Also, the combination of traits in the double crossover tells us which gene is in the middle. When a chromatid undergoes a double crossover, it separates the gene in the middle from the other two genes at either end:

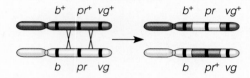

$b^+\quad pr^+\quad vg^+$ $b^+\quad pr\quad vg^+$

$b\quad pr\quad vg$ $b\quad pr^+\quad vg$

In the double crossover categories, the recessive purple-eye allele is separated from the other two recessive alleles. When mated to a homozygous recessive fly in the test cross, this yields flies with gray bodies, purple eyes, and normal wings, or with black bodies, red eyes, and vestigial wings. This observation indicates that the gene for eye color lies between the genes for body color and wing shape.

Step 4. *Calculate the map distance between pairs of genes.* To do this, we must regroup the data according to pairs of genes. From the parental generation, we know that the dominant alleles are initially linked to each other, as are the recessive alleles. This allows us to group pairs of genes into parental and nonparental combinations. The parental combinations are composed of a pair of dominant or a pair of recessive genes, whereas nonparental combinations have one dominant and one recessive gene. After we have regrouped the data in this way, the map distance between two genes can be calculated:

Parental Offspring	Total	Nonparental Offspring	Total
Gray body, red eyes		Gray body, purple eyes	
(411 + 61)	472	(30 + 2)	32
Black body, purple eyes		Black body, red eyes	
(412 + 60)	472	(28 + 1)	29
	944		61
Gray body, normal wings		Gray body, vestigial wings	
(411 + 2)	413	(30 + 61)	91
Black body, vestigial wings		Black body, normal wings	
(412 + 1)	413	(28 + 60)	88
	826		179
Red eyes, normal wings		Red eyes, vestigial wings	
(411 + 28)	439	(61 + 1)	62
Purple eyes, vestigial wings		Purple eyes, normal wings	
(412 + 30)	442	(60 + 2)	62
	881		124

Map distance between body color and eye color:

$$\text{Map distance} = \frac{61}{944 + 61} \times 100 = 6.1 \text{ mu}$$

Map distance between body color and wing shape:

$$\text{Map distance} = \frac{179}{826 + 179} \times 100 = 17.8 \text{ mu}$$

Map distance between eye color and wing shape:

$$\text{Map distance} = \frac{124}{881 + 124} \times 100 = 12.3 \text{ mu}$$

Step 5. *Construct the map.* Based on the map unit calculation, the body color and wing shape genes are farthest apart. The eye color gene must lie in the middle. As mentioned earlier, this order of genes is also confirmed by the pattern of traits found in the double crossovers. To construct the map, we use the distances between the genes that are closest together:

In our example, we have placed the body color gene first and the wing shape gene last. Our data also are consistent with a map in which the wing shape gene comes first and the body color gene comes last. In detailed genetic maps, the locations of genes are mapped relative to the centromere. This topic is discussed later in this chapter.

Interference can influence the number of double crossovers that occur in a short region

In Chapter 2, we introduced the product rule to determine the probability that two independent events will both occur. The product rule allows us to predict the expected likelihood of a double crossover provided we know the individual probabilities of each single crossover. Let's reconsider the data of the trihybrid test cross just described to see if the frequency of double crossovers is what we would expect based on the product rule. If we multiply the likelihood of a single crossover between *b* and *pr* (6.1%) times the likelihood of a single crossover between *pr* and *vg* (12.3%), then the product rule predicts

Expected likelihood of a double crossover = $0.061 \times 0.123 = 0.0075 = 0.75\%$

Based on a total of 1005 offspring produced,

Expected number of offspring due to a double crossover = $1005 \times 0.0075 = 7.5$

In other words, we would expect about 7 or 8 offspring to be produced as a result of a double crossover. The observed number of offspring was only 3 (namely, 2 with gray bodies, purple eyes, and normal wings, and 1 with a black body, red eyes, and vestigial wings). This lower than expected value is not due to random sampling error. Instead, it is due to a common genetic phenomenon known as **positive interference**. When a crossover occurs in one region of a chromosome, it often decreases the probability that another crossover will occur nearby. In other words, the first crossover interferes with the ability to form a second crossover in the immediate vicinity. To provide interference with a quantitative value, we first calculate the *coefficient of coincidence* (*C*):

$$C = \frac{\text{Observed number of double crossovers}}{\text{Expected number of double crossovers}}$$

Interference (*I*) is expressed as

$$I = 1 - C$$

For the data of the trihybrid test cross, the observed number of crossovers is 3 and the expected number is 7.5, and so the coefficient of coincidence equals 3/7.5 = 0.40. In other words, only 40% of the expected number of double crossovers were actually observed. The value for interference equals $1 - 0.4 = 0.60$ or 60%. This means that 60% of the expected number of crossovers were prevented from occurring. Since *I* has a positive value, this is positive interference. Rarely, the outcome of a test cross yields a negative value for interference. A negative interference value suggests that a first crossover enhances the rate of a second crossover in a nearby region. Although the molecular mechanisms that cause interference are not entirely understood, most organisms regulate the number of crossovers so that very few occur per chromosome. The reasons for positive and negative interference will require further research.

GENETIC MAPPING IN HAPLOID EUKARYOTES

Before ending our discussion of genetic mapping, it is interesting to consider some pioneering studies that involved the genetic mapping of haploid organisms. It may be surprising to you that certain species of lower eukaryotes, particularly unicellular algae and fungi, which spend the greatest part of their life cycle in the haploid state, have also been used in mapping studies. The sac fungi (ascomycetes) have been particularly useful to geneticists because of their unique style of sexual reproduction. In fact, much of our earliest understanding of genetic recombination came from the genetic analyses of fungi.

Fungi may be unicellular or multicellular organisms. Fungal cells are typically haploid (1n) and can reproduce asexually. In addition, fungi can also reproduce sexually by the fusion of two haploid cells to create a diploid zygote (2n) (Figure 5-11). The diploid zygote can then proceed through meiosis to produce four haploid cells, which are called **spores**. This group of four spores is known as a **tetrad** (not to be confused with a tetrad of four sister chromatids). In some species, meiosis is followed by a mitotic division to produce eight cells, known as an **octad**. The cells of a tetrad or octad are contained within a sac known as an **ascus** (plural, *asci*). In other words, the products of a single meiotic division are contained within one sac. This is a key feature that is useful to geneticists, and it dramatically differs from sexual reproduction in animals and plants. For example, in animals, oogenesis produces a single functional egg, and spermatogenesis occurs in the testes, where the resulting sperm become mixed with millions of other sperm.

Using a microscope, researchers can dissect asci and study the traits of each haploid spore. In this way, these organisms offer a unique opportunity for geneticists to identify and study all of the cells that are derived from a single meiotic division. In this section, we will consider how the analysis of asci can be used to map genes in fungi.

In fungal asci, tetrads and octads of spores can be ordered or unordered

The arrangement of spores within an ascus varies from species to species (Figure 5-12a). In some cases, the ascus provides enough space for the tetrads or octads of spores to randomly mix together. This is known as an *unordered tetrad* or *octad*. These occur in fungal species such as *S. cerevisiae* and *A. nidulans* and also in certain unicellular algae (*Chlamydomonas rheinhardii*). By comparison, other species of fungi produce a very tight ascus that prevents spores from randomly moving around. This can create a *linear tetrad or octad*. Figure 5-12b illustrates how a linear octad is formed in *N. crassa*. In this example, spores which carry the *A* allele have orange pigmentation, while spores having the *a* (albino) allele are white.

A key feature of linear tetrads or octads is that the position and order of spores within the ascus reflects their relationship to each other as they were produced by meiosis and mitosis. This idea is schematically shown in Figure 5-12b. After the original diploid cell has undergone chromosome replication, the first meiotic division produces two cells that are arranged next to each other within the sac. The second meiotic division then produces four cells that are also arranged in a straight row. Due to the tight enclosure of the sac around the cells, each pair of daughter cells is forced to lie next to each other in a linear fashion. Likewise, when each of these four cells divides by mitosis, each pair of daughter cells is located next to each other.

Linear tetrad analysis can be used to map the distance between a gene and the centromere

In the case of species that make linear tetrads and octads, experimenters can analyze the phenotypes of the spores within the asci and map the distance between a single gene and the centromere. Since the location of the centromere can be seen under the microscope, the mapping of a gene relative to the centromere provides a way to correlate a gene's location with the cytological characteristics of a chromosome. This approach has been extensively exploited in *N. crassa*.

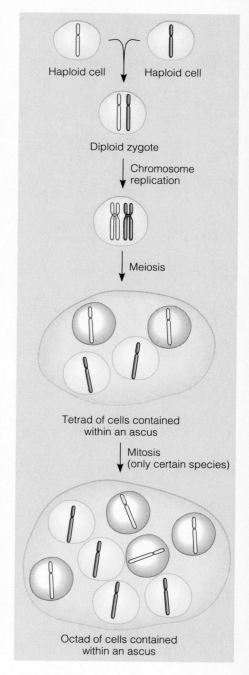

Haploid cell   Haploid cell

Diploid zygote

Chromosome replication

Meiosis

Tetrad of cells contained within an ascus

Mitosis (only certain species)

Octad of cells contained within an ascus

FIGURE 5-11

Sexual reproduction in ascomycetes.
For simplicity, this diagram shows each haploid cell as having only one chromosome per haploid set. However, fungal species actually contain several chromosomes per haploid set.

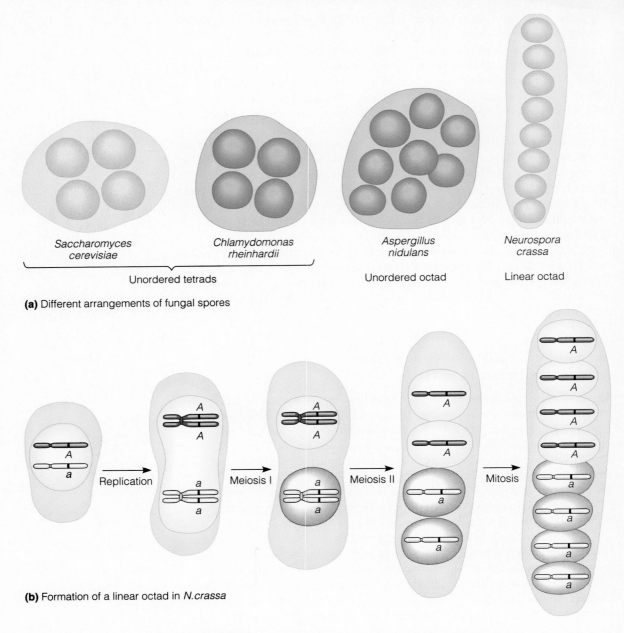

FIGURE **5-12**

Arrangement of spores within asci of different fungal species. (a) *Saccharomyces cerevisiae* and *Chlamydomonas rheinhardii* (an alga) produce unordered tetrads, *Aspergillus nidulans* produces an unordered octad, and *Neurospora crassa* produces a linear octad. **(b)** Linear octads are produced in *Neurospora* by meiosis and mitosis in such a way that the eight resulting cells are arranged linearly.

Figure 5-13 compares the arrangement of cells within a *Neurospora* ascus depending on whether or not a crossover has occurred between two homologues that differ at a gene with alleles *A* (orange pigmentation) and *a* (albino, which results in a white phenotype). In Figure 5-13a, a crossover has not occurred, and so the octad contains a linear arrangement of four haploid cells carrying the *A* allele, which are adjacent to four haploid cells that contain the *a* allele. This 4:4 arrangement of spores within the ascus is called a *first-division segregation (FDS) pattern* or an *M1 pattern*.

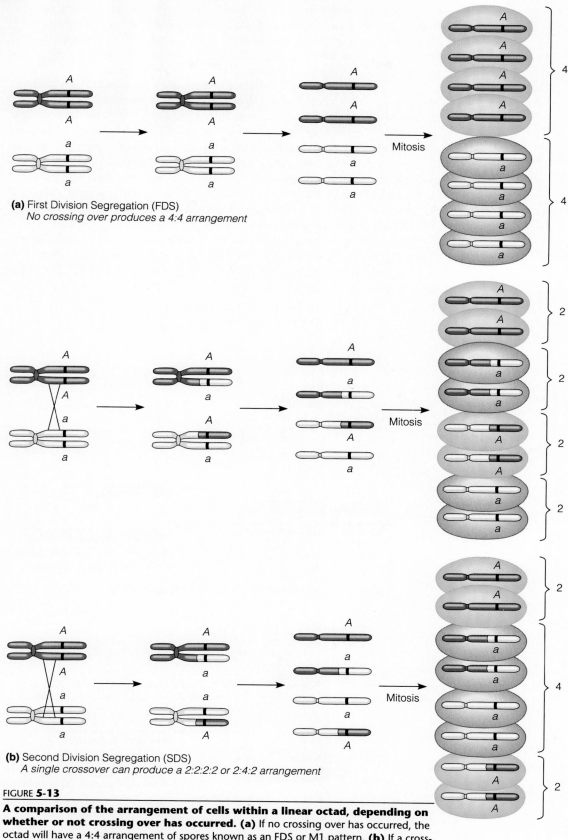

(a) First Division Segregation (FDS)
No crossing over produces a 4:4 arrangement

(b) Second Division Segregation (SDS)
A single crossover can produce a 2:2:2:2 or 2:4:2 arrangement

FIGURE **5-13**

A comparison of the arrangement of cells within a linear octad, depending on whether or not crossing over has occurred. **(a)** If no crossing over has occurred, the octad will have a 4:4 arrangement of spores known as an FDS or M1 pattern. **(b)** If a crossover has occurred between the centromere and gene of interest, a 2:2:2:2 or 2:4:2 pattern, known as an SDS or M2 pattern, is observed.

In contrast, as shown in Figure 5-13b, if a crossover occurs between the centromere and the gene of interest, the linear octad will deviate from the 4:4 pattern. Depending on the relative locations of the two chromatids that participated in the crossover, the ascus will contain a 2:2:2:2 or 2:4:2 pattern (see Figure 5-13b). These are called *second-division segregation (SDS) patterns* or *M2 patterns*.

Since a pattern of second division segregation is a result of crossing over, the percentage of M2 asci can be used to calculate the map distance between the centromere and the gene of interest. To understand why this is possible, let's consider the pattern of movement of a crossover site, or **chiasma** (plural, *chiasmata*). As shown in Figure 5-14, a chiasma forms at a particular site on a chromosome and then moves away from the centromere and toward the terminal end of the chromosome. As shown here, a crossover will only separate a gene from its original centromere if it begins in the region between the centromere and that gene. Therefore, the chances of getting a 2:2:2:2 or 2:4:2 pattern depend on the distance between the gene of interest and the centromere.

To determine the map distance between the centromere and a gene, the experimenter must count the number of SDS asci and the total number of asci. In SDS asci, only half of the spores are actually the product of a crossover. Therefore, the map distance is calculated as

$$\text{Map distance} = \frac{(1/2)(\text{Number of SDS asci})}{\text{Total number of asci}} \times 100$$

FIGURE **5-14**

The movement of chiasmata during crossing over. (a) If the chiasma initially forms between the centromere and the gene of interest, the gene will be separated from its original centromere. **(b)** If the chiasma initiates outside of this region, the gene remains attached to its original centromere.

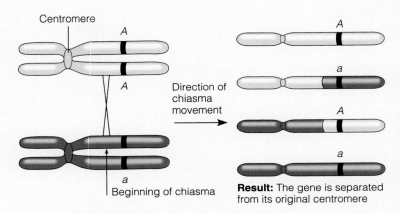

(a) Chiasma begins between centromere and gene of interest

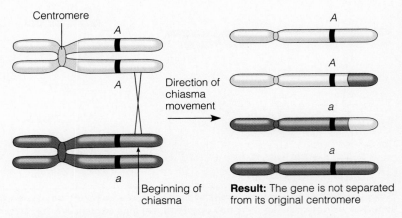

(b) Chiasma does not begin between centromere and gene of interest

Unordered tetrad analysis can be used to map genes in dihybrid crosses

Unordered tetrads contain a group of spores that are randomly arranged and the product of meiosis. An experimenter can conduct a dihybrid cross, remove the spores from each asci, and determine the phenotypes of the spores. This analysis can determine if two genes are linked or assort independently. If two genes are linked, a tetrad analysis can also be used to compute map distance.

Figure 5-15 illustrates the possible outcomes starting with a diploid yeast zygote that has the genotype *ura+ura-2 arg+arg-3*. *Ura+* and *arg+* are normal alleles required for uracil and arginine biosynthesis. *Ura-2* and *arg-3* are defective alleles that result in yeast strains that require uracil and arginine in the growth medium. This diploid cell was produced from the fusion of two haploid cells that were *ura+arg+* and *ura-2arg-3*. After the diploid cell has completed meiosis, there are three distinct possible combinations of four haploid cells. One possibility is that the tetrad will contain four spores with the parental combinations of alleles. This ascus is said to have the *parental ditype (PD)*. Alternatively, an ascus with a *nonparental ditype (NPD)* contains four cells with nonparental genotypes. Finally, it is possible to have a ascus that has two parental cells and two nonparental cells. This is called a *tetratype (T)*.

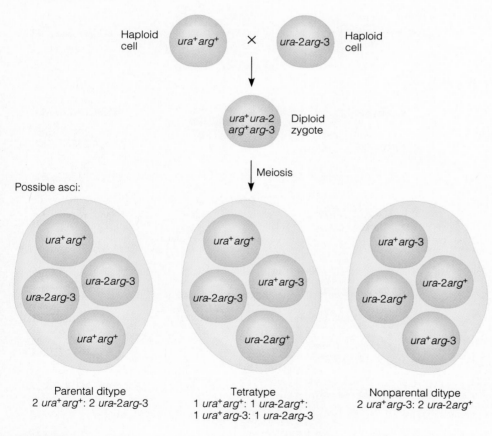

Parental ditype
2 *ura+arg+*: 2 *ura-2arg-3*

Tetratype
1 *ura+arg+*: 1 *ura-2arg+*:
1 *ura+arg-3*: 1 *ura-2arg-3*

Nonparental ditype
2 *ura+arg-3*: 2 *ura-2arg+*

FIGURE **5-15**

The assortment of two genes in an unordered tetrad. If the tetrad contains 100% parental cells, this ascus has the parental ditype. If it contains 50% parental and 50% recombinant cells, it is a tetratype. Finally, an ascus with 100% recombinant cells is called a nonparental ditype. This figure does not illustrate the chromosomal locations of the alleles. In this type of experiment, the goal is to determine whether the two genes are linked on the same chromosome.

When two genes assort independently, the number of asci having a parental ditype is expected to equal the number having a nonparental ditype, thus yielding 50% recombinant spores. For linked genes, Figure 5-16 illustrates the relationship between crossing over and the type of ascus that will result. If no crossing over occurs in the region between the two genes, then the parental ditype will be created. A single crossover event will produce a tetratype. Double crossovers can yield either a parental ditype, tetratype, or nonparental ditype depending on the combination of chromatids that are involved. A nonparental ditype is produced when a double crossover involves all four chromatids. A tetratype will result from a three-chromatid crossover. Finally, a double crossover between the same two chromosomes will produce the parental ditype.

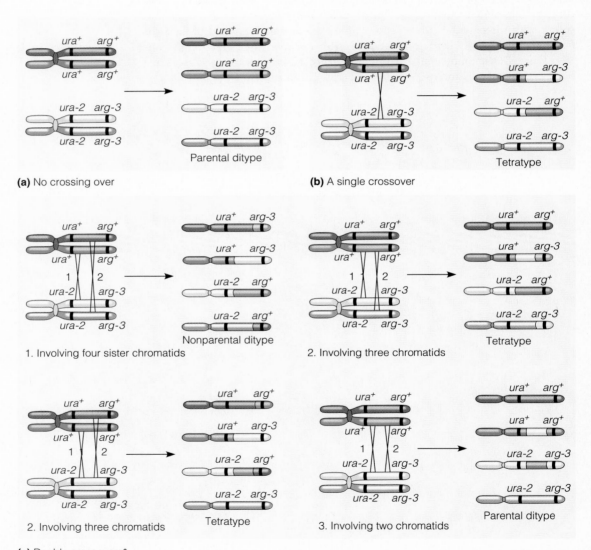

(a) No crossing over

(b) A single crossover

1. Involving four sister chromatids

2. Involving three chromatids

2. Involving three chromatids

3. Involving two chromatids

(c) Double crossovers*

* Since chiasma move from the centromere toward the chromosome end, crossover labeled 1 exchanges first and then crossover labeled 2 exchanges subsequently.

FIGURE 5-16

Relationship between crossing over and the production of the parental ditype, tetratype, and nonparental ditype for two linked genes.

The data from a tetrad analysis can be used to calculate the map distance between two linked genes. As in conventional mapping, the map distance is calculated as the percentage of offspring that carry recombinant chromosomes. As mentioned, a tetratype contains 50% recombinant chromosomes, a nonparental ditype 100%. Therefore, the map distance is computed as

$$\text{Map distance} = \frac{\text{Nonparental ditypes} + (1/2)(\text{Tetratypes})}{\text{Total number of asci}} \times 100$$

Over short map distances, this calculation provides a fairly reliable measure of distance. However, it does not adequately account for double crossovers. When two genes are far apart on the same chromosome, the calculated map distance using this equation underestimates the actual map distance due to multiple crossing over. Fortunately, a particular strength of tetrad analysis is that we can derive another equation that accounts for double crossovers and thereby provides a more accurate value for map distance. To begin this derivation, let's consider a more precise way to calculate map distance:

$$\text{Map distance} =$$
$$\frac{\text{Single crossover tetrads} + (2)(\text{Double crossover tetrads})}{\text{Total number of asci}} \times 0.5 \times 100$$

This equation includes the number of single and double crossovers in the computation of map distance. The total number of crossovers equals the number of single crossovers plus two times the number of double crossovers. Overall, the tetrads that contain single and double crossovers also contain 50% nonrecombinant chromosomes. To calculate map distance, therefore, we divide the total number of crossovers by the total number of asci and multiply by 50%.

Next, we need to relate this equation to the number of parental ditypes, nonparental ditypes, and tetratypes that are obtained by experimentation. To derive this relationship, we must consider the types of tetrads that are produced from no crossing over, a single crossover, and double crossovers. To do so, let's take another look at Figure 5-16. As shown there, the parental ditype and tetratype are ambiguous. The parental ditype can be derived from no crossovers or a double crossover; the tetratype can be derived from a single crossover or a double crossover. However, the nonparental ditype is unambiguous, since it can only be produced from a double crossover. We can use this observation as a way to determine the actual number of single and double crossovers. As seen in Figure 5-16, 1/4 of all the double crossovers are nonparental ditypes. Therefore, the total number of double crossovers equals 4 times the number of NPD.

Next, we need to know the number of single crossovers. A single crossover will yield a tetratype, but double crossovers can also yield a tetratype. Therefore, the total number of tetratypes overestimates the true number of single crossovers. Fortunately, we can compensate for this overestimation. Since there are two types of tetratypes that are due to a double crossover, the actual number of tetratypes arising from a double crossover should equal 2NPD. Therefore, the true number of single crossovers is calculated as $T - 2\text{NPD}$.

Now we have accurate measures of both single and double crossovers. The number of single crossovers equals $T - 2\text{NPD}$, and the number of double crossovers equals 4NPD. We can substitute these values into our previous equation:

$$\text{Map distance} = \frac{(T - 2\text{NPD}) + (2)(4\text{NPD})}{\text{Total number of asci}} \times 0.5 \times 100$$

$$= \frac{T + 6\text{NPD}}{\text{Total number of asci}} \times 0.5 \times 100$$

This equation provides a more accurate measure of map distance, since it considers both single and double crossovers.

Linkage refers to the phenomenon that many different genes may be located on the same chromosome. Chromosomes are sometimes called *linkage groups* because they contain a group of linked genes. Linkage affects the pattern of inheritance, because closely linked genes do not assort independently during gamete formation. This produces a greater percentage of offspring that display parental phenotypes. Nevertheless, nonparental offspring can be produced as a result of *crossing over*.

The likelihood of crossing over depends on the *distance* between two genes. If two genes are far apart from each other on the same chromosome, it is more likely that crossing over will occur between them. Therefore, when two genes are widely separated, a substantial percentage of recombinant offspring will be obtained from a *test cross*. However, the percentage of recombinant offspring cannot exceed a value of 50%, even when two genes are more that 50 map units apart on the same chromosome. The relationship between the percentage of recombinant offspring and the linear distance between genes is the basis for *genetic mapping*.

Experimentally, the phenomenon of linkage was deduced from genetic crosses. Bateson and Punnett were the first scientists to notice that certain genes do not assort independently. Morgan conducted crosses involving X-linked traits in fruit flies and correctly proposed that linkage is due to the location of particular genes on the same chromosome. He also hypothesized that recombinant phenotypes occur because of crossing over during meiosis. Morgan realized that the likelihood of crossing over depends on the distance between two genes. The proposal that genetic recombination is due to crossing over was confirmed cytologically by the studies of Creighton and McClintock, which showed that the production of recombinant offspring correlates with the production of recombinant chromosomes.

Genetic mapping is the determination of gene order and distance along chromosomes. In this chapter, we have considered how test crosses are conducted as a method to map genes. Sturtevant was the first person to understand that the percentage of recombinant offspring in a test cross could be used as a measure of the relative distance between two genes. Map distance is computed as the number of recombinant offspring divided by the total number of offspring. This approach can be readily applied to map genes using dihybrid and trihybrid test crosses. Genetic mapping is most accurate when map distances are calculated between closely linked genes. As the map distance approaches 50 map units (mu) and above, the percentage of recombinant offspring is not a reliable measure of map distance. In Chapter 20, several molecular methods of genetic mapping will be described.

This chapter ended with a discussion of gene mapping methods in fungi. A group of fungi known as the ascomycetes have been extensively used in genetic studies, because they produce all the products of a single meiosis within an ascus. For fungi such as *Neurospora* that make a linear ascus, the spores are arranged in a manner that reflects their relationship to each other during meiosis (and mitosis). Linear asci can be analyzed to map the location of a single gene relative to the centromere. Fungal species that produce unordered asci have also been used in mapping studies. In this case, dihybrid crosses are made, and the distance between the two genes can be computed by determining the proportions of parental ditypes, tetratypes, and nonparental ditypes.

In the next chapter, we will consider the linkage of genes within bacterial chromosomes and bacteriophages. Although bacteria normally reproduce asexually, they still can transfer genetic material by various different mechanisms. As we will see, these mechanisms also provide a way to map genes along the bacterial chromosome.

PROBLEM SETS

SOLVED PROBLEMS

1. In the garden pea, orange pods (*orp*) are recessive to normal pods (*Orp*), and sensitivity to pea mosaic virus (*mo*) is recessive to resistance to the virus (*Mo*). A plant with orange pods and sensitivity to the virus is crossed to a true-breeding plant with normal pods and resistance to the virus. The F_1 plants were then test crossed to plants with orange pods and sensitivity to the virus. The following results were obtained:

160	orange pods/virus sensitive
165	normal pods/virus resistant
36	orange pods/virus resistant
39	normal pods/virus sensitive
400	total

A. Conduct a chi square analysis to see if these genes are linked.
B. If they are linked, calculate the map distance between the two genes.

Answer:

A. Chi square analysis:

 1. Our hypothesis is that the genes are not linked.
 2. Calculate the predicted number of offspring based on the hypothesis. The test cross is

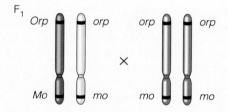

The predicted outcome of this cross under our hypothesis is a 1:1:1:1 ratio of plants with the four possible phenotypes. In other words, 1/4 should have the phenotype orange pods/virus sensitive, 1/4 should have normal pods/virus resistant, 1/4 should have orange pods/virus resistant, and 1/4 should have normal pods/virus sensitive. Since there were a total of 400 offspring, our hypothesis predicts 100 offspring in each category.
 3. Calculate the chi square:

$$\chi^2 = \frac{(O_1 - E_1)^2}{E_1} + \frac{(O_2 - E_2)^2}{E_2} + \frac{(O_3 - E_3)^2}{E_3} + \frac{(O_4 - E_4)^2}{E_4}$$

$$= \frac{(160 - 100)^2}{100} + \frac{(165 - 100)^2}{100} + \frac{(36 - 100)^2}{100} + \frac{(39 - 100)^2}{100}$$

$$= 36 + 42.3 + 41.0 + 37.2$$

$$= 156.5$$

 4. Interpret the chi square value. The calculated chi square value is quite large. This indicates that the deviation between observed and expected values is very high. According to Table 2-1, such a large deviation is expected to occur by chance alone less than 1%

of time. Therefore, we reject the hypothesis that the genes are assorting independently. Instead, we conclude the two genes are linked.

B. Calculate the map distance:

$$\text{Map distance} = \frac{\text{Number of nonparental offspring}}{\text{Total number of offspring}} \times 100$$

$$= \frac{36 + 39}{36 + 39 + 160 + 165} \times 100$$

$$= 18.8 \text{ mu}$$

The genes are approximately 18.8 map units apart.

2. Two recessive disorders in mice, droopy ear and flaky tail, are caused by genes that are located 6 mu apart on chromosome 3. A true-breeding mouse with normal ears (*De*) and a flaky tail (*ft*) is crossed to a true-breeding mouse with droopy ears (*de*) and a normal tail (*Ft*). The F_1 offspring are then crossed to mice with droopy ears and flaky tails. If this test cross produces 100 offspring, what is the expected outcome?

Answer: The test cross is

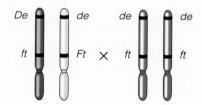

The parental offspring are

Dede	*ftft*	normal ears, flaky tail
dede	*Ftft*	droopy ears, normal tail

The recombinant offspring are

dede	*ftft*	droopy ears, flaky tail
Dede	*Ftft*	normal ears, normal tail

Since the two genes are located 6 mu apart on the same chromosome, 6% of the offspring will be recombinants. Therefore, the expected outcome for 100 offspring is

3	droopy ears, flaky tail
3	normal ears, normal tail
47	normal ears, flaky tail
47	droopy ears, normal tail

3. The following X-linked recessive traits are found in fruit flies: vermillion eyes are recessive to normal (red) eyes; miniature wings are recessive to normal (long) wings; and sable body is recessive to normal (gray) body. A cross was made between wild-type males and females with vermillion eyes, miniature wings, and sable bodies. The heterozygous females from this cross, which had a wild-type phenotype, were then crossed to males with vermillion eyes, miniature wings, and sable bodies. The following outcome was obtained:

Males and Females:

1320 vermillion eyes/miniature wings/sable body
1346 red eyes/long wings/gray body
102 vermillion eyes/miniature wings/gray body
90 red eyes/long wings/sable body
42 vermillion eyes/long wings/gray body
48 red eyes/miniature wings/sable body
2 vermillion eyes/long wings/sable body
1 red eyes/miniature wings/gray body

A. Calculate the map distance between the three genes.
B. Is positive interference occurring?

Answer:

A. The first step is to determine the order of the three genes. We can do this by evaluating the pattern of inheritance in the double crossovers. As discussed in the chapter, the double crossover group occurs with the lowest frequency. Thus, the double crossovers are vermillion eyes/long wings/sable body and red eyes/miniature wings/gray body. Compared with the parental combinations of alleles (vermillion eyes/miniature wings/sable body and red eyes/long wings/gray body), the gene for wing length has been reassorted. Two flies have long wings associated with vermillion eyes and sable body, and one fly has miniature wings associated with red eyes and gray body. Taken together, these results indicate that the wing length gene is found in between the eye color and body color genes:

Eye color——wing length——body color

We now calculate the distance between eye color and wing length, and between wing length and body color. To do this, we rearrange the data according to gene pairs:

vermillion eyes/miniature wings = 1320 + 102 = 1422
red eyes/long wings = 1346 + 90 = 1436
vermillion eyes/long wings = 42 + 2 = 44
red eyes/miniature wings = 48 + 1 = 49

The recombinants are vermillion eyes/long wings and red eyes/miniature wings. The map distance between these two genes is

$$(44 + 49)/(1422 + 1436 + 44 + 49) \times 100 = 3.2 \text{ mu}$$

Likewise, the other gene pair is wing length and body color:

miniature wings/sable body = 1320 + 48 = 1368
long wings/gray body = 1346 + 42 = 1388
miniature wings/gray body = 102 + 1 = 103
long wings/sable body = 90 + 2 = 92

The recombinants are miniature wings/gray body and long wings/sable body. The map distance between these two genes is

$$(103 + 92)/(1368 + 1388 + 103 + 92) \times 100 = 6.6 \text{ mu}$$

With these data, we can produce the following genetic map:

```
    ver           min            sab
    |←—— 3.1 ——→| |←——— 6.6 ———→|
```

B. To calculate the interference value, we must first calculate the coefficient of coincidence:

$$C = \frac{\text{Observed number of double crossovers}}{\text{Expected number of double crossovers}}$$

Based on our mapping in part A, the percentage of single crossovers equals 3.2% (0.032) and 6.6% (0.066). The expected number of double crossovers equals 0.032 times 0.066, which is 0.002, or 0.2%. There were a total of 2951 offspring produced. If we multiply 2951 times 0.002, we get 5.9, which is the expected number of double crossovers. The observed number was 3. Therefore:

$$C = 3/5.9 = 0.51$$
$$I = 1 - C = 1 - 0.51 = 0.49$$

In other words, approximately 49% of the double crossovers did not occur due to interference.

4. In *Saccharomyces cerevisiae*, a cross is made involving two genes, *ura-2* (uracil requiring) and *arg-3* (arginine-requiring) by mating a haploid *ura-2 arg+* strain to a *ura+ arg-3* strain. From the analysis of 400 individual tetrads, the following results were obtained:

2 spores: *ura-2arg+* + 2 spores: *ura+arg-3* = 231
2 spores: *ura-2arg-3* + 2 spores: *ura+arg+* = 3
1 spore: *ura-2arg-3* + 1 spore: *ura-2arg+*
+ 1 spore: *ura+arg-3* + 1 spore: *ura+arg+* = 166

A. Explain how an experimenter would determine the genotypes of the spores.
B. What is the map distance between these two genes?

Answer:

A. In this cross, an experimenter would dissect each ascus, and then grow each spore individually on a medium that contained uracil and arginine. On this medium, the spores will divide by mitosis to produce many haploid cells that are genetically identical to the original spore. Some of these cells can then be streaked onto two types of media: one that lacks uracil and another that lacks arginine. If the cells cannot grow on either medium, the original spore was *ura-2 arg-3*. If the cells can grow on both media, the original spore was *ura+ arg+*. If the cells can only grow on the medium lacking uracil, but not on that lacking arginine, the original spore was *ura+ arg-3*. Finally, if the cells can grow on the medium lacking arginine, but not that lacking uracil, the original spore was *ura-2 arg+*.

B. The first step is to determine which asci are parental ditypes, nonparental ditypes, and tetratypes. A parental ditype will contain a 2:2 combination of spores having the same genotypes as the original haploid parents. The top combination of 231 asci are the parental ditypes. The nonparental ditypes are those containing a 2:2 combination of genotypes that are unlike the parentals. The middle combination of 3 asci fit this description. Finally, the tetratypes contain a 1:1:1:1 arrangement of genotypes, half of which have a parental genotype and half of which do not. There are 166 tetratypes in this case. First, let's compute the map distance using

$$\text{Map distance} = \frac{\text{Nonparental ditypes} + (1/2)(\text{Tetratypes})}{\text{Total number of asci}} \times 100$$

$$= \frac{3 + (1/2)(166)}{400} \times 100$$

$$= 21.5 \text{ mu}$$

If we use the more accurate equation:

$$\text{Map distance} = \frac{(T + 6NPD)}{\text{Total number of asci}} \times 0.5 \times 100$$

$$= \frac{166 + 6(3)}{400} \times 0.5 \times 100$$

$$= 23.0 \text{ mu}$$

CONCEPTUAL QUESTIONS

1. What is the difference in meaning between the terms *genetic recombination* and *crossing over*?

2. When true-breeding mice with brown fur and short tails (*BBtt*) are crossed to true-breeding mice with white fur and long tails (*bbTT*), all the F_1 offspring have brown fur and long tails. The F_1 offspring are crossed to mice with white fur and short tails. What are the possible phenotypes and genotypes of the F_2 offspring? Which F_2 offspring are recombinant and which are nonrecombinant? What will be the ratios of the F_2 offspring if independent assortment is taking place? How will the ratios be affected by linkage?

3. Figure 5-2 shows the first experimental results that indicated linkage between two different genes. Conduct a chi square analysis to confirm that the genes are really linked and the data could not be explained by independent assortment.

4. When applying a chi square approach in a linkage problem, explain why an independent assortment hypothesis is used.

5. In the tomato, red fruit (*R*) is dominant over yellow fruit (*r*), and yellow flowers (*Wf*) are dominant over white flowers (*wf*). A cross is made between true-breeding plants with red fruit and yellow flowers and plants with yellow fruit and white flowers. The F_1 generation plants are then crossed to plants with yellow fruit and white flowers. The following results are obtained:

> 333 red fruit/yellow flowers
> 64 red fruit/white flowers
> 58 yellow fruit/yellow flowers
> 350 yellow fruit/white flowers

Calculate the map distance between the two genes.

6. Two genes are located on the same chromosome and are known to be 12 map units apart. An *AABB* individual is crossed to an *aabb* individual to produce *AaBb* offspring. The *AaBb* offspring are then crossed to *aabb* individuals.

A. If this cross produces 1000 offspring, what are the predicted numbers of offspring with each of the four genotypes: *AaBb*, *Aabb*, *aaBb*, and *aabb*?
B. What would be the predicted numbers of offspring with these four genotypes if the parental generation had been *AAbb* and *aaBB* instead of *AABB* and *aabb*?

7. Two genes, designated *A* and *B*, are located 10 map units from each other. A third gene, designated *C*, is located 15 map units from *B* and 5 map units from *A*. A parental generation consists of *AAbbCC* and *aaBBcc* individuals. The F_1 heterozygotes are then test crossed to *aabbcc* individuals. What percentage of offspring would you expect with the following genotypes?

A. *AaBbCc*
B. *aaBbCc*
C. *Aabbcc*

How would positive interference affect these predicted values?

8. Two different traits affecting pod characteristics in garden pea plants are encoded by genes found on chromosome 5. Narrow pod is recessive to normal pod; yellow pod is recessive to green pod. A true-breeding plant with narrow, green pods was crossed to a true-breeding plant with normal, yellow pods. The F_1 offspring were then test crossed to plants with narrow, yellow pods. The following results were obtained:

> 144 narrow green pods
> 150 normal yellow pods
> 11 narrow yellow pods
> 9 normal green pods

How far apart are these two genes?

9. What is mitotic recombination? A heterozygous individual (*Bb*) with brown eyes has one eye with a small patch of blue. Provide two or more explanations for how the blue patch may have occurred.

10. Mitotic recombination can occasionally produce a twin spot. Let's suppose an animal species can be heterozygous for two genes that govern fur color and length: one gene affects pigmentation, with dark pigmentation (*A*) dominant to albino (*a*); long hair (*L*) is dominant to short hair (*l*). The two genes are linked on the same chromosome. Let's assume an animal is *AaLl*; *A* is linked to *l*, and *a* is linked to *L*. What is the phenotype of this animal? Draw the chromosomes labeled with these alleles, and explain how mitotic recombination could produce a twin spot with one spot having albino pigmentation and long fur, the other having dark, short hair.

11. Two genes in tomatoes are 61 map units apart; normal fruit (*F*) is dominant to fasciated fruit, and normal numbers of leaves (*Lf*) is dominant to leafy (*lf*). A true-breeding plant with normal leaves and fruit is crossed to a leafy plant with fasciated fruit. The F_1 offspring are then crossed to leafy plants with fasciated fruit. If this cross produces 600 offspring, what are the expected numbers of plants in each of the four possible categories: normal leaves/normal fruit, normal leaves/fasciated fruit, leafy/normal fruit, and leafy/fasciated fruit?

12. Two recessive traits in mice are crinkly tail and soft coat. A mouse with a crinkly tail and soft coat was crossed to a true-breeding normal mouse. All the F_1 offspring were normal. The F_1 mice were then crossed to mice with crinkly tails and soft coats. The following results were obtained:

> 104 crinkly tail/soft coat
> 102 crinkly tail/normal coat
> 97 normal tail/normal coat
> 99 normal tail/soft coat

A. Conduct a chi square analysis to determine if these two genes are linked.
B. If the genes are linked, calculate the map distance between them.

13. In the tomato, three genes are linked on the same chromosome. Tall is dominant to dwarf, fruit skin that is smooth is dominant to skin that is peachy, and fruit with a normal tomato shape is dominant to oblate shape. A plant that is true-breeding for the dominant traits is crossed to a dwarf plant with peachy, oblate fruit. The F_1 plants are then test crossed to dwarf plants with peachy, oblate fruit. The following results are obtained:

> 151 tall/smooth/normal
> 33 tall/smooth/oblate
> 11 tall/peach/oblate

2 tall/peach/normal
155 dwarf/peach/oblate
29 dwarf/peach/normal
12 dwarf/smooth/normal
0 dwarf/smooth/oblate

A. Construct a genetic map that describes the order of these three genes and the distances between them.
B. Is positive interference occurring?

14. In *Drosophila*, two genes are 18 mu apart on the same autosome. Dumpy wings are recessive to normal wings, and short legs are recessive to normal legs. A true-breeding fly with normal wings and short legs is crossed to a fly with dumpy wings and normal legs. The F_1 generation is then test crossed to flies with dumpy wings and short legs. For the F_2 generation:

A. What proportion will have normal wings and normal legs?
B. What proportion will have dumpy wings and short legs?

15. A crossover has occurred in the tetrad shown here:

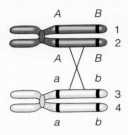

If a second crossover occurs in same region between these two genes, which two chromatids would be involved in order to produce the following outcomes?

A. 100% recombinants
B. 0% recombinants
C. 50% recombinants

16. A crossover has occurred in the tetrad shown here:

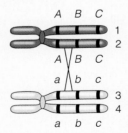

What would be the outcome of this single crossover event? If a second crossover occurs somewhere between *A* and *C*, explain which two chromatids it would involve and where it would occur to produce the following types of chromosomes:

A. *AbC*, *aBc*, *ABC*, and *abc*
B. *Abc*, *aBC*, *Abc*, and *aBC*
C. *Abc*, *aBC*, *ABc*, and *abC*
D. *ABC*, *ABC*, *abc*, and *abc*

17. A trait in garden peas involves the curling of leaves. A dihybrid cross was made of a plant with yellow pods and curling leaves to a wild-type plant with green pods and normal leaves. All the F_1 offspring had green pods and normal leaves. The F_1 plants were then crossed to plants with yellow pods and normal leaves. The following results were obtained:

117 green pods/normal leaves
115 yellow pods/curling leaves
78 green pods/curling leaves
80 yellow pods/normal leaves

A. Conduct a chi square analysis to determine if these two genes are linked.
B. If they are linked, calculate the map distance between the two genes. How accurate do think this distance is?

18. Though we often think of genes in terms of the phenotypes they produce (e.g., curly leaves, flaky tail, brown eyes), the molecular function of most genes is to encode proteins. Many cellular proteins function as enzymes. The table that follows describes the map distances between six different genes that encode six different enzymes. Construct a genetic map that describes the locations of all six genes.

Ada, adenosine deaminase
Hao-1, hydroxyacid oxidase-1
Hdc, histidine decarboxylase
Odc-2, ornithine decarboxylase-2
Sdh-1, sorbitol dehydrogenase-1
Ass-1, arginosuccinate synthetase-1

Map distances between two genes:

	Ada	Hao-1	Hdc	Odc-2	Sdh-1	Ass-1
Ada		14		8	28	
Hao-1	14		9		14	
Hdc		9		15	5	
Odc-2	8		15			63
Sdh-1	28	14	5			43
Ass-1				63	43	

19. In mice, the gene that encodes the enzyme inosine triphosphatase is 12 mu from the gene that encodes the enzyme ornithine decarboxylase. Let's suppose you have isolated a strain of mice homozygous for a defective inosine triphosphatase gene that does not produce any of this enzyme and is also homozygous defective for the ornithine decarboxylase gene. In other words, this strain of mice cannot make either enzyme. You cross this homozygous recessive strain to a normal strain of mice to produce heterozygotes. The heterozygotes are then backcrossed to the strain that cannot produce either enzyme. What is the probability of obtaining a mouse that cannot make either enzyme?

20. In the garden pea, several different genes affect pod characteristics. A gene affecting pod color (green is dominant to yellow) is approximately 7 mu away from a gene affecting pod width (wide is dominant to narrow). Both of these genes are located on chromosome 5. A third gene, located on chromosome 4, affects pod length (long is dominant to short). A true-breeding wild-type plant (green, wide, long pods) is crossed to a plant with yellow, narrow, short pods. The F_1 offspring are then test crossed to plants with yellow, narrow, short pods. If the test cross produces 800 offspring, what are the expected numbers of the eight possible phenotypic combinations?

21. Describe the unique features of ascomycetes that lend themselves to genetic analysis.

22. In fungi, what is the difference between a tetrad and an octad? What cellular process occurs in an octad that does not occur in a tetrad?

23. Explain the difference between an unordered versus a linear octad.

24. *His-5* and *lys-1* are alleles found in baker's yeast that require histidine and lysine for growth, respectively. A cross is made between two haploid yeast that are *his-5 lys-1* and *his+ lys+*. From the analysis of 818 individual tetrads, the following numbers of tetrads were obtained:

2 spores: *his-5/lys+* + 2 spores: *his+/lys-1* = 4

2 spores: *his-5/lys-1* + 2 spores: *his+/lys+* = 502

1 spore: *his-5/lys-1* + 1 spore: *his-5/lys+*

+ 1 spore: *his+/lys-1* + 1 spore: *his+/lys+* = 312

A. Compute the map distance between these two genes using the method of calculation that considers double crossovers and the one that does not. Which method gives a higher value? Explain why.

B. What is the frequency of single crossovers between these two genes?

C. Based on your answer to part B, how many NPDs are expected from this cross? Explain your answer. Is positive interference occurring?

25. On chromosome 4 in *Neurospora*, the allele *pyr-1* results in a pyrimidine requirement for growth. A cross is made between a *pyr-1* and a *pyr+* (wild-type) strain, and the following results are obtained:

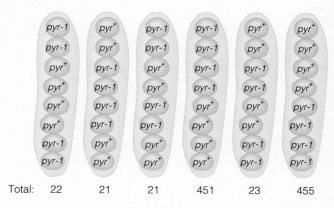

| Total: | 22 | 21 | 21 | 451 | 23 | 455 |

What is the distance between the *pyr-1* gene and the centromere?

26. On chromosome 3 in *Neurospora*, the *pro-1* gene is located approximately 9.8 mu from the centromere. Let's suppose a cross was made between a *pro-1* and a *pro+* strain and 1000 asci were analyzed.

A. What are the six types of asci that can be produced?

B. What are the expected numbers of each type of asci?

27. In *Neurospora*, a cross is made between a wild-type and an albino mutant strain (which produce orange and white spores, respectively). Draw two different ways that an octad might look if it was displaying second division segregation.

EXPERIMENTAL QUESTIONS

1. Explain the rationale behind a test cross. Is it necessary for one of the parents to be homozygous recessive for the genes of interest? In the heterozygous parent of a test cross, must all the dominant alleles be linked on the same chromosome and all the recessive alleles be linked on the homologue?

2. In Morgan's trihybrid test cross of Figure 5-3, he realized that crossing over was more frequent between the eye color and wing shape genes than between the body color and eye color genes. Explain how he determined this.

3. In Experiment 5A, the researchers followed the inheritance pattern of chromosomes that were abnormal at both ends to correlate genetic recombination with the physical exchange of chromosome pieces. Is it necessary to use a chromosome that is abnormal at both ends, or could they have had a strain with two abnormal chromosome 9s, one

with a knob at one end and its homologue with a translocation at the other end?

4. In an experiment similar to Experiment 5A, McClintock and Creighton observed on rare occasions a plant with recombinant characteristics but without recombinant chromosomes. How could this occur?

5. In Experiment 5B, describe which dihybrid cross data Sturtevant used to construct his genetic map.

6. In Experiment 5B, the percentage of crossovers between *y* and *m* was 37.5 (but the map distance is 57.6), and the crossover percentage between *w* and *m* was 45.2 (but the map distance is 56.6). Explain why the percentage of recombinant offspring was such an inaccurate measure of map distance.

QUESTIONS FOR STUDENT DISCUSSION/COLLABORATION

1. In mice, a dominant gene that causes a short tail is located on chromosome 2. On chromosome 3, a recessive gene causing droopy ears is 6 mu away from another recessive gene that causes a flaky tail. A recessive gene that causes a jerker (uncoordinated) phenotype is located on chromosome 4. A jerker mouse with droopy ears and a short, flaky tail is crossed to a normal mouse. All the F_1 generation mice are phenotypically normal, except they have short tails. These F_1 mice are then test crossed to jerker mice with droopy ears and long, flaky tails. If this cross produced 400 offspring, what would be the proportions of the sixteen possible phenotypic categories?

2. In Chapter 3, we discussed the idea that the X- and Y-chromosomes have a few genes in common. These genes are inherited in a pseudoau-tosomal pattern. With this phenomenon in mind, discuss whether or not the X- and Y-chromosomes are really distinct linkage groups.

3. Mendel studied seven traits in pea plants, and the garden pea happens to have seven different chromosomes. It has been pointed out that Mendel was very lucky not to have conducted crosses involving two traits governed by genes that are closely linked on the same chromosome, since the results would have confounded his theory about independent assortment. It has even been suggested that, perhaps, Mendel did not publish data involving traits that were linked! An article by Blixt, S. (1975) *Nature 256*, p. 206, considers this issue. Look up this article and discuss why Mendel did not find linkage.

CHAPTER 6

GENETIC TRANSFER AND MAPPING IN BACTERIA AND BACTERIOPHAGES

Thus far, our attention in Part II of this text has focused mostly on genetic analyses of eukaryotic species such as fungi, plants, and animals. As we have seen, these organisms are amenable to genetic studies for two reasons. First, allelic differences, such as white versus red eyes in *Drosophila* and tall versus dwarf pea plants, provide readily discernable traits among different individuals. Second, since eukaryotic species reproduce sexually, crosses can be made, and the pattern of transmission of traits from parent to offspring can be analyzed. The ability to follow allelic differences in a genetic cross is a basic tool in genetic examination of eukaryotic species.

In this chapter, we will turn our attention to the genetic analysis of bacteria. Like their eukaryotic counterparts, bacteria often possess allelic differences that affect their cellular traits. Common allelic variations among bacteria that are readily discernable involve traits such as sensitivity to antibiotics and varying nutrient requirements. Throughout this chapter, we will consider interesting experiments that genetically examine bacterial strains with allelic differences affecting such traits. However, compared with eukaryotes, one striking difference in prokaryotic species is their mode of reproduction. Since bacteria reproduce asexually, crosses are not used in genetic analyses of bacterial species. Instead, researchers rely on a similar phenomenon called **genetic transfer**. In this process, a segment of bacterial chromosomal DNA is transferred from one bacterium to another. As we will learn, there are several ways that bacterial species can transfer genetic material. Researchers have used genetic transfer to map the locations of genes along the single, circular bacterial chromosome.

GENETIC TRANSFER AND MAPPING IN BACTERIA

INTRAGENIC MAPPING IN BACTERIOPHAGES

In the second part of this chapter, we will examine **bacteriophages** (also known as **phages**), which are viruses that infect bacteria. Bacteriophages contain their own genetic material, which governs the traits of the phage. As we will see, the genetic analysis of phages can yield a highly detailed genetic map of a small chromosomal region. These types of analyses have provided researchers with insights regarding the structure and function of genes.

GENETIC TRANSFER AND MAPPING IN BACTERIA

Genetic transfer is a process whereby genetic material from one bacterium is transferred to another bacterium. Like sexual reproduction in eukaryotes, genetic transfer in bacteria is thought to enhance the genetic diversity of bacterial species. For example, a bacterial cell carrying a gene that provides antibiotic resistance may transfer this gene to another bacterial cell, allowing that bacterial cell to survive exposure to the antibiotic.

There are three naturally occurring ways that bacteria can transfer genetic material. The first route, known as **conjugation**, involves direct physical interaction between two bacterial cells. One bacterium acts as donor and transfers genetic material to a recipient cell. A second means of transfer is called **transduction**. In this situation, a virus that infects bacteria, a bacteriophage, transfers bacterial genetic material from one bacterium to another. The last mode of genetic transfer is **transformation**. In this case, genetic material is released into the environment when a bacterial cell dies. This material then binds to a living bacterial cell, which can take it up. In this section, we will describe these three systems of genetic transfer in greater detail. We will also learn how genetic transfer between bacterial cells has provided unique ways to accurately map bacterial genes.

Bacteria can transfer genetic material during conjugation

The natural ability of some bacteria to transfer genetic material between each other was first recognized by Joshua Lederberg and Edward Tatum while working at Yale University in 1946. They were studying strains of *Escherichia coli* (*E. coli*) that had different nutritional requirements for growth. The general methods for growing bacteria in a laboratory are described in the appendix. As shown in the experiment in Figure 6-1, they studied one strain, designated $B^-M^-P^+T^+$, which required one vitamin, biotin (B), and one amino acid, methionine (M), in order to grow. This strain did not require the amino acids phenylalanine (P) or threonine (T) for growth. Another strain, designated $B^+M^+P^-T^-$, had just the opposite requirements. It needed phenylalanine and threonine for growth, but not biotin and methionine. These differences in nutritional requirements correspond to variations in the genetic material of the two strains. The first strain had two defective genes encoding enzymes necessary for biotin and methionine synthesis. The second strain contained two defective genes required to make phenylalanine and threonine.

Figure 6-1 compares the results when the two strains were mixed together and when they were not mixed. Without mixing, about one billion (10^9) $B^-M^-P^+T^+$ cells were applied to plates, and no colonies were observed to grow. This result is expected, since the plates do not contain biotin and methionine. Likewise, when 10^9 $B^+M^+P^-T^-$ cells were plated, no colonies were observed, because phenylalanine and threonine were also missing from this growth medium. However, when $B^-M^-P^+T^+$ and $B^+M^+P^-T^-$ cells were mixed together and then 10^9 cells plated, approximately 100 cells were observed to grow and divide to form visible bacterial colonies. Since growth occurred, the genotype of the cells within these colonies must have been $B^+M^+P^+T^+$. To explain these results, Lederberg and Tatum hypothesized that some genetic material was being transferred between the two strains. One possibility is that the genetic material providing the ability to syn-

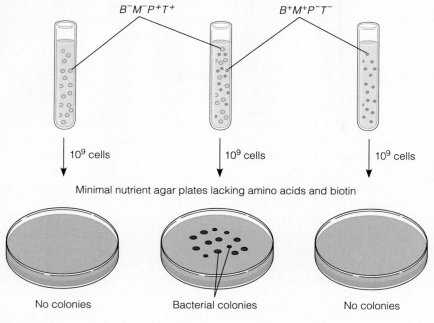

$B^-M^-P^+T^+$ $B^+M^+P^-T^-$

10^9 cells 10^9 cells 10^9 cells

Minimal nutrient agar plates lacking amino acids and biotin

No colonies   Bacterial colonies   No colonies

FIGURE **6-1**

Experiment of Lederberg and Tatum demonstrating genetic transfer during conjugation in E. coli. When the $B^-M^-P^+T^+$ or $B^+M^+P^-T^-$ strains were plated on a minimal medium lacking amino acids and biotin, they were unable to grow. However, if they were mixed together and then plated, many colonies were observed. These colonies are due to the transfer of genetic material between these two strains by conjugation.

thesize phenylalanine and threonine (P^+T^+) was transferred to the $B^+M^+P^-T^-$ strain. Alternatively, the ability to synthesize biotin and methionine (B^+M^+) may have been transferred to the $B^-M^-P^+T^+$ cells. The results of this experiment cannot distinguish between these two possibilities.

Bernard Davis at Cornell subsequently conducted experiments showing that the two strains of bacteria must make physical contact with each other to transfer genetic material. The apparatus he used, known as a **U-tube**, is shown in Figure 6-2. At the bottom of the U-tube is a filter with pores small enough to allow the passage

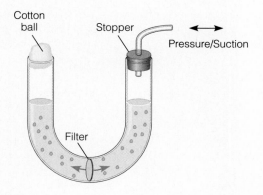

Cotton ball   Stopper
                              Pressure/Suction

Filter

FIGURE **6-2**

A U-tube apparatus like that used by Bernard Davis. A conceptual diagram of the U-tube. The fluid in the tube is forced through the filter by alternating suction and pressure. However, the pores in the filter are too small for the passage of bacteria.

of genetic material (i.e., DNA molecules), but too small to permit the passage of bacterial cells. On one side of the filter, Davis added a bacterial strain with a certain combination of nutritional requirements (the $B^-M^-P^+T^+$ strain), on the other side a different bacterial strain (the $B^+M^+P^-T^-$ strain). The application of pressure or suction promoted the movement of liquid through the filter. Since the bacteria were too large to fit through the pores, the movement of liquid did not allow the two types of bacterial strains to mix with each other. However, genetic material (which could have been released from a bacterium) could pass through the filter.

After incubation in a U-tube, bacteria from either side of the tube were placed on plates that could select for the growth of strains that were $B^+M^+P^+T^+$. These selective plates lacked biotin, methionine, phenylalanine, and threonine but contained all other nutrients essential for growth. In this case, no bacterial colonies grew on the plates. Thus, without physical contact, the two bacterial strains did not transfer genetic material to one another.

The term **conjugation** is now used to describe the natural process of genetic transfer between bacterial cells that requires direct cell-to-cell contact. Figure 6-3 is a micrograph that shows two bacteria making physical contact during conjugation. Many, but not all, species of bacteria can conjugate. During the late 1940s and the 1950s, scientists continued to explore this phenomenon of genetic transfer between bacterial cells. Working independently, Joshua and Esther Lederberg, William Hayes, and Luca Cavalli-Sforza discovered that only certain bacterial strains can act as donors of genetic material. For example, only about 5% of natural isolates of *E. coli* can act as donor strains. Research studies showed that a strain incapable of acting as a donor could subsequently be converted to a donor strain after being mixed with another donor strain. Hayes correctly proposed that donor strains contain a fertility factor that can be transferred to conjugation-defective strains to make them conjugation proficient.

We now know that donor strains usually contain a small circular segment of genetic material known as an **F factor** (for Fertility factor) in addition to their circular chromosome. Strains of bacteria that contain an F factor are designated F^+, strains without F factors are F^-. The more general term **plasmid** is used to describe circular pieces of DNA that exist independently of the chromosome. Plasmids, and their uses in molecular cloning, will be examined in detail in Chapter 19.

The process of bacterial conjugation is shown in Figure 6-4. It has been extensively studied in many bacterial species, particularly *E. coli*. F^+ strains of *E. coli* produce structures on their cell surface known as **sex pili** (singular, *pilus*; see Figure 6-3).

FIGURE **6-3**

Two *E. coli* cells in the act of conjugation. This is a micrograph (a photograph of cells under a microscope) showing two conjugating *E. coli* cells. The cell on the right is F^+, and the one on the left is F^-. The two cells make contact with each other via sex pili that are made by the F^+ cell.

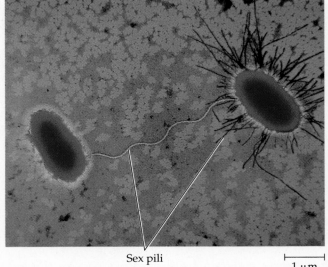

Sex pili

1 μm

The pili act as attachment sites that promote the binding of bacteria to each other. In this way, an F^+ strain makes physical contact with an F^- strain.

To begin conjugation, one of the strands in the double-stranded F factor DNA is cut at a location known as the **origin of transfer**. This cut DNA strand is then transferred in a linear manner to the F^- cell. The other strand of the F factor DNA remains in the donor cell. Although DNA replication is examined in Chapter 11, let's briefly consider how DNA replication coincides with conjugation so that we can understand the conjugation process. DNA replication in the donor cell restores the F factor DNA to its original double-stranded condition. Similarly, after the recipient cell receives a single strand of the F factor DNA, it is replicated in the recipient cell to become double stranded. Finally, the two ends of the linear F factor DNA in the recipient cell ligate to each other to become a circular F factor. As seen in Figure 6-4, the result of conjugation is that the recipient cell has acquired an F factor, converting it from an F^- to an F^+ cell. The genetic composition of the donor strain has not been changed.

Hfr Strains contain an F^+ factor integrated into the bacterial chromosome; during conjugation, the bacterial chromosome is transferred from the donor to the recipient strain in a linear fashion

An important advance in our understanding of bacterial genetics occurred when Luca Cavalli-Sforza discovered a strain of *E. coli* that was very efficient at transferring many chromosomal genes to recipient F^- strains. Cavalli-Sforza designated this bacterial strain an **Hfr strain** (for High frequency of recombination). It was originally derived from an F^+ strain. In an Hfr strain, an F factor has become integrated into the bacterial chromosome (Figure 6-5a). The term **episome** is used to describe a segment of bacterial DNA that can exist as an F factor and also integrate into the chromosome.

Hayes, who independently isolated another Hfr strain, demonstrated that conjugation between an Hfr strain and an F^- strain involves the transfer of the bacterial chromosome from the Hfr strain to the F^- cell (Figure 6-5b). A region within the integrated F^+ factor known as the origin of transfer determines the starting point and direction of this transfer process. One of the DNA strands is cut at the origin of transfer. This cut, or nicked, site is the starting point that will enter the F^- recipient cell. Behind this starting point, the rest of the bacterial chromosome enters the F^- cell in a linear manner. This transfer process occurs in conjunction with chromosomal replication, so that the Hfr cell retains its original chromosomal composition. It generally takes about 1.5 to 2 hours for the entire Hfr chromosome to be passed into the F^- cell. Since most matings do not last that long, usually only a portion of the Hfr chromosome is transmitted to the F^- cell.

Once inside the F^- cell, the chromosomal material from the Hfr cell can swap, or recombine, with the homologous region of the recipient cell's chromosome. The net result is that the recipient cell has received a new segment of chromosomal DNA from the Hfr cell. As illustrated in Figure 6-5b, this recombination may provide the recipient cell with a new combination of alleles. Prior to mating, the recipient strain was *lac⁻* (unable to metabolize lactose) and *pro⁻* (unable to synthesize proline). If mating occurred for a short time, the recipient cell received a short segment of chromosomal DNA from the donor. In this case, the recipient cell has become *lac⁺* but remains *pro⁻*. If the mating is prolonged, the recipient cell will receive a longer segment of chromosomal DNA from the donor. After a longer mating, the recipient becomes *lac⁺* and *pro⁺*. As noted in Figure 6-5b, an important feature of Hfr mating is that the bacterial chromosome is transferred linearly to the recipient strain. In this example, *lac⁺* is always transferred first and *pro⁺* is transferred later.

In any particular Hfr strain, the origin of transfer has a specific orientation that either promotes a counterclockwise or clockwise transfer of genes. Also,

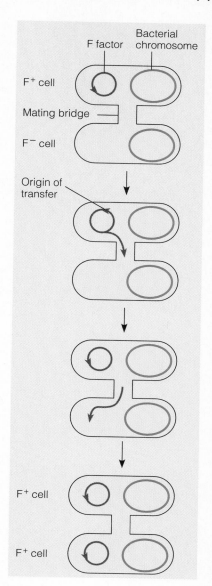

FIGURE **6-4**

The transfer of a fertility factor during bacterial conjugation. The origin of transfer is the site on the F factor that is transferred first. This transfer coincides with F factor replication. The net result is that the recipient F^- cell has received a copy of the F factor from the F^+ donor cell.

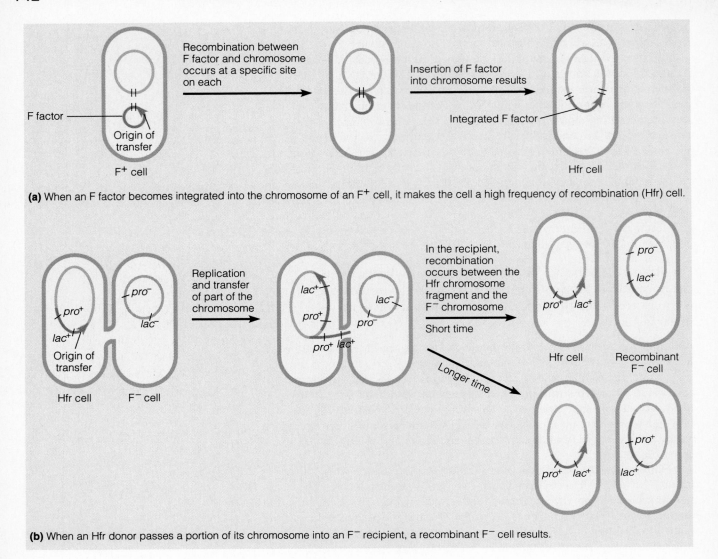

(a) When an F factor becomes integrated into the chromosome of an F⁺ cell, it makes the cell a high frequency of recombination (Hfr) cell.

(b) When an Hfr donor passes a portion of its chromosome into an F⁻ recipient, a recombinant F⁻ cell results.

FIGURE **6-5**

The transfer of the *E. coli* chromosome occurs in a linear manner during an Hfr-mediated conjugation. (a) An Hfr cell is created when an F factor integrates into the bacterial chromosome. **(b)** The transfer of the bacterial chromosome begins at the origin of replication and then proceeds around the circular chromosome. After a segment of chromosome has been transferred to the F⁻ recipient cell, it recombines with the recipient cell's chromosome. If mating occurs for a short period, only a short segment of the chromosome is transferred. If mating is prolonged, a longer segment of the bacterial chromosome is transferred.

GENES ⟶ TRAITS: The *F⁻* recipient cell was originally *lac⁻* (unable to metabolize lactose) and *pro⁻* (unable to synthesize proline). If mating occurred for a short period of time, the recipient cell acquired *lac⁺*, allowing it to metabolize lactose. If mating occurred for a longer period of time, the recipient cell also acquired *pro⁺*, enabling it to synthesize proline.

among different Hfr strains, the origin of transfer may be located in different regions of the chromosome. Therefore, the order of gene transfer depends on the location and orientation of the origin of transfer. For example, another Hfr strain could have its origin of transfer next to *pro⁺* and transfer *pro⁺* first and then *lac⁺*.

Wollman and Jacob used Hfr matings to map genes along the *E. coli* chromosome in the 1950s

The first genetic mapping experiments in bacteria were carried out by Elie Wollman and François Jacob during the 1950s while working at the Pasteur Institute in Paris. At the time of their studies, there was not much information about the organization of bacterial genes along the chromosome. A few key advances during the 1940s and early 1950s made Wollman and Jacob realize that the process of genetic transfer could be used to map the order of genes in *E. coli*. Most importantly, the discovery of conjugation by Lederberg and Tatum, and the identification of Hfr strains by Cavalli-Sforza and Hayes, made it clear that bacteria can transfer genes from donor to recipient cells in a linear fashion. In addition, Wollman and Jacob were aware of previous microbiological studies concerning bacteriophages that bind to *E. coli* cells and subsequently infect them. These studies showed that bacteriophages can be sheared from the surface of *E. coli* cells if they are spun in a blender. In this treatment, the bacteriophages are detached from the surface of the bacterial cells, but the bacteria themselves remain healthy and viable. Wollman and Jacob reasoned that a blender treatment could also be used to separate bacterial cells that were in the act of conjugation without killing them. This technique is known as an **interrupted mating**.

The rationale behind Wollman and Jacob's mapping strategy is that the time it takes for genes to enter a donor cell is directly related to their order along the bacterial chromosome. Because the Hfr chromosome is transferred linearly, they realized that interruptions of mating at different times would lead to various lengths of the Hfr chromosome being transferred to the F^- recipient cell. If two bacterial cells had mated for a short period of time, only a small segment of the Hfr chromosome would be transferred to the recipient bacterium. However, if the bacterial cells were allowed to mate for a longer period of time before being interrupted, a longer segment of the Hfr chromosome could be transferred. By determining which genes were transferred during short matings and which required longer mating times, Wollman and Jacob were able to deduce the order of particular genes along the *E. coli* chromosome.

For this experiment, Wollman and Jacob began with two *E. coli* strains. The donor (Hfr) strain had the following genetic composition:

T^+:   able to synthesize threonine, an essential amino acid for growth

L^+:   able to synthesize leucine, an essential amino acid for growth

Az^s: sensitive to killing by azide (a toxic chemical)

$T1^s$:   sensitive to infection (i.e., a plaque formation) by bacteriophage T1

lac^+:   able to metabolize lactose, and use it for growth

gal^+:   able to metabolize galactose, and use it for growth

str^s: sensitive to killing by streptomycin (an antibiotic)

The recipient (F^-) strain had the opposite genotype: T^- L^- Az^r $T1^r$ lac^- gal^- str^r (r = resistant). Before the experiment, Wollman and Jacob already knew that the T^+ gene was transferred first, followed by the L^+ gene, and both were transferred relatively soon (5–10 minutes) after mating. Their main goal in this experiment was to determine the times at which the other genes (Az^s $T1^s$ lac^+ gal^+) were transferred to the recipient strain. The transfer of the str^s gene was not examined because streptomycin was used to kill the donor strain following conjugation.

THE HYPOTHESIS

The chromosome of the donor strain in an Hfr mating is transferred in a linear manner to the recipient strain. The order of genes along the chromosome can be deduced by determining the time various genes take to enter the recipient strain.

Starting materials: The two *E. coli* strains already described, one Hfr strain ($T^+ L^+ Az^s T1^s lac^+ gal^+ Str^s$) and one F⁻ ($T^- L^- Az^r T1^r lac^- gal^- Str^r$).

Experimental Level

Conceptual Level

1. Mix together a large number of Hfr donor and F⁻ recipient cells.

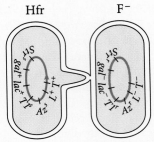

Hfr   F⁻

In this conceptual example, the cells have been incubated about 20 minutes.

2. After different periods of time, take a sample of cells and interrupt conjugation in a blender.

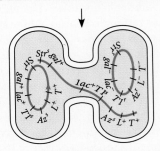

Separate by blending; donor DNA recombines with recipient cell chromosome.

3. Plate the cells on minimal media lacking threonine and leucine, but containing streptomycin. Note: The general methods for growing bacteria in a laboratory are described in the appendix.

Minimal solid growth medium and streptomycin

Overnight growth

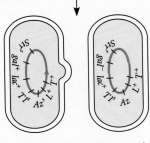

Cannot survive on minimal plates with streptomycin

Can survive on minimal plates with streptomycin

Surviving colonies

Sterile loop

Additional tests

4. Pick each surviving colony, which would have to be $T^+L^+Str^r$, and test to see if it is sensitive to killing by azide, sensitive to infection by T1 bacteriophage, and if it is able to metabolize lactose or galactose.

Plaques

No growth

+Azide

+T1 phage

Bacterial growth

+Lactose

No growth

+Galactose

The conclusion is that the colony that was picked contained cells with a genotype of $T^+L^+Az^sT1^slac^+gal^-Str^r$.

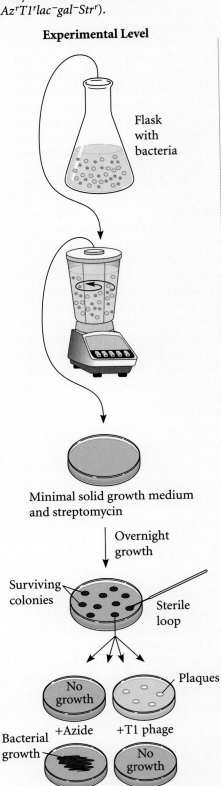

Minutes that Bacterial Cells were Allowed to Mate Before Blender Treatment	Percent of Surviving Bacterial Colonies with the Following Genotypes:				
	T^+L^+	Az^s	$T1^s$	lac^+	gal^+
5	—*	—	—	—	—
10	100	12	3	0	0
15	100	70	31	0	0
20	100	88	71	12	0
25	100	92	80	28	0.6
30	100	90	75	36	5
40	100	90	75	38	20
50	100	91	78	42	27
60	100	91	78	42	27

* There were no surviving colonies after 5 minutes of mating. Data from: Jacob, F., and Wollman, E. L. (1961) *Sexuality and the Genetics of Bacteria.* Academic Press, New York.

Before discussing the conclusions of this experiment, it is helpful to describe how Wollman and Jacob monitored gene transfer. To determine if particular genes had been transferred after mating, they took the mated cells and first plated them on growth media that lacked threonine and leucine but contained streptomycin. On these plates, the original donor and recipient strains could not grow, because the donor strain was streptomycin sensitive and the recipient strain required threonine and leucine. However, mated cells in which the donor transferred chromosomal DNA carrying the T^+ and L^+ genes would be able to grow.

To determine the order of gene transfer of the Az^s, $T1^s$, lac^+, and gal^+ genes, Wollman and Jacob picked colonies from the first plates and restreaked them on plates that contained azide or bacteriophage T1, or on minimal plates that contained lactose or galactose as the sole source of energy for growth. The plates were incubated overnight to observe the formation of visible bacterial growth. Whether or not the bacteria could grow depended on their genotypes. For example, a cell that is Az^s cannot grow on plates containing azide, and a cell that is lac^- cannot grow on minimal plates containing lactose as the carbon source for growth. By comparison, a cell that is Az^r and lac^+ can grow on both types of plates.

Now let's discuss the data. After the first plating, all survivors would be cells in which the T^+ and L^+ alleles had been transferred to the F^- recipient strain (which is already streptomycin resistant). As seen in the data, five minutes was not sufficient time to transfer the T^+ and L^+ alleles, since no surviving colonies were observed. After 10 minutes or longer, however, surviving bacterial colonies with the $T^+ L^+$ genotype were obtained. To determine the order of the remaining genes (Az^s, $T1^s$, lac^+, and gal^+), each surviving colony was tested to see if it was sensitive to killing by azide, sensitive to infection by T1 bacteriophage, able to utilize lactose for growth, or able to utilize galactose for growth. A consistent pattern emerged from the data. The gene that conferred sensitivity to azide (Az^s) was transferred first, followed by $T1^s$, lac^+, and finally gal^+. From these data, as well as those from other experiments, Wollman and Jacob constructed a genetic map that described the order of these genes along the *E. coli* chromosome:

```
|----------------------------------------------------------|
   T   L    Az    T1          lac              gal
```

This work provided the first method for bacterial geneticists to map the order of genes along the bacterial chromosome. Throughout the course of their studies, Wollman and Jacob identified several different Hfr strains in which the origin of transfer had been integrated at different places along the bacterial chromosome. When they compared the order of genes among different Hfr strains, their results were consistent with the idea that the *E. coli* chromosome is circular (see solved problem 2).

A detailed genetic map of the *E. coli* chromosome has been obtained from many conjugation studies

Conjugation experiments have been used to map more than 1000 genes along the circular *E. coli* chromosome. A map of the *E. coli* chromosome is shown in Figure 6-6. This simplified map only shows the locations of 12 genes. Since the chromosome is circular, we must arbitrarily assign a starting point on the map. The gene *thrA* is at the location of 0 minutes. The *E. coli* genetic map is 100 minutes long, which is approximately the time that it takes to transfer the complete chromosome during an Hfr mating.

As shown in Figure 6-6, we scale genetic maps from bacterial conjugation studies in units of **minutes**. This unit refers to the relative time it takes for genes to first enter an F⁻ recipient strain during a conjugation experiment. The distance between two genes is determined by comparing their times of entry during a conjugation experiment. As shown in Figure 6-7, the time of entry is found by conducting mating experiments at different time intervals before interruption. We compute the time of entry by extrapolating the time back to the origin. In this experiment, the time of entry of the *lacZ* gene was approximately 16 minutes, and that of the *galE* gene was 25 minutes. Therefore, these two genes are approximately 9 minutes apart from each other along the *E. coli* chromosome.

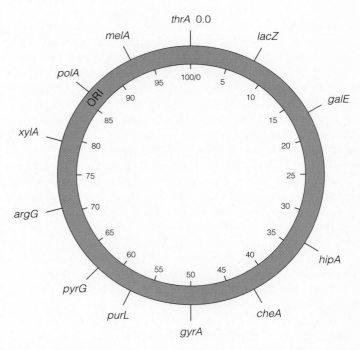

FIGURE **6-6**

A simplified genetic map of the *E. coli* chromosome indicating the positions of several genes. *E. coli* has a circular chromosome with about 2000 different genes. This map shows the locations of 12 of them. The map is scaled in units of minutes, and it proceeds in a clockwise direction. The starting point on the map is the gene *thrA*.

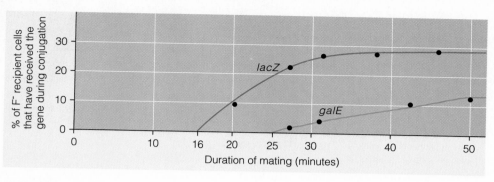

FIGURE **6-7**

Time course of an interrupted _E. coli_ conjugation experiment. By extrapolating the data back to the origin, the approximate time of entry of the _lacZ_ gene is found to be 16 minutes, that of the _galE_ gene 25 minutes. Therefore, the distance between these two genes is 9 minutes. Data from: Wollman, E. L., Jacob, F., and Hayes, W. (1956) _Cold Spring Harbor Symposia on Quantitative Biology 21_, 141.

Bacteriophages can also transfer genetic material from one bacterial cell to another in the process of transduction

Before we discuss the ability of bacteriophages to transfer genetic material between bacterial cells, let's consider some general features of a phage's life cycle. Bacteriophages are composed of genetic material that is surrounded by a protein coat. As shown in Figure 6-8, certain types of bacteriophages can bind to the surface of a bacterium and inject their genetic material into the bacterial cytoplasm. Depending on

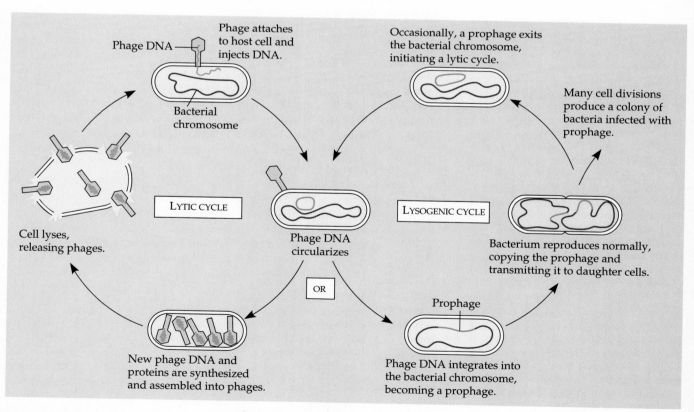

FIGURE **6-8**

The lytic and lysogenic life cycles of bacteriophages.

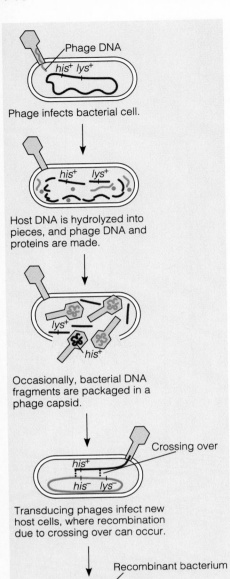

Phage infects bacterial cell.

Host DNA is hydrolyzed into pieces, and phage DNA and proteins are made.

Occasionally, bacterial DNA fragments are packaged in a phage capsid.

Transducing phages infect new host cells, where recombination due to crossing over can occur.

The recombinant bacterium has a genotype (his^+ lys^-) that is different from the recipient bacterial cell (his^- lys^-).

FIGURE 6-9

Transduction in bacteria.

GENES ⟶ TRAITS: The transducing phage introduced DNA into a new host cell that was originally his^- lys^- (unable to synthesize histidine and lysine). During transduction, it received a segment of bacterial chromosomal DNA that carried his^+. Following recombination, the host cell's genotype was changed to his^+, so that it now could synthesize histidine.

its type and the growth conditions, the phage may follow a **lytic cycle** or a **lysogenic cycle**. Some phages can follow both cycles. During the lytic cycle, the bacteriophage directs the synthesis of many copies of the phage genetic material and coat proteins (see left side of Figure 6-8). These components then assemble to make new phages. When synthesis and assembly is completed, the bacterial host cell is lysed (broken apart), releasing the newly made phages into the environment.

In other cases, a bacteriophage will infect a bacterium and follow the lysogenic cycle (see right side of Figure 6-8). A bacteriophage that usually exists in the lysogenic life cycle is called a **temperate phage**. Under most conditions, temperate phages do not produce new phages and will not kill the host bacterial cell. During the lysogenic cycle, a phage integrates its genetic material into the chromosome of the bacterium. This integrated phage DNA is known as a **prophage**. A prophage can exist in a dormant state for a long time, during which no new bacteriophages are made. When a bacterium containing a lysogenic prophage divides to produce two daughter cells, it will copy the prophage's genetic material along with its own chromosome. Therefore, both daughter cells will inherit the prophage. At some later time, a prophage may become activated to excise itself from the bacterial chromosome and enter the lytic cycle. Then, once again, it will promote the synthesis of new phages and eventually lyse the host cell.

With a general understanding of bacteriophage life cycles, we will now examine the ability of phages to transfer genetic material between bacteria. This process is called **transduction**. Examples of phages that can transfer bacterial chromosomal DNA from one bacterium to another are the P22 and P1 phages that infect the bacterial species *Salmonella typhimurium* and *E. coli*, respectively. The P22 and P1 phages can follow either the lytic or lysogenic cycle. While in the lytic cycle, the phage directs the synthesis of phage DNA and coat proteins, which then assemble to make new phages. During this process, the bacterial chromosome becomes fragmented into small pieces of DNA (Figure 6-9).

Occasionally, a mistake can happen in which a piece of bacterial DNA assembles with the coat proteins. This creates a phage that contains bacterial chromosomal DNA. When phage synthesis is completed, the bacterial cell is lysed and releases the newly made phage into the environment. Following release, this abnormal phage can still bind to a living bacterial cell and introduce its genetic material into the bacterium. This DNA fragment (which was derived from the bacterial chromosomal DNA) can then recombine with the bacterial chromosome inside the infected cell. In the example of Figure 6-9, any piece of the bacterial chromosomal DNA can be incorporated into the phage. This type of transduction is called **generalized transduction**.

Transduction was first discovered by Joshua Lederberg and Norton Zinder in 1952 while working at the University of Wisconsin in Madison. In an experimental strategy similar to that of Figure 6-1, they mixed together two strains of the bacterium *S. typhimurium*. One strain, designated LA-22, was *phe⁻*, *trp⁻*, *met⁺*, and *his⁺* (unable to synthesize phenylalanine or tryptophan but able to synthesize methionine and histidine); the other strain, LA-2, was *phe⁺*, *trp⁺*, *met⁻*, and *his⁻* (able to synthesize phenylalanine and tryptophan, but unable to synthesize methionine or histidine). When they placed this mixture of cells on plates with minimal growth media lacking these four amino acids, approximately one cell in 100,000 was observed to grow. The genotype of the surviving bacterial cells must have been *phe⁺ trp⁺ met⁺ his⁺*. Therefore, Lederberg and Zinder concluded that genetic material had been transferred between the two strains.

A novel result occurred when Lederberg and Zinder repeated this experiment using a U-tube apparatus as previously described in Figure 6-2. They placed the LA-22 strain (*phe⁻ trp⁻ met⁺ his⁺*) on one side of the filter and LA-2 (*phe⁺ trp⁺ met⁻ his⁻*) on the other. They removed samples from either side of the tube and plated the cells on minimal plates lacking the four amino acids. Surprisingly, they obtained colonies from the side of the tube that contained LA-22 but not from the side that contained LA-2. From these results, they concluded that some filterable agent was being transferred from LA-2 to LA-22 that converted LA-22 to a *phe⁺ trp⁺ met⁺ his⁺* genotype. In other words, some filterable agent carrying the *phe⁺* and *trp⁺* genes was being produced on the side of the tube containing LA-2, traversing the filter, and then being taken up by the LA-22 strain. By conducting this type of experiment using filters with different pore sizes, Lederberg and Zinder found that the filterable agent was slightly less than 0.1 μm in diameter, a size much smaller than a bacterium. They correctly concluded that the filterable agent in these experiments was a bacteriophage.

Cotransduction can be used to map genes that are within 2 minutes of each other

During transduction, there is a maximum size of the DNA segment that can be packaged by bacteriophages. P1 cannot package pieces that are greater than 2 to 2.5% of the entire *E. coli* chromosome, and P22 cannot package pieces that are greater than 1% of the length of the *S. typhimurium* chromosome. If two genes are close together along the chromosome, a bacteriophage may package a single piece of the chromosome that carries both genes and transfer that piece to another bacterium. This phenomenon is called **cotransduction**. The likelihood that two genes will be cotransduced depends on how close together they lie. If two genes are far apart along the bacterial chromosome, they will never be cotransduced, because the bacteriophage cannot physically package a DNA fragment that is larger than 1 to 2.5% of the bacterial chromosome. In genetic mapping studies, cotransduction is only used to determine the order and distance between genes that lie fairly close to each other.

To map genes using cotransduction, a researcher selects for the transduction of one gene and then monitors whether or not a second gene is cotransduced along with it. As an example, let's consider a donor strain of *E. coli* that is *arg⁺ met⁺ strˢ* (able to synthesize arginine and methionine, but sensitive to killing by streptomycin) and a recipient strain that is *arg⁻ met⁻ strʳ* (see Figure 6-10). The donor strain is infected with phage P1. Some of the *E. coli* cells are lysed by P1, and this P1 lysate is mixed with the recipient cells. After allowing sufficient time for transduction, the recipient cells are plated on a minimal medium that contains arginine and streptomycin but not methionine. Therefore, these plates select for the growth of cells in which the *met⁺* gene has been transferred to the recipient strain, but do not select for the growth of cells in which the *arg⁺* gene has been transferred, since the plates are supplied with arginine.

Nevertheless, the *arg⁺* gene may have been cotransduced with the *met⁺* gene if the two genes are close together. To determine this, a sample of cells from each bacterial colony can be picked up with a needle and restreaked on minimal plates that lack both amino acids. If the colony can grow, it must have also obtained the *arg⁺* gene during transduction. In other words, cotransduction of both the *arg⁺* and *met⁺* genes has occurred. Alternatively, if the restreaked colony does not grow, it must have only received the *met⁺* gene during transduction. Data from this type of experiment are shown at the bottom of Figure 6-10. These data indicate a cotransduction frequency of $21/(21 + 29) = 0.42$.

In 1966, T. T. Wu at Harvard University derived a mathematical expression that relates cotransduction frequency with map distance obtained from conjugation experiments. This equation is

$$\text{Cotransduction frequency} = (1 - d/L)^3$$

where

d = the distance between two genes in minutes

L = the size of chromosomal pieces (in minutes) that the phage carries during transduction.

(For P1 transduction, this size is approximately 2% of the bacterial chromosome, which equals 2 minutes.)

FIGURE 6-10

The steps in a cotransduction experiment. The donor strain is *arg⁺ met⁺ strˢ* (able to synthesize arginine and methionine but sensitive to streptomycin), and the recipient strain is *arg⁻ met⁻ strʳ*. The donor strain is infected with phage P1. Some of the *E. coli* cells are lysed by P1, and this P1 lysate is mixed with the recipient cells. P1 phages in this lysate may carry fragments of the donor cell's chromosome, and the P1 phage may inject that DNA into the recipient cells. To identify recipient cells that have received the *met⁺* gene from the donor strain, the recipient cells are plated on minimal medium that contains arginine and streptomycin but not methionine. To determine if the *arg⁺* gene has also been cotransduced, cells from each bacterial colony are sampled with a needle and placed on minimal plates that lack both amino acids. If the cells can grow, cotransduction of the *arg⁺* and *met⁺* genes has occurred. Alternatively, if the restreaked colony does not grow, it must have received only the *met⁺* gene during transduction.

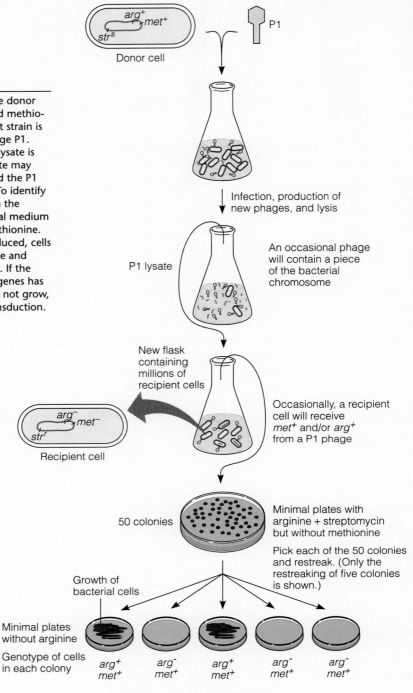

Selected gene	Nonselected gene	Number of colonies that grew on		Cotransduction frequency
		minimal +arginine	minimal −arginine	
met⁺	arg⁺	50	21	0.42

This equation assumes that the bacteriophage randomly packages pieces of the bacterial chromosome that are similar in size. Depending on the type of phage used in a transduction experiment, this assumption may not always be valid. Nevertheless, this equation has been fairly reliable in computing map distance for P1 transduction experiments in *E. coli*. We can use this equation to estimate the distance between the two genes described in Figure 6-10:

$$0.42 = (1 - d/2)^3$$

$$(1 - d/2) = \sqrt[3]{0.42}$$

$$1 - d/2 = 0.75$$

$$d/2 = 0.25$$

$$d = 0.5 \text{ minutes}$$

This equation tells us that the distance between the *met*+ and *arg*+ genes is approximately 0.5 minutes.

Genetic mapping strategies in bacteria often involve data from both conjugation and transduction experiments. Conjugation is commonly used to determine the relative order and distance of genes, particularly those far apart along the chromosome. By comparison, transduction experiments can provide very accurate mapping data for genes that are fairly close together.

Bacteria can also transfer genetic material by transformation

A third natural mechanism for the transfer of genetic material from one bacterium to another is known as **transformation**. This process was first discovered by Frederick Griffith in 1928 while working with strains of *Streptococcus pneumoniae*. During bacterial transformation, a living bacterial cell will take up a fragment of DNA released from a dead bacterium. This DNA fragment may then recombine with the bacterial chromosome, producing a bacterium with genetic material that it has received from the dead bacterium. (In Chapter 9, we will consider how bacterial transformation was used as an experimental tool to demonstrate that DNA is the genetic material.)

Since the initial studies of Griffith, we have learned much about the events that occur in transformation. Only certain bacterial cells, known as **competent cells**, can be transformed by extracellular DNA. Temperature, ionic conditions, and nutrients in the growth media can greatly influence whether or not a bacterium will be competent to take up genetic material from its environment.

In recent years, geneticists have unraveled some of the steps that occur when competent bacterial cells are transformed by genetic material in their environment. Figure 6-11 describes these steps. First, a large fragment of genetic material binds to the surface of the bacterial cell. Before entering the cell, however, this large piece of chromosomal DNA must be cut into smaller fragments. This cutting is accomplished by a bacterial enzyme known as an *endonuclease*, which makes occasional random cuts in the long piece of chromosomal DNA. At this stage, the DNA fragments are composed of double-stranded DNA.

The next step is for the DNA fragment to begin its entry into the bacterial cytoplasm. For this to occur, the double-stranded DNA interacts with proteins in the bacterial membrane. In some bacterial species, one of the DNA strands is

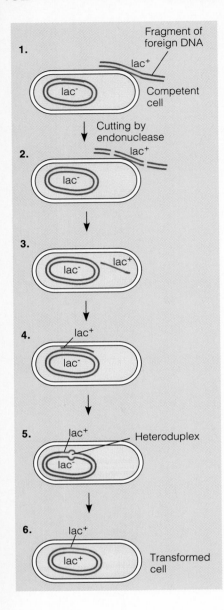

FIGURE **6-11**

The steps of bacterial transformation. (1) A large piece of genetic material containing the *lac*+ gene binds to the surface of a competent bacterium. **(2)** The genetic material is digested to smaller fragments by a cellular endonuclease. **(3)** One of the DNA strands is degraded, and the other, which contains the *lac*+ gene, enters the bacterial cell. **(4)** The newly imported DNA strand aligns with the matching section of the bacterial chromosome, in this case where the *lac*− gene is located. **(5)** The imported strand is substituted for one of the chromosomal strands. The resulting temporary mismatch creates a heteroduplex. **(6)** The heteroduplex is repaired.

GENES ⟶ TRAITS: Bacterial transformation can also lead to new genetic traits for the recipient cell. The original cell was *lac*− (unable to metabolize lactose). Following transformation, it became *lac*+. This result would transform the original bacterial cell into a cell that could grow on minimal plates that contain lactose. Before transformation, the original *lac*− cell would not have been able to grow on lactose minimal plates.

degraded and the other strand enters the bacterial cytoplasm. The single-stranded DNA then aligns itself with the homologous location on the bacterial chromosome. In the example shown in Figure 6-11, the foreign DNA carries a functional *lac*+ gene. This strand of DNA aligns itself with a nonfunctional (mutant) *lac*− gene already present within the bacterial chromosome. The foreign DNA then recombines with one of the strands in the bacterial chromosome of the competent cell. In other words, the foreign DNA replaces one of the chromosomal strands of DNA, which is subsequently degraded. During recombination, alignment of the *lac*− and the *lac*+ alleles results in a region of mismatch called a **heteroduplex**. However, the heteroduplex is only a temporary situation. DNA repair enzymes in the recipient cell recognize the heteroduplex and repair it. In this case, the heteroduplex has been repaired by eliminating the mutation that caused the *lac*− phenotype, thereby creating a *lac*+ gene. In this example, the recipient cell has been transformed from a *lac*− strain to a *lac*+ strain.

Transformation has also been used to map many bacterial genes. Such gene mapping is conducted similarly to the cotransduction experiments described earlier. If two genes are close together, the cotransformation frequency is expected to be high, whereas genes that are far apart will have a cotransformation frequency that is very low or even zero. Like cotransduction, genetic mapping via transformation is only used to map genes that are relatively close together.

INTRAGENIC MAPPING IN BACTERIOPHAGES

Biologists do not consider viruses to be living, because they rely on a host cell for their existence and proliferation. Nevertheless, we can think of viruses as having traits, because they have unique biological structures and functions. Each type of virus has its own genetic material, which contains many viral genes. In this section, we will focus our attention on a bacteriophage called T4. Its genetic material contains several dozen different viral genes. These genes encode viral proteins that carry out a variety of functions. For example, some of the genes encode proteins needed for the synthesis of new viruses and the lysis of the host cell. Other genes encode the viral coat proteins that are found in the head, shaft, tail baseplate, and tail fibers. Figure 6-12 illustrates the locations of several coat proteins that make up the structure of T4. For example, five different proteins, designated by the numbers 34–38, assemble to form the tail fibers (see inset to Figure 6-12). The expression of T4 genes to make proteins 34–38 provides the bacteriophage with the trait of having tail fibers, enabling it to attach itself to the surface of a bacterium.

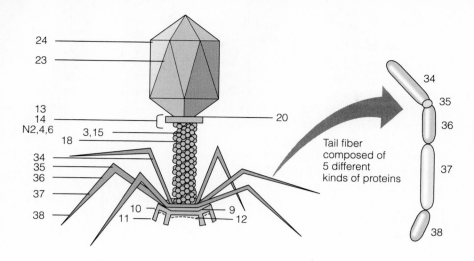

FIGURE **6-12**

Structure of the T4 virus, with labeled proteins.

GENES ⟶ TRAITS: The inset to this figure shows the five different proteins, coded for by five different genes, that make up the tail fiber. The expression of these genes provides the bacteriophage with the trait of having tail fibers, enabling it to attach itself to the surface of a bacterium.

The study of viral genes has been instrumental in our basic understanding of how the genetic material works. During the early 1950s, Seymour Benzer at Purdue University embarked on a ten-year study that focused on the function of viral genes in the T4 bacteriophage. In this section, we will examine some of his pivotal results, which advanced our knowledge of gene function. We will also explore how he conducted a detailed type of genetic mapping known as **intragenic** or **fine structure mapping**. The difference between intragenic and intergenic mapping is shown here:

Intergenic mapping

Intragenic mapping (also known as fine structure mapping)

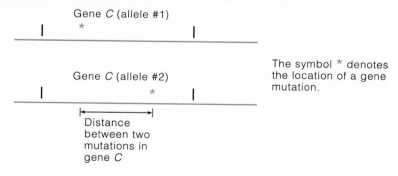

The symbol * denotes the location of a gene mutation.

We have considered intergenic mapping earlier in this chapter and in Chapter 5. In that type of mapping, the goal is to determine the distance between two different genes. As shown here, determining the distance between genes *A* and *B* is an example of intergenic mapping. By comparison, intragenic mapping seeks to ascertain distances within the same gene. For example, in a population, gene *C* may exist as two different mutant alleles. One allele may be due to a mutation near the beginning of the gene, the second to a mutation near the end. The goal of intragenic mapping is to determine the distance between the two mutations that occur in the same gene. In this section, we will explore the pioneering studies that showed the feasibility of intragenic mapping.

Lytic bacteriophages produce viral plaques; mutations in viral genes can alter plaque morphology

As they progress through the lytic cycle, bacteriophages ultimately produce new phages, which are released when the bacterial cell lyses (refer back to Figure 6-8). In the laboratory, researchers can visually observe the consequences of bacterial cell lysis in the following way (Figure 6-13). A sample of bacterial cells and lytic bacteriophages are mixed together. This mixture of cells and phages is then poured onto petri plates that contain nutrient agar for bacterial cell growth. Bacterial cells, which are not infected by a bacteriophage, will rapidly grow and divide to produce a "lawn" of bacteria. This lawn of bacteria is opaque (i.e., you cannot see through it to the underlying agar). In the experiment shown in Figure 6-13, seven bacterial cells have been infected by bacteriophages, and these infected cells are found at random locations in the lawn of uninfected bacteria. The infected cells will lyse and release newly made bacteriophages. These bacteriophages will then infect the nearby bacteria within the lawn. These cells will eventually lyse and also release newly made phages. Over time, these repeated cycles of infection and lysis will produce, around each site of an original phage infection, an observable clear area, or **plaque**, where the bacteria have been lysed (Figure 6-13).

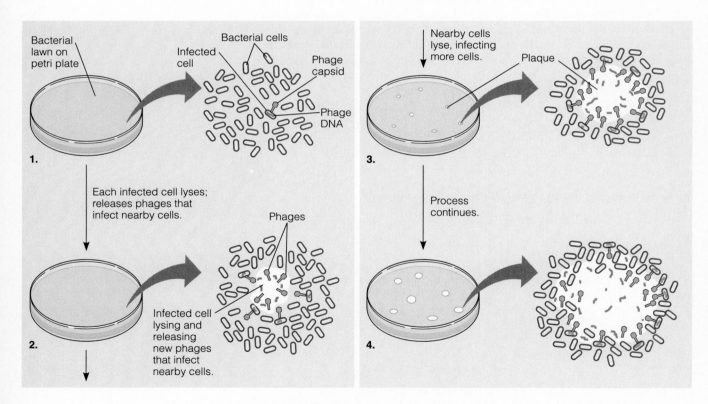

FIGURE **6-13**

The formation of phage plaques on a lawn of bacteria in the laboratory. In this experiment, bacterial cells are mixed with a small number of lytic bacteriophages. In the figure, seven bacterial cells are initially infected by phages. The cells are poured onto petri plates containing nutrient agar for bacterial cell growth. Bacterial cells will rapidly grow and divide to produce an opaque lawn of densely packed bacteria. The seven infected cells will lyse and release newly made bacteriophages. These bacteriophages will then infect the nearby bacteria within the lawn. Likewise, these newly infected cells will lyse and release new phages. By this repeated process, the area of cell lysis will create a clear zone known as a viral plaque.

A genetic analysis of any organism requires strains with allelic differences. Since bacteriophages are microscopic, it is rather difficult for geneticists to analyze mutations that affect phage morphology. However, some mutations in the bacteriophage's genetic material can alter the ability of the phage to cause plaque formation. Therefore, we can view the morphology of plaques as a trait of the bacteriophage. Since plaques are visible with the naked eye, mutations affecting this trait lend themselves to a much easier genetic analysis. An example is a rapid-lysis mutant of bacteriophage T4, which tends to form unusually large plaques (Figure 6-14). These types of mutants were first identified by A. D. Hershey in 1946 while working at Cold Spring Harbor, New York. The plaques are large because the mutant phages lyse the bacterial cells more rapidly than do the wild-type phages. Rapid-lysis mutants form large, clearly defined plaques, as opposed to wild-type bacteriophages that produce smaller, fuzzy-edged plaques. Mutations in several different bacteriophage genes can produce a rapid-lysis phenotype.

Seymour Benzer studied one category of T4 phage mutants, designated rII (r stands for rapid lysis). In a bacterial strain called *E. coli B*, rII phages produce abnormally large plaques. Nevertheless, *E. coli B* strains produce low yields of rII phages, because the rII phages lyse the bacterial cells so quickly they do not have sufficient time to produce many new phages. To help study this phage, Benzer wanted to obtain large quantities of it. Therefore, to improve his yield of rII phage, he decided to test its yield in other bacterial strains.

On the day Benzer decided to do this, he happened to be teaching a phage genetics class at Purdue University. For that class, he was growing two *E. coli* strains designated *E. coli K12S* and *E. coli K12(λ)*. He was growing these two strains to teach his class about the lysogenic cycle. *E. coli K12(λ)* has DNA from another phage, called lambda, integrated into its chromosome, whereas *E. coli K12S* does not. To see if the use of these strains might improve phage yield, *E. coli B*, *E. coli K12S*, and *E. coli K12(λ)* were infected with the rII and wild-type T4 phage strains. As expected, the wild-type phage could infect all three bacterial strains. However, the rII mutant strains behaved quite differently. In *E. coli B*, the rII strains produced large plaques that had poor yields of bacteriophage. In *E. coli K12S*, the rII mutants produced normal plaques that gave good yields of phage. Surprisingly, in *E. coli K12(λ)*, the rII mutants were unable to produce plaques at all, for reasons that were not understood. Nevertheless, as we will see later, this fortuitous observation was a critical feature that allowed intragenic mapping in this bacteriophage.

Wild-type plaque   Plaque caused by a rapid-lysis mutant

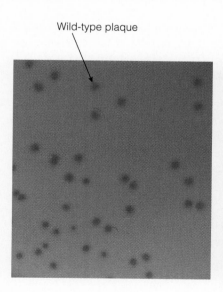 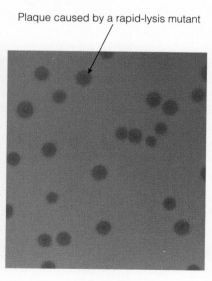

Plaques caused by wild-type bacteriophages

Plaques caused by rapid-lysis bacteriophage strains

FIGURE **6-14**

A comparison of plaques produced by the wild-type T4 bacteriophage and a rapid-lysis mutant.

GENES ⟶ TRAITS: The plaques on the left side are caused by the infection and lysis of *E. coli* cells by the wild-type T4 phage. On the right, a mutation in a phage gene, called a rapid-lysis mutation, causes the phage to lyse *E. coli* cells more quickly. A phage carrying a rapid-lysis mutant allele yields much larger plaques.

A complementation test can reveal if mutations are in the same gene or in different genes

When conducting a genetic analysis for any trait, researchers may identify many different mutations that affect the outcome of the trait. There are two possible ways that different mutations can affect a single trait. One possibility is that the mutations are in the same gene. In other words, two or more mutations can be alleles of each other. Alternatively, two mutations can be in different genes yet affect the same trait. Epistasis is an example of this phenomenon.

In his experiments Benzer was interested in a single trait, the ability to form plaques. He had isolated many rII mutant strains that could form large plaques in *E. coli B* but could not form plaques in *E. coli K12(λ)*. To determine the numbers of genes affecting plaque formation, he needed to know if all the rII mutations were in the same gene or if they involved mutations in different genes. To accomplish this, he conducted **complementation** experiments. In this type of approach, the goal is to determine if two different mutations (which affect the same trait) are in the same gene or in two different genes. The possible outcomes of complementation experiments involving mutations that affect plaque formation are shown in Figure 6-15.

As shown here, distinct rII mutations in two different strains of T4 bacteriophage, designated strain 1 and 2, prevented plaque formation in *E. coli K12(λ)*. To conduct this complementation experiment, bacterial cells were coinfected with the two different strains of phage. Two distinct outcomes are possible. If each rII mutation was in a different phage gene (e.g., gene *A* and gene *B*), a bacterial cell that is coinfected by both types of phages will have two mutant genes, but also two wild-type genes. If the mutant phage genes behave in a recessive fashion, the doubly infected cell will have a wild-type phenotype. In other words, coinfected cells will be lysed in the same manner as the wild-type strain. Therefore, this coinfection should be able to produce plaques in *E. coli K12(λ)*. This is called complementation, because the defective genes in each rII strain are complemented by the corresponding wild-type genes.

If, however, the two rII phage strains possess mutations in the same gene, they will not complement each other. For example, if both rII phages contain deleterious mutations in gene *A*, an *E. coli K12(λ)* cell that is coinfected by both phages will not be able to make plaques, because it cannot make a wild-type gene *A* product. This is called **noncomplementation**.

By carefully considering the pattern of complementation and noncomplementation, Benzer found that the rII mutations occurred in two different genes, which were termed *rIIA* and *rIIB*. The identification of two distinct genes affecting plaque formation was a necessary step that preceded his intragenic mapping analysis, which will be described next. Benzer coined the term **cistron** to refer to the smallest genetic unit that gives a negative complementation test. In other words, if two mutations occur within the same cistron, they cannot complement each other. Since these studies of the 1950s, it has become clear that a cistron is equivalent to a gene. In recent decades, the term gene has gained wide popularity while the term cistron is not commonly used. Nevertheless, it would be correct to refer to gene *A* as cistron *A* or gene *B* as cistron *B*.

Intragenic maps were constructed using data from a recombinational analysis of mutants within the rII region

As we have seen in Figure 6-15, the ability of a coinfection to produce many viral plaques is due to complementation. Noncomplementation occurs when two different strains have mutations in the same gene. However, at an extremely low rate, two noncomplementing strains of viruses can produce an occasional viral plaque if intragenic recombination has taken place. For example, Figure 6-16 shows a

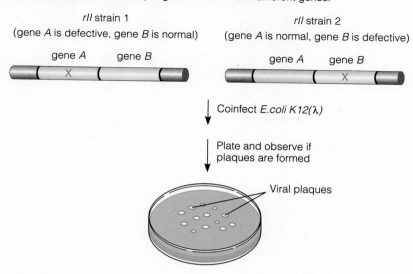

(a) Complementation: The phage mutations are in different genes.

rII strain 1
(gene *A* is defective, gene *B* is normal)

rII strain 2
(gene *A* is normal, gene *B* is defective)

gene *A*   gene *B*

gene *A*   gene *B*

Coinfect *E.coli K12(λ)*

Plate and observe if plaques are formed

Viral plaques

Complementation occurs since the coinfected cell is able to make normal products of gene *A* and gene *B*. The coinfected bacterial cell will produce viral particles that lyse the cell, resulting in the appearance of clear plaques.

(b) Noncomplementation: The phage mutations are in the same gene.

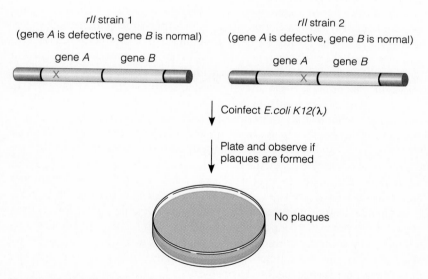

rII strain 1
(gene *A* is defective, gene *B* is normal)

rII strain 2
(gene *A* is defective, gene *B* is normal)

gene *A*   gene *B*

gene *A*   gene *B*

Coinfect *E.coli K12(λ)*

Plate and observe if plaques are formed

No plaques

No complementation occurs since the coinfected cell is unable to make the normal product of gene *A*. The coinfected cell will not produce viral particles, thus no bacterial cell lysis and plaque formation occurs.

FIGURE **6-15**

A comparison of complementation and noncomplementation. Two different T4 phage strains (designated #1 and #2) that carry rII mutations were coinfected into *E. coli K12(λ)*. If the rII mutations are in different genes (say, gene *A* and gene *B*), a coinfected cell will have two mutant genes, but also two wild-type genes. Doubly infected cells with a wild-type copy of each gene can produce new phages and form plaques. This result is called complementation, because the defective genes in each rII strain are complemented by the corresponding wild-type genes. If the two rII phage strains possess mutations in the same gene, noncomplementation will occur.

coinfection experiment between two phage strains that both contain rII mutations in gene *A*. These mutations are located at different places within the same gene. Rarely, a crossover occurs in the very short region between each mutation. This produces a gene *A* with two mutations and also a wild-type gene *A*. Since this event has produced a wild-type gene *A*, the normal product of gene *A* will be produced and new phages can be made in *E. coli K12(λ)*, resulting in the formation of viral plaques.

Even though intragenic recombination can take place, an experimental obstacle makes it difficult to measure. Since the distance between two mutations within the same gene is very short, the likelihood is extremely small that a crossover will occur

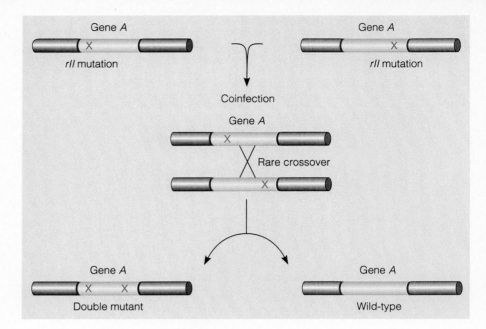

FIGURE **6-16**

Intragenic recombination. Following coinfection, a rare crossover has occurred between the sites of the two mutations. This produces a wild-type phage with no mutations and a doubly mutant phage with both mutations.

in this region to produce recombinant chromosomes—perhaps less than one chance in 100,000. In conventional crosses involving eukaryotic organisms such as *Drosophila* or the pea plant, such a low rate of recombination is very difficult to measure experimentally, since it would involve the analysis of hundreds of thousands or even millions of offspring. The accidental discovery that rII mutations behave differently in *E. coli B*, *E. coli K12S*, and *E. coli K12(λ)* provided an experimental system to detect recombinants that occur at a very low rate, even less than one in 1,000,000 times! In Benzer's own words, "I dropped everything else and embarked on this project."

Figure 6-17 describes the general strategy for intragenic mapping of rII phage mutations. In steps 1–3, bacteriophages from two different noncomplementing rII phage mutants (here, r103 and r104) were mixed together in equal amounts, and then infected into *E. coli B*. In this strain, the rII mutants could grow and propagate. When these two different mutants coinfected the same cell, intragenic recombination may occur, producing wild-type phages and double-mutant phages. However, these intragenic recombinants were produced at a very low rate. Following coinfection and lysis of *E. coli B*, a new population of phages was isolated in step 4. This population was expected to mostly contain nonrecombinant phages. However, due to intragenic recombination, it would also contain a very low percentage of wild-type phages and doubly mutant phages (refer back to Figure 6-16).

To determine the relative amounts of parental and recombinant phages, Benzer took advantage of the observation that rII phages cannot grow in *E. coli K12(λ)*. In step 5, he took his reisolated population of phages and used some of them to infect *E. coli B* and some to infect *E. coli K12(λ)*. After plating in step 6, the *E. coli B* infection was used to determine the total number of phages, since rII mutants as well as wild-type phages can produce plaques in this strain. The overwhelming majority of these phages were expected to be nonrecombinant phages. The *E. coli K12(λ)* infection was used to determine the number of rare intragenic recombinants that produce wild-type phages.

Steps 4–6 of Figure 6-17 illustrate the great advantage of this experimental system in detecting a low percentage of recombinants. In the laboratory, phage preparations containing several billion phages per milliliter are readily made. Among billions

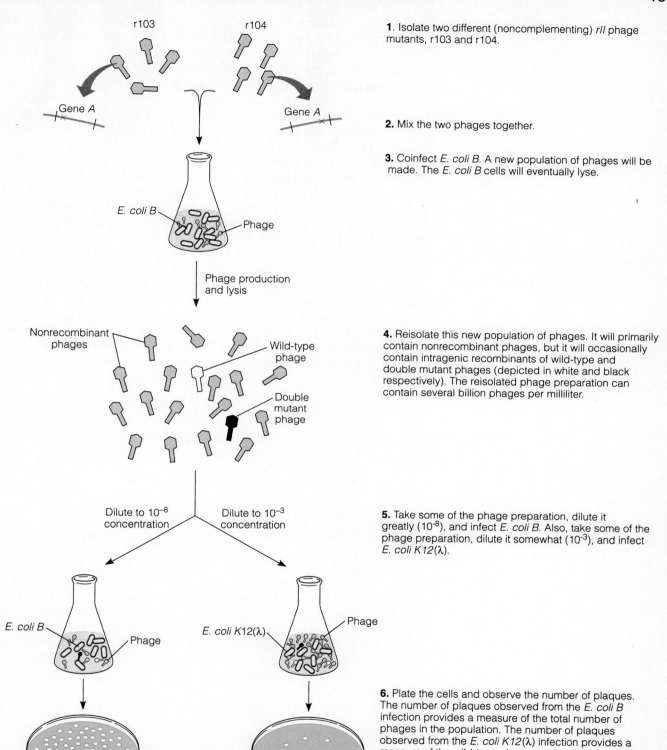

1. Isolate two different (noncomplementing) *rII* phage mutants, r103 and r104.

2. Mix the two phages together.

3. Coinfect *E. coli B*. A new population of phages will be made. The *E. coli B* cells will eventually lyse.

4. Reisolate this new population of phages. It will primarily contain nonrecombinant phages, but it will occasionally contain intragenic recombinants of wild-type and double mutant phages (depicted in white and black respectively). The reisolated phage preparation can contain several billion phages per milliliter.

5. Take some of the phage preparation, dilute it greatly (10^{-8}), and infect *E. coli B*. Also, take some of the phage preparation, dilute it somewhat (10^{-3}), and infect *E. coli K12(λ)*.

6. Plate the cells and observe the number of plaques. The number of plaques observed from the *E. coli B* infection provides a measure of the total number of phages in the population. The number of plaques observed from the *E. coli K12(λ)* infection provides a measure of the wild-type phage produced by intragenic recombination.

FIGURE **6-17**

Benzer's method of intragenic mapping in the rII region.

of phages, a low percentage (e.g., one in every 100,000) may be wild-type phages arising from intragenic recombination. The wild-type recombinants can produce plaques in *E. coli K12(λ)*, whereas the rII mutant strains cannot. In other words, only the tiny fraction of wild-type recombinants would produce plaques in *E. coli K12(λ)*.

We can determine the frequency of recombinant phages by comparing the number of recombinant (wild-type) phages and the total number of phages. As shown in Figure 6-17, the total number of phages can be deduced from the number of plaques obtained from the infection of *E. coli B*. In this experiment, the phage preparation was diluted by 10^{-8} (1 : 100,000,000), and 1 ml was used to infect *E. coli B*. Since this plate produced 66 plaques, the total number of phages in the original preparation was $66 \times 10^8 = 6.6 \times 10^9$, or 6.6 billion phages per milliliter. By comparison, the phage preparation used to infect *E. coli K12(λ)* was only diluted by 10^{-3} (1 in 1000). This plate produced 11 plaques. Therefore, the number of wild-type phages was 11×10^3, which equals 11,000 wild-type phages per milliliter.

As we have already seen in this and previous chapters, genetic mapping distance is computed by dividing the number of recombinants by the total population (non-recombinants and recombinants). In this experiment, intragenic recombination produces an equal amount of two types of recombinants: wild-type phages and doubly mutant phages. Only the wild-type phages are detected in the infection of *E. coli K12(λ)*. Therefore, to obtain the total number of recombinants, the number of wild-type phages must be multiplied by two. With all this information, we can compute the frequency of recombinants using the experimental approach described in Figure 6-17:

$$\text{Frequency of recombinants} = \frac{2\,(\text{Wild-type plaques obtained in } E.\ coli\ K12(\lambda))}{\text{Total number of plaques obtained in } E.\ coli\ B}$$

We can use this equation to calculate the frequency of recombinants obtained in the experiment described in Figure 6-17:

$$\text{Frequency of recombinants} = \frac{2\,(11 \times 10^3)}{6.6 \times 10^9}$$

$$= 3.3 \times 10^{-6} = 0.0000033$$

Thus, there were approximately 3.3 recombinants per 1,000,000 phages.

The frequency of recombinants provides a measure of map distance. In eukaryotic mapping studies, we compute the map distance by multiplying the frequency of recombinants by 100 to give a value in map units (also known as centimorgans). Similarly, in these experiments, the frequency of recombinants can provide a measure of map distance along the bacteriophage chromosome. In this case, the map distance is between two mutations within the same gene. Like intergenic mapping, the frequency of intragenic recombinants is correlated with the distance between the two mutations; the farther apart they are, the higher the frequency of recombinants. If two mutations happen to be located at exactly the same site within a gene, they would not be able to produce any wild-type recombinants, and so the map distance would be zero. These are known as **homoallelic** mutations.

Deletion mapping can be used to localize many rII mutations to specific regions in the *rIIA* or *rIIB* genes

Now that we have seen the general approach to intragenic mapping, let's consider a method to efficiently map hundreds of rII mutations within the two genes designated *rIIA* and *rIIB*. As you may have realized, the coinfection experiments described in Figure 6-17 are quite similar to Sturtevant's strategy of making dihybrid crosses to map genes along the *Drosophila* chromosome (refer back to Experiment 5B). Similarly, Benzer wanted to "cross" or coinfect different rII mutants in order to map the sites of the mutations within the *rIIA* and *rIIB* genes. During the course of his work, he obtained hundreds of different rII mutant strains that he wanted to map. However, making all of the pairwise combinations would have been an overwhelming task. Instead, Benzer used an approach known as **deletion mapping** as a first step in localizing his rII mutations to a fairly short region within gene *A* or gene *B*.

Figure 6-18 describes the general strategy used in deletion mapping. This approach is easier to understand if we use an example. Let's suppose that the goal is to know the approximate location of an rII mutation, such as r103. To do so, *E. coli K12(λ)* would be coinfected with r103 and a deletion strain. Each deletion strain is a T4 bacteriophage that is missing a known segment of the *rIIA* and/or *rIIB* gene. If the deleted region includes the same region that contains the r103 mutation, it will be impossible for a coinfection to produce intragenic wild-type recombinants. However, if a deletion strain recombines with r103 to produce a wild-type phage, the deleted region did not contain the r103 mutation. In the example shown in Figure 6-18, the r103 strain produced wild-type recombinants when coinfected with deletion strains PB242, A105, and 638. However, coinfection of r103 with PT1, J3, 1241, and 1272 did not produce intragenic wild-type recombinants. Since coinfection with PB242 produced recombinants and PT1 did not, the r103 mutation must be located in the region that is missing in PT1 but not missing in PB242. This region is called A4 (A refers to the *rIIA* gene). In other words, the r103 mutation is located somewhere within the A4 region, but not in the other six regions (A1, A2, A3, A5, A6, and B).

As described in Figure 6-18, this first step in the deletion mapping strategy localized an rII mutation to one of seven regions; six of these were in *rIIA* and one was in *rIIB*. Other deletion strains were used to eventually localize each rII mutation to one of 47 short regions; 36 were in *rIIA*, 11 in *rIIB*. At this point, pairwise coinfections were made between mutant strains that had been localized to the same region by deletion mapping. For example, 24 mutations were deletion mapped to a region called A5d. Pairwise coinfection experiments were conducted among this group of 24 mutants to precisely map their locations relative to each other in the A5d region. Similarly, all of the mutants in each of the 46 other groups were mapped by pairwise coinfections. In this way, a fine structure map was constructed depicting the locations of hundreds of different rII mutations (Figure 6-19). As seen here, certain locations contained an abnormally high percentage of mutations compared with other sites. These were termed **hot spots** for mutation.

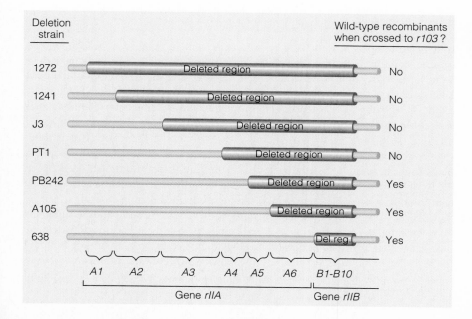

FIGURE **6-18**

The use of deletion strains to localize rII mutants to short regions within the *rIIA* or *rIIB* gene. The deleted regions are shown in gray.

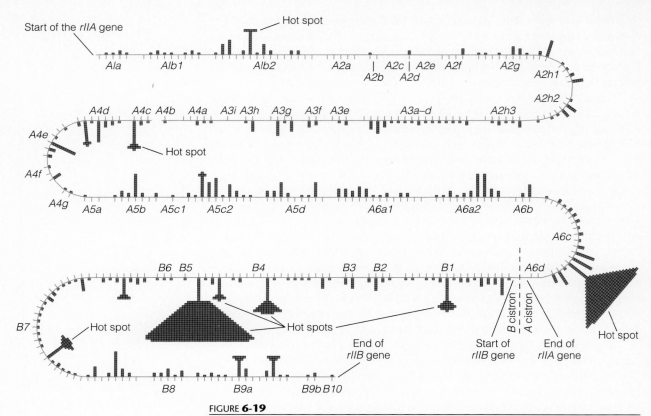

FIGURE 6-19

The outcome of intragenic mapping of many rII mutations. The purple line represents the linear sequence of the *rIIA* and *rIIB* genes, which are found within the T4 phage's genetic material. Each small red box attached to the purple line symbolizes a mutation that was mapped by intragenic mapping. In many cases, several mutations mapped to the same site. In this figure, mutations at the same site form columns of boxes. Hot spots contain a large number of mutations and are represented as a triangular group of boxes attached to a column of boxes. A hot spot contains many mutations at exactly the same site within the *rIIA* or *rIIB* gene.

Intragenic mapping experiments provided insight into the relationship between traits and molecular genetics

Intragenic mapping studies were a pivotal achievement in our early understanding of gene structure. Since the time of Mendel, geneticists had considered a gene to be a unit of heredity that provided an organism with its inherited traits. In the early 1960s, however, the molecular nature of the gene was not understood. Since it was a unit of heredity, some scientists envisioned a gene as being a particle-like entity that could not be further subdivided into additional parts. However, intragenic mapping studies revealed, convincingly, that this is not the case. These studies showed that mutations can occur at many different sites within a single gene. Furthermore, intragenic crossing over can recombine these mutations, resulting in wild-type genes. Therefore, rather than being an indivisible particle, a gene must be composed of a large structure, which can be subdivided during crossing over.

Benzer's results were published in the late 1950s and early 1960s, around the same time that the structure of DNA was being elucidated by Watson and Crick. We now know that a gene is a segment of DNA that is composed of smaller building blocks called *nucleotides*. We will describe the structures of nucleotides and DNA in Chapter 9. A typical gene is a linear sequence of several hundred to several thousand nucleotides. As the genetic map of Figure 6-19 indicates, mutations can occur at many sites along the linear structure of a gene; intragenic crossing over can recombine mutations that are located at different sites within the same gene. We will discuss the molecular structure and function of genes throughout Part IV of this text.

This chapter has been concerned with *genetic transfer* and mapping in prokaryotes. There are three mechanisms for the transfer of genetic material from one bacterial cell to another. During *conjugation*, bacteria make direct physical contact with each other. Donor cells can be either F^+ cells or Hfr cells. An F^+ cell contains a circular piece of genetic material, a *plasmid* called an F (fertility) *factor* that is transferred to the recipient (F^-) cell during mating. By comparison, an *Hfr* (for High frequency of recombination) bacterium transfers chromosomal DNA to the recipient cell. A second route of genetic transfer is *transduction*. In this case, a bacteriophage accidentally packages a portion of the bacterial chromosome, which is then transferred to another bacterium upon infection. Finally, a third mechanism of genetic transfer is *transformation*. For this to occur, a dead bacterial cell must release its genetic material into the environment. A living bacterial cell in a *competent* state subsequently imports the DNA, and then genetic recombination causes the imported DNA to replace segments of genetic material along the bacterial chromosome.

This chapter concluded with a description of *intragenic mapping* studies in T4 bacteriophages. Benzer constructed a *fine structure* genetic map of two bacteriophage genes, *rIIA* and *rIIB*. In the early 1960s, his fine structure map provided important insights into the molecular nature of the gene. It revealed that a gene is not an indivisible particle. Rather, his results indicated that mutations can occur at many sites along the linear structure of a gene and that intragenic crossing over can recombine mutations that are located at different sites within the same gene. As we will learn in Chapters 9 and 12, these results are consistent with the molecular structure of genes.

Research studies of genetic transfer in bacteria have provided a unique experimental strategy for mapping the linear order and relative locations of bacterial genes. Conjugation has been used to map the locations of many genes along the bacterial chromosome. In this approach, map distances are determined by the number of minutes it takes for a gene to enter a recipient cell during conjugation. In addition, cotransduction and cotransformation experiments have been commonly used to accurately map bacterial genes that are relatively close to each other on the chromosome.

As mentioned in the conceptual summary, intragenic mapping studies of T4 bacteriophages have provided great insight into the molecular nature of genes. A fine structure genetic map was constructed of two bacteriophage genes, *rIIA* and *rIIB*. This was accomplished using a coinfection strategy, in which the very low percentage of intragenic recombinants yielding wild-type phages could be identified by their unique ability to infect a strain of bacteria known as *E. coli K12(λ)*. This approach illustrates the power of phage genetics compared with the analysis of offspring in eukaryotic species. It is technically easy to isolate huge numbers of phages. The selective inability of rII mutants to lyse *E. coli K12(λ)* allowed detection of the very few intragenic recombinants, which produced wild-type phages that could lyse *E. coli K12(λ)*. In this way, the very short distance that occurs between mutations within a single gene could be determined.

PROBLEM SETS

SOLVED PROBLEMS

1. In *E. coli*, the gene *bioD* encodes an enzyme involved in biotin synthesis, and *galK* encodes an enzyme involved in galactose utilization. An *E. coli* strain that contained wild-type versions of both genes was infected with P1, and then a P1 lysate was obtained. This lysate was used to transduce a strain that was *bioD⁻* and *galK⁻*. The cells were plated on minimal plates containing galactose as the sole carbon source for growth to select for transduction of the *galK* gene. These plates also were supplemented with biotin. The colonies were then restreaked on plates that lacked biotin to see if the *bioD* gene had been cotransduced. The following results were obtained:

Selected Gene	Non-selected Gene	Number of Colonies that Grew on		Cotrans-duction Frequency
		Galactose + Biotin	Galactose − Biotin	
galK	bioD	80	10	0.125

How far apart are these two genes?

Answer: We can use the cotransduction frequency to calculate the distance between the two genes (in minutes) using Wu's equation:

$$\text{Contransduction frequency} = (1 - d/2)^3$$

$$0.125 = (1 - d/2)^3$$

$$\sqrt[3]{0.125} = 1 - d/2$$

$$1 - d/2 = 0.5$$

$$d/2 = 1 - 0.5$$

$$d = 1.0 \text{ minute}$$

The two genes are approximately 1 minute apart on the *E. coli* chromosome.

2. By conducting mating experiments between a single Hfr strain and a recipient strain, Wollman and Jacob mapped the order of many bacterial genes. Throughout the course of their studies, they identified several different Hfr strains in which the *F⁺* factor DNA had been integrated at different places along the bacterial chromosome. A sample of their experimental results is shown in the following:

Hfr Strain	Origin	Order of transfer of several different bacterial genes
		First ↓ Last ↓
H	O	T L Az T1 pro lac gal Str M
1	O	L T M Str gal lac pro T1 Az
2	O	pro T1 Az L T M Str gal lac
3	O	lac pro T1 Az L T M Str gal
4	O	M Str gal lac pro T1 Az L T
5	O	M T L Az T1 pro lac gal Str
6	O	M T L Az T1 pro lac gal Str
7	O	T1 Az L T M Str gal lac pro

Data from: Jacob, F., and Wollman, E. L. (1961) *Sexuality and the Genetics of Bacteria*. Academic Press, New York.

A. Explain how these results are consistent with the idea that the bacterial chromosome is circular.

B. Draw a map that shows the order of genes and the locations of the origins of transfer among these different Hfr strains.

Answer:

A. In comparing the data among different Hfr strains, the order of the nine genes was always the same or the reverse of the same order. For example, HfrH and Hfr1 have the same order of genes but are reversed relative to each other. In addition, the Hfr strains showed an overlapping pattern of transfer with regard to the origin. For example, Hfr1 and Hfr2 had the same order of genes, but Hfr1 began with *L* and ended with *Az* while Hfr2 began with *pro* and ended with *lac*. From these findings, Wollman and Jacob concluded that the segment of DNA that was the origin of transfer had been inserted at different points within a circular *E. coli* chromosome in different Hfr strains. They also concluded that the origin can be inserted in either orientation, so that the direction of gene transfer can be clockwise or counterclockwise around the circular bacterial chromosome.

B. A genetic map that is consistent with these results is shown here:

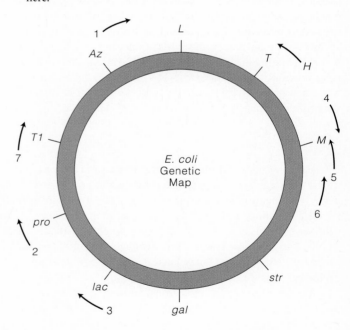

This map describes the order and locations of the nine genes and the predicted site where the origin had been inserted in the eight different Hfr strains. The direction of chromosome transfer is also noted.

CONCEPTUAL QUESTIONS

1. The terms conjugation, transduction, and transformation are used to describe three different natural forms of genetic transfer between bacterial cells. Briefly discuss the similarities and differences between these processes.

2. Conjugation is sometimes called "bacterial mating." Is it a form of sexual reproduction? Explain.

3. If you mix together an equal amount of F^+ and F^- cells, how do you expect the proportions to change over time? In other words, do you expect an increase in the relative proportions of F^+ or of F^- cells? Explain your answer.

4. What is the difference between an F^+ and an Hfr strain? Which type of strain do you expect to be able to transfer many bacterial genes to recipient cells?

5. What is the role of the origin of transfer during F^+- and Hfr-mediated conjugation? What is the significance of the direction of transfer in Hfr-mediated conjugation?

6. An Hfr strain that is $hisE^+$ and $pheA^+$ was mated to a strain that is $hisE^-$ and $pheA^-$. The mating was interrupted and the percentage of recombinants for each gene was determined by streaking on minimal plates that lacked either histidine or phenylalanine. The following results were obtained:

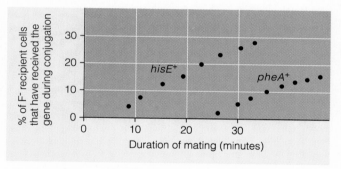

A. Determine the map distance (in minutes) between these two genes.

B. In a previous experiment, it was found that $hisE$ is 4 minutes away from the gene $pabB$. $PheA$ was shown to be 17 minutes from this gene. Draw a genetic map describing the locations of all three genes.

7. Refer to solved problem 2 and in your own words, explain how the analysis of Hfr matings involving several different Hfr strains made it possible to deduce that the *E. coli* chromosome is circular. Could the circularity of the *E. coli* chromosome have been deduced if Wollman and Jacob only had two Hfr strains with origins in opposite directions?

8. Briefly describe the lytic and lysogenic cycles of bacteriophages. In your answer, explain what a prophage is.

9. What is cotransduction? What determines the likelihood that two genes will be cotransduced?

10. If two bacterial genes are 0.6 minutes apart on the bacterial chromosome, what frequency of cotransductants would you expect to observe in a P1 transduction experiment?

11. In an experiment involving P1 transduction, the cotransduction frequency was 0.53. How far apart are the two genes?

12. In a cotransduction experiment, the transfer of one gene is selected for and the presence of the second gene is then determined. If zero out of 1000 P1 transductants that carry the first gene also carry the second gene, what would you conclude about the minimum distance between the two genes?

13. Describe the steps that occur during bacterial transformation. What is a competent cell? What factors may determine whether a cell will be competent?

14. Which bacterial genetic transfer process does not require recombination with the bacterial chromosome?

15. What is fine structure mapping? How does it differ from the conventional gene mapping techniques discussed in Chapter 5?

16. What does the term complementation mean? If two different mutations that produce the same phenotype can complement each other in a genetic cross, what can you conclude about the locations of each mutation?

17. Why do rII plaque morphology mutants lend themselves to an easy genetic analysis?

18. Explain how Benzer's results indicated that a gene is not an indivisible unit.

EXPERIMENTAL QUESTIONS

1. In the experiment of Figure 6-1, a $B^-M^-P^+T^+$ cell could become $B^+M^+P^+T^+$ by a (rare) double mutation that converts the B^-M^- genetic material into B^+M^+. Likewise, a $B^+M^+P^-T^-$ cell could become $B^+M^+P^+T^+$ by double mutations that convert the P^-T^- genetic material into P^+T^+. From the results of Figure 6-1, how do you know that the occurrence of 100 $B^+M^+P^+T^+$ colonies is not due to these types of rare double mutations?

2. In the experiment of Figure 6-1, Lederberg and Tatum could not discern whether B^+M^+ genetic material was transferred to the

$B^-M^-P^+T^+$ strain or if P^+T^+ genetic material was transferred to the $B^+M^+P^-T^-$ strain. Let's suppose that one strain is streptomycin resistant (say, $B^+M^+P^-T^-$) while the other strain is sensitive to streptomycin. Describe an experiment that could determine whether the B^+M^+ genetic material was transferred to the $B^-M^-P^+T^+$ strain or if P^+T^+ genetic material was transferred to the $B^+M^+P^-T^-$ strain.

3. Explain how a U-tube apparatus can distinguish between genetic transfer involving conjugation and that from transduction. Do you think a U-tube could be used to distinguish between transduction and transformation?

4. The following table presents data from Benzer's original complementation experiments.

Phage Mixture	Complementation
r47 and r51	Yes
r47 and r102	Yes
r103 and r104	No
r101 and r102	Yes
r101 and r103	No
r101 and r105	No
r105 and r106	No
r103 and r51	Yes
r104 and r47	No
r51 and r102	No

Data adapted from: Benzer, S. (1955) *Proc. Natl. Acad. Sci. USA* 41, 344–354.

Explain which mutations are in one phage gene (*rIIA*) and which are in the other gene (*rIIB*). The strain designated r47 is an rII mutation in gene *rIIA*.

5. What is an interrupted mating experiment? What type of experimental information can be obtained from this type of study? Why is it necessary to interrupt mating?

6. In a conjugation experiment, what is meant by the term *time of entry*? How is it determined experimentally?

7. If you took a pipette tip and removed a phage plaque from a petri plate, what would it contain?

8. Phages with rII mutations cannot produce plaques in *E. coli* K12(λ), but wild-type phages can. From an experimental point of view, explain why this observation is so significant.

9. In the experimental strategy described in Figure 6-17, explain why it was necessary to dilute the *E. coli* B infection much more that the *E. coli* K12(λ) infection.

QUESTIONS FOR STUDENT DISCUSSION/COLLABORATION

1. Discuss the advantages of the genetic analysis of bacteria and bacteriophages. Make a list of the types of allelic differences among bacteria and phages that are suitable for genetic analyses.

2. Complementation occurs when alleles in two different genes produce the same phenotype. What other examples of complementation have we encountered in previous chapters of this text?

CHAPTER 7

NON-MENDELIAN
INHERITANCE

In Chapters 2 through 6, we have focused on the inheritance patterns of many different genes in a variety of species. In the case of eukaryotes, a Mendelian pattern of inheritance refers to the inheritance of genes located on chromosomes in the cell nucleus that directly influence the outcome of an offspring's traits. To predict phenotype, we must consider several factors. These include the dominant/recessive relationships of alleles, gene interactions that may affect the expression of a single trait, and the roles that gender and the environment play in influencing the individual's phenotype. Once these factors are understood, the phenotypes of offspring can be predicted from their genotypes.

Most genes in eukaryotic species follow a Mendelian pattern of inheritance. However, there are many that do not. In this chapter, we will examine several interesting and even unusual types of inheritance patterns that deviate from a Mendelian pattern. In the first two sections of this chapter, we will consider two important examples of non-Mendelian inheritance called the maternal effect and epigenetic inheritance. Even though these inheritance patterns involve genes on chromosomes within the cell nucleus, the genotype of the offspring does not directly govern their phenotype in the ways predicted by Mendel. We will see how the timing of gene expression and gene inactivation can cause a non-Mendelian pattern of inheritance.

In the third section, we will examine a deviation from Mendelian inheritance that arises because some genetic material is not located in the cell nucleus. There are certain cellular organelles, such as mitochondria and chloroplasts, that contain their own genetic material. We will survey the inheritance of organellar genes and a few other examples in which traits are influenced by genetic material outside the cell nucleus.

MATERNAL EFFECT

EPIGENETIC INHERITANCE

EXTRANUCLEAR INHERITANCE

MATERNAL EFFECT

We will begin by considering genes that have a **maternal effect**. This term refers to an inheritance pattern for certain **nuclear genes** (i.e., genes located on chromosomes that are found in the cell nucleus) in which the genotype of the mother directly determines the phenotypic traits of her offspring. Surprisingly, for maternal effect genes, the genotypes of the father and offspring themselves do not affect the phenotype of the offspring. We will see that this phenomenon is explained by the accumulation of gene products that the mother provides to her developing eggs.

The genotype of the mother determines the phenotype of the offspring for traits that are governed by maternal effect genes

The first example of a maternal effect gene was studied in the 1920s by A. E. Boycott and his colleagues at the University of London. They were interested in the morphological features of the water snail, *Limnea peregra*. In this species, the shell and internal organs can be arranged in either a right-handed (dextral) or left-handed (sinistral) direction. The dextral orientation is more common and is dominant to the sinistral orientation. Whether a snail's body plan curves in a dextral or sinistral direction depends on the cleavage pattern of the egg immediately following fertilization.

Figure 7-1 describes the results of a genetic analysis carried out by Boycott. In this experiment, he began with two different true-breeding strains of snails with either a dextral or sinistral morphology. Many combinations of crosses produced re-

FIGURE **7-1**

Experiment showing the inheritance pattern of snail coiling. In this experiment, *D* (dextral) is dominant to *d* (sinistral). The genotype of the mother determines the phenotype of the offspring. This phenomenon is known as the maternal effect. In this case, a *DD* or *Dd* mother will produce dextral offspring and a *dd* mother will produce sinistral offspring. The genotypes of the father and offspring themselves do not affect the offspring's phenotype.

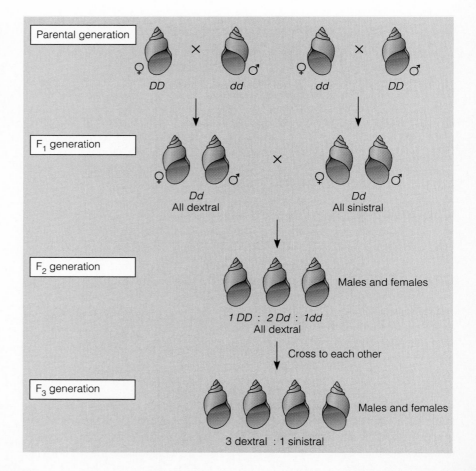

sults that could not be explained by a Mendelian pattern of inheritance. First, Boycott conducted a **reciprocal cross**. In this type of cross, the experimenter examines the phenotypic traits of the offspring relative to the genders and phenotypes of the parents. When a dextral female was crossed to a sinistral male, all the offspring were dextral. However, in the reciprocal cross, where a sinistral female was crossed to a dextral male, the opposite result was obtained. Taken together, these results contradict a Mendelian pattern of inheritance.

The idea that snail coiling is due to a maternal effect gene that exists in dextral (D) and sinistral (d) alleles was first proposed in 1924 by Alfred Sturtevant at Columbia University. His conclusions were drawn from the inheritance patterns of the F_2 and F_3 generations (see Figure 7-1). In the experiment shown in Figure 7-1, the genotype of the F_1 generation is expected to be heterozygous (Dd). When these F_1 individuals are crossed to each other, a genotypic ratio of 1 DD : 2 Dd : 1 dd is predicted for the F_2 generation. Since the D allele is dominant to the d allele, a 3:1 phenotypic ratio of dextral to sinistral snails would be produced according to a Mendelian pattern of inheritance. Instead of this predicted phenotypic ratio, however, the F_2 generation is composed of all dextral snails. This incongruity with Mendelian inheritance is due to the maternal effect. The phenotype of the offspring depends solely on the genotype of the mother. The F_1 mothers are Dd; the D allele in the mothers is dominant to the d allele and causes the offspring to be dextral, even if the offspring's genotype is dd. When the members of the F_2 generation are crossed, the F_3 generation exhibits a 3:1 ratio of dextral to sinistral snails. This is due to the genotypes of the F_2 females that are the mothers of the F_3 generation. The ratio of genotypes for the F_2 females is 1 DD : 2 Dd : 1 dd. The DD and Dd females produce dextral offspring, while the dd females produce sinistral offspring. This explains the 3:1 ratio of dextral and sinistral offspring in the F_3 generation.

Female gametes receive gene products from the mother that affect early developmental stages of the organism

The non-Mendelian inheritance pattern of maternal effect genes can be explained by the process of oogenesis in female animals (Figure 7-2a). As an animal oocyte matures, surrounding maternal cells called *nurse cells* provide it with nutrients. The nurse cells are diploid cells as opposed to the oocyte, which will become haploid (refer back to Figure 3-12). In the example of Figure 7-2a, the nurse cells are heterozygous for the snail coiling maternal effect gene, with the alleles designated D and d. Depending on the outcome of meiosis, the haploid egg may receive the D allele or the d allele, but not both. Within the diploid nurse cells, however, both genes are activated to produce their gene products (mRNA and proteins). These gene products are then transported into the oocyte. As shown here, the egg has received both the D allele gene product and the d allele gene product, though the egg itself only carries the d allele. These gene products persist for a significant time after the egg has been fertilized and begins its embryonic development. In this way, the gene products of the nurse cells, which reflect the genotype of the mother, influence the early developmental stages of the embryo.

Now that we have an understanding of the relationship between oogenesis and maternal effect genes, let's reconsider the experiment of Figure 7-1. As shown in Figure 7-2b, a female snail that is DD will transmit only the D gene products to the eggs. During the early stages of embryonic development, these gene products will cause the egg cleavage to occur in a way that promotes a right-handed body plan. A heterozygous female will transmit both D and d gene products. Since the D allele is dominant, the maternal effect will also be a right-handed body plan. Finally, a dd mother will only contribute d gene products that promote a left-handed body plan, even if the egg is fertilized by a sperm carrying a D allele. The sperm's genotype is irrelevant, because the expression of the sperm's gene occurs too late.

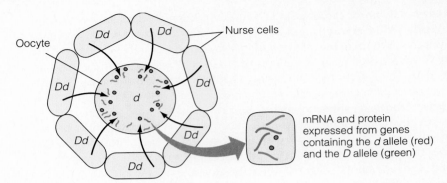

Oocyte

Nurse cells

mRNA and protein
expressed from genes
containing the *d* allele (red)
and the *D* allele (green)

(a) Transfer of gene products from nurse cells to oocyte

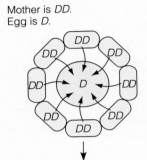

Mother is *DD*.
Egg is *D*.

All offspring are dextral
because the egg received
the gene products of the *D* allele

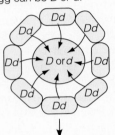

Mother is *Dd*.
Egg can be *D* or *d*.

All offspring are dextral
because the egg received
the gene products of the
dominant *D* allele

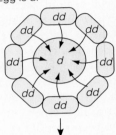

Mother is *dd*.
Egg is *d*.

All offspring are sinistral
because the egg only
received the gene products
of the *d* allele

(b) Maternal effect in snail coiling

FIGURE **7-2**

The mechanism of maternal effect in snail coiling. (a) Transfer of gene products
from nurse cells to an oocyte. The nurse cells are heterozygous (*Dd*). Both the *D* and *d*
alleles are activated in the nurse cells to produce *D* and *d* gene products (mRNA and/or
proteins). These products are transported into the cytoplasm of the oocyte, where they
accumulate to significant amounts. **(b)** Explanation of the maternal effect in snail coiling.

GENES ⟶ TRAITS: If the nurse cells are *DD* or *Dd*, they will transfer the *D* gene
product to the oocyte and thereby cause the resulting offspring to be dextral. If the
nurse cells are *dd*, only the *d* gene product will be transferred to the oocyte, so that
the resulting offspring will be sinistral.

Since these early studies, researchers have found that maternal effect genes
encode RNA and proteins that play important roles in the early steps of embryogenesis. Maternal effect genes often play a role in cell division, cleavage
pattern, and body axis orientation. Therefore, defective alleles in maternal effect
genes tend to have a dramatic effect on the phenotype of the individual. Abnormal maternal effect alleles usually alter major features of morphology, often
with dire consequences.

Our understanding of maternal effect genes has been greatly aided by their
identification in experimental organisms such as *Drosophila melanogaster* and
Caenorhabditis elegans (a nematode that is often studied by developmental geneticists). In experimental organisms with a short generation time, geneticists have successfully searched for mutant alleles that prevent the normal process of embryonic
development. In *Drosophila*, for example, geneticists have identified several dozen
maternal effect genes with dramatic effects on the early stages of development. The
pattern of development of a *Drosophila* embryo occurs along axes, such as the an-

teroposterior axis and the dorso-ventral axis. The proper development of each axis requires a distinct set of maternal gene products. Mutant alleles of maternal effect genes often lead to abnormalities in the anteroposterior or the dorso-ventral pattern of development. Chapter 24 will examine the relationships among the actions of several maternal effect genes during embryonic development.

EPIGENETIC INHERITANCE

As we have just seen, events during oogenesis can cause the inheritance pattern of traits to deviate from a Mendelian pattern. Likewise, **epigenetic inheritance** is a pattern in which a modification occurs to a nuclear gene or chromosome that alters gene expression, but the expression is not permanently changed over the course of many generations. As we will see, epigenetic inheritance patterns are caused by DNA and chromosomal modifications that occur during oogenesis, spermatogenesis, or early stages of embryogenesis. Once they occur, epigenetic changes alter the expression of particular genes in a way that may be fixed during an individual's lifetime. Therefore, epigenetic changes can affect the phenotype of the individual. However, epigenetic modifications are not permanent over the course of many generations. For example, a gene may undergo an epigenetic change that inactivates it for the lifetime of an individual. However, when this individual makes gametes, the gene may become activated and remain operative during the lifetime of an offspring who inherits the active gene.

In this section, we will discuss two examples of epigenetic inheritance called dosage compensation and genomic imprinting. The purpose of **dosage compensation** is to offset differences in the number of active sex chromosomes. One of the sex chromosomes is altered so that males and females will have similar levels of gene expression, even though they do not possess the same complement of sex chromosomes. In mammals, dosage compensation occurs during the early stages of embryonic development. By comparison, **genomic imprinting** occurs prior to fertilization; it involves a change in a single gene or chromosome during gamete formation. Depending on whether the modification occurs during spermatogenesis or oogenesis, imprinting governs whether an offspring will express a gene that has been inherited from its mother or father.

Dosage compensation is necessary to ensure genetic equality between the sexes

Dosage compensation refers to the phenomenon that the level of expression of many genes on the sex chromosomes (particularly the X-chromosome) is similar in both genders even though males and females have a different complement of sex chromosomes. This term was coined in 1932 by Hermann Muller at the University of Texas to explain the effects of eye color mutations in *Drosophila*. Muller observed that female flies homozygous for certain X-linked eye color alleles had a similar phenotype to hemizygous males. For example, an X-linked gene conferring an apricot eye color produces a very similar phenotype in homozygous females and hemizygous males. In contrast, a female that has one copy of the apricot allele and a deletion of the apricot gene on the other X-chromosome has eyes of paler color. Therefore, one copy of the allele in the female is not equivalent to one copy of the allele in the male. Instead, two copies of the allele in the female produce a phenotype that is similar to one copy in the male. In other words, the difference in gene *dosage* (two copies in females versus one copy in males) is being compensated for at the level of gene expression. In *Drosophila*, however, dosage compensation does not occur for all eye color alleles. As described previously in Experiment 4A, the eosin eye color allele exhibits a gene dosage effect. The reasons why some alleles show dosage compensation, and others do not, is not understood.

Since these early studies, dosage compensation has been studied extensively in mammals, *Drosophila*, and *C. elegans*. Depending on the species, dosage compensation occurs via different mechanisms. Table 7-1 describes how X-chromosome compensation is accomplished in different species. Mammals equalize the expression of X-linked genes by turning off one X-chromosome in the somatic cells of females. This process is known as **X-inactivation**. In *C. elegans*, the XX hermaphrodite diminishes the expression of X-linked genes in a different manner. The level of expression of genes on both X-chromosomes is decreased to approximately 50% of that in the male. In *Drosophila*, the male accomplishes dosage compensation by doubling the expression of most X-linked genes.

Even though dosage compensation is widespread among animal species, it is not universal. Certain species, such as birds and butterflies, may not compensate for differences in the number of sex chromosomes. Also, as we will see later in this section, not all genes on the female X-chromosome are inactivated in mammals.

Dosage compensation occurs in female mammals by the random inactivation of one X-chromosome

In 1961, Mary Lyon at the Medical Research Council in England proposed that dosage compensation in mammals occurs by the inactivation of a single X-chromosome in females. This proposal brought together two lines of study. The first type of evidence came from cytological studies. In 1949, Murray Barr and Ewart Bertram at the University of Western Ontario identified a highly condensed structure in the interphase nuclei of somatic cells in female cats that was not found in male cats. This structure became known as the **Barr body** (Figure 7-3a). In 1960, Susumu Ohno at the City of Hope Medical Center in California proposed that the Barr body is actually a highly condensed X-chromosome.

TABLE 7–1　Mechanisms of Dosage Compensation among Different Species

Species	Sex Chromosomes in Females	Sex Chromosomes in Males	Mechanism of Compensation
Mammals	XX	XY	One of the X-chromosomes in the somatic cells of females is inactivated. In certain species, the paternal X-chromosome is inactivated; in other species, such as humans, either the maternal or paternal X-chromosome is randomly inactivated throughout the female's body.
Marsupials	XX	XY	The paternally derived X-chromosome is inactivated in the somatic cells of females.
Drosophila melanogaster	XX	XY	The level of expression of genes on the X-chromosome in males is increased twofold.
Caenorhabditis elegans	XX*	X0	The level of expression of genes on both X-chromosomes in hermaphrodites is decreased to 50% levels.

*In nematodes, an XX individual is a hermaphrodite, not a female.

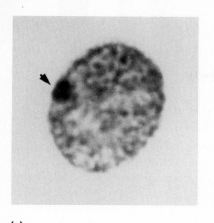

(a)   (b)

FIGURE **7-3**

X-chromosome inactivation in the cells of female mammals. (a) Nucleus from a human female cell. The arrow denotes the Barr body. **(b)** A calico cat.

GENES —→ TRAITS: The pattern of black and orange fur on this cat is due to random X-inactivation during embryonic development. The orange patches of fur are due to the inactivation of the X-chromosome that carries a black allele; the black patches are due to the inactivation of the X-chromosome that carries the orange allele. Like all calicos, this cat is a heterozygous female.

In addition to this cytological evidence, Lyon was also familiar with mammalian mutations in which the coat color had a variegated pattern. An example is shown in Figure 7-3b, which illustrates a calico cat. Calico cats are females that are heterozygous for an X-linked gene that can occur as an orange or a black allele. (The white underside is due to a dominant allele in a different gene.) The orange and black patches are randomly distributed in different female individuals. Similar kinds of mutations have also been identified in the mouse. Based on these genetic observations and the cytological data, Lyon suggested that these results can be explained by X-inactivation in the cells of female mammals.

The mechanism of X-inactivation, also known as the **Lyon hypothesis**, is illustrated in Figure 7-4. This example involves a white and black variegated coat color found in certain strains of mice. As shown here, a female mouse has inherited an X-chromosome from her mother that carries an allele conferring white coat color (X^b); the X-chromosome from her father carries a black coat color allele (X^B). Initially, both X-chromosomes are active. However, at an early stage of embryonic development, one of the two X-chromosomes is randomly inactivated in each somatic cell. For example, one embryonic cell may have the X^B chromosome inactivated. As the embryo continues to grow and mature, this embryonic cell will divide and may eventually give rise to billions of cells in the adult animal. The epithelial (skin) cells that are derived from this embryonic cell will produce a patch of white fur, because the X^B chromosome has been inactivated. Alternatively, another embryonic cell may have the other X-chromosome inactivated (i.e., X^b). The epithelial cells derived from this embryonic cell will produce a patch of black fur. Since the primary event of X-inactivation is a random process that occurs at an early stage of development, the result is an animal with patches of white fur and other patches of black fur. This is the basis for the variegated phenotype.

During inactivation, the chromosomal DNA becomes highly compacted so that most of the genes on the inactivated X-chromosome cannot be expressed.

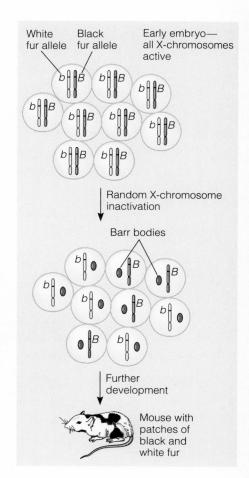

FIGURE **7-4**

The mechanism of X-chromosome inactivation.

GENES —→ TRAITS: The top of this figure represents a mass of several cells that compose the early embryo. Initially, both X-chromosomes are active. At an early stage of embryonic development, random inactivation of one X-chromosome occurs in each cell. This inactivation pattern is maintained as the embryo matures into an adult. As noted in this schematic diagram, the pattern of X-inactivation in the embryo parallels the pattern of white and black fur found in the adult mouse.

When cell division occurs and the inactivated X-chromosome is replicated, both copies remain highly compacted and inactive. Likewise, during subsequent cell divisions, X-inactivation is passed along to all future somatic cells.

Davidson, Nitowsky, and Childs provided evidence that X-inactivation in mammals occurs randomly in embryonic cells

According to the Lyon hypothesis, each somatic cell of female mammals will express the genes on one of the X-chromosomes, but not both. If an adult female is heterozygous for an X-linked gene, only one of two alleles will be expressed in any given cell. In 1963, Ronald Davidson, Harold Nitowsky, and Barton Childs at Johns Hopkins University set out to test the Lyon hypothesis at the cellular level. To do so, they analyzed the expression of a human X-linked gene that encodes an enzyme involved with sugar metabolism known as glucose-6-phosphate dehydrogenase (G-6-PD).

Prior to the experiment described here, biochemists had found that individuals vary with regard to the G-6-PD enzyme. This variation can be detected when the enzyme is subjected to gel electrophoresis (see the appendix for a description of gel electrophoresis). One G-6-PD allele encodes a G-6-PD enzyme that migrates very quickly during gel electrophoresis (the "fast" enzyme), whereas another G-6-PD allele produces an enzyme that migrates more slowly (the "slow" enzyme). As shown here, a sample of red blood cells from heterozygous females produces both types of enzymes, whereas hemizygous males produce either the fast or slow type:

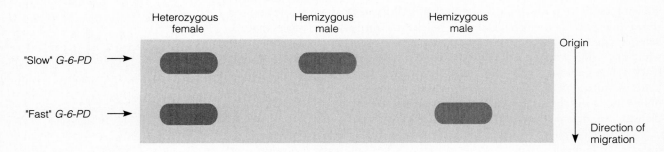

The difference in migration between the fast and slow G-6-PD enzymes is due to minor differences in the structures of these enzymes. These minor differences do not significantly affect G-6-PD function, but they do enable geneticists to distinguish the proteins encoded by the two X-linked alleles.

Davidson, Nitowsky, and Childs tested the Lyon hypothesis using cell culturing techniques. They removed small samples of epithelial cells from a heterozygous female and grew them in the laboratory. As described in the appendix, human epithelial cells can be sparsely plated on growth media. A single cell will grow and divide to produce a colony of cells. This is called a **clone** of cells, because all the cells within the colony have been derived from a single cell. Davidson, Nitowsky, and Childs reasoned that all of the cells within a clone would only express one of the two G-6-PD alleles if the Lyon hypothesis was correct.

THE HYPOTHESIS

According to the Lyon hypothesis, a female who is heterozygous for the fast and slow G-6-PD alleles should express only one of the two alleles in any particular somatic cell and its descendants, but not both.

Starting material: Small skin samples taken from a woman who was heterozygous for the fast and slow alleles of *G-6-PD*.

Experimental Level

Conceptual Level

1. Mince the tissue to separate the individual cells.

2. Grow the cells in a liquid growth medium and then plate (sparsely) onto solid growth medium. The cells then divide to form a clone of many cells.

3. Take nine isolated clones and grow in liquid cultures. (Only three are shown here.)

4. Take cells from the liquid cultures, prepare protein lysates, and subject to gel electrophoresis. (This technique is described in the appendix.)

Note: As a control, prepare protein lysates from cells in step 1, and subject the lysate to gel electrophoresis. This control sample is not from a clone. It is a mixture of cells derived from a woman's skin sample.

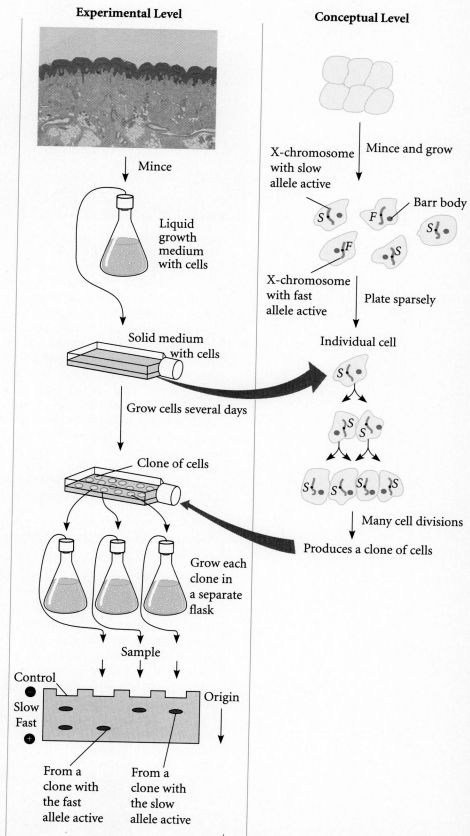

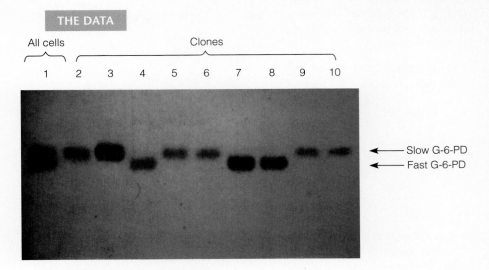

THE DATA

All cells

Clones

1   2   3   4   5   6   7   8   9   10

← Slow G-6-PD
← Fast G-6-PD

INTERPRETING THE DATA

In the data shown here, lane 1 contains a mixture of epithelial cells from a heterozygous woman who produces both types of G-6-PD enzymes. Bands corresponding to the fast and slow enzymes are observed in this lane. As described in steps 2–4, this mixture of epithelial cells was also used to generate nine clones, which are shown in lanes 2–10. Each clone was a population of cells independently derived from a single epithelial cell. Since the epithelial cells were biopsied from an adult female, the Lyon hypothesis predicts that each epithelial cell would already have one of its X-chromosomes permanently inactivated and would pass this trait to its progeny cells. For example, suppose that an epithelial cell had inactivated the X-chromosome that encoded the fast G-6-PD. If this cell was allowed to form a clone of cells on a plate, all of the cells in this clonal population would be expected to have the same X-chromosome inactivated (namely, the X-chromosome encoding the fast G-6-PD). Therefore, this clone of cells should only express the slow G-6-PD. As shown in the data, all nine clones expressed either the fast or slow G-6-PD, but not both. These results are consistent with the hypothesis that X-inactivation has already occurred in any given epithelial cell and that this trait is passed to all of its progeny cells.

X-inactivation in mammals depends on the Xic locus and the *XIST* gene

Since the 1960s, researchers have been interested in the genetic control of X-inactivation. They have found that mammalian cells possess the ability to count their X-chromosomes and allow only one of them to remain active. In normal females, two X-chromosomes are counted and one is inactivated; in males, one X-chromosome is counted and none inactivated. On occasion, an abnormal female mammal is born with three X-chromosomes instead of two. The cells in this individual's body count three X-chromosomes and inactivate two of them.

Although the genetic control of inactivation is not entirely understood, a site on the X-chromosome called the **X-inactivation center (Xic)** appears to play a critical role. Eeva Therman and Klaus Patau at the University of Wisconsin in Madison identified Xic from its key role in X-inactivation. The counting of human X-chromosomes is accomplished by counting the number of Xics. There is an absolute requirement for a Xic on each X-chromosome for inactivation to occur. Therman and Patau found that if one of the X-chromosomes is missing its Xic due to a chromosome mutation, neither X-chromosome will be inactivated. This is a lethal condition for a human female embryo.

The process of X-inactivation can be divided into three phases: *initiation*, *spreading*, and *maintenance* (Figure 7-5). During initiation, one of the X-chromo-

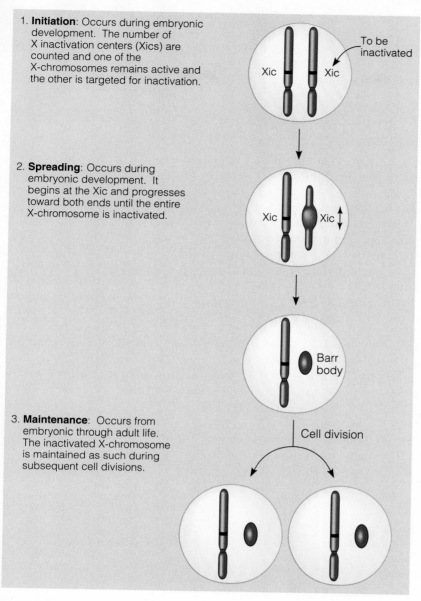

1. **Initiation**: Occurs during embryonic development. The number of X inactivation centers (Xics) are counted and one of the X-chromosomes remains active and the other is targeted for inactivation.

2. **Spreading**: Occurs during embryonic development. It begins at the Xic and progresses toward both ends until the entire X-chromosome is inactivated.

3. **Maintenance**: Occurs from embryonic through adult life. The inactivated X-chromosome is maintained as such during subsequent cell divisions.

To be inactivated

Barr body

Cell division

FIGURE **7-5**

The function of the Xic locus during X-chromosome inactivation.

somes is targeted to remain active and the other is inactivated. This is followed by the spreading phase, in which the chosen X-chromosome is inactivated. It is called the spreading phase because inactivation is thought to begin at the Xic and spread in both directions along the X-chromosome. Both initiation and spreading occur during embryonic development. Subsequently, the inactivated X-chromosome is maintained as such during future cell divisions.

In 1991, a gene that is only expressed on the inactivated X-chromosome was discovered and named **XIST** (for X-inactive specific transcript). Researchers were excited to discover that the *XIST* gene is located within the Xic region of the X-chromosome, suggesting that it is required for X-inactivation (Figure 7-6).

A few genes on the inactivated X-chromosome are expressed in the somatic cells of adult female mammals. These genes are said to escape the effects of X-inactivation. *XIST* is an example of a gene that is expressed from the highly condensed

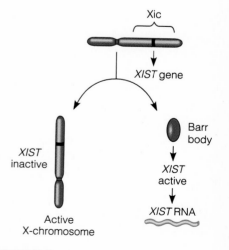

FIGURE **7-6**

Location of the *XIST* gene within the Xic region of the X-chromosome. The *XIST* gene is transcribed into RNA from the inactive X-chromosome but not from the active X-chromosome.

Barr body. In addition, researchers have identified a few other human X-linked genes that escape X-inactivation. Among these are the pseudoautosomal genes that are also found on the Y-chromosome (see Chapter 3). Dosage compensation is not necessary for X-linked pseudoautosomal genes, since they are located on both the X- and Y-chromosomes.

The expression of an imprinted gene depends on the gender of the parent from which the gene was inherited

As we have just seen, dosage compensation changes the level of expression of many genes located on the sex chromosomes. We now turn to another epigenetic phenomenon, known as imprinting. The term imprinting implies a type of marking process that has a memory. For example, newly hatched birds identify marks on their parents, allowing them to distinguish their parents from other individuals. The term genomic (or genetic) imprinting refers to an analogous situation in which a segment of DNA is marked, and that mark is retained and recognized throughout the life of the organism inheriting the marked DNA. Imprinted genes follow a non-Mendelian pattern of inheritance, because the marking process causes the offspring to distinguish between maternally and paternally inherited alleles. Depending on how the genes are marked, the offspring will express one of the two alleles, but not both.

Let's look at an example of imprinting. A mouse gene, *Igf-2*, encodes a growth hormone called insulin-like growth factor 2. A functional *Igf-2* gene is necessary for normal size. A mutant allele of this gene, designated *Igf-2m*, yields defective insulin-like growth factor 2. This may cause a mouse to be a dwarf, but the dwarfism depends on whether the mutant allele is inherited from the male or female parent:

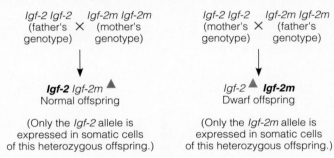

▲ denotes an allele that has been "imprinted" during oogenesis so that it is inactive in the offspring.

In the mouse, imprinting of the *Igf-2* gene occurs so that the maternal allele is not expressed. Only the paternal allele is expressed. On the left side, an offspring has inherited the *Igf-2* allele from its father and the *Igf-2m* allele from its mother. Due to imprinting in the mother, only the *Igf-2* allele is expressed in the offspring, which grow to normal size. Alternatively, in the reciprocal cross on the right side, an individual has inherited the *Igf-2m* allele from its father and the *Igf-2* allele from its mother. In this case, only the *Igf-2m* allele is expressed in the offspring, which has a dwarf phenotype. As shown here, both offspring have the exact same genotype; they are heterozygous for the *Igf-2* alleles (i.e., *Igf-2 Igf-2m*). They are phenotypically different, however, because the allele that is expressed in their somatic cells depends on the parents from which the alleles were inherited.

At the cellular level, imprinting is an epigenetic process that can be divided into three stages: the *establishment* of the imprint during gametogenesis; the *maintenance* of the imprint during embryogenesis and in adult somatic cells; and the *erasure and reestablishment* of the imprint in the germ cells. These stages are described in Figure 7-7. This example also considers the imprinting of the *Igf-2* gene. The two mice shown here have inherited the *Igf-2* allele from their father and the *Igf-2m* allele from their mother. Due to imprinting, both mice express the *Igf-2* allele in their somatic cells. In other words, the imprint that inactivates the *Igf-2m* allele is maintained as each mouse grows from a fertilized egg into an adult. However, the imprint is maintained only in the somatic cells. In the germ cells (i.e., sperm and eggs), the imprint is erased; it may be reestablished, depending on the gender. The female mouse on the left will establish the imprint on both alleles. Therefore, the eggs of this mouse will transmit inactive alleles to her offspring. The imprint will remain erased in the sperm cells of the male mouse. Thus, this mouse will transmit active alleles to his offspring.

As seen in Figure 7-7, genomic imprinting is permanent in the somatic cells of an animal, but the marking of alleles can be altered from generation to generation. For example, the female mouse on the left has an active copy of the *Igf-2* allele but

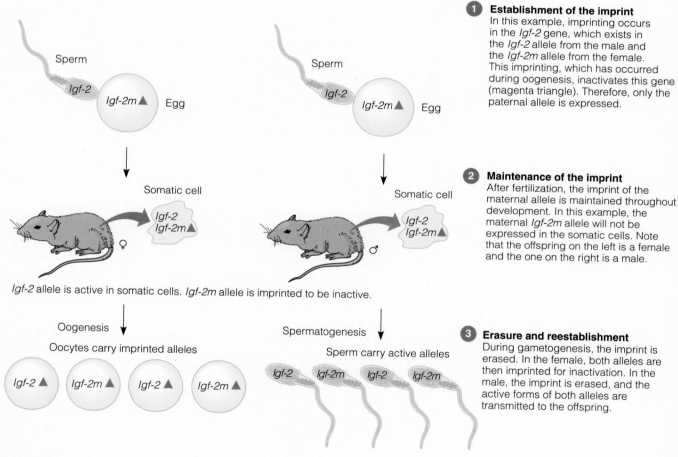

1 **Establishment of the imprint**
In this example, imprinting occurs in the *Igf-2* gene, which exists in the *Igf-2* allele from the male and the *Igf-2m* allele from the female. This imprinting, which has occurred during oogenesis, inactivates this gene (magenta triangle). Therefore, only the paternal allele is expressed.

2 **Maintenance of the imprint**
After fertilization, the imprint of the maternal allele is maintained throughout development. In this example, the maternal *Igf-2m* allele will not be expressed in the somatic cells. Note that the offspring on the left is a female and the one on the right is a male.

3 **Erasure and reestablishment**
During gametogenesis, the imprint is erased. In the female, both alleles are then imprinted for inactivation. In the male, the imprint is erased, and the active forms of both alleles are transmitted to the offspring.

FIGURE **7-7**

Genomic imprinting during gametogenesis. This example involves a mouse gene *Igf-2*, which is found in two alleles designated *Igf-2* and *Igf-2m*. The left side shows a female mouse that was produced from a sperm carrying the *Igf-2* allele and an egg carrying the *Igf-2m* allele. In the somatic cells of this female animal, the *Igf-2* allele is active. However, when this female produces eggs, both alleles will be imprinted. Therefore, the eggs of this female will contain either an inactive *Igf-2* allele or an inactive *Igf-2m* allele. The right side of this figure shows a male mouse that was also produced from a sperm carrying the *Igf-2* allele and an egg carrying the *Igf-2m* allele. In the somatic cells of this male animal, the *Igf-2* allele is active. However, when this male produces sperm, the imprinting is erased. Therefore, the sperm from this male will contain either an active *Igf-2* allele or an active *Igf-2m* allele.

may transmit an inactive copy of this allele to her offspring. The male mouse has an imprinted copy of the *Igf-2m* allele, but can transmit a nonimprinted copy of this allele to his offspring.

Genomic imprinting occurs in several species including numerous insects, plants, and mammals. Imprinting may involve a single gene, a part of a chromosome, an entire chromosome, or even all the chromosomes from one parent. Helen Crouse at Columbia University discovered the first example of imprinting, which involved an entire chromosome in the house fly, *Sciara coprophilia*. In this species, the fly normally inherits three sex chromosomes (rather than two as in most other species); one X-chromosome is inherited from the female, two from the male. In male flies, both paternal X-chromosomes are lost during embryogenesis. In female flies, only one of the paternal X-chromosomes is lost. In both genders, the maternally inherited X-chromosome is never lost. These results indicate that the maternal X-chromosome is marked so as to promote its retention.

Genomic imprinting can also be correlated with the process of X-inactivation described previously. In certain species, imprinting is involved in the choice of the X-chromosome for inactivation. In marsupials, the paternal X-chromosome is always inactivated in the somatic cells of the individual. In placental mammals, X-inactivation of the paternal X-chromosome occurs in the extraembryonic tissue (e.g., the placenta), while X-inactivation occurs randomly between the maternal and paternal X-chromosomes in the embryo itself.

The imprinting of genes and chromosomes is a molecular marking process that may involve DNA methylation

As we have seen, genomic imprinting must involve a marking process. A particular gene or chromosome must be marked dissimilarly during spermatogenesis versus oogenesis. After fertilization takes place, this differential marking affects the expression of particular genes. In mammals, a modification known as DNA methylation is thought to be a common method of marking imprinted genes.

As described in Chapter 11, DNA methylation involves the attachment of a methyl group ($-CH_3$) to the DNA. Imprinted genes are postulated to be differentially methylated during oogenesis versus spermatogenesis. This idea is shown in Figure 7-8. In this example, a male offspring and female offspring have inherited an imprinted allele from their mother and a nonimprinted allele from their father. This pattern of imprinting is maintained in the somatic cells of both offspring. However, when the male offspring makes gametes, the imprinting is erased during early spermatogenesis so that the male can pass either active allele to its offspring. In the female, erasure of the imprint also occurs during early oogenesis, but then *de novo* (new) methylation occurs in both genes. Therefore, the female offspring transmits inactive alleles of this gene to her offspring.

Genomic imprinting is a fairly new area of research. Thus far, imprinting has been identified in several mammalian genes (Table 7-2). In some cases, the female alleles are active in the offspring, whereas in other cases the male alleles are active. The biological significance of imprinting is still a matter of speculation.

Imprinting plays a role in the inheritance of certain human diseases. One example concerns two different inherited diseases known as **Prader–Willi syndrome (PWS)** and **Angelman syndrome (AS)**. PWS is characterized by reduced motor function, obesity, and mental deficiencies. AS patients are hyperactive, have unusual seizures and repetitive symmetrical muscle movements, and show mental deficiencies. Most commonly, PWS and AS involve a small deletion in human chromosome 15. If this deletion is inherited from the maternal parent, it leads to Angelman syndrome; paternal inheritance leads to Prader–Willi syndrome (Figure 7-9). To explain these syndromes, researchers have proposed that this region contains two closely linked but distinct genes that are maternally or

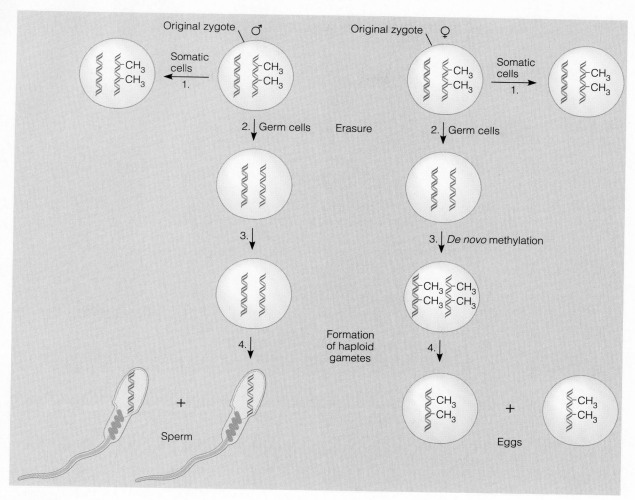

FIGURE **7-8**

DNA methylation in the imprinting process. In this example, a male and a female mouse have inherited an imprinted (methylated) allele and an active allele from their mother and father, respectively. **(1)** Maintenance methylation retains the imprinting in somatic cells during embryogenesis and in adulthood. **(2)** Demethylation occurs in cells that are destined to become gametes. **(3)** *De novo* methylation occurs only in cells that are destined to become eggs. **(4)** Haploid male gametes transmit an active allele, whereas haploid female gametes transmit an inactive allele.

TABLE **7–2 Examples of Mammalian Genes and Inherited Human Diseases that Involve Imprinted Genes**

Gene	Allele Expressed	Function
WT1	Maternal	Wilms tumor suppressor gene—suppresses uncontrollable cell growth
INS	Paternal	Insulin—hormone involved in cell growth and metabolism
Igf-2	Paternal	Insulin-like growth factor II—similar to insulin
Igf-2r	Maternal	Receptor for insulin-like growth factor II
H19	Maternal	Unknown
Snrpn	Paternal	Splicing factor
Gabrb	Maternal	Neurotransmitter receptor

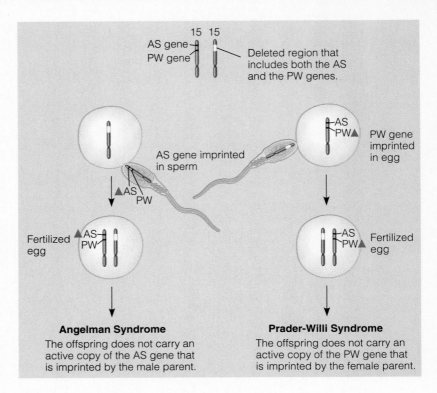

Angelman Syndrome
The offspring does not carry an
active copy of the AS gene that
is imprinted by the male parent.

Prader-Willi Syndrome
The offspring does not carry an
active copy of the PW gene that
is imprinted by the female parent.

FIGURE **7-9**

**The role of imprinting in the development of Angelman and Prader–Willi
syndromes.**

GENES ⟶ TRAITS: A small region on chromosome 15 is thought to contain two different genes called the AS gene and PW gene in this figure. The AS gene is imprinted
during spermatogenesis, and the PW gene is imprinted during oogenesis. The imprinted genes are inactive in the offspring that inherit them. If a chromosome 15
deletion is inherited from the maternal parent, it leads to Angelman syndrome. This
phenotype occurs because the offspring does not inherit an active copy of the AS
gene that is imprinted by the male parent. Alternatively, the chromosome 15 deletion
may be inherited from the male parent, leading to Prader–Willi syndrome. The phenotype of this syndrome occurs because the offspring does not inherit an active copy of
the PW gene, which is imprinted by the female parent.

paternally imprinted. One gene in this region is maternally expressed; if it is
deleted from the maternally inherited chromosome 15, AS will result. A different
gene on chromosome 15 is paternally expressed; if it is deleted from the paternally
inherited chromosome 15, PWS will occur.

EXTRANUCLEAR INHERITANCE

Thus far, we have considered several types of non-Mendelian inheritance patterns. These include maternal effect genes, dosage compensation, and genomic
imprinting. All of these inheritance patterns involve genes found on chromosomes in the cell nucleus. Another cause of non-Mendelian inheritance patterns
involves genes that are not located in the cell nucleus. In eukaryotic species, the
most biologically important example of extranuclear inheritance concerns genetic material in cellular organelles. Since these organelles are found within the
cytoplasm of the cells, the inheritance of organellar genetic material is called
extranuclear inheritance (the prefix *extra-* means outside of) or **cytoplasmic
inheritance.** In addition to the cell nucleus where the linear chromosomes reside,
the mitochondria and plastids (e.g., chloroplasts) contain their own genetic ma-

terial. This genetic material contains genes that encode proteins that function within these organelles.

In this section, we will examine the genetic composition of mitochondria and plastids, and explore the pattern of transmission of these organelles from parent to offspring. We will also examine a few other examples of inheritance patterns that cannot be explained by the transmission of nuclear genes.

Mitochondria and chloroplasts contain multiple copies of circular chromosomes; each chromosome carries several genes

In 1951, the Japanese researcher Y. Chiba was the first to suggest that chloroplasts contain their own DNA. He based his conclusion on the staining properties of a DNA-specific dye known as Feulgen. During the 1960s, researchers developed techniques to separate organellar DNA from nuclear DNA and thereby characterize the chromosomes of mitochondria and chloroplasts. In addition, electron microscopy studies provided interesting insights into the organization and composition of mitochondrial and chloroplast chromosomes. More recently, the advent of molecular genetic techniques in the 1970s and 1980s has allowed researchers to characterize organellar DNA at the molecular level. From these types of studies, the chromosomes of mitochondria and chloroplasts were found to resemble smaller versions of bacterial chromosomes. (The organization and structure of bacterial chromosomes will be described in Chapter 10.)

The genetic material of mitochondria and chloroplasts is located inside the organelle in a region known as the **nucleoid** (Figure 7-10). The genome is composed of a single circular chromosome, although a nucleoid contains several copies of this chromosome. In addition, a mitochondrion or chloroplast often contains more than one nucleoid. In mice, for example, each mitochondrion contains 1–3 nucleoids, with each nucleoid containing 2–6 copies of the circular mitochondrial genome. On average, each mouse mitochondrion contains 5–6 copies of the mitochondrial genome. However, this number is variable and depends on the type of cell and the stage of development. In comparison, the plastids of algae and more complex plants tend to have more nucleoids per organelle. Table 7-3 describes the genetic composition of mitochondria and chloroplasts for a few selected species.

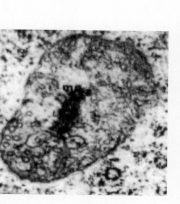

(a) Mitochondrial nucleoids   **(b)** Chloroplast nucleoids

FIGURE **7-10**

Nucleoids within (a) mitochondria and (b) chloroplasts.
The mitochondrial and chloroplast chromosomes are found within the nucleoid region of the organelle.

TABLE **7–3** Genetic Composition of Mitochondria and Chloroplasts from Selected Species			
Species	Organelle	Nucleoids per Organelle	Total Number of Chromosomes per Organelle
Tetrahymena	Mitochondrion	1	6–8
Mouse	Mitochondrion	1–3	5–6
Chlamydomonas	Chloroplast	5–6	~80
Euglena	Chloroplast	20–34	100–300
Higher plants	Chloroplast	12–25	~60

Data from: Gillham, N. W. (1994) *Organelle Genes and Genomes.* Oxford University Press, New York.

Among different species, the size of mitochondrial and plastid genomes varies greatly. For example, there is a 400-fold variation in the size of the mitochondrial chromosomes. In general, the mitochondrial genomes of animal species tend to be fairly small; those of fungi, algae, and protists are intermediate in size; and those of plant cells tend to be fairly large. Among algae and more complex plants, there is substantial variation in the size of plastid chromosomes.

Figure 7-11 illustrates the genome of a human mitochondrion; this genetic material is called **mtDNA**. Each copy of the mitochondrial chromosome contains a circular DNA molecule that is only 17,000 (bp) in length. This size is less than 1% of a typical bacterial chromosome. The mtDNA carries relatively few genes. There are genes that encode ribosomal RNA and transfer RNA, which are necessary for the synthesis of proteins inside the mitochondrion. In addition, there are thirteen genes encoding polypeptides that function within the mitochondrion. The primary functional role of mitochondria is to provide cells with the large bulk of their ATP (adenosine triphosphate), which is used as an energy source to drive cellular reactions. These thirteen polypeptides function in a process known as oxidative phosphorylation, which enables the mitochondrion to synthesize ATP. However, mitochondria require many additional proteins to carry out oxidative phosphorylation and other mitochondrial functions. Most mitochondrial proteins are encoded by genes within the cell nucleus. When these nuclear genes are expressed, the

FIGURE **7-11**

A genetic map of the human mitochondrial genome. This diagram illustrates the locations of many genes along the circular mitochondrial chromosome. The genes shown in dark orange are genes that encode tRNAs. For example, *tRNA^Arg* encodes the tRNA that carries arginine. The genes that encode ribosomal RNA are shown in pale orange. The remaining genes encode polypeptides that function within the mitochondrion.

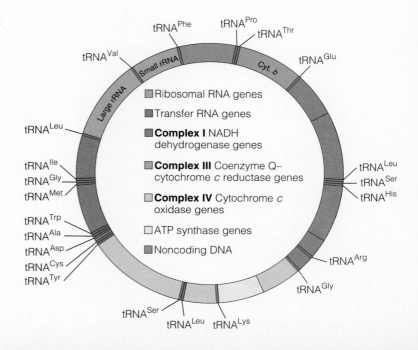

mitochondrial proteins are first synthesized outside of the mitochondrion in the cytosol of the cell. After synthesis, nuclearly encoded mitochondrial proteins are then transported into the mitochondria.

Chloroplast genomes tend to be larger than mitochondrial genomes, and they have a correspondingly greater number of genes. A typical chloroplast genome is approximately 100,000 to 200,000 base pairs in length, which is about ten times larger than the mitochondrial genome of animal cells. Figure 7-12 shows the chloroplast genome, called **cpDNA**, of the tobacco plant. It is a circular DNA molecule that contains 156,000 base pairs. It carries between 110 and 120 different genes. These include genes that encode ribosomal RNAs, transfer RNAs, and many proteins required for photosynthesis. As with mitochondria, many chloroplast proteins are in fact encoded by genes found in the plant cell nucleus. These proteins contain chloroplast-targeting signals that direct them to the chloroplasts.

Extranuclear inheritance produces non-Mendelian results in reciprocal crosses

In a diploid eukaryotic species, the genes within the nucleus obey a Mendelian pattern of inheritance, because the homologous pairs of chromosomes segregate during gamete formation. Except for sex-linked traits, offspring inherit one copy of each gene from both the maternal and paternal parents. By comparison, the inheritance pattern of extranuclear genetic material does not usually display a Mendelian pattern. This is because mitochondria and plastids do not typically segregate into the gametes in the same way as nuclear chromosomes.

In 1909, Carl Correns at the University of Leipzig in Berlin discovered a trait that showed a non-Mendelian pattern of inheritance. This observation was obtained by making reciprocal crosses.

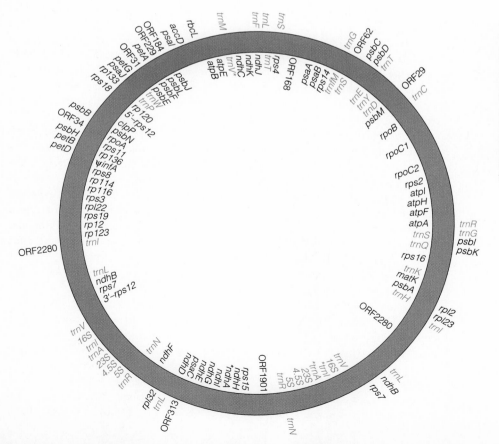

FIGURE **7-12**

A genetic map of the tobacco chloroplast genome. This diagram illustrates the locations of many genes along the circular chloroplast chromosome. The genes shown in blue encode tRNAs. The genes that encode ribosomal RNA are shown in red. The remaining genes shown in black encode polypeptides that function within the chloroplast. The genes designated ORF (open reading frame) encode polypeptides with unknown functions.

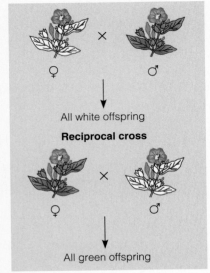

(a)

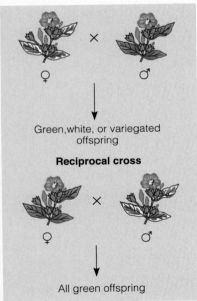

(b)

FIGURE **7-13**
The reciprocal crosses of four-o'clock plants by Carl Correns. (a) A pair of crosses between white-leaved and green-leaved plants. **(b)** A pair of crosses between variegated and green-leaved plants.

GENES ⟶ TRAITS: In this example, the white phenotype is due to chloroplasts that carry a mutant allele that prevents green pigmentation. The variegated phenotype is due to a mixture of chloroplasts, some of which carry the normal (green) allele and some of which carry the white allele. In the crosses shown here, the parent providing the eggs determines the phenotypes of the offspring. This is due to maternal inheritance. The egg contains the chloroplasts that are inherited by the offspring. If the plant providing the eggs has a white phenotype, all the offspring will inherit chloroplasts that carry the mutant allele and will have white leaves. If the plant providing the eggs is green, all the offspring will inherit normal chloroplasts and will be green. If the plant providing the eggs is variegated, it can transmit normal chloroplasts, mutant chloroplasts, or a mixture of these to the offspring. This will produce green, white, or variegated offspring, respectively.

He followed the inheritance of a pigmentation trait in *Mirabilis jalapa* (the four-o'clock plant). Leaves could be either green, white, or variegated (with both green and white sectors). As shown in Figure 7-13, Correns discovered that the pigmentation of the offspring depended solely on the maternal parent and not at all on the paternal parent. If the female parent had white pigmentation, all the offspring had white leaves. Similarly, if the female was green, all the offspring were green. When the female was variegated, the offspring could be either green, white, or variegated.

The pattern of inheritance observed by Correns is a type of extrachromosomal inheritance called **maternal inheritance** (not to be confused with maternal effect). In this example, maternal inheritance occurs because the plastids, which are the site of green pigment synthesis, are only inherited through the cytoplasm of the egg. In contrast, the pollen grains of *Mirabilis jalapa* do not transmit plastids to the resulting offspring.

Another early example of maternal inheritance involved a trait in corn plants that affects gamete fertility. In a planting of corn at Arequipa, Peru, in the early 1930s, the geneticist R. Emerson of Cornell University discovered a plant with sterile pollen—an unusual trait called *male sterility*. However, the female gametes from the male-sterile plant could be fertilized by pollen from a different plant. In other words, this strain of corn had no signs of female sterility.

Subsequent studies showed that the male-sterility trait was maternally inherited, suggesting an extranuclear mode of inheritance. In 1933, Marcus Rhoades of Columbia University conducted an extensive genetic analysis of the inheritance of the mutant allele that caused pollen sterility. At that time, it was already known that corn contains ten different chromosomes. Several genes had been mapped to each of these ten linkage groups. By comparing the inheritance of known nuclear genes with the inheritance of the male-sterility trait, Rhoades discovered that the male sterility allele was not linked to any of the ten corn chromosomes. This study provided convincing evidence that the male-sterility trait is controlled by genetic material outside of the nucleus. We now know this trait is encoded in the mitochondrial genome.

Pollen sterility is of great agricultural importance. Many crop plants such as corn, beets, and sugar cane are grown commercially from hybrid seed produced by cross-fertilization. Since hybrid seed production involves controlled cross-pollination, the use of male-sterile plants has great practical value. In the case of corn, for example, a cross can be made between a male-sterile strain (which makes fertile eggs) and a male-fertile strain; the seeds are collected from the male-sterile strain. Unlike Mendel's cross-pollination experiments with pea plants, where he had to remove the pollen-bearing structures, it is unnecessary to remove the corn tassels (which bear pollen) when one of the strains produces sterile pollen. Since detasseling is an expensive undertaking, male-sterile strains can save seed producers a substantial amount of money.

Studies in yeast and *Chlamydomonas* provided genetic evidence for extranuclear inheritance of mitochondria and chloroplasts

The early studies of Correns, Rhoades, and others indicated that some traits, such as plant pigmentation and pollen sterility, are inherited in a non-Mendelian manner. However, these studies did not definitively determine that maternal inheritance was due to genetic material within the subcellular organelles. Further progress in the investigation of extranuclear inheritance was provided by detailed genetic analyses of eukaryotic microorganisms such as yeast and algae. In these species, researchers isolated and identified mutant phenotypes that specifically affected the chloroplasts or mitochondria.

During the late 1940s and 1950s, yeasts and molds became model eukaryotic organisms for investigating the inheritance of mitochondria. Because mitochondria produce energy for cells in the form of ATP, mutations that yield defective mitochondria are expected to make cells grow much more slowly. In 1949, the French geneticist Boris Ephrussi and his colleagues identified mutations in *S. cerevisiae* that had such a phenotype. These mutants were called **petites** to describe their formation of small colonies on agar plates (as opposed to wild-type strains that formed larger colonies, which were termed **grande**). Biochemical and physiological evidence indicated that petite mutants had defective mitochondria. The researchers found that petite mutants could not grow at all when the cells only had an energy source requiring the metabolic activity of mitochondria, but they could form small colonies when grown on sugars that are metabolized by the glycolytic pathway, which occurs outside of the mitochondria.

Genetic analyses showed that petite mutants are inherited in different ways. *Segregational petite mutants* were shown to have mutations in genes located in the nucleus. These mutations affect genes encoding proteins necessary for mitochondrial function. Many nuclear genes encode proteins that are synthesized in the cytoplasm and are then taken up by the mitochondria, where they perform their functions. Segregational petites get their name because they segregate in a Mendelian manner during meiosis (Figure 7-14a). By comparison, the second category of petite mutants, known as *vegetative petite mutants*, do not segregate in a Mendelian manner. This is because vegetative petites involve mutations in the mitochondrial genome itself.

Ephrussi and colleagues identified two types of vegetative petites, called *neutral petites* and *suppressive petites*. They crossed the vegetative petites to a wild-type strain, revealing a non-Mendelian pattern of inheritance (Figure 7-14b). In a cross between a wild-type strain and a neutral petite, all four resulting haploid progeny were wild-type. This contradicts the normal 2:2 ratio expected for the segregation of Mendelian traits (see the discussion of tetrad analysis in Chapter 5). By comparison, a cross between a wild-type strain and a suppressive petite yielded all petite colonies. Thus, vegetative petites are defective in mitochondrial function and show a non-Mendelian pattern of inheritance, together providing compelling evidence that the mitochondrion has its own genetic material.

Since these initial studies, researchers have found that neutral petites lack most of their mitochondrial DNA, whereas suppressive petites usually lack small segments of the mitochondrial genetic material. When two yeast cells are mated, the resulting progeny inherit mitochondria from both parents. For example, in a cross between a wild-type and a neutral petite strain, the progeny would inherit both types of mitochondria. Since wild-type mitochondria are inherited, the cells display a

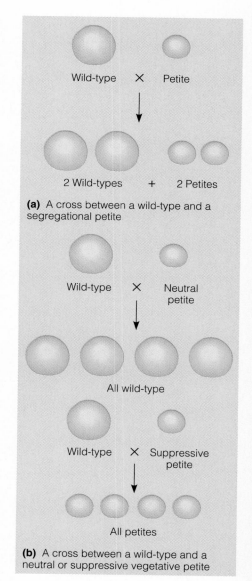

(a) A cross between a wild-type and a segregational petite

(b) A cross between a wild-type and a neutral or suppressive vegetative petite

FIGURE **7-14**

Transmission of the petite trait in *Saccharomyces cerevisiae*. (a) A wild-type strain crossed to a segregational petite. **(b)** A wild-type strain crossed to a neutral and to a suppressive vegetative petite.

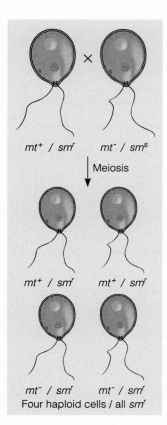

mt^+ / sm^r mt^- / sm^s

Meiosis

mt^+ / sm^r mt^+ / sm^r

mt^- / sm^r mt^- / sm^r
Four haploid cells / all sm^r

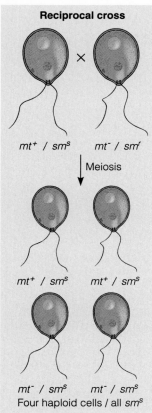

Reciprocal cross

mt^+ / sm^s mt^- / sm^r

Meiosis

mt^+ / sm^s mt^+ / sm^s

mt^- / sm^s mt^- / sm^s
Four haploid cells / all sm^s

FIGURE 7-15

Chloroplast inheritance in *Chlamydomonas*. Mt^+ and mt^- indicate the two mating types of the organism. Sm^r indicates streptomycin resistance, whereas sm^s indicates sensitivity to this antibiotic.

normal phenotype. The inheritance pattern of suppressive petites is more difficult to explain, since the offspring inherit both normal and suppressive petite mitochondria. One possibility is that the suppressive petite mitochondria replicate more rapidly, so that the wild-type mitochondria are not maintained in the cytoplasm for many doublings. Alternatively, genetic exchanges between the genomes of normal and suppressive petites may ultimately produce a defective population of mitochondria.

Let's now turn our attention to the inheritance of chloroplasts, which are found in eukaryotic species capable of photosynthesis (namely, algae and higher plants). The unicellular alga *Chlamydomonas reinhardtii* has been used as a model organism to investigate the inheritance of chloroplasts. This organism contains a single chloroplast that occupies approximately 40% of the cell volume. Genetic studies of chloroplast inheritance began in 1954, when Ruth Sager at Rockefeller University identified a mutation that provided *Chlamydomonas* with resistance to the antibiotic streptomycin (sm^r). By comparison, most strains are sensitive to killing by streptomycin (sm^s). As shown in Figure 7-15, Sager isolated a sm^r mutant that was not inherited in a Mendelian manner. Like yeast, *Chlamydomonas* is an organism that can be found in two mating types, designated mt^+ and mt^-. Sager and her colleagues discovered that the sm^r trait was inherited from the mt^+ parent but not from the mt^- parent. In subsequent studies, they mapped several genes, including the sm^r gene, to a single chromosome that was not inherited in Mendelian manner. This linkage group is the chloroplast chromosome.

The pattern of inheritance of mitochondria and plastids varies from species to species and depends on the gender of the parent

The inheritance of traits via genetic material within mitochondria and plastids is now a well-established phenomenon, which geneticists have investigated in many different species. In heterogamous species, the female gamete tends to be large and provides most of the cytoplasm to the resulting zygote, while the male gamete is small and often provides little more than a nucleus. Therefore, mitochondria and plastids are most often inherited from the mother. However, this is not always the case. Table 7-4 describes the inheritance patterns of mitochondria and plastids in several selected species.

In species where maternal inheritance is generally observed, the male parent may, on rare occasions, provide mitochondria to the zygote. This is called **paternal leakage**. It is fairly common in many species that normally exhibit maternal inheritance of their organelles. In the mouse, for example, approximately 1–4 paternal mitochondria are inherited for every 100,000 maternal mitochondria per generation of offspring. This means that most mouse zygotes do not inherit any paternal mitochondria, but a rare zygote may inherit a mitochondrion from the sperm.

A few rare human diseases are caused by mitochondrial mutations

The human mitochondrial genome has thirteen genes that encode polypeptides necessary for the synthesis of ATP. In addition, the mtDNA has genes that encode ribosomal RNA and transfer RNA molecules. Human mtDNA is maternally inherited, since it is transmitted from mother to offspring via the cytoplasm of the egg. Therefore, the transmission of human mitochondrial diseases follows a strict

TABLE 7–4 Transmission of Organelles among Different Species

Species	Organelle	Transmission
Mammals	Mitochondria	Maternal inheritance
Mussels	Mitochondria	Biparental
S. cerevisiae	Mitochondria	Biparental
Molds	Mitochondria	Usually maternal inheritance, paternal inheritance has been found in the genus, *Allomyces*
Chlamydomonas	Mitochondria	Inherited from the parent with the mt^- mating type
Chlamydomonas	Chloroplasts	Inherited from the parent with the mt^+ mating type
Plants		
Angiosperms	Mitochondria and plastids	Often maternally inherited, although biparental inheritance is not uncommon among many species
Gymnosperms	Mitochondria and plastids	Usually paternal inheritance

Data from: Gillham, N. W. (1994) *Organelle Genes and Genomes.* Oxford University Press, New York.

maternal inheritance pattern. Table 7-5 describes several mitochondrial diseases that have been discovered in humans.

Human diseases involving mutations in mitochondrial genes are usually chronic degenerative disorders that affect the brain, heart, muscle, kidney, and endocrine glands. For example, Leber's hereditary optic neuropathy (LHON) is a disease that affects the optic nerve. It may lead to the progressive loss of vision in one or both eyes. LHON can be caused by a defective mutation in one of several different mitochondrial genes. However, it is still unclear how a defect in these mitochondrial genes produces the symptoms of this disease.

TABLE 7–5 Examples of Human Mitochondrial Diseases

Disease	Mitochondrial Gene Mutated
Leber's hereditary optic neuropathy	A mutation in one of several mitochondrial genes that encode respiratory proteins, including *ND1, ND2, CO1, ND4, ND5, ND6,* or *cytb*
Neurogenic muscle weakness	A mutation in the *ATPase6* gene that encodes a subunit of the mitochondrial ATP-synthetase, which is required for ATP synthesis
Mitochondrial encephalomyopathy, lactic acidosis, and stroke-like episodes	A mutation in genes that encode tRNAs for leucine and lysine
Mitochondrial myopathy	A mutation in a gene that encodes a tRNA for leucine
Maternal myopathy and cardiomyopathy	A mutation in a gene that encodes a tRNA for leucine
Myoclonic epilepsy and ragged-red muscle fibers	A mutation in a gene that encodes a tRNA for lysine

Data from: Wallace, D. C. (1993) Mitochondrial diseases: Genotype versus phenotype. *Trends Genet. 9,* 128–133.

Extranuclear genomes of mitochondria and chloroplasts evolved from an endosymbiotic relationship between bacteria and primitive eukaryotic cells

The idea that the nucleus, mitochondria, and chloroplasts all contain their own separate genomes may at first seem puzzling. It would appear simpler to have all the genetic material in one place in the cell. The underlying reason that there are distinct genomes for mitochondria and chloroplasts can be traced back to their evolutionary origin, which is thought to involve a symbiotic association.

A *symbiotic relationship* occurs when two different species live together in direct contact with each other. The symbiont is the smaller of the two species. The term **endosymbiosis** describes a symbiotic relationship in which the symbiont actually lives inside (*endo-*, inside) the larger of the two species. In 1883, Andreas Schimper at the University of Bonn in Germany proposed that plastids were descended from an endosymbiotic relationship between cyanobacteria and eukaryotic cells. This idea, now known as the **endosymbiosis theory**, proposed that the ancient origin of plastids was initiated when a bacterium took up residence within a primordial eukaryotic cell (Figure 7-16). Over the course of evolution, the characteristics of the intracellular bacterial cell gradually changed to those of a plastid. In 1922, Ivan Wallin at the University of Colorado in Boulder also proposed an endosymbiotic origin for mitochondria.

In spite of these early hypotheses, the question of endosymbiosis was largely ignored until researchers in the 1950s discovered that chloroplasts and mitochondria contain their own genetic material. The issue of endosymbiosis was resurrected (and hotly debated) in 1970 when Lynn Margulis published a book entitled *Origin of Eukaryotic Cells*. During the 1970s and 1980s, the advent of molecular genetic techniques allowed researchers to analyze genes from chloroplasts, mitochondria, bacteria, and eukaryotic nuclear genomes. They found that genes in chloroplasts and mitochondria are very similar to bacterial genes but not as similar to those

FIGURE **7-16**

The endosymbiotic origin of mitochondria and chloroplasts. According to the endosymbiotic theory, plastids descended from an endosymbiotic relationship between cyanobacteria and eukaryotic cells. This arose when a bacterium took up residence within a primordial eukaryotic cell. Over the course of evolution, the intracellular bacterial cell gradually changed its characteristics, eventually becoming a plastid. Similarly, mitochondria were derived from an endosymbiotic relationship between purple bacteria and eukaryotic cells.

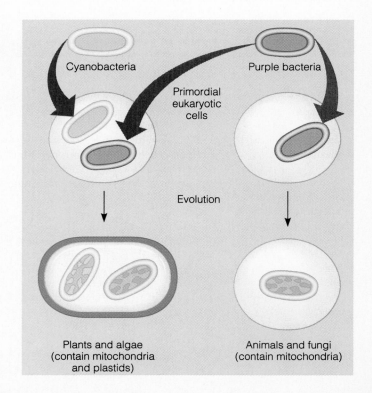

found within the nucleus of eukaryotic cells. This observation provided strong support for the endosymbiotic origin of mitochondria and chloroplasts.

Symbiosis occurs because the relationship is beneficial to one or both species. In the case of the endosymbiosis theory, the relationship provided eukaryotic cells with useful cellular characteristics. Plastids were derived from *cyanobacteria*, a prokaryotic species that is capable of photosynthesis. The ability to carry out photosynthesis is beneficial to plant cells, providing them with the ability to use the energy from sunlight. It is less clear how the relationship would have been beneficial to a cyanobacterium. By comparison, mitochondria are thought to have been derived from a different species of bacteria, known as *purple bacteria*. In this case, the endosymbiotic relationship enabled eukaryotic cells to synthesize greater amounts of ATP.

During the evolution of eukaryotic species, most genes that were originally found in the genome of the primordial cyanobacteria and purple bacteria have been transferred from the organelles to the nucleus. In other words, genes have been removed from the mitochondrial and chloroplast chromosomes and relocated to the nuclear chromosomes. This has occurred many times throughout evolution, so that modern mitochondria and chloroplasts have lost most of the genes that are still found in present-day purple bacteria and cyanobacteria.

The molecular mechanism that underlies gene transfer between organelles is a relatively new topic of genetic research. While this mechanism is not entirely understood, the direction of transfer is well established. During evolution, gene transfer has occurred primarily from the organelles to the nucleus. Transfer of genes from the nucleus to the organelles almost never occurs, although there is one example of a nuclear gene in plants that has been transferred to the mitochondrial genome. This unidirectional gene transfer from organelles to the nucleus explains why the organellar genomes now contain relatively few genes. In addition, gene transfer can occur between organelles. It can happen between two mitochondria, between two chloroplasts, and between a chloroplast and mitochondrion. Overall, the transfer of genetic material between the nucleus, chloroplasts, and mitochondria is an established phenomenon, although its biological benefits remain unclear.

Eukaryotic cells occasionally contain symbiotic infective particles

Other unusual endosymbiotic relationships have been identified in eukaryotic organisms. There are a few rare cases where infectious particles establish a symbiotic relationship with their host. In some cases, research indicates that symbiotic infectious particles are bacteria or viruses that exist within the cytoplasm of eukaryotic cells. While examples of symbiotic infectious particles are relatively rare, they have provided some interesting and even bizarre examples of the extranuclear inheritance of traits.

In the early 1940s, Tracy Sonneborn at Indiana University studied a trait in the protozoan *Paramecia aurelia* known as the **killer trait**. Killer paramecia secrete a substance called *paramecin*, which kills some but not all strains of paramecia. The killer strains themselves are resistant to paramecin. Sonneborn found that killer strains contain particles in their cytoplasm known as *kappa particles*. Each kappa particle is 0.4 μm long and has its own DNA. Genes within the kappa particle encode the paramecin toxin. In addition, other kappa particle genes provide the killer paramecia with resistance to paramecin. When nonkiller strains are mixed with a cell extract derived from killer paramecia, the kappa particles within the extract can infect the nonkiller strains and convert them into killer strains. In other words, the extranuclear particle that determines the killer trait is infectious.

Infectious particles have also been identified in fruit flies. The French researcher P. l'Heritier identified strains of *Drosophila melanogaster* that are highly sensitive to killing by CO_2. Reciprocal crosses between CO_2-sensitive and normal flies revealed that the trait is inherited in a non-Mendelian manner. Furthermore, cell extracts from a sensitive fly can infect a normal fly and make it sensitive to CO_2.

Another example of an infectious particle in fruit flies involves a trait known as sex ratio. It was discovered by Chana Malogolowkin at Columbia University and D. Poulson at Yale University. This trait was found in one strain of *Drosophila willistoni* where most of the offspring of female flies were daughters; nearly all the male offspring died. The sex ratio trait is transmitted from mother to offspring. The rare surviving males do not transmit this trait to their male or female offspring. This result indicates a maternal inheritance pattern for the sex ratio trait. The agent in the cytoplasm of female flies was found to be a symbiotic microorganism. Its presence is usually lethal to males but not to females. This infective agent can be extracted from the tissues of adult females and used to infect the females of a normal strain of flies.

CONCEPTUAL SUMMARY

In this chapter, we have considered inheritance patterns that differ from Mendelian inheritance. In some cases, genes that are physically located on nuclear chromosomes fail to follow a Mendelian pattern of inheritance. For example, genes that display a *maternal effect* are expressed in the mother's nurse cells during oogenesis and affect the phenotype of the offspring. *Epigenetic inheritance* describes another class of examples in which *nuclear genes* do not follow a Mendelian pattern of inheritance. During *dosage compensation*, the level of expression of genes on the sex chromosomes is altered. In mammals, dosage compensation occurs via the *inactivation* of one X-chromosome in the female. According to the *Lyon hypothesis*, one of the two X-chromosomes in females is randomly inactivated during embryonic development to produce a transcriptionally inactive *Barr body*. The active and inactive X-chromosomes are then inherited during later stages of development to create a mosaic animal that is random patchwork in which a single type of X-chromosome is active in each cell. A second example of epigenetic inheritance is known as *genomic imprinting*. In this case, a gene or even an entire chromosome is marked in such as way as to be inherited in an inactive form. DNA methylation is thought to be important in this marking process.

Other examples of non-Mendelian inheritance are due to the existence of genetic material outside the nucleus. This is known as *extranuclear* or *cytoplasmic inheritance*. The two most important examples of extranuclear inheritance are due to genetic material within mitochondria and chloroplasts. In many cases, extranuclear genetic material follows a *maternal inheritance* pattern. This is because the oocyte is large and, in most species, is more likely than the sperm to transmit organelles to the offspring. However, some species display biparental and even paternal patterns of organelle inheritance. Mitochondria and plastids contain genetic material due to their *endosymbiotic* origin. Mitochondria were derived from purple bacteria, plastids from cyanobacteria. During evolution, many genes have been transferred from the organellar genomes to the nuclear genomes, so that modern mitochondria and chloroplast genomes are small and contain only a few genes. Other examples of extranuclear inheritance include symbiotic *infective particles*.

EXPERIMENTAL SUMMARY

The primary way in which researchers have identified non-Mendelian patterns of inheritance is by making a series of genetic crosses and then analyzing the phenotypes of the offspring. In many cases, reciprocal crosses have yielded results that deviate from Mendelian inheritance. For genes with a maternal effect, reciprocal crosses reveal that the genotype of the mother governs the phenotype of the offspring. In genomic imprinting, reciprocal crosses indicate that one parent transmits an active form of a gene while the other parent transmits a marked copy that is inactive. Genetic crosses are also consistent with the phenomenon of X-inactivation. For example, the trait of variegated coat pattern only occurs in heterozygous female animals.

Similarly, the outcomes of reciprocal crosses are usually different for extra-chromosomal inheritance. In most species, organelles are inherited from the maternal parent, although many exceptions are known. The inheritance of the genetic material within mitochondria and chloroplasts, first identified in studies of yeast and algae, is thus non-Mendelian.

Researchers have also investigated non-Mendelian inheritance at the cellular and molecular levels. The cellular explanation of maternal effect genes is that the gene products are transferred to the developing oocyte by the nurse cells. In Chapter 24, we will examine methods that demonstrate the accumulation of maternal effect gene products within the oocyte. Genomic imprinting is also being investigated at the molecular level. Researchers exploring the biochemistry of imprinted genes have identified DNA methylation as a likely key to marking. Further experimentation is needed to elucidate the mechanisms of erasure and *de novo* methylation that occur during gametogenesis.

Likewise, researchers have explored the cellular and molecular mechanisms of X-inactivation. Cytological examination of female mammalian cells revealed the presence of a highly condensed X-chromosome, called the Barr body. A study of G-6-PD expression at the cellular level confirmed that one X-chromosome is permanently inactivated in the somatic cell lineages of females. A site on the X-chromosome called Xic is required for the proper counting of X-chromosomes. Current research is focused on the role of the *XIST* gene in the X-inactivation mechanism.

Finally, molecular tools have been used to dissect the genetic composition of organellar chromosomes. Some of these methods, which include DNA cloning and sequencing, will be described in Chapter 19. These techniques have enabled researchers to identify the genes located in mtDNA and cpDNA.

PROBLEM SETS

SOLVED PROBLEMS

1. A maternal effect gene in *Drosophila*, called *torso*, is found in a recessive allele that prevents the correct development of anterior- and posteriormost structures. A wild-type male is crossed to a female of unknown genotype. This mating produces 100% larva that are missing their anterior- and posteriormost structures and therefore die during early development. What is the genotype and phenotype of the female fly in this cross? What are the genotypes and phenotypes of the female fly's parents?

Answer: Since this cross is producing 100% abnormal offspring, the female fly must be homozygous for the abnormal *torso* allele. Even so, the female fly must be phenotypically normal in order to reproduce. This female fly would have had a mother that was heterozygous for a normal and abnormal *torso* allele and a father that was either heterozygous, or homozygous for the abnormal *torso* allele:

$$Torso^+ \; Torso^- \quad \times \quad Torso^+ \; Torso^- \text{ or } Torso^- \; Torso^-$$
female   male

↓

$$Torso^- \; Torso^-$$
(female that is mother
of 100% abnormal offspring)

This female fly is phenotypically normal because her mother was heterozygous and provided the gene products of the *torso*⁺ allele from the nurse cells. However, this homozygous female will only produce abnormal offspring, because she cannot provide them with normal *torso*⁺ gene products.

2. An individual with Angelman syndrome produces an offspring with Prader–Willi syndrome. Why does this occur? What are the genders of the parent with Angelman syndrome and the offspring with Prader–Willi syndrome?

Answer: These two different syndromes are most commonly caused by a small deletion in chromosome 15. In addition, genomic imprinting plays a role, since genes in this deleted region are differentially imprinted depending on gender. If this deletion is inherited from the paternal parent, the offspring develops Prader–Willi syndrome. Therefore, in this problem, the individual with Angelman syndrome must have been a male, since he produced a child with Prader–Willi syndrome. The child could be either a male or female.

3. In yeast, a haploid petite mutant also carries a mutant gene that requires the amino acid histidine for growth. The *petite/his⁻* strain is crossed to a wild-type strain to yield the following tetrad:

2 cells: *petite/his⁻*
2 cells: *petite/his⁺*

Explain the inheritance of the *petite* and *his⁻* mutations.

Answer: The *his⁻* and *his⁺* alleles are segregating in a 2:2 ratio. This indicates a nuclear pattern of inheritance. By comparison, all four cells in this tetrad have a petite phenotype. This is a suppressive petite that arises from a mitochondrial mutation.

4. Let's suppose that you are a horticulturist that has recently identified an interesting plant with variegated leaves. How would you determine if this trait is nuclearly or cytoplasmically inherited?

Answer: Make reciprocal crosses involving normal and variegated strains. In many species, plastid genomes are inherited maternally, although this is not always the case. In addition, a significant percentage of paternal leakage may occur. Nevertheless, when reciprocal crosses yield different outcomes, an organellar mode of inheritance is usually at work. On rare occasions, it is important to be mindful of the possibility of a nuclear gene affecting plastid development.

Conceptual Questions

1. Define the term epigenetic, and describe two examples.

2. Describe the pattern of inheritance that maternal effect genes follow. Explain how the maternal effect occurs at the cellular level. What are the expected functional roles of the proteins that are encoded by maternal effect genes?

3. A maternal effect gene exists in a dominant *N* (normal) allele and a recessive *n* (abnormal) allele. What would be the ratios of genotypes and phenotypes for the offspring of the following crosses:

    A. *nn* female × *NN* male
    B. *NN* female × *nn* male
    C. *Nn* female × *Nn* male

4. A *Drosophila* embryo dies during early embryogenesis due to a recessive maternal effect allele called *bicoid*. The wild-type allele is designated *bicoid⁺*. What are the genotypes and phenotypes of the embryo's mother and maternal grandparents?

5. With regard to the numbers of sex chromosomes, explain why dosage compensation is necessary.

6. What is a Barr body? How is its structure different from that of other chromosomes in the cell? How does the structure of a Barr body affect the level of X-linked gene expression?

7. Among different species, describe three distinct strategies for accomplishing dosage compensation.

8. Describe when X-inactivation occurs, and how this leads to phenotypic results at the organismal level. In your answer, you should explain why X-inactivation causes results such as variegated coat patterns in mammals. Why do two different calico cats have their patches of orange and black fur in different places? Explain whether or not a variegated coat pattern could occur in marsupials due to X-inactivation.

9. Describe the molecular process of X-inactivation. This description should include the three phases of inactivation and the role of the Xic locus. Explain what happens to X-chromosomes during embryogenesis, in adult somatic cells, and during oogenesis.

10. On rare occasions, an abnormal human male is born who is somewhat feminized compared with normal males. Microscopic examination of the cells of one such individual reveals that he has a single Barr body in each cell. What is the chromosomal composition of this individual?

11. When does the erasure and reestablishment phase of imprinting occur? Explain why it is necessary to erase an imprint and then reestablish it in order to always maintain imprinting from the same gender of parent.

12. In what types of cells would you expect *de novo* methylation to occur? In what cell types would it not occur?

13. On rare occasions, people are born with a condition known as **uniparental disomy**. It happens when an individual inherits both copies of a chromosome from one parent and no copies from the other parent. This occurs when two abnormal gametes happen to complement each other to produce a diploid zygote. For example, an abnormal sperm that lacks chromosome 15 could fertilize an egg that contains two copies of chromosome 15. In this situation, the individual would be said to have *maternal uniparental disomy-15*, because both copies of chromosome 15 come from the mother. Alternatively, an abnormal sperm with two copies of chromosome 15 could fertilize an egg with no copies. This is known as *paternal uniparental disomy-15*.

If a female is born with paternal uniparental disomy-15, would you expect her to be phenotypically normal, have Angelman syndrome, or have Prader–Willi syndrome? Explain. Would you expect her to produce normal offspring or offspring affected with AS or PWS?

14. What is extranuclear inheritance? Describe three examples.

15. What is a reciprocal cross? Let's suppose that a gene is found in a wild-type allele and a recessive mutant allele. What would be the expected outcomes of reciprocal crosses if a true-breeding normal individual was crossed to a true-breeding individual carrying the mutant allele? What would be the results if the gene was maternally inherited?

16. Does extranuclear inheritance always follow a maternal inheritance pattern? Why or why not?

17. A variegated trait in plants is analyzed using reciprocal crosses. The following results are obtained:

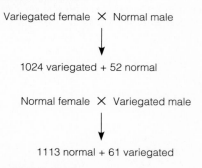

Explain this pattern of inheritance.

18. What is the phenotype of a petite mutant? Where can a petite mutation occur—in nuclear genes, extranuclear genetic material, or both? What is the difference between a neutral and suppressive petite?

19. Extranuclear inheritance often correlates with maternal inheritance. Even so, paternal leakage is not uncommon. Explain what paternal leakage is. If a cross produced 200 offspring and the rate of mitochondrial paternal leakage was 3%, how many offspring would be expected to contain paternal mitochondria?

20. Discuss the structure and organization of the mitochondrial and chloroplast genomes. How large are they, how many genes do they contain, and how many copies of the genome are there per organelle?

21. Explain the likely evolutionary origin of mitochondrial and chloroplast genomes. How have the sizes of the mitochondrial and chloroplast genomes changed since their origin? How has this occurred?

EXPERIMENTAL QUESTIONS

1. Figure 7-1 describes an example of a maternal effect gene. Explain how Sturtevant deduced a maternal effect gene based on the F_2 and F_3 generations. Why weren't the results of F_1 generation crosses sufficient to propose a maternal effect gene?

2. Discuss the types of experimental observations that Mary Lyon had to consider in proposing her hypothesis concerning X-inactivation. In your own words, explain how these observations were consistent with her hypothesis.

3. In Experiment 7A, why does a clone of cells only produce one type of G-6-PD enzyme? What would you expect to happen if a clone was derived from an early embryonic cell?

4. In Experiment 7A, why does the initial sample of tissue produce both forms of G-6-PD?

5. Let's suppose that you have two strains of petunias: one has pink flowers, and the other has flowers that are variegated with pink and white pigmentation. Considering the role of plastids, why do you think this has occurred? Describe the type of genetic analysis you would follow to determine the mode of inheritance of this variegated flower phenotype.

6. Sager and her colleagues discovered that the mode of inheritance of streptomycin resistance in *Chlamydomonas* could be altered if the mt^+ cells were exposed to UV irradiation prior to mating. This exposure dramatically increased the frequency of biparental inheritance. What would be the expected outcome of a cross between an mt^+ sm^r and an mt^- sm^s strain in the absence of UV irradiation? How would the result differ if the mt^+ strain was exposed to UV light?

QUESTIONS FOR STUDENT DISCUSSION/COLLABORATION

1. Recessive maternal effect genes are identified in flies (for example) when a phenotypically normal mother cannot produce any normal offspring. Since all the offspring are dead, this female fly cannot be used to produce a strain of heterozygous flies that could be used in future studies. How would you identify heterozygous individuals that are carrying a recessive maternal effect allele? How would you maintain this strain of flies in a laboratory over many generations?

2. What is an infective particle? Discuss the similarities and differences between infective particles and organelles such as mitochondria and chloroplasts. Do you think the existence of infective particles supports the endosymbiotic theory of the origin of mitochondria and chloroplasts?

VARIATION IN CHROMOSOME STRUCTURE AND NUMBER

The term **genetic variation** refers to genetic differences between members of the same species, or those between different species. Throughout Chapters 2 to 7, we have focused primarily on variation in specific genes. This is called *allelic variation*. Allelic differences are due to small changes (i.e., mutations) that occur within a particular gene. In this chapter, our emphasis will shift to larger types of genetic changes that affect the chromosomal composition of eukaryotic organisms. As we will see, these changes may affect the expression of many genes and lead to interesting changes in phenotypes.

In the first section of this chapter, we will examine how the structure of a eukaryotic chromosome can be modified, either by altering the total amount of the genetic material or by rearranging the order of genes along a chromosome. A substantial change in chromosome structure, possibly affecting more than a single gene, is referred to as a **chromosome mutation**. In most cases, a mutant chromosome can be detected microscopically; its structure will differ visually from a normal chromosome. In this chapter, we will examine how chromosome mutations occur, how they are transmitted from parent to offspring, and how they affect the phenotype of the individual who inherits them.

The rest of the chapter will be concerned with changes in the total number of chromosomes. A change in chromosome number is called a **genome mutation**. This type of mutation comprises two subtypes: changes in the number of sets of chromosomes, and changes in the numbers of individual chromosomes within a set. Natural variation in the number of sets of chromosomes is relatively common among different species, particularly in the plant kingdom. Within the same species, changes in chromosome number within a set can also occur, but such alterations usually are detrimental. In this chapter, we will explore how variation in chromosome number occurs, and describe examples where it has significant phenotypic consequences.

VARIATION IN CHROMOSOME STRUCTURE

We will begin our discussion of chromosome variation by considering several ways in which the structures of eukaryotic chromosomes can be altered. As discussed in Chapters 3 and 5, the chromosomes found in the nuclei of eukaryotic cells are long, linear molecules that carry hundreds or even thousands of genes. In this section, we will examine how the composition of a chromosome can be changed. As you will see, segments of a chromosome can be lost, duplicated, or rearranged in a new way.

The study of chromosomal variation is important for several reasons. First, geneticists have discovered that variations in chromosome structure can have major effects on the phenotype of an organism. For example, we now know that several human genetic diseases are caused by changes in chromosome structure. In other cases, an individual possessing a chromosomal rearrangement may have a normal phenotype yet have a high probability of producing offspring with genetic abnormalities. Later in this section, we will examine irregularities in gamete formation that account for this phenomenon. Finally, changes in chromosome structure have been an important force in the evolution of new species. This topic will be explored in greater detail in Chapter 27.

In this section, we will examine the cellular mechanisms that underlie changes in chromosome structure. We will explore unusual events during meiosis that affect how altered chromosomes are transmitted from parents to offspring. Also, we will consider many examples where chromosomal alterations affect an organism's phenotypic characteristics.

Natural variation exists in chromosome structure

Before we begin to discuss how chromosome structure can be altered, we need to have a reference point for a normal set of chromosomes. To determine what the normal chromosomes of a species look like, a *cytogeneticist* microscopically examines the chromosomes from several members of the species. In most cases, two phenotypically normal individuals of the same species will have the same number and types of chromosomes.

To determine the chromosomal composition of a species, the chromosomes in actively dividing cells are karyotyped. The procedure for making a karyotype was described in Chapter 3 (Figure 3-1). A **karyotype** is a micrograph in which all the chromosomes within a single cell have been arranged in a standard fashion. Figure 8-1a shows normal karyotypes for three species: humans, fruit flies, and corn. As seen here, humans contain 46 chromosomes (23 pairs), fruit flies have 8 chromosomes (4 pairs), and corn has 20 chromosomes (10 pairs). Except for the sex chromosomes, which differ between males and females, most members of the same species have very similar karyotypes. By comparison, distantly related species (e.g., humans versus fruit flies) have very different karyotypes. Nevertheless, a karyotype of 46 chromosomes is normal for humans, as is a karyotype of 8 chromosomes for fruit flies.

As seen in Figure 8-1a, the chromosomes from any given species can vary considerably in size and shape. Cytogeneticists have various ways to classify and identify chromosomes. The three most commonly used features are size, location of the centromere, and banding patterns that are revealed when the chromosomes are treated with stains. By convention, the chromosomes are numbered according to their size, with the largest chromosomes having the smallest numbers. For example, human chromosomes 1, 2, and 3 are relatively large, whereas 21 and 22 are the smallest (see Figure 8-1a). An exception to the numbering system are the sex chromosomes, which are designated with letters (for humans, X and Y).

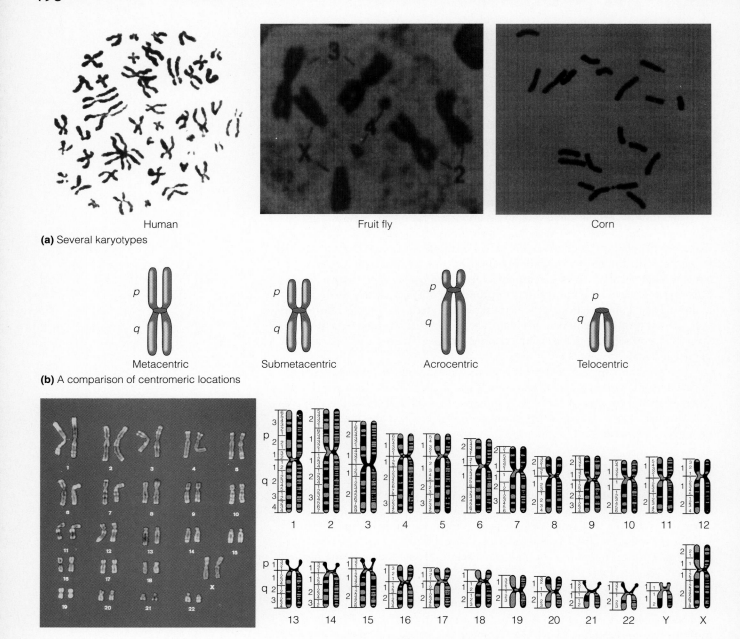

(a) Several karyotypes

Human

Fruit fly

Corn

(b) A comparison of centromeric locations

Metacentric   Submetacentric   Acrocentric   Telocentric

(c) Giemsa staining of human chromosomes

(d) Conventional numbering system of G-bands in human chromosomes

FIGURE **8-1**

Features of normal chromosomes.
(a) Karyotypes of humans, fruit flies, and corn.
(b) A comparison of centromeric locations.
Centromeres can be either metacentric (in the middle), submetacentric (slightly off center), acrocentric (near one end), or telocentric (at the end). **(c)** Human chromosomes that have been stained with Giemsa. **(d)** The conventional numbering of bands in Giemsa-stained human chromosomes. The banding patterns of chromatids change as the chromatids condense. The left side of each chromosome shows the banding pattern of a chromatid in mid-metaphase, and the right side shows a chromatid as it would appear in prophase.

Another distinguishing feature of eukaryotic chromosomes is the location of the centromere. As shown in Figure 8-1b, chromosomes are classified as **metacentric**, **submetacentric**, **acrocentric**, and **telocentric**. Because the centromere is not exactly in the center of a chromosome, each chromosome has a long arm (designated with the letter q) and a short arm (designated with the letter p). In the case of telocentric chromosomes, the short arm may be almost nonexistent. When preparing a karyotype, the chromosomes are aligned with the short arms on top and the long arms on the bottom.

Since different chromosomes often have similar sizes and centromeric locations (e.g., compare human chromosomes 8, 9, and 10), geneticists must use additional methods to accurately identify each type of chromosome. For detailed identification, chromosomes are treated with stains to produce characteristic banding patterns. Figure 8-1c shows a human karyotype in which the chromosomes have been treated with a chemical dye called *Giemsa stain*. In some chromosomal regions, the stain binds heavily and produces a dark band; in other regions, the stain hardly binds at all and produces a light band. As shown here, the alternating banding pattern, known as **G bands**, is a unique feature for each chromosome. Figure 8-1d shows the conventional numbering system that is used to designate each band along a Giemsa stained chromosome.

The banding pattern of eukaryotic chromosomes is useful in several ways. First, individual chromosomes can be distinguished from each other, even if they have similar sizes and centromeric locations. For example, compare the differences in banding patterns between human chromosomes 8 and 9 (Figure 8-1c, d). These differences permit us to distinguish these two chromosomes even though their sizes and centromeric locations are very similar. Banding patterns are also used to detect changes in chromosome structure. As discussed next, chromosomal rearrangements or changes in the total amount of genetic material are more easily detected when viewing banded chromosomes. Also, chromosome banding may reveal evolutionary relationships among the chromosomes of closely related species. This topic is discussed in Chapter 27.

Mutations can alter chromosome structure

Now that we understand that normal chromosomes can come in a variety of shapes and sizes, we can consider how the structures of normal chromosomes can be modified. As mentioned earlier, a chromosome mutation is a substantial change in the structure of a chromosome. There are two primary ways that the structure of chromosomes can be altered. First, the total amount of genetic material within a single chromosome can be increased or decreased significantly. Second, the genetic material in one or more chromosomes may be rearranged without affecting the total amount of material. As shown in Figure 8-2, these alterations are categorized as deficiencies, duplications, inversions, and translocations.

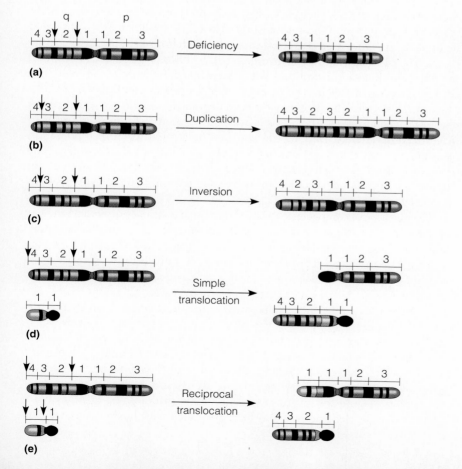

(a)

(b)

(c)

(d)

(e)

FIGURE **8-2**

Types of changes in chromosome structure. The larger chromosome shown in purple is human chromosome 1. The smaller chromosome is human chromosome 21. **(a)** A deficiency occurs which removes the q2 region. **(b)** A duplication occurs which doubles the q2-q3 region. **(c)** An inversion occurs that inverts the q2-q3 region. **(d)** The q2-q4 region of chromosome 1 is translocated to chromosome 21. **(e)** The q2-q4 region of chromosome 1 is exchanged with the q1 region of chromosome 21.

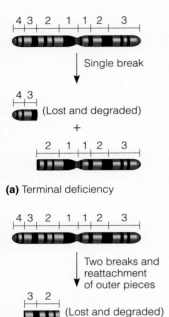

(a) Terminal deficiency

(b) Interstitial deficiency

FIGURE **8-3**

Production of terminal and interstitial deficiencies. This illustration shows the production of deletions in human chromosome 1.

Deficiencies and duplications are changes in the total amount of genetic material within a single chromosome. In Figure 8-2, human chromosomes are labeled according to their normal banding patterns. When a **deficiency** occurs, a segment of chromosomal material is missing. In other words, the affected chromosome is deficient in a significant amount of genetic material. Deficiencies are also referred to as **deletions**. In contrast, a **duplication** is the opposite situation—an amount of genetic material is repeated compared with the normal parent chromosome.

Inversions and translocations are chromosomal rearrangements. An **inversion** involves a change in the direction of the genetic material along a single chromosome. For example, in Figure 8-2c, a segment of one chromosome has been inverted, so that the order of four G-bands is opposite to that of the parent chromosomes. A **translocation** occurs when one segment of a chromosome breaks off and becomes attached to a different chromosome. When a single piece of chromosome is attached to another chromosome, this is called a **simple translocation**. In other cases, two different types of chromosomes can exchange pieces, thereby producing two abnormal chromosomes carrying translocations. This situation is referred to as a **reciprocal translocation**.

Figure 8-2 illustrates the four common ways that the structure of chromosomes can be changed. Throughout the rest of this section, we will consider how these changes occur, how the changes are detected experimentally, and how they affect the phenotypes of the individuals who inherit them.

The loss of genetic material in a deficiency tends to be detrimental to an organism

A chromosomal deficiency occurs when a chromosome breaks in one or more places and a fragment of the chromosome is lost. This phenomenon is shown in Figure 8-3. In Figure 8-3a, a normal chromosome has broken into two separate pieces. The piece without the centromere will be lost and degraded. Therefore, this event produces a chromosome with a **terminal deficiency**. In Figure 8-3b, a chromosome has broken in two places to produce three chromosomal fragments. The central fragment has become lost (and degraded), and the two outer pieces have reattached to each other. This process has created a chromosome with an **interstitial deficiency**. Deficiencies can also be created when recombination takes place at incorrect locations between two homologous chromosomes. The products of this type of aberrant recombination event are one chromosome with a deficiency and another chromosome with a duplication. This process will be examined later in this chapter.

The phenotypic consequences of a chromosomal deficiency depend on the size of the deletion and whether it includes genes (or portions of genes) vital to the development of the organism. When deletions have a phenotypic effect, they are usually detrimental. Larger deletions tend to be more harmful, because more genes are missing. Many examples are known in which deficiencies have significant phenotypic influences. In humans, for example, a genetic disease known as *cri-du-chat syndrome* involves a deficiency in a segment of the short arm of human chromosome 5 (Figure 8-4a). Individuals who carry a single copy of this abnormal chromosome (along with a normal chromosome 5) display an array of abnormalities including severe mental retardation, facial anomalies, and an unusual catlike cry (the meaning of the French name for the syndrome) (Figure 8-4b). Some other human genetic diseases, such as Angelman syndrome and Prader–Willi syndrome (see Chapter 7, Figure 7-9), are due to a deficiency in chromosome 15. Many types of chromosomal deletions have also been associated with human cancers. This topic will be discussed in Chapter 23.

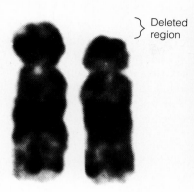

} Deleted region

(a) Chromosome 5

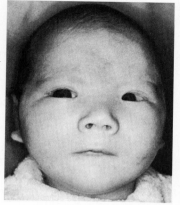

(b) A child with cri-du-chat syndrome

FIGURE **8-4**

Cri-du-chat syndrome. (a) Chromosome 5 from the karyotype of an individual with this disorder. A section of the short arm of the chromosome is missing. **(b)** An affected individual.

GENES ⟶ TRAITS: Compared with a normal individual who has two copies of each gene on chromosome 5, an individual with cri-du-chat syndrome only has one copy of the genes that are located within the missing segment. This genetic imbalance (one versus two copies of many genes on chromosome 5) causes the phenotypic characteristics of this disorder.

Deficiencies can be detected using cytological, genetic, and molecular techniques

A variety of experimental techniques are used to detect a deletion within a chromosome. These include microscopic, genetic, and molecular methods described in Chapter 19. Microscopic techniques, which are relatively quick and easy, can detect chromosomal deficiencies large enough to be seen with a light microscope. Cri-du-chat and Prader–Willi syndromes are chromosomal deletions easily detectable via light microscopy. Fairly small deletions, which may involve only one or a few genes, are often difficult to detect with simple microscopy. However, more complicated microscopy techniques, such as *in situ* hybridization, have enhanced our ability to detect small deletions. This technique will be described in Chapter 20.

When a mutation causes a phenotypic effect, genetic analysis can sometimes show that it is due to a deletion. As will be described in Chapter 17, deletions in particular genes can be identified by their inability to revert back to the wild-type allele. In addition, deletions can sometimes be revealed by a phenomenon known as **pseudodominance**. This occurs when one copy of a gene has been deleted from a chromosome and the remaining copy of a recessive allele on the homologous chromosome is phenotypically expressed. Under these conditions, the individual is hemizygous for the recessive allele.

Duplications tend to be less harmful than deletions

Duplications result in extra genetic material. They are usually caused by abnormal events during recombination. Under normal circumstances, crossing over occurs at analogous sites between homologous chromosomes. On rare occasions, the crossover may occur at two different sites on the homologues (Figure 8-5). This results in one chromatid with an internal duplication and another chromatid with a deletion. In Figure 8-5, the chromosome with the extra genetic material carries a **gene duplication**, because the number of copies of gene *C* has been increased from one to two. In most cases, gene duplications happen as rare, sporadic events during the evolution of species. Later in this section, we will consider how multiple copies of genes can evolve to produce a family of genes with specialized functions.

Like deletions, the phenotypic consequences of duplications tend to be correlated with size. Duplications are more likely to have phenotypic effects if they involve a large piece of the chromosome. In general, small duplications are less likely to have harmful effects than are deletions of comparable size. This observation suggests that having one copy of a gene is more harmful than having three copies. In humans, relatively few well-defined syndromes are known to be caused by small chromosomal duplications.

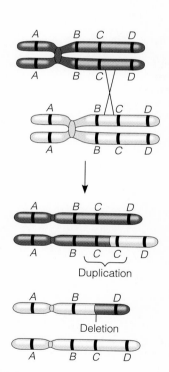

Duplication

Deletion

FIGURE **8-5**

Abnormal crossing over, leading to a duplication and a deficiency. A crossover has occurred at sites between genes *C* and *D* in one chromatid and between *B* and *C* in another chromatid. After crossing over is completed, one chromatid contains a duplication and the other contains a deletion.

In 1936, Bridges found that gene duplications produced the bar and ultra-bar eyes phenotype in *Drosophila*

Early insight into the causes of gene duplications came from studies in *Drosophila* involving a trait that affects the number of facets in the eye. In 1914, Sabra Colby Tice of Columbia University found a fly that had a reduced number of facets. This trait was called bar eyes. A genetic analysis of the trait revealed that it is an X-linked trait that shows incomplete dominance. Females homozygous for the *bar* allele have a more severe phenotype than heterozygous females that have one *bar* allele and one wild-type allele:

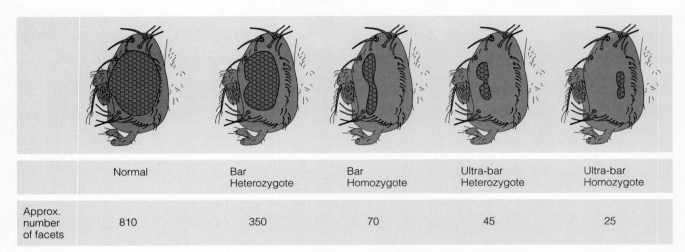

	Normal	Bar Heterozygote	Bar Homozygote	Ultra-bar Heterozygote	Ultra-bar Homozygote
Approx. number of facets	810	350	70	45	25

In 1920, Charles Zeleny at the University of Illinois identified some rare mutants in a stock of flies that was originally homozygous for the *bar* allele. Some rare flies had eyes that were phenotypically normal (*bar revertants*), and other flies had eyes with even fewer facets than the *bar* homozygote. Zeleny called these flies *ultrabar* (also known as *double-bar*). This trait also shows incomplete dominance, since a female fly that is homozygous for the *ultra-bar* allele has fewer facets than a heterozygous female fly carrying one *ultra-bar* allele and one normal allele.

Based on the results of genetic crosses, Alfred Sturtevant and Thomas Hunt Morgan of Columbia University initially suggested that the ultra-bar phenotype was due to a gene duplication in which two copies of the *bar* allele were on the same X-chromosome. Since it was (erroneously) thought that an ultra-bar female fly had two copies of the *bar* allele, yet displayed a more intense phenotype than a homozygous bar female that also had two copies of the *bar* allele, the term **position effect** was coined. This term refers to changes in phenotype that are caused by the positions of genes on a chromosome. In this case, it was suggested that having two *bar* alleles on the same chromosome produced a different phenotype compared with having two copies of the *bar* allele on separate X-chromosomes.

To gain further insight into the cause of the bar and ultra-bar phenotypes, Calvin Bridges at the California Institute of Technology investigated the bar/ultra-bar phenomenon at the cytological level. In cells of the *Drosophila* salivary gland, the chromosomes are easy to study under the microscope, because they replicate many times to form gigantic **polytene chromosomes**. The structure of polytene chromosomes will be described later in this chapter. (If you are not familiar with polytene chromosomes, you may wish to read pp. 215–216 before continuing with this experiment.)

Because polytene chromosomes are so large, their banding pattern is very easy to see in great detail. It is thus possible to detect very small changes in chromosome structure. In the experiment described here, which was conducted in 1936, Bridges investigated changes in chromosome structure associated with the bar and ultra-bar phenotypes.

THE HYPOTHESIS

The rare formation of the ultra-bar phenotype is due to a duplication of the *bar* allele. Information concerning the nature of the bar and ultra-bar phenotypes may be revealed by a cytological examination of polytene chromosomes.

TESTING THE HYPOTHESIS

Starting materials: A strain of flies homozygous for the bar allele and a homozygous normal strain.

1. Within the strain of homozygous bar-eyed flies, identify rare flies that have normal eyes (bar revertants) or ultra-bar eyes.

2. Dissect the salivary glands from the larva of a normal strain, a homozygous bar strain, a bar-revertant strain, and an ultra-bar strain.

3. Prepare the salivary cells for the microscopic examination of the polytene chromosomes. Note: The microscopic examination of chromosomes is described in Chapter 3. It involves gently breaking open the cells, staining the chromosomes with dyes, and squashing the preparation on a microscope slide underneath a coverslip.

4. View the banding patterns of the polytene chromosomes under the microscope.

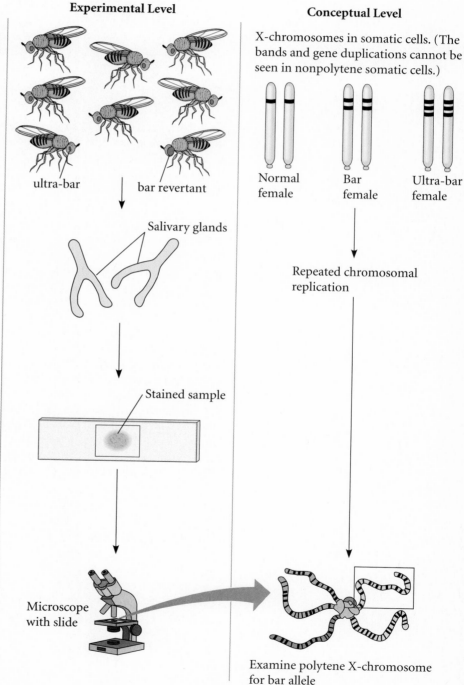

Experimental Level

ultra-bar   bar revertant

Salivary glands

Stained sample

Microscope with slide

Conceptual Level

X-chromosomes in somatic cells. (The bands and gene duplications cannot be seen in nonpolytene somatic cells.)

Normal female   Bar female   Ultra-bar female

Repeated chromosomal replication

Examine polytene X-chromosome for bar allele

THE DATA

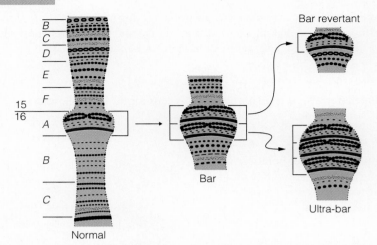

Note: This is a drawing of a short segment of a polytene chromosome that corresponds to the region of the X-chromosome where the *bar* allele is located. This *bar* allele is found within the region designated 16A.

INTERPRETING THE DATA

The cytological examination of region 16A provided direct insight into the nature of the bar and ultra-bar phenotypes. Bridges found that the bar phenotype itself is caused by a duplication of the genetic material in the 16A region of the chromosome. Furthermore, homozygous *bar* stocks can occasionally produce phenotypically normal flies (bar revertants) and ultra-bar flies by altering gene number. The polytene chromosomes of bar revertants showed that the 16A region had returned to the wild-type banding pattern. In other words, the duplicated region was returned to a single copy. By comparison, the opposite situation occurred for the ultra-bar phenotype. In this case, an additional duplication occurred, so that there were three copies of the 16A region. As stated by Bridges, "The Bar-eye reduction [in the number of facets] is thus seen to be interpretable as the effect of increasing the action of certain genes by doubling or triplicating their number."

The mechanism for the formation of the *bar* allele can be explained by unequal crossing over as described previously in Figure 8-5. The formation of *ultra-bar* and *bar-revertant* alleles can likewise be explained by unequal crossing over. As shown here, an unequal crossover between two X-chromosomes carrying the duplicated *bar* allele can result in one chromosome carrying one copy (*bar revertant*) and the other chromosome carrying three copies (*ultra-bar*):

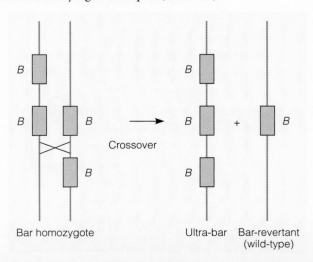

The gene duplication and triplication seen with *bar* and *ultra-bar* alleles are also associated with position effects. A female that is homozygous for the *bar* allele has four copies of this region, and a female that is heterozygous for the *ultra-bar* allele (three copies) and the wild-type allele (one copy) also has four copies. However, a female that is heterozygous for the *ultra-bar* and normal alleles has fewer facets (45) than does a *bar* homozygote (70). This is a position effect—the positioning of three copies next to each other on an X-chromosome increases the severity of the defect. As will be discussed in Chapter 17, position effects are caused by the influences of chromosome structure or genetic regulatory regions on the level of gene expression. In this case, each of the three genes in the *ultra-bar* allele is expressed at a higher level than it is when there are just one or two copies on the X-chromosome.

Duplications provide additional material for gene evolution, sometimes leading to the development of gene families

In contrast to the gene duplication that causes the bar phenotype in *Drosophila*, the majority of small chromosomal duplications have no phenotypic effect. Nevertheless, they are still very important, because they provide raw material for the addition of more genes into a species' chromosomes. Over the course of many generations, this can lead to the formation of a **gene family** consisting of two or more genes that are similar to each other. As shown in Figure 8-6, the members of a gene family are derived from the same ancestral gene. Over time, two copies of an ancestral gene can accumulate different mutations. Therefore, after many generations, the two genes will be similar but not identical. During evolution, this type of event can occur several times, creating a family of many similar genes.

When two or more genes are derived from a single ancestral gene, the genes are said to be **homologous**. Homologous genes make up a gene family. A well-studied example of a gene family is shown in Figure 8-7, which illustrates the evolution of the globin gene family found in humans. The globin genes encode polypeptides that are subunits of proteins that function in oxygen binding. For example, hemoglobin is found in red blood cells; its function is to carry oxygen throughout the body. The globin gene family is composed of fourteen homologous genes that were originally derived from a single ancestral globin gene. According to an evolutionary analysis, the ancestral globin gene duplicated about 800 million years ago. Since that time, additional duplication events and chromosomal rearrangements have occurred to produce the current number of fourteen genes on three different human chromosomes.

Gene families have been important in the evolution of traits. Even though all of the globin polypeptides are subunits of proteins that play a role in oxygen binding,

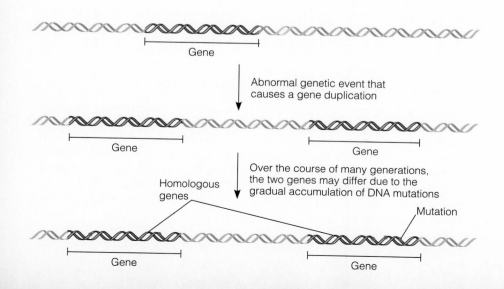

Gene

Abnormal genetic event that
causes a gene duplication

Gene   Gene

Homologous
genes

Over the course of many generations,
the two genes may differ due to the
gradual accumulation of DNA mutations

Mutation

Gene   Gene

FIGURE 8-6

Gene duplication and the evolution of homologous genes. An abnormal crossover event, like the one described in Figure 8-5, leads to a gene duplication. Over time, each gene accumulates different mutations.

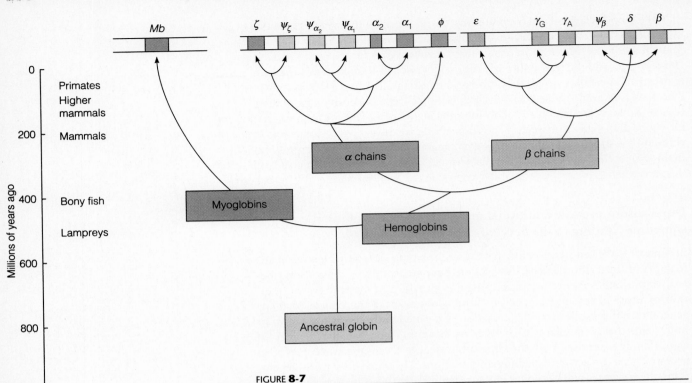

FIGURE **8-7**

The evolution of the globin gene family in humans. The globin gene family evolved from a single ancestral globin gene. The first gene duplication produced two genes that accumulated mutations and became the genes encoding myoglobin and the family of hemoglobins. The hemoglobins evolved from a primordial hemoglobin gene that duplicated to produce the α-chains and β-chains. Further duplications of primordial α-chain and β-chain genes produced several homologous genes on chromosomes 16 and 11, respectively. The four genes shown in gray are nonfunctional pseudogenes.

the accumulation of different mutations in the various family members has created globins that are more specialized in their function. For example, *myoglobin* is better at binding and storing oxygen, whereas the *hemoglobins* are better at binding and transporting oxygen via the red blood cells. Also, these different globin genes are expressed during different stages of human development. The ε- and ζ-globin genes are expressed very early in embryonic life, while the γ-globin gene exhibits its maximal expression during the second and third trimesters of gestation. Following birth, the γ-globin gene is turned off and the β-globin gene is turned on. These differences in the expression of the globin genes reflect the differences in the oxygen transport needs of humans during the embryonic, fetal, and postpartum stages of life.

Inversions often occur without phenotypic consequences

We now turn our attention to changes in chromosome structure that involve a rearrangement in the genetic material. A chromosome with an inversion contains a segment that has been flipped to the opposite orientation. Geneticists classify inversions according to the location of the centromere. If the centromere lies within the inverted region of the chromosome, the inverted region is known as a **pericentric inversion**. Alternatively, if the centromere is found outside the inverted region, the inverted region is called a **paracentric inversion**. The two types of inversions are illustrated in Figure 8-8.

When a chromosome contains an inversion, the total amount of genetic material remains the same as in a normal chromosome. For this reason, the great majority of inversions do not have any phenotypic consequences in the individuals who carry them. In rare cases, however, an inversion can alter the phenotype of an individual. Whether or not this occurs is related to the boundaries of the inverted segment. When an inversion occurs, the chromosome is broken in two places and the ends flip around to produce the inversion. If either breakpoint occurs within a vital gene, the function of the gene is expected to be disrupted, possibly producing a phenotypic effect. For example, some people with *hemophilia* (type A) have inherited an X-linked inversion that has inactivated a gene for Factor VIII (a blood clotting protein).

In other cases, an inversion (or translocation) may reposition a gene on a chromosome in a way that alters its normal level of expression. Like the *bar* gene duplication discussed in Experiment 8A, this is a position effect. In Chapter 17, we will consider an example in which a position effect decreases the expression of a gene in *Drosophila* that influences red versus white eye pigmentation. In Chapter 23, we will examine how a position effect can lead to an abnormal increase in gene expression and cause cancer.

Since inversions seem like an unusual genetic phenomenon, it is perhaps surprising that they are found in human populations in quite significant numbers. About 2% of the human population carry inversions that are detectable with a light microscope. In most cases, these individuals are phenotypically normal and live their lives without knowing they contain this abnormality. In a few cases, however, an individual with an inversion chromosome may produce (one or more) offspring with genetic abnormalities. This event may prompt a physician to request a microscopic examination of the individual's chromosomes. In this way, a phenotypically normal individual may discover that they have a chromosome with an inversion. Next, we will examine why an individual carrying an inversion may produce offspring with phenotypic abnormalities.

Inversion heterozygotes may produce abnormal gametes due to crossing over

An individual who carries one copy of a normal chromosome and one copy of an inverted chromosome is known as an **inversion heterozygote**. Such an individual, though possibly phenotypically normal, may have a high probability of producing abnormal gametes. The frequency of abnormal gametes depends on the size of the inverted segment. An inversion heterozygote with a fairly large inverted segment may produce a sizable number of abnormal gametes (perhaps 1/3 or even higher).

The underlying cause of abnormal gametes is crossing over within the inverted region. During meiosis I of gamete formation, pairs of homologous sister chromatids synapse with each other. Figure 8-9 illustrates how this occurs in an inversion heterozygote. For the normal chromosome and inversion chromosome to synapse properly, a loop, termed an **inversion loop**, must form to permit the homologous genes on both chromosomes to align next to each other despite the inverted sequence. If a crossover occurs within the inversion loop, highly abnormal chromosomes will be produced.

The consequences of this type of crossover depend on whether the inversion is pericentric or paracentric. Figure 8-9a describes a crossover in the inversion loop when one of the homologues has a pericentric inversion. This is a single crossover that involves only two of the four sister chromatids. Following the completion of mitosis, this single crossover yields two abnormal chromosomes. Both of these abnormal chromosomes have a segment that is deleted and a different segment that is duplicated. In this example, one of the abnormal chromosomes is missing genes *A*, *B*, and *C* and has extra copies of genes *H* and *I*. The other abnormal

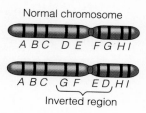

Normal chromosome

A B C   D E   F G H I

A B C   G F   E D   H I
Inverted region

(a) Pericentric inversion

A   E D   C B   F G H I
Inverted region

(b) Paracentric inversion

FIGURE **8-8**

Types of inversions. A pericentric inversion **(a)** includes the centromere, whereas a paracentric inversion **(b)** does not.

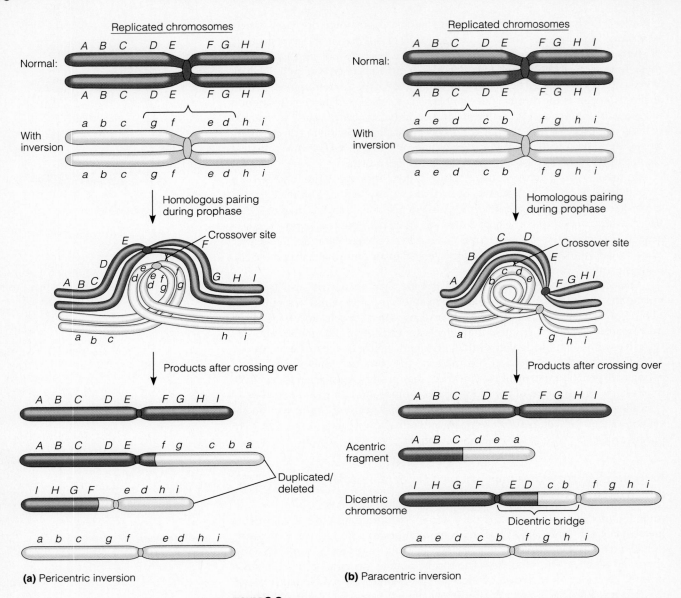

(a) Pericentric inversion

(b) Paracentric inversion

FIGURE **8-9**

The consequences of crossing over in the inversion loop. (a) Crossover within a pericentric inversion. **(b)** Crossover within a paracentric inversion.

chromosome has the opposite situation; it is missing genes *H* and *I* and has an extra copy of genes *A*, *B*, and *C*. If these abnormal chromosomes are passed on to offspring, they may produce phenotypic abnormalities depending on the amount and nature of the duplicated/deleted genetic material.

The outcome of a crossover involving a paracentric inversion is shown in Figure 8-9b. A single crossover has occurred between two homologous sister chromatids; the other two sister chromatids have not participated in a crossover. This single crossover event produces a very strange outcome. One chromosome is produced that contains two centromeres. This is called a **dicentric chromosome**; the region of chromosome connecting the two centromeres is called a **dicentric bridge**. The crossover also produces a piece of chromosome without any centromere. This **acentric** (without a centromere) fragment will be lost and degraded in subsequent cell divisions. The dicentric chromosome is a temporary condition. When the two centromeres try to move toward opposite poles during anaphase, the dicentric bridge will be forced to break at some random location. Therefore, the net result of this crossover is to produce two normal chromosomes (namely, the ones that did not cross over) and two chromosomes that contain deletions. The chromosomes with deletions are the result of the snapping in two of the dicentric chromosome.

Translocations can be balanced or unbalanced; unbalanced translocations usually have detrimental phenotypic effects

Another type of chromosomal rearrangement is a translocation. As mentioned at the beginning of this section, reciprocal translocations involve an exchange of chromosomal fragments between two different chromosomes. As shown in Figure 8-10, a reciprocal translocation can be produced when two nonhomologous chromosomes cross over with each other. This type of rare aberrant event results in a rearrangement of the genetic material, though not a change in the total amount of genetic material. For this reason, a reciprocal translocation is also called a **balanced translocation**. Like inversions, balanced translocations are usually without any phenotypic consequences, because the individual has a normal amount of genetic material. In a few cases, balanced translocations can result in position effects in a fashion similar to inversions.

The inheritance of a simple translocation, in which one fragment of a chromosome is attached to another chromosome, can result in an unbalanced translocation. Unbalanced translocations are generally associated with phenotypic abnormalities or even lethality. An inherited human syndrome known as *familial Down syndrome* provides an example of this phenomenon. In this condition, a large segment of chromosome 21 is translocated to chromosome 14. In addition, this individual already has two normal copies of chromosome 21. This results in an imbalance in genetic material, since the individual has three copies of the genes that are found on a large segment of chromosome 21. As will be discussed later in this chapter, imbalances among many genes often lead to phenotypic anomalies. In familial Down syndrome (Figure 8-11), the person exhibits characteristics very similar to those of an individual with the more common form of Down syndrome, which is due to three entire copies of chromosome 21. This common form of Down syndrome will be described later in this chapter.

Individuals with balanced translocations may produce abnormal gametes due to the segregation of chromosomes

Individuals who carry balanced translocations have a greater risk of producing gametes with unbalanced combinations of chromosomes. Whether or not this occurs depends on the segregation pattern during meiosis I. This idea is illustrated in Figure 8-12. In this example, the parent carries a balanced, reciprocal translocation and is likely to be phenotypically normal. During gamete formation, the homologous chromosomes attempt to synapse with each other. Because of the translocations, the pairing of homologous regions leads to the formation of an unusual structure that contains four pairs of sister chromatids, termed a **translocation cross**.

To understand the segregation of translocated chromosomes, pay close attention to the centromeres, which are numbered in the figure. During anaphase I of meiosis, each daughter cell will receive one of the two pairs of sister chromatids. For these translocated chromosomes, the expected segregation is due to the segregation of the centromeres; each daughter should receive a centromere located on chromosome 1 and a centromere located on chromosome 2. This can occur in two ways. One possibility is **alternate segregation**. As shown in Figure 8-12a, this occurs when the chromosomes on opposite sides of the translocation cross segregate into

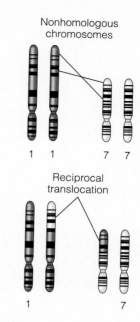

FIGURE **8-10**
Formation of a reciprocal translocation. A crossover has occurred between chromosome 1 and chromosome 7. This crossover yields two chromosomes that carry translocations.

FIGURE **8-11**
An individual affected with familial Down syndrome. This individual carries a simple translocation in which a large segment of chromosome 21 has been translocated to chromosome 14. In addition, the individual also carries two normal copies of chromosome 21.

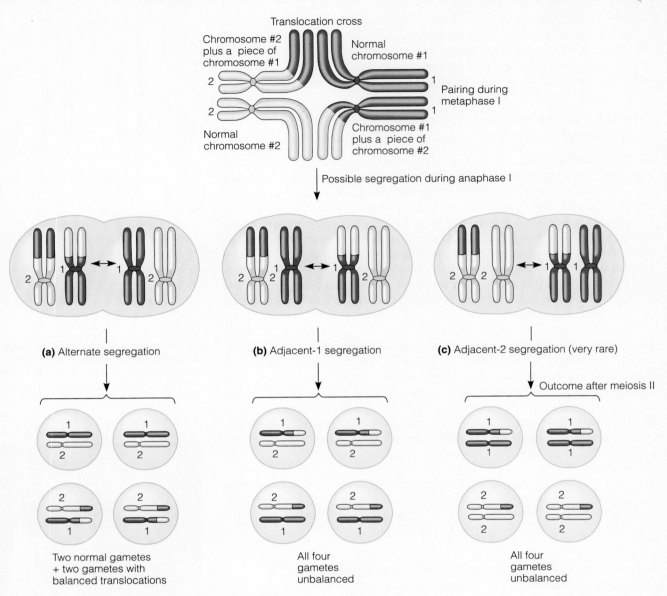

FIGURE **8-12**

Meiotic segregation of a reciprocal translocation. Follow the numbered centromeres through each process.

the same cell. One daughter cell receives two normal chromosomes, and the other cell gets two translocated chromosomes. Following meiosis II, four gametes are produced: two are normal and two have reciprocal (balanced) translocations.

Another possible segregation pattern is called **adjacent-1 segregation** (Figure 8-12b). This occurs when adjacent chromosomes (one of each type of centromere) segregate into the same cell. When this occurs, each daughter cell receives one normal chromosome and one translocated chromosome. This produces four gametes, all of which are genetically unbalanced because part of one chromosome has been deleted and part of another has been duplicated. If the gametes produced from adjacent-1 segregation unite with a normal gamete, the resulting zygote is expected to be abnormal genetically, and possibly phenotypically.

Alternate and adjacent-1 segregation are the likely outcomes when an individual carries a reciprocal translocation. Depending on the sizes of the translocated segments, both types may be equally likely to occur. In many cases, the gametes from adjacent-1 segregation cannot produce viable offspring, thereby lowering the fertility of the parent. On very rare occasions, **adjacent-2 segregation** can occur (Figure 8-12c). In this case, the centromeres do not segregate as they should. One daughter cell has received both copies of the centromere on chromosome 1, the other both copies of the centromere on chromosome 2. This rare segregation pattern also yields four abnormal gametes that contain an unbalanced combination of chromosomes.

VARIATION IN CHROMOSOME NUMBER

As we have seen in the preceding section, chromosome structure can be altered in a variety of ways. The total number of chromosomes, too, can vary. Eukaryotic species typically contain several chromosomes that are inherited as one or more sets. Variations in chromosome number can be categorized in two ways: variation in the number of sets of chromosomes, and variation in the number of particular chromosomes within a set.

Organisms that are **euploid** have a chromosome number that is an exact multiple of a chromosome set. In *Drosophila melanogaster*, a normal individual possesses eight chromosomes. The species is diploid, having two sets of four chromosomes each (Figure 8-13a). We can also recognize that a normal fruit fly is euploid because eight chromosomes divided by four chromosomes per set equals two exact sets. On rare occasions, an abnormal fruit fly can be produced with twelve chromosomes, containing three sets of four chromosomes each. This alteration in euploidy produces a **triploid** fruit fly. Such a fly is also euploid, since it contains exactly three sets of chromosomes. Organisms with three or more sets of chromosomes are also called **polyploid**. Geneticists use the letter *n* to represent a

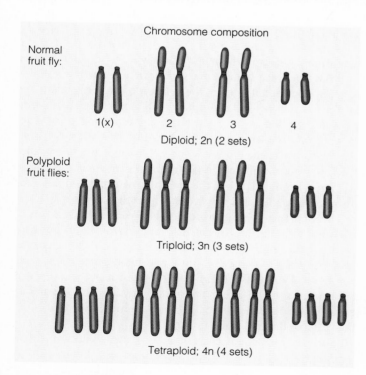

(a) Variations in euploidy

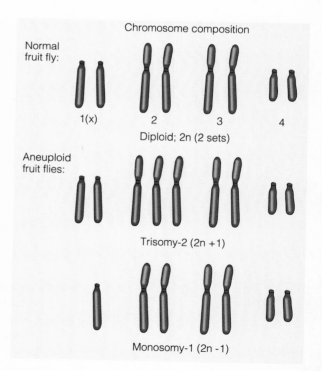

(b) Variations in aneuploidy

FIGURE **8-13**

Types of variation in chromosome number.

set of chromosomes. A diploid organism is referred to as $2n$, a triploid organism as $3n$, a **tetraploid** organism as $4n$, and so forth.

A second way in which chromosome number can vary is by **aneuploidy** (literally, not euploidy). This term refers to a variation that involves an alteration in the number of particular chromosomes, so that the total number of chromosomes is not an exact multiple of a set (not an exact multiple of n). For example, an abnormal fruit fly could contain nine chromosomes instead of eight because it had three copies of chromosome 2 instead of the normal two copies (Figure 8-13b). Such an animal is said to have *trisomy-2* or to be trisomic. Instead of being perfectly diploid (i.e., $2n$), a trisomic animal is $2n + 1$. By comparison, a fruit fly could be lacking a copy of chromosome 1 and contain a total of seven chromosomes ($2n - 1$). This animal would have *monosomy-1*.

In this section, we will examine euploid and aneuploid variation in many eukaryotic species. We will learn that euploid variation among different species occurs occasionally in animals and quite frequently in plants. We will see how natural variation in the number of sets of chromosomes affects phenotypic variation. By comparison, aneuploidy is generally regarded as an abnormal condition. We will consider several examples in which aneuploidy has a negative impact on an organism's phenotype.

Aneuploidy causes an imbalance in gene expression that is often detrimental to the phenotype of the individual

The phenotype of every organism is influenced by thousands of different genes. Humans, for example, possess approximately 100,000 different genes. Many genes are only expressed in certain cell types or during particular stages of development. To produce a normal individual, intricate coordination has to occur in the expression of thousands of genes. In the case of humans and many other species, this process works correctly when there are two copies of each chromosome per cell. In other words, when a human is diploid, the balance of gene expression among many different genes usually produces a human with a normal phenotype.

Aneuploidy commonly causes an abnormal phenotype. To understand why, consider the relationship between gene expression and chromosome number (Figure 8-14). For many (but not all) genes, the amount of expression is correlated with the number of genes per cell. Compared with a diploid cell, if a gene is carried on a chromosome that is present in three copies instead of two, approximately 150% of the normal amount of gene product will be made. Alternatively, if only one copy of that gene is present due to a monosomy abnormality, only 50% of the gene product will be made. Therefore, in trisomic and monosomic individuals, there is an imbalance in the level of gene expression between the majority of chromosomes found in pairs versus those that are not.

At first glance, the difference in gene expression between euploid and aneuploid individuals may not seem terribly drastic. Keep in mind, however, that a eukaryotic chromosome carries hundreds or even thousands of different genes. Therefore, when an organism is trisomic or monosomic, many gene products will occur in excessive or deficient amounts. This imbalance among many genes appears to underlie the abnormal phenotypic effects that aneuploidy frequently causes. In most cases, these effects are detrimental and produce an individual that is less likely to survive than a euploid individual.

The first person to recognize the harmful effects of aneuploidy was Albert Blakeslee at Cold Spring Harbor. In the 1920s, Blakeslee and his colleagues identified many examples of alterations in chromosome number in the Jimson weed (*Datura stramonium*). Figure 8-15 shows the differences in the capsule morphology of the normal and trisomic strains. (The capsule is the dry fruit of the Jimson weed.) All 12 trisomies had capsules that were morphologically different from those of the normal diploid strain. These aneuploid plants also had many other morphologically

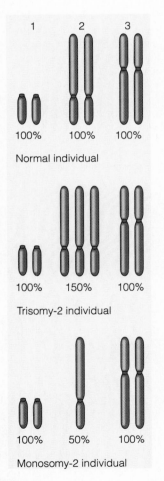

FIGURE 8-14

Imbalance of gene products in trisomic and monosomic individuals. Aneuploidy of chromosome 2 (i.e., trisomy and monosomy) leads to an imbalance in the amount of genes products from chromosome 2 compared with the amounts from chromosomes 1 and 3.

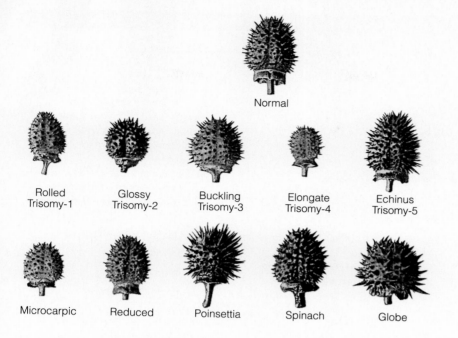

Normal

Rolled
Trisomy-1

Glossy
Trisomy-2

Buckling
Trisomy-3

Elongate
Trisomy-4

Echinus
Trisomy-5

Microcarpic

Reduced

Poinsettia

Spinach

Globe

FIGURE **8-15**

The effects of trisomy on the phenotype of the capsules of the *Datura* plant.

GENES ⟶ TRAITS: As described in Figure 8-14, trisomy leads to an imbalance in the copy number of genes on the trisomic chromosomes (here, 150% of normal) versus those on the other chromosomes (100%). This drawing compares the morphology of the capsule of the *Datura* plant in normal individuals and those carrying trisomy of each of the 12 types of chromosomes in this species. As seen here, trisomy of each chromosome causes changes in the phenotype of the capsule. This is due to changes in the balance of gene expression.

distinguishable traits, including changes in leaf shape, size, and so forth. In many cases, the observed changes in chromosome number produced detrimental traits. For example, Blakeslee noted that the cocklebur plant (trisomy-6) is "weak and lopping with the leaves narrow and twisted."

Aneuploidy in humans causes abnormal phenotypes

One important reason that geneticists are so interested in aneuploidy is its relationship to certain inherited disorders in humans. Even though most people are born with a normal number of chromosomes (i.e., 46), alterations in chromosome number occur fairly frequently during gamete formation. About 5–10% of all fertilized human eggs result in an embryo with an abnormality in chromosome number. In most cases, these abnormal embryos do not develop properly and result in a spontaneous abortion very early in pregnancy. Approximately 50% of all spontaneous abortions are due to alterations in chromosome number.

In some cases, an abnormality in chromosome number produces an offspring that can survive. Several human disorders involve abnormalities in chromosome number. The most common are trisomies of chromosomes 21, 18, or 13, or abnormalities in the number of the sex chromosomes (Table 8-1). Most of the known trisomies involve chromosomes that are relatively small (i.e., chromosomes 21, 18, and 13; refer back to Figure 8-1). Trisomies of the other human autosomes and monosomies of the autosomes usually produce a lethal phenotype that causes early spontaneous abortion.

Variation in the number of X-chromosomes, unlike the case for most other large chromosomes, is often nonlethal. The survival of trisomy-X individuals is explained by X-inactivation (see Chapter 7). In an individual with more than one X-chromosome, all additional X-chromosomes are converted to Barr bodies in the somatic cells of adult tissues. In an individual with trisomy-X, for example, two out of the three X-chromosomes are converted to inactive Barr bodies. Unlike the autosomes, the normal level of expression for X-linked genes is from a single X-chromosome. In other words, the correct level of mammalian gene expression results from two copies of each autosomal gene and one copy of each X-linked gene. This explains how the expression of X-linked genes in males (XY) can be maintained at the same levels as in females (XX). It may also explain why

TABLE 8–1 Aneuploid Conditions in Humans

Condition	Frequency	Syndrome	Characteristics
Autosomal			
Trisomy-21	1/800	Down	Mental retardation, abnormal pattern of palm creases, slanted eyes, flattened face, short stature
Trisomy-18	1/6,000	Edward	Mental and physical retardation, facial abnormalities, extreme muscle tone
Trisomy-13	1/15,000	Patau	Mental and physical retardation, wide variety of defects in organs, large triangular nose
Sex chromosomal			
XXY	1/1,000 (males)	Klinefelter	Sexual immaturity, no sperm, breast swelling
XYY	1/1,000 (males)	Jacobs	Tall
XXX	1/1,500 (females)	Superfemale	Tall and thin, menstrual irregularity
X0	1/5,000 (females)	Turner	Short stature, webbed neck, sexually undeveloped

monosomy-X and trisomy-X are not lethal conditions. The phenotypic effects noted in Table 8-1 involving sex chromosomal abnormalities may be due to the expression of X-linked genes prior to embryonic X-inactivation, or due to the imbalance in the expression of pseudoautosomal genes.

Some human abnormalities in chromosome number are influenced by the age of the parents. Older parents are more likely to produce children with abnormalities in chromosome number. For example, the incidence of Down syndrome rises with the age of the parents, particularly that of the mother (Figure 8-16). This syndrome was first described by the English physician John Langdon Down in 1866. The association between maternal age and Down syndrome was later discovered by L. S. Penrose in 1933, even before the chromosomal basis for the disorder was identified by the French scientist Jerome Lejeune in 1959. Down syndrome is most commonly caused by nondisjunction at meiosis I in the oocyte. The mechanism of nondisjunction will be described later in this chapter.

Different theories have been proposed to explain the relationship between maternal age and Down syndrome. One popular idea suggests that it may due to

FIGURE **8-16**

The incidence of Down syndrome births according to the age of the mother.

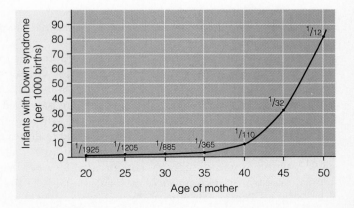

the age of the oocytes. Human primary oocytes are produced within the ovary of the female fetus prior to birth and are arrested at prophase I of meiosis I until the time of ovulation. Therefore, as a women ages, her primary oocytes have been in prophase I for a progressively longer period of time. This added length of time may contribute to an increased frequency of nondisjunction. About 5% of the time, Down syndrome is due to an extra paternal chromosome. Prenatal tests can determine if a fetus has Down syndrome or some other genetic abnormalities. The topic of fetal testing will be discussed in Chapter 23.

Variations in euploidy occur naturally in a few animal species

We now turn our attention to changes in the number of sets of chromosomes, referred to as variations in euploidy. Most species of animals are diploid. In many cases, changes in euploidy are not well tolerated. For example, polyploidy in mammals is generally a lethal condition. In less complex vertebrates and invertebrates, however, a few examples of naturally occurring variations in euploidy occur. Male bees (drones) contain a single set of chromosomes; they are produced from unfertilized eggs. By comparison, female bees are diploid. Geneticists often refer to male bees as being "haploid," although in a strict genetic sense, the term haploid describes gametes that contain half the genetic material of the parent's somatic cells. An organism with a single set of chromosomes within its somatic cells is more accurately called **monoploid.**

A few examples of polyploid animals have been discovered. In some cases, animals that are morphologically very similar to each other can be found as a diploid species as well as a separate polyploid species. This situation occurs among certain amphibians and reptiles. Figure 8-17 presents photographs of a diploid and a tetraploid frog. As you can see, they look fairly similar to each other. Their differences in euploidy can only be revealed by an examination of the chromosome number in the somatic cells of the animals.

Variations in euploidy can occur in certain tissues within an animal; polytene chromosomes are an extreme example of repeated chromosome doubling

Thus far, we have considered variations in chromosome number that occur at fertilization, so that all the somatic cells of an individual contain this variation. In many animals, certain tissues of the body will display normal variations in the number of sets of chromosomes. In particular, diploid animals sometimes produce tissues that are polyploid. The cells of the human liver, for example, can vary to a great degree in their ploidy. Liver cells can be triploid, tetraploid, and even octaploid ($8n$). This phenomenon is known as **endopolyploidy.** The biological significance of endopolyploidy is not completely understood. One possibility is that the increase in chromosome number in certain cells may enhance their ability to produce specific gene products that are needed in great abundance.

An unusual example of natural variation in the ploidy of somatic cells occurs in *Drosophila* and a few other insects. Within certain tissues, such as the salivary glands, the chromosomes undergo repeated rounds of chromosome replication without cellular division. For example, in the salivary gland cells of *Drosophila*, the pairs of chromosomes double approximately nine times ($2^9 = 512$). Figure 8-18a illustrates how repeated rounds of chromosomal replication produce a bundle of chromosomes that lie together in a parallel fashion. This bundle of many chromatids lying side-by-side is a polytene chromosome. They were first observed by E. G. Balbiani in 1881. In the early 1930s, Theophilus Painter and his colleagues at

(a) Hyla chrysocelis

(b) Hyla versicolor

FIGURE **8-17**

Differences in euploidy in two closely related frog species. The frog in **(a)** is diploid, whereas the frog in **(b)** is tetraploid.

GENES → TRAITS: A doubling of the chromosomal composition from diploid to tetraploid only has a minor effect on the phenotypes of these two species. At the level of gene expression, this observation suggests that the copy number of each gene (two versus four) does not critically affect the phenotype of these two species.

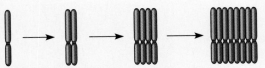

(a) Repeated chromosome replication produces polytene chromosome

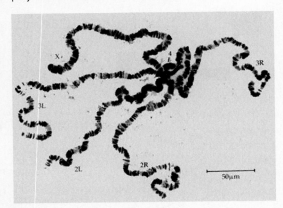

(b) A polytene chromosome

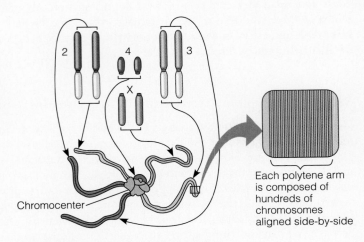

Chromocenter

Each polytene arm is composed of hundreds of chromosomes aligned side-by-side

(c) Composition polytene chromosome from regular *Drosophila* chromosomes

FIGURE **8-18**

Polytene chromosomes in *Drosophila*. (a) A schematic illustration of the production of polytene chromosomes. Several rounds of repeated replication result in a bundle of sister chromatids that lie side by side. Both homologues also lie next to each other. This replication does not occur in highly condensed, heterochromatic DNA near the centromere. **(b)** A photograph of a polytene chromosome. **(c)** This drawing shows the relationship between the four pairs of chromosomes and the formation of a polytene chromosome in the salivary gland. The heterochromatic regions of the chromosomes aggregate at the chromocenter, and the arms of the chromosomes project outward.

the University of Texas recognized that the size and morphology of polytene chromosomes provided geneticists with unique opportunities to study chromosome structure and gene organization.

Figure 8-18b shows a photograph of a polytene chromosome observed under a light microscope. Prior to the formation of polytene chromosomes, *Drosophila* cells contain eight chromosomes (two sets of four chromosomes each). In the salivary gland cells, the homologous chromosomes synapse with each other and replicate to form a polytene structure. During this process, the four types of chromosomes aggregate to form a single structure with several polytene arms. The central point where the chromosomes aggregate is known as the **chromocenter**. Each of the four types of chromosome are attached to the chromocenter near their centromeres. The X and Y and chromosome 4 are telocentric, and chromosomes 2 and 3 are metacentric. Therefore, chromosomes 2 and 3 have two arms that radiate from the chromocenter while the X and Y and chromosome 4 have a single arm projecting from the chromocenter (see Figure 8-18c).

Because of their considerable size, polytene chromosomes lend themselves to an easy microscopic examination. Ordinarily, we use light microscopy to visualize the highly condensed metaphase chromosomes seen during mitosis or meiosis (as in Figure 8-1). Since polytene chromosomes are so large, we can even see them during interphase, when normal chromosomes are not visible. In fact, a polytene chromosome during interphase is actually 100–200 times as large as the average metaphase chromosome. As shown in Figure 8-18b, polytene chromosomes exhibit a characteristic banding pattern. Each dark band is known as a **chromomere**. The structure of the genetic material within a dark band is more compact than in the interband region. More than 95% of the DNA is found within the chromomeres. The banding patterns of polytene chromosomes are much more detailed than those observed in metaphase chromosomes. Cytogeneticists have identified approximately 5000 bands along polytene chromosomes. At one time, each chromomere was thought to correspond to one gene. However, this idea was found to be incorrect.

Polytene chromosomes have allowed geneticists to study the organization and functioning of interphase chromosomes in great detail. When a gene is deleted or duplicated in a mutation, researchers can map the change by observing abnormalities in the structure of polytene chromosomes. Experiment 8A presented an example of this approach. In addition, since polytene chromosomes can be observed during interphase, the expression of particular genes can be correlated with changes in the compaction of certain bands in the polytene chromosome. The phenomenon of polytene chromosome "puffing" and its relationship to gene expression will be discussed in Chapter 12.

Variations in euploidy are common in plants; this variation has allowed the development of many agricultural crops

We now turn our attention to the variations of euploidy that occur in plants. In contrast to animals, plants commonly exhibit polyploidy. Among ferns and flowering plants, about 30–35% of the species are polyploid. Polyploidy is also important in agriculture. Many of the fruits and grains we eat are produced from polyploid plants. For example, the species of wheat that we use to make bread, *Triticum aestivum*, is hexaploid (6*n*). In many instances, polyploid strains of plants display outstanding agricultural characteristics. They are often larger in size and more robust. These traits are clearly advantageous in the production of food. In addition, polyploids tend to exhibit a greater adaptability, which allows them to withstand harsher environmental conditions. Also, polyploid ornamental plants often produce larger flowers than their diploid counterparts (Figure 8-19).

Polyploids having an odd number of chromosome sets, such as triploids or pentaploids, are usually sterile. The sterility arises because they produce highly aneuploid gametes. To understand why aneuploidy occurs, consider what happens during anaphase I in a triploid organism (Figure 8-20). Since there are three copies of each replicated chromosome, they cannot be divided equally between two daughter cells. For each type of chromosome, a daughter cell randomly gets one or two copies. For example, one daughter cell might receive one copy of chromosome 1, two copies of chromosome 2, two copies of chromosome 3, one copy of chromosome 4, and so forth. For a triploid species containing many different chromosomes in a set, it is very unlikely that any gamete will be euploid. If we assume that a daughter cell will receive either one copy *or* two copies of each kind of chromosome, the probability that a gamete will end up being perfectly haploid or diploid is $(1/2)^{N-1}$, where N is the number of chromosomes in a set. As an example, in a triploid species containing 20 chromosomes per set, the probability of producing a haploid or diploid gamete is 1 in 524,288. Thus, a gamete is almost certain to contain one copy of some chromosomes and two copies of the other chromosomes. This high probability of aneuploidy underlies the reason for triploid sterility.

Sterility can be an agriculturally useful trait, since it may result in a seedless fruit. For example, seedless watermelons and bananas are triploid varieties. The domestic banana, which is triploid, was originally derived from a normal diploid species and has been asexually propagated by humans via cuttings. The small black spots in the center of a domestic banana are degenerate seeds. In the case of

(a) Diploid

(b) Tetraploid

FIGURE **8-19**

Differences in euploidy in two closely related petunia species. The flower in **(a)** is diploid, whereas the one in **(b)** is tetraploid.

GENES ⟶ TRAITS: The doubling of chromosome number from diploid to tetraploid affects the phenotype of the individual. In the case of many plant species, such as the petunia, a tetraploid individual is larger and more robust than its diploid counterpart. This suggests that having four copies of each gene is somewhat better than having two copies of each gene. This phenomenon in plants is rather different than the situation in animals. Tetraploidy in animals may have little effect (as in Figure 8-17), but it is more common for polyploidy in animals to be detrimental. Geneticists do not understand why polyploidy is often beneficial in plants but not in animals.

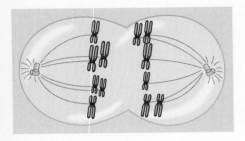

FIGURE 8-20

Schematic representation of anaphase I in a triploid organism containing three sets of four chromosomes each. In this example, the homologous chromosomes (three each) do not evenly separate during anaphase. Each cell receives one copy of some chromosomes and two copies of other chromosomes. This produces aneuploid gametes.

flowers, the seedless phenotype can also be beneficial. Seed producers such as Burpee are developing triploid varieties of flowering plants. The triploid marigold is sterile and unable to set seed. According to Burpee, "they bloom and bloom, unweakened by seed bearing."

NATURAL AND EXPERIMENTAL WAYS TO PRODUCE VARIATIONS IN CHROMOSOME NUMBER

As we have seen, variations in chromosome number are fairly widespread and usually have a significant impact on the phenotypes of plants and animals. For these reasons, researchers have wanted to understand the cellular mechanisms that cause variations in chromosome number. In some cases, a change in chromosome number is the result of nondisjunction. The term **nondisjunction** refers to the event in which the chromosomes do not separate properly during anaphase. As we will see, it may be caused by an improper separation of homologous pairs in a tetrad, or a failure of the centromere to split during meiosis II or mitosis.

Meiotic nondisjunction can produce gametes that have too many or too few chromosomes. If such a gamete fuses with a normal gamete during fertilization, the resulting individual will have an abnormal chromosomal composition in all of the cells of the organism. An abnormal nondisjunction event also may occur after fertilization in one of the somatic cells of the body. This second mechanism is known as **mitotic nondisjunction**. When this occurs during embryonic stages of development, it may lead to a patch of tissue in the organism that has an abnormal chromosomal compostion.

Finally, a third common way in which the chromosome composition of an organism can vary is by **interspecies matings** to produce an alloploid organism. An **alloploid** organism contains sets of chromosomes from two (or more) different species. This term refers to the occurrence of chromosome sets (ploidy) from different (*allo-*) species.

In this section, we will examine these three mechanisms in greater detail. Also, in the past few decades, researchers have devised several methods to manipulate chromosome number in experimentally and agriculturally important species. As we will learn, the experimental manipulation of chromosome number has had an important impact on genetic research and agriculture.

Meiotic nondisjunction can produce aneuploidy; complete nondisjunction can produce polyploidy

The phenomenon of meiotic nondisjunction was first described by Calvin Bridges, who compared inheritance patterns in *Drosophila* with the cytological presence of certain chromosomes (see Chapter 3). As shown in Figure 8-21, a nondisjunction event can occur during meiosis I or meiosis II. If it occurs during anaphase of meiosis I (Figure 8-21a), an entire tetrad will migrate into one of the two daughter cells. All the gametes produced from this event will be abnormal. A second possibility is that nondisjunction can occur during anaphase of meiosis II (Figure 8-21b). If this occurs, the net result will be two normal and two abnormal gametes. In this scenario, the individual produced from an abnormal gamete will be aneuploid (namely, either monosomic or trisomic for one chromosome).

In rare cases, all the chromosomes can undergo nondisjunction and migrate to one of the daughter cells. The net result of *complete nondisjunction* is a diploid gamete and a gamete without chromosomes. While the gamete without chromosomes is nonviable, the diploid gamete might participate in fertilization with a normal haploid gamete to produce a triploid individual. Therefore, complete nondisjunction can produce individuals that are polyploid.

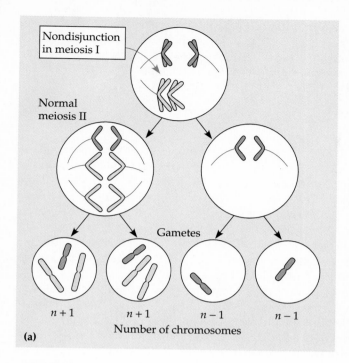

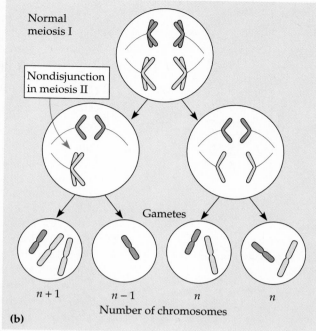

FIGURE **8-21**

Nondisjunction during meiosis I and II. The pair of chromosomes shown in purple are behaving properly during meiosis I and II, so that each gamete receives one copy of this chromosome. The pair of chromosomes shown in yellow are not disjoining correctly. In **(a)**, nondisjunction occurred in meiosis I, so that a gamete receives either two copies of the orange chromosome or zero copies. In **(b)**, nondisjunction occurred during meiosis II, so that one gamete has two yellow chromosomes and another gamete has zero. The remaining two gametes are normal.

Mitotic nondisjunction or chromosome loss can produce a patch of tissue with an altered chromosome number

Abnormalities in chromosome number occasionally occur after fertilization takes place. In this case, the abnormal event happens during mitosis rather than meiosis. One possibility is that the sister chromatids could separate improperly, so that one daughter cell would have three copies of that chromosome while the other daughter cell would only have one (Figure 8-22a). Alternatively, the sister chromatids could separate during anaphase of mitosis but one of the chromosomes could be improperly attached to the spindle, so that it would not migrate to a pole (Figure 8-22b). If this happens, a chromosome will be degraded if it is left outside of the nucleus when the nuclear membrane re-forms. In this case, one of the daughter cells would have two copies of that chromosome, while the other would have only one.

When genetic abnormalities occur after fertilization, part of the organism will contain cells that are genetically different from the rest of the organism. This condition is referred to as **mosaicism**. The size and location of the mosaic region will depend on the timing and location of the original abnormal event. If a genetic alteration happens very early in the embryonic development of an organism, the abnormal cell will be the precursor for a large section of the organism. In the most extreme case, an abnormality could take place at the first mitotic division. As an example, consider a fertilized *Drosophila* egg that is XX. One of the X-chromosomes may be lost during the first mitotic division, producing one daughter cell that is X0

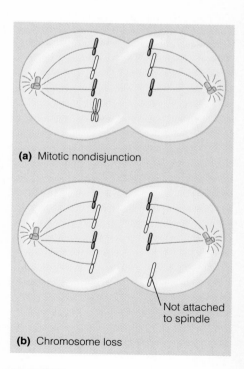

(a) Mitotic nondisjunction

(b) Chromosome loss

Not attached to spindle

FIGURE **8-22**

Nondisjunction and chromosome loss during mitosis in somatic cells. (a) Mitotic nondisjunction produces a trisomic and a monosomic daughter cell. **(b)** Chromosome loss produces a normal and a monosomic daughter cell.

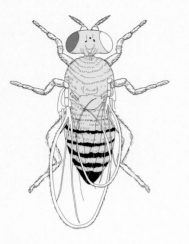

FIGURE **8-23**

A bilateral gynandromorph of
Drosophila melanogaster.

GENES ⟶ TRAITS: In *Drosophila*, the ratio
between genes on the X-chromosome and
genes on the autosomes determines gender. This fly began as an XX female. One
X-chromosome carried the recessive white-
eye and miniature wing alleles, while the
other X-chromosome carried the wild-type
alleles. The X-chromosome carrying the wild-
type alleles was lost from one of the cells
during the first mitotic division, producing
one XX cell and one X0 cell. The XX cell
became the precursor for one side of the fly,
which developed as female because the ratio
of genes on the X-chromosomes to the ratio
of genes on the autosomes is 1. The X0 cell
became the precursor for the other side of
the fly, which developed as male, with a
white eye and a miniature wing.

and one that is XX. As described in Chapter 3, XX flies develop into females while
X0 flies develop into males. Therefore, in this example, one-half of the organism
will become male and one-half will become female. This peculiar and rare individual is referred to as a *bilateral gynandromorph* (Figure 8-23).

Changes in euploidy can occur by autopolyploidy, alloploidy, and allopolyploidy

Different mechanisms account for changes in the number of chromosome sets among
natural populations of plants and animals (Figure 8-24). As mentioned earlier, complete nondisjunction can produce an individual with one or more extra sets of chromosomes. This individual is known as an **autopolyploid** (Figure 8-24a). The terms
auto- (meaning self) and polyploid (meaning many sets of chromosomes) refer to the
increase in the number of chromosome sets occurring within a single species.

A much more common mechanism for change in chromosome number,
called **alloploidy**, is a result of interspecies matings (Figure 8-24b). This event is
most likely to occur between species that are close evolutionary relatives. For example, closely related species of grasses may interbreed to produce alloploids. As
shown in Figure 8-24c, the **allopolyploid** contains a combination of both autopolyploidy and alloploidy. In this case, the allopolyploid contains two complete
sets of chromosomes from two different species for a total of four sets. As mentioned earlier in this chapter, domesticated wheat is a hexaploid species. It is actually an allopolyploid that contains two sets of chromosomes from three
different, but closely related, wild species.

Experimental treatments can promote polyploidy

Because polyploid plants often exhibit desirable traits, the development of polyploids is of considerable interest among plant breeders. Experimental studies on the
ability of environmental agents to promote polyploidy began in the early 1900s.
Since that time, various agents have been shown to promote nondisjunction and
thereby lead to polyploidy. These include abrupt temperature changes during the
initial stages of seedling growth, and the treatment of plants with chemical agents
that interfere with the formation of the spindle apparatus.

The drug *colchicine* is commonly used to promote polyploidy. Once inside
the cell, colchicine binds to tubulin (a protein found in the spindle apparatus)
and thereby interferes with normal chromosome segregation during mitosis or
meiosis. In 1937, Alfred Blakeslee and Amos Avery applied colchicine to plant tissue and, at high doses, were able to cause complete mitotic nondisjunction. As
shown in Figure 8-25, this strategy can be used to produce new polyploid strains
of plants. Colchicine can be applied to seeds, young embryos, or rapidly growing
regions of a plant. This application may produce aneuploidy (usually an undesirable outcome), but it often produces polyploid cells, which may grow faster than
the surrounding diploid tissue. If the parent plant is diploid, colchicine may cause
complete mitotic disjunction, yielding tetraploid ($4n$) cells. As the tetraploid cells
continue to divide, they generate a portion of the plant that is often morphologically distinguishable from the remainder. For example, a polyploid stem may
have a larger diameter and produce larger leaves and flowers. Since individual
plants can be propagated asexually from pieces of plant tissue (i.e., cuttings), the
polyploid portion of the plant can be removed, treated with the proper growth
hormones, and grown as a separate plant. Alternatively, the polyploid region of a
plant may have flowers that produce polyploid seeds; a tetraploid flower will produce diploid pollen and eggs, which will combine to produce tetraploid offspring.
In this way, the use of colchicine provides a straightforward method to produce
autopolyploid strains of plants.

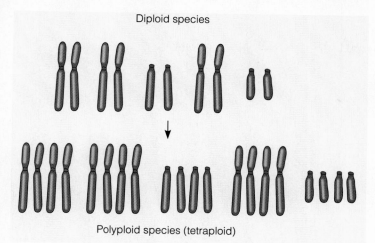

Diploid species

Polyploid species (tetraploid)

(a) Autopolyploidy

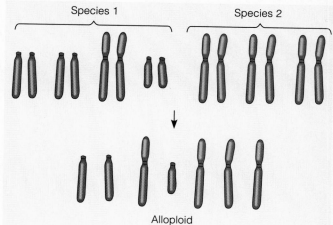

Species 1   Species 2

Alloploid

(b) Alloploidy

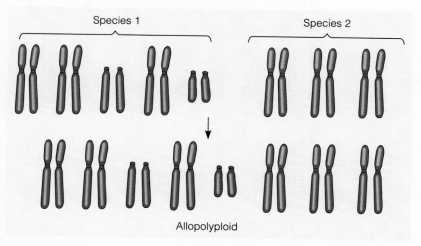

Species 1   Species 2

Allopolyploid

(c) Allopolyploidy

FIGURE **8-24**

A comparison of autopolyploidy, alloploidy, and allopolyploidy.

Alloploids are often sterile, but allopolyploids are more likely to be fertile

Geneticists are interested in the production of alloploids and allopolyploids as ways to generate interspecies hybrids with desirable traits. For example, if one species of grass can withstand hot temperatures and a closely related species is adapted to survive through cold winters, a plant breeder may hope to produce an interspecies hybrid that combines both qualities (i.e., good growth in the heat and survival through the winter). Such an alloploid would be desirable in climates with both hot summers and cold winters.

An important determinant of success in producing an interspecies hybrid is the degree of similarity of the different species' chromosomes. In two very closely related species, the number of chromosomes might be identical or very similar. Figure 8-26 shows a karyotype of an interspecies hybrid between the roan antelope (*Hippotragus equinus*) and the sable antelope (*H. niger*). As seen here, these two closely related species have the same number of chromosomes. Moreover, the sizes and banding patterns of the chromosomes show that they correspond to one

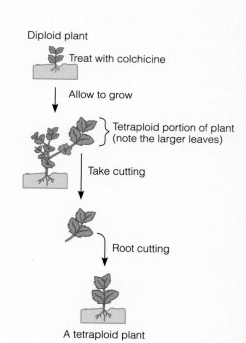

Diploid plant

Treat with colchicine

Allow to grow

Tetraploid portion of plant (note the larger leaves)

Take cutting

Root cutting

A tetraploid plant

FIGURE **8-25**

Use of colchicine to promote polyploidy in plants. Colchicine interferes with the mitotic spindle apparatus and promotes nondisjunction. If complete nondisjunction occurs in a diploid cell, one daughter cell will become tetraploid and the other cell will not receive any chromosomes (and die). The tetraploid cell may continue to divide and produce a segment of the plant with more robust characteristics. This segment may be cut from the rest of the plant and rooted. In this way, a tetraploid plant can be obtained.

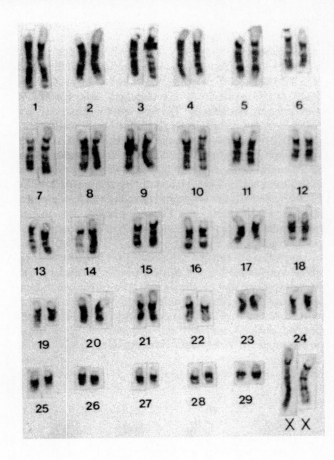

FIGURE **8-26**

The karyotype of a hybrid animal produced from two closely related antelope species. In each chromosome pair in this karyotype, one chromosome was inherited from the roan antelope and the other from the sable antelope. For chromosomes with slightly different banding patterns between the two species, the roan chromosomes are shown on the left side of each pair.

another. For example, chromosome 1 from both species is fairly large and has very similar banding patterns. This chromosome derived from both species likely carries many of the same genes. Analogous chromosomes from related species are called **homeologous** chromosomes (not to be confused with homologous).

The degree of similarity between the homeologous chromosomes also greatly influences whether an alloploid is fertile. This phenomenon was first recognized by the Russian cytogeneticist G. D. Karpechenko in 1928. He crossed a radish (*Raphanus*) and a cabbage (*Brassica*), both of which are diploid and contain 18 chromosomes. Each of these organisms produces haploid gametes containing 9 chromosomes each. Therefore, the alloploid produced from this interspecies mating contains 18 chromosomes. Since the radish and cabbage are not closely related, the nine *Raphanus* chromosomes are distinctly different from the nine *Brassica* chromosomes. During meiosis I, the radish and cabbage chromosomes cannot synapse with each other, resulting in a high degree of aneuploidy (Figure 8-27a). Therefore, the radish/cabbage hybrid is sterile.

Karpechenko discovered a way to overcome the sterility of the radish/cabbage hybrid. The idea is to produce an allopolyploid (rather than an alloploid) with a diploid number of chromosomes from each of the two species. In the example here, a radish/cabbage allopolyploid would contain 36 chromosomes instead of 18. During metaphase I, the homologous chromosomes from each of the two species can synapse properly (Figure 8-27b). When anaphase I occurs, the pairs of synapsed chromosomes can disjoin equally to produce gametes with 18 chromosomes each (a haploid set from the radish and a haploid set from the cabbage). These gametes with 18 chromosomes can combine with each other to produce an allopolyploid containing 36 chromosomes. In this way, the allopolyploid is a fertile organism. Un-

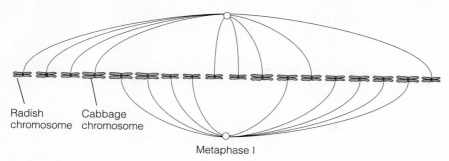

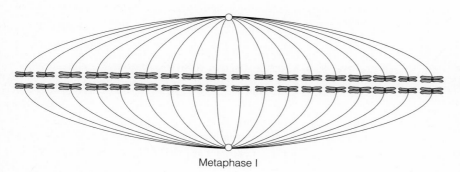

Radish
chromosome

Cabbage
chromosome

Metaphase I

(a) Alloploid with a monoploid set from each species

Metaphase I

(b) Allopolyploid with a diploid set from each species

FIGURE **8-27**

**A comparison of metaphase I in an allo-
ploid and an allopolyploid.** The red chromosomes are from the radish, and the blue
are from the cabbage. **(a)** In the alloploid,
the radish and cabbage chromosomes have
not synapsed and are randomly aligned during metaphase. **(b)** In the allopolyploid, the
homologous radish and homologous cabbage
chromosomes are properly aligned along the
metaphase plate.

fortunately, it is not agriculturally useful, because its leaves are like the radish and
its roots are like the cabbage!

Modern plant breeders employ several different strategies to produce allopolyploids. One is to start with two different tetraploid species. Since these make
diploid gametes, the hybrid from this cross would contain a diploid set from each
species. Alternatively, if an alloploid containing a monoploid set from each species
already exists, a second approach is to create an allopolyploid by using agents such
as colchicine.

Cell fusion techniques can be used to make hybrid plants

Thus far, we have discussed several mechanisms that produce variations in chromosome number. Some of these processes can occur naturally and have figured
prominently in speciation and evolution. In addition, agricultural geneticists can
administer treatments such as colchicine to promote nondisjunction and thereby
obtain useful strains of organisms. More recently, researchers are devising cellular
approaches to produce hybrids with altered chromosomal composition. As described here, these cellular approaches have important applications in research
and agriculture.

In the technique known as **cell fusion**, individual cells are mixed together and
made to fuse. As will be described in Chapter 20, this method can be used in gene
mapping. In agriculture, cell fusion can create new strains of plants. Cell fusion may
allow the crossing of two plants that normally cannot interbreed. There are several
reasons that interbreeding cannot occur. For example, one of the parents may be
sterile. Also, two distantly related species may not be able to cross-pollinate. Similarly, within a single species there may exist "incompatibility alleles" that prevent
certain strains from cross-pollinating or self-fertilizing. In many cases, cell fusion
techniques have successfully circumvented such interbreeding problems.

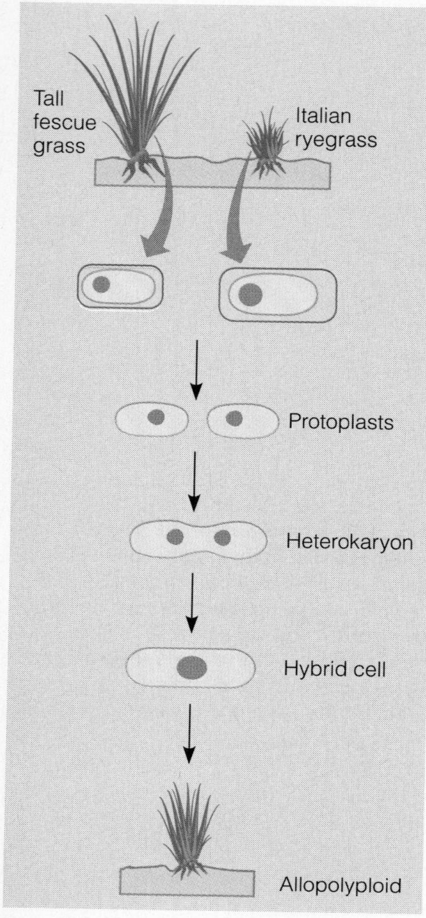

FIGURE **8-28**

The technique of cell fusion. This technique is shown here with cells from tall fescue grass and Italian ryegrass. The resulting strain is an allopolyploid.

GENES → TRAITS: The allopolyploid contains two copies of genes from each parent. In this case, the allopolyploid displays characteristics that are intermediate between the tall fescue grass and the Italian ryegrass.

Figure 8-28 illustrates the use of cell fusion to produce a hybrid grass. The parent cells were derived from tall fescue grass (*Festuca arundinacea*) and Italian ryegrass (*Lolium multiflorum*). Prior to fusion, the plant cells from these two species were treated with agents that gently digest the cell wall without rupturing the plasma membrane. A plant cell without a cell wall is called a **protoplast**. The protoplasts are mixed together and treated with agents that promote fusion. Immediately after this takes place, a cell containing two separate nuclei is formed. This cell, known as a **heterokaryon**, will spontaneously go through a nuclear fusion process to create a **hybrid cell** with a single nucleus. After nuclear fusion, the hybrid cells can be grown on laboratory media and eventually regenerate an entire plant. The allopolyploid shown in Figure 8-28 has phenotypic characteristics that are intermediate between tall fescue grass and Italian ryegrass. This type of approach is sometimes used by agricultural geneticists to produce intraspecies and interspecies hybrids.

Monoploids produced in agricultural and genetic research can be used to create homozygous and hybrid strains

Plant breeders sometimes wish to have diploid strains of crop plants that are homozygous for all of their genes. One true-breeding strain can then be crossed to a different true-breeding strain to produce an F_1 hybrid that is heterozygous for many genes. Such hybrids are often more vigorous than the corresponding homozygous strains. This phenomenon, known as **hybrid vigor** or **heterosis**, will be described in greater detail in Chapter 25.

Seed companies often use this strategy to produce hybrid seed for many crops, such as corn and alfalfa. To achieve this goal, the companies must have homozygous parental strains that can be crossed to each other to produce the hybrid seed. One way to obtain these homozygous strains involves inbreeding over many generations. This may be accomplished after several rounds of self-fertilization. As you might imagine, this can be a rather time-consuming endeavor.

As an alternative, the production of monoploids can be used as part of an experimental strategy to develop homozygous diploid strains of plants. In 1966, Sipra Guha and Satish Maheswari at the University of Delhi produced monoploid plants directly from pollen (containing the haploid male gamete). Monoploids produced experimentally have been used to improve many agricultural crops such as wheat, rice, corn, barley, and potato. Figure 8-29 describes an experimental technique called **anther culture**, which can be used to produce a diploid strain homozygous for all of its genes. It involves alternation between monoploid and diploid generations. The parental plant is diploid but not homozygous for all of its genes. The anthers from this diploid plant are collected, and the haploid pollen grains within them are induced to begin development by a cold shock treatment. After several weeks, monoploid plantlets will emerge, and these can be grown on agar media in a laboratory. Eventually, the plantlets can be transferred to small pots. After the plantlets grow to a reasonable size, a section of the monoploid plantlet can be treated with colchicine to convert it to diploid tissue. A cutting from this diploid section can then be used to generate a separate plant. This diploid plant is expected to be homozygous for all of its genes, because it was produced by the chromosomal doubling of a monoploid strain.

In certain animal species, monoploids can be produced by experimental treatments that induce the eggs to begin development without fertilization by sperm. This process is known as **parthenogenesis**. In many cases, however, the haploid zygote will only develop for a short period of time before it dies. Nevertheless, a short phase of development can be useful to research scientists. For example, the zebrafish (*Brachydanio rerio*), a common aquarium fish, has gained recent popularity among researchers interested in vertebrate development. The haploid egg can be induced to begin development by exposure to sperm rendered biologically inactive by UV-irradiation.

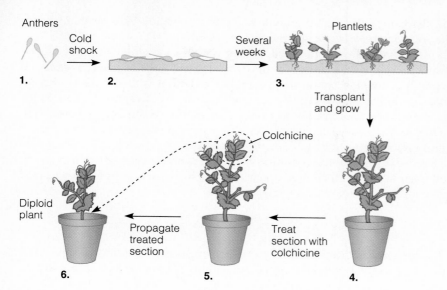

FIGURE **8-29**

The experimental production of monoploids. In the technique of anther culture, **(1)** the anthers are collected, and **(2)** the pollen within them is induced to begin development by a cold shock treatment. **(3)** After several weeks, monoploid plantlets will emerge, and these can be grown on an agar medium. Eventually, the plantlets can be transferred to small pots. **(4)** After the plantlets have grown to a reasonable size, **(5)** a section of the monoploid plantlet can be treated with colchicine to convert it to diploid tissue. **(6)** A cutting from this diploid section can then be used to generate a diploid plant.

CONCEPTUAL SUMMARY

We classify variations in chromosome structure as *deletions* (or *deficiencies*), *duplications*, *inversions*, and *translocations*. Deletions and duplications involve changes in the total amount of genetic material; inversions and translocations are genetic rearrangements. Deletions tend to be detrimental to the phenotype of the individual, although this depends on the size and location of the deletion. In some cases, a deletion heterozygote will exhibit *pseudodominance* of (normally recessive) genes located in the corresponding region of the nondeleted chromosome. In comparison, duplications tend to be less harmful. Gene duplications provide the raw material for the evolution of *gene families* in which a group of genes encode proteins that carry out similar yet specialized functions.

Inversions and translocations are chromosomal rearrangements; they often do not affect the phenotype of the individual who carries them. *Reciprocal translocations* are sometimes called *balanced translocations*, because the individual has a normal amount of genetic material. In a few cases, inversions and translocations can affect phenotype because a breakpoint is within a vital gene or the rearrangement results in a *position effect* that alters gene expression or regulation. Inversions and translocations are often associated with fertility problems and a higher probability of producing abnormal offspring. Even though an individual carrying an inversion or balanced translocation may be phenotypically normal, crossing over and chromosomal segregation may lead to the production of gametes that do not have a balanced amount of genetic material. Such gametes are likely to produce phenotypic abnormality or even lethality.

Chromosome number is another critical factor in determining the phenotype of an organism. In the condition known as *aneuploidy*, an individual may have an extra chromosome (trisomy) or may be missing a chromosome (monosomy). This usually is detrimental to the phenotype. For example, various human genetic diseases, such as Down syndrome (trisomy-21), are due to irregularities in chromosome number. Likewise, in plants, aneuploidy also significantly affects the phenotype of an individual.

Variations in the number of sets of chromosomes are also common, particularly in the plant kingdom. This has dramatically influenced agriculture. Many of our modern crops are *polyploid* plants. These frequently exhibit characteristics superior to those of their diploid counterparts. Polyploids with an odd number of chromosome sets are usually sterile, and these are useful in producing seedless varieties of plants. In animals, some cells of the body may be *endopolyploid*—that is, they have more sets of chromosomes than other somatic cells. An extreme example are the *polytene chromosomes* found in *Drosophila*.

Changes in the chromosome number can arise via several different mechanisms. In natural populations, the most important of these are *nondisjunction* and *interspecies matings*. Nondisjunction produces gametes with alterations in chromosome number, which can lead to aneuploidy or even polyploidy. Also, nondisjunction or chromosome loss can occur in cells after fertilization to produce an individual that is a genetic *mosaic*. Another way to produce new combinations of chromosomes is by interspecies matings. These can produce *alloploids*, which have one set of chromosomes from each species, or *allopolyploids*, which have two or more sets from each species. Allopolyploids are more likely to be fertile when the chromosomes are found in homologous pairs.

EXPERIMENTAL SUMMARY

The primary experimental strategy for studying variation in chromosome structure and number is the cytological examination of chromosomes. A karyotype is a photographic representation of the chromosomes from actively dividing eukaryotic cells. When preparing a karyotype, the chromosomes are usually stained with a dye such as Giemsa, which gives individual chromosomes their own characteristic banding pattern. By observing the banding patterns and number of chromosomes under a microscope, a cytogeneticist can determine if an individual carries a change in chromosome structure or number. In Chapter 20, we will learn about methods such as *in situ* hybridization, which can detect relatively small changes in chromosome structure (e.g., small deletions) that are difficult to visualize by normal light microscopy.

Experimental methods are also available to cause changes in chromosome number. For example, drugs such as colchicine can alter chromosome number by promoting nondisjunction. In addition, laboratory methods can produce hybrids by cell fusion techniques. This can be useful in making fertile interspecies hybrids. Also, monoploid plants and animals can be produced by the experimental activation of haploid gametes.

PROBLEM SETS

SOLVED PROBLEMS

1. Describe how a gene family is produced.

Answer: A gene family is produced when a single gene is copied one or more times by a gene duplication event. This duplication occurs by an abnormal crossing over, which produces a chromosome with a deficiency and another chromosome with a gene duplication:

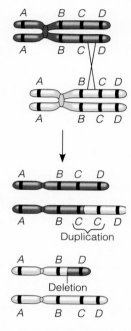

Over time, this type of duplication may occur several times to produce many copies of a particular gene. In addition, translocations may move the duplicated genes to other chromosomes, so that the members of the gene family may be dispersed among several different chromosomes. Eventually, each member of a gene family will accumulate mutations, which may subtly alter their function.

2. An inversion heterozygote has the following inverted chromosome:

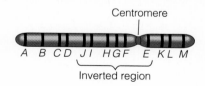

What is the result if a crossover occurs between genes *F* and *G* on one inverted and one normal chromosome?

Answer: The resulting product is four chromosomes. One chromosome is normal, one is an inverted chromosome, and two chromosomes have duplications and deficiencies. The two duplicated/deficient chromosomes are shown here:

3. In humans, the number of chromosomes per set equals 23. Even though the following conditions are lethal, what would be the total number of chromosomes for the following individuals?

- **A.** Trisomy-22
- **B.** Monosomy-11
- **C.** Triploid individual
- **D.** Tetrasomy-21

Answer:

- **A.** 47 (the diploid number, 46, plus 1)
- **B.** 45 (the diploid number, 46, minus 1)
- **C.** 69 (3 times 23)
- **D.** 48 (the diploid number, 46, plus 2)

4. A diploid species with 44 chromosomes (i.e., 22/set) is crossed to another diploid species with 38 chromosomes (i.e., 19/set). What would be the number of chromosomes in an alloploid and allopolyploid produced from this cross? Would you expect the offspring to be sterile or fertile?

Answer: An alloploid would have 22 + 19 = 41 chromosomes. This individual would likely be sterile, because the chromosomes would not have homologous or homeologous partners to pair with during meiosis. This yields aneuploidy, which usually causes sterility. An allopolyploid would have 44 + 38 = 82 chromosomes. Since each chromosome would have a homologous partner, the allopolyploid would be more likely to be fertile.

CONCEPTUAL QUESTIONS

1. Which changes in chromosome structure cause a change in the total amount of genetic material, and which do not?

2. Explain why small deletions and duplications are less likely to have a detrimental effect on an individual's phenotype than large ones. If a small deletion happens to have a phenotypic effect, what would you conclude about the genes in this region?

3. How does a chromosomal duplication occur?

4. What is a gene family? How are gene families produced over time? With regard to gene function, what is the biological significance of a gene family?

5. Following a gene duplication, two genes will accumulate different mutations, causing them to have slightly different sequences. In Figure 8-7, which two genes would you expect to have more similar sequences, α_1 and α_2 or Ψ_{α_1} and α_2? Explain your answer.

6. Two chromosomes have the following order of genes:

  Normal: *A B C* centromere *D E F G H I*
  Abnormal: *A B G F E D* centromere *C H I*

Is this a pericentric or paracentric inversion? Draw a sketch showing how these two chromosomes would pair during prophase I of meiosis.

7. An inversion heterozygote has the following inverted chromosome:

What would be the products if a crossover occurred between genes *H* and *I* on one inverted and one normal chromosome?

8. An inversion heterozygote has the following inverted chromosome:

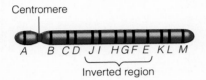

What would be the products if a crossover occurred between genes *H* and *I* on one inverted and one normal chromosome?

9. Explain why inversions and reciprocal translocations do not usually cause a phenotypic effect. In a few cases, however, they do. Explain how.

10. An individual has the following reciprocal translocation:

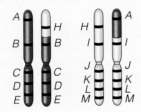

What would be the outcome of alternate and adjacent-1 segregation?

11. A phenotypically normal individual has the following combinations of abnormal chromosomes:

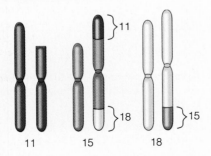

The normal chromosomes are shown on the left of each pair. Suggest a series of events (breaks, translocations, crossovers, etc.) that may have produced this combination of chromosomes.

12. Two phenotypically normal parents produce an abnormal child in which chromosome 5 is missing part of its long arm but has a piece of chromosome 7 attached to it. The child also has one normal copy of chromosome 5 and two normal copies of chromosome 7. With regard to chromosomes 5 and 7, what do you think are the chromosomal compositions of the parents?

13. In the segregation of centromeres, why is adjacent-2 segregation less frequent than alternate or adjacent-1 segregation?

14. A diploid fruit fly has eight chromosomes. How many total chromosomes would be found in the following flies?

  A. Tetraploid
  B. Trisomy-2
  C. Monosomy-3
  D. $3n$
  E. $4n + 1$

15. A person is born with one X-chromosome, zero Y-chromosomes, trisomy-21, and two copies of the other chromosomes. How many chromosomes does this person have altogether? Explain whether this person is euploid or aneuploid.

16. Two phenotypically normal parents produce two children with familial Down syndrome. With regard to chromosomes 14 and 21, what are the chromosomal compositions of the parents?

17. Aneuploidy is typically detrimental, whereas polyploidy is sometimes beneficial, particularly in plants. Discuss why you think this is the case.

18. Explain how aneuploidy, deletions, and duplications cause genetic imbalances. Why do you think that deletions and monosomies are more detrimental than duplications and trisomies?

19. Female fruit flies homozygous for the X-linked white-eye allele are crossed to males with red eyes. On very rare occasions, an offspring is a male with red eyes. These rare offspring are *not* due to a new mutation in one of the mother's X-chromosomes that converted the white-eye allele into a red-eye allele. Explain how this red-eyed male arose.

20. Aneuploidy occurs rather frequently during human gamete formation, though very few people are born with this genetic composition. Why not?

21. A cytogeneticist has collected blood samples from members of the same butterfly species. Some of the butterflies were located in Canada, while others were found in Mexico. Upon karyotyping, the cytogeneticist discovered that chromosome 5 of the Canadian butterflies had a large inversion compared with the Mexican butterflies. The Canadian butterflies were inversion homozygotes, whereas the Mexican butterflies had two normal copies of chromosome 5.

  A. Explain whether a mating between the Canadian and Mexican butterflies would produce phenotypically normal offspring.

B. Explain whether the offspring of a cross between Canadian and Mexican butterflies would be fertile.

22. Why do you think that human trisomy-13, -18, -21 can survive but the other trisomies are lethal? Even though X-chromosomes are large, aneuploidies of this chromosome are also tolerated. Explain why.

23. A zookeeper has collected a male and female lizard that look like they belong to the same species. They mate with each other and produce phenotypically normal offspring. However, the offspring are sterile. Suggest one or more explanations for their sterility.

24. What is endopolyploidy? What is its biological significance?

25. What is a genetic mosaic? How is it produced?

26. Explain how polytene chromosomes are produced and how they form a six-armed structure.

27. Describe some of the advantages of polyploid plants. What are the consequences of having an odd number of chromosome sets?

28. Describe three naturally occurring ways that the chromosome number can change.

29. Meiotic nondisjunction is much more likely than mitotic nondisjunction. Based on this observation, would you conclude that meiotic nondisjunction is usually due to nondisjunction during meiosis I or meiosis II? Explain your reasoning.

30. A woman who is heterozygous, *Bb*, has brown eyes. In one of her eyes, however, there is a patch of blue color. Give three different explanations for how this might have occurred.

31. What is an alloploid? What factor determines the fertility of an alloploid? Why are allopolyploids more likely to be fertile?

32. What are homeologous chromosomes?

EXPERIMENTAL QUESTIONS

1. Describe five or more experimental ways to show that a mutation is actually a deletion.

2. With regard to gene mapping, explain the profound experimental advantage that polytene chromosomes offer.

3. Describe how colchicine can be used to alter chromosome number.

4. Describe the steps you would take to produce a tetraploid plant that is homozygous for all of its genes.

5. In agriculture, what is the primary purpose of anther culture?

6. What are some experimental advantages of cell fusion techniques as opposed to interbreeding approaches?

7. It is an exciting time to be a plant breeder, because so many options are available for the development of new types of agriculturally useful plants. Let's suppose that you wish to develop a seedless tomato that could grow in a very hot climate, and is resistant to a viral pathogen that commonly infects tomato plants. At your disposal, you have a seed-bearing tomato strain that is heat resistant and produces great-tasting tomatoes. You also have a wild strain of tomato plants (which have lousy-tasting tomatoes) that is resistant to the viral pathogen. Suggest one or more strategies that you might follow to produce a great-tasting, seedless tomato that is resistant to heat and the viral pathogen.

QUESTIONS FOR STUDENT DISCUSSION/COLLABORATION

1. A chromosome involved in a reciprocal translocation also has an inversion. Make a drawing that shows how these chromosomes will pair during metaphase I.

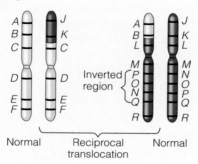

2. Besides the ones mentioned in this text, look for other examples of variations in euploidy. Perhaps you might look in more advanced texts concerning population genetics, ecology, or the like. Discuss the phenotypic consequences of these changes.

3. Cell biology texts often discuss cellular proteins encoded by genes that are members of a gene family. Examples of such proteins include myosins and glucose transporters. Take a look through a cell biology text and identify some proteins encoded by members of gene families. Discuss the importance of gene families at the cellular level.

4. Discuss how variation in chromosome number has been useful in agriculture.

Part III

Molecular Structure and Replication of the Genetic Material

CHAPTER 9

MOLECULAR STRUCTURE OF DNA AND RNA

IDENTIFICATION OF DNA AS THE GENETIC MATERIAL

NUCLEIC ACID STRUCTURE

In Chapters 2 through 8, we focused on the relationships among the inheritance of genes, chromosomes, and the outcome of an organism's traits. In this chapter, we will shift our attention to **molecular genetics**, the study of DNA structure and function at the molecular level. An exciting goal of molecular genetics is to use our knowledge of DNA structure to understand how DNA functions as the genetic material. By studying the structure of DNA, researchers have determined the molecular organization of many genes. This information, in turn, has helped us understand how the expression of genes at the molecular level governs the outcome of an individual's inherited traits.

The past several decades have seen a virtual explosion in techniques and approaches to investigate and even to alter the genetic material. These advances have greatly increased our understanding of molecular genetics as well as transmission and population genetics. In addition, molecular genetic technology is widely used in supporting disciplines such as biochemistry, cell biology, and microbiology.

To a large extent, our understanding of genetics comes from our knowledge of the molecular structure of DNA (deoxyribonucleic acid) and RNA (ribonucleic acid). In this chapter, we will begin by describing two classic experiments that showed that DNA is the genetic material. We will then survey the basic structural features of DNA and RNA that underlie their function.

IDENTIFICATION OF DNA AS THE GENETIC MATERIAL

In his classic experiments of the 1860s, Gregor Mendel studied several different traits in pea plants. By conducting the appropriate crosses, he showed that traits are inherited in an unaltered way as they pass from parent to offspring. This implies that living organisms contain a genetic material that governs an individual's traits, a substance that is transferred during the process of reproduction.

To fulfill its role, the genetic material must meet several criteria:

1. The genetic material must contain the information necessary to construct an entire organism. In other words, it must provide the blueprint that determines all of the inherited traits that an organism will have.
2. During reproduction, the genetic material must usually pass unaltered from parent to offspring.
3. Since a parent can produce many offspring, the genetic material must be copied. Also, in multicellular organisms, each cell must be supplied with a copy of the genetic material.
4. Within any species, a significant amount of phenotypic variability occurs. For example, Mendel studied several traits in pea plants that were variable among different plants. These included height (i.e., tall vs. dwarf) and seed texture (i.e., round vs. wrinkled). Therefore, the genetic material must also have variation that can account for the known phenotypic variation within each species.

Along with Mendel's work, the experiments of many other geneticists in the early 1900s were consistent with these four properties. However, the experimental study of genetic crosses cannot by itself identify the chemical nature of the genetic material.

As discussed in Chapter 3, August Weismann and Carl Nägeli championed the idea that there is a chemical substance within living cells that is responsible for the transmission of traits from parents to offspring. In the early 1900s, the chromosome theory of inheritance was developed, and experimentation demonstrated that the chromosomes are the carriers of the genetic material. Nevertheless, the story was not complete, because chromosomes contain several types of macromolecules, including DNA, RNA, and proteins. Therefore, further research was needed to precisely identify the genetic material. In this section, we will examine two innovative experimental approaches showing that DNA is the genetic material.

Early experiments with pneumococcus bacteria suggested that DNA is the genetic material

Some early work in microbiology was important in developing an experimental strategy to identify the genetic material. In the late 1920s, Frederick Griffith studied a type of bacterium known then as pneumococci and now classified as *Diplococcus pneumoniae*. Certain strains of *D. pneumoniae* secrete a polysaccharide capsule, whereas other strains do not. When streaked on petri plates containing solid growth media, capsule-secreting strains have a smooth colony morphology, whereas those strains unable to secrete a capsule have a rough appearance.

When comparing two different smooth strains of *D. pneumoniae*, researchers found that the chemical composition of their capsules can differ significantly. Using biochemical techniques, these different types of smooth strains were classified. For example, a type II smooth strain and a type III smooth strain both make a capsule, but their capsules differ from each other biochemically. This idea is schematically shown in Figure 9-1. Rare mutations can occasionally convert a smooth bacterium into a rough bacterium, and vice versa. These infrequent interconversions are type-specific. For example, a type III smooth strain can mutate to become a rough strain. On rare occasions, a bacterium from this rough strain may mutate back to become a smooth strain. In this case, it can only mutate to become a type III smooth strain, never a type II smooth strain.

Griffith carried out experiments that involved the injection of live and/or heat-killed bacteria into mice; he then observed whether or not the bacteria caused

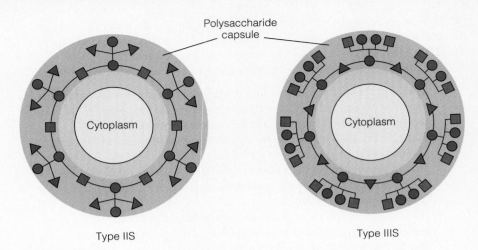

Type IIS   Type IIIS

FIGURE **9-1**

Schematic of biochemical differences between the capsules of two strains of pneumococci. The capsule is composed of sugar units linked to form a polysaccharide material. This figure is a schematic diagram illustrating how the types of sugars (represented by circles, squares, and triangles) and their arrangement to form a polysaccharide capsule can differ among different strains. This figure is not meant to represent the actual polysaccharide capsule, which is much more complex than shown here. Instead, this figure emphasizes that the structures of the polysaccharide capsule are different in type IIS and type IIIS bacteria.

GENES $\longrightarrow$ TRAITS: The chemical composition of the polysaccharide coat is governed by genes within the bacterial cell. These genes encode enzymes that synthesize the polysaccharide capsule. The genes within type IIS bacteria encode enzymes that synthesize the polysaccharide capsule depicted on the left. The genes within type IIIS bacteria encode enzymes that function somewhat differently to synthesize the capsule shown on the right.

a lethal infection. He was working with two strains of *D. pneumoniae*, a type IIIS (S for smooth) and a type IIR (R for rough). When injected into a live mouse, the type IIIS bacteria proliferated within the mouse's bloodstream and ultimately killed the mouse (Figure 9-2a). In this process, the capsule is very important in preventing the mouse's defenses from inhibiting the bacteria. Following the death of the mouse, Griffith found many type IIIS bacteria within the mouse's blood. In contrast, when type IIR bacteria were injected into a mouse, they did not cause a lethal infection (Figure 9-2b). To verify that the proliferation of the smooth bacteria was causing the death of the mouse, Griffith killed the smooth bacteria with heat treatment before injecting them into the mouse. In this case, the mouse survived (Figure 9-2c).

The critical and unexpected result was obtained in the experiment outlined in Figure 9-2d. In this experiment, live type IIR bacteria were mixed with heat-killed type IIIS bacteria. As shown here, the mouse died. Upon analysis, the dead mouse was found to contain living type IIIS bacteria! The interpretation of these data is that something from the dead type IIIS bacteria was transforming the type IIR bacteria into type IIIS. This process, known as **transformation**, was also described in Chapter 6 (pp. 151–152).

At this point, let's look at what Griffith's observations mean in genetic terms. The ability to form a capsule is a *trait* of pneumococci. Among different strains, there is *variation* in the type of capsule that is made and whether a capsule is made. In addition, this trait is *inherited* from one generation of bacteria to the next. Taken together, these observations are consistent with the idea that the formation of a capsule is governed by the bacteria's genetic material. Significantly, Griffith's experi-

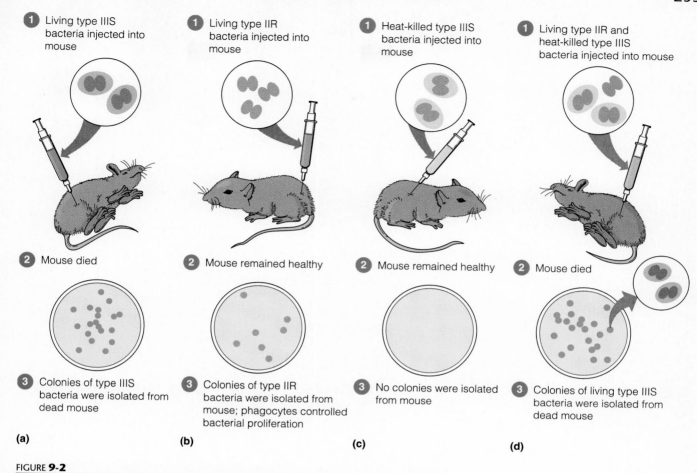

(1) Living type IIIS bacteria injected into mouse

(2) Mouse died

(3) Colonies of type IIIS bacteria were isolated from dead mouse

(a)

(1) Living type IIR bacteria injected into mouse

(2) Mouse remained healthy

(3) Colonies of type IIR bacteria were isolated from mouse; phagocytes controlled bacterial proliferation

(b)

(1) Heat-killed type IIIS bacteria injected into mouse

(2) Mouse remained healthy

(3) No colonies were isolated from mouse

(c)

(1) Living type IIR and heat-killed type IIIS bacteria injected into mouse

(2) Mouse died

(3) Colonies of living type IIIS bacteria were isolated from dead mouse

(d)

FIGURE **9-2**

Griffith's experiments on genetic transformation in pneumococcus.

ments showed that some genetic material from the dead bacteria was being transferred to the living bacteria and providing them with a new trait. However, Griffith did not know what the transforming substance was.

Important scientific discoveries often take place when researchers recognize that someone else's experimental observations can be used to address a particular scientific question. In the 1940s, Oswald Avery, Colin MacLeod, and Maclyn McCarty at the Rockefeller Institute realized that Griffith's observations could be used as part of an experimental strategy to identify the genetic material. They asked the question, "What substance is being transferred from the dead type IIIS bacteria to the live type IIR?" To answer this question, they incorporated some additional biochemical techniques in their experimental methods. Their approach is outlined in Figure 9-3.

At the time of these experiments, researchers already knew that DNA, RNA, and proteins are major constituents of living cells. To separate these components, and to determine if any of them was the genetic material, Avery, MacLeod, and McCarty used established biochemical purification procedures and prepared from the dead bacteria an extract that was responsible for converting the type IIR bacteria into type IIIS. They used the term *transformation principle* to describe this extract; it contained a purified preparation of DNA. As outlined in Figure 9-3, when this extract was mixed with type IIR bacteria, some of the bacteria were converted to type IIIS. However, if no extract was added, no type IIIS bacterial colonies were observed on the petri plates.

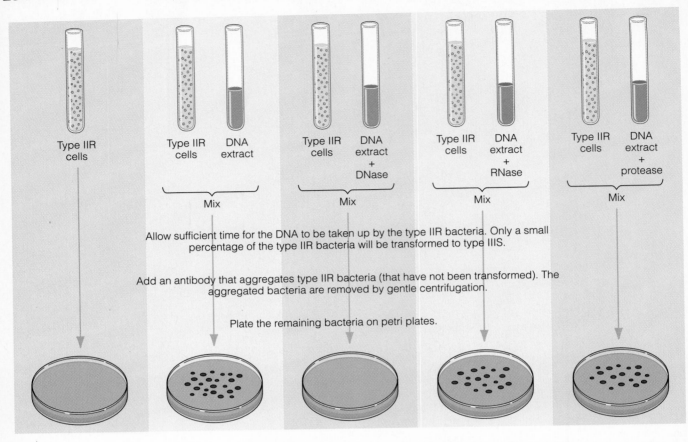

FIGURE 9-3

Experimental protocol used by Avery, MacLeod, and McCarty to identify the transformation principle. As noted here, antibodies are molecules that can specifically recognize the molecular structure of macromolecules. In this case, the antibodies recognize the cell surface of type IIR bacteria. When the antibodies recognize the type IIR bacteria, they cause them to clump together. The clumped bacteria can be removed by a gentle centrifugation step. Only the bacteria that have not been recognized by the antibody (namely, the type IIIS bacteria) remain in the supernatant.

In the experiment outlined in Figure 9-3, a biochemist might point out that the DNA extract may not be 100% pure. In fact, any purified extract might contain small traces of some other substances. Therefore, one can argue that a small amount of contaminating material in the DNA extract might actually be the genetic material. The most likely contaminating substances in this case would be protein or RNA. To further verify that the DNA in the extract was indeed responsible for the transformation, Avery, MacLeod, and McCarty treated samples of the DNA extract with enzymes that digest DNA (called *DNase*), RNA (*RNase*), or protein (*protease*). When the DNA extracts were treated with RNase or protease, they still converted type IIR bacteria into type IIIS. These results indicated that contaminating RNA or protein in the extract was not acting as the genetic material. However, when the extract was treated with DNase, its ability to convert type IIR into type IIIS was lost. These results indicated that the degradation of the DNA in the extract by DNase prevented conversion of type IIR to type IIIS. This interpretation is consistent with the hypothesis that DNA is the genetic material. A more elegant way of saying this is that the transforming principle is DNA.

Hershey and Chase provided evidence that the genetic material injected into the bacterial cytoplasm is T2 phage DNA

A second experimental approach indicating that DNA is the genetic material came from the studies of A. D. Hershey and Martha Chase at Cold Spring Harbor. Their research centered on the study of a virus known as T2. This virus infects *Escherichia coli* bacterial cells and is therefore known as a **bacteriophage** or simply a **phage**. As shown in Figure 9-4, the external structure of the T2 phage, known as the *phage coat*, contains a *capsid*, *sheath*, *baseplate*, and *tail fibers*. Biochemically, the phage coat is composed entirely of protein, which includes several different polypeptides. DNA is found inside the T2 capsid. From a molecular point of view, this virus is rather simple, since it is only composed of two types of macromolecules: DNA and proteins.

Although the viral genetic material contains the blueprint to make new viruses, a virus itself cannot synthesize new viruses. Instead, a virus must introduce its genetic material into the cytoplasm of a living cell. In the case of T2, this first involves the attachment of its tail to the bacterial cell wall and the subsequent injection of its genetic material into the cytoplasm of the cell (Figure 9-5). The phage coat remains attached on the outside of the bacterium and does not enter the cell. After the entry of the viral genetic material, the bacterial cytoplasm provides all the synthetic machinery necessary to make viral proteins and DNA. The viral proteins and DNA assemble to make new viruses, which are subsequently released from the cell by **lysis** (i.e., cell breakage).

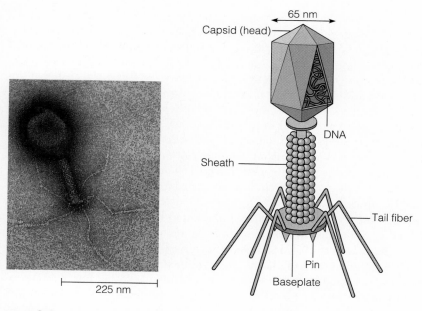

FIGURE **9-4**

Structure of the T2 bacteriophage. The T2 bacteriophage is composed of a phage coat and genetic material inside the phage head. The phage coat is divided into regions called the capsid, sheath, baseplate, and tail fibers. These components are composed of proteins. The genetic material is composed of DNA.

GENES ⟶ TRAITS: The genetic material of a bacteriophage contains many genes, which provide the blueprint for making new viruses. When the bacteriophage injects its genetic material into a bacterium, these genes are activated and direct the host cell to make new bacteriophages as described in Figure 9-5.

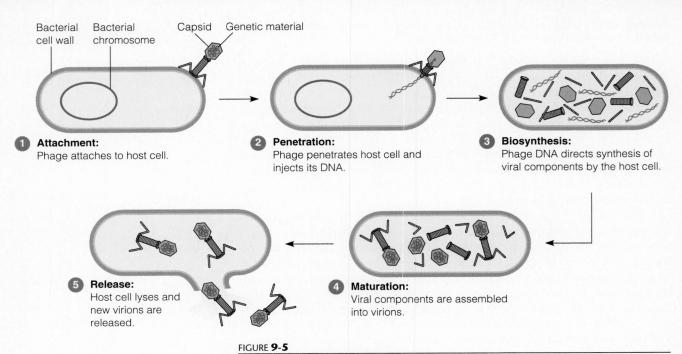

FIGURE **9-5**

Life cycle of the T2 bacteriophage.

To verify that DNA is the genetic material, Hershey and Chase devised a method to separate the phage coat that is attached to the outside of the bacterium from the genetic material that is injected into the cytoplasm. They were aware of microscopy experiments by T. F. Anderson showing that the T2 phage attaches itself to the outside of a bacterium by its tail. Hershey and Chase reasoned that this is a fairly precarious attachment that could be disrupted by subjecting the bacteria to high shear forces such as those produced in a blender. As described in this experiment, their method was to expose bacteria to T2 phage, allowing sufficient time for the viruses to attach to bacteria and inject their genetic material. They then sheared the phage coats from the surface of the bacteria by a blender treatment. In this way, the phage's genetic material, which had been injected into the cytoplasm of the bacterial cell, could be separated from the rest of the phage coat, which was sheared away.

Before discussing this experiment, it should be mentioned that radioisotopes were used to distinguish between DNA and proteins. Hershey and Chase obtained T2 phages that were radiolabeled with either ^{35}S (a radioisotope of sulfur) or ^{32}P (a radioisotope of phosphorus). Sulfur atoms are found in proteins but not in DNA, whereas phosphorus atoms are found in DNA but not in viral proteins. Therefore, ^{35}S and ^{32}P were used in this experiment to specifically label proteins and DNA, respectively.

THE HYPOTHESIS

This experiment tests the hypothesis that bacteriophage T2 injects DNA rather than protein into the bacterial cytoplasm, where it functions as the viral genetic material.

TESTING THE HYPOTHESIS

Starting materials: The starting materials are *E. coli* cells and two preparations of T2 phage. One preparation is labeled with ^{35}S to label the phage proteins; the other preparation is labeled with ^{32}P to label the phage DNA.

Experimental Level   **Conceptual Level**

1. Grow bacterial cells. Divide into two flasks.

2. Into one flask, add ^{35}S-labeled phage; in the second flask, add ^{32}P-labeled phage.

3. Allow infection to occur.

4. Spin solutions in blenders for different lengths of time to shear the empty phages off the bacterial cells.

5. Centrifuge at 10,000 rpm.

6. Note: The heavy bacterial cells sediment to the pellet, while the lighter phages remain in the supernatant. (See appendix for an explanation of centrifugation.)

7. Count the amount of radioisotope in the supernatant with a scintillation counter (see the appendix). Compare it with the starting amount.

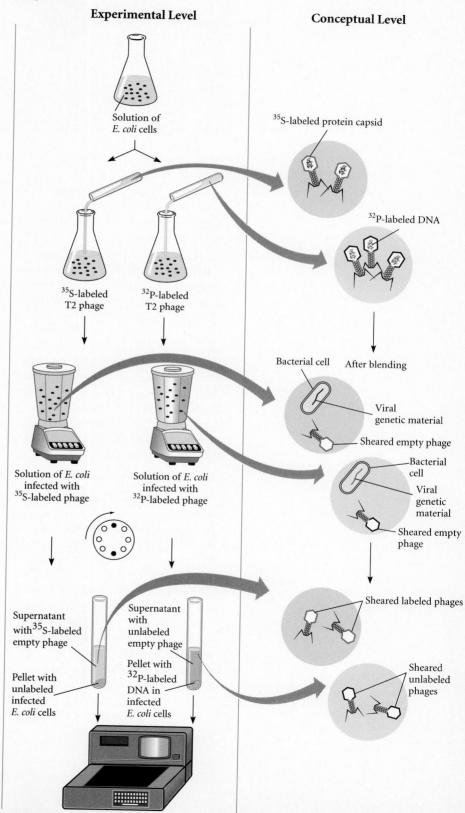

Solution of *E. coli* cells

^{35}S-labeled T2 phage ^{32}P-labeled T2 phage

^{35}S-labeled protein capsid

^{32}P-labeled DNA

Bacterial cell   After blending

Viral genetic material

Sheared empty phage

Bacterial cell

Viral genetic material

Sheared empty phage

Solution of *E. coli* infected with ^{35}S-labeled phage

Solution of *E. coli* infected with ^{32}P-labeled phage

Supernatant with ^{35}S-labeled empty phage

Supernatant with unlabeled empty phage

Pellet with unlabeled infected *E. coli* cells

Pellet with ^{32}P-labeled DNA in infected *E. coli* cells

Sheared labeled phages

Sheared unlabeled phages

THE DATA

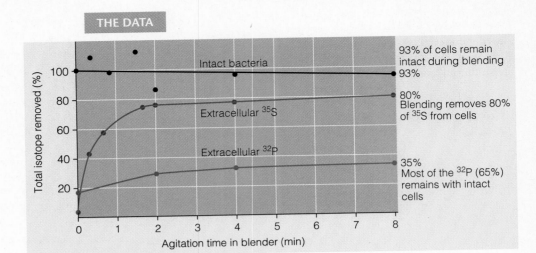

INTERPRETING THE DATA

Following infection, most of the phage protein was sheared from the bacterial cells and ended up in the supernatant. This indicated that the empty phages contain primarily protein. Furthermore, less than 40% of the DNA was found in the supernatant following shearing. Therefore, most of the DNA was located within the bacterial cells in the pellet. These results are consistent with the idea that the DNA is injected into the bacterial cytoplasm during infection. This is the expected result if DNA is the genetic material.

By themselves, the results described in Experiment 9A were not conclusive evidence that DNA is the genetic material. For example, you may have noticed that less that 100% of the viral protein was found in the supernatant. Therefore, some of the phage protein could have been introduced into the bacterial cells (and could function as the genetic material). Nevertheless, the results of Hershey and Chase leaned toward the conclusion that DNA is the genetic material rather than protein. Overall, their studies of the T2 phage, published in 1952, were quite influential in promoting the belief that DNA is the genetic material.

RNA functions as the genetic material in some viruses

We now know that bacteria, protozoa, fungi, algae, plants, and animals all use DNA as their genetic material. As mentioned, viruses also have their own genetic material. Hershey and Chase concluded from their experiments that this genetic material is DNA. In the case of T2 bacteriophage, that is the correct conclusion. However, many viruses use RNA, rather than DNA, as their genetic material. In 1956, A. Gierer and G. Schramm at the Max Planck Institute in Germany isolated RNA from the tobacco mosaic virus (TMV), which infects plant cells. When this purified RNA was applied to plant tissue, the plants developed the same types of lesions that occurred when they were exposed to intact TMV viruses. Gierer and Schramm correctly concluded that the viral genome of tobacco mosaic virus is composed of RNA. Since that time, many other viruses have been found to contain RNA as their genetic material. Table 9-1 compares the genetic compositions of several different types of viruses.

NUCLEIC ACID STRUCTURE

Geneticists, biochemists, and biophysicists have been interested in the molecular structure of nucleic acids for many decades. Both DNA and RNA are large macro-

TABLE 9–1 Examples of DNA- and RNA-containing viruses		
Virus	Host	Nucleic Acid
Tomato bushy stunt virus	Tomato	RNA
Tobacco mosaic virus	Tobacco	RNA
Influenza virus	Humans	RNA
HIV	Humans	RNA
f2	*E. coli*	RNA
QB	*E. coli*	RNA
Cauliflower mosaic virus	Cauliflower	DNA
Herpes virus	Humans	DNA
SV40	Primates	DNA
Epstein–Barr virus	Humans	DNA
T2	*E. coli*	DNA
M13	*E. coli*	DNA

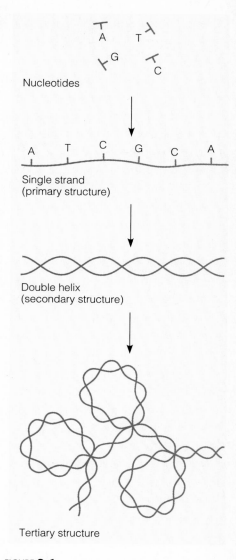

Nucleotides

Single strand
(primary structure)

Double helix
(secondary structure)

Tertiary structure

FIGURE **9-6**

Levels of nucleic acid structure.

molecules composed of smaller building blocks. To fully appreciate their structures, we need to consider several levels of complexity (Figure 9-6):

1. **Nucleotides** form the repeating structural unit of nucleic acids.
2. Nucleotides are linked together to form a **strand** of DNA or RNA. The linear sequence of nucleotides within a strand is known as the **primary structure** of DNA or RNA.
3. Two strands of DNA (and sometimes RNA) can interact with each other to form a **double helix**. The double helix is an example of a regular repeating **secondary structure**. As we will see, the DNA double helix can adopt several secondary structures, known as A-DNA, B-DNA, and Z-DNA.
4. The folding and bending of secondary structures into a final three-dimensional structure is known as the **tertiary structure**. Within living cells, DNA is associated with a wide variety of proteins that influence its final tertiary structure. Chapter 10 will be devoted to the roles of these proteins and the three-dimensional structure of DNA found within chromosomes.

In this section, we will first examine the structure of individual nucleotides and then progress through the primary, secondary, and tertiary structures of DNA and RNA. Along the way, we will also review some of the pivotal experiments that led to the discovery of the double helix.

Nucleotides are the building blocks of nucleic acids

The nucleotide is the repeating structural unit of DNA and RNA. A **nucleotide** has three components: a phosphate group, a pentose sugar, and a nitrogenous base. As shown in Figure 9-7, nucleotides can vary with regard to the sugar and the nitrogenous base. There are two types of sugars, ribose and deoxyribose. The five different bases are subdivided into two categories, the **purines** and the **pyrimidines**. The purine bases, *adenine* (A) and *guanine* (G), contain a double-ring structure; the pyrimidine bases, *cytosine* (C), *thymine* (T), and *uracil* (U), contain a single-ring structure. The chemical differences between nucleotides are important in distinguishing DNA and RNA (see Figure 9-8). For example, the sugar in DNA is always *deoxyribose*; in RNA, it is *ribose*. Second, the base thymine is not found in RNA. Rather, uracil is found in RNA instead of thymine. Adenine, cytosine, and guanine occur in both DNA and RNA.

The terminology used to describe nucleic acid units is based on three structural features: the type of base, the type of sugar, and the number of phosphate groups. When a base is attached to only a sugar, we call this pair a **nucleoside**. For

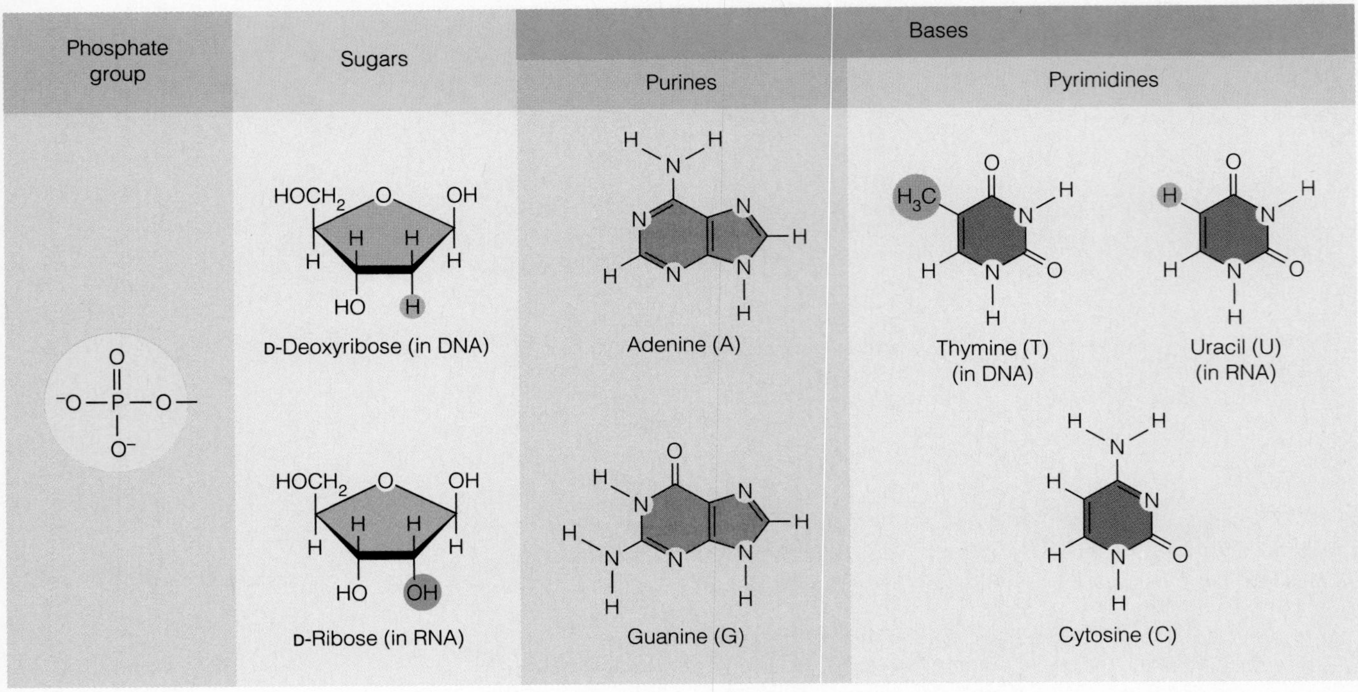

FIGURE 9-7

The components of nucleotides. The three building blocks of a nucleotide are one or more phosphate groups, a sugar, and a base. The bases are categorized as purines (adenine and guanine) and pyrimidines (thymine, uracil, and cytosine).

example, if adenine is attached to ribose, this is called *adenosine* (Figure 9-9). Likewise, nucleosides containing guanine, thymine, cytosine, or uracil are called *guanosine*, *thymidine*, *cytidine*, and *uridine*, respectively. The covalent attachment of one or more phosphate molecules to a nucleoside creates a **nucleotide**. If a nucleotide contains adenine, ribose, and three phosphate groups, it is called *adenosine triphosphate* (see Figure 9-9), abbreviated ATP. Or if it contains guanosine, ribose,

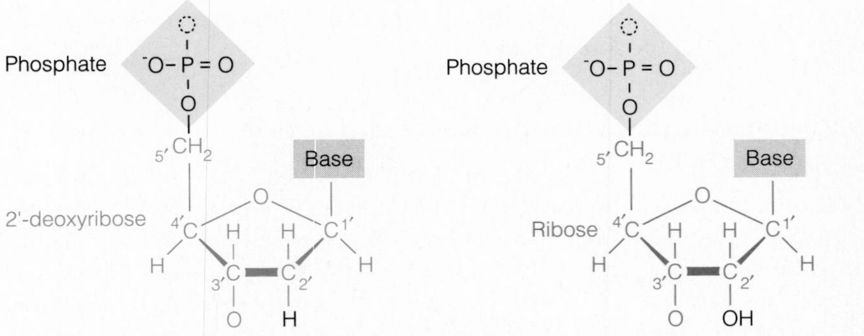

(a) Repeating unit of deoxyribonucleic acid (DNA)

(b) Repeating unit of ribonucleic acid (RNA)

FIGURE 9-8

The structure of nucleotides found in (a) DNA and (b) RNA. DNA contains deoxyribose as its sugar, and the bases A, T, G, and C. RNA contains ribose as its sugar, and the bases A, U, G, and C.

In a DNA or RNA strand, the oxygen on the 3'-carbon is linked to the phosphorus atom of phosphate in the adjacent nucleotide. This bond formation removes the hydrogen attached to the 3'-oxygen atom, and it removes an oxygen (shown with a dashed line) attached to phosphorus.

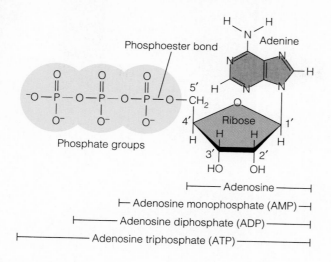

FIGURE **9-9**

A comparison between the structures of an adenine-containing nucleoside and nucleotides.

and three phosphate groups, it is *guanosine triphosphate*, or GTP. If a nucleotide contains adenine, ribose, and one phosphate, it is *adenosine monophosphate* (AMP). A nucleotide can be composed of adenine, deoxyribose, and three phosphate groups. This nucleotide is referred to as *deoxyadenosine triphosphate*.

The locations of the attachment sites of the base and phosphate to the sugar molecule are important to the nucleotide's function. Figures 9-8 and 9-9 illustrate the conventional numbering of the carbon atoms in a pentose sugar. In the sugar ring, carbon atoms are numbered in a clockwise direction, beginning with the carbon atom adjacent to the ring oxygen atom. The fifth carbon is outside of the ring structure. In a single nucleotide, the base is always attached to the number 1 carbon atom and the phosphate group(s) are attached at the number 5 position. As we will discuss in the next section, the —OH group attached to the number 3 carbon is important in allowing nucleotides to form covalent linkages with each other.

Nucleotides are linked together to form a strand

A **strand** of DNA or RNA contains nucleotides that are covalently attached to each other in a linear fashion. Figure 9-10 depicts a short strand of DNA with six nucleotides. A few structural features here are worth noting. First, the linkage involves an ester bond between a phosphate group on one nucleotide and the sugar molecule on the other nucleotide. Another way of viewing this linkage is to notice that a phosphate group is connecting *two* sugar molecules together. For this reason, the linkage in DNA or RNA strands is referred to as a **phosphodiester linkage**. The phosphates and sugar molecules form the **backbone** of a DNA or RNA strand. The bases project from the backbone. The backbone is negatively charged due to the negative charge on the phosphate groups.

A second important structural feature is the orientation of the nucleotides. As mentioned, the carbon atoms in a sugar molecule are numbered in a particular way. A phosphodiester linkage involves a phosphate attachment to the number 5 carbon in one nucleotide and to the number 3 carbon in the other. In a strand, all sugar molecules are oriented in the same direction. As shown in Figure 9-10, all of the number 5 carbons in every sugar molecule are above the number 3 carbons. Therefore, a strand has a **directionality** based on the orientation of the sugar molecules within that strand. In Figure 9-10, the direction of the strand is 5′ to 3′ when going from top to bottom.

A critical aspect regarding DNA and RNA structure is that *a strand contains a specific sequence of bases*. In Figure 9-10, the sequence of bases is thymine–adenine–

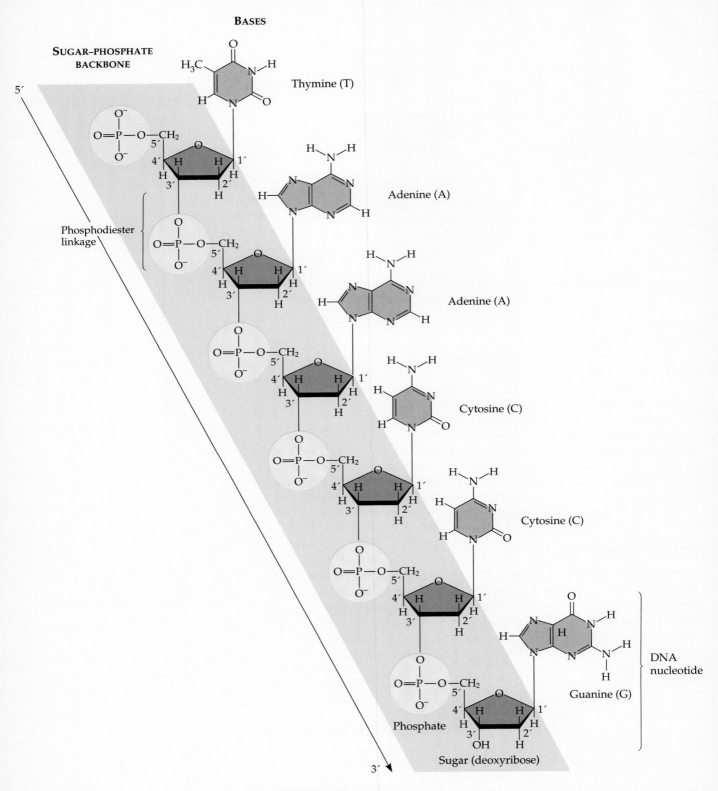

FIGURE **9-10**

A short strand of DNA containing six nucleotides. Nucleotides are covalently linked together to form a strand of DNA.

adenine–cytosine–cytosine–guanine. This sequence is abbreviated TAACCG. Furthermore, to show the directionality, the strand can be abbreviated 5′-TAACCG-3′. The nucleotides within a strand are covalently attached to each other so that the sequence of bases cannot shuffle around and become rearranged. Therefore, the sequence of bases in a DNA strand will remain the same over time, except in the rare cases when mutations occur. As we will see throughout this text, the sequence of bases within DNA and RNA is the defining feature that allows them to carry information.

A few key events led to the discovery of the double helix structure

A discovery of paramount importance in molecular genetics was made during the 1950s by James Watson and Francis Crick. At that time, DNA was already known to be composed of nucleotides. However, it was not understood how the nucleotides are bonded together to form the primary and secondary structure of DNA. Watson and Crick committed themselves to the goal of elucidating the structure of DNA, because they felt that this knowledge was needed to understand the functioning of genes. Others, like Rosalind Franklin and Maurice Wilkins, shared this view. Before we describe the characteristics of the double helix, let's consider the events that provided the scientific framework for Watson and Crick's breakthrough.

In the early 1950s, Linus Pauling proposed that proteins can fold into secondary structural regions known as an α-helices (Figure 9-11a). To discover this, Pauling built large models by linking together simpler ball-and-stick units (Figure 9-11b). In this

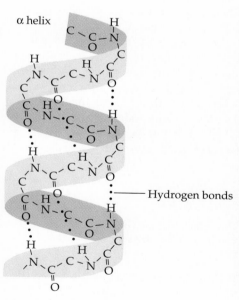

(a) An α helix

(b) Linus Pauling and his model

FIGURE **9-11**

Linus Pauling and the α-helix protein structure. (a) An α-helix is a secondary structure found in proteins. This structure emphasizes the polypeptide backbone (shown as a green ribbon), which is composed of amino acids linked together in a linear fashion. Hydrogen bonding between hydrogen and oxygen atoms stabilizes the helical conformation. **(b)** A photograph of Linus Pauling with a ball and stick model.

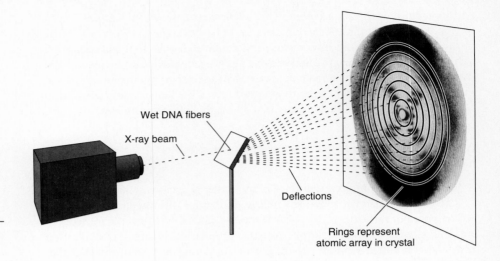

Wet DNA fibers

X-ray beam

Deflections

Rings represent
atomic array in crystal

FIGURE **9-12**

**A diffraction pattern of wet DNA
fibers in the B-DNA conformation.**

way, he could visualize if atoms fit together properly in a complicated three-dimensional structure. As we will see, Watson and Crick used a similar approach to solve the structure of the DNA double helix. Interestingly, Watson and Crick were well aware that Pauling might figure out the structure of DNA before they could. This provided a stimulating rivalry between the researchers.

A second important development that led to the elucidation of the double helix was X-ray diffraction data. When a purified substance, such as DNA, is subjected to X-rays, it will produce a well-defined diffraction pattern if the molecule is organized into a regular structural pattern (e.g., a helix). An interpretation of the diffraction pattern (using mathematical theory) can ultimately provide information concerning the structure of the molecule. Rosalind Franklin, working in the laboratory of Maurice Wilkins at King's College in London, used X-ray diffraction to study wet DNA fibers. The diffraction pattern of Rosalind Franklin's DNA fibers is shown in Figure 9-12. This pattern suggested several structural features of DNA. First, it was consistent with a helical structure. Second, the diameter of the helical structure was too wide to be only a single-stranded helix. Finally, the diffraction pattern indicated that the helix contains about ten nucleotides per complete turn. As we will discuss, these observations were instrumental in working out the structure of DNA.

EXPERIMENT 9B

**Chargaff and his colleagues found that DNA has a striking
biochemical composition in which the amount of A equals T
and the amount of G equals C**

A third piece of information that led to the discovery of the double helix structure came from the studies of Erwin Chargaff at Columbia University. Chargaff and his colleagues pioneered many of the biochemical techniques for the isolation, purification, and measurement of nucleic acids from living cells. This is not a trivial undertaking, because the biochemical composition of living cells is very complex. At the time of Chargaff's work, it was already known that the building blocks of DNA are nucleotides containing the bases adenine, thymine, cytosine, or guanine. In the experiment described here, Chargaff analyzed the base composition within DNA that was isolated from many different species. He expected that the results might provide important clues concerning the structure of DNA.

THE HYPOTHESIS

An analysis of the base composition of DNA in different organisms may reveal important features about the structure of DNA.

TESTING THE HYPOTHESIS

Starting material: The following types of cells were obtained: *Escherichia coli*, *Diplococcus pneumoniae* (type III), yeast, turtle red blood cells, salmon sperm cells, chicken red blood cells, and human liver cells.

Experimental Level   **Conceptual Level**

1. For each type of cell, extract the chromosomal material. This can be done in a variety of ways including the use of high salt, detergent, or mild alkali treatment. Note: The chromosomes contain both DNA and protein.

2. Remove the protein. This can be done in several ways including chloroform extraction or treatment with proteases.

3. Hydrolyze the DNA to remove the bases. A common way to do this is by strong acid treatment.

4. Separate the bases by chromatography. Paper chromatography provides an easy way to separate the four types of bases. (The technique of chromatography is described in the appendix.)

5. Extract bands from paper into solutions and determine the amounts of each base by spectroscopy. Each base will absorb light at a particular wavelength. By examining the absorption profile of a sample of base, it is then possible to calculate the amount of the base. (Spectroscopy is described in the appendix.)

6. Compare the results for cells from different organisms.

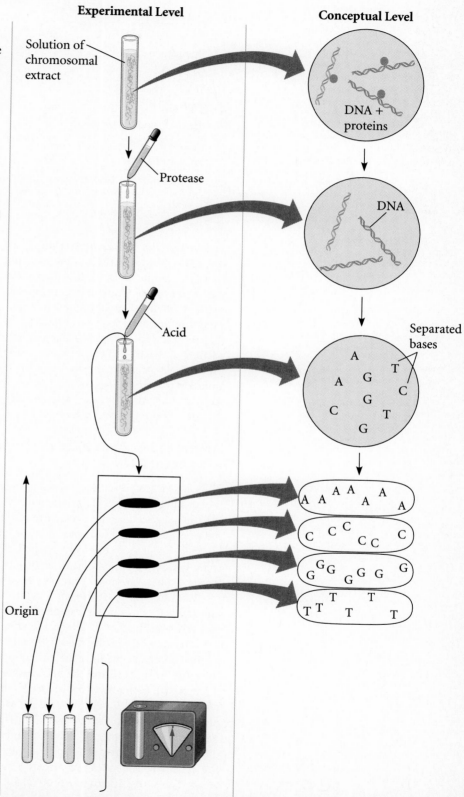

Base Content in the DNA from a Variety of Organisms*

Organism	% of Bases (based on molarity)			
	Adenine	Thymine	Guanine	Cytosine
Escherichia coli	26.0	23.9	24.9	25.2
Pneumococcus (type III)	29.8	31.6	20.5	18.0
Yeast	31.7	32.6	18.3	17.4
Turtle (RBC)	28.7	27.9	22.0	21.3
Salmon (sperm)	29.7	29.1	20.8	20.4
Chicken (RBC)	28.0	28.4	22.0	21.6
Human (liver)	30.3	30.3	19.5	19.9

*Data selected from Tables IX, X, and XII in Chargaff, E., and Davidson, J. N., eds (1955) *The Nucleic Acids: Chemistry and Biology,* pp. 307–368. Academic Press, New York.

The data shown here are only a small sampling of Chargaff's results. Between the late 1940s and early 1950s, Chargaff published many papers concerned with the chemical composition of DNA from biological sources. Hundreds of measurements were made. The compelling observation was that the amount of adenine was similar to thymine and the amount of cytosine was similar to guanine. These results are not sufficient to propose a model for the secondary structure of DNA. However, they provided the important clue that DNA is structured so that each molecule of adenine interacts with thymine, and each molecule of guanine interacts with cytosine. A DNA structure in which A binds to T, and G to C, would explain the equal amounts of A and T, and G and C observed in Chargaff's experiments. As we will see, this observation became crucial evidence that Watson and Crick used to deduce the structure of the double helix.

Watson and Crick deduced the double helical structure of DNA

Thus far, we have discussed several key pieces of information used in determining the structure of DNA. In particular, the crystallography work of Franklin and Wilkins suggested a helical structure with two strands, with ten bases per turn. And the work of Chargaff indicated that the amount of A equals T, and the amount of G equals C. Furthermore, Watson and Crick were familiar with Pauling's success in using ball-and-stick models to deduce the secondary structure of proteins. With these key observations, they set out to solve the secondary structure of DNA.

Watson and Crick assumed that DNA is composed of nucleotides that are linked together in a linear fashion. They also assumed that the chemical linkage between two nucleotides is always the same. With these ideas in mind, Watson and Crick tried to build ball-and-stick models that incorporated all the known experimental observations. Since the diffraction pattern suggested that there must be two (or more) strands within the helix, a critical question was how two strands could interact with each other. As discussed in the book *The Double Helix*, in an early attempt at model building, they considered the possibility that the negatively charged phosphate groups together with magnesium ions were promoting an interaction between the backbones of DNA strands (Figure 9-13). However, more detailed diffraction data were not consistent with this model.

Since the Mg^{++} idea was incorrect, it was back to the drawing board (or back to the ball-and-stick units) for Watson and Crick. During this time, Rosalind

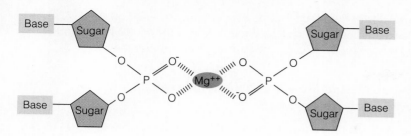

FIGURE **9-13**

An incorrect hypothesis for the structure of the DNA double helix. This illustration shows an early hypothesis of Watson and Crick's suggesting that two DNA strands interact by a crosslink between the negatively charged phosphate groups in the backbone and divalent Mg^{++} cations.

Franklin had produced even clearer X-ray diffraction patterns, which provided greater detail concerning the relative locations of the bases and backbone of DNA. In their model building, the emphasis shifted to models containing the two backbones on the outside of the model, with the bases projecting toward each other. A major breakthrough came when the researchers considered that a two-strand interaction could be promoted by hydrogen bonding between the nitrogenous bases. At first, a structure was considered in which the bases form hydrogen bonds with the identical base in the opposite strand (i.e., A to A, T to T, G to G, and C to C). However, the model building revealed that the bases could not fit together this way. The final hurdle was overcome when it was realized that the hydrogen bonding of adenine to thymine was structurally similar to that of cytosine to guanine. With an interaction between A and T, and between G and C, the ball-and-stick models showed that the two strands would fit together properly. This ball-and-stick model, shown in Figure 9-14, was consistent with all the known data regarding DNA structure. For

(a) Watson and Crick

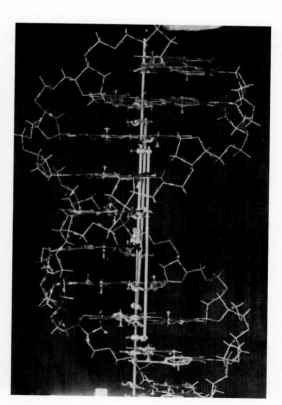

(b) Original model of DNA double helix

FIGURE **9-14**

Watson and Crick and their model of the DNA double helix. (a) James Watson is shown here on the left and Francis Crick on the right. **(b)** The molecular model they originally proposed for the double helix, in which A hydrogen bonds to T, and G hydrogen bonds with C.

their work, Watson, Crick, and Wilkins were awarded the Nobel prize in 1962. As will be discussed in Chapters 11 and 12, the structure of the double helix immediately suggested ways that DNA can replicate and also store information.

The molecular structure of the DNA double helix has several key features

The general structural features of the double helix are shown in Figure 9-15. In a DNA double helix, two DNA strands are twisted together around a common axis to form a structure that resembles a circular staircase. There are ten *base pairs* within a complete twist. This double-stranded structure is stabilized by *hydrogen bonding* between the base pairs. A distinguishing feature of this hydrogen bonding is that an adenine base in one strand always hydrogen bonds with a thymine base in the opposite strand. Or a guanine base always hydrogen bonds with a cytosine. This **A–T/G–C rule** (also known as Chargaff's rule) explained the earlier data of Chargaff showing that the DNA from many organisms contains equal amounts of A and T, and equal amounts of G and C (see Experiment 9B). The A–T/G–C rule indicates that purines (A and G) always bond with pyrimidines (T and C). As noted in Figure 9-15, there are three hydrogen bonds between G and C but only two

Key Features

1. Two strands of DNA form a right-handed double helix.

2. The bases in opposite strands hydrogen bond according to the A/T and G/C rule.

3. The two strands are antiparallel with regard to their 5′ to 3′ directionality.

4. There are ~10.5 nucleotides in each strand per complete 360° turn of the helix.

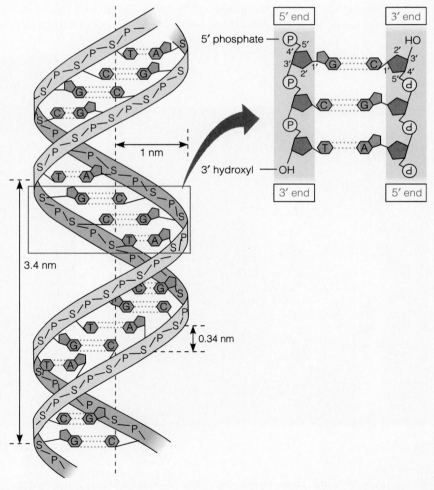

FIGURE **9-15**

Key features of the structure of the double helix.

between A and T. For this reason, DNA sequences that have a high proportion of G and C tend to form more stable double-stranded structures.

The A–T/G–C rule implies that base sequences within two DNA strands are **complementary** to each other. We can predict the sequence in one DNA strand if the sequence in the opposite strand is known. For example, if one strand has the sequence of ATGGCGGATTT, then the opposite strand must be TACCGCCTAAA. In genetic terms, we would say that these two sequences are complementary to each other, or that the two sequences exhibit complementarity.

Another issue to consider in a double-stranded structure is the orientation of the two backbones. When going from the top of Figure 9-15 to the bottom, one strand is running in the 5′ to 3′ direction while the other strand is 3′ to 5′. This opposite orientation of the two DNA strands is referred to as an **antiparallel** arrangement. An antiparallel structure was initially proposed in the models of Watson and Crick.

Figure 9-16a is a schematic model that emphasizes certain molecular features of DNA structure. The DNA backbone is what forms a helical structure. It is twisted in a right-handed direction. The black line depicts a **central axis** around which the helical backbones are twisted. Counting the bases, if you move past ten base pairs, you have gone 360° around the backbone. In addition, the bases in this model are depicted as flat rectangular structures that are hydrogen bonding in pairs. (The hydrogen bonds are the dashed lines.) Although the bases are not actually rectangular, they do form flattened planar structures (see Figure 9-16a). Within DNA, the bases are oriented so that the flattened regions are facing each other. This is referred to as **base stacking**. In other words, if you think of the bases as flat plates, these plates are stacked on top of each other in the double-stranded DNA structure. Along with hydrogen bonding, base stacking is a structural feature that stabilizes the double helix.

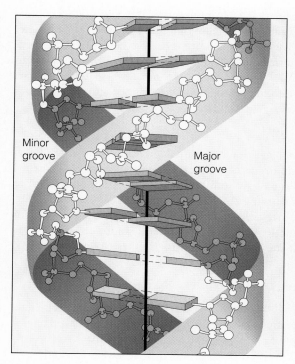

(a) Ball and stick model of DNA

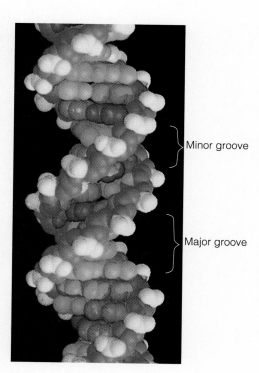

(b) Space-filling model of DNA

FIGURE **9-16**

Two models of the double helix. (a) Ball-and-stick model of the double helix. The ribose–phosphate backbone is shown in detail, while the bases are depicted as flattened rectangles. **(b)** Space-filling model of the double helix.

Figure 9-16b is a space-filling model for DNA in which the atoms are represented by spheres. This model emphasizes what the surface of DNA is like. Note that the blue and yellow backbone (composed of sugar and phosphate groups, respectively) is on the outermost surface. In a living cell, the backbone has the most direct contact with water. In contrast, the bases are more internally located within the double-stranded structure. Biochemists use the term **grooves** to describe the indentations where the atoms of the bases are in contact with the surrounding water. In the helical structure of DNA, there are actually two grooves as you go around the structure. These are called the **major groove** and the **minor groove**. As will be discussed in later chapters, certain proteins can bind within these grooves and interact with a particular sequence of bases, thereby regulating gene expression.

DNA can form alternative types of double helices

The DNA double helix can form different types of secondary structure. Figure 9-17 compares the structures of A-DNA, B-DNA, and Z-DNA. The highly detailed structures shown here were deduced by X-ray crystallography on short segments of DNA. B-DNA is the predominant form of DNA found in living cells. However, under certain conditions, the two strands of DNA can twist into A-DNA and Z-DNA, which differ significantly from B-DNA. A- and B-DNA are right-handed helices; Z-DNA has the opposite, left-handed, orientation. In addition, the helical backbone in Z-DNA appears to slightly zig-zag as it winds itself around in the double helical structure. The numbers of base pairs per 360° turn are 10.9, 10.0, and 12.0 in A-, B-, and Z-DNA, respectively. In B-DNA, the bases tend to be centrally located and the hydrogen bonds between base pairs occur relatively perpendicular to the central axis. In contrast, the bases in A-DNA and Z-DNA are substantially tilted relative to the central axis.

The ability of the predominant B-DNA to adopt A-DNA and Z-DNA conformations depends on the conditions. In crystallization studies, A-DNA occurs under conditions of low humidity. The ability of a double helix to adopt a Z-DNA conformation depends on various factors. At high ionic strength (i.e., high salt concentration), formation of a Z-DNA conformation is favored by a sequence of bases that alternates between purines and pyrimidines. One such sequence is

5′–GCGCGCGCG–3′
3′–CGCGCGCGC–5′

At lower ionic strength, the methylation of cytosine bases can favor Z-DNA formation. Cytosine **methylation** occurs when a cellular enzyme attaches a methyl group ($-CH_3$) to the cytosine base. In addition, negative supercoiling (a topic to be discussed in Chapter 10) favors the Z-DNA conformation. An important question is the functional significance of A- and Z-DNA. There is little evidence to indicate that A-DNA is biologically important but several studies in recent years have suggested a possible role for Z-DNA in the process of transcription.

DNA can form a triple helix, called triplex DNA

A surprising discovery made by Alexander Rich, David Davies, and Gary Felsenfeld during the late 1950s was that DNA can form a triple helical structure called **triplex DNA**. This triplex was formed *in vitro* using pieces of DNA that were made synthetically. Although this result was interesting, it seemed to have little, if any, biological relevance.

About thirty years later, interest in triplex DNA was renewed by the discovery that triplex DNA can form by the mixture of natural double-stranded DNA and a third short strand that is synthetically made. The synthetic strand binds into the major groove of the naturally occurring double-stranded DNA (Figure 9-18). As shown here, an interesting feature of triplex DNA formation is that it is *sequence specific*. In other words, the synthetic third strand only incorporates itself into a

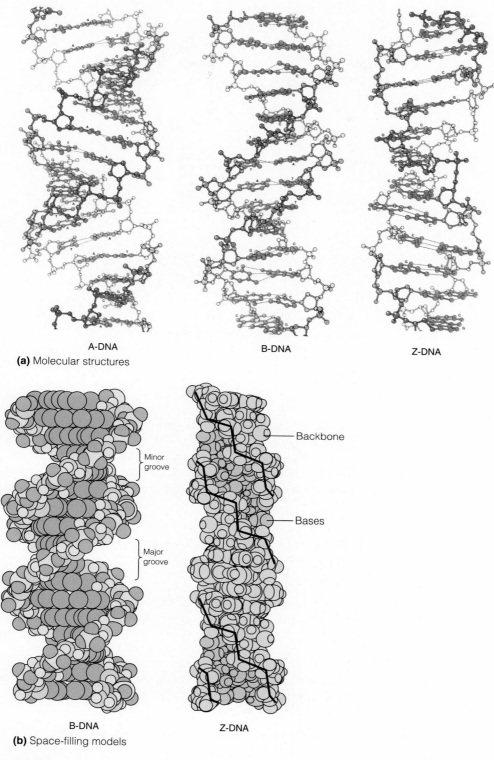

(a) Molecular structures

A-DNA   B-DNA   Z-DNA

(b) Space-filling models

B-DNA   Z-DNA

FIGURE **9-17**

Comparison of the structures of A-DNA, B-DNA, and Z-DNA. (a) The highly detailed structures shown here were deduced by X-ray crystallography performed on short segments of DNA, rather than the less detailed structures obtained from DNA wet fibers. The diffraction pattern obtained from the crystallization of short segments of DNA provides much greater detail concerning the exact placement of atoms within a double helical structure. Alexander Rich at MIT, Richard Dickerson then at the California Institute of Technology, and their colleagues were the first researchers to crystallize a short piece of DNA. **(b)** Space-filling models of the B-DNA and Z-DNA structures.

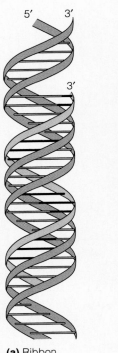

(a) Ribbon model

(b) Example sequence of bases

FIGURE **9-18**
_____

The structure of triplex DNA. Within triplex DNA, the biological DNA is double-stranded due to hydrogen bonding between A and T, and between G and C. The third, synthetic, strand binds within the major groove of this double-stranded structure according to the rule T to A–T, and C to G–C.

triple helix due to specific interactions between the synthetic DNA and the biological DNA. The pairing rules are that a thymine in the synthetic DNA will bind near an A–T pair in the biological DNA, and that a cytosine in the synthetic DNA will bind near a G–C pair. It is not yet clear if triplex DNA has any biological significance, although it has been hypothesized to play a role in genetic recombination (see Chapter 18).

The three-dimensional structure of DNA within chromosomes requires additional folding and the association with proteins

The double helical structure of DNA is its secondary structure. To fit within a living cell, the long double helical structure of chromosomal DNA must be greatly compacted into a tertiary conformation. The double helix becomes greatly twisted and folded with the aid of DNA-binding proteins. Figure 9-19 depicts the relationship between the DNA double helix and the compaction that occurs within a eukaryotic chromosome. Chapter 10 will be devoted to the topic of chromosome structure and organization.

RNA molecules are also composed of strands that can fold into secondary and tertiary structures

Let's now turn our attention to RNA structure, which bears many similarities to DNA structure. The primary structure of an RNA strand is much like a DNA strand (Figure 9-20). Strands of RNA are typically several hundred or several thousand nucleotides in length which is much shorter than chromosomal DNA. When RNA is made during transcription, the DNA is used as a template to make a copy

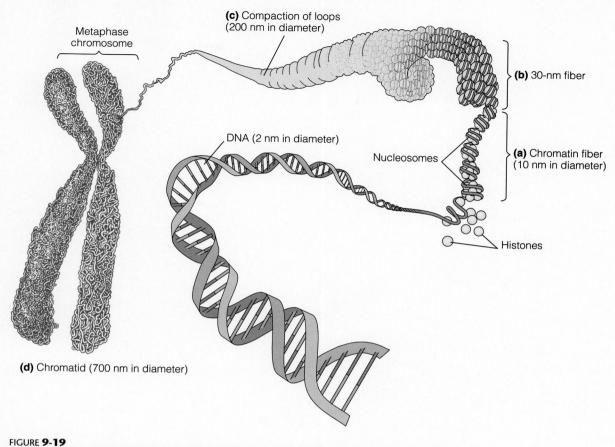

Metaphase chromosome

(c) Compaction of loops (200 nm in diameter)

(b) 30-nm fiber

DNA (2 nm in diameter)

Nucleosomes

(a) Chromatin fiber (10 nm in diameter)

Histones

(d) Chromatid (700 nm in diameter)

FIGURE **9-19**

The steps in eukaryotic chromosomal compaction leading to the metaphase chromosome. The DNA double helix is wound around histone proteins and then further compacted to form a highly condensed metaphase chromosome. The levels of DNA compaction will be described in greater detail in Chapter 10.

of single-stranded RNA. In most cases, only one of the two DNA strands is used as a template for RNA synthesis. Therefore, both complementary strands of RNA are not usually made. Nevertheless, relatively short sequences within one RNA molecule or between two separate RNA molecules can be complementary to each other. These sequences can form short double-stranded regions.

The secondary structure of RNA molecules is due to the ability of complementary regions to form base pairs between A and U and between G and C. This base pairing allows short regions to form a double helix. RNA double helices are antiparallel and typically are the right-handed A-form with 11–12 base pairs per turn. As shown in Figure 9-21, different types of secondary structural patterns are possible. These include bulges, internal loops, multibranched junctions, and stem-loops (also called hairpins). Note that these structures contain regions of complementarity punctuated by regions of noncomplementarity. In Figure 9-21, the complementary regions are held together by connecting hydrogen bonds, while the noncomplementary regions have their bases projecting away from the double-stranded region.

There are many factors that contribute to the tertiary structure of RNA molecules. Within the RNA molecule itself, these include the base-paired double-stranded helices, stacking between bases, and hydrogen bonding between bases and backbone regions. In addition, interactions with ions, small molecules,

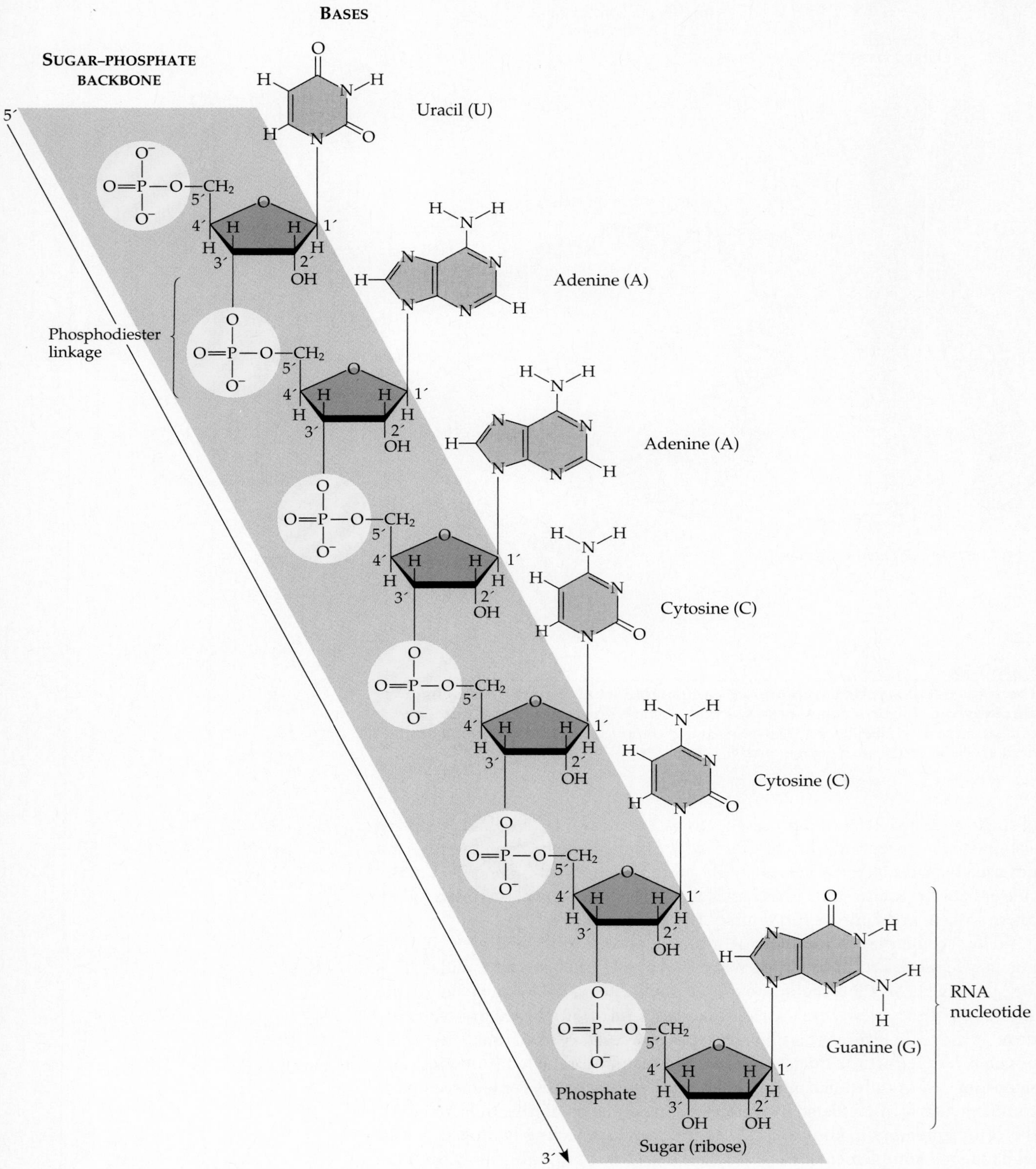

FIGURE **9-20**

A strand of RNA. This structure is very similar to a DNA strand (see Figure 9-10), except that the sugar is ribose instead of deoxyribose and uracil is substituted for thymine.

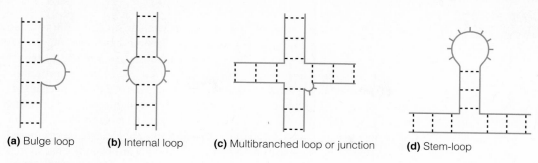

(a) Bulge loop   **(b)** Internal loop   **(c)** Multibranched loop or junction   **(d)** Stem-loop

FIGURE **9-21**

Possible secondary structures of RNA molecules. The double-stranded regions are depicted by connecting hydrogen bonds. Loops are noncomplementary regions that are not hydrogen bonded with complementary bases. Double-stranded RNA structures can form within a single RNA molecule or between two separate RNA molecules. This figure is modified from Jaeger et al. (1993) *Ann. Rev. Biochem. 62*, 255–287.

and large proteins may influence RNA tertiary structure. Figure 9-22 depicts the tertiary structure of a transfer RNA molecule known as tRNA[phe] (i.e., a tRNA molecule that carries the amino acid phenylalanine). As will be discussed in Chapter 11, tRNAs function during the translation of mRNA into an amino acid sequence. tRNA[phe] was the first naturally occurring RNA to have its structure solved. Note that this RNA molecule contains several double-stranded and single-stranded regions. These regions are folded to produce the three-dimensional structure.

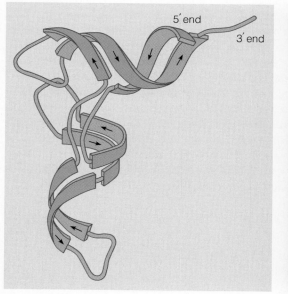

(a) Ribbon model

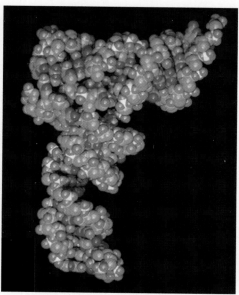

(b) Space-filling model

FIGURE **9-22**

The tertiary structure of tRNA[phe], the transfer RNA molecule that carries phenylalanine. (a) Here the double-stranded regions of the molecule are shown as antiparallel ribbons. **(b)** A space-filling model of tRNA.

CONCEPTUAL SUMMARY

The molecular structure of nucleic acids underlies their function. *Nucleotides*, which are composed of a sugar, phosphate, and nitrogenous base, form the repeating structural unit of nucleic acids. The *primary structure* is a *strand* that contains a linear sequence of nucleotides. The formation of *secondary structure* occurs because strands of DNA (and RNA) can form *complementary* regions due to hydrogen bonding between adenine and thymine (or uracil), and between guanine and cytosine. *Base stacking* also stabilizes a double-stranded structure. The most common form of DNA is a right-handed double helix; the *backbones* in the *double helix* are composed of *sugar–phosphate linkages*, with the bases projecting inward from the backbones and hydrogen bonding with each other. The two strands are *antiparallel*. In DNA, different helical conformations have been identified, including A-DNA, B-DNA, and Z-DNA. B-DNA is the predominant form found in living cells, but short regions of Z-DNA may play an important functional role. Within chromosomes, DNA is folded into a *tertiary conformation* with the aid of proteins. The structure of chromosomes will be discussed in the next chapter. It is also common for short segments within RNA to form double helical structures such as hairpins, bulges, loops, and junctions. As we will see in later chapters, RNA secondary structures can also play many important functional roles. The final tertiary structure of RNA is dictated by several factors including double helical regions, *base stacking*, hydrogen bonding between bases and backbone, and interactions with other molecules.

EXPERIMENTAL SUMMARY

Two different experimental approaches were used to show that DNA is the genetic material. Avery, MacLeod, and McCarty took advantage of Griffith's observations regarding transformation in pneumococci. They purified DNA from type IIIS strains and used that purified DNA to transform type IIR strains into type IIIS. Furthermore, DNase prevented the transformation, whereas RNase and protease did not. These observations indicated that DNA is the transforming substance that provides type IIR bacteria with a new trait. Hershey and Chase showed that T2 bacteriophage injects DNA into bacterial cells, showing that DNA is the genetic material of this virus. Taken together, these two studies were instrumental in convincing scientists that the genetic material is composed of DNA.

In the 1950s, several experimental observations enabled Watson and Crick to solve the secondary structure of DNA. The crystallography work of Franklin and Wilkins indicated a helical structure with two strands, with ten bases per turn. And the work of Chargaff indicated that the amount of A equals T, and the amount of G equals C. With these observations, Watson and Crick followed Pauling's strategy of using ball-and-stick models to visualize the structure of macromolecules. In this way, they deduced the secondary structure of DNA.

PROBLEM SETS

SOLVED PROBLEMS

1. A naturally occurring DNA has the following sequence:

5′–A–A–G–G–A–A–A–A–G–G–G–A–G–G–A–G–A–G–3′
3′–T–T–C–C–T–T–T–T–C–C–C–T–C–C–T–C–T–C–5′

What sequence of DNA molecule could form triplex DNA with this double helix?

Answer: 5′–T–T–C–C–T–T–T–T–C–C–C–T–C–C–T–C–T–C–3′

2. As we will discuss in future chapters, the formation of stem-loop structures within RNA can be crucial in many functional ways. For example, the formation of a stem-loop at the beginning of mRNA can influence the rate at which that mRNA is translated into protein. A hypothetical sequence at the beginning of an mRNA molecule is

5′-A<u>UUUGC</u>CC<u>UAGCAAAC</u>GUAG<u>CAAAC</u>G. . .rest of the coding
sequence

Using two out of the three underlined sequences, draw two possible models for potential stem-loop structures at the 5′ end of this mRNA.

Answer:

CONCEPTUAL QUESTIONS

1. What is the meaning of the term genetic material?

2. After the DNA from type IIIS bacteria is exposed to type IIR bacteria, list all of the steps that you think must occur for the bacteria to start making a type IIIS capsule.

3. What are the building blocks of a nucleotide? With regard to the 5′ and 3′ positions on a sugar molecule, how are nucleotides linked together to form a strand of DNA?

4. Draw the structure of guanine, guanosine, and deoxyguanosine triphosphate.

5. Draw the structure of a phosphodiester linkage.

6. Describe how bases interact with each other in the double helix. This discussion should address the issues of complementarity, hydrogen bonding, and base stacking.

7. If one DNA strand is 5′–GGCATTACACTAGGCCT–3′, what is the complementary strand?

8. What is meant by the term DNA sequence?

9. Make a side-by-side drawing of two DNA helices, one with 10 base pairs per 360° turn and the other with 15 base pairs per 360° turn.

10. Discuss the differences in the structural features of A-DNA, B-DNA, and Z-DNA.

11. What parts of a nucleotide (namely, phosphate, sugar, and/or bases) occupy the major and minor grooves of double-stranded DNA? If a DNA-binding protein does not recognize a specific nucleotide sequence, do you expect that it recognizes the major groove, the minor groove, or the DNA backbone? Explain.

12. List the structural differences between DNA and RNA.

13. Draw the structure of deoxyribose, and number the carbon atoms. Describe the numbering of the carbon atoms in deoxyribose with regard to the directionality of a DNA strand. In a DNA double helix, what does the term antiparallel mean?

14. Write out a sequence of an RNA molecule that could form a stem-loop with 24 nucleotides in the stem and 16 nucleotides in the loop.

15. Compare the structural features of a double-stranded RNA stem structure with those of a DNA double helix.

16. Which of the following DNA double helices would be more difficult to separate into single-stranded molecules?

A.

GGCGTACCAGCGCAT
CCGCATGGTCGCGTA

B.

ATACGATTTACGAGA
TATGCTAAATGCTCT

Explain your choice.

17. What structural feature allows DNA to store information?

18. Discuss the structural significance of complementarity in DNA and in RNA.

19. An organism has a G + C content of 64% in its DNA. What are the percentages of A, T, G, and C?

20. Some viruses contain single- or double-stranded RNA as their genetic material. If a virus contains the following amounts of nucleotides, would you conclude that its genetic material is single-stranded or double-stranded: A = 15%, U = 29%, G = 28%, and C = 28%?

21. Let's suppose that you have recently identified an organism that was scraped off of an asteroid that hit the earth. (Fortunately, no one was injured.) When you analyze this organism, you discover that its DNA is a triple helix, composed of six different nucleotides: A, T, G, C, X, and Y. You measure the chemical composition of the bases and find the following amounts of these six bases: A = 24%, T = 23%, G = 11%, C = 12%, X = 21%, Y = 9%. What rules would you propose govern triplex DNA formation in this organism? Note: There is more than one possibility.

22. Upon further analysis of the DNA described in problem 21, you discover that the triplex DNA in this alien organism is composed of a double helix, with the third helix wound within the major groove (just like the DNA in Figure 9-18). How would you propose that this DNA is able to replicate itself? In your answer, be specific about the base pairing rules within the double helix, and which part of the triplex DNA would be replicated first.

23. A DNA-binding protein recognizes the following double-stranded sequence:

$$5'–GCCCGGGC–3'$$
$$3'–CGGGCCCG–5'$$

This type of double-stranded structure could also occur within the stem region of an RNA molecule. Discuss the structural differences between RNA and DNA that might prevent this DNA-binding protein from recognizing a double-stranded RNA molecule.

24. Within a protein, certain amino acids are positively charged (e.g., lysine and arginine), some are negatively charged (e.g., glutamate and aspartate), some are polar but uncharged, and some are nonpolar. If you knew that a DNA-binding protein was recognizing the DNA backbone rather than base sequences, which amino acids in the protein would be good candidates for interacting with the DNA?

EXPERIMENTAL QUESTIONS

1. In the experiment described in Figure 9-3, list several possible reasons why only a small percentage of the type IIR bacteria were converted to type IIIS.

2. Another interesting trait that some bacteria exhibit is resistance to killing by antibiotics. For example, certain strains of bacteria are resistant to tetracycline, whereas other strains are sensitive. Describe an experiment that you would carry out to demonstrate that tetracycline resistance is an inherited trait encoded by the DNA of the resistant strain.

3. In Experiment 9A, give possible explanations why some of the DNA is in the supernatant.

4. Plot the results of Experiment 9A if the radioactivity in the pellet, rather than in the supernatant, had been measured.

5. In Experiment 9A, why were ^{32}P and ^{35}S chosen as radioisotopes to label the phages?

6. List some possible reasons that less than 100% of the phage protein is removed from the bacterial cells during the shearing process.

7. It is possible to specifically label DNA or RNA by providing bacteria with radiolabeled thymine or uracil, respectively. With this type of tool, design an experiment to show whether a newly identified bacteriophage contains DNA or RNA as its genetic material. Describe your expected results depending on whether the genetic material is DNA or RNA.

8. The type of model building that was used by Pauling, Watson, and Crick involved the use of small ball-and-stick units. Now we can do model building on a computer screen. Even though you may not be familiar with this approach, discuss some potential advantages computers might provide in molecular model building.

9. In Chargaff's experiment (Experiment 9B), what is the purpose of paper chromatography?

10. Would Chargaff's experiments have been convincing if they had been done on only one species? Discuss.

QUESTIONS FOR STUDENT DISCUSSION/COLLABORATION

1. Try to propose structures for a genetic material that are substantially different from the double helix. Remember that the genetic material must have a way to store information and a way to be faithfully replicated.

2. How might you provide evidence that DNA is the genetic material in mice?

CHAPTER 10

CHROMOSOME ORGANIZATION AND MOLECULAR STRUCTURE

VIRAL GENOMES

PROKARYOTIC CHROMOSOMES

EUKARYOTIC CHROMOSOMES

Chromosomes are the structures that contain the genetic material. They are composed of a complex between DNA and various proteins. The term **genome** describes all the types of genes and DNA sequences that an organism can possess. For prokaryotes, the genome is typically a single circular chromosome. For eukaryotes, the nuclear genome refers to one complete set of chromosomes that reside in the cell nucleus. In other words, the haploid complement of chromosomes is considered a nuclear genome. As discussed in Chapter 7, eukaryotes possess a mitochondrial genome, and plants also have a chloroplast genome. Unless otherwise noted, the term eukaryotic genome refers to the nuclear genome.

The primary function of the genetic material is to store the information needed to produce the characteristics of an organism. As we have seen in Chapter 9, the sequence of bases in a DNA molecule can store information. To fulfill their role at the molecular level, chromosomal sequences facilitate four important processes. DNA sequences are necessary for (1) the synthesis of RNA and cellular proteins, (2) the proper segregation of chromosomes, (3) the replication of chromosomes, and (4) the compaction of chromosomes so they can fit within living cells. In this chapter, we will examine the general organization of DNA sequences within viral, bacterial, and eukaryotic chromosomes. In addition, the molecular mechanisms that account for the packaging of chromosomal DNA in viruses, bacteria, and eukaryotic cells will be described. We will begin by considering the comparatively simple genomes of viruses.

VIRAL GENOMES

Viruses are small infectious particles that contain nucleic acid as their genetic material, surrounded by a *capsid* of proteins (Figure 10-1a). In addition, some bacteriophages contain a sheath, base plate, and tail fibers, and certain eukaryotic viruses also have an *envelope* consisting of a membrane embedded with spike proteins (Figure 10-1b). By themselves, viruses are not cellular organisms. They do not contain energy-producing enzymes, ribosomes, or other cellular organelles. Instead, viruses rely on their **host cells** (i.e., the cells they infect) for replication. In general, most viruses exhibit a limited **host range**, which is the spectrum of host cell types that a virus can infect. Many viruses can infect only specific types of cells of only one host species. Depending on the life cycle of the virus, the host cell may or may not be destroyed during the process of viral replication and release. In this section, we will consider the genetic composition of viruses and describe how viral genomes are packaged into virus particles.

Viral genomes are relatively small and can be composed of RNA or DNA

The **viral genome** is the genetic material packaged within the capsid of virus particles. The term **viral chromosome** is also used to describe the viral genome. Surprisingly, the nucleic acid composition of viral genomes varies markedly among different types of viruses. Table 10-1 describes the genome characteristics of a few selected viruses. The genome can be DNA or RNA, but not both. In some cases it is single-stranded, whereas in others it is double-stranded. Depending on the virus, the genome can be linear or circular.

As shown in Table 10-1, viral genomes vary in size from a few thousand to more than a hundred thousand nucleotides in length. For example, the genomes of some simple viruses, such as Qβ, are only a few thousand nucleotides in length and contain only a few genes. Other viruses, particularly those with a complex structure, can contain many more genes. The T even phages (T2, T4, etc.), discussed in Chapters 6 and 9, are examples of more complex viruses.

FIGURE **10-1**

General structure of viruses. (a) The simplest viruses contain a nucleic acid molecule (DNA or RNA) surrounded by a protein capsid. **(b)** Other viruses also contain an envelope composed of a lipid bilayer and spike proteins. The lipid bilayer is obtained from the host cell when the virus buds through the plasma membrane.

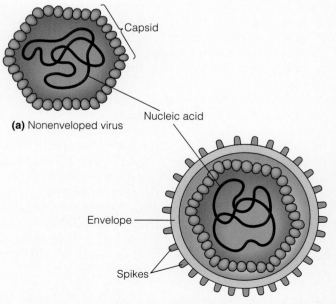

Capsid

(a) Nonenveloped virus

Nucleic acid

Envelope

Spikes

(b) Enveloped virus with spikes

TABLE 10–1 Characteristics of selected viral genomes

Virus	Host	Type of Nucleic Acid	Size (Kilobases)*	Genes
Parvovirus	Mammals	ssDNA	5.0	5
Phage fd	E. coli	ssDNA	6.4	10
Lambda	E. coli	dsDNA	48.5	36
T4	E. coli	dsDNA	165.0	>190
Qß	E. coli	ssRNA	4.2	4
TMV	Many plants	ssRNA	6.4	4
Influenza virus	Mammals	ssRNA	13.5	12
Reovirus	Animals and plants	dsRNA	23.0	22

* A Kilobase equals 1000 bases.

Viral genomes are packaged into the capsid in an assembly process

In an infected cell, the life cycle of the virus eventually leads to the synthesis of viral nucleic acids and proteins. Newly synthesized viral chromosomes and capsid proteins must then come together and assemble to make mature virus particles. Viruses with a simple structure may **self-assemble**. This means that the nucleic acid and capsid proteins spontaneously bind to each other to form a mature virus. The structure of one self-assembling virus, the tobacco mosaic virus is shown in Figure 10-2. As shown here, the proteins assemble around the RNA genome, which becomes trapped inside the hollow capsid. This assembly process can occur *in vitro* if purified capsid proteins and RNA are mixed together.

Some viruses, such as T2 bacteriophage, have more complicated structures that do not self-assemble. The correct assembly of certain viruses requires the help of proteins that are not found within the mature virus particle itself. When virus assembly requires the participation of noncapsid proteins, the process is called **directed assembly**, as these noncapsid proteins are necessary to direct the proper assembly of the virus. The noncapsid proteins required for directed assembly usually carry out two main functions. First, some proteins, called scaffolding proteins, catalyze the assembly process and are transiently associated with the capsid. However, as viral assembly nears completion, the scaffolding proteins are expelled from the mature virus. Second, some noncapsid proteins act as proteases that specifically cleave viral capsid proteins. This cleavage produces a capsid protein that is slightly smaller and able to assemble correctly. For many viruses, the cleavage of capsid proteins into smaller units is an important event that precedes viral assembly.

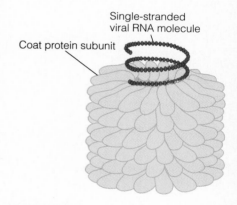

FIGURE **10-2**

Structure of the tobacco mosaic virus. A complete TMV virus is composed of a coiled RNA molecule surrounded by 2130 identical protein subunits. Only a portion of the TMV virus is shown here. Several layers of proteins have been omitted from the upper part of this illustration to reveal the RNA genome, which is normally trapped inside the protein coat.

PROKARYOTIC CHROMOSOMES

Let's now turn our attention to the chromosomes found within bacterial species. Inside a bacterial cell, the chromosome is highly compacted and found within a region of the cell known as the **nucleoid**. Although bacteria usually contain a single type of chromosome, more than one copy of that chromosome is likely to be found within one bacterial cell. It is common for bacteria to have one to four identical chromosomes per cell. However, the number of copies varies depending on the bacterial species and the growth conditions. As shown in Figure 10-3, each chromosome occupies a distinct nucleoid region within the cell. Unlike the eukaryotic nucleus, the bacterial nucleoid is not a separate cellular compartment bounded by a membrane. Rather, the DNA in the nucleoid is in direct contact with the cytoplasm of the cell. In this section, we will study two important features of bacterial chromosomes. First, the organization of DNA sequences along the chromosome will be examined.

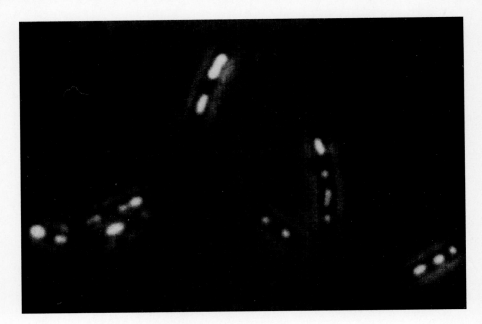

FIGURE **10-3**

The localization of nucleoids within the bacterium *Bacillus subtilis*. The nucleoids are fluorescently labeled and are seen as bright, oval-shaped regions within the bacterial cytoplasm. Note that there are two or more nucleoids within each cell.

Second, we will consider the mechanisms that cause the chromosome to become a compacted structure within the nucleoid of the bacterium.

Bacterial chromosomes contain circular DNA molecules in which a few thousand gene sequences are interspersed with other functionally important sequences

Bacterial chromosomal DNA is usually a circular molecule a few million nucleotides in length. For example, the chromosome of *Escherichia coli* has approximately 4.7 million nucleotide pairs (i.e., base pairs), and the *Hemophilus influenzae* chromosome has roughly 1.2 million. A typical bacterial chromosome contains a few thousand different genes. These genes are interspersed throughout the entire chromosome (Figure 10-4). **Structural gene sequences** (nucleotide sequences that code for proteins) account for the majority of bacterial DNA.

Other sequences within chromosomal DNA influence DNA replication, gene expression (i.e., the transcription of genes), and chromosome structure. For example, bacterial chromosomes have one **origin of replication**, which is a few hundred nucleotides in length. This nucleotide sequence functions as an initiation site for the assembly of several proteins that are required for DNA replication. Also, a variety of short (less than 10 nucleotides) repetitive sequences have been identified in many bacterial species; these sequences are found in multiple copies and are interspersed throughout the bacterial chromosome. Short repetitive sequences may play a role in a variety of genetic processes including DNA folding, gene expression, and genetic recombination. Figure 10-4 summarizes the key features of sequence organization within bacterial chromosomes.

The formation of chromosomal loops helps make the bacterial chromosome more compact

To fit within the bacterial cell, the chromosomal DNA must be compacted about 1000-fold. Part of this compaction process involves the formation of **loop domains** within the bacterial chromosome (Figure 10-5). As its name suggests, a loop domain is a segment of chromosomal DNA that is folded into a structure that resembles a loop. DNA-binding proteins are thought to play an important role in

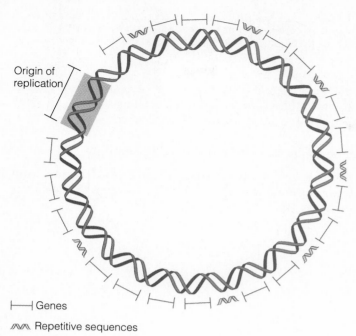

Origin of replication

Key features:

- Most, but not all, bacterial species contain circular chromosomal DNA.
- A typical chromosome is several million base pairs in length.
- Several thousand different genes are interspersed throughout the chromosome.
- One origin of replication is required to initiate DNA replication.
- Short repetitive sequences may be interspersed throughout the chromosome.

⊢—⊣ Genes

〜〜 Repetitive sequences

FIGURE **10-4**

Organization of sequences in prokaryotic chromosomal DNA.

holding the loops in place. The number of loops varies according to the size of bacterial chromosome and the species. In *E. coli*, there are approximately 100 loop domains with about 40,000 base pairs of DNA in each loop. This looped structure compacts the circular chromosome about 10-fold.

DNA supercoiling further compacts the bacterial chromosome

A phenomenon called **DNA supercoiling** is a second important way to compact the bacterial chromosome (Figure 10-6). Because DNA is a long thin molecule, its conformation can be dramatically changed by twisting forces. This is similar to the effects of twisting a rubber band. If twisted in one direction, a rubber band will eventually coil itself into a compact structure as it absorbs the energy applied by the twisting motion. Likewise, the DNA double helix can be subjected to twisting forces that will cause it to change its secondary conformation (i.e., the B-DNA conformation described in Chapter 9). Since the DNA double helices already coil around each other, the formation of additional coils due to twisting forces is referred to as supercoiling.

Figure 10-7 illustrates how twisting can affect DNA structure. In part (a), a double-stranded DNA molecule containing five complete turns is anchored between two plates. In this hypothetical example, the ends of the DNA molecule cannot rotate freely. Both *underwinding* and *overwinding* of the DNA double helix can induce supercoiling of the helix. Since B-DNA is a right-handed helix, underwinding is a left-handed twisting motion and overwinding is a right-handed twist. At the top of Figure 10-7, one of the plates is given a turn in the direction that tends to unwind the helix. As the helix absorbs this force, two things can happen. The underwinding motion can cause fewer turns (part b) or cause a supercoil to form (part c). This is referred to as *negative supercoiling*. At the bottom of Figure 10-7, the DNA is twisted in the direction that overwinds the double helix; it is given a right-handed turn. This can lead to either more

(a) Circular chromosomal DNA

Compaction

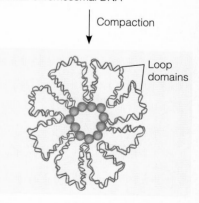

Loop domains

(b) Looped chromosomal DNA with associated proteins

FIGURE **10-5**

The formation of loop domains within the bacterial chromosome. As a way to compact the large circular chromosomal DNA **(a)**, it is organized into smaller looped chromosomal DNA with loop domains and associated proteins **(b)**.

(a) Looped chromosomal DNA

Compaction

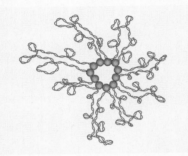

(b) Supercoiled and looped DNA

FIGURE **10-6**

DNA supercoiling leads to compaction of the looped chromosomal DNA. The looped chromosomal DNA **(a)** becomes much more compacted **(b)** due to supercoiling within the loops.

turns (part d) or *positive supercoiling* (part e). The DNA conformations shown in Figure 10-7a, c, and e differ with regard to supercoiling. These three DNA conformations are referred to as **topoisomers** of each other.

Chromosome function is influenced by DNA supercoiling

The chromosomal DNA in living bacteria is negatively supercoiled. There is about one negative supercoil per 40 turns of the double helix. There are several impor-

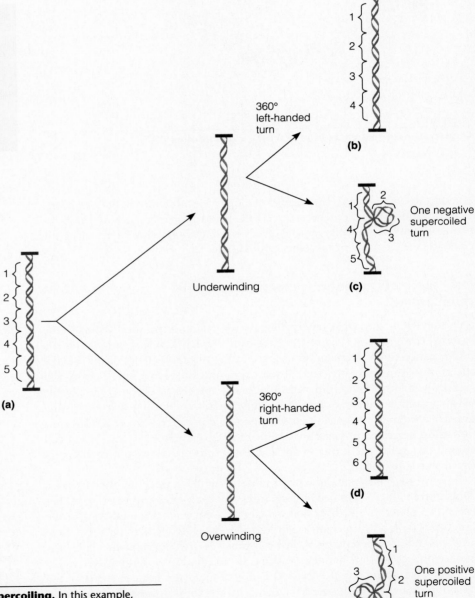

360°
left-handed
turn

Underwinding

(b)

One negative
supercoiled
turn

(c)

360°
right-handed
turn

Overwinding

(d)

One positive
supercoiled
turn

(e)

(a)

FIGURE **10-7**

Schematic representation of DNA supercoiling. In this example, the DNA in **(a)** is anchored between two plates and given a twist as noted by the arrows. A left-handed twist (underwinding) tends to produce either fewer turns **(b)** or a negative supercoil **(c)**. A right-handed twist (overwinding) produces more turns **(d)** or a positive supercoil **(e)**.

tant consequences of negative supercoiling. As already mentioned, the supercoil-ing of chromosomal DNA makes it much more compact (Figure 10-6). Therefore, supercoiling helps decrease the size of the bacterial chromosome. In addition, negative supercoiling also has a significant effect on DNA function. To understand this effect, remember that negative supercoiling is due to an underwinding force on the DNA. Therefore, negative supercoiling creates tension on the DNA strands, which may be released by DNA strand separation (Figure 10-8). Although most of the chromosomal DNA is negatively supercoiled and compact, the force of negative supercoiling may promote DNA strand separation in small regions. This enhances genetic activities, such as replication and transcription, that require the DNA strands to be separated.

Two bacterial enzymes are primarily responsible for the control of supercoil-ing within living bacteria. In 1976, Martin Gellert and collaborators discovered the enzyme **DNA gyrase** (also known as topoisomerase II). This enzyme introduces negative supercoils into DNA using energy from ATP (Figure 10-9). Gyrase can also relax positive supercoils when they occur. A second enzyme, **topoisomerase I**, can relax negative supercoils. The competing actions of DNA gyrase and topoisomerase I govern the overall supercoiling of the bacterial DNA.

The ability of gyrase to introduce negative supercoils into DNA appears to be important in bacteria but not in eukaryotic cells. For this reason, much research has been aimed at identifying drugs that will specifically block bacterial gyrase function as a way to cure or alleviate diseases caused by bacteria. Two main classes, quinolones and coumarins, inhibit gyrase and thereby block bacterial cell growth. This finding has been the basis for the production of many drugs with important antibacterial applications.

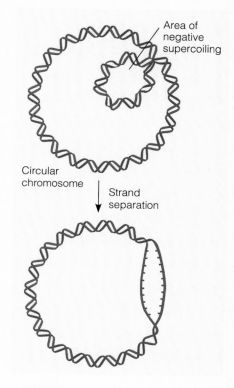

FIGURE **10-8**

Negative supercoiling promotes strand separation.

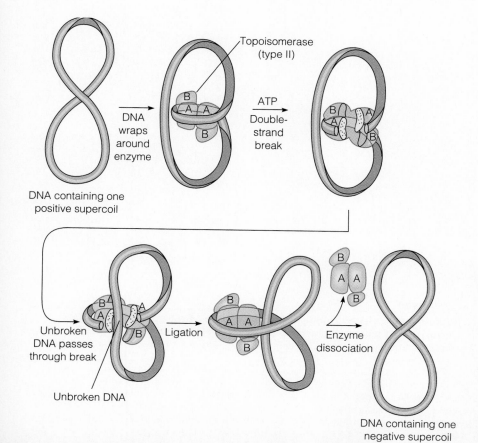

FIGURE **10-9**

The action of DNA gyrase. DNA gyrase, also known as topoisomerase II, is composed of two A and two B subunits. In the first step, the A subunits bind and cleave the DNA. The unbroken segment of DNA then passes through the break, and the break is repaired. The B subunits capture the energy from ATP hydrolysis to catalyze this process. The result is that two negative turns have been introduced into the DNA molecule. In this example, the DNA originally contained one positive supercoil and now it has one net negative supercoil.

EUKARYOTIC CHROMOSOMES

Eukaryotic species contain one or more sets of chromosomes; each set is composed of several different linear chromosomes (see Figure 8-1, p. 198). Humans, for example, have two sets of 23 chromosomes each, for a total of 46. It is not surprising that the total amount of DNA in eukaryotic species is much greater than in simpler bacteria. This enables eukaryotic genomes to encode many more genes than their bacterial counterparts. In this section, we will examine the sizes of eukaryotic genomes and the organization of DNA sequences along the length of eukaryotic chromosomes. We will consider several techniques used to analyze the composition of sequences that are found in chromosomes.

A distinguishing feature of eukaryotic cells is that their chromosomes are located within a separate cellular compartment, known as the **nucleus**. To fit within the nucleus, the length of DNA must be compacted by a remarkable amount. As in bacterial chromosomes, this is accomplished by the binding of the DNA to many different cellular proteins. The term **chromatin** is used to describe the DNA–protein complex that is found within eukaryotic chromosomes. Chromatin is a dynamic structure that can change its shape and composition during the life of a cell. In this section, we will examine the mechanisms that account for the compaction of eukaryotic chromosomes during different stages of the cell cycle.

The sizes of eukaryotic genomes vary substantially

When comparing different eukaryotic species, there is often a dramatic variation in genome size (Figure 10-10a; note that this is a log scale). In general, this variation

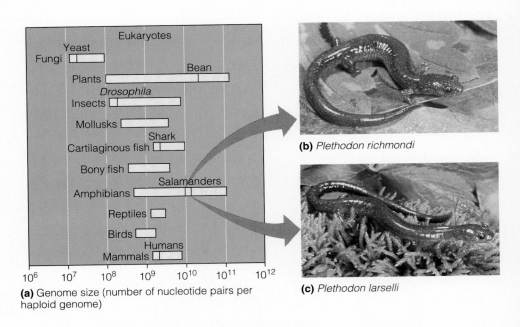

(a) Genome size (number of nucleotide pairs per haploid genome)

(b) *Plethodon richmondi*

(c) *Plethodon larselli*

FIGURE **10-10**

Haploid genome sizes among eukaryotic species. (a) Ranges of genome sizes among different species of eukaryotes. **(b)** A species of salamander, *Plethodon richmondi* and **(c)** a close relative, *P. larselli*. The genome of *P. larselli* is over two times as large as that of *P. richmondi*.

GENES ⟶ TRAITS: The two species of salamander shown here have very similar morphological features, even though the genome of *P. larselli* is over twice as large as that of *P. richmondi*. However, the genome of *P. larselli* is not expected to contain more genes. Rather, the additional DNA is due to the accumulation of short DNA sequences that do not code for genes and are present in many copies.

is not related to the complexity of the species. For example, Figure 10-10b and c compare two closely related species of salamander, *Plethodon richmondi* and *P. larselli*. The genome of *P. larselli* is over two times as large as that of *P. richmondi*. However, the genome of *P. larselli* is not expected to contain more genes. Rather, the additional DNA is due to the accumulation of short DNA sequences that are present in many copies. In some species, these **repetitive sequences** can accumulate to enormous levels. Highly repetitive sequences do not encode proteins, and their function remains a matter of controversy and great interest. The structure and significance of highly repetitive DNA will be discussed later in this chapter.

Eukaryotic chromosomes contain many functionally important sequences including genes, origins of replication, centromeres, and telomeres

Each eukaryotic chromosome contains a long, linear DNA molecule (Figure 10-11). Three types of DNA sequences are required for chromosomal replication and segregation: origins of replication, centromeres, and telomeres. **Origins of replication** are nucleotide sequences within chromosomal DNA that are necessary to initiate DNA replication. Unlike bacterial chromosomes, which only contain one, eukaryotic chromosomes contain many origins, interspersed approximately every 100,000 base pairs (bp). The function of origins will be discussed in greater detail in Chapter 11. **Centromeres** are DNA sequences required for the proper segregation of

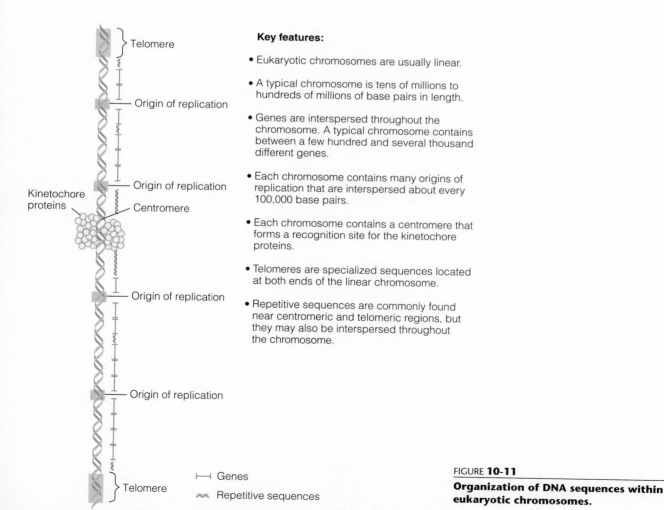

Key features:

- Eukaryotic chromosomes are usually linear.

- A typical chromosome is tens of millions to hundreds of millions of base pairs in length.

- Genes are interspersed throughout the chromosome. A typical chromosome contains between a few hundred and several thousand different genes.

- Each chromosome contains many origins of replication that are interspersed about every 100,000 base pairs.

- Each chromosome contains a centromere that forms a recognition site for the kinetochore proteins.

- Telomeres are specialized sequences located at both ends of the linear chromosome.

- Repetitive sequences are commonly found near centromeric and telomeric regions, but they may also be interspersed throughout the chromosome.

⊢⊣ Genes

〰 Repetitive sequences

FIGURE **10-11**

Organization of DNA sequences within eukaryotic chromosomes.

chromosomes during mitosis and meiosis (see Chapter 3). Each eukaryotic chromosome contains a single centromere. The centromere serves as an attachment site for a group of proteins that form the **kinetochore**. During mitosis and meiosis, the kinetochore is linked to the spindle apparatus and thereby assures the proper segregation of the chromosomes to each daughter cell. Finally, at the ends of linear chromosomes are found specialized sequences known as **telomeres**. An unusual form of DNA replication occurs at the telomere to ensure that eukaryotic chromosomes do not become shortened with each round of DNA replication (see Chapter 11).

Genes are located between the centromeric and telomeric regions along the entire eukaryotic chromosome. A single chromosome is expected to contain from a few hundred to several thousand different genes. In simpler eukaryotes such as yeast, genes are relatively small and primarily contain the nucleotide sequences that code for the amino acid sequences within proteins (i.e., structural genes). In more complex eukaryotes such as mammals and higher plants, the genes tend to be much longer due to the presence of introns (noncoding intervening sequences; see Chapter 12). Introns range in size from less than 100 nucleotides to more than 10,000. Therefore, the presence of large introns can greatly increase the lengths of eukaryotic genes.

The genome of eukaryotes contains sequences that are unique, moderately repetitive, and highly repetitive

The term **sequence complexity** refers to the number of times a particular base sequence appears throughout the genome. **Unique** or **nonrepetitive sequences** are those found only once or a few times within the genome. Unique sequences are usually sequences for genes. The sequence of a typical eukaryotic gene is usually a few thousand to tens of thousand nucleotides in length. The vast majority of proteins in eukaryotic cells are encoded by genes that are present in one or a few copies. In the case of humans, unique sequences make up roughly 60–70% of the entire genome.

Moderately repetitive sequences are those found a few hundred to several thousand times in the genome. In a few cases, moderately repetitive sequences are multiple copies of the same gene. For example, the genes that encode ribosomal RNA (rRNA) are found in many copies. Ribosomal RNA is necessary for the functioning of ribosomes. The cell needs a large amount of rRNA for its cellular ribosomes; one way to accomplish this is by having multiple copies of the genes that encode rRNA. Likewise, the histone genes are also found in multiple copies, because a large amount of histone proteins are needed for chromatin. In addition, other types of functionally important sequences can be moderately repetitive. For example, multiple copies of origins of replication are found within eukaryotic chromosomes. Other moderately repetitive sequences may play a role in the regulation of gene transcription and translation.

Highly repetitive sequences are those that are found tens of thousands or even millions of times throughout the genome. Each copy of a highly repetitive sequence is relatively short, ranging from a few nucleotides to several hundred in length. A widely studied example is the *Alu family* of sequences found in humans and other primates. The Alu sequence is approximately 300 base pairs in length. This sequence derives its name from the observation that it contains a site for cleavage by a restriction enzyme known as AluI. (The function of restriction enzymes will be described in Chapter 19.) The Alu sequence is present in 500,000–1,000,000 copies in the human genome. It represents about 5–6% of the total human DNA and occurs approximately every 5000 to 6000 bases. Evolutionary studies suggest that the Alu sequence arose 65 million years ago from a section of a single ancestral gene known as the 7SL RNA gene. Since that time, the Alu sequence has been copied and inserted into the human genome to achieve the modern number of more than 500,000 copies. The mechanism for the proliferation of Alu sequences will be described in Chapter 18.

Some highly repetitive sequences, like the Alu family, are interspersed throughout the genome. However, other highly repetitive sequences are clustered together in a **tandem array** or **tandem repeats**. In a tandem array, a very short nucleotide sequence is repeated many times in a row. In *Drosophila*, for example, 19% of the chromosomal DNA is highly repetitive DNA found in tandem arrays. An example is shown here:

A A T A T A A T A T A A T A T A A T A T A A T A T A T A A T A T
T T A T A T T A T A T T A T A T T A T A T T A T A T A T T A T A

In this particular tandem array, two related sequences, AATAT and AATATAT, are repeated many times. Tandem arrays of highly repetitive sequences are commonly found in centromeric regions of chromosomes and can be quite long, sometimes more than 1,000,000 base pairs in length!

Whether highly repetitive sequences play any significant functional role is controversial. Some experiments in *Drosophila* indicate that highly repetitive sequences may be important in the proper segregation of chromosomes during meiosis. It is not yet clear if highly repetitive DNA plays the same role in other species. The sequences within highly repetitive DNA vary greatly from species to species. In fact, as noted earlier, the amount of highly repetitive DNA can vary a great deal even among closely related species (as shown in Figure 10-10b).

Highly repetitive DNA can be separated from the rest of the chromosomal DNA by equilibrium density centrifugation

As we will discuss later in this chapter, centromeric and telomeric regions tend to be highly compacted, or **heterochromatic**. During the 1960s and 1970s, many scientists were interested in the relationship between heterochromatic regions and repetitive sequences. At that time, cytological evidence suggested that highly repetitive sequences are localized to heterochromatic regions. To gain further insight into this question, biochemical methods were developed to detect and isolate highly repetitive DNA. *Drosophila* was considered an ideal organism to study, because a large proportion (20%) of its genome is heterochromatic. Therefore, if it is correct that heterochromatic DNA is composed of highly repetitive sequences, it follows that *Drosophila* has a large amount of highly repetitive DNA.

When repetitive DNA is in a tandem array, its base composition may be quite different from rest of the chromosomal DNA. For example, the tandem array shown above is 100% A/T, whereas the average base composition of nonrepetitive chromosomal DNA in *Drosophila* is approximately 60% A/T and 40% G/C. One way to distinguish an A/T pair from a G/C pair is by their relative densities. An A/T pair is slightly less dense than a G/C pair. Therefore, a DNA fragment with a high proportion of A/T pairs will be of lighter density than a fragment with a high percentage of G/C pairs. The experiment described here was conducted by the Australian group of W. Peacock, D. Brutlag, E. Goldring, R. Appels, C. Hinton, and D. Lindsley in 1974. As we will see, they took advantage of the experimental observation that molecules with different densities can be separated from each other via equilibrium density centrifugation (see the appendix for a description of density centrifugation).

THE HYPOTHESIS

Highly repetitive DNA may have a base composition that is significantly different from the rest of the chromosomal DNA. If so, it may be possible to separate repetitive DNA from the rest of the chromosomal DNA by equilibrium density centrifugation.

TESTING THE HYPOTHESIS

Starting material: Nuclei isolated from *Drosophila melanogaster*.

Experimental Level   **Conceptual Level**

1. Extract the DNA from the nuclei. This involves treatment with detergent (to dissolve the nuclear membrane) and then addition of phenol (to remove the protein). The DNA remains in the aqueous phase, while most of the protein goes into the phenol phase. During this procedure, the chromosomal DNA breaks up into small fragments.

2. Load the aqueous phase, which contains the DNA fragments, onto a CsCl density gradient. (Note: Density gradient centrifugation is described in the appendix.)

3. Centrifuge for 18 hours until the DNA fragments reach their equilibrium density.

4. Collect fractions along the gradient.

5. Determine the amount of DNA in each fraction by using a spectrophotometer. DNA absorbs light in the UV range. (The use of a spectrophotometer is described in the appendix.)

6. Plot the amount of DNA in each fraction.

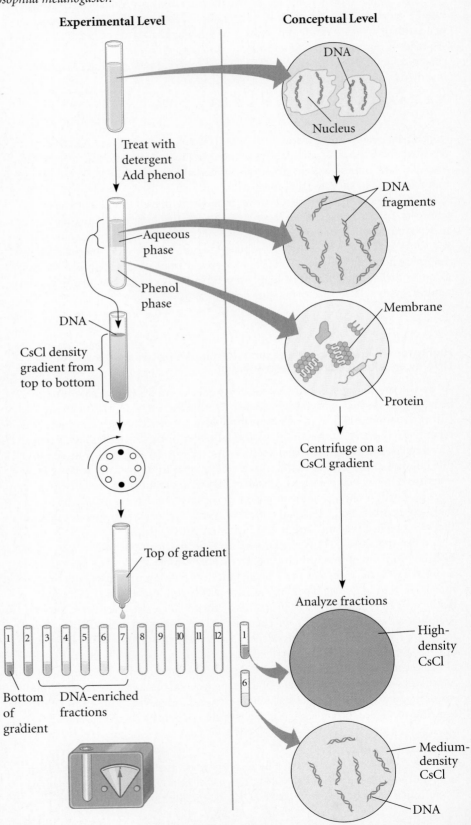

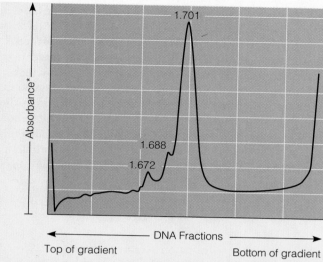

Top of gradient

Bottom of gradient

←————— DNA Fractions —————→

*Higher absorbance means a higher amount of DNA in the fractions.

As shown in these data, most of the DNA is found in a fraction of density 1.701 g/cm³. This peak contains about 80% of the *Drosophila* DNA. It corresponds to the chromosomal DNA that is not heterochromatic. We now know that this peak contains the DNA pieces derived from the nonrepetitive chromosomal DNA. Its average base composition is approximately 60% A/T and 40% G/C. In addition, the data show two other notable peaks of DNA, found in fractions that have a lighter density (namely, 1.672 and 1.688 g/cm³). These peaks of DNA are referred to as **satellite DNA**, since they are separated from the main peak. In this experiment, the two peaks of satellite DNA have a base composition that is significantly lighter than the rest of the chromosomal DNA. Other experiments have been carried out to identify satellite DNA in *Drosophila*. By changing the DNA preparation techniques and the centrifugation conditions, several more satellite DNAs have been identified in this organism.

The experiment described here demonstrated that some of the DNA in *Drosophila* has a base composition that is significantly different from rest of the chromosomal DNA. Furthermore, density centrifugation provided a way to biochemically separate the satellite DNA from the rest of the DNA. This enabled researchers to clone the satellite DNA and subject it to DNA sequencing. The sequencing of satellite DNA confirmed that it is composed of highly repetitive sequences. The satellite with a density of 1.672 g/cm³ has been sequenced; it is the highly repetitive DNA shown on page 269. Since this repeat is 100% A/T, its density is lighter than the main band of chromosomal DNA, which is 60% A/T and 40% G/C and has a density of 1.701 g/cm³.

Sequence complexity can be evaluated in a renaturation experiment

A second approach that has proved useful in understanding genome complexity has come from renaturation studies. These kinds of experiments were first carried out by Roy Britten and David Kohne in the late 1960s. In a renaturation study, the DNA is broken up into pieces containing several hundred nucleotides. The double-stranded DNA is then "melted" (separated) into single-stranded pieces by heat treatment (Figure 10-12a). When the temperature is lowered, the pieces of DNA that are complementary can renature with each other to form double-stranded molecules.

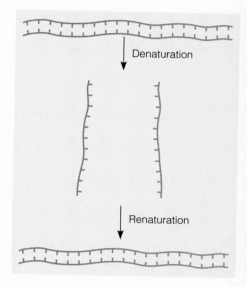

(a) Renaturation of DNA strands

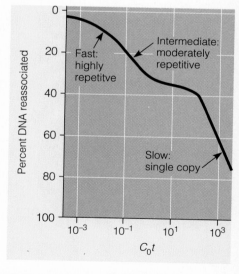

(b) Human chromosomal DNA $C_0 t$ curve

FIGURE **10-12**

Renaturation and DNA sequence complexity. (a) Denaturation and renaturation (or reassociation) of DNA strands. **(b)** A $C_0 t$ curve for human chromosomal DNA.

The rate of renaturation of complementary DNA strands provides a way to distinguish unique, moderately repetitive, and highly repetitive sequences. For a given category of DNA sequences, the renaturation rate will depend on the concentration of its complementary partner. Highly repetitive DNA sequences will renature much faster, because there are many copies of the complementary sequences. In contrast, unique sequences, such as those found within genes, will take longer to renature because of the added time it takes for the unique sequences to find each other.

The renaturation of two DNA strands is a bimolecular reaction that involves the collision of two complementary DNA strands. Its rate is proportional to the product of the concentrations of both strands. If C is the concentration of single-stranded DNA, then for any DNA derived from a double-stranded fragment, the concentration of one DNA strand (denoted C_1) equals the concentration of its complementary partner (denoted C_2). Letting $C = C_1 = C_2$, we see the rate of renaturation is represented by the second-order equation

$$\frac{-dC}{dt} = kC^2$$

(This is a called a second-order equation because the rate depends on the concentration of both reactants—i.e., C_1 and C_2. In this case, this product is simplified to C^2 because $C_1 = C_2$.)

This equation says that a change in concentration of a single DNA strand ($-dC$) with respect to time (dt) equals a rate constant (k) times the concentration of the single-stranded molecule squared (C^2). This equation can then be integrated to determine how the concentration of the single-stranded DNA will change from time zero to a later time:

$$\frac{C}{C_0} = \frac{1}{1 + k_2 C_0 t}$$

Where C = the concentration of single-strand DNA at a later time, t

C_0 = the concentration of single-stranded DNA at time zero

k_2 = the second-order rate constant for renaturation

C/C_0 is the fraction of DNA still in single-stranded form after a given length of time. For example, if C/C_0 equals 0.4 after a certain period of time, 40% of the DNA is still in single-stranded form while 60% has renatured into double-stranded form.

A renaturation experiment can provide quantitative information about the complexity of DNA sequences within chromosomal DNA. In the experiment shown in Figure 10-12b, human DNA was sheared into small pieces (each about 600 bp in length), subjected to heat, and then allowed to renature at a lower temperature. The rates of renaturation for the DNA pieces can be represented in a plot of C/C_0 versus $C_0 t$. This is referred to as a $C_0 t$ curve (called a "cot" curve). A small amount of the DNA renatures very rapidly. This is the highly repetitive DNA. Some DNA renatures at a moderate rate, but most of the DNA renatures fairly slowly. From these data, the relative amounts of highly repetitive, moderately repetitive, and unique DNA sequences can be approximated. As seen in Figure 10-12b, 60-70% of human DNA is unique DNA sequences.

Eukaryotic chromatin must be compacted to fit within the cell

We now turn our attention to ways that eukaryotic chromosomes are folded in order to fit within a living cell. A typical eukaryotic chromosome contains a single, linear double-stranded DNA molecule that may be hundreds of millions of base pairs in length. If all the haploid human DNA from a single cell was stretched from end to end, it would be over 1 m in length. By comparison, most eukaryotic cells are only 10–100 µm in diameter, and the cell nucleus is only about 2–4 µm in diameter. Therefore, the DNA in a eukaryotic cell must be folded and packaged by a staggering amount to fit inside the nucleus.

The compaction of linear DNA within eukaryotic chromosomes is accomplished through mechanisms that involve interactions between DNA and several different proteins. In recent years, it has become increasingly clear that the proteins bound to chromosomal DNA are subject to change during the life of the cell. These changes in protein composition, in turn, affect the degree of compaction of the chromatin. Chromosomes are very dynamic structures, which alternate between tight and loose compaction states in response to changes in protein composition.

Linear DNA wraps around histone proteins to form nucleosomes, the repeating structural unit of chromatin

The repeating structural unit within eukaryotic chromatin is the **nucleosome**. As proposed by Roger Kornberg in 1974, the nucleosome is composed of double-stranded DNA wrapped around an octamer of **histone proteins** (Figure 10-13a). Each octamer contains eight histone subunits (two copies each of four different histone proteins). The DNA lies on the surface and makes 1.8 negative superhelical turns around the histone octamer. The amount of DNA that is required to wrap around the histone octamer is 145 base pairs. The entire nucleosome is about 11 nm in diameter.

The chromatin of eukaryotic cells contains a repeating pattern in which the nucleosomes are connected by linker regions of DNA that vary in length from 20 to 100 bp, depending on the species and cell type. It has been suggested that the overall structure of connected nucleosomes resembles beads on a string. This structure shortens the length of the DNA molecule about sevenfold.

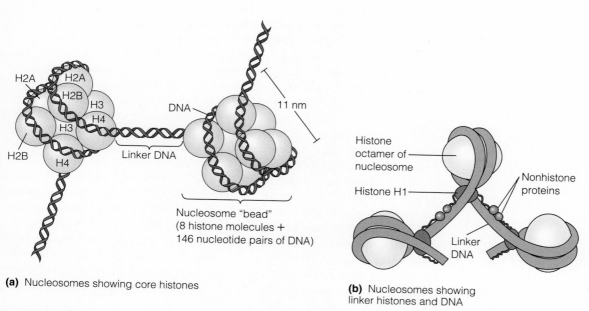

(a) Nucleosomes showing core histones

(b) Nucleosomes showing linker histones and DNA

FIGURE **10-13**

Nucleosome structure. (a) 145 base pairs of DNA are wrapped around an octamer of core histones. **(b)** The linker DNA connects adjacent nucleosomes. The linker histone H1 and nonhistone proteins also bind to this linker region.

overall structure of connected nucleosomes resembles beads on a string. This structure shortens the length of the DNA molecule about sevenfold.

Histone proteins are very basic proteins because they contain a large number of positively charged lysine and arginine amino acids. The arginine residues, in particular, play a major role in binding to the DNA. Arginine residues within the histone proteins form electrostatic and hydrogen bonding interactions within the phosphate groups along the DNA backbone. The octamer of histones contains two molecules each of four different histone proteins: H2A, H2B, H3, and H4. These are called the *core histones*. Another histone, H1, is found in most eukaryotic cells and is called the *linker histone*. It binds to the DNA on one side of a nucleosome and to the linker region of DNA (see Figure 10-13b). As will be discussed later, H1 is thought to promote the compaction of nucleosomes with each other. In addition, *nonhistone proteins* are bound to this linker region. The linker histones are less tightly bound to the DNA than are the core histones.

EXPERIMENT 10B

Noll showed that the repeating nucleosome structure contains 200 bp of DNA by treating chromatin with a nuclease that cuts the DNA in the linker region

The model of nucleosome structure shown in Figure 10-13 is now firmly established. It was originally proposed by Roger Kornberg in 1974 while working at the MRC Laboratory of Molecular Biology in Cambridge, England. He based his proposal on several observations. Biochemical experiments had shown that chromatin contains a ratio of one molecule of each of the four core histones (namely, H2A, H2B, H3, and H4) per 100 base pairs of DNA. Approximately one H1 protein was found per 200 bp of DNA. In addition, purified core histone proteins were observed to bind to each other via specific pairwise interactions. X-ray diffraction studies showed that chromatin is composed of a repeating pattern of smaller units. And finally, electron microscopy of chromatin fibers revealed a diameter of approximately 10 nm. Taken together, these observations led Kornberg to propose a model in which the DNA double helix is wrapped around an octamer of core histone proteins. Including the linker region, this would involve about 200 bp of DNA.

At the time of Kornberg's proposal, an experiment by Dean Hewish and Leigh Burgoyne in Australia suggested that the chromatin in rat liver nuclei is digested by cellular nucleases to yield DNA pieces in multiples of 200 bp. In these experiments, however, the sizes of the DNA fragments were determined by their rate of movement during centrifugation, and the results were regarded as only approximate. As described here, Markus Noll, who also was working at the MRC, decided to test Kornberg's model by a similar approach in which he digested chromatin with DNaseI and accurately determined the molecular weight of the DNA fragments by gel electrophoresis. The rationale behind this experiment is that the linker region of DNA will be more accessible to enzymes, such as DNaseI, that can digest DNA. In other words, DNaseI is more likely to make cuts in the linker region than in the 145 bp region that is tightly bound to the core histones. If this is correct, incubation with DNaseI is expected to make cuts in the linker region and thereby produce DNA pieces that are approximately 200 bp in length. (Note: The size of the DNA fragments may vary somewhat, since the linker region is not of constant length and because the cut within the linker region may occur at different sites.)

THE HYPOTHESIS

This experiment seeks to test the beads-on-a-string model for chromatin structure. According to this model, DNaseI should preferentially cut the DNA in the linker region, and thereby produce DNA pieces that are about 200 bp in length.

TESTING THE HYPOTHESIS

Starting material: Nuclei from rat liver cells.

Experimental Level

Conceptual Level

1. Incubate the nuclei with low, medium, and high concentrations of DNaseI. The conceptual level illustrates a low DNaseI concentration.

2. Extract the DNA (see step 1 of Experiment 10A).

3. Load the DNA into a well of an agarose gel and run the gel to separate the DNA pieces according to size. On this gel, also load DNA fragments of known molecular weight (Marker lane).

4. Visualize the DNA fragments by staining the DNA with ethidium bromide. This is a dye that binds to DNA and is fluorescent when excited by UV light.

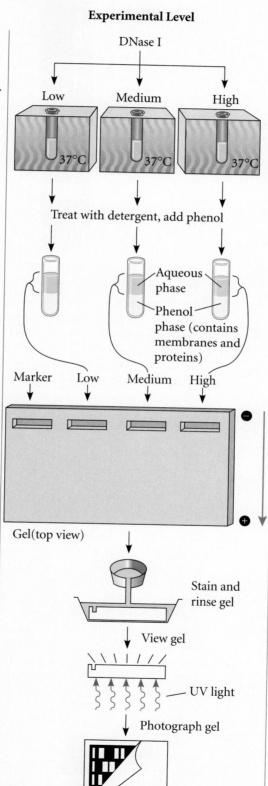

DNase I

Low   Medium   High

37°C   37°C   37°C

Treat with detergent, add phenol

Aqueous phase

Phenol phase (contains membranes and proteins)

Marker   Low   Medium   High

Gel(top view)

Stain and rinse gel

View gel

UV light

Photograph gel

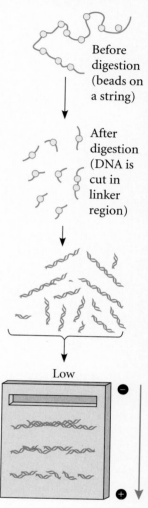

Before digestion (beads on a string)

After digestion (DNA is cut in linker region)

Low

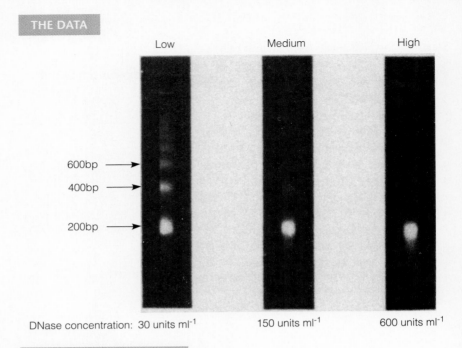

DNase concentration: 30 units ml⁻¹   150 units ml⁻¹   600 units ml⁻¹

At high DNaseI concentrations, all the chromosomal DNA is digested into fragments of approximately 200 base pairs in length. This is the result predicted by the beads-on-a-string model. Furthermore, at lower DNaseI concentrations, longer pieces were observed, and these were always roughly in multiples of 200 bp (400, 600, etc.). These longer pieces, occurring in multiples of 200 bp, may be explained by occasional uncut linker regions at lower DNaseI concentrations. For example, a DNA piece might contain two nucleosomes and be 400 bp in length. If two consecutive linker regions were not cut, this would produce a piece with three nucleosomes containing about 600 bp of DNA. And so on. Taken together, these results strongly supported the nucleosome model for chromatin structure.

Nucleosomes become closely associated to form a 30 nm fiber

In eukaryotic chromatin, nucleosomes associate with each other to form an even more compact structure, 30 nm in diameter. Evidence for the packaging of nucleosomes was obtained in the microscopy studies of F. Thoma and colleagues in Switzerland. As shown in Figure 10-14, chromatin samples were treated with or without solutions of moderate salt concentration (100 mM NaCl) and then observed with an electron microscope. Moderate salt concentrations are expected to remove the H1 histone but not the core histones, because the H1 histone is more loosely attached to the DNA. At moderate salt concentrations (Figure 10-14a), the chromatin exhibited the classic beads-on-a-string morphology. At low salt concentrations (when H1 is expected to remain bound to the DNA), these "beads" associated with each other into a more compact conformation (Figure 10-14b). These results suggest that the nucleosomes are packaged into more compact units. Furthermore, the results are consistent with a key role for H1 in the packaging of nucleosomes into more compact structures.

The experiment of Figure 10-14 and other experiments have established that nucleosome units are organized into a more compact structure that is 30 nm in diameter, known as the **30 nm fiber** (Figure 10-15a). The 30 nm fiber shortens the total length of DNA another sevenfold. The structure of the 30 nm fiber has proved difficult to determine, because the conformation of the DNA may be substantially altered when it is extracted from living cells. An early model by Thoma, known as the solenoid model, suggested a helical structure in which nucleosome contacts produce a

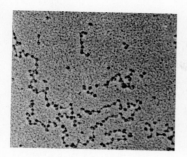
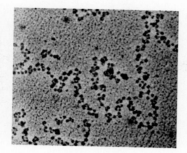

(a) At moderate salt concentration

(b) At low salt concentration

FIGURE **10-14**

The nucleosome structure of eukaryotic chromatin as viewed by electron microscopy. The chromatin in **(a)** has been treated with moderate salt concentrations to remove the linker histone H1. It exhibits the classic beads-on-a-string morphology. The chromatin in **(b)** has been incubated at lower salt concentrations and shows a more compact morphology. From Thoma, F., and Koller, Th. (1977) Influence of histone H1 on chromatin structure. *Cell 12*, 101–107.

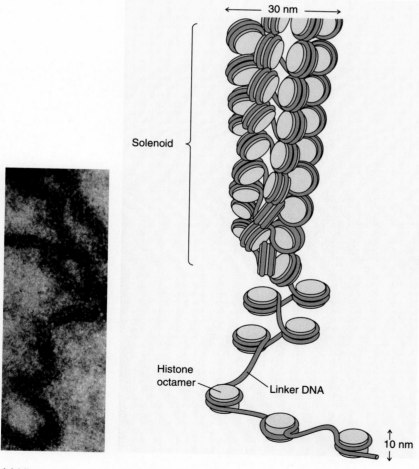
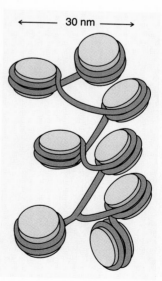

(a) Micrograph of a 30 nm fiber

(b) Solenoid model

(c) 3D Zig-zag model

FIGURE **10-15**

The 30 nm fiber. (a) A photomicrograph of the 30 nm fiber. **(b)** In the solenoid model, the nucleosomes are packed in a spiral configuration containing six nucleosomes per turn. **(c)** In the 3D zig-zag model, the linker DNA forms a more irregular structure and there is less contact between adjacent nucleosomes. The 3D zig-zag model is consistent with more recent data regarding chromatin conformation.

symmetrically compact structure within the 30 nm fiber (Figure 10-15b). However, more recent data suggest that the 30 nm fiber does not form such a regular structure. Instead, a newer model based on cryoelectron microscopy (i.e., electron microscopy at low temperature) has been proposed by Rachel Horowitz and Christopher Woodcock at the University of Massachusetts in Amherst and their colleagues (Figure 10-15c). According to their model, linker regions within the 30 nm structure are variably bent and twisted, and there is little face-to-face contact between nucleosomes. The 30 nm fiber forms an asymmetric 3D zig-zag of nucleosomes.

Chromosomes are further compacted by folding the 30 nm fiber into radial loop domains

Thus far, we have described two mechanisms that compact eukaryotic DNA. These involve the wrapping of DNA within nucleosomes and the arrangement of nucleosomes to form a 30 nm fiber. Taken together, these two events shorten the DNA about 50-fold. A third mechanism that compacts the DNA even further involves the formation of **radial loop domains**, similar to those described for the bacterial chromosome. In this process, chromatin is organized into loops, often 50,000 to 100,000 base pairs in size, that are anchored to a **nuclear scaffold** (also known as the **nuclear matrix**). The nuclear scaffold is composed of many cellular proteins that bind to each other and to the DNA. These scaffolding proteins come together

(a)

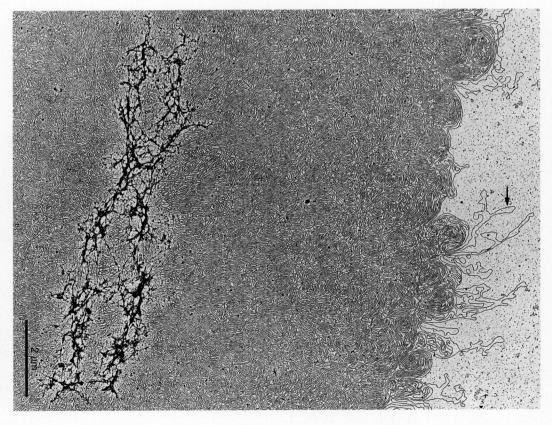

(b)

FIGURE **10-16**

The importance of histones and scaffolding proteins in the compaction of eukaryotic chromosomes. (a) A metaphase chromosome. **(b)** A metaphase chromosome following treatment with high salt concentration to remove the histone proteins. The arrow points to an elongated strand of DNA.

to form a fairly rigid structure that provides a support for the attachment of the radial loop domains.

Microscopy experiments have shown that the proteins making up the nuclear scaffold are critical to the three-dimensional structure of chromosomes. This point is emphasized in the experiment shown in Figure 10-16. At the left is a human metaphase chromosome. In this condition, the loops of DNA are in a relatively compact configuration and are attached to the scaffold. If this chromosome is treated with a very high concentration of salt to remove both the core and linker histones, the highly compact configuration is lost, but the elongated loops remain attached to the nuclear scaffold. An arrow points to an elongated DNA strand emanating from the darkly staining nuclear scaffold. It is remarkable that the nuclear scaffold retains the shape of the original metaphase chromosome even though the DNA strands have become greatly elongated. These results illustrate the importance of both the nuclear scaffold and the histones in determining chromosome structure.

The proteins making up the nuclear scaffold have not been entirely delineated, but they are a diverse group that includes nuclear lamins (described later), histone H1, and topoisomerase II as well as a network of other poorly identified proteins. The DNA sequences that bind to the nuclear scaffold are known as **scaffold-attached regions (SARs)** or **matrix-attached regions (MARs)**. These regions are a few hundred nucleotides in length and have a high percentage of A and T residues (Figure 10-17). Within SARs, it is common to find many sequences of three or more As or Ts (i.e., AAA or TTT) in a row. This characteristic of the SAR sequence may be important in the binding of the SARs to the scaffolding proteins. In recent years, evidence has accumulated suggesting that two SARs may flank a region of DNA that contains one or more coordinately regulated genes (i.e., genes that are turned on the same cell type or at the same time in development). The relationship between radial loops and gene expression will be discussed in Chapter 16.

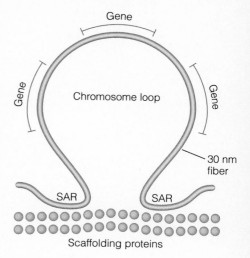

FIGURE **10-17**

The relationship between scaffold-attached regions (SARs) and gene sequences in eukaryotic cells. The SARs, which contain a high percentage of A and T residues, are anchored to the nuclear scaffold to create chromosome loops. This causes a greater compaction of eukaryotic chromosomal DNA.

Chromosome structure changes during the cell cycle

Eukaryotic cells progress through a series of stages known as the **cell cycle** (Figure 10-18). The level of compaction of eukaryotic chromosomes varies considerably throughout the cell cycle. Changes in the composition of scaffolding proteins and other DNA-binding proteins occur at different stages of the cell cycle and cause dramatic differences in chromosome structure. For example, during replication, the

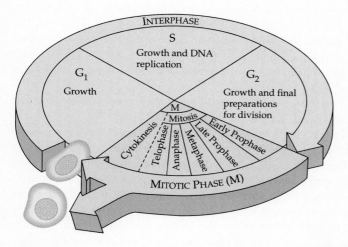

FIGURE **10-18**

The eukaryotic cell cycle.

DNA must be in a relatively loose conformation so that the strands can unwind and be copied. In contrast, after the chromosomes have been completely replicated, they are condensed into very compact structures during mitosis. This change in chromosome structure is caused by changes in the proteins that are bound to the DNA

Even in nondividing cells, the structures of chromosomal regions show considerable variability. Some regions are rather tightly compacted, whereas other regions have a more extended conformation. This variation has important consequences for the transcription of genes into RNA.

Changes in the compaction of eukaryotic chromosomes are particularly dramatic during M phase. The chromosomes become highly condensed in preparation for cell division. Figure 10-19 depicts the progression of DNA compaction leading to the highly compacted chromosome that is found during metaphase (a part of M phase). The DNA is first wrapped into nucleosomes, which are further coiled into a 30 nm structure. For metaphase chromosomes, the 30 nm structure adopts a larger-diameter structure due to further folding of the 30 nm fiber. This highly compacted structure forms the radial loops that are anchored to the nuclear scaffold.

Highly condensed chromosomes undergo little gene transcription, because it is difficult for transcription enzymes (namely, RNA polymerases) to gain access to the compacted DNA. Therefore, most transcriptional activity ceases during M phase, although a few specific genes may be transcribed. M phase is usually a short period of the cell cycle.

During interphase, the chromosomes are less compacted and are located in the nucleus

The interphase of the cell cycle includes the G_1, S, and G_2 phases (see Figure 10-18). During interphase, eukaryotic chromosomes are found within the cell nucleus (Figure 10-20). The cell nucleus is a double-membrane-bounded organelle. Along the inner nuclear membrane are found a collection of proteins known as *nuclear lamins*. The lining of the lamins along the inner nuclear membrane is termed the *nuclear lamina*. Rather than floating freely inside the nucleus, the chromosomes are attached to the lamina at specific sites. In addition, nuclear pores, which connect the inner and outer nuclear membrane, allow molecules to pass into and out of the nucleus.

During interphase, the chromatin forms looped domains that are attached to the nuclear scaffold. However, interphase chromosomes are substantially less compacted than the tightly condensed metaphase chromosomes. In addition, the compaction of the chromosomes is much less uniform. This variability can be seen with a light microscope and was first described by the German cytologist E. Heitz in the late 1920s. He used the term **heterochromatin** to describe the tightly compacted regions of chromosomes. In general, these regions of the chromosome are transcriptionally inactive. By comparison, the less condensed regions, known as **euchromatin**, reflect areas that are capable of gene transcription.

Figure 10-21 illustrates the distribution of euchromatin and heterochromatin in a typical eukaryotic chromosome during interphase. The chromosome contains regions of both heterochromatin and euchromatin. Heterochromatin is most abundant in the centromeric regions of the chromosomes and, to a lesser extent, in the telomeric regions. The term **constitutive heterochromatin** refers to chromosomal regions that are always heterochromatic and permanently inactive with regard to transcription. Constitutive heterochromatin usually contains highly repetitive DNA sequences, such as tandem repeats, rather than gene sequences. **Facultative heterochromatin** refers to the conversion of euchromatin to heterochromatin. An example of this phenomenon occurs in female mammals,

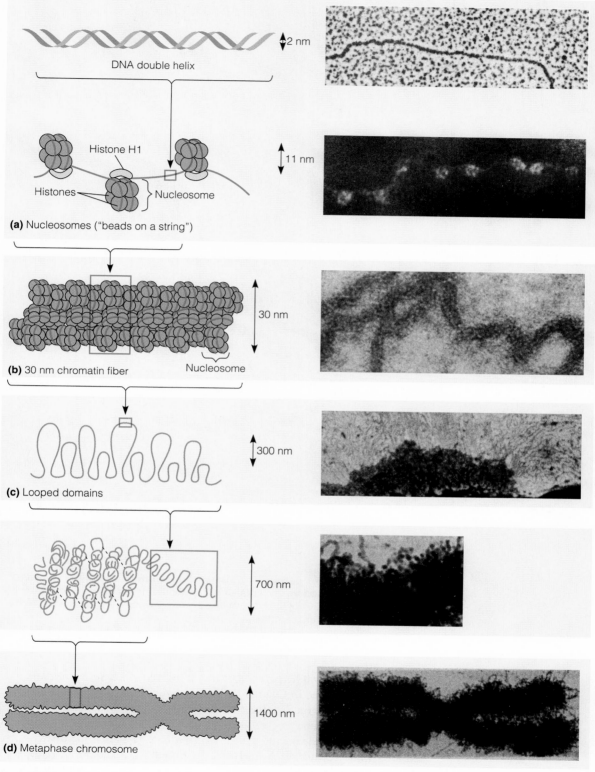

(a) Nucleosomes ("beads on a string")

DNA double helix

Histone H1

Histones

Nucleosome

2 nm

11 nm

(b) 30 nm chromatin fiber

Nucleosome

30 nm

(c) Looped domains

300 nm

700 nm

(d) Metaphase chromosome

1400 nm

FIGURE **10-19**

The steps in eukaryotic chromosomal compaction leading to the metaphase chromosome.

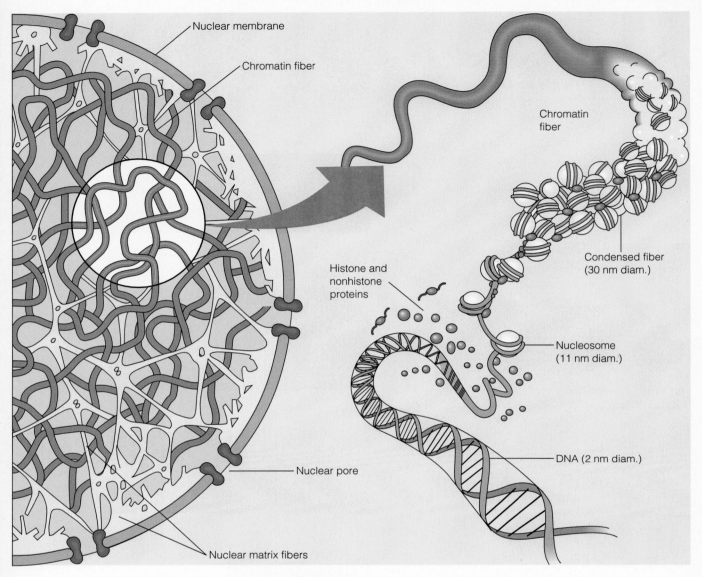

Nuclear membrane

Chromatin fiber

Chromatin
fiber

Condensed fiber
(30 nm diam.)

Histone and
nonhistone
proteins

Nucleosome
(11 nm diam.)

DNA (2 nm diam.)

Nuclear pore

Nuclear matrix fibers

FIGURE **10-20**

The eukaryotic nucleus during interphase, showing the steps in eukaryotic chromosomal compaction. The DNA is first compacted into an 11 nm fiber, and then into a 30 nm fiber. The 30 nm fiber (which may be further compacted) is anchored to the nuclear matrix, which is attached to the nuclear membrane.

in which one of the two X-chromosomes is converted to a heterochromatic **Barr body**. As was discussed in Chapter 7, the Barr body is transcriptionally inactive, and so only the genes on the other (euchromatic) X-chromosome can function.

Interphase is a time of great genetic activity. Gene transcription primarily involves genes located within euchromatic regions. The great majority of proteins are synthesized during the interphase period of a cell's life. In addition, during G_1, a cell may make the commitment to divide. If this happens, the DNA is replicated during S phase. The process of DNA replication is the topic of the following chapter.

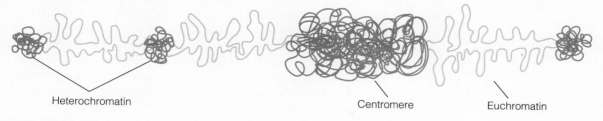

Heterochromatin   Centromere   Euchromatin

FIGURE **10-21**

Chromatin structure during interphase. Heterochromatic regions are more highly condensed and tend to be localized in centromeric and telomeric regions.

The chromosomal location and function of DNA sequences are central to our understanding of genetics. *Viruses* contain relatively small *genomes* that are composed of DNA or RNA. The *viral genome* is packaged into mature virus particles in an *assembly* process. In prokaryotes, several thousand different genes are interspersed throughout a circular chromosome; a single *origin of replication* is also present. Highly repetitive sequences, which may play a role in chromosome structure and gene function, are found throughout the chromosome. In comparison, eukaryotes have many linear chromosomes. A single *centromere* sequence is found on each chromosome; this functions as a recognition site for a *kinetochore* complex. At the ends of the linear chromosomes are *telomeres*. Hundreds or even thousands of different genes are located between the centromeres and telomeres. An unusual feature of many eukaryotic species is that their DNA contains an abundance of *highly repetitive sequences*. The functional importance of these sequences remains unclear and will be discussed further in Chapter 18.

A second important issue that underlies chromosomal structure and function is the amount of compaction that occurs within chromosomal DNA. In bacteria, two main events, chromosomal looping and *DNA supercoiling*, are responsible for the folding of the circular chromosome into a compact structure within a *nucleoid*. Certain DNA-binding proteins are likely involved in this process. In eukaryotic *chromatin*, the DNA is first folded into *nucleosomes* that contain the DNA double helix wrapped around *histone proteins*. The nucleosomes are then compacted into a 30 nm structure. The *30 nm fiber* is further organized into *radial loop domains* as they are anchored to the *nuclear scaffold*. During interphase, eukaryotic chromosomes are found within the cell nucleus and contain fairly condensed *heterochromatic* regions and less condensed *euchromatic* regions. During cell division, the nuclear membrane is fragmented and the chromosomes become highly compacted.

Our understanding of genome complexity and organization has been aided by various cytological, genetic, biochemical, and molecular techniques. In this chapter, we have considered equilibrium density centrifugation and renaturation experiments as techniques to probe the complexity of genomes. In Chapters 5 and 6, we examined ways to genetically map the locations of genes along eukaryotic and bacterial chromosomes. Chapter 20 will explore the use of cytological techniques such as *in situ* hybridization to identify particular sequences within an intact chromosome. In addition, Chapter 20 will describe molecular methods in which segments of a genome are cloned and analyzed molecularly.

The second topic in this chapter was the compaction of chromosomes to fit within living cells. This phenomenon can also be examined via several different approaches. As seen throughout this chapter, the microscopic analysis of chromosomes provides key insight into their structure. In addition, chromatin can be analyzed biochemically by many different methods. For example, the repeating nucleosome structure was verified by DNaseI digestion (as described in Experiment 10B). More recently, the detailed structure of nucleosomes has been resolved by crystallization studies. Currently, much research is aimed at identifying the proteins that are bound to chromosomal DNA and elucidating the roles that these proteins play in determining chromatin structure.

PROBLEM SETS

SOLVED PROBLEMS

1. Here is a C_0t curve for a hypothetical eukaryotic species:

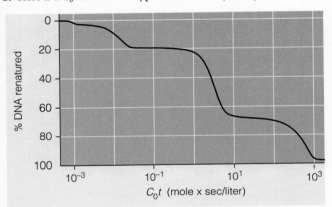

Estimate the amount of highly repetitive DNA, moderately repetitive DNA, and unique DNA.

Answer: About 20% is highly repetitive and renatures quickly, about 50% is moderately repetitive, and about 30% is unique and renatures very slowly.

2. Let's suppose that a DNA molecule is given a left-handed twist. How does this affect the structure and function of the DNA?

Answer: A left-handed twist is negative supercoiling. Negative supercoiling makes the bacterial chromosome more compact. It also promotes DNA functions that involve strand separation. These include gene transcription and DNA replication.

CONCEPTUAL QUESTIONS

1. In viral replication, what is the difference between self-assembly and directed assembly?

2. What is a bacterial nucleoid? With regard to cellular membranes, what is the difference between a bacterial nucleoid and a eukaryotic nucleus?

3. Describe the two main mechanisms by which the bacterial DNA becomes compacted.

4. Draw a DNA molecule. In your drawing, explain the function of DNA gyrase.

5. How are two topoisomers different from each other? How are they the same?

6. What is the function of a centromere? At what stage of the cell cycle would you expect the centromere to be the most important?

7. Describe the characteristics of highly repetitive DNA.

8. The Alu family of highly repetitive sequences in primates is interspersed throughout the genome. Would you expect Alu sequences to form a satellite peak during equilibrium density gradient centrifugation?

9. Do satellite DNAs that sediment at different equilibrium densities have the same or different DNA sequences? Explain.

10. Explain how a renaturation experiment can provide quantitative information about genome sequence complexity.

11. Describe the structures of a nucleosome and a 30 nm fiber.

12. Beginning with the G_1 phase of the cell cycle, describe the level of compaction of the eukaryotic chromosome. How does the level of compaction change as the cell progresses through the cell cycle? Why is it necessary to further compact the chromatin during mitosis?

13. If you assume that the average length of linker DNA is 50 base pairs, approximately how many nucleosomes are there in the haploid human genome, which contains 3 billion base pairs?

14. Draw the binding between the nuclear scaffold and SARs. Be specific about the types of DNA sequences in the SARs that are binding to proteins in the scaffold.

15. Compare heterochromatin and euchromatin. What are the differences between them?

16. Compare the structure and cell localization of chromosomes during interphase and M phase.

17. What types of genetic activities occur during interphase? Explain why these activities cannot occur during M phase.

EXPERIMENTAL QUESTIONS

1. Here are the results of an equilibrium density centrifugation experiment carried out on two different species (a mammal and an amphibian):

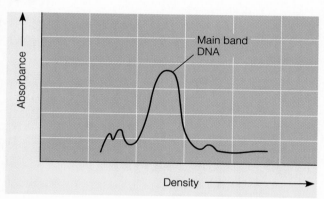

(a) Mammal

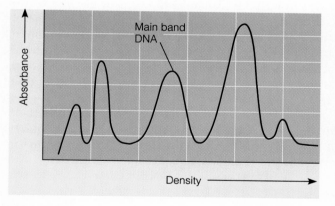

(b) Amphibian

The amphibian has 11 times as much DNA as the mammal. Discuss why the amphibian has so much more DNA.

2. Let's suppose that you have isolated DNA from a cell and have viewed it under a microscope. It looks supercoiled. What experiment would you perform to determine if it is positively or negatively supercoiled? In your answer, you should describe your expected results. You may assume that you have purified topoisomerases at your disposal.

3. We seem to know more about the structure of eukaryotic chromosomal DNA than that of bacterial DNA. Discuss several experimental procedures that have yielded important information concerning the compaction of eukaryotic chromatin.

4. Is it possible for repetitive DNA to sediment at the same density as the main band DNA? Explain.

5. If an organism is shown to have two different satellite DNAs by equilibrium density gradient centrifugation, what does this tell you about the types of sequences within the satellites? In other words, do the two satellites have the same or different repeated sequences? Explain your answer.

6. An organism contains 20% highly repetitive DNA, 10% moderately repetitive DNA, and 70% unique sequences. Draw the expected C_0t curve that would be obtained from this organism.

7. Let's suppose you have isolated chromatin from some bizarre eukaryote with a linker region that is usually 300 to 350 base pairs in length. The nucleosome structure is the same as in other eukaryotes. If you digested this eukaryotic organism's chromatin with a high concentration of DNaseI, what would be your expected results?

8. If you were given a sample of chromosomal DNA and were asked to determine if it is prokaryotic or eukaryotic, what experiment would you perform and what would be your expected results?

QUESTIONS FOR STUDENT DISCUSSION/COLLABORATION

1. Based on your reading of this chapter, discuss whether or not you feel that DNA sequences are organized or disorganized along bacterial and eukaryotic chromosomes.

2. Bacterial and eukaryotic chromosomes are very compact. Discuss the advantages and disadvantages of having a compact structure.

3. The prevalence of highly repetitive sequences seems rather strange to many geneticists. Do they seem strange to you? Why or why not? Discuss whether or not you think they have an important function.

4. Discuss and make a list of the similarities and differences between prokaryotic and eukaryotic chromosomes.

CHAPTER 11

DNA REPLICATION

STRUCTURAL OVERVIEW OF DNA REPLICATION

PROKARYOTIC DNA REPLICATION

EUKARYOTIC DNA REPLICATION

As we have seen throughout Chapters 2 to 8, genetic material is transmitted from parent to offspring and from cell to cell. For this to occur, the genetic material must be copied. During this process, known as **DNA replication**, the original DNA strands are used as templates for the synthesis of new DNA strands. This chapter will begin with a consideration of the structural features of the double helix that pertain to the replication process. Then we will examine how chromosomes are replicated within living cells: where does DNA replication begin, how does it proceed, and where does it end? At the molecular level, it is rather remarkable that the replication of chromosomal DNA occurs very quickly, very accurately, and at the appropriate time in the life of the cell. For this to happen, many cellular proteins play vital roles. In this chapter, we will examine the mechanism of DNA replication and consider the functions of several proteins involved in the process.

STRUCTURAL OVERVIEW OF DNA REPLICATION

We begin by recalling a few important structural features of the double helix from Chapter 9, since they bear directly on the replication process. The double helix is composed of two DNA strands, and the individual building blocks of each strand are nucleotides. The nucleotides contain one of four bases: adenine, thymine, guanine, or cytosine. The double-stranded structure is held together by hydrogen bonding between the bases in opposite strands. A critical feature of the double helix structure is that it will only fit together if adenine hydrogen bonds with thymine and guanine hydrogen bonds with cytosine. This rule, known as the A–T/G–C rule or Chargaff's rule, is the basis for the complementarity of the base sequences in double-stranded DNA.

Another feature worth noting is that the strands within a double helix have an antiparallel alignment. This directionality is determined by the orientation of sugar residues within the sugar–phosphate backbone. If one strand is running in the 5′ to 3′ direction, the complementary strand is running in the 3′ to 5′ direction. The issue of directionality will be important later in this chapter, when we consider the function of the enzymes that synthesize new DNA strands. In this section, we will consider how the structure of the DNA double helix allows it to be replicated.

The complementarity of base sequences provides the basis for DNA replication

As shown in Figure 11-1a, DNA replication relies on the complementarity of DNA strands according to the A–T/G–C rule. During the replication process, the two

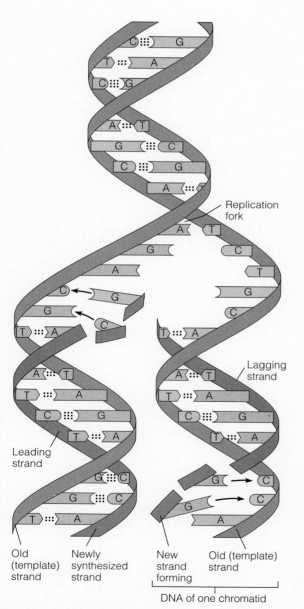

(a) The mechanism of DNA replication

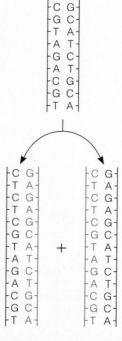

(b) The products of replication

FIGURE **11-1**

The structural basis for DNA replication. (a) The mechanism of DNA replication as originally proposed by Watson and Crick. As discussed later in this chapter, the synthesis in the leading strand occurs in the direction toward the replication fork, while the synthesis of the lagging strand occurs in small segments away from the replication fork. **(b)** DNA replication produces two copies of DNA with the same sequence as the original DNA molecule.

complementary strands of DNA come apart and serve as **template strands** for the synthesis of two new strands of DNA. After the template strands have separated, individual nucleotides can hydrogen bond with the template strands. This hydrogen bonding must obey the A–T/G–C rule. To complete the replication process, a covalent bond is formed between the phosphate on one nucleotide and the sugar on the previous nucleotide. The two newly made strands are referred to as the **daughter strands**, while the original strands are called the **parental strands**. Note that the base sequences are identical in both double-stranded molecules after replication (Figure 11-1b). Therefore, DNA can be replicated so that both copies retain the same information (i.e., the same DNA sequence) as the original molecule.

PARENT
CELL

FIRST
REPLICATION

SECOND
REPLICATION

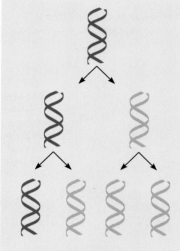

(a) Conservative model: The parental double helix remains intact and a second, all-new copy is made.

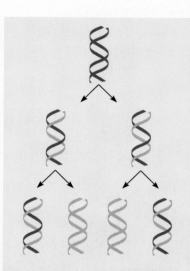

(b) Semiconservative model: The two strands of the parental molecule separate, and each functions as a template for synthesis of a new complementary strand.

PARENT
CELL

FIRST
REPLICATION

SECOND
REPLICATION

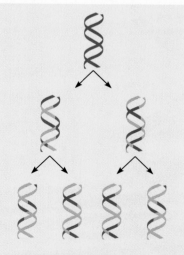

(c) Dispersive model: Each strand of *both* daughter molecules would contain a mixture of old and newly synthesized parts.

FIGURE **11-2**

Three possible models for DNA replication. The two original parental DNA strands are shown in dark blue, and the newly made strands after one or two generations are shown in light blue.

In the 1950s, three different models were proposed that described the net result of DNA replication

Scientists in the late 1950s had considered three different mechanisms that described what the net result of DNA replication would be. These mechanisms are shown in Figure 11-2. The first is referred to as a **conservative** model. According to this hypothesis, both strands of parental DNA remain together following DNA replication. Therefore, the original arrangement of parental strands is completely conserved, and the two daughter strands are also together following replication. The second is called a **semiconservative** model. As seen in Figure 11-2b, the double-stranded DNA is "half" conserved following the replication process. In other words, the double-stranded DNA contains one parental strand and one daughter strand. The third, **dispersive**, model proposes that segments of parental DNA and newly made DNA are interspersed in both strands following the replication process. As we will see, only one of the three models shown in Figure 11-2 is actually correct.

The work of Meselson and Stahl in 1958 supported the semiconservative model of DNA replication

EXPERIMENT 11A

To identify the correct model of those shown in Figure 11-2, Matthew Meselson and Franklin Stahl at the California Institute of Technology experimentally distinguished newly made daughter strands from the original parental strands. The technique they used is called *heavy isotope labeling*. Nitrogen, which is found within the bases of DNA, occurs in a light (^{14}N) form and a heavy (^{15}N) form. Meselson and Stahl reasoned that if the parental DNA was made within cells that only have ^{15}N, all the parental DNA would be heavy. Prior to DNA replication, the cells could then be provided with only ^{14}N so that all their newly made DNA would be made using ^{14}N. Therefore, the daughter DNA strands would be light while the original parental DNA strands would be heavy.

Meselson and Stahl could then analyze the labeling of DNA strands when the DNA was subjected to centrifugation. (The procedure of gradient centrifugation is described in the appendix.) If both DNA strands contained ^{14}N, the DNA would have a light density. If one strand contained ^{14}N and the other strand contained ^{15}N, the DNA would be half-heavy. Alternatively, if both strands contained ^{15}N, the DNA would be heavy and sediment closer to the bottom of the centrifuge tube.

THE HYPOTHESIS

One of the three models shown in Figure 11-2 may accurately describe the net result of DNA replication. The purpose of this experiment is to determine which of the three models is correct.

TESTING THE HYPOTHESIS

Starting material: A strain of *E. coli* that has been grown for many generations in the presence of ^{15}N. All of the nitrogen in the DNA is labeled with ^{15}N.

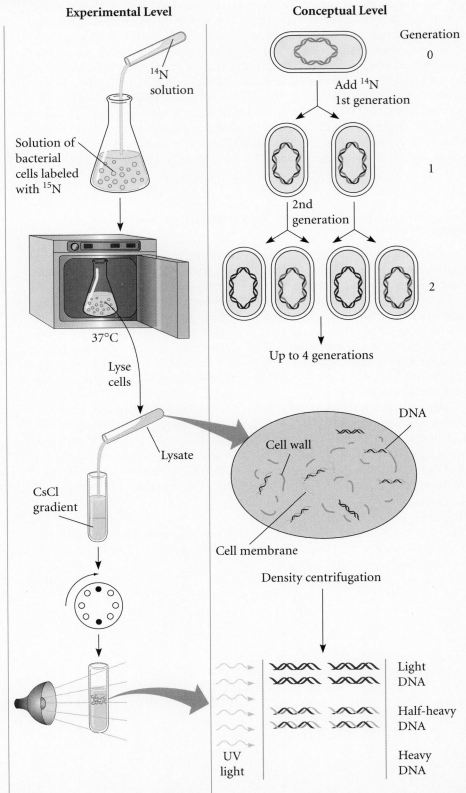

Experimental Level

Conceptual Level

1. Add an excess of ^{14}N containing compounds to the bacterial cells so that all the newly made DNA will contain ^{14}N.

2. Incubate the cells for various lengths of time.

3. Lyse the cells by the addition of lysozyme and detergent, that disrupt the bacterial cell wall and cell membrane respectively.

4. Load the lysate onto a CsCl gradient that contains ethidium bromide. (Note: The average density of DNA is around 1.7 gm/cm^3, which is well isolated from other cellular macromolecules.)

5. Centrifuge the gradients until the DNA molecules reach their equilibrium densities.

6. DNA within the gradient can be observed with a UV light since the DNA is now stained with ethidium bromide, a UV sensitive dye.

^{14}N solution

Solution of bacterial cells labeled with ^{15}N

37°C

Lyse cells

Lysate

CsCl gradient

Generation 0

Add ^{14}N
1st generation

1

2nd generation

2

Up to 4 generations

DNA

Cell wall

Cell membrane

Density centrifugation

UV light

Light DNA

Half-heavy DNA

Heavy DNA

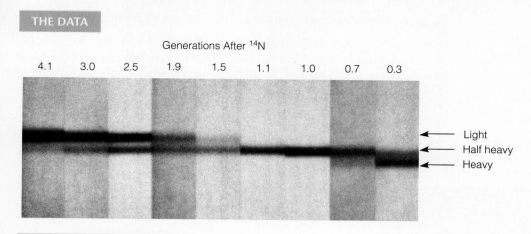

After one complete generation (i.e., one round of DNA replication), all the DNA sedimented at a density that was half-heavy. These results are consistent with both the semiconservative and dispersive models. The conservative model had predicted two separate DNA types: a light type and a heavy type. Since all the DNA had sedimented as a single band, this model was now disproved. According to the semiconservative model, the replicated DNA would contain one original strand (a heavy strand) and a newly made daughter strand (a light strand). Likewise, in a dispersive model, there should have been all half-heavy DNA after one generation as well. To determine which of these two remaining models was correct, therefore, Meselson and Stahl had to investigate future generations.

After approximately two generations (precisely, 1.9 generations), there was a mixture of light DNA and half-heavy DNA. This observation was also consistent with the semiconservative mode of DNA replication, because some DNA molecules should contain all light DNA while other molecules should be half-heavy (see Figure 11-2b). In a dispersive model, however, the DNA after two generations would have been 1/4-heavy. This is because a dispersive model predicts that the heavy nitrogen would be evenly dispersed among four strands, each strand thus containing 1/4 heavy nitrogen and 3/4 light nitrogen (see Figure 11-2c). However, this result was not obtained. Instead, the results of the Meselson and Stahl experiment provided compelling evidence in favor of the semiconservative model for DNA replication.

PROKARYOTIC DNA REPLICATION

Thus far in this chapter, we have considered how complementarity underlies the ability of DNA to be copied. In addition, the experiments of Meselson and Stahl showed that DNA replication results in two double helices, each one containing an original parental strand and a newly made daughter strand. We will now turn our attention to how DNA replication actually occurs within living cells. Much research has focused on DNA replication in the bacterium *E. coli*. The results of these studies have provided the foundation for our current understanding of DNA replication. The replication of the bacterial chromosome is a stepwise process in which many cellular proteins participate. In this section, we will follow this process from beginning to end.

Bacterial chromosomes contain a single origin of replication

Figure 11-3 presents an overview of the process of bacterial chromosome replication. The site on the bacterial chromosome where DNA synthesis begins is known as the origin of replication. Bacterial chromosomes have a single origin of replication. The synthesis of new daughter strands is initiated within the origin and proceeds **bidirectionally** (in both directions) around the bacterial chromosome. This means that two **replication forks** move in opposite directions outward from

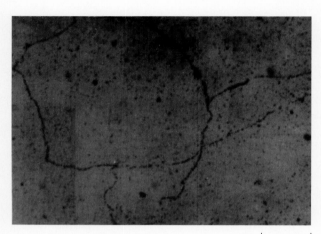

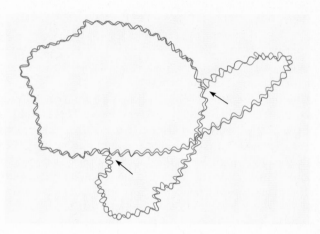

(a) A replicating *E. coli* chromosome

0.25 μm

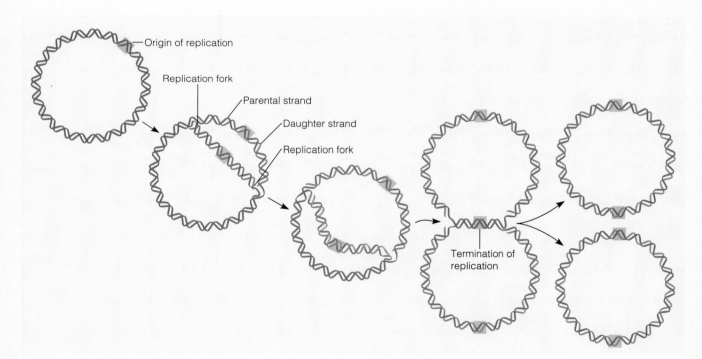

(b) The process of replication

FIGURE **11-3**

The process of bacterial chromosome replication. (a) A replicating *E. coli* chromosome visualized by autoradiography. The diagram at the right shows the locations of the two replication forks (at the arrows). The chromosome is about one-third replicated. New strands are shown in blue. **(b)** An overview of the process of bacterial chromosome replication.

the origin. Eventually, these replication forks meet each other on the opposite side of the bacterial chromosome to complete the replication process.

Replication is initiated by the binding of DnaA protein to the origin of replication

Considerable research has focused on the origin of replication in *E. coli*. This origin has been named *oriC* (for *ori*gin of *c*hromosomal replication; Figure 11-4). There are three types of DNA sequences within *oriC* that are functionally important: an A/T-rich region, dnaA box sequences, and GATC methylation sites. The third functional sequence, GATC methylation sites, will be discussed later when we consider the regulation of replication.

DNA replication is initiated by the binding of *DnaA proteins* to sequences within the origin known as **dnaA box sequences**. The dnaA box serves as a recognition site for the binding of the DnaA protein. As shown in Figure 11-5, several DnaA proteins bind to the four dnaA boxes in *oriC* to initiate DNA replication. The DnaA protein has several important functions in DNA replication. First, it binds specifically to the dnaA boxes in *oriC*, thereby recognizing this region as the single origin of replication. Next, DnaA proteins cause the A/T-rich region to denature into separate single-stranded regions. Since A/T base pairs form only two hydrogen bonds while G/C base pairs form three, it is easier to separate two DNA strands in an A/T-rich region.

Following denaturation of the A/T-rich region, the DnaA proteins, with the help of the DnaC protein, recruit **DNA helicase** (also known as DnaB helicase) to bind to this site. DNA helicases begin strand separation within the *oriC* region and continue to separate the DNA strands beyond the origin. These enzymes use the energy from ATP hydrolysis to catalyze the separation of the double-stranded parental DNA (see Figure 11-5). DNA helicases bind to single-stranded DNA and travel along the DNA in a 5′ to 3′ direction to keep the replication fork moving. When a helicase encounters a double-stranded region, it breaks the hydrogen bonds between the two strands, thereby generating two single strands. As shown in Figure 11-5, the action of DNA helicases promotes the movement of the two replication forks outward from *oriC* in opposite directions.

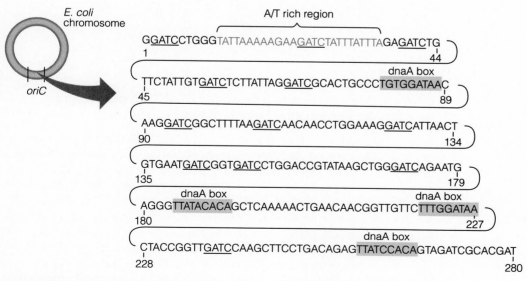

FIGURE **11-4**

The sequence of *oriC* in *E. coli*. The GATC methylation sites are underlined. The dnaA boxes are highlighted in orange, and the A/T-rich region is highlighted in blue. This sequence only depicts one DNA strand; the complementary strand in *oriC* is not shown.

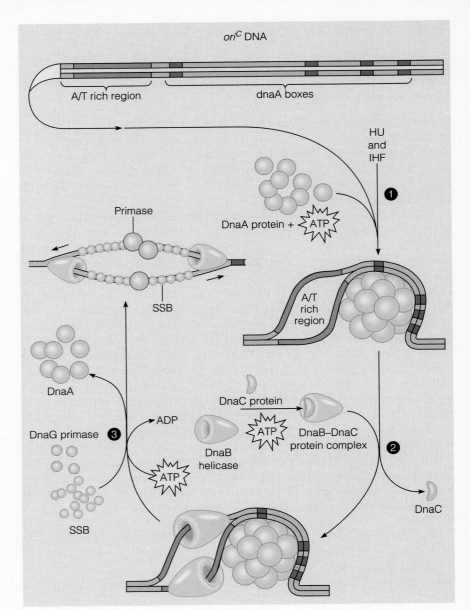

FIGURE **11-5**

The events that occur at *oriC* to initiate the DNA replication process. (1) To initiate DNA replication, DnaA proteins bind to the four dnaA boxes, which causes the DNA strands to separate. Additional proteins called HU and IHF facilitate this process. **(2)** DnaA and DnaC proteins then recruit helicase into this region. Helicase is composed of six subunits, which form a ring on two DNA strands and migrate in the 5′ to 3′ direction. **(3)** As shown here, the movement of helicase serves to separate the DNA strands beyond the *oriC* region. Later, primase (also called DnaG primase) and SSB (single-strand binding protein) bind to the single-stranded DNA, and DnaA proteins are released.

DNA strand separation and the synthesis of RNA primers are necessary before daughter strands of DNA can be made

Figure 11-6 provides an overview of the molecular events that occur as a replication fork moves around the bacterial chromosome. The action of DNA helicase is believed to generate positive supercoiling ahead of the replication fork. An enzyme known as **topoisomerase** (i.e., DNA gyrase; see Chapter 10) travels ahead of the helicase enzyme and alleviates this positive supercoiling.

After the two parental strands have been separated and the supercoiling relaxed, they must be kept that way until the complementary daughter strands have been made. The function of the **single-strand binding protein** is to bind to both of the single strands of parental DNA and prevent them from re-forming a double helix. In this way, the bases within the parental strands are kept in an exposed condition that enables them to hydrogen bond with single nucleotides.

The next event in DNA replication involves the synthesis of a short strand of RNA (rather than DNA) called an *RNA primer*. This strand of RNA is synthesized by the linkage of ribonucleotides via an enzyme known as **primase**. These RNA

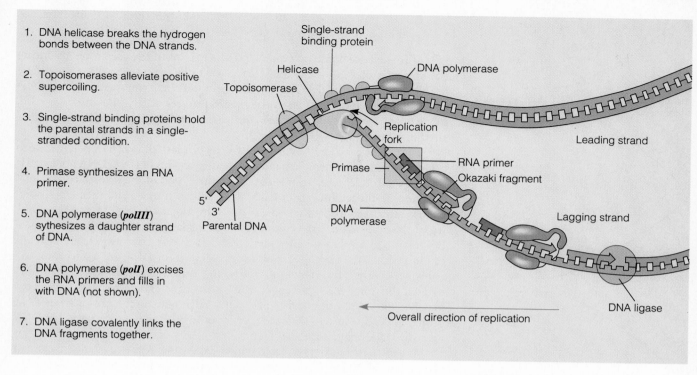

1. DNA helicase breaks the hydrogen bonds between the DNA strands.

2. Topoisomerases alleviate positive supercoiling.

3. Single-strand binding proteins hold the parental strands in a single-stranded condition.

4. Primase synthesizes an RNA primer.

5. DNA polymerase (*polIII*) sythesizes a daughter strand of DNA.

6. DNA polymerase (*polI*) excises the RNA primers and fills in with DNA (not shown).

7. DNA ligase covalently links the DNA fragments together.

FIGURE **11-6**
Enzymology of DNA replication.

strands start, or prime, the process of DNA replication (Figure 11-6). At a later stage in DNA replication, the RNA primers are removed.

DNA polymerases link nucleotides to synthesize the daughter strands

The enzymes known as **DNA polymerases** are responsible for covalently attaching nucleotides together to make new daughter strands. In *E. coli*, there are three distinct proteins that function as DNA polymerases, designated *polI*, *polII*, and *polIII*. *PolI* and *polIII* are involved in DNA replication, and *polII* is important in DNA repair. The action of *polIII* is shown in Figure 11-7. In DNA replication, there are two important features of DNA polymerases that may seem unusual:

1. DNA polymerases cannot link together the first two nucleotides in a new strand. Rather, they can only elongate a strand starting with a primer (see Figure 11-8a).

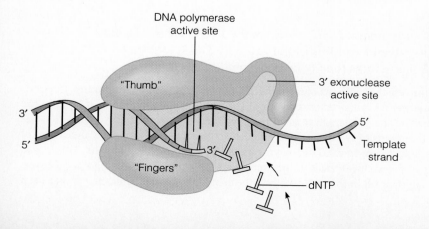

FIGURE **11-7**

The action of DNA polymerase (*polIII*). DNA polymerase slides along the template strand, and it synthesizes new strands by connecting deoxynucleoside triphosphates (dNTPs) in a 5′ to 3′ direction. The structure of DNA polymerase resembles a hand that is wrapped around the template strand. In this regard, the movement of DNA polymerase along the template strand is similar to a hand that is sliding along a rope.

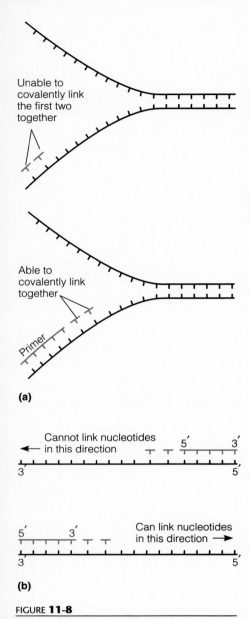

Unable to covalently link the first two together

Able to covalently link together

Primer

(a)

Cannot link nucleotides in this direction ←

5′   3′

3′   5′

5′   3′

3′   5′

Can link nucleotides in this direction →

(b)

FIGURE **11-8**

Unusual features of DNA polymerase function. (a) DNA polymerase can only elongate a strand starting with a short RNA primer. **(b)** DNA polymerase can only attach nucleotides in a 5′ to 3′ direction. Note the template strand is in the opposite, 3′ to 5′, direction.

2. DNA polymerases can attach nucleotides only in the 5′ to 3′ direction, not in the 3′ to 5′ direction (Figure 11-8b).

At first glance, these two characteristics may seem to pose problems during the synthesis of DNA. Let's now turn back to Figure 11-6 and see how these problems are overcome. As shown there, additional enzymes within the replication fork compensate for these two unusual features of DNA polymerases. Also, the two new DNA strands (namely, the leading and lagging strands) are synthesized in different ways.

The synthesis of an RNA primer by RNA primase allows DNA polymerase to begin the synthesis of complementary daughter strands of DNA. DNA polymerase catalyzes the attachment of nucleotides to the 3′ end of the primer, in a 5′ to 3′ direction. In the **leading strand**, one RNA primer is made and then DNA polymerase can attach nucleotides in a 5′ to 3′ direction as it slides toward the opening of the replication fork (Figure 11-6). In the **lagging strand**, the synthesis of DNA is also in a 5′ to 3′ manner, but it occurs in the direction away from the replication fork. In the lagging strand, short segments of DNA are made; these are known as **Okazaki fragments**, after Reiji and Tuneko Okazaki, who initially discovered them in 1968. The length of Okazaki fragments in bacteria is approximately 1000–2000 nucleotides. Each Okazaki fragment contains a short RNA primer at the 5′ end, and then the remainder of the fragment is a strand of DNA made by DNA polymerase (*polIII*).

As shown in Figure 11-9, DNA polymerases catalyze the covalent attachment between the phosphate in one nucleotide and the sugar in the previous nucleotide. Prior to this event, the nucleotide that is about to be attached to the growing strand is a nucleoside triphosphate. It contains three phosphate groups attached at the 5′–OH group of the sugar. The nucleoside triphosphate first hydrogen bonds to the template strand according to the A–T/G–C rule. Next, the 3′–OH group on the previous nucleotide reacts with the innermost phosphate group on the incoming nucleotide to form a covalent bond. The formation of this covalent bond causes the newly made strand to grow in the 5′ to 3′ direction. As shown in Figure 11-9, pyrophosphate (PO_4—PO_4) is released.

Not only does DNA polymerase have to catalyze the covalent attachment of nucleotides, it has to do this very quickly. In *E. coli*, DNA polymerase attaches approximately 850 nucleotides per second! DNA polymerase can catalyze the synthesis of the daughter strands so quickly because it is a **processive enzyme**. This means that it does not dissociate from the growing strand after it has catalyzed the covalent joining of two nucleotides. Rather, as depicted in Figure 11-7, it remains clamped to the DNA template strand and slides along the template as it catalyzes the synthesis of the daughter strand. Certain proteins, known as processivity factors, are responsible for this clamping phenomenon. For example, in *E. coli*, processivity factors called β (beta), τ (tau), and γ (gamma), which are subunits of *polIII*, are important in clamping the enzyme to the DNA template strand.

DNA polymerase (*polI*) removes the RNA primers in the lagging strand and fills in the gap with DNA, and then DNA ligase links the Okazaki fragments

To complete the synthesis of Okazaki fragments within the lagging strand, three additional events must occur: removal of the RNA primers, synthesis of DNA in the area where the primers have been removed, and the covalent attachment of adjacent fragments of DNA. In *E. coli*, the RNA primers are removed by the action of DNA polymerase (*polI*). This enzyme has a 5′ to 3′ exonuclease activity. This means that *polI* digests away the RNA primers in a 5′ to 3′ direction, leaving a vacant area. *PolI* can then fill in this region by synthesizing DNA there. After the gap has been

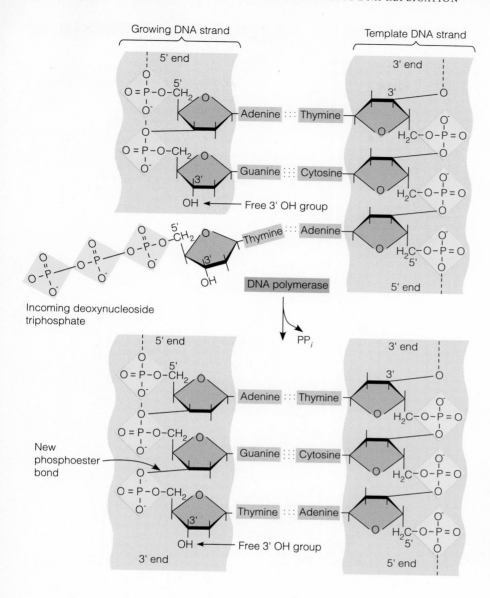

FIGURE **11-9**

The enzymatic action of DNA polymerase. DNA polymerase catalyzes the formation of a covalent bond between the 3′–OH group on the previous nucleotide and the innermost 5′–phosphate group on the incoming nucleotide. Pyrophosphate (PP_i) is released.

completely filled in, there is still a bond missing between the last nucleotide added by *polI* and the first nucleotide in the next DNA fragment. An enzyme known as **DNA ligase** catalyzes a covalent bond between these adjacent DNA fragments to complete the process (see Figure 11-6). This completes the replication process in the lagging strand. Table 11-1 gives a functional summary of the proteins involved in *E. coli* DNA replication.

Replication is terminated when the replication forks meet at the terminus sequences

On the opposite side from *oriC*, the *E. coli* chromosome contains a pair of **termination sequences** called *ter* sequences. A protein known as the *ter*-binding protein binds to the *ter* sequences and stops the movement of the replication forks. As shown in Figure 11-10, one of the *ter* sequences stops the clockwise-moving fork while the other *ter* sequence stops the counterclockwise-moving fork. Therefore, DNA replication ends at the *terminus* sequences and DNA ligase covalently links all

TABLE 11–1 Proteins involved in *E. coli* DNA replication

Common Name	Function
DnaA protein	Binds to dnaA boxes within the origin to initiate DNA replication
Helicase (DnaB)	Separates double-stranded DNA
DnaC protein	Aids DnaA in the recruitment of helicase to the origin
Topoisomerase (gyrase)	Removes positive supercoils
Primase	Synthesizes short RNA primers
DNA polymerase (*polIII*)	Synthesizes DNA in the leading and lagging strands
DNA polymerase (*polI*)	Removes RNA primers, fills in gaps
DNA ligase	Covalently attaches adjacent DNA fragments
ter binding protein	Binds to *ter* sequences and prevents the advance of the replication fork

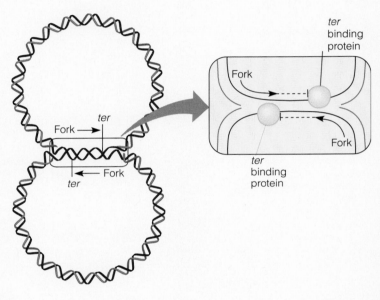

FIGURE **11-10**

The termination of DNA replication. A site in the bacterial chromosome, shown in the rectangle, contains two *ter* sequences. As shown in the inset, the binding of the two *ter*-binding proteins, one in each strand, prevents the replication forks from proceeding past the *ter* sequences.

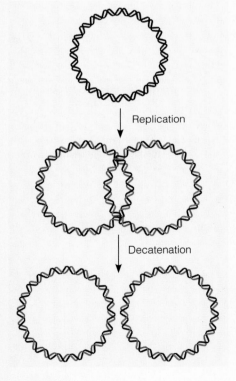

FIGURE **11-11**

Separation of catenanes. When DNA replication is complete, the two circular chromosomes may be interlocked. These catenanes can be separated by the action of topoisomerases.

four DNA strands, creating two circular double-stranded molecules. Curiously, the *ter* sequences are arranged so that the forks would pass each other to reach the *ter* sequence that stops its own advancement (see the inset in Figure 11-10). Since this could not occur, one explanation is that only one *ter* sequence is required to stop the advancement of one replication fork, and that the other fork ends its synthesis of DNA when it reaches the halted replication fork.

After DNA replication is completed, one last problem may exist. Bacterial DNA replication often results in two intertwined DNA molecules (Figure 11-11). Interlocked circular molecules are known as **catenanes**. Fortunately, catenanes are only transient structures in DNA replication. Topoisomerases introduce a temporary break into the DNA strands and then rejoin them after the strands have become unlocked. This allows the catenanes to be separated into individual circular molecules.

Certain enzymes of DNA replication bind to each other to form a complex

Figure 11-12 provides a more three-dimensional view of the DNA replication process. DNA helicase, primase, and several accessory proteins form a multiprotein complex known as a **primosome**. This complex leads the way at the replication fork. The primosome tracks along the DNA processively, separating the parental strands and synthesizing RNA primers at regular intervals along the lagging strand. By acting within a primosome, the actions of helicase and primase can be better coordinated.

Behind the primosome are the other proteins involved in DNA replication. In this drawing, it is suggested that two DNA polymerases (*polIII*) act in concert to replicate the leading and lagging strands. In other words, two DNA polymerases associate together to form a dimeric replicative DNA polymerase that moves as a unit. For this to occur, the lagging strand is looped around one of the DNA polymerase enzymes so that it can synthesize DNA in a 5′ to 3′ direction, yet move

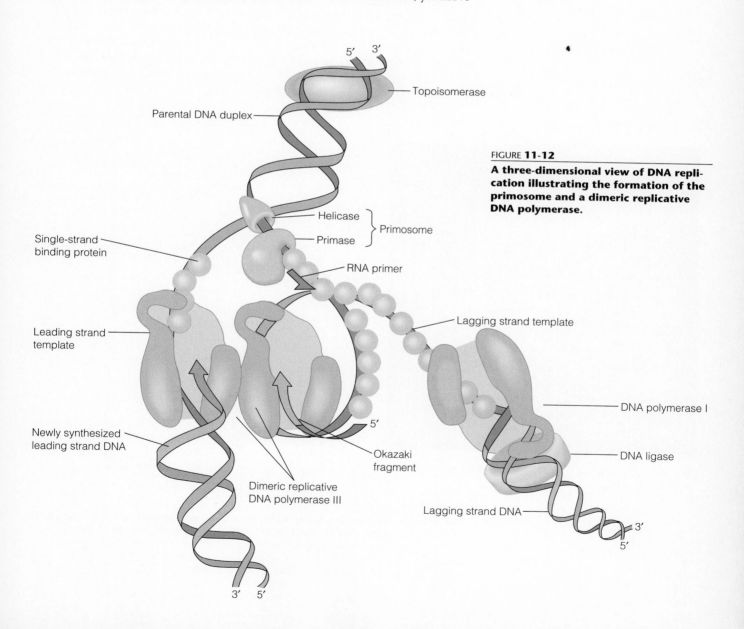

FIGURE 11-12

A three-dimensional view of DNA replication illustrating the formation of the primosome and a dimeric replicative DNA polymerase.

toward the opening of the replication fork. Research also suggests that the DNA polymerases may be associated with the primosome, so that DNA helicase, primase, and the DNA polymerases would move as a single unit along the replication fork. This putative complex has been given the name **replisome**.

The fidelity of DNA replication is ensured by proofreading mechanisms

With replication occurring so rapidly, one might imagine that mistakes could happen in which the wrong nucleotide could be incorporated into the growing daughter strand. Although mistakes can happen during DNA replication, they are extraordinarily rare. In the case of DNA synthesis via *polIII*, only one mistake per 100,000,000 nucleotides is made. Therefore, DNA synthesis occurs with relatively few errors. Another way of saying this is that DNA replication exhibits a high degree of **fidelity**.

There are several reasons that fidelity is high. First, the hydrogen bonding between G/C and A/T is much more stable than that between mismatched pairs. However, this accounts for only part of the fidelity, since mismatching due to stability considerations would account for one mistake per 1000 nucleotides. Two additional characteristics of DNA polymerases contribute to their fidelity of DNA replication. The active site of DNA polymerase is such that it will preferentially catalyze the attachment of nucleotides when the correct bases are

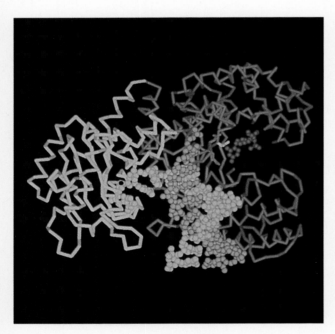

(a) Crystal graphic structure of DNA polymerase I bound to DNA while proofreading

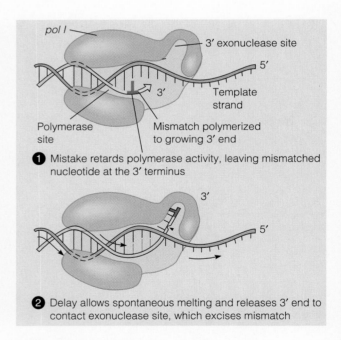

(b) The proofreading function of DNA polymerase

FIGURE **11-13**

The structure and proofreading function of DNA polymerase. (a) This molecular model shows a portion of DNA polymerase I (shown in yellow and blue stick form) as it is synthesizing and proofreading DNA. The DNA is shown as balls: the template strand is green, the growing strand is light blue, and a mispaired cytosine in the 3′ exonuclease site is shown in red. **(b)** A schematic diagram of proofreading.

hydrogen bonding in opposite strands. In other words, DNA polymerase is unlikely to catalyze bond formation between adjacent nucleotides if a mismatched base pair is formed. This induced-fit phenomenon decreases the error rate to one in 100,000 to 1,000,000.

Another way that DNA polymerase decreases the error rate is by the enzymatic removal of mismatched nucleotides. As shown in Figure 11-13, DNA polymerase can identify a mismatched nucleotide and remove it from the daughter strand. This occurs by cleavage of the bond between adjacent nucleotides at the 3′ end of the newly made strand. This process is referred to as **exonuclease** cleavage (the prefix *exo-* indicates that DNA polymerase digests away nucleotides at the end of a DNA strand). The ability to remove mismatched bases by this mechanism has been called the **proofreading function** of DNA polymerase.

Prokaryotic DNA replication is coordinated with cell division

Bacterial cells can divide into two daughter cells at an amazing rate. Under optimal conditions, certain bacteria such as *E. coli* can divide every 20 to 30 minutes. It is important that DNA replication only take place when a cell is about to divide. If DNA replication occurs too frequently, then there will be too many copies of the bacterial chromosome per cell. Alternatively, if DNA replication does not occur frequently enough, a daughter cell will be left without a chromosome. Therefore, cell division in bacterial cells must coordinate with DNA replication.

Bacterial cells regulate the DNA replication process by controlling the initiation of replication at the origin. This control has been extensively studied in *E. coli*. In this bacterium, several different mechanisms have been proposed to account for the control of DNA replication. In general, the regulation aims to prevent the premature initiation of DNA replication at *oriC*. Two different mechanisms are described here.

First, the initiation of replication is coupled to cell division by the action of the **DnaA** protein (Figure 11-14). As discussed previously in this chapter, the DnaA protein binds to dnaA boxes within the origin of replication and opens the DNA helix at the A/T-rich region within the origin. To initiate DNA replication, the concentration of the DnaA protein must become high enough to bind all the dnaA boxes. Immediately following DNA replication, there are twice as many dnaA boxes, and so there is insufficient DnaA protein to reinitiate a second round of replication. Also, some of the DnaA protein may be unavailable for DNA binding because it becomes attached to the cell membrane during cell division. Since it takes time to accumulate newly made DnaA protein, DNA replication cannot occur until the daughter cells have had time to grow.

Another way to regulate DNA replication involves the *GATC sites* within *oriC*. These sites can be methylated by an enzyme known as *dam* methylase. The *dam* methylase recognizes the GATC sequence, binds there, and attaches a methyl group onto the adenine base (see Figure 11-15a). DNA methylation within *oriC* helps regulate the replication process. Prior to DNA replication, these sites are methylated in both strands. This full methylation of the GATC sites is necessary for DNA replication to be initiated at the origin. Following DNA replication, the newly made strands are not methylated, since adenine rather than methyladenine is found in the daughter strands (see Figure 11-15b). The hemimethylated (i.e., half-methylated) DNA cannot initiate replication until after it has become fully methylated. Since it takes several minutes for the *dam* methylase to methylate all the GATC sequences within this region, DNA replication will not occur again too quickly.

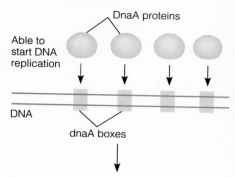

To begin DNA replication:

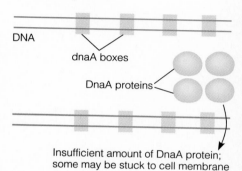

After DNA replication:

FIGURE 11-14

The amount of DnaA protein provides a way to regulate DNA replication. To begin replication, there must be enough DnaA protein to bind to all the dnaA boxes. Immediately after DNA replication, there is insufficient DnaA protein to reinitiate a second (premature) round of DNA replication. This is because there are twice as many dnaA boxes after DNA replication, and because some DnaA protein may be stuck to the cell membrane after cell division.

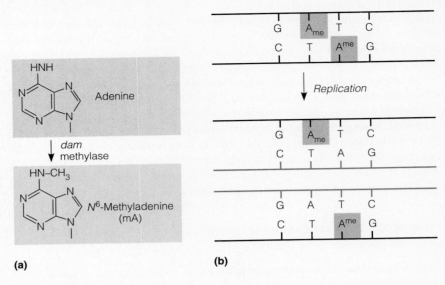

FIGURE 11-15

Methylation of GATC sites in *oriC*. (a) The action of *dam* methylase, which covalently attaches a methyl group to adenine. **(b)** Prior to DNA replication, the action of *dam* methylase causes both adenines within the GATC sites to be methylated. After DNA replication, only the adenines in the original strands are methylated. These are called hemimethylated double helices. Several minutes will pass before *dam* methylase will methylate these unmethylated adenines.

Arthur Kornberg's work provided a way to measure DNA replication *in vitro*

Much of our understanding of the process of prokaryotic DNA replication has come from thousands of experiments in which DNA replication has been studied *in vitro*. A common experimental strategy is to purify proteins from cell extracts and to determine their roles in the replication process. In other words, purified proteins (see Table 11-1), nucleotides, template DNA, and so forth can be mixed together in a test tube, and the synthesis of new DNA strands will occur. While at the Washington University and later at Stanford, this approach was pioneered by Arthur Kornberg and his colleagues.

Although we will not discuss the procedures for purifying replication proteins, Experiment 11B illustrates an early success in measuring DNA replication *in vitro*. In 1958, not long after the discovery of the double helix, Kornberg and colleagues set out to develop methods for measuring DNA synthesis. They had to base their initial work on hypotheses about the chemical structure of the precursor nucleotides that are involved in DNA replication. Kornberg correctly hypothesized that nucleoside triphosphates are the precursors for DNA synthesis. Also, he knew that nucleoside triphosphates are soluble in an acidic solution, whereas long strands of DNA are not. Rather, DNA strands will precipitate out of solution at an acidic pH. This precipitation event provides a method to separate nucleotides (e.g., nucleoside triphosphates) from strands of DNA.

THE HYPOTHESIS

DNA synthesis can occur *in vitro* if all the necessary components are present.

TESTING THE HYPOTHESIS

Starting material: An extract of proteins from *E. coli*.

1. Mix together the extract of *E. coli* proteins, template DNA that is not radiolabeled, and ^{32}P-radiolabeled deoxynucleoside triphosphates. This is expected to be a complete system that contains everything necessary for DNA synthesis. As a control, a second sample is made in which the template DNA was omitted from the mixture.

2. Incubate the mixture for 30 minutes at 37°C.

3. Add perchloric acid to precipitate DNA. It does not precipitate free nucleotides.

4. Centrifuge the tube. (10,000 × *g*, 3 minutes) Note: The radiolabeled nucleoside triphosphates that have not been incorporated into DNA will remain in the supernatant.

5. Collect the pellet, which contains precipitated DNA and proteins. (The control pellet is not expected to contain DNA.)

6. Count the amount of radioactivity in the pellet using a scintillation counter. (See the appendix.)

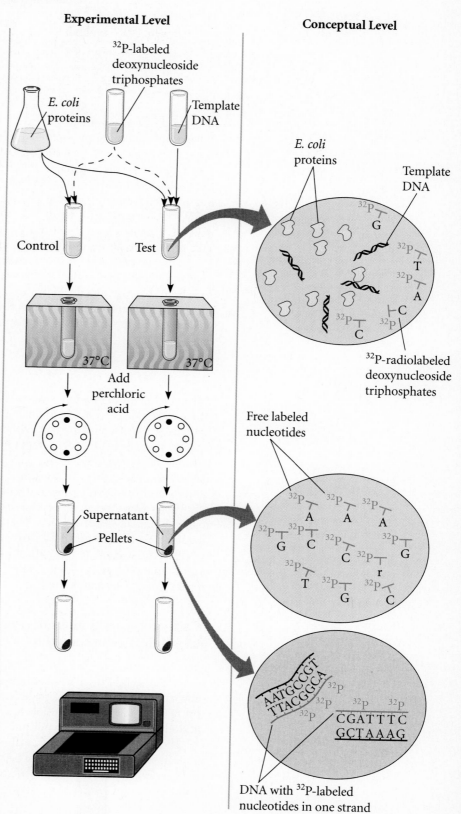

Experimental Level

Conceptual Level

THE DATA

Conditions	Amount of Radiolabeled DNA*
Complete system	3300
Template DNA omitted	0.0

*Calculated in picomoles of ^{32}P-labeled DNA.
From Lehman, I. R., Bessman, M. J., Simms, E. S., and Kornberg, A. (1958) Enzymatic synthesis of deoxyribonucleic acid. I. Preparation of substrates and partial purification of an enzyme from *Escherichia coli*. *J. Biol. Chem.* 233, 163–170.

INTERPRETING THE DATA

When the *E. coli* proteins were mixed with nonlabeled template DNA and radiolabeled deoxynucleoside triphosphates, an acid-precipitable and radiolabeled product was formed. This product must be newly synthesized DNA strands rather than individual nucleotides, since the nucleotides are not acid precipitable. As a control, if nonlabeled template DNA was omitted from the assay, no radiolabeled DNA was made. This is the expected result, since the template DNA is necessary to make new daughter strands. Taken together, these results indicate that this technique can be used to measure the synthesis of DNA *in vitro*.

In the late 1950s, the *in vitro* approach provided an exciting starting point to unravel the complexities of the DNA replication process. Since that time, researchers have been able to purify the individual proteins described in Table 11-1 and determine their roles in the replication process. This approach still continues, particularly as we try to understand the added complexities of eukaryotic DNA replication.

EUKARYOTIC DNA REPLICATION

Eukaryotic DNA replication is not as well understood as prokaryotic replication. Much research has been carried out in a variety of experimental organisms, particularly yeast and mammals. Many of these studies have found extensive similarities between the general features of DNA replication in prokaryotes and eukaryotes. For example, the types of prokaryotic enzymes described in Table 11-1 have also been identified in eukaryotes. DNA helicases, primases, DNA polymerases, single-strand binding proteins, DNA ligases, and topoisomerases have all been found in eukaryotic species. Nevertheless, at the molecular level, eukaryotic DNA replication appears to be substantially more complex. These complexities are related to several features of eukaryotic cells. In particular, eukaryotic cells have larger, linear chromosomes; the chromatin is more tightly packed within nucleosomes; and cell cycle regulation is much more complicated. This section will emphasize some of the unique features of eukaryotic DNA replication.

Initiation occurs at multiple origins of replication on linear eukaryotic chromosomes

Since eukaryotes have long linear chromosomes, they require multiple origins of replication so that the DNA can be replicated in a reasonable length of time. As shown in Figure 11-16, DNA replication proceeds bidirectionally from many origins of replications. The multiple replication forks eventually make contact with each other to complete the replication process.

The molecular features of eukaryotic origins of replication may have some similarities to the origins found in bacteria. At the molecular level, eukaryotic origins have been the most extensively studied in the yeast *Saccharomyces cerevisiae*. In this organism, several replication origins have been identified and sequenced.

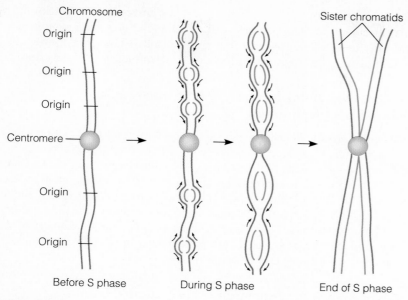

FIGURE **11-16**

The replication of eukaryotic chromosomes. At the beginning of the S phase of the cell cycle, eukaryotic chromosome replication begins from multiple origins of replication. As the S phase continues, the replication forks move bidirectionally to replicate the DNA. By the end of the S phase, all the replication forks have merged. The net result is two sister chromatids that are attached to each other at the centromere.

They have been named *ARS elements* (autonomously replicating sequence). ARS elements are necessary to initiate chromosome replication *in vivo*. There are two common sequence features among different ARS elements. First, they have a higher percentage of A and T bases than the rest of the chromosomal DNA. Second, the sequence (A or T)TTTAT(A or G)TTT(A or T) is found in all ARS elements. Similar to the dnaA boxes found in bacteria, this sequence may function as a recognition site for an initiator protein. Consistent with this idea is the observation that a complex of six proteins, termed *ORC* (origin recognition complex), appears to bind to this region of the ARS element. Current research is underway to elucidate the mechanism of action of ORC.

Eukaryotes contain several different DNA polymerases

In mammalian cells, there are at least five different DNA polymerases (Table 11-2). Four of these, designated α (alpha), β (beta), δ (delta), and ε (epsilon), function within the cell nucleus, whereas γ (gamma) functions in the mitochondria to replicate mitochondrial DNA. In the nucleus, DNA polymerases β and ε play a role in

TABLE **11–2 Eukaryotic DNA polymerases**

Mammalian Designation	α	β	δ	ε	γ
Analogous enzyme in yeast	*polI*		*polIII*	*polII*	
Cell location	Nucleus	Nucleus	Nucleus	Nucleus	Mitochondrial
Function	Chromosomal replication	DNA repair	Chromosomal replication	DNA repair	Mitochondria DNA replication

repairing damaged DNA. DNA polymerases α and δ are responsible for chromosomal DNA replication. DNA polymerase α contains a subunit that functions as a primase, whereas δ lacks any primase activity. DNA polymerase α may be responsible for synthesizing the lagging strand, while ε may synthesize the leading strand.

Nucleosomes containing new histone proteins are quickly formed after DNA replication

The DNA within eukaryotic cells is wrapped around histone proteins to form a nucleosome structure. As DNA replication occurs, the histone octamers remain attached to one of the strands of the parental DNA. Since replication doubles the amount of DNA, the cell must synthesize more histone proteins to accommodate this increase. Like DNA synthesis, the synthesis of histones occurs during the S phase of the cell cycle. These histones are assembled into octamer structures and associate with the newly made DNA very near the replication fork. Following DNA replication, each daughter strand contains a random mixture of original histone octamers and newly assembled histone octamers (Figure 11-17).

The ends of eukaryotic chromosomes are replicated by a telomerase

Linear eukaryotic chromosomes contain telomeres at both ends. The term **telomere** refers to the complex of telomeric sequences within the DNA and the special proteins that are bound to these sequences. As shown in Figure 11-18, telomeric sequences consist of a moderately repetitive tandem array and a 3′ overhang region that is 12 to 16 nucleotides in length.

The tandem array that occurs within the telomere has been studied in a wide variety of eukaryotic organisms. A common feature is that the telomeric sequence con-

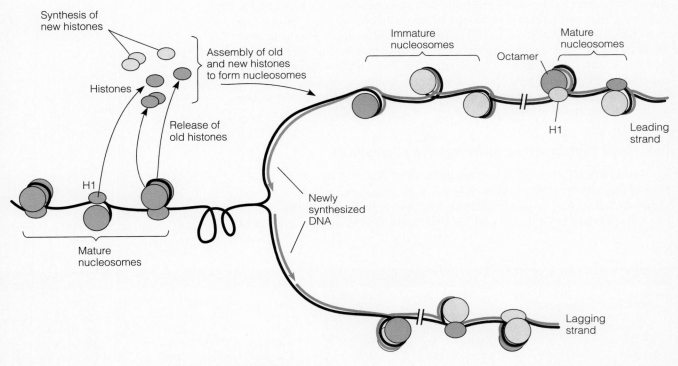

FIGURE **11-17**

Newly assembled histone octamers form nucleosomes near the replication fork.

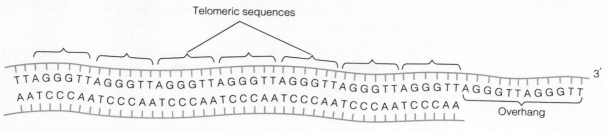

Telomeric sequences

TTAGGGTTAGGGTTAGGGTTAGGGTTAGGGTTAGGGTTAGGGTTAGGGTTAGGGTTAGGGTT 3'
AATCCCAATCCCAATCCCAATCCCAATCCCAATCCCAATCCCAA

Overhang

FIGURE **11-18**

General structure of telomeric sequences. The sequence consists of a tandemly repeated sequence and a 12- to 16-nucleotide overhang.

tains several guanine nucleotides and often many thymine nucleotides (Table 11-3). Depending on the species and the cell type, this sequence can be tandemly repeated between several and several hundred times in the telomere region.

Telomeres have a specialized form of DNA replication. As discussed previously in this chapter, DNA polymerases only synthesize DNA in a 5′ to 3′ direction, and they cannot start DNA synthesis at the first two nucleotides (a primer is required here). These two features of DNA polymerase function pose a problem at the ends of linear chromosomes. In particular, one of the strands shown in Figure 11-19 cannot be replicated at the very end by DNA polymerase. Upstream from this point, there is no place on the parental DNA strand for the primer to bind. Therefore, if this problem were not solved, the linear chromosome would become progressively shorter with each round of DNA replication.

TABLE **11-3 Telomeric sequences within selected organisms**

Organism	Telomeric Repeat Sequence
Mammals	AGGGTT
Slime molds	AGGGTT
Filamentous fungi	AGGGTT
Tetrahymena	GGGGTT
Paramecium	GGG[GT]TT
Higher plants	AGGGTTT
Baker's yeast	G(1–3)T

From Blackburn, E. H. (1991) Telomeres. *Trends Biochem. Sci. 16*, 378–381.

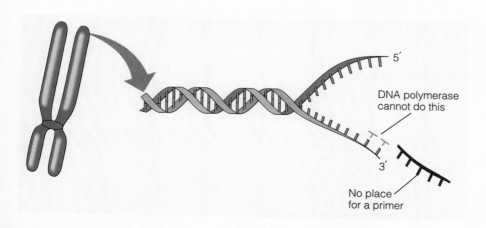

No place for a primer

DNA polymerase cannot do this

FIGURE **11-19**

The replication problem at the ends of linear chromosomes. DNA polymerase cannot synthesize a DNA strand that is complementary to the 3′ end, because there is no primer upstream from this site.

To prevent the loss of genetic information due to chromosome shortening, eukaryotic cells attach additional DNA sequences to the ends of telomeres. This specialized mechanism involves the attachment of telomeric sequences using an enzyme known as **telomerase**. This enzyme recognizes telomeric sequences at the ends of eukaryotic chromosomes and synthesizes additional numbers of telomeric repeat sequences. Figure 11-20 shows the rather interesting mechanism by which telomerase works. The telomerase enzyme contains both protein and RNA. The RNA part of telomerase contains a sequence that is complementary to the DNA sequence found in the telomeric repeat. This allows telomerase to bind to the end of the telomere. Following binding, the RNA sequence beyond the binding site can also function as a template allowing the attachment of a six-nucleotide sequence to the end of the DNA strand. This is called *polymerization*, since

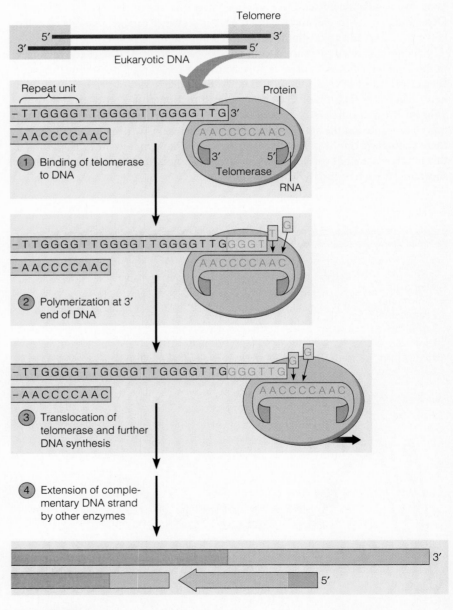

FIGURE **11-20**

The enzymatic action of telomerase.

it is analogous to the function of DNA polymerase. The telomerase can then move (i.e., translocate) to the new end of this DNA strand and attach another six nucleotides to the end. And so on. This binding–polymerization–translocation cycle can occur many times and thereby greatly lengthen one of the DNA strands in the telomeric region. The complementary strand is synthesized by DNA polymerase (see step 4 in Figure 11-20). In this way, the progressive shortening of eukaryotic chromosomes is prevented.

The structural basis for *DNA replication* is the double-stranded helix of DNA, in which the A–T/G–C rule is obeyed. This complementarity between strands allows DNA to be replicated in a *semiconservative* fashion. DNA replication begins at a specific site within the DNA, known as the origin of replication. Circular bacterial chromosomes have a single origin of replication. Specific proteins, such as the *DnaA* protein found in *E. coli*, recognize the origin and initiate the DNA replication process. The synthesis of DNA strands proceeds *bidirectionally* from the origin. This synthesis requires various proteins. *Helicase* breaks the hydrogen bonding between the *parental strands*, *topoisomerase* removes positive supercoils, and *single-strand binding proteins* hold the parental strands in a single-stranded state. *Primase* synthesizes RNA primers, which are necessary for *DNA polymerase* to elongate new *daughter strands* in a 5′ to 3′ direction. In the *leading strand*, synthesis of the daughter strand is continuous. In the *lagging strand*, the synthesis of DNA occurs in short *Okazaki fragments*. The RNA primers are removed, DNA polymerase fills in the gaps, and *DNA ligase* covalently attaches the fragments together. DNA replication is terminated when both *replication forks* reach the *ter* sequences, and a possible *catenane* structure is resolved.

Coordination of cell division and DNA replication is accomplished via the cellular regulation of replication. In bacteria, the control of DNA replication occurs at the origin. Several mechanisms can prevent premature DNA replication. Two examples are the availability of the DnaA protein and the methylation of GATC sites.

DNA replication in eukaryotes has several unique features. For example, eukaryotic chromosomes contain multiple origins of replication. Also, since eukaryotic chromosomes are linear, a specialized mechanism exists for the replication of the ends of the chromosomes within the *telomere*. An enzyme known as *telomerase* attaches telomeric repeat sequences. This prevents chromosome shortening with each round of DNA replication. Also, since eukaryotic chromosomes exist in a nucleosome structure, it is necessary to assemble the DNA and histone octamers following replication. This assembly occurs in the vicinity of the replication fork, and replicated chromosomes contain a random mixture of new and old histone octamers.

The semiconservative mechanism of DNA replication was initially proposed based on the double helix structure determined by Watson and Crick. Nevertheless, researchers needed to experimentally demonstrate that this mechanism was correct. The isotope labeling experiments of Meselson and Stahl were consistent with a semiconservative mode of replication and were inconsistent with conservative and dispersive models.

Once the semiconservative model was established, researchers focused their attention on the details of DNA replication within living cells. Arthur Kornberg and his colleagues developed methods to synthesize bacterial DNA *in vitro*. This allowed researchers to identify the components required for DNA replication. Many of these components were found to be enzymes, such as DNA polymerase, primase, helicase, topoisomerase, and ligase. Also, an analysis of DNA revealed two sequences, *oriC* and *ter*, that are necessary for the initiation and termination of DNA replication. In eukaryotes, similar findings have been obtained, but there are more enzymes and the sequences of the origins of replications are more complex. These added complexities have made it more difficult to analyze DNA replication in eukaryotes, although much progress has been made. As we will see in Chapter 23, the regulation of DNA replication is an important topic and an area of intense research, because it may underlie the proliferation of cancer cells.

PROBLEM SETS

SOLVED PROBLEMS

1. Describe three ways to account for the high fidelity of DNA replication. Discuss the quantitative contributions of each of the three ways.

Answer: First: A/T and G/C pairs are preferred in the double helix structure. This provides fidelity to around one mistake per 1000.

Second: Induced fit by DNA polymerase prevents covalent bond formation unless the proper nucleotides are in place. This increases fidelity another 100- to 1000-fold, to about one error in 100,000 to 1,000,000.

Third: Exonuclease proofreading increases fidelity another 100- to 1000-fold, to about one error per 100,000,000 nucleotides added.

2. What do you think would happen if the *ter* sequences were deleted from the bacterial DNA?

Answer: Instead of meeting at the *ter* sequences, the two replication forks would meet somewhere else. This would depend on how fast they are moving. For example, if the counterclockwise-moving fork was advancing faster than the clockwise-moving fork, they would meet closer to where the clockwise-moving fork started. In fact, researchers have actually conducted this experiment. Interestingly, *E. coli* without the *ter* sequences seemed to survive just fine.

CONCEPTUAL QUESTIONS

1. What are the key structural features of the DNA molecule that underlie its ability to be faithfully replicated?

2. With regard to DNA replication, define the term bidirectionality.

3. One way that bacterial cells regulate DNA replication is by GATC methylation sites within the origin of replication. Would this mechanism work if the DNA was conservatively (rather than semiconservatively) replicated?

4. The chromosome of *E. coli* contains 4.7 million base pairs. How long will it take to replicate its DNA? Assuming that *polIII* is the primary enzyme involved, and this enzyme can actively proofread during DNA synthesis, how many base pair mistakes will be made in one round of DNA replication in a bacterial population containing 1000 bacteria?

5. Here are two strands of DNA:

———————————— DNA polymerase→

———————————————————————

Template strand

The one on the bottom is a template strand, and the one on the top is being synthesized by DNA polymerase in the direction shown by the arrow. Label the 5′ and 3′ ends of the top and bottom strands.

6. A DNA strand has the following sequence:

5′–GATCCCGATCCGCATACATTTACCAGATCACCACC–3′

In what direction would DNA polymerase slide along this strand? If this strand was used as a template by DNA polymerase, what would

QUESTIONS FOR STUDENT DISCUSSION/COLLABORATION

be the sequence of the newly made strand? Indicate the 5' and 3' ends of the newly made strand.

7. Discuss the three types of sequences within bacterial origins of replication that are functionally important.

8. What is DNA methylation? Why is DNA in a hemimethylated condition immediately after DNA replication? What are the functional consequences of methylation in the regulation of DNA replication?

9. Describe the three important functions of the DnaA protein.

10. If a strain of bacteria was making too much DnaA protein, how would you expect this to affect its ability to regulate DNA replication? With regard to the number of chromosomes per cell, how might this strain differ from a normal bacterial strain?

11. Draw a picture that describes how helicase works.

12. What is an Okazaki fragment? In which strand of DNA are Okazaki fragments found? Why are they necessary?

13. Discuss the similarities and differences in the synthesis of DNA in the lagging and leading strands. What is the advantage of a primosome and a replisome as opposed to having all of the replication enzymes functioning independently of each other?

14. Explain the proofreading function of DNA polymerase.

15. What is a processive enzyme? Explain why this is an important feature of DNA polymerase.

16. Aside from DNA polymerase, which enzymes involved in DNA replication do you think are processive enzymes?

17. How and when are nucleosomes assembled during DNA replication?

18. Why is it important for living organisms to regulate DNA replication?

19. What enzymatic features of DNA polymerase function require a specialized form of DNA replication at the telomere? Compared with DNA polymerase, how is telomerase different in its ability to synthesis a DNA strand? What does telomerase use as its template for the synthesis of a DNA strand? How does the use of this template result in a telomere sequence that is tandemly repetitive?

20. As shown in Figure 11-20, telomerase attaches additional DNA, six nucleotides at a time, to the ends of eukaryotic chromosomes. However, it only works in one DNA strand. Describe how the opposite strand is replicated.

EXPERIMENTAL QUESTIONS

1. What would be the expected results if the Meselson and Stahl experiment (see Experiment 11A) was carried out for four or five generations?

2. What would be the expected results of the Meselson and Stahl experiment after three generations if the mechanism of DNA replication were dispersive?

3. Figure 11-3 shows an autoradiograph of a replicating bacterial chromosome. If you analyzed many replicating chromosomes, what types of information could you learn about the mechanism of DNA replication?

4. Experiment 11B described a method for determining the amount of DNA that is made during replication. Let's suppose that you can pu-

rify all the proteins required for DNA replication. You then want to "reconstitute" DNA synthesis by mixing together all the purified components that are necessary to synthesize a complementary strand of DNA. If you started with single-stranded DNA as a template, what additional proteins and molecules would you have to add for DNA polymerization to occur? What additional proteins would be necessary if you started with a double-stranded DNA molecule?

5. Using the reconstitution strategy described in question 4, how would you determine if a purified protein functions as a helicase?

6. Using the reconstitution strategy described in question 4, what components would you have to add to measure the ability of telomerase to synthesize DNA? Be specific about the type of template DNA that you would add to your mixture.

QUESTIONS FOR STUDENT DISCUSSION/COLLABORATION

1. The complementarity of double-stranded DNA is the underlying reason that DNA can be faithfully copied. Propose alternative chemical structures that could be faithfully copied.

2. The technique described in Experiment 11B makes it possible to measure DNA synthesis *in vitro*. Let's suppose that you have purified the following enzymes: DNA polymerases (*polI* and *polIII*), helicase, ligase, primase, single-strand binding protein, and topoisomerase. You also have the following reagents available:

A. Radiolabeled nucleotides (labeled with ^{32}P, a radioisotope of phosphorus)
B. Nonlabeled double-stranded DNA

C. Nonlabeled single-stranded DNA
D. An RNA primer that binds to one end of the nonlabeled single-stranded DNA

With these reagents, how could you show that helicase is necessary for strand separation, and primase is necessary for the synthesis of an RNA primer? Note: In this question, think about conditions where helicase or primase would be necessary to allow DNA replication and other conditions where they would be unnecessary.

3. DNA replication is fast, virtually error-free, and coordinated with cell division. Discuss which of these three features you think is the most important.

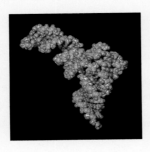

Part IV
Molecular Properties of Genes

CHAPTER 12

GENERAL PROPERTIES OF GENE STRUCTURE AND FUNCTION

THE RELATIONSHIP AMONG GENES, PROTEINS, AND TRAITS

AN OVERVIEW OF GENE EXPRESSION

Over the past few decades, the molecular investigation of gene structure and function has exploded. Molecular genetics has become one of the most important research areas for biological scientists. This chapter will provide an overview of gene structure and function. In the following chapters of Part IV, different molecular aspects of gene structure and function will be described in greater detail.

Chromosomal DNA contains thousands of different genes. Each **gene** is considered to be a single unit of heredity. In other words, the function of a gene can influence the traits of an organism. In humans, for example, there are genes that affect many traits; eye color, hair texture, and sugar metabolism are a few examples. During reproduction, genes are transmitted from parent to offspring to provide the blueprint (**genotype**) that determines the individual's traits (**phenotype**).

Within chromosomes, DNA sequences store genetic information. This genetic information is accessed during the process of **gene expression**. Most genes provide the code for the amino acid sequence within a particular polypeptide or protein. These are referred to as **structural genes**. The functioning of proteins within living cells and organisms influences the outcome of traits. In this chapter, we will first consider gene function at the cellular and organismal level, in which genes affect phenotypes, and then we will examine gene function at the molecular level, in which gene structure allows the expression of stored genetic information.

THE RELATIONSHIP AMONG GENES, PROTEINS, AND TRAITS

Genetics texts often focus on DNA's role as the blueprint for making an organism. To understand how that blueprint operates, it is helpful to consider the chemical composition of living organisms. A living cell arises from the construction and organization of chemical substances (Figure 12-1). **Small organic molecules** such as *nucleotides*, *amino acids*, and *glucose* are produced from the linkage of atoms via chemical bonds. These organic molecules are essential for cell vitality in several ways. First, the breakage of chemical bonds during the degradation of small molecules provides energy to drive cellular processes. Second, organic molecules serve as the building blocks for the synthesis of **macromolecules** as shown in Figure 12-1. The four types of biological macromolecules are *nucleic acids* (DNA and RNA), *proteins*, *carbohydrates* (such as cellulose), and *lipids* (which are found in cell membranes).

The formation of cells and cellular structures relies on the interactions of macromolecules to form **supramolecular structures**. For example, cell membranes are formed from the association of lipids to form a bilayer. This bilayer also includes some proteins and carbohydrates. Another example is the eukaryotic chromosome. It is composed of deoxyribonucleic acid (DNA) along with a variety of proteins. As a general theme, the formation of supramolecular structures is derived from the interactions among different macromolecules. These supramolecular structures are organized to make cell organelles and other cellular structures, which in turn are arranged to make a complete living cell.

The characteristics of living cells determine the phenotypic, or outwardly appearing, characteristics of an organism. Though our naked eyes cannot usually discern a single living cell, when we observe the phenotype of an organism, we are really examining the phenotypes of its cells. For example, the color of a person's eyes is determined by characteristics of the cells that constitute his or her iris. Likewise, the shape of an insect's wing is determined by the geometry and organization of the cells that form the wing. Thus, we can understand the genetic basis of a phenotype if we can understand the characteristics of cells. In this section, we will examine how cellular proteins are the main determinants of cell structure and function.

Cellular proteins are primarily responsible for the characteristics of living cells

To a great extent, the characteristics of a cell depend on the types of proteins that it makes. Proteins can perform a variety of functions (see Table 12-1). Some proteins are important in determining the shape and structure of a given cell. For example, the protein *tubulin* assembles into large cytoskeletal structures known as *microtubules*, which provide the cell with internal structure and organization. Some proteins are inserted into membranes and aid in the transport of ions and small molecules across the membrane. An example is sodium channels, which transport sodium ions into nerve cells. Another interesting category of proteins are those that function as biological motors, such as *myosin*, which is involved in the contractile properties of muscle cells. Within multicellular organisms, certain proteins function in cell-to-cell recognition and signaling. For example, hormones such as *insulin* are secreted by endocrine cells and bind to the insulin receptor protein found within the plasma membrane of target cells.

A key category of proteins are **enzymes**, which function to accelerate chemical reactions within the cell. Some enzymes assist in the breaking down of molecules or macromolecules into smaller units. These are known as *catabolic enzymes* and are important in generating cellular energy. In contrast, *anabolic enzymes* function in

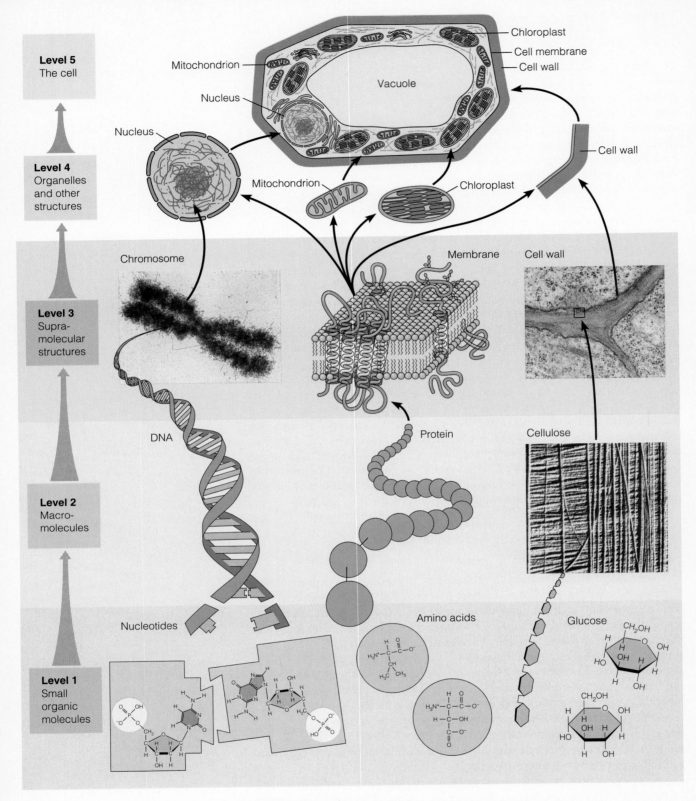

FIGURE **12-1**

The levels of structural organization within living cells. Small organic molecules are linked together to form larger macromolecules. Macromolecules then associate with each other to form supramolecular structures. These supramolecular structures are organized into cell organelles and other structures to form a living cell.

TABLE 12–1 Functions of Selected Cellular Proteins	
Function	Examples
Cell shape & organization	Tubulin: forms cytoskeletal structures known as microtubules. Ankyrin: anchors cytoskeletal proteins to the plasma membrane.
Transport	Lactose permease: transports lactose across the bacterial cell membrane. Sodium channels: transports sodium ions across the nerve cell membrane. Hemoglobin: transports oxygen within red blood cells.
Movement	Myosin: involved in muscle cell contraction. Kinesin: involved in the movement of chromosomes during mitosis and meiosis.
Cell signaling	Insulin: a hormone that influences target cell metabolism and growth. Epidermal growth factor: a growth factor that promotes cell division.
Cell surface recognition	Insulin receptor: recognizes insulin and initiates a cell response. Integrins: bind to large extracellular proteins.
Enzymes	Hexokinase: phosphorylates glucose during the first step in glycolysis. Beta-galactosidase: cleaves lactose into glucose and galactose. Glycogen synthetase: uses glucose as building blocks to synthesize a large carbohydrate known as glycogen. Acyl transferase: links together fatty acids and glycerol phosphate during the synthesis of phospholipids. RNA polymerase: uses ribonucleotides as building blocks to synthesize RNA. DNA polymerase: uses deoxyribonucleotides as building blocks to synthesize DNA.

the synthesis of molecules and macromolecules. Several anabolic enzymes are listed in Table 12-1, including DNA polymerase, which is required for the synthesis of DNA from nucleotide building blocks. Throughout the cell, the synthesis of molecules and macromolecules relies on enzymes and accessory proteins. Ultimately, then, the construction of a cell greatly depends on its anabolic enzymes, since these are required to synthesize all cellular macromolecules.

DNA stores the information for the synthesis of many different cellular proteins

As we have just seen, cellular proteins are critically important as active participants in cell structure and function. The primary role of the genetic material (i.e., DNA) is to store the information needed for the synthesis of all the proteins that a cell can make. In other words, the main function of the genetic blueprint is to encode the production of cellular proteins in the correct cell, at the proper time, and in suitable amounts. This is an extremely complicated task, since living cells make thousands of different kinds of proteins. Genetic analyses have shown that a typical bacterium can make a few thousand different proteins, and estimates among eukaryotes range from tens of thousands to more than one hundred thousand.

In living cells, the DNA is found within chromosomes. The DNA of an average human chromosome is an extraordinarily long linear (double-stranded) structure that contains well over one hundred million nucleotides. Along the immense length of such a chromosome, the genetic information is parceled into shorter segments known as genes. Most genes code for the sequence of amino acids within proteins. Typical genes vary from a few hundred to several thousand nucleotides in length. An average-sized human chromosome is expected to contain several thousand different genes. In most eukaryotes, functional genes are interspersed with sizable intergenic regions that do not code for proteins.

DNA's ability to store information is based on its molecular structure. To synthesize proteins, a cell must access the genetic information stored within the DNA. This occurs by a molecular process known as gene expression. We will examine the events of gene expression later in this chapter.

Archibald Garrod proposed that some genes code for the production of a single enzyme

The idea that a relationship exists between genes and the production of proteins was first suggested at the beginning of the 20th century by Archibald Garrod, a British physician. Prior to Garrod's studies, biochemists had studied many metabolic pathways within living cells. These pathways consist of a series of metabolic conversions of one molecule to another, each step catalyzed by an enzyme. Each type of enzyme is a distinctly different protein that catalyzes a particular chemical reaction. For example, Figure 12-2 illustrates part of the metabolic pathway for the degradation of *phenylalanine*, an amino acid commonly found in human diets. The enzyme *phenylalanine hydroxylase* catalyzes the conversion of phenylalanine to *tyrosine*. A different enzyme, *tyrosine aminotransferase*, converts tyrosine into *p-hydroxyphenylpyruvic acid*. In all of the steps shown in Figure 12-2, a specific enzyme catalyzes a single type of chemical reaction.

Garrod studied patients who had defects in their ability to metabolize certain compounds. He was particularly interested in the inherited disease known as **alkaptonuria**. In this disorder, the patient's body accumulates abnormal levels of *homogentisic acid* (alkapton) which is secreted in the urine. Garrod proposed that this defect is due to a missing enzyme, namely *homogentisic acid oxidase* (see Figure 12-2).

Furthermore, Garrod already knew that alkaptonuria is an inherited trait that follows a recessive pattern of inheritance; thus, an individual with alkaptonuria must have inherited the gene that causes this disorder from each parent. From these observations, Garrod proposed that a relationship exists between the inheritance of the trait and the inheritance of a defective enzyme. Namely, if an individual inherited the defective gene from both parents, he or she would not produce any normal enzyme and would thereby be unable to metabolize homogentisic acid. Garrod described alkaptonuria as an **inborn error of metabolism**. This hypothesis was the first suggestion that a connection exists between the function of genes and the production of enzymes. At the turn of the century, this was a particularly insightful idea, since the structure and function of the genetic material were completely unknown.

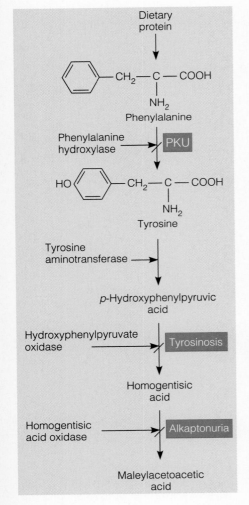

FIGURE **12-2**

The metabolic pathway of phenylalanine metabolism and related genetic diseases. This diagram shows part of the metabolic pathway of phenylalanine breakdown. This pathway consists of enzymes that successively convert one molecule to another. Certain human genetic diseases are caused when enzymes in this pathway are missing.

GENES ⟶ TRAITS: When a person inherits two defective copies of the gene that encodes homogentisic acid oxidase, he or she cannot convert homogentisic acid into maleylacetoacetic acid. Such a person accumulates large amounts of homogentisic acid in their urine and has other symptoms of the disease known as alkaptonuria. Similarly, if a person has two mutant alleles of the gene encoding phenylalanine hydroxylase, he or she is unable to synthesize the enzyme phenylalanine hydroxylase and has the disease called phenylketonuria (PKU).

Beadle and Tatum's experiments with *Neurospora* led them to propose the one gene–one enzyme theory

In the early 1940s, George Beadle and Edward Tatum at Stanford University were also interested in the relationship among genes, enzymes, and traits. Their important contribution to genetics was their formulation of an experimental system for investigating the relationship between genes and the production of particular enzymes. Consistent with the ideas of Garrod, the underlying assumption behind their approach was that a relationship exists between genes and the production of enzymes. However, the quantitative nature of this relationship was very unclear. In particular, they wondered whether one gene was controlling the production of one enzyme or whether one gene was controlling the synthesis of many enzymes involved in a complex biochemical pathway.

At the time of their studies, many geneticists were also trying to understand the nature of the gene by studying morphological traits. However, Beadle and Tatum realized that morphological traits are "likely to be based on systems of biochemical reactions so complex as to make analysis exceedingly difficult." Therefore, they turned their genetic studies to the analysis of simple nutritional requirements in the common bread mold *Neurospora crassa*.

Neurospora can be easily grown in the laboratory and has few nutritional requirements. The minimum requirements for growth are a carbon source (namely, sugar), inorganic salts, and one vitamin known as biotin. Otherwise, *Neurospora* has many different cellular enzymes that can synthesize all the small molecules, such as amino acids and many vitamins, that are essential for growth.

Beadle and Tatum wanted to understand how enzymes are controlled by genes. They reasoned that a mutation in a gene, causing a defect in an enzyme needed for the cellular synthesis of an essential molecule, would prevent that mutant strain from growing on minimal medium (which only contains a carbon source, inorganic salts, and biotin). For example, if a *Neurospora* mutant strain could not make the vitamin pantothenic acid, it would be unable to grow on minimal medium. However, it would grow if the growth medium was supplemented with pantothenic acid.

A second important characteristic of *Neurospora* is that it is easy to determine if a mutant strain has a single mutation rather than several. Since Beadle and Tatum were interested in the quantitative relationship between genes and enzymes, they wanted to make sure that a mutant strain contained only one defective mutation. In *Neurospora*, it is easy to determine this by mating the mutant strain to the wild-type strain. This is a type of backcross. *Neurospora* is haploid (monoploid), meaning that it has only one copy of each gene. Prior to mating, *Neurospora* produces haploid "sexual" spores that can combine to produce a diploid cell. The diploid stage of the *Neurospora* life cycle can then proceed (via meiosis and a single mitosis) to produce an ascus containing eight haploid spores (see Chapter 5, p. 123 for a description of an ascus). When a mutant strain is backcrossed to a normal strain, the ascus will contain four mutant spores and four normal spores if the original mutant strain only contained a single mutation. For their studies, Beadle and Tatum analyzed mutant *Neurospora* strains that on backcrossing yielded a 4:4 ratio of mutant to normal spores.

THE HYPOTHESIS

Genes are responsible for encoding proteins that function as enzymes. The purpose of this experiment was to understand the quantitative relationship between genes and the production of enzymes. (Does a gene control a complicated biochemical pathway, or does a gene control the synthesis of a single enzyme?)

Starting material: A normal strain of *Neurospora*.

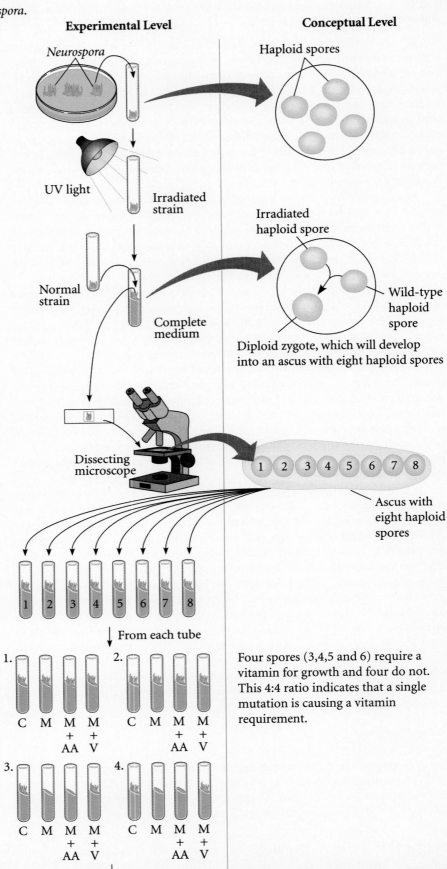

Experimental Level

Conceptual Level

1. Isolate haploid spores from a normal strain.

2. Irradiate spores with X-rays or UV light to cause mutations in genes that were previously normal.

3. Mate the irradiated strains with a normal strain. (Note: This is a backcross). Then allow them to grow on complete medium.

4. Following meiosis, use a dissecting microscope to dissect the haploid spores from an ascus. Grow each on complete medium.

5. From each complete medium tube, restreak some cells onto four tubes with complete media (C), minimal media (M), or minimal media supplemented with all amino acids (M+AA) or all vitamins (M+V).

Neurospora

Haploid spores

UV light

Irradiated strain

Irradiated haploid spore

Normal strain

Wild-type haploid spore

Complete medium

Diploid zygote, which will develop into an ascus with eight haploid spores

Dissecting microscope

1 2 3 4 5 6 7 8

Ascus with eight haploid spores

1 2 3 4 5 6 7 8

From each tube

1. C M M+AA M+V

2. C M M+AA M+V

3. C M M+AA M+V

4. C M M+AA M+V

Four spores (3,4,5 and 6) require a vitamin for growth and four do not. This 4:4 ratio indicates that a single mutation is causing a vitamin requirement.

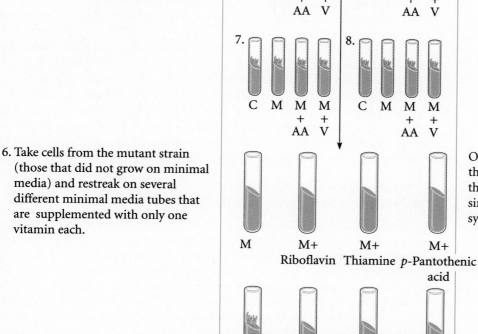

6. Take cells from the mutant strain (those that did not grow on minimal media) and restreak on several different minimal media tubes that are supplemented with only one vitamin each.

Only one vitamin, pyridoxine, satisfies the vitamin requirement for growth in this mutant strain. This suggests that a single gene involved in pyridoxine synthesis is defective.

7. Observe where growth occurs.

THE DATA

Single Vitamin Added to Minimal Media	Growth*
None	No
Riboflavin	No
Thiamin	No
Pantothenic acid	No
Pyridoxine	Yes
Niacin	No
Folic acid	No
p-Aminobenzoic acid	No

*These data are from a strain shown in step 5 that could not grow on minimal media but could grow on media supplemented with all of the vitamins.

INTERPRETING THE DATA

The table shown here only describes the data from a single mutant strain. In their first studies, published in 1941, Beadle and Tatum analyzed more than 2000 irradiated strains; they found 3 strains that were unable to grow on minimal medium and gave a 4:4 ratio in the backcross. In each of these cases, the mutant strain only required a single vitamin to restore its growth on minimal medium. One of these strains required pyridoxine for growth (as in the table), the second strain required thiamine, and the third strain required *p*-aminobenzoic acid. In the normal strain, these vitamins are synthesized by cellular enzymes. In the mutant strains, a genetic defect in one gene prevented the synthesis of a functional enzyme required for the synthesis of a particular vitamin. Therefore, Beadle and Tatum concluded that a single gene controlled the synthesis of a single enzyme. This was referred to as the **one gene–one enzyme theory**.

In the early 1940s, Beadle and Tatum did not know that DNA is the genetic material. Therefore, they did not understand the molecular relationship between DNA, mRNA, and polypeptides. Nevertheless, they had the insight to devise a simple experimental system that allowed geneticists to appreciate the relationship between genes and enzymes. In later decades, as the biochemical structures of enzymes became clearer, some enzymes were found to be composed of two or more polypeptide subunits. Therefore, the one gene–one enzyme theory had to be modified to the one gene–one polypeptide theory.

Geneticists have identified many examples in which the relationship between genes and traits is understood

As mentioned earlier in this chapter, the functions of cellular proteins are of central importance in determining the phenotypic characteristics of an organism. Since a living cell makes thousands of different proteins, it is often difficult to understand the contribution that any specific protein makes to an organism's phenotype. To overcome this problem, different scientific approaches can help elucidate the functional role that a particular protein plays. These include biochemical, cellular, and genetic analyses. In a **biochemical analysis**, the experiment seeks to determine the molecular function of a protein. This often involves the purification and characterization of the protein *in vitro*. In a **cellular** (or **cytological**) **analysis**, a researcher may attempt to understand protein function by observing its location and function within living cells.

A particularly powerful approach to discern protein function is a **genetic analysis**. In this strategy, the aim is to compare individuals who are different from each other with regard to a particular gene. This method is possible because mutations can alter the DNA sequence of a gene and thereby produce alternative versions of a gene, which are known as **alleles**. In many cases, a mutation within a gene will alter the function of the protein that the gene encodes, or it may change the amount of protein that is made.

A common approach in genetic analyses is to compare the phenotypes of individuals who have normal copies of a gene with those of individuals in which the gene has been rendered inactive by a mutation that alters its DNA sequence. In other words, this type of genetic analysis involves a comparison between wild-type individuals carrying the normal allele and those individuals carrying a defective allele of the same gene. Figure 12-3 illustrates how this approach can clarify the relationship between an organism's genotype and phenotype at the organismal, cellular, and molecular levels. This example considers a trait that was described in Chapter 3, namely, the white eye and red eye phenotype in *Drosophila* that was originally described by Morgan. At the organismal level, this diagram shows two female fruit flies that differ in eye color. The genotype of the red-eyed fly is $X^{w+}X^{w+}$, while the white-eyed fly is $X^{w}X^{w}$. The phenotypic differences can be explained by differences in events that occur within the eye cells of these flies.

The eye of the fruit fly is composed of several types of cells. Some of them are pigment cells that produce eye color. In the pigment cells of the red-eyed fly, the X-chromosomes contain the red-eyed allele (w^{+}), which codes for a protein that is located in the cell membrane. In the 1990s, the cloning and DNA sequencing of this

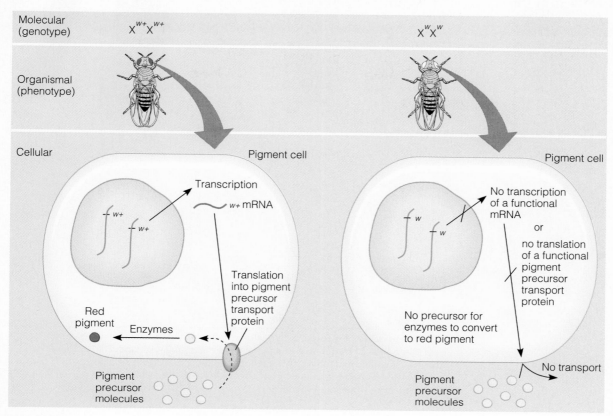

FIGURE **12-3**

A comparison of genotype and phenotype at the molecular, organismal, and cellular levels.

GENES ⟶ TRAITS: The wild-type allele encodes a protein that transports a colorless precursor into the pigment cells of the eye. This precursor is converted to a red pigment, so that the eyes become red. The mutant *w* allele encodes a defective transporter, so that little or none of the precursor is transported into the pigment cells. Therefore, the X^wX^w homozygous fly has colorless (white) eyes.

gene suggested that the encoded protein transports colorless pigment precursor molecules into the cell. In the pigment cells of the red eye, the X^{w+} genes are transcribed and translated to produce this transport protein. This allows the cells to take up the pigment precursor molecules. Once inside, additional enzymes (that are encoded by different genes) convert the colorless precursors into red pigment. Therefore, the eyes of this fly appear red. In contrast, the white allele (*w*) of this gene is an example of a defective or **loss of function allele**. In the pigment cells of the white-eyed fly, this gene is not expressed properly, so that little or no functional transport protein is made. Without this transport protein, the pigment precursor molecules are not transported, and so the red pigment cannot be made. Because of this defect, the eyes of the fly remain white.

AN OVERVIEW OF GENE EXPRESSION

The previous section familiarized you with the relationship between the production of cellular proteins and their impact on the phenotype of an organism. In this section, we will examine how the genetic material acts as a storage unit, containing the instructions to synthesize proteins with a defined amino acid sequence. At the molecular level, a multistage process is required for a gene to be expressed and

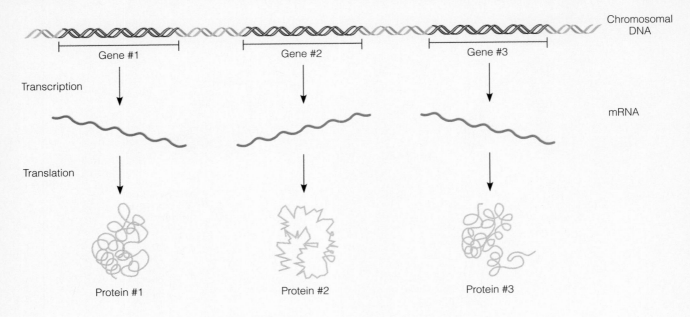

FIGURE **12-4**

Gene expression. The expression of a gene is a two-step process. During transcription, one of the DNA strands is used as a template to make an RNA strand. During translation, the RNA strand is used to specify the sequence of amino acids within a polypeptide. In this diagram, there are three different genes, which encode three polypeptides having their own distinctive structures and functions.

produce a cellular protein (Figure 12-4). The first stage involves the copying of a gene sequence into an RNA sequence called **messenger RNA (mRNA)**. This event is known as **transcription**. In the second stage, an mRNA sequence is **translated** into the amino acid sequence of a polypeptide. The translation of RNA into a polypeptide sequence requires the aid of a **ribosome** and **transfer RNAs (tRNA)**. After translation is completed, the sequence and chemistry of the amino acid side chains within a polypeptide ultimately determine how that polypeptide will fold (and possibly associate with other polypeptides) to make a functional protein.

Figure 12-4 illustrates the process of gene expression for three different genes. Note that the three genes in this figure are separate genes that encode three polypeptides with distinct amino acid sequences. As schematically illustrated in this figure, the differences among the amino acid sequences cause these three proteins to have different molecular structures. The molecular structure of each protein, in turn, will determine its function. At the cellular and organismal levels, the functions of proteins within living cells are expected to influence the outcome of traits.

Genes are organized into functional groups of base sequences

At the molecular level, a gene is a relatively short segment of DNA located within a much larger chromosome. A typical gene is a few hundred to many thousand nucleotides in length. The defining feature of a gene is its nucleotide base sequence. Within a complete gene, there are several different types of base sequences, which perform different roles during the process of gene expression. We will give a general discussion of these sequences in this chapter, deferring a more detailed consideration of their roles to later chapters.

Figure 12-5 shows a common organization of sequences within a bacterial structural gene. Different types of nucleotide sequences perform their function

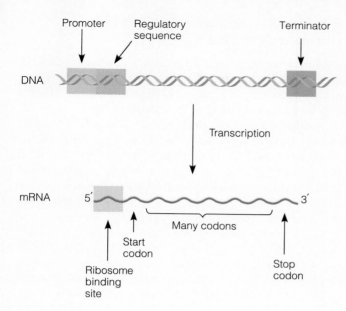

DNA:

1. Promoter: site for RNA polymerase binding; signals the beginning of transcription.

2. Terminator: signals the end of transcription.

3. Regulatory sequences: site for the binding of regulatory proteins; the role of regulatory proteins is to influence gene expression. In eukaryotes, regulatory sequences can be found in a variety of locations.

mRNA:

4. Ribosomal binding site: site for ribosome binding; translation begins near this site in the mRNA. In eukaryotes, the ribosome scans the mRNA for a start codon.

5. Start codon: specifies the first amino acid in a protein sequence, usually a methionine.

6. Codons: a three-nucleotide sequence within the mRNA that specifies a particular amino acid. The sequence of codons within the mRNA determines the sequence of amino acids within a polypeptide.

FIGURE **12-5**

Organization of sequences within a bacterial gene and its mRNA transcript.

during specific stages of gene expression. For example, the promoter and terminator are nucleotide base sequences that function during gene transcription. Specifically, the **promoter** sequence provides a signal to begin transcription, and the **terminator** signals the end of transcription. Therefore, these two sequences cause RNA synthesis to occur within a defined location. As shown in Figure 12-5, the DNA is transcribed into RNA in a region that spans from the end of the promoter through the end of the terminator. The sequence of an entire gene is usually considered to be the region from the promoter to the terminator site.

A second category of sequences are those involved in the regulation of gene expression. These are termed **regulatory sequences**, which act as binding sites for genetic regulatory proteins. The binding of regulatory proteins to a regulatory sequence is expected to affect gene expression. For example, certain transcription factors bind to specific genetic regulatory sequences and thereby influence positively or negatively the rate of transcription. A variety of gene regulation mechanisms will be considered in Chapters 15 and 16.

A third category of sequences ensures that the nucleotide sequence within an mRNA will be translated into the amino acid sequence of a polypeptide. These sequences actually function within the mRNA itself as it is being used in the translation process. In bacteria, a short sequence within the mRNA known as the **ribosomal-binding site** provides a location for the ribosomes to recognize and begin the process of translation. In bacteria, this ribosomal-binding site is also called the Shine–Dalgarno sequence after the researchers who discovered, in 1974, that it is complementary to a sequence in ribosomal RNA (see Chapter 14, Figure 14-11).

During polypeptide synthesis, the sequence of nucleotides within the mRNA is read as groups of three nucleotides, known as **codons**. The first codon, which is very close to the ribosomal-binding site, is known as the **start codon**. This is followed by many more codons that dictate the sequence of amino acids within the synthesized polypeptide. The relationship between the codons and the genetic code will be discussed later in this chapter. Finally, a **stop codon** signals the end of translation.

During gene transcription, the information within DNA is accessed to produce a molecule of RNA

Transcription accesses the information stored within the DNA by making an RNA copy of a DNA strand. Figure 12-6 is a simplified overview of the transcription process. To begin transcription, an enzyme known as **RNA polymerase** binds to the promoter region. (As will be discussed in Chapter 13, a variety of transcription factors ensure that RNA polymerase recognizes the promoter sequence.) Following binding of RNA polymerase, the DNA begins to unwind. Only one of the two DNA strands will serve as a **template strand** for the synthesis of a complementary RNA strand. For the synthesis to occur, RNA polymerase slides along the template strand and catalyzes the covalent attachment of nucleotides that have hydrogen bonded to the template strand due to their complementarity. Note that these nucleotides are ribonucleotides, not deoxyribonucleotides. During transcription, the bases of the ribonucleotides hydrogen bond to the coding strand of DNA according to the following rules: U_{RNA} to A_{DNA}; A_{RNA} to T_{DNA}; G_{RNA} to C_{DNA}; and C_{RNA} to G_{DNA}. Like DNA replication, the synthesis of RNA during transcription occurs in the 5′ to 3′ direction.

RNA synthesis stops after the termination sequence has been copied. When this occurs, the completed RNA, as well as the RNA polymerase, dissociate from the template strand to finish the transcription process. As shown in Figure 12-6, the

FIGURE **12-6**

Gene transcription. RNA polymerase binds to the promoter site and slides along the DNA, making a complementary copy of RNA. Transcription is ended when polymerase reaches the termination sequence.

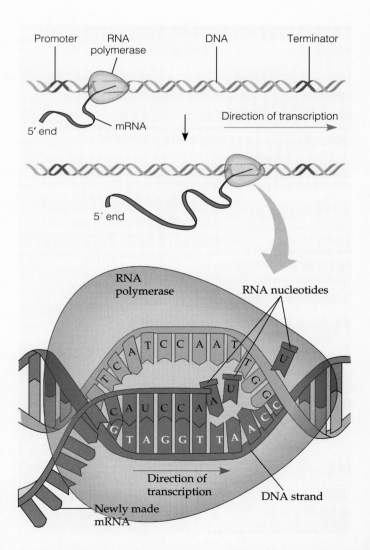

structure of the DNA has not been altered during the transcription process. Rather, the DNA nucleotide sequence has only been accessed to make a copy of RNA. Therefore, the same DNA can continue to store genetic information.

Eukaryotic pre-mRNA transcripts are often spliced to form functional mRNAs

In bacteria, the mRNA is ready for translation immediately after it is made. By comparison, the transcription of eukaryotic genes produces a precursor mRNA, called pre-mRNA. The pre-mRNA for eukaryotic genes must be modified before it can be properly translated. One important modification is **splicing**, in which certain pieces of the RNA are removed and the remaining pieces are covalently attached to each other. This event is diagrammed in Figure 12-7.

The gene shown in Figure 12-7 contains three **exons** and two **introns**. The exons (coding regions) contain the coding information for the sequence of amino acids within a polypeptide. The introns (noncoding regions) are sequences that intervene between the exons. Both the introns and exons are transcribed into RNA. During splicing, the introns are excised from the RNA, and the remaining exons are attached to each other. The process of RNA splicing is quite common among eukaryotes. In addition, other types of RNA modifications are required prior to its translation within the cytoplasm. These topics will be covered in greater detail in Chapter 13.

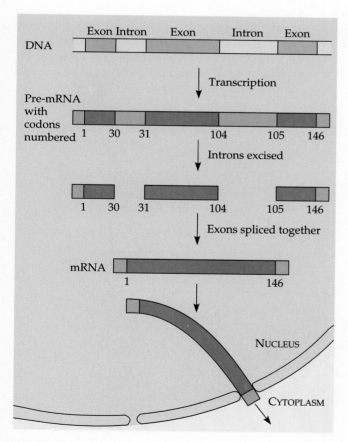

FIGURE **12-7**

The splicing of RNA due to the occurrence of introns within a eukaryotic gene.
Introns, or intervening sequences, are found interspersed within a gene. Following transcription, the introns (depicted in light red) are spliced out of the RNA, and the remaining exons (depicted in dark red) are covalently attached to each other.

During translation, the genetic code within mRNA is used to make a polypeptide with a specific amino acid sequence

For structural genes, the nucleotide sequence of the mRNA is translated into an amino acid sequence. The ability of the mRNA to be translated into a specific sequence of amino acids relies on the **genetic code**. The sequence of bases within an mRNA molecule provides coded information that is read in groups of three nucleotides (Figure 12-8). Each group of three nucleotides is known as a codon. Depending on its sequence of three bases, a codon can have two different functions. The sequence of three bases in most codons specifies a particular amino acid. For example, the codon CCC specifies the amino acid proline, the codon AUG the amino acid methionine. As mentioned, a few codons are used to end the process of translation. These are known as termination or stop codons.

The details of the genetic code are shown in Table 12-2. Since there are four types of bases in mRNA (A, U, G, and C) and three nucleotides within a codon, there are $4^3 = 64$ different codons. However, there are only 20 different amino acids. Because there are more codons than amino acids, the genetic code is **degenerate**. This means that more that one codon specifies the same amino acid. For example, the codons GGU, GGC, GGA, and GGG all specify the amino acid glycine. In many instances, the third base in the codon is the degenerate base. The third base is sometimes referred to as the **wobble base**. This term is derived from the idea that the complementary base in the tRNA can "wobble" a bit during the recognition of the third base in the codon in mRNA. The significance of the wobble base will be discussed in greater detail in Chapter 14.

From the analysis of many different species, including bacteria, protozoa, fungi, plants, and animals, researchers have found that the genetic code is nearly universal. Only a few rare exceptions to the genetic code have been noted. These are described in Table 12-3. For example, as was discussed in Chapter 7, the eukaryotic

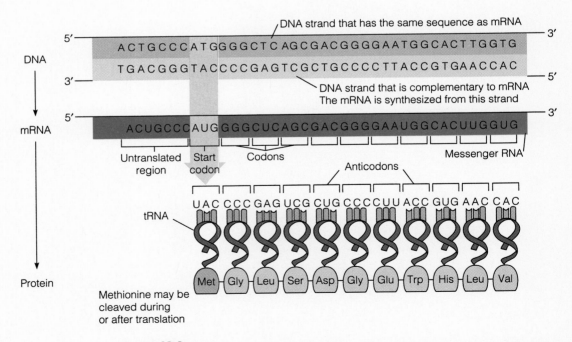

FIGURE **12-8**

The relationships among DNA coding sequence, mRNA codons, tRNA anticodons, and amino acids in a polypeptide. The sequence of nucleotides within DNA is transcribed to make a complementary sequence of nucleotides within mRNA. This sequence of nucleotides in mRNA is translated into a sequence of amino acids within a polypeptide. tRNA molecules act as intermediates in this translation process.

TABLE 12–2 The Genetic Code

		Second position			
First position	U	C	A	G	Third position
U	UUU ⎤ UUC ⎦ Phe UUA ⎤ UUG ⎦ Leu	UCU ⎤ UCC ⎥ Ser UCA ⎥ UCG ⎦	UAU ⎤ Tyr UAC ⎦ UAA Stop UAG Stop	UGU ⎤ Cys UGC ⎦ UGA Stop UGG Trp	U C A G
C	CUU ⎤ CUC ⎥ Leu CUA ⎥ CUG ⎦	CCU ⎤ CCC ⎥ Pro CCA ⎥ CCG ⎦	CAU ⎤ His CAC ⎦ CAA ⎤ Gln CAG ⎦	CGU ⎤ CGC ⎥ Arg CGA ⎥ CGG ⎦	U C A G
A	AUU ⎤ AUC ⎥ Ile AUA ⎦ AUG Met/start	ACU ⎤ ACC ⎥ Thr ACA ⎥ ACG ⎦	AAU ⎤ Asn AAC ⎦ AAA ⎤ Lys AAG ⎦	AGU ⎤ Ser AGC ⎦ AGA ⎤ Arg AGG ⎦	U C A G
G	GUU ⎤ GUC ⎥ Val GUA ⎥ GUG ⎦	GCU ⎤ GCC ⎥ Ala GCA ⎥ GCG ⎦	GAU ⎤ Asp GAC ⎦ GAA ⎤ Glu GAG ⎦	GGU ⎤ GGC ⎥ Gly GGA ⎥ GGG ⎦	U C A G

organelles known as mitochondria have their own DNA, which encodes a few structural genes. In mammals, the mitochondrial genetic code contains differences such as AUA = methionine and UGA = tryptophan. Also, in mitochondria and certain ciliated protozoa, AGA and AGG specify stop codons instead of arginine. Aside from the few examples listed in Table 12-3, the genetic code shown in Table 12-2 has been found to be **universal** in all organisms studied thus far.

For translation of mRNA to occur at an efficient rate within a living cell, several additional cellular components are required. These include ribosomes, transfer RNAs

TABLE 12–3 Exceptions to the Genetic Code

Codon	Universal Meaning	Exception
AUA	Isoleucine	Methionine in yeast and mammalian mitochondria
UGA	Stop	Tryptophan in mammalian mitochondria
CUU, CUA, CUC, CUG	Leucine	Threonine in yeast mitochondria
AGA, AGG	Arginine	Stop codon in ciliated protozoa and in yeast and mammalian mitochondria
CGG	Arginine	Tryptophan in plant mitochondria
UAA, UAG	Stop	Glutamine in ciliated protozoa

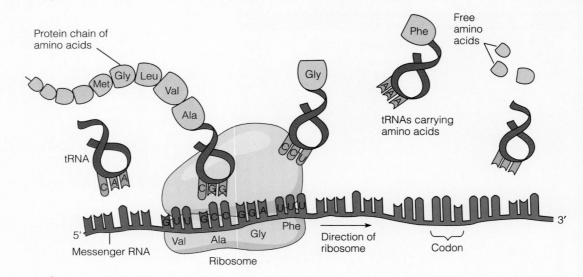

FIGURE **12-9**

mRNA translation. The ribosome slides along the mRNA and catalyzes the synthesis of a polypeptide. For synthesis to occur, the anticodon in the tRNA must bind to a complementary codon in the mRNA, carrying with it the proper amino acid.

(tRNAs), protein factors, and small molecules. We will discuss the details of translation in Chapter 14. Figure 12-9 shows a simplified overview of the translation process.

Certain sequences within the mRNA are necessary to ensure that the correct amino acid sequence will be synthesized. As already noted, in prokaryotes a ribosomal-binding site is required so that the ribosome has a signal to begin translation at the correct location. In eukaryotes, the ribosome scans the mRNA in search of a start codon, which is the starting point for translation. The start codon is usually AUG. This codon specifies that the first amino acid in the polypeptide will be a methionine.

The synthesis of a polypeptide involves interactions among the mRNA, the ribosome, and tRNA molecules. The ribosome slides along the mRNA in the 5′ to 3′ direction. As the ribosome slides over a codon of the mRNA, a tRNA with a complementary **anticodon** will bind to the codon. At its other end, the tRNA carries an amino acid. Therefore, the tRNA has two important roles: its anticodon binds specifically to a codon in the mRNA, and it carries the correct amino acid at its other end. In this way, the tRNA is the intermediate molecule that acts as the translator during polypeptide synthesis. Beyond the start codon in mRNA, there are usually several hundred or even several thousand codons that specify the amino acid sequence of the polypeptide to be made. The ribosome slides past each codon in a stepwise manner, as tRNA molecules bind to the codons via their anticodon sequences and carry with them the proper amino acid. As this is occurring, the amino acids are detached from the tRNA molecules and are linked to each other to form a polypeptide chain. Finally, a stop codon is reached, signaling the end of translation. At this point, the ribosome, mRNA, and polypeptide dissociate from each other.

EXPERIMENT 12B

Nirenberg and Ochoa independently deciphered the genetic code

During the early 1960s, the genetic code was deciphered by two different research groups, headed by Marshall Nirenberg at the National Institutes of Health (NIH) and by Severo Ochoa at the New York University School of Medicine. They accomplished this by synthesizing polypeptides in a cell extract. Prior to their work, several laboratories had already determined that extracts from bacterial cells are able to synthesize polypeptides. In these earlier studies, the proteins made in a cell extract were specified by the mRNAs that were already present in the extract. It was then found that the addition of DNase to the extract prevented protein synthesis. DNase is expected to digest the DNA found within the extract. Since DNA acts as the template to make mRNA, a DNase-treated extract cannot make new mRNA. Therefore, after a short period of time, a DNase-treated extract becomes depleted of mRNA, because mRNA has a very short half-life (a few minutes). In this way, the extract becomes unable to translate new polypeptides.

An important advance occurred when Nirenberg and colleagues were able to add RNA back to DNase-treated extracts and thereby regain polypeptide synthesis. Furthermore, if radiolabeled amino acids were added to these extracts, the synthesized polypeptides would be radiolabeled and easy to detect. This approach enabled Nirenberg and Ochoa, working independently, to decipher the genetic code. They did this by making RNA molecules of a known base composition, adding them to the DNase-treated cell extract, and then analyzing the amino acid composition of the resultant polypeptides. For example, if an RNA molecule consisted of only a string of adenine-containing nucleotides (e.g., 5′–AAAAAAAAAAAAAAAAAAAA–3′), Nirenberg and Ochoa could add this polyA RNA to an extract and ask the question, "What amino acids are specified by codons that only contain adenine nucleotides?"

Before discussing the details of this experiment, let's consider how the synthetic RNA molecules were made. To synthesize RNA, an enzyme known as *polynucleotide phosphorylase* was used. This enzyme randomly links ribonucleotides together to make a polymer of RNA. The experimenter can control the amounts of the ribonucleotides that are added during the RNA synthesis. For example, if only uracil-containing triphosphates, UTPs, are added, then a polyU RNA (e.g., 5′–UUUUU-UUUUUUUUUU–3′) will be made. If ribonucleotides containing two different bases, such as uracil and guanine, are added, then the phosphorylase will make a random polymer containing both nucleotides (e.g., 5′–UGGGUGUUUUGUGUG–3′). Furthermore, the experimenter can control the amounts of the nucleotides that are added. For example, if 70% G and 30% U are mixed together, the predicted amounts of the codons within the random polymer are as follows:

Codon Possibilities	Percentage in the Random Polymer		
GGG	$0.7 \times 0.7 \times 0.7 = 0.34 =$	34%
GGU	$0.7 \times 0.7 \times 0.3 = 0.15 =$	15%
GUU	$0.7 \times 0.3 \times 0.3 = 0.06 =$	6%
UUU	$0.3 \times 0.3 \times 0.3 = 0.03 =$	3%
UGG	$0.3 \times 0.7 \times 0.7 = 0.15 =$	15%
UUG	$0.3 \times 0.3 \times 0.7 = 0.06 =$	6%
UGU	$0.3 \times 0.7 \times 0.3 = 0.06 =$	6%
GUG	$0.7 \times 0.3 \times 0.7 = 0.15 =$	15%
		100%

Therefore, by controlling the amounts of the nucleotides in the phosphorylase reaction, the relative amounts of the possible codons can be predicted. The experiment described here, performed by Marshall Nirenberg and J. Heinrich Matthaei in 1961, used this approach to decipher the genetic code.

THE HYPOTHESIS

The sequence of amino acids within a polypeptide is determined by the sequence of codons within mRNA. The purpose of this experiment is to decipher the genetic code.

Starting material: A bacterial cell extract that has been treated with DNase.

Experimental Level	Conceptual Level

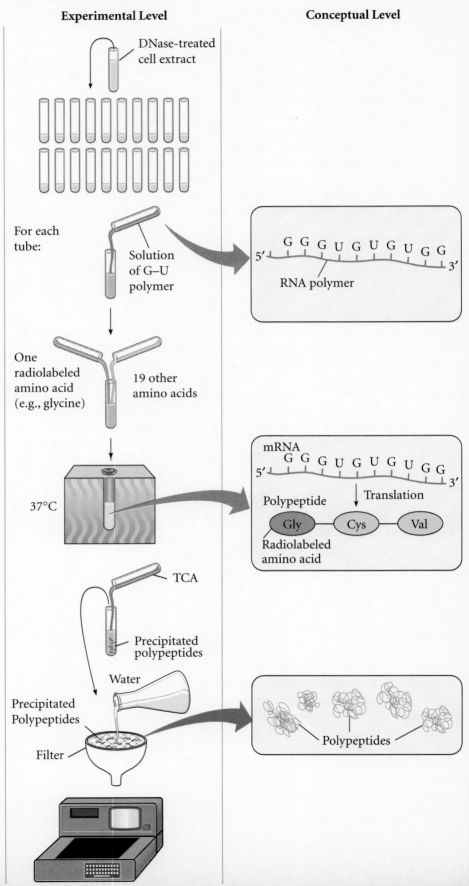

1. Add the DNase-treated cell extract to each of 20 tubes.

DNase-treated cell extract

2. To each tube, add a random RNA polymer of G and U made via polynucleotide phosphorylase using 70% G and 30% U.

For each tube:

Solution of G–U polymer

$5'$ —G G G U G U G U G G— $3'$

RNA polymer

3. Add a different radiolabeled amino acid to each tube, and add the other 19 non-radiolabeled amino acids.

One radiolabeled amino acid (e.g., glycine)   19 other amino acids

mRNA

$5'$ —G G G U G U G U G G— $3'$

Polypeptide   Translation

Gly — Cys — Val

Radiolabeled amino acid

4. Incubate for 60 minutes to allow translation to occur.

37°C

5. Add 15% trichloroacetic acid (TCA). Note: This precipitates polypeptides but not amino acids.

TCA

Precipitated polypeptides

6. Place the precipitate onto a filter and wash.

Water

Precipitated Polypeptides

Filter

Polypeptides

7. Count the radioactivity on the filter in a scintillation counter (see the appendix for a description).

8. Calculate the amount of radiolabeled amino acids in the precipitated polypeptides.

Radiolabeled Amino Acid Added	Relative Amount of Radiolabeled Amino Acid Incorporated into Translated Polypeptides (% of total)
Alanine	0
Arginine	0
Asparagine	0
Aspartic acid	0
Cysteine	6
Glutamic acid	0
Glutamine	0
Glycine	49
Histidine	0
Isoleucine	0
Leucine	6
Lysine	0
Methionine	0
Phenylalanine	3
Proline	0
Serine	0
Threonine	0
Tryptophan	15
Tyrosine	0
Valine	21

This table represents adapted data using the protocol described (see Nirenberg and Matthaei (1961) *PNAS* 47, 1588–1602). The counts per minute derived in step 7 can be used to calculate the amount of the radiolabeled amino acid that was incorporated into a polypeptide. In the table, this amount is expressed as the percentage of the total amount found in all 20 tubes.

INTERPRETING THE DATA

According to the calculation described at the beginning of this experiment, there should be the following proportions of codons: 34% GGG, 15% GGU, 6% GUU, 3% UUU, 15% UGG, 6% UUG, 6% UGU, and 15% GUG. In the data given here, the value of 49% for glycine is due to two codons: GGG (34%) and GGU (15%) (see Table 12-2, p. 327). The 6% cysteine is due to UGU. And so on. It is important to realize that the genetic code was not deciphered in a single experiment such as the one described here. Rather, by comparing many different RNA polymers, it became possible to determine the specific code between codons and amino acids. In their first experiments, Nirenberg and Matthaei showed that a random polymer containing only uracil produced a polypeptide containing only phenylalanine. From this result, they inferred that UUU specifies phenylalanine. This idea is consistent with the data shown in the table. In the random G and U polymer, it is expected that 3% of the codons will be UUU. Likewise, 3% of the amino acids within the polypeptides were found to be phenylalanine.

We can highlight a few interesting points about these experiments. First, when considering their approach, keep in mind that these experiments were carried out in the early 1960s, before it was possible to determine the sequence of nucleotides within DNA or RNA. Therefore, the researchers had to rely on measurements of base compositions and amino acid compositions rather than on sequencing data. Second, it may have occurred to you that a random G and U polymer, like the one described in this experiment, does not contain a methionine start codon. Therefore, it may seem strange that any polypeptide was made at all. In these experiments, a large amount of synthetic RNA was added to the cell extract. Even though the RNA did not contain the types of signals (namely, ribosomal-binding site and start codon) required for efficient polypeptide translation, detectable levels of translation were still obtained in their extracts.

The amino acid sequences of polypeptides determine the structure and function of proteins

Now that we understand how genes are transcribed and translated, let's consider the structure and function of the gene product. Following gene transcription and mRNA translation, the net result is a polypeptide with a defined amino acid sequence. This sequence is known as the **primary structure** of a polypeptide. Figure 12-10 shows the primary structure of an enzyme called *lysozyme*, a relatively small polypeptide containing 129 amino acids. The primary structure of a typical polypeptide may be several hundred or even several thousand amino acids in length.

Within a living cell, a newly made polypeptide does not remain in a long linear state for a significant length of time. Rather, it rapidly folds itself into a more compact three-dimensional structure. The progression from the primary sequence of a polypeptide to the three-dimensional structure of a protein is dictated by the amino acid sequence within the polypeptide. In particular, the chemical properties of the amino acid side chains play a central role in determining the folding pattern of a protein.

As shown in Figure 12-11, there are 20 amino acids found within polypeptide chains. Each amino acid contains a different **side chain**, which has its own particular chemical properties. For example, aliphatic and aromatic amino acids are *nonpolar*, which means that they are less likely to associate with water. These *hydrophobic* (meaning water-fearing) amino acids are often buried within the interior of a folded protein. In contrast, the *polar* and *charged* amino acids are *hydrophilic* (meaning water-loving) and are more likely to be on the surface of a protein, where they can favorably interact with the surrounding water. Overall, the collective contributions of the amino acid side chains within the primary structure of polypeptides ultimately determine the unique structure of a particular protein.

This folding process occurs in multiple stages. The first stage involves the formation of regular, repeating shapes known as **secondary structures**. The two common types of secondary structure are the **α-helix** and the **β-sheet** (see Figure 12-12). A single polypeptide may have some regions that fold into an α-helix and other regions that fold into a β-sheet. Because of the geometry of secondary structures, certain amino acids (e.g., glutamic acid, alanine, and methionine) are good candidates to form an α-helix, whereas others (e.g., valine, isoleucine, and tyrosine) are more likely to be found in a β-sheet conformation. Secondary structures within polypeptides are primarily stabilized by the formation of hydrogen bonds.

The short regions of secondary structure within a polypeptide are folded relative to each other to make the **tertiary structure** of a polypeptide. As shown in Figure 12-12, α-helical regions and β-sheet regions are connected by irregularly

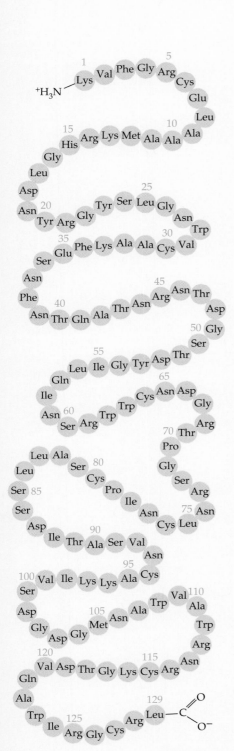

FIGURE **12-10**

An example of polypeptide primary structure. This is the amino acid sequence of the enzyme lysozyme, which contains 129 amino acids in its primary structure. As you may have noticed, the first amino acid is not methionine; instead, it is lysine. The first methionine residue in this polypeptide sequence is removed after translation is completed. The removal of the first methionine occurs in many (but not all) polypeptides.

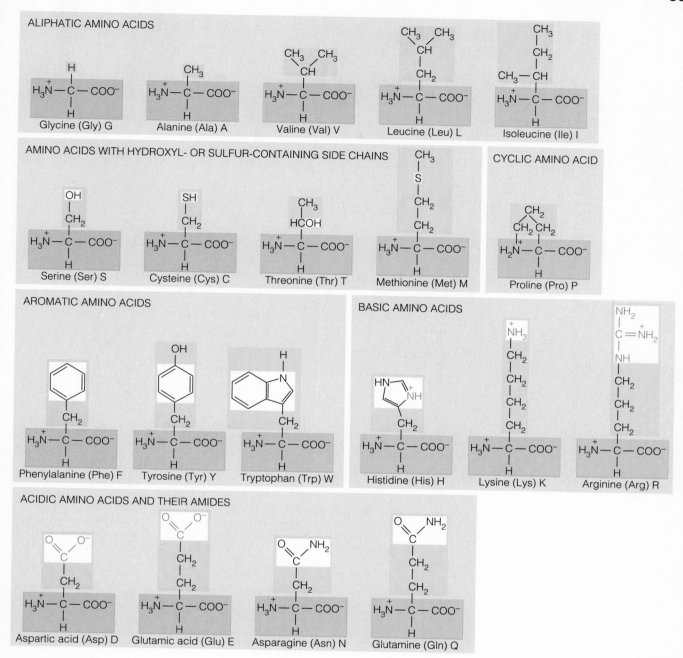

FIGURE **12-11**

The 20 amino acids that are found within polypeptides.

shaped segments to determine the three-dimensional or tertiary structure of the polypeptide. The folding of a polypeptide into its secondary and then tertiary conformation can occur spontaneously, because it is a thermodynamically favorable process. Various factors influence the folding process. These include the tendency of hydrophobic amino acids to avoid water, ionic interactions among charged amino acids, hydrogen bonding among amino acids in the folded polypeptide, and weak bonding known as van der Waals interactions.

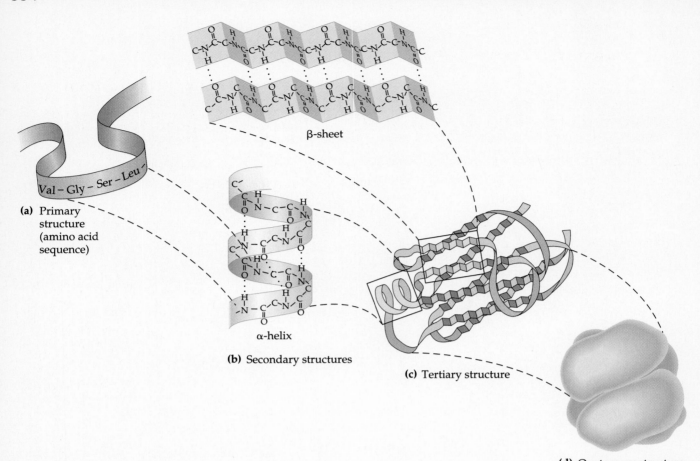

β-sheet

Val – Gly – Ser – Leu

(a) Primary
structure
(amino acid
sequence)

α-helix

(b) Secondary structures

(c) Tertiary structure

(d) Quaternary structure

FIGURE **12-12**

Levels of structures formed in polypeptides. (a) The primary structure of a polypeptide is its amino acid sequence. **(b)** The primary structure may fold into a secondary structure; the two types of secondary structures are called α-helices and β-sheets. **(c)** Both of these secondary structures can be found within the tertiary structure of a polypeptide. **(d)** Some polypeptides associate with each other to form a protein with a quaternary structure.

Functional proteins can be composed of single or multiple polypeptide chains

Proteins can be made up of different numbers of polypeptide chains. Certain proteins are composed of a single polypeptide, whereas others are composed of two or more. In cases where a protein is composed of a single polypeptide, the tertiary structure of the polypeptide is identical to the three-dimensional structure of the final protein. For example, the lactose permease protein described in Table 12-1 functions as a monomer, containing a single polypeptide. Other proteins, however, are composed of two or more polypeptides that associate with each other to make a functional protein with a **quaternary structure** (see Figure 12-12d). An example of a protein with a quaternary structure is hemoglobin, which carries oxygen within the red blood cells. Hemoglobin is composed of four polypeptide chains: two α-chains and two β-chains. (Note: A polypeptide is sometimes called a chain, because it is composed of a chain of amino acids). The α- and β-chains of hemoglobin are encoded by different genes. After the α-globin gene and β-globin gene have been expressed, the folded polypeptides associate with each other to form a protein with four polypeptides. The individual polypeptides are called **subunits** of the protein. Each subunit has its own tertiary structure. The association of multiple subunits is the quaternary structure of a protein.

An appreciation of the molecular characteristics of gene structure and function is central to our understanding of genetics. Early ideas of Garrod and of Beadle and Tatum suggested that there was a link between *genes* and the production of *enzymes*. Since their time, the molecular nature of that link has become clear. Segments of DNA within the chromosomes are organized into units called genes. Most genes are *structural genes*, which means that they provide the code for the amino acid sequence within a polypeptide. The function of most genes, therefore, is to store the information for the synthesis of a particular polypeptide—ultimately, for the synthesis of a specific protein. It is the action of proteins within living cells and organisms that directly determines the outcome of an organism's traits.

During *gene expression* in eukaryotes, DNA is *transcribed* into a *pre-mRNA*, which is subsequently spliced and finally *translated* into an amino acid sequence. The translation process employs a *genetic code*, which is nearly *universal* in all species. The use of this code involves the recognition of three-base *codons* within the mRNA by complementary *anticodons* in *tRNA* molecules. The tRNAs carry the correct amino acid, allowing the synthesis of a polypeptide with a defined amino acid sequence. This sequence is known as a polypeptide's *primary structure*. The sequence and chemistry of the amino acid *side chains* within a polypeptide cause it to fold into *secondary* and *tertiary conformations*. Some proteins also exhibit a *quaternary structure* if they are composed of two or more *subunits*. The final three-dimensional structure of a protein determines its functional role.

The first biochemical insights into the function of genes came from the experimental observations of Garrod. By making the connection between inherited defects in metabolism and the absence of a specific functional enzyme, he proposed that the function of genes is related to the production of enzymes. Later, Beadle and Tatum analyzed mutant genes in *Neurospora* and concluded that a single gene encodes a single enzyme.

Since these early studies, an enormous amount of research has probed the molecular structure of genes. Many of the techniques used to examine gene structure and function will be described in Chapters 13 to 21. As you will learn, there are a diverse array of approaches to determine how genes are expressed at the molecular and cellular levels. In this chapter, we examined the experimental strategy used by Nirenberg and Ochoa to establish the relationship between the mRNA sequence and the polypeptide sequence. By synthesizing RNAs *in vitro* with a known nucleotide composition, they were able to correlate the predicted ratio of codons with the observed ratio of amino acids within a polypeptide chain. This enabled them to decipher the genetic code by determining the relationship between a codon sequence and the amino acid it specifies.

PROBLEM SETS

SOLVED PROBLEMS

1. What is a gene? Describe the function of a gene at the molecular and organismal levels.

Answer: A gene is a unit of heredity that is responsible for determining an organism's traits. At the molecular level, a gene is a sequence of nucleotides within a much larger chromosome. Most genes are structural genes that encode the amino acid sequence within a protein. To function, a gene must be expressed. This requires the transcription of the DNA sequence into an mRNA sequence, and the translation of the mRNA sequence into a polypeptide sequence. As will be discussed in Chapters 13 and 14, the transcript and polypeptide may also be subject to many types of modifications. At the organismal level, the functioning of a gene influences an organism's phenotype. This occurs because proteins (which are the products of gene expression) affect the characteristics of cells. When we observe phenotypic characteristics of an organism, they are due to the attributes of its cells.

2. What is the difference between the primary, secondary, tertiary, and quaternary structure of a protein?

Answer: Primary structure: the sequence of amino acids within a polypeptide.

Secondary structure: a repeating pattern of structure within a polypeptide sequence. The two types of secondary structure are the α-helix and β-sheet.

Tertiary structure: the folding of secondary structures relative to each other to produce a three-dimensional arrangement of structures. The tertiary structure of a polypeptide comprises secondary structures connected by random coils.

Quaternary structure: the association of two or more polypeptides to make a functional protein. Each polypeptide chain is considered to be a subunit of the protein.

CONCEPTUAL QUESTIONS

1. Explain the nature of the genetic defect that causes alkaptonuria.

2. Some people consider proteins to be the "active participants" in biology while DNA is considered a "passive participant." Discuss whether or not you think that this is an accurate statement.

3. The photographs shown here illustrate organisms that are wild-type versus those that carry a defective allele for a particular gene. Make one or more guesses about the function of the protein that is encoded by the defective gene.

(a)   **(b)**

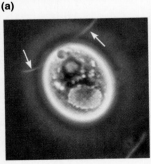

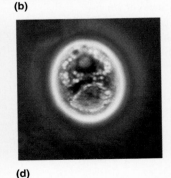

(c)   **(d)**

4. Discuss how the usual English definitions of the words *transcription* and *translation* are related to the two genetic processes they describe.

5. For each of the following, how would transcription be affected if it was the only sequence missing from a structural gene?

A. Promoter
B. Terminator
C. Ribosomal-binding site
D. Start codon
E. Stop codon
F. Regulatory sequence

6. For each of the sequences in problem 5, how would translation be affected if it was the only sequence missing from a structural gene?

7. What is the function of a ribosomal-binding site?

8. What is the specific function of a start codon?

9. A template strand of DNA has the following sequence:

3′–AATTGGCGTTCTAATCTTAC–5′

What would be the sequence of an mRNA made from this template?

10. After RNA transcription is completed, how has the sequence of DNA been altered?

11. Draw and describe the events of splicing. What types of chemical bonds are broken and formed during this process?

12. Let's suppose a gene has three exons and two introns. What would be the consequences if one of the introns was accidentally not spliced out? How would this affect translation?

13. An mRNA has the following sequence:

5′–GGCGAUGGGCAAUAAACCGGGCCAGUAAGC–3′

Identify the start codon and determine the complete amino acid sequence that would be translated from this mRNA.

14. What is an anticodon? Where is it found, and what is its function?

15. A tRNA that carries tryptophan has an anticodon sequence of ACC. A mutation alters the anticodon sequence to AAC. How will this mutation affect the polypeptide sequences of proteins that are translated within the cell?

16. What does it mean when we say that the genetic code is degenerate?

17. Discuss the universality of the genetic code.

18. What determines the three-dimensional structure of a polypeptide? Describe the steps in the folding process.

19. What is meant by the term protein subunit?

EXPERIMENTAL QUESTIONS

1. Describe the experimental observations that led Garrod to propose the idea of inborn errors of metabolism.

2. In Experiment 12A, what is the purpose of using X-rays or UV light? Explain your answer with regard to gene structure and function.

3. In Experiment 12B, what would be the predicted amounts of amino acids incorporated into polypeptides if the RNA was a random polymer containing 50% C and 50% G?

4. Discuss how the data in Experiment 12A were consistent with the one gene–one enzyme hypothesis.

5. In the table shown in Experiment 12A, explain why the mutant can grow in the presence of pantothenic acid but cannot grow in its absence.

6. In Experiment 12B, why was it necessary for the cell extract to be DNase treated?

QUESTIONS FOR STUDENT DISCUSSION/COLLABORATION

1. Suppose you are on a debate team and the topic is the following: Which is more important in determining traits, genes or proteins? Pick the side that you favor, and discuss your argument.

2. Discuss how the phenotype of an organism is determined at the cellular level. See if you can think of examples, similar to the *Drosophila* example of red eyes versus white eyes, where you can describe phenotypic differences between two individuals at the cellular level.

3. The lives of many living organisms begin with a fertilized egg. What kinds of macromolecules are necessary to allow the egg to divide and continue on with life? If you wanted to start life in a test tube by mixing together macromolecules, what macromolecules would you need?

CHAPTER 13

GENE TRANSCRIPTION
AND RNA MODIFICATION

OVERVIEW OF
TRANSCRIPTION

TRANSCRIPTION
IN PROKARYOTES

TRANSCRIPTION
IN EUKARYOTES

RNA MODIFICATION

The chromosomal DNA of living organisms contains thousands of different genes. As was outlined in Chapter 12, the information within those genes is accessed during gene expression. In this chapter, we will reconsider the first step in gene expression, transcription, in much greater detail. In particular, we will emphasize a molecular understanding of gene transcription and the modification of RNA molecules. For example, it is important to understand the sequences within genes that play a role in transcription. In this chapter we will examine these sequences. In addition, we will discuss the roles that different proteins play in the process of transcription. These proteins include RNA polymerase, transcription factors, and termination factors. Finally, we will consider the modifications that are necessary to produce functional RNA molecules.

Gene transcription is one of the hottest topics in molecular biology. The transcription of genes is of central importance in a variety of research areas, including developmental biology, cancer biology, and biotechnology. In this chapter, we will focus on the basic components that are required to transcribe genes, in both prokaryotic and eukaryotic cells. In Chapters 15 and 16, we will examine the ability to turn genes on and off by transcriptional regulation.

OVERVIEW OF TRANSCRIPTION

The word transcription means the act or process of making a copy. In genetics, this term refers to the copying of a DNA sequence within a gene into an RNA sequence. DNA is a double-stranded molecule in which the two strands have complementary sequences according to the A–T/G–C rule. During transcription, one of the DNA strands is used as a template to make a complementary copy of RNA. Except for the substitution of uracil in RNA for thymine in DNA, the base sequence in the RNA transcript is complementary to the **template strand** and identical to the DNA sequence in the opposite strand, known as the **coding strand**. In this introductory section, we will consider the general steps in the transcription process and the types of RNA transcripts that can be made.

The three stages of transcription are initiation, elongation, and termination

Transcription occurs in three stages: **initiation**, **synthesis** of the RNA transcript (also called **elongation**), and **termination**. These steps involve protein–DNA interactions in which proteins such as RNA polymerase interact with DNA sequences. As was discussed in the previous chapter, a gene contains two sequences that define the beginning and ending of transcription. These are known as the promoter and terminator, respectively. As shown in Figure 13-1, the initiation stage in the transcription process

FIGURE **13-1**

Stages of transcription.

GENES ⟶ TRAITS: The ability of genes to produce an organism's traits relies on the molecular process of gene expression. Transcription is the first step in gene expression. During transcription, the gene's sequence within the DNA is used as a template to make a complementary copy of RNA. In Chapter 14, we will examine how the sequence in mRNA is translated into a polypeptide chain. After polypeptides are made within a living cell, they fold into functional proteins that govern an organism's traits. For example, once the gene encoding phenylalanine hydroxylase is transcribed and then translated, the enzyme phenylalanine hydroxylase is made, and the cells can metabolize phenylalanine. As was described in Chapter 12 (see Figure 12-2), individuals who possess two defective copies of this gene cannot synthesize this enzyme and thus cannot metabolize phenylalanine, a disease known as phenylketonuria (PKU).

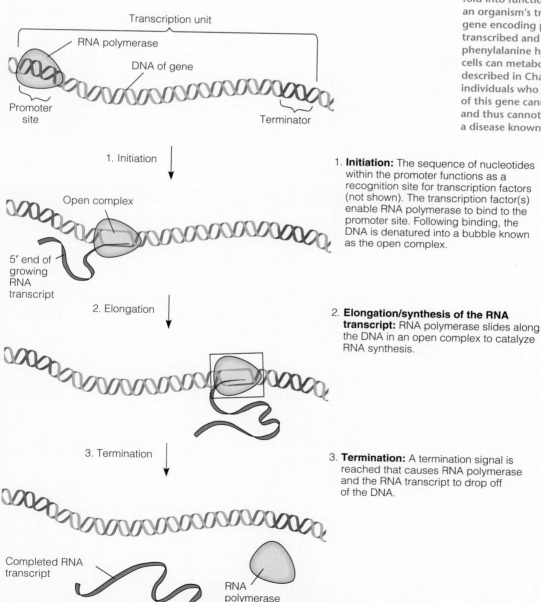

1. **Initiation:** The sequence of nucleotides within the promoter functions as a recognition site for transcription factors (not shown). The transcription factor(s) enable RNA polymerase to bind to the promoter site. Following binding, the DNA is denatured into a bubble known as the open complex.

2. **Elongation/synthesis of the RNA transcript:** RNA polymerase slides along the DNA in an open complex to catalyze RNA synthesis.

3. **Termination:** A termination signal is reached that causes RNA polymerase and the RNA transcript to drop off of the DNA.

is a recognition step. The sequence of bases within the promoter region is recognized by proteins known as **transcription factors**. The specific binding of transcription factors to the promoter sequence identifies the starting site for transcription.

After the binding of transcription factors and RNA polymerase to the promoter region, the DNA strands must be separated for transcription to occur. One of the two strands can then be used as a template for the synthesis of a complementary strand of RNA. This synthesis occurs as RNA polymerase slides along the DNA, creating a small bubblelike structure known as the **open complex**. Eventually, the RNA polymerase reaches a termination sequence, which causes it and the newly made RNA transcript to dissociate from the DNA. This event is known as transcriptional termination.

RNA transcripts have different functions

Once they are made, RNA transcripts play several different functional roles (Table 13-1). A **structural gene** is transcribed, producing an RNA transcript known as messenger RNA (mRNA). The function of mRNA is to specify the amino acid sequence of a polypeptide. As will be described in the following chapter, the process of translation uses the codon sequence in mRNA to synthesize a polypeptide with a defined amino acid sequence. Well over 90% of all genes are structural genes.

In addition, several other genes encode RNAs that are never translated. As described in Table 13-1, the RNA transcripts from nonstructural genes have various important cellular functions. In some cases, the RNA transcript becomes part of a complex that contains both protein subunits and one or more RNA molecules. Examples of protein–RNA complexes include ribosomes, signal recognition particles, spliceosomes, and certain enzymes such as *RNase P*. The function of spliceosomes and RNase P will be described later in this chapter.

TABLE 13–1 Functions of RNA Molecules

Type of RNA	Description
mRNA	Messenger RNA (mRNA) encodes the sequence of amino acids within a polypeptide.
tRNA	Transfer RNA (tRNA) is necessary in the translation of mRNA.
rRNA	Ribosomal RNA (rRNA) is necessary in the translation of mRNA. rRNAs are components of ribosomes, which are composed of both rRNAs and protein subunits. The structure and function of ribosomes will be examined in Chapter 14.
Other small RNAs	
7S RNA	7S RNA is necessary for targeting proteins to the endoplasmic reticulum. 7S RNA is a component of a complex known as signal recognition particle (SRP), which is composed of 7S RNA and six different protein subunits.
RNA of RNase P	RNase P is an enzyme that is necessary for processing bacterial tRNA molecules. The RNA is the catalytic component of this enzyme. RNase P is composed of a 350–410 nucleotide RNA and one protein subunit.
snRNA	Small nuclear RNA (snRNA) is necessary in the splicing of eukaryotic pre-mRNA. snRNAs are components of a spliceosome, which is composed of both snRNAs and protein subunits. The structure and function of spliceosomes are examined later in this chapter.
snoRNA	Small nucleolar RNA (snoRNA) is necessary in the processing of eukaryotic rRNA transcripts. snoRNAs are also associated with protein subunits. In eukaryotes, snoRNAs are found in the nucleolus, where rRNA processing and ribosome assembly occurs.

TRANSCRIPTION IN PROKARYOTES

Our molecular understanding of gene transcription initially came from studies involving bacteria and bacteriophages. Several early investigations focused on the production of viral RNA after bacteriophage infection. The first suggestion that RNA is derived from the transcription of DNA was made in the 1950s by Eliot Volkin and his colleagues at the Oak Ridge National Laboratory. They discovered that T2 bacteriophage infection of *Escherichia coli* resulted in the synthesis of RNA having a base composition similar to the DNA of the bacteriophage but different from the DNA of *E. coli*. These results were consistent with the idea that the bacteriophage DNA was being used as a template in the synthesis of bacteriophage RNA.

In 1960, Matthew Meselson at Cal Tech and François Jacob at the Pasteur Institute in Paris found that proteins are synthesized on ribosomes. One year later, Jacob and Jacques Monod proposed that a certain type of RNA acts as a genetic messenger (from the DNA to the ribosome) to provide the information for protein synthesis. They hypothesized that this RNA, which they called messenger RNA, is transcribed from the sequence within the DNA and then directs the synthesis of a particular polypeptide. In the early 1960s this was a remarkable proposal, considering that it was made even before the actual isolation and characterization of the mRNA molecules *in vitro*. The mRNA hypothesis was confirmed by Sydney Brenner at Cambridge in collaboration with Jacob and Meselson. They found that when a virus infects a bacterial cell, a virus-specific RNA is made that rapidly associates with preexisting ribosomes in the cell.

Since these pioneering studies of the 1950s and 1960s, we have learned a great deal about the molecular features of bacterial gene transcription. In this section, we will examine the three steps in the bacterial gene transcription process as they occur at the molecular level.

Prokaryotic transcription is initiated when RNA polymerase holoenzyme binds at a promoter sequence

The type of DNA sequence known as the promoter gets its name from the idea that it "promotes" gene expression. More precisely, this sequence of nucleotides directs the precise location for the initiation of RNA transcription. A promoter is located just upstream from the site where transcription of a gene actually begins. When we describe the bases in a promoter sequence, by convention we number them relative to the transcription start site. In this system, the first base used as a template for RNA transcription is denoted +1. The bases preceding this are numbered in a negative direction. There is no base numbered 0. Therefore, the promoter region is labeled with negative numbers that describe the number of bases preceding the beginning of transcription. The numbering of the promoter region is shown in Figure 13-2.

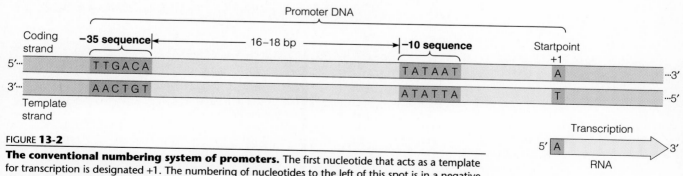

FIGURE **13-2**

The conventional numbering system of promoters. The first nucleotide that acts as a template for transcription is designated +1. The numbering of nucleotides to the left of this spot is in a negative direction, while the numbering to the right is in a positive direction. For example, the nucleotide that is immediately to the left of the +1 nucleotide is numbered −1, and the nucleotide to the right of the +1 nucleotide is numbered +2. There is no zero nucleotide in this numbering system. In many bacterial promoters, sequence elements at the −35 and −10 regions play a key role in promoting transcription.

Although the promoter may encompass a region that is several dozen nucleotides in length, there are short *sequence elements* within the promoter region that are particularly critical for promoter recognition. By comparing the sequence of DNA bases within many promoters, researchers have learned that certain sequences of bases are necessary to create a functional promoter. In many bacterial promoters, two sequence elements are important. These are located at approximately the −35 and −10 sites in the promoter region (Figure 13-2). The sequence at the −35 region is 5′–TTGACA–3′, and the one at the −10 region is 5′–TATAAT–3′. The TATAAT sequence is sometimes called the Pribnow box after David Pribnow, who discovered it in 1975 while working at Harvard University.

The sequences at the −35 and −10 sites can vary among different genes. For example, Figure 13-3 illustrates the sequences found in several different prokaryotic promoters. The most commonly occurring bases within a sequence element form the **consensus sequence**. It is oftentimes the sequence that is most efficiently recognized. For many bacterial genes, there is a good correlation between the rate of RNA transcription and the degree to which the −35 and −10 regions agree with their consensus sequences.

The enzyme that catalyzes the synthesis of RNA is **RNA polymerase**. It consists of multiple polypeptide subunits. In *E. coli*, the *core enzyme* is composed of four subunits, $\alpha_2\beta\beta'$. The association of a fifth subunit, known as **sigma factor**, with the core enzyme is referred to as the RNA polymerase *holoenzyme*.

The different subunits within the holoenzyme play distinct functional roles. The two α subunits are important in the proper assembly of the holoenzyme and in the process of binding to DNA. The β and β' subunits are critical in the catalytic

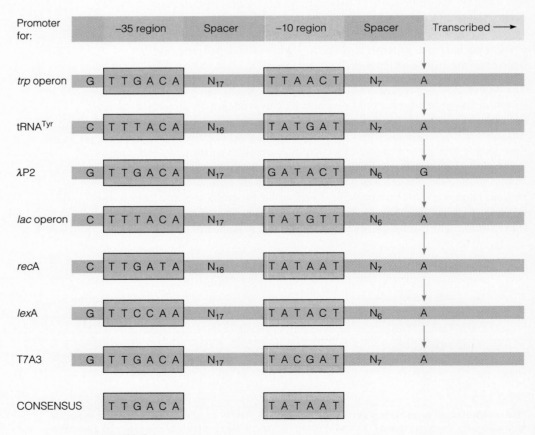

Promoter for:	−35 region	Spacer	−10 region	Spacer	Transcribed ⟶
trp operon	G T T G A C A	N_{17}	T T A A C T	N_7	A
tRNA^{Tyr}	C T T T A C A	N_{16}	T A T G A T	N_7	A
λP2	G T T G A C A	N_{17}	G A T A C T	N_6	G
lac operon	C T T T A C A	N_{17}	T A T G T T	N_6	A
*rec*A	C T T G A T A	N_{16}	T A T A A T	N_7	A
*lex*A	G T T C C A A	N_{17}	T A T A C T	N_6	A
T7A3	G T T G A C A	N_{17}	T A C G A T	N_7	A
CONSENSUS	T T G A C A		T A T A A T		

FIGURE **13-3**

Examples of −35 and −10 sequences within a variety of bacterial promoters.
This figure shows the −35 and −10 sequences for seven different bacterial and bacteriophage promoters. The consensus sequence is shown at the bottom.

synthesis of RNA. Overall, the core enzyme is necessary for RNA synthesis, and the primary role of the sigma factor is to recognize the promoter. Proteins, such as sigma factor, that influence the ability of RNA polymerase to transcribe a gene are known as **transcription factors**.

After RNA polymerase holoenzyme is assembled into its five subunits, it binds loosely to the DNA. It then scans along the DNA much as a train rolls down the tracks. When it encounters a promoter region, sigma factor recognizes the promoter elements at both the −35 and −10 positions. A region within the sigma factor protein that contains a **helix–turn–helix structure** is involved in a tighter binding to the DNA. Alpha-helices within the protein can fit into the major groove of the DNA double helix and form hydrogen bonds with the bases. This phenomenon is shown in Figure 13-4. Hydrogen bonding occurs between nucleotides in the −35 and −10 regions of the promoter and amino acid side chains in the helix–turn–helix portion of the sigma factor.

As shown in Figure 13-5, the process of transcription is initiated when the sigma factor within the holoenzyme has bound to the promoter region to form the **closed complex**. For transcription to begin, the double-stranded DNA must then be unwound into an open complex. This unwinding first occurs in the vicinity of the TATAAT box. The TATAAT box in the −10 region contains only A/T base pairs. As you may recall from Chapter 9, A/T base pairs form only two hydrogen bonds, whereas G/C pairs form three. Therefore, it is easier to separate DNA in an A/T-rich region, since fewer hydrogen bonds must be broken. The formation of the open complex completes the initiation stage of transcription. The core enzyme may now slide down the DNA to synthesize a strand of RNA. During the synthesis stage, sigma factor is no longer necessary, and it dissociates from the holoenzyme (Figure 13-5).

The RNA transcript is synthesized during the elongation stage

After the initiation stage of transcription is completed, the RNA transcript is actually made in the elongation stage of transcription. During the synthesis of the RNA transcript, RNA polymerase moves along the DNA, causing it to unwind (see Figure 13-6). The DNA strand that is used as a template for RNA synthesis is called the template or noncoding strand. The opposite DNA strand is called the coding strand; it has the same sequence as the RNA transcript except that T in the DNA corresponds to U in the RNA. Within a given gene, only the template strand is used for RNA synthesis while the coding strand is never used. As it moves down the DNA, the open complex formed by the action of RNA polymerase is approximately 17 base pairs long. On average, the rate of RNA synthesis is about 43 nucleotides per second! Behind the open complex, the DNA rewinds back into a double helix, thereby forcing the RNA behind the open complex to dissociate from the DNA.

As described in Figure 13-6 and also shown in Figure 12-6 (p. 324), the chemistry of transcription by RNA polymerase is similar to the synthesis of DNA via DNA polymerase that was discussed in Chapter 11. RNA polymerase always connects nucleotides in the 5′ to 3′ direction. During this process, RNA polymerase catalyzes the formation of a bond between the 5′–phosphate group on one nucleotide and the 3′–OH group on the previous nucleotide. The complementarity rule is similar to the A–T/G–C rule, except that uracil substitutes for thymine in the RNA. In other words, RNA synthesis obeys an A_{RNA}–T_{DNA}/U_{RNA}–A_{DNA}/G_{RNA}–C_{DNA}/C_{RNA}–G_{DNA} rule.

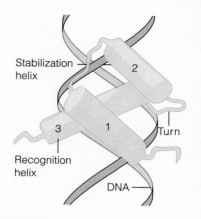

FIGURE **13-4**

The binding of a transcription factor protein to the DNA double helix. In this example, the protein contains a helix–turn–helix motif. The two α-helices of the protein (labeled 2 and 3) can fit within the major groove of the DNA. Amino acids within the α-helices hydrogen bond with the nitrogenous bases within the DNA.

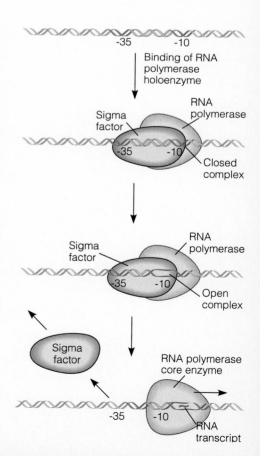

FIGURE **13-5**

The initiation stage of transcription in bacteria. The sigma factor subunit of the RNA polymerase holoenzyme recognizes the −35 and −10 regions of the promoter. The DNA unwinds in the −10 region to form an open complex. Sigma factor then dissociates from the holoenzyme, and the RNA polymerase core enzyme can proceed down the DNA to transcribe RNA.

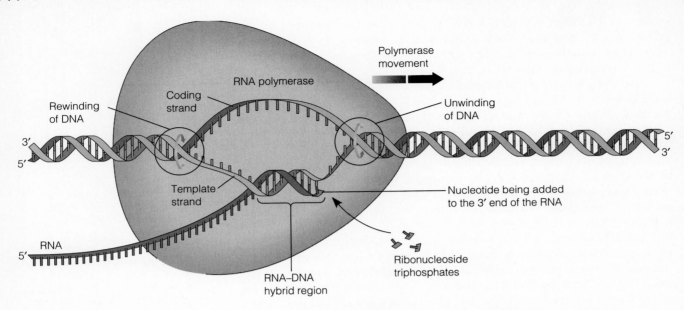

1. RNA polymerase slides along the DNA, creating an open complex as it moves.

2. The DNA strand known as the template strand is used to make a complementary copy of RNA as an RNA–DNA hybrid.

3. The RNA is synthesized in a 5′ to 3′ direction using ribonucleoside triphosphates as precursors. Pyrophosphate is released (not shown).

4. The complementarity rule is the same as the A–T and G–C rule except that U is substituted for T in the RNA.

FIGURE 13-6
Synthesis of the RNA transcript.

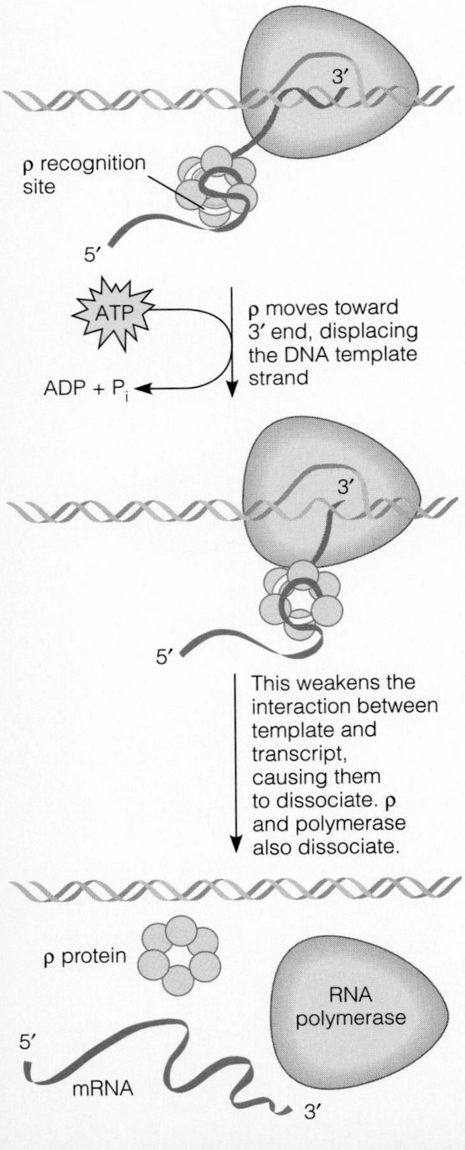

Transcription is terminated by either an RNA-binding protein or an intrinsic terminator

The end of RNA synthesis is referred to as termination. Prior to termination, the hydrogen bonding between the DNA and RNA within the open complex is of central importance in preventing dissociation of the RNA polymerase from the template strand. Termination occurs when this short RNA–DNA hybrid region is forced to separate, thereby releasing the newly made RNA transcript as well as the RNA polymerase. In *E. coli*, two different mechanisms for termination have been identified. For certain genes, a protein known as ρ (rho) is responsible for terminating transcription, a mechanism called ρ-dependent termination. For other genes, termination does not require ρ. This is referred to as ρ-independent termination. Both mechanisms will be described here.

In **ρ-dependent termination**, a sequence near the 3′ end of the newly made RNA acts as a recognition site for the binding of the ρ protein (Figure 13-7). Rho protein functions as a helicase, an enzyme that can separate RNA–DNA hybrid regions. After the ρ recognition site is synthesized in the RNA, ρ binds to the RNA and moves in the direction of the RNA polymerase. A **termination sequence** near the 3′ end of the gene causes RNA polymerase to pause in its synthesis of RNA. This allows ρ to catch up with RNA polymerase and break the hydrogen bonds between the DNA and RNA within the open complex. When this occurs, the completed RNA strand is separated from the DNA along with the RNA polymerase.

ρ recognition site

5′

ATP

ADP + P$_i$

ρ moves toward 3′ end, displacing the DNA template strand

This weakens the interaction between template and transcript, causing them to dissociate. ρ and polymerase also dissociate.

ρ protein

RNA polymerase

5′

mRNA

3′

FIGURE 13-7
ρ-dependent termination.

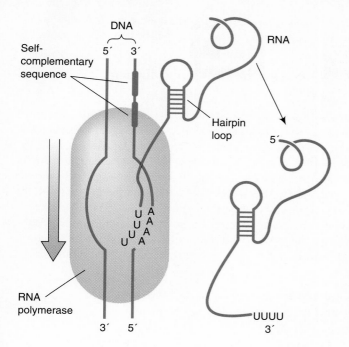

DNA
Self-complementary sequence
5′ 3′
RNA
Hairpin loop
5′
U U U A A A A
RNA polymerase
3′ 5′
UUUU 3′

FIGURE **13-8**

ρ-**independent or intrinsic termination.** When RNA polymerase reaches the end of the gene, it transcribes a uracil-rich sequence. Soon after this uracil-rich sequence is transcribed, a stem-loop forms just upstream from the open complex. The formation of this stem-loop causes RNA polymerase to pause in its synthesis of the transcript. While it is pausing, the RNA–DNA hybrid region is a uracil-rich sequence. Since U–A hydrogen bonds are relatively weak interactions, the transcript and RNA polymerase dissociate from the DNA.

Rho-independent termination is facilitated by two sequence elements within the RNA (Figure 13-8). One element is a uracil-rich sequence located at the 3′ end of the RNA. The second sequence is slightly upstream from the uracil-rich sequence, near the 3′ end; it promotes the formation of a stem-loop (also called a hairpin) structure. As you may recall from Chapter 9, a stem-loop structure can form due to complementary sequences within the RNA. These RNA stem-loops can form almost immediately after they are synthesized.

In the sequence of events described in Figure 13-8, the formation of the stem-loop near the 3′ end of the RNA causes the RNA polymerase to pause in its synthesis of RNA. At the time RNA polymerase pauses, the uracil-rich sequence in the RNA transcript is bound to the DNA template strand. As previously mentioned, the hydrogen bonding of the RNA to the DNA keeps the RNA polymerase clamped onto the DNA. However, the binding of this uracil-rich sequence to the DNA template strand is thought to be unstable, causing the RNA transcript to spontaneously fall off the DNA and terminate further transcription. Because this process does not require another protein to remove the RNA transcript from the DNA, the sequence elements that cause this type of termination are referred to as **intrinsic terminators**.

TRANSCRIPTION IN EUKARYOTES

Many of the basic features of gene transcription are very similar in prokaryotic and eukaryotic species. Much of our understanding of transcription has come from studies in *Saccharomyces cerevisiae* (baker's yeast) and higher eukaryotic species such as mammals. In general, gene transcription in eukaryotes is more complex than in their prokaryotic counterparts. Eukaryotic cells are larger and contain a variety of compartments known as organelles. This added level of cellular complexity dictates that eukaryotes contain many more genes encoding cellular proteins. In addition, higher eukaryotic species are multicellular, being composed of many different cell types. Multicellularity adds the requirement that genes be transcribed in the correct type of cell and during the proper stage of development. Therefore, in any given species, the transcription of the thousands of different genes that an organism possesses requires the appropriate timing and coordination. In this section, we will examine some features of gene transcription that are unique to eukaryotes. In addition, the regulation of eukaryotic gene transcription will be covered in Chapter 16.

Three eukaryotic RNA polymerases transcribe different types of genes

The genetic material within the nucleus of a eukaryotic cell is transcribed by three different RNA polymerase enzymes, designated I, II, and III. Each of the three RNA polymerases transcribes different categories of genes. *RNA polymerase I* transcribes all of the genes that encode ribosomal RNA (rRNA) except for the 5S rRNA. *RNA polymerase II* transcribes all of the structural genes and also certain snRNA genes. RNA polymerase II is responsible for the synthesis of all mRNA. *RNA polymerase III* transcribes all of the tRNA genes and the 5S rRNA gene.

All three RNA polymerases have many subunits. They contain two large subunits that are similar to the β and β′ subunits of bacterial RNA polymerase. Eukaryotic RNA polymerases also contain eight or more additional subunits.

Transcription of eukaryotic genes is initiated when RNA polymerase and transcription factors bind to a promoter sequence

In eukaryotes, the promoter sequence is more variable and often more complex than that found in prokaryotes. It is usually composed of a **core promoter** and **regulatory elements**. This chapter will focus primarily on the structure of the core promoter region (regulatory elements will be covered in Chapter 16). The core promoter is obligatory for transcription to take place. Similar to the function of a bacterial promoter, the core promoter provides the initial binding site for **general transcription factors** and RNA polymerase.

Figure 13-9 shows common patterns of sequences found within the promoters of higher eukaryotes that are recognized by the three types of RNA polymerases. Short DNA sequences, which are usually called **elements** or **boxes**, are important for general transcription factors and RNA polymerase to assemble at the promoter region. For example, Figure 13-9b describes a common pattern of sequences found within the core promoter region that is recognized by RNA polymerase II. In the vicinity of the −25 position, a TATAAAA sequence called the **TATA box** is usually found. In addition, two other promoter elements are commonly found upstream from the TATA box. A sequence GGCCAATCT, called the **CAAT box**, and a sequence GGGCGG, called the **GC box**, may be located in the −50 to −100 region of structural genes in eukaryotes. However, the number and locations of GC boxes and CAAT boxes vary considerably among different eukaryotic structural genes.

The role of the TATA box differs from that of the CAAT and GC boxes. The TATA box is important in determining the precise starting point for transcription. If it is missing from the core promoter, the transcription starting point becomes undefined, and transcription may start at a variety of different locations. In contrast, the CAAT and GC boxes function as recognition sites that are bound by transcription factors and recruit RNA polymerase into the general vicinity of the promoter region.

Several general transcription factors are found in eukaryotes. These proteins interact in a fairly complicated way to recruit RNA polymerase to the core promoter region. Many general transcription factors have been identified, and their roles in transcriptional initiation have been studied. At present, the exact series of steps that occur during initiation is not entirely understood. Nevertheless, enough information is available to propose models that describe the events leading to the formation of the open complex.

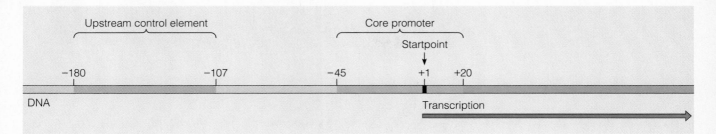

(a) Promoter for RNA polymerase I

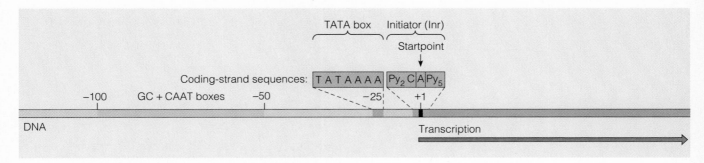

(b) General promoter for RNA polymerase II

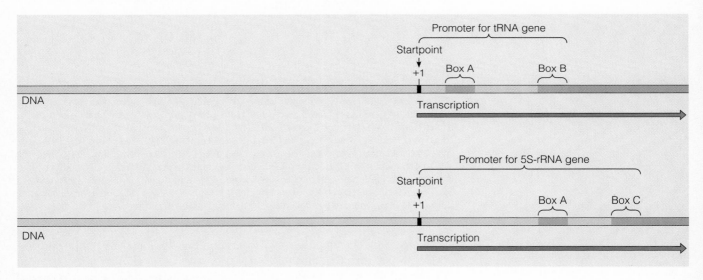

(c) Two types of promoters for RNA polymerase III

FIGURE **13-9**

A common pattern found within the promoter of many eukaryotic genes recognized by RNA polymerases. (a) Promoter for RNA polymerase I. **(b)** Core promoter for RNA polymerase II. The startpoint usually occurs at adenine; there are two pyrimidines and a cytosine that precede this adenine, and five pyrimidines that follow it. **(c)** Two types of promoters for RNA polymerase III. The binding of RNA polymerase III is facilitated by 10bp sequences called BoxA, BoxB, and BoxC.

Figure 13-10 is a model that describes the assembly of transcription factors and RNA polymerase II at the TATA box. As shown here, a series of interactions leads to the formation of the open complex. Transcription Factor IID (TFIID) first binds to the TATA box and thereby plays a critical role in the recognition of the

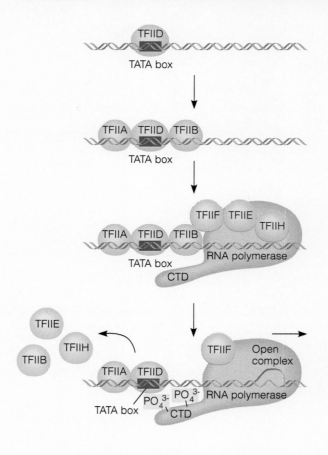

1. TFIID binds to the TATA box. TFIID is a complex of proteins that includes the TATA-binding protein (TBP) and several TBP-associated factors (TAFs).

2. TFIIA and TFIIB bind to TFIID.

3. TFIIB acts as a bridge to bind RNA polymerase II. Note that other transcription factors such as TFIIE, TFIIF, and TFIIH are also bound to RNA polymerase II.

4. The phosphorylation of the CTD within RNA polymerase II causes it to dissociate from TFIIB. TFIIH promotes the denaturation of DNA to form an open complex. TFIIB, TFIIE, and TFIIH dissociate from the complex.

5. RNA polymerase II is now free to move down the DNA.

FIGURE **13-10**

Steps leading to the formation of the open complex. (1) TFIID first binds to the TATA box. A subunit of TFIID, known as the TATA-binding protein, recognizes the TATA box sequence. **(2)** TFIIA and TFIIB then bind to TFIID. **(3)** TFIIB promotes the binding of RNA polymerase II. Transcription factors TFIIE, TFIIF, and TFIIH are also bound to RNA polymerase to form the closed complex. **(4)** To form the open complex, TFIIH hydrolyzes ATP and phosphorylates a region in RNA polymerase II known as the carboxy terminal domain (CTD). RNA polymerase II is released from TFIIB. TFIIH also functions as a helicase that breaks the hydrogen bonding between the double-stranded DNA and thereby promotes the formation of the open complex. After the open complex has formed, TFIIB, TFIIE, and TFIIH dissociate, and **(5)** RNA polymerase II proceeds to the synthesis stage of transcription.

promoter. TFIID is composed of several subunits including TATA-binding protein (TBP), which directly binds to the TATA box, and several other proteins called TBP-associated factors (TAFs). After TFIID binds to the TATA box, it associates with TFIIA and TFIIB. TFIIB promotes the binding of RNA polymerase II. Transcription Factors IIE, IIF, and IIH are also bound to RNA polymerase, forming the closed complex.

To form the open complex, TFIIH hydrolyzes ATP and phosphorylates a domain in RNA polymerase II known as the carboxy terminal domain (CTD). These reactions release the contact between RNA polymerase II and TFIIB. TFIIH also functions as a helicase that breaks the hydrogen bonding between the double-stranded DNA and thereby promotes the formation of the open complex. After the open complex has formed, TFIIB, TFIIE, and TFIIH dissociate. RNA polymerase II is then free to proceed to the synthesis stage of transcription.

In vitro, when researchers mix together TBP, TFIIB, TFIIE, TFIIF, TFIIH, RNA polymerase, and a DNA sequence containing a TATA box, the DNA is transcribed into RNA. Therefore, these components are referred to as the **basal transcription**

apparatus. In a living cell, however, additional components regulate transcription and allow it to proceed at a reasonable rate. As shown in Figure 13-10, TAFs (which are part of TFIID) and TFIIA are normally involved in the initiation of transcription. A multiprotein complex called *mediator* also binds to the CTD of RNA polymerase and regulates transcription. In addition, as we will discuss in Chapter 16, regulatory transcription factors bind to sequence elements in the vicinity of the promoter and influence the assembly of the basal transcription apparatus.

The nucleosome structure is disrupted so that the RNA transcript can be synthesized

The chemistry of the synthesis of RNA in eukaryotes is identical to that in prokaryotes, as was described in Figure 13-6. In eukaryotes, however, the DNA is packaged into a nucleosome structure. Since RNA polymerase is a very large enzyme compared with a nucleosome, the tight wrapping of DNA within a nucleosome is expected to inhibit the ability of RNA polymerase to transcribe the DNA. To circumvent this problem, the nucleosome structure is significantly perturbed during transcription. Some studies suggest that the histones are completely displaced, whereas others suggest that they are loosened but remain attached to the DNA.

Different models have been proposed to explain how the histones might be partially or completely removed from the DNA during transcription. One possibility is that there are transcription factors that function in the removal of the histone proteins. Alternatively, it has been hypothesized that the movement of RNA polymerase along the DNA causes positive supercoiling ahead of the open complex and negative supercoiling behind it. Since the nucleosome structure is negatively supercoiled, positive supercoiling occurring in front of RNA polymerase would disrupt the nucleosome structure. Along these same lines, negative supercoiling occurring behind RNA polymerase would favor the re-formation of a tight nucleosome structure.

RNA MODIFICATION

During the 1960s and 1970s, studies in bacteria established the physical structure of the gene. The analysis of bacterial genes showed that the sequence of DNA within the coding strand corresponds to the sequence of nucleotides in the mRNA. During translation, the sequence of codons in the mRNA is then read, providing the instructions for the correct amino acid sequence in a polypeptide. The one-to-one correspondence between the sequence of codons in the DNA coding strand and the amino acid sequence of the polypeptide has been termed the **colinearity** of gene expression. Based on these early results in bacteria, researchers concluded that modification of mRNA transcripts is unnecessary for gene expression.

The situation, however, rapidly changed in the late 1970s, when the tools became available to study eukaryotic genes at the molecular level. The scientific community was astonished by the discovery that eukaryotic structural genes are not always colinear with their functional mRNAs. Instead, the coding sequences within many eukaryotic genes are separated by DNA sequences that are not translated into protein. As was mentioned in Chapter 12, the coding sequences are called exons, and the sequences that interrupt them are called intervening sequences or introns. During transcription, an RNA is made corresponding to the entire gene sequence. Subsequently, the sequences in the RNA that correspond to the gene introns are cut out, while the RNA sequences derived from the exons are connected, or spliced together. This process is called **RNA splicing**. Since the 1970s, it has become apparent that splicing is a common genetic phenomenon in eukaryotic species. Splicing occurs occasionally in prokaryotes as well.

Aside from splicing, research has also shown that RNA transcripts can be modified in several other ways. Table 13-2 describes the general types of RNA modifications. For example, rRNAs and tRNAs are synthesized as long transcripts that are cleaved into smaller functional pieces. In addition, most eukaryotic mRNAs have a cap attached to their 5′ end and a tail attached at their 3′ end. In this section, we will examine the molecular mechanisms that account for several types of RNA modifications.

TABLE 13–2 Modifications that May Occur to RNAs

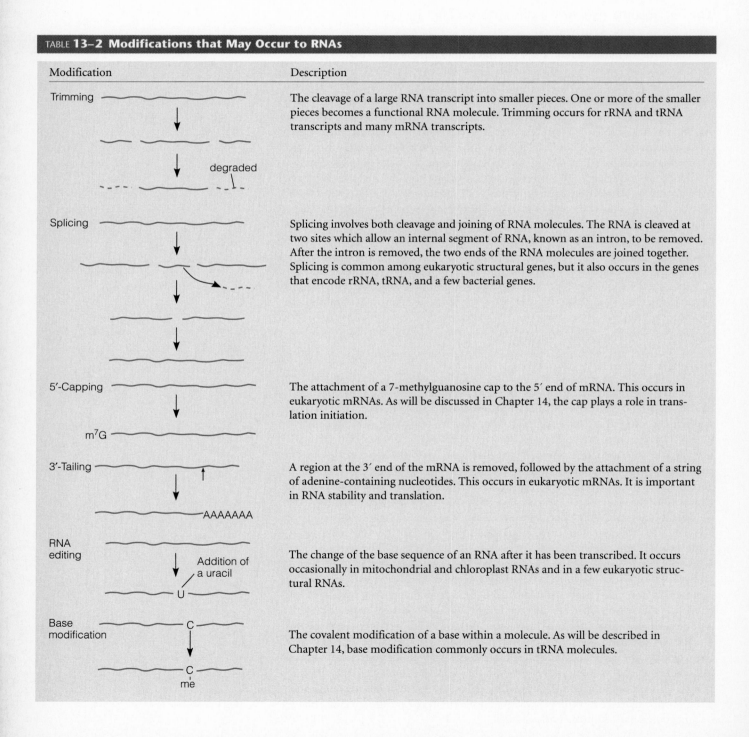

Modification	Description
Trimming	The cleavage of a large RNA transcript into smaller pieces. One or more of the smaller pieces becomes a functional RNA molecule. Trimming occurs for rRNA and tRNA transcripts and many mRNA transcripts.
Splicing	Splicing involves both cleavage and joining of RNA molecules. The RNA is cleaved at two sites which allow an internal segment of RNA, known as an intron, to be removed. After the intron is removed, the two ends of the RNA molecules are joined together. Splicing is common among eukaryotic structural genes, but it also occurs in the genes that encode rRNA, tRNA, and a few bacterial genes.
5′-Capping	The attachment of a 7-methylguanosine cap to the 5′ end of mRNA. This occurs in eukaryotic mRNAs. As will be discussed in Chapter 14, the cap plays a role in translation initiation.
3′-Tailing	A region at the 3′ end of the mRNA is removed, followed by the attachment of a string of adenine-containing nucleotides. This occurs in eukaryotic mRNAs. It is important in RNA stability and translation.
RNA editing	The change of the base sequence of an RNA after it has been transcribed. It occurs occasionally in mitochondrial and chloroplast RNAs and in a few eukaryotic structural RNAs.
Base modification	The covalent modification of a base within a molecule. As will be described in Chapter 14, base modification commonly occurs in tRNA molecules.

Some large RNA transcripts are processed to smaller functional transcripts by enzymatic cleavage

In many cases, the RNA transcript initially made during gene transcription is processed into smaller functional RNA molecules. This involves cleavage of the RNA to make smaller pieces. As an example, Figure 13-11 shows the processing of mammalian ribosomal RNA. The ribosomal RNA gene is transcribed by RNA polymerase I to make a long primary transcript, known as 45S rRNA. (The term 45S refers to the sedimentation characteristics of this transcript in Svedberg units.) Following the synthesis of the 45S rRNA, cleavage occurs at several points as shown in Figure 13-11. Three of the fragments, termed 5.8S, 18S, and 28S rRNA, are functional rRNA molecules that bind to proteins within the ribosomes. In eukaryotes, the processing of 45S rRNA and the assembly of ribosomal subunits occur in a structure within the cell nucleus known as the **nucleolus**. The biosynthesis of ribosomes will be described further in Chapter 14.

The production of tRNA molecules also requires processing. Like ribosomal RNA, tRNAs are synthesized as large precursor RNAs that must be cleaved at both the 5′ and 3′ ends to produce mature functional tRNAs (i.e., ones that bind to amino acids). This processing event has been studied extensively in *E. coli*. Figure 13-12 shows the processing of a precursor tRNA that carries tyrosine (tRNA^{Tyr}). Interestingly, the cleavage occurs differently at the 5′ end and the 3′ end. At the 5′ end, the precursor tRNA is recognized by an enzyme known as **RNase P**. This enzyme is an endonuclease that cuts the precursor tRNA. The action of RNase P produces the correct 5′ end of the mature tRNA. At the 3′ end, two different enzymes trim the tRNA precursor. First, an endonuclease cleaves the precursor RNA to remove a 170-nucleotide segment. Next, an exonuclease, designated *RNase D*, binds to the 3′ end and digests the RNA in the 3′ to 5′ direction. When it reaches a CCA sequence, the exonuclease stops digesting the precursor RNA molecule. Therefore, all tRNAs in

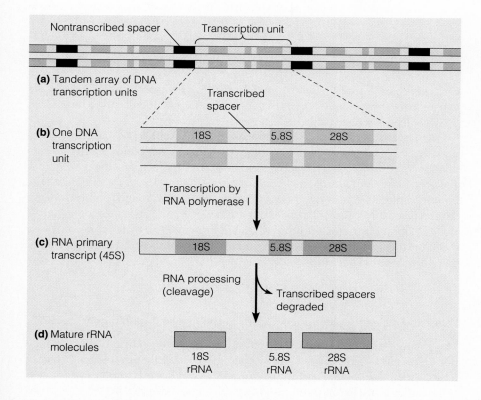

FIGURE 13-11

The processing of ribosomal RNA in eukaryotes. The large ribosomal RNA gene is transcribed into a long 45S rRNA primary transcript. This transcript is cleaved to produce 5.8S, 18S, and 28S rRNA molecules, which become associated with protein subunits in the ribosome. This processing occurs within the nucleus of the cell.

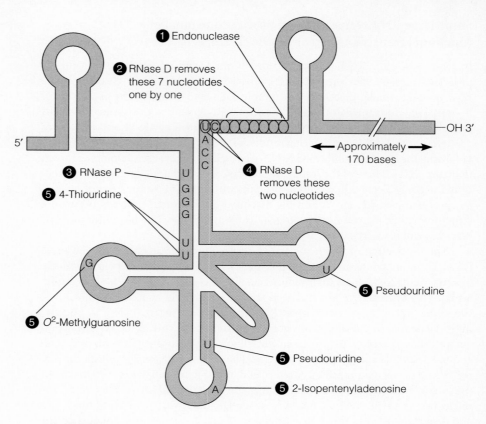

FIGURE **13-12**

The processing of tRNA molecules. Near the 3' end of the tRNA, an endonuclease makes a cut (step 1) and then RNase D removes seven nucleotides at the 3' end (step 2). In step 3, RNase P makes a cut near the 5' end. After this cut, RNase D removes two additional nucleotides at the 3' end (step 4). In addition to these cleavage steps, several bases within the tRNA molecule are modified to other bases (steps 5).

E. coli have a CCA sequence at their 3' ends. Finally, certain bases in tRNA molecules may be covalently modified to alter their structure. The modification of bases in tRNAs will be described further in Chapter 14.

As researchers studied tRNA processing, they found certain enzymatic features that were very unusual and exciting, changing the way biologists view the actions of enzymes. RNase P has been found to be an enzyme that contains both RNA and protein subunits. Surprisingly, Sidney Altman and colleagues at Yale University found that the RNA portion of this enzyme contains the catalytic ability to cleave the precursor tRNA. RNase P is an example of a **ribozyme**, which means that its catalytic ability is due to the action of RNA, not protein. Prior to the study of RNase P and the identification of self-splicing RNAs, biochemists had staunchly believed that only proteins could function as enzymes.

EXPERIMENT 13A

Leder and colleagues identified introns in the mouse β-globin gene in 1978

The discovery that tRNA and rRNA transcripts are processed to a smaller form did not seem unusual to geneticists and biochemists, because the enzymatic cleavage of RNA was similar to the cleavage that can occur for other macromolecules such as DNA and proteins. In sharp contrast, when splicing was detected in the 1970s, it was a novel concept. Splicing involves cleavage at two sites; an intron is removed; and—in a unique step—the remaining fragments are hooked back together again.

Eukaryotic introns were first detected by comparing the base sequence of genes and mRNAs during viral infection of mammalian cells by adenovirus. This research was carried out by two groups: Philip Sharp and colleagues at MIT, and Richard Roberts and colleagues at Cold Spring Harbor. This pioneering observation led to the next question: "Are introns a peculiar phenomenon that only occurs in viral genes, or are they found in chromosomal genes as well?"

In the late 1970s, several research groups, including those of Pierre Chambon in Strasbourg, France, Bert O'Malley at Baylor College of Medicine, and Phillip Leder at the National Institutes of Health (NIH), investigated the presence of introns in eukaryotic structural genes. The experiment described here is that of Leder, which used electron microscopy to identify introns in the β-globin gene.

In this experiment, Leder and his colleagues isolated the mRNA for the mouse β-globin gene. Beta-globin is a polypeptide that is a subunit of hemoglobin, the protein that carries oxygen in red blood cells. Prior to this work, the β-globin gene had been cloned. To detect introns within the cloned gene, Leder used a strategy involving **hybridization**. In this approach, the double-stranded DNA of the cloned β-globin gene was first denatured and mixed with mRNA for β-globin. Since this mRNA is complementary to the anticoding strand of the DNA, the anticoding strand and the mRNA will specifically bind to each other. This event is called hybridization. Later, when the DNA is allowed to renature, the binding of the mRNA to the anticoding strand of DNA prevents the two strands of DNA from forming a double helix. In the absence of any introns, the single-stranded DNA will form a loop. Since the RNA is displacing one of the DNA strands, this structure is known as an RNA displacement loop or **R loop**, as shown here:

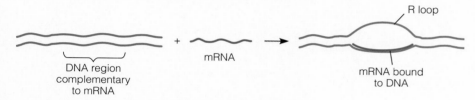

Leder and colleagues realized that a different type of R loop structure would be formed if the DNA contains an intron. For example, when mRNA hybridizes to a gene containing one intron, two single-stranded R loops will form that are separated by a double-stranded DNA region:

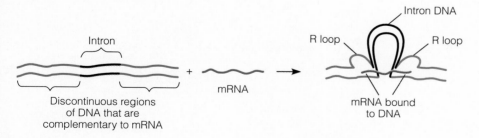

The intervening double-stranded region occurs because an intron has been spliced out of the mRNA, so that the mRNA cannot hybridize to this segment of the gene. As we will see, Leder and colleagues used electron microscopy to observe these kinds of structures in their experiment.

THE HYPOTHESIS

The β-globin gene from the mouse contains introns.

TESTING THE HYPOTHESIS

Starting material: A cloned fragment of chromosomal DNA that contains mouse β-globin gene.

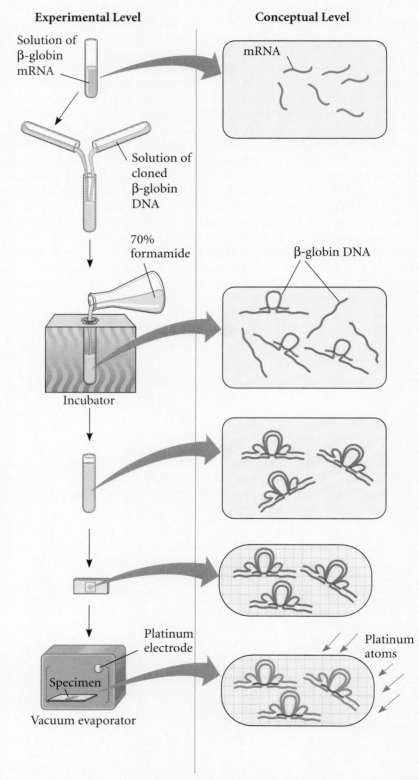

Experimental Level

Conceptual Level

1. Isolate mature mRNA for the mouse β-globin gene. Note: Globin mRNA is abundant in reticulocytes, which are immature red blood cells.

Solution of β-globin mRNA

mRNA

2. Mix the mRNA and cloned DNA together.

Solution of cloned β-globin DNA

70% formamide

β-globin DNA

3. Separate the double-stranded DNA and allow the mRNA to hybridize. This is done using 70% formamide, 52°C, for 16 h.

Incubator

4. Dilute the sample to decrease the formamide concentration. This allows the DNA to reform a double-stranded structure. Note: The DNA cannot form a double-stranded structure in regions where the mRNA has already hybridized.

5. Spread the sample onto a microscopy grid.

6. Stain with uranyl acetate and shadow with heavy metal. Note: The technique of electron microscopy is described in the appendix.

Platinum electrode

Platinum atoms

Specimen

Vacuum evaporator

7. View the sample under the electron microscope.

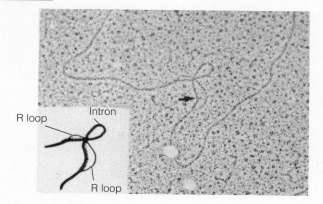

R loop

Intron

R loop

As seen in the electron micrograph, the mRNA hybridized to the DNA to form two R loops separated by a double-stranded DNA region. These data are consistent with the idea that the DNA of the β-globin gene contains an intron. Similar results were obtained by Chambon and O'Malley for other structural genes. Since these early discoveries, introns have been found in many eukaryotic genes. The occurrence and biological significance of introns will be discussed later in this chapter.

DNA sequencing methods developed in the late 1970s have permitted an easier and more precise way of detecting introns. As will be described in Chapter 19, researchers can clone a fragment of chromosomal DNA that contains a particular gene. This is called a **genomic clone**. In addition, mRNA can be used as a starting material to make a copy of DNA known as **complementary DNA** or **cDNA** (see Chapter 19). The cDNA will *not* contain introns, because the introns have been previously removed during RNA splicing. In contrast, if a gene contains introns, a genomic clone for a eukaryotic gene will also contain introns. Therefore, a comparison of the DNA sequences of genomic and cDNA clones can provide direct evidence that a particular gene contains introns.

Three different splicing mechanisms can remove introns

Since the 1970s, the investigations of many research groups have shown that most structural genes among higher eukaryotes contain one or more introns. Less commonly, introns can occasionally occur within tRNA and rRNA genes. At the molecular level, three different RNA splicing mechanisms have been identified (Figure 13-13). In all three cases, splicing leads to removal of the intron RNA and the covalent connection of the exon RNA by a phosphodiester linkage.

The chemistry of splicing among **Group I and II introns** is called **self-splicing**. It is given this name because the splicing event does not require the aid of other enzymes. Instead, the RNA functions as its own ribozyme. Group I introns that occur within the rRNA of *Tetrahymena* (a protozoan) have been studied extensively by Thomas Cech at the University of Colorado, Boulder. In this organism, the splicing process involves the binding of a single guanosine to a site within the intron (Figure 13-13a). This leads to the cleavage of RNA at the 3′ end of exon 1. Next, the bond between a different guanine nucleotide (that is found within the intron sequence) and the 5′ end of exon 2 is cleaved. This event allows the 3′ end of exon 1 to then form a phosphodiester bond with the 5′ end of exon 2. The intron RNA is subsequently degraded. In this example, the RNA molecule functions as its own ribozyme, since it splices itself without the aid of a catalytic protein.

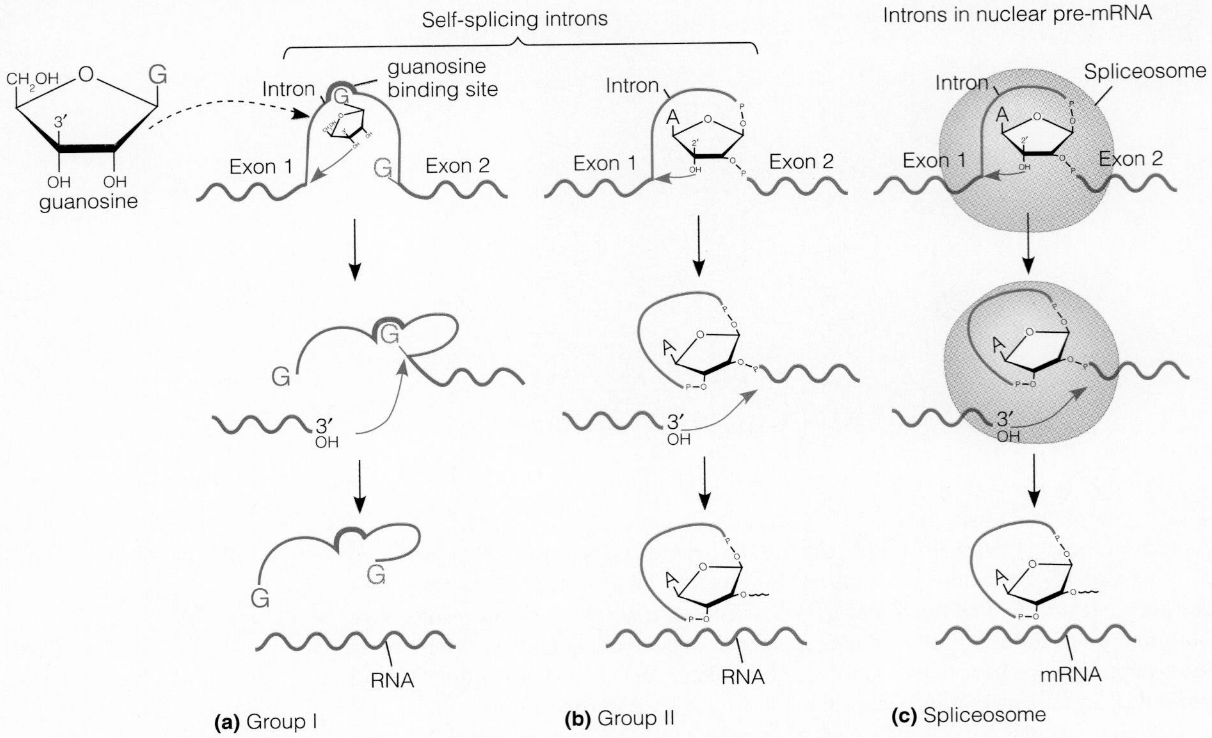

FIGURE 13-13

Mechanisms of RNA splicing. Group I and II introns are self-splicing. **(a)** The splicing of Group I introns involves the binding of a free guanosine to a site within the intron, leading to the cleavage of RNA at the 3′ end of exon 1. The bond between a different guanine nucleotide (in the intron strand) and the 5′ end of exon 2 is cleaved. The 3′ end of exon 1 then forms a phosphodiester bond with the 5′ end of exon 2. **(b)** In Group II introns, a similar splicing mechanism occurs, except that the 2′–OH group on an adenine nucleotide (already within the intron) begins the catalytic process. **(c)** Pre-mRNA splicing requires the aid of a multicomponent structure known as the spliceosome. The mechanism of pre-mRNA splicing is described later in Figure 13-17.

In Group II introns, a similar splicing mechanism occurs, except the 2′–OH group on ribose found in an adenine nucleotide (already within the intron strand) begins the catalytic process (Figure 13-13b). Experimentally, Group I and II self-splicing can occur *in vitro* without the addition of any proteins. However, in a living cell (i.e., *in vivo*), proteins known as **maturases** often enhance the rate of splicing of Group I and II introns.

In eukaryotes, the transcription of structural genes produces a long transcript known as **pre-mRNA**, which is located within the nucleus. These large RNA transcripts are also known as **heterogenous nuclear RNA** or **hnRNA**. This pre-mRNA is usually altered by splicing and other modifications before it exits the nucleus. Unlike the Group I and II introns, which may undergo self-splicing, pre-mRNA splicing requires the aid of a multicomponent structure known as the **spliceosome**. The mechanism of pre-mRNA splicing shown in Figure 13-13c will be described later in this chapter.

Table 13-3 describes the occurrence of introns among the genes of different species. The biological significance of Group I and II introns is not understood. By comparison, pre-mRNA splicing is a widespread phenomenon among higher eukaryotes. In mammals and higher plants, most structural genes have at least one intron that can be located anywhere within the gene. The functional significance of eukaryotic introns will be discussed later in this chapter.

TABLE 13–3 Occurrence of Introns

Type of Intron	Mechanism	Occurrence
Pre-mRNA	Spliceosome	Very commonly found in structural genes within the nucleus of eukaryotes.
Group I	Self-splicing	Found in rRNA genes within the nucleus of Tetrahymena and other lower eukaryotes. Found in a few structural, tRNA, and rRNA genes within the mitochondrial DNA (fungi and plants) and chloroplast DNA. Found very rarely in tRNA genes within bacteria.
Group II	Self-splicing	Found in a few structural, tRNA, and rRNA genes within the mitochondrial DNA (fungi and plants) and in chloroplast DNA.

Most eukaryotic pre-mRNAs are trimmed, capped, tailed, and spliced

Before a pre-mRNA can be translated, several modifications must occur. These are summarized in Figure 13-14. A segment of RNA at the 5′ or 3′ end of the pre-mRNA may be removed by enzymatic cleavage. This process is known as **trimming**. Most mRNAs then have a 7-methylguanosine covalently attached to the 5′ end. Such an attachment of this nucleotide to mRNA is known as **capping**; the cap is attached by a

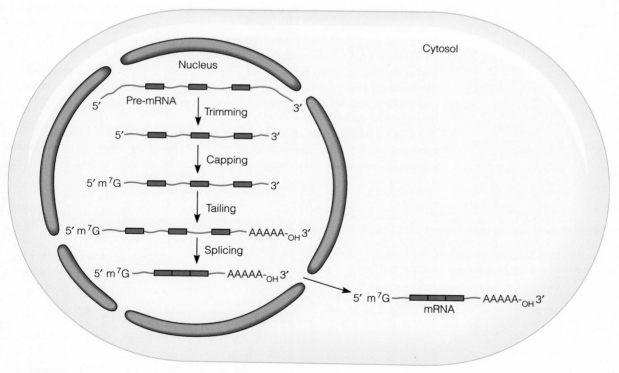

FIGURE **13-14**

Common modifications to eukaryotic pre-mRNA. The pre-mRNA may be cleaved at its 5′ or 3′ end to trim it to a shorter molecule. The 5′ end is then capped with a 7-methylguanosine group, and a polyA tail is attached to the 3′ end. If the pre-mRNA contains introns, these are removed prior to the exit from the nucleus. After all these modifications have occurred, the mRNA is referred to as **mature mRNA** or simply mRNA.

capping enzyme. As will be discussed in Chapter 14, this cap structure plays an important role during the early stages of translation.

At the 3′ end, most mRNAs have a string of adenine nucleotides, referred to as a polyA **tail**. The polyA tail appears to be important in the stability of mRNA and in the translation of polypeptides. However, it is not absolutely required for translation; some mRNAs without polyA tails are translated efficiently. The steps required to synthesize a polyA tail are shown in Figure 13-15. To acquire a polyA tail, the pre-mRNA needs to contain a **polyadenylation sequence** near its 3′ end. In higher eukaryotes, the consensus sequence is AAUAAA. This sequence is downstream (toward the 3′ end) from the stop codon in the pre-mRNA. An endonuclease recognizes the polyadenylation sequence and cuts the pre-mRNA several nucleotides beyond the 3′ end of the AAUAAA site. The fragment beyond the 3′ cut is degraded. Next, an enzyme known as *polyA-polymerase* attaches many adenosine nucleotides. The length of the polyA tail varies among different mRNAs, from a few dozen to several hundred adenine nucleotides.

Besides trimming, capping, and tailing, a pre-mRNA may also be spliced before it is ready to exit the nucleus (Figure 13-14). The mechanism of pre-mRNA splicing will be described next.

Pre-mRNA splicing occurs by the action of a spliceosome

As noted previously, the spliceosome is a large complex that splices pre-mRNA. It is composed of several subunits known as snRNPs (pronounced "snurps"). Each snRNP contains *small nuclear RNA* and a set of *proteins*. During splicing, the subunits of a spliceosome carry out several functions. First, spliceosome subunits bind to an intron sequence and precisely recognize the intron–exon boundaries. In addition, the spliceosome must hold the pre-mRNA in the correct configuration so that splicing can occur. And finally, the spliceosome catalyzes the chemical reactions that cause the introns to be removed and the exons to be covalently linked.

Intron RNA is recognized by the sequence of nucleotides within the intron and at the intron–exon boundaries. The consensus sequences for the splicing of mammalian pre-mRNA are shown in Figure 13-16. These sequences serve as recognition sites for the binding of the spliceosome. The most highly conserved bases are shown in bold.

The molecular mechanism of pre-mRNA splicing is depicted in Figure 13-17. First, the snRNP designated U1 binds to the 5′ end of the intron at its boundary with the 3′ end of the preceding exon, known as the *splice donor* site. The binding

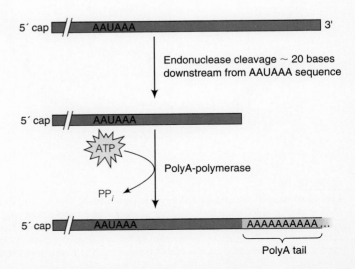

FIGURE **13-15**

Attachment of a polyA tail. First, an endonuclease cuts the RNA 11 to 30 nucleotides after the AAUAAA polyadenylation signal making the RNA shorter at its 3′ end. Adenine-containing nucleotides are then attached, one at a time, to the 3′ end by the enzyme polyA-polymerase.

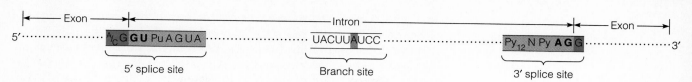

FIGURE 13-16

Consensus sequences for pre-mRNA splicing in higher eukaryotes. Consensus sequences exist at the intron–exon boundaries, and at a branch site that is found within the intron itself. The adenine nucleotide shown in green in this figure corresponds to the boxed adenine nucleotide in Figure 13-17. The nucleotides shown in bold are highly conserved. Designations: A/C = A or C, Pu = purine, Py = pyrimidine, N = any of the four bases.

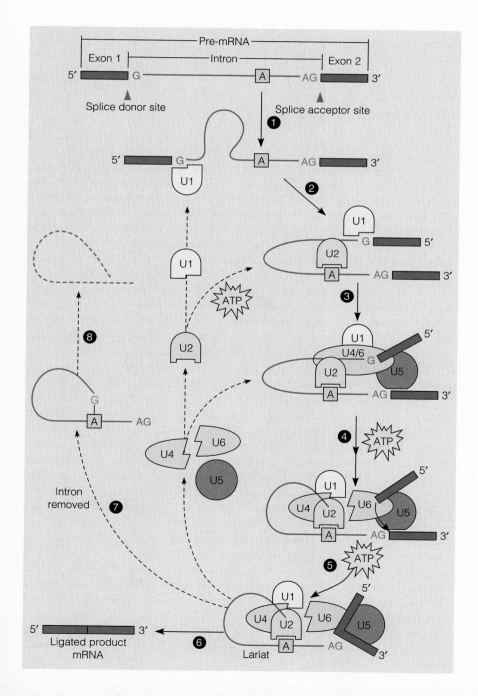

FIGURE 13-17

Splicing of pre-mRNA via a spliceosome. **(1)** First, U1 binds to the 5′ intron–exon boundary, followed by **(2)** the binding of U2 to form the A complex. The formation of the A complex depends on ATP. **(3)** Three more snRNPs (U4, U5, and U6) then assemble to form a complete spliceosome. The splicing reactions are transesterifications. **(4)** In the first reaction, an ester bond in the RNA is broken at the 3′ end of exon 1, and an ester bond forms between the 5′ end of the intron and an adenine nucleotide within the intron. Note that this yields a circular intron structure, known as a lariat. In the second reaction, **(5)** an ester bond at the 5′ end of exon 2 is broken, and an ester bond forms between exon 1 and exon 2. **(6)** This connects the exons. **(7)** The intron (now in a lariat configuration) is released and **(8)** will be degraded within the nucleus.

of U1 snRNP creates a *commitment complex* that is destined to proceed through the splicing cycle. This is followed by the binding of U2 to the branch site of the intron. Although not shown in this figure, U2 also recognizes the *splice acceptor site* (namely, the 5′ end of the next exon); this step in spliceosome assembly is called the *A complex*. The formation of the A complex requires ATP. Three more snRNPs (U4, U5, and U6) then assemble to form a complete spliceosome.

The enzymology of splicing involves two transesterification reactions. In a transesterification, one ester bond is broken and, at the same time, another ester bond is formed. During the first step in splicing, an ester bond in the RNA is broken at the 3′ end of exon 1 (see Figure 13-16). An ester bond forms between the 5′ end of the intron and a second site within the intron. Note that this makes a circular intron structure, known as a **lariat**. During the second step, an ester bond at the 5′ end of exon 2 is broken and an ester bond is formed between exon 1 and exon 2. At this point, the mRNA has been spliced, and it is released from the spliceosome. The intron, now in a lariat configuration, is also released and will be degraded within the nucleus. The snRNPs can then be recycled to splice more introns.

Alternative splicing allows different proteins to be made from the same structural gene

When it was first discovered, the phenomenon of splicing seemed like a rather wasteful process. During transcription, energy is used to synthesize intron sequences. Likewise, energy is also used to remove introns via a large spliceosome complex. This observation troubled many geneticists, because evolution tends to select against wasteful processes. Therefore, instead of simply viewing splicing as a wasteful process, many geneticists expected to find that pre-mRNA splicing has one or more important biological roles. In recent years, one very important biological advantage has become apparent. This is **alternative splicing**, which refers to the fact that a pre-mRNA can be spliced in more than one way.

To understand the biological effects of alternative splicing, remember that the sequence of amino acids within a polypeptide determines the structure and function of a protein. Alternative splicing produces two (or more) polypeptides with minor differences in their amino acid sequences, leading to subtle changes in their functions. In most cases, the alternative versions of the protein will have similar functions, because much of their amino acid sequences are identical to each other. Nevertheless, alternative splicing produces small differences in amino acid sequences that will provide each polypeptide with its own unique characteristics.

The biological advantage of alternative splicing is that two (or more) different polypeptide sequences can be derived from a single gene. This allows an organism to carry fewer genes in its genome. Figure 13-18 considers an example of alternative splicing for a gene that encodes a protein known as α-tropomyosin. This protein functions in the regulation of cell contraction. It is located along the thin filaments found in smooth muscle cells (e.g., in the uterus and small intestine) and striated muscle cells (e.g., in cardiac and skeletal muscle). Alpha-tropomyosin is synthesized in many types of nonmuscle cells as well. Within a multicellular organism, different types of cells must regulate their contractibility in subtly different ways. One way that this may be accomplished is to produce different forms of α-tropomyosin by alternative splicing.

Figure 13-18 describes the intron–exon structure of the rat α-tropomyosin gene and the alternative ways that the pre-mRNA can be spliced. The alternatively spliced versions of α-tropomyosin mRNA produce α-tropomyosin proteins that differ slightly from each other in their structure and function. All of the alternatively spliced versions of this mRNA contain exons 1, 4–6, and 9–10. Presumably, these exons (shown in blue) encode polypeptide segments of the α-tropomyosin protein

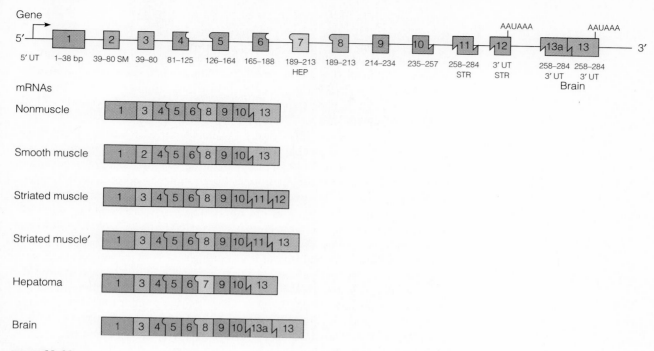

FIGURE **13-18**

Alternative ways that the rat α-tropomyosin pre-mRNA can be spliced. The top part of this figure depicts the gene structure of the rat α-tropomyosin gene. Exons are shown as colored boxes, and introns are illustrated as connecting black lines. The lower part of the figure describes the final mRNA products in different types of cells. The regions designated with a UT are untranslated portions of the mRNA.

GENES ⟶ TRAITS: Alpha-tropomyosin functions in the regulation of cell contraction in muscle and nonmuscle cells. Alternative splicing of the α-tropomyosin gene provides a way to vary contractibility in different types of cells by modifying the function of α-tropomyosin. As shown here, the alternatively spliced versions of α-tropomyosin mRNA produce α-tropomyosin proteins that differ slightly from each other in their structure (i.e., amino acid sequence). Presumably, these alternatively spliced versions of α-tropomyosin vary in function to meet the needs of the cell type in which they are found. For example, the sequence of exons 1–2–4–5–6–8–9–10–13 produces an α-tropomyosin protein that functions suitably in smooth muscle cells. Overall, alternative splicing affects the traits of an organism by allowing a single gene to encode several versions of a protein, each optimally suited to the cell type in which it is made.

that are necessary for its general structure and function. However, the other exons are variable in different α-tropomyosin mRNAs. These alternative exons subtly change the function of α-tropomyosin to meet the needs of the cell type in which they are found. The specific pattern of splicing is usually uniform in any given cell. For example, the α-tropomyosin mRNA found in smooth muscle contains exon 2, which is not present in other cell types. Presumably, exon 2 encodes a segment of the α-tropomyosin protein that slightly alters its function to make it suitable for smooth muscle cells. In another example, the splicing pattern of α-tropomyosin mRNA that contains exons 1, 3–6, 8–10, and 13 occurs in nonmuscle cells but not in smooth muscle, striated muscle, or brain cells.

Interestingly, certain abnormalities are associated with alternative splicing. For example, a *hepatoma* is a disease condition in which a liver cell becomes cancerous. While normal liver cells do not transcribe the α-tropomyosin gene at a significant rate, a laboratory hepatoma cell line has been shown to produce an alternatively spliced version of α-tropomyosin that contains exon 7. We will consider the molecular mechanism for the regulation of alternative splicing in Chapter 16.

The nucleotide sequence of RNA can be modified by RNA editing

The term **RNA editing** refers to a change in the nucleotide sequence of an RNA molecule that involves additions or deletions of particular bases, or a conversion of one type of base to another (e.g., a cytosine to a uracil). In the case of mRNAs, editing can have various effects, such as generating start codons, stop codons, and additional coding sequences.

The phenomenon of RNA editing was first identified in the mid-1980s in trypanosomes, the protozoa that cause sleeping sickness. As with the discovery of RNA splicing, the initial finding of RNA editing was met with great skepticism. Since that time, however, RNA editing has been shown to occur in various organisms and in a variety of ways, although its functional significance remains unclear. Over the next decade or so, it will be interesting to see if RNA editing is a widespread occurrence or if it only occurs in a few selected cases. Table 13-4 describes several examples where RNA editing has been found.

The molecular mechanisms of RNA editing have been the subject of many investigations. In the specific case of trypanosomes, the editing process involves the nucleotides of a **guide RNA** that directs the addition of one or more uracil nucleotides into the mRNA. As shown in Figure 13-19, the guide mRNA has two important characteristics: (1) its 5′ anchor is complementary to the mRNA that is to be edited; and (2) it has uracil nucleotides at its 3′ end. During the editing process, the 5′ anchor binds to the mRNA. The mRNA is cleaved at a defined location, and the 3′ end of the guide RNA becomes covalently attached to the mRNA. The guide RNA is then cleaved near its 3′ end, but in the example shown in Figure 13-19, it leaves a uracil nucleotide behind. Finally, the two pieces of mRNA are rejoined. Note, however, that an extra uracil nucleotide has been inserted into the mRNA. This uracil nucleotide came from the 3′ end of the guide RNA.

A different mechanism occurs for mammalian RNA editing that involves changes of one type of base to another. For example, an mRNA that encodes a protein called *apolipoprotein B* is edited so that a single C is changed to a U. This converts a glutamine codon (CAA) to a stop codon (UAA) and thereby results in a shorter apolipoprotein, termed B48. In this case, therefore, RNA editing produces an apolipoprotein B with an altered structure. Therefore, RNA editing can produce two proteins from the same gene, much like the phenomenon of alternative splicing described earlier in this chapter.

The apolipoprotein B mRNA editing process occurs within the nucleus of the cell at the same time that the pre-mRNA is being spliced. The C to U conversion is catalyzed by a complex known as an **editosome**, which recognizes a small region adjacent to the editing site and then modifies the cytosine base. The term editosome has also been used to describe the protein machinery that may be necessary for editing via guide RNA.

TABLE 13–4 Examples of RNA Editing

Organism	Type of Editing	Found in
Trypanosomes (protozoa)	Primarily additions, but occasionally deletions of uracil nucleotides	Many mitochondrial mRNAs
Flowering plants	C-to-U conversion	Many mitochondrial and chloroplast mRNAs
Slime mold	C additions	Many mitochondrial mRNAs
Mammals	C-to-U conversion	Apolipoprotein B mRNA
	A-to-G conversion	Glutamate receptor mRNA
	Pyrimidine conversions	tRNAs

Adapted from Gray and Covello, *FASEB J. 7*, 64–71.

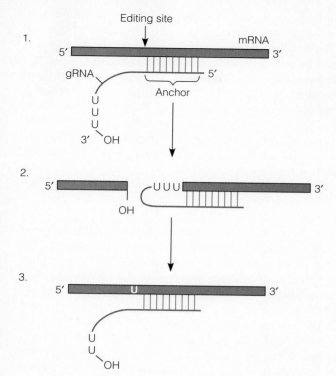

Editing site

1.

5'   mRNA   3'

gRNA

Anchor

5'

U
U
U
3'   OH

2.

5'   3'

OH

U U U

3.

5'   U   3'

U
U
U
OH

FIGURE **13-19**

RNA editing in trypanosomes. The 5' end of the guide RNA (the anchor) is complementary to the mRNA. From step 1 to 2, the mRNA is cleaved at the editing site, and the 3' end of the guide RNA is covalently attached to the mRNA. From step 2 to 3, the guide RNA is cleaved, leaving behind a uracil residue, and the ends of the mRNA are rejoined.

CONCEPTUAL SUMMARY

Transcription is the process of RNA synthesis. Different types of RNA transcripts can be made. These include mRNA that specifies the sequence of amino acids within a polypeptide, and tRNA and rRNA, which are necessary to translate mRNA. In addition, several other small RNA molecules, such as snRNA, snoRNA, 7S RNA, and the RNA component of *RNase P*, have various functions in the cell.

The synthesis of RNA during gene transcription is a three-step process: *initiation*, *elongation*, and *termination*. During the initiation stage, specific sequences within the promoter region are recognized by proteins that bind to DNA. In the case of many bacterial promoters, the −35 and −10 sequences are recognized by the *sigma factor* that is part of the *RNA polymerase* holoenzyme. To begin the *synthesis* of a transcript, the DNA must be converted to an *open complex* by denaturing a small double-stranded region. At this point, the RNA polymerase can slide down the DNA, synthesizing a complementary RNA molecule. At the end of the gene, a termination signal is found. In bacteria, there are two different ways that termination signals can promote transcriptional termination. In ρ-*dependent termination*, a protein known as ρ is responsible for dissociating the RNA transcript and RNA polymerase. In ρ-*independent* or *intrinsic termination*, the formation of a stem-loop within the RNA causes RNA polymerase to stall and eventually fall off the DNA.

Transcription in eukaryotes is similar to, but more complex, than that in prokaryotes. In eukaryotes, there are three RNA polymerases, designated I, II, and III, that transcribe different types of genes. The initiation of transcription of *structural genes* requires several *transcription factors* that recognize sequence elements such as *CAAT* and *GC boxes* and assemble at a *TATA box* in the −25 region. Transcription is initiated by a complicated assembly process. During elongation, the histone proteins must be completely or partially removed so that RNA polymerase can transcribe the template DNA.

Following transcription, a newly made RNA transcript may be modified in various ways. For example, rRNA and tRNA transcripts are processed to smaller forms by endo- and exonucleases. Some genes contain introns that must be removed from the RNA via *splicing*. *Group I* and *Group II* introns are *self-splicing*, which means that they catalyze their own removal. These types of introns are found in the rRNA genes of *Tetrahymena* and other simpler eukaryotes, in organelle DNA, and occasionally in bacterial genes. In contrast, introns are commonly found in eukaryotic *pre-mRNA*, where they are removed by a multimeric complex known as a *spliceosome*. Besides splicing, eukaryotic pre-mRNA is also *trimmed*, *capped* at its 5′ end, and a polyA *tail* may be attached to its 3′ end. It is particularly interesting that many eukaryotic pre-mRNAs can be *alternatively spliced*. The net result of alternative splicing is that a single gene can produce more than one type of polypeptide. Similarly, the sequence of bases within RNA may be modified by *RNA editing*. In Chapter 14, we will examine the next step of gene expression, in which an mRNA transcript is translated into a polypeptide sequence.

EXPERIMENTAL SUMMARY

Experimentally, there are several methods for studying the transcription of genes at the molecular level. Most of these techniques will be described in Chapter 19. For example, the quantity of an RNA transcript can be determined by Northern blotting, in which RNA transcripts are identified through a DNA–RNA hybridization technique. This method can also be used to detect alternative splicing of mRNAs.

Our understanding of the sequences required for transcription, such as promoter sequences, has come from the isolation of mutants in which the promoter sequence was altered. More recently, the technique of site-directed mutagenesis (also described in Chapter 19) has been developed to alter gene sequences. Researchers now can examine how an alteration in a gene sequence affects genetic processes such as transcription.

The binding and assembly of proteins during transcription can be examined in a variety of ways. In Chapter 11, we considered how Arthur Kornberg developed an *in vitro* system to study DNA replication. Similarly, researchers have succeeded in developing *in vitro* systems to study transcription. In this way, they have identified the basal transcriptional apparatus needed for transcription to occur.

In this chapter, we have also considered how hybridization and electron microscopy can be used to detect introns. In this approach, the hybridization of mRNA to a DNA sequence blocks the DNA's ability to re-form double-stranded regions. Using an electron microscope to examine the pattern of single-stranded DNA and double-stranded DNA regions, the hybridization of RNA to DNA allows us to determine if a gene has introns. As mentioned at the end of Experiment 13A, DNA sequencing methods that will be described in Chapter 19 allow an easier and more precise way of detecting introns. A comparison of the DNA sequences of genomic and cDNA clones can determine if a particular gene contains introns.

SOLVED PROBLEMS

1. Describe the important events that occur during the three stages of gene transcription in prokaryotes. What proteins play critical roles in the three stages?

Answer: The three stages are initiation, elongation, and termination:

Initiation: RNA polymerase holoenzyme scans along the DNA until sigma factor recognizes the consensus sequence of a promoter. Sigma factor binds tightly to this sequence, forming a closed complex. The DNA is then denatured to form a bubblelike structure known as the open complex.

Elongation: RNA polymerase core enzyme slides along the DNA, synthesizing RNA as it goes. The α-subunits of RNA polymerase keep the enzyme bound to the DNA, while the β-subunits are responsible for the catalytic synthesis of RNA. During elongation, RNA is made according to the A–U/G–C rule, with nucleotides being added in the 5′ to 3′ direction.

Termination: The RNA polymerase eventually reaches a sequence at the end of the gene that signals the cessation of transcription. In ρ-independent termination, the properties of the termination sequences in the RNA are sufficient to cause termination. In ρ-dependent termination, the ρ protein recognizes a sequence within the RNA, binds there, and travels toward RNA polymerase. When the formation of a stem-loop structure causes RNA polymerase to pause, ρ catches up and knocks it off the DNA.

2. What is the difference between a structural gene and a nonstructural gene?

Answer: Structural genes encode mRNA that is translated into a polypeptide sequence. Nonstructural genes encode RNAs that are never translated. Examples of nonstructural genes include tRNA and rRNA, which function during translation; 7S RNA, which is part of SRP; the RNA of RNase P; snoRNA, which is involved in rRNA trimming; and snRNA, which is part of pre-mRNA spliceosomes. In many cases, the RNA from nonstructural genes becomes part of a complex composed of RNA molecules and protein subunits.

CONCEPTUAL QUESTIONS

1. In prokaryotes, what event marks the end of the initiation stage of transcription?

2. What is the meaning of the term consensus sequence? Give an example. Describe the locations of consensus sequences within bacterial promoters. What are their functions?

3. What is the consensus sequence of the following six DNA molecules?

G G C A T T G A C T
G C C A T T G T C A
C G C A T A G T C A
G G A A A T G G G A
G G C T T T G T C A
G G C A T A G T C A

4. A mutation within a gene sequence changes the start codon to a stop codon. How will this mutation affect the transcription of this gene?

5. What is the subunit composition of bacterial RNA polymerase holoenzyme? What are the functional roles of the different subunits?

6. At the molecular level, describe how sigma factor recognizes bacterial promoters. Be specific about the structure of sigma factor and the type of chemical bonding.

7. Let's suppose that a DNA mutation changes the consensus sequence at the −35 location so that sigma factor is no longer able to bind there. Explain how a mutation would prevent sigma factor from binding to the DNA. Look at Figure 13-3 and describe two specific base substitutions that you think would inhibit the binding of sigma factor. Explain why you think your base substitutions would have this effect.

8. What is the complementarity rule that governs the synthesis of an RNA molecule during transcription? An RNA transcript has the following sequence:

5′–GGCAUGCAUUACGGCAUCACACUAGGGAUC–3′

What is the sequence of the template and coding strands of the DNA that encodes this RNA? On which side (5′ or 3′) of the template strand is the promoter located?

9. Describe the movement of the open complex down the RNA.

10. Describe what happens to the chemical bonding interactions when transcriptional termination occurs. Be specific about the type of chemical bonding.

11. Discuss the differences between ρ-dependent and ρ-independent termination.

12. For each of the following RNA polymerases, if it was missing from a eukaryotic cell, what types of genes would not be transcribed?

  A. RNA polymerase I
  B. RNA polymerase II
  C. RNA polymerase III

13. What sequence elements are found within the core promoter of eukaryotes? Describe their locations and specific functions.

14. For each of the following transcription factors, how would eukaryotic transcriptional initiation be affected if it was missing?

  A. TFIIB
  B. TFIID
  C. TFIIH

15. Discuss how the binding of DNA to histones in a nucleosome structure may affect the process of transcription. Do you think that the nucleosome structure would affect the ability of transcription factors to recognize the promoter sequence in eukaryotic genes? Explain.

16. A eukaryotic structural gene contains two introns and three exons: Exon 1–Intron 1–Exon 2–Intron 2–Exon 3. The splice donor site of exon 2 has been eliminated by a small deletion in the gene. Describe how this gene would be spliced. Indicate which introns and exons will be found in the mRNA after splicing occurs.

17. Describe the processing events that occur during the production of tRNA.

18. Describe the structure and function of a spliceosome. Speculate on why you think the spliceosome subunits contain snRNA. In other words, what do you think is the functional role(s) of snRNA during splicing?

19. What is the unique feature of ribozyme function? Give two examples described in this chapter.

20. What does it mean to say that gene expression is colinear? Give examples of genes that are colinear and genes that are not.

21. What is meant by the term self-splicing? What types of introns are self-splicing?

22. What types of modification occur to pre-mRNA?

23. What is a transesterification reaction? Draw the chemical bonds that are broken and re-formed.

24. What is alternative splicing? What is its biological significance?

EXPERIMENTAL QUESTIONS

1. A research group has sequenced the cDNA and genomic DNA from a particular gene. As will be described in Chapter 19, cDNA is derived from mRNA, so that it does not contain introns. Here are DNA sequences derived from the cDNA and genomic DNA:

cDNA:

5′–ATTGCATCCAGCGTATACTATCTCGGGCCCAATTAA
TGCCAGCGGCCAGACTATCACCCAACTAGGTTACCTAC
TAGTATATCCCATATACTAGCATATATTTTACCCATAA
TTTGTGTGTGGGTATACAGTATAATCATATA–3′

genomic DNA:

5′–ATTGCATCCAGCGTATACTATCTCGGGCCCAATTAA
TGCCAGCGGCCAGACTATCACCCAACTAGGTAAGTA
CCCCCCAGGTTTACACAGTCATACCATACATACAAAAA
TCGCAGTTACCCCAAAAAAACCTAGTACTTATCCATAC
CCCACATTTTACTATTAATCCTCTTTCTTTCTAGGTTACCT
ACTAGTATATCCCATATACTAGCATATATTTTACCCATA
ATTTGTGTGTGGGTATACAGTATAATCATATA–3′

Write out the sequence of the intron. Does the intron contain the normal consensus splice site sequences? Underline these splice site sequences, and indicate whether or not they fit the consensus sequence.

2. What is an R loop?

3. In an R loop experiment, to which strand of DNA does the mRNA bind, the coding strand or template strand?

4. If a gene contains three introns, draw what it would look like in an R loop experiment.

QUESTIONS FOR STUDENT DISCUSSION/COLLABORATION

1. Based on your knowledge of introns and pre-mRNA splicing, discuss whether or not you think alternative splicing fully explains the existence of introns. Can you think of other possible reasons to explain the existence of introns?

2. Discuss the types of RNA transcripts and the functional roles that they play. Why do you think that some RNAs form complexes with protein subunits?

CHAPTER 14

TRANSLATION OF mRNA

The synthesis of cellular proteins occurs via the translation of the codons within mRNA into a sequence of amino acids within a polypeptide. The general steps that occur in this process have already been outlined in Chapter 12. This chapter reexamines translation with an eye toward the molecular level. During the past few decades, the concerted efforts of geneticists, cell biologists, and biochemists have profoundly advanced our understanding of translation. Even so, many questions remain unanswered, and this topic continues to be an exciting area of research investigation.

In this chapter, we will discuss the current state of knowledge regarding the molecular features of mRNA translation. A variety of cellular components play important roles in translation. These include many different proteins, RNAs, and small molecules. In the first section of this chapter, we will begin by examining the structure and function of tRNA molecules, which act as the translators of the genetic information within mRNA. Later, we will consider the composition of ribosomes, and then examine the stages of translation in prokaryotic and eukaryotic cells.

tRNA STRUCTURE AND FUNCTION

RIBOSOME STRUCTURE AND ASSEMBLY

STAGES OF TRANSLATION

tRNA STRUCTURE AND FUNCTION

Biochemical studies of protein synthesis and tRNA molecules began in the 1950s. As work progressed on protein translation, it became evident that RNA molecules are involved in the incorporation of amino acids into growing polypeptides. Francis Crick and Mahlon Hoagland were the first to propose that the position of an amino acid within a polypeptide chain is determined by the bonding between the mRNA and a tRNA carrying a specific amino acid. This idea, known as the **adaptor hypothesis**, suggested that the tRNA plays a direct role in the recognition of the codons within the mRNA. In particular, this hypothesis proposed that a tRNA has two functions: recognizing a three-base codon sequence in mRNA, and carrying an amino acid that is specific for that codon. In this section, we will begin by examining the general function of tRNA molecules and describe an experiment that was critical in supporting the adaptor hypothesis. We will then explore some of the important structural features that underlie tRNA function.

The function of a tRNA depends on the specificity between the amino acid it carries and its anticodon

tRNA molecules recognize the codons within an mRNA and carry the correct amino acids to the site of polypeptide synthesis. During mRNA–tRNA recognition, the anticodon in a tRNA molecule binds to a codon in mRNA due to their complementary sequences (Figure 14-1). Importantly, the anticodon in the tRNA corresponds to the amino acid that it carries. For example, if the anticodon in the tRNA is 3′–AAG–5′, it is complementary to a 5′–UUC–3′ codon. According to the genetic code described in Chapter 12, the UUC codon specifies phenylalanine. Therefore, the tRNA with a 3′-AAG-5′ anticodon must carry a phenylalanine. As another example, if the tRNA has a 3′–GGC–5′ anticodon, it is complementary to a 5′–CCG–3′ codon, which specifies proline. This tRNA must carry proline.

As was discussed in Chapter 12, the genetic code has 3 stop codons and 61 different codons that specify the 20 amino acids. Therefore, to synthesize proteins, a cell must produce many different tRNA molecules that have specific anticodon sequences. To do so, the chromosomal DNA contains many distinct tRNA genes that encode tRNA molecules with different sequences. According to the adaptor hypothesis, the anticodon in a tRNA always specifies the type of amino acid that it carries. Due to this specificity, tRNA molecules are named according to the type of amino acid that they bear. For example, a tRNA that carries an phenylalanine is described as tRNA^{phe}, while a tRNA that carries proline is tRNA^{pro}.

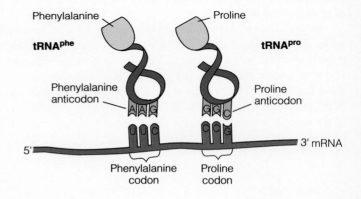

FIGURE **14-1**

Recognition between tRNAs and mRNA. The anticodon in the tRNA binds to a complementary sequence in the mRNA. At its other end, the tRNA carries the aminoacid that corresponds to the codon in the mRNA via the genetic code.

Chapeville and colleagues tested the adaptor hypothesis of tRNA function

In 1962, François Chapeville, in collaboration with several colleagues at the Rockefeller Institute, Johns Hopkins University, and Purdue University, conducted an experiment that was aimed at testing the adaptor hypothesis. The technical strategy was similar to that of the Nirenberg and Ochoa experiments, which deciphered the genetic code (see Experiment 12B). In this approach, a translation system was used in which polypeptides were synthesized *in vitro* by isolating cell extracts that contained the components necessary for translation. These components include ribosomes, tRNAs, and other translation factors. An *in vitro* translation system can be used to investigate the role of specific factors by adding a particular mRNA template and varying individual components required for translation.

According to the adaptor hypothesis, the amino acid attached to a tRNA is not directly involved in codon recognition. Chapeville reasoned that if this were true, the alteration of an amino acid already attached to a tRNA should cause that altered amino acid to be incorporated into the polypeptide instead of the normal amino acid. For example, consider a tRNA^{cys} that carries the amino acid cysteine. If the attached cysteine is changed to an alanine, then this tRNA^{cys} will insert an alanine into a polypeptide where it would normally put a cysteine. Fortunately, Chapeville could carry out this strategy, because he had a reagent, known as Raney nickel, that can chemically convert cysteine to alanine.

An elegant aspect of the experimental design was the choice of the mRNA template. Chapeville and his colleagues synthesized an mRNA template that only contained U and G. Therefore, this template could only contain the following codons (refer back to the genetic code in Table 12-2):

UUU = phenylalanine	GUU = valine
UUG = leucine	GUG = valine
UGU = cysteine	GGU = glycine
UGG = tryptophan	GGG = glycine

Among the eight possible codons, there is one cysteine codon, but there are no alanine codons that can be formed from a polyUG template.

THE HYPOTHESIS

Codon recognition is dictated only by the tRNA; the chemical structure of the amino acid attached to the tRNA does not play a role.

TESTING THE HYPOTHESIS

Starting material: A bacterial cell extract containing the components required for translation, such as ribosomes, tRNAs, etc.

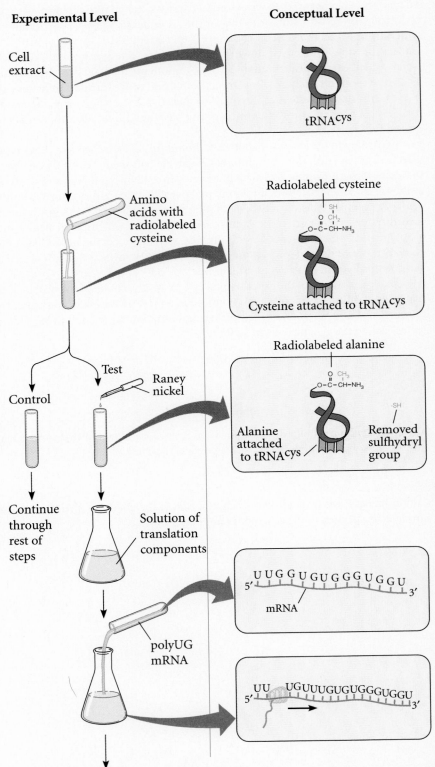

Experimental Level

Conceptual Level

1. Isolate an extract containing tRNAs. Note: This drawing only emphasizes tRNA^{cys}, even though the extract contains all types of tRNAs. When the extract is isolated, a substantial proportion of the tRNAs do not have an attached amino acid. The extract also contains enzymes that attach amino acids to tRNAs. (These enzymes will be described later in the chapter.)

Cell extract

tRNA^{cys}

2. Add amino acids, including radiolabeled cysteine, to the extract. An enzyme within the extract will specifically attach the radiolabeled cysteine to tRNA^{cys}. The other tRNAs will have unlabeled amino acids attached to them.

Amino acids with radiolabeled cysteine

Radiolabeled cysteine

Cysteine attached to tRNA^{cys}

3. In one tube, treat the tRNAs with Raney nickel. This removes the —SH group from cysteine, converting it to alanine. In the control tube, do not add Raney nickel.

Control   Test   Raney nickel

Radiolabeled alanine

Alanine attached to tRNA^{cys} Removed sulfhydryl group

4. Add the tRNAs to the other components that are necessary for translation: ribosomes, translation factors, and so forth.

Continue through rest of steps

Solution of translation components

UUGGUGUGGGUGGU
5'   3'

mRNA

5. Add polyUG mRNA as a template. As noted, polyUG contains one cysteine codon but no alanine codons.

polyUG mRNA

UU UGUUUGUGUGGGUGGU
5'   3'

6. Allow translation to proceed.

7. Isolate the newly made polypeptides by precipitating them with trichloroacetic acid and then isolating the precipitated polypeptides on a filter.

8. Hydrolyze the polypeptides to their individual amino acids by treatment with a solution containing concentrated hydrochloric acid.

9. Run the sample over a column that separates cysteine and alanine. (See the appendix for a description of column chromatography.) Separate into fractions. Note: Cysteine runs through the column more quickly and comes out in fraction 3. Alanine comes out later, in fraction 7.

10. Determine the amount of radioactivity in the fractions that contain alanine and cysteine.

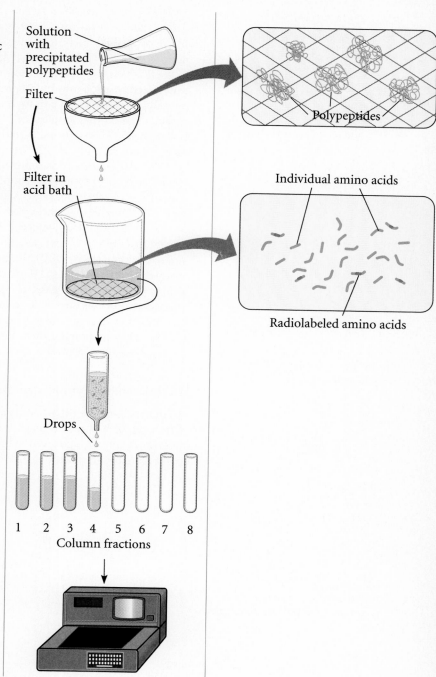

THE DATA

	Amount of Radiolabeled Amino Acids Incorporated into Polypeptide (cpm)*		
Conditions	Cysteine	Alanine	Total
Control, untreated tRNA	2835	83	2918
Raney nickel-treated tRNA	990	2020	3010

*Cpm is the counts per minute of radioactivity in the sample.
Adapted from Chapeville *et al* (1962) *PNAS 48*, 1086–1092.

INTERPRETING THE DATA

In the control sample, nearly all the radioactivity was found in the fraction containing cysteine. This was expected, since the only radiolabeled amino acid added to the tRNAs was cysteine. The low radioactivity (83 counts per minute) in the alanine fraction probably represents contamination of this fraction by a small amount of cysteine. By comparison, when the tRNAs were treated with Raney nickel, a substantial amount of radiolabeled alanine became incorporated into polypeptides. This occurred even though the mRNA template did not contain any alanine codons.

These results are consistent with the explanation that a tRNA^{cys} carrying alanine instead of cysteine incorporated alanine into the synthesized polypeptide. Thus, these observations indicate that the codons in mRNA are identified directly by the tRNA, with the attached amino acid playing no role in codon recognition.

Note that the Raney nickel–treated sample still had 990 cpm of cysteine incorporated into polypeptides. This is about one-third of the total amount of radioactivity (namely, 990/3010). In other experiments conducted in this study, the researchers showed that the Raney nickel did not react with about one-third of the tRNA^{cys}. Therefore, this proportion of the Raney nickel–treated tRNA^{cys} would still carry cysteine. This observation was consistent with the data shown here. Overall, the results of this experiment supported the adaptor hypothesis, indicating that tRNAs act as adaptors to carry the correct amino acid to the ribosome based on their anticodon sequence.

tRNAs share common structural features

To understand how tRNAs act as carriers of the correct amino acids during translation, researchers have examined the structural characteristics of these molecules in great detail. Though a cell makes many different tRNAs, all tRNAs share some common structural features. As originally proposed by Robert Holley at Cornell,

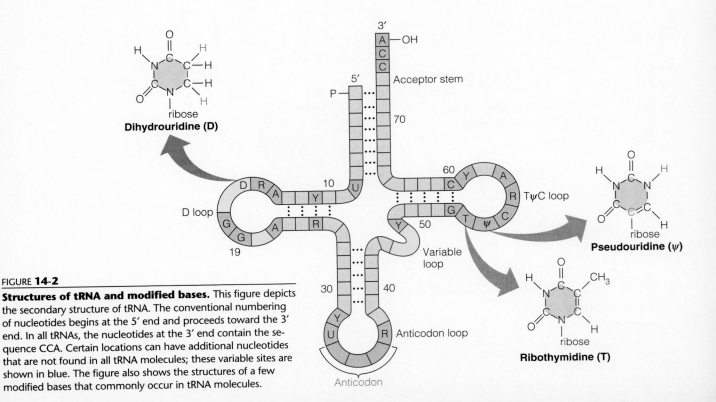

FIGURE **14-2**

Structures of tRNA and modified bases. This figure depicts the secondary structure of tRNA. The conventional numbering of nucleotides begins at the 5′ end and proceeds toward the 3′ end. In all tRNAs, the nucleotides at the 3′ end contain the sequence CCA. Certain locations can have additional nucleotides that are not found in all tRNA molecules; these variable sites are shown in blue. The figure also shows the structures of a few modified bases that commonly occur in tRNA molecules.

the secondary structure of tRNAs exhibits a cloverleaf pattern. There are three stem-loop structures, a variable region, an acceptor stem, and a 3′ single-stranded region (Figure 14-2). A conventional numbering system for the nucleotides within a tRNA molecule begins at the 5′ end and proceeds toward the 3′ end. Among different types of tRNA molecules, there are three variable regions that can differ in the number of nucleotides they contain. The anticodon is located in the second stem-loop region.

The actual three-dimensional or tertiary structure of tRNA molecules involves additional folding of the secondary structure. In the tertiary structure of tRNA, the stem-loop regions are folded into a much more compact molecule (see Figure 9-22). The ability of RNA molecules to form stem-loop structures and the tertiary folding of tRNA molecules was also described in Chapter 9.

Interestingly, in addition to the normal A, U, G, and C nucleotides, tRNA molecules commonly contain modified nucleotides within their primary structures. For example, Figure 14-2 illustrates a tRNA that contains several modified bases. Among many different species, researchers have found more than 60 different nucleotide modifications that can occur in tRNA molecules. The significance of modified bases in codon recognition will be described later in this chapter.

Aminoacyl-tRNA synthetases charge tRNAs by attaching an appropriate amino acid

To function correctly, each type of tRNA must have the appropriate amino acid attached to it. The enzymes that catalyze the attachment of amino acids to tRNA molecules are known as **aminoacyl-tRNA synthetases**. In most species, there are 20 different aminoacyl-tRNA synthetase enzymes, one for each of the 20 distinct amino acids. Each aminoacyl-tRNA synthetase is named for the specific amino acid it attaches to tRNA. For example, alanyl-tRNA synthetase recognizes a tRNA with an alanine anticodon (i.e., tRNA^{ala}) and attaches an alanine to it.

Aminoacyl-tRNA synthetases actually catalyze a two-step chemical reaction involving three different molecules: an amino acid, a tRNA molecule, and ATP. As shown in Figure 14-3, the first step of the reaction involves the activation of an amino acid by the covalent attachment of an AMP molecule. Pyrophosphate is released. During the second step, the activated amino acid is attached to the ribose located at the 3′ end of the tRNA molecule in the acceptor stem; AMP is released. At this stage, the tRNA is called a **charged tRNA**. In a charged tRNA molecule, the amino acid is attached to the 3′ end of the tRNA by an ester bond.

The ability of the aminoacyl-tRNA synthetases to recognize tRNAs has sometimes been called the second genetic code. This recognition process is necessary to maintain the fidelity of genetic information. The frequency of error for aminoacyl-tRNA synthetases is less than 10^{-5}. In other words, the wrong amino acid will be attached to a tRNA less than once in 100,000 times! As you might expect, the anticodon region of the tRNA is usually important for recognition by the correct aminoacyl-tRNA synthetase. In studies of *Escherichia coli* synthetases, 17 of the 20 types of aminoacyl-tRNA synthetases recognize the anticodon region of the tRNA. However, other regions of the tRNA also appear to be important recognition sites. These include the acceptor stem and bases in the stem-loop regions (see Figure 14-4).

As mentioned earlier in this chapter, tRNA molecules frequently contain bases within their structure that have been chemically modified. These modified bases can have important effects on tRNA function. For example, modified bases within tRNA molecules affect the rate of translation and the recognition of tRNAs by aminoacyl-tRNA synthetases. Positions 34 and 37 contain the largest variety of modified nucleotides; position 34 is the first base in the anticodon that matches the third base in the codon of mRNA. As discussed next, a modified base at position 34 can have important effects on codon–anticodon recognition.

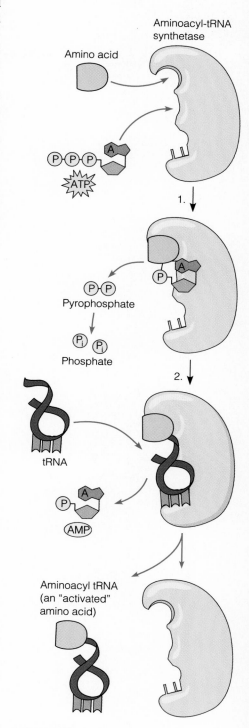

FIGURE **14-3**

Catalytic function of aminoacyl-tRNA synthetase. Aminoacyl-tRNA synthetase contains binding sites for ATP, a specific amino acid, and a particular tRNA. In the first step, the enzyme catalyzes the covalent attachment of AMP to an amino acid, yielding an activated amino acid. In the second step, the activated amino acid is attached to a tRNA.

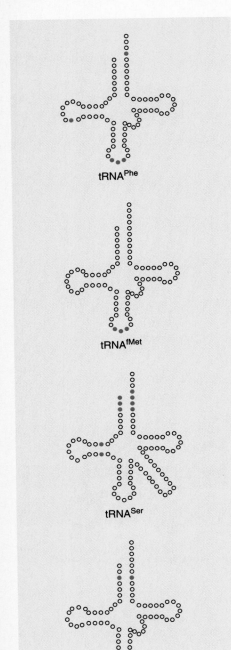

tRNA^Phe

tRNA^fMet

tRNA^Ser

tRNA^Ala

FIGURE **14-4**

Important recognition sites in *E. coli* tRNA molecules. Different aminoacyl-tRNA synthetases found in *E. coli* recognize different sites in tRNAs. The bases critical for recognition are depicted in red.

Mismatches that follow the wobble rule can occur at the third position in codon–anticodon pairing

After considering the structure and function of tRNA molecules, it is interesting to reexamine some of the subtle features of the genetic code. As was discussed in Chapter 12, the genetic code is degenerate. This means that more than one codon can specify the same amino acid. With the exception of serine, arginine, and leucine, this degeneracy always occurs at the third position in the codon. For example, valine is specified by GUU, GUC, GUA, and GUG. In all four cases, the first two bases are G and U. The third base, however, can be U, C, A, or G. To explain this pattern of degeneracy, Francis Crick proposed that it is due to "wobble" at the third position in the codon–anticodon recognition process. According to the **wobble hypothesis**, the first two positions pair strictly according to the A–U/G–C rule. However, the third position can tolerate certain types of mismatches (Figure 14-5). This proposal suggests that the base at the third position can actually move a bit so that hydrogen bonding can occur between the codon and anticodon.

Because of the wobble rules, a single type of tRNA can recognize more than one codon. For example, a tRNA with an anticodon sequence of 3′–AAG–5′ can recognize a 5′–UUC–3′ and a 5′–UUU–3′ codon. The 5′–UUC–3′ codon is a perfect match with this tRNA. The 5′–UUU–3′ codon is mismatched according to the

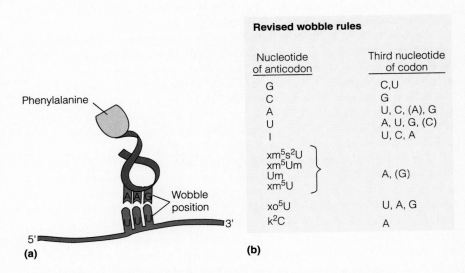

Revised wobble rules

Nucleotide of anticodon	Third nucleotide of codon
G	C, U
C	G
A	U, C, (A), G
U	A, U, G, (C)
I	U, C, A
xm⁵s²U xm⁵Um Um xm⁵U	A, (G)
xo⁵U	U, A, G
k²C	A

(a)   (b)

FIGURE **14-5**

Wobble position and base-pairing rules. (a) The wobble position occurs between the first base (meaning the first base in the 5′ to 3′ direction) in the anticodon and the third base in the mRNA codon. **(b)** The revised wobble rules are slightly different from those originally proposed by Crick. The revised rules shown here are from Yokoyama, S., and Nishimura, S. (1995) *tRNA Structure, Biosynthesis, and Function*, Soll, D., and RajBhandary, U. L., eds., p. 209, ASM Press, Washington, D.C. The standard bases found in RNA are G, C, A, and U. In addition, the structures of bases in tRNAs may be modified. Some modified bases that may occur in the wobble position in tRNA are I = inosine; xm⁵s²U = 5-methyl-2-thiouridine; xm⁵Um = 5-methyl–2′–O-methyluridine; Um = 2′–O-methyluridine; xm⁵U = 5-methyluridine; xo⁵U = 5-hydroxyuridine; k²C = lysidine (a cytosine derivative). The mRNA bases in parentheses are recognized very poorly by the tRNA.

standard RNA–RNA hybridization rules (namely, G in the anticodon is mismatched to U in the codon), but the two can fit according to the wobble rules described in Figure 14-5. Likewise, the modification of the wobble base to an inosine can allow a tRNA to recognize three different codons. At the cellular level, the ability of a single tRNA to recognize more than one codon makes it unnecessary for a cell to make 61 different tRNA molecules corresponding to the 61 different possible codons.

RIBOSOME STRUCTURE AND ASSEMBLY

In the previous section, we examined how the structure and function of tRNA molecules are important in translation. According to the adaptor hypothesis, tRNAs bind to mRNA due to complementarity between the anticodons and codons. Concurrently, the tRNA molecules have the correct amino acid attached to their 3′ ends.

To synthesize a polypeptide, additional events must occur. In particular, the bond between the 3′ end of the tRNA and the amino acid must be broken, and a peptide bond must be formed between the adjacent amino acids. To facilitate these events, translation occurs on the surface of a large macromolecular complex known as the ribosome. The ribosome can be thought of as the macromolecular arena where translation takes place.

In this section, we will begin by examining the biochemical compositions of ribosomes in prokaryotic and eukaryotic cells. We will then examine the key functional sites on ribosomes for the translation process.

Prokaryotic and eukaryotic ribosomes are assembled from rRNA and proteins

Prokaryotic cells have one type of ribosome, which is found within the bacterial cytoplasm. By comparison, eukaryotic cells are compartmentalized into cellular organelles that are bounded by membranes. Eukaryotic cells contain biochemically distinct ribosomes in different cellular locations. The most abundant type of ribosome functions in the cytosol, which is the region of the eukaryotic cell that is inside the plasma membrane but outside of the organelles. Besides the cytosolic ribosomes, all eukaryotic cells have ribosomes within an organelle known as the mitochondrion. In addition, plant cells have ribosomes in their chloroplasts. The compositions of mitochondrial and chloroplast ribosomes are quite different from that of the cytosolic ribosomes. Unless otherwise noted, the term "eukaryotic ribosome" refers to the ribosomes in the cytosol, not to those found within organelles. Likewise, the description of eukaryotic translation refers to translation via cytosolic ribosomes.

Each ribosome is composed of structures called the large and small subunits (see Figure 14-6). This term is perhaps misleading, since each ribosomal subunit itself is formed from the assembly of many different proteins and rRNA molecules. In prokaryotic ribosomes, the 30S subunit is formed from the assembly of 21 different ribosomal proteins and one rRNA molecule; the 50S subunit contains 34 different proteins and two different rRNA molecules (see Figure 14-6a). The designations 30S and 50S refer to the rate that these subunits sediment when subjected to a centrifugal force. This rate is described as a sedimentation coefficient in Svedberg units (S), in honor of T. Svedberg, who invented the ultracentrifuge in the 1920s (see the appendix). Together, the 30S and 50S subunits form a 70S ribosome. (Note: Svedberg units don't add up!) In prokaryotes, the ribosomal proteins and rRNA molecules are synthesized in the cytoplasm, and the ribosomal subunits are assembled there.

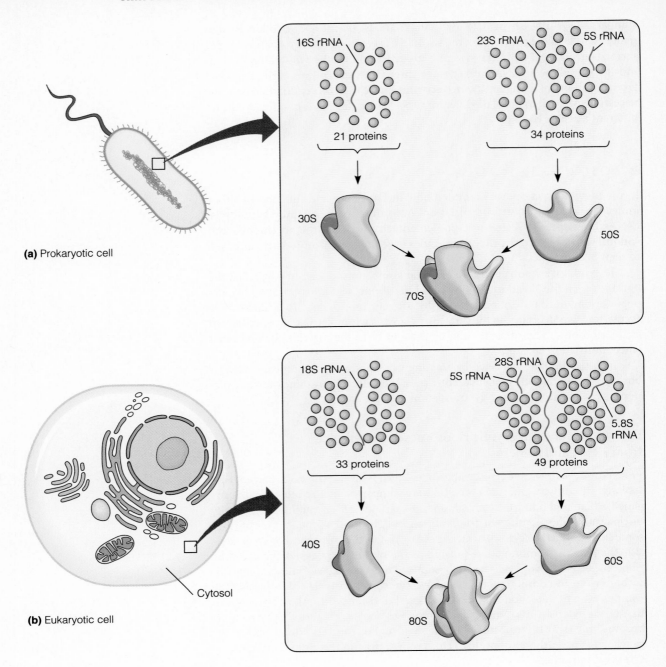

FIGURE 14-6

Composition of prokaryotic and eukaryotic ribosomes. (a) Prokaryotic ribosomes contain two subunits: 30S and 50S. The 30S subunit contains 21 different proteins and 16S rRNA; the 50S subunit contains 34 proteins and 5S and 23S rRNAs. **(b)** Eukaryotic ribosomes found in the cytosol contain two subunits: 40S and 60S. The 40S subunit contains 33 proteins and 18S rRNA; the 60S subunit contains 49 proteins and 5S, 5.8S, and 28S rRNAs. The 40S and 60S subunits assemble in the nucleolus and then are exported into the cytosol.

By comparison, the synthesis of eukaryotic rRNA occurs within the nucleus, and the ribosomal proteins are made in the cytosol. The 40S subunit is composed of 33 proteins and an 18S rRNA; the 60S subunit is made up of 49 proteins and 5S, 5.8S, and 28S rRNAs (Figure 14-6b). The assembly of the rRNAs and ribosomal proteins to make the 40S and 60S subunits occurs within a specialized region of the nucleus known as the **nucleolus**. The 40S and 60S are then exported into the cytosol, where they associate to form an 80S ribosome during translation.

Ribosomes were first visualized by electron microscopy in the mid-1950s, and their protein synthesizing capability was demonstrated a few years later. A micrograph of prokaryotic ribosomes is shown in Figure 14-7a, and their general structure is described in Figure 14-7b. The ribosome is a fairly complicated macromolecular structure. Certain regions of the ribosomal subunits have names that describe their geometrical shape. For example, a region on the 50S subunit is called the central protuberance, since it protrudes from the center of this subunit. The large subunit also contains an indentation called the valley. Similarly, the 30S subunit has a cleft. In the 70S ribosome, the cleft and valley provide an opening through the ribosome. This opening is occupied by mRNA during the translation process.

Components of ribosomal subunits form functional sites for translation

To understand the structure and function of the ribosome at the molecular level, researchers must determine the locations and functional roles of the individual ribosomal proteins and rRNAs. In recent years, many advances have been made toward a molecular understanding of ribosomes. Figure 14-8a is a more refined structural model of the prokaryotic ribosome. This model shows the locations of many ribosomal proteins and rRNA molecules.

During prokaryotic translation, the mRNA lies in a space between the 30S and 50S subunits. As the polypeptide is being synthesized, it exits through a hole within the 50S subunit. As depicted in Figure 14-8b, ribosomes also contain discrete sites where tRNA binds and the polypeptide is synthesized. In 1963, James Watson was the first to propose a two-site model for tRNA binding to the ribosome. As shown here, these sites are known as the **peptidyl site (P site)** and **aminoacyl site (A site)**. Twenty years later, Knud Nierhaus and Hans-Jorg Rheinberger at the Max Planck Institute for Molecular Genetics in Berlin proposed a three-site model. This model incorporated the observation that uncharged tRNA molecules can bind to a site on the ribosome that is distinct from the P and A sites. This third site is now known as the **exit site (E site)**. In the next section, we will examine the roles of the A, P, and E sites during the three stages of translation.

FIGURE **14-7**

Ribosomal structure. (a) Electron micrograph of ribosomes purified from *E. coli* bacterial cells. **(b)** A model depicting the three-dimensional structures of ribosomal subunits and the 70S ribosome. The mRNA is threaded through the space between the 30S and 50S subunits. This space also contains binding sites for tRNA molecules. Note: the previous drawing in Figure 14-6a gives a different view of the ribosome structure. If the drawings in Figure 14-6a were rotated 90° to the left and 90° toward the reader, they would match the drawings shown in Figure 14-7.

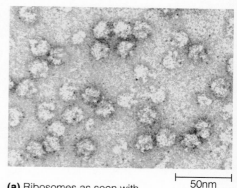

(a) Ribosomes as seen with electron microscopy

50nm

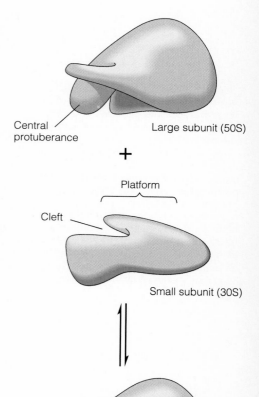

Central protuberance

Large subunit (50S)

+

Platform

Cleft

Small subunit (30S)

70S ribosome

(b) Prokaryotic ribosome and its subunits

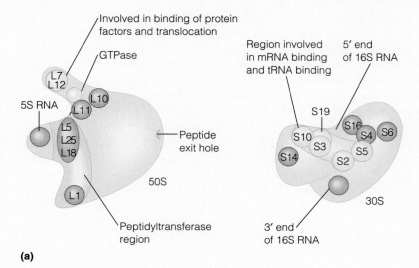

(a)

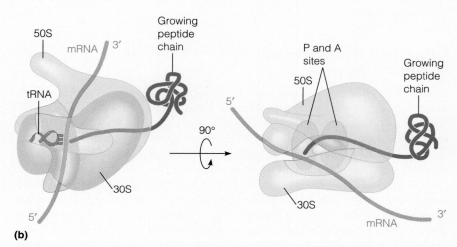

(b)

FIGURE **14-8**

A model depicting the location of several ribosomal proteins and rRNAs within the small and large ribosomal subunits. The yellow regions depict sites where mRNA and tRNA molecules are known to interact with the ribosome. The purple regions show where the synthesis of the polypeptide chain occurs. **(a)** S and L designate proteins of the small and large subunits, respectively. For example, S10 is protein #10 of the small subunit. The locations of certain rRNAs are also indicated. **(b)** The interaction of mRNA and tRNA molecules with the ribosomal subunits during translation. The location of the E site is not shown.

STAGES OF TRANSLATION

Like transcription, the process of translation can be viewed as occurring in three stages: **initiation**, **elongation**, and **termination**. Figure 14-9 presents an overview of these stages. During initiation, the mRNA, initiator tRNA (i.e., the tRNA bound to the start codon), and ribosomal subunits assemble to form a complex. After the initiation complex is formed, the ribosome slides along the mRNA in the 5′ to 3′ direction, moving over the codons. As the ribosome moves, tRNA molecules sequentially bind to the mRNA and ribosome, bringing with them the appropriate amino acids. In this way, the amino acids are linked in the order dictated by the codon sequence in the mRNA. Finally, a stop codon is reached, signaling the end of translation. At this point, disassembly occurs and the newly made polypeptide is released. In this section, we will examine the components that are required for the translation process and consider their functional roles during the three stages of translation.

Many components are required for translation to occur

Protein synthesis is a vital cellular process. Much cellular energy and many proteins are devoted to translation. As already described in this chapter, there are three

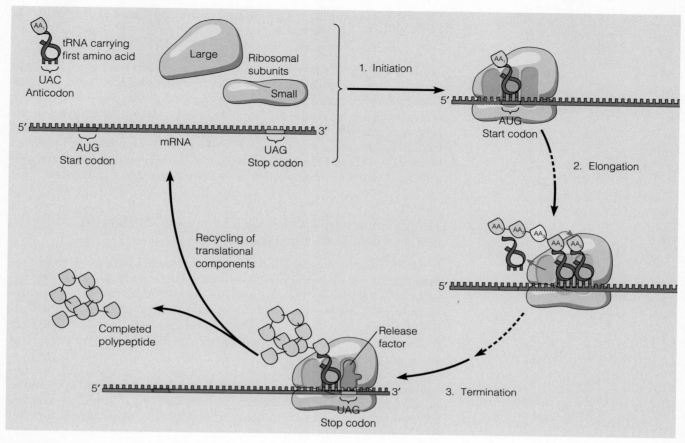

FIGURE **14-9**

Overview of the stages of translation. **(1)** The initiation stage involves the assembly of the ribosomal subunits, mRNA, and the tRNA carrying the first amino acid. **(2)** During elongation, the ribosome slides along the mRNA and synthesizes a polypeptide chain. **(3)** Translation ends when a stop codon is reached and the polypeptide is released from the ribosome. Note: in this figure, and succeeding figures in this chapter, the ribosomes are drawn schematically to emphasize different aspects of the translation process. The actual structures of ribosomes have been described in Figures 14-6 through 14-8.

GENES ⟶ TRAITS: The ability of genes to produce an organism's traits relies on the molecular process of gene expression. During translation, the codon sequence within mRNA (which is derived from a gene sequence during transcription) is translated into a polypeptide sequence. After polypeptides are made within a living cell, they function as proteins to govern an organism's traits. For example, once the globin polypeptide is made, it functions within the hemoglobin protein and provides red blood cells with the ability to carry oxygen, a vital trait for survival. Translation allows functional proteins to be made within living cells.

components that are centrally important in the synthesis of polypeptides. An overview of their functions is described here:

1. **mRNA** is necessary to specify the amino acid sequence within a polypeptide. The sequence of codons within mRNA specifies the sequence of amino acids within a polypeptide. In addition, the structure of the 5′ end of the mRNA helps initiate translation at the correct start codon. A stop codon is also required to signal the end of translation.

2. **tRNA** molecules are necessary to decipher the genetic information coded within the mRNA. Due to the action of the aminoacyl-tRNA synthetases, each tRNA carries with it the amino acid that corresponds to its anticodon. During translation, the anticodon in the tRNA binds to the codon in the mRNA.

3. **Ribosomes** are large macromolecular structures that act as the catalytic site for polypeptide synthesis. The ribosome allows the mRNA and tRNAs to be correctly positioned as the polypeptide is made.

Besides these components, many small molecules and proteins participate in the three stages of translation. For example, molecules such as ATP and GTP provide important sources of energy. These molecules are necessary because the molecular interactions between mRNA and tRNAs, and also the synthesis of the polypeptide chain, utilize a significant amount of energy. In addition, many cellular proteins function at specific stages of translation. Their functions are described in Table 14-1. The individual roles that many of these molecules and proteins play will also be described as we progress through the three stages of translation.

The initiation stage involves the binding of mRNA and the initiator tRNA to the ribosomal subunits

Initiation, the first stage of translation, involves the binding of mRNA and the first tRNA to the ribosomal subunits. A specific tRNA functions as the initiator tRNA, which recognizes the start codon in the mRNA. In bacteria, the initiator tRNA carries a methionine that has been covalently modified to N-formylmethionine; this tRNA is called tRNA^{fmet}. In this modification, a formyl group (—CHO) is attached to the nitrogen atom in methionine after the methionine has been attached to the tRNA.

Figure 14-10 outlines the steps that occur during the initiation stage of translation in prokaryotes. During translational initiation, the mRNA, tRNA^{fmet}, and ribosomal subunits associate with each other to form an initiation complex. The formation of this complex requires the participation of three initiation factors: IF1, IF2, and IF3. IF1 and IF3 are necessary to dissociate the 50S and 30S ribosomal subunits so that the 30S subunit can reinitiate with another mRNA molecule.

TABLE 14–1 Comparison of Translational Protein Factors in Prokaryotes and Eukaryotes

Prokaryotic Factor	Eukaryotic Factor	Function
Initiation Factors		
IF1		
IF2	eIF2	Involved in forming initiation
IF3	eIF3, eIF4C	complex
	CBPI	Involved in cap binding
	eIF4A, eIF4B, eIF4F	Involved in search for first AUG
	eIF5	Helps dissociate eIF2, eIF3, eIF4C
	eIF6	Helps dissociate 60S subunit from inactive ribosomes
Elongation Factors		
EF-Tu	eEFIα	Delivery of aminoacyl tRNA to ribosomes
EF-Ts	eEF1βγ	Aids in recycling factor above
EF-G	eEF2	Translocation factor
Release Factors		
RF1	eRF	Release of complete
RF2		polypeptide chain
RF3		

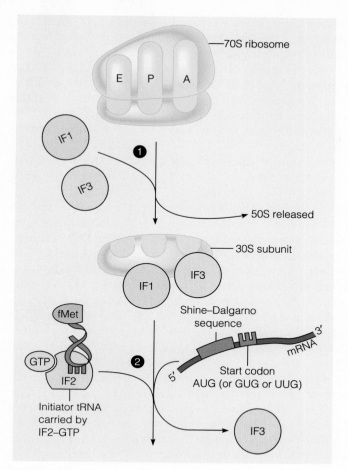

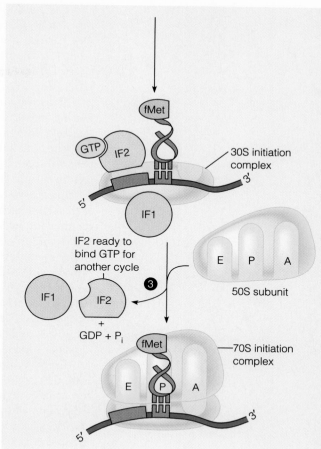

FIGURE **14-10**

Translation initiation in bacteria. (1) IF1 and IF3 associate with the 30S subunit.
(2) The initiator tRNA/IF2, mRNA, and the 30S subunit then associate with each other.
IF3 is released. Note that the codon in the mRNA and the anticodon in the tRNA recognize
each other. **(3)** IF2 and IF1 are released, and the 50S subunit associates to form a 70S
initiation complex. This marks the end of the initiation stage.

The binding of the mRNA to the 30S ribosomal subunit relies on the complementarity between a short region of the mRNA and rRNA within the 30S subunit.
A sequence within bacterial mRNAs, known as the **ribosomal-binding site** or
Shine–Dalgarno sequence, is involved in the binding of the mRNA to the 30S subunit. As shown in Figure 14-11, the Shine–Dalgarno sequence within the mRNA is
complementary to a short sequence within the 16S rRNA. This complementarity
promotes the hydrogen bonding of the mRNA to the 30S subunit.

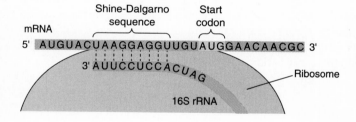

FIGURE **14-11**

**The locations of the Shine–Dalgarno sequence and the start
codon in prokaryotic mRNA.** The Shine–Dalgarno sequence is complementary to the 16S rRNA. It hydrogen bonds with the 16S rRNA to
promote initiation. The start codon is usually a few nucleotides downstream from the Shine–Dalgarno sequence. In bacteria, the start codon
is usually AUG. Occasionally, the start codon can be GUG (valine) or
UUG (leucine). However, in these rare cases, the GUG and UUG start
codons are still recognized by tRNA^fmet, so that a formylmethionine
occurs as the first amino acid instead of valine or leucine.

The binding of the tRNA^{fmet} to the 30S subunit and the mRNA requires the function of IF2. In most cases, the tRNA^{fmet} binds to the first start codon (namely, AUG), which is typically a few nucleotides downstream from the Shine–Dalgarno sequence. The start codon is usually AUG, but in some cases it can be GUG or UUG. Even when the start codon is GUG (which normally encodes valine) or UUG (which normally encodes leucine), the first amino acid in the polypeptide is still a formylmethionine, because only a tRNA^{fmet} can initiate translation. During translation, the formyl group or the entire formylmethionine may be removed from the polypeptide.

The initiation stage of prokaryotic translation is completed when the 50S ribosomal subunit associates with the 30S subunit. For this to occur, the initiation factors must be released from the 30S subunit.

In eukaryotes, the assembly of the initiation complex bears many similarities to that in prokaryotes. However, as described in Figure 14-12, additional factors are required for the initiation process. Note that *eukaryotic Initiation Factors* are designated eIF to distinguish them from prokaryotic initiation factors. The initiator tRNA in eukaryotes carries methionine rather than formylmethionine (as in prokaryotes). A eukaryotic initiation factor, eIF2, binds directly to tRNA^{met} to recruit it to the 40S subunit. Also, several initiation factors (CBPI [cap-binding protein], eIF4A, eIF4B, eIF4F, and others) bind to the mRNA. These initiation factors recognize the 7-methylguanosine cap structure, unwind any secondary structure in the mRNA so that it can bind to the ribosome, and aid in the identification of a start codon (see Table 14-1).

In eukaryotes, the identification of the correct AUG start codon differs greatly from that in prokaryotes. Most eukaryotic mRNAs contain a 7-methylguanosine cap structure at their 5′ end. The recognition of this cap structure is key to the initial binding of mRNA to the ribosome. After this occurs, the next step is to locate an AUG start codon that is somewhere downstream from the 5′ cap structure. In 1978, while at New York University, Marilyn Kozak proposed that the ribosome accomplishes this task by scanning along the mRNA in the 3′ direction in search of an AUG start codon. In many, but not all, cases, the ribosome uses the first AUG codon that it encounters as a start codon. When a start codon is identified, the 60S subunit assembles with the aid of eIF5.

By analyzing the sequences of many eukaryotic mRNAs, researchers have found that not all AUG codons near the 5′ end of mRNA can function as start codons. In some cases, the scanning ribosome passes over the first AUG codon and chooses an AUG farther down the mRNA. As it turns out, the sequence of nucleotides around the AUG codon plays an important role in determining whether or not it will be selected as the start codon by a scanning ribosome. The consensus sequence for optimal start codon recognition is shown here:

Start Codon

G   C   C   (A/G)   C   C   A   U   G   G
-6 -5 -4   -3   -2 -1 +1 +2 +3 +4

Aside from an AUG codon itself, a guanosine at the +4 position and a purine, preferably an adenine, at the −3 position are the most important for start codon selection. These rules for optimal translation initiation are called Kozak's rules. Presumably, this consensus sequence is recognized by a ribosomal component that facilitates the initiation of translation.

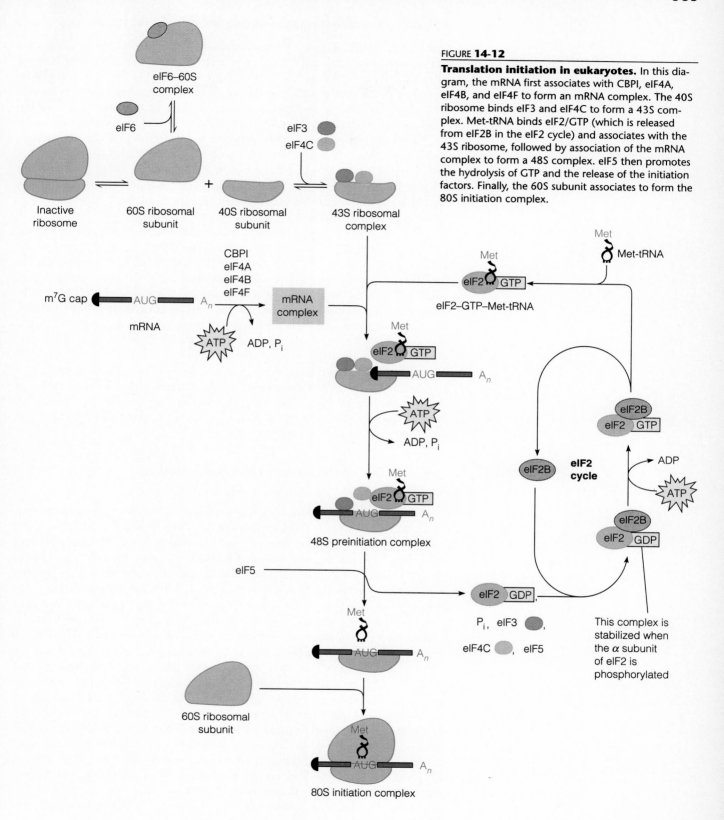

FIGURE **14-12**

Translation initiation in eukaryotes. In this diagram, the mRNA first associates with CBPI, eIF4A, eIF4B, and eIF4F to form an mRNA complex. The 40S ribosome binds eIF3 and eIF4C to form a 43S complex. Met-tRNA binds eIF2/GTP (which is released from eIF2B in the eIF2 cycle) and associates with the 43S ribosome, followed by association of the mRNA complex to form a 48S complex. eIF5 then promotes the hydrolysis of GTP and the release of the initiation factors. Finally, the 60S subunit associates to form the 80S initiation complex.

Polypeptide synthesis occurs during the elongation stage

During the elongation stage of translation, amino acids are added, one at a time, to the polypeptide chain. The addition of each amino acid occurs via the series of steps outlined in Figure 14-13. These steps are referred to as the **elongation cycle**. Even though this cycle is rather complex, it occurs at a remarkable rate. Under normal cellular conditions, a polypeptide chain can elongate at a rate of 15–18 amino acids per second in bacteria and 6 amino acids per second in eukaryotes!

In the elongation cycle, a tRNA brings a new amino acid to the ribosome so that it can be attached to the end of the growing polypeptide chain. At the upper left of Figure 14-13, a short polypeptide is attached to the tRNA located at the P site of the ribosome. A new tRNA carrying a single amino acid binds to the A site. This binding occurs because the anticodon in the tRNA is complementary to the codon in the mRNA.

The formation of a peptide bond between the amino acid at the A site and the growing polypeptide chain occurs via a **peptidyl transfer** reaction. This term means that the poly*peptide* is removed from the tRNA in the P site and *transfer*red to the amino acid at the A site. This transfer is accompanied by the formation of a peptide bond between the amino acid at the A site and the polypeptide chain, lengthening the chain by one amino acid. The peptidyl transfer reaction is catalyzed by an enzyme complex known as **peptidyltransferase**, which is composed of several proteins and rRNA. Interestingly, researchers have speculated that the rRNA may act as a ribozyme to catalyze this chemical reaction.

After the peptidyl transfer reaction is complete, the ribosome moves, or translocates, to the next codon in the mRNA. This moves the tRNAs at the P and A sites to the E and P sites, respectively. Also note that the next codon in the mRNA is now exposed in the unoccupied A site. Finally, the uncharged tRNA exits the E site, and the cycle can begin all over again.

Termination occurs when a stop codon is reached in the mRNA

The final stage of translation, known as termination, occurs when a stop codon is reached in the mRNA. In most species, the three stop codons are UAG, UAA, and UGA. These codons are sometimes referred to as **amber** (UAG), **ocher** (UGA), and **opal** (UAA). (Note: The term amber, or brown stone, is the English translation of the name Bernstein, a graduate student at the California Institute of Technology who was involved in the discovery of the UAG codon. In keeping with this tradition, the lighthearted names ocher and opal were given to the other two stop codons.) The stop codons, also known as nonsense codons, are not recognized by a tRNA with a complementary sequence. Instead, they are recognized directly by proteins known as **release factors**. In bacteria, there are two different release factors, designated RF1 and RF2, that recognize stop codons. RF1 recognizes UAA and UAG, and RF2 recognizes UGA and UAA. A third release factor, RF3, is also required. In eukaryotes, a single release factor, eRF, recognizes all three stop codons.

Figure 14-14 illustrates the termination stage of translation. At the top of this figure, the completed polypeptide chain is attached to a tRNA in the P site. A stop codon is located at the A site. In the next step, RF1 or RF2 binds to the stop codon at the A site and RF3 binds at a different location on the ribosome. After RF1 (or RF2) and RF3 have bound, the bond between the polypeptide and the tRNA is hydrolyzed. The polypeptide is then released from the ribosome. The final step in translational termination is marked by the disassembly of tRNA, ribosomal subunits, and the release factors.

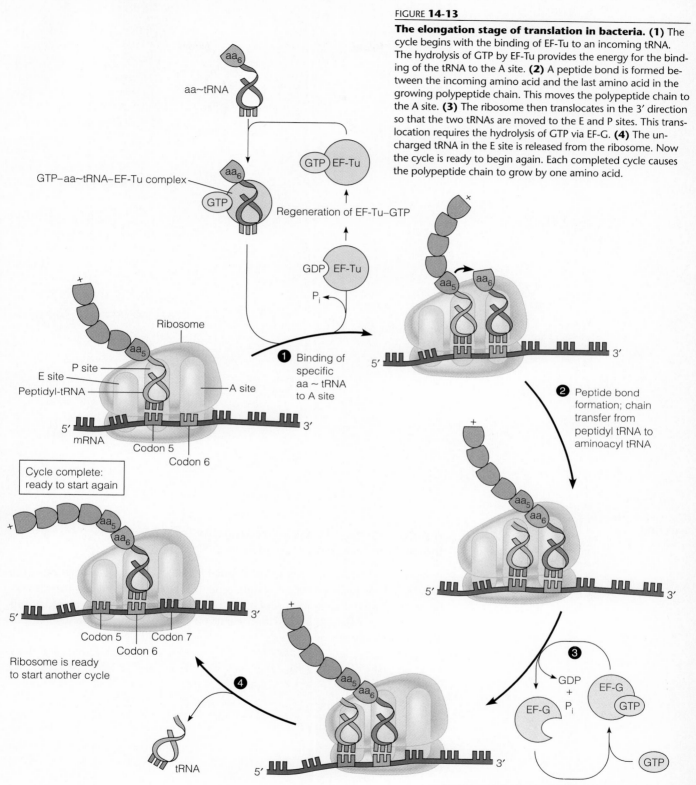

FIGURE **14-13**

The elongation stage of translation in bacteria. (1) The cycle begins with the binding of EF-Tu to an incoming tRNA. The hydrolysis of GTP by EF-Tu provides the energy for the binding of the tRNA to the A site. **(2)** A peptide bond is formed between the incoming amino acid and the last amino acid in the growing polypeptide chain. This moves the polypeptide chain to the A site. **(3)** The ribosome then translocates in the 3' direction so that the two tRNAs are moved to the E and P sites. This translocation requires the hydrolysis of GTP via EF-G. **(4)** The uncharged tRNA in the E site is released from the ribosome. Now the cycle is ready to begin again. Each completed cycle causes the polypeptide chain to grow by one amino acid.

aa~tRNA

GTP–aa~tRNA–EF-Tu complex

GTP EF-Tu

Regeneration of EF-Tu–GTP

GDP EF-Tu

P_i

❶ Binding of specific aa ~ tRNA to A site

Ribosome

E site

P site

Peptidyl-tRNA

A site

5' mRNA 3'

Codon 5

Codon 6

Cycle complete: ready to start again

Codon 5 Codon 7

Codon 6

Ribosome is ready to start another cycle

❹

tRNA

❷ Peptide bond formation; chain transfer from peptidyl tRNA to aminoacyl tRNA

5' 3'

5' 3'

❸

GDP + P_i

EF-G

EF-G GTP

GTP

5' 3'

Translocation of peptidyl tRNA from A site to P site. Ribosome moves one codon to the right, and the now uncharged tRNA moves from P site to E site

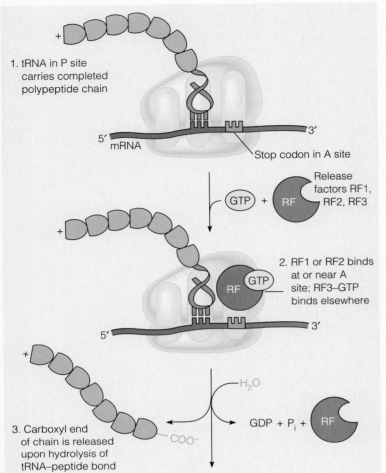

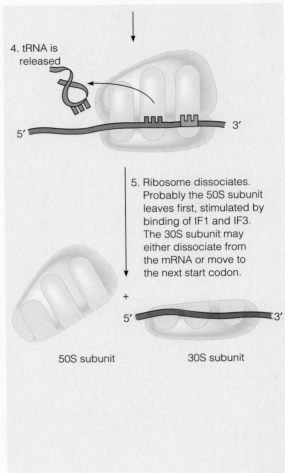

FIGURE **14-14**

The termination of translation in bacteria.

A polypeptide chain has a directionality from its amino terminus to its carboxy terminus

Polypeptide synthesis has a directionality that parallels the 5′ to 3′ orientation of the mRNA. During each cycle of elongation, a **peptide bond** is formed between the carboxyl group in the last amino acid of the polypeptide chain and the amino group in the amino acid being added. This occurs via a condensation reaction that releases a water molecule:

Last peptide bond formed in the growing chain of amino acids

Figure 14-15 compares the sequence of a very short polypeptide with the mRNA that encodes it. The first amino acid is said to be at the **N-terminal** or **amino terminal end** of the polypeptide. The term N-terminal refers to the presence of a nitrogen atom (N) at this end. By comparison, the last amino acid in a completed polypeptide is located at the **C-terminal** or **carboxyl terminal end**. A carboxyl group (COOH) is always found at this site in the polypeptide chain.

Bacterial translation can begin before transcription is completed

While most of our knowledge concerning transcription and translation has come from genetic and biochemical studies, electron microscopy has also been an important tool in elucidating the mechanisms of transcription and translation. As described earlier in this chapter, electron microscopy (EM) has been a critical technique in facilitating our understanding of ribosome structure. In addition, EM can be used to visualize genetic processes such as translation.

The first success in the EM observation of gene expression was achieved in 1967 by Oscar Miller Jr. and his colleagues at the Oak Ridge National Laboratory in Tennessee. Figure 14-16 shows an EM photograph of a bacterial gene in action! Prior to this experiment, biochemical and genetic studies had suggested that the translation of a bacterial structural gene begins before the mRNA transcript is completed. In other words, as soon as an mRNA strand is long enough, a ribosome will attach to the 5′ end and begin translation—even before RNA polymerase has reached the transcriptional termination site within the gene. This is termed the *coupling* between transcription and translation in bacterial cells.

As labeled in Figure 14-16, only one of the DNA strands is used as a template for the synthesis of mRNA transcripts. Several different RNA polymerase enzymes have recognized this gene and begun to transcribe it. Since the transcripts on the right side are longer than those on the left, Miller concluded that transcription was proceeding from left to right in the micrograph. This EM image also shows the process of translation. Relatively small mRNA transcripts, near the left side of the figure, had a few ribosomes attached to them. As the transcripts became longer, additional ribosomes

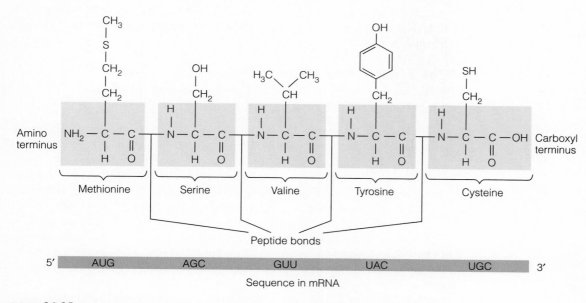

FIGURE **14-15**

The directionality of polypeptide synthesis. The first amino acid in a polypeptide chain (usually methionine) is located at the amino terminus, the last amino acid at the carboxy terminus. Thus, the directionality of amino acids in a polypeptide chain is from the amino to carboxyl terminus, which corresponds to the 5′ to 3′ orientation of codons in mRNA.

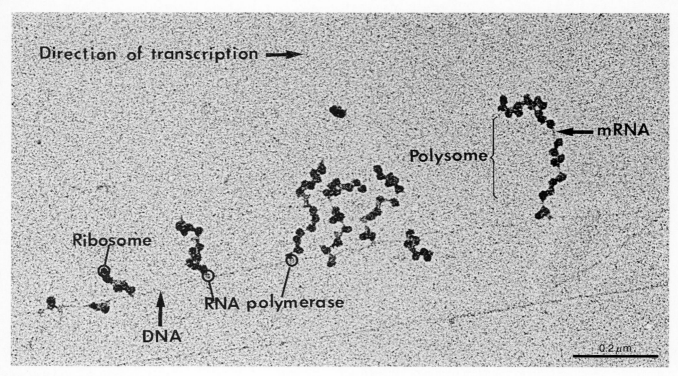

FIGURE **14-16**

Coupling between transcription and translation in bacteria. An electron micrograph showing the simultaneous transcription and translation processes. The DNA is transcribed by many RNA polymerases that move along the DNA from left to right. Note that the RNA transcripts are getting longer as you go from left to right. Ribosomes attach to the mRNA, even before transcription is completed. The binding of many ribosomes to the mRNA is called a polyribosome. Several polyribosomes are seen here.

were attached to them. The term **polyribosome** or **polysome** is used to describe an mRNA transcript that has many bound ribosomes in the act of translation. In this early study, the nascent polypeptide chains were too small for researchers to observe. In later studies, as EM techniques became more refined, the polypeptide chains emerging from the ribosome were also visible.

When they were published, in 1970, these micrographs were remarkably similar to the diagrams presented in genetics texts, illustrations that were based on biochemical and genetic studies of bacterial gene expression. These EM studies left little doubt that gene expression occurs via the transcription of DNA and the translation of mRNA.

The sorting of eukaryotic proteins can occur posttranslationally or cotranslationally; sorting signals are contained within the amino acid sequence of a protein

Unlike bacteria, eukaryotic cells cannot couple their transcription and translation, because the two processes occur in different cellular compartments. Transcription occurs in the nucleus, translation in the cytosol. The compartmentalization of eukaryotic cells also has other consequences for translation. Since they are compartmentalized into many different membrane-bound organelles, eukaryotic cells must sort their proteins into the correct compartment. In general, any particular protein is only meant to function in a single compartment. For example, the enzyme F_0F_1-ATP-synthetase functions in the mitochondrion and is not found in other locations in the cell. Cytoskeletal proteins such as tubulin and actin are found in the cytosol but not within the lumen of cellular organelles. To fulfill its function, each type of protein must be sorted into the correct cellular compartment. In eukaryotes, this sorting can occur during translation (termed **cotranslational sorting**) or after translation is completed (termed **posttranslational sorting**).

Most proteins begin their synthesis on ribosomes in the cytosol. Translation is completed in the cytosol for those proteins destined for the cytosol, nucleus, mitochondrion, chloroplast, or peroxisome (Figure 14-17). The uptake of proteins

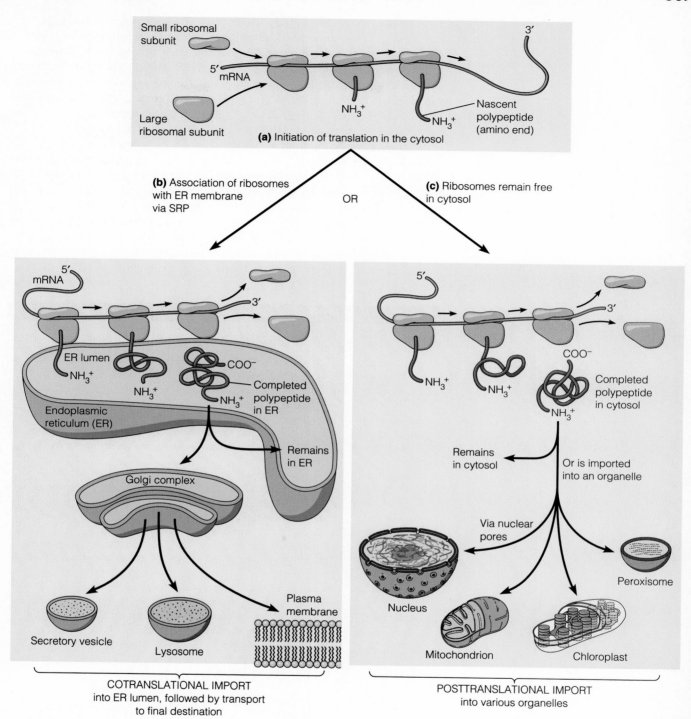

FIGURE **14-17**

Cotranslational and posttranslational protein sorting in eukaryotic cells.
(a) Nuclear genes begin their translation in the cytosol. (b) Some polypeptides contain an amino acid sequence at their amino terminal end that is recognized by the signal recognition particle (SRP). SRP directs these proteins to the ER, where they are sorted to the ER, Golgi, lysosome, plasma membrane, or are secreted. (c) Other proteins destined for the nucleus, mitochondrion, chloroplast, or peroxisome are completely synthesized in the cytosol and then are sorted post-translationally.

into the nucleus, mitochondrion, chloroplast, and peroxisome then occurs post-translationally. By comparison, the synthesis of other eukaryotic proteins begins in the cytosol and then is temporarily halted until the ribosome has become bound to the membrane of the endoplasmic reticulum (ER). After the ribosome has bound to the ER membrane, translation resumes, and the polypeptide is synthesized into the ER lumen or ER membrane. This event is termed **cotranslational import**. Proteins that are destined for the ER, Golgi, lysosome, plasma membrane, or secretion are first directed to the ER via this type of cotranslational import mechanism.

Genetics texts commonly focus attention on the importance of the amino acid sequence in dictating the structure and function of a protein. Even so, another vital function of the polypeptide sequence in eukaryotes is protein sorting. In eukaryotic proteins, there are short stretches of amino acid sequences that direct the protein to its correct cellular location. These short amino acid sequences are called **sorting** or **traffic signals**. Table 14-2 provides a general description of the sorting signals contained within eukaryotic proteins. Each traffic signal is recognized by specific cellular components that facilitate the sorting of the protein to its correct compartment, either cotranslationally or posttranslationally.

TABLE 14–2 Sorting Signals in Eukaryotic Proteins

Type of Signal	Description
Mitochondrial-targeting signal	Usually a short sequence at the amino terminus of a protein that contains several positively charged residues (i.e., lysine and arginine). This signal folds into an α-helix in which the positive charges are on one face of the helix.
Nuclear-localization signal	Can be located almost anywhere in the polypeptide sequence. The signal is four to eight amino acids in length and contains several positively charged residues and usually one or more prolines.
Peroxisomal-targeting signal	A specific sequence of three amino acids, serine–lysine–leucine, which is usually located near the carboxyl terminus of the protein.
SRP signal (sorting to the ER)	A sequence of ~20 amino acids near the amino terminus that is composed of mostly nonpolar amino acids.
ER retention signal	A sequence of four amino acids, lysine–aspartic acid–glutamic acid–leucine, which is located at the carboxyl terminus of the protein.
Golgi retention signal	A sequence of 20 hydrophobic amino acids that forms a transmembrane domain flanked by positively charged residues.
Lysosomal-targeting signal	A patch of amino acids within the polypeptide sequence. Positively charged residues within this patch are thought to play an important role. This patch causes lysosomal proteins to be covalently modified to contain a mannose-6-phosphate residue that directs the protein to the lysosome.

Destination of a Cellular Protein	Type of Signal the Protein Contains within Its Amino Acid Sequence
Cytosol	No signal required.
Mitochondrion	Mitochondrial-targeting signal.
Nucleus	Nuclear-localization signal.
Peroxisome	Peroxisomal-localization signal.
ER	SRP signal and an ER retention signal.
Golgi	SRP signal and a Golgi retention signal.
Lysosome	SRP signal and a lysosomal targeting signal.
Plasma membrane	SRP signal and the protein contains a hydrophobic transmembrane domain that anchors it in the membrane.
Secretion	SRP signal. No additional signal required.

The translation of mRNA into a polypeptide sequence requires many cellular components, including *ribosomes*, *tRNAs*, *mRNA*, and small effector molecules. According to the *adaptor hypothesis*, the anticodons in tRNA molecules act as the translators of the genetic code within mRNA. Prior to translation, enzymes known as *aminoacyl-tRNA synthetases* recognize tRNAs and attach the correct amino acid to their 3′ end. During translation, the anticodon in a tRNA binds to a codon in the mRNA due to their complementary sequences. Concurrently, the tRNA carries the correct amino acid, which is then added to the growing polypeptide chain. At the *wobble* position, the first base in the anticodon of the tRNA binds to the third base in the codon of the mRNA. In many cases, the wobble position can accommodate more than one type of base. In addition, modified bases in the tRNA can often be recognized by different bases in the mRNA.

The ribosomes are large macromolecular structures that provide a site for translation. Each ribosome is composed of one small and one large subunit; each subunit contains several proteins and rRNAs that assemble together. In eukaryotes, this assembly occurs within the *nucleolus*; the ribosomal subunits are then exported into the cytosol, where translation occurs.

The translation process occurs in three stages, known as *initiation*, *elongation*, and *termination*. During initiation, the ribosomal subunits, mRNA, and initiator tRNA assemble together. In prokaryotes, the *ribosomal-binding site* is located a few nucleotides upstream from the start codon. In eukaryotes, the ribosome binds near the 5′ end of the mRNA and then scans along the mRNA until it finds a start codon. After the start codon has been identified, translation can proceed to the elongation phase, in which the polypeptide is made. During elongation, there are three sites on the ribosome for tRNA binding. An appropriate tRNA first binds to the *A site* when its anticodon recognizes the codon in the mRNA. A peptide bond then forms between the amino acid attached to the tRNA at the A site and the growing polypeptide chain attached to a tRNA in the *P site*. The tRNA at the P site can then exit the ribosome from the *E site*. Finally, during the termination stage, the stop codon in the mRNA is reached. At this point, termination factors bind to the ribosome and cause the disassembly of the ribosomal subunits, tRNA, mRNA, and completed polypeptide chain.

Some interesting differences are found between prokaryotic and eukaryotic translation due to cell compartmentalization. Since bacteria contain a single intracellular compartment, translation of mRNA can begin before transcription is completed. This is referred to as coupling between transcription and translation. Coupling does not occur in eukaryotes, because the cells are compartmentalized; transcription takes place in the nucleus, and translation occurs in the cytosol. Compartmentalization also requires eukaryotic cells to sort their proteins into the correct locations. This occurs via the presence of *sorting signals* within the translated polypeptides. The targeting of cytosolic, nuclear, mitochondrial, chloroplast, and peroxisomal proteins occurs *posttranslationally*, whereas proteins destined for the ER, Golgi, lysosomes, plasma membrane, or secretion are initially targeted to the ER in a *cotranslational* manner.

Our molecular understanding of translation has come from several lines of experimentation, including biochemistry, genetics, and microscopy. The structure and composition of tRNAs, mRNAs, and ribosomes have been biochemically investigated for many decades. From such studies, we know the structures of tRNAs, mRNAs, and even the ribosomes, which are highly complex. Electron microscopy has aided our elucidation of ribosome structure and has also provided a method to demonstrate that transcription and translation are coupled in bacteria.

The steps of translation have also been investigated at the molecular level. By modifying the amino acid attached to tRNA, Chapeville showed that the anticodon sequence in the tRNA is the determining factor during translation. This experimental observation supported the adaptor hypothesis proposed by Crick. We have also learned a great deal about the translation process by examining genetic sequences. As was described in Chapter 12, the genetic code in mRNA was unraveled by the experiments of Nirenberg and Ochoa. In addition, geneticists have identified the functional roles of sequences such as the Shine–Dalgarno sequence in prokaryotes, and Kozak's rules in eukaryotes, by analyzing mutations that interfere with translation.

PROBLEM SETS

SOLVED PROBLEMS

1. The first amino acid within a certain bacterial polypeptide chain is methionine. Its start codon is GUG, which codes for valine. Why isn't the first amino acid formylmethionine or valine?

Answer: The first amino acid in a polypeptide chain is carried by the initiator tRNA, which always carries formylmethionine. This occurs even when the start codon is GUG (valine) or UUG (leucine). After polypeptide synthesis occurs, the formyl group is commonly removed, leaving a methionine as the first amino acid in the polypeptide.

2. A tRNA has the anticodon sequence 3′–CAG–5′. What amino acid does it carry?

Answer: Since the anticodon is 3′–CAG–5′, it would be complementary to a codon with the sequence 5′–GUC–3′. According to the genetic code, this codon specifies the amino acid valine. Therefore, this tRNA must carry valine at its acceptor site.

CONCEPTUAL QUESTIONS

1. List the components required for translation. Describe the relative sizes of these different components. In other words, which components are small molecules, macromolecules, or assemblies of macromolecules?

2. Describe the components of eukaryotic ribosomal subunits and where assembly of the subunits occurs within living cells.

3. In your own opinion, what structural feature of the ribosome do you find the most interesting and/or unusual? Explain your opinion.

4. If a tRNA molecule carries a glutamic acid, what are the two possible anticodon sequences that it could contain? Be specific about the 5′ and 3′ ends.

5. A tRNA has an anticodon sequence 3′–GGU–5′. What amino acid does it carry?

6. If a tRNA has an anticodon sequence 3′–CCI–5′, what codon(s) can it recognize?

7. Describe the anticodon of a single tRNA that could recognize the codons 5′–AAC–3′ and 5′–AAU–3′. How would this tRNA need to modified for it to also recognize 5′–AAA–3′?

8. Describe the structural features that all tRNA molecules have in common.

9. In the tertiary structure of tRNA, where is the anticodon region relative to the attachment site for the amino acid? Are they adjacent to each other?

10. What is an aminoacyl-tRNA synthetase? Why has the function of this type of enzyme been described as the second genetic code?

11. What is an activated amino acid?

12. Discuss the significance of modified bases within tRNA molecules.

13. How and when does formylmethionine become attached to the initiator tRNA in bacteria?

14. Is it necessary for a cell to make 61 different tRNA molecules, corresponding to the 61 codons for amino acids? Explain your answer.

15. What are the three stages of translation? Discuss the main events that occur during these three stages.

16. Describe the sequences in prokaryotic mRNA that promote recognition by the 30S subunit.

17. For each of the following initiation factors, how would eukaryotic initiation of translation be affected if it was missing?

    A. eIF2
    B. eIF4A
    C. eIF5

18. How does a eukaryotic ribosome select its start codon? Describe the sequences in eukaryotic mRNA that provide an optimal context for a start codon.

19. Which of the following sequences would be a good context for eukaryotic ribosomal initiation?

GACGCCAUGG

GCCUCCAUGC

GCCAUCAAGG

GCCACCAUGG

Rank them in order from best to worst.

20. Explain the functional roles of the A, P, and E sites during translation.

21. An mRNA has the following sequence: 5′–AUG UAC UAU GGG GCG UAA–3′. Draw the molecular structure of the peptide that would be encoded by this mRNA. Be specific about the amino and carboxyl terminal ends.

22. For eukaryotic proteins, what is the function of a sorting signal? Describe an example.

23. A mutation causes a protein to contain an SRP signal, an ER retention signal, and a mitochondrial-targeting signal. What cellular compartment do you think this protein would go to? Explain.

24. What is the difference between posttranslational and cotranslational protein sorting? What cellular components are required for cotranslational sorting?

EXPERIMENTAL QUESTIONS

1. Discuss how the elucidation of the structure of the ribosome can help us to understand its function.

2. In Experiment 14A, explain why a polyUG mRNA template was used.

3. In Experiment 14A, would you radiolabel the cysteine with the isotope ^{14}C or ^{35}S? Explain your choice.

4. An experimenter has a chemical reagent that modifies threonine to another amino acid. Following the protocol described in Experiment 14A, an mRNA is made that is composed of 50% C and 50% A, which are arranged randomly in the mRNA. The amino acid composition of the resultant polypeptides is 12.5% lysine, 12.5% asparagine, 25% serine, 12.5% glutamine, 12.5% histidine, and 25% proline. One of the amino acids present in this polypeptide is due to the modification of threonine. Which amino acid is it? Based on the structure of the amino acid side chains, explain how the structure of threonine has been modified.

5. As described in Experiment 14A, polypeptides can be translated *in vitro*. Would a prokaryotic mRNA be translated *in vitro* by eukaryotic ribosomes? Would a eukaryotic mRNA be translated *in vitro* by prokaryotic ribosomes? Why or why not?

6. Antibodies are molecules that bind specifically to cellular macromolecules such as proteins. Antibodies have been made that are specific for many of the proteins found within ribosomal subunits. The antibodies can be radiolabeled, fluorescently labeled, or labeled with atoms of a heavy metal. Describe how you might use these antibodies to study the structure of ribosomes.

7. Figure 14-16 shows an electron micrograph of a bacterial gene as it is being transcribed and translated. In this figure, label the 5′ and 3′ ends of the DNA and RNA strands. Place an arrow where you think the start codons are found in the mRNA transcripts.

8. As mentioned in question 6, antibodies can be labeled with heavy metals, such as gold particles, so that they can be visualized under the EM. Let's suppose you repeated the experiment of Miller shown in Figure 14-16 but added a gold-labeled antibody that recognizes RNA polymerase. Draw what you think the micrographs would look like. Note: The gold-labeled antibodies look like black, perfectly round spheres when viewed under the EM.

QUESTIONS FOR STUDENT DISCUSSION/COLLABORATION

1. Discuss why you think the ribosome needs to contain so many proteins and rRNA molecules. Does it seem like a waste of cellular energy to make so large a structure so that translation can occur?

2. Discuss and make a list of the similarities and differences in the events that occur during the initiation, elongation, and termination stages of transcription (Chapter 13) and translation (this chapter).

3. Which events during translation involve molecular recognition of a nucleotide base sequence within RNA? Which events involve recognition between different protein molecules?

CHAPTER 15

GENE REGULATION
IN BACTERIA AND
BACTERIOPHAGES

TRANSCRIPTIONAL
REGULATION

TRANSLATIONAL AND
POSTTRANSLATIONAL
REGULATION

GENE REGULATION IN
THE BACTERIOPHAGE
LIFE CYCLE

Chromosomes of bacteria, such as *Escherichia coli*, contain a few thousand different genes. Some of these genes are regulated; others are not. The term **gene regulation** means that the level of gene expression can vary under different conditions. By comparison, unregulated genes have essentially constant levels of expression in all conditions over time. These are called **constitutive genes**. Frequently, constitutive genes encode proteins that are always necessary for the survival of the bacterium. In contrast, the majority of genes are **regulated** so that the proteins they encode can be produced at the proper times and in the proper amounts.

The benefit of regulating genes is that the encoded proteins will only be produced when they are required. Therefore, gene regulation allows a cell to avoid wasting its valuable cellular energy making proteins it does not need. From the viewpoint of natural selection, this enables a bacterium to compete as efficiently as possible for a limited amount of available resources. This is particularly important since bacteria find themselves in an environment that is frequently changing with regard to temperature, nutrients, and many other factors. A few common processes that are regulated at the genetic level are listed here:

1. *Metabolism:* Some proteins function in the metabolism of small molecules. For example, certain enzymes are needed for a bacterium to metabolize a particular sugar. These enzymes are only required when the bacterium is exposed to that sugar in its environment.
2. *Response to environmental stress:* Certain proteins help a bacterium to survive environmental stress (e.g., osmotic shock, heat shock). These proteins are only required when the bacterium is confronted with the stress.

3. *Cell division:* Some proteins are needed for cell division. These are only necessary when the bacterial cell is getting ready to divide.

The expression of **structural genes**, which encode distinct polypeptides, ultimately leads to the production of functional cellular proteins. As we have seen in Chapters 12 to 14, gene expression is a multistep process that proceeds from transcription to translation, and it may involve posttranslational effects on protein structure and function. As shown in Figure 15-1, gene regulation can occur at any of these levels of gene expression. In this chapter, we will examine the molecular mechanisms that account for these types of gene regulation. We will also consider how gene regulation affects the phenotype of the bacterium.

TRANSCRIPTIONAL REGULATION

In bacteria, the most common way to regulate gene expression is by influencing the rate at which transcription is initiated. Although we frequently refer to genes as being "turned on or off," it is more accurate to say that the level of gene expression is increased or decreased. At the level of transcription, this means that the rate of RNA synthesis can be increased or decreased.

In most cases, transcriptional regulation involves the actions of regulatory proteins that can bind to the DNA and affect the rate of transcription of one or more nearby genes. There are two common types of regulatory proteins. A **repressor** is a regulatory protein that binds to the DNA and inhibits transcription, whereas **activators** are proteins that increase the rate of transcription. The term **negative control** refers to transcriptional regulation by repressor proteins, **positive control** to regulation by activator proteins.

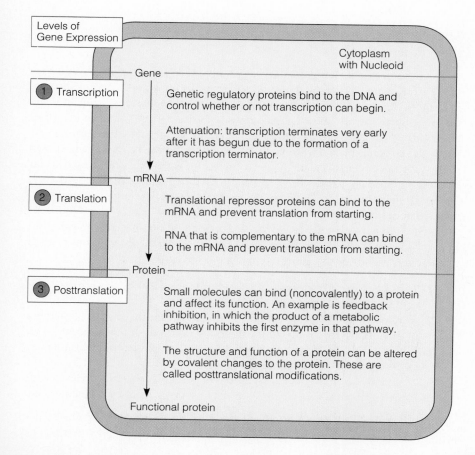

FIGURE **15-1**

Common points of genetic regulation in bacteria.

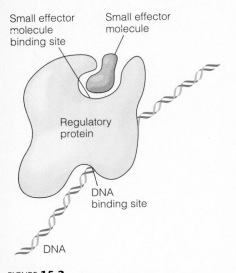

Small effector molecule binding site

Small effector molecule

Regulatory protein

DNA binding site

DNA

FIGURE **15-2**

Binding sites on a genetic regulatory protein. A regulatory protein has two binding sites, one for a small effector molecule and one for DNA. The binding of the small effector molecule will change the conformation of the regulatory protein and thereby influence whether the protein can bind to the DNA.

In conjunction with regulatory proteins, small effector molecules often play a critical role in transcriptional regulation. However, small effector molecules do not bind directly to the DNA to alter transcription. Rather, an effector molecule exerts its effects by binding to a regulatory protein such as an activator or repressor. The binding of the effector molecule causes a conformational change in the regulatory protein and thereby influences whether or not the protein can bind to the DNA. As shown in Figure 15-2, genetic regulatory proteins that respond to small effector molecules have two functional domains. One domain is a site where the protein binds to the DNA; the other domain is the binding site for the effector molecule.

Small effector molecules are given names that describe how they affect transcription when they are present. (The regulatory proteins are given names that describe how they affect transcription when they are *bound* to the DNA.) An **inducer** is a small molecule that will cause transcription to increase. An inducer may accomplish this in two ways: it could bind to an activator protein and cause it to bind to the DNA; or it could bind to a repressor protein and prevent it from binding to the DNA. In either case, the transcription rate is increased. Genes that are regulated in this manner are called **inducible**.

Alternatively, the presence of a small effector molecule may inhibit transcription. This can also occur in two ways. A **corepressor** is a small molecule that binds to a repressor protein, thereby causing the protein to bind to the DNA. An **inhibitor** binds to an activator protein and prevents it from binding to the DNA. Both corepressors and inhibitors act to reduce the rate of transcription. Therefore, the genes they regulate are **repressible**.

Unfortunately, this terminology can be confusing, since a repressible system could involve an activator protein or an inducible system could involve a repressor protein. In this section, we will examine several examples where genes are regulated by the actions of genetic regulatory proteins that influence the rate of transcription.

The phenomenon of enzyme adaptation is due to the synthesis of cellular proteins

To a significant extent, our understanding of molecular genetics can be traced back to the creative minds of Jacques Monod and François Jacob at the Pasteur Institute in France. Their research into genes and gene regulation stemmed from an interest in the phenomenon known as **enzyme adaptation**, which had been identified at the turn of the 20th century. Enzyme adaptation refers to the observation that a particular enzyme appears within a living cell only after the cell has been exposed to the substrate for that enzyme. When a bacterium is not exposed to a particular substance, it does not make the enzymes needed to metabolize that substance. However, when the bacterium is exposed to the substance, it will synthesize the enzymes that can metabolize it.

In the 1950s, Jacob and Monod focused their attention on lactose metabolism in *E. coli* to investigate this problem. Some key experimental observations that led to an understanding of this genetic system are listed here:

1. The exposure of bacterial cells to lactose increased the levels of lactose-utilizing enzymes by 1000- to 10,000-fold.
2. Antibody and labeling techniques revealed that the increase in the activity of these enzymes was due to the increased synthesis of the enzymes.
3. The removal of lactose from the environment caused an abrupt termination in the synthesis of the enzymes.
4. Mutations that prevented the synthesis of particular enzymes involved with lactose utilization showed that each enzyme was encoded by a separate gene.

These critical observations indicated that enzyme adaptation is due to the synthesis of specific cellular proteins in response to lactose in the environment. In the following sections, we will learn how Jacob and Monod discovered that this phenomenon is due to the interactions between genetic regulatory proteins and small effector molecules. In other words, we will see that enzyme adaptation is due to the transcriptional regulation of genes.

The *lac* operon encodes proteins that are involved in lactose metabolism

In bacteria, it is common for a few structural genes to be arranged together in a regulatory unit that is under the transcriptional control of a single promoter. This arrangement is known as an **operon**. The biological advantage of an operon organization is that it allows a bacterium to coordinately regulate a group of genes that encode proteins with a common function. The key feature of an operon is that the expression of the structural genes occurs as a single unit.

An operon contains several different regions. For transcription to take place, an operon is flanked by a **promoter** to signal the beginning of transcription and a **terminator** to signal the end of transcription. Two or more structural genes are found between these two sequences. To control the ability of RNA polymerase to transcribe an operon, an additional DNA sequence, known as the **operator site**, is present.

Figure 15-3 shows the organization of the genes within the lactose operon in *E. coli*. There are actually two separate transcriptional units. The first unit, known as the *lac* operon, contains a promoter and three structural genes, *lacZ*, *lacY*, and *lacA*. *LacZ* encodes the enzyme β-galactosidase, which enzymatically cleaves lactose and lactose analogues. As a side reaction, β-galactosidase also converts a small amount of lactose into *allolactose*, a related sugar. As we will see later, allolactose acts as a small effector molecule to regulate the *lac* operon. The *lacY* gene encodes lactose permease, a membrane protein required for the active transport of lactose into the cytoplasm of the bacterium. The *lacA* gene encodes a transacetylase enzyme that can covalently modify lactose and lactose analogues. Although the functional necessity of the transacetylase remains unclear, the acetylation of nonmetabolizable lactose analogues may prevent their toxic buildup within the bacterial cytoplasm.

In the vicinity of the *lac* promoter are two regulatory sites, designated the operator and the CAP site. The **operator** is a sequence of nucleotides that provides a binding site for a repressor protein, and the **CAP site** is a DNA sequence recognized by an activator protein called the catabolite activator protein (CAP).

A second transcriptional unit involved in genetic regulation is the *lacI* gene. The *lacI* gene encodes the *lac* repressor, a protein that is important for the regulation of the *lac* operon. The *lac* repressor functions as a tetramer, composed of four identical subunits. The *lacI* gene has its own promoter, the *i* promoter, which is constitutively expressed at fairly low levels. The amount of *lac* repressor made is

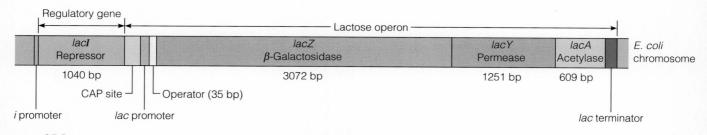

FIGURE **15-3**

Organization of the *lac* operon and other genes involved with lactose metabolism in *E. coli*. The CAP site is the binding site for CAP; the operator provides a binding site for the *lac* repressor. The *i* promoter is responsible for the transcription of the *lacI* gene, and the *lac* promoter (*lacP*) is responsible for the transcription of the *lacZ*, *lacY*, and *lacA* genes as a single unit.

approximately ten tetramer proteins per cell. Only a small amount of the *lac* repressor protein is needed to repress the *lac* operon.

The *lac* operon is regulated by a repressor protein

The *lac* operon can be transcriptionally regulated in more than one way. The first method that we will examine is an inducible, negative control mechanism. As shown in Figure 15-4, this form of regulation involves the *lac* repressor protein, which binds to the sequence of nucleotides found within the *lac* operator site. Once bound, the *lac* repressor prevents RNA polymerase from sliding past the operator site and transcribing the *lacZ*, *lacY*, and *lacA* genes (Figure 15-4a).

The ability of the *lac* repressor to bind to the operator site depends on whether or not allolactose (a product of lactose) is bound to it. The repressor protein functions as a tetramer; each of the four subunits has a single binding site for the inducer (namely, allolactose). When allolactose binds to the repressor, a conformational change occurs that prevents the *lac* repressor from binding to the operator site. Under these conditions, RNA polymerase is now free to transcribe the operon (Figure 15-4b). In genetic terms, we would say that the operon has been **induced**.

To better appreciate this form of regulation at the cellular level, let's consider the process as it occurs over time. Figure 15-5 illustrates the effects of external lactose on the regulation of the *lac* operon. In the absence of lactose (stage 1), no inducer is available to bind to the *lac* repressor. Therefore, the *lac* repressor binds to the operator site and inhibits transcription. In reality, the repressor does not completely inhibit transcription, so that a very small amount of β-galactosidase, lactose permease, and transacetylase are made. However, the levels are too low for the bacterium to

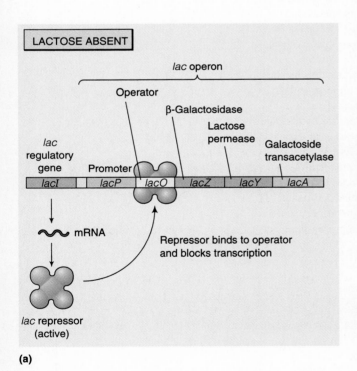

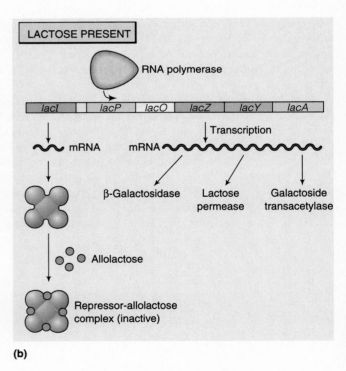

(a)

(b)

FIGURE **15-4**

Mechanism of induction of the *lac* operon. (a) In the absence of the inducer allolactose, the repressor protein is tightly bound to the operator site, thereby inhibiting the ability of RNA polymerase to transcribe the operon. **(b)** When there is sufficient allolactose, it binds to the repressor. This alters the conformation of the repressor protein in such a way as to prevent it from binding to the operator site. Therefore, RNA polymerase can transcribe the operon. (The CAP site is not shown in this drawing.)

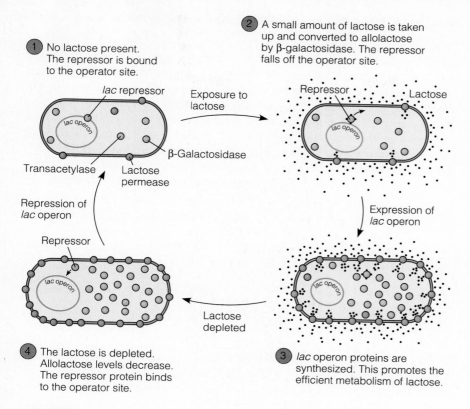

① No lactose present. The repressor is bound to the operator site.

lac repressor

lac operon

Transacetylase   Lactose permease   β-Galactosidase

Exposure to lactose

② A small amount of lactose is taken up and converted to allolactose by β-galactosidase. The repressor falls off the operator site.

Repressor   Lactose

lac operon

Expression of *lac* operon

③ *lac* operon proteins are synthesized. This promotes the efficient metabolism of lactose.

lac operon

Lactose depleted

Repression of *lac* operon

Repressor

lac operon

④ The lactose is depleted. Allolactose levels decrease. The repressor protein binds to the operator site.

FIGURE 15-5

The cycle of *lac* operon induction and repression.

GENES ⟶ TRAITS: The genes and the genetic regulation of the *lac* operon provide the bacterium with the trait of being able to metabolize lactose when it is present in the environment. When lactose is present, the genes of the *lac* operon are induced, and the bacterial cell can efficiently metabolize this sugar. When lactose is absent, these genes are repressed, so that the bacterium does not waste its energy expressing these genes.

readily utilize lactose. When the bacterium is exposed to lactose (stage 2), a small amount can be transported into the cytoplasm via the lactose permease, and β-galactosidase will convert some of it to allolactose. As this occurs, the cytoplasmic level of allolactose gradually rises; eventually, allolactose binds to the *lac* repressor. This prevents the repressor from binding to the *lac* operator site and thereby allows transcription of the *lacZ*, *lacY*, and *lacA* genes to occur (stage 3).

To understand how the induction process is shut off in a lactose-depleted environment (stage 4), let's consider the interaction between allolactose and the *lac* repressor. Allolactose does not bind irreversibly to the *lac* repressor. Rather, the *lac* repressor has a measurable affinity for binding and releasing allolactose. When the concentration of allolactose is above the affinity for the repressor protein, allolactose will bind to the *lac* repressor. In contrast, when the concentration of allolactose is below its affinity for the repressor, it will not be bound. With these ideas in mind, we can see how the *lac* operon shuts down when lactose is depleted from the environment. Since the cell is metabolizing its internal lactose, the cytoplasmic levels of lactose and allolactose will fall. Once the allolactose concentration within the cell drops below its affinity for the *lac* repressor, the unoccupied repressor will again bind to the *lac* operator site and cause repression. The proteins encoded by the *lac* operon have a finite half life, and they will eventually be degraded. This returns the cell to stage 1.

Jacob, Monod, and Pardee investigated the function of the *lacI* gene

Now that we have an understanding of the *lac* operon, let's look back at one of the approaches that was used in the 1950s to elucidate the regulation of the *lac* operon. In their studies of lactose metabolism in bacteria, Jacob and Monod, along with their colleague, Arthur Pardee, had identified a few rare mutant strains of bacteria that had abnormal lactose adaptation. One type of mutant, which involved a defect in the *lacI* gene (designated *lacI*⁻), resulted in the constitutive expression of the *lac* operon even in the absence of lactose. They were able to map genetically *lacI*⁻ mutations using conjugation methods described in Chapter 6 (see Experiment 6A). The *lacI*⁻ mutations mapped very close to the *lac* operon. From these observations, Jacob, Monod, and Pardee proposed two different possible functions of the *lacI* gene:

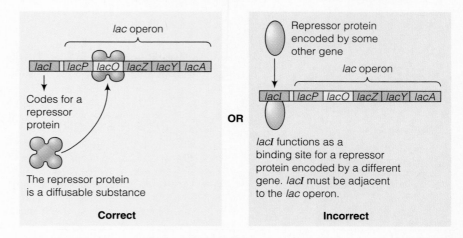

One hypothesis was that the *lacI* gene encodes a repressor protein that binds to the operon and inhibits transcription. (As we have already learned, this is the correct hypothesis.) Alternatively, since the *lacI* gene is physically next to the operon, a second possibility would be that the *lacI* gene is functioning as a site for the binding of a repressor protein produced by some other unknown gene. In other words, a second (incorrect) possibility would be that a *lacI* mutation prevents a repressor protein from binding to an operator site adjacent to the *lac* operon.

Jacob, Monod, and Pardee used a genetic approach to distinguish between these two alternative possibilities. Using bacterial conjugation methods, they introduced different portions of the *lac* operon into bacterial strains that carried *lacI* mutations. Even though we discussed bacterial mating in Chapter 6, it is necessary to briefly review this process to understand this experiment. In general, bacteria are haploid and reproduce asexually by binary fission. On occasion, however, bacteria can mate with each other and transfer circular segments of DNA known as F factors. Sometimes, an F factor also carries genes that were originally found within the bacterial chromosome. These types of F factors are called F′ factors.

In their studies, Jacob, Monod, and Pardee identified F′ factors that carried portions of the *lac* operon. These F′ factors can be transferred from donor to recipient strains during bacterial conjugation. For example, let's consider an F′ factor that carries the *lacI* gene. A bacterium that has received this F′ factor would actually have two copies of the *lacI* gene: one copy of the *lacI* gene on the chromosome and a second copy on the F′ factor. A strain of bacteria containing F′ factor genes is called a **merozygote**, or partial diploid.

The production of merozygotes was instrumental in allowing Jacob, Monod, and Pardee to elucidate the function of the *lacI* gene. There are two key points. First, the two *lacI* genes in a merozygote may be different alleles. For example, the *lacI* gene on the chromosome may be a defective *lacI*⁻ allele while the *lacI* gene on the F′ factor may be normal. Second, the genes on the F′ factor and the genes on the bacterial chromosome are not physically adjacent to each other. According to the correct hypothesis shown on p. 400, a *lacI* gene on an F′ factor should be able to produce a repressor protein that could diffuse within the cell and eventually bind to the operator site of the *lac* operon located on the chromosome. In contrast, the incorrect hypothesis suggests that the *lacI* gene provides a binding site for a repressor protein. In this case, it must be physically adjacent to the *lac* operon to exert its effects. If the *lacI* gene functioned as a binding site for a repressor protein, a merozygote containing a normal *lacI* gene on an F′ factor would not be able to affect the expression of the *lac* operon caused by a chromosomal mutation in the *lacI* gene. In this way, the approach of making merozygotes enabled Jacob, Monod, and Pardee to distinguish between these two possibilities.

THE HYPOTHESIS

This experiment was done to determine whether the *lacI* gene codes for a protein that regulates the *lac* operon *or* whether the *lacI* gene serves as a repressor protein binding site and must be physically next to the *lac* operon to regulate it. The rationale is that if the *lacI* gene encodes a regulatory protein, then the *lacI* gene itself does not have to be physically next to the *lac* operon to repress it; the protein produced from the *lacI* gene (namely, the *lac* repressor) can diffuse throughout the cell and bind to an operator site regardless of the physical location of the *lacI* gene.

TESTING THE HYPOTHESIS

Starting material: The genotype of the recipient strain was *lacI*⁻ *lacZ*⁺ *lacY*⁺ *lacA*⁺. The donor strain had an F′ factor that was *lacI*⁺ *lacZ*⁺ *lacY*⁺ *lacA*⁺.

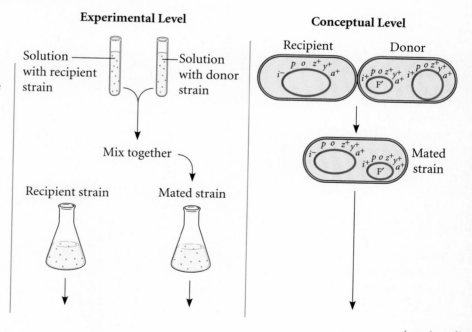

Experimental Level

Conceptual Level

1. Mate the donor strain to the recipient strain. The mated strain contains a *lacI*⁻ mutation on the chromosome and a normal *lacI* gene on the F′ factor.

2. Grow an unmated recipient strain and a mated strain separately.

(continued)

3. Divide each strain into two tubes.

4. In one of the two tubes, add lactose.

5. Incubate the cells long enough to allow *lac* operon induction.

6. Burst the cells with a sonicator. This allows β-galactosidase to escape from the cells.

7. Add β-*o*-nitrophenylgalactoside. This is a colorless compound. β-galactosidase will cleave the compound to produce galactose and *o*-nitrophenol (O-NP). *O*-nitrophenol has a yellow color. The deeper the yellow color, the more β-galactosidase was produced.

8. Incubate the sonicated cells to allow β-galactosidase time to cleave β-*o*-nitrophenylgalactoside (β-ONPG).

9. Measure the yellow color produced with a spectrophotometer. (See the appendix for a description of spectrophotometry.)

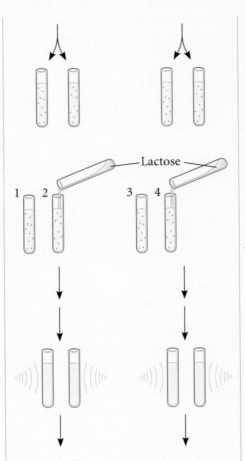

Lactose

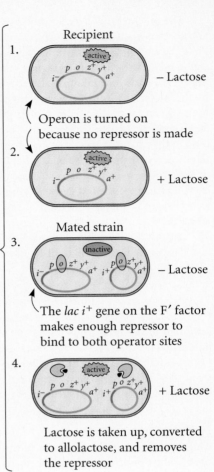

Recipient

1. Operon is turned on because no repressor is made

   − Lactose

2. + Lactose

   Mated strain

3. The *lac* i^+ gene on the F′ factor makes enough repressor to bind to both operator sites

   − Lactose

4. Lactose is taken up, converted to allolactose, and removes the repressor

   + Lactose

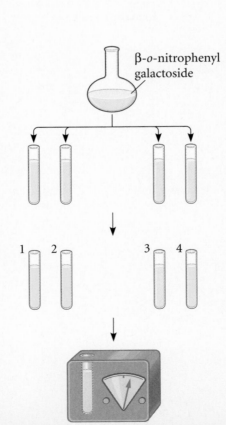

β-*o*-nitrophenyl galactoside

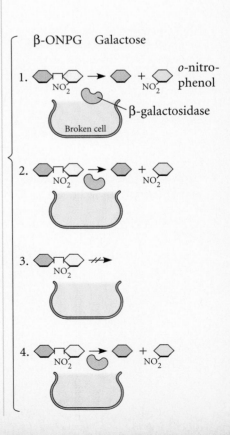

β-ONPG   Galactose

1. *o*-nitro-phenol

   β-galactosidase

   Broken cell

2.

3.

4.

THE DATA

Strain	Addition of Lactose	Amount of ß-Galactosidase (percentage of parent strain)
Recipient	No	100%
Recipient	Yes	100%
Mated	No	<1%
Mated	Yes	220%

This experiment is adapted from: Pardee, A. B., Jacob, F., and Monod, F. (1959) *J. Mol. Biol. 1,* 165. In their original studies, negative mutations in the *lacZ* gene were included in combination with the *lacI* mutations.

INTERPRETING THE DATA

In the recipient strain, the amount of yellow color produced was the same in the presence or absence of lactose. This is expected, since the expression of β-galactosidase in the *lacI⁻* recipient strain was already known to be constitutive. In other words, the presence of lactose was no longer needed to induce the operon due to a defective *lacI* gene. In the mated strain, however, a different result was obtained. In the absence of lactose, the *lac* operons were repressed, even the operon on the chromosome. Since the normal *lacI* gene on the F′ factor was not physically located next to the chromosomal *lac* operon, this result is consistent with the hypothesis that the *lacI* gene codes for a repressor protein that can diffuse throughout the cell and bind to any *lac* operon. In the presence of lactose, the operon is induced in the mated strain.

The interactions between regulatory proteins and DNA sequences illustrated in this experiment have led to the definition of two genetic terms. A **trans effect** is a form of genetic regulation that can occur even though two DNA segments are not physically adjacent. The action of the *lac* repressor is a trans effect. In contrast, a **cis effect** or a **cis-acting element** is a DNA segment that must be adjacent to the gene(s) that it regulates. The *lac* operator site is an example of a cis-acting element. A trans effect is mediated by genes that encode regulatory proteins, whereas a cis effect is mediated by DNA sequences that bind regulatory proteins.

The *lac* operon is also regulated by an activator protein

The *lac* operon can be transcriptionally regulated in a second way, known as **catabolite repression** (as we shall see, a somewhat imprecise term). This form of transcriptional regulation is influenced by the presence of the sugar glucose, which is a catabolite (it is broken down—i.e., catabolized—inside the cell). The presence of glucose ultimately leads to the repression of the *lac* operon.

Glucose, however, is not itself the small effector molecule that binds directly to a genetic regulatory protein. Instead, this form of regulation involves a small effector molecule, **cyclic AMP (cAMP)**, which is produced from ATP via an enzyme known as **adenylate cyclase**. cAMP binds to an activator protein known as the **catabolite activator protein (CAP)** or the **cAMP receptor protein (CRP)**.

As shown in Figure 15-6, the cAMP–CAP complex is an example of genetic regulation that is inducible and under positive control. When cAMP binds to CAP, the cAMP–CAP complex binds to the CAP site near the *lac* promoter and increases the rate of transcription. The term catabolite repression may seem puzzling, since this regulation involves the action of an inducer and activator protein rather than a repressor. The term was coined before the action of the cAMP–CAP complex was understood. At that time, the primary observation was that glucose (a catabolite) inhibited (repressed) lactose metabolism. We now know that when a bacterium is exposed to glucose, the intracellular concentration of cAMP decreases because the presence of glucose inhibits

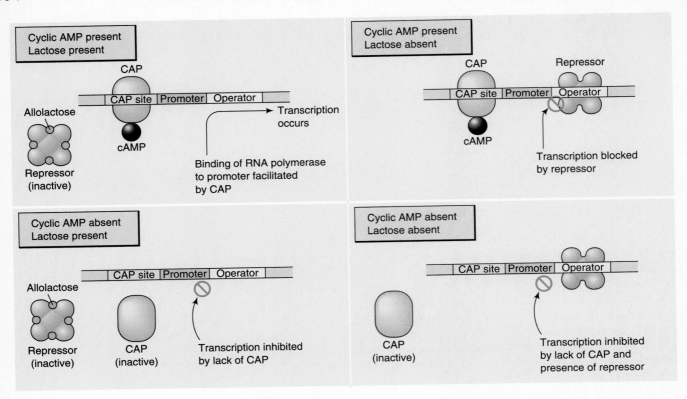

FIGURE **15-6**

The roles of the *lac* repressor and catabolite activator protein (CAP) in the regulation of the *lac* operon. Under conditions of low glucose, cAMP is made. This small effector molecule binds to CAP, causing it to change its conformation and bind to the region adjacent to the *lac* promoter. The binding of the CAP–cAMP complex to this region facilitates the binding of RNA polymerase to the promoter. This leads to an enhanced rate of transcription of the *lac* operon. By comparison, the *lac* repressor protein binds to the operator and inhibits the operon when lactose levels are low. This figure illustrates how the *lac* operon will be regulated depending on the levels of cAMP and lactose within the cell.

GENES ⟶ TRAITS: The mechanism of catabolite repression provides the bacterium with the trait of being able to choose between two sugars. When exposed to both glucose and lactose, the bacterium chooses glucose first. After the glucose is used up, it then will express the genes that are necessary for lactose metabolism. This trait allows the bacterium to more efficiently utilize sugars from its environment.

adenylate cyclase, an enzyme needed for cAMP synthesis. When this happens, cAMP is no longer available to bind to the CAP. Therefore, the unoccupied CAP does not bind to the CAP site. This causes the transcription rate to decrease.

Catabolite repression enables the bacterium to utilize two sugars more efficiently. When exposed to both glucose and lactose, *E. coli* cells will first utilize glucose, and catabolite repression will prevent the utilization of lactose. This is advantageous to the bacterium, because it does not have to express all the genes that are necessary for both glucose and lactose metabolism. If the glucose is used up, however, catabolite repression will be alleviated and the bacterium will then express the *lac* operon. The sequential use of two sugars by a bacterium is known as **diauxic growth**. It is a common phenomenon among many bacterial species. When glucose is one of the two sugars available, it is typical that the bacterium will metabolize glucose first, and then a second sugar after the depletion of glucose. Among *E. coli* and related species, diauxic growth is regulated by intracellular cAMP levels and the CAP.

In recent years, more detailed studies have revealed that the *lac* operon has three operator sites for the *lac* repressor

Our traditional view of the regulation of the *lac* operon has been modified as we have gained a greater molecular understanding of the process. In particular, detailed genetic and crystallographic studies have shown recently that the binding of the *lac* repressor is more complex than originally realized. The site in the *lac* operon that is commonly called the operator site was first identified by mutations that prevented *lac* repressor binding. These mutations, called *lacO^C* mutants, resulted in the constitutive expression of the *lac* operon even in strains that make a normal *lac* repressor protein. *LacO^C* mutations were found to be localized in the *lac* operator site (which is now known as O_1). This led to the simple view that a single operator site was bound by the *lac* repressor to inhibit transcription (as in Figure 15-4).

In the late 1970s and 1980s, however, two additional operator sites were identified. As shown at the top of Figure 15-7, these sites are called O_2 and O_3. O_1 is the operator site that is next to the promoter. O_2 is located downstream (3′ direction) in the *lacZ* coding sequence, and O_3 is located slightly upstream (5′ direction) from the CAP-binding site. The O_2 and O_3 operators were initially called pseudo-operators, because the absence of either one of them only decreased repression by two- to threefold. However, studies by Benno Müller-Hill and his colleagues at the Institute for Genetics in Köln, Germany, revealed a surprising result. As shown in Figure 15-7, if both O_2 and O_3 are missing, repression is dramatically reduced even when O_1 is present. When O_1 is missing, repression is nearly abolished. This result supported the hypothesis that the *lac* repressor must bind to two out of three operators to cause maximal repression. According to this view, the *lac* repressor can bind to O_1 and O_2, or to O_1 and O_3, but not to O_2 and O_3. Since these operator sites are a fair distance away from each other, it was proposed that the binding of the *lac* repressor to two operator sites causes the DNA to form a loop (Figure 15-8a).

The proposal that the *lac* repressor binds to two operator sites has been confirmed by crystallographic studies. X-ray crystallography can provide a detailed picture of the molecular structure of macromolecules. To study protein–DNA interactions using crystallography, one must purify a large number of complexes that contain the DNA-binding protein and a fragment of DNA that is bound to the protein. These complexes are then subjected to conditions that will cause them to crystallize with each other, much like the formation of a NaCl crystal. When subjected to X-rays, a crystal will produce a pattern on an X-ray film. A computer analysis of the X-ray pattern can reveal information about the arrangement of atoms within the crystal.

In 1996, the *lac* repressor was crystallized by Mitchell Lewis and his colleagues at the University of Pennsylvania and the Oregon Health Sciences University. The crystal structure of the *lac* repressor has provided exciting insights into its mechanism of action. As mentioned earlier in this chapter, the *lac* repressor is a tetramer of four identical subunits. The crystal structure revealed that each dimer within the tetramer recognizes one operator site. Figure 15-8b is a molecular model illustrating the binding of the *lac* repressor to the O_1 and O_3 sites. The amino acid side chains in the protein interact directly with bases in the major groove of the DNA helix. This is how genetic regulatory proteins can recognize specific DNA sequences. Each dimer within the tetramer is recognizing a single operator site. The association of two dimers to form a tetramer forces the two operator sites close to each other. For this to occur, a loop must form in the DNA. The formation of this loop is expected to inhibit the ability of RNA polymerase to transcribe this region.

Figure 15-8b also shows the binding of the cAMP–CAP complex to the CAP site. A particularly striking observation is that the binding of the cAMP–CAP complex to

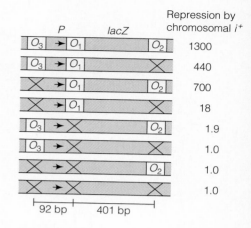

FIGURE **15-7**

The identification of three *lac* operator sites. The top of this figure shows the locations of three *lac* operator sites, designated O_1, O_2, and O_3. O_1 is analogous to the *lac* operator shown in previous figures. When all three operators are present, the repression of the *lac* operon is 1300-fold; this means there is 1/1300th the level of expression than there is when lactose is present. This figure also shows the amount of repression when one or more operator sites are removed. (The removal of an operator site is designated with a ×.) A repression value of 1.0 indicates that no repression is occurring. In other words, a value of 1.0 indicates constitutive expression.

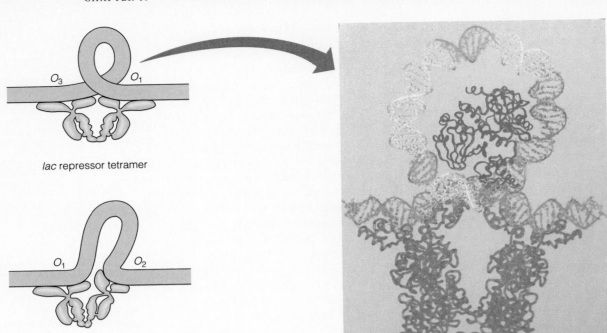

lac repressor tetramer

lac repressor tetramer

(a) Possible DNA loops caused by the binding of the *lac* repressor

(b) Proposed model of the *lac* repressor binding to O_1 and O_3 based on crystallography studies

FIGURE **15-8**

The binding and structure of the *lac* repressor. (a) The binding of the *lac* repressor protein to the O_1 and O_2 or to the O_1 and O_3 operator sites. **(b)** A molecular model for the binding of the *lac* repressor to O_1 and O_3. Each repressor dimer binds to one operator site, so that the repressor tetramer brings the two operator sites together. This causes the formation of a DNA loop in the intervening region. Note that the DNA loop contains the −35 and −10 regions that are recognized by the sigma factor of RNA polymerase. This loop also contains the binding site for the CAP–cAMP complex, which is shown here. Note: The figure shown in part (b) is from the cover of SCIENCE 271, March 1, 1996. The article is Lewis, M., Chang, G., Horton, N. C., Kercher, M. A., Pace, H. C., Schumacher, M. A., Brennan, R. G., and Lu, P. (1996) Crystal structure of the lactose operon repressor and its complexes with DNA and inducer. *Science 271*, 1247–1254.

the DNA causes a 90° bend in the DNA structure. When the repressor is active (i.e., not bound to allolactose), the cAMP–CAP complex facilitates the binding of the *lac* repressor to the O_1 and O_3 sites. When the repressor is inactive, this bending also appears to be important in the ability of RNA polymerase to initiate transcription slightly downstream from the bend.

The *ara* operon can be regulated positively or negatively by the same regulatory protein

Now that we have considered the regulation of *lac* operon, let's compare its regulation with that of other genes in the bacterial chromosome. Another operon in *E. coli* involved in sugar metabolism is the *ara* (arabinose) operon. The sugar arabinose is a constituent of the cell walls of a few types of plants. As shown in Figure 15-9, the *ara* operon contains three structural genes, *araB*, *araA*, and *araD*, encoding three enzymes involved in arabinose metabolism. The actions of the three enzymes metabolize arabinose into D-xylulose-5-phosphate.

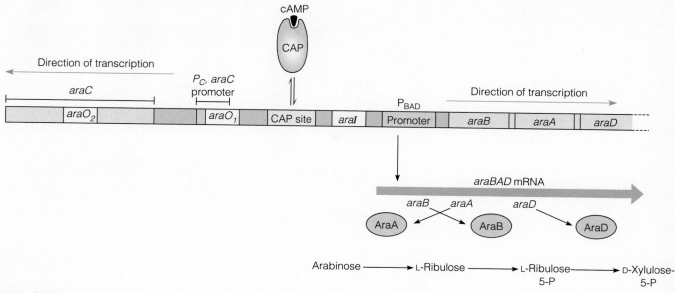

Organization of the *ara* operon and other genes involved in arabinose metabolism. *AraC* encodes a genetic regulatory protein called the araC protein. This protein can bind to three different regulatory sites, called *araO₁*, *araO₂*, and *araI*. P_C is the promoter for the *araC* gene; P_{BAD} is the promoter for the *ara* operon, which contains three genes (*araB*, *araA*, and *araD*) encoding proteins involved in arabinose metabolism.

Like the *lac* operon, the *ara* operon contains a single promoter, designated P_{BAD}. This operon also contains a CAP site for the binding of the catabolite activator protein. The *araC* gene, which has its own promoter (P_C), is adjacent to the *ara* operon. *AraC* encodes a regulatory protein, called the araC protein, that can bind to operator sites designated *araI*, *araO₁*, and *araO₂*.

As we have seen with the *lac* operon, some regulatory proteins such as the *lac* repressor inhibit transcription whereas others such as CAP turn on transcription. The araC protein is rather interesting and unusual because it can act as either a negative or positive regulator of transcription, depending on whether or not arabinose is present. Figure 15-10 illustrates the possible actions of the araC protein. In the absence of arabinose, the araC protein binds to the *araI*, *araO₁*, and *araO₂* operator sites. An araC protein dimer is bound to *araO₁*, while monomers are bound at *araO₂* and *araI*. The binding of the araC protein to the *araO₁* site activates the transcription of the *araC* gene. In other words, the araC protein is a positive regulator of the *araC* gene.

In contrast, the araC proteins bound at *araO₂* and *araI* repress the arabinose operon. As shown in Figure 15-10a, the araC proteins at *araO₂* and *araI* can bind to each other by causing a loop in the DNA, as originally proposed in 1990 by Robert Schleif and his colleagues at Johns Hopkins University. This DNA loop prevents RNA polymerase from transcribing the *ara* operon. Therefore, in absence of arabinose, the *ara* operon is turned off.

Figure 15-10b illustrates the activation of the *ara* operon in the presence of arabinose. When arabinose is bound to the araC protein, the interaction between the araC proteins at the *araO₂* and *araI* sites is broken. This breaks the DNA loop. In addition, a second araC protein binds at the *araI* site. This araC dimer at the *araI* operator site activates transcription. This activation can occur in conjunction with the activation of the *ara* operon by CAP and cAMP if glucose levels are low. When the *ara* operon is activated, the bacterial cell can efficiently metabolize arabinose.

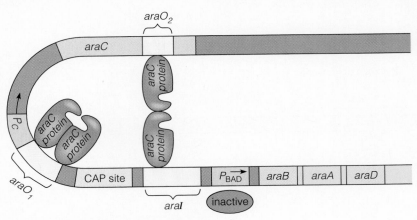

(a) Operon inhibited in the absence of arabinose

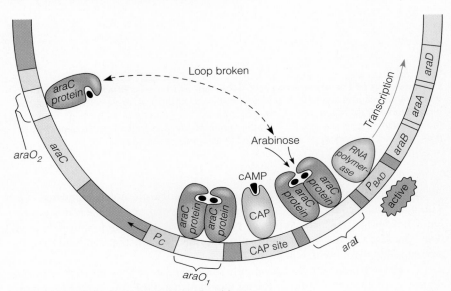

(b) Operon activated in the presence of arabinose

FIGURE **15-10**

DNA looping and unlooping via the araC protein. (a) In the absence of arabinose, an *araC* protein binds to the *araI* operator site and another to the *araO₂* site. These two araC proteins interact to promote a loop in the DNA. This loop prevents RNA polymerase from transcribing the *ara* operon from the P_{BAD} promoter. **(b)** In the presence of arabinose, an araC dimer is bound to the *araI* operator site. The interaction between the araC proteins bound at *araI* and *araO₂* no longer occurs. This causes the DNA loop to be broken. Under these conditions, RNA polymerase can transcribe the *ara* operon.

The *trp* operon is regulated by a repressor protein and also by attenuation

The *trp* operon (pronounced "trip") encodes enzymes that are needed for the biosynthesis of the amino acid tryptophan (Figure 15-11). The *trpE*, *trpD*, *trpC*, *trpB*, and *trpA* genes encode enzymes involved in tryptophan biosynthesis. The *trpL* and *trpR* genes are involved in regulating the *trp* operon in two different ways. The *trpR* gene encodes the *trp* repressor protein. When tryptophan levels within the cell are very low, the *trp* repressor cannot bind to the operator site. Under these conditions, RNA polymerase transcribes the operon (Figure 15-11a). In this way, the cell expresses the genes that encode the synthesis of tryptophan. When the tryptophan levels within the cell become high, tryptophan acts as a corepressor to bind to the *trp* repressor protein. This causes a conformational change in the *trp* repressor, which allows it to bind to the *trp* operator site (Figure 15-11b). This inhibits the ability of RNA polymerase to transcribe the operon. Therefore, when there is a high level of tryptophan within the cell (i.e., when the cell does not need to make more tryptophan), the *trp* operon is turned off.

After the action of the *trp* repressor was elucidated, Charles Yanofsky and coworkers at Stanford made a few unexpected observations. Two *trp* operon mu-

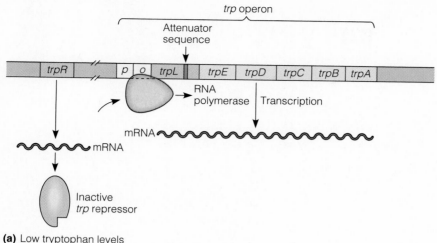

(a) Low tryptophan levels

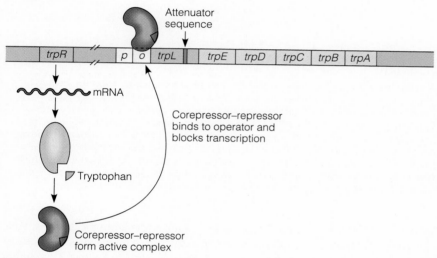

(b) High tryptophan levels

FIGURE **15-11**

Organization of the *trp* operon and regulation via the *trp* repressor protein. **(a)** When tryptophan levels are low, tryptophan does not bind to the repressor protein; the latter thus cannot bind to the operator site. Under these conditions, RNA polymerase can transcribe the operon, which leads to the expression of the *trpE, trpD, trpC, trpB,* and *trpA* genes. These genes encode enzymes involved in tryptophan biosynthesis. **(b)** When tryptophan levels are high, tryptophan acts as a corepressor that binds to the *trp* repressor protein. The tryptophan–*trp* repressor complex then binds to the operator site to inhibit transcription.

tations were identified in which a region including the *trpL* gene was missing from the operon. These mutations actually had higher levels of expression of the other genes in the *trp* operon. In addition, other mutant strains were found that lacked the *trp* repressor protein. Surprisingly, these mutant strains still could inhibit expression of the *trp* operon in the presence of tryptophan. As is often the case, unusual observations can lead people into interesting avenues of study. By pursuing this research further, Yanofsky discovered a second regulatory mechanism in the *trp* operon, called **attenuation**. This mechanism is mediated by the *trpL* gene.

Attenuation can occur only in bacteria that normally couple the processes of transcription and translation (as was described in Figure 14-16). During attenuation, transcription actually begins, but it is terminated before the entire mRNA is made. A segment of DNA, termed the **attenuator**, is important in facilitating this termination. When attenuation occurs, the mRNA encoding the *trp* operon is made as only a short piece that terminates shortly past the *trpL* region. Since this short mRNA does not encode the genes required for tryptophan biosynthesis, this mechanism can prevent the expression of these genes.

The segment of the *trp* operon immediately downstream from the operator site plays a critical role during attenuation. The first coding sequence in the *trp* operon is the *trpL* region, which encodes a short peptide called the *Leader* peptide. As

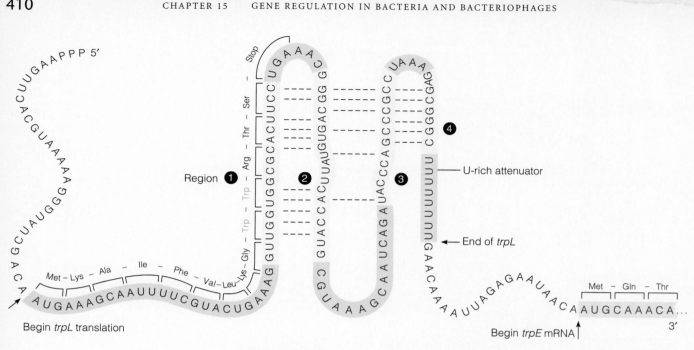

FIGURE **15-12**

Sequence of the *trpL* mRNA produced during attenuation. A second method of regulation of the *trp* operon is attenuation. At high tryptophan levels, it may occasionally happen that the *trp* repressor protein does not bind to the operator site to inhibit transcription. If so, a short mRNA is made that encodes the *trpL* region. As shown here, this mRNA has several regions that are complementary to each other. The hydrogen bonding between yellow regions 1 and 2, 2 and 3, and 3 and 4 is also shown. The last U in the purple attenuator sequence is the last nucleotide that would be transcribed during attenuation. At low tryptophan concentrations, however, transcription would occur beyond the end of *trpL* and proceed through the *trpE* gene and the rest of the *trp* operon.

shown in Figure 15-12, two features are key in the attenuation mechanism. First, there are two tryptophan codons within the sequence that encodes the *trp* leader peptide. As we will see later, these two codons provide a way to sense whether or not there is sufficient tryptophan for the bacterium to translate its proteins. Second, the RNA that is transcribed from this region can form stem-loop structures. The type of stem-loop structure that forms underlies attenuation.

Different combinations of stem-loop structures are possible due to interactions among four sequences within the RNA transcript (see the yellow regions in Figure 15-12). Region 2 is complementary to region 1 and also to region 3. Region 3 is complementary to region 2 as well as to region 4. Therefore, several stem-loop structures are possible. Even so, keep in mind that a particular segment of RNA can only participate in the formation of one stem-loop structure. For example, if region 2 forms a stem-loop with region 1 it cannot (at the same time) form a stem-loop with region 3. Alternatively, if region 2 forms a stem-loop with region 3, then region 3 cannot form a stem-loop with region 4. Though several stem-loop structures are possible, the 3–4 stem-loop structure is unique: it can act as a transcriptional terminator. The 3–4 stem-loop and the U-rich attenuator sequence act as an intrinsic terminator (i.e., ρ-independent terminator as described in Chapter 13). Therefore, the formation of the 3–4 stem-loop causes RNA polymerase to pause and the U-rich sequence dissociates from the DNA. This terminates transcription at the 3′ end of the *trpL* gene. By comparison, if region 3 forms a stem-loop with region 2, transcription will not be terminated because a 3–4 stem-loop cannot form.

Conditions that favor the formation of the 3–4 stem-loop ultimately rely on the translation of the *trpL* gene. As shown in Figure 15-13, there are three possi-

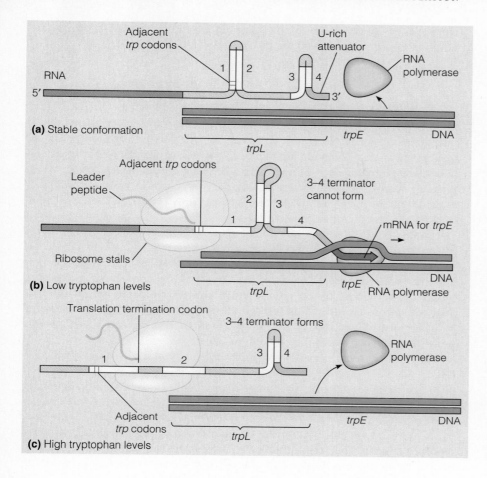

(a) Stable conformation

(b) Low tryptophan levels

(c) High tryptophan levels

FIGURE **15-13**

Possible stem-loop structures formed from *trpL* mRNA under different conditions of translation. (a) When translation is not coupled with transcription, the most stable form of the mRNA is when region 1 hydrogen bonds to region 2 and region 3 hydrogen bonds to region 4. A terminator stem-loop forms, and transcription will be terminated just past the *trpL* gene. **(b)** Coupled transcription and translation have occurred under conditions where the tryptophan concentration is low. The ribosome pauses at the *trp* codons in the *trpL* gene because of insufficient amounts of tRNA^{Trp}. This pause blocks region 1 of the mRNA, so that region 2 can only hydrogen bond with region 3. Since region 3 is already hydrogen bonded to region 2, the 3–4 stem-loop structure cannot form. Transcriptional termination does not occur, and RNA polymerase transcribes the rest of the operon. **(c)** Coupled transcription and translation occurs under conditions where there is a sufficient amount of tryptophan in the cell. Translation of the *trpL* gene progresses to its stop codon, where the ribosome pauses. This blocks region 2 from hydrogen bonding with any region and thereby enables region 3 to hydrogen bond with region 4. This terminates transcription at the end of the *trpL* gene.

ble scenarios. In Figure 15-13a, translation is not coupled with transcription. Since it is the most stable form of the mRNA, region 1 rapidly hydrogen bonds to region 2, and region 3 is left to hydrogen bond to region 4. Therefore, the terminator stem-loop forms, and transcription will be terminated just past the *trpL* gene. In Figure 15-13b, coupled transcription and translation occur under conditions where the tryptophan concentration is low. When tryptophan levels are low, the cell cannot make a sufficient amount of tRNA^{Trp}. As we see in Figure 15-13b, the ribosome pauses at the *trp* codons in the *trpL* gene, because it is waiting for the cell to provide tRNA^{Trp}. This pause occurs in such a way that the ribosome shields region 1 of the mRNA. This sterically prevents region 1 from hydrogen bonding to region 2. As an alternative, region 2 hydrogen bonds to region 3. Therefore, since region 3 is already hydrogen bonded to region 2, the 3–4 stem-loop structure cannot form. Under these conditions, transcriptional termination does not occur, and RNA polymerase transcribes the rest of the operon. This ultimately enables the bacterium to make more tryptophan.

Finally, in Figure 15-13c, coupled transcription and translation occur under conditions where there is sufficient tryptophan in the cell. In this case, translation of the *trpL* gene progresses to its stop codon, where the ribosome pauses. The pausing at the stop codon prevents region 2 from hydrogen bonding with any region and thereby enables region 3 to hydrogen bond with region 4. As in Figure 15-13a, this terminates transcription. Of course, keep in mind that the *trpL* gene contains two tryptophan codons. For the ribosome to smoothly progress to the *trpL* stop codon, there must have been enough tRNA^{Trp} in the cell to translate this gene. It follows that the bacterium must have a sufficient amount of tryptophan.

Since the cell does not need to synthesize more tryptophan, the rest of the transcription of the operon is terminated.

Repressible operons usually encode anabolic enzymes, and inducible operons encode catabolic enzymes

In the preceding parts of this chapter, we have seen that bacterial genes can be transcriptionally regulated in a positive or negative way—often times both. The *lac* operon and arabinose operon are inducible systems regulated by sugar molecules that activate transcription of these operons. By comparison, the *trp* operon is a repressible operon regulated by a corepressor, tryptophan, that binds to the repressor and turns the operon off. In addition, an abundance of tRNA^{Trp} in the cytoplasm can turn the operon off via attenuation.

In studying the genetic regulation of many operons, geneticists have noticed a general trend concerning inducible versus repressible regulation. When the genes in an operon encode proteins that function in the breakdown (i.e., catabolism) of a substance, they are usually regulated in an inducible manner. The substance to be broken down (or a related compound) often acts as the inducer. For example, allolactose and arabinose act as inducers of the *lac* and *ara* operons, respectively. An inducible form of regulation allows the bacterium to phenotypically express the appropriate genes only when they are needed to catabolize these sugars.

In contrast, other enzymes are important for synthesizing small molecules (i.e., anabolism). The genes that encode these anabolic enzymes tend to be regulated by a repressible mechanism. The inhibitor or corepressor is commonly the small molecule that is the product of the enzymes' biosynthetic activity. For example, tryptophan is produced by the sequential action of several enzymes that are encoded by the *trp* operon. Tryptophan itself acts as a corepressor that can bind to *trp* repressor protein when the intracellular levels of tryptophan become relatively high. This mechanism turns off the genes required for tryptophan biosynthesis when enough of this amino acid has been made. Therefore, genetic regulation via repression provides the bacterium with a way to prevent the overproduction of the product of an anabolic pathway.

TRANSLATIONAL AND POSTTRANSLATIONAL REGULATION

Though genetic regulation in bacteria is exercised predominantly at transcription, there are many examples where regulation is exerted at a later stage in gene expression. In some cases, specialized mechanisms have evolved to regulate the translation of certain mRNAs. As was described in Chapter 14, the translation of mRNA occurs in three stages: initiation, elongation, and termination. Genetic regulation of translation is usually aimed at preventing the initiation step. In this section, we will examine a couple of ways that bacteria can regulate the initiation of translation.

The net result of translation is the synthesis of a protein. It is the activities of proteins within living cells that ultimately determine an organism's traits. Therefore, any modification that alters protein function can be considered to affect gene expression. The term **posttranslational regulation** refers to the control of the functioning of proteins that are already present in the cell rather than regulation of transcription or translation. Posttranslational control can either activate or inhibit the function of a protein. Compared with transcriptional or translational control, posttranslational control can be relatively fast, occurring in a matter of seconds, which is an important advantage. In contrast, transcriptional and translational control typically require several minutes or even hours to take effect, since these two mechanisms involve the synthesis and turnover of mRNA and polypeptides. In this section, we will also examine ways that protein function can be regulated post-translationally.

Repressor proteins can inhibit the initiation of translation

One common way to exert translational control involves the binding of a regulatory protein to the mRNA. A **translational regulatory protein** recognizes sequences within the mRNA, much as transcription factors recognize DNA sequences. In most cases, translational regulatory proteins act to inhibit translation. These are known as **translational repressors**. When a translational repressor protein binds to the mRNA, it can inhibit translational initiation in one of two ways. One possibility is that it can bind in the vicinity of the Shine–Dalgarno sequence and/or the start codon and thereby sterically block the ribosome's ability to initiate translation in this region. Alternatively, the repressor protein may bind outside of the Shine–Dalgarno/start codon region, but stabilize an mRNA secondary structure that prevents initiation.

The synthesis of antisense RNA can also regulate translation

A second way to regulate translation is via the synthesis of **antisense RNA**. This term refers to an RNA strand that is complementary to a strand of mRNA. The mRNA strand has the same sequence as the DNA sense strand (also known as the coding strand). To understand this form of genetic regulation, let's consider a trait known as osmoregulation, which is essential for the survival of most bacteria. Osmoregulation refers to the ability to control the amount of water inside the cell. Since the solute concentrations in the external environment may rapidly change between hypotonic and hypertonic conditions, bacteria must have an osmoregulation mechanism to maintain their internal cell volume. Otherwise, bacteria would be susceptible to the harmful effects of lysis or shrinking.

In *E. coli*, an outer membrane protein encoded by the *ompF* gene is important in osmoregulation. At low osmolarity, the OmpF protein is preferentially produced, whereas at high osmolarity its synthesis is decreased. The expression of another gene, known as *micF*, is responsible for inhibiting the expression of the *ompF* gene at high osmolarity. As shown in Figure 15-14, the inhibition occurs because the *micF* RNA is complementary to the *ompF* mRNA. When the *micF* gene is transcribed, its RNA product binds to the *ompF* mRNA via hydrogen bonding between their complementary regions. The binding of the *micF* RNA to the *ompF* mRNA prevents the *ompF* mRNA from being translated. The RNA transcribed from the *micF* gene is called antisense RNA because it is complementary to the *ompF* RNA, which is a sense strand of mRNA that encodes a polypeptide. The *micF* RNA does not encode a polypeptide.

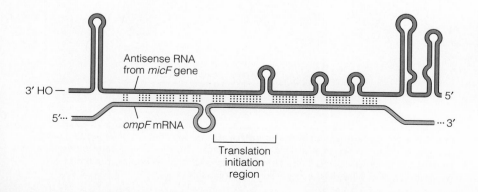

FIGURE **15-14**

The double-stranded RNA structure formed between the *micF* antisense RNA and the *ompF* mRNA. This double-stranded structure cannot be translated, because the ribosome must bind to a single-stranded ribosomal-binding site to begin translation.

Posttranslational regulation can occur via feedback inhibition and covalent modifications

A common mechanism to regulate the activity of metabolic enzymes is **feedback inhibition**. The synthesis of many cellular molecules such as amino acids, vitamins, and nucleotides occurs via the action of a series of enzymes that convert precursor molecules to a particular product. The final product in a metabolic pathway often can inhibit an enzyme that acts early in the pathway. This idea is schematically shown in Figure 15-15, in which the metabolic product inhibits enzyme 1 of the pathway.

In Figure 15-15, enzyme 1 is an example of an **allosteric enzyme**. It contains two different binding sites. The *catalytic site* is responsible for the binding of the substrate and its conversion to intermediate A. The second site is a *regulatory site*. This site binds the product of the metabolic pathway. When bound to the regulatory site, the product inhibits the catalytic ability of enzyme 1.

To appreciate feedback inhibition at the cellular level, we can consider the relationship between the product concentration and the regulatory site on enzyme 1. As the product is made within the cell, its concentration will gradually increase. Once the product concentration has reached a level that is similar to its affinity for enzyme 1, it becomes likely that the product will bind to the regulatory site on enzyme 1 and inhibit its function. This blocks the further synthesis of intermediate A. In this way, the net result is that the product of a metabolic pathway inhibits the further synthesis of more product. Under these conditions, the concentration of the product has reached a level that is sufficient for the purpose of the bacterium.

A second strategy to control the function of proteins is by the covalent modification of their structure. Certain types of modifications are involved primarily in the assembly and construction of a functional protein. These alterations include proteolytic processing, disulfide bond formation, and the attachment of prosthetic

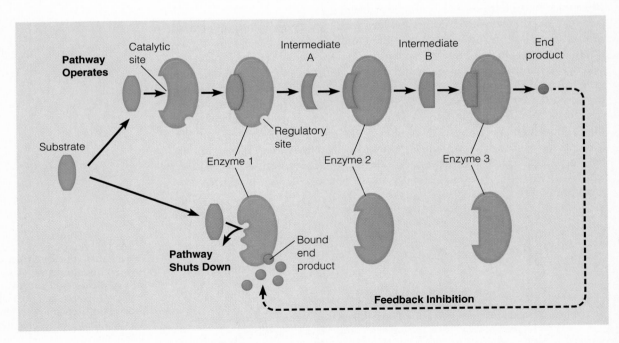

FIGURE **15-15**

Feedback inhibition in a metabolic pathway. The substrate is converted to a product by the sequential action of three different enzymes. Enzyme 1 has a catalytic site that recognizes the substrate, and it also has a regulatory site that recognizes the product. When the product binds to the regulatory site, it inhibits enzyme 1.

groups, sugars, or lipids. These are typically irreversible changes required to produce a functional protein. In contrast, other types of modifications, such as phosphorylation ($-PO_4$), acetylation ($-COCH_3$), and methylation ($-CH_3$), are often reversible modifications that transiently affect the function of a protein.

GENE REGULATION IN THE BACTERIOPHAGE LIFE CYCLE

Viruses are small particles that contain genetic material surrounded by a protein coat. They can infect a living cell and then propagate themselves by utilizing the energy and metabolic machinery of the host cell. For this to occur, the genetic material of viruses orchestrates an intricate series of steps, involving the expression of many viral genes. During the past several decades, the reproduction of viruses has presented an interesting and challenging problem for geneticists to investigate. The study of bacteriophages, which are viruses that infect bacteria, has greatly advanced our basic knowledge of how genetic regulatory proteins work. In addition, the study of viruses has been instrumental in our ability to devise medical strategies aimed at combating viral disease. For example, our knowledge of the life cycles of human viruses has led to the development of drugs that are used to inhibit viral growth. For example, azidothymidine (AZT), which is used to combat HIV (human immunodeficiency virus), suppresses the production of viral DNA by inhibiting a viral gene product involved in viral DNA synthesis.

In this section, we will focus on the function of bacteriophage genes that encode genetic regulatory proteins. The structural genes of bacteriophages are often in an operon arrangement. This enables the genes within an operon to be controlled by regulatory proteins that bind to operator sites and influence the function of nearby promoters. Like bacterial operons, phage operons can be controlled by repressor proteins or activator proteins. To understand how this works, we will carefully examine the two life cycles of a virus called phage λ (lambda), which was discovered in the 1940s by Andre Lwoff and his colleagues at the Pasteur Institute in Paris. Since its discovery, phage λ has been investigated extensively and has provided geneticists with a model on which to base our understanding of viral proliferation.

Phage λ can follow a lytic or lysogenic life cycle

Phage λ can bind to the surface of a bacterium and inject its genetic material into the bacterial cytoplasm. After this occurs, the phage will "choose" between two alternative life cycles, known as the lytic cycle and the lysogenic cycle (refer back to Figure 6-8). During the lytic cycle, the genetic instructions of the bacteriophage direct the synthesis of many copies of the phage genetic material and coat proteins. These are then assembled to make new phages. When synthesis and assembly is completed, the bacterial host cell is lysed, and the newly made phages are released into the environment.

Alternatively, phage λ can infect a bacterium and not progress to the lytic cycle. Instead, this phage can act as a **temperate phage** that will usually not produce new phages and will not kill the bacterial cell that acts as its host. Rather, the phage follows a lysogenic cycle. During the lysogenic cycle, phage λ integrates its genetic material into the chromosome of the bacterium. This integrated phage DNA is known as a **prophage**. A prophage can exist in a dormant state for a long time, during which no new bacteriophages are made. When a bacterium containing a lysogenic prophage divides to produce two daughter cells, it will copy the prophage's genetic material along with its own chromosome. Therefore, both daughter cells will inherit the prophage. At some later time, a prophage may become activated to excise itself from the bacterial chromosome and enter the lytic cycle. This process is called induction. During induction, the phage will promote the synthesis of new phages and, eventually, lyse the host cell.

Figure 15-16 shows the genome of phage λ. Inside the virion head, phage λ DNA is linear. After injection into the bacterium, the two ends of the DNA become covalently attached to each other to form a circular piece of DNA. The organization of the genes within this circular structure reflects the two alternative life cycles of this virus. The genes in the top center of the figure are transcribed very soon after infection. This occurs at the beginning of either life cycle. As we will discuss later, the expression pattern of these early genes will determine whether the lytic or lysogenic cycle prevails.

If the lysogenic cycle prevails, the integrase (*int*) gene will be subsequently turned on. The integrase gene encodes an enzyme that integrates the λ DNA into the bacterial chromosome. The mechanism of phage λ integration will be described in Chapter 18.

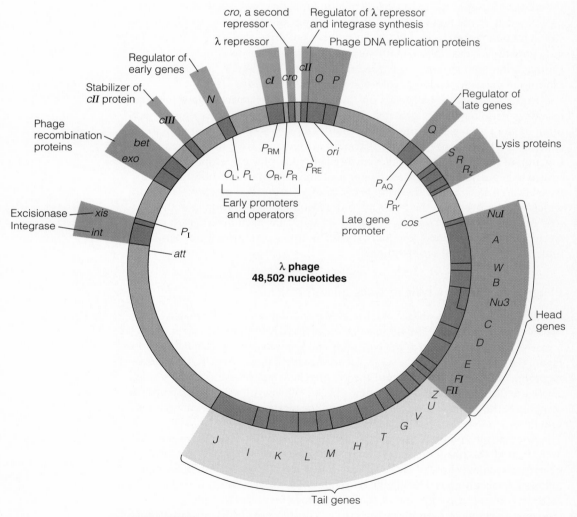

FIGURE **15-16**

The genome of phage λ. The genes shaded in red encode genetic regulatory proteins that determine whether the lysogenic or lytic cycle prevails. Genes shaded in green encode proteins that are necessary for the lysogenic cycle. The genes shaded in blue and gray encode proteins that are required for the lytic cycle.

If the lysogenic cycle is not chosen, the genes on the right side of the figure will be transcribed. This operon contains many genes that are necessary for λ assembly and release. These genes encode replication proteins, coat proteins, proteins involved in coat assembly, proteins involved in packaging the DNA into the phage head, and enzymes that will cause the bacterium to lyse.

During the lytic cycle, the λ repressor and integrase proteins are made

Now that we have an understanding of the phage λ life cycles and genome organization, let's examine how the decision is made between the lytic and lysogenic cycles. This choice depends on the actions of several genetic regulatory proteins. Our molecular understanding of the phage λ life cycles represents an extraordinary accomplishment in the field of genetic regulation. This process is quite detailed, since it involves a series of intricate steps in which regulatory proteins bind to several different sites in the λ genome. To simplify these events, we will begin by describing the steps that occur when the lysogenic cycle prevails. In the following section, we will examine the steps that occur when the lytic cycle is followed.

Soon after infection, two promoters, designated P_L and P_R, are used for transcription. This initiates a competition between the lytic and lysogenic cycles (see Figure 15-17). Initially, the transcription from P_R and P_L results in the synthesis of two short RNA transcripts that only encode two proteins called the *cro protein* and the *N protein*, both of which are genetic regulatory proteins. We will consider the function of the cro protein later in this chapter. The N protein is a genetic regulatory protein with an interesting function that we have not yet considered. Its function, known as **antitermination**, is to prevent transcriptional termination. The N protein binds to three sites, designated t_L, t_{R1}, and t_{R2}. When it binds to the t_{R1} and t_{R2} sites, the transcript from P_R is extended to include *cII*, *O*, *P*, and *Q*. The *cII* gene encodes an activator protein, the *O* and *P* genes encode enzymes needed for the initiation of λ DNA synthesis, and the *Q* gene encodes a regulatory protein required for the lytic cycle. When the N protein binds to t_L, the transcript from P_L is extended to include *int*, *xis*, and *cIII*. The cIII protein helps to stabilize the cII activator protein.

As shown in Figure 15-17a, if the cII–cIII complex accumulates to sufficient levels, the lysogenic cycle is favored. Once it is made, the cII protein activates two different promoters in the λ genome. When the cII protein binds to the promoter P_{RE}, it turns on the transcription of *cI*, a gene that encodes the λ repressor. The cII protein also activates the *int* gene by binding to the promoter P_I. The λ repressor and integrase protein play central roles in promoting the lysogenic life cycle. When the λ repressor is made in sufficient quantities, it binds to operator sites that are adjacent to P_R and P_L. When the λ repressor is bound to these operator sites, it inhibits the expression of the genes that are required for lytic infection.

As you are looking at Figure 15-17a, you may have noticed that the binding of the λ repressor to O_R will inhibit the expression of *cII*. This may seem counterintuitive since the cII protein was initially required to activate the *cI* gene (which encodes the λ repressor). You may be thinking that the inhibition of the *cII* gene will eventually prevent the expression of the λ repressor protein. This does not occur during the lysogenic cycle because the *cI* gene actually has two promoters, P_{RE} (which is activated by the cII protein) and also a second promoter called P_{RM}. Transcription from P_{RE} occurs early in the lysogenic cycle. P_{RE} gets its name because the use of this promoter results in the expression of the λ *Repressor* during the *Establishment of the lysogenic cycle. The transcript made from the use of P_{RE} is very stable and quickly leads to a buildup of the λ repressor protein. This causes an abrupt inhibition of the lytic cycle because the the binding of the λ repressor protein to O_R blocks the P_R promoter. Later in the lysogenic cycle, it is no longer necessary to make a large amount of the λ repressor. At this point, the use of the P_{RM} promoter is sufficient to make enough *Repressor* protein to *Maintain the lysogenic cycle.

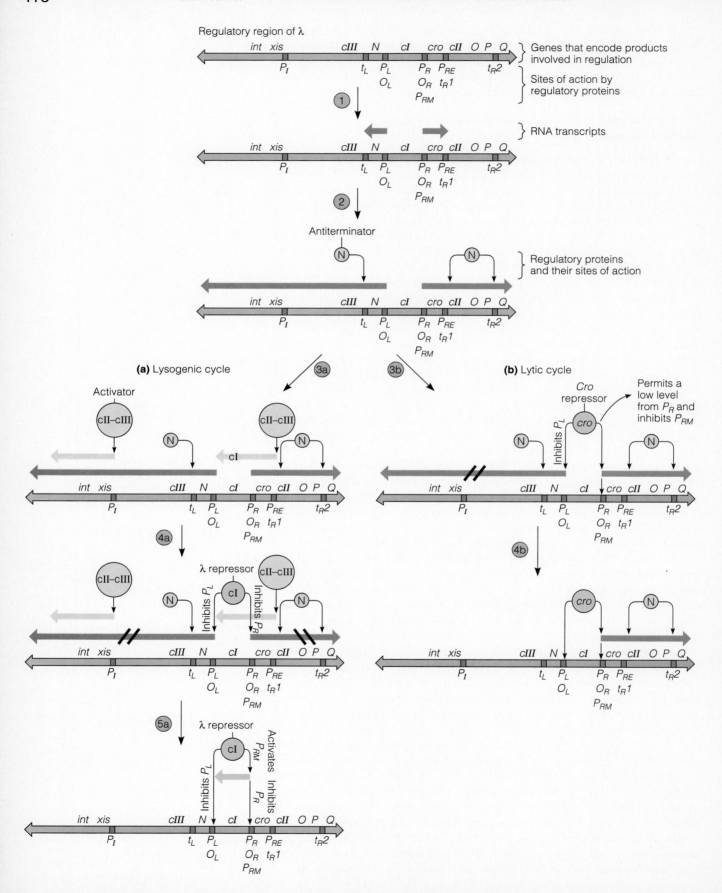

FIGURE **15-17**

The sequence of events that occur during the (a) lysogenic and (b) lytic cycles of phage lambda. The top part of this figure shows the region of the phage λ genome that regulates the choice between the lytic and lysogenic cycles. **(1)** Immediately after infection, P_L and P_R are used to make two short mRNAs. These mRNAs only encode two early proteins, designated N and cro. **(2)** The N protein binds to three sites (t_L, t_{R1}, and t_{R2}) to prevent mRNA termination. This allows the transcription of the delayed early genes, which include *cIII*, *cII*, *O*, *P*, and *Q*. (Note: *O*, *P*, and *Q* are only necessary for the lytic cycle.) **(a) The lysogenic cycle. (3a)** The cII–cIII complex binds to P_{RE}. This activates the transcription of the *cI* gene, which encodes the λ repressor. The cII–cIII complex also binds to P_I to activate the transcription of the *int* gene. **(4a)** The λ repressor binds to O_R and O_L to inhibit transcription from P_R and P_L. This prevents the lytic cycle. The integrase protein catalyzes the integration of the λ DNA into the *E. coli* chromosome. **(5a)** Later in the lysogenic cycle, the *cI* gene is transcribed from the P_{RM} promoter. This results in a low but steady synthesis of the λ repressor, which is necessary to further maintain the lysogenic state. **(b) The lytic cycle. (3b)** The cro protein binds to O_R and blocks transcription from P_{RM}. This prevents the synthesis of the λ repressor. Transcription from P_R is still allowed to occur at a low level, which leads to a buildup of the O, P, and Q proteins. **(4b)** The O and P proteins catalyze the DNA replication of additional λ DNA. The Q protein stimulates transcription from $P_{R'}$ (not show in this figure, but see Figure 15–16). This leads to the synthesis of many proteins that are necessary to make new λ phages and to lyse the cell.

GENES ⟶ TRAITS: The ability to choose between two alternative life cycles can be viewed as a trait of this bacteriophage. As described here, the choice between the two life cycles depends on the pattern of gene regulation.

The lytic cycle depends on the action of the cro protein

As we have just seen, the λ repressor protein binds to O_R and prevents the expression of the operons needed for the lytic cycle. For the lytic cycle to occur, the λ repressor must be prevented from inhibiting P_R (Figure 15-17b). This is the role of the cro protein. If the activity of the cro protein exceeds the activity of the cII protein, the lytic cycle prevails. As was mentioned, an early step in the expression of λ genes is the transcription from P_R to produce the cro protein. If the concentration of the cro protein builds to sufficient levels, it will bind to two operator regions, O_R and O_L. The binding of cro to O_L inhibits transcription from P_L; the binding of cro to O_R has several effects. When the cro protein binds to O_R, it inhibits transcription from P_{RM} in the leftward direction (see Figure 15-17b). This inhibition prevents the expression of cI which encodes the λ repressor; the λ repressor is needed to maintain the progression to lysogeny. Therefore, the λ repressor cannot successfully shut down transcription from P_R.

The binding of the cro protein to O_R also allows a low level of transcription from P_R in the rightward direction. This enables the transcription and translation of the O, P, and Q genes. The O and P proteins are necessary for the replication of the λ DNA. The Q protein is a regulatory protein that activates another promoter, designated $P_{R'}$ (not to be confused with P_R). The $P_{R'}$ promoter controls a very large operon that encodes the proteins necessary for the phage coat, the assembly of the coat proteins, the packaging of the λ DNA, and the lysis of the bacterial cell (see Figure 15-16). These proteins are made toward the end of the lytic cycle. The expression of these late genes leads to the synthesis and assembly of many new λ phages that are released from the bacterial cell when it lyses.

Cellular proteases influence the choice between the lytic and lysogenic cycle

As you may have noticed from Figure 15-17, the first two steps of the lysogenic and lytic cycles are identical. Whether the lysogenic or lytic cycle prevails depends on the steps that occur after the early genes are transcribed. In particular, the activity of the cII protein plays a key role in directing λ to the lysogenic or lytic cycle. A critical physiological issue is that the cII protein is easily degraded by cellular proteases that are produced by *E. coli*. Whether or not these proteases are made depends on the environmental conditions. If the growth conditions are very favorable (e.g., a rich growth medium), the intracellular protease levels are relatively high, and the cII protein tends to be degraded. When cII protein is present at a low level, it cannot bind to P_{RE} and activate the λ repressor gene which is required for the lysogenic cycle. Instead, the cro protein slowly accumulates to sufficient levels (as in Figure 15-17b). The binding of the cro protein to O_R prevents transcription of the λ repressor gene from P_{RM} and, at the same time, allows the lytic cycle to proceed. In this way, environmental conditions that are favorable for growth promote the lytic cycle. This makes sense because a sufficient supply of nutrients is necessary to synthesize new bacteriophages.

Alternatively, starvation conditions favor the lysogenic cycle. When nutrients are limiting, cellular proteases are relatively inactive. Under these conditions, the cII protein will build up much more quickly than the cro protein. Therefore, the cII protein will turn on P_{RE} and lead to the transcription of the λ repressor (as in Figure 15-17a). This event favors the lysogenic cycle. It is advantageous to favor lysogeny under starvation conditions, because there may not be sufficient nutrients available for the production of new λ phages.

After lysogeny has been established, certain environmental conditions will favor induction to the lytic cycle. For example, exposure to UV light will promote induction. This also is caused by the activation of cellular proteases. In this case, a cellular protein known as RecA (a protein ordinarily involved in facilitating recombination between DNA molecules) senses the DNA damage and is activated to become a cellular protease. RecA protein cleaves the λ repressor and thereby inactivates it. This allows transcription from P_R and eventually leads to the accumulation of the cro protein. This favors the lytic cycle. Under these conditions, it may be advantageous for λ to make new phages and lyse the cell, because the exposure to UV light may have already damaged the bacterium and prevented further bacterial growth and division.

The O_R region provides a genetic switch between the lytic and lysogenic cycles

Before we end this section on the λ life cycles, it is interesting to consider how the O_R region acts as a genetic switch between the lytic and lysogenic life cycles. Depending on the binding of genetic regulatory proteins to this region, the switch can be turned to favor the lytic or lysogenic cycle. To understand how this switch works, we need to take a closer look at the O_R region (Figure 15-18).

The O_R region contains three operator sites, designated O_{R1}, O_{R2}, and O_{R3}. The λ repressor protein or the cro protein can bind to any or all of these sites. The binding of these two proteins at these sites governs the switch between the lysogenic and lytic cycles. There are two critical issues that influence this binding event. The first is the relative affinities that the regulatory proteins have for these operator sites. The second is the concentrations of the λ repressor protein and the cro protein within the cell.

Let's first consider how an increasing concentration of the λ repressor protein can switch on the lysogenic cycle and switch off the lytic cycle (Figure 15-18a). This protein was first isolated by Mark Ptashne and his colleagues while at Harvard University in 1967. Their studies showed that λ repressor binds with highest affinity to O_{R1}, followed by O_{R2} and O_{R3}. As the concentration of the λ repressor builds within the cell, a dimer of the λ repressor protein will first bind to O_{R1}, because it has the highest affinity for this site. Next, a second λ repressor dimer will bind to O_{R2}. This occurs very rapidly, because the binding of the first dimer to O_{R1} favors the binding of a second dimer to O_{R2}. This is called a *cooperative interaction*. The binding of the λ repressor to O_{R1} and O_{R2} inhibits transcription from P_R and thereby switches off the lytic cycle.

Early in the lysogenic cycle, the λ repressor protein concentration may become so high that it will occupy O_{R3}. Eventually, however, the λ repressor concentration will begin to drop, because the inhibition of P_R will decrease the synthesis of cII, which activates the λ repressor gene (from P_{RE}). As the λ repressor concentration gradually falls, it will first be removed from O_{R3}. This allows transcription from P_{RM}. In fact, the

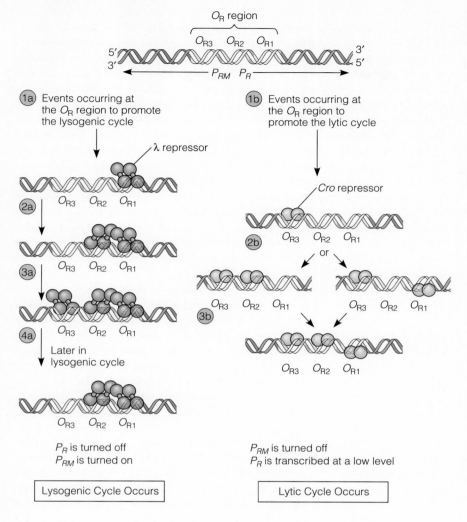

1a) Events occurring at the O_R region to promote the lysogenic cycle

λ repressor

2a) O_{R3} O_{R2} O_{R1}

3a) O_{R3} O_{R2} O_{R1}

4a) O_{R3} O_{R2} O_{R1}

Later in lysogenic cycle

O_{R3} O_{R2} O_{R1}

P_R is turned off
P_{RM} is turned on

Lysogenic Cycle Occurs

1b) Events occurring at the O_R region to promote the lytic cycle

Cro repressor

2b) O_{R3} O_{R2} O_{R1}

or

3b) O_{R3} O_{R2} O_{R1} O_{R3} O_{R2} O_{R1}

O_{R3} O_{R2} O_{R1}

P_{RM} is turned off
P_R is transcribed at a low level

Lytic Cycle Occurs

O_R region

O_{R3} O_{R2} O_{R1}

5'
3'   3'
                      5'

P_{RM} P_R

FIGURE **15-18**

The O_R region, the genetic switch between the lysogenic and lytic life cycles. **(a)** Events at the O_R region that promote the lysogenic life cycle. The λ repressor protein contains two globular domains connected by a short link. A dimer binds much more tightly to each operator site than a monomer would. **(1a)** The λ repressor protein dimer will first bind to O_{R1}, because it has the highest affinity for this site. **(2a)** A second λ repressor dimer will bind to O_{R2}. This occurs very rapidly, because the binding of the first dimer to O_{R1} favors the binding of a second dimer to O_{R2}. The binding of the λ repressor to O_{R1} and O_{R2} inhibits transcription from P_R and thereby switches off the lytic cycle. **(3a)** Early in the lysogenic cycle, the λ repressor protein concentration may become so high that it will occupy O_{R3}. **(4a)** Later, the λ repressor concentration will begin to drop, because the inhibition of P_R will decrease the synthesis of cII protein, which activates the λ repressor gene (from P_{RE}). As the λ repressor concentration gradually falls, it will first be removed from O_{R3}. This allows transcription from P_{RM} and maintains the lysogenic cycle. **(b)** Events at the O_R region that promote the lytic cycle. The cro repressor protein is a small globular protein that also binds to each operator site as a dimer. **(1b)** The cro protein has its highest affinity for O_{R3}, and so it binds there first. This blocks transcription from P_{RM} and thereby switches off lysogeny. **(2b)** The cro protein has a similar affinity for O_{R2} and O_{R1}, and so it may occupy either of these sites next. **(3b)** Later in the lytic cycle, the cro protein concentration will continue to rise, so that eventually it will bind to both O_{R2} and O_{R1}. This turns down the expression from P_R, which is not needed in the later stages of the lytic cycle.

term λ repressor is somewhat misleading, because the binding of the λ repressor at only O_{R1} and O_{R2} acts as an activator of P_{RM}. The ability of the λ repressor to activate its own transcription allows the switch to the lysogenic cycle to be maintained.

In the lytic cycle (Figure 15-18b), the binding of the cro protein controls the switch. The cro protein has its highest affinity for O_{R3} and has a similar affinity for O_{R2} and O_{R1}. Under conditions that favor the lytic cycle, the cro protein accumulates, and a cro dimer first binds to O_{R3}. This blocks transcription from P_{RM} and thereby switches off lysogeny. Later in the lytic cycle, the cro protein concentration will continue to rise, so that eventually it will bind to O_{R2} and O_{R1}. This turns down expression from P_R, which is not needed in the later stages of the lytic cycle.

Genetic switches, like the one just described for phage λ, represent an important form of genetic regulation. As we have just seen, a genetic switch can be used to control two alternative life cycles of a bacteriophage. In addition, genetic switches are also important in the developmental pathways for bacteria and eukaryotes. For example, certain species of bacteria can grow in a vegetative state when nutrients are abundant but will sporulate when conditions are unfavorable for growth. The choice between sporulation and vegetative growth involves genetic switches. Likewise, genetic switches operate in the developmental pathways in eukaryotes. As will be described in Chapter 24, they are key events in the initiation of cell differentiation during development. Studies of the phage λ life cycle have provided fundamental information with which to understand how these other switches can operate at the molecular level.

Genes can be *regulated* in different ways. In prokaryotic cells, some genes are organized into multigene units called *operons*. An operon typically consists of a *promoter*, an *operator site*, several *structural genes,* and a terminator. The organization of an operon allows two or more genes to be coordinately regulated as a single unit. For transcriptional regulation, an important concept is that the binding of regulatory proteins to the operon can greatly influence the rate of transcription. Furthermore, the binding of small effector molecules to regulatory proteins can influence whether or not the regulatory proteins can bind to the DNA. Some regulatory proteins that exert *negative control*, such as the *lac* repressor and *trp* repressor, inhibit transcription. Others, such as the *catabolite activator protein (CAP)*, exert *positive control* by enhancing transcription. The araC protein is interesting because it can act as both a *repressor* or an *activator* depending on the presence or absence of arabinose. Experiments involving the use of F factors showed that regulatory proteins are synthesized and can diffuse within the cell to ultimately bind to a distant operator site and cause repression. This action of a regulatory protein is called a *trans effect*. By comparison, operators are *cis regulators* by providing a binding site for genetic regulatory proteins.

Our understanding of transcriptional regulation has been greatly aided by studying the life cycles of viruses. In this chapter, we have considered the ability of phage λ to choose between the lytic and lysogenic cycle. This choice is determined by the actions of several genetic regulatory proteins that can bind to the λ DNA and influence the transcription of nearby genes. The O_R region provides a genetic switch for the lytic or lysogenic cycle. If the λ repressor controls the switch, the lysogenic cycle is favored, whereas binding of the cro protein to this switch favors the lytic cycle. Cellular proteases, which are influenced by environmental conditions, play a key role in the choice between the lytic and lysogenic cycles.

In addition to direct transcriptional regulation, gene expression in bacteria can be affected at later stages in the expression process. During *attenuation* of the *trp* operon, for example, transcription actually begins, but the length of the transcript is determined by the translation of the *trpL* gene. This form of regulation is unique to bacteria, because they couple the processes of transcription and translation. The translation of a complete mRNA transcript can also be regulated. One form of translational regulation involves the binding of *translational repressor* proteins that prevent translational initiation. A second way involves the binding of *antisense RNA* to the mRNA to block translation. Finally, the function of proteins that have already been made can be influenced *posttranslationally. Feedback inhibition* involves the noncovalent binding of molecules to an *allosteric* site on a protein. This inhibition blocks the function of an enzyme that is required during an early step in a metabolic pathway. In addition, the function of proteins can be controlled by irreversible and reversible covalent modifications.

Overall, it is clear that genes can be regulated in a variety of different ways. If you are a scientist trying to understand the genetic regulation of a particular gene, your research can be a complicated, frustrating, and challenging task. At the same time, it can be a great deal of fun.

Early insights into the molecular mechanisms of gene expression came from experiments involving lactose metabolism in *E. coli*. Jacob and Monod examined the phenomenon of enzyme adaptation and concluded that the proteins involved in lactose metabolism were made after the bacterium was exposed to lactose. This observation led them to investigate the genes involved in this regulation process. By constructing merozygotes, they discovered that the *lacI* gene encodes a repressor protein that is bound to the operator site in the absence of lactose. More recently, studies by Müller-Hill showed that there are actually three operator sites. The crystal structure of the *lac* repressor tetramer bound at two out of three sites has been determined. Similarly, biophysical studies have determined the structure of CAP, a positive regulator of the *lac* operon.

A different form of genetic regulation, known as attenuation, was discovered by Yanofsky and colleagues. They identified mutant strains of bacteria that were missing the *trp* repressor, yet still could repress the *trp* operon when tryptophan levels were high. Additional mutations in the *trpL* region destroyed the ability to attenuate transcription. An analysis of the DNA sequence in this region revealed the ability of the mRNA to form alternative stem-loop structures. The 3–4 stem-loop structure forms a transcriptional terminator and thereby prevents the further synthesis of the *trp* operon.

An analysis of genetic sequences also revealed a posttranscriptional method of gene regulation involving antisense RNA. By comparing the sequence of the *ompF* mRNA and the *micF* RNA, researchers discovered that the two were complementary to each other. This explains how the synthesis of the *micF* RNA inhibits the translation of the *ompF* mRNA. When the (antisense) *micF* RNA hybridizes to the *ompF* mRNA, the latter cannot be recognized by the ribosomes.

We concluded this chapter with a description of gene regulation in bacteriophage λ. Many of the same kinds of experimental approaches were used to elucidate the molecular mechanisms that underlie the choice between the lysogenic and lytic cycles. In particular, many researchers have studied how mutations in particular genes in the λ genome favor the lysogenic or lytic cycles. In addition, biochemical and biophysical studies have analyzed the detailed interactions between λ regulatory proteins and the operator sites that they recognize.

PROBLEM SETS

SOLVED PROBLEMS

1. Researchers have identified mutations in the promoter region of the *lacI* gene that make it more difficult for the *lac* operon to be induced. These are called *lacI^Q* mutants, because a greater quantity of *lac* repressor protein is made. Explain why an increased transcription of the *lacI* gene makes it more difficult to induce the *lac* operon.

Answer: An increase in the amount of *lac* repressor protein will make it easier to repress the *lac* operon. When the cells become exposed to lactose, allolactose levels will slowly rise. Some of the allolactose will bind to the *lac* repressor protein. If there are many more *lac* repressor proteins found within the cell, it will take more allolactose to ensure that there are no unoccupied repressor proteins that can repress the operon.

2. Explain how the pausing of the ribosome in the presence or absence of tryptophan affects the formation of a terminator stem-loop.

Answer: The key issue is the location where the ribosome stalls. In the absence of tryptophan, it will stall over the *trp* codons in the *trpL* mRNA. Stalling at this site will also shield region 1 in the attenuator region. Since region 1 is unavailable to hydrogen bond with region 2, region 2 will hydrogen bond with region 3. Therefore, region 3 cannot form a terminator stem-loop with region 4. Alternatively, if there is sufficient tryptophan in the cell, the ribosome will stall over the stop codon in *trpL*. In this case, the ribosome will shield region 2. Therefore, region 3 and 4 will hydrogen bond with each other to form a terminator stem-loop. This will abruptly halt the continued transcription of the *trp* operon.

3. What are the key proteins that cellular proteases affect to choose the lytic or lysogenic cycle?

Answer: After infection, the key protein affected by cellular proteases is cII. If protease levels are high, as under good growth conditions, cII will be degraded. This promotes the lytic cycle. Under starvation conditions, the protease levels are low. This prevents the degradation of cII and thereby promotes the lysogenic cycle. After lysogeny has been established, the key protein affected by cellular proteases is the λ repressor. Agents such as UV light activate cellular proteases that digest the λ repressor. This permits induction of the lytic cycle.

CONCEPTUAL QUESTIONS

1. What is the difference between a constitutive gene and a regulated gene?

2. In general, why is it important to regulate genes? Discuss examples of situations in which it would be advantageous for a bacterial cell to regulate genes.

3. What is an operon? Describe its general organization.

4. In the *lac* operon, how would gene expression be affected if one of the following segments was missing?

   A. *lac* operon promoter
   B. Operator site
   C. *lacA* gene

5. The presence of a small effector molecule inhibits transcription. What combination(s) of effector molecule (inducer, inhibitor, or corepressor) and regulatory protein (activator or repressor protein) could explain this event?

6. If a gene is repressible and under positive control, describe what kind of effector molecule and regulatory protein are involved. Explain how the binding of the effector molecule affects the regulatory protein.

7. Some mutations have a cis effect, whereas others have a trans effect. Explain the molecular differences between cis and trans mutations. Which type of mutation (cis or trans) can be complemented in a merozygote experiment?

8. What is enzyme adaptation? From a genetic point of view, how does it occur?

9. Explain the meaning of the terms inducible and repressible.

10. If an abnormal repressor protein could still bind allolactose but the binding of allolactose did not alter the conformation of the repressor protein, how would this affect the expression of the *lac* operon?

11. What is diauxic growth? Explain the roles of cAMP and the catabolite activator protein in this process.

12. Describe the function of the araC protein. How does it positively and negatively regulate the arabinose operon?

13. What is meant by the term attenuation? Is it an example of gene regulation at the level of transcription or translation? Explain your answer.

14. As described in Figure 15-12, there are four regions within the *trpL* gene that can form stem-loop structures. Let's suppose that we have previously isolated mutations that disrupt the ability of a particular region to form a stem-loop structure with a complementary region. For example, a region 1 mutant cannot form a 1–2 stem-loop structure but it can still form a 2–3 or 3–4 structure. Likewise, a region 4 mutant can form a 1–2 or 2–3 stem-loop but not a 3–4 stem-loop. Under the following conditions, would attenuation occur?

   A. Region 1 is mutant, tryptophan is high, and translation is occurring.
   B. Region 2 is mutant, tryptophan is low, and translation is occurring.
   C. Region 3 is mutant, tryptophan is high, and translation is not occurring.

   D. Region 4 is mutant, tryptophan is low, and translation is not occurring.

15. Translational control is usually aimed at preventing the initiation of translation. With regard to cellular efficiency, why do you think this is the case?

16. What is a translational repressor protein? What does it bind to? How does it affect translational initiation?

17. What is antisense RNA? How does it affect the translation of a complementary mRNA?

18. What is feedback inhibition? At the cellular level, explain why feedback inhibition is useful.

19. A species of bacteria can synthesize the amino acid histidine so that it does not require histidine in its growth medium. A key enzyme, which we will call histidine synthetase, is necessary for histidine biosynthesis. When these bacteria are given histidine in their growth media, they stop synthesizing histidine intracellularly. Based on this observation alone, propose three different regulatory mechanisms to explain why histidine biosynthesis ceases when histidine is in the growth medium. To explore this phenomenon further, you measure the amount of intracellular histidine synthetase protein when cells are grown in the presence and absence of histidine. In both conditions, the amount of this protein is identical. Which mechanism of regulation would be consistent with this observation?

20. Describe three posttranslational modifications. How can reversible modifications modulate protein function?

21. Using three examples, describe how allosteric sites are important in the function of genetic regulatory proteins.

22. What are key features that distinguish the lytic and lysogenic cycles?

23. With regard to promoting the lytic or lysogenic cycle, what would happen if the following genes were missing from the λ genome?

   A. *cro*
   B. *cI*
   C. *cII*
   D. *int*
   E. *cI* and *cro*

24. How do the λ repressor and the cro protein affect the transcription from P_R and P_{RM}? Explain where these proteins are binding to cause their effects.

25. In your own words, explain why it is necessary for the *cI* gene to have two promoters. What would happen if it only had P_{RE}?

26. A mutation in P_R causes its transcription rate to be increased tenfold. Do you think that this mutation would favor the lytic or lysogenic cycle? Explain your answer.

27. When an *E. coli* bacterium already has a λ prophage integrated into its chromosome, another λ phage cannot usually infect the cell and establish the lysogenic or lytic cycle. Based on your understanding of the genetic regulation of the λ life cycles, why do you think the other phage would be unsuccessful?

EXPERIMENTAL QUESTIONS

1. In Experiment 15A, the optical density values were twice as high for the mated strain as for parent strain. Why was this result obtained?

2. A parent strain has a defective *lac* operator site that results in the constitutive expression of the *lac* operon. Outline an experiment you would carry out to demonstrate that the operator site must be physically adjacent to the genes that it influences. Based on your knowledge of the *lac* operon, describe the results you would expect to get.

3. Let's suppose you have isolated a mutant strain of *E. coli* in which the *lac* operon is constitutively expressed. To understand the nature of this defect, you mate this parent strain so that it now contains an F′ factor with a normal *lac* operon. You then compare the parent and mated strain with regard to their ß-galactosidase activities in the presence and absence of lactose. You obtain the following results:

Strain	Addition of Lactose	Amount of ß-Galactosidase (percentage of parent strain)
Parent	No	100
Parent	Yes	100
Mated	No	100
Mated	Yes	200

Explain the nature of the defect in the parent strain.

4. In Experiment 15A, a *lacI*⁻ mutant was mated to a strain that had a functional *lacI* gene on an F factor. The results of this experiment were important in determining the action of the *lac* repressor protein. What results would you expect if you used the same approach to investigate the regulation of the arabinose operon via *araC*? In other words, what would be the level of expression of the arabinose operon in an *araC*⁻ strain in the presence and absence of arabinose, and how would the level of expression change when a functional *araC* gene was introduced into the strain via mating?

5. The crystallization of the CAP–DNA complex revealed a 90° turn in the DNA. Think about the process of transcription (which was described in Chapter 13). Propose a model that explains why the 90° bend promotes transcription.

6. A segment of DNA that contains a loop is more compact than the same DNA segment without a loop. Therefore, when these two alternative structures (looped versus unlooped) are electrophoresed through a gel, the looped structure will migrate more quickly to the bottom of the gel, since it can more easily penetrate the gel matrix. Let's suppose that a mutant *E. coli* strain has been identified in which the araC protein represses the arabinose operon, even in the presence of arabinose. In the experiment shown here, mutant or normal araC protein was mixed with a segment of DNA containing the *ara* operon in the absence or presence of arabinose, and then run on a gel:

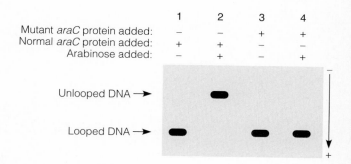

Describe the defect in this mutant araC protein.

QUESTIONS FOR STUDENT DISCUSSION/COLLABORATION

1. Discuss the advantages and disadvantages of genetic regulation at the different levels described in Figure 15-1.

2. As you look at Figure 15-8, discuss possible "molecular ways" that the cAMP–CAP complex and *lac* repressor may influence RNA polymerase function. In other words, try to explain how the bending and looping in DNA may affect the ability of RNA polymerase to initiate transcription.

3. Certain environmental conditions such as UV light are known to activate lysogenic λ prophages and cause them to progress into the lytic cycle. UV light initially causes the repressor protein to be proteolytically degraded. Make a flow diagram that describes the subsequent events that would lead to the lytic cycle. Note: The *xis* gene codes for an enzyme that is necessary to excise the λ prophage from the *E. coli* chromosome. The integrase enzyme is also necessary to excise the λ prophage.

CHAPTER 16

GENE REGULATION IN EUKARYOTES

REGULATION OF DNA AND CHROMATIN STRUCTURE

REGULATION OF TRANSCRIPTION

REGULATION OF RNA PROCESSING AND TRANSLATION

Eukaryotic organisms, a category that includes protozoa, algae, fungi, plants, and animals, have many reasons for regulating their genes. Like their prokaryotic counterparts described in Chapter 15, eukaryotic cells need to adapt to changes in their environment. For example, eukaryotic cells can respond to changes in nutrient availability by enzyme adaptation much as prokaryotic cells do. Eukaryotic cells can also respond to environmental stresses such as UV radiation by inducing genes that provide protection against this harmful environmental agent. An example is the ability of sunbathers to develop a tan. The tanning response protects a person's cells against the damaging effects of ultraviolet rays.

Among plants and animals, multicellularity and a more complex cell structure also demand a much greater level of gene regulation. The life cycle of higher eukaryotic organisms involves the progression through several developmental stages to achieve a mature organism. Some genes are only expressed during early stages of development (e.g., the embryonic stage), whereas others are expressed in the adult. In addition, complex eukaryotic species are composed of many different tissues that contain a variety of cell types. Gene regulation is necessary to ensure the differences among distinct cell types. It is amazing that the various cells within a multicellular organism usually contain the same genetic material, yet, phenotypically, may look quite different. For example, the appearance of a human nerve cell is about as similar to a muscle cell as an amoeba is to a paramecium. In spite of these phenotypic differences, a human nerve and muscle cell actually contain the same 46 chromosomes of the human genome. Nerve and muscle cells look so different because of gene regulation rather than differences in DNA content. There are many genes that are expressed in the nerve cell and not the muscle cell, and vice versa.

The molecular mechanisms that underlie gene regulation in eukaryotes bear many similarities to the ways that bacteria regulate their genes. Figure 16-1 describes some common points of genetic control where eukaryotic gene expression can be affected. Regulation of gene transcription is an important form of control. In addition, research in the past few decades has revealed that eukaryotic organisms frequently regulate gene expression at points other than transcription. In this chapter, we will discuss some well-studied examples where genes are regulated at many of these control points.

Eukaryotic species frequently regulate gene expression posttranslationally. For example, small effector molecules can regulate the activity of certain enzymes by

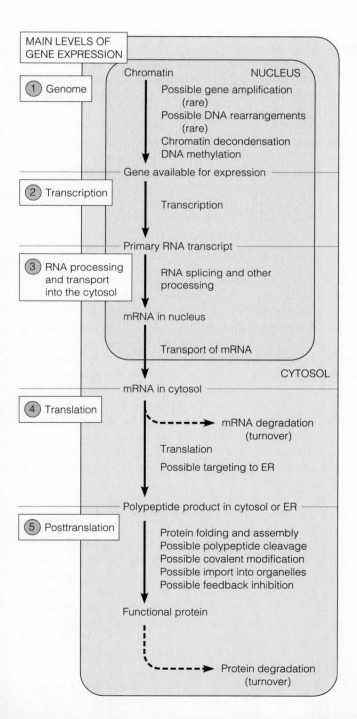

FIGURE **16-1**

Points of genetic control in eukaryotes.

feedback inhibition. In addition, the covalent posttranslational modification of proteins is a common way for eukaryotes to regulate the activities of cellular proteins. The mechanisms of posttranslational regulation of proteins were described in Chapter 15.

REGULATION OF DNA AND CHROMATIN STRUCTURE

Since genes are segments of DNA, it is not surprising that alterations in DNA structure can affect gene expression. This can occur in several ways. One possibility is that the number of copies of a gene can be increased. This is called **gene amplification**. Although this is an uncommon way to regulate gene expression, we will learn that it is occasionally used in eukaryotes. Another change in DNA that affects gene expression is the rearrangement of the DNA structure. In Chapter 18, we will see that **gene rearrangement** occurs in the DNA of immune system cells to generate a diverse array of antibody proteins. A third, more common way to alter DNA structure is **DNA methylation**, a topic also discussed in Chapters 7 and 11. In this section, we will examine how DNA methylation occurs and how it can regulate genes in a tissue-specific manner.

Since DNA is a component of chromosomes, the three-dimensional packing of chromatin is an important parameter affecting gene expression. If the chromatin is very tightly packed or in a **closed conformation**, transcription may be difficult or impossible. By comparison, chromatin that is in an **open conformation** is more easily accessible to transcription factors and RNA polymerase, so that transcription can take place. In this section, we will consider how conversion between the open and closed conformations can regulate the expression of genes.

The number of copies of a gene can be increased through DNA amplification

While the majority of the genome remains constant in all of the cells of a eukaryotic organism, alterations in the amount of DNA occasionally occur during the normal course of eukaryotic development. For example, the oocytes of animals need to contain a large number of ribosomes to synthesize the vast amount of proteins that are stored in the egg. In the oocytes of the South African clawed toad (*Xenopus laevis*), for example, approximately 10^{12} ribosomes are required per oocyte. To construct so many ribosomes, a large amount of ribosomal RNA must be transcribed. To accomplish this task, the number of rRNA genes is increased through gene amplification. In this case, the mechanism involves an unusual DNA replication event in which the rRNA genes are replicated to form many copies of circular DNA molecules called **minichromosomes** (Figure 16-2). Each circular DNA molecule contains 1 to 20 copies of the rRNA gene. These minichromosomes accumulate within the nucleoli of the oocyte.

In laboratory cell lines, researchers can also select for the amplification of particular genes. For example, the drug methotrexate specifically inhibits an enzyme known as dihydrofolate reductase (DHFR). This enzyme, which catalyzes a step required in nucleotide biosynthesis, is essential for cell growth. Therefore, treating cells with methotrexate inhibits their growth. In a laboratory, cells can be exposed to this drug to select for cells that have become resistant to methotrexate inhibition. Methotrexate-resistant cells were determined to contain an increased number of genes that encode DHFR. This increase in the number of genes allows the cells to synthesize more DHFR and thereby overcome the inhibition from methotrexate. By analyzing the DNA of methotrexate-resistant cells, researchers have found that the increase in DHFR genes is due to circular minichromosomes or tandem repeats of the DHFR gene (Figure 16-3).

In addition to these cases of gene amplification, a select number of genes are regulated by the rearrangement of DNA. A common example of this situation oc-

Nucleoli

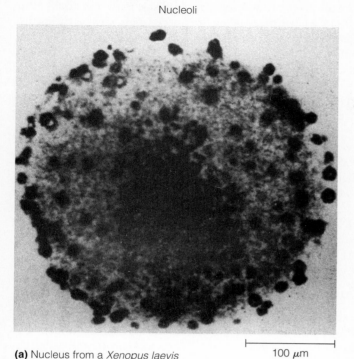

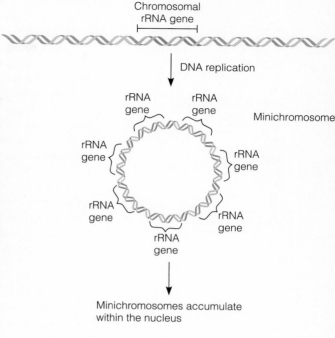

Minichromosome

Minichromosomes accumulate within the nucleus

(a) Nucleus from a *Xenopus laevis* oocyte   100 μm

(b) Process of amplification of rRNA genes in *Xenopus laevis* oocytes

FIGURE **16-2**

The amplification of the DNA that encodes rRNA genes within *Xenopus laevis* oocytes. (a) Micrograph of a *Xenopus laevis* oocyte stained to show the nucleoli, in which minichromosomes accumulate. **(b)** The rRNA genes are specifically replicated to generate minichromosomes, small circular DNA molecules that contain 1 to 20 copies of the genes encoding rRNAs.

curs during the development of the cells in the immune system. To generate a diverse array of antibodies to ward off infection, the cells of the immune system rearrange the DNA within the genes that encode antibodies. This occurs by a genetic recombination mechanism that will be discussed in detail in Chapter 18.

DNA methylation inhibits gene transcription

Another way that DNA structure can be modified is by the covalent attachment of methyl groups. DNA methylation is common in some eukaryotic species but certainly not all. For example, yeast and *Drosophila* have little or no detectable methylation of their DNA, whereas DNA methylation in vertebrates and plants is relatively abundant. In mammals, approximately 2 to 7% of the DNA is methylated. As shown in Figure 16-4, eukaryotic DNA methylation occurs on cytosine residues at the number 5 position of the cytosine base. The sequence that is methylated is

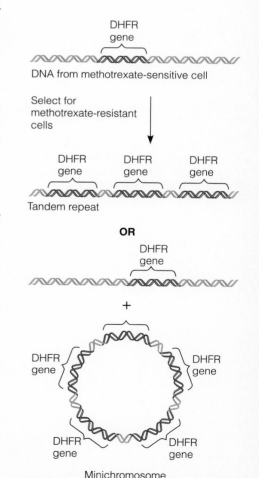

FIGURE **16-3**

Selection for methotrexate resistance by amplification of the dihydrofolate reductase gene. The conversion of normal methotrexate-sensitive cells to methotrexate-resistant cells involves an increase in the number of copies of the gene encoding dihydrofolate reductase (DHFR). This gene can be duplicated many times to form tandem repeats within the chromosome. Alternatively, a single DHFR gene remains in the chromosome but this gene is duplicated many times to form a minichromosome that is independent of the chromosomal DNA. The formation of tandem repeats or minichromosomes is an abnormal, relatively rare event. In this case, however, the presence of methotrexate selects for the survival of cells in which this event has occurred.

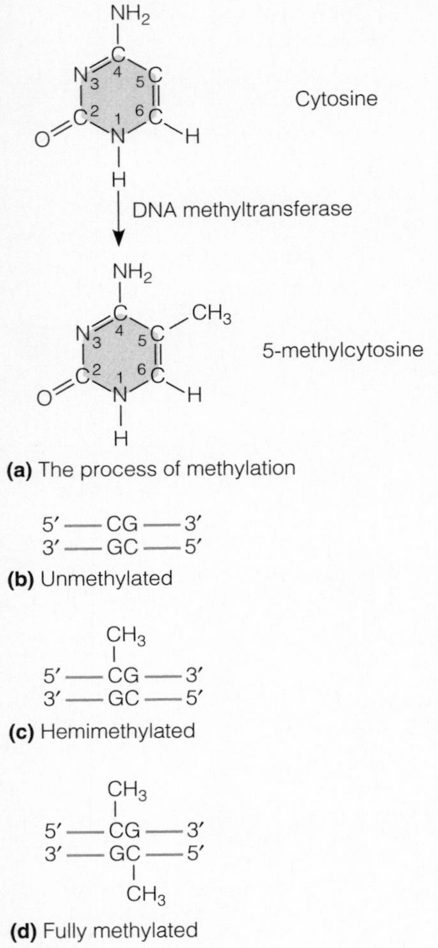

FIGURE 16-4

DNA methylation on cytosine residues.
(a) Methylation occurs via an enzyme known as DNA methylase or DNA methyltransferase, which attaches a methyl group to the number 5 carbon on cytosine. A CG sequence can be **(b)** unmethylated, **(c)** hemimethylated, or **(d)** fully methylated.

5′–CG–3′. Note that this sequence also contains a cytosine in the opposite strand. When one strand is methylated, this is called *hemimethylation*, whereas the methylation of the cytosines in both strands is termed *full methylation*.

DNA methylation inhibits the transcription of eukaryotic genes, particularly when it occurs in the vicinity of the promoter. The methylation pattern may affect the ability of transcription factors and RNA polymerase to initiate transcription. In vertebrates, **housekeeping genes** typically encode proteins that are required in most cells of a multicellular organism. By comparison, other genes are highly regulated and may only be expressed in a particular cell type—**tissue-specific genes**—and/or at a particular stage of development. Many housekeeping genes contain **CG islands** near their promoters. These CG islands, which are commonly 1000 to 2000 nucleotides in length, contain many unmethylated CG sequences. Therefore, housekeeping genes are not inhibited and tend to be expressed in most cell types. By comparison, many genes that are only expressed in certain cell types may also contain CG islands, but evidence is accumulating that DNA methylation may play an important role in the expression of these tissue-specific genes. The methylation of genes in particular tissues may prevent them from being expressed in the wrong cell type or at the wrong time.

DNA methylation is inheritable

Methylated DNA sequences are inherited during cell division. Experimentally, if fully methylated DNA is introduced into a vertebrate cell, the DNA will remain fully methylated even in subsequently produced daughter cells. However, if the same sequence of nonmethylated DNA is introduced into a cell, it will remain nonmethylated in the daughter cells. These observations indicate that the pattern of methylation is retained following DNA replication and, therefore, is inherited in daughter cells of future generations.

Figure 16-5 illustrates a molecular model that explains how methylation can be passed from parent to daughter cell. This model was originally suggested by Arthur Riggs at the City of Hope Research Institute in California and Robin Holliday and J. E. Pugh at the National Institute for Medical Research in London. In the parent cell, the DNA becomes methylated in both strands by *de novo* methylation (i.e., new methylation). When this cell divides, it will replicate its DNA and synthesize complementary daughter strands. Initially, the newly made daughter strands contain nonmethylated cytosines. This hemimethylated DNA is efficiently recognized by DNA methylase (also called DNA methyltransferase), which makes it fully methylated. This process is called *maintenance methylation*, since it preserves the methylated condition in future generations of cells. Overall, maintenance methylation appears to be an efficient process that routinely occurs within vertebrate and plant cells. By comparison, *de novo* methylation and demethylation are infrequent and highly regulated events. According to this view, the initial methylation or demethylation of a specific gene can be regulated at a specific time or stage of development. Once methylation has occurred, it can then be transmitted from parent to daughter cell via maintenance methylation.

As an example, let's consider DNA methylation within the context of vertebrate development. Suppose embryonic muscle cells have a gene that becomes fully methylated early in development due to *de novo* methylation. According to the model of Figure 16-5, this gene will be maintenance methylated in all of the muscle cells derived from these embryonic cells. Furthermore, since methylation inhibits transcription, all of these muscle cells will not transcribe this gene. In contrast, this same gene may not be methylated in embryonic nerve cells. Therefore, it will be transcribed in the nerve cells of the organism. Along these lines, developmental geneticists are eager to determine how variations in DNA methylation patterns may be an important mechanism in the establishment of tissue characteristics during early stages of vertebrate development. Additional research will be necessary to understand how specific genes may be targeted for *de*

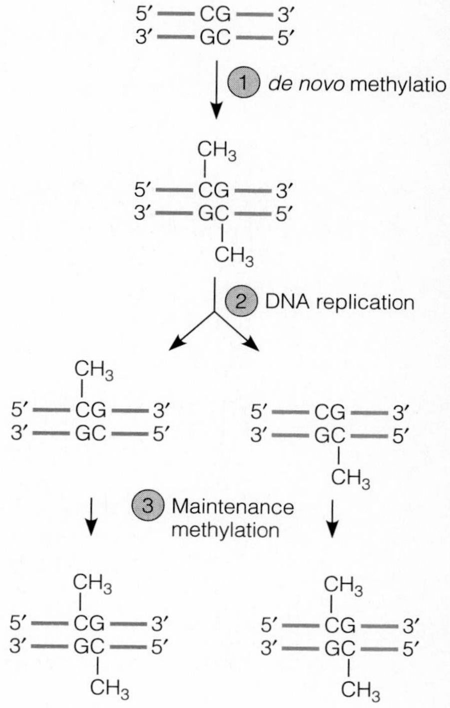

FIGURE **16-5**

A molecular model for the inheritance of DNA methylation. (1) The DNA initially becomes methylated *de novo*. The *de novo* methylation is a rare, highly regulated event. **(2)** Once this occurs, DNA replication produces hemimethylated DNA molecules, **(3)** which are then fully methylated by a DNA methylase. Maintenance methylation is a routine event that is expected to occur for all hemimethylated DNA.

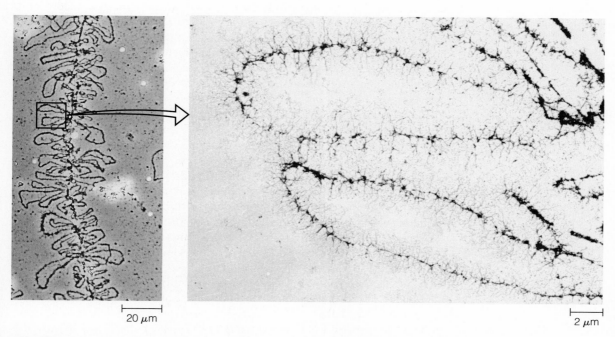

novo methylation or demethylation during different developmental stages or in specific cell types.

Gene accessibility can be controlled by changes in chromatin structure

Chromatin is composed of DNA and proteins that are organized into a compact structure that fits inside the nucleus of the cell. The DNA is wound around histone proteins to form nucleosomes that are 11 nm in diameter (refer back to Figure 10-13). This 11 nm fiber is condensed further to a 30 nm fiber, which is the predominant form of euchromatin found within the nucleus during interphase, when gene expression primarily occurs. Nevertheless, chromatin is a very dynamic structure that can alternate between highly condensed and highly extended conformations. The dynamic nature of chromatin is important in regulating gene transcription.

Variations in the degree of chromatin packing occur along the length of eukaryotic chromosomes during interphase. Tightly packed chromatin in a closed conformation cannot be transcribed. During gene activation, such chromatin must be converted to an open conformation that is less tightly packed than a 30 nm fiber. In certain cells, researchers can microscopically observe the decondensation of the 30 nm fiber when transcription is occurring. Figure 16-6 shows photomicrographs of a chromosome from an amphibian oocyte, the genes of which are actively being transcribed. This chromosome does not form a uniform, compact 30 nm fiber. Instead, many decondensed loops radiate from the central axis of the chromosome. These

FIGURE **16-6**

Lampbrush chromosomes in amphibian oocytes. The loops that radiate from a chromosome are regions of DNA containing genes that are being actively transcribed.

loops are regions of DNA in which the genes are being actively transcribed. These chromosomes have been named *lambrush chromosomes*, because their feathery appearance resembles the brushes that were once used to clean kerosene lamps.

Weintraub and Groudine used DNaseI sensitivity to study changes in chromatin structure during transcription of the β-globin gene in 1976

To understand the interconversions between open and closed chromatin conformations, we need methods to evaluate the degree of chromatin packing as it occurs in living cells. An important method developed during the 1970s was the use of DNaseI to monitor DNA conformation. As you may recall from Experiment 10B, DNaseI is an endonuclease that cleaves DNA. DNaseI is much more likely to cleave DNA in an open conformation than that in a closed conformation, because a looser conformation allows greater accessibility of DNaseI to the DNA. Regions of DNA that are highly susceptible to DNaseI digestion are called **DNaseI hypersensitive sites**.

While working at Princeton University, Harold Weintraub and Mark Groudine used DNaseI sensitivity as a tool to evaluate differences in chromatin structure that occur when a gene is actively transcribed. As you may recall from Chapter 12, humans possess several different genes that encode globin polypeptides. These polypeptides are the subunits of the oxygen-carrying protein hemoglobin. Prior to this work, it had been well established that globin genes are specifically expressed in red blood cells but not in other cell types such as brain cells and fibroblasts. Weintraub and Groudine asked the question, "Is there a difference in the chromatin packing of globin genes in cells that can actively transcribe the globin genes compared with that in cells in which the globin genes are turned off?" To answer this question, they used DNaseI sensitivity to compare the degree of globin gene packing in red blood cells versus that in brain cells and fibroblasts.

Since the globin genes are but a small part of the total chromosomal DNA, having a way to specifically monitor the digestion of the globin gene was a vital aspect of Weintraub and Groudine's experimental protocol. They accomplished this by using a cloned fragment of DNA (probe) that was complementary to the β-globin gene. This fragment was hybridized to the chromosomal DNA to determine specifically if the chromosomal globin gene was intact. If the chromosomal globin gene had been digested by DNaseI, it would not hybridize to the probe DNA because the corresponding chromosomal DNA would have been digested. However, if the chromosomal globin gene had not been digested by DNaseI, it could hybridize to the probe (Figure 16-7).

In Weintraub and Groudine's experiments, the cloned fragment of the globin gene was radiolabeled to allow detection of its presence. Following hybridization, the samples were then exposed to another enzyme, known as S1 nuclease. This enzyme digests DNA strands but, interestingly, only cuts DNA when it is single-stranded, not when it is double-stranded. As shown in Figure 16-7, S1 nuclease would digest the radiolabeled DNA probe if it had not hybridized with the complementary chromosomal strand, but it would be unable to do so if the radiolabeled probe and chromosomal strand had formed a double-stranded structure.

After S1 digestion, Weintraub and Groudine reasoned that their sample would contain a radiolabeled DNA probe if the chromosomal DNA was in a closed conformation. This is because DNaseI would not digest the chromosomal gene, and therefore, the β-globin gene would be available to hybridize to the radiolabeled DNA probe. However, if the chromosomal globin gene was in an open conformation, DNaseI would digest the chromosomal DNA, preventing it from hybridizing with the radiolabeled DNA probe. In this situation, the radiolabeled DNA strand would be digested by S1 nuclease. Therefore, the susceptibility of the radiolabeled DNA to S1 nuclease digestion allowed them to evaluate whether the chromosomal globin gene was in an open or closed conformation.

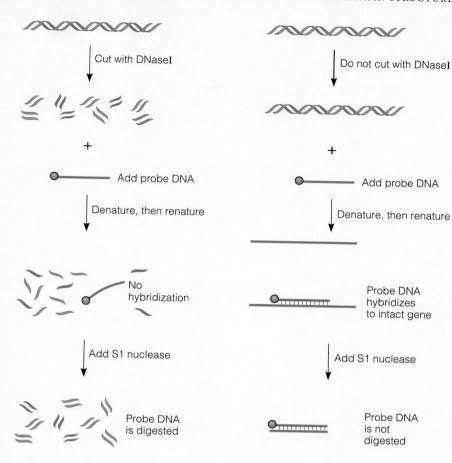

FIGURE **16-7**

The use of DNaseI and S1 nuclease to probe the chromatin structure of the β-globin gene. On the left side, the DNA has been digested into small pieces using DNaseI. These small pieces cannot hybridize with the probe DNA. When S1 nuclease is added, single-stranded probe DNA is also digested. On the right side, the DNA is not cut with DNaseI. The probe DNA can hybridize with a strand of chromosomal DNA. When SI nuclease is added, it will digest only the single-stranded regions of DNA, not the double-stranded region where the probe DNA and chromosomal DNA are bound to each other.

THE HYPOTHESIS

Changes in chromatin structure occur when globin genes are transcriptionally active.

TESTING THE HYPOTHESIS

Starting material: Nuclei were isolated from three different cell types in chicken: reticulocytes (immature red blood cells), brain cells, and fibroblasts. The globin genes are expressed in red blood cells but not in brain cells and fibroblasts.

1. Treat all three types of nuclei with the same amount of DNaseI.

Experimental Level

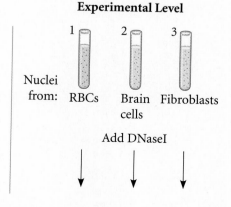

Conceptual Level

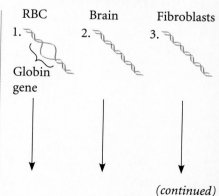

(continued)

(continued)

2. Extract the DNA from the nuclei. This involves lysing the nuclei with detergent, removing the protein by treatment with a phenol–chloroform mixture, and then precipitating the DNA by adding ethanol. The precipitated DNA forms a pellet at the bottom of a test tube following centrifugation. The DNA at the bottom of the tube can then be resuspended in an appropriate solution for the next step.

3. Subject the DNA to sound waves (i.e., sonication) to break the DNA into fragments of an average length of 500 bp.

4. Add a radiolabeled DNA probe that is complementary to the β-globin gene.

5. Denature the DNA into single strands by treatment with high temperature. Then cool it to 65°C to allow the complementary DNA strands to hybridize with each other.

6. Reduce the temperature and divide each sample into two tubes. Into one tube of each sample add S1 nuclease. Omit S1 nuclease from the second tube.

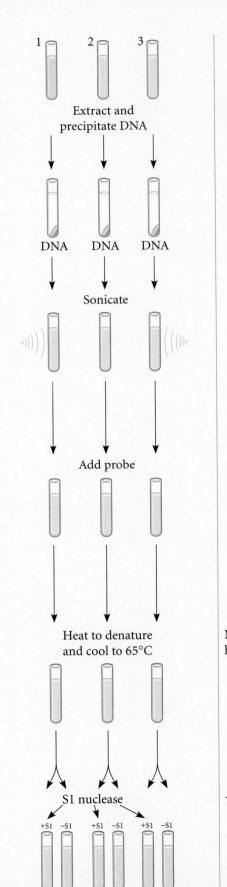

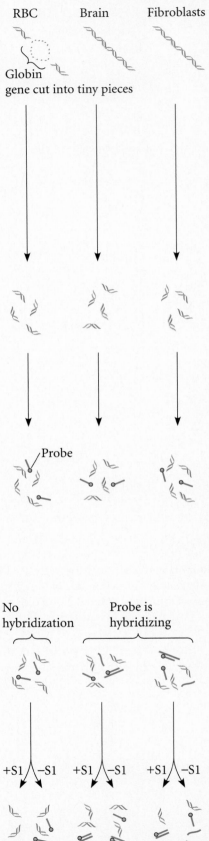

7. Precipitate the double-stranded DNA with trichloroacetic acid.

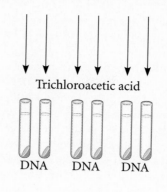

Trichloroacetic acid

DNA   DNA   DNA

Remove pellet and count radioactivity

Single and double-stranded DNA is in the pellet in the absence of S1; only double-stranded DNA is in the pellet in the presence of S1.

8. Count the amount of radioactivity. (The technique of scintillation counting is described in the appendix.) The amount of radioactivity in the presence of S1 nuclease divided by the amount of radioactivity in the absence of S1 nuclease provides a measure of the percentage of radiolabeled DNA that has hybridized to the chromosomal DNA.

THE DATA

Source of Nuclei	% Hybridization of DNA Probe
Reticulocytes	25%
Brain cells	>94%
Fibroblasts	>94%

Adapted from: Weintraub, H., and Groudine, M. (1976) *Science 193*, 848–856.

INTERPRETING THE DATA

In red blood cells, a much smaller percentage of the radiolabeled DNA probe hybridized to the chromosomal DNA than in brain cells and fibroblasts. These results were interpreted to mean that DNaseI is digesting the globin gene in the chromatin of red blood cells into small fragments that are too small to hybridize to the radiolabeled DNA probe. In other words, the globin genes in red blood cells are DNaseI hypersensitive. By comparison, the globin genes in brain cells and fibroblasts are relatively resistant to DNaseI digestion. Since the globin genes are expressed in red blood cells but not in brain cells and fibroblasts, these results are consistent with the hypothesis that the globin gene is less tightly packed when it is being expressed. In cells where the globin genes should not be expressed (e.g., brain cells and fibroblasts), the chromatin containing the β-globin gene is tightly compacted. However, in red blood cells where this gene is expressed, the chromatin is more loosely packed so that transcription of the globin genes can occur. This phenomenon provides one way to regulate globin gene expression among different cell types.

Several mechanisms have been proposed for the regulation of chromatin conformation

In recent years, geneticists have been trying to identify the proteins and DNA sequences that regulate the interconversion between the closed and open conformations of chromatin. From a biochemical viewpoint, eukaryotic chromosomes are large, complex macromolecules. Therefore, it has been technically difficult to definitively identify all of the factors that influence chromatin structure within the nucleus of a living cell. Nevertheless, many different factors have been proposed to influence chromatin structure and thereby alter the transcriptional activity of genes. Some of these are listed in Table 16-1.

One proposed mechanism for altering chromatin structure involves the posttranslational modification of proteins that are bound to the DNA. For example, the modification of core histone proteins may be involved in chromatin decondensation. The histones within actively transcribed genes are covalently modified so that the lysine residues are acetylated. This acetylation eliminates the positive charge on the lysine residue and may loosen the tight interaction between the negatively charged DNA and the positively charged histone protein.

Besides acetylation, another suggested mechanism involves various proteins that may have important functions in altering chromatin structure during gene activation or repression. Histone H1, which is the linker histone, appears to play a critical role in the formation of the 30 nm fiber. Therefore, the regulation of histone H1 activity may contribute to the interconversion between the 30 nm diameter fiber and looser chromatin conformations. In addition, other nonhistone proteins may be important in regulating chromatin structure. A diverse group of DNA-binding proteins known as the **high mobility group proteins** are abundant within the nucleus of eukaryotic cells. While their role remains unclear, their relative abundance and ability to bind to DNA is consistent with the idea that they play an important role in chromatin structure.

Chromatin packing and nucleosome location are altered during globin gene expression

Since the early studies of Weintraub and Groudine described in Experiment 16A, more detailed information has been gathered from molecular research in globin gene expression. Before discussing these studies, let's briefly consider globin gene organization and expression. Although the family of globin genes are expressed in red blood cells, individual members are expressed at different stages of development. For example, β-globin is expressed in adult red blood cells, whereas γ-globin is expressed in fetal red blood cells. As shown in Figure 16-8a, several of the globin genes are located adjacent to each other on the same chromosome.

Worldwide studies have been conducted that focus on inherited defects in hemoglobin composition. These disorders are known as *hemoglobinopathies*. An in-

TABLE **16–1 Factors that May Regulate Chromatin Structure**	
Factor	Effect
Acetylation of core histones	The acetylation removes the positive charge on histones and may loosen the tight binding between DNA and the core histones.
Histone H1	The regulation of the linker histone may be involved in the interconversion of the 11 nm and 30 nm conformations.
High mobility group proteins	This broad category of DNA binding proteins may alter chromatin structure.
Transcription factors	Transcription factor proteins may influence the degree of chromatin packing or the position of nucleosomes.

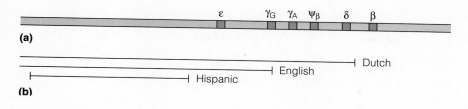

(a)

(b)

FIGURE 16-8

Globin genes and hemoglobinopathies. (a) The organization of globin genes on human chromosome 11. **(b)** These lines depict the segments of DNA that are deleted in certain hemoglobinopathies found in Dutch, English, and Hispanic populations.

GENE $\longrightarrow$ TRAITS: When the segment of DNA, depicted by the lines, is deleted, as has been found in some Dutch, English, and Hispanic people, the β-globin gene is not expressed even though it is still present. This is because the locus control region (LCR), which is necessary for the expression of the globin genes, is missing. Without the LCR, β-globin gene is not expressed, and this causes a blood disorder known as hemoglobinopathy.

triguing observation among certain patients is that they cannot synthesize β-globin even though the DNA sequence of the β-globin gene is perfectly normal. As shown in Figure 16-8b, this type of hemoglobinopathy has been found in Dutch, English, and Hispanic populations. It involves a DNA deletion that occurs upstream from the β-globin gene, although the β-globin gene itself is intact. Nevertheless, the β-globin gene is turned off in these patients. This unexpected finding prompted further investigations into how the globin genes are regulated.

Since the initial studies of hemoglobinopathies, a DNA region upstream from the β-globin gene has been identified as necessary for globin gene expression. This region, known as the **locus control region (LCR)**, is involved in the regulation of chromatin opening and closing. It is missing in certain persons with hemoglobinopathies. As hypothesized in Figure 16-9, this DNA locus provides recognition sites for proteins that promote the general opening of the entire region. This gives RNA polymerase and gene-specific transcription factors access to the region. Not only does the locus control region affect the transcription of β-globin, it also influences the other globin genes in this region. Additional research will be necessary to determine if locus control regions are commonly involved with the regulation of chromatin packing for many eukaryotic genes.

Aside from the degree of chromatin packing, a second structural issue to consider is the position of nucleosomes. In chromatin, the nucleosomes are usually positioned at regular intervals along the DNA. The position of a nucleosome may greatly influence whether or not a gene can be transcribed. For example, if the TATA box is tightly bound to the histone core, it may be inaccessible to general transcription factors and RNA polymerase. Nucleosomes have been shown to change position in cells that normally express a particular gene but not in cells where the gene is inactive. For example, in fibroblasts that do not express the β-globin gene, nucleosomes are positioned at regular intervals from nucleotides −3000 to +1500 (see Figure 16-10a). However, in red blood cells that can express the β-globin gene, a disruption in nucleosome positioning occurs in the region from nucleotide −500 to +200 (see Figure 16-10b). This disruption may be an important first step in gene activation.

FIGURE 16-9

The function of the globin locus control region. In red blood cells that express globin genes, proteins may bind to the locus control region (LCR) and promote the general opening of the region, so that all of the globin genes are accessible to transcription factors and RNA polymerase.

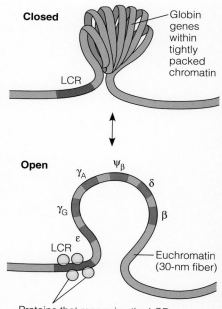

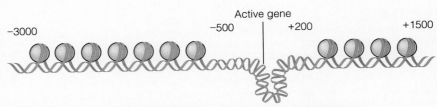

FIGURE **16-10**

Changes in nucleosome position during the activation of the β-globin gene.
(a) In fibroblasts, which do not express β-globin, nucleosome positioning is uninterrupted. **(b)** In red blood cells that can express this gene, disruption occurs in the positions of nucleosomes in the −500 to +200 region. (Position +1 is the beginning of transcription.) It is not yet clear whether the histones are removed, partially displaced, and/or other proteins are bound to this region.

(a) Nucleosomes in DNA of fibroblasts, which do not express the β-globin gene

(b) Nucleosomes in DNA of reticulocytes, which do express the β-globin gene

REGULATION OF TRANSCRIPTION

In this text, we have frequently used the term transcription factor to describe proteins that influence the transcription of a given gene. In addition to the general transcription factors necessary for the basal level of transcription, eukaryotic cells also possess an interesting array of **regulatory transcription factors** that serve to regulate the transcription of nearby genes. As mentioned in the previous section, some proteins may affect transcription by altering the degree of chromatin packing or the position of nucleosomes. Alternatively, many regulatory transcription factors exert their effects by directly influencing RNA polymerase and/or general transcription factors.

Regulatory transcription factors commonly recognize a particular DNA sequence or consensus sequence. These sequences are analogous to the operator sites found near bacterial promoters. In eukaryotes, DNA sequences, recognized by regulatory transcription factors, are known as **response elements** or **control elements**. When a regulatory transcription factor binds to a response element, it affects the transcription of an associated gene (Figure 16-11). For example, the binding of regulatory transcription factors may facilitate the binding of general transcription factors and RNA polymerase in the vicinity of the promoter. This would enhance the rate of transcription (Figure 16-11a). Alternatively, regulatory transcription factors may act as repressors by preventing transcription from occurring (Figure 16-11b). In this section, we will examine several features of regulatory transcription factor function. We will begin by considering the structural features of transcription factor proteins and their response elements. We will then examine two well-studied examples that illustrate how the function of a transcription factor is modulated within a living cell.

Structural features of regulatory transcription factors allow them to bind to DNA

Genes that encode general and regulatory transcription factor proteins have been identified and sequenced from a wide variety of eukaryotic species including yeast, plants, and animals. Several different families of related transcription factors have been discovered. In recent years, the molecular structures of transcription factor proteins have become an area of intense research. Transcription factor proteins contain regions, called **domains**, that have specific functions. For example, one domain of a transcription factor may have a DNA-binding function, while another may provide a binding site for a small effector molecule. When a domain with a very similar structure or amino acid sequence is found in many different proteins, it is sometimes called a **motif**.

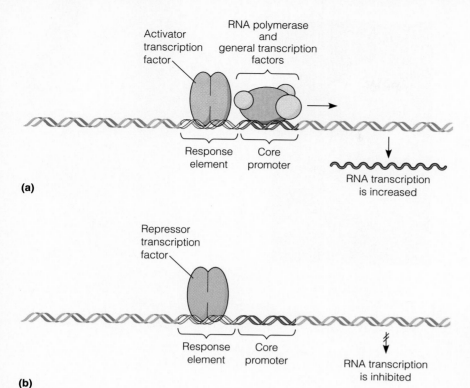

(a)

(b)

FIGURE **16-11**

Overview of transcriptional regulation by regulatory transcription factors.
These factors can act as either **(a)** an activator to increase the rate of transcription or **(b)** a repressor to decrease that rate.

Figure 16-12 depicts several different domain structures found in transcription factor proteins. The protein secondary structure known as an α-helix is frequently important in the recognition of the DNA double helix. In helix-turn–helix and helix–loop–helix motifs, an α-helix called the *recognition helix* makes contact with and recognizes a base sequence along the major groove of the DNA (Figure 16-12a, b). As discussed in Chapter 9, the major groove is a region of the DNA double helix where the bases contact the surrounding water in the cell. Hydrogen bonding between an α-helix and nucleotide bases is one way that a transcription factor can bind to a specific DNA sequence. Similarly, a zinc finger motif is composed of one α-helix and two β-sheet structures that are held together by a zinc (Zn^{++}) metal ion (see Figure 16-12c). The zinc finger also can recognize DNA sequences within the major groove.

A second interesting feature of certain motifs is that they promote protein dimerization. The leucine zipper (see Figure 16-12d) and helix–loop–helix motifs (see Figure 16-12e) mediate protein dimerization. For example, Figure 16-12d depicts the dimerization and DNA binding of two proteins that have leucine zippers. Alternating leucine residues in both proteins interact ("zip up"), resulting in protein dimerization. In some cases, two identical transcription factor proteins will come together to form a **homodimer** or two different transcription factors can form a **heterodimer**. As discussed later in this chapter, the dimerization of transcription factors can be an important way to modulate their function.

Regulatory transcription factors recognize response elements that function as enhancers or silencers

When the binding of a regulatory transcription factor to a response element increases transcription, the response element is known as an **enhancer**. Enhancers can stimulate or **up-regulate** transcription 10- to 1000-fold. Alternatively, response elements that serve to inhibit transcription are called **silencers**. This is called **down-regulation**.

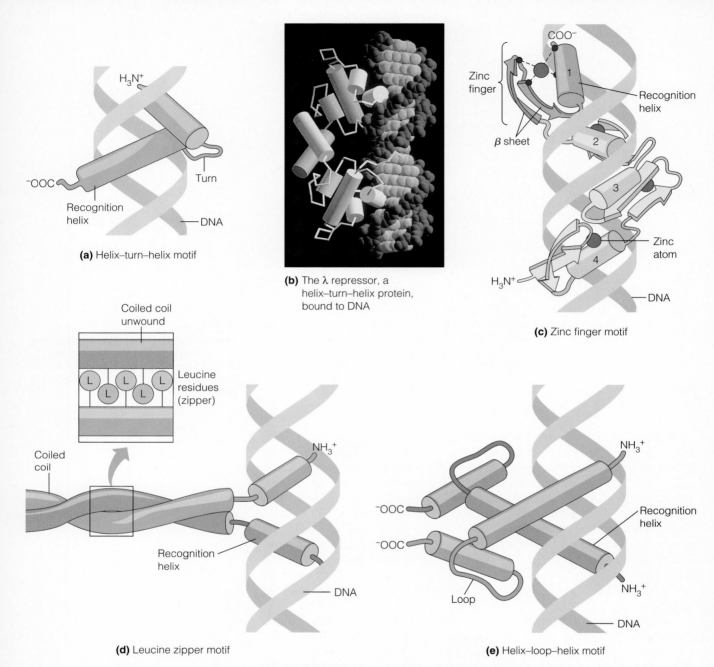

(a) Helix–turn–helix motif

(b) The λ repressor, a helix–turn–helix protein, bound to DNA

(c) Zinc finger motif

(d) Leucine zipper motif

(e) Helix–loop–helix motif

FIGURE **16-12**

Structural domains found in transcription factor proteins. Certain types of protein secondary structure are found in many different transcription factors. In this figure, α-helices are shown as cylinders and β-sheets as flattened arrows. **(a)** Helix–turn–helix motif: Two α-helices are connected by a turn. The α-helices lie in the major groove of the DNA. **(b)** The lambda repressor: This is a specific example of a helix–turn–helix protein binding to DNA. **(c)** Zinc finger motif: Each zinc finger is composed of one α-helix and two antiparallel β-sheets. A zinc atom, shown in red, holds the zinc finger together. This illustration shows four zinc fingers in a row. **(d)** Leucine zipper motif: The leucine zipper promotes the dimerization of two transcription factors, which are shown in purple and green. Two α-helices (a coiled coil) are intertwined due to the leucine residues (see inset). **(e)** Helix–loop–helix motif: A short α-helix is connected to a longer α-helix by a loop. In this illustration, a dimer is formed from the interactions of two helix–loop–helix motifs, and the longer helices are binding to the DNA.

Many response elements are **orientation independent** or **bidirectional**. This means that the response element can function in the forward or reverse orientation. For example, if the forward orientation of an enhancer is

5′-GATA-3′

3′-CTAT-5′

then this enhancer will also bind to a regulatory transcription factor and enhance transcription even when it is oriented in the reverse direction:

5′-TATC-3′

3′-ATAG-5′

There is also striking variation in the location of response elements relative to a gene's promoter. In general, response elements are located in a region within a few hundred nucleotides upstream from the promoter site. Nevertheless, response elements are sometimes found several thousand nucleotides away from the actual promoter site. In some cases, response elements are located downstream from the promoter site and may even be found within introns! As you may imagine, the variation in response element orientation and location profoundly complicates the efforts of geneticists to identify the response elements that affect the expression of any given gene.

Different mechanisms have been proposed to explain how a regulatory transcription factor can bind to a response element and thereby affect gene transcription. Indeed, more than one mechanism may be involved. As mentioned, some regulatory transcription factors may influence transcription by altering chromatin packing or nucleosome positioning. This alteration can affect whether or not RNA polymerase can gain access to the promoter site. Alternatively, many regulatory transcription factors affect the ability of RNA polymerase and/or general transcription factors to function at the core promoter site. Since the response element may be a substantial distance away from the promoter site, these types of regulatory transcription factors can influence RNA polymerase by a looping mechanism. Figure 16-13 presents this idea schematically.

According to this view, the regulatory transcription factor bound at the response element can physically influence events at the core promoter due to looping in the DNA. When the regulatory transcription factor enhances the rate of transcription, this is known as **transactivation**. The prefix trans- implies that the **activator** can bind to an enhancer site that may be a substantial distance away from the core promoter. As noted in Figure 16-13, a transactivating regulatory factor can exert this effect by recognizing and recruiting general transcription factors and RNA polymerase to the core promoter region.

The domain in an activator that stimulates transcription is called the *transactivating domain*. Like the DNA-binding domains and dimerization domains described in Figure 16-12, transactivating domains have unique structural features. Most transactivating domains fall into one of three categories: acidic, glutamine-rich, or proline-rich. Each category is characterized by the prevalence of a particular type of amino acid. For example, a yeast activator known as GAL4 has an acidic transactivating domain, which is 49 amino acids in length and contains 11 acidic (glutamic acid and aspartic acid) residues. As their name suggests, glutamine-rich and proline-rich domains have a high proportion (typically 20–25%) of glutamine and proline, respectively.

Alternatively, other regulatory transcription factors may inhibit transcription by **transinhibition.** Transcription factors that cause transinhibition are also called **repressors.** They can inhibit transcription in a variety of ways. For example, repressors may physically interact with general transcription factors or RNA

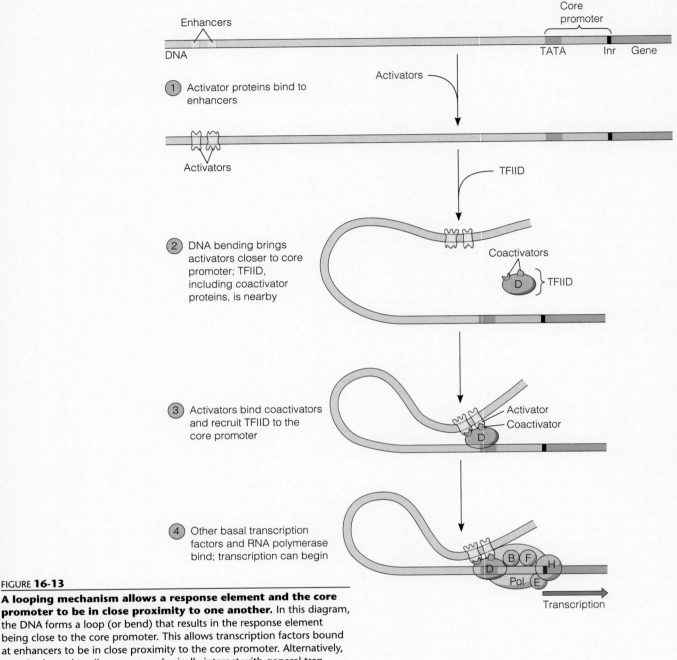

FIGURE **16-13**

**A looping mechanism allows a response element and the core
promoter to be in close proximity to one another.** In this diagram,
the DNA forms a loop (or bend) that results in the response element
being close to the core promoter. This allows transcription factors bound
at enhancers to be in close proximity to the core promoter. Alternatively,
proteins bound to silencers may physically interact with general transcription factors and RNA polymerase bound to the core promoter.

polymerase and prevent them from functioning. In addition, certain repressors
exert their effects by inhibiting the function of transactivators.

The function of regulatory transcription factor proteins can be modulated in three ways

The function of regulatory transcription factor proteins can be affected directly in
three common ways. These are (1) the binding of an effector molecule, (2) protein–protein interactions, and (3) covalent modification of the transcription factor

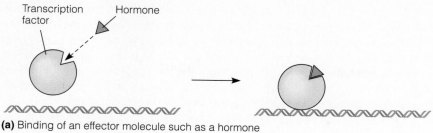

Transcription factor   Hormone

(a) Binding of an effector molecule such as a hormone

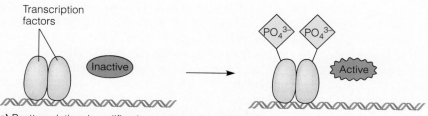

Transcription factor   Transcription factor

(b) Protein–protein interaction

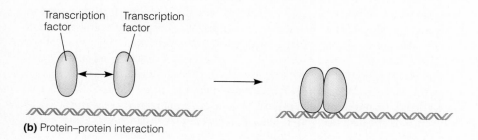

Transcription factors

Inactive   Active

(c) Posttranslational modification such as phosphorylation

FIGURE **16-14**

Common ways to modulate the function of regulatory transcription factors.
(a) The binding of an effector molecule such as a hormone may influence the ability of a transcription factor to bind to the DNA.
(b) Protein–protein interactions among transcription factor proteins may influence their functions. **(c)** Posttranslational modifications such as phosphorylation may alter transcription factor function.

itself. Figure 16-14 depicts these three means of modulating regulatory transcription factor function. Usually, one or more of these modulating effects are important in determining whether a transcription factor can bind to the DNA and/or influence transcription by RNA polymerase. For example, an effector molecule may bind to a regulatory transcription factor and promote its binding to the DNA (Figure 16-14a). Later in this chapter, we will see that steroid hormones function in this manner. Another important way is via protein–protein interactions (Figure 16-14b). The formation of homodimers and heterodimers is a fairly common means of controlling transcription. Finally, the function of regulatory transcription factors can be affected by covalent modifications such as the attachment of a phosphate group (Figure 16-14c). As discussed later, the phosphorylation of activators can alter their ability to stimulate transcription.

Steroid hormones exert their effects by binding to a regulatory transcription factor and controlling the transcription of nearby genes

Thus far in this section, we have considered the general properties of transcription factor structure, the characteristics of the response elements they recognize, and the three ways that regulatory transcription factors can be modulated. We now will turn to specific examples that illustrate how regulatory transcription factors function within living cells. Our first example is a category that respond to *steroid hormones*. This type of regulatory transcription factor is known as a **steroid receptor**, because the steroid hormone binds directly to the protein.

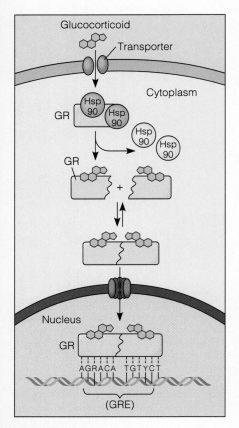

FIGURE **16-15**

The action of glucocorticoid hormones.
Once inside the cell, the hormone binds to the glucocorticoid receptor, releasing it from a protein known as Hsp90. Two glucocorticoid receptors then form a dimer and travel into the nucleus, where the dimer binds to glucocorticoid response elements (GREs) that are next to particular genes. The binding of the glucocorticoid receptor to the GRE activates the transcription of the adjacent gene.

GENES ⟶ TRAITS: Glucocorticoid hormones are produced by the endocrine glands in response to eating and activity. They enable the body to regulate its metabolism properly. When glucocorticoids are produced, they are taken into cells and bind to the glucocorticoid receptors. This eventually leads to the activation of genes that encode proteins involved in the synthesis of glucose, the breakdown of proteins, and the mobilization of fats.

The ultimate action of a steroid hormone is to affect gene transcription. Steroid hormones are synthesized by endocrine glands of animals and secreted into the bloodstream. The hormones are then taken up by cells that can respond to the hormones in different ways. For example, *glucocorticoid hormones* influence nutrient metabolism in most body cells by promoting glucose utilization, fat mobilization, and protein breakdown. Other steroid hormones, such as estrogen and testosterone, are called *gonadocorticoids* because they influence the growth and function of the gonads.

Figure 16-15 shows the stepwise action of a glucocorticoid hormone. In this example, the hormone is transported into the cytosol of a cell. Once inside, there are glucocorticoid receptors that can specifically bind the hormone. Prior to hormone binding, the glucocorticoid receptor is complexed with proteins known as heat shock proteins, an example being Hsp90. After the hormone binds to the glucocorticoid receptor, Hsp90 is released. This exposes a nuclear localization signal within the receptor that allows it to travel into the nucleus through the nuclear pore. Two glucocorticoid receptors form a homodimer once inside the nucleus. The glucocorticoid receptor homodimer binds to glucocorticoid *response elements* (GREs) that are next to particular genes. The GREs function as enhancers. The binding of the glucocorticoid receptor homodimer to GREs activates the transcription of the adjacent gene, eventually leading to the synthesis of the encoded protein.

Animal cells usually have a large number of glucocorticoid receptors within the cytoplasm. Since GREs are located near several different genes, the uptake of many hormone molecules can activate many glucocorticoid receptors and thereby up-regulate many different genes. For example, if ten different genes within the nucleus contain GREs, all ten will be activated when the hormone is present. For this reason, a cell can respond to the presence of the hormone in a very complex way. Glucocorticoid hormones stimulate many genes that encode proteins involved in several different cellular processes, including the synthesis of glucose, the breakdown of proteins, and the mobilization of fats.

The CREB protein is an example of a regulatory transcription factor modulated by protein–protein interaction and covalent modification

As we have just seen, some signaling molecules such as steroid hormones bind directly to regulatory transcription factors to alter their function. This enables a cell to respond to a hormone by up-regulating a particular set of genes. Most extracellular signaling molecules, however, do not enter the cell and bind directly to transcription factors. Instead, most signaling molecules must bind to receptors in the plasma membrane. This binding activates the receptor and leads to the synthesis of an intracellular signal that causes a cellular response. One type of cellular response is to affect the transcription of particular genes within the cell.

As our second example of regulatory transcription factor function within living cells, we will examine the **cAMP response element–binding (CREB) protein**. The CREB protein is a regulatory transcription factor that becomes activated in response to specific cell-signaling molecules. It recognizes a response element with the consensus sequence 5′–TGACGTCA–3′. This response element, which is found near many different genes, has been termed a cAMP response element (CRE).

Figure 16-16a shows the steps leading to the activation of the CREB protein. A wide variety of hormones, growth factors, neurotransmitters, and other signaling molecules can bind to plasma membrane receptors to initiate an intracellular response. In this case, the response involves the production of a second messenger, known as cAMP. (The extracellular signaling molecule itself is considered the primary messenger.) When the signaling molecule binds to the receptor, it activates a G protein that subsequently activates an enzyme known as adenylate cyclase. The activated

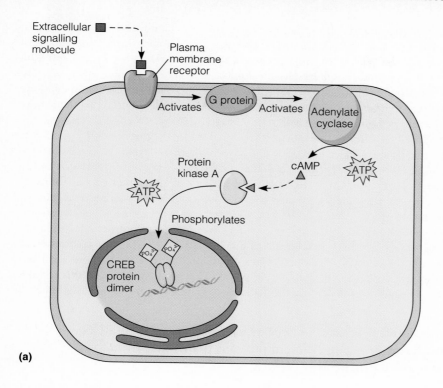

(a)

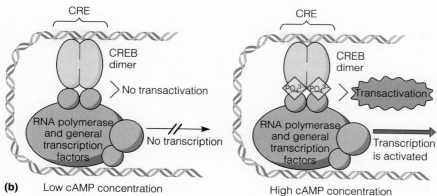

(b)

FIGURE **16-16**

The activity of the CREB protein.
(a) When an extracellular signal binds to a receptor in the plasma membrane, adenylate cyclase is activated, leading to synthesis of cAMP. cAMP activates protein kinase A, which travels into the nucleus and phosphorylates the CREB protein. Once phosphorylated, the CREB protein acts as a transcriptional activator. **(b)** The CREB protein is inactive prior to phosphorylation. However, when cAMP activates protein kinase A, the CREB protein is phosphorylated, and it can then transactivate transcription.

adenylate cyclase catalyzes the synthesis of cAMP. The cAMP molecule then activates a second enzyme, protein kinase A. This enzyme can phosphorylate several different cellular proteins, including the CREB protein. When phosphorylated, the CREB protein stimulates transcription. In contrast, the unphosphorylated CREB protein can still bind to CREs but does not transactivate RNA polymerase (Figure 16-16b).

REGULATION OF RNA PROCESSING AND TRANSLATION

In eukaryotic species, gene expression is commonly regulated at the RNA level. The function of RNA can be controlled in three general ways. One way is pre-mRNA processing. Following transcription, a pre-mRNA transcript is processed before it becomes a functional mRNA. These processing events, which include trimming, splicing, capping, polyA tailing, and RNA editing, were described in Chapter 13. We will see shortly how alternative splicing is regulated at the RNA level.

Another strategy for regulating gene expression is to influence the concentration of mRNA. As we have seen in this chapter, this can be done by regulating the rate of transcription. When the transcription of a gene is increased, a higher concentration of the corresponding RNA results. In addition, RNA concentration is greatly affected by the stability or half-life of a particular RNA. Factors that increase RNA stability are expected to raise the concentration of that RNA molecule. Later in this chapter, we will examine how sequences within mRNA molecules greatly affect their stability.

Finally, a third way to regulate RNA function is to control the ability of mRNAs to be translated. As we have seen in Chapter 14, mRNA translation relies on the translational machinery (e.g., ribosomes, tRNA, etc.). The rate of translation can be regulated by influencing the functional activity of ribosomes or the ability of mRNA to be translated. In eukaryotes, one important strategy for regulating translation is to alter the rate of translation via the ribosomal machinery. This occurs primarily in two ways. One is to directly affect the function of translational initiation factors. This has the general effect of increasing or decreasing the translation of many mRNAs within the cell. Second, RNA-binding proteins can prevent ribosomes from initiating the translation process for specific mRNAs. In this section, we will examine both of these mechanisms for regulating mRNA translation.

Alternative splicing regulates which exons occur in an RNA transcript, allowing different proteins to be made from the same structural gene

In Chapter 13, we examined the phenomenon of alternative splicing, although we did not consider how this process is regulated. During alternative splicing, a premRNA can be spliced in more than one way, leading to different combinations of exons in the resulting mRNAs. This produces two (or more) proteins that have specialized differences in structures and functions (refer back to Figure 13-18).

Alternative splicing is not a random event. Rather, the specific pattern of splicing is regulated in any given cell. The molecular mechanism for the regulation of alternative splicing is not entirely understood, although recent evidence indicates that it is due to variations in the concentrations of proteins known as **splicing factors**. Certain splicing factors play a key role in the choice of particular splice sites. One such category of splicing factors are the *SR proteins*. These splicing factors contain a domain at their carboxy terminal end that is rich in serine (S) and arginine (R). They also contain an RNA-binding domain at their amino terminal end.

In some cases, SR proteins modulate the ability of general splicing factors to choose the 5′ splice site. The RNA-binding domain of the SR protein recognizes a region in the exon such as the 5′ splice junction. It is hypothesized that the SR protein then recruits the general splicing factors into this region by using its SR (serine-arginine) domain. The net effect of SR protein binding and splicing factor recruitment is to promote exon recognition and inclusion of the recognized exon in the final mRNA product. Alternative splicing in different tissues may occur because each cell type has its own characteristic concentration of one or more types of SR proteins. These differences in SR protein concentration may play a key role in alternative splicing decisions.

The stability of mRNA influences mRNA concentration

In eukaryotes, the stability of mRNAs can vary considerably. Certain mRNAs have very short half-lives (namely, several minutes), whereas others can persist for several days. In some cases, the stability of an mRNA can be regulated so that its half-life is shortened or lengthened. A change in the stability of mRNA can greatly influence the cellular concentration of that mRNA molecule. In this way, factors that influence RNA stability can dramatically affect gene expression.

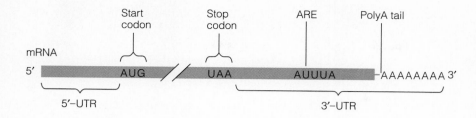

FIGURE **16-17**

The location of AU-rich elements (AREs) within mRNAs. One or more AREs are commonly found within the 3'–UTRs (untranslated regions) of mRNAs with short half-lives. The 5'–UTR is the untranslated region of the mRNA that precedes the start codon.

Various factors can play a role in mRNA stability. One important structural feature is the length of the polyA tail. As you may recall from Chapter 13, most newly made mRNAs contain a polyA tail that is approximately 200 nucleotides in length. This polyA tail is recognized by the **polyA-binding protein**. As an mRNA ages, its polyA tail tends to be shortened by the action of cellular exonucleases. Once it becomes less than 10 to 30 adenosines in length, the polyA-binding protein cannot bind, and the mRNA is rapidly degraded by exo- and endonucleases.

Certain mRNAs, particularly those with short half-lives, contain sequences that act as destabilizing elements. While these destabilizing elements can be located anywhere within the mRNA, they are most commonly located at the 3' end between the stop codon and the polyA tail. This region of the mRNA is known as the *3'–untranslated region (3'–UTR)*. An example of a destabilizing element is the **AU-rich element** (**ARE**) that is found in many short-lived mRNAs (Figure 16-17). This element, which contains the consensus sequence AUUUA, is recognized by cellular proteins that bind to the ARE and thereby influence whether or not the mRNA is rapidly degraded.

Phosphorylation of ribosomal initiation factors can alter the rate of translation

Modulation of translational initiation factors is widely used to control fundamental cellular processes. Under certain conditions, it is advantageous for a cell to stop synthesizing proteins. For example, if a cell is infected by a virus, it is vital to inhibit protein synthesis so that the virus cannot manufacture viral proteins. Likewise, if critical nutrients are in short supply, it is beneficial for a cell to conserve its resources by inhibiting protein synthesis.

As was discussed in Chapter 14, several eukaryotic initiation factors are required to initiate translation. The phosphorylation of many different initiation factors has been found to affect translation. Two factors, eIF2 and eIF4F, appear to play a central role in controlling the initiation of translation. The functions of these two translational initiation factors are modulated by phosphorylation in opposite ways. When the α-subunit of eIF2 (known as eIF2α) is phosphorylated, translation is inhibited, whereas the phosphorylation of eIF4F increases the rate of translation.

Figure 16-18 shows the events leading to translational inhibition by eIF2α. A variety of conditions can lead to a shutdown of protein synthesis, including viral infection, nutrient deprivation, heat shock, and the presence of toxic heavy metals. These conditions promote the activation of protein kinases known as *eIF2α protein kinases*. Several eIF2α protein kinases have been identified. Once activated, eIF2α protein kinase can phosphorylate eIF2α. The phosphorylation of eIF2α causes it to bind tightly to another initiation factor subunit, known as eIF2B. Functional eIF2B is necessary so that eIF2 can promote the binding of the initiator tRNA^{Met} to the 40S subunit (also see Figure 14-12, p. 383). However, when the phosphorylated eIF2α binds to eIF2B, it prevents eIF2B from functioning. Therefore, the initiator tRNA^{Met} does not bind to the 40S subunit, and translation is inhibited.

A second important way to control translation is via the eIF4F translation factor that modulates the binding of mRNA to the ribosomal initiation complex. The

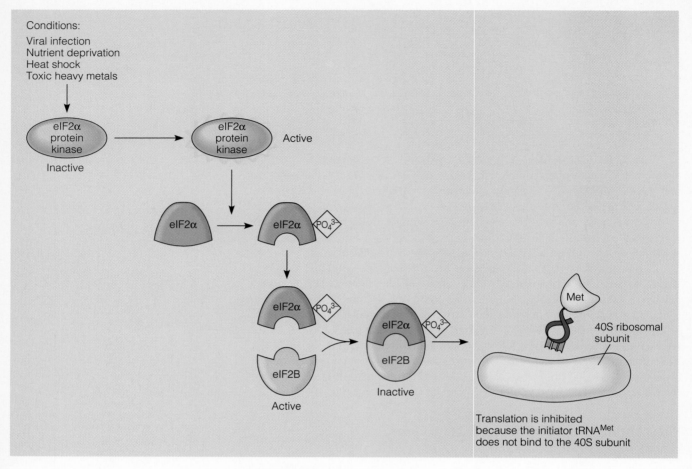

FIGURE **16-18**

The pathway that leads to the phosphorylation of eIF2α (eukaryotic initiation factor) and the inhibition of translation.

function of eIF4F is stimulated by phosphorylation. A variety of conditions have been shown to cause eIF4F to become phosphorylated. These include the presence of growth factors, insulin, and other signaling molecules that promote cell proliferation. Conversely, conditions such as heat shock and viral infection decrease the level of eIF4F phosphorylation and thereby inhibit translation.

The regulation of iron assimilation is an example of the regulatory effect of RNA-binding proteins on translation

As we have just seen, the phosphorylation of ribosomal proteins can modulate the translation of mRNA. Since the ribosomes are necessary to translate all of a cell's mRNA, this form of regulation affects many mRNAs. By comparison, particular mRNAs are sometimes regulated by RNA-binding proteins that directly affect translational initiation or RNA stability. The regulation of iron assimilation provides a well-studied example in which both of these phenomena occur. Before discussing the translational control, it is interesting to consider the biology of iron metabolism.

Iron is an essential element for the survival of living organisms, since it is required for the function of many different enzymes. The pathway whereby mammalian cells take up iron is depicted in Figure 16-19. Iron ingested by an animal is absorbed into the bloodstream and becomes bound to a protein known as **transferrin**, a carrier of iron through the bloodstream. The transferrin–Fe^{++} com-

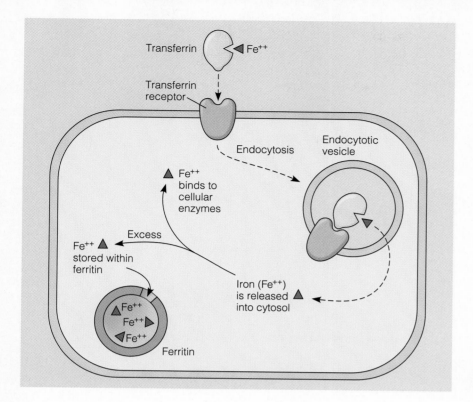

FIGURE **16-19**

The uptake of iron (Fe^{++}) into mammalian cells.

plex is recognized by a **transferrin receptor** on the surface of cells; the complex binds to the receptor and then is transported into the cytosol by endocytosis. Once inside, the iron is then released from transferrin. At this stage, it may bind to cellular enzymes that require iron for their activity. Alternatively, if there is an overabundance of iron, the excess iron is stored within a hollow, spherical protein known as **ferritin**. The storage of excess iron within ferritin prevents the toxic buildup of too much iron within the cell.

Since iron is a vital yet potentially toxic substance, mammalian cells have evolved an interesting way to regulate iron assimilation. The two mRNAs that encode ferritin and the transferrin receptor are both influenced by an RNA-binding protein known as the **iron regulatory protein (IRP)**. This protein binds to a regulatory element within the mRNA known as the **iron regulatory element (IRE)**. The ferritin mRNA has an IRE in its 5′–UTR. When IRP binds to this IRE, it inhibits the translation of the ferritin mRNA (see Figure 16-20a). However, when iron is abundant in the cytosol, the iron binds directly to IRP and prevents it from binding to the IRE. Under these conditions, the ferritin mRNA is translated to make more ferritin protein. The synthesis of ferritin prevents the toxic buildup of iron within the cytosol.

The transferrin receptor mRNA also contains iron response elements. However, the IREs in the transferrin receptor mRNA are located in the 3′–UTR. When IRP binds to these IREs, it does not inhibit translation. Instead, it increases the stability of the mRNA by blocking the action of RNA-degrading enzymes. This leads to increased amounts of transferrin receptor mRNA within the cell when the cytosolic levels of iron are very low (Figure 16-20b). Under these conditions, more transferrin receptor is made. This promotes the uptake of iron when in short supply. In contrast, when iron is abundant within the cytosol, IRP is removed from the transferrin receptor mRNA, and the mRNA becomes rapidly degraded. This leads to a decrease in the amount of transferrin receptor and thereby helps to prevent the assimilation of too much iron into the cell.

Low iron

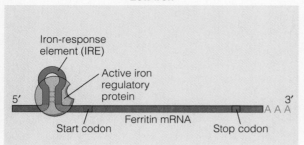

High iron

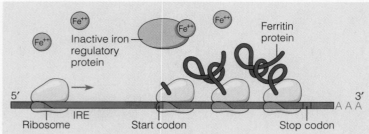

Regulatory protein binds IRE;
ferritin synthesis low

Regulatory protein cannot bind IRE;
ferritin synthesis high

(a) Iron increases translation initiation from ferritin mRNA

Low iron

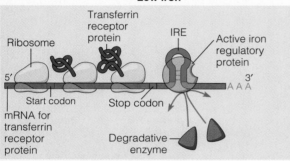

High iron

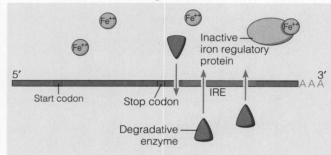

Regulatory protein binds IRE;
mRNA protected;
transferrin receptor synthesis high

Regulatory protein cannot bind IRE;
mRNA degraded;
transferrin receptor synthesis low

(b) Iron decreases stability of transferrin receptor mRNA

FIGURE **16-20**

The regulation of iron assimilation genes by IREs and IRP (iron regulatory elements and protein). (a) The binding of the iron regulatory protein (IRP) to the iron regulatory element (IRE) in the 5′–UTR of ferritin mRNA inhibits translation. When Fe^{++} binds to IRP, the protein is removed from the ferritin mRNA, so that translation can proceed. **(b)** The binding of IRP to IREs in the 3′–UTR of the transferrin receptor mRNA. In this case, the binding of IRP enhances the stability of the mRNA and leads to a higher concentration of this mRNA. Therefore, more transferrin receptor protein is made. When Fe^{++} levels are high and this ion binds to IRP, the IRP dissociates from the IREs, and the transferrin receptor mRNA is rapidly degraded.

GENES ⟶ TRAITS: This form of translational control allows cells to utilize iron appropriately. When the cellular concentration of iron is low, the translation of the transferrin receptor is increased, thereby enhancing the ability of cells to take up more iron. Also, the translation of ferritin mRNA is inhibited, which is not needed to bind excess iron. By comparison, when the cellular concentration of iron is high, the translation of ferritin mRNA is enhanced. This leads to the synthesis of ferritin, an iron storage protein that prevents the toxic buildup of iron. Also, when iron is high, the transferrin receptor mRNA is degraded, which decreases further uptake of iron.

CONCEPTUAL SUMMARY

In this chapter, we have surveyed a wide variety of mechanisms that allow eukaryotes to regulate gene expression. The DNA itself can be altered by *gene amplification*, *rearrangement*, or *methylation*. DNA methylation in vertebrates and higher plants appears to be an important mechanism to turn genes off in a *tissue-specific* or developmentally specific manner. In addition, the degree of chromatin packing is an important parameter that affects gene expression. For transcription to occur, the chromatin cannot be in a *closed conformation*. Instead, it must be in a loosely packed or *open conformation*. In the case of globin genes, a segment of DNA called the *locus control region* plays a role in regulating the packing of DNA in this region.

A wide array of *regulatory transcription factors* have been identified. These vary in their structures and modes of action. Some transcription factors bind to DNA and alter chromatin packing or nucleosome positioning. Other transcription factors recognize *response elements* in the vicinity of genes and influence the events that occur at the core promoter. *Enhancers* are response elements that *up-regulate* gene expression, whereas *silencers* have the opposite effect of *down-regulating* transcription. Response elements can be *bidirectional* and function at a fairly large distance away from the core promoter site. In this chapter, we have considered the molecular mechanism of two regulatory transcription factors that respond to cell hormones. The glucocorticoid receptor is a regulatory transcription factor that binds glucocorticoid hormone directly. Once the hormone is bound, it activates several cellular genes by binding to glucocorticoid response elements (GREs) that are next to genes. The *CREB protein* responds to intracellular levels of cAMP. The CREB protein binds to response elements known as CREs. When CREB becomes phosphorylated, it stimulates transcription by *transactivation*.

Gene regulation can also occur at the RNA level. For example, an RNA transcript can be regulated by alternative RNA splicing. This RNA processing event influences the type of protein that is made from the mRNA transcript. The stability of RNA can also be affected by particular sequences within the mRNA. Factors that promote RNA stability lead to a higher concentration of that RNA transcript. In addition, the rate of mRNA translation can be controlled in two ways. First, the translational machinery can be regulated by affecting the activity of translational initiation factors. Two particular factors, eIF2α and eIF4F, are modulated by phosphorylation in opposite ways. When eIF2α is phosphorylated, translation is inhibited, whereas the phosphorylation of eIF4F increases the rate of translation. Second, specific RNA-binding proteins can affect the ability of certain mRNAs to be translated.

The regulation of gene expression in eukaryotes has been studied via many different techniques. In this chapter, we have considered a few approaches to studying this phenomenon. We saw, in one example, that Weintraub and Groudine were able to detect changes in the degree of DNA packing of the globin genes by using a DNaseI sensitivity assay. This approach exploits the fact that tightly packed chromatin is less susceptible to DNaseI digestion than is DNA in an open conformation. Furthermore, the identification of people who possess intact β-globin genes yet have hemoglobinopathies has suggested that a region near globin genes, known as the locus control region, is involved in controlling the degree of chromatin packing.

Other studies have focused on the identification of regulatory transcription factors that influence the expression of particular genes. By comparing the structures of many different transcription factors, researchers have found that they tend to have common domains that act to either bind DNA or interact with effector molecules. The sequences of many response elements have also been determined. In a few cases, the combined efforts of many research groups have elucidated the detailed molecular mechanisms for the regulation of particular genes. For example, in this chapter, we have examined how the glucocorticoid receptor and CREB protein regulate genes at the level of transcription. We have also considered how IRP regulates ferritin and transferrin receptor expression at the level of translation.

PROBLEM SETS

SOLVED PROBLEMS

1. Describe how tight packing of chromatin in a closed conformation may prevent gene transcription.

Answer: There are several possible ways that the tight packing of chromatin physically inhibits transcription. First, it may prevent general transcription factors and/or RNA polymerase from binding to the major groove of the DNA. Second, it may prevent RNA polymerase from forming an open complex, which is necessary to begin transcription. Third, it could prevent looping in the DNA that may be necessary to activate transcription.

2. What are the two alternative ways that IRP can affect gene expression at the RNA level?

Answer: The ferritin mRNA has an IRE in its 5′–UTR. When IRP binds to this IRE, it inhibits the translation of the ferritin mRNA. This decreases the amount of ferritin protein, which is not needed when iron levels are low. However, when iron is abundant in the cytosol, the iron binds directly to IRP and prevents it from binding to the IRE. This allows the ferritin mRNA to be translated, producing more ferritin protein. IREs are also located in the transferrin receptor mRNA in the 3′–UTR. When IRP binds to these IREs, it increases the stability of the mRNA. This leads to an increase in the amount of transferrin receptor mRNA within the cell when the cytosolic levels of iron are very low. Under these conditions, more transferrin receptor is made to promote the uptake of iron, which is in short supply. When the iron is found in abundance within the cytosol, IRP is removed from the mRNA, and the mRNA becomes rapidly degraded. This leads to a decrease in the amount of transferrin receptor.

3. Eukaryotic transcription factors are often orientation independent and can function in a variety of locations. Explain the meaning of this statement.

Answer: Orientation independence means that the response element can function in the forward or reverse direction. In addition, response elements can function at a variety of locations that may be upstream or downstream from the promoter. DNA looping is thought to bring the regulatory transcription factors bound at the response elements and the general transcription factors bound at the core promoter in close proximity with one another.

CONCEPTUAL QUESTIONS

1. Discuss the common points of control in eukaryotic gene regulation.

2. What is meant by the term DNA amplification? Give two examples.

3. What is DNA methylation? How is it passed from a mother to a daughter cell?

4. Let's suppose that a vertebrate organism carries a mutation that causes some cells that would normally differentiate into nerve cells to differentiate into muscle cells. A molecular analysis of this mutation revealed that it was in a gene that encodes a methylase. Explain how an alteration in a methylase could produce this phenotype.

5. What is a CG island? Where would you expect one to be located? How does the methylation of CG islands affect gene expression?

6. What is the predominant form of chromatin in the eukaryotic nucleus? What must happen to its structure for transcription to take place? Discuss what is believed to cause this change in chromatin structure.

7. Explain how the acetylation of core histones may loosen chromatin packing.

8. Explain how the position of nucleosomes can affect transcription.

9. Explain the function of a locus control region.

10. Let's suppose you have found a patient with a hemoglobinopathy in which the globin genes have been translocated to a different chromosome. The sequence of the β-globin gene is perfectly normal in this patient with a translocation, yet it is not expressed. Propose a reasonable explanation for this observation.

11. Discuss the general ways that regulatory transcription factors can influence transcription.

12. Explain how the binding of a regulatory transcription factor can occur in a sequence-specific manner. Why is it functionally important for regulatory transcription factors to recognize particular DNA sequences?

13. Discuss the structure and function of response elements. Where are they located relative to the core promoter?

14. What is meant by the term transcription factor modulation? List three general ways that this can occur.

15. What do the terms transactivation and transinhibition mean? Explain how they occur at the molecular level.

16. Explain what must happen in order for the glucocorticoid receptor to bind to a GRE.

17. Let's suppose a mutation in the glucocorticoid receptor does not prevent the binding of the glucocorticoid hormone to the protein, but does prevent the ability of the receptor to activate transcription. Make a list of all the possible defects that may explain why transcription cannot be activated.

18. Explain how phosphorylation affects the function of the CREB protein.

19. A particular drug inhibits the protein kinase that is responsible for phosphorylating the CREB protein. How would this drug affect the following events?

    A. The ability of CREB to bind to CREs
    B. The ability of extracellular hormone to enhance cAMP levels
    C. The ability of CREB to stimulate transcription
    D. The ability of CREB to dimerize

20. What is the function of an SR protein? Explain how SR proteins can regulate the tissue-specific splicing of mRNAs.

21. What conditions lead to the phosphorylation of eIF2α? Discuss why a cell would want to shut down translation when these conditions occur.

22. What is the relationship between mRNA stability and mRNA concentration? What factors affect mRNA stability?

23. Describe how the binding of the iron regulatory factor affects the mRNAs for ferritin and the transferrin receptor. How does iron influence this process?

EXPERIMENTAL QUESTIONS

1. In Figure 16-3, it was shown how methotrexate resistance is due to the presence of minichromosomes or tandem repeats of the dihydrofolate reductase gene, a gene that has been cloned. Based on experimental tools described elsewhere (namely, *in situ* hybridization, Chapter 20), describe the types of experiments that you think would have been done to come to this conclusion.

2. What is a DNaseI hypersensitive site?

3. With regard to DNaseI sensitivity, what do you expect will happen when a gene is converted from an inactive to a transcriptionally active state?

4. A laboratory cell line has suffered a mutation that causes the inability of the CREB protein to activate transcription in the presence of adrenalin. Make a list of three or more mutations that may cause this effect. Pick two of your possible types of mutations and describe how

you would experimentally distinguish between them. You may wish to review the section on Detection of Genes and Gene Products in Chapter 19 when preparing your answer.

5. In Experiment 16A, explain how the use of S1 nuclease makes it possible to determine whether or not the β-globin gene is DNaseI hypersensitive.

6. Explain how investigations of hemoglobinopathies ultimately led to studies that identified a locus control region.

7. Certain hormones, such as adrenalin, can increase the levels of cAMP within cells. Let's suppose you can pretreat cells with or without adrenalin and then prepare a cell extract that contains the CREB protein. You then analyze the ability of the CREB protein to bind to a DNA fragment containing a CRE using a gel retardation assay (see Chapter 19). Describe what the expected results would be.

QUESTIONS FOR STUDENT DISCUSSION/COLLABORATION

1. Explain how DNA methylation could be used to regulate gene expression in a tissue-specific way. When and where would *de novo* methylation occur, and when would demethylation occur? What would occur in the germ line?

2. Enhancers can be almost anywhere and affect the transcription of a gene. Let's suppose you have a gene cloned on a piece of DNA, and the DNA fragment is 50,000 base pairs in length. Using cloning methods described in Chapter 19, you can cut out short segments from this 50,000 bp fragment and then reintroduce the smaller fragments into a cell that can express the gene. You would like to know if there are any

enhancers within the 50,000 bp region that may affect the expression of the gene. Discuss the *most efficient* strategy you can think of to trim your 50,000 bp fragment and thereby locate enhancers. You can assume that the coding sequence of the gene is in the center of the 50,000 bp fragment and that you can trim the 50,000 bp fragment into any size piece you want using molecular techniques described in Chapter 19.

3. How are regulatory transcription factors and regulatory splicing factors similar in their mechanism of action? In your discussion, consider the domain structures of both types of proteins. How are they different?

CHAPTER 17

GENE MUTATION
AND DNA REPAIR

CONSEQUENCES
OF MUTATIONS

OCCURRENCE
AND CAUSES
OF MUTATION

DNA REPAIR

As we have seen throughout this text, the function of DNA is to store the information for the synthesis of cellular proteins. A key aspect of the gene expression process is that the DNA itself normally does not change. This allows DNA to function as a permanent storage unit. However, on relatively rare occasions, a **mutation** can occur. The term mutation refers to a heritable change in the genetic material. This means that the structure of DNA has been changed permanently and this alteration can be inherited by daughter cells from a mother cell following cell division. If a mutation occurs in the germ line cells that produce gametes, it may also be passed from parent to offspring.

The topic of mutation is centrally important in all fields of genetics, including molecular genetics, Mendelian inheritance, and population genetics. Mutations provide the allelic variation that we have discussed throughout this text. For example, phenotypic differences such as tall versus dwarf pea plants are due to mutations that alter the expression of particular genes. Mutations can have both beneficial and detrimental effects. On the positive side, mutations are essential to the continuity of life. They provide the variation that enables species to change and adapt to their environments. On the negative side, however, new mutations are much more likely to be harmful rather than beneficial to the individual. For example, many inherited human diseases result from mutated genes. In addition, diseases such as skin and lung cancer can be caused by environmental agents that are known to cause DNA mutations. For these and many other reasons, understanding the molecular nature of mutations is a deeply compelling area of research. In this chapter, we will consider the nature of mutations and their consequences on gene expression at the molecular level.

Since most mutations are harmful, organisms have developed several ways to repair damaged DNA. DNA repair systems reverse DNA damage before it results in a mutation that could potentially cause cell or organism lethality. DNA repair systems have been studied extensively in many organisms, particularly *Escherichia coli*, yeast, and mammals. A variety of systems exist that repair different types of DNA lesions. In this chapter, we will examine the ways that several of these DNA repair systems operate. First, though, we will discuss mutation and its consequences.

CONSEQUENCES OF MUTATION

To understand why DNA mutations are beneficial or detrimental, we must appreciate how changes in DNA structure can ultimately affect DNA function. Much of our understanding of mutation has come from the study of experimental organisms, such as bacteria, yeast, and *Drosophila*. Researchers can expose these organisms to environmental agents that cause mutation and then study the consequences of the induced mutations. In addition, since these organisms have a short generation time, researchers can investigate the effects of mutation when they are passed from parent to offspring.

Changes in chromosome structure are referred to as **chromosome mutations**, and changes in chromosome number are called **genome mutations**. Both chromosome and genome mutations can usually be seen with the aid of a light microscope. As discussed in Chapter 8, they are important occurrences within natural populations of many eukaryotic organisms. By comparison, **single gene mutations** are relatively small changes in DNA structure that occur within a particular gene. In this chapter, we will be concerned primarily with the ways that mutations may affect the molecular and phenotypic expression of single genes. We will also consider how the timing of mutations during an organism's development has important consequences.

Gene mutations are molecular changes in the DNA sequence of a gene

A gene mutation occurs when the sequence of the DNA within a gene is altered in a permanent way. A gene mutation can change the base sequence within a gene, or it can involve a removal or addition of one or more nucleotides.

A **point mutation** is a change in a single base pair within the DNA. For example, the DNA sequence shown here has been altered by a **base substitution**:

$$5'-AACGCTAGATC-3' \quad \rightarrow \quad 5'-AACGCGAGATC-3'$$
$$3'-TTGCGATCTAG-3' \quad \quad \quad 3'-TTGCGCTCTAG-5'$$

A change of a pyrimidine to another pyrimidine (C $\rightarrow$ T) or a purine to another purine (A $\rightarrow$ G) is called a **transition**. This type of mutation is more common than a **transversion**, in which a purine is interchanged with a pyrimidine. The example just shown is a transversion (T $\rightarrow$ G change), not a transition.

Besides base substitutions, a short sequence of DNA may be deleted from or added to the chromosomal DNA:

$$5'-AACGCTAGATC-3' \quad \rightarrow \quad 5'-AACGCTC-3'$$
$$3'-TTGCGATCTAG-3' \quad \quad \quad 3'-TTGCGAG-5' \quad \text{(deletion of 4 base pairs)}$$

$$5'-AACGCTAGATC-3' \quad \rightarrow \quad 5'-AACAGTCGCTAGATC-3' \quad \text{(addition of}$$
$$3'-TTGCGATCTAG-3' \quad \quad \quad 3'-TTGTCAGCGATCTAG-5' \quad \text{4 base pairs)}$$

As we will see next, small deletions or additions to the sequence of a gene can significantly affect its function.

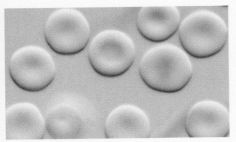

(a) Normal red blood cells   10 μm

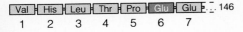

| Val | His | Leu | Thr | Pro | Glu | Glu |·· 146
| 1 | 2 | 3 | 4 | 5 | 6 | 7 |

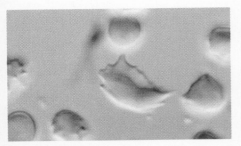

(b) Sickled red blood cells   10 μm

| Val | His | Leu | Thr | Pro | Val | Glu |·· 146
| 1 | 2 | 3 | 4 | 5 | 6 | 7 |

FIGURE **17-1**

Missense mutation in sickle cell anemia.
(a) Normal red blood cells and the normal
amino acid sequence of the β-globin gene.
(b) Sickled red blood cells and the missense
mutation in the β-globin gene.

GENES ⟶ TRAITS: A missense mutation
alters the structure of β-globin, which is a
subunit of hemoglobin, the oxygen-carrying
protein in the red blood cells. When an individual is homozygous for this allele, this missense mutation causes the red blood cells to
sickle under conditions of low oxygen tension. The sickling phenomenon is a description of the trait at the cellular level. At the
organism level, the sickled cells clog the
capillaries, thereby causing painful crises
and symptoms of anemia.

Gene mutations can alter the coding sequence within a gene

The occurrence of a mutation within the coding sequence of a structural gene can have various effects on the amino acid sequence of the polypeptide encoded by a gene. Table 17-1 describes the possible effects of point mutations. **Silent mutations** are those that do not alter the amino acid sequence of the polypeptide even though the nucleotide sequence has changed. Since the genetic code is degenerate, silent mutations can occur in the wobble base so that the type of amino acid is not changed. In contrast, **missense mutations** are base substitutions in which an amino acid change does occur. An example of a missense mutation occurs in the human disease known as sickle cell anemia, which was discussed in Chapter 4. This disease involves a mutation in the β-globin gene. In the most common form of this disease, a missense mutation alters the polypeptide sequence so that the sixth amino acid is changed from a glutamic acid to valine. This single amino acid substitution alters the structure and function of the hemoglobin protein. One consequence of this alteration is that the red blood cells sickle under conditions of low oxygen. A photo of this is shown in Figure 17-1.

Nonsense mutations involve a change from a normal codon to a termination codon. This causes the polypeptide to be terminated earlier than expected, producing a truncated polypeptide. Finally, **frameshift mutations** involve the addition or deletion of nucleotides in multiples of one or two. Since the codons are read in multiples of three, this shifts the reading frame so that a completely different amino acid sequence occurs downstream from the mutation.

Except for silent mutations, most new mutations are likely to produce polypeptides that have reduced rather than better function. For example, nonsense mutations will produce polypeptides that are substantially shorter and, therefore, unlikely to function properly. Likewise, frameshift mutations dramatically alter the amino acid sequence of polypeptides and are thereby likely to disrupt function. Missense mutations are less likely to alter function, since they involve a change of a single amino acid within polypeptides that typically contain hundreds of amino acids. When a missense mutation has no detectable effect on protein function, it is referred to as a **neutral mutation**. Silent mutations are also considered neutral mutations.

Mutations can occasionally produce a polypeptide that has an enhanced ability to function. While these favorable mutations are relatively rare, they may result in an organism with a greater likelihood to survive and reproduce. If this is the case, natural selection may cause such a favorable mutation to increase in frequency within a population. This topic will be discussed later in this chapter and also in Chapter 26.

Gene mutations are also given names that describe how they affect the wild-type genotype and phenotype

Thus far, we have introduced several genetic terms that describe the molecular effects of mutations. Genetic terms are also used to describe the effects of mutations relative to a wild-type genotype. In a natural population, the **wild-type** is the most common genotype. For example, the most common form of the β-globin gene is called the wild-type allele. For some genes, there may be multiple alleles that are prevalent in a population, so that there is not a single wild-type allele.

A **forward mutation** changes the wild-type genotype into some new variation. For example, in the sickle cell allele of the β-globin gene, the sixth amino acid is changed from a glutamic acid to a valine. The mutation is "forward" in an evolutionary sense, since it has changed from the prevalent genotype in the population. If a forward mutation is beneficial, it may move evolution forward; otherwise, it will probably be eliminated from the population. A **reverse mutation** has the opposite effect. For example, if the sickle cell allele mutated back to the wild-type allele (valine to glutamic acid), this would be a reverse mutation or a **reversion**.

TABLE 17–1 Consequences of Point Mutations within the Coding Sequence

Type of Mutation	Change in DNA	
None	None	5′–A–T–G–A–C–C–G–A–C–C–C–G–A–A–A–G–G–G–A–C–C–3′* – Met – Thr – Asp – Pro – Lys – Gly – Thr –
Silent	Base substitution	5′–A–T–G–A–C–C–G–A–C–C–C–C–A–A–A–G–G–G–A–C–C–3′ – Met – Thr – Asp – Pro – Lys – Gly – Thr –
Missense	Base substitution	5′–A–T–G–C–C–C–G–A–C–C–C–G–A–A–A–G–G–G–A–C–C–3′ – Met – Pro – Asp – Pro – Lys – Gly – Thr –
Nonsense	Base substitution	5′–A–T–G–A–C–C–G–A–C–C–C–G–T–A–A–G–G–G–A–C–C–3′ – Met – Thr – Asp – Pro – STOP!
Frameshift	Addition/ substitution	5′–A–T–G–A–C–C–G–A–C–G–C–C–G–A–A–A–G–G–G–A–C–C– – Met – Thr – Asp – Ala – Glu – Arg – Asp –

*DNA sequence in the coding strand. Note that this sequence is the same as the mRNA sequence except that the RNA contains uracil (U) instead of thymine (T).

Another way to describe a mutation is based on its influence on the wild-type phenotype. When a mutation alters the phenotypic characteristics of an organism, it is said to be a **variant**. Variants are often characterized by their differential ability to survive. As mentioned, a neutral mutation does not alter protein function, so it does not affect survival. A **deleterious mutation** will decrease the chances of survival. The extreme example of deleterious mutation is a **lethal mutation**, which results in death to the cell or organism. On the other hand, a **beneficial mutation** will enhance the survival of the organism. In some cases, an allele may be either deleterious or beneficial depending on the genotype and/or the environmental conditions. An example is the sickle cell allele. In the homozygous state, the sickle cell allele lessens the chances of survival. However, when an individual is heterozygous for the sickle cell allele and wild-type allele, this increases the chances of survival due to malarial resistance. Finally, some mutations are called **conditional mutants** because they only affect the phenotype under a defined set of conditions. Geneticists often study conditional mutants in microorganisms; a common example are temperature-sensitive (*ts*) mutants. A bacterium that has a *ts* mutation will grow normally in one temperature range (i.e., the permissive temperature) but will exhibit defective growth at a different temperature range (i.e., the nonpermissive temperature). For example, an *E. coli* strain carrying a *ts* mutation may be able to grow at 37°C but not at 42°C, whereas the wild-type strain can grow at either temperature.

A second mutation will sometimes affect the phenotypic expression of a first mutation. For example, a forward mutation may cause an organism to grow very slowly. A second mutation at another site in the organism's DNA may restore the normal growth rate, converting the original mutant back to the wild-type condition. Geneticists call these second site mutations **suppressors** or **suppressor mutations**. This name reflects that a suppressor mutation acts to suppress the phenotypic effects of another mutation. A suppressor mutation differs from a reversion, because it occurs at a DNA site that is distinct from that of the first mutation.

Suppressor mutations are classified according to their relative locations with regard to the mutation they suppress. When the second mutant site is within the same gene as the first mutation, it is termed an **intragenic suppressor**. Alternatively, a suppressor mutation can be in a different gene from the first mutation; this is called an **intergenic suppressor**.

There are two general types of intergenic suppressors: those that involve an ability to defy the genetic code and those that involve a mutant structural gene. A common example of the first type are suppressor tRNA genes, which have been identified in microorganisms. Suppressor tRNA genes have a change in the anticodon region that causes the tRNA to behave contrary to the genetic code. For example, nonsense suppressors are mutant tRNAs that recognize a stop codon and put an amino acid into the growing polypeptide chain. This type of mutant tRNA can suppress a nonsense mutation in another gene.

A second type of intergenic suppressor mutation are those that occur within structural genes. These suppressor mutations usually involve a change in the expression of one gene that compensates for a defective mutation affecting another gene. For example, a first mutation may cause one protein to be partially or completely defective. An intergenic suppressor mutation in a different structural gene might overcome this defect by altering the structure of a second protein so that it could take over the functional role that the first protein cannot perform. Alternatively, intergenic suppressors may involve proteins that participate in a common cellular function. When a first mutation decreases the activity of a protein, a suppressor mutation could enhance the function of the second protein involved in this common function and thereby overcome the defect in the first protein. Interestingly, intergenic suppressors sometimes involve mutations in genetic regulatory proteins such as transcription factors. When a first mutation causes a protein to be defective, a suppressor mutation may occur in a gene that encodes a transcription factor. The mutant transcription factor transcriptionally activates other genes that can compensate for the loss of function mutation in the first gene.

Gene mutations can occur outside of the coding sequence and still influence gene expression

Thus far in this chapter, we have focused our attention primarily on mutations in the coding regions of genes and their effects on gene expression. In Chapters 12 through 16, we learned how various sequences outside of the coding sequence play important roles during the process of gene expression. A mutation can occur within noncoding sequences and thereby affect gene expression (Table 17-2). For example, a mutation may alter the sequence within the core promoter of a gene. If the mutant promoter sequence becomes more like the consensus sequence, the mutation may increase the rate of transcription. This is called an *up-promoter mutation*. In contrast, a *down-promoter mutation* occurs when a mutation causes the promoter to become less like the consensus sequence, decreasing its affinity for regulatory factors and decreasing the transcription rate.

In Chapter 15, we considered how mutations can affect regulatory sequences. For example, mutations in the *lac* operator site, called *lacO^C* mutants, prevent the binding of the *lac* repressor protein. This causes the *lac* operon to be constitutively

TABLE 17-2 Possible Consequences of Gene Mutations Outside of the Coding Sequence

Sequence	Effect of Mutation
Promoter	May increase or decrease the rate of transcription
Response element/ operator site	May disrupt the ability of the gene to be properly regulated
5′UTR/3′UTR	May alter the ability of mRNA to be translated; may alter mRNA stability
Splice recognition sequence	May alter the ability of pre–mRNA to be properly spliced

expressed even in the absence of lactose. Bacteria strains with *lacO^C* mutations are at a selective disadvantage compared with wild-type *E. coli* strains, because they waste their energy expressing the *lac* operon even when these proteins are not needed. As noted in Table 17-2, mutations can also occur in other noncoding regions of a gene and alter gene expression in a way that may affect phenotype. For example, mutations in eukaryotic genes can alter splice junctions and affect the order and/or number of exons that are contained within mRNA. In addition, mutations that affect the untranslated regions of mRNA (i.e., 5′– and 3′–UTRs) may affect gene expression if they alter the stability of mRNA or its ability to be translated.

DNA sequences known as trinucleotide repeats may cause mutation

Researchers have discovered several human genetic diseases caused by an unusual form of mutation known as **trinucleotide repeat expansion** (**TNRE**). These diseases include fragile X syndrome (FRAXA), FRAXE mental retardation, myotonic muscular dystrophy (DM), spinal and bulbar muscular atrophy (SBMA), Huntington disease (HD), and spinocerebellar ataxia (SCA1) (Table 17-3).

There are two particularly unusual features that TNRE disorders have in common. First, the severity of the disease tends to worsen in future generations. This phenomenon is called **anticipation**. A second perplexing feature of TNRE disorders is that the severity of the disease depends on whether the disease is inherited from the mother or father. In the case of Huntington disease, TNRE is likely to occur if inheritance occurs from the father. In contrast, myotonic muscular dystrophy is more likely to get worse if it is inherited from the mother. Overall, TNRE is a newly discovered form of mutation that is receiving a lot of attention by the research community. It poses many challenging questions in molecular genetics. TNRE also makes it particularly difficult for genetic counselors to advise couples as to the severity of these diseases if they are passed to their children.

To understand why TNRE occurs, we need to take a molecular look at the expansion process. As the name suggests, trinucleotide repeat expansion involves increased repetition of a trinucleotide sequence. In normal individuals, certain genes and chromosomal locations contain regions where trinucleotide sequences are repeated in tandem. These sequences are transmitted normally from parent to offspring

TABLE 17–3 TNRE Disorders

	SBMA	HD	SCA1	FRAXA	FRAXE	DM
Repeat sequence	CAG	CAG	CAG	CGG	GCC	CTG
# of repeats in normal individuals	11–33	6–37	19–36	6–50	6–25	5–35
# of repeats in affected individuals	36–62	27–121	43–81	>200	>200	>200
Pattern of inheritance	X–linked	Autosomal dominant	Autosomal dominant	X–linked	X–linked	Autosomal dominant
Disease symptoms	Neurodegenerative ————————————→			X–breakage	Mental retardation	Muscle disease
Anticipation*	—	Male	Male	Female	—	Female

*Indicates the parent in which TNRE occurs most prevalently.

Taken from Bates, G., and Lehrach, H. (1994) Trinucleotide repeat expansions and human genetic disease. *BioEssays* 16, 277–284.

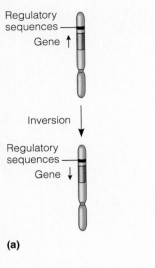

Regulatory sequences

Gene ↑

Inversion

Regulatory sequences

Gene ↓

(a)

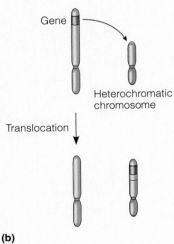

Gene

Heterochromatic chromosome

Translocation

(b)

FIGURE **17-2**

Causes of position effects. (a) A chromosomal inversion has repositioned the promoter for the coding sequence of a gene next to a regulatory sequence from another gene. Now the regulatory sequences are near the gene promoter. **(b)** A translocation has moved a gene from a euchromatic to a heterochromatic chromosome. This type of position effect prevents the expression of the relocated gene.

without mutation. However, in persons with TNRE disorders, the length of a trinucleotide repeat has increased above a certain critical size and becomes prone to frequent expansion. This phenomenon is depicted in the following, where the trinucleotide repeat of CAG has expanded from 11 tandem copies to 18 copies:

$$-CAGCAGCAGCAGCAGCAGCAGCAGCAGCAGCAG- \quad n = 11$$
$$\downarrow$$
$$-CAGCAGCAGCAGCAGCAGCAGCAGCAGCAGCAGCAGCAG$$
$$CAGCAGCAGCAGCAG- \quad n = 18$$

The cause of TNRE is not well understood. It has been speculated that the trinucleotide repeat produces alterations in DNA structure, such as stem-loop formation, and this may lead to errors in DNA replication. However, future research will be necessary to understand the underlying mechanism that causes TNRE. Nevertheless, it is well established that TNRE within certain genes alters the expression of the gene and thereby produces the disease symptoms described in Table 17-3.

Changes in chromosome structure can affect the expression of a gene

The mutations that we have previously considered in this chapter have been small changes in the DNA sequence of particular genes. However, a change in chromosome structure can also be associated with an alteration in the expression of a specific gene. Figure 17-2 depicts a schematic example in which a piece of one chromosome has been inverted or translocated to a different chromosome. Quite commonly, an inversion or translocation has no obvious phenotypic consequence. However, Alfred Sturtevant recognized as early as 1925 that chromosomal rearrangements in *Drosophila melanogaster* can influence phenotypic expression (namely, eye morphology). In some cases, a chromosomal rearrangement may affect a gene because the chromosomal breakpoint actually occurs within the gene itself. In other cases, a gene may be left intact, but its expression may be altered when it is moved to a new location. When this occurs, the change in gene location is said to have a **position effect**.

There are two common reasons for position effects. One possibility is that a gene may be moved next to regulatory sequences (i.e., silencers or enhancers) that influence the expression of the relocated gene (Figure 17-2a). Alternatively, a chromosomal rearrangement may reposition a gene from a euchromatic region to a chromosome that is very highly condensed (heterochromatic). When the gene is moved to a heterochromatic region, its expression may be turned off (Figure 17-2b). This second type of position effect may produce a variegated phenotype in which the expression of the gene is variable. For genes that affect pigmentation, this produces a mottled appearance rather than an even color. Figure 17-3 shows a position effect that alters eye color in *Drosophila*. Figure 17-3a depicts a normal red-eyed fruit fly, and Figure 17-3b shows a mutant fly in which a gene affecting eye color has been relocated to a heterochromatic chromosome. The variegated appearance of the eye occurs because the degree of heterochromatinization varies across different regions of the eye. In cells where heterochromatinization has turned off the eye color gene, a white phenotype occurs, while other cells allow this same region to remain euchromatic and produce a red phenotype.

Mutations can occur in germ line or somatic cells

In this section, we have considered many different ways that mutation can affect gene expression. For multicellular organisms, the timing of mutation also plays an important role. A mutation can occur very early in life, such as in a gamete or a fertilized egg, or it may occur later in life, such as in embryonic or adult stages. The exact time when mutations occur can be important with regard to the severity of the genetic effect and their ability to be passed from parent to offspring.

Geneticists classify the cells of sexually reproducing organisms into two types: the germ line and the somatic cells. The term **germ line** refers to the cells that give rise to the gametes such as eggs and sperm. A germ line mutation can occur directly in a sperm or egg cell, or it can occur in a precursor cell that produces the gametes. If a mutant gamete participates in fertilization, all of the cells of the resulting offspring will contain the mutation (Figure 17-4a). Likewise, when an individual who has inherited a germ line mutation produces gametes, the mutation may be passed along to future generations of offspring.

The **somatic** cells comprise all cells of the body excluding the germ line cells. Examples include muscle cells, nerve cells, and so forth. Mutations can also occur within somatic cells at early or late stages of development. Figure 17-4b illustrates the consequences of a mutation that occurs during the embryonic stage. In this example, a mutation has occurred within a single embryonic cell. As the embryo grows, this single cell will be the precursor for many cells of the adult organism. Therefore, in the adult, a patch of tissue will contain the mutation. The size of the patch will depend on the timing of the mutation. In general, the earlier the mutation occurs during development, the larger the patch. An individual who has somatic regions that are genotypically different from each other is called a **genetic mosaic**.

Figure 17-5 illustrates an individual who has had a somatic mutation occur during an early stage of development. In this case, the person has a patch of gray hair

(a) Wild r-type red eye (normal allele)

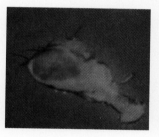

(b) Variegated eye

FIGURE **17-3**

A position effect that alters eye color in *Drosophila*. (a) A normal red eye. **(b)** An eye in which an eye-color gene has been relocated to a heterochromatic chromosome. This inactivates the gene and produces a variegated phenotype.

GENES ⟶ TRAITS: The variegated trait of eye color occurs because the degree of heterochromatinization varies throughout different regions of the eye. In some cells, heterochromatinization occurs and turns off the eye color gene, thereby leading to the white phenotype. In other patches of cells, the region containing the eye color allele remains euchromatic, yielding a red phenotype.

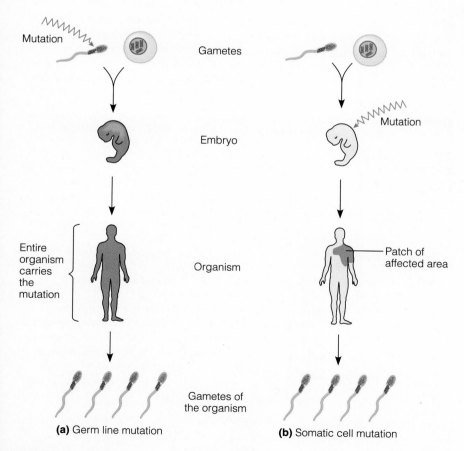

Mutation

Gametes

Mutation

Embryo

Entire organism carries the mutation

Organism

Patch of affected area

Gametes of the organism

(a) Germ line mutation

(b) Somatic cell mutation

FIGURE **17-4**

The effects of somatic versus germ line mutations.

FIGURE **17-5**

Example of a somatic mutation.

GENES → TRAITS: This individual has a patch of gray hair, because a somatic mutation occurred in a single cell during embryonic development that prevented pigmentation of the hair. This cell continued to divide to produce a patch of gray hair.

while the rest of the hair is pigmented. Presumably, this individual initially had a single mutation occur in a embryonic cell that ultimately gave rise to a patch of scalp that produced the gray hair. Although a patch of gray hair is not a particularly harmful phenotypic effect, mutations during early stages of life can be quite harmful, especially if they disrupt essential developmental processes. Therefore, even though it is prudent to avoid environmental agents that cause mutations during all stages of life, the possibility of somatic mutations is a rather compelling reason to avoid them during the very early stages of life such as fetal development, infancy, and early childhood.

OCCURRENCE AND CAUSES OF MUTATION

As we have seen, mutations can have a wide variety of effects on the phenotypic expression of genes. For this reason, geneticists have spent a great deal of effort identifying the causes of mutations. This has been a truly challenging task, since various agents can alter the structure of DNA and thereby cause mutation. Geneticists categorize the cause of mutation in one of two ways. **Spontaneous mutations** are changes in DNA structure that result from abnormalities in biological processes, whereas **induced mutations** are caused by environmental agents (Table 17-4).

Many causes of spontaneous mutations are examined in other chapters throughout this text. As discussed in Chapter 8, abnormalities in crossing over can produce chromosome mutations such as deletions, duplications, translocations, and inversions. Aberrant segregation of chromosomes during meiosis can cause genome mutations (i.e., changes in chromosome number). In Chapter 11, we discovered that DNA polymerase can make a mistake during DNA replication by putting the wrong base in a newly synthesized daughter strand. Errors in DNA replication are usually infrequent except in certain viruses, such as HIV, that have relatively high rates of spontaneous mutations. In addition, normal metabolic processes may produce chemicals within the cell that can react directly with the DNA and alter its structure. Also, as will be described in Chapter 18, transposable genetic elements can alter gene sequences

TABLE **17–4 Causes of Mutations**

Common Causes of Mutations	Description
Spontaneous	
Aberrant recombination	Abnormal crossing over may cause deletions, duplications, translocations, and inversions (Chapter 8).
Aberrant segregation	Abnormal chromosomal segregation may cause aneuploidy or polyploidy (Chapter 8).
Errors in DNA replication	A mistake by DNA polymerase may cause a point mutation (Chapter 11).
Toxic metabolic products	The products of normal metabolic processes may be chemically reactive agents that can alter the structure of DNA.
Transposable elements	Transposable elements can insert themselves into the sequence of a gene (Chapter 18).
Induced	
Chemical agents	Chemical substances may cause changes in the structure of DNA. Many examples of chemical mutagens will be described in this chapter.
Physical agents	Physical phenomena such as UV light and X-rays damage the DNA. These will also be described in this chapter.

by inserting themselves into genes. Overall, a distinguishing feature of spontaneous mutations is that their underlying cause originates within the cell.

By comparison, the cause of induced mutations originates outside of the cell. Induced mutations are produced by environmental agents that enter the cell and then alter the structure of DNA. Agents that are known to alter the structure of DNA are called **mutagens**. Mutagens can be chemical substances or physical agents that ultimately lead to changes in DNA structure.

In this section, we will begin by examining the random nature of spontaneous mutations and some general features of the mutation rate. We will then explore several mechanisms by which mutagens can alter the structure of DNA. Laboratory tests that can identify potential mutagens will then be described.

Spontaneous mutations are random events

For many centuries, biologists have wondered whether genotypic changes occur purposefully as a result of environmental conditions or whether they are spontaneous events that may occur randomly in any gene of any individual. The question of whether such mutations are spontaneous occurrences or causally related to environmental conditions has an interesting history. In the 19th century, Jean Baptiste Lamarck proposed that physiological events (e.g., use and disuse) determine whether traits are passed along to offspring. For example, his theory suggested that an individual who practiced and became adept at a physical activity, such as the long jump, would pass that quality on to their offspring. Alternatively, Charles Darwin proposed that genetic variation occurs as a matter of chance and that natural selection results in the differential survival of organisms that are better adapted to their environments. According to this view, those individuals who happen to contain beneficial mutations will be more likely to survive and pass these genes to their offspring. These opposing theories of the 19th century were tested in bacterial studies in the 1940s and 1950s. Two of these studies are described here.

Salvadore Luria and Max Delbruck, while working at Indiana and Vanderbilt Universities, were interested in the ability of bacteria to become resistant to infection by a bacteriophage called T1. When a population of *E. coli* cells is exposed to T1, a small percentage of bacteria become resistant to T1 infection and pass this trait to their progeny. Luria and Delbruck were interested in whether such resistance, called Ton^R (*T on*e Resistance), is due to the occurrence of spontaneous mutations or whether it is a physiological adaptation that occurs at a low rate within the bacterial population.

According to the physiological adaptation theory, the rate of adaptation should be a relatively constant value and would depend on the exposure to the bacteriophage. Therefore, when comparing different populations of bacteria, the number of Ton^R bacteria should be an essentially constant proportion of the total population. In contrast, the spontaneous mutation theory depends on the timing of mutation. If a Ton^R mutation occurs early within the proliferation of a bacterial population, many Ton^R bacteria will be found within that population. However, if it occurs much later in population growth, then fewer Ton^R bacteria will be observed. In general, a spontaneous mutation theory predicts a much greater fluctuation in the number of Ton^R bacteria among different populations. This test, therefore, has become known as the *fluctuation test*.

To distinguish between the physiological adaptation and spontaneous mutation theories, Luria and Delbruck inoculated 20 individual tubes and one large flask with *E. coli* cells and grew them in the absence of T1 phage. The flask was grown to produce a very large population of cells, while each individual culture was grown to a smaller population of approximately 20 million cells. They then plated the individual cultures onto media containing T1 phage. Likewise, 10 subsamples, each consisting of 20 million bacteria, were removed from the large flask and plated onto media with T1 phage.

The results of the Luria and Delbruck experiment are shown in Figure 17-6. Within the smaller individual cultures, a great fluctuation was observed in the number of Ton^R mutants. These results are consistent with a spontaneous mutation theory in which the timing of a mutation during the growth of a culture greatly affects the number of mutant cells. For example, in tube #14 there were many Ton^R bacteria. Luria and Delbruck reasoned that a mutation occurred randomly in one bacterium at an early stage of the population growth, before the bacteria were exposed to T1 on plates. This mutant bacterium then divided to produce many daughter cells that inherited the Ton^R trait. In other tubes, such as #1 and #3, this spontaneous mutation did not occur, and so none of the bacteria had a Ton^R phenotype. By comparison, the cells plated from the large flask tended toward a relatively constant and intermediate number of Ton^R bacteria. Since the large growth flask had so many cells, several independent Ton^R mutations were likely to have occurred during different stages of its growth. In a single flask, however, these independent events would be mixed together to give an average value of Ton^R cells.

Randomly occurring mutations can give an organism a survival advantage

Joshua Lederberg and Ester Lederberg were also interested in the relationship between mutation and the environmental conditions that select for mutation. At the

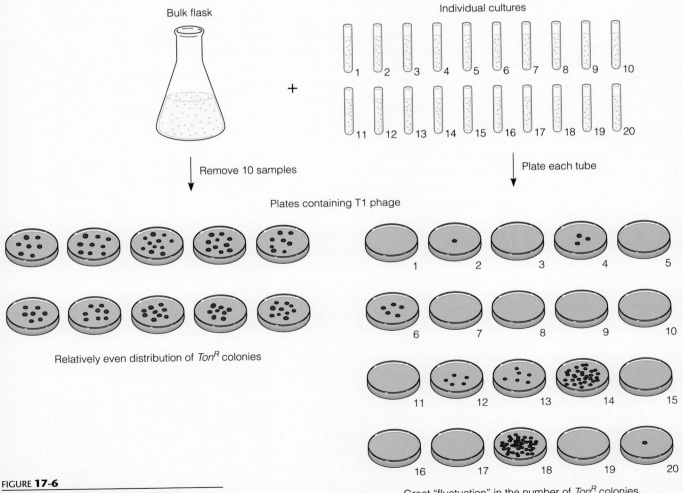

FIGURE **17-6**

The Luria–Delbruck fluctuation test.

time of their studies, some scientists still held the belief that selective conditions could promote the formation of specific mutations allowing the organism to survive. Similar to the Lamarck theory of adaptive mutations, these scientists thought that if bacteria were exposed to an antibiotic, for example, the presence of the antibiotic would actually cause gene mutations that confer antibiotic resistance. This theory, which was similar to the adaptation theory, was known as the *directed mutation theory*. In contrast, the *random mutation theory*, which was consistent with a Darwinian viewpoint, proposed that mutations occur at random. According to this theory, environmental factors that affect survival simply select for the survival of those individuals that happen to possess a beneficial mutation.

An experimental approach to distinguish between these two possibilities was developed by the Lederbergs in the 1950s while working at the University of Wisconsin in Madison. They used a technique known as **replica plating**. As shown in Figure 17-7, they plated a large number of bacteria onto a master plate that did not contain any selective agent (namely, T1). A sterile piece of velvet cloth was lightly touched to this plate in order to pick up a few bacterial cells from each colony. This replica was then transferred to secondary plates that contained an agent that selected for the growth of bacterial cells with a particular genotype.

In the example shown in Figure 17-7, the secondary plates contained T1 bacteriophages. On these plates, only those mutant cells that are *Ton^R* could grow. On the secondary plates, a few colonies were observed. Furthermore, they occupied the same location on each plate. These results indicated that the mutations conferring *Ton^R* occurred randomly while the cells were growing on the (nonselective) master plate; the presence of the bacteriophage in the secondary plates simply selected for the growth of previously occurring *Ton^R* mutants. These results supported the random mutation theory. In contrast, the directed mutation theory would have predicted that *Ton^R* bacterial mutants would occur after the cells were transferred to the secondary plates. If that had been the case, they would not be expected to arise in identical locations on different secondary plates.

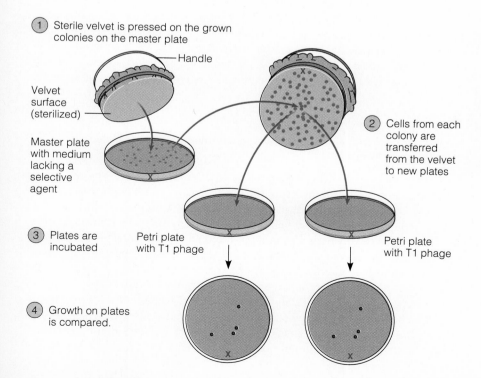

1. Sterile velvet is pressed on the grown colonies on the master plate

Handle

Velvet surface (sterilized)

Master plate with medium lacking a selective agent

2. Cells from each colony are transferred from the velvet to new plates

3. Plates are incubated

Petri plate with T1 phage

Petri plate with T1 phage

4. Growth on plates is compared.

FIGURE **17-7**

Replica plating. Bacteria were first plated on a master plate under nonselective conditions. A sterile velvet cloth was used to make a replica of the master plate. This replica was gently pressed onto two secondary plates that contained a selective agent. In this case, the two secondary plates contained T1 bacteriophage. Only those mutant cells that are *Ton^R* could grow to form visible colonies. Note: the red x indicates the alignment of the filter and the plates.

The mutation rate is a measure of new mutations per generation; the mutation frequency is the relative occurrence of a mutation within a population

Since mutations occur spontaneously among populations of living organisms, geneticists have been greatly interested in learning how prevalent they are. The term **mutation rate** is the likelihood that a gene will be altered by a new mutation. It is commonly expressed as the number of new mutations in a given gene per generation. In unicellular organisms such as bacteria, the spontaneous mutation rate for a particular gene is in the range of 1 in 100,000 to 1 in 1,000,000,000, or 10^{-5} to 10^{-9} per cell generation. In more complex eukaryotes, the rate of germ line mutations can be expressed as the number of new mutations per generation. These numbers tell us that it is very unlikely that a particular gene will mutate due to natural causes. However, the mutation rate is not a constant number. The presence of mutagens within the environment can increase the rate of induced mutations to a much higher value than the spontaneous mutation rate. In addition, mutation rates vary substantially from species to species and even within different strains of the same species. One explanation for this variation is the many different causes of mutations (see Table 17-4).

As we have learned more about mutation rate, it has been necessary to modify the view that mutations are a totally random process. Within the same individual, different genes can vary widely with regard to their individual mutation rates. Some genes mutate at a much higher rate than other genes. This is because some genes are larger than others (providing a greater chance for mutation) and their locations within the chromosome may be more susceptible to mutation. Even within a single gene, there are usually **hot spots**, which are select regions of a gene that are more likely to mutate than are other regions (refer back to Figure 6-19).

Before we end our discussion of mutation rate, it is helpful to distinguish the rate of new mutation from the concept of **mutation frequency**. The mutation frequency for a gene is the number of mutant genes divided by the total number of genes within the population. If 1,000,000 bacteria were plated and 10 were found to be mutant, the mutation frequency would be 1 in 100,000 or 10^{-5}. As described earlier in the chapter, Luria and Delbruck showed that among the bacteria in the 20 tubes, the timing of mutations influenced the mutation frequency within any particular tube. Some tubes had a high frequency of mutation, while others did not. The mutation frequency is an important genetic concept, particularly in the field of population genetics. As will be discussed in Chapter 26, the mutation or allelic frequency depends on a variety of factors, including the mutation rate and the forces of natural selection.

EXPERIMENT 17A

X-rays were the first environmental agent shown to cause induced mutations, by Müller in 1927

Geneticists have used allelic variation (e.g., red versus white eyes) as an important tool to explore the mechanisms of inheritance. In natural populations, allelic variation can occur as a result of spontaneous mutation. As we have just discussed, though, the rate of spontaneous mutations for any given gene tends to be very low, in the range of one in a million. From an experimental viewpoint, this low rate of spontaneous mutation presents an obstacle to the study of allelic variation among individuals. For example, it is not an easy task to observe millions of fruit flies in search of rare spontaneous mutations that cause a phenotypic variation. As you may recall from Chapter 3, Thomas Hunt Morgan spent two years before he obtained his first white-eye mutation (see Experiment 3A). In response to this obstacle, many geneticists have sought to identify environmental agents that promote mutation and thereby make it easier to study mutation in experimental organisms.

In 1927, long before the structure of DNA was known, Hermann Müller at the University of Texas devised an approach that demonstrated that X-rays can cause mutation. This work was carried out using *Drosophila melanogaster* as an experimental organism. Müller reasoned that a mutagenic agent might cause some genes to become defective. His experimental approach focused on the ability of a mutagen to cause defects in X-linked genes that result in a recessive, lethal phenotype. In particular, the goal of Müller's experiment was to determine whether X-rays increase the production of X-linked recessive mutations.

To determine if X-rays increase the rate of recessive X-linked lethal mutations, Müller wanted to have an easy way to detect the occurrence of such mutations. He cleverly realized that he had available a laboratory strain of fruit flies that could make this possible. In particular, he conducted his crosses in such a way that a female fly that inherited a new mutation causing a recessive X-linked lethal allele would not be able to produce any male offspring. This made it very easy for him to detect lethal mutation; he only had to count the number of female flies that could not produce sons. In flies that had not been exposed to X-rays, nearly all females were able to produce sons. In contrast, if X-rays were acting as a mutagenic agent, Müller hypothesized that exposure to X-rays would increase the numbers of females unable to produce sons.

To understand Müller's crosses, we need to take a closer look at a peculiar version of one of the X-chromosomes in a strain of flies that he used in his crosses. This X-chromosome, designated ClB, had three important genetic alterations:

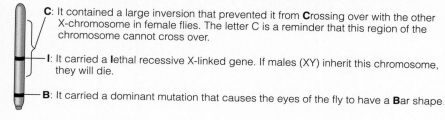

C: It contained a large inversion that prevented it from **C**rossing over with the other X-chromosome in female flies. The letter C is a reminder that this region of the chromosome cannot cross over.

l: It carried a **l**ethal recessive X-linked gene. If males (XY) inherit this chromosome, they will die.

B: It carried a dominant mutation that causes the eyes of the fly to have a **B**ar shape.

Note: C and B are uppercase because they are inherited in a dominant manner, while l is lowercase because it is a recessive allele.

A female fly that has one copy of this X-chromosome would have bar-shaped eyes. Even though this X-chromosome has a lethal allele, a female fly can survive if the corresponding gene on the other X-chromosome is a normal allele. In Müller's experiments, the goal was to determine if the normal X-chromosome (not the ClB chromosome) had obtained a lethal mutation in any gene except for the same lethal gene on the ClB chromosome. If a recessive lethal mutation occurred on the normal X-chromosome, this female could survive because it would be heterozygous for recessive lethal mutations in two different genes. However, since each X-chromosome would have a lethal mutation, this female would not be able to produce any living sons:

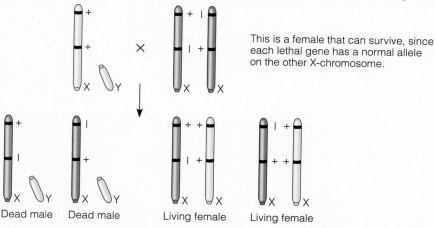

This is a female that can survive, since each lethal gene has a normal allele on the other X-chromosome.

Dead male   Dead male   Living female   Living female

THE HYPOTHESIS

The exposure of flies to X-rays will increase the rate of mutation.

TESTING THE HYPOTHESIS

Starting material: The female flies contain one normal X-chromosome and on ClB X-chromosome. The male flies contain a normal X-chromosome.

Experimental Level   **Conceptual Level**

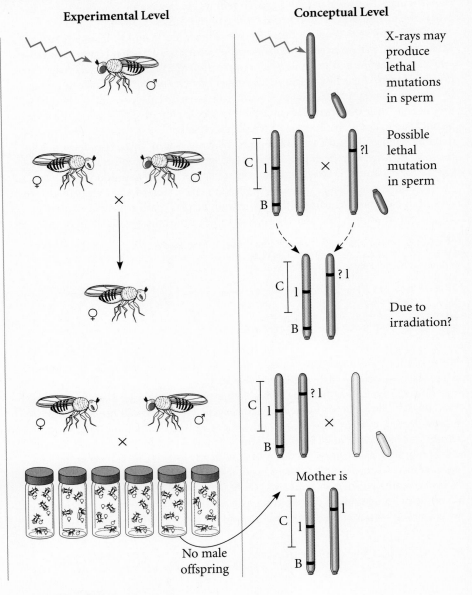

1. Expose male flies to X-rays. Also, have a control group that is not exposed to X-rays.

X-rays may produce lethal mutations in sperm

2. Mate the male flies to female flies carrying one normal X-chromosome and one ClB X-chromosome.

Possible lethal mutation in sperm

3. Save about 1000 daughters with bar eyes. Note: These females contain a ClB X-chromosome from their mothers and an X-chromosome from their fathers that may (or may not) have a recessive lethal mutation.

Due to irradiation?

4. Mate each bar-eyed daughter with normal (nonirradiated) males. Note: This is done in (1000) individual tubes. (Only six are shown.)

5. Count the number of crosses that do not contain any male offspring. These crosses indicate that the bar-eyed female parent contained an X-linked recessive mutation on the non-ClB X-chromosome.

Mother is

No male offspring

THE DATA

Treatment	Number of ClB Daughters Crossed to Normal Males*	Number of Tubes Containing Any Offspring**	Number of Tubes with Female Offspring but Lacking Male Offspring
Control	1011	947	1
X-ray treated	1015	783	91

*See Step 4. **The reason why these values are less than the numbers of mated ClB daughters is because some crosses did not produce living offspring.

The data are from Muller, H. J. (1928) *Verh. V. Intern. Kongr. Vererb. 1.* Described in: Timofeeff-Ressovsky, N. W. (1934) The experimental production of mutation. *Biol. Rev.* 9, 411–457.

In the absence of X-ray treatment, only one cross in approximately 1000 was unable to produce male offspring. This means that the spontaneous rate for any X-linked lethal mutation was relatively low. By comparison, X-ray treatment (of the fathers) of the ClB females produced 91 crosses without male offspring. Since these females inherited their non-ClB chromosome from irradiated fathers, these results indicate that X-rays greatly increase the rate of X-linked recessive lethal mutations. This conclusion has been confirmed in many subsequent studies, which have shown that the increase in mutation rate is correlated with the amount of exposure to X-rays.

Mutagens alter DNA structure in different ways

Since this pioneering study of Müller, researchers have found that an enormous array of agents can act as mutagens to permanently alter the structure of DNA. We often hear in the news media that we should avoid these agents in our foods and living environment. We even use products such as sunscreens that help us avoid the mutagenic effects of ultraviolet (UV) rays. The public is concerned about mutagens for two important reasons. First, mutagenic agents are often involved in the development of human cancers. This topic will be discussed in Chapter 23. In addition, since most new mutations are deleterious, people want to avoid mutagens to prevent gene mutations that may have harmful effects in their offspring.

Mutagenic agents are usually classified as **chemical** or **physical** mutagens. Examples of both types of agents are listed in Table 17-5. In other cases, chemicals that are not mutagenic can be altered to a mutagenically active form after they have been ingested into the body. Cellular enzymes such as oxidases have been shown to activate some mutagens. Certain foods contain chemicals that act as antioxidants. Many scientists are investigating whether antioxidants can counteract the effects of mutagens and thereby lower the cancer rate.

Mutagens can alter the structure of DNA in various ways. Some mutagens act by covalently modifying the structure of nucleotides. For example, **nitrous acid** replaces amino groups with keto groups ($-NH_2$ to $=O$). This can change cytosine to uracil, and adenine to hypoxanthine. When this mutated DNA replicates, the modified bases do not pair with the appropriate nucleotides in the newly made strand. Instead, uracil pairs with adenine, and hypoxanthine pairs with cytosine (see Figure 17-8). Other chemical mutagens can also disrupt the appropriate pairing between nucleotides by alkylating bases within the DNA. During alkylation, methyl or ethyl groups are covalently attached to the bases. Examples of alkylating agents include **nitrogen mustards** and **ethyl methanesulfonate (EMS)**.

TABLE 17-5 Examples of Mutagens

Mutagen	Effect(s) on DNA structure
Chemical	
Nitrous acid	Deaminates bases
Hydroxylamine	Hydroxylates cytosine
Nitrogen mustard	Alkylating agent
Ethyl methanesulfonate	Alkylating agent
Proflavin	Interchelates within DNA helix
5-Bromouracil	Base analogue
2-Aminopurine	Base analogue
Physical	
UV light	Promotes pyrimidine dimer formation
X–rays	Causes base deletions, single nicks in DNA strands, cross–linking, and chromosomal breaks

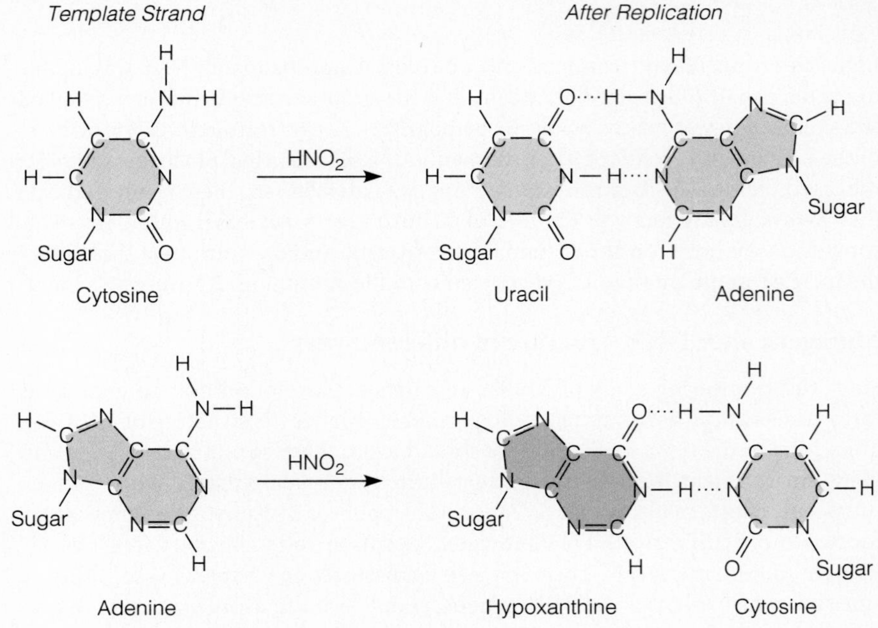

Template Strand

After Replication

Cytosine

$\xrightarrow{HNO_2}$

Uracil   Adenine

Adenine

$\xrightarrow{HNO_2}$

Hypoxanthine   Cytosine

FIGURE **17-8**

Mispairing of modified bases. During DNA replication, uracil will pair with adenine, and hypoxanthine will pair with cytosine. This will create mutations in the newly replicated strand during DNA replication.

Some mutagens exert their effects by directly interfering with the DNA replication process. For example, **acridine dyes**, such as **proflavin**, contain flat planar structures that insert themselves into the double helix, thereby distorting the helical structure. When DNA containing these mutagens is replicated, single nucleotide additions and/or deletions can be incorporated into the newly made daughter strands.

Compounds such as **5-bromouracil (5BU)** and **2-aminopurine** are nucleotide base analogues that become incorporated into daughter strands during DNA replication. 5-Bromouracil is a thymine analogue that can be incorporated into DNA instead of thymine. Like thymine, 5BU can base pair with adenine. However, at a relatively high rate, it can also base pair with guanine (Figure 17-9a). When this occurs during DNA replication, 5BU causes a mutation in which an A–T base pair is changed to a G–5BU base pair (Figure 17-9b). This is a transition, since the adenine has been changed to a guanine, both of which are purines. During the next round of DNA replication, the template strand containing the guanosine base will create

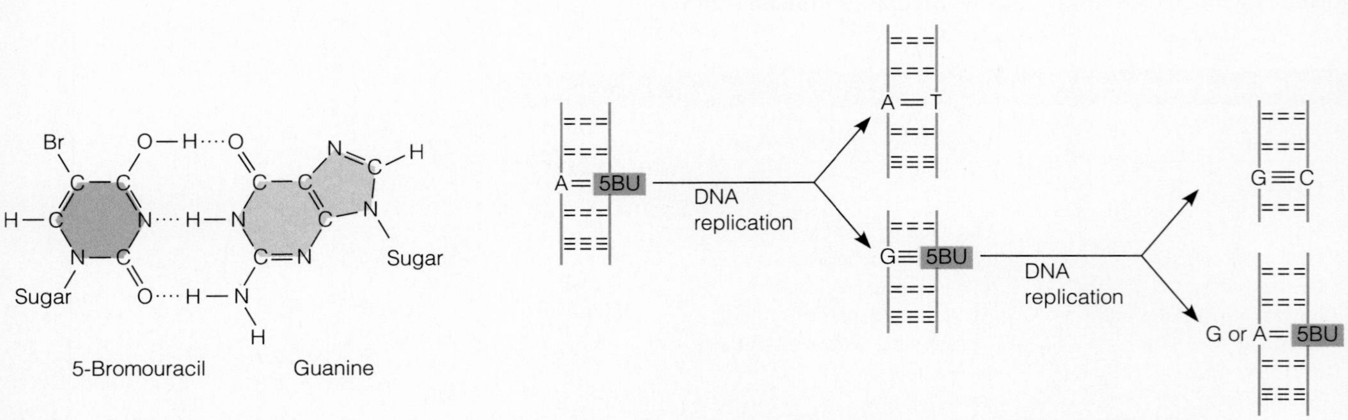

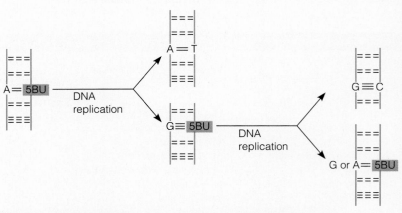

5-Bromouracil   Guanine

(a) Base pairing of 5BU with guanine

(b) How 5BU causes a mutation in a base pair during DNA replication

FIGURE **17-9**

Structure and effects of the mispairing of 5-bromouracil with guanine.

a G–C base pair. In this way, 5-bromouracil can promote a change of an A–T base pair into a G–C base pair.

DNA molecules are also sensitive to physical agents such as radiation. In particular, radiation of short wavelength and high energy is known to alter DNA structure. Ionizing radiation includes X-rays and gamma rays. This type of radiation can penetrate deeply into biological materials, where it creates chemically reactive molecules known as *free radicals*. These molecules can alter the structure of DNA in a variety of ways. Exposure to high doses of ionizing radiation can cause base deletions, single nicks in DNA strands, crosslinking, and even chromosomal breaks. Nonionizing radiation, such as UV light, contains less energy, and so it only penetrates the surface of material such as the skin. Nevertheless, UV light is known to cause DNA mutation. For example, as shown in Figure 17-10, UV light causes the formation of crosslinked **thymine dimers**. A thymine dimer within a DNA strand may cause a mutation when that DNA strand is replicated.

Testing methods can determine if an agent is a mutagen

To determine if an agent is mutagenic, researchers utilize testing methods that can monitor whether or not an agent increases the rate of mutation. Many different kinds of tests have been used to evaluate mutagenicity. One commonly used test is the **Ames test**, which was developed by Bruce Ames at UC Berkeley in the 1970s. This test uses strains of a bacterium, *Salmonella typhimurium*, that cannot synthesize the amino acid histidine. These strains contain a deleterious mutation within a gene that encodes an enzyme required for histidine biosynthesis. Therefore, the bacteria cannot grow on petri plates unless histidine has been added to the growth medium. However, a second mutation may occur that restores the ability to synthesize histidine. In other words, a second mutation can cause a reversion back to the wild-type condition. The Ames test monitors the rate at which this second mutation occurs and thereby indicates whether an agent increases the mutation rate above the spontaneous rate.

Figure 17-11 outlines the steps in the Ames test. The suspected mutagen is mixed with a rat liver extract and a bacterial strain that can't synthesize histidine. A mutagen may require activation by cellular enzymes; the rat liver extract provides a mixture of enzymes that may activate a mutagen. This step improves the ability of the test to identify agents that may cause mutation in mammals. After the incubation period, a large number of bacteria are then plated on a minimal

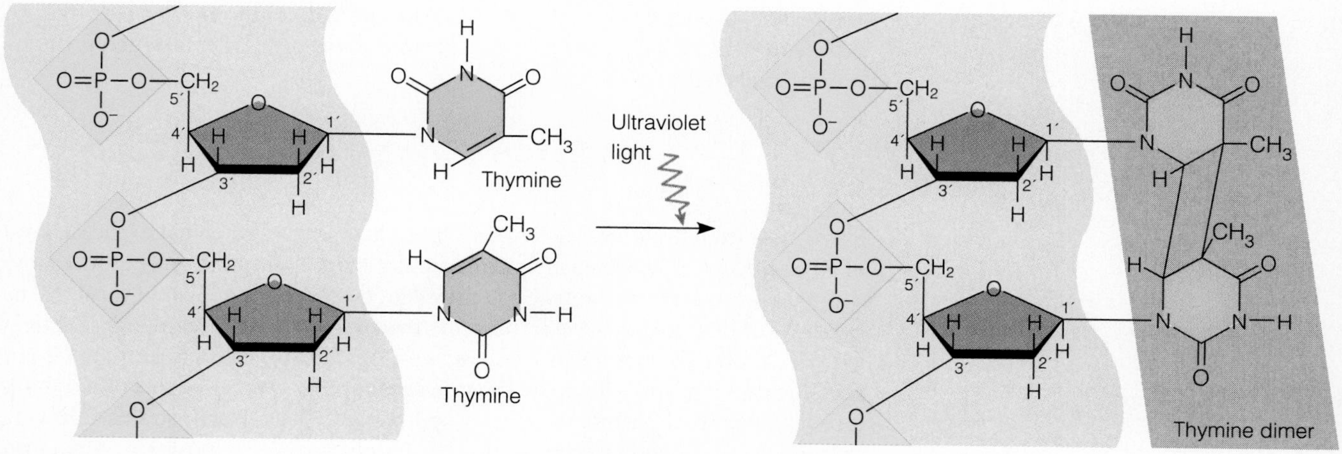

FIGURE **17-10**

Formation and structure of a thymine dimer.

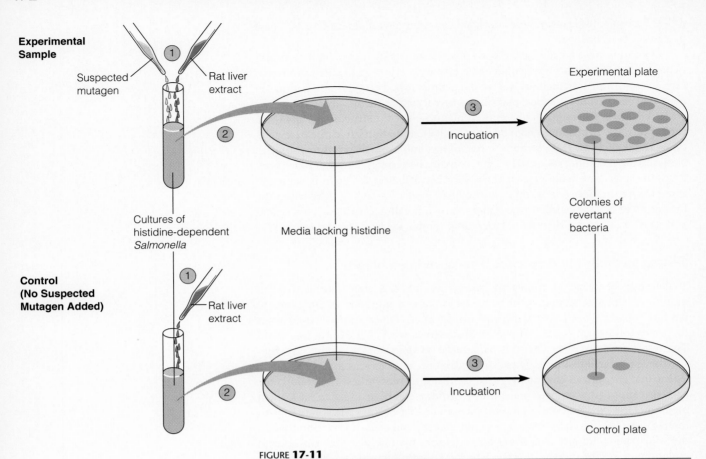

FIGURE **17-11**
The Ames test for mutagenicity.

growth medium that does not contain histidine. The *Salmonella* strain is not expected to grow on these plates. However, if a mutation has occurred that allows a bacterium to synthesize histidine, it can grow on these plates to form a visible bacterial colony. To estimate the mutation rate, the colonies that grow on the minimal media are counted and compared with the total number of bacterial cells that were originally streaked on the plate. For example, if 10,000,000 bacteria were plated and 10 growing colonies were observed, the rate of mutation is 10 out of 10,000,000; this equals 1 in 10^6, or 10^{-6}. As a control, bacteria that have not been exposed to the mutagen are also tested, since a low level of spontaneous mutations is expected to occur.

DNA REPAIR

Since most mutations are deleterious, DNA repair systems are vital to the survival of all organisms. If DNA repair systems did not exist, environmentally induced and spontaneous mutations would be so prevalent that few species would survive. The necessity of DNA repair systems becomes evident when they are missing. Bacteria contain several different DNA repair systems. Yet when even a single system is absent, the bacteria have a much higher rate of mutation. In fact, the rate of mutation is so high that these bacterial strains are sometimes called mutator strains. Likewise, in humans, an individual who is defective in only a single DNA repair system may manifest various disease symptoms, including a higher risk of skin cancer. This increased risk is due to the inability to repair UV-induced mutations.

TABLE 17-6 Common Types of DNA Repair Systems

System	Description
Direct repair	An enzyme recognizes an incorrect alteration in DNA structure and directly converts it back to a correct structure.
Excision repair	An abnormal nucleotide or base is first recognized and removed from the DNA. A segment of DNA in this region is excised, and then the complementary DNA strand is used as a template to synthesize a normal DNA strand.
Mismatch repair	Similar to excision repair except that the DNA defect is a base pair mismatch in the DNA, not an abnormal nucleotide. The mismatch is recognized, and a segment of DNA in this region is removed. The parental strand is used to synthesize a normal daughter strand of DNA.
Recombinational repair	Occurs when a DNA mutation causes a gap in synthesis during DNA replication. The gap is exchanged between the abnormal DNA and the corresponding region in the normal replicated double helix. After this occurs, it is possible to fill in the gap using the complementary DNA strand in the normal DNA strand.

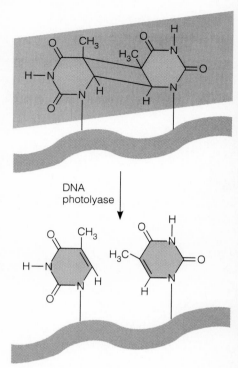

(a) Repair of a thymine dimer

Living cells contain several DNA repair systems that can fix different types of DNA alterations (Table 17-6). Each repair system is composed of one or more proteins that play specific roles in the repair mechanism. In most cases, DNA repair is a multistep process. First, one or more proteins in the DNA repair system detect an irregularity in DNA structure. Next, the abnormality is removed by the action of DNA repair enzymes. In this section, we will examine several different DNA repair systems that have been characterized in bacteria, yeast, and mammals. Their diverse ways of repairing DNA underscore the extreme necessity for the structure of DNA to be maintained properly.

Damaged bases can be directly repaired

In a few cases, the covalent modification of nucleotides by mutagens can be reversed by specific cellular enzymes. As discussed earlier in this chapter, UV light causes the formation of thymine dimers. In the early 1960s, it was found that the yeast enzyme *photolyase* can repair thymine dimers by splitting the dimers, restoring DNA to its original condition (Figure 17-12a). This process directly restores the structure of DNA.

An enzyme known as O⁶-alkylguanine alkyltransferase can remove the methyl or ethyl groups from guanine bases that have been mutagenized by agents such as nitrogen mustards and ethyl methanesulfonate. This enzyme is called a transferase because it transfers the methyl or ethyl group from the base to a cysteine side chain within the alkyltransferase protein (Figure 17-12b). Surprisingly, this permanently inactivates alkyltransferase—it can only be used once!

Excision repair systems remove damaged nucleotides or bases from DNA

An important general process for DNA repair is **nucleotide excision repair (NER)**. This type of system can repair many different types of DNA damage, including UV-induced damage (namely, thymine dimers), chemically modified bases, missing bases, and certain types of crosslinks. In NER, several nucleotides in the damaged strand are removed from the DNA, and the intact strand is used as a template for resynthesis of a normal complementary strand. NER is found in both eukaryotes

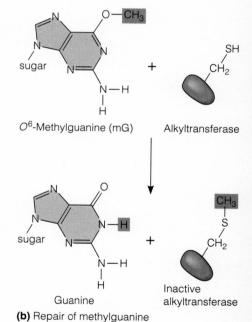

O^6-Methylguanine (mG)   Alkyltransferase

Guanine   Inactive alkyltransferase

(b) Repair of methylguanine

FIGURE **17-12**

Direct repair of damaged bases in DNA.
(a) The repair of thymine dimers by photolyase. **(b)** The repair of methylguanine by transfer of the methyl group to alkyltransferase.

and prokaryotes, although its molecular mechanism is better understood in prokaryotic species.

In *E. coli*, the NER system requires four key proteins, designated UvrA, UvrB, UvrC, and UvrD, plus the help of DNA polymerase and DNA ligase. UvrA, B, C, and D recognize and remove a short segment of a damaged DNA strand. They are named Uvr, because they are involved in *Ultraviolet* light *repair* of pyrimidine dimers, although the UvrA–D proteins are also important in repairing chemically damaged DNA.

Figure 17-13 outlines the steps involved in the *E. coli* NER system. A protein trimer consisting of two UvrA molecules and one UvrB molecule tracks along the DNA in search of damaged DNA. Such DNA will have a distorted double helix, which is sensed by the UvrA–B complex. When a damaged segment is identified, the two UvrA proteins are released and UvrC binds to the site. Together, UvrB and UvrC make incisions in the damaged strand on both sides of the damaged site. UvrB probably makes the 3′ cut and UvrC the 5′ cut. After this incision process, UvrC is released and UvrD binds. UvrD is a helicase that unravels the DNA so that the short damaged segment can be released or excised from the DNA. Following the excision of the damaged DNA, DNA polymerase fills in the gap using the undamaged strand as a template. Finally, DNA ligase makes the final covalent connection between the newly made DNA and the original DNA strand.

A second type of excision repair system involves the function of an enzyme known as **DNA-N-glycosylase**. This enzyme can recognize an abnormal base and cleave the bond between it and the sugar in the DNA backbone (Figure 17-14). Depending on the organism, this repair system can eliminate abnormal bases such as uracil, N- and 3-methyladenine, 7-methylguanine, and pyrimidine dimers.

Figure 17-14 illustrates the steps involved in DNA repair via DNA-N-glycosylase. In this example, the DNA contains a uracil residue in its sequence. This could have happened by the action of a chemical mutagen (e.g., nitrous acid) that deaminates cytosine to produce uracil. N-Glycosylase recognizes a uracil within the DNA and cleaves the bond between the sugar and base. This releases the uracil base and leaves behind an apyrimidinic (AP) nucleotide. This abnormal nucleotide is recognized by a second enzyme, AP-endonuclease, which makes a cut on the 5′ side. DNA polymerase (which has a 5′ to 3′ exonuclease activity) removes the abnormal region and, at the same time, replaces it with normal nucleotides. This process is called **nick translation** (although nick replication would be a more accurate term). Finally, DNA ligase closes the nick.

Several human diseases have been shown to involve inherited defects in genes involved in nucleotide excision repair. These include xeroderma pigmentosum (XP), Cockayne's syndrome (CS), and PIBIDS. (PIBIDS is an acronym for a syndrome with symptoms that include *p*hotosensitivity, *i*chthyosis [a skin abnormality], *b*rittle hair, *i*mpaired intelligence, *d*ecreased fertility, and *s*hort stature.) A common characteristic in all three syndromes is an increased sensitivity to sunlight because of an inability to repair UV-induced lesions. Figure 17-15 shows a photograph of an individual with xeroderma pigmentosum. These individuals have pigmentation abnormalities, many premalignant lesions, and a high predisposition to skin cancer. They may also develop early degeneration of the nervous system.

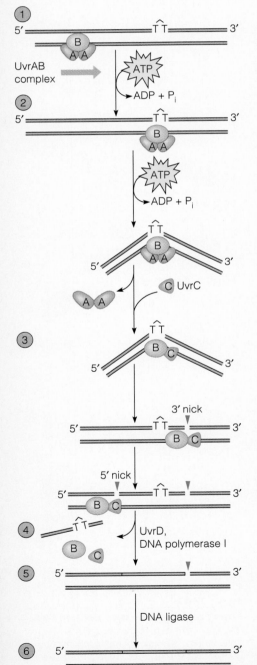

FIGURE **17-13**

Nucleotide excision repair in *E. coli*. (1) UvrA/UvrB trimer scan along the DNA. (2) When the trimer encounters a thymine dimer in the DNA, the UvrA dimer is released. (3) UvrC binds to UvrB. The proteins make two cuts in the damaged DNA strand. One cut is eight nucleotides 5′ to the site, the other approximately five nucleotides 3′ from the site. (4) UvrC is released and UvrD binds to the site. UvrD is a helicase that causes the damaged strand to be removed. The action of UvrD and strand removal are not shown in this figure. (5) UvrD and UvrB are released. DNA polymerase resynthesizes a complementary strand. (6) DNA ligase makes the final connection.

Genetic analyses of patients with XP, CS, and PIBIDS have revealed that these syndromes result from defects in a variety of different genes that encode NER proteins. For example, xeroderma pigmentosum can be caused by defects in seven different NER genes. In all cases, individuals have a defective nucleotide excision repair pathway. In recent years, several human NER genes have been successfully cloned and sequenced. Although more research is needed to understand completely the mechanisms of DNA repair, the identification of NER genes has helped unravel the complexities of NER pathways in human cells.

Mismatch repair systems recognize and correct a base pair mismatch

Thus far, we have considered several DNA repair systems that recognize abnormal nucleotide structures within DNA, including thymine dimers, alkylated bases, and the presence of uracil in the DNA. Another type of abnormality that should not occur in DNA is a **base mismatch**. As described in Chapter 9, the structure of the DNA double helix obeys the A–T/G–C rule of base pairing. During the normal course of DNA replication, however, an incorrect nucleotide may be added to the growing strand by mistake. This creates a mismatch between a nucleotide in the parental and newly made strand. There are various DNA repair mechanisms that recognize and remove this mismatch. For example, as described in Chapter 11, DNA polymerase has a 3′ to 5′ proofreading ability that can detect mismatches and remove them. However, if this proofreading ability fails, cells contain several additional DNA repair systems that can detect base mismatches and fix them.

An interesting DNA repair system is the **methyl-directed mismatch repair** system that has been studied extensively in *E. coli*. This system involves the participation of several proteins that detect the mismatch and specifically remove the segment from the newly made daughter strand. Keep in mind that the newly made daughter strand contains the incorrect base, while the parental strand is normal. Therefore, an important aspect of methyl-directed mismatch repair is that it repairs specifically the newly made strand rather than the original template strand.

As shown in Figure 17-16, three proteins, designated MutL, MutH, and MutS, detect the mismatch and direct the removal of the mismatched base from the newly made strand. These proteins are named Mut because their absence leads to a much higher *mut*ation rate than that in normal strains of *E. coli*. A key characteristic of the MutH protein is that it can distinguish between the parental DNA strand and the daughter strand. It can do this because the MutH protein recognizes a methylated DNA sequence but does not recognize nonmethylated DNA sequences. Immediately after DNA replication, the parental strand is methylated, but it takes a short time before a newly made daughter strand is methylated (refer back to Figure 11-15). Therefore, the MutH protein specifically recognizes the parental DNA strand rather than the daughter strand.

The role of MutS is to locate mismatches. MutS forms a complex with MutL. When the MutS–MutL complex has located a DNA mismatch, it interacts with MutH by a looping mechanism (see Figure 17-16). This stimulates MutH to make a cut in the nonmethylated DNA strand. After the strand is cut, an exonuclease digests the nonmethylated DNA strand in the direction of the mismatch and proceeds

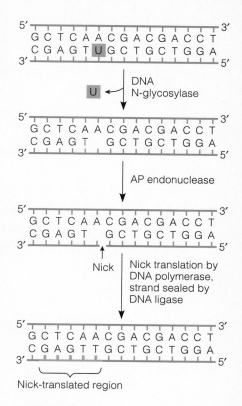

FIGURE **17-14**

Mechanism of DNA repair by DNA-N-glycosylase and AP-endonuclease.

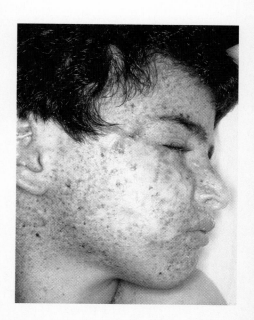

FIGURE **17-15**

An individual affected with xeroderma pigmentosum.

GENES ⟶ TRAITS: This disease involves a defect in genes that are involved with nucleotide excision repair. Xeroderma pigmentosum can be caused by defects in seven different NER genes. Affected individuals have an increased sensitivity to sunlight because of an inability to repair UV-induced DNA lesions. In addition, they may also have pigmentation abnormalities, many premalignant lesions, and a high predisposition to skin cancer.

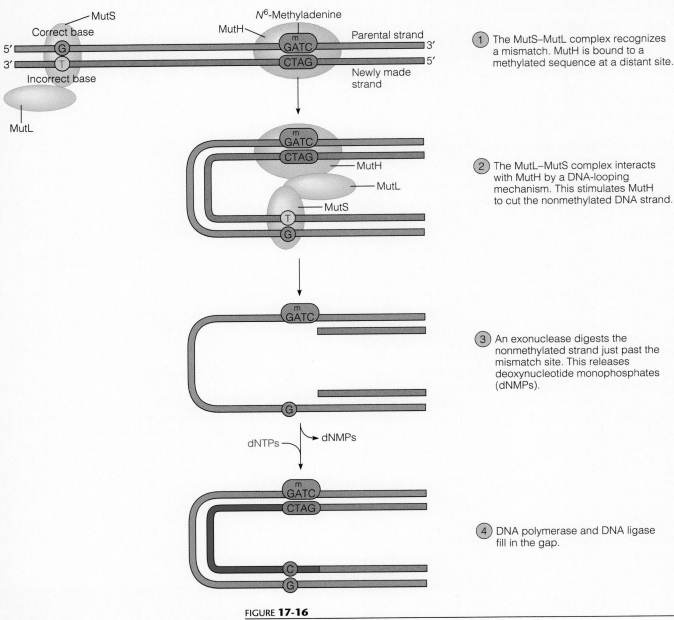

FIGURE **17-16**

Methyl-directed mismatch repair in _E. coli_.

just beyond the mismatch site. This leaves a gap in the daughter strand, which is repaired by DNA polymerase and DNA ligase. As illustrated in Figure 17-16, the net result is that the mismatch has been corrected by removing the incorrect region in the daughter strand and then resynthesizing the correct sequence using the parental DNA as a template.

Damaged DNA can be repaired by recombination

Certain types of DNA damage can halt the progression of DNA replication. For example, a thymine dimer inhibits DNA replication, because it prevents base pairing in a newly made daughter strand. For this reason, a gap will be created in the region that cannot be replicated. DNA gaps are particularly harmful, since they may cause chromosomal breaks and further mutations. To avoid these consequences, the unreplicated gap may be repaired by genetic recombination. We will describe **recombinational repair** in this chapter, and we will examine the molecular mechanisms of genetic recombination in greater detail in Chapter 18.

DNA replication produces a genetically identical pair of double-stranded DNA molecules (see Chapter 9). Recombinational repair occurs while the two DNA copies are being made. To understand this process, we must pay close attention to the relationship between newly made DNA copies during replication. In Figure 17-17, the strands in the two double helices are labeled A, B, C, and D. Since DNA replication makes identical copies of genetic material, strands A and C have the same DNA sequence. Strands B and D also have the same sequence and are complementary to A and C. In Figure 17-17, a thymine dimer had previously occurred in the parental strand (which is now labeled strand D). The thymines in a thymine dimer cannot hydrogen bond properly to incoming nucleotides during DNA replication. Therefore, a thymine dimer disrupts the DNA replication process by creating a gap in the newly made strand after the replication fork has passed. Recombinational repair fixes this gap. This occurs in a two-step process. First, the gap is replaced with the same region in the original DNA strand: a short segment of strand A replaces the corresponding segment of strand C. This removes the gap in strand C, but it creates a gap in strand A. The next step is to fill in the gap in strand A, using the normal strand B as a template. The net result is that neither of the replicated DNA double helices contain gaps, although the thymine dimer is still present in one of the double helices.

Certain bacteria, such as *E. coli*, also have an alternative mechanism to fill in gaps that occur during the replication of damaged DNA. This mechanism, known as the **SOS response**, occurs under extreme conditions that promote DNA damage. Causes of the SOS response include high doses of UV light and other types of mutagens. When the SOS response occurs, the gaps in replicating DNA are filled in by a mechanism that does not involve recombination. When the gaps are filled in during the SOS response, a high error rate of mutation occurs. Even though this is a harmful result, it nevertheless allows the bacteria to survive under conditions of extreme environmental stress. Furthermore, the high rate of mutation may provide genetic variability within the bacterial population, so that certain cells may become resistant to the harsh conditions.

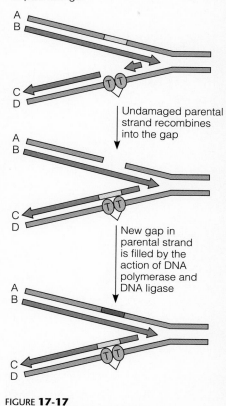

Replication stops at the thymine dimer, but resumes after the dimer is passed, creating a gap. The parental strand complementary to the region containing the dimer is shown in pale orange.

Undamaged parental strand recombines into the gap

New gap in parental strand is filled by the action of DNA polymerase and DNA ligase

FIGURE **17-17**

DNA repair by recombination.

Actively transcribing DNA is repaired more efficiently than nontranscribed DNA

Before we end our discussion of DNA repair systems, it is interesting to mention that not all DNA is repaired at the same rate. In the 1980s, Philip Hanawalt and colleagues at Stanford conducted experiments showing that actively transcribing genes in eukaryotes and prokaryotes are more efficiently repaired following radiation damage than is nontranscribed DNA. The targeting of DNA repair enzymes to actively transcribing genes may have several biological advantages. First, active genes are loosely packed and may be more susceptible to DNA damage. Likewise, the process of transcription itself may cause DNA damage. In addition, DNA regions that contain actively transcribing genes are more likely to be important for survival than nontranscribed regions. This may be particularly true in terminally differentiated cells, which no longer divide. In nondividing cells, gene transcription (rather than DNA replication) is of utmost importance.

In eukaryotes, the mechanism that couples DNA repair and transcription is not completely understood. Certain general transcription factors, such as TFII-H, play a role in both transcription and excision repair. In *E. coli*, a protein

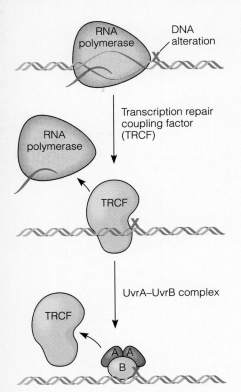

FIGURE **17-18**

Targeting of the nucleotide excision repair system to an actively transcribing gene containing a DNA alteration.

known as **transcription–repair coupling factor** (**TRCF**) is responsible for targeting the excision repair system to actively transcribing genes having damaged DNA. Figure 17-18 presents a model to explain how this works. In this scenario, RNA polymerase is actively transcribing a gene until it encounters a DNA lesion and becomes stalled. TRCF binds to the stalled RNA polymerase and displaces it from the damaged site. TRCF recruits the excision repair system to the damaged site by recognizing the UvrA–UvrB complex. After the UvrA–UvrB complex has bound to this site, TRCF is released from the damaged site. At this stage, the UvrA–B complex can begin excision repair (as previously described in Figure 17-13).

CONCEPTUAL SUMMARY

Mutations are heritable changes in the genetic material. They can result from small changes in the genetic material (*single gene mutations*), rearrangements in chromosome structure (*chromosome mutations*), or changes in chromosome number (*genome mutations*). Mutations can have various consequences on gene expression. If a mutation occurs within the coding sequence, it may alter the amino acid sequence of the encoded polypeptide. *Missense mutations* cause single amino acid substitutions, *nonsense mutations* result in truncated polypeptides, and *frameshift mutations* produce changes in the amino acid reading frame. In general, these types of mutations are more likely to be detrimental rather than beneficial to protein structure and function. Mutations can also occur outside of the coding sequence and significantly affect gene expression. For example, mutations within promoters, response elements, splicing sequences, and untranslated sequences can have major consequences on the level of gene expression. If a mutation occurs within the *germ line*, it will occur in all the cells of an organism. Furthermore, a germ line mutation can be passed from parent to offspring. By comparison, *somatic* mutations only affect a patch of the organism. The size of the patch depends on how early during development the mutation has occurred.

Mutations are random events, although *hot spots* for higher rates of mutation can occur at certain locations within genes. A variety of agents, known as *mutagens*, are known to induce mutations. These include many different types of *chemical* and *physical* agents. In addition, *spontaneous mutations* can arise during normal cellular conditions. Mutagens can alter DNA structure by modifying bases, removing bases, causing errors in DNA replication, and producing breaks in DNA strands. Testing methods such as the *Ames test* can determine whether an agent is mutagenic.

Due to the prevalence of mutagens in the environment, living cells have evolved DNA repair systems that can recognize and repair different types of DNA lesions. A few types of DNA repair systems can recognize altered bases and repair them directly. Alternatively, *nucleotide excision repair* systems recognize alterations in DNA structure and excise the abnormal region. Excision repair mechanisms are found in all organisms and represent a prominent, general system for DNA repair. Excision repair is particularly important in repairing actively transcribing genes. Certain proteins, such as the *transcription–repair coupling factor* of *E. coli*, can target the excision repair system to a damaged gene that is being actively transcribed. Finally, living organisms also possess *mismatch repair* and *recombinational repair* systems that operate after or during the DNA replication process.

Experimentally, there are many ways to study the occurrence, causes, and consequences of mutation. Luria and Delbruck conducted a fluctuation test to show that mutations occur randomly in a population. Similarly, the Lederbergs' used replica plating to demonstrate that mutations occur randomly, and that a selective agent such as T1 phage simply selects for the growth of organisms that happen to have incurred a mutation providing T1 resistance. However, there are hot spots where mutations may occur more frequently. In Chapter 6, the work of Benzer showed that a few sites within two bacteriophage genes were more likely to mutate than were other sites.

Müller was the first scientist to establish that environment agents such as X-rays can cause mutation. His work showed that X-rays dramatically increased the likelihood of recessive, X-linked lethal alleles. Since these studies, many mutagens have been discovered by biochemists, microbiologists, and geneticists. Testing methods, such as the Ames test, can ascertain whether or not an agent is a mutagen.

PROBLEM SETS

SOLVED PROBLEMS

1. There are mutant tRNAs that act as nonsense and missense suppressors. At the molecular level, explain how you think these suppressors work.

Answer: A suppressor is a second site mutation that suppresses the phenotypic effects of a first mutation. Intergenic suppressor mutations in tRNA genes can act as nonsense or missense suppressors. For example, let's suppose a first mutation puts a stop codon into a structural gene. A second mutation in a tRNA gene can occur to alter the anticodon region of the tRNA so that the anticodon recognizes a stop codon but inserts an amino acid at this site instead of terminating translation. A missense suppressor is a mutation in a tRNA gene that changes the anticodon so that it puts in the wrong amino acid at a normal codon that is not a stop codon. These mutant tRNAs are termed missense tRNAs. For example, a tRNA that normally recognizes glutamic acid may incur a mutation that changes its anticodon sequence so that it recognizes a glycine codon instead. Like nonsense suppressors, missense suppressors can be produced by mutations in the anticodon region of tRNAs so that the tRNA recognizes an incorrect codon. Alternatively, missense suppressors can also be produced by mutations in aminoacyl-tRNA synthetases that cause them to attach the incorrect amino acid to a tRNA.

2. If the rate of mutation is 10^{-5}, how many new mutations would you expect in a population of one million bacteria?

Answer: If we multiply the mutation rate times the number of bacteria ($10^{-5} \times 10^{6}$), we obtain a value of 10 new mutations in this population. This answer is correct, but it is an oversimplification of mutation rate. For any given gene, the mutation rate is based on a probability that an event will occur. Therefore, when we consider a particular population of bacteria, we should be aware that the actual rate of new mutation will vary. Even though the rate may be 10^{-5}, we would not be surprised if a population of 1,000,000 bacteria had 9 or 11 new mutations instead of the expected number of 10. We would be surprised if it had 5000 new mutations, since this value would deviate much too far from our expected number.

3. During the Ames test, why is the potential mutagen mixed with a rat liver extract?

Answer: The rat liver extract contains enzymes that may chemically modify a nonmutagenic substance and thereby convert it into a mutagenic substance. This step makes it possible to identify chemicals that may become mutagenic after they have been ingested into our bodies.

CONCEPTUAL QUESTIONS

1. For each of the following mutations, is it a transition, transversion, addition, or deletion? The original DNA strand is 5′–GGACTA-GATAC–3′ (Note: Only one strand is shown):

A. 5′–GAACTAGATAC–3′
B. 5′–GGACTAGAGAC–3′
C. 5′–GGACTAGTAC–3′
D. 5′–GGAGTAGATAC–3′

2. Discuss the differences between gene, chromosome, and genome mutations.

3. Draw and explain how alkylating agents alter the structure of DNA.

4. A gene mutation changes an A–T base pair to a G–C pair. This causes a gene to encode a truncated protein that is nonfunctional. An organism that carries this mutation cannot survive at high temperatures. Make a list of all the genetic terms that could be used to describe this type of mutation.

5. What does a suppressor mutation suppress? What is the difference between an intragenic and intergenic suppressor?

6. How would each of the following types of mutations affect the amount of functional protein that is expressed from a gene?

A. Nonsense
B. Missense
C. Up-promoter mutation
D. Mutation that affects splicing

7. Explain how a mutagen can interfere with DNA replication to cause a mutation. Give two examples.

8. What type of mutation (transition, transversion, and/or frameshift) would you expect each of the following mutagens to cause?

A. Nitrous acid
B. 5-Bromouracil
C. Proflavin

9. Explain what happens to the sequence of DNA during trinucleotide repeat expansion. If someone was mildly affected with a TNRE disorder, what issues would be important when considering to have offspring?

10. Nonsense suppressors tend to be very inefficient at their job of allowing readthrough of a stop codon. How would it affect the cell if they were efficient at their job?

11. During methyl-directed mismatch repair, why is it necessary to distinguish between the template strand and the newly made daughter strand? How is this accomplished?

12. Are each of the following mutations silent, missense, nonsense, or frameshift? The original DNA strand is 5′–ATGGGACTAGATACC–3′ (Note: Only the coding strand is shown; the first codon is methionine):

A. 5′–ATGGGTCTAGATACC–3′
B. 5′–ATGCGACTAGATACC–3′
C. 5′–ATGGGACTAGTTACC–3′
D. 5′–ATGGGACTAAGATACC–3′

13. In Chapters 12 through 16, we discussed many sequences that are outside of the coding sequence but are important for gene expression. Look up two of these sequences and write them out. Explain how a mutation could change these sequences and thereby alter gene expression.

14. Distinguish between spontaneous and induced mutations. Which are more harmful? Which are avoidable?

15. Are mutations random events? Explain your answer.

16. Give an example of a mutagen that can change cytosine to uracil. Which DNA repair system(s) would be able to repair this defect?

17. If a mutagen causes bases to be removed from nucleotides within DNA, what repair system would fix this damage?

18. Explain two ways that a chromosomal rearrangement can cause a position effect.

19. How would nucleotide excision repair be affected in each case if one of the following proteins was missing? Describe the condition of the DNA that had been repaired in the absence of the protein.

A. UvrA
B. UvrC
C. UvrD
D. DNA polymerase

20. When does recombinational repair occur? What would happen if a mutation occurred in exactly the same location in both parental DNA strands?

21. When DNA-N-glycosylase recognizes thymine dimers, it only detects the thymine located on the 5′ side of the thymine dimer as being abnormal. Draw and explain the steps whereby a thymine dimer would be removed by the consecutive actions of DNA-N-glycosylase, AP-endonuclease, and DNA polymerase.

22. Discuss the relationship between transcription and DNA repair. Why is it beneficial to repair actively transcribed DNA more efficiently than nontranscribed DNA?

23. Discuss the consequences of a germ line versus a somatic cell mutation.

EXPERIMENTAL QUESTIONS

1. From an experimental point of view, is it better to use haploid or diploid organisms for mutagen testing? Consider both the Ames test and Experiment 17A when preparing your answer.

2. During an Ames test, bacteria were exposed to a potential mutagen. Also, as a control, another sample of bacteria was not exposed to the mutagen. In both cases, 10,000,000 bacteria were plated and the following results were obtained:

No mutagen: 17 colonies
With mutagen: 2017 colonies

Calculate the mutation frequency in the presence and absence of the mutagen. How much does the mutagen increase the frequency of mutation?

3. In the Ames test, there are several *Salmonella* strains that contain different types of mutations within the gene that encodes an enzyme necessary for histidine biosynthesis. These mutations include transversions, transitions, and frameshifts. Why do you think it would be informative to test a mutagen with these different types of strains?

4. In Müller's experiment with ClB chromosomes, is the experiment measuring the mutation rate within a single gene? Explain. If we divide 91 by 783, we obtain a mutation rate of 11.6%. In your own words, explain what this value means.

5. Suggest ways that you could modify Müller's approach so that you could measure the mutation rate within a single gene.

6. In Experiment 17A, why was it useful for Müller's strains to be crossover deficient? How would his data have been affected if the ClB chromosome could have crossed over with the chromosome containing the X-ray-induced mutations?

7. Explain how the technique of replica plating supports a random mutation theory but conflicts with the adaptation theory. Outline how you would use this technique to show that antibiotic resistance is due to random mutations.

8. In 1964, Richard Boyce and Paul Howard-Flanders conducted an experiment that provided biochemical evidence that thymine dimers are removed from the DNA by a DNA repair system (see Boyce, R. P., and Howard-Flanders, P. (1964) *Proc. Natl. Acad. Sci. USA 51*, 293–300). In their studies, bacterial DNA was radiolabeled so that the amount of radioactivity reflected the amount of thymine dimers. The DNA was then subjected to UV light, causing the formation of thymine dimers. When radioactivity was found in the soluble fraction, thymine dimers had been excised from the DNA by a DNA repair system. But when the radioactivity was in the insoluble fraction, the thymine dimers had been retained within the DNA. The following table illustrates some of their results involving a normal strain of *E. coli* and a second strain that was very sensitive to killing by UV light.

Strain	Treatment	Radioactivity in the Insoluble Fraction (cpm)	Radioactivity in the Soluble Fraction (cpm)
Normal	No UV	<100	<40
Normal	UV-treated, incubated 2h at 37°	357	940
Mutant	No UV	<100	<40
Mutant	UV-treated, incubated 2h at 37°	890	<40

Adapted from: Boyce, R. P., and Howard-Flanders, P. (1964) *Proc. Natl. Acad. Sci. USA 51*, 293–300.

Explain the results found in this table. Why is the mutant strain sensitive to UV light?

QUESTIONS FOR STUDENT DISCUSSION/COLLABORATION

1. In *E. coli*, a variety of mutator strains have been identified in which the spontaneous rate of mutation is much higher than normal strains. Make a list of the types of abnormalities that could cause a strain of bacteria to become a mutator strain. Which abnormalities do you think would give the highest rate of spontaneous mutation?

2. Discuss the times in a person's life when it would be most important to avoid mutagens. Which parts of a person's body should be most protected from mutagens?

3. A large amount of research is aimed at studying mutation. However, there is not an infinite amount of research dollars. Where would you put your money for mutation research:

Testing of potential mutagens
Investigating molecular effects of mutagens
Investigating DNA repair mechanisms
Or some other place?

CHAPTER 18

RECOMBINATION AND TRANSPOSITION AT THE MOLECULAR LEVEL

SISTER CHROMATID
EXCHANGE AND
HOMOLOGOUS
RECOMBINATION

SITE-SPECIFIC
RECOMBINATION

TRANSPOSITION

Genetic recombination is the process in which chromosomes are broken and then rejoined to form a new genetic combination, different from the original. A major category of genetic recombination is homologous recombination, which is an essential feature of all organisms. As its name suggests, **homologous recombination** occurs between DNA segments that are homologous to each other. This process enhances genetic diversity, helps to maintain genome integrity (i.e., DNA repair), and ensures the proper segregation of chromosomes.

In this chapter, we will also examine ways that nonhomologous segments of DNA may recombine with each other. During **site-specific recombination**, nonhomologous DNA segments are recombined at specific sites. This type of recombination occurs within genes that encode antibody polypeptides and also occurs when certain viruses integrate their genomes into host cell DNA. Finally, we will end this chapter with a description of an unusual form of genetic recombination known as **transposition**. As we will learn, small segments of DNA called **transposons** can move themselves to multiple locations within the host's chromosomal DNA.

From a molecular viewpoint, homologous recombination, site-specific recombination, and transposition all are important mechanisms for DNA rearrangement. These processes involve a series of steps that direct the breakage and rejoining of DNA fragments. Various cellular proteins are necessary for these steps to occur properly. The past few decades have seen many exciting advances in our understanding of genetic recombination at the molecular level. In the first part of this chapter, we will consider the general concepts of recombination and examine several models that explain how recombination occurs.

SISTER CHROMATID EXCHANGE AND HOMOLOGOUS RECOMBINATION

As was described in Chapters 3 and 5, chromosomes that have similar or identical sequences frequently participate in crossing over during meiosis I and occasionally during mitosis. As you may recall, crossing over involves the alignment of a pair of homologous chromosomes, followed by the breakage of two chromosomes at analogous locations, and the subsequent exchange of the corresponding segments (refer back to Figure 3-9). When crossing over occurs between sister chromatids, it is called **sister chromatid exchange (SCE)**. Since sister chromatids are genetically identical to each other, SCE does not produce a new combination of alleles (Figure 18-1a). Therefore, it is not considered a form of recombination. By comparison, it is also common for homologous chromosomes to cross over. This is considered homologous recombination, because two similar (but not identical) homologues have exchanged genetic material. As shown in Figure 18-1b, homologous recombination may produce a new combination of alleles in the resulting chromosomes. **Recombinant chromosomes** contain a combination of alleles not found in the **parental chromosomes**.

In this section, we will begin with an experimental approach to detect sister chromatid exchange, overcoming the obstacle that the exchanged chromosomes are genetically identical to each other. We will then focus our attention on the molecular mechanisms that underlie homologous recombination.

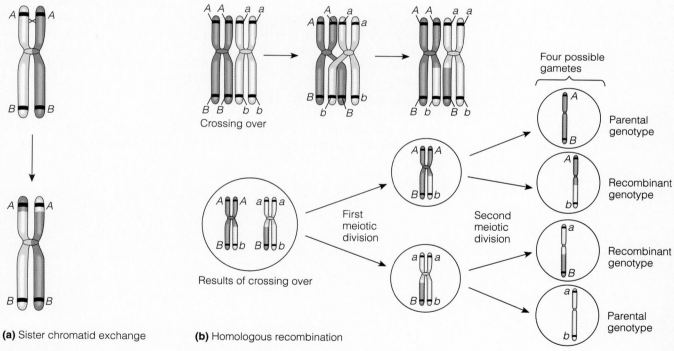

(a) Sister chromatid exchange

(b) Homologous recombination

FIGURE **18-1**

Crossing over between eukaryotic chromosomes. (a) Sister chromatid exchange occurs when genetically identical chromatids cross over. **(b)** Homologous recombination occurs when homologous chromosomes cross over. Homologous recombination may lead to a new combination of alleles, which is called a recombinant (or nonparental) genotype.

GENES → TRAITS: Homologous recombination is particularly important when we consider the relationships between multiple genes and multiple traits. For example, if one chromosome carried alleles for brown eyes and brown hair and its homologue carried alleles for blue eyes and blond hair, homologous recombination could produce recombinant chromosomes that carry alleles for brown eyes and blond hair, or for blue eyes and brown hair. Therefore, new combinations of two or more traits can arise when homologous recombination takes place.

Perry and Wolff produced harlequin chromosomes to reveal recombination between sister chromatids

Our understanding of crossing over and genetic recombination has come from a variety of experimental approaches including genetic, biochemical, and cytological analyses. Chromosomal staining methods have made it possible to visualize the genetic exchange between eukaryotic chromosomes. In the 1970s, the Russian cytogeneticist A. F. Zakharov and colleagues spent much effort developing methods that improved our ability to identify chromosomes. They made the interesting observation that chromosomes labeled with the nucleotide analogue 5-bromo-deoxyuridine (BrdU) become more fluorescent when stained with Giemsa and then visualized microscopically. As we will see in the present experiment, Paul Perry at the University of California in San Francisco and Sheldon Wolff in Edinburgh, U.K., extended this approach to differentially stain sister chromatids and microscopically identify sister chromatid exchanges.

Before we consider the experiment of Perry and Wolff, let's examine how their staining procedure allowed them to accurately discern the two sister chromatids. In their approach, eukaryotic cells were grown in a laboratory and exposed to BrdU for two rounds of DNA replication. After the second round of DNA replication, one of the sister chromatids contained one normal strand and one BrdU-labeled strand. The other sister chromatid had two BrdU-labeled strands (Figure 18-2). When treated with two dyes, Hoechst 33258 and Giemsa, the sister chromatid containing two strands with BrdU stains very weakly and appears light, whereas the sister chromatid with only one strand containing BrdU stains much more strongly and appears very dark. In this way, the two sister chromatids can be distinguished microscopically. Chromosomes stained in this way have been referred to as *harlequin chromosomes*, because they are reminiscent of a harlequin character's costume with its variegated pattern of light and dark patches.

A chromosomal staining method will make it possible to identify recombination events where genetically identical pieces of sister chromatids are exchanged with each other.

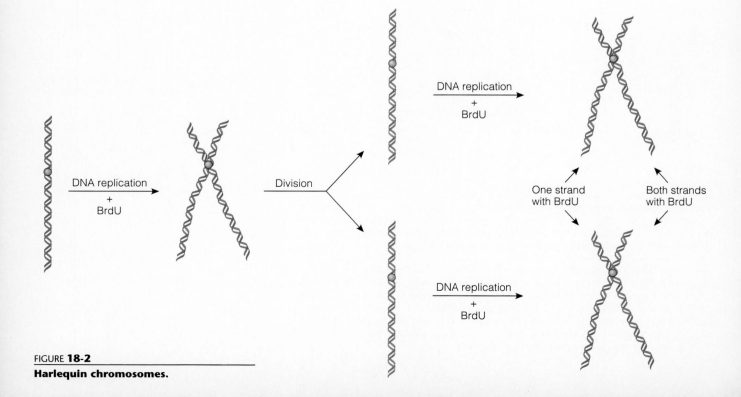

FIGURE **18-2**
Harlequin chromosomes.

TESTING THE HYPOTHESIS

Starting material: A laboratory cell line of Chinese hamster ovary (CHO) cells.

1. Expose CHO cells to BrdU for two cell generations (approximately 24 hours).

2. Near the end of the growth, expose the cells to colcemid. This prevents the cells from completing mitosis following the second round of DNA replication.

3. Add 0.075 M KCl to spread the chromosomes and then methanol/acetic acid to fix the cells.

4. Stain with Hoechst 33258, rinse, and later stain with Giemsa. Note: This refinement in the staining procedure greatly improved the ability to discern the sister chromatids.

5. View under a microscope.

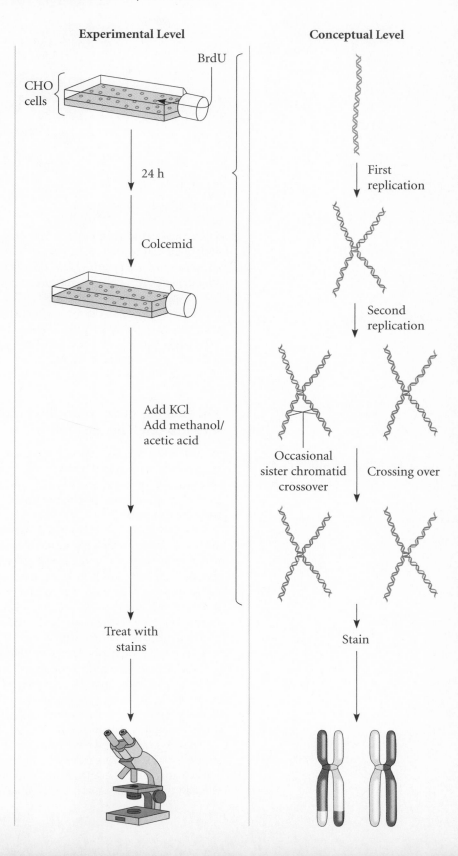

Experimental Level

BrdU

CHO cells

24 h

Colcemid

Add KCl
Add methanol/ acetic acid

Treat with stains

Conceptual Level

First replication

Second replication

Occasional sister chromatid crossover

Crossing over

Stain

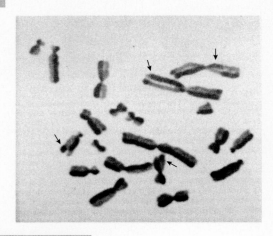

INTERPRETING THE DATA

The arrows depict regions where crossing over has taken place. In this study, Perry and Wolff found that SCEs occurred at a frequency of approximately 0.67 per chromosome. This method has provided an accurate (and dramatic) way to visualize genetic exchange between eukaryotic chromosomes.

Many subsequent studies have utilized the harlequin staining method to study the effects of agents that may influence the frequency of genetic exchanges. Researchers have found that DNA damage caused by radiation and chemical mutagens tends to increase the level of genetic exchange. When cells are exposed to these types of mutagens, the technique of harlequin staining has revealed a substantial increase in the frequency of SCEs.

The Holliday model describes a molecular mechanism for the recombination process

We will now turn our attention to genetic exchange that occurs between homologous chromosomes. Perhaps it is surprising that the first molecular model of homologous recombination did not come from a biochemical analysis of DNA or from electron microscopy studies. Instead, it was deduced from the outcome of genetic crosses in fungi.

As was discussed in Chapter 5, geneticists have learned a great deal from the analysis of fungal asci, because an ascus contains the products of a single meiosis. When two haploid fungi that differ at a single gene are crossed to each other, it is expected that the ascus will contain an equal proportion of each genotype. For example, if a pigmented strain of *Neurospora* producing orange spores is crossed to an albino strain producing white spores, the octad should contain four orange spores and four white spores. As early as 1934, however, H. Zickler noticed that unequal proportions of the spores sometimes occurred within asci. He occasionally observed octads with six orange spores and two white spores, or six white spores and two orange spores.

Zickler used the term **gene conversion** to describe this phenomenon. It occurred at too high a rate to be explained by new mutations. Subsequent studies in the 1950s by several researchers confirmed this phenomenon in yeast and *Neurospora*. When gene conversion occurs, one allele is converted to the allele on the homologous chromosome.

Based on studies involving gene conversion, Robin Holliday at the National Institute for Medical Research in London proposed a model in 1964 to explain the molecular steps that occur during homologous recombination. In this chapter, we will first consider the steps in the Holliday model and then move on to more recent models. Later in this chapter, we will examine how the Holliday model can explain the phenomenon of gene conversion.

The Holliday model is shown in Figure 18-3a. At the beginning of the process depicted in this figure, two homologous chromosomes are aligned with each other. According to the Holliday model, in step 1 a break occurs at identical sites in one strand of both parental chromosomes. During step 2, the strands then invade the opposite helices and base pair with the complementary strands. In step 3, this event is followed by a covalent linkage to create a *Holliday junction*.

As shown in step 4, the cross in the Holliday junction can migrate in a lateral direction. As it does so, a DNA strand in one helix is being swapped for a DNA strand in the other helix. This process is called **branch migration**, because the branch connecting the two double helices migrates laterally. Since the DNA sequences in the homologous chromosomes are similar but not identical, the swapping of the DNA strands during branch migration may produce regions in the double-stranded DNA that are called **heteroduplexes**. A heteroduplex is a DNA double helix that contains mismatches. In other words, since the DNA strands in this region are from homologous chromosomes, their sequences are not perfectly complementary, yielding mismatches.

During step 5, the Holliday structure may or may not make a 180° turn. This is called *isomerization*, because the two structures shown at step 5 are structural *isomers* of each other. This means that they are chemically identical except for the relative locations of certain segments.

The final steps in the recombination process are called **resolution**, since they involve the breakage and rejoining of two DNA strands to create two separate chromosomes. In other words, the entangled DNA strands become resolved into two separate structures. Resolution can occur in two ways. Steps 6A–8A are the result without isomerization, whereas steps 6B–8B will occur if isomerization has taken place at step 5. As shown in steps 6A–8A, breakage can occur in the same two strands that were also broken in step 1. If this occurs, the strands are rejoined to produce a nonrecombinant pair of chromosomes. The only difference between this nonrecombinant pair and the original parental pair is the short heteroduplex region. Alternatively, as shown in steps 6B–8B, the resolution phase can involve breakage of the two DNA strands that were *not* broken during step 1. In this case, the rejoining of the corresponding strands produces two recombinant chromosomes.

The Holliday model can account for the general properties of recombinant chromosomes that are formed during eukaryotic meiosis. As was mentioned, the original model was based on the results of crosses in fungi where the products of meiosis are contained within a single ascus. Nevertheless, molecular research in many other organisms have supported the central tenets of the Holliday model. A particularly convincing piece of evidence came from electron microscopy studies in which recombination structures could be visualized. Figure 18-3b shows an electron micrograph of two DNA fragments that are in the process of recombination. This structure can be equated to those found at step 5 of Figure 18-3a. They have been referred to as chi (χ) forms because their shape is similar to the Greek letter χ.

More recent models have refined the molecular steps of homologous recombination

As more detailed studies of genetic recombination have become available, certain steps in the Holliday model have been reconsidered. In particular, more recent models have modified the initiation phase of recombination. It is now known that the first step need not involve nicks at identical sites in two corresponding strands. A DNA helix with a break in both strands or DNA molecules with a single nick have been shown to participate in genetic recombination. Therefore, newer models have tried to incorporate these experimental observations. A model proposed by Matthew

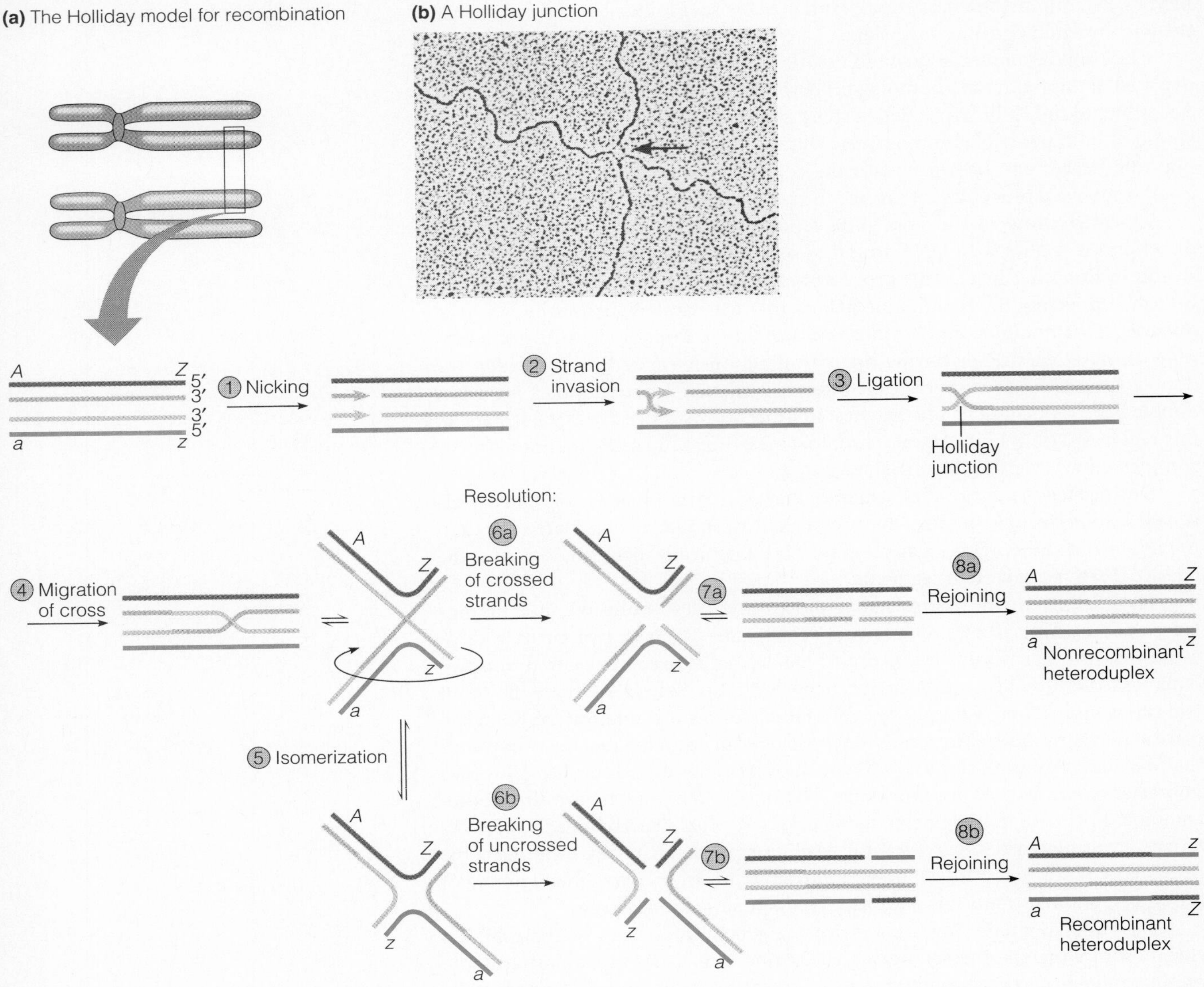

(a) The Holliday model for recombination

(b) A Holliday junction

FIGURE **18-3**

The Holliday model for homologous recombination. (a) The Holliday model for homologous recombination. Adapted from: Holliday, R. (1964) A mechanism for gene conversion in fungi. *Genet. Res. 5*, 282–303. **(b)** A Holliday structure viewed via electron microscopy.

Meselson (Harvard University) and Charles Radding (Yale University) hypothesizes that a single nick in one DNA strand initiates recombination. A second model, proposed by Jack Szostak, Terry Orr-Weaver, Rodney Rothstein, and Franklin Stahl, suggests that a double-stranded break initiates the recombination process. This is called the double-stranded break model. In Figure 18-4, the Holliday model is compared with the Meselson–Radding and double-stranded break models.

A few important differences from the Holliday model are worth noting in these newer models. As mentioned, the pattern of breakage of one or more DNA strands is different among the three models. Unlike the Holliday model, the newer models propose that a short region of strand degradation occurs. This occurs via the action of nucleases that can degrade a DNA strand over a short distance. (This occurs in step 3 in the Meselson–Radding model and step 1 in the double-stranded

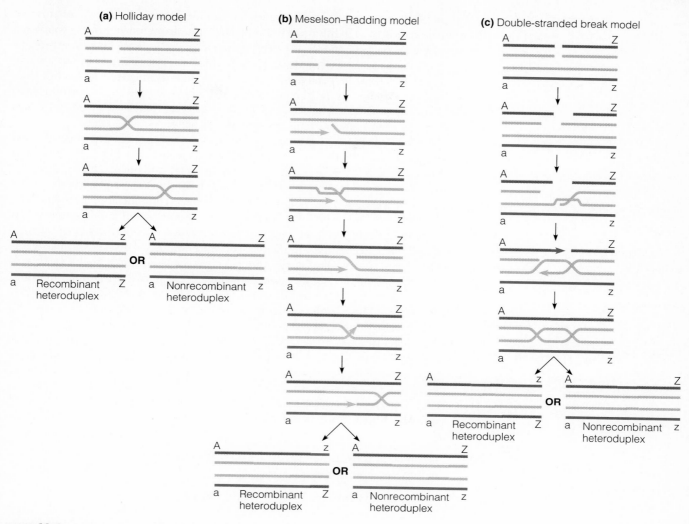

(a) Holliday model

Recombinant heteroduplex   OR   Nonrecombinant heteroduplex

(b) Meselson–Radding model

Recombinant heteroduplex   OR   Nonrecombinant heteroduplex

(c) Double-stranded break model

Recombinant heteroduplex   OR   Nonrecombinant heteroduplex

FIGURE **18-4**

A comparison of (a) the Holliday model with (b) the Meselson–Radding model and (c) the double-stranded break model. Note: The figure shown is from Kowalczy-kowski, S. C., Dixon, D. A., Eggleston, A. K., Lauder, S. C., and Rehrauer, W. M. (1994) Biochemistry of homologous recombination in *Escherichia coli. Microbiol. Rev. 58,* 401–465.

break model.) Since DNA strand degradation takes place, these two models also require the synthesis of new DNA. This DNA synthesis occurs in the relatively short gaps where a DNA strand is missing. For this reason, the DNA synthesis is called **DNA gap repair synthesis.** The blue arrowheads in Figure 18-4b and c indicate the areas where DNA gap repair synthesis occurs. In the Meselson–Radding model, gap repair synthesis occurs in one strand; in the double-stranded break model, it occurs in two strands.

Various proteins are necessary to facilitate homologous recombination

The homologous recombination process requires the participation of many proteins that catalyze different steps in the recombination pathway. Though homologous recombination takes places in all organisms, the enzymology of this process is best understood in *Escherichia coli.* Figure 18-5 presents a more complete version of the double-stranded break model that includes some of the *E. coli* proteins that play critical roles in this process. RecBCD is a multicomponent protein composed of the RecB, RecC, and RecD proteins. (The term Rec indicates that these proteins are involved with *re*combination.) The RecBCD complex plays an important role in the initiation of recombination involving double-stranded breaks. RecBCD recognizes a double-stranded break within DNA and catalyzes DNA unwinding and strand degradation. The action of RecBCD produces single-stranded DNA ends that can participate in strand invasion and exchange.

The RecA protein can bind to the single-stranded ends of DNA molecules that are generated from the activity of RecBCD. A large number of RecA proteins bind to single-stranded DNA, forming a structure called a filament. During **synapsis**, this filament makes contact with the unbroken chromosome. Initially, this contact is most likely to occur at nonhomologous regions. The contact point slides along the DNA until it reaches a homologous region. Some current models suggest that a homologous region is recognized by the formation of triplex DNA (see Chapter 9), although this is not firmly established. One of the original strands is quickly displaced, and the invading single-stranded DNA forms a normal double helix with the other strand. RecA proteins mediate the movement of the invading strand and the displacement of the complementary strand. In step 3 of Figure 18-5, this displaced strand invades the vacant region of the broken chromosome.

Proteins that bind specifically to Holliday junctions have also been identified. In the double-stranded break model, DNA gap repair synthesis creates a double Holliday junction (see step 4). RecG and RuvAB proteins specifically bind to Holliday junctions. Either of these proteins can catalyze the branch migration of Holliday junctions. RuvC protein also recognizes Holliday junctions. It is called a **resolvase**, because it is an endonuclease that makes the final cuts in the DNA during the resolution phase of recombination (refer back to Figure 18-3).

Gene conversion may result from DNA gap repair synthesis or DNA mismatch repair

As was mentioned earlier in the chapter, genetic recombination can lead to an event where two different alleles become two identical alleles. Since one of the alleles has been converted to the other, this process is known as gene conversion. The Holliday model, as well as newer models, can account for the phenomenon of gene conversion.

There are two possible ways that gene conversion can occur. One way is via DNA gap repair synthesis. Figure 18-6 illustrates how gap repair synthesis can lead to gene conversion according to the double-stranded break model. The top chromosome, which carries the *a* allele, has suffered a double-stranded break. A gap is created by the digestion of the DNA in the double helix. This digestion eliminates the *a* allele. The two template strands used in gap repair synthesis are from the other double helix. This helix carries the *A* allele. After gap repair synthesis takes place, the top chromosome will contain the *A* allele.

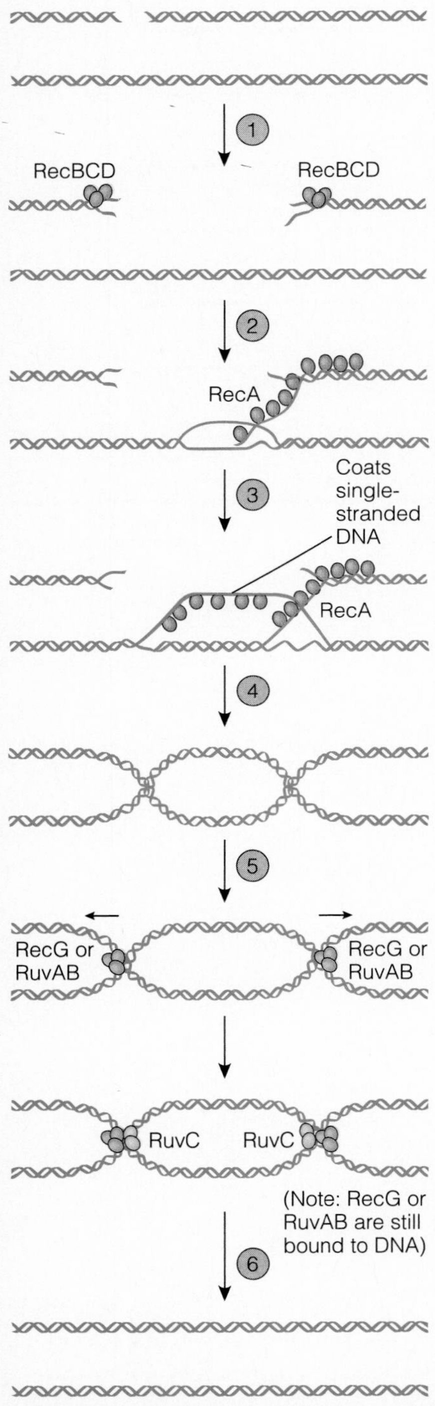

FIGURE **18-5**

Proteins that facilitate homologous recombination in *E. coli* according to the double-stranded break model. (1) RecBCD recognizes a double-stranded break in DNA. It unwinds the ends of the DNA and promotes strand degradation. **(2)** RecA proteins bind to the single-stranded DNA to form a filament. In this step, the single-stranded DNA first aligns with its homologue. Then single-stranded DNA invades the double helix and displaces the other strand. This displacement is called strand exchange. **(3)** The strand that was displaced in step 2 synapses with the broken chromosome. **(4)** DNA gap repair synthesis (via DNA polymerase and ligase) creates a double Holliday junction. **(5)** Branch migration occurs. This process is facilitated by RecG or RuvAB. **(6)** RuvC catalyzes the cleavage of the Holliday junction that occurs during the resolution step.

FIGURE **18-6**

Gene conversion by gap repair synthesis in the double-stranded break model.
A gene found in two alleles, designated *A* and *a*, is located near the double-stranded break site. In this example, **(1)** both DNA strands corresponding to the *a* allele are digested away. **(2)** A complementary DNA strand corresponding to the *A* allele migrates to this region and **(3, 4)** provides the template to synthesize a double-stranded region that contains the *A* allele. Following **(5)** resolution, both DNA double helices carry the *A* allele.

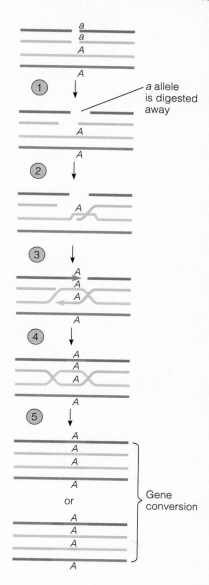

A second mechanism to account for gene conversion is DNA mismatch repair, a topic that was described in Chapter 17. To understand how this works, let's take a closer look at the heteroduplex structure that is formed during homologous recombination (refer back to Figure 18-3 or Figure 18-4). A heteroduplex contains a DNA strand from each of the two original parental chromosomes. It is possible that the two parental chromosomes may contain an allelic difference within this region. In other words, this short region may contain DNA sequence differences. If this is the case, the heteroduplex region that is formed after branch migration will contain an area of base mismatch. Gene conversion occurs when recombinant chromosomes are repaired to produce the same allele.

The DNA repair of a heteroduplex to cause gene conversion is shown in Figure 18-7. The two parental chromosomes contained an allelic difference in their

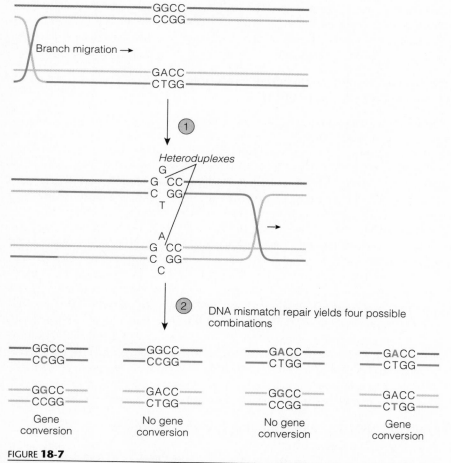

FIGURE **18-7**

Gene conversion by mismatch DNA repair. (1) A branch migrates past a homologous region that contains slightly different DNA sequences. This produces two heteroduplexes—DNA double helices with mismatches. **(2)** The mismatches can be repaired in four possible ways by the mismatch DNA repair system described in Chapter 17. Two of these ways result in gene conversion. The repaired base is shown in red.

DNA sequences as shown at the top of the figure. During recombination, branch migration has occurred across this region, thereby creating two heteroduplexes with base mismatches. As was described in Chapter 17, DNA mismatches will be recognized by DNA repair systems and repaired to a DNA double helix that obeys the A–T/G–C rule. These two mismatches can be repaired in four possible ways. As shown here, two possibilities will produce no gene conversion, whereas the other two will lead to gene conversion.

SITE-SPECIFIC RECOMBINATION

Thus far in this chapter, we have examined recombination between segments of DNA that are homologous or identical to each other. Site-specific recombination is another mechanism where DNA fragments can recombine to make new genetic combinations. During this type of recombination, two DNA segments with little or no homology align themselves at specific sites. The sites are relatively short DNA sequences (a dozen or so nucleotides in length) that provide a specific location where recombination will occur. Chromosome breakage and reunion occurs at these defined sites to create a recombinant chromosome. These sites are recognized by specialized enzymes that catalyze the breakage and rejoining of DNA fragments within the sites.

Certain viruses use site-specific recombination to insert their viral chromosome into their host cell's chromosome. This process has been examined carefully in bacteriophage λ. In addition, mammalian genes that encode antibody polypeptides are rearranged by site-specific recombination, which enables the generation of a diverse array of antibodies. In this section, we will consider both types of mechanisms.

The integration of viral genomes can occur by site-specific recombination

The life cycle of some viruses involves the integration of viral DNA into host cell DNA. Certain bacteriophages, for example, can integrate their viral DNA into the bacterial chromosome, creating a **prophage**. This prophage can exist in a latent, or **lysogenic**, state for many generations. The integration of phage DNA is well understood for bacteriophage λ, which infects *E. coli*. The life cycles of phage λ were described in Chapter 15. The integration occurs by a mechanism involving site-specific recombination. The steps leading to integration of the viral DNA are outlined in Figure 18-8.

Integration of the λ DNA into the *E. coli* chromosome requires sequences known as **attachment sites**. As shown at the top of Figure 18-8, the attachment site sequences are identical in the λ DNA and the *E. coli* chromosome. An enzyme known as **integrase** is encoded by a gene in the λ DNA. Several molecules of integrase recognize the attachment site sequences and bring them close together. Integrase then makes staggered cuts in both the λ and *E. coli* attachment sites. The strands are then exchanged, and the ends are ligated together. In this way, the phage DNA is integrated into the host cell chromosome. As a prophage, the λ DNA may remain latent for many generations. Certain conditions (e.g., when the bacterium is exposed to UV light) may act to stimulate the excision of the prophage from the host DNA, thereby reactivating the virus. Excision also requires integrase, which catalyzes the reverse reaction, as well as a second protein known as **excisionase**.

Antibody diversity in the immune system is produced by site-specific recombination

As we have just seen, viruses can integrate their DNA into the host cell chromosome by a site-specific recombinational event. This process requires a viral enzyme, integrase, that recognizes the sites and catalyzes the recombination reaction. A similar process occurs in certain cells of the immune system. The DNA sequences within antibody genes are rearranged by enzymes that recognize specific sites within those

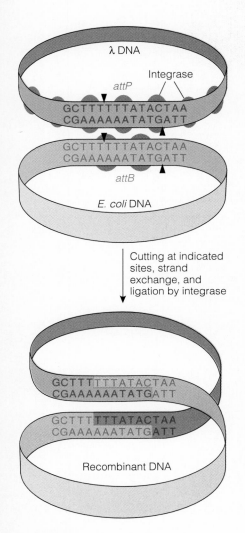

FIGURE 18-8

The integration of λ DNA into the *E. coli* chromosome. The site designated *attP* is the sequence (i.e., attachment site) in the λ DNA that attaches to the *attB* site in the *E. coli* chromosome. As noted here, the *attP* and *attB* sequences are identical to each other and thereby provide recognition sites for site-specific recombination.

genes and catalyze the breakage and reunion of DNA segments. Before we discuss the details of this mechanism, it is interesting to consider the biology of antibodies.

Antibodies or **immunoglobulins (Igs)** are proteins produced by the B-cells of the immune system. Their function is to recognize foreign substances (namely, viruses, bacteria, etc.) and target them for destruction. Antibodies recognize sites known as epitopes within the structures of foreign substances (also known as antigens). The recognition between an antibody and antigen is very specific, with each type of antibody recognizing a single epitope. Within the immune system, each B-cell produces a single type of antibody. However, our bodies have millions of B-cells. Site-specific recombination allows each B-cell to produce an antibody with a different amino acid sequence. These differences in the amino acid sequences of antibody proteins enable them to recognize different epitopes. In this way, the immune system can identify an impressive variety of substances as being foreign antigens and thereby target them for destruction.

From a genetic viewpoint, the production of millions of different antibodies poses an interesting problem. If a distinct gene were needed to produce each different antibody polypeptide, the DNA would need to contain millions of different antibody genes. By comparison, consider that the entire human genome contains only about 100,000 different genes. To generate millions of antibody molecules with different polypeptide sequences, an unusual mechanism has evolved in which the DNA is cut and reconnected by site-specific recombination. We might call this "DNA splicing," although the term splicing is usually reserved for cutting and rejoining of RNA molecules. With this mechanism, only a few large antibody precursor genes are needed to produce millions of different antibodies. These precursor genes are spliced in many different ways to produce a vast array of polypeptides with differing amino acid sequences.

Figure 18-9 shows the site-specific recombination of an antibody precursor gene. Antibodies are tetrameric proteins composed of two heavy polypeptide chains and two light chains. One type of light chain is the κ (kappa) light chain, which is a component of a class of antibodies known as immunoglobulin G (IgG). The organization of this precursor gene of the κ light chain is shown at the top of Figure 18-9. At the left side, there are 300 regions known as variable (V) sequences or domains. In addition, there are four different joining (J) sequences and a single constant (C) sequence. Each variable domain or joining domain encodes a different amino acid sequence.

During the differentiation of B-cells, the κ light precursor gene is cut and rejoined so that one variable sequence becomes adjacent to a joining sequence. This recombination event is catalyzed by an enzyme known as **V(D)J recombinase**. At the 3′ end of every V domain and the 5′ end of the J domains is located a **recombination signal sequence** that functions as a location for site-specific recombination between the V and J regions. These signal sequences are recognized by a V(D)J recombinase that carries out the recombination reaction. Following transcription, the fused VJ region is contained within a pre-mRNA transcript that is also spliced to connect the J and C regions. After this has occurred, a B-cell will only produce the particular κ light chain encoded by the specific VJ fusion domain and the constant domain.

The recombination process within immunoglobulin genes produces an enormous diversity in polypeptides. Even though it occurs at specific junctions within the antibody gene, the recombination is random with regard to the particular V and J segments that can be joined. Any of the 300 different variable sequences can be spliced next to any of the four joining sequences. This amounts to 1200 possible combinations.

The heavy-chain polypeptides are produced by a similar recombination mechanism. In this case, there are about 500 variable segments and four joining segments. In addition, there are also 12 diversity (D) segments, which are found between the variable and joining segments. The recombination first involves the connection of a D and J segment, followed by the connection of a V and DJ segment. The number of heavy-chain possibilities is $500 \times 12 \times 4 = 24,000$. Since any light chain–heavy

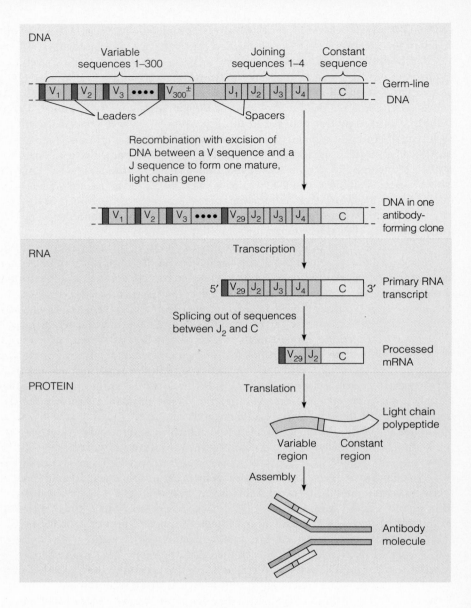

FIGURE **18-9**

Site-specific recombination within the gene that encodes the κ light chain for immunoglobulin G (IgG) proteins.

chain combination is possible, this yields $1200 \times 24{,}000 = 28{,}800{,}000$ possible antibody molecules from the splicing of only two precursor genes!

TRANSPOSITION

The last form of recombination that we will discuss is transposition. In some ways, transposition resembles the site-specific recombination that we examined for phage λ. In that case, a segment of λ DNA was able to integrate itself into the *E. coli* chromosome. Transposition also involves the integration of small segments of DNA into the chromosome. Transposition, though, can occur at many different locations within the genome. The DNA segments that transpose themselves are known as **transposable elements (TEs)**. TEs have sometimes been referred to as "jumping genes," because they are inherently mobile.

Transposable elements were first identified by Barbara McClintock in the early 1950s from her classical studies with corn plants. Since that time, geneticists have discovered many different types of TEs in species as diverse as bacteria, fungi, plants, and animals. The advent of molecular technology has allowed scientists to understand more about the characteristics of TEs that enable them to be mobile. In this section, we will examine the characteristics of transposable elements and explore mechanisms that explain how TEs move. We will also discuss the biological significance of TEs and their uses as experimental tools.

McClintock found that chromosomes of corn plants contain loci that can move

Barbara McClintock began her scientific career as a student at Cornell University. Her interests quickly became focused on the structure and function of the chromosomes of corn plants, an interest that continued for the rest of her life. She spent countless hours examining corn chromosomes under the microscope. She was technically gifted and, in addition, had a theoretical mind that could propose ideas that conflicted with conventional wisdom.

During her long career as a scientist, McClintock identified many unusual features of corn chromosomes. She noticed that one strain of corn had the strange characteristic that a particular chromosome, number 9, tended to break at a fairly high rate at the same site. McClintock termed this a **mutable site** or **locus**. This observation initiated a six-year study concerned with highly unstable chromosomal locations. In 1951, at the end of her study, McClintock proposed that these mutable sites are actually locations where transposable elements have been inserted into the chromosomes. At the time of McClintock's studies, such an idea was entirely unorthodox.

McClintock focused her efforts on the relationship between a mutable locus and its phenotypic effects on corn kernels. There were several genes on a single chromosome that could be used as chromosomal markers in her crosses. All of these genes affected the phenotype of corn kernels. Each gene had (at least) one dominant and one recessive allele that could be detected easily by examining the kernels. A few of these alleles are described here. The chromosome also contained a mutable locus that McClintock termed *Ds* (for dissociation) since the locus was known to frequently cause chromosomal breaks. In the chromosome shown here, the *Ds* locus is located next to several genes affecting kernel traits.

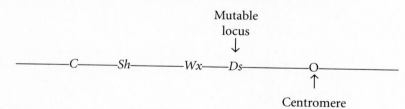

C = normal kernel color (dark red); c = colorless kernel; C^I = a dominant allele that also causes colorless kernel

Sh = normal endosperm; sh = shrunken endosperm
   Note: The endosperm is the storage material in the kernel that is used by the plant embryo to provide energy for growth.

Wx = normal starch in the endosperm; wx = (waxy), no starch in the endosperm

McClintock conducted crosses demonstrating that the *Ds* locus could be transposed to different locations within the corn genome. Using the chromosome shown here, it was straightforward to determine if *Ds* had moved, because the movement of *Ds* occasionally causes chromosome breakage. Before discussing her crosses, it is important to mention that the endosperm of a kernel is triploid: it is derived from two maternal haploid nuclei and one paternal (i.e., pollen) haploid nucleus. McClintock produced a cross with the following genotype:

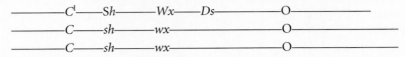

This kernel is expected to be colorless, because the C^I allele is dominant and causes a colorless phenotype. However, as the kernel develops, some chromosomes may break at the *Ds* locus and lose the distal part of this chromosome:

Within a single kernel, this will produce a patch of cells having a different phenotype. As shown in Figure 18-10, these patches will be red, waxy, and shrunken.

By analyzing many kernels, McClintock was able to identify cases in which *Ds* had moved to a new location. For example, if *Ds* had inserted between *Sh* and *Wx*, a break at *Ds* would produce the following combination:

This genotype would produce patches on the kernel that are red and shrunken but not waxy. In this way, McClintock identified 20 independent cases in which the *Ds* element had moved to a new location within this chromosome. Overall, the results from many crosses were consistent with the idea that *Ds* can transpose itself throughout the corn genome.

McClintock also found that a second locus, termed **Ac** (for activator), was necessary for *Ds* transposition to occur. It is now known that the *Ac* locus contains a gene that encodes the enzyme transposase, which is necessary for the *Ds* region to move. We will discuss the function of transposase later in this chapter. Some strains of McClintock's corn contained this activator locus, whereas others did not.

In one cross, McClintock noticed a particularly exciting and unusual event. She started with a plant that had the following genotype:

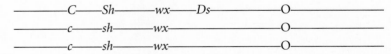

In this cross, the kernels are expected to be red. If the strain contains the *Ac* locus, breakage will occur occasionally at the *Ds* locus to produce colorless patches. Among 4000 kernels, she noticed one kernel with the opposite phenotype. It had a colorless background with red patches. This suggested that the background genotype was recessive, *c*, and it was mutable to become *C*! In this kernel, the *Ds* element had moved into the *C* gene:

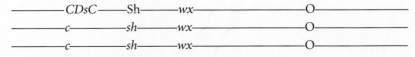

Furthermore, the red patch phenotype required the *Ac* locus.

McClintock postulated that the colorless phenotype was due to a transposition of *Ds* into the *C* gene. When *Ds* is located within the *C* gene, she proposed, it inac-

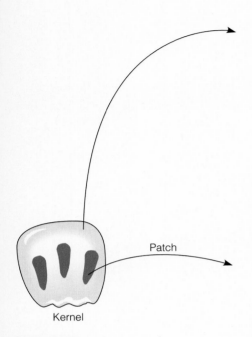

FIGURE **18-10**

The sectoring trait in corn kernels.

GENES ⟶ TRAITS: This kernel is expected to be colorless, because the C^I allele is dominant. On occasion, though, the movement of *Ds* within the developing kernel may cause a chromosome break, thereby losing the C^I allele. As such a cell continues to divide, it will produce a patch of daughter cells that are red, waxy, and shrunken. Therefore, this sectoring trait arises from the loss of genes that occurs when the movement of *Ds* causes chromosome breakage.

tivated the *C* gene, producing the recessive colorless phenotype. However, when *Ds* occasionally transposed out of the gene, the *C* allele would be restored and a red patch would result. (Note: In this case, the formation of patches, or *sectoring*, is due to the movement of *Ds* out of its original location but not due to chromosome breakage.) According to this hypothesis, the red phenotype should be associated with two observations. First, *Ds* should have moved to a new location. Second, the restored *C* allele should no longer be mutable.

THE HYPOTHESIS

The transposition of the *Ds* element into the normal *C* gene prevents kernel pigmentation. When the *Ds* element transposes back out of the *C* gene, the normal *C* allele is restored and the kernel becomes red.

TESTING THE HYPOTHESIS

Starting material: The male pollen of the corn plant had a chromosome in which the *Ds* element had moved into the *C* gene. The male plant also contains the *Ac* locus.

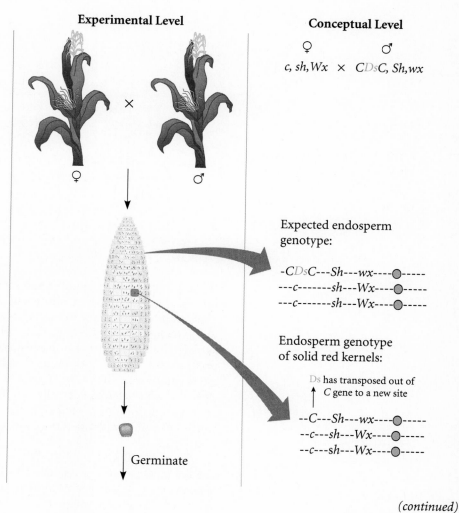

Experimental Level

Conceptual Level

1. Cross the male with a female that is *c, sh, Wx,* (no *Ds*) homozygous.

♀   ♂

c, sh, Wx × *CDsC, Sh, wx*

♀   ♂

2. In this cross, most kernels will have a white background with red sectoring. On rare occasions, *Ds* may have transposed during male gamete formation, producing a completely red kernel.

Expected endosperm genotype:

-*CDsC*---*Sh*---*wx*----◯-----
---*c*-------*sh*---*Wx*----◯-----
---*c*-------*sh*---*Wx*----◯-----

3. Identify those occasional kernels that are completely red.

Endosperm genotype of solid red kernels:

Ds has transposed out of
↑ *C* gene to a new site

--*C*---*Sh*---*wx*----◯-----
--*c*---*sh*---*Wx*----◯-----
--*c*---*sh*---*Wx*----◯-----

4. Germinate the solid red kernels.

Germinate

(continued)

(continued)

5. Conduct crosses to determine the location of *Ds* in the plants derived from the solid red kernels.
   Note: This can be done using the chromosomal markers and the strategy that was described at the beginning of this experiment. When observing the results of these crosses, determine whether the *C* gene is still mutable.
   (Is red sectoring occurring?)

Make crosses and
observe kernels

THE DATA

Strain	Kernel Phenotype	Location of Ds	Mutability?
From parent cross (see Step 2)	White background with red sectoring	Within the *C* gene	Yes, red sectoring occurred in strains containing *Ac*.
From red kernels (see Step 3)	Red kernels	*Ds* had moved out of the *C* gene to another location.	No, the C gene was stable; no sectoring was observed.

INTERPRETING THE DATA

In the parent strain, previous results showed that the *Ds* locus had transposed into the *C* gene. This parent strain had red sectoring on a colorless background. This result is consistent with the idea that the red sectoring is due to the removal of the *Ds* locus from the *C* gene, thereby restoring a normal *C* gene. Solid red kernels occasionally arose from this parental strain. By conducting the appropriate crosses, McClintock found that in the progeny of the red kernels, the *Ds* locus had moved out of the *C* gene to another location. In addition, the "restored" *C* gene behaved normally. In other words, it was no longer a highly mutable gene exhibiting a sectoring phenotype. Taken together, the results are consistent with the idea that the *Ds* locus can move around the corn genome by transposition.

When McClintock published these results in 1951, they were met with great skepticism. Most geneticists of this time were unable to accept the idea that the genetic material was susceptible to frequent rearrangement. Instead, they wanted to believe that the genetic material was always very stable and permanent in its structure. It was not until several decades later that the scientific community accepted the existence of transposable elements. Much like Gregor Mendel and Charles Darwin, Barbara McClintock was clearly ahead of her time. She was awarded the Nobel prize in 1983, more than 30 years after her original discovery.

Transposable elements and retroelements move via one of three transposition pathways

Since the pioneering studies of Barbara McClintock, many different transposable elements have been found in bacteria, fungi, plant, and animal cells. Three general types of transposition pathways have been identified (Figure 18-11). In **simple** or **conservative transposition**, the TE is removed from its original site and transferred to a new target site. By comparison, **replicative transposition** involves the replication of the TE and insertion of the newly made copy into a second site. In this case, one of the TEs remains in its original location and the other is inserted at another location. A third category of elements move via an RNA intermediate. These are known as **retroelements**, **retrotransposons**, or **retroposons**. Like replicative transposons, retroelements increase in number during retrotranspositional events.

Each type of transposable element has a characteristic pattern of DNA sequences

As researchers have studied TEs from many species, they have found that DNA sequences within transposable elements are organized in several different ways. Figure 18-12 describes a few ways that TEs are organized, although many variations are possible. The simplest TEs, which are commonly found in bacteria, are known as **insertion sequences**. As shown in Figure 18-12a, insertion sequences

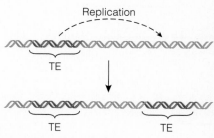

(a) Simple transposition

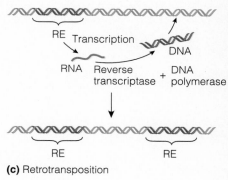

(b) Replicative transposition

(c) Retrotransposition

FIGURE **18-11**

Three mechanisms of transposition.

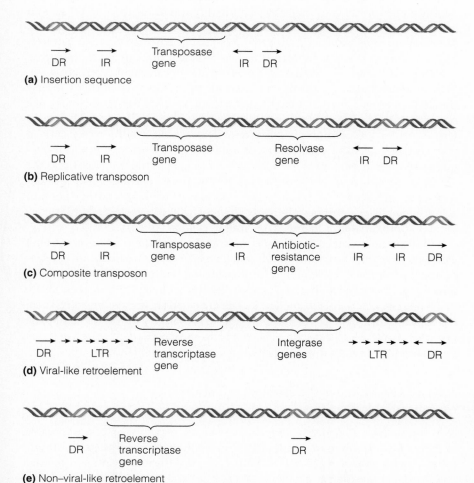

(a) Insertion sequence

(b) Replicative transposon

(c) Composite transposon

(d) Viral-like retroelement

(e) Non–viral-like retroelement

FIGURE **18-12**

Common organizations of transposable elements. Direct repeats (DRs) are found within the host DNA. Inverted repeats (IRs) are at the ends of most transposable elements. Long terminal repeats (LTRs) are regions containing a large number of tandem repeats.

have two important characteristics. First, both ends of the insertion sequence contain **inverted repeats (IRs)**. Inverted repeats are DNA sequences that are identical (or very similar) but run in opposite directions, such as the following:

$$5'\text{–CTGACTCTT–}3' \qquad \text{and} \qquad 5'\text{–AAGAGTCAG–}3'$$
$$3'\text{–GACTGAGAA–}5' \qquad\qquad 3'\text{–TTCTCAGTC–}5'$$

In addition, insertion sequences may contain a central region that encodes the enzyme **transposase**, which catalyzes the transposition event. In the case of replicative transposons, the enzyme resolvase is also encoded (see Figure 18-12b).

Composite transposons contain additional genes that are not necessary for transposition *per se*. Composite transposons commonly contain genes that confer a selective advantage to the organism. Composite transposons are prevalent in bacteria, where they often contain genes that provide resistance to antibiotics or toxic heavy metals. Composite transposons contain such a resistance gene that is sandwiched between two TEs. For example, the composite transposon shown in Figure 18-12c contains two insertion sequences flanking a gene that confers antibiotic resistance. During transposition of a composite transposon, only the two inverted repeats at the ends of the transposon are involved in the transpositional event. Any time insertion sequences are found at both ends of a gene, they create a composite transposon.

The organization of retroelements can be quite variable, and they are categorized based on their evolutionary relationship to retroviral sequences. Retroviruses are RNA viruses that make a DNA copy that integrates into the host's genome. The viral-like retroelements contain **long terminal repeat sequences (LTRs)** at both ends of the element (see Figure 18-12d). These elements tend to be fairly large and show striking similarities to known retroviruses. In particular, viral-like retroelements encode virally related proteins such as **reverse transcriptase** and integrase.

By comparison, non-viral-like retroelements appear less like retroviruses in their sequence, although some similarity, such as the occurrence of a reverse transcriptase gene, may be present (see Figure 18-12e). Many non-viral-like retroelements, however, do not share any sequence similarity with known viruses. Instead, some non-viral-like retroelements are evolutionarily derived from normal eukaryotic genes. For example, the Alu family of repetitive sequences found in humans is derived from a single ancestral gene known as the 7SL RNA gene (a component of the signal recognition particle, which was described in Chapter 14). This gene sequence has been copied by retroposition to achieve the current number of greater than 500,000 copies. Incredibly, it constitutes approximately 5 to 6% of the human genome!

Transposable elements are considered to be **complete** (or **autonomous**) **elements** when they contain all the information necessary for transposition or retroposition to take place. However, TEs are often **incomplete** (or **nonautonomous**). An incomplete element typically lacks a gene such as transposase or reverse transcriptase that is necessary for transposition. The *Ds* element described in Experiment 18B is an incomplete element that lacks a transposase gene. The *Ac* locus provides a transposase gene that enables *Ds* to transpose. Therefore, incomplete TEs such as *Ds* can transpose when the transposase gene is present at another region in the genome.

Transposase catalyzes the excision and insertion of transposable elements

Now that we have an understanding of the sequence organization of transposable elements, we can examine the steps of the transposition process. The enzyme transposase catalyzes the removal of a TE from its original site in the chromosome and its subsequent insertion at another location. A general scheme for simple or conservative transposition is shown in Figure 18-13. Transposase proteins recognize the

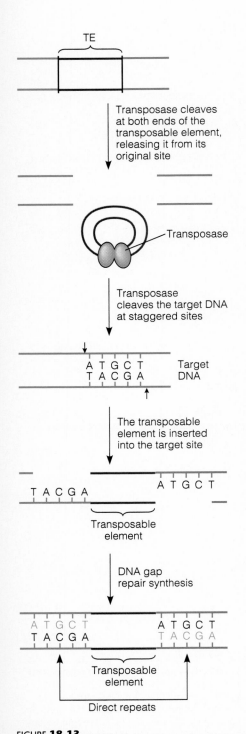

TE

Transposase cleaves at both ends of the transposable element, releasing it from its original site

Transposase

Transposase cleaves the target DNA at staggered sites

A T G C T
T A C G A   Target DNA

The transposable element is inserted into the target site

T A C G A   A T G C T
Transposable element

DNA gap repair synthesis

A T G C T   A T G C T
T A C G A   T A C G A
Transposable element

Direct repeats

FIGURE 18-13
Conservative transposition.

inverted repeat sequences at the ends of the TE and bring them close together. The DNA is cleaved at the ends of the TE, excising it from its original site within the chromosome. This excision event is followed by an insertion process in which transposase cleaves the target DNA sequence at staggered recognition sites. The TE is then ligated to the target DNA.

As noted in Figure 18-13, the ligation of the transposable element into its new site initially leaves short gaps in the target DNA. Notice that the DNA sequences in these gaps are complementary to each other (in this case, ATGCT and TACGA). Therefore, when they are filled in by DNA gap repair synthesis, the DNA base pair sequences at both ends of the TE are identical. These two sequences are called **direct repeats (DRs)**, because they are in the same *direction* and they are *repeat*ed at both ends of the element. The occurrence of direct repeats is a common feature of all TEs (refer back to Figure 18-12).

Replicative transposition requires both transposase and resolvase

Replicative transposition has been studied in several bacterial transposons and in bacteriophage μ (mu), which behaves like a transposon. The net result of replicative transposition is that a transposable element occurs at a new site and the original TE remains in its original location. Figure 18-14 describes a model for replicative transposition between two circular DNA molecules. One DNA molecule already has a TE, whereas the other does not. In this mechanism, transposase initially makes two cuts in the TE and in the target DNA. Note that this differs from conservative transposition, in which the transposase makes four cuts in the DNA and thereby removes the TE from its original site (compare Figures 18-13 and 18-14). In replicative transposition, one strand of the TE is left at its original location.

Following ligation, both the host DNA and the transposable element have a long gap. Gap repair DNA synthesis copies the host DNA gap as well as the TE. This creates two copies of the TE within a large circular molecule known as a *cointegrant*. The enzyme resolvase catalyzes homologous recombination within the TEs so that the cointegrant can be resolved into two separate DNA molecules. One of these molecules contains the TE in its original location, and the other has a TE at a new location.

In the preceding discussion, we have seen that replicative transposons are duplicated when they transpose, whereas conservative TEs are simply removed and placed in a new location. Thus, you might think that conservative TEs would have difficulty increasing in number. Interestingly, this is not the case. Conservative TEs can easily increase in number, because transposition often occurs during DNA replication. For conservative TEs, an increase in number can happen in the following way. After a replication fork has passed a region containing a TE, there will be two TEs behind the fork (i.e., one in each of the replicated regions). One of these TEs could then transpose from its original location into a region ahead of the replication fork. After the replication fork has passed this second region and DNA replication is completed, there will be two TEs in one of the chromosomes, and one TE in the other chromosome. In this way, conservative TEs can increase in number. We will discuss the biological significance of transposon proliferation later in this chapter.

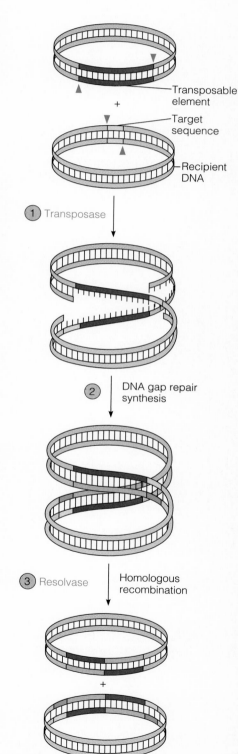

Transposable element

Target sequence

Recipient DNA

1 Transposase

2 DNA gap repair synthesis

3 Resolvase — Homologous recombination

FIGURE **18-14**

Replicative transposition. (1) The transposase catalyzes the movement of a DNA strand carrying the transposable element to a new recipient site. **(2)** DNA gap repair synthesis at the previous and new sites produces two double-stranded elements within a large circular molecule known as a cointegrant. **(3)** Similar to the resolution step of homologous recombination described in Figure 18-3, a resolvase separates the cointegrant into two separate structures, which each contain a TE.

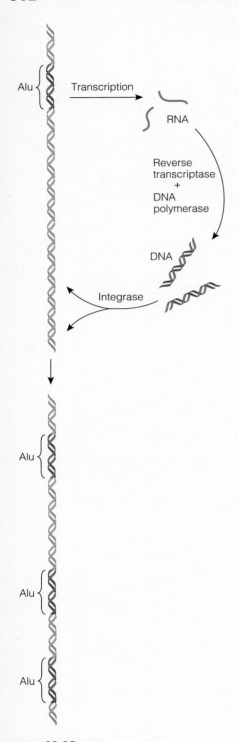

FIGURE **18-15**
The mechanism of retrotransposition.

Retroelements utilize reverse transcriptase and integrase for retrotransposition

Thus far, we have considered how DNA elements (i.e., transposons) can move throughout the genome. By comparison, retroelements utilize an RNA intermediate in their transposition mechanism. As shown in Figure 18-15, the movement of retroelements also requires two key enzymes, reverse transcriptase and integrase. In this example, the host cell already contains a retroelement known as the Alu sequence within its genome. This retroelement, which behaves like a gene, is transcribed into RNA. Reverse transcriptase uses this RNA as a template to synthesize a complementary strand of DNA. The single-stranded DNA is then used as template by DNA polymerase to make double-stranded DNA. The ends of the double-stranded DNA are then recognized by integrase, which catalyzes the insertion of the DNA into the host chromosomal DNA. The integration of retroelements can occur at many locations within the genome. Furthermore, since a single retroelement can be copied into many RNA transcripts, retroelements may accumulate rapidly within a genome.

Transposable elements may have important influences on mutation and evolution

Over the past few decades, researchers have found that transposable elements probably occur in the genomes of all species. Table 18-1 describes a few TEs that have been studied in great detail. As discussed in Chapter 10, the genomes of eukaryotic species typically contain moderately and highly repetitive sequences. In some cases, these repetitive sequences are due to the proliferation of TEs. In mammals, for example, **LINEs** are *l*ong *in*terspersed *e*lement*s* that are usually 1 to 5 kbp in length and found in 20,000 to 100,000 copies per genome. **SINEs** are *s*hort *in*terspersed *e*lements that are less than 500 bp in length. A specific example of a SINE is the Alu sequence, present in 500,000 to 1,000,000 copies in the human genome.

The biological significance of transposons in the evolution of prokaryotic and eukaryotic species remains a matter of intense debate. According to the **selfish DNA theory**, TEs exist because they contain the characteristics that allow them to multiply within the host cell DNA. In other words, they resemble parasites in the sense that they inhabit the host without offering any selective advantage. They can proliferate within the host as long as they do not harm the host to the extent that they significantly disrupt survival.

Alternatively, other geneticists have argued that most transpositional events are deleterious and, therefore, TEs would be eliminated from the genome if they did not also offer a compensating advantage. Several potential advantages have been suggested. For example, TEs may cause greater genetic variability by promoting recombination. In addition, bacterial TEs often carry an antibiotic resistance gene that provides the organism with a survival advantage. Researchers have also suggested that transposition may cause the insertion of exons into the coding sequences of structural genes. This phenomenon, called **exon shuffling**, may lead to the evolution of genes with more diverse functions.

While this controversy remains unresolved, it is clear that transposable elements can rapidly enter the genome of an organism and proliferate quickly. In *Drosophila melanogaster*, for example, a TE known as the P element likely was introduced into this species in the 1950s. Laboratory stocks of *D. melanogaster* collected prior to this time do not contain P elements. Remarkably, in the last 50 years, the P element has expanded throughout *D. melanogaster* populations worldwide. The only strains without the P element are laboratory strains collected prior to the 1950s. This observation underscores the surprising ability of TEs to infiltrate a population of organisms.

TABLE 18–1 Examples of Transposable Elements

Element	Type	(Approximate) Length (bp)	Description
Prokaryotic			
IS1	Insertion sequence	768	An insertion sequence that is commonly found in 5–8 copies in *E. coli*.
Mu	Replicative transposon	36,000	Mu is a true virus that can insert itself anywhere in the *E. coli* chromosome. Its name, Mu, is derived from its ability to insert into genes and *mu*tate them.
Tn10	Composite transposon	9,300	One of many different bacterial transposons that carries antibiotic resistance. Tn10 carries a gene that confers tetracycline resistance.
Tn951	Composite transposon	16,600	A composite transposon that provides bacteria with genes that allow them to metabolize lactose.
Yeast			
Ty elements	Viral-like retroelements	6,200	A retroelement found in *S. cerevisiae* at about 35 copies per genome.
Drosophila			
P elements	Transposon	500–3,000	A conservative transposon that may be found in 30–50 copies in P strains of *Drosophila*. It is absent from M strains.
Copia-like elements	Viral-like retroelements	5,000-8,000	There are several families of copia-like elements found in *Drosophila,* which vary slightly in their lengths and sequences. Typically, each family of copia elements is found at about 5 to 100 copies per genome.
Humans			
Alu sequences	Nonviral retroelement	300	As discussed in Chapter 10, Alu sequences are abundantly interspersed throughout the human genome.
L1	Viral-like retroelement	6,500	A human LINE found in 50,000–100,000 copies in the human genome.
Plants			
Ac/Ds	Transposon	4,500	*Ac* is an autonomous transposable element found in corn and other plant species. It carries a transposase gene. *Ds* is a nonautonomous version that is missing the transposase gene.

Transposable elements have a variety of effects on chromosome structure and gene expression (Table 18-2). Since many of these outcomes are likely to be harmful, transposition is usually a highly regulated phenomenon that only occurs in a few individuals under certain conditions. Agents such as radiation, chemical mutagens, and hormones stimulate the movement of TEs. When it is not carefully regulated, transposition is likely to be potently detrimental. For example, in *D. melanogaster*, if M strain females (which lack P elements) are crossed with P strain males (which contain numerous P elements), the hybrid offspring contain a variety of abnormalities, which include a high rate of mutation and chromosome breakage. This deleterious outcome, which is called **hybrid dysgenesis**, occurs because the P elements can transpose freely.

Transposons have become important tools in molecular biology

The unique and unusual features of transposons have made them an important experimental tool in molecular biology. For example, the introduction of a transposon into a cell is a convenient way to abolish the expression of particular genes. If a transposon "hops" into a gene, it is likely to inactivate the gene's function. This phenomenon can be utilized to clone a particular gene in an approach

TABLE 18–2 Possible Consequences of Transposition

Consequence	Cause
Chromosome structure	
Chromosomal breakage	Excision of a TE.
Chromosomal rearrangements	Homologous recombination between TEs located at different positions in the genome.
Gene expression	
Mutation	Incorrect excision of TEs.
Gene inactivation	Insertion of a TE into a gene.
Alteration in gene regulation	Transposition of a gene next to regulatory sequences, or of regulatory sequences next to a gene.
Alteration in the exon content of a gene	Insertion of exons into the coding sequence of a gene via TEs. This phenomenon is called exon shuffling.
Gene duplications	Creation of a composite transposon that transposes to another site in the genome.

known as **transposon tagging**. In this strategy, researchers use transposons in an attempt to clone novel genes.

The first example of transposon tagging involved an X-linked gene in *Drosophila* that affects eye color. As was described in Chapters 3 and 4, this *Drosophila* X-linked gene can exist in the wild-type (red) allele and a loss of function allele that causes a white-eyed phenotype. In 1981, Paul Bingham at Research Triangle Park in North Carolina, in collaboration with Robert Levis and Gerald Rubin at the Carnegie Institution of Washington, used transposon tagging to clone this gene (Figure 18-16).

Prior to their cloning work, a wild-type strain of *Drosophila* had been characterized that carried a transposable element called *copia*. From this red-eyed strain, a white-eyed strain was obtained in which the *copia* element had transposed to a region on the X-chromosome that corresponded to where the eye-color gene mapped. The researchers reasoned that the white-eyed phenotype could be due to the insertion of the *copia* element into the wild-type gene, thereby inactivating it.

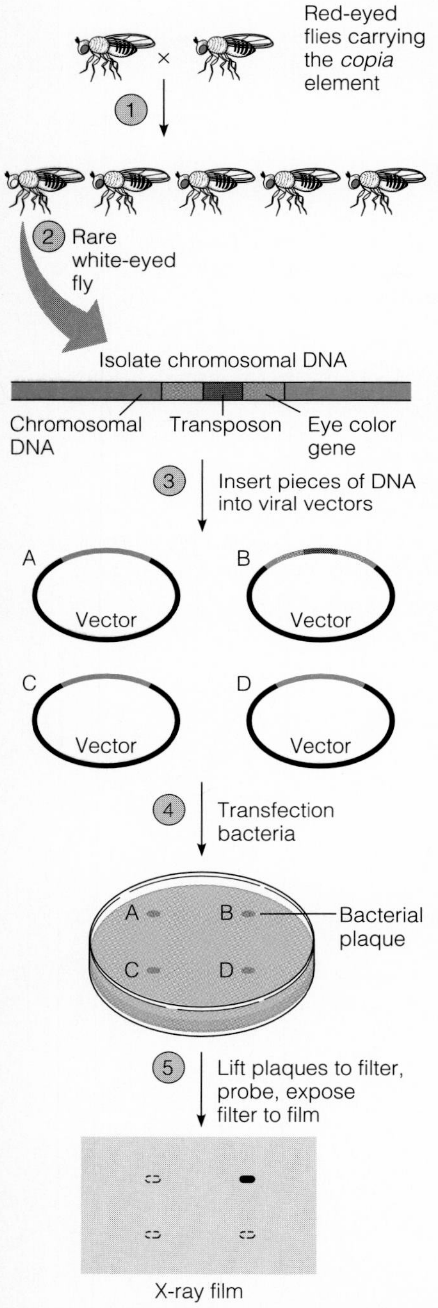

FIGURE **18-16**

The procedure of transposon tagging. A transposon, known as *copia*, was introduced into a *Drosophila* strain with red eyes. **(1)** On rare occasions, a white-eyed fly was produced from this red-eyed strain. In this case, the phenotype has occurred because the *copia* transposon has inserted itself into the normal eye color gene and thereby inactivated it. **(2)** Chromosomal DNA was isolated from the white-eyed fly, **(3, 4)** digested with a restriction enzyme, and cloned into a vector, creating a DNA library as described in Chapter 19. **(5)** The DNA library is probed with a radiolabeled fragment complementary to the *copia* element. A plaque carrying the TE shows up as a dark spot on an X-ray film. This plaque may also contain a version of the eye-color gene that was disrupted by the TE.

GENES ⟶ TRAITS: A white-eyed fly may be produced from a red-eyed fly due to the insertion of a transposable element into the gene that confers eye color. As was discussed in Chapter 4 (see also Figure 12-3), the normal eye color gene encodes a protein that is necessary for red pigment production. When a TE inserts into this gene, it disrupts the coding sequence and thereby causes the gene to produce a nonfunctional protein. Therefore, no red pigment can be made, and a white-eyed phenotype results. In many cases, transposons affect the phenotypes of organisms by inactivating individual genes.

To clone the eye-color gene, chromosomal DNA from this white-eyed strain was isolated, digested with restriction enzymes, and cloned into viral vectors. This procedure of creating a DNA library will be described in Chapter 19. A DNA library is a collection of vectors that contain different pieces of chromosomal DNA. If a transposon has "jumped" into the eye-color gene, vectors that contain this gene will also contain the transposon sequence. In other words, the presence of the transposon tags the eye-color gene. Therefore, a radiolabeled fragment of DNA that is complementary to the transposon sequence can be used as a probe to identify plaques that also contain the eye-color gene. The method of using a probe to screen a DNA library will also be described in Chapter 19. In the example of Figure 18-16, the method of transposon tagging was successful at cloning an eye-color gene in *Drosophila*.

Genetic recombination occurs when segments of DNA are broken and reconnected to form new combinations. During *homologous recombination*, homologous DNA regions become aligned and exchange DNA segments. This enhances genetic variability by producing *recombinant chromosomes* with new combinations of alleles.

At the molecular level, several models have attempted to explain the steps in homologous recombination. The Holliday model was the first example of a molecular explanation for the recombination process. This model was able to account for the phenomenon of *gene conversion*, in which one allele of a gene is converted to another allele. More recently, other models have more accurately described certain steps in the recombination pathway. In addition, much progress has been made toward identifying the many proteins that play important roles in recombination.

A second mechanism for recombination is known as *site-specific recombination*, because the breakage and rejoining of the DNA segments occurs at particular DNA sequences. Site-specific recombination is responsible for the integration of certain bacteriophages such as λ into the host genome. In addition, site-specific recombination within *immunoglobulin* genes is important in generating an astounding diversity in *antibody* polypeptides. During this process, regions in precursor antibody genes known as V, D, and J are connected to each other via recombination. This mechanism enables the production of a very diverse array of antibody proteins.

Transposition is a third way that DNA segments can rearrange. Short segments of DNA known as *transposable elements (TEs)* possess characteristics that enable them to be mobile. During *conservative* transposition, a "cut and paste" mechanism occurs where the TE is cut out of its original site and ligated to a new site. *Transposase* catalyzes this reaction. By comparison, in *replicative* transposition a TE is duplicated, with one TE remaining in its original location and a new TE located at a new site. Finally, retroposition occurs via an RNA intermediate. In this case, the *retroelement* is transcribed to RNA and then copied to DNA by *reverse transcriptase*. This DNA is then integrated into the chromosome by *integrase*. From an evolutionary viewpoint, all mechanisms of transposition can cause many different types of mutations.

Crossing over, which may occur between sister chromatids or homologous chromosomes, can be detected by several techniques. As described in Experiment 18A, sister chromatid exchange can be observed using staining methods that distinguish the sister chromatids within a pair. As was discussed in Chapter 5, homologous recombination can also be identified from the outcome of dihybrid crosses involving linked genes, where the frequency of recombinant offspring is a measure of the frequency of crossing over. At the molecular level, researchers have studied homologous recombination by identifying and characterizing the proteins and intermediates that facilitate the process. As shown in Figure 18-3, Holliday junctions are a type of intermediate involved in homologous recombination.

Likewise, researchers have elucidated the mechanisms of site-specific recombination and transposition using genetic and molecular techniques. In Experiment 18B, we learned that McClintock followed genetic crosses in corn to provide compelling evidence for the existence of transposable elements. Since that time, researchers have studied many different TEs. The sequencing of TEs has revealed certain patterns of DNA sequences that are required for transposition and retroposition. Researchers have also identified proteins that are necessary to promote the movement of TEs. In addition, molecular biologists have taken advantage of our knowledge of transposition and now routinely use transposons to inactivate and clone genes via transposon tagging.

PROBLEM SETS

SOLVED PROBLEMS

1. Make a drawing that describes a hypothetical mechanism in which a RecA filament forms triplex DNA with the homologous chromosome and displaces the complementary strand.

Answer:

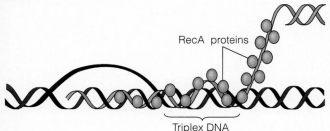

RecA proteins

Triplex DNA

2. What is DNA gap repair synthesis? What role does it play in the Meselson–Radding model of genetic recombination?

Answer: DNA gap repair synthesis occurs when a single-stranded DNA gap is produced within DNA. This gap is repaired by the action of DNA polymerase and DNA ligase. In the Meselson–Radding model, the invading strand leaves its DNA duplex to invade the unbroken chromosome. This leaves a gap in the broken strand, which is filled in via DNA gap repair synthesis. Note that this gap cannot be filled in by the invasion of the displaced strand (see Figure 18-4b, step 3), because a segment of the displaced strand is degraded.

CONCEPTUAL QUESTIONS

1. Describe the similarities and differences between sister chromatid exchange and homologous recombination. Would you expect the same types of proteins to be involved in both processes? Explain.

2. Is homologous recombination an example of mutation? Explain.

3. What are recombinant chromosomes? How do they differ from the original parental chromosomes from which they are derived?

4. During homologous recombination (Figure 18-3), the resolution steps can produce recombinant or nonrecombinant chromosomes. Explain how this can occur.

5. What is gene conversion?

6. Make a list of the differences among the Holliday model, the Meselson–Radding model, and the double-stranded break model.

7. In recombinant chromosomes, where is gene conversion likely to have taken place: near the breakpoint, or far away from the breakpoint? Explain.

8. What are the events that RecA protein facilitates?

9. According to the double-stranded break model, does gene conversion necessarily involve DNA mismatch repair? Explain.

10. What type of DNA structure is recognized by RecG, RuvAB, and RuvC? Do you think these proteins recognize DNA sequences? Be specific about what type(s) of molecular recognition these proteins can perform.

11. Briefly describe the three ways that antibody diversity is produced.

12. Describe the function of V(D)J recombinase.

13. Describe the role that integrase plays during the insertion of λ DNA into the host chromosome.

14. If you were examining a sequence of chromosomal DNA, what characteristics would cause you to believe that the DNA contained a transposable element?

15. According to the current model for replicative transposition, does the transposable element replicate before or after it transposes? Explain your answer.

16. Make a drawing that describes how a composite transposon can transpose by a conservative mechanism. In your drawing, remember that a composite transposon actually has four inverted repeats. With regard to the inverted repeats, make it clear where the cutting and ligation occurs.

17. Why does transposition always produce direct repeats in the host DNA?

18. Which types of transposable elements have the greatest potential for proliferation: conservative TEs, replicative TEs, or retroelements? Explain your choice.

19. Do you consider transposable elements to be mutagens? From Table 18-2, pick two ways that TEs can cause mutation, and make molecular drawings that explain how the mutation occurs.

EXPERIMENTAL QUESTIONS

1. With the harlequin staining technique, one sister chromatid appears to fluoresce more brightly than the other. Why?

2. In the data shown here, harlequin staining was used to determine the rate of SCEs in the presence of a suspected mutagen.

Frequency of SCEs/Chromosome

No mutagen	0.67
With suspected mutagen	14.7

Would you conclude that this substance is a mutagen?

3. In the experiment described in the previous problem, at what point would you need to add the mutagen: before the first round of DNA replication, after the first round but before the second round, or after the second round?

4. Briefly explain how McClintock determined that *Ds* was occasionally moving from one chromosomal location to another. Discuss the type of data she examined to arrive at this conclusion.

5. In the table of Experiment 18B, is the solid red phenotype due to chromosome breakage or the excision of a transposable element? Explain how you have arrived at your conclusion.

6. In your own words, explain the term transposon tagging.

7. Tumor suppressor genes are normal human genes that prevent uncontrollable cell growth. Starting with a normal laboratory human cell line, describe how you could use transposon tagging to identify tumor suppressor genes.

QUESTIONS FOR STUDENT DISCUSSION/COLLABORATION

1. Make a list of the similarities and differences among homologous recombination, site-specific recombination, and transposition.

2. If *no* genetic recombination of any kind could occur, what would be the harmful and beneficial consequences?

3. Based on your current knowledge of genetics, discuss whether or not you think the selfish DNA theory is correct.

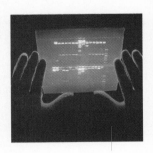

Part V
Genetic Technologies

CHAPTER 19

RECOMBINANT
DNA TECHNOLOGY

GENE CLONING

DETECTION OF GENES
AND GENE PRODUCTS

ANALYSIS AND
ALTERATION OF
DNA SEQUENCES

Recombinant **DNA technology** is the use of *in vitro* molecular techniques to isolate and manipulate fragments of DNA. In the early 1970s, the first successes in making recombinant DNA molecules were accomplished independently by two groups at Stanford University: David Jackson, Robert Symons, and Paul Berg; and Peter Lobban and A. Dale Kaiser. Both groups were able to isolate and purify pieces of DNA in a test tube, and then covalently link together two or more DNA fragments. In other words, they constructed recombinant DNA molecules. Shortly thereafter, it became possible to introduce such recombinant DNA molecules into living cells. Once inside a host cell, the recombinant molecules can be replicated to produce many identical copies of a gene. This achievement ushered in the era of **gene cloning**.

Recombinant DNA technology and gene cloning have enabled geneticists to probe relationships between gene sequences and phenotypic consequences and, thereby, have been fundamental to our understanding of gene structure and function. Most researchers in molecular genetics are familiar with recombinant DNA technology and apply it frequently in their work. Significant practical applications of recombinant DNA technology also have been developed. These include advances such as **gene therapy**, screening for human diseases, recombinant vaccines, and the production of **transgenic** plants and animals in agriculture (in which the cloned gene from one species is transferred to some other species). In this chapter, we will focus primarily on the use of recombinant DNA technology as a way to further our understanding of gene structure and function. We will examine how molecular techniques make it possible to clone genes, detect genes, and analyze DNA sequences. In Chapter 21, we will consider some modern applications of recombinant DNA technology.

GENE CLONING

The term gene cloning refers to the phenomenon of isolating and making many copies of a gene. The laboratory methods that are necessary to clone a gene were devised during the early 1970s. Since then, many technical advances have enabled gene cloning to become a widely used procedure among scientists, including geneticists, biochemists, plant biologists, microbiologists, evolutionary biologists, and biotechnologists. In this section, we will examine the reagents and procedures that are used to clone genes.

Cloning experiments usually involve two kinds of DNA molecules: chromosomal DNA and vector DNA

Let's begin our discussion of gene cloning by considering a recombinant DNA technology in which a gene is removed from its native site within a chromosome and inserted into a smaller segment of DNA known as a **vector**. When introduced into a living cell, a vector can replicate and produce many identical copies of the inserted gene.

If a scientist wants to clone a particular gene, the source of the gene is the chromosomal DNA of the species that carries the gene. For example, if the goal is to clone the rat β-globin gene, this gene is found within the chromosomal DNA of rat cells (see pp. 205–206 for a description of globin gene function). In this case, therefore, the rat's chromosomal DNA is one type of DNA that is needed in a cloning experiment. To prepare chromosomal DNA, an experimenter first obtains some cellular tissue from the organism of interest. To clone the rat β-globin gene, a researcher would obtain some tissue from a rat. The preparation of chromosomal DNA then involves the breaking open of cells and the extraction and purification of the DNA using biochemical techniques such as chromatography and centrifugation (see the appendix).

A second type of DNA is used in a cloning experiment—a small, specialized DNA segment known as a vector. The purpose of vector DNA is to act as a carrier of the DNA segment that is to be cloned. (The term vector comes from a Latin term meaning carrier.) In cloning experiments, a vector may carry a small segment of chromosomal DNA, perhaps only a single gene. By comparison, a chromosome carries many more genes, perhaps a few thousand. Like a chromosome, a vector is replicated when it resides within a living cell; a cell that harbors a vector is called the **host cell**. When a vector is replicated within a host cell, the DNA that it carries is also replicated.

The vectors commonly used in gene cloning experiments were derived originally from two natural sources. Some vectors are **plasmids**, which are small circular pieces of DNA. Plasmids are found naturally in many strains of bacteria and occasionally in eukaryotic cells. Many naturally occurring plasmids carry genes that confer resistance to antibiotics and other toxic substances. These plasmids are called **R factors**. Many of the plasmids used in modern cloning experiments were derived from R factors.

Plasmids also contain a DNA sequence, known as an origin of replication, that is recognized by the replication enzymes of the host cell. It is the sequence of the origin of replication that determines the host cell specificity of a vector. Some plasmids have origins of replications with a broad host range. Such a plasmid can replicate in the cells of many different species. Alternatively, many vectors used in cloning experiments have a limited host cell specificity. In cloning experiments, researchers must choose a vector that will replicate in the appropriate cell types for their experiments. If researchers want a cloned gene to be propagated in *Escherichia coli*, the vector they employ must have an origin of replication that is recognized by this species of bacterium.

Commercially available plasmids have been genetically engineered for effective use in cloning experiments. They contain unique sites where geneticists who are cloning genes insert pieces of DNA. Another useful feature of cloning vectors is that they often contain resistance genes that provide host bacteria with the ability to grow in the presence of a toxic substance. Such a gene is called a **selectable marker**, since the expression of the gene selects for the growth of the bacterial cells. Many selectable markers are genes that confer antibiotic resistance to the host cell. For example, the gene amp^R encodes an enzyme known as β-lactamase. This enzyme degrades ampicillin, an antibiotic that normally kills bacteria. Bacteria containing the amp^R gene can grow on media containing ampicillin, because they can degrade it. In a cloning experiment where the amp^R gene is found within the plasmid, the growth of cells in the presence of ampicillin identifies bacteria that contain the plasmid.

A second type of vector used in cloning experiments is a **viral vector**. As discussed in Chapter 6, viruses can infect living cells and propagate themselves by taking control of the host cell's metabolic machinery. When a chromosomal gene is inserted into a viral genome, the gene will be replicated whenever the viral DNA is replicated. Therefore, viruses can be used as vectors to carry other pieces of DNA. When a virus is used as a vector, the researcher analyzes viral plaques rather than bacterial colonies. The characteristics of viral plaques are described in Chapter 6 (see Figure 6-13).

Molecular biologists use hundreds of different vectors in cloning experiments. It is beyond the scope of this text to mention more than a few. Table 19-1

TABLE 19–1 Vectors Used in Cloning Experiments

Example	Type	Description
pBR322	Plasmid	This is one of the first vectors used by molecular geneticists. It is used to clone small segments of DNA and propagate them in *E. coli*.
YEp24	Plasmid	This is an example of a **shuttle vector** that can replicate in two different species, *E. coli* and *Saccharomyces cerevisiae*. It carries the origin of replication from pBR322 and another origin of replication that enables its replication in yeast. Using a shuttle vector, a researcher can propagate this plasmid in two unrelated species.
λgt11	Viral	This vector is derived from the bacteriophage λ, which was described in Chapter 15. λgt11 also contains a promoter from the *lac* operon. When fragments of DNA are cloned next to this promoter, the DNA is expressed in *E. coli*. This is an example of an **expression vector.** An expression vector is designed to clone the coding sequence of genes so that they will be transcribed and translated correctly.
M13mp18	Viral	This virus produces single-stranded DNA as a natural part of its life cycle. For this reason, it is commonly used as a vector to determine the sequence of cloned DNA fragments, as described later in this chapter.
SV40	Viral	This virus naturally infects mammalian cells. Genetically altered derivatives of the SV40 viral DNA are used as vectors for the cloning and expression of genes in laboratory grown mammalian cells.
Baculovirus	Viral	This virus naturally infects insect cells. In a laboratory, insect cells can be grown in liquid media. Unlike most other types of cells, insect cells often express large amounts of proteins that are encoded by cloned genes. When researchers want to make a large amount of a protein, they can clone the gene that encodes the protein into baculovirus and then purify the protein from insect cells.

provides a general description of several different types of vectors that are commonly used to clone small segments of DNA. In addition, two other types of vectors, called **cosmids** and **YACs**, are used to clone large pieces of DNA. These vectors will be described in Chapter 20. Also, vectors designed to introduce genes into plants and animals will be discussed in Chapter 21.

Enzymes are used to cut DNA into pieces and join the pieces together

A key step in a cloning experiment is the insertion of chromosomal DNA into a plasmid or viral vector. This requires the cutting and pasting of DNA fragments. To cut DNA, researchers use enzymes known as **restriction endonucleases** or **restriction enzymes**. The restriction enzymes used in cloning experiments bind to a specific base sequence and then cleave the DNA backbone at two defined locations, one in each strand. Figure 19-1 shows the action of a restriction endonuclease. Restriction enzymes were discovered in the 1960s and 1970s by Werner Arber at the University of Geneva, and Hamilton Smith and Daniel Nathans at Johns Hopkins University. They are made naturally by many different species of bacteria. Restriction endonucleases protect bacterial cells from invasion by foreign DNA, particularly that of bacteriophages. Researchers originally isolated and purified restriction enzymes from these bacterial species and now use them in their cloning experiments.

Currently, several hundred different restriction enzymes from various bacterial species have been identified and are available commercially to molecular biologists. Table 19-2 gives a few examples. Restriction enzymes usually recognize sequences that are palindromic. This means that the sequence is identical when read in the opposite direction in the complementary strand. For example, the sequence recognized by *Eco*RI is 5′–GAATTC–3′ in the top strand. Read in the opposite direction in the bottom strand, this sequence is also 5′–GAATTC–3′.

Restriction enzymes are useful in cloning because many digest DNA into fragments with "sticky ends." This means, as shown in Figure 19-1, that these DNA fragments will hydrogen bond to each other due to their complementary sequences. As shown there, the complementary sequences promote interactions between two different pieces of DNA. The ends of two different DNA pieces will hydrogen bond to each other because of their compatible sticky ends.

The hydrogen bonding between the sticky ends of DNA fragments promotes temporary interactions between two DNA fragments. However, this interaction is not stable, because it involves only a few hydrogen bonds between complementary bases. To establish a permanent connection between two DNA fragments, the sugar–phosphate backbones within the DNA strands must be covalently linked together. This linkage is catalyzed by an enzyme known as **DNA ligase**. Figure 19-1 illustrates DNA ligase catalyzing covalent bond formation in both DNA strands after the sticky ends have hydrogen bonded with each other.

Gene cloning involves the insertion of DNA fragments into vectors; the vectors are propagated within host cells

Now that we are familiar with the materials, let's outline the general strategy that is followed in a typical cloning experiment. The procedure shown in Figure 19-2 seeks to clone a chromosomal gene into a plasmid vector that already carries the *amp^R* gene. To begin this experiment, the chromosomal DNA is isolated and digested with a restriction enzyme. This enzyme will cut the chromosomes into many small fragments. The plasmid DNA is also cut at one site with the same restriction enzyme. The digested chromosomal DNA and plasmid DNA are mixed together and incubated under conditions that promote the binding of complementary sticky ends.

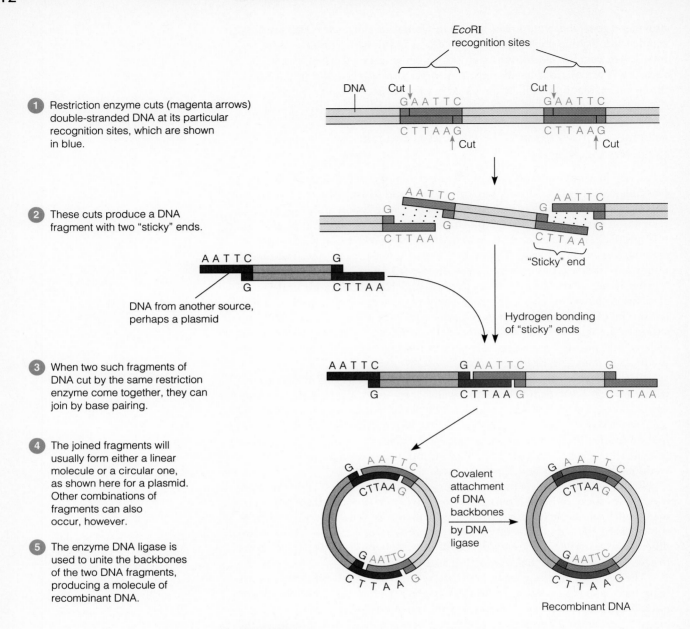

1 Restriction enzyme cuts (magenta arrows) double-stranded DNA at its particular recognition sites, which are shown in blue.

2 These cuts produce a DNA fragment with two "sticky" ends.

3 When two such fragments of DNA cut by the same restriction enzyme come together, they can join by base pairing.

4 The joined fragments will usually form either a linear molecule or a circular one, as shown here for a plasmid. Other combinations of fragments can also occur, however.

5 The enzyme DNA ligase is used to unite the backbones of the two DNA fragments, producing a molecule of recombinant DNA.

FIGURE 19-1

The action of a restriction enzyme and the production of recombinant DNA.
The restriction enzyme, *Eco*RI, binds to a specific sequence, in this case 5'–GAATTC–3'. It then cleaves the DNA backbone, producing DNA fragments. The single-stranded ends of the DNA fragments can hydrogen bond with each other, because they have complementary sequences. The enzyme DNA ligase can then catalyze the formation of covalent bonds in the DNA backbones of the fragments.

DNA ligase is then added to catalyze the covalent linkage between DNA fragments. In some cases, the two ends of the vector will simply ligate back together, restoring the vector to its original circular structure. This is called a **recircularized vector**. In other cases, a fragment of chromosomal DNA may become ligated to both ends of the vector; thus, a segment of chromosomal DNA has been inserted into the vector. The vector containing a piece of chromosomal DNA is referred to as a **hybrid vector**.

TABLE 19–2 Some Restriction Enzymes Used in Gene Cloning

Restriction Enzyme	Bacterial Source	Sequence Recognized*
BamHI	Bacillus amyloliquefaciens H	↓ 5′–G–G–A–T–C–C–3′ 3′–C–C–T–A–G–G–5′ ↑
ClaI	Caryophanon latum	↓ 5′–A–T–C–G–A–T–3′ 3′–T–A–G–C–T–A–5′ ↑
EcoRI	E. coli RY13	↓ 5′–G–A–A–T–T–C–3′ 3′–C–T–T–A–A–G–5′ ↑
PstI	Providencia stuartii	↓ 5′–C–T–G–C–A–G–3′ 3′–G–A–C–G–T–C–5′ ↑
SacI	Streptomyces achromonogenes	↓ 5′–G–A–G–C–T–C–3′ 3′–C–T–C–G–A–G–5′ ↑

*The arrows show the locations in the upper and lower DNA strands where the restriction enzymes cleave the DNA backbone.

Following ligation, the DNA is introduced into bacterial cells treated with agents that render them permeable to DNA molecules. This step is called **transformation**, when a plasmid vector is used, or **transfection**, when a viral vector is introduced into a host cell. In the experiment shown here, a plasmid is introduced into bacterial cells that were originally sensitive to ampicillin. The bacteria are then streaked on plates containing bacterial growth media and ampicillin. If a bacterium has taken up a plasmid carrying the amp^R gene, it will continue to divide and form a bacterial colony containing tens of millions of cells. Since each cell within a single colony is derived from the same original cell, all the cells within a colony contain the same type of plasmid DNA.

In the experiment shown in Figure 19-2, the experimenter can distinguish between bacterial colonies that contain a recircularized vector versus those with a hybrid vector carrying a piece of chromosomal DNA. As shown here, the chromosomal DNA has been inserted into a region of the vector that contains the *lacZ* gene, which encodes the enzyme β-galactosidase. The insertion of chromosomal DNA into the vector disrupts the *lacZ* gene. By comparison, a recircularized vector has a functional *lacZ* gene. The functionality of *lacZ* can be determined by providing the growth media with a colorless compound, Xgal, which is cleaved by β-galactosidase into a blue dye. Bacteria grown in the presence of Xgal and IPTG (an inducer of the *lacZ* gene) will form blue colonies if they have a functional β-galactosidase, and white colonies if they do not. In this experiment, therefore, bacterial colonies containing recircularized vectors will form blue colonies, while colonies containing hybrid vectors will be white.

In Figure 19-2, the bacterial cells that make up a white colony contain a hybrid vector, which carries a segment of chromosomal DNA. This example illustrates a hybrid vector that carries a human gene; the segment containing the human gene is shown in black. The net result of gene cloning is to produce an

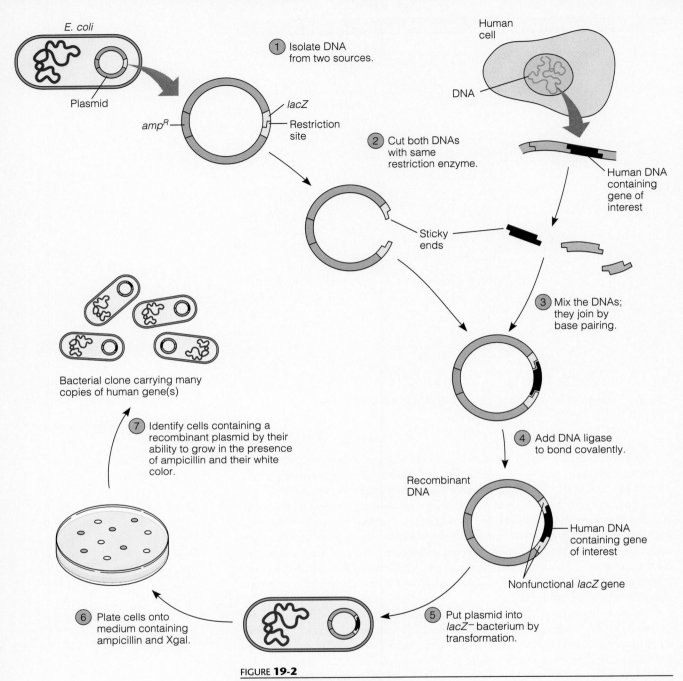

FIGURE **19-2**

The steps in gene cloning.

enormous number of copies of a gene within a hybrid vector. During transformation, a single bacterial cell usually takes up a single copy of a hybrid vector. However, two subsequent events lead to the amplification of the cloned gene. First, since the vector has an origin of replication, the bacterial host cell replicates the hybrid vector to produce many copies per cell. Second, the bacterial cells divide approximately every 30 minutes. Following overnight growth, a population of many millions of bacteria will be obtained. Each of these bacterial cells will contain many copies of the cloned gene. For example, a bacterial colony may comprise 10,000,000 cells, with each cell containing 50 copies of the hybrid vector. Therefore, this bacterial colony would contain 500 million copies of the cloned gene!

The preceding description has meant to acquaint you with the steps required to clone a gene. A misleading aspect of Figure 19-2 is that the digestion of the chromosomal DNA with restriction enzymes seems to yield only a few DNA fragments, one of which contains the gene of interest. In an actual cloning experiment, however, the digestion of the chromosomal DNA with a restriction enzyme produces tens of thousands of different pieces of chromosomal DNA, not just a single piece of chromosomal DNA that happens to be the gene we want to clone. Later in this chapter, we will consider how to use probes to identify bacterial colonies containing the specific gene that a researcher wants to clone.

Recombinant DNA technology can also be used to clone fragments of DNA that do not code for genes. For example, sequences such as telomeres, centromeres, and highly repetitive sequences have been cloned by this procedure. In addition, RNA can be used as a starting material to clone DNA. The enzyme reverse transcriptase can use RNA as a template to make a complementary strand of DNA. This RNA–DNA hybrid is then exposed to the enzyme RNaseH, which digests specifically the RNA in RNA–DNA hybrids. The action of RNaseH leaves behind short segments of RNA that are used as primers for DNA polymerase to produce the other strand of DNA. In this way, double-stranded DNA can be made from single-stranded RNA. This double-stranded DNA can then be cloned into a vector using the methods already described in this chapter. In this case, the DNA is referred to as **complementary DNA** or **cDNA**.

In the first gene cloning experiment, Cohen, Chang, Boyer, and Helling inserted a bacterial *kanamycin^R* gene into a plasmid vector

Now that we are familiar with the basic procedures followed in a cloning experiment, let's consider the first successful attempt at creating a recombinant DNA molecule and propagating that molecule in bacterial cells. This was accomplished in 1973 in a collaboration between Stanley Cohen and Annie Chang at Stanford University and Herbert Boyer and Robert Helling at the University of California at San Francisco. Several prior important discoveries led to their ability to clone a gene. In 1970, H. Gobind Khorana at the Massachusetts Institute of Technology found that DNA ligase could covalently link DNA fragments together. In 1972, Janet Mertz, Ronald Davis, and Vittorio Sgaramella at Stanford discovered that the digestion of DNA with the restriction enzyme *Eco*RI produced sticky ends that enabled the DNA fragments to hydrogen bond. With these two observations in mind, Cohen, Chang, Boyer, and Helling realized that it might be possible to create recombinant DNA molecules by digesting DNA with *Eco*RI, allowing the fragments to hydrogen bond with each other, and then covalently linking these fragments together with DNA ligase.

One last condition was necessary for Cohen, Chang, Boyer, and Helling to succeed in cloning a gene. They needed to identify a vector that could independently replicate itself once inside a host cell. They chose a plasmid vector as their vehicle for gene cloning. In their collection of plasmids at Stanford, Cohen, Chang, Boyer, and Helling found one small plasmid, designated pSC101, that had a single *Eco*RI site and carried a gene for tetracycline resistance:

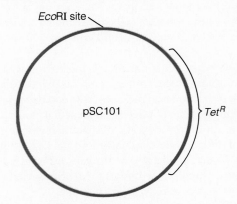

They reasoned that it should be possible to open up this plasmid with *Eco*RI and then insert another piece of DNA into this *Eco*RI site via the hydrogen bonding of sticky *Eco*RI ends.

As a source of a gene to insert into pSC101, they obtained a second plasmid, which they called pSC102:

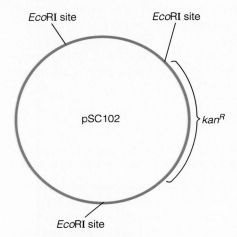

This plasmid carried a gene, *kan^R*, which provides resistance to the antibiotic kanamycin. However, this second plasmid was cut at three locations by *Eco*RI, yielding three DNA fragments. One of these DNA fragments would be expected to carry the kanamycin resistance gene (unless, unluckily, *Eco*RI happened to cut in the middle of the *kan^R* gene). Their goal in this experiment was to clone the *kan^R* gene by inserting the DNA fragment carrying this gene into the *Eco*RI site of pSC101, and then introducing the hybrid plasmid into bacterial cells.

THE HYPOTHESIS

A piece of DNA carrying a gene can be inserted into a plasmid vector using recombinant DNA techniques. If this recombinant plasmid is introduced into a bacterial host cell, it will be replicated and transmitted to daughter cells, producing many copies of the recombinant plasmid.

Starting material: Three different strains of *E. coli*: One strain that did not carry any plasmid and two strains that carried pSC101 or pSC102.

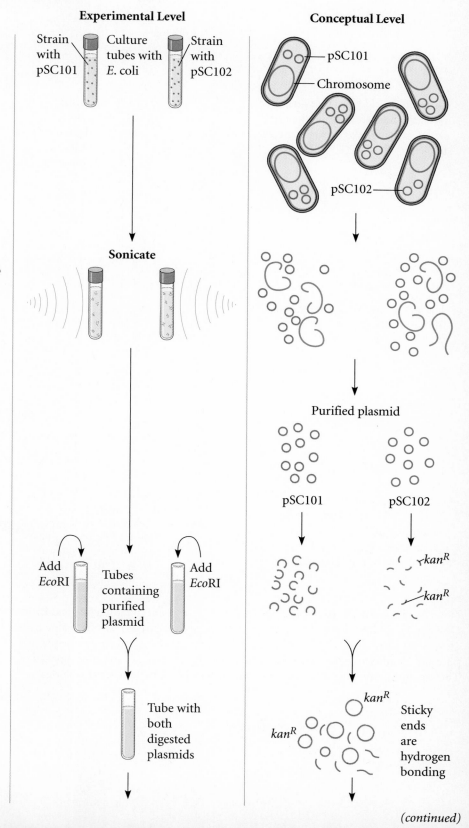

Experimental Level

Conceptual Level

1. Isolate and purify the two types of plasmid DNA:

Grow the bacterial cells containing the plasmids.

Break open the cells. One way to break open cells is to subject them to harsh sound waves, or sonication.

Isolate the plasmid DNA by density gradient centrifugation (see the appendix for a description of this procedure).

2. Digest the plasmid DNAs with *Eco*RI.

3. Mix together the two samples.

Strain with pSC101   Culture tubes with *E. coli*   Strain with pSC102

pSC101

Chromosome

pSC102

Sonicate

Purified plasmid

pSC101   pSC102

Add *Eco*RI   Tubes containing purified plasmid   Add *Eco*RI

kan^R

kan^R

Tube with both digested plasmids

kan^R

kan^R

Sticky ends are hydrogen bonding

(continued)

(continued)

4. Add DNA ligase.

5. Grow an *E. coli* strain that does not carry a plasmid. Treat the cells with CaCl$_2$ to make them permeable to DNA.

6. Add the ligated DNA samples to the bacterial cells.

7. Plate the cells on growth media containing both tetracycline and kanamycin. Grow overnight to allow the growth of visible bacterial colonies.

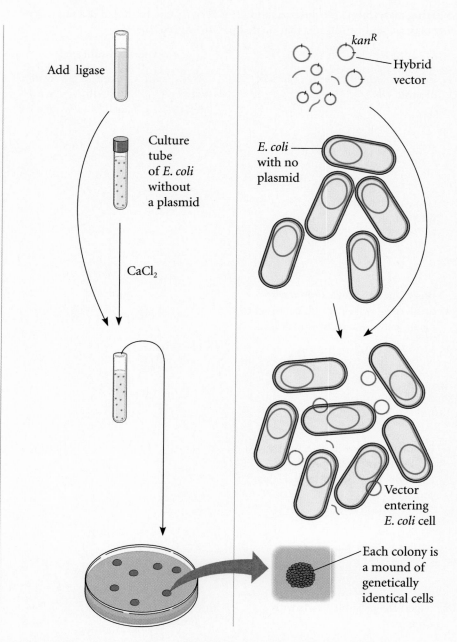

Add ligase

Culture tube of *E. coli* without a plasmid

CaCl$_2$

kan^R

Hybrid vector

E. coli with no plasmid

Vector entering *E. coli* cell

Each colony is a mound of genetically identical cells

Note: At this point, the bacterial colonies could be resistant to tetracycline and kanamycin for two different reasons. One possibility (the one they hoped for) is that the kanamycin-resistance gene had been inserted into pSC101. However, a second possibility is that a single bacterial cell had taken up one pSC101 plasmid and one pSC102 plasmid. In other words, a bacterial cell could have two types of plasmids instead of a single recombinant plasmid. The remaining steps are intended to distinguish between these two possibilities.

8. Pick four colonies from the plates and grow the colonies in liquid culture containing radiolabeled compounds.

9. To isolate the radiolabeled plasmid DNA, break open the cells and subject the DNA to cesium chloride density gradient centrifugation. As described in the appendix, collect fractions and count them in a scintillation counter. (See data shown on the left.)

10. Also, take some of the plasmid DNA, digest it with EcoRI, and subject it to gel electrophoresis. (See data shown on the right.)

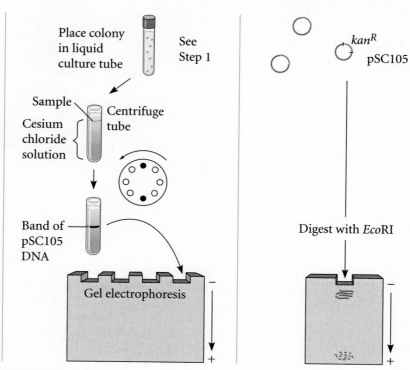

Place colony in liquid culture tube — See Step 1

Sample — Centrifuge tube

Cesium chloride solution

Band of pSC105 DNA

Gel electrophoresis

kan^R pSC105

Digest with EcoRI

Note: As a control, the procedures in Steps 8–10 were also conducted on bacteria carrying the pSC101 and pSC102 plasmids.

THE DATA

Results from Step 9:

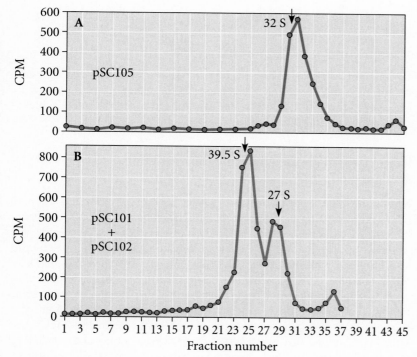

A — pSC105 — 32 S

CPM

B — pSC101 + pSC102 — 39.5 S — 27 S

CPM

Fraction number

Results from Step 10:

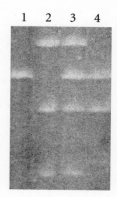

1 2 3 4

Let's first consider the results of density gradient centrifugation. As shown in the control experiment, the pSC102 plasmid is larger than pSC101 and sediments at a density of 39.5S, whereas the latter plasmid has a density of 27S. (Note: The letter S refers to Svedberg units, a unit of centrifugation. See the appendix.) In this control experiment, the pSC101 and pSC102 plasmids were mixed together and then subjected to density gradient centrifugation. This yielded two peaks at 27S and 39.5S. The top figure shows the results for pSC105 obtained from a bacterial colony that was resistant to both tetracycline and kanamycin. In this case, there is a plasmid with an intermediate density of 32S rather than a mixture of two plasmids. These results indicate that this bacterial colony contained a recombinant plasmid, not a mixture of pSC101 and pSC102.

This conclusion is confirmed in the gel electrophoresis analysis. When digested with *Eco*RI, pSC101 yielded a single band, whereas pSC102 yielded three bands (see lanes 1 and 2). In lane 3, the experimenters had mixed together samples of pSC101 and pSC102 (isolated in step 1) and digested them with *Eco*RI. As expected, this produced four bands. Lane 4 shows the plasmid DNA, pSC105, from a bacterial colony that was tetracycline and kanamycin resistant. This plasmid showed the pSC101 band plus one other band that was found in pSC102. The results of lane 4 are consistent with the idea that the pSC105 plasmid is formed by the insertion of one fragment from pSC102 into the single *Eco*RI site of pSC101. This recombinant plasmid is shown here:

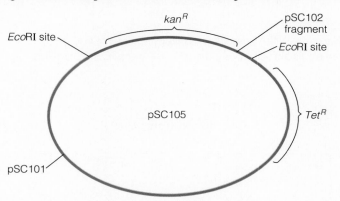

This experiment showed the scientific community that it is possible to create recombinant DNA molecules and then propagate the recombinant molecules in bacterial cells. In other words, it is possible to clone genes. This hallmark achievement ushered in the era of gene cloning.

Restriction mapping is used to locate the restriction sites within a vector

As we have seen, DNA or gene cloning involves the digestion of vector and chromosomal DNA with restriction enzymes and the subsequent ligation of DNA fragments into vectors. In this type of procedure, the locations of restriction enzyme sites are crucial. In the vector, for example, it is desirable to have a unique restriction site for the insertion of chromosomal DNA. After successfully cloning a large chromosomal DNA insert, a researcher may wish to use the hybrid vector further to obtain clones having smaller pieces of the chromosomal DNA. The process of making smaller clones from a larger one is called **subcloning**. To make a subclone, the hybrid vector DNA must be cut with one or more restriction enzymes to produce smaller fragments of the insert DNA, which are then inserted into a new vector. Overall, cloning and subcloning methods require knowledge of the locations of restriction enzyme sites in vectors and hybrid vectors.

A common approach to examining the locations of restriction sites is known as **restriction mapping**. Figure 19-3 outlines the restriction mapping of a bacterial plasmid, pBR322 (described in Table 19-1). To begin this experiment, (1) the small circular plasmid DNA is isolated and purified from host cells. (2) Samples of the purified DNA are then placed in separate test tubes that (3) contain a particular

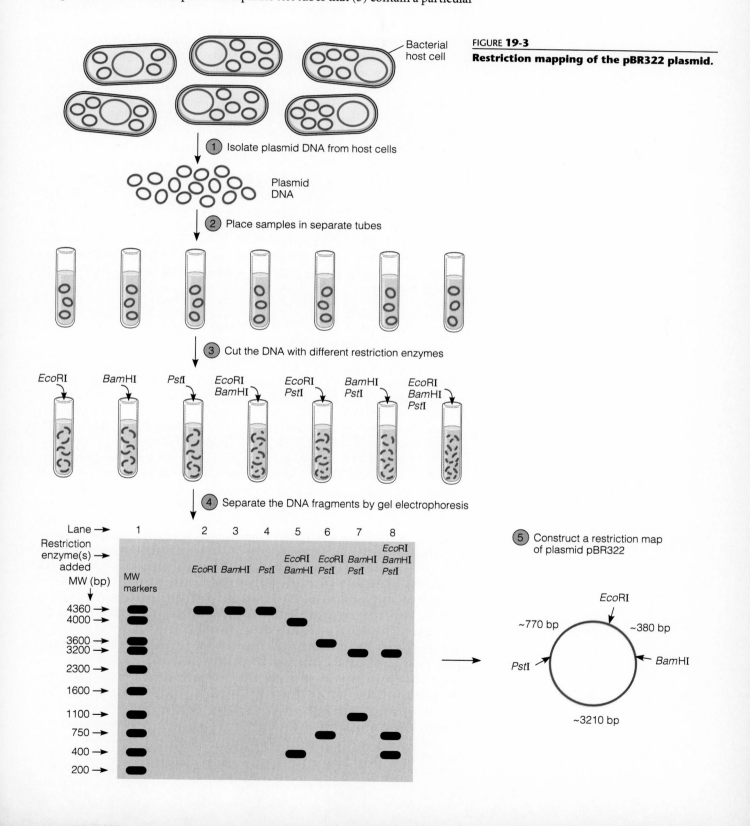

FIGURE **19-3**

Restriction mapping of the pBR322 plasmid.

restriction enzyme or combination of enzymes. The plasmid DNA is incubated with the restriction enzymes long enough for digestion to occur. (4) The DNA fragments are then separated by gel electrophoresis. In lane 1 of this experiment, a different DNA sample, known as molecular weight markers, is also subjected to gel electrophoresis. This sample contains a mixture of DNA fragments with molecular weights that are known from previous experiments. (These markers are purchased from commercial sources or can be prepared in the laboratory.) To determine the sizes of the fragments obtained by digesting pBR322, the fragments in lanes 2–8 are compared with the known markers in lane 1.

The restriction map shown in (5) was deduced by comparing the sizes of fragments obtained from the single, double, and triple digestions. The starting plasmid is a circular molecule 4363 base pairs in length. A single digestion with any of the three enzymes yields a single linear fragment of size 4363 bp. This means that *Eco*RI, *Bam*HI, and *Pst*I cut the plasmid at a single site.

The double digestion with *Eco*RI and *Bam*HI yields two fragments, of about 380 bp and 3980 bp. This result indicates that the *Eco*RI and *Bam*HI sites are approximately 380 bp apart in one direction along the circle and 3980 bp apart along the circle in the opposite direction. Likewise, the pairwise combinations of *Eco*RI–*Pst*I and *Bam*HI–*Pst*I indicate how far apart these sites are along the circular plasmid.

Finally, the triple digestion confirms the locations of the single sites for *Eco*RI, *Bam*HI, and *Pst*I. Taken together, these results provide a map of the restriction sites within this plasmid. A similar approach can be used on a hybrid vector to determine the locations of restriction sites within a fragment of DNA that has been inserted into a plasmid.

Polymerase chain reaction (PCR) can also be used to make many copies of DNA

In our previous discussions of gene cloning, the DNA of interest was inserted into a vector, which then was introduced into a host cell. The replication of the vector within the host cell, and the proliferation of the host cells, led to the production of many copies of the DNA. Another way to copy DNA, without the aid of vectors and host cells, is a technique called **polymerase chain reaction (PCR)**, which was developed by Kary Mullis in 1985 while at the Cetus Corporation.

The PCR method is outlined in Figure 19-4. In this example, the starting material contains a sample of DNA, known as the **template DNA**. The goal of PCR is to make many copies of the template DNA. Several reagents are added to facilitate the synthesis of DNA. These include a sample containing many copies of two oligonucleotide primers (which are complementary to sequences within the DNA fragment), deoxynucleotides, and a thermostable form of DNA polymerase called *taq* **polymerase**. A thermostable form of DNA polymerase is necessary, because PCR involves heating steps that would inactivate most other natural forms of DNA polymerase (which are thermolabile).

To make copies of the DNA, this double-stranded template is denatured by heat treatment and then the oligonucleotide primers bind to the template DNA as the temperature is lowered. The binding of the primers to the DNA is called **annealing**. Once the primers have annealed, *taq* polymerase will catalyze the synthesis of complementary DNA strands. This doubles the amount of the template DNA. The sequential process of denaturation–annealing–synthesis is then iterated repeatedly to double the amount of template DNA many times over. This method is called a chain reaction, because the products of each previous reaction (i.e., newly made DNA strands) are used as reactants (i.e., as template strands) in subsequent reactions.

A thermal reactor that automates the timing of each cycle, known as a **thermocycler**, is used to carry out PCR. The experimenter mixes the DNA sample, an excess amount of primers, *taq* polymerase, and deoxynucleotides together in a single tube. The tube is placed in a thermocycler, and the experimenter

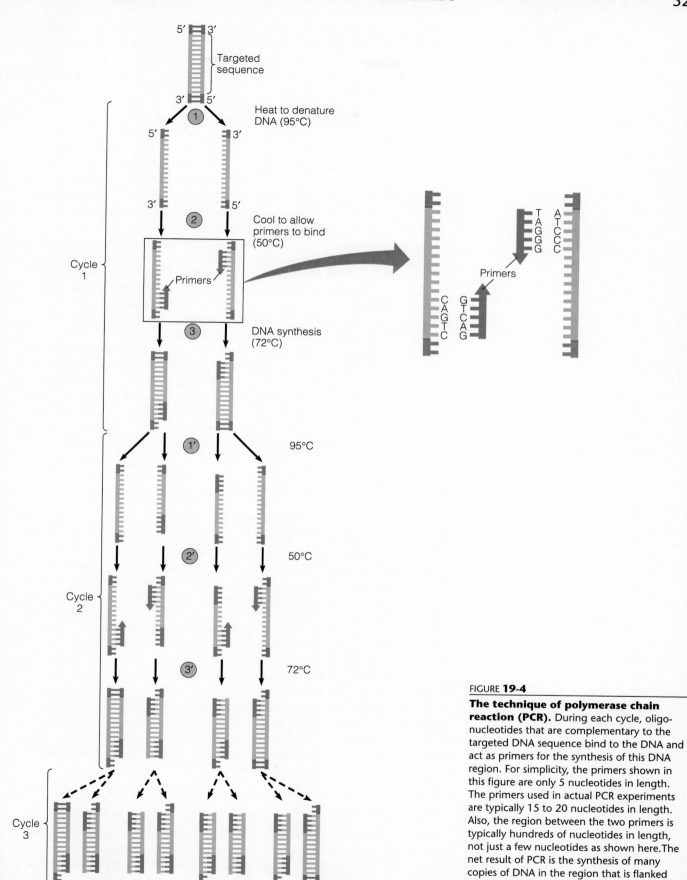

FIGURE **19-4**

The technique of polymerase chain reaction (PCR). During each cycle, oligo-nucleotides that are complementary to the targeted DNA sequence bind to the DNA and act as primers for the synthesis of this DNA region. For simplicity, the primers shown in this figure are only 5 nucleotides in length. The primers used in actual PCR experiments are typically 15 to 20 nucleotides in length. Also, the region between the two primers is typically hundreds of nucleotides in length, not just a few nucleotides as shown here. The net result of PCR is the synthesis of many copies of DNA in the region that is flanked by the two primers.

sets the machine to operate within a defined temperature range and number of cycles. During every cycle, the thermocycler increases the temperature to denature the DNA strands and then lowers the temperature to allow annealing and DNA synthesis to take place. After a few minutes, the cycle is repeated by increasing and then lowering the temperature. A typical PCR run is likely to involve 20 to 30 cycles of replication and takes several hours to complete. The PCR technique can amplify the amount of DNA by a staggering amount. After 30 cycles of amplification, a DNA sample will increase 2^{30}-fold, which is approximately a billion-fold.

The PCR reaction shown in Figure 19-4 seeks to amplify a particular DNA segment. As shown here, the sequences of the PCR primers are complementary to two specific sequences within the template DNA. Therefore, the two primers bind to these sites and the intervening region is replicated. To conduct this type of PCR experiment, the researcher must have knowledge about the sequence of the template DNA in order to design oligonucleotide primers complementary to two sites in the template sequence. When specific primers can be constructed, PCR can amplify a specific region of DNA from a complex mixture of template DNA. For example, if a researcher uses two primers that anneal to the human β-globin gene, PCR can specifically amplify the β-globin gene from a DNA sample that contains all of the human chromosomes!

Alternatively, PCR can be used to amplify a sample of chromosomal DNA semi-specifically or nonspecifically. As will be discussed in Chapter 21, this approach is used in DNA fingerprinting analysis. In a semispecific PCR experiment, the primers recognize a repetitive DNA sequence found at several sites within the genome. When chromosomal DNA is used as a template, this will amplify many different DNA fragments. In a nonspecific approach, a mixture of short PCR primers with many different random sequences is used. These primers will anneal randomly throughout the genome and amplify most of the chromosomal DNA. Nonspecific DNA amplification is used to increase the total amount of DNA in very small samples, such as blood stains found at crime scenes. This topic will be discussed in Chapter 21.

DETECTION OF GENES AND GENE PRODUCTS

The advent of gene cloning since the 1970s has enabled scientists to investigate gene structure and function at the molecular level. However, unlike the first gene cloning experiment in which a single gene was removed from one plasmid (pSC102) and cloned into another plasmid (pSC101), molecular geneticists usually want to study genes that are originally within the chromosomes of living species. This presents a problem, because chromosomal DNA contains thousands to hundreds of thousands of different genes. For this reason, researchers need methods to identify specifically a gene within a mixture of many other genes or DNA fragments. The term **gene detection** refers to methods that distinguish one particular gene among a mixture of thousands or even millions of other genes. We will see that the molecular identification of genes has advanced greatly our understanding of how genes function. Also, later in this section, we will learn how gene detection methods are used to clone genes.

When studying gene expression at the molecular level, scientists also need techniques for the identification of gene products, such as the RNA that is transcribed from a particular gene or the protein that is encoded by an mRNA. In this section, we will consider the methodology and uses of a few common detection strategies.

Blotting methods can be used to detect DNA sequences and gene products

The technique of **Southern blotting** can detect the presence of a particular gene sequence within a mixture of many chromosomal DNA fragments. It was developed by E. M. Southern in 1975. Southern blotting has several uses. It can determine the copy number of a gene within the genome of an organism. For example, Southern blotting has revealed that rRNA genes are found in multiple copies within a genome whereas many other structural genes are unique. In the study of human genetic diseases, Southern blotting can also detect small gene deletions that cannot be distinguished under the light microscope (see Solved Problem 2 at the end of the chapter). A common use of Southern blotting is to identify gene families. As was discussed in Chapter 8, a gene family is a group of two or more genes derived from the same ancestral gene. The members of a gene family have similar but not identical DNA sequences; they are called homologous genes. As we will see, Southern blotting can distinguish the homologous members of a gene family. Similarly, Southern blotting can identify homologous genes among different species.

Prior to a Southern blotting experiment, a gene of interest, or a fragment of a gene, has been cloned. This cloned DNA is labeled (e.g., radiolabeled) *in vitro*, and then the labeled DNA is used as a **probe** to detect the presence of the gene within a mixture of many DNA fragments. The basis for a Southern blotting experiment is that two DNA fragments will bind to each other only if they have complementary sequences. In the experiment, the labeled strands from the cloned gene will pair specifically with complementary DNA strands, even if the complementary strands are found within a mixture of many other DNA pieces.

Figure 19-5 shows the Southern blotting procedure. The goal of this experiment is to determine if the chromosomal DNA contains a base sequence complementary to a previously cloned gene. To begin the experiment, the chromosomal

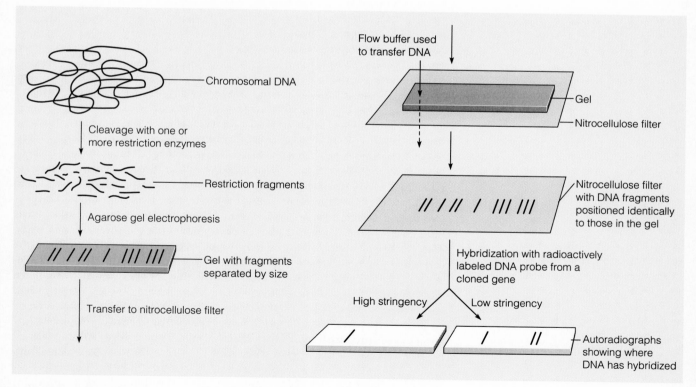

FIGURE **19-5**

The technique of Southern blotting. When the probe DNA is hybridized to the mixture of chromosomal DNA fragments at high temperature and/or high ionic strength (i.e., high stringency), the probe only recognizes a single band. When hybridization is performed at lower temperature and/or lower ionic strength (i.e., low stringency), this band and two additional bands hybridize with the radiolabeled probe.

DNA is isolated and digested with a restriction enzyme. Since the restriction enzyme cuts the chromosomal DNA at many different sites within the chromosomes, this step produces thousands of DNA pieces of different sizes. The chromosomal pieces are loaded onto a gel that separates them according to their size. The DNA pieces within the gel are then extracted by transferring (i.e., blotting) onto a nitrocellulose (or nylon) filter. The DNA fragments are then tightly fixed to that filter, and denatured. At this point, the filter contains many unlabeled DNA fragments that have been denatured and separated according to size.

The next step is to determine if any of these unlabeled fragments from the chromosomal DNA contain sequences complementary to the gene that has been cloned. To detect the cloned DNA, it must be labeled. A common labeling method is to incorporate the radioisotope ^{32}P into the cloned DNA, which labels the phosphate group in the DNA backbone. The filter, which has the unlabeled chromosomal DNA attached to it, is submerged in a solution containing the radiolabeled DNA. If the radiolabeled DNA and a fragment of chromosomal DNA are complementary, they will hydrogen bond to each other. Any unbound radiolabeled DNA is then washed away, and the filter is exposed to X-ray film. Locations where radiolabeled DNA has bound will appear as dark bands on the X-ray film.

An important variable in the Southern blot procedure is the temperature and ionic strength of the hybridization and wash steps. If these steps are done at very high temperatures (namely, at 65°C) or at high salt (NaCl) concentrations, then the probe DNA and chromosomal fragment must be very complementary—nearly a perfect match—to hybridize. This condition is called *high stringency*. Conditions of high stringency are used to detect a match between the cloned gene and a chromosomal DNA fragment. However, if the temperature and/or ionic strength are lower, genes that are similar but not necessarily identical may hybridize to the probe. This is called *low stringency*. Conditions of low stringency are used to detect homologous genes with DNA sequences that are similar but not identical to the cloned gene being used as a probe. In the results shown in Figure 19-5, conditions of high stringency reveal that the gene of interest is found only in a single copy in the genome. At low stringency, however, two other bands are detected. These results suggest that this gene is a member of a gene family composed of three distinct members.

Let's now turn our attention to the technique known as **Northern blotting**, which is used to identify a specific RNA within a mixture of many RNA molecules. (Note: Even though Southern blotting was named after E. M. Southern, Northern blotting was not named after anyone called Northern! It was originally termed reverse-Southern blotting and later Northern blotting.) Northern blotting is used to investigate the transcription of genes at the molecular level. This method can determine if a specific gene is transcribed in a particular cell type (e.g., nerve versus muscle cells) or at a particular stage of development (e.g., fetal versus adult cells). Also, Northern blotting can reveal if a pre-mRNA transcript is alternatively spliced into two or more mRNAs of different sizes.

From a technical viewpoint, Northern blotting is rather similar to Southern blotting with a few important differences. In Northern blotting, RNA (rather than chromosomal DNA) is extracted and purified from living cells. This RNA can be isolated from a particular cell type or during a particular stage of development. Any given cell will produce hundreds or thousands of different types of RNA molecules, because cells express many genes at any given time. After the RNA is extracted from cells and purified, it is loaded onto a gel that separates the RNA transcripts according to their size. The RNAs within the gel are then blotted onto a nitrocellulose (or nylon) filter and probed with a radiolabeled fragment of DNA from a cloned gene. Using this method, RNAs that are complementary to the radiolabeled DNA fragment are detected as dark bands on an X-ray film.

For structural genes, the net result of gene expression is the synthesis of proteins. A particular protein within a mixture of many different protein molecules can be detected by a third detection procedure, **Western blotting**. (Here again, Western blotting was not named after anyone called Western!) Similar to Northern blotting, Western blotting can determine if a specific protein is made in a particular cell type or at a particular stage of development. Technically, this procedure is also rather similar to Southern and Northern blotting. In a Western blotting experiment, proteins are extracted from living cells. These proteins can be isolated from different cell types or from the same cell type during different stages of development. As with RNA, any given cell will produce many different proteins at any time, because it is expressing many structural genes. After the proteins have been extracted from the cells, they are loaded onto a gel that separates them by molecular weight. To perform the separation step, the proteins are first dissolved in sodium dodecyl sulfate (SDS), a detergent that denatures proteins and coats them with negative charges. The negatively charged proteins are then separated in a gel made of polyacrylamide. This method of separating proteins is called SDS-PAGE (*poly*acrylamide *gel* *e*lectrophoresis).

Following SDS-PAGE, the proteins within the gel are blotted onto a nitrocellulose (or nylon) filter. The next step is to use a probe that will recognize a specific protein of interest. An important difference between Western blotting and either Southern or Northern blotting is the use of an antibody, rather than a cloned gene, as a probe. Antibodies bind to sites on molecules known as **epitopes** or antigenic sites within **antigens**. In the case of proteins, an antigenic site is a short sequence of amino acids. Since the amino acid sequence is a unique feature of each protein, any given antibody will specifically recognize a particular protein. This is called the primary antibody.

After the primary antibody has been given sufficient time to recognize the protein of interest, any unbound primary antibody is washed away and a secondary antibody is added. A **secondary antibody** is an antibody that binds to the primary antibody. Secondary antibodies, which may be radiolabeled, are used for convenience, as labeled secondary antibodies are available commercially. In general, it is easier for researchers to purchase these antibodies rather than to label their own primary antibodies. In a Western blotting experiment, the labeling of the secondary antibody provides a way to detect the protein of interest on a gel. For example, if the secondary antibody is radiolabeled, the protein that is recognized by the primary antibody will show up as a dark band on the X-ray film.

A DNA library is a collection of many cloned fragments of DNA; probing methods are necessary to detect a particular gene within a DNA library

Now that we understand some methods used to detect genes and gene products, let's reexamine the steps required to clone a gene (refer back to Figure 19-2). Detection methods are also needed to identify desirable hybrid clones. In a typical cloning experiment, the treatment of the chromosomal DNA with restriction enzymes yields tens of thousands of different DNA fragments. Therefore, after the DNA fragments are ligated individually to vectors, the researcher has a collection of hybrid vectors, with each vector containing a particular fragment of chromosomal DNA. This collection of hybrid vectors is known as a **DNA library** (see Figure 19-6). When the starting material is chromosomal DNA, the library is called a **genomic library**. Similarly, it is also common for researchers to make a **cDNA library** that contains hybrid vectors with cDNA inserts.

In most cloning experiments, the ultimate goal is to clone a specific gene. For example, suppose that a geneticist wishes to clone the rat β-globin gene. To begin the cloning experiment, rat chromosomal DNA would be isolated from rat cells. This chromosomal DNA would be digested with a restriction enzyme, yielding thousands of DNA fragments. The chromosomal fragments would then be ligated

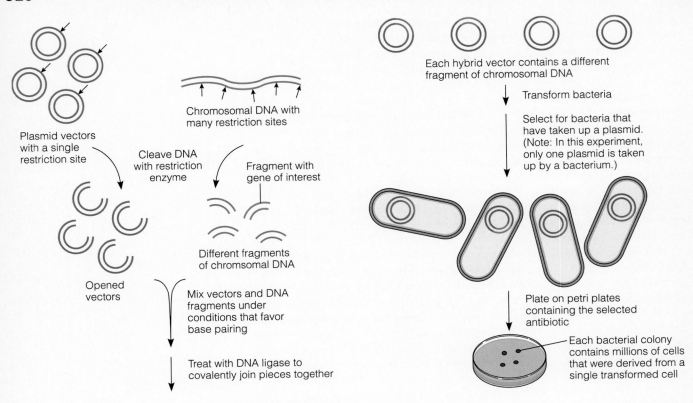

Plasmid vectors with a single restriction site

Cleave DNA with restriction enzyme

Chromosomal DNA with many restriction sites

Fragment with gene of interest

Opened vectors

Different fragments of chromsomal DNA

Mix vectors and DNA fragments under conditions that favor base pairing

Treat with DNA ligase to covalently join pieces together

Each hybrid vector contains a different fragment of chromosomal DNA

Transform bacteria

Select for bacteria that have taken up a plasmid. (Note: In this experiment, only one plasmid is taken up by a bacterium.)

Plate on petri plates containing the selected antibiotic

Each bacterial colony contains millions of cells that were derived from a single transformed cell

FIGURE **19-6**

The construction of a DNA library. The digestion of chromosomal DNA produces many fragments. The fragment containing the gene of interest is highlighted in red. Following ligation, each vector contains a different piece of chromosomal DNA.

to vector DNA and transformed into bacterial cells. Unfortunately, only a small percentage, perhaps one in a few thousand, of the hybrid vectors would actually contain the rat β-globin gene. For this reason, researchers must have some way to identify those rare bacterial colonies that happen to contain the cloned gene of interest, in this case, a colony that contains the rat β-globin gene.

Figure 19-7 describes a method for using a DNA probe to identify a bacterial colony that has the rat β-globin gene. This procedure is referred to as **colony hybridization**. The master plate shown at the top of Figure 19-7 has many bacterial colonies. Each bacterial colony is composed of bacterial cells containing a hybrid vector with a different piece of rat chromosomal DNA. The goal is to identify a colony that contains the rat β-globin gene. To do so, a nitrocellulose filter is laid gently onto the master plate containing many bacterial colonies. After the filter is lifted, some cells from each colony are attached to it. In this way, the nitrocellulose filter paper contains a replica of the colonies on the master plate.

The cells on the filter are then permeabilized and the DNA within the cells is denatured and fixed to the nitrocellulose filter. The filter is then submerged in a solution containing a radiolabeled DNA probe. In this case, the probe is a DNA strand that is complementary to the rat β-globin gene. The probe is given time to hybridize to the DNA on the filter. If and only if a bacterial colony contains the rat β-globin gene, the probe will hybridize to the DNA in this colony. Most bacterial colonies are not expected to contain this gene. The unbound probe is then washed away, and the filter is placed next to X-ray film. If the DNA within a bacterial colony did hybridize to the probe, a dark spot will appear on the film in the corresponding location. Therefore, a dark spot identifies a bacterial colony that contains

FIGURE **19-7**

A colony hybridization experiment. (1) A replica of the master plate is blotted onto a nitrocellulose filter. **(2)** The filter is treated with detergent (SDS) to permeabilize the bacteria. **(3)** The filter is then treated with sodium hydroxide (NaOH) to separate the DNA into single strands. **(4)** Radioactively labeled probes are added. Probes are single-stranded DNA with base sequences complementary to that of the gene of interest. **(5)** The probe will hybridize with the desired gene from the bacterial cell DNA. **(6)** The filter is washed to remove unbound probe and then exposed to X-ray film. **(7)** Finally, the developed film is compared with the replica of the master plate to identify bacterial colonies carrying the gene of interest.

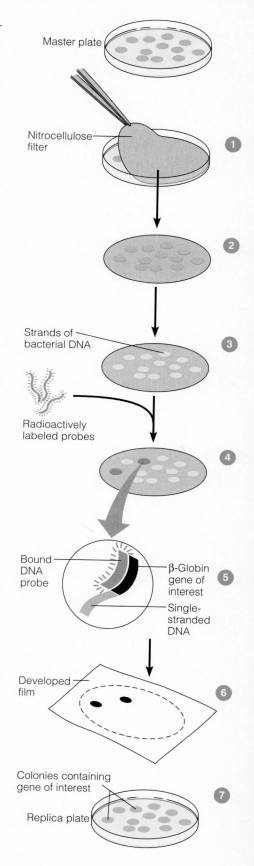

the rat β-globin gene. Following the identification of "hot" colonies, the experimenter can go back to the master plate (which contains living cells) to identify and grow bacteria containing the cloned gene.

There are several different strategies a scientist might use to obtain a probe for a gene cloning experiment. In some cases, the gene of interest may have already been cloned. A piece of the cloned gene then can be used as a radiolabeled probe. However, in many circumstances, a scientist may attempt to clone a gene that has never been cloned before. In such a case, one strategy is to use a probe that likely has a sequence similar to that of the gene of interest. For example, if the goal is to clone the β-globin gene from a South American rodent, it is expected that this gene is similar to the rat β-globin gene, which has already been cloned. Therefore, the cloned rat β-globin gene can be used as a DNA probe to "fish out" the β-globin gene from this species. In other cloning experiments, a scientist may be looking for a novel type of gene that no one else has ever cloned before from any species. In that case, a useful approach is to obtain antibodies that recognize the polypeptide encoded by the gene. Like DNA probes, radiolabeled antibodies can be used as probes in a colony hybridization experiment.

Techniques can be used to detect the binding of proteins to DNA sequences

In addition to detecting the presence of genes and gene products using blotting techniques, researchers often want to study the binding of proteins to specific sites on a DNA molecule. For example, the molecular investigation of transcription factors requires methods that can identify interactions between transcription factor proteins and specific DNA sequences. A technically simple, widely used method for identifying this type of interaction is the **gel retardation assay** (also known as the **band shift assay**). This technique was used originally to study rRNA–protein interactions and quickly became popular after its success in studying protein–DNA interactions in the *lac* operon. Now it is commonly used as a technique to detect interactions between eukaryotic transcription factors and DNA response elements.

The technical basis for a gel retardation assay is that the binding of a protein to a DNA fragment will retard the fragment's ability to move within a polyacrylamide or agarose gel. During electrophoresis, DNA fragments migrate through a gel by wriggling through the gel matrix in a wormlike fashion. Smaller fragments of DNA migrate more quickly through a gel matrix than do larger fragments. As you might expect, therefore, the binding of a protein to a DNA fragment will retard the DNA's rate of movement through the gel matrix, because a protein–DNA complex has a higher mass. When comparing a DNA fragment and a protein–DNA complex after electrophoresis, the complex will produce a band closer to the top of the gel. The protein–DNA complex is shifted to a higher band than the DNA alone, because the complex migrates more slowly to the bottom of the gel (see Figure 19-8).

A gel retardation assay must be carried out under nondenaturing conditions. This means that the buffers and gel may not allow the unfolding of proteins and the

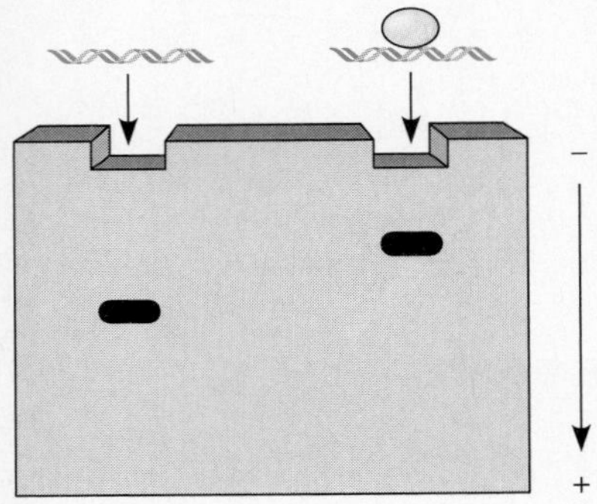

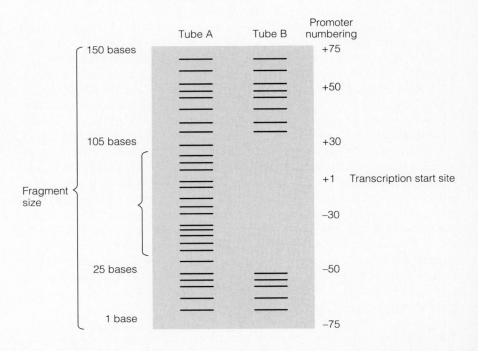

FIGURE **19-8**

The results of a gel retardation assay. The binding of protein to a fragment of DNA retards its rate of movement through a gel.

DNA double helix. This is necessary so that the proteins and DNA retain their proper structure and thereby can bind to each other. The nondenaturing conditions of a gel retardation assay differ from the more common SDS–gel electrophoresis, in which the proteins are denatured by the detergent SDS.

A second method for studying protein–DNA interactions is **DNA footprinting**. This technique was described originally by David Galas and Albert Schmitz in 1978 while working at the University of Geneva in Switzerland. A DNA footprinting experiment attempts to identify one or more regions of DNA that interact with a DNA-binding protein. In their original study, Galas and Smith identified a site in the *lac* operon, known as the *lac* operator site, that is bound by a DNA-binding protein called the *lac* repressor (see Chapter 15).

To understand the basis of a DNA footprinting experiment, we need to consider the molecular interactions among three things: a fragment of DNA, DNA-binding proteins, and agents that can alter DNA structure. As an example, let's examine the binding of RNA polymerase to a bacterial promoter, a topic discussed in Chapter 15. When the RNA polymerase holoenzyme binds to the promoter to form a closed complex, it binds tightly to the −35 and −10 promoter region, but the protein covers up an even larger region of the DNA. Thus, holoenzyme bound at the promoter prevents other molecules from gaining access to this region of the DNA. The enzyme DNaseI, which can cleave covalent bonds in the DNA backbone, is used as a reagent to determine if a DNA region has a protein bound to it. Galas and Schmitz reasoned that DNaseI cannot cleave the DNA at locations where a protein is bound. In the foregoing example, it is expected that RNA polymerase holoenzyme will bind to a promoter and protect this DNA region from DNaseI cleavage.

Figure 19-9 shows the results of a DNA footprinting experiment. In this experiment, a sample of many, identical cloned DNA fragments which are 150 bp in length were radiolabeled at only one end. The fragment sample was then divided into two tubes; tube A did not contain any holoenzyme, whereas tube B contained RNA polymerase holoenzyme. DNaseI was then added to both tubes. The tubes were incubated long enough for DNaseI to cleave the DNA at a single site in each DNA strand. Each tube contained many 150 bp DNA fragments, and the cutting in any DNA strand by DNaseI occurs randomly. Therefore, the DNaseI treatment should produce a mixture

FIGURE **19-9**

A DNA footprinting experiment. Both tubes contained 150 bp fragments of DNA that were incubated with DNaseI. Tube B also contained RNA polymerase holoenzyme. The binding of RNA polymerase holoenzyme protected a region of about 80 nucleotides (namely, the −50 region to the +30 region) from DNaseI digestion. Note: The promoter numbering convention shown here is the same as that described previously in Figure 13-2.

of many smaller DNA fragments. A key point, however, is that DNaseI will not cleave the DNA in a region where RNA polymerase holoenzyme is bound. After DNaseI treatment, the DNA fragments within the two tubes were separated by gel electrophoresis and DNA fragments containing the labeled end were detected by autoradiography.

In the absence of RNA polymerase holoenzyme (tube A), DNaseI should cleave the 150 bp fragments randomly at any single location. Therefore, a continuous range of sizes occurs (see Figure 19-9). In contrast, DNaseI should not be able to cleave the DNA in the region where RNA polymerase holoenzyme is bound. Looking at the gel lane from tube B, there are no bands in the size range from 25 to 105 nucleotides. These bands are missing because DNaseI cannot cleave the DNA within the region where the holoenzyme is bound. These results occur because the middle portion of the 150 bp fragment contains a promoter sequence that binds the RNA polymerase holoenzyme. The right side of the figure numbers the bases according to their position within the gene. (The +1 site is where transcription begins.) As seen here, RNA polymerase covers up a fairly large region of about 80 nucleotides, from the −50 region to the +30 region. As was discussed in Chapter 15, the promoter sequence is found at the −35 and −10 regions. The observation that RNA polymerase shields an 80 nucleotide region has been obtained for many other bacterial promoters.

As illustrated in this experiment, DNA footprinting can identify the DNA region that interacts with a DNA-binding protein. In addition to RNA polymerase–promoter binding, DNA footprinting has been used to identify the sites of binding for many other types of DNA-binding proteins, such as eukaryotic transcription factors and histones. This technique has greatly facilitated our understanding of protein–DNA interactions.

ANALYSIS AND ALTERATION OF DNA SEQUENCES

As we have seen throughout this text, our knowledge of genetics can be largely attributed to our understanding of DNA structure and function. The feature that underlies all aspects of inherited traits is the DNA sequence. For this reason, analyzing and altering DNA sequences is a powerful approach in understanding genetics. In this last section, we will begin by examining a technique called **DNA sequencing**. This method enables researchers to determine the base sequence of DNA found in genes and other chromosomal regions. It is one of the most important tools for exploring genetics at the molecular level. DNA sequencing is practiced by scientists around the world. The amount of scientific information that is contained within experimentally determined DNA sequences has become enormous. In the next chapter, we will learn how researchers can determine the complete DNA sequence of entire genomes. In Chapter 22, we will consider how computers play an essential role in the storage and analysis of genetic sequences.

Not only can researchers determine DNA sequences, another technique, known as **site-directed mutagenesis**, allows scientists to change the sequence of cloned DNA segments. At the end of this section, we will examine how site-directed mutagenesis is conducted, and how it provides information regarding the function of genes.

The dideoxy method of DNA sequencing is based on our knowledge of DNA replication

Molecular geneticists often seek to determine nucleotide sequences as a first step toward understanding the function and expression of genes. For example, the investigation of genetic sequences has been vital in our understanding of promoters, regulatory elements, and the genetic code itself. Likewise, an examination of sequences has facilitated our understanding of origins of replication, centromeres,

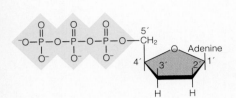

2′,3′-Dideoxyadenosine triphosphate

FIGURE **19-10**
The structure of a dideoxynucleotide.
Note that the 3′ group is a hydrogen rather than an —OH group. For this reason, another nucleotide cannot be attached at the 3′ position.

telomeres, and transposable elements. In this section, we will examine the technique of DNA sequencing used to determine the base sequence within a DNA strand.

During the 1970s, two methods for DNA sequencing were devised. One method, developed by Alan Maxam and Walter Gilbert while at Harvard University, involves the base-specific cleavage of DNA. Another method, developed by Frederick Sanger and colleagues at the Medical Research Council (MRC) in Cambridge, England, is known as **dideoxy sequencing**. Since it has become the more popular method of DNA sequencing, we will consider the dideoxy method here.

The dideoxy procedure of DNA sequencing is based on our knowledge of DNA replication with a clever twist. As was described in Chapter 11, DNA polymerase connects adjacent deoxynucleotides by catalyzing a covalent linkage between the 5′–phosphate on one nucleotide and the 3′–OH group on the previous nucleotide (refer back to Figure 11-9). Chemists, though, can synthesize deoxynucleotides that are missing the —OH group at the 3′ position. These synthetic nucleotides are called **dideoxynucleotides**. (Note: The prefix "dideoxy-" refers to the fact that there are two (di) removed (de) oxygens (oxy) compared with ribose; ribose has —OH groups at both the 2′ and 3′ positions.) Sanger reasoned that if a dideoxynucleotide is added to a growing DNA strand, the strand can no longer grow, because the dideoxynucleotide is missing the 3′–OH group (Figure 19-10). The incorporation of a dideoxynucleotide into a growing strand is therefore referred to as **chain termination**.

Before describing the steps of this DNA sequencing protocol, we need to become acquainted with the DNA segments that are used in a sequencing experiment. Prior to DNA sequencing, the segment of DNA to be sequenced must be obtained in large amounts. This is accomplished using gene cloning or PCR techniques, which were described earlier in this chapter. In Figure 19-11, the segment of DNA to be sequenced (which we will call the **target DNA**) was cloned into a vector at a defined location. In many DNA sequencing experiments, the target DNA is cloned into the vector at a site adjacent to a *primer annealing site*. The aim of the experiment is to determine the base sequence of the target DNA that has been inserted next to the primer annealing site. In the experiment shown in Figure 19-11, the vector DNA is from a virus called M13. The target DNA is cloned into the double-stranded form of the viral DNA. After cloning, the viral DNA is introduced into a host cell and it will produce single-stranded DNA as part of its life cycle.

With these ideas in mind, we can now describe the steps involved in DNA sequencing (see Figure 19-11). First, a sample containing many copies of the single-stranded DNA is mixed with primers that will bind to the primer annealing site. This annealing process is identical to hybridization, since the primer and primer annealing site are complementary to each other. All four types of deoxynucleotides and DNA polymerase are then added to the annealed DNA fragments, and the mixture is divided into four separate tubes. In addition to the four deoxynucleotides, each of the four tubes has a low concentration of a different dideoxynucleotide. The tubes are then incubated to allow DNA polymerase to make strands complementary to the target DNA sequence. In the third tube, which in this example contains ddGTP, chain termination can occasionally occur at the sixth or thirteenth position of the newly synthesized DNA strand if a ddG becomes incorporated at either of these sites. Note that the complementary C base is found at the sixth and thirteenth position in the target DNA. Therefore, in this tube, we would expect to make some DNA strands that will terminate at the sixth or thirteenth positions. Likewise, in the first tube, a ddATP can only cause chain termination at the second, seventh, eighth, or eleventh positions because a T is found at the corresponding positions in the target strand.

Within the four tubes, mixtures of DNA strands of different lengths have been made. These DNA strands can be separated according to their lengths by running them on an acrylamide gel. The shorter strands move to the bottom of the gel more quickly than the longer strands. To detect the newly made DNA strands, a small amount of

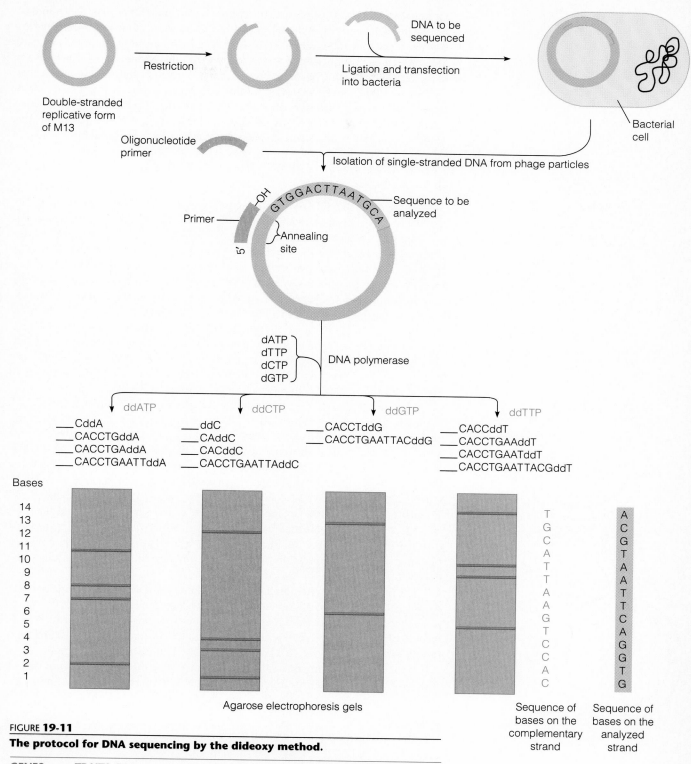

FIGURE **19-11**

The protocol for DNA sequencing by the dideoxy method.

GENES → TRAITS: DNA sequencing can often reveal important information regarding the relationship between genes and traits. For example, DNA sequencing has been used to compare the β-globin gene sequence from normal individuals versus that from people affected with sickle cell anemia. DNA sequencing has shown that the β-globin gene from affected individuals has a missense mutation that changes codon 6 from a glutamic acid codon to a valine codon. This amino acid substitution alters the function of β-globin and leads to the symptoms of sickle cell anemia.

radiolabeled deoxynucleotides are added to each reaction. This enables the strands to be visualized as bands when the gel is exposed to X-ray film. In Figure 19-11, the DNA strands in the four tubes were run in separate lanes on an acrylamide gel. Since we know which dideoxynucleotide was added to each tube, we also know which base is at the very end of each DNA strand separated on this gel. Therefore, it becomes possible to read the DNA sequence by reading which base is at the end of every DNA strand and matching this sequence with the length of the strand. Reading the base sequence, from bottom to top, is much like climbing a ladder of bands. Therefore, the sequence obtained by this method is referred to as a **sequencing ladder**.

In vitro site-directed mutagenesis is a technique to alter DNA sequences

As we have seen, the dideoxy technique provides a way to determine the base sequence of DNA. To understand how the genetic material functions, researchers often analyze mutations that alter the normal DNA sequence and thereby affect the expression of genes and the outcome of traits. For example, geneticists have discovered that many inherited human diseases, such as sickle cell anemia and hemophilia, involve mutations within specific genes. These mutations provide insight into the function of the genes in normal individuals. Hemophilia, for example, involves deleterious mutations in genes that normally encode blood clotting factors.

Since the analysis of mutations can provide important information about normal genetic processes, researchers often wish to obtain mutant organisms. Mutations can arise spontaneously; Mendel's pea plants are a classic example of allelic strains with different phenotypes that arose from spontaneous mutations. In addition, experimental organisms can be treated with mutagens that increase the rate of mutations.

More recently, researchers have developed molecular techniques to make mutations within cloned genes or other DNA segments. One widely used method, known as *in vitro* site-directed mutagenesis, allows a researcher to produce a mutation at a specific site within a cloned DNA segment. With this technique, a DNA sequence can be altered in a specific way. For example, if a DNA sequence is 5′–AAATTTCTTTAAA–3′, a researcher can use site-directed mutagenesis to change it to 5′–AAATTTGTTTAAA–3′; in this case, the researcher deliberately changed the seventh base from a C to a G. Since the sequence of DNA has been altered at a specific site, this approach is called site-directed or site-specific mutagenesis. The site-directed mutant can then be introduced into a living organism to see how the mutation affects the expression of a gene, the function of a protein, and the phenotype of an organism.

The first successful attempts at site-directed mutagenesis involved changes in the sequences of viral genomes. These studies were conducted in the 1970s. Since that time, researchers have devised methods for the mutagenesis of cloned DNA segments. A protocol for the site-directed mutagenesis of DNA that has been cloned into a viral vector was developed by Mark Zoller and Michael Smith. It is shown in Figure 19-12.

Prior to this experiment, a DNA fragment must be inserted into a viral vector such as M13 that synthesizes single-stranded DNA as a part of its natural life cycle.

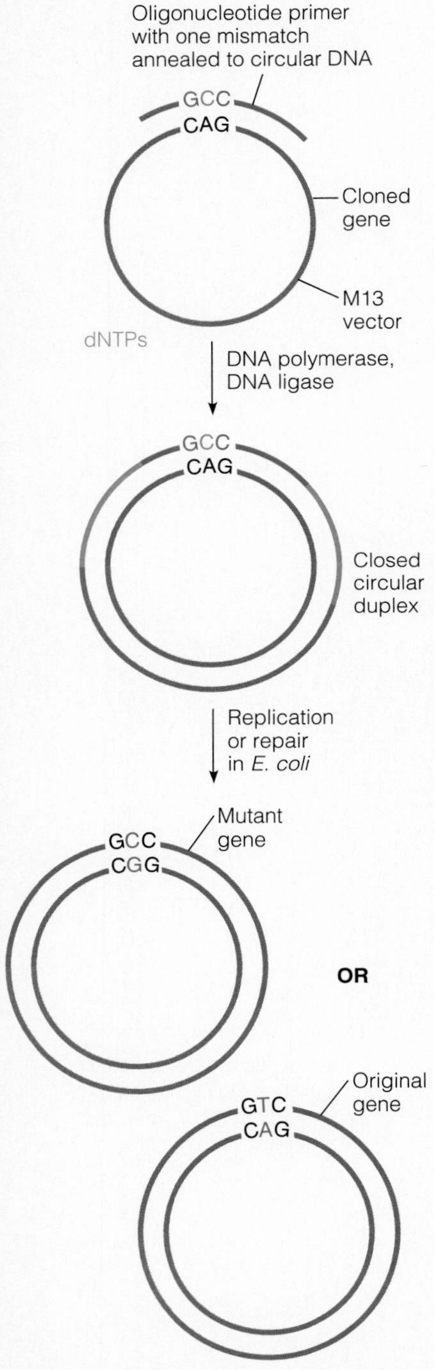

FIGURE 19-12

The method of site-directed mutagenesis.

GENES ⟶ TRAITS: To examine the relationship between genes and traits, researchers can alter gene sequences via site-directed mutagenesis. The altered gene can then be introduced into a living organism to examine how the mutation affects the organism's traits. For example, a researcher could introduce a nonsense mutation into the middle of the *lacY* gene in the *lac* operon. If this site-directed mutant was introduced into an *Escherichia coli* bacterium that did not a have a normal copy of the *lacY* gene, the bacterium would be unable to utilize lactose. These results indicate that a functional *lacY* gene is necessary for bacteria to have the trait of lactose utilization.

This single-stranded DNA can be isolated and used in an *in vitro* site-directed mutagenesis experiment. As in PCR, this single-stranded DNA is referred to as the template DNA, because it is used as a template to synthesize a complementary strand.

To begin the experiment, an oligonucleotide primer is allowed to hybridize or anneal to the template DNA. The primer, typically 20 or so nucleotides in length, is synthesized chemically. (A shorter version is shown in Figure 19-12 for simplicity.) The base sequence of the primer is determined by the scientist. The primer has two important characteristics. First, most of the sequence of the primer is complementary to the site in the DNA where the mutation is to be made. However, a second feature is that the primer contains a region of mismatch where the primer and template DNA are not complementary. The mutation will occur in this mismatched region. For this reason, site-directed mutagenesis is sometimes referred to as oligonucleotide-directed mutagenesis.

After the primer and template have annealed, the complementary strand is synthesized by adding deoxynucleoside triphosphates (dNTPs), DNA polymerase, and DNA ligase. This yields a double-stranded molecule that contains a mismatch only at the desired location. This double-stranded DNA is then transfected into a bacterial cell. Within the cell, the DNA mismatch will likely be repaired (see Chapter 17). Depending on which base is replaced, this may produce the mutant sequence or the original sequence. Clones containing the desired mutation can be identified by DNA sequencing and used for further studies.

After a site-directed mutation has been made within a cloned gene, its consequences are analyzed by introducing the mutant gene into a living organism. As described earlier in this chapter, hybrid vectors containing cloned genes can be introduced into bacterial cells. Following transformation or transfection, a researcher can study the differences in function between the mutant and wild-type genes and the proteins they encode. Similarly, mutant genes made via site-directed mutagenesis can be introduced into plants and animals as will be described in Chapter 21.

Recombinant DNA technology, the development of which began in the early 1970s, has revolutionized our understanding of molecular genetics. It is now possible to use *restriction enzymes* to cut DNA fragments out of their native sites within a chromosome. The DNA fragments can then be ligated to a *vector*, which propagates within a living *host cell*. This technology is known as DNA cloning, or if the fragment of DNA contains a gene, as *gene cloning*. In addition, *PCR* can produce many copies of a DNA fragment. DNA cloning has enabled researchers to study the structure and function of genes at the molecular level. Tens of thousands of different genes have been cloned from hundreds of different species. In addition, gene cloning has many practical applications, which will be described in Chapter 21.

In some cases, a researcher needs to detect a specific gene or gene product. For example, *gene detection* procedures are usually needed to identify particular cloned genes in a *DNA library*. In addition, detection of gene products can provide information regarding the expression pattern of a gene in particular cell types or during specific stages of development. Three common methods of detection are *Southern blotting*, *Northern blotting*, and *Western blotting*. In Southern blotting, a DNA *probe* is used to detect the presence of a gene or other DNA sequence within a mixture of DNA fragments. Under conditions of high stringency, this technique can determine if a species has a particular gene; under conditions of low stringency, if a gene is a member of a certain family. The Northern blotting procedure is used to detect the transcription of RNA from a specific gene. In this technique, a DNA probe is used to detect RNA. Both Southern and Northern blotting rely on sequence homology between the DNA probe and the DNA or RNA in the sample. They are called hybridization techniques, because the labeled DNA probe forms a hybrid with a specific molecule in the sample. Finally, Western blotting is used to detect the protein product from a particular gene. In this technique, an antibody is used as a probe because antibodies bind to proteins very specifically. Besides detection methods, researchers can use *gel retardation* and *DNA footprinting* as a way to study protein–DNA interactions.

Perhaps the greatest advance of recombinant DNA technology is that it has enabled researchers to determine the nucleotide base sequence of DNA. In the *dideoxy* method of *DNA sequencing*, dideoxynucleotides are used that terminate the action of DNA polymerase at specific locations in the growing DNA strands. When the terminated DNA strands are separated on the basis of their size by gel electrophoresis, this creates a *ladder* of DNA fragments that are terminated at defined locations. The DNA sequence can then be determined by reading the bases along the ladder. In addition, researchers can answer questions concerning the functional importance of DNA sequences by intentionally altering the DNA sequence via *site-directed mutagenesis*. In this procedure, a mutagenic DNA primer, which is synthesized chemically, is used to alter the DNA sequence at a particular site in a cloned fragment of DNA.

PROBLEM SETS

SOLVED PROBLEMS

1. A DNA strand has the sequence 3′–ATACGACTAGTCGGGAC-CATATC–5′. If the primer in a dideoxy sequencing reaction anneals just to the left of this sequence, draw what the sequencing ladder would look like.

Answer:

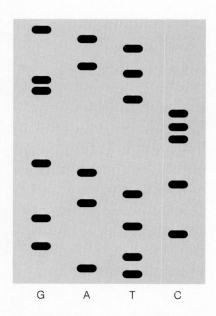

2. The human genetic disease PKU involves a defect in a gene that encodes the enzyme phenylalanine hydroxylase. It is inherited as a recessive autosomal disorder. Using the normal phenylalanine hydroxylase gene as a probe, a Southern blot was carried out on a PKU patient, one of their parents, and a normal unrelated person. The following results were obtained:

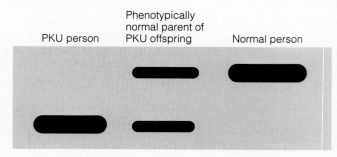

Suggest an explanation for these results.

Answer: In this person, the PKU defect is caused by a small deletion within the PKU gene. The parent is heterozygous for the normal gene and the deletion. The PKU-affected person only carries the deletion, which runs at a lower molecular weight than the normal gene.

CONCEPTUAL QUESTIONS

1. Discuss three important advances that have resulted from gene cloning.

2. What is a restriction enzyme? What structure does it recognize? What type of chemical bond does it cleave? Be as specific as possible.

3. Use your imagination and try to draw the active site of a restriction enzyme such as *Eco*RI with its substrate bound at the active site.

4. Write a sequence that is 20 nucleotides long and is palindromic.

5. What is cDNA? In eukaryotes, how would cDNA differ from genomic DNA?

6. Explain and draw the structural feature of a dideoxynucleotide that causes chain termination.

EXPERIMENTAL QUESTIONS

1. What is the functional significance of sticky ends in a cloning experiment? What types of chemical bonds make the ends sticky?

2. Describe the important features of cloning vectors. Explain the purpose of selectable marker genes in cloning experiments.

3. How does gene cloning produce many copies of a gene?

4. In your own words, describe the series of steps necessary to clone a gene. Your answer should include the use of a probe to identify a bacterial colony that contains the cloned gene of interest.

5. Let's suppose that you have recently cloned a gene, which we will call gene *X*. You use this cloned gene in a Southern blot experiment under conditions of low and high stringency and obtain the following results:

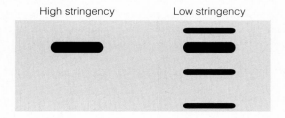

What do these results mean?

6. What is a DNA library? Do you think that this is an appropriate name?

7. Some vectors used in cloning experiments contain bacterial promoters that are adjacent to unique cloning sites. This makes it possible to insert a gene sequence next to the bacterial promoter and express the gene in bacterial cells. These are called *expression vectors*. If you wanted to express a eukaryotic protein in bacterial cells, would you clone genomic DNA or cDNA into the expression vector? Explain your choice.

8. In a Southern blot experiment, how might the results be different if the hybridization is conducted under conditions of high stringency compared with results under conditions of low stringency?

9. Southern and Northern blotting depend on the phenomenon of hybridization. In these two techniques, explain why hybridization occurs. What member of the hybrid is labeled?

10. In Southern, Northern, and Western blotting, what is the purpose of gel electrophoresis?

11. What is the purpose of a Northern blot experiment? What types of information can it tell you about the transcription of a gene?

12. If you wanted to know if a protein was made during a particular stage of development, what technique would you choose?

13. Explain the basis for using an antibody as a probe in a Western blot experiment. Why is a secondary antibody used? What does the secondary antibody recognize?

14. Starting with pig cells and a probe that is the human β-globin gene, describe how you would clone the β-globin gene from pigs. You may assume that you have available all the materials needed in a cloning experiment.

15. In Experiment 19A, explain how density gradient centrifugation was used to show that a hybrid vector had been produced. Are the electrophoresis results consistent with the idea that pSC105 is smaller than pSC102 but larger than pSC101? Explain.

16. A cloned gene fragment contains a response element that is recognized by a regulatory transcription factor. Previous experiments have shown that the presence of a hormone results in transcriptional activation by this transcription factor. To study this effect, you conduct a gel retardation assay and obtain the following results.

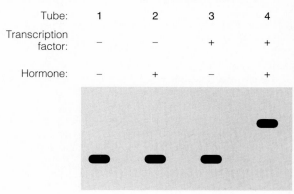

Explain the action of the hormone.

17. From a technical viewpoint, describe why "polymerase chain reaction" has been given its name.

18. Describe the rationale behind a gel retardation assay.

19. Certain hormones, such as adrenalin, can increase the levels of cAMP within cells. Let's suppose you can pretreat cells with or without adrenalin and then prepare a cell extract that contains the CREB protein (see Chapter 16 for a description of the CREB protein). You then use a gel retardation assay to analyze the ability of the CREB protein to bind to a DNA fragment containing a CRE. Describe what the expected results would be.

20. Explain the rationale behind a DNA footprinting experiment.

21. The following are the results from a DNA footprinting experiment using a 325 bp DNA fragment with or without a protein called protein X (Note: Protein X was only added to tube B):

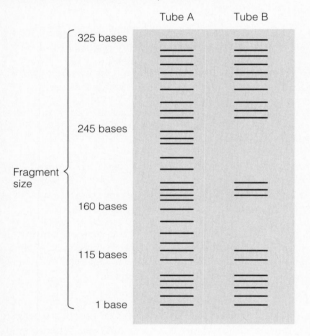

Based on these results, make a drawing that describes the binding of protein X to the DNA fragment.

22. Besides RNA polymerase, what other types of proteins are known to bind to DNA and could be studied by DNA footprinting?

23. In a footprinting experiment, why is it necessary to radiolabel only one end of a DNA fragment? What results would have been obtained in Figure 19-9 if both ends were radiolabeled?

24. In this chapter, we discussed DNA footprinting, which uses DNaseI, as a way to study protein–DNA interactions. In Experiment 10B, DNaseI was used to investigate chromatin structure. How are these approaches similar and different?

25. In a DNA sequencing experiment, a primer is 20 nucleotides in length and anneals immediately adjacent to the target DNA. The third base in the template strand of the target DNA is a guanine. How many nucleotides are there in the fastest running band in the ddC lane (i.e.,

the band on a sequencing film that is in the ddC lane and is at the bottom of the film)?

26. The following is a photograph of a DNA sequencing film from a dideoxy sequencing experiment:

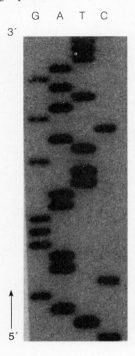

Start at the bottom of this film and read the sequence. Also, which end of this sequence is the 5′ and which is the 3′? To figure this out, you may want to think about the directionality of the primer shown in Figure 19-11.

27. In a Northern blot experiment, mRNA was isolated from four types of cells. The blot was probed with a DNA fragment containing gene X. The results are as follows:

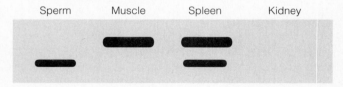

Based on these results, what would you conclude about the expression of gene X in these cells?

28. The following are the results from a Western blot experiment in which proteins were isolated from red blood cells, spleen cells, and muscle cells and then probed with an antibody that recognizes a protein known as myosin:

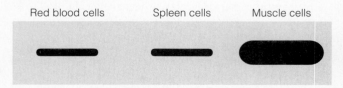

Explain what you think these results mean.

29. DNA sequencing can help us to identify mutations within genes. The following is an experiment in which a normal gene and a mutant gene have been sequenced:

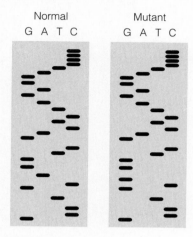

Normal
G A T C

Mutant
G A T C

Locate and describe the mutation.

30. Let's suppose you want to use site-directed mutagenesis to investigate a DNA sequence that functions as a response element for hormone binding. From previous work, you have narrowed down the response element to a sequence of DNA that is 20 base pairs in length with the following sequence:

5′–GGACTGACTTATCCATCGGT–3′
3′–CCTGACTGAATAGGTAGCCA–5′

As a strategy to pinpoint the actual response element sequence, you decide to make 10 different site-directed mutants and then analyze their effects by a gel retardation assay. What mutations would you make? What results would you expect to obtain?

31. Site-directed mutagenesis can also be used to explore the structure and function of proteins. For example, changes can be made to the coding sequence of a gene to determine how alterations in the amino acid sequence affect the function of a protein. Let's suppose that you are interested in the functional importance of one glutamic acid residue within a protein you are studying. By site-directed mutagenesis, you make mutant proteins in which this glutamic acid codon has been changed to other codons. You then test the encoded mutant proteins for functionality. The results are as follows:

	Functionality
Normal protein	100%
Mutant proteins containing:	
Tyrosine	5%
Phenylalanine	3%
Aspartic acid	94%
Glycine	4%

From these results, what would you conclude about the functional significance of the glutamic acid residue within the protein?

32. Part of a gene sequence is shown here:

5′– ATG CCC GAT GGC ATC GAT TTT CCT ATC –3′
     Met Pro Asp Gly Ile Asp Phe Pro Ile

Let's suppose you wanted to change the glycine residue to a glutamic acid. To do so, you make an oligonucleotide 20 bases in length to direct the mutation. Describe the sequence of the oligonucleotide that could accomplish this goal. Note: The mutagenic base(s) should be in the middle of the oligonucleotide.

QUESTIONS FOR STUDENT DISCUSSION/COLLABORATION

1. Discuss and make a list of some of the reasons that it would be informative for a geneticist to determine the amount of a gene product. Use specific examples of known genes (e.g., β-globin and other genes) when making your list.

2. Make a list of all the possible genetic questions that could be answered using site-directed mutagenesis.

GENOME ANALYSIS

The term **genome** refers to the total genetic composition of an organism. For example, the nuclear genome of humans is composed of 22 different autosomes and two sex chromosomes. In addition, humans have a mitochondrial genome that is composed of a single circular chromosome.

As genetic technology has progressed over the past few decades, researchers have gained increasing ability to analyze the composition of genomes as a whole unit. The term **genomics** refers the molecular analysis of the entire genome of a species. Genome analysis is a molecular dissection process applied to a complete set of chromosomes. During genome analysis, segments of chromosomes are cloned and analyzed in progressively smaller pieces, the locations of which are known on the intact chromosomes. This is the mapping phase of genome analysis. The mapping of the genome ultimately progresses to the determination of the complete DNA sequence. A complete DNA sequence provides the most detailed description of an organism's genome at the molecular level.

Currently, efforts are under way to map and sequence the genomes of many prokaryotes and eukaryotes. These include several bacterial species, *Caenorhabditis elegans* (a nematode), *Drosophila melanogaster*, *Arabidopsis thaliana* (a small plant), mice, and humans. In 1995, a team of researchers headed by Craig Venter at the Institute for Genomic Research in Gaithersburg, Maryland, and Hamilton Smith at Johns Hopkins obtained the first complete DNA sequence of the bacterial genome from *Hemophilus influenzae*. Its genome is composed of a single circular chromosome 1.83 million base pairs (bp) in length and contains approximately 1743 genes (Figure 20-1). In 1996, the first entire DNA sequence of a eukaryote, *Saccharomyces cerevisiae* (baker's yeast), was completed. This work was carried out by a European-

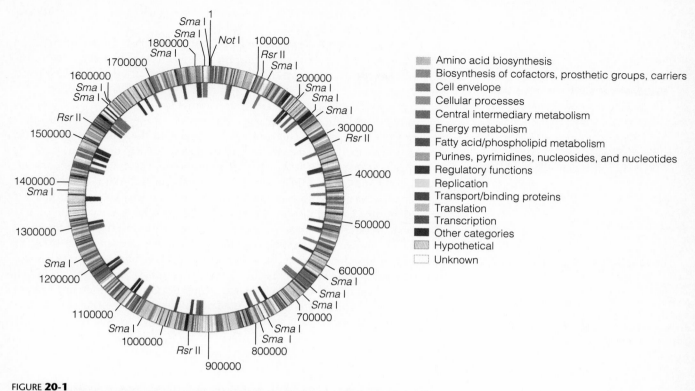

FIGURE **20-1**

A complete map of the genome of the bacterium *Hemophilus influenzae.* On the outside of the circular chromosome, the numbering of nucleotides begins at the top of the circle and proceeds clockwise in 100,000 nucleotide increments. The map also shows the locations of restriction enzyme sites for *Not*I, *Sma*I, and *Rsr*II. As shown in the key, the general categories of genes are color coded.

GENES ⟶ TRAITS: By mapping and DNA sequencing the entire genome of a species, researchers can identify all the genes it possesses. It is the expression of these many genes that ultimately determines an organism's inherited traits.

led consortium of more than 100 laboratories, including some in the United States, Japan, and Canada. The overall coordinator was Andre Goffeau of the University of Louvain in Belgium. The yeast genome contains 16 linear chromosomes; these 16 chromosomes have a combined length of about 12 million bp, which contain approximately 6000 genes.

A large portion of this chapter will be devoted to experimental strategies aimed at mapping a genome. We will examine three different mapping approaches: cytogenetic mapping, linkage mapping, and physical mapping. Before we discuss these approaches in detail, it is helpful to get an overview by comparing them.

Cytogenetic (also called cytological) mapping relies on the localization of gene sequences within chromosomes that are viewed microscopically. In most cases, each chromosome has a characteristic banding pattern, and genes are mapped cytologically relative to a band location. This type of mapping is more refined in organisms with polytene chromosomes, such as *Drosophila*, because their polytene chromosomes have a very detailed banding pattern (see Chapter 8).

By comparison, in Chapter 5, we considered how genetic crosses are conducted to map the relative locations of genes within a chromosome. Such genetic studies, which are called linkage mapping, use the frequency of genetic recombination between different genes to determine their relative order along a chromosome. In eukaryotes, linkage analysis involves crosses among organisms who are heterozygous for two or more genes. The number of recombinant offspring

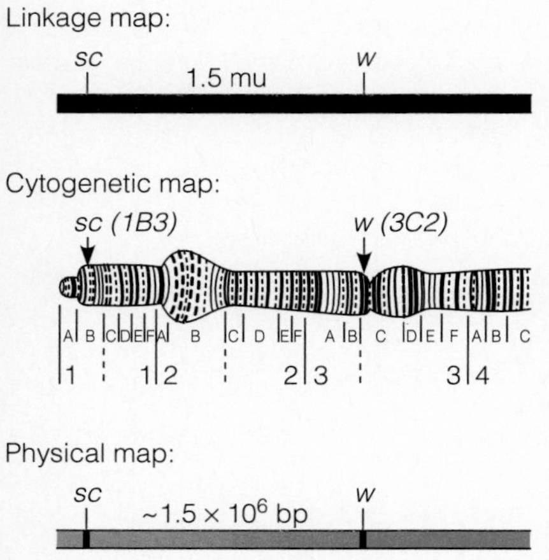

Linkage map:

Cytogenetic map:

Physical map:

FIGURE **20-2**

A comparison of linkage, cytological, and physical maps. Each of these maps shows the distance between the *sc* and *w* genes in *Drosophila melanogaster*.

provides a relative measure of the distance between genes, which is computed in map units (or centimorgans).

Finally, genes can be physically mapped. In this approach, DNA cloning techniques are used to determine the location of and distance between genes. In a physical map, the distances are computed in the number of nucleotide base pairs between genes.

Figure 20-2 compares the map distance between two genes, *sc* (scute, an abnormality in bristle formation) and *w* (white eye) in *Drosophila melanogaster* according to linkage, cytogenetic, and physical maps. In the linkage map, genetic crosses indicate that the two genes are approximately 1.5 map units (mu) apart. In the cytogenetic map, analysis of polytene chromosomes has shown that the *sc* gene is located at band 1B3 and the *w* gene is located at band 3C2. With regard to physical mapping, the two genes are approximately 1.5×10^6 base pairs apart from each other along the X-chromosome. Correlations between linkage, cytogenetic, and physical maps often vary from species to species and from one region of the chromosome to another. For example, a distance of 1 mu may correspond to 10^6 bp in one region of the chromosome, but other regions may recombine at a much lower rate so that a distance of 1 mu may be a much longer physical segment of DNA.

In this chapter, we will explore several techniques aimed at producing cytogenetic, linkage, and physical maps. At the end of this chapter, we will consider an ongoing worldwide project to analyze the human genome.

CYTOGENETIC MAPPING

Cytogenetic mapping provides a way to determine the locations of genes by microscopically examining the chromosomes. It is more commonly used in eukaryotes, which have much larger chromosomes. Cytologically, eukaryotic chromosomes can be distinguished by their size, centromeric location, and banding patterns. By treating chromosomal preparations with particular dyes, a discrete banding pattern is obtained for each chromosome. Cytogeneticists use this banding pattern as a way to describe locations along each chromosome. This topic was discussed in Chapter 8; you may wish to review Figure 8-1.

Cytogenetic mapping attempts to determine the location of a particular gene relative to a banding pattern along a chromosome. For example, the human gene that encodes phenylalanine hydroxylase, the enzyme that is defective in people with phenylketonuria (PKU), is located on chromosome 12, at the band designated q24. Cytogenetic mapping is often used as a first step in the localization of genes in plants and animals. However, since they rely on light microscopy, cytological studies have a fairly crude limit of resolution. In most species, cytogenetic mapping is only accurate within limits of approximately 5,000,000 bp along a chromosome. In species that have polytene chromosomes, such as *Drosophila*, the resolution is much better.

A common strategy used by geneticists is to roughly locate a gene by cytogenetic analysis and then determine its location more precisely by the physical mapping methods described later in this chapter. In this section, we will consider several different approaches used to cytologically map genes along eukaryotic chromosomes.

In situ hybridization can localize genes along particular chromosomes

In situ **hybridization** can detect the location of a gene at a particular site within an intact chromosome. This method is used widely to map cytologically the locations of genes or other DNA sequences within large eukaryotic chromosomes. The term *in situ* (from the Latin for "in place") indicates the procedure is conducted on chromosomes that are being held in place (i.e., adhered to a surface).

To map a gene via *in situ* hybridization, researchers use a probe to detect the location of the gene within a set of chromosomes. Most often, the gene of interest has been cloned previously as described in Chapter 19. The DNA of the cloned gene is used as a probe to find where the same gene is located within a set of chromosomes. Because the cloned gene, which is a very small piece of DNA relative to a chromosome, will hybridize only to its complementary sequence on a particular chromosome, this technique provides an accurate location for any cloned gene. For example, let's consider the gene that causes the white-eye phenotype in *Drosophila*. This gene has already been cloned. If a piece of this cloned DNA is mixed with intact *Drosophila* chromosomes, it will bind only to the X-chromosome, at the location corresponding to the site of the white-eye gene.

The most common method of *in situ* hybridization uses fluorescently labeled DNA probes. This specialized form of *in situ* hybridization is referred to as **fluorescence *in situ* hybridization** (**FISH**). Figure 20-3 describes the steps of the FISH procedure. The cells are prepared using a technique that keeps the chromosome intact. A DNA probe is then added. For example, the added DNA probe might be complementary to a specific gene. In this case, the goal of a FISH experiment is to determine the location of the gene within a particular set of chromosomes.

To detect where the probe has bound to a chromosome, the probe is subsequently tagged with a fluorescent molecule (step 5). A *fluorescent molecule* is one that absorbs light at a particular wavelength and then emits light at a longer wavelength. To detect the light emitted by a fluorescent probe, a fluorescence microscope is used. Such a microscope contains filters that allow the passage of light only within a defined wavelength range. The sample is illuminated at the absorption wavelength of the fluorescent molecule and then viewed through a filter that transmits light within the emission wavelength of the molecule. Since the fluorescent probe binds only to a specific sequence, it is seen as a colorfully glowing region against a colorless, dark background. Because of this background, the slightest amount of color is readily detectable. This makes fluorescence microscopy a very sensitive technique; it is widely used in genetics and cell biology. FISH has also gained widespread use in research and clinical applications.

Figure 20-4 illustrates the results of an experiment involving six different DNA probes. The six probes were cloned pieces of DNA corresponding to six different fragments of DNA from human chromosome 5. In this experiment, each probe was labeled with a different combination of fluorescent molecules. This enabled researchers to distinguish the probes when they bound to their corresponding locations on chromosome 5. In this experiment, computer imaging methods were used

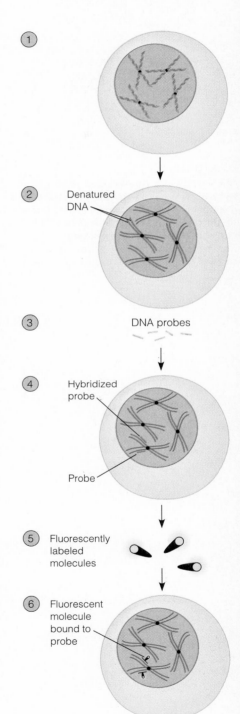

FIGURE 20-3

The technique of fluorescence *in situ* hybridization (FISH). (1) The cells are treated with agents that make them swell and are then fixed on slides. **(2)** The chromosomal DNA is denatured to single-stranded regions. **(3)** The single-stranded probe is added. **(4)** The DNA probe hybridizes to the denatured chromosomal DNA at specific sites that are complementary to the probe. **(5)** A fluorescently labeled molecule that binds to the probe DNA is added. (The probe DNA has been chemically modified to allow the fluorescent label to attach specifically to probe DNA.) **(6)** The preparation is viewed under a fluorescence microscope. Note that the chromosomes are highly condensed metaphase chromosomes that have already replicated. Therefore, each X-shaped chromosome actually contains two copies of a particular chromosome. These are sister chromatids. Since the sister chromatids are identical, a probe that recognizes one sister chromatid will also bind to the other. In some cases, the sister chromatids are so close to each other that the fluorescence looks like a single colored spot. In other cases, you can see two distinguishable adjacent spots from the probe binding to both sister chromatids.

Results of FISH experiment using six DNA probes that bind to human chromosome 5.

Probe #:	Color:
1	white
2	orange
3	green
4	red
5	yellow
6	pink

FIGURE **20-4**

The results of a fluorescence *in situ* hybridization experiment. In this experiment, six different probes were used to locate six different sites along chromosome 5. The colors are due to computer imaging of fluorescence emission; they are not the actual colors of the fluorescent labels.

to assign each fluorescently labeled probe a different color. In this way, fluorescence *in situ* hybridization discerns the sites along chromosome 5 corresponding to the six different probes. Referring to the probes by number, probe 1 (white) labels a site at the end of chromosome 5. Probe 2 (orange) is located closest to probe 1, followed by probe 3 (green), probe 4 (red), and then probe 5 (yellow). Probe 6 (pink) is located at the opposite end of chromosome 5. In a visual, colorful way, this technique was used here to determine the order of and relative distances between several specific sites along a single chromosome.

Somatic cell hybrids can be used to map human genes

Another cytological method of locating genes uses **cell hybrids,** which are made by fusing together two different cell types. Cell hybrids are used most commonly to map the locations of human genes to specific chromosomes. A **somatic cell hybrid** is produced in the laboratory by fusing together two different types of somatic cells. This method was first developed in 1960 by Georges Barsky and his colleagues in Paris, who were able to fuse two different mouse cell lines.

It soon became apparent that somatic cell hybrids could also be made between different species. In particular, human cells could be fused with rodent (rat or mouse) cells. A peculiar and useful characteristic of human–rodent hybrids is that they randomly discard the human chromosomes (Figure 20-5). This results in hybrid cell lines that retain only a few human chromosomes. For example, one hybrid cell line may retain human chromosomes 2, 14, and 17 while a different cell line may retain human chromosomes 1, 4, 5, 17, and 18. The human chromosomal composition of different hybrid lines can be determined by light microscopy.

The variation in the composition of human chromosomes retained among different somatic cell hybrids makes it possible to deduce the locations of human genes on particular chromosomes. If a hybrid cell is missing certain human chromosomes, it will be missing the genes that are located on those chromosomes. Therefore, a geneticist can analyze a collection of different somatic cell hybrids to see if they contain a particular human gene or gene product (i.e., polypeptide or protein encoded by a gene). Those hybrids that do contain the gene or gene product of interest must have retained the human chromosome that carries that gene. In contrast, those hybrids that do not contain the human gene or do not make the gene product have discarded that human chromosome.

Table 20-1 considers a hypothetical analysis of a group of human–mouse hybrids that contain differing combinations of human chromosomes. In this example, the human gene we wish to map encodes an enzyme that is distinguishable from any mouse enzymes; the human gene product, an enzyme, can be detected biochemically. In the data of Table 20-1, the experimenter has analyzed several different hybrid cell lines to see if they are producing this human enzyme. Some of them do, some do not. Likewise, the same hybrid cell lines have been subjected to a cytological examination to determine which human chromosomes they contain. By comparing the two sets of data, the researcher can deduce which human chromosome carries the gene that encodes this enzyme. In this example, the presence of the gene product is always correlated with the presence of chromosome 9. When this chromosome is missing from a hybrid, the human enzyme is not made. Therefore, these results indicate the gene that codes for this enzyme maps to human chromosome 9.

In the example of Table 20-1, a gene was mapped to a particular chromosome. In some cases, this method can map genes to more specific regions of human chromosomes. For example, some somatic cell hybrids contain human chromosomes that have suffered a large deletion. Such abnormalities can be detected cytologically. If the absence of a gene product is correlated with the lack of a chromosomal segment, the gene can be mapped to the deleted segment. Simi-

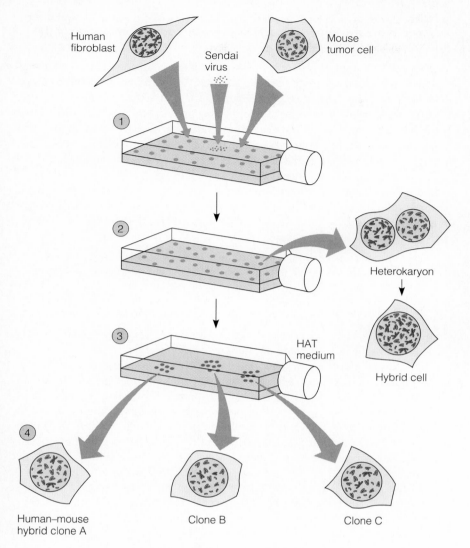

FIGURE **20-5**

Producing human–mouse somatic cell hybrids. (1) Two different cell lines are grown in a laboratory and mixed together. Certain agents, such as chemicals and viruses, can cause the cells to fuse with each other. For example, a virus called Sendai virus can promote cell fusion. **(2)** Initially, after fusion, the cells contain two separate nuclei and are known as heterokaryons. **(3)** The nuclei then fuse with each other to form a hybrid cell that can grow on selective media such as HAT medium (HAT is an acronym for the chemicals hypoxanthine, aminopterin, and thymidine, which are contained within this medium.) The presence of aminopterin requires the cells to synthesize nucleotides by a pathway involving the enzymes thymidine kinase (TK) and hypoxanthine guanine phosphoribosyl transferase (HGPRT). If the human cell is TK+ and HGPRT− while the rodent cell is TK− and HGPRT+, this medium will only allow human–rodent cell hybrids (which are TK+ and HGPRT+) to grow. **(4)** During subsequent cell divisions, it is likely that human chromosomes will be randomly lost, yielding hybrid lines that contain varying collections of only a few human chromosomes.

TABLE **20–1 Gene Mapping by Somatic Cell Hybrid Analysis**

Hybrid Line	Human Chromosomes in the Hybrids																							Gene Product in the Hybrids
	1	2	3	4	5	6	7	8	9	10	11	12	13	14	15	16	17	18	19	20	21	22	X	
1		+							+			+					+				+			+
2			+								+						+							−
3			+											+			+		+					−
4					+		+		+		+					+		+						+
5	+								+	+			+				+							+
6		+															+							−

larly, some human–rodent hybrid cells contain abnormal chromosomes in which a piece of a human chromosome has been translocated to a rodent chromosome. If the presence of this abnormal chromosome is correlated with the synthesis of a human gene product, the gene maps to this translocated piece of the human chromosome.

In 1936, Demerec and Hoover were able to detect microscopically small changes in the structure of *Drosophila* polytene chromosomes and use this information to map genes

During the 1930s, many geneticists became excited by the potential usefulness of polytene chromosomes in cytogenetic analyses. The structure of polytene chromosomes was described in Chapter 8 (see Figure 8-18). Much of this excitement was generated by the pioneering studies of Theophilus Painter at the University of Texas. He began his study of polytene chromosomes in the fall of 1932. Painter was interested in the relationship between chromosome structure and the location of genes. In the 1930s, molecular techniques for analyzing gene and chromosome structure had not yet been invented. Instead, the scientists of this time had to rely on genetic and light microscopic analyses to elucidate the relationships between chromosome structure and gene organization. Unfortunately, the ability to view normal metaphase chromosomes in most eukaryotic species was severely limited. In his own words, Painter wrote:

> Ever since the formulation of the chromosome theory of heredity, cytologists and geneticists alike have dreamed of the day when someone would find somewhere an organism in which the chromosomes were so large that it would be possible to see qualitative differences along their lengths corresponding to the different genes which we know must reside there.

In his search for such an organism, Painter decided to investigate the large structures found in the salivary gland of *Drosophila melanogaster*.

Painter and his colleagues determined that polytene chromosomes exhibit a reproducible banding pattern. After the banding pattern of polytene chromosomes from normal flies had been resolved, Painter wanted to correlate the structure of polytene chromosomes with that of normal metaphase chromosomes (found in other cells of the fruit fly). He first examined the salivary cells of a fruit fly known to have a break in its X-chromosome. In this fly, one of the arms of the polytene chromosome was different from that of normal flies. This arm of the polytene chromosome, Painter deduced, must correlate with the X-chromosome.

In this experiment, we will consider another very early genetic study conducted by M. Demerec and Margaret Hoover at Cold Spring Harbor in 1936. It illustrates the precision that can be achieved in mapping the locations of deletions relative to the bands on polytene chromosomes. Prior to this study, genetic crosses had produced a map of the X-chromosome that showed the *y* (yellow body) gene at the end, followed by *ac* (achete, an allele that affects bristle formation), *sc* (scute), and *svr* (an allele that *s*uppresses *ver*million eye color), as shown here:

Polytene arm of X-chromosome

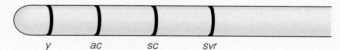

y   ac   sc   svr

In addition, two different mutant strains that carried deletions had been identified in genetic crosses. One strain, designated 260-1, was missing *y*, *ac*, and *sc*, but not *svr*. The other mutant, designated 260-2, was missing *y* and *ac*, but not *sc* and *svr*. In addition, the routine observation of polytene chromosomes in normal flies identified a third mutant strain, designated 260-5. This strain was not missing any of the four genes, but in polytene chromosomes it was possible to see that the very tip of its X-chromosome was missing. In the experiment described here, Demerec and Hoover used these three strains to cytologically map the positions of the *y*, *ac*, and *sc* genes along the polytene chromosome.

An examination of polytene chromosomes in mutant flies may reveal structural alterations that make it possible to determine the locations of particular genes.

TESTING THE HYPOTHESIS

Starting material: Three strains of *Drosophila* flies, 260-1, 260-2, and 260-5. Previous genetic crosses showed that 260-1 is missing the *y*, *ac*, and *sc* genes, but not the *svr* gene. 260-2 is only missing *y* and *ac*. 260-5 is not missing any of these genes.

| **Experimental Level** | **Conceptual Level** |

1. Carry out the appropriate crosses to produce female flies with one normal X-chromosome and one abnormal chromosome containing the 260-1, 260-2, or 260-5 deletion.

2. Dissect the salivary gland from the larva.

3. Cytologically examine the polytene chromosomes in the wild-type and deletion heterozygotes by staining the cells with an acidic solution containing a dye such as orcein or carmine.

4. Make "squashes" of the salivary cells—squash the cells under a coverslip on a slide.

5. View the polytene chromosomes under a light microscope.

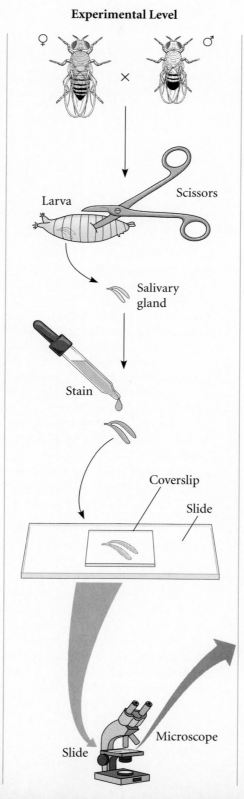

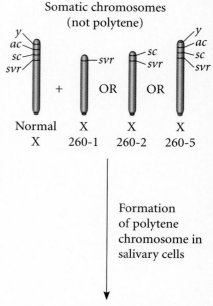

Somatic chromosomes (not polytene)

Formation of polytene chromosome in salivary cells

X-arm of a polytene chromosome

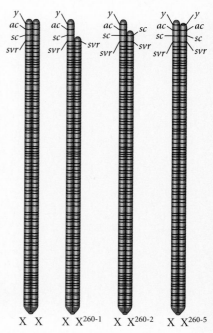

The left side of each polytene chromosome is composed of about 500 copies of the normal X-chromosome. The right side is composed of about 500 copies of the other X-chromosome.

THE DATA

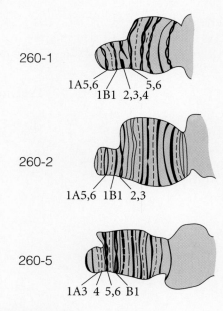

260-1

1A5,6   5,6
1B1 2,3,4

260-2

1A5,6 1B1 2,3

260-5

1A3 4 5,6 B1

INTERPRETING THE DATA

Strain 260-1, which was known to be missing the *y*, *ac*, and *sc* genes, lacked all of the 1A bands and the 1B1–1B4 bands, but not the 1B5 band. This deletion localizes these three genes to the 1A1–1B4 region of the polytene chromosome. Strain 260-2, which is only missing the *y* and *ac* genes, was found to lack all of the 1A bands and the 1B1 and 1B2 bands; this strain has the 1B3–1B4 bands. Since 260-2 has the *sc* gene but 260-1 does not, these results indicate that the *sc* gene is somewhere in the 1B3–1B4 bands. (Later studies placed it in the 1B3 band.) The 260-5 strain is not missing any of these genes, but it lacks the 1A1–1A4 bands. These results tell us that the *y* and *ac* genes must be somewhere in the 1A5–1B2 region.

This early gene mapping experiment illustrates the detailed information that can be provided by analyzing polytene chromosomes. Since that time, it has even become possible to map small deletions that involve a single band in a polytene chromosome. This has enabled *Drosophila* geneticists to construct very detailed maps that correlate linkage map data (from crosses) with cytogenetic data (from the analysis of polytene chromosomes). This is one of the reasons that *Drosophila* is one of the favorite organisms of geneticists.

GENETIC LINKAGE MAPPING USING MOLECULAR MARKERS

Genetic linkage mapping attempts to determine the distance between sites that are located along the same chromosome. The linkage mapping described in Chapters 5 and 6 used allelic differences between genes to map the relative locations of those genes along a chromosome. As an alternative to gene mapping, geneticists have realized that fragments of DNA (which need not encode genes) can be used as genetic markers along a chromosome. A **molecular marker** is a segment of DNA that is found at a specific site in the genome and has properties that enable it to be uniquely recognized using molecular tools such as gel electrophoresis. As with alleles, the characteristics of molecular markers vary from individual to individual. Therefore, the distances between linked molecular markers can be determined from the outcomes of crosses. We will see that, for many species, it is much easier to identify many molecular markers rather than identify many allelic differences among individuals. For this reason, geneticists have increasingly turned to molecular markers as points of reference along genetic maps.

There are many different types of DNA fragments that can be used as markers along a chromosome. In this section, we will focus on one particular type of molecular marker, called **restriction fragment length polymorphisms (RFLPs)**. As shown in Figure 20-6, an RFLP is a type of genetic variation that affects the lengths of DNA fragments produced when chromosomes are digested with restriction enzymes. The three individuals shown here differ: the one on the left is homozygous for a short RFLP, the one in the middle is homozygous for a long RFLP, and the one on the right is heterozygous.

In this section, we will begin with an explanation of RFLPs and how they are detected via molecular techniques. We will then consider experiments that examine how the distance between two or more RFLPs along the same chromosome can be determined. Finally, we will consider how RFLPs can be used to map and clone genes.

Restriction fragment length polymorphisms arise from differences in the numbers and locations of restriction enzyme sites among individuals

As we discussed in Chapter 19, restriction enzymes recognize specific DNA sequences and cleave the DNA at those sequences. Along a very long chromosome, a particular restriction enzyme will recognize many sites. For example, a commonly used restriction enzyme, *Eco*RI, recognizes GAATTC. Simply by chance, this six nucleotide sequence is expected to occur (on average) every 4^{-6} or once in every 4096 nucleotides. A chromosome composed of millions of nucleotides contains many *Eco*RI sites, which will be randomly distributed along the chromosome. Therefore, *Eco*RI will digest a chromosome into many smaller pieces of different lengths.

As we have seen throughout this text, genetic variation arises from differences in DNA sequences. For example, two alleles of the same gene differ from each other because they have different DNA sequences. The same is true with regard to restriction enzyme sites. An *Eco*RI site, GAATTC, may mutate to become GAAGTC. This mutant site will no longer be recognized by *Eco*RI. Likewise, when comparing DNA sequences within a population, each individual will display different patterns of restriction enzyme sites along their chromosomes. Therefore, when the DNA fragments from a restriction digestion are separated by gel electrophoresis, there will be differences in the lengths of certain DNA fragments between two individuals. We say that there is **polymorphism** in the population with regard to the locations of particular restriction enzyme sites.

The identification of a restriction fragment length polymorphism (RFLP) is shown in Figure 20-7. This simplified example considers a short region of chromosomal DNA from three people. Since humans are diploid, each individual has two copies of this region. In this experiment, the DNA has been isolated and cut by *Eco*RI. The DNA from individual #1 is cut at six sites by *Eco*RI, whereas individual #2 has only five locations where the DNA is cut. In individual #2, an *Eco*RI site is missing because the sequence of DNA in this region has been changed (from GAATTC), so that it is not recognized by *Eco*RI. Compared with individuals #1 and #2, individual #3 is a heterozygote; one chromosome is missing the *Eco*RI site, the other is not.

When subjected to gel electrophoresis, the change in the DNA sequence at a single *Eco*RI restriction site produces a variation in the sizes of homologous DNA fragments. The term polymorphism (meaning many forms) refers to the idea that the individuals within a population differ with regard to these particular DNA fragments. As noted in Figure 20-7, not all restriction sites are polymorphic. The three individuals share many DNA fragments that are identical in size. When a DNA segment is identical among all members of a population, it is said to be **monomorphic** (meaning one form). By comparison, if there is genetic variation in a population for a certain DNA region, it is said to be polymorphic.

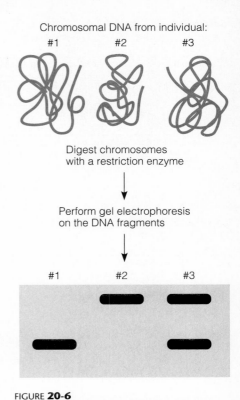

Chromosomal DNA from individual:

#1   #2   #3

Digest chromosomes
with a restriction enzyme

Perform gel electrophoresis
on the DNA fragments

#1   #2   #3

FIGURE **20-6**

Restriction fragment length polymorphisms (RFLPs).

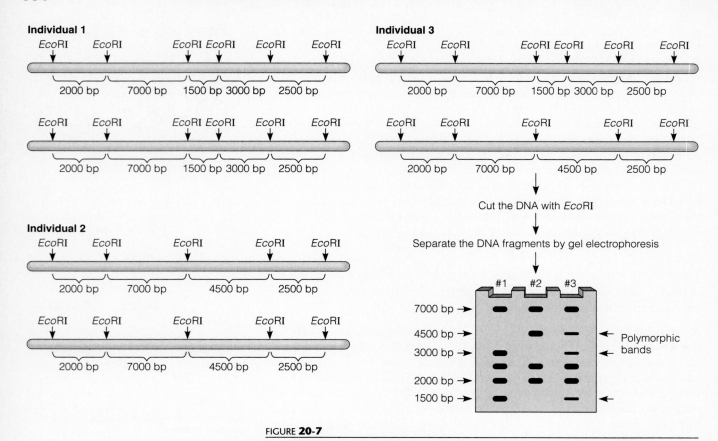

FIGURE **20-7**
An RFLP analysis of chromosomal DNA from three different individuals.

As a practical rule of thumb, a DNA segment is considered monomorphic when over 99% of the individuals in the population have identical sequences at that segment. In Figure 20-7, the polymorphism affects the lengths of particular DNA fragments generated by restriction enzyme digestion. Therefore, when the length of a particular DNA fragment varies within a population, this is called a restriction fragment length polymorphism.

The preceding discussion was an overview of the phenomenon of RFLPs. The experiment of Figure 20-7 is a technical oversimplification, because it only considers a very short region of chromosomal DNA. In an actual RFLP analysis, DNA samples containing all of the chromosomal DNA would be isolated from these three individuals. The digestion of the chromosomal DNA with *Eco*RI would then yield so many fragments that the results would be very difficult to analyze. To circumvent this problem, Southern blotting is used to identify specific RFLPs.

Figure 20-8 illustrates such an experiment that reconsiders the chromosomes from the three individuals described in Figure 20-7. A cloned piece of DNA that corresponds to the region near the fourth *Eco*RI site is used as a radiolabeled probe in a Southern blot of the chromosomal DNA. When the blot is exposed to X-ray film, we will see only the DNA band that can hybridize to the radiolabeled probe. As shown in Figure 20-8, individual #1 shows one band of 3000 bp and individual #2 has a band that is 4500 bp long. As we discussed in regard to Figure 20-7, this difference is due to the absence of an *Eco*RI site in individual #2.

Individual #3 has both bands because of heterozygosity for this *Eco*RI site. Since RFLPs are analyzed at the molecular level, they are always inherited in this codominant manner. In other words, a heterozygote will have two bands of different lengths, whereas homozygotes will only display one band, because the RFLP procedure detects the molecular products of each chromosome.

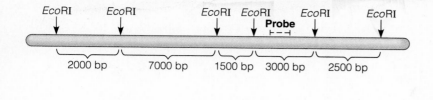

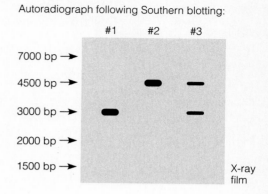

FIGURE **20-8**

Southern blot hybridization of a specific RFLP. The left side of the figure shows the region of the chromosome that was used as a probe in a Southern blot hybridization. The radiolabeled probe is a DNA segment located just to the right of the fourth EcoRI site. In this experiment, a Southern blot was conducted on the chromosomal DNA from the same three individuals described in Figure 20-7.

The distance between two linked RFLPs can be determined

Now that we understand what an RFLP is, let's consider how they are mapped and used as chromosomal markers. In the genetic mapping experiments described in Chapter 5, we learned how the results of genetic crosses are used to map the distance between two genes. In that type of analysis, the proportions of offspring with recombinant phenotypes were used to calculate the map distance between genes. Likewise, we can map the distance between two RFLPs by making crosses and analyzing the offspring. In an RFLP analysis, however, we do not look at the phenotypic characteristics of the offspring (e.g., white eyes or miniature wings). Instead, we look at their DNA bands on a gel. To do this, DNA is obtained from a tissue sample, such as a small sample of blood.

Figure 20-9 presents an example of how RFLP mapping is done. Let's suppose there are two RFLPs that are located at different sites in the genome. One RFLP is detected with probe #1 and yields either a 4500 bp band or a 6500 bp band. A second RFLP is detected with probe #2 and yields either a 2000 bp band or a 1500 bp band. In the experiment of Figure 20-9, the goal is to determine whether or not the 4500/6500 and 2000/1500 sites are linked along the same chromosome and, if so, to find the map distance between them. (1) The researcher begins with two strains that are homozygous: 4500 and 2000 versus 6500 and 1500. (2) These strains are crossed to each other to (3) produce a heterozygote. (4) The heterozygote can then be crossed to a homozygote that only carries the 4500 and 2000 bands. (5) The results of this cross depend on whether the RFLPs are closely linked or not (see the right side of Figure 20-9). The parental combinations are 6500, 1500, 4500, and 2000 bp or 4500 and 2000 bp. The recombinant categories are 6500, 4500, and 2000 bp or 4500, 2000, and 1500 bp. If the RFLPs are not linked, we would expect a 1:1:1:1 ratio of the four types among the offspring. If the RFLPs are linked, we would expect a higher percentage of offspring with a parental combination of RFLPs.

Step (6) of Figure 20-9 describes the results obtained from an analysis of 100 offspring. If the RFLPs assort independently, we would expect equal numbers of parental and recombinant offspring. In fact, there are many more parental offspring, indicating that the RFLPs are closely linked.

In an actual analysis of RFLP data, researchers decide on the likelihood of linkage between two RFLPs by using a statistical test called the **lod** (*l*ogarithm of *od*ds) **score method**. This method was devised by Newton Morton in 1955. Although the theoretical basis of this approach is beyond the scope of this text, computer programs analyze pooled data from a large number of pedigrees or crosses. The programs determine the probability that two markers exhibit a certain degree of linkage (i.e., are

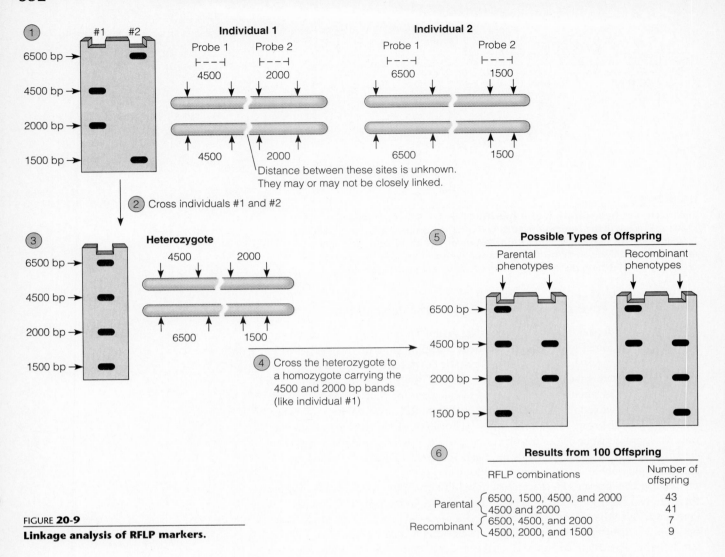

FIGURE **20-9**

Linkage analysis of RFLP markers.

within a particular number of map units apart), and also the probability that the data would have been obtained if the two markers are unlinked. The lod score is then calculated as

$$\text{lod score} = \log_{10} \frac{\text{probability of a certain degree of linkage}}{\text{probability of independent assortment}}$$

For example, if the lod score equals +3, $\log_{10}$ of 3 equals 1000, and so there is a 1000-fold greater probability that the two markers are linked than they are assorting independently. Traditionally, geneticists accept that two markers are linked if the lod score is +3 or higher.

Figure 20-9 illustrates the general type of approach that researchers can follow to map RFLP markers. If a lod score suggested that these two markers were linked, we can divide the recombinant offspring (7 + 9 = 16) by the total number (100) to obtain a map distance of 16 map units.

An RFLP map describes the locations of many different RFLPs throughout the genome; functional genes can be located within an RFLP map

As described in the preceding section, the map distance between two RFLPs can be determined by analyzing the transmission of RFLP markers from parents to offspring. Such an analysis can be conducted on many different RFLPs to determine their relative locations throughout a genome. RFLPs are quite common in virtually all species, so that geneticists easily can map the locations of many RFLPs within a genome. A genetic map composed of many RFLP markers is called an **RFLP map**. In 1980, David Botstein at the Massachusetts Institute of Technology and his colleagues suggested that the construction of RFLP maps would be beneficial in the physical mapping of eukaryotic genomes. Since then, RFLP maps have become increasingly useful in the genetic analysis of plants and mammals, where the genomes are large and relatively few genes have been mapped. Figure 20-10 shows an early

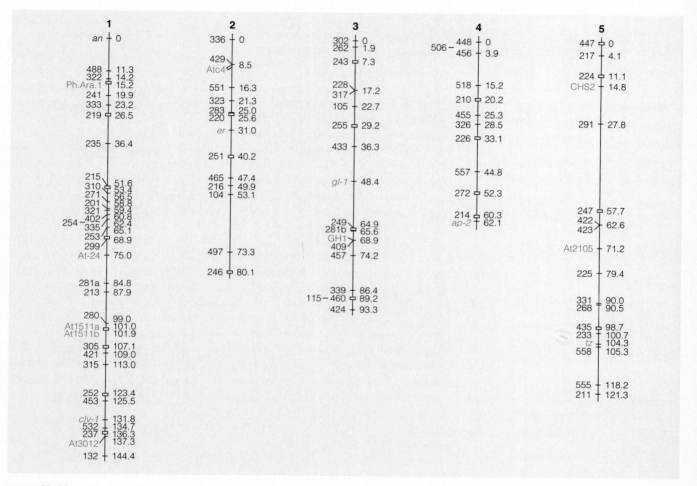

FIGURE **20-10**

An RFLP linkage map of *Arabidopsis thaliana*. This plant has five different chromosomes. The left side of each chromosome describes the locations of RFLP markers. For example, 219 and 235 are RFLPs located 9.9 mu apart on chromosome 1. The numbers along the right side of each chromosome are the map distances in map units. The top marker at the end of each chromosome was arbitrarily assigned as the starting point (zero) for each chromosome. In addition, the map shows the locations of a few known genes (shown in red): *Ph.Ara.*1 = phytochrome, *At*-24 = nitrate reductase, *At*1511a/b = small RNA found in seeds, *clv*-1 = clavata-1, *At*3012 = alcohol dehydrogenase, *Atc*4 = actin, *er* = erecta, *gl*-1 = glabra-1, *GH*1 = acetolactate synthase, *ap*-2 = apelata-2, *CHS*2 = chalcone synthase, *At*2105 = 12S seed storage protein, and *tz* = thiazole-requiring allele.

GENES ⟶ TRAITS: As a first step in mapping the locations of an organism's genes, researchers may initially determine the sites of RFLPs along the chromosomes. In this case, researchers determined the locations of many of these sites along the *Arabidopsis* chromosomes. By mapping these RFLP sites, it becomes easier to locate genes within the *Arabidopsis* genome. The identification of genes helps researchers to elucidate the relationship between genes and traits.

RFLP map of the plant *Arabidopsis thaliana*, (a small plant in the mustard family), which is one of the favorite organisms of plant molecular geneticists. Many RFLPs have been mapped to different locations along the five *Arabidopsis* chromosomes.

An RFLP map, like the one shown in Figure 20-10, can be used to locate functional genes within the genome. In the analysis of human disease-causing genes and in the genetic analysis of agriculturally important crops, RFLPs are gaining widespread use as a way to locate genes along particular chromosomes. As an example, Solved Problem 3 at the end of this chapter illustrates how RFLP analysis can be used to identify a gene that confers herbicide resistance.

In 1978, Kan and Dozy used RFLP analysis to follow the inheritance of human disease-causing alleles

As we have just seen, RFLP analysis can be used to locate particular genes relative to an RFLP marker. This method may also be applied to let prospective parents know if it is likely that they are heterozygous for a recessive allele of a gene that may cause a genetic disease. For example, a couple may want to know if they are heterozygous carriers of the recessive allele that causes sickle cell anemia (see Chapter 4, p. 86, for a description of this disease). This information would allow them to predict the likelihood of having children affected with the disorder. This genetic testing method is particularly useful for diseases in which the phenotype of the heterozygote may not differ from that of the normal homozygote.

More modern techniques are now used to distinguish sickle cell homozygotes and heterozygotes (see Chapter 23). Nevertheless, Kan and Dozy's experiment is a classic study, which confirmed that RFLP markers can be used to predict the likelihood that an individual carries a mutant gene. In 1978, Yuet Kan and Andree Dozy at the University of California at San Francisco conducted an RFLP analysis in the region of the β-globin gene. The normal β-globin gene encodes the β-globin polypeptide; this is a subunit of normal hemoglobin, designated hemoglobin A (Hb^A). A mutant form of the β-globin gene encodes an abnormal polypeptide, which results in the formation of hemoglobin S (Hb^S). Individuals who are homozygous for this mutant gene develop sickle cell anemia. Kan and Dozy set out to determine if RFLP differences could be detected between the normal and mutant forms of the β-globin gene. In other words, they wanted to discover if the restriction digestion of chromosomal DNA in the vicinity of the β-globin gene was different in the Hb^A and Hb^S forms of the gene.

THE HYPOTHESIS

A mutant allele may be closely linked to a particular RFLP. People who carry the mutant allele can be identified by the presence of this RFLP.

TESTING THE HYPOTHESIS

Starting materials: The β-globin gene had been cloned previously. This clone was used as a radiolabeled probe. The researchers needed blood samples from several individuals, some of whom were known to be affected with sickle cell anemia.

Experimental Level

Conceptual Level

1. Take blood samples from 73 people. (Note: The *Hb* genotype was already known from a biochemical analysis of the subjects' hemoglobin.)

Take blood sample that contains white blood cells

Leukocytes

DNA in the nuclei

2. Isolate chromosomal DNA from the cells. This is similar to plasmid isolation. It involves breaking the cells and then using chromatography or centrifugation to purify the DNA. See Experiment 19A.

Tube with purified chromosomal DNA

Long chromosomal DNA

3. Digest the chromosomal DNA with the restriction enzyme *Hpa*I.

Add *Hpa*I

DNA fragments following digestion

4. Use a radiolabeled DNA probe of the β-globin gene to conduct a Southern blot experiment. (Note: The procedure of Southern blotting is described in Chapter 19.)

See Figure 19.5 for a description of Southern blotting

Radiolabeled probe recognizing a DNA fragment

X-ray film

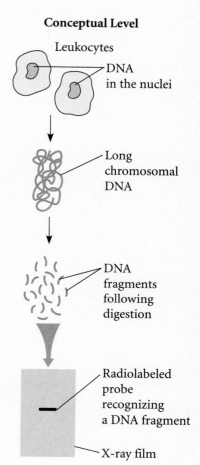

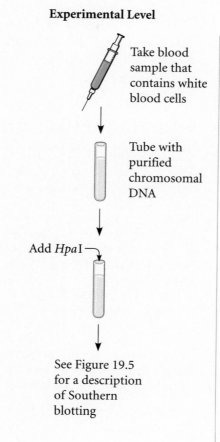

THE DATA

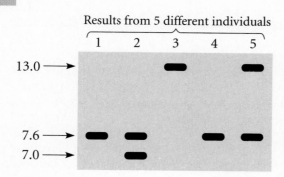

Results from 5 different individuals

1   2   3   4   5

13.0 →

7.6 →
7.0 →

Summary of Data from 73 People

| Hemoglobin Composition | HpaI β-Globin RFLP | | | | | | Frequency of the 13 kbp Fragment |
	7.6/7.6	7.6/7.0	7.0/13	7.6/13	13/13	Total	
African*							
AA	8	6	0	1	0	15	3%
AS	5	1	1	9	0	16	31%
SS	0	0	0	4	11	15	87%
Caucasian							
AA	12	0	0	0	0	12	0%
Asian							
AA	15	0	0	0	0	15 / 73	0%

*Racial origin of the individuals used in this study.

Data from: Kan, Y. W., and Dozy, A. M. (1978) Polymorphism of DNA sequence adjacent to human β-globin structural gene: Relationship to sickle mutation. *Proc. Natl. Acad. Sci. USA 75,* 5631–5635.

INTERPRETING THE DATA

In this study, the genotypes of the 73 individuals were already known from a biochemical analysis of their hemoglobin characteristics. The goal of this research was to see if molecular genetic tests could also distinguish individuals carrying the mutant gene from those carrying the normal gene. Using the β-globin gene as a probe, three different RFLPs were found among a population of 73 individuals. These RFLPs produced fragments of molecular weights 7000 bp, 7600 bp, and 13,000 bp (or 7.6, 7.0, and 13 kbp [kilobase pairs], respectively).

With regard to the Hb^A and Hb^S alleles, the occurrence of these RFLPs was not random. Instead, the 13 kbp RFLP was usually found in persons who were known to have at least one copy of the Hb^S allele. For example, in approximately 87% of the homozygotes ($Hb^S Hb^S$), the Hb^S allele was associated with the 13 kbp fragment. This observation is consistent with the diagram shown here:

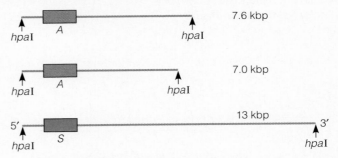

In most people with hemoglobin A, the site where the normal β-globin gene is located has a distance of 7.6 or 7.0 kbp between the *hpa*I sites. For people carrying the Hb^S allele, the mutant β-globin gene is found in a chromosomal region where the *hpa*I sites are usually 13 kbp apart.

In most but not all cases, the Hb^S allele was found to be closely linked to an RFLP of 13 kbp. This type of information can be used as a predictive tool in human genetic analysis. For example, if an individual was found to be heterozygous, 7.6/13, it would be fairly likely that the individual is a heterozygous, Hb^A/Hb^S, carrier of the abnormal gene. By comparison, if an RFLP analysis revealed that an individual was 7.6/7.6, they would more likely be homozygous for the normal Hb^A allele.

Since these initial studies of the β-globin gene, RFLP markers have been determined to be linked closely to many other alleles associated with human genetic diseases. The analysis of RFLP markers has been used frequently as a tool to predict if individuals are heterozygous carriers of recessive alleles that cause human genetic diseases. This approach has also been used in prenatal testing for a variety of genetic abnormalities.

PHYSICAL MAPPING

We now turn our attention to methods aimed at establishing a physical map of a species' genome. Physical mapping usually requires the cloning of many pieces of chromosomal DNA. The cloned DNA fragments are then characterized by size (i.e., length in nucleotide base pairs), the gene(s) they contain, and their relative locations along a chromosome. In recent years, physical mapping studies have led to the DNA sequencing of entire genomes.

As was mentioned in Chapter 10, eukaryotic genomes are very large; the human genome, for example, is approximately 3 billion bp in length and the *Drosophila* genome is roughly 180 million bp. When making a physical map of a genome, researchers must characterize many DNA clones that contain much smaller pieces of the genome. For any given species, the physical mapping of the entire genome is a massive undertaking carried out as a collaborative project by many genome research laboratories. In this section, we will examine the general strategies used in creating a physical map of a species' genome. We will also consider how physical mapping information can be used to clone genes.

Specific chromosomes can be isolated by pulsed-field gel electrophoresis or chromosome sorting

Eukaryotic genomes usually are composed of many different chromosomes. A first step in the dissection of a large genome involves techniques that separate individual chromosomes or large pieces of chromosomes. One technique that can accomplish this goal is **pulsed-field gel electrophoresis (PFGE)**, developed by David Schwartz and Charles Cantor at Columbia University. Through this method, entire small chromosomes or large pieces of chromosomes are subjected to electrophoresis. As its name suggests, pulsed-field gel electrophoresis differs from conventional electrophoresis; the DNA is driven through the gel using alternating pulses of electrical current that are at different angles to each other. This can be achieved using electrophoresis devices that have two sets of electrodes. The alternating pulses can achieve a good separation of DNA fragments that vary in length from 50,000 up to 10 million bp. Therefore, this method is used to separate large fragments of DNA or even small chromosomes.

For PFGE, individual chromosomes or large fragments of chromosomes must be isolated very carefully. To avoid mechanical breakage of the chromosomes, cells are embedded in agarose blocks (also called plugs). The agarose protects the chromosomal DNA from mechanical breakage. Once inside the plug, the cells are lysed, and if desired, the chromosomal DNA can be subjected to a restriction digestion while still in the plug.

PFGE can be performed in conjunction with restriction digestion to obtain large DNA fragments. This can be achieved by using a restriction enzyme that cuts very infrequently. To separate the large pieces of DNA, the plug is inserted into a well in an agarose gel and then subjected to PFGE. Figure 20-11a illustrates a separation of yeast chromosomes that have not been subjected to restriction digestion. Figure 20-11b shows fragments of DNA that have been obtained by digesting a bacterial chromosome with three different restriction enzymes that cut the bacterial chromosome infrequently.

PFGE has two important uses in physical mapping studies. First, it can be used to purify large pieces of DNA for the construction of libraries. For example, the DNA containing a particular yeast chromosome can be isolated from a gel slice and used

kbp
914 —

550 —

214 —

(a) Each band is a different yeast chromosome

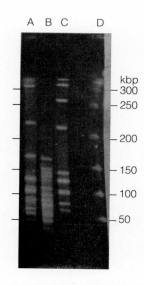

A B C   D

kbp
— 300
— 250

— 200

— 150

— 100

— 50

(b) Each band is a fragment of a bacterial chromosome

FIGURE **20-11**

Pulsed-field gel electrophoresis (PFGE). **(a)** Yeast chromosomes that have not been digested with a restriction enzyme. The numbers at the left indicate the size of the DNA fragments in kilobase pairs (kbp). **(b)** The *Hemophilus influenzae* chromosome digested with *Eag*I (lane A), *Nae*I (lane B), and *Sma*I (lane C). Lane D shows molecular weight markers.

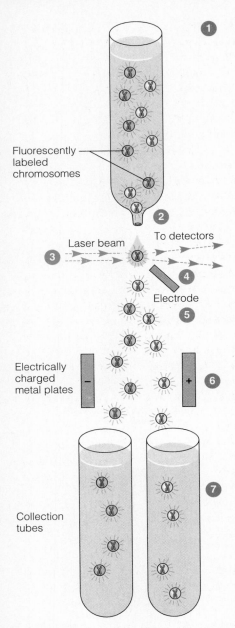

FIGURE **20-12**

Chromosome sorting. (1) A mixture of chromosomes is treated with two fluorescent stains, Hoechst 33258 and chromomycin A₃, to label the chromosomes. Each type of chromosome is already known to have its own unique level of fluorescence. **(2)** The chromosome mixture leaves the nozzle in droplets with one chromosome per droplet. **(3)** A laser beam strikes each droplet. **(4)** A fluorescence detector identifies fluorescent chromosomes by the intensity and wavelength of light emitted by the chromosomes. **(5)** An electrode negatively charges a droplet carrying the desired chromosome 5. **(6)** As droplets pass between the electrically charged plates, the droplets with a negative charge move closer to the positive plate. **(7)** In this way, the chromosomes are separated as they fall into different collection tubes.

to construct a chromosome-specific library. In addition, pulsed-field gels can be used to coarsely map the locations of genes. If an experimenter has cloned two different genes, these clones can be labeled and used as probes in a Southern blot of a pulsed-field gel. If both probes bind to the same band, then they are located on the same chromosome in the genome or on the same large piece of a chromosome.

The chromosomes of bacteria, and some simpler eukaryotes such as yeast and protozoa, are small enough to be resolved by PFGE. However, the chromosomes of more complex eukaryotes (e.g., mammals and plants) are too large to be resolved by this procedure. Instead, individual chromosomes can be isolated by **chromosome sorting**, a technique pioneered by scientists at the Los Alamos National Research Laboratory (Figure 20-12).

As shown in the figure, the starting material is a mixture of chromosomes. The chromosomes from these cells are stained with two fluorescent dyes: Hoechst 33258, which binds preferentially to AT-rich DNA, and chromomycin A₃, which binds to GC-rich DNA. Each chromosome will have its own characteristic numbers of AT- and GC-rich regions, so that each chromosome will have a distinctive level of fluorescence.

The stained chromosomes pass through a laser beam that excites the fluorescent dyes. The stream containing the chromosomes is separated into individual droplets, which are given an electric charge if the fluorescence detector indicates that they contain the desired chromosome. In this example, the goal is to separate human chromosome 5 from the rest of the chromosomes. If the fluorescence intensity in the droplet indicates that the drop contains human chromosome 5, the droplet is given an electric charge and deflected away from the main stream and into a separate tube. In this way, a particular chromosome can be separated from a mixture of many different chromosomes. This device can separate chromosomes at the amazing rate of 1000–2000 per second!

A physical map of a chromosome is constructed by creating a contiguous series of clones from a chromosome-specific library

As described in the preceding section, both pulsed-field gel electrophoresis and chromosome sorting can be used to obtain a sample that contains a single type of chromosome. The sample of DNA can then be digested into many pieces with a restriction enzyme; these fragments are then cloned into vectors to create a chromosome-specific library. Such a library contains a collection of hybrid vectors with different pieces of chromosomal DNA.

The next step in physical mapping studies is to determine the relative locations of the cloned chromosomal pieces as they would occur in an intact chromosome. In other words, the members of the library must be organized according to their actual locations along a chromosome. To determine a complete physical map of a chromosome, researchers need a series of clones that contain overlapping pieces of chromosomal DNA. Such a collection of clones, known as a **contig**, contains a contiguous region of a chromosome that is found as overlapping regions within a group of vectors (Figure 20-13). As will be discussed later, cloning vectors known as YACs and cosmids commonly are used in the construction of a contig.

Different experimental strategies can be used to align the members of a contig. The general approach is to identify adjacent members that contain overlapping regions. Southern blotting (described in Chapter 19) can determine if two different clones contain an overlapping region. For example, the DNA from clone #1 in Figure 20-13 could be radiolabeled *in vitro* and then used as a probe in a Southern blot of the other clones shown in Figure 20-13. Clone #1 would hybridize to clone #2, because they share identical DNA sequences in the overlapping region. Similarly, clone #2 could be used as a probe to show that it will hybridize to clone #1 and clone #3. By conducting Southern blots between many combinations of clones, re-

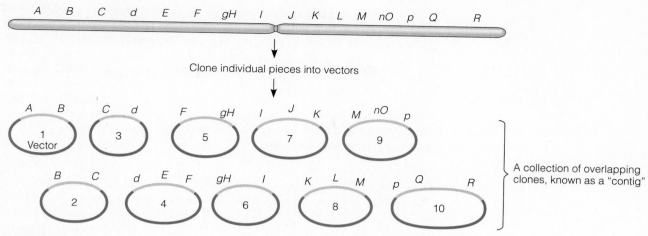

FIGURE **20-13**

The construction of a contig. Pieces of chromosomal DNA are cloned into vectors. The numbers denote the order of the members of the contig. For example, #3 is between #2 and #4. Note that the chromosome is labeled with letters that designate the locations of genes. The members of the contig have overlapping regions. For example, clone #3 ends with gene *d* and clone #4 begins with gene *d*; clone #4 ends with gene *F* and clone #5 begins with gene *F*.

searchers can determine which clones have common overlapping regions and thereby order them as they would occur along the chromosome.

An ultimate goal of physical mapping procedures is to obtain a complete contig for each type of chromosome within a full set. For example, in the case of humans, a complete physical map will require a contig for each of the 22 autosomes and for the X- and Y-chromosomes. At the end of this chapter, we will examine some physical mapping strategies that have been used in the Human Genome Project.

A contig represents a physical map of a chromosome. Geneticists can also correlate cloned DNA fragments in a contig with locations along a chromosome obtained from linkage or cytogenetic mapping. This can be accomplished by identifying members of the contig with inserts that have already been mapped by linkage or cytological methods. For example, a member of a contig may contain a gene that previously has been mapped by genetic linkage. Figure 20-14 considers a situation in which two members of a contig carry genes already mapped by linkage analysis to be approximately 1.5 mu apart on chromosome 11. In this example, clone #2 has an insert that corresponds to gene *A*, while clone #7 has an insert that corresponds to gene *B*. Since a contig is composed of overlapping members, a researcher can line up the contig along chromosome 11 starting with gene *A* and gene *B* as reference points. In this example, genes *A* and *B* serve as markers that identify the location of specific members of the contig.

The identification of markers makes it possible to line up a contig along a particular chromosome. Unfortunately, for most eukaryotic species, few genes have

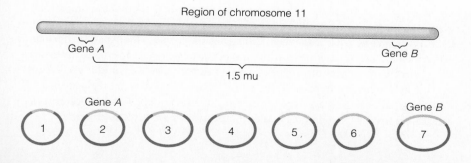

FIGURE **20-14**

The use of genetic markers to align a contig. In this example, gene *A* and gene *B* had been mapped previously to specific regions of chromosome 11. Gene *A* was found within the insert of clone #2, gene *B* within the insert of clone #7. This made it possible to align the contig using gene *A* and gene *B* as genetic markers (i.e., reference points) along chromosome 11.

been genetically mapped and are thereby available as markers. As an alternative, molecular markers such as RFLPs may be used to correlate the locations in a linkage map with the members of a contig. **Sequence-tagged sites (STSs)** may also be used. Each STS is a short segment of DNA, usually between 100 and 400 bp long, with a base sequence known to be unique within the entire genome. The identification of STSs will be described later in this chapter.

YAC cloning vectors are used to make contigs of eukaryotic chromosomes

For large eukaryotic genomes, it is much easier to create contigs when the cloning vector can accept chromosomal DNA inserts of very large size. In general, most plasmid and viral vectors can accommodate inserts only a few thousand to perhaps tens of thousands of nucleotides in length. If a plasmid or viral genome has a DNA insert that is too large, it will have difficulty with DNA replication and is likely to suffer deletions in the insert.

By comparison, another cloning vector, known as a **yeast artificial chromosome (YAC)**, can reliably contain inserted DNA fragments of much larger size. This type of cloning vector was developed in 1987 by David Burke, Georges Carle, and Maynard Olson at Washington University in St. Louis. An insert within a YAC can be several hundred thousand to perhaps 2 million nucleotides in length. For a small human chromosome, a few hundred YACs would be sufficient to create a contig with fragments that span the entire length of the chromosome. By comparison, it would take thousands or even tens of thousands of hybrid plasmids to create such a contig.

At the molecular level, YACs have structural similarities to normal eukaryotic chromosomes yet have characteristics that make them suitable for cloning. The general structure of a YAC is shown at the top left of Figure 20-15. The YAC contains two telomeres (*TEL*), a centromere (*CEN*), a bacterial origin of replication (*ORI*), a yeast origin of replication (known as an *ARS* for autonomous replication sequence), selectable markers, and unique cloning sites that are recognized by a single restriction enzyme. Without an insert, the circular form of this vector can replicate in *E. coli*. After a large fragment has been inserted, the linear form of the vector can replicate in yeast.

In the experiment shown in Figure 20-15, the chromosomal DNA is digested with the restriction enzyme *Eco*RI at a low concentration so that only some of the restriction sites are cut. This partial digestion will result in only occasional cleavage of the chromosomal DNA to yield very large DNA fragments. (In other words, most restriction sites that are normally recognized by the enzyme will not be cleaved, because there is not enough of the restriction enzyme.) The circular YAC vector is also digested with *Eco*RI and a second restriction enzyme, *Bam*HI, to yield two arms of the YAC. The YAC arms are then ligated to the large fragments of chromosomal DNA and transformed into yeast cells. When ligation occurs in the desired way, a large piece of chromosomal DNA becomes ligated to both arms of the YAC. Since each arm contains a different selectable marker, it is possible to select for the growth of yeast cells that carry a YAC construct having both arms.

YAC cloning vectors are very useful in the construction of contigs that span long segments of chromosomes. They are commonly the first step in creating a rough physical map of a genome. While this is an important step in physical mapping, the large insert sizes of YACs make them difficult to use in gene cloning and sequencing experiments. Therefore, libraries containing hybrid vectors with smaller insert sizes are needed. Most commonly, a type of cloning vector called a **cosmid** is used. A cosmid is a hybrid between a plasmid vector and phage λ; its DNA can replicate in a cell like a plasmid or be packaged into a protein coat like a phage. Cosmid vectors typically can accept DNA fragments tens of thousands of base pairs in length.

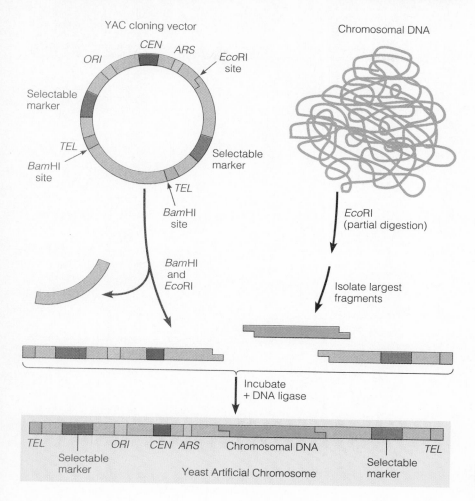

YAC cloning vector

CEN

ORI   *ARS*

*Eco*RI
site

Selectable
marker

TEL

*Bam*HI
site

Selectable
marker

TEL

*Bam*HI
site

*Bam*HI
and
*Eco*RI

Chromosomal DNA

*Eco*RI
(partial digestion)

Isolate largest
fragments

Incubate
+ DNA ligase

TEL   *ORI*   *CEN* *ARS*   Chromosomal DNA   *TEL*

Selectable
marker

Yeast Artificial Chromosome

Selectable
marker

FIGURE **20-15**

The use of YAC vectors in DNA cloning.

Positional cloning can be achieved using chromosome walking

The creation of a contig bears many similarities to a gene cloning strategy known as **positional cloning**. This term refers to a strategy in which a gene is cloned based on its mapped position along a chromosome. This approach has been successful in the cloning of many human genes, particularly those that cause genetic diseases when mutated. These include genes involved in cystic fibrosis, Huntington disease, and Duchenne muscular dystrophy.

A common method used in positional cloning is known as **chromosome walking**. To initiate this type of experiment, a gene's position relative to a marker must be known from mapping studies. For example, a gene may be known to be fairly close to a previously mapped gene or RFLP marker. This provides a starting point to molecularly "walk" toward the gene of interest.

Figure 20-16 considers a chromosome walk in which the goal is to locate a gene that we will call gene *A*. In this example, previous genetic mapping studies have revealed that gene *A* is relatively close to another gene, called gene *B*, that has been previously cloned. Gene *A* and gene *B* have been deduced from genetic crosses to be approximately 1 mu apart. To begin this chromosome walk, a cloned DNA fragment of gene *B* can be used as a starting point to walk to gene *A*.

To walk from gene *B* to gene *A*, a series of library screening methods are followed. In this example, the starting materials are a cosmid library and a clone containing

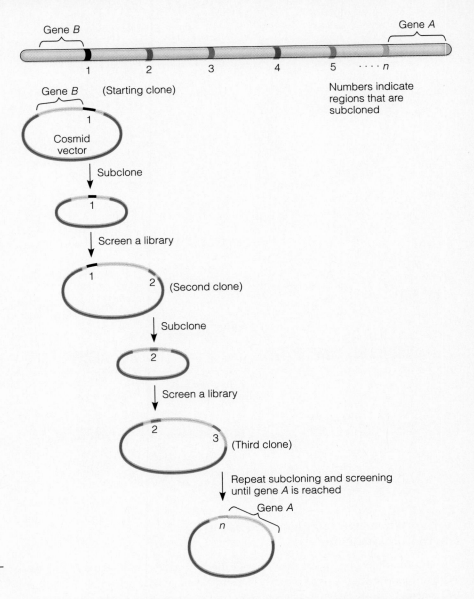

FIGURE **20-16**

The technique of chromosome walking.

gene *B*. A small piece of DNA from the gene *B* clone is inserted into another vector. This is called **subcloning**. The subclone is radiolabeled and used as a probe to screen a cosmid library. This will enable the researchers to identify a second clone that is closer to gene *A*. A subclone from this second clone is then used to screen the library a second time. This will allow the researchers to identify a clone that is even closer to gene *A*. This repeated pattern of subcloning and library screening is used to walk toward gene *A*. The term chromosome walking is an appropriate description of this technique, because each clone takes you a step closer to the gene of interest.

The number of steps required to reach the gene of interest depends on the distance between the starting and ending points. If the two points are 1 mu apart, they are expected to be approximately 1 million bp apart, although the correlation between map units and physical distances can be quite variable. In a typical walking experiment, each clone might have an average insert size of 50,000 bp. Therefore, it will take more than 20 walking steps to reach the gene of interest. If chromosome walking is done in conjunction with DNA sequencing (see Chapter 19, pp. 531–534), this can be a laborious undertaking spanning several years. For this reason, researchers are eager to locate starting points in a chromosome walking experiment that are as close as possible to the gene of interest.

It may have occurred to you that the subcloning steps of chromosome walking are unnecessary if a researcher previously has created a contig in the region of the chromosome where the gene is located. In this case, the experimenter may analyze the members of the contig, rather than following the extremely tedious strategy described in Figure 20-16. This desire to more easily clone genes via a positional strategy has provided a strong impetus for the creation of contigs that can be shared by many researchers.

THE HUMAN GENOME PROJECT

The **Human Genome Project** aims to create a detailed map of the human genome. It is the largest internationally coordinated undertaking in the history of biological research. Discussions among scientists to undertake this project began in the mid 1980s. In 1988, the National Institutes of Health in Bethesda, Maryland, established an Office of Human Genome Research, with James Watson as its first director. The Human Genome Project officially began on October 1, 1990. The goals of the human genome project are the following:

1. *To obtain a genetic linkage map of the human genome.* This involves the identification of thousands of genetic markers and their localization along the autosomes and sex chromosomes.

2. *To obtain a physical map of the human genome.* This requires the cloning of many segments of chromosomal DNA into YACs and cosmids.

3. *To obtain the DNA sequence of the entire human genome.* This long-term goal will require the DNA sequencing of the entire genome, which is approximately 3 billion nucleotides in length. If the entire human genome was typed in a textbook like this, it would be nearly 1 million pages long!

4. *To develop technology for the management of human genome information.* The amount of information that will be obtained from this project is staggering, to say the least. The Human Genome Project will develop user-friendly tools to provide scientists easy access to up-to-date information obtained from the project. The Human Genome Project will also develop analytical tools to interpret genomic information.

5. *To analyze the genomes of other model organisms.* These include bacterial species (e.g., *Escherichia coli* and *Bacillus subtilis*), *Drosophila melanogaster* (fruit fly), *Caenorhabditis elegans* (a nematode), *Arabidopsis thaliana* (a simple plant), and *Mus musculus* (mouse).

6. *To develop programs focused on understanding and addressing the ethical, legal, and social implications of the results obtained from the Human Genome Project.* The Human Genome Project will attempt to identify the major genetic issues that will affect the members of our society and to develop policies to address these issues. For example, some people are worried that their medical insurance companies may discriminate against them if it is found that they carry a disease-causing or otherwise deleterious gene.

7. *To develop technological advances in genetic methodologies.* Some of the efforts of the Human Genome Project are expected to involve improvements in molecular genetic technology such as gene cloning, contig construction, DNA sequencing, and so forth. The Project is also aimed at developing computer technology for data processing, storage, and the analysis of sequence information (see Chapter 22).

A great benefit expected from the characterization of the human genome is the ability to identify our genes. Mutations in many different genes are known to be correlated with human diseases, which include cancer, heart disease, and many other abnormalities. The potential identification of genes that are mutated in the development of diseases is a strong motivation for the Human Genome Project. A detailed genetic and physical map will make it much easier for researchers to locate such genes. Furthermore, a complete DNA sequence of the human genome will provide researchers with insight into the types of proteins encoded by these genes. The cloning and sequencing of disease-causing alleles is expected to play an increasingly important role in the diagnosis and treatment of disease. In the rest of this section, we will examine methods that are used to analyze the human genome and the types of important information this analysis is expected to yield.

The genetic linkage mapping of the human genome analyzes the transmission of genetic markers in large human pedigrees

As was described in Chapter 5, the general strategy for constructing a genetic linkage map is to make crosses between individuals who are heterozygous for two or more genes. The percentages of offspring exhibiting various combinations of traits will indicate if two genes are assorting independently. If they are not, the frequency of recombinant offspring provides a measure of map distance between the two genes. In humans, a similar approach can be taken, except that the numbers of recombinant offspring must be determined by analyzing very large pedigrees rather than making crosses.

To make a highly refined genetic map of a genome, many different polymorphic sites must be identified and their transmission followed from parent to offspring over many generations. In the case of humans, however, relatively few polymorphic genes are available for this type of analysis. Therefore, linkage mapping in humans often focuses on the transmission of genetic markers that are not actually genes. For example, RFLPs provide many additional markers within the human genome.

Nevertheless, because the human genome is extremely large, and since the Human Genome Project seeks to produce a very detailed genetic map, the search for many more genetic markers has been undertaken. Fortunately, a straightforward method of identifying genetic markers has been discovered. These newer markers are short, simple sequences called **microsatellites** that are abundantly interspersed throughout the human genome and are quite variable in length among different individuals. The most common microsatellite encountered in humans is the sequence $(CA)_n$, where n may range from 5 to more than 50. (In other words, this dinucleotide sequence can be tandemly repeated 5 to 50 or more times.) These sequences are called microsatellites to distinguish them from the much larger tandemly repeated sequences called satellites (see Chapter 10).

The $(CA)_n$ microsatellite is found, on average, about every 10,000 bases in the human genome. An initial goal of the Human Genome Project is to identify many different pieces of human DNA that contain this microsatellite sequence. This is accomplished by using a complementary DNA probe (namely, $(GT)_n$) to screen genomic DNA libraries. Researchers have identified thousands of different DNA segments that contain $(CA)_n$ microsatellites, located at many distinct sites within the human genome.

In most cases, the $(CA)_n$ microsatellite is flanked by unique DNA sequences. Using primers complementary to these associated unique DNA sequences, a particular microsatellite can be amplified specifically by PCR (which is described in Chapter 19; see Figure 19-4). In other words, the PCR primers will only amplify a particular microsatellite, but not the thousands of others that are interspersed throughout the genome (Figure 20-17).

When a pair of PCR primers amplify a single DNA sequence within the haploid human genome, the amplified region is called a sequence-tagged site (STS).

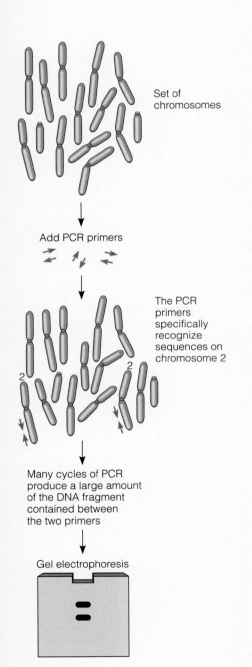

Set of chromosomes

Add PCR primers

The PCR primers specifically recognize sequences on chromosome 2

Many cycles of PCR produce a large amount of the DNA fragment contained between the two primers

Gel electrophoresis

FIGURE 20-17

Identifying a microsatellite using PCR primers.

Since humans are diploid, however, there will usually be two copies of a given STS in an individual. When an STS contains a microsatellite, the two PCR products will be identical if the (CA)$_n$ region is the same length in both copies (i.e., if the individual is homozygous for the microsatellite). However, if an individual has two copies that vary in the number of repeats in the microsatellite sequence, the two PCR products obtained will be slightly different in length (as in Figure 20-17). Like RFLPs, microsatellites have length polymorphisms that allow researchers to follow their transmission from parent to offspring. A key difference between RFLPs and microsatellites is that RFLPs are generated by restriction enzyme digestion, which is technically more difficult than the PCR used to produce microsatellites. A more detailed description of how STSs are made will be given later in this chapter.

In the Human Genome Project, PCR amplification of particular microsatellites provides an important strategy in the genetic analysis of human pedigrees. This idea is shown in Figure 20-18. A unique segment of DNA containing a microsatellite has been characterized. Using PCR primers complementary to the unique flanking segments, two individuals and their three offspring can be tested for the inheritance of this microsatellite. A small sample of cells is obtained from each individual and subjected to PCR amplification as described in Chapter 19. The amplified PCR products are then analyzed by high-resolution gel electrophoresis, which can detect small differences in the lengths of DNA fragments. The father's PCR products were 154 and 150 bp in length, while the mother's were 146 and 140 bp. Their first offspring inherited the 154 bp product from the father and the 146 bp from the mother, the second inherited the 150 bp from the father and the 146 bp from the mother, and the third inherited the 150 bp from the father and the 140 bp from the mother. As shown in the figure, the transmission of polymorphic microsatellites is relatively easy to follow from generation to generation.

The simple pedigree analysis shown in Figure 20-18 illustrates the general method used to follow the transmission of a single microsatellite that is polymorphic in length. In genetic linkage studies, the goal is to follow the transmission of many different microsatellites to determine which are linked along the same chromosome and which are not. Those that are not linked will independently assort from generation to generation. Those that are linked tend to be transmitted to the same offspring. In a large pedigree, it is possible to identify cases where linked microsatellites have segregated due to crossing over. As discussed in Chapter 5, the frequency of crossing over provides a measure of the map distance, in this case between different microsatellites. This approach can help researchers to obtain a finely detailed genetic linkage map of the human chromosomes.

The human genome project requires the physical mapping of each chromosome

The construction of contigs for each human chromosome is being carried out by many laboratories participating in the Human Genome Project. Since this is a collaborative effort, the various groups need common genetic markers to use as landmarks in the human genome. Furthermore, since the human genome is so large, many unique landmarks are required. Maynard Olson at the University of Washington proposed the use of sequence-tagged sites (STSs) as unique landmarks for the human genome.

In Figures 20-17 and 20-18, we considered STSs that contain microsatellite sequences. Such STSs commonly are used as molecular markers in genetic linkage analysis. In addition, STSs that do not contain microsatellite sequences are useful in physical mapping studies. An initial physical mapping goal of the Human

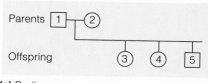

(a) Pedigree

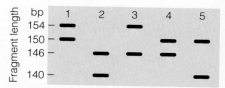

(b) Electrophoretic gel of PCR products for the polymorphic STSs found in the family in (a).

FIGURE **20-18**

Inheritance pattern of microsatellites in a human pedigree.

Genome Project was to produce a series of unique STSs spaced roughly 100,000 bases apart along the contig map of each human chromosome. To identify an STS (Figure 20-19), one begins by sequencing a relatively short region of chromosomal DNA, perhaps 400 bp in length. Within this region, the DNA sequence is used to design two PCR primers 200 to 400 bp apart:

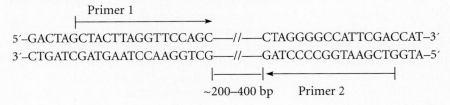

Primer 1

5′–GACTAGCTACTTAGGTTCCAGC——//——CTAGGGGCCATTCGACCAT–3′
3′–CTGATCGATGAATCCAAGGTCG——//——GATCCCCGGTAAGCTGGTA–5′

~200–400 bp   Primer 2

The primer sequences are complementary to specific regions within the DNA segment; they usually are chosen by computer programs that scan the DNA sequence and identify regions with optimal sequences for PCR. The PCR primers, which are about 20 nucleotides in length, are made synthetically. The two primers are then used in a PCR experiment that includes a sample of human DNA containing all the chromosomes. If the PCR produces a single product of the correct size, a unique STS has been identified; the PCR primers are uniquely recognizing this single site in the human genome. Alternatively, the PCR primers may sometimes bind to multiple sites in the genome, because the primer sequences are not unique within the DNA. These products generally are not useful in physical mapping studies.

Using the approach described in Figure 20-19, about half the time you get a unique STS and about half you get multiple PCR products. The sequences of PCR primers that yield unique STSs are entered into a computer database (see Chapter 22 for a description of databases). Any researcher can access this database to look up PCR primers that yield unique STSs. Therefore, researchers worldwide can use this database to make the same STSs, and have the same landmarks in their physical mapping studies.

Now that we understand what an STS is, let's consider how they can be used to align a contig. As a simple example, let's suppose that ten different STSs have been identified by sequencing short segments of DNA from a chromosome 6–specific DNA library (Table 20-2). The ten STSs are found along chromosome 6, but their order is not known. A researcher has also made seven different YAC clones from a chromosome 6–specific library. The goal in the experiment of Table 20-2 is to deduce the order of the STSs and the YACs. To accomplish this goal, each YAC clone is tested to see if it will produce a PCR product, using the pairs of primers specific for each of the ten STSs. If a YAC clone produces a PCR product (designated by a + sign), the corresponding STS must be located somewhere within the YAC's insert. In the example shown here, each YAC tests positive for one or more STSs. As shown at the bottom of Table 20-2, the pattern of positive and negative results can be used to deduce the order of STSs and YACs. This result illustrates a common strategy for aligning YACs using STS markers.

A goal of mapping studies is to correlate cytogenetic, linkage, and physical maps

To correlate locations within cytogenetic, linkage, and physical maps, researchers need techniques that can compare the relative locations between different maps.

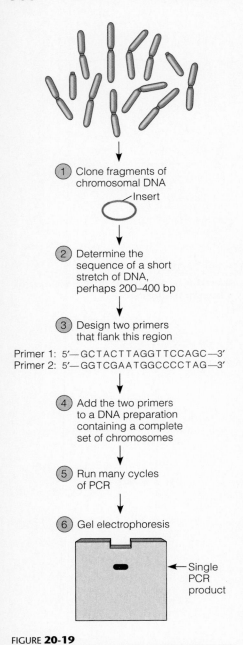

1. Clone fragments of chromosomal DNA

Insert

2. Determine the sequence of a short stretch of DNA, perhaps 200–400 bp

3. Design two primers that flank this region

Primer 1: 5′—GCTACTTAGGTTCCAGC—3′
Primer 2: 5′—GGTCGAATGGCCCCTAG—3′

4. Add the two primers to a DNA preparation containing a complete set of chromosomes

5. Run many cycles of PCR

6. Gel electrophoresis

Single PCR product

FIGURE 20-19
Identification of a sequence-tagged site (STS).

TABLE 20–2 Alignment of STSs and YACs

	STSs									
YACs	1	2	3	4	5	6	7	8	9	10
1	+	−	−	+	−	−	−	−	−	−
2	+	−	−	−	−	+	−	−	−	−
3	−	+	+	−	−	−	+	−	−	−
4	−	−	−	−	+	+	−	−	−	−
5	−	−	+	−	−	−	−	−	+	−
6	−	−	−	+	−	−	−	+	+	+

Deduced Outcome

	STSs[a]									
YAC	5	6	1	4	10	8	9	3	2	7
4	←	→								
2		←	→							
1			←	→						
6				←	→					
5						←	→			
3							←	→		

[a]The relative order of 10 & 8, and 2 & 7 cannot be determined from this analysis.

One common method is *in situ* hybridization, which was described earlier in this chapter. A YAC or cosmid clone can be hybridized to intact chromosomes to see where the clone is located. This provides a way to correlate a cytogenetic map with a physical map. Also, when a cloned gene is used as a probe, a linkage map can be correlated with a cytogenetic map. When a genetically mapped gene can be localized to a particular chromosomal region in somatic cell hybrids, a linkage map can be correlated with a cytogenetic map.

Finally, a more recent approach for correlating linkage and physical maps involves the use of microsatellites. As discussed earlier in this chapter, some STSs also contain microsatellite $(CA)_n$ sequences. These are termed **polymorphic STSs** (see Figures 20-17 and 20-18). A nice feature of polymorphic STSs is that they can be used in both physical and genetic linkage analyses to correlate locations on a genetic linkage map with locations on a physical map.

Figure 20-20 illustrates a comparison between cytogenetic, linkage, and physical maps of chromosome 16. Actually, this is a fairly crude map of chromosome 16. A much more detailed map is now available, although it would take well over ten pages of this textbook to print it!

The duration of the Human Genome Project is estimated at 15 years, though an accurate time frame is quite difficult to predict. The total estimated cost is anticipated to be $3 billion. While many argue that this is wise investment of research dollars, other scientists have asserted that this effort will drain precious funds away from other important research projects. In spite of these objections, the Human Genome Project continues to move ahead. During the 1990s, it has become clear that genome research will play an increasingly important role in the genetic analysis of humans and other organisms.

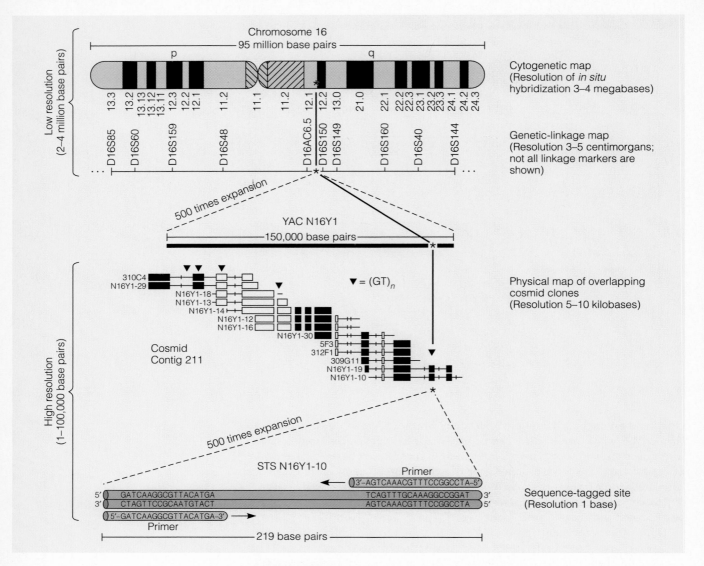

FIGURE **20-20**

A correlation of cytogenetic, linkage, and physical maps of human chromosome 16. The top part of the figure shows the cytological map of human chromosome 16 according to its G-banding pattern. A very simple linkage map of molecular markers (D16S85, D16S60, etc.) is aligned below the cytological map. A correlation between the linkage map and a segment of the physical map is shown below the linkage map. A YAC clone designated YAC N16Y1 is located between markers D16AC6.5 and D16S150 on the linkage map. Pieces of DNA from this YAC were subcloned into cosmid vectors and the cosmids (N16Y1-29, N16Y1-18, etc.) were aligned relative to each other. One of the cosmids (N16Y1-10) was sequenced, and this sequence was used to generate an STS shown at the bottom of the figure.

CONCEPTUAL
AND EXPERIMENTAL
SUMMARY

In this chapter, we have considered a variety of methods aimed at the molecular analysis of entire *genomes*. This approach, termed *genomics*, begins with the mapping of the genome and progresses ultimately to a complete DNA sequence. As we have seen, the three main strategies to map the locations of genes and molecular markers are cytogenetic, linkage, and physical mapping.

Cytogenetic mapping relies on light microscopy to locate genes or other DNA markers along intact chromosomes. The most commonly used method of cytogenetic mapping is *fluorescence in situ hybridization (FISH)*, which can locate genes with remarkable sensitivity. In this technique, a small DNA probe is hybridized to intact chromosomes to determine where that DNA sequence is located in the genome. A second method that is used primarily to map human genes involves the use of *somatic cell hybrids*. Such hybrids vary with regard to their human chromosomal content, so that the presence of particular human chromosomes or fragments of human chromosomes can be correlated with the presence of certain gene products. Finally, cytogenetic mapping is particularly advanced in species that have very large polytene chromosomes, such as *Drosophila*. In this organism, very small changes in chromosome structure can be correlated with the loss of particular genes by analyzing alterations in the banding pattern of the polytene chromosomes.

Linkage mapping relies on the outcome of crosses to map the relative locations of genes along chromosomes. In Chapter 5, we studied linkage mapping that involved crosses between individuals heterozygous for two or more genes. In this chapter, we have considered how *molecular markers*, such as *restriction fragment length polymorphisms (RFLPs)*, can be mapped by a similar approach. Compared with conventional linkage analysis, molecular mapping uses techniques that examine the composition of an individual's chromosomes rather than relying on their external phenotypes. It is now common for geneticists to construct *RFLP maps* of entire genomes. These maps can be used to determine the locations of functional genes by correlating the inheritance of the gene with the inheritance of a particular RFLP.

Physical mapping relies on the molecular analysis of an organism's genome via DNA cloning methods. As a start, it is common to isolate individual chromosomes or large fragments of chromosomes by *PFGE* or *chromosome sorting*. A chromosome-specific library can then be made using *yeast artificial chromosome (YAC)* cloning vectors, the large DNA inserts of which can be aligned to make a *contig*. A more refined contig is made by cloning smaller fragments of DNA into *cosmid* vectors. The methods of constructing a contig are similar in theory to *positional cloning*. To clone a gene by this method, the gene is first mapped to a region of the chromosome and a clone from that region is used as a starting point in a *chromosome walk* toward the gene of interest. This is accomplished by sequential hybridization methods. Positional cloning can be accomplished more easily if a contig has already been constructed in the region where the gene is located.

Finally, this chapter discussed the *Human Genome Project*. The techniques used to map the human genome are analogous to those described earlier in this chapter, although the large size of the human genome has required the development of some specialized methods. In particular, the mapping of the human genome has used many *sequence-tagged sites (STSs)* as physical markers of the human genome. Many of these STSs are also *microsatellites*, which vary in their lengths much as RFLPs do. Both STSs and microsatellites are detected by PCR methods. Currently, much progress has been made toward the cytogenetic, linkage, and physical mapping of the human genome and the correlation of these three maps. The Human Genome Project is expected to yield a great deal of information about our genes and will aid in the identification of many genes that play a role in human diseases.

PROBLEM SETS

SOLVED PROBLEMS

1. The human genetic disease phenylketonuria (PKU) involves a defect in the production of the enzyme phenylalanine hydroxylase. To map this gene to a particular human chromosome, a normal human cell line was fused with rodent cells. Several hybrid lines were then examined with regard to their chromosomal composition and ability to produce human phenylalanine hydroxylase. The results in the table below were obtained. On which human chromosome does this gene map?

Answer: The strategy in this experiment is to correlate the presence of the gene product with the presence of a particular human chromosome. Likewise, the absence of the same chromosome should correlate with the absence of the gene product. In the data shown here, the gene product (namely, phenylalanine hydroxylase) is always made when human chromosome 12 is present but is never made when this chromosome is not found in the hybrid line. Therefore, these results indicate that human chromosome 12 carries the gene that encodes phenylalanine hydroxylase.

2. A RFLP marker is located 1 million bases away from a gene of interest. Your goal is to start at this RFLP marker and walk to this gene. The average insert size in the library is 55,000 bp and the average overlap at each end is 5000 bp. Approximately how many steps will it take to get there?

Answer: Each step is only 50,000 bp (i.e., 55,000 minus 5000 bp), because you have to subtract the overlap between adjacent fragments, which is 5000 bp. Therefore, it will take about 20 steps to go 1 million bp.

3. When many RFLPs have been mapped within the genome of a plant, an RFLP analysis can be used to map a herbicide resistance gene. For example, let's suppose that an agricultural geneticist has two strains, one that is herbicide resistant and one that is herbicide sensitive. The two strains differ with regard to many RFLPs. The sensitive and resistant strains are crossed, and the F_1 offspring are allowed to self-fertilize. The F_2 offspring are then analyzed with regard to their herbicide sensitivity and RFLP markers. The following results are obtained:

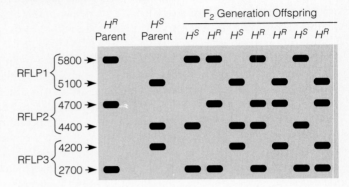

To which RFLP might the herbicide resistance gene be linked?

Answer: The aim of this experiment is to correlate the presence of a particular RFLP with the herbicide-resistant phenotype. As shown in the data, the herbicide-resistant parent and all the herbicide-resistant offspring have an RFLP 4700 bp in length. In an actual experiment, a more thorough lod score analysis would be conducted to determine if linkage is considered likely. If so, the 4700 bp RFLP may either contain the gene that confers herbicide resistance or, as is more likely, the two may be linked. If the 4700 bp RFLP has already been mapped to a particular site in the plant's genome, the herbicide-resistance gene also maps to the same site or very close to it. This information may then be used by the plant breeder when making future crosses to produce herbicide-resistant plant strains.

Data for Question 1, above

Hybrid Line	Human Chromosomes in the Hybrids 1 2 3 4 5 6 7 8 9 10 11 12 13 14 15 16 17 18 19 20 21 22 X	Human phenylalanine Hydroxylase in the Hybrids
1	+ + + + +	+
2	+ + +	−
3	+ + + +	−
4	+ + + + + + +	+
5	+ + + + + + +	+
6	+ +	−

CONCEPTUAL AND EXPERIMENTAL QUESTIONS

1. In your own words, describe the distinction between genetic, cytological, and physical mapping.

2. A person with a rare genetic disease has a sample of her chromosomes subjected to *in situ* hybridization using a probe that is known to recognize band p11 on chromosome 7. Even though her chromosomes look cytologically normal, the probe does not bind to them. How would you explain these results? How would you use this information to positionally clone the gene that is defective in this person?

3. Discuss the goals of the Human Genome Project.

4. Cystic fibrosis is due to a defect in a protein that transports Cl^- ions across the plasma membrane of lung and intestinal cells. To map the location of this gene, a normal human cell line was fused with rodent cells. Several hybrid lines were then examined with regard to their chromosomal composition and ability to produce this Cl^--transporting human protein. The results in the table below were obtained.

A. On which human chromosome does this gene map?
B. How would these results be different if the human cell line used in this experiment was from a patient with cystic fibrosis or from a heterozygous carrier of the disease?

5. A human genetic disease known as Tay–Sachs disease involves a defect in the production of the enzyme hexosaminidase A. To map this gene to a particular human chromosome, a normal human line was fused with rodent cells. Several hybrid lines were then examined with regard to their chromosomal composition and ability to produce human hexosaminidase A. The results in the table below were obtained. Where does this gene map?

6. What is a contig? Explain how you would determine that two clones in a contig are overlapping.

7. Contigs are often made using YAC and cosmid vectors. What are the experimental advantages and disadvantages of these two types of vectors? Which type of contig would you make first, a YAC or cosmid contig? Explain.

8. In general terms, what is a polymorphism? Explain the molecular basis for a restriction fragment length polymorphism (RFLP). How is an RFLP detected experimentally? Why are they useful in physical mapping studies? How can they be used to clone a particular gene?

9. Describe the molecular features of a YAC cloning vector. What is the primary advantage of a YAC compared with plasmid or viral vectors?

10. Let's suppose there are two different RFLPs in a species. RFLP 1 is found in 4500 bp and 5200 bp lengths; RFLP 2 can be 2100 bp or 3200 bp. A homozygote, 4500 bp and 2100 bp, is crossed to a homozygote, 5200 bp and 3200 bp. The F_1 offspring are then crossed to a homozygote containing the 4500 bp and 2100 bp fragments. The following results are obtained:

5200, 4500, and 2100	40 offspring
5200, 4500, 3200, and 2100	98 offspring
4500 and 2100	97 offspring
4500, 2100, and 3200	37 offspring

Conduct a chi square analysis to determine if these two RFLPs are linked. If they are linked, calculate the map distance between them.

11. Explain how a detailed RFLP map of a species' genome can be helpful in mapping the locations of functional genes. Describe an experimental strategy you would follow to map a gene to a nearby RFLP.

12. At the molecular level, what is a microsatellite? How do researchers synthesize them in a laboratory? How are they used in linkage and physical mapping studies? How are microsatellites similar to and different from RFLPs?

13. Five RFLPs designated 1A and 1B, 2A and 2B, 3A and 3B, 4A and 4B, and 5A and 5B are known to map along chromosome 4 of corn. A plant breeder has obtained a strain of corn that carries a pesticide resistance gene known (from previous experiments) to map somewhere along chromosome 4. She crosses this pesticide-resistant strain, which is homozygous for RFLPs 1A, 2B, 3A, 4A, and 5A to a

Data for Question 4, above

Hybrid Line	Human Chromosomes in the Hybrids																							Gene Product in the Hybrids
	1	2	3	4	5	6	7	8	9	10	11	12	13	14	15	16	17	18	19	20	21	22	X	
1		+					+					+					+					+		+
2			+						+		+						+							–
3		+							+						+		+		+					–
4				+			+				+				+		+							+
5	+						+			+			+				+							+
6		+															+							–

Data for Question 5, above

Hybrid Line	Human Chromosomes in the Hybrids*																			Hexosaminidase A in the Hybrids
	1	2p	2q	4	4p	7q	11	11q	13	15p	15q	16	17	17p	17q	18p	18q	20	X	
1		+					+		+	+					+				+	–
2			+					+			+				+					+
3		+										+		+		+				–
4				+	+		+	+	+		+		+		+					+
5	+				+		+	+	+					+						–
6		+									+				+					+

*q designates the long arm of the chromosome; p designates the short arm.

pesticide-sensitive strain homozygous for 1B, 2A, 3B, 4B, and 5B. The F$_1$ generation plants are allowed to self-hybridize and produce the following F$_2$ plants:

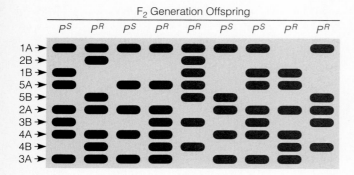

F$_2$ Generation Offspring

Based on these results, which RFLP does the pesticide resistance gene map closest to?

14. A woman has been married to two different men and produced five children. This group is analyzed with regard to three different sequence-tagged sites (STSs): STS-1 is 146 and 122 bp; STS-2 is 102 and 88 bp, and STS-3 is 188 and 204 bp. The mother is homozygous for all three STSs: STS-1 = 122, STS-2 = 88, and STS-3 = 188. Father 1 is homozygous for STS-1 = 122 and STS-2 = 102, and heterozygous for STS-3 = 188/204. Father 2 is heterozygous for STS-1 = 122/146 and STS-2 = 88/102, and homozygous for STS-3 =204. The five children have the following results:

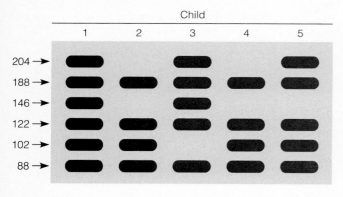

Child

Which children can you definitely assign to father 1 and father 2?

15. Why are sequence-tagged sites (STSs) important in large genome projects?

16. A researcher used primers to nine different STSs to test their presence along five different YAC clones. The following results were obtained:

Alignment of STSs and YACs

	STSs								
YACs	1	2	3	4	5	6	7	8	9
1	–	–	–	–	–	+	+	–	+
2	+	–	–	–	+	–	–	+	–
3	–	–	–	+	–	+	–	–	–
4	–	+	+	–	–	–	–	+	–
5	–	–	+	–	–	–	–	–	+

Make a contig map describing the alignment of the five YACs.

17. Explain how somatic cell hybrids can be used to map the locations of human genes. How much information can you get from this type of experiment?

18. Describe the technique of *in situ* hybridization. Explain how it can be used to map genes.

19. The cells from a malignant tumor were subjected to *in situ* hybridization using a probe that recognizes a unique sequence on chromosome 14. The probe hybridized exactly once within each cell. Explain these results and speculate on their significance with regard to the malignant characteristics of these cells.

20. In an *in situ* hybridization experiment, what is the relationship between the sequence of the probe DNA and the site on the chromosomal DNA where the probe binds?

21. In Experiment 20B, explain the conclusion that the 13 kbp fragment is more closely linked to *Hb^S* than are the 7.0 and 7.6 fragments. Is it always linked? Why or why not? How is this type of information useful?

22. Describe how you would clone a gene by positional cloning. Explain how a (previously made) contig would make this task much easier.

23. Describe two experimental strategies to obtain a purified preparation of a particular type of chromosome. Why are these techniques useful in genome analysis?

QUESTIONS FOR STUDENT DISCUSSION/COLLABORATION

1. How is it possible to obtain an RFLP linkage map? What kind of experiments would you conduct to correlate the RFLP linkage with the positions of known, previously cloned genes? Discuss the uses of RFLPs in genetic analyses.

2. What is a molecular marker? Give two examples. Discuss why it is easier to locate and map many molecular markers rather than functional genes.

3. Which goals of the Human Genome Project do you think are the most important? Why? Discuss the types of ethical problems that might arise as a result of identifying all of our genes.

MOLECULAR GENETICS
AND BIOTECHNOLOGY

Biotechnology is defined broadly as technologies that involve the use of living organisms, or products from living organisms, as a way to benefit humans. Biotechnology is not a new topic. It began about 12,000 years ago when humans began to domesticate animals and plants for the production of food. Since that time, many species of microorganisms, plants, and animals have become routinely used for human benefit.

More recently, the term biotechnology has become associated with molecular genetics. Since the 1970s, molecular genetic tools have provided new, improved ways to make use of living organisms to benefit humans. As was discussed in Chapter 19, recombinant DNA techniques can be used to genetically engineer microorganisms. In addition, recombinant methods enable the introduction of genetic material into plants and animals. As will be discussed in this chapter, an organism that has integrated recombinant DNA into its genome is called **transgenic**.

In the 1980s, court rulings made it possible to patent recombinant microorganisms as well as transgenic plants and animals. This has provided great economic stimulus to the growth of many biotechnology industries. In this chapter, we will examine how molecular techniques have expanded our knowledge of the genetic characteristics of commercially important species. We will also discuss examples in which recombinant microorganisms and transgenic plants and animals have been given characteristics that are useful to humans. These include recombinant bacteria that make human insulin, transgenic tomatoes with a longer shelf life, and transgenic livestock that produce human proteins in their milk.

In the last two sections of this chapter, we will learn how molecular genetics and biotechnology can be applied directly to humans. Molecular genetics has

**USES OF MICROORGANISMS
IN BIOTECHNOLOGY**

**NEW METHODS FOR
GENETICALLY
MANIPULATING PLANTS
AND ANIMALS**

**APPLICATIONS OF
TRANSGENIC PLANTS
AND ANIMALS**

DNA FINGERPRINTING

HUMAN GENE THERAPY

spawned technologies based on the analysis of DNA among different individuals. DNA fingerprinting is now a common method to analyze the DNA from an individual. Like conventional fingerprinting, this technique can be utilized for identification purposes. It also can be applied to analyze possible genetic relationships between individuals. In the last section of this chapter, we will learn how certain inherited diseases are now being treated via gene therapy. In this relatively new approach, cloned genes are introduced into individuals with genetic diseases in an attempt to compensate for mutant genes.

USES OF MICROORGANISMS IN BIOTECHNOLOGY

Microorganisms are used to benefit humans in various ways (Table 21-1). In this section, we will examine how molecular genetic tools have become increasingly important in influencing and improving our use of microorganisms. Such tools can produce recombinant microorganisms with genes that have been manipulated *in vitro*. This approach can improve strains of microorganisms currently in use, and it has even yielded strains that make products not normally produced by microorganisms. For example, several human genes have been introduced into bacteria to produce medically important products such as insulin and human growth hormone.

Overall, the use of recombinant microorganisms is an area of great research interest and potential. As we will discuss in this section, several recombinant strains are in current use, making profitable products. However, in some areas of biotechnology, the commercial use of recombinant strains has proceeded slowly. This is particularly true for applications in which recombinant microorganisms may be used to produce food products or where they are released into the environment. In such cases, safety concerns and negative public perceptions have slowed the commercial use of recombinant microorganisms. Nevertheless, molecular genetic research continues, and many biotechnologists expect an expanding use of recombinant microbes in the future.

TABLE 21-1 Common Uses of Microorganisms

Application	Examples
Production of medicinal agents	Antibiotics
	Synthesis of human insulin in recombinant *E. coli*
Food fermentation	Cheese, yogurt, vinegar, wine, beer
Biological control	Control of plant diseases, insect pests, and weeds
	Symbiotic nitrogen fixation
	Prevent frost formation
Bioremediation	Cleanup of environmental pollutants such as petroleum hydrocarbons and recalcitrant synthetics

EXPERIMENT 21A

Scientists at Genentech engineered the production of somatostatin, the first human peptide hormone produced by recombinant bacteria

During the 1970s, geneticists became aware of the great potential of recombinant DNA technology to produce therapeutic agents to treat certain human diseases. Healthy individuals possess many different genes that encode short peptide and longer polypeptide hormones. Diseases can result when an individual is unable to produce these hormones. This can occur because the individual may have inherited a defective gene or because the cells that produce the hormone have been damaged.

In 1976, Robert Swanson and Herbert Boyer formed Genentech Inc. The aspiration of this company was to engineer bacteria to synthesize useful products, particularly peptide and polypeptide hormones. Their first contract was with Keiichi Itakura and Arthur Riggs at the City of Hope National Medical Center in Duarte, California. Their intent was to engineer a bacterial strain that would produce somatostatin, a human hormone that functions to inhibit the secretion of a number of other hormones, including growth hormone, insulin, and glucagon. Somatostatin was chosen because it is very small (it only contains 14 amino acids) and can be detected easily via radioimmunoassay (see the appendix).

Before discussing the details of this experiment, let's consider the researchers' approach to constructing the somatostatin gene. To express somatostatin in bacteria, the coding sequence for somatostatin must be inserted next to a bacterial promoter that is contained within a plasmid. In this experiment, the researchers did not clone the human somatostatin gene and insert it into a bacterial plasmid. Instead, they chemically synthesized short oligonucleotides that would hydrogen bond with each other to form the coding sequence for somatostatin:

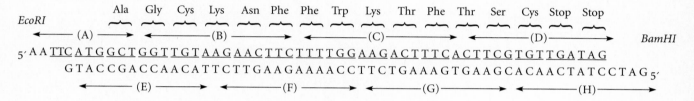

As shown here, eight separate oligonucleotides (labeled A through H) were synthesized chemically. Due to base complementarity within their sequences, the oligonucleotides hydrogen bonded to each other, forming a longer double-stranded DNA fragment with several important characteristics. First, its single-stranded ends allow it to be inserted into *Eco*RI and *Bam*HI restriction sites within plasmid DNA. The middle of this DNA fragment encodes the amino acid sequence of the somatostatin peptide hormone.

In addition, an extra methionine was added at the amino terminal end. This methionine provided a link between somatostatin and a bacterial protein (namely, β-galactosidase). This was necessary because the researchers learned, during the course of their experiments, that somatostatin made in bacteria is degraded rapidly by cellular proteases. To prevent this from happening, they linked the somatostatin sequence to the bacterial gene encoding β-galactosidase. When this linked gene is expressed in bacteria, a fusion protein is made between somatostatin and β-galactosidase. The fusion protein is not rapidly degraded. The researchers could then separate somatostatin from β-galactosidase by treatment with cyanogen bromide (CNBr), which cleaves polypeptides at the carboxyl terminal side of methionine:

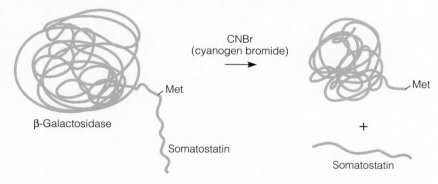

THE HYPOTHESIS

It is possible to produce human somatostatin in a recombinant bacterium.

TESTING THE HYPOTHESIS

Starting material: A normal *E. coli* strain that was unable to synthesize somatostatin.

Experimental Level   **Conceptual Level**

1. Chemically synthesize eight oligonucleotides. When they are mixed together, the complementary oligonucleotides will hybridize to each other.

2. Treat with DNA ligase to covalently link the oligonucleotides, which will form the molecule shown at the right.

3. Using recombinant techniques described in Chapter 19 (Figure 19-2), insert this fragment into a plasmid by digesting the plasmid at unique sites for *Eco*RI and *Bam*HI.

   Note: In their first attempt, the somatostatin fragment was expressed using a bacterial promoter, but the short peptide was degraded rapidly. As an alternative, the somatostatin sequence was cloned in-frame with the bacterial enzyme β-galactosidase. This plasmid encodes a fusion protein between β-galactosidase and somatostatin. As a control, a plasmid in which β-galactosidase is in the wrong orientation was also constructed. This plasmid does not encode a correct fusion protein.

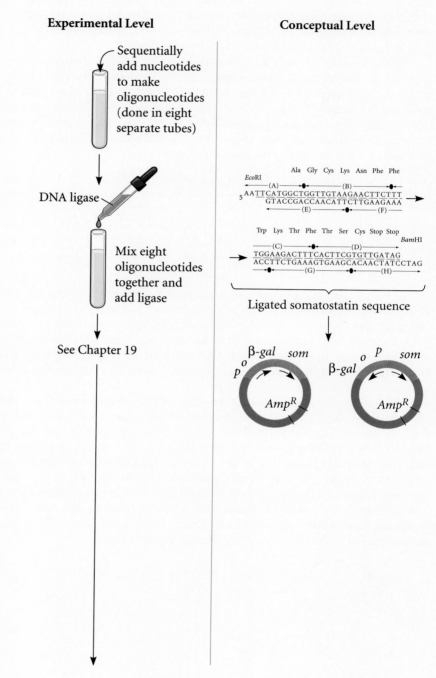

4. Transform the plasmids into *E. coli* by treatment with CaCl₂. The transformed cells are spread on plates containing ampicillin. Grow overnight to obtain bacterial colonies that are ampicillin resistant because they contain the plasmid.

5. Grow these recombinant bacteria and induce the somatostatin gene with isopropyl thiodigalactoside (IPTG).

   Note: The *lac* promoter is controlled by the Lac repressor (see Chapter 15). IPTG is an inducer that activates transcription of the β-galactosidase gene by removing the Lac repressor.

6. Place in a tube and centrifuge to obtain a bacterial cell pellet.

7. Resuspend the pellet in 70% formic acid and cyanogen bromide (5 mg/ml).

   Note: This breaks open the cells and cleaves polypeptides at methionine residues.

8. Determine the amount of somatostatin using a radioimmune assay. (See the appendix for a description of this procedure.)

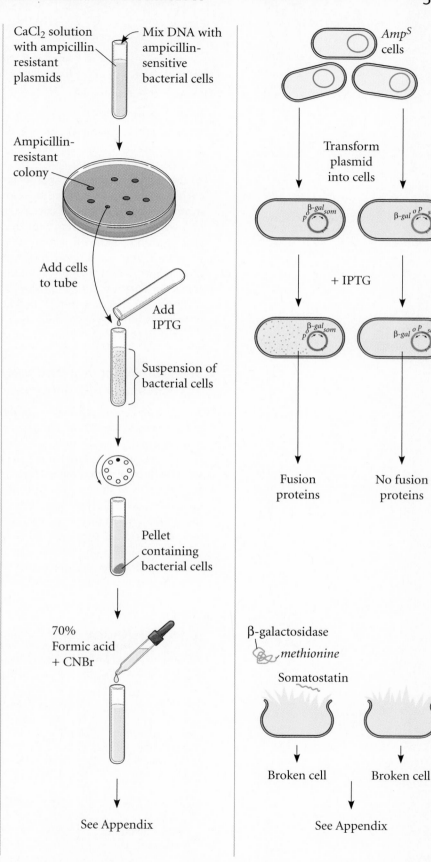

THE DATA

Plasmid Strain	Amount of Somatostatin (detected by RIA) (picograms of Somatostatin/milligram of bacterial proteins)
Correct orientation	8–320*
Incorrect orientation	< 0.1–0.4

*The amount of somatostatin was determined in several independent experiments.

INTERPRETING THE DATA

As shown in these data, recombinant bacteria carrying the somatostatin gene in the correct orientation produced this hormone. The amount of somatostatin varied from 8 to 320 pg per mg of bacterial proteins. This variability could be attributed to several factors, including protein degradation, incomplete cyanogen bromide cleavage, and unknown genetic changes in the plasmids during bacterial cell growth. In spite of this variability, the exciting result was the production of a human hormone in recombinant bacteria. By comparison, the plasmid with the incorrect orientation did not produce any significant amount of the hormone. This study was the first demonstration that recombinant bacteria could make products encoded by human genes. At the time, this was a major breakthrough that catalyzed the growth of the biotechnology industry!

Many important medicines are produced by recombinant microorganisms

Since the pioneering studies described in Experiment 21A, recombinant DNA technology has developed bacterial strains that synthesize several other human peptides and proteins. A few examples are described in Table 21-2.

In 1982, the U.S. Food and Drug Administration gave approval to human insulin made by recombinant bacteria. In normal individuals, insulin is produced by the beta-cells of the pancreas. Insulin functions to regulate several physiological processes, particularly the uptake of glucose into fat and muscle cells. Persons with insulin-dependent diabetes cannot synthesize an adequate amount of insulin, due to a defect in their beta-cells. Today, these people can purchase genetically engineered human insulin to treat their disease. Prior to 1982, insulin was isolated from the pancreases removed from cattle. Unfortunately, in some cases, diabetic individuals became allergic to cow insulin. These allergic patients had to use very ex-

TABLE 21–2 Examples of Medical Agents Produced by Recombinant Microorganisms

Drug	Action	Treatment
Insulin	A hormone involved in the uptake of glucose	For diabetic patients
Tissue plasminogen activator (TPA)	Dissolves blood clots	For heart attack victims and other arterial occlusions
Superoxide dismutase	Antioxidant	For heart attack victims to minimize damage to cardiac tissue
Factor VIII	Blood clotting factor	For certain types of hemophilia patients
Renin inhibitor	Lowers blood pressure	For hypertension
Erythropoietin	Stimulates the synthesis of red blood cells	For anemia

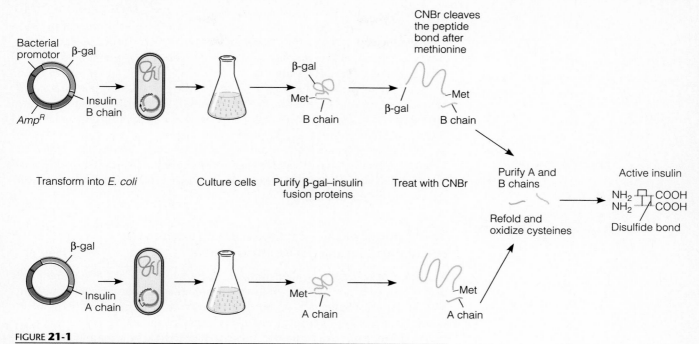

FIGURE **21-1**

The use of bacteria to make human insulin.

GENES ⟶ TRAITS: The synthesis of human insulin is not a trait that bacteria normally possess. However, genetic engineers can introduce the genetic sequences that encode the A and B chains of human insulin via recombinant DNA technology, yielding bacteria that make these polypeptides as fusion proteins with β-galactosidase. CNBr treatment releases the A and B polypeptides, which are then purified and oxidized to form functional human insulin.

pensive combinations of insulin from human cadavers and other animals. Now, of course, they can use human insulin made by recombinant bacteria.

As shown in Figure 21-1, insulin is a hormone composed of two polypeptide chains, called the A and B chains. To make this hormone using bacteria, the coding sequences of the A and B chains are placed next to the coding sequence of a native *E. coli* protein, β-galactosidase. This creates a fusion protein comprising β-galactosidase and the A or B chain. As with somatostatin, this step is necessary because the A and B chains are degraded rapidly when expressed in bacterial cells by themselves. The fusion proteins, however, are not. After the fusion proteins are expressed in bacteria, they can be purified and then treated with cyanogen bromide (CNBr) to separate β-galactosidase from the A or B chain. The A and B chains are then purified and mixed together under conditions in which they will fold and associate with each other to make a functional insulin hormone.

Bacterial species can be used as biological control agents

The term **biological control** refers to the use of microorganisms or their products to alleviate plant diseases or damage from environmental conditions (e.g., frost damage). During the past 20 years, interest in the biological control of plant diseases and insect pests as an alternative to chemical pesticides has increased. Biological control agents can prevent disease in several ways. In some cases, nonpathogenic microorganisms are used to compete effectively against pathogenic strains for nutrients or space. In other cases, microorganisms may produce a toxin that inhibits other pathogenic microorganisms or insects without harming the plant.

Biological control can also involve the use of microorganisms living in the field. A successful example is the use of *Agrobacterium radiobacter* to prevent crown gall disease caused by *Agrobacterium tumefaciens*. *A. radiobacter* produces agrocin 84, an antibiotic that kills *A. tumefaciens*. *A. radiobacter* contains genes that confer resistance to the agrocin 84 it produces.

Molecular geneticists have determined that the genes responsible for agrocin 84 synthesis and resistance are located on a plasmid. Unfortunately, this plasmid occasionally is transferred from *A. radiobacter* to *A. tumefaciens* during interspecies conjugation. When this occurs, *A. tumefaciens* can gain resistance to agrocin 84. To prevent this from happening, researchers have identified *A. radiobacter* strains in which this plasmid has been altered genetically to prevent its transfer during conjugation. This conjugation-deficient strain is now used commercially worldwide to prevent crown gall disease.

The release of recombinant microorganisms into the environment is sometimes controversial

As we have seen, genetically altered strains can have commercial applications in the field. Whether or not a microorganism is recombinant has become an important issue in the use of biological control agents that are released into the environment. If DNA is altered *in vitro* using molecular techniques and then reintroduced back into a microorganism, that microorganism is considered to be recombinant. In contrast, alterations in the genetic characteristics of a microorganism, such as the acquisition of naturally occurring plasmids or mutagenesis by chemical agents and radiation, are not considered to create recombinant strains.

Knowledge from molecular genetic research is used to develop both nonrecombinant and recombinant strains with desirable characteristics. Each year, many new strains of nonrecombinant microorganisms are analyzed in field tests for the biological control of plant diseases and insect pests. By comparison, the release of recombinant microorganisms and their use in field tests has proceeded much more slowly. This slow progress is related to increased levels of governmental regulation and, in some cases, to negative public perception of recombinant microorganisms.

As an example of the controversial nature of this topic, let's consider the first field test of a recombinant bacterium. It involved the use of a genetically engineered strain of *Pseudomonas syringae* to control frost damage. Experiments by Steven Lindow at the University of California, Berkeley, and his colleagues showed that the formation of ice on the surface of plants is enhanced by the presence of certain bacterial species. These *Ice*+ species synthesize cellular proteins that promote ice nucleation (i.e., the initiation of ice crystals). A way to prevent this from occurring is to use strains that cannot make ice-nucleation proteins (*Ice*− species). When applied to the surface of plants, an *Ice*− strain can compete with and thereby reduce the proliferation of *Ice*+ bacteria. Using recombinant DNA technology, an *Ice*− strain of *P. syringae* was constructed in the early 1980s.

Lindow sought approval for field tests of an *Ice*− recombinant strain in Tulelake, California. For several years, these tests were delayed because of a lawsuit from the Foundation on Economic Trends in Washington, D.C. During that time, Lindow made great efforts to ensure the safety of this project by studying the local environment. He also consulted with local townspeople where the field test was to take place. Initially, the idea was well received by the local residents. Unfortunately, however, another company tested similar bacteria on the roof of an Oakland facility. This experiment had not been approved by the Environmental Protection Agency (EPA). The media reported this incident, and it caused many townspeople to become apprehensive about the release of recombinant bacteria in Tulelake. Nevertheless, in 1987, approval was finally granted for the field testing of the recombinant *P. syringae* (Figure 21-2).

FIGURE **21-2**

The release of recombinant micro-organisms in a field test.

In the first test on several thousand strawberry plants, the plants were ripped out by vandals. In a second field test, the ability of *Ice⁻* bacteria to protect potato plants was tested. Again, some (but not all) of the plants were destroyed by vandals. The results of this field experiment showed that the *Ice⁻* bacteria did protect the potato plants from frost damage. In addition, soil sampling showed that the recombinant bacteria were contained at the field site and did not proliferate into surrounding areas. Even so, the release of recombinant microorganisms into the environment remains controversial. Since this first test, relatively few recombinant strains have been released.

Microorganisms can reduce environmental pollutants

The term **bioremediation** refers to the use of microorganisms to decrease pollutants in the environment. As its name suggests, this is a biological remedy for pollution. During bioremediation, enzymes produced by a microorganism modify a toxic pollutant by altering or transforming its structure. This event is called **biotransformation**. In many cases, biotransformation results in **biodegradation**, in which the toxic pollutant is degraded, yielding less complex, nontoxic metabolites. Alternatively, biotransformations without biodegradation can also occur. For example, toxic heavy metals can often be rendered less toxic by oxidation or reduction reactions carried out by microorganisms. Another way to alter the toxicity of organic pollutants is by promoting polymerization. In many cases, polymerized toxic compounds are less likely to leach from the soil and, therefore, are less environmentally toxic than their parent compounds.

Since the turn of the century, microorganisms have been used in the treatment and degradation of sewage. More recently, the field of bioremediation has expanded into the treatment of hazardous and refractory chemical wastes (i.e., chemicals that are difficult to degrade), usually associated with chemical and industrial activity. These pollutants include petroleum hydrocarbons, halogenated organic compounds, pesticides, herbicides, and organic solvents. Many new applications that use microorganisms to degrade these pollutants are being tested. The field of bioremediation has been fostered, to a large extent, by better knowledge of how pollutants are degraded by microorganisms, the identification of new and useful strains of microbes, and the ability to enhance bioremediation through genetic engineering.

Molecular genetic technology is key in identifying genes that encode enzymes involved in bioremediation. The characterization of the relevant genes greatly enhances our understanding of how microbes can modify toxic pollutants. In addition, recombinant strains created in the laboratory can be more efficient at degrading certain types of pollutants.

In 1980, in a landmark case (*Diamond v. Chakrabarty*), the U.S. Supreme Court ruled that a live, recombinant microorganism is patentable as a "manufacture or composition of matter." The first recombinant microorganism to be patented was an "oil-eating" bacterium, which contained a laboratory-constructed plasmid. This strain can oxidize the hydrocarbons commonly found in petroleum. It grew faster on crude oil than did any of the natural isolates that were tested. However, it has not been a commercial success, because this recombinant strain only metabolizes a limited number of toxic compounds; the number of compounds actually present in crude oil is over 3,000. Unfortunately, the recombinant strain did not degrade many higher-molecular-weight compounds, which tend to persist in the environment.

Currently, bioremediation should be considered a developing industry. This field will need well-trained molecular geneticists to conduct research aimed at elucidating the mechanisms whereby microorganisms degrade toxic pollutants. In the future, recombinant microorganisms may provide an effective way to decrease the levels of toxic chemicals within our environment. However, this approach will require careful studies to demonstrate that recombinant organisms are effective at reducing pollutants and safe when released into the environment.

NEW METHODS FOR GENETICALLY MANIPULATING PLANTS AND ANIMALS

As mentioned at the beginning of this chapter, transgenic organisms contain recombinant DNA that has been integrated into their genome. The production of transgenic plants and animals is a relatively new, exciting area of biotechnology. In recent years, a few transgenic species have reached the stage of commercialization. Many researchers believe that this technology holds great promise for innovations in agricultural quality and productivity. However, the degree to which this potential may be realized will depend, in part, on the public's concern about the release and consumption of transgenic species.

In some cases, transgenic organisms have been made in which the cloned gene from one species (the **transgene**) is transferred to some other species. A dramatic example of this is shown in Figure 21-3. In the experiment shown here, the gene that encodes the human growth hormone was introduced into the genome of a mouse. The larger mouse shown on the right is a transgenic mouse that expresses the human growth hormone gene.

In this section, we will begin by examining the two mechanisms whereby cloned DNA becomes integrated into the chromosomal DNA of animal and plant cells. We will then explore the current techniques used to create transgenic animals and plants.

The integration of a cloned gene into a chromosome can result in gene addition or gene replacement

In Chapter 19, we considered methods used to clone genes. As described there, a common approach is to insert a chromosomal gene into a vector and then propagate the vector in living microorganisms such as bacteria or yeast cells. Cloned genes can also

FIGURE **21-3**

A comparison between a normal mouse and a transgenic mouse.

GENES ⟶ TRAITS: The transgenic mouse (on the right) carries a gene encoding human growth hormone. This introduction of the human gene into the mouse's genome causes it to grow larger.

be introduced into plant and animal cells. However, to be inherited stably from generation to generation, the cloned gene must become integrated into one (or more) of the chromosomes that reside in the cell nucleus. The integration of cloned DNA into a chromosome occurs by recombination (which was described in Chapter 18).

Figure 21-4 illustrates the two ways that a cloned gene can integrate into a chromosome. In this example, the cloned gene has been rendered functionally inactive by a mutation. If the mutant gene recombines with the normal chromosomal gene by homologous recombination, then the mutant gene will replace the normal gene within the chromosome (Figure 21-4a). This is called **gene replacement**. In this case, it becomes possible to study how the loss of the normal gene affects the organism.

Alternatively, the mutant gene may recombine at another location within the genome by nonhomologous recombination (Figure 21-4b). When this occurs, both the normal and mutant genes will be present. This process is called **gene addition**.

Molecular biologists can produce mice that contain gene replacements

In bacteria and yeast, which have relatively small genomes, homologous recombination between cloned genes and the host cell chromosome occurs at a relatively high rate, so that gene replacement is commonly achieved. This is useful when a researcher or biotechnologist wants to make a gene mutation *in vitro* and then compare the effects of the normal and mutant genes on the phenotype of the organism. However, in more complex eukaryotes with very large genomes, the introduction of cloned genes into cells is much more likely to result in gene addition rather than gene replacement. For example, when a mutant gene is introduced into a mouse cell, it will undergo homologous recombination only 0.1% of the time. Almost always (namely, 99.9%), gene addition will occur.

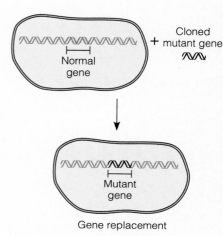

Gene replacement

(a) Homologous recombination

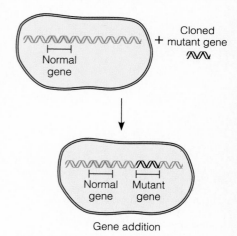

Gene addition

(b) Nonhomologous recombination

FIGURE **21-4**

The introduction of a cloned gene into a cell can lead to gene replacement or gene addition. (a) If the gene undergoes homologous recombination and replaces the normal chromosomal gene, this is gene replacement. **(b)** Alternatively, the cloned gene may recombine nonhomologously at some other chromosomal location, leading to gene addition.

To produce mice with gene replacements, molecular biologists have devised laboratory tricks to preferentially select cells in which homologous recombination has occurred. This approach is shown in Figure 21-5. As shown here, the cloned gene is altered using two selectable marker genes. These selectable markers influence whether or not mouse embryonic cells can grow in the presence of certain drugs. First, the normal cloned gene, which is called the target gene, is inactivated by inserting a neomycin resistance gene (called Neo^R) into the center of its coding sequence. Neo^R provides cells with resistance to neomycin. Next, a

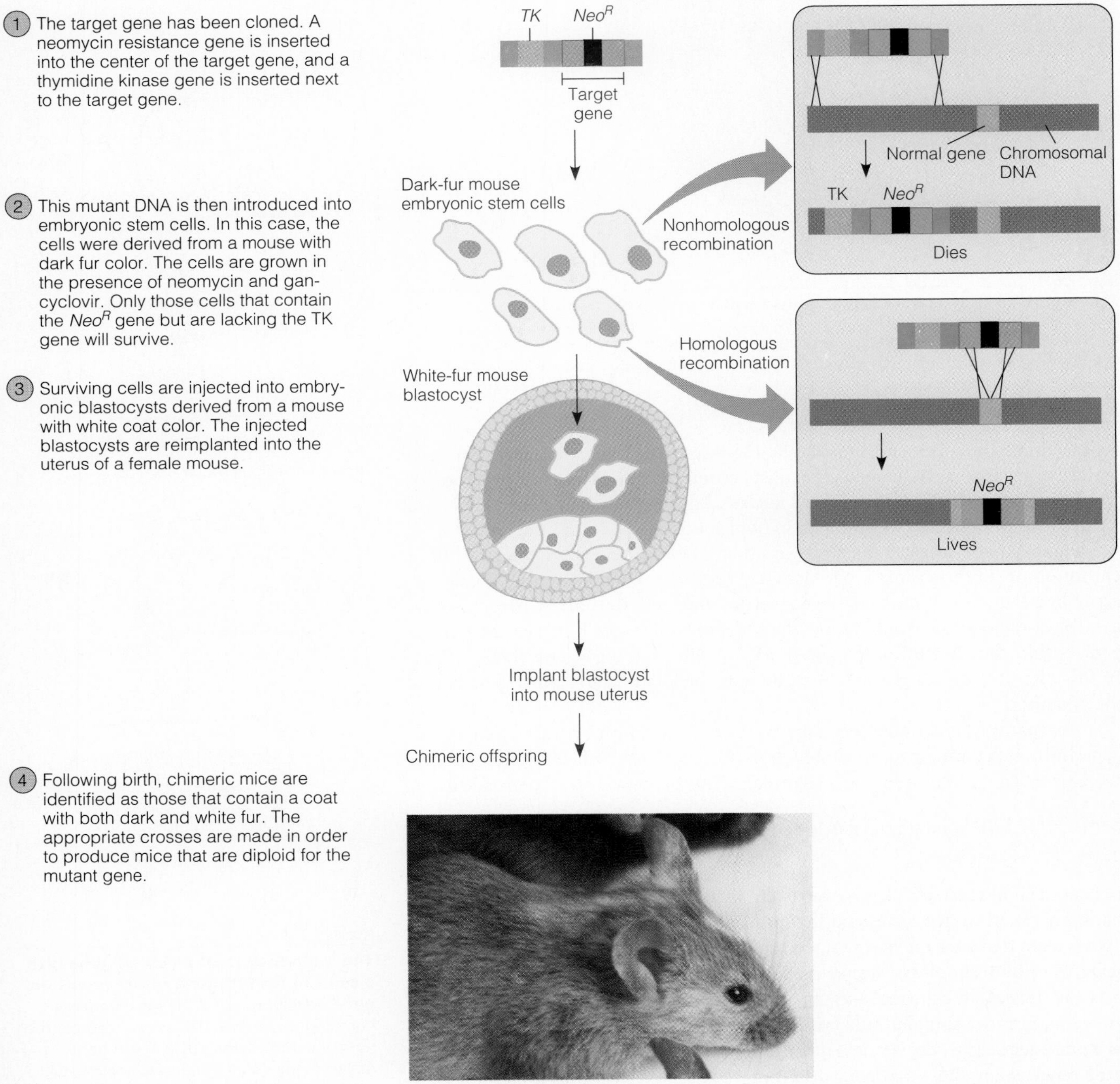

1. The target gene has been cloned. A neomycin resistance gene is inserted into the center of the target gene, and a thymidine kinase gene is inserted next to the target gene.

2. This mutant DNA is then introduced into embryonic stem cells. In this case, the cells were derived from a mouse with dark fur color. The cells are grown in the presence of neomycin and gancyclovir. Only those cells that contain the Neo^R gene but are lacking the TK gene will survive.

3. Surviving cells are injected into embryonic blastocysts derived from a mouse with white coat color. The injected blastocysts are reimplanted into the uterus of a female mouse.

4. Following birth, chimeric mice are identified as those that contain a coat with both dark and white fur. The appropriate crosses are made in order to produce mice that are diploid for the mutant gene.

TK Neo^R

Target gene

Dark-fur mouse embryonic stem cells

Nonhomologous recombination

Normal gene   Chromosomal DNA

TK Neo^R

Dies

White-fur mouse blastocyst

Homologous recombination

Neo^R

Lives

Implant blastocyst into mouse uterus

Chimeric offspring

FIGURE **21-5**

Producing a gene replacement in mice. The bottom shows a photograph of a chimeric mouse. Note the patches of black and white fur.

thymidine kinase gene, designated *TK*, is inserted adjacent to the target gene but not within the target gene itself. This gene renders cells sensitive to killing by a drug called gancyclovir.

After the target gene has been modified *in vitro*, it is introduced into mouse embryonic cells that can be grown in the laboratory. When the cells are grown in the presence of neomycin and gancyclovir, most nonhomologous recombinants will be killed, since they will also carry the *TK* gene. In contrast, homologous recombinants only contain the *Neo^R* gene, and so they will be resistant to both drugs. The surviving embryonic cells can then be injected into a blastocyst, an early embryo that is obtained from a pregnant mouse. In the example shown in Figure 21-5, the embryonic cells are from a mouse with dark fur and the blastocysts are from a mouse with white fur. The embryonic cells can mix with the blastocyst cells to create a **chimera**. This is an organism that contains cells from two different individuals. To identify chimeras, the injected blastocysts are reimplanted into a female mouse and allowed to develop. When this mouse gives birth, chimeras are identified easily, because they contain patches of white and dark coat color (see Figure 21-5).

Chimeric animals that contain a single gene replacement can then be mated to other mice to produce offspring that carry the mutant gene. Since mice are diploid, it is usually necessary to make two or more subsequent crosses to create a strain of mice that contain both copies of the mutant target gene. Later in this chapter, we will consider some of the applications of gene replacements in research and medicine.

Agrobacterium tumefaciens can be used to make transgenic plants

As we have just seen, the introduction of cloned genes into embryonic cells can produce transgenic animals. The production of transgenic plants is somewhat easier, because plant cells are totipotent, which means that an entire organism can be regenerated from somatic cells. Therefore, a transgenic plant can be made by the introduction of cloned genes into somatic tissue, such as the tissue of a leaf. After the cells of a leaf have become transgenic, an entire plant can be regenerated by the treatment of the leaf with plant growth hormones that cause it to form roots and shoots.

Molecular biologists can use the bacterium *Agrobacterium tumefaciens*, which naturally infects plant cells, to produce transgenic plants. A plasmid from the bacterium, known as the **Ti plasmid** (*T*umor-*i*nducing plasmid), induces tumor formation after a plant has been infected (see Figure 21-6a). A segment of the plasmid DNA, known as **T-DNA** (for transferred DNA), is transferred from the bacterium to the infected plant cells. The T-DNA from the Ti plasmid becomes integrated into the chromosomal DNA of the plant cell by recombination. After this occurs, genes within the T-DNA that encode plant growth hormones cause uncontrolled plant cell growth. This produces a cancerous plant growth known as a *crown gall tumor* (see Figure 21-6b).

Since *A. tumefaciens* inserts its T-DNA into the chromosomal DNA of plant cells, it can be used as a vector to introduce cloned genes into plants. Molecular geneticists have been able to modify the Ti plasmid to make this an efficient process. The T-DNA genes that cause tumorigenesis have been identified. Fortunately for genetic engineers, when these genes are deleted, the T-DNA is still taken up into plant cells and integrated within the plant chromosomal DNA. However, a crown gall tumor does not form. In addition, geneticists have inserted selectable marker genes into the T-DNA to allow selection of plant cells that have taken up the T-DNA. A gene that provides resistance to the antibiotic kanamycin is a commonly used selectable marker. Finally, the Ti plasmids used in cloning experiments have been modified to contain unique restriction sites for the convenient insertion of any gene.

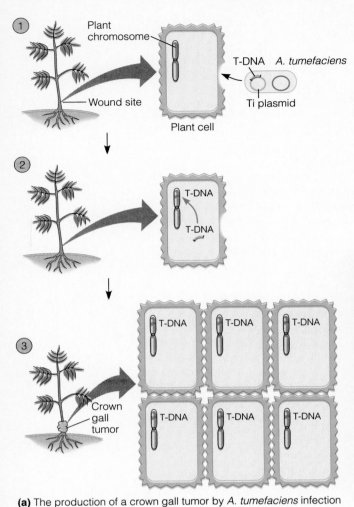

(1) *Agrobacterium tumefaciens* is found within the soil. A wound on the plant enables the bacterium to infect the plant cells.

(2) During infection, the T-DNA within the Ti plasmid is transferred to the plant cell. The T-DNA becomes integrated into the plant cell's DNA. Genes within the T-DNA promote uncontrolled plant cell growth.

(3) The growth of the plant cells produces a crown gall tumor.

(a) The production of a crown gall tumor by *A. tumefaciens* infection

(b) A crown gall on a pecan tree

FIGURE **21-6**

***Agrobacterium tumefaciens* infecting a pecan tree and causing a crown gall tumor.**

Figure 21-7 shows the general strategy for producing transgenic plants via T-DNA–mediated gene transfer. A desired gene is cloned into a genetically engineered Ti plasmid and then transformed into *A. tumefaciens*. Plant cells are exposed to the transformed *A. tumefaciens*. After allowing time for infection, the plant cells are grown in a liquid culture (step 2) using a growth medium that contains kanamycin and carbenicillin. Carbenicillin kills *A. tumefaciens*, and kanamycin kills any plant cells that have not taken up the T-DNA. Therefore, the only surviving cells are those plant cells that have integrated the T-DNA into their genome. Since the T-DNA also contains the cloned gene of interest, the selected plant cells are expected to have received this cloned gene as well. The medium also contains the plant growth hormones necessary for the regeneration of entire plants. These plants can then be analyzed to verify that they are transgenic plants containing the cloned gene.

Other methods are available for introducing genes into plant cells. For example, in an approach known as **biolistic gene transfer** (i.e., biological ballistics), plant cells are bombarded with high-velocity microprojectiles coated with DNA. When fired upon by this "DNA gun," the microprojectiles penetrate the cell wall and membrane and thereby enter the plant cell. Alternatively, DNA can enter plant cells by **microinjection** (i.e., by use of microscopic-sized needles) or may be targeted electrophoretically into cells by **electroporation** (i.e., by use of electrical current that creates transient pores in the plasma membrane through which DNA can enter a cell).

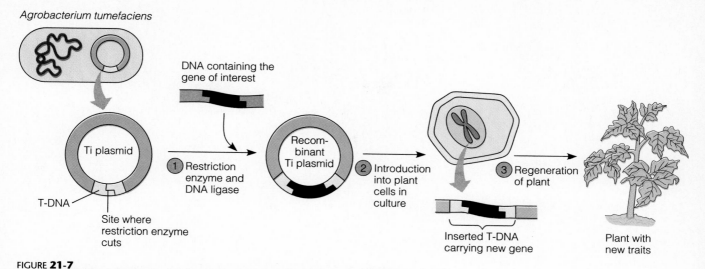

FIGURE **21-7**

The transfer of genes into plants using the Ti plasmid from *A. tumefaciens* as a vector.

Since the rigid plant cell wall is a difficult barrier for DNA entry, other approaches involve the use of protoplasts, which are plant cells that have had their cell walls removed. DNA can be introduced into protoplasts using a variety of methods, including treatment with polyethylene glycol and calcium phosphate.

The production of transgenic plants has become routine practice for many agriculturally important plant species. These include alfalfa, corn, cotton, soybean, tobacco, and tomato. Some of the applications of transgenic plants will be described in the next section.

Researchers have succeeded in cloning mammals from somatic cells

We now turn our attention to cloning as a way to genetically manipulate plants and animals. The term cloning has many different meanings. In Chapter 19, we discussed gene cloning, which involves methods that produce many copies of a gene. The cloning of an entire organism is a different matter. **Organismal cloning** refers to methods that produce two or more genetically identical individuals. By accident, this happens in nature; identical twins are genetic clones that began from the same fertilized egg. Similarly, researchers can take mammalian embryos at an early stage of development (e.g., two-cell to eight-cell stage), separate the cells, implant them into the uterus, and obtain multiple births of genetically identical individuals.

In the case of plants, cloning is an easier undertaking. Plants can be cloned from somatic cells. In most cases, it is easy to take a cutting from a plant, expose it to growth hormones, and obtain a separate plant that is genetically identical to the original. However, until recently, this approach has not been possible with mammals. For several decades, scientists believed that the somatic cells of mammals have incurred irreversible genetic changes that render them unsuitable for cloning. However, this hypothesis has proved to be incorrect. In 1997, Ian Wilmut at the Roslin Institute near Edinburgh and his colleagues created clones of sheep using somatic cells. As you may have heard, they named the first cloned lamb Dolly.

Figure 21-8 illustrates how Dolly was created. The researchers removed mammary cells from an adult female sheep and grew them in the laboratory. The researchers then extracted the nucleus from a sheep oocyte and fused a diploid mammary cell with the enucleated oocyte cell. Fusion was promoted by electrical

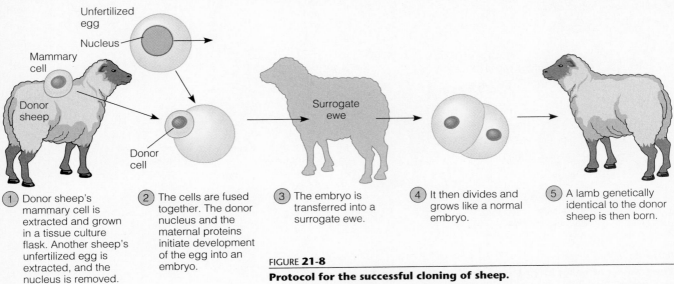

① Donor sheep's mammary cell is extracted and grown in a tissue culture flask. Another sheep's unfertilized egg is extracted, and the nucleus is removed.

② The cells are fused together. The donor nucleus and the maternal proteins initiate development of the egg into an embryo.

③ The embryo is transferred into a surrogate ewe.

④ It then divides and grows like a normal embryo.

⑤ A lamb genetically identical to the donor sheep is then born.

FIGURE **21-8**

Protocol for the successful cloning of sheep.

GENES → TRAITS: Dolly is a genetically identical copy of the sheep that donated a mammary cell to create Dolly. Dolly and the donor sheep are genetically identical in the same way that identical twins are; they carry the same set of genes and look remarkably similar.

pulses. After fusion, the zygote was implanted into the uterus of an adult sheep. One hundred and forty-eight days later, Dolly was born.

Mammalian cloning is still at an early stage of development. Nevertheless, the breakthrough of creating Dolly has shown that it is possible. This provides the potential for many practical applications. With regard to livestock, cloning would enable farmers to use the somatic cells from their best individuals to create genetically homogeneous herds. This could be advantageous with regard to agricultural yield, although such a genetically homogeneous herd may be more susceptible to rare diseases.

Aside from practical uses, however, people have become concerned with the possibility of human cloning. This prospect has raised serious ethical questions. In the future, society will have to make decisions regarding the application of cloning to humans.

APPLICATIONS OF TRANSGENIC PLANTS
AND ANIMALS

In this section, we will examine applications arising from the ability of scientists to make transgenic plants and animals. In research, transgenic species can provide important information regarding the functional roles of genes. In addition, transgenic animals as model organisms are becoming invaluable tools for investigating the mechanisms and treatment of human diseases.

A particularly exciting avenue of research is the use of transgenic species in agriculture. For centuries, agriculture has relied on selective breeding programs to produce plants and animals with desirable characteristics. Traditional mating strategies yield offspring carrying genes that will provide them with worthwhile traits. For agriculturally important species, this often means the production of strains that are larger, have disease resistance, and yield high-quality food. It is now technologically possible to complement traditional breeding strategies with modern molecular genetic approaches. In this section, we will discuss the current and potential uses of transgenic organisms in agriculture.

Gene replacements in mice can be used to understand gene function and human disease

Earlier, in Figure 21-5, we considered how a researcher can replace a normal mouse gene with one that has been inactivated by the insertion of an antibiotic resistance gene. When a mouse is homozygous for the inactivated gene, this is called a **gene knockout**. Both copies of the normal gene have been replaced by the inactive mutant gene. In other words, the function of the normal gene has been knocked out. Gene replacements and gene knockouts have become powerful tools for understanding gene function.

In some cases, gene knockouts have shown that the function of a gene is critical within a particular tissue or during a specific stage of development. In many cases, however, a gene knockout produces no detectable phenotypic effect at all. This has led geneticists to conclude that mammalian genomes have a fair amount of **gene redundancy**. This means that when one type of gene is inactivated, another gene with a similar function may be able to compensate for the inactive gene.

A particularly exciting avenue of gene replacement research is its application in the study of human disease. For example, the disease cystic fibrosis (CF) is one of the most common and severe inherited human disorders. We will describe its symptoms later in this chapter. In humans, the defective gene that causes CF has been identified. Using molecular techniques, the homologous gene in mice was also identified. CF gene knockouts have produced mice with disease symptoms resembling those found in humans (namely, respiratory and digestive abnormalities). Therefore, these mice can be used as model organisms to study this human disease. Furthermore, these mice models have been used to test the effects of various therapies in the treatment of the disease.

Biotechnology holds promise in producing transgenic livestock

The technology for creating transgenic mice has been extended to other animals, and much research is under way to develop transgenic species of livestock, including fish, sheep, pigs, goats, and cattle. The ability to modify the characteristics of livestock via the introduction of cloned genes is an exciting prospect.

A novel avenue of research involves the production of medically important proteins in the mammary glands of livestock. This approach is sometimes called **molecular pharming**. As shown in Table 21-3, several human proteins have been successfully produced in the milk of domestic livestock. Compared with the production of proteins in bacteria, one advantage is that certain proteins are more likely to function properly when expressed in mammals. This may be due to post-translational modifications (e.g., attachment of carbohydrate groups) that occur in

TABLE 21–3 Proteins that Can be Produced in the Milk of Domestic Animals

Protein	Host	Use
Lactoferrin	Cattle	Used as an iron supplement in infant formula
Tissue plasminogen activator (TPA)	Goat	Dissolves blood clots
α-1-Antitrypsin	Sheep	Treatment of emphysema
Factor IX	Sheep	Treatment of certain inherited forms of hemophilia
Insulin-like growth factor I	Cattle	Treatment of diabetes

eukaryotes but not in bacteria. In addition, certain proteins may be degraded rapidly or folded improperly when expressed in bacteria. Therefore, as an alternative to recombinant bacteria, the expression of human proteins in the milk of transgenic livestock is an area of great interest.

To introduce a human gene into an animal so that the encoded protein will be secreted into its milk, the strategy is to clone the gene next to a *milk-specific promoter*. As was discussed in Chapter 16, eukaryotic genes often are expressed in a tissue-specific fashion. In mammals, certain genes are expressed specifically within the mammary gland so that their protein product will be secreted into the milk. Examples of milk-specific genes include genes that encode the milk proteins such as β-lactoglobulin, casein, and whey acidic protein. In order to express a human gene into a domestic animal's milk, the promoter and regulatory sequences for these milk-specific genes are linked to the coding sequences for the human gene (Figure 21-9). In this way, the protein encoded by the human gene will be expressed within the mammary gland and secreted into the milk. The milk can then be obtained from the animal, and the human protein isolated.

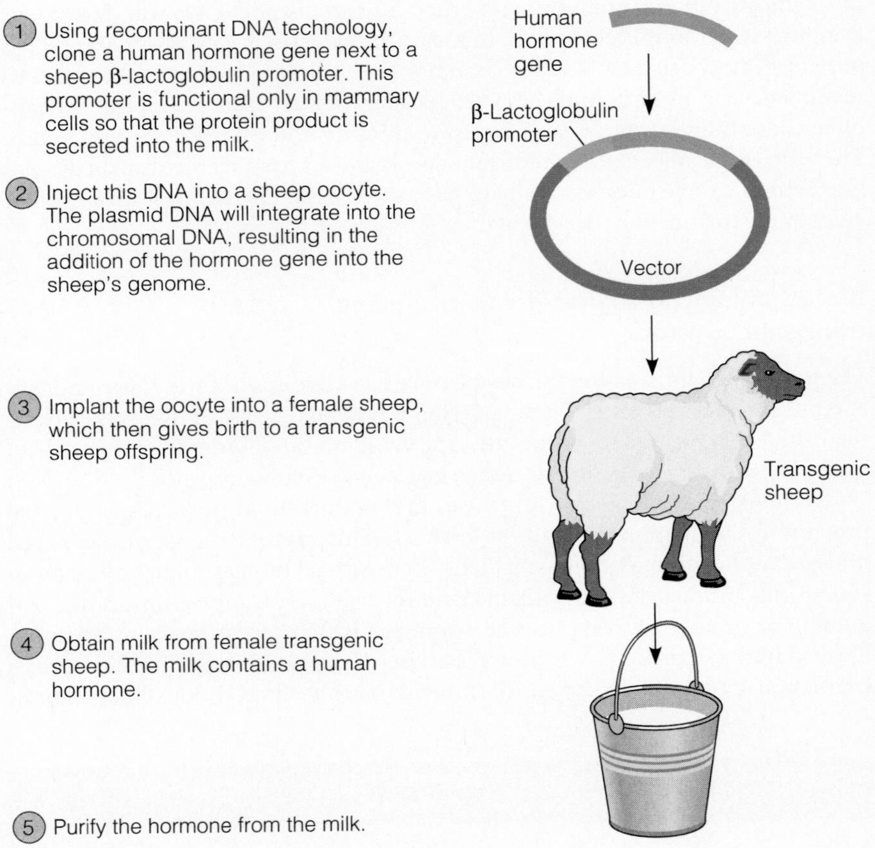

1. Using recombinant DNA technology, clone a human hormone gene next to a sheep β-lactoglobulin promoter. This promoter is functional only in mammary cells so that the protein product is secreted into the milk.

2. Inject this DNA into a sheep oocyte. The plasmid DNA will integrate into the chromosomal DNA, resulting in the addition of the hormone gene into the sheep's genome.

3. Implant the oocyte into a female sheep, which then gives birth to a transgenic sheep offspring.

4. Obtain milk from female transgenic sheep. The milk contains a human hormone.

5. Purify the hormone from the milk.

FIGURE **21-9**

Strategy for expressing human genes in a domestic animal's milk. The β-lactoglobulin promoter is normally expressed in mammary cells, whereas the human hormone gene is not. To express the human hormone gene in milk, the promoter from the milk-specific gene is linked to the coding sequence of the hormone gene. In addition to the promoter, a short signal sequence may also be necessary so that the protein will be secreted from the mammary cells and into the milk.

GENES → TRAITS: By using genetic engineering, researchers can give sheep the trait of producing a human hormone in their milk. This hormone can be purified from the milk and used to treat humans.

Transgenic plants can be given characteristics that are agriculturally useful

Various traits can be modified in transgenic plants. Frequently, transgenic research has sought to produce plant strains resistant to insects, disease, and herbicides. For example, transgenic plants highly tolerant of particular herbicides have been made. The Monsanto company has produced transgenic plant strains tolerant of glyphosate, the active agent in the herbicide Roundup™. Compared with nontransgenics, these plants grow quite well in the presence of glyphosate-containing herbicides (Figure 21-10). Another important approach is to make plant strains that are disease resistant. In many cases, virus-resistant plants have been developed by introducing a gene that encodes a viral coat protein. When the plant cells express the viral coat protein, they become resistant to infection by that pathogenic virus.

Transgenic plants are already on the market. One example is the FlavrSavr tomato (Figure 21-11), which was developed by Calgene Inc. (in California). This transgenic tomato plant has been given a gene that encodes an antisense RNA. This antisense RNA is complementary to the mRNA that codes for the enzyme polygalacturonase, which is involved in ripening. The antisense RNA thus binds to this mRNA, preventing it from being translated. In this way, the antisense RNA silences the expression of the polygalacturonase gene. The practical advantage of this transgenic tomato is improved shelf life; these tomatoes do not spoil (i.e., become overripe) as quickly as traditional tomatoes. Therefore, to improve their flavor, these tomatoes can be allowed to ripen on the vine for a longer period of time. This is an important consideration in the $5 billion annual U.S. tomato market. By comparison, other commercial tomatoes commonly are picked while they are green and ripened later.

FIGURE **21-11**

The genetically engineered FlavrSavr tomato.

GENES → TRAITS: This transgenic tomato plant has been engineered genetically to contain an artificial gene that encodes an antisense RNA to the gene encoding polygalacturonase. The antisense RNA will bind to the mRNA that encodes polygalacturonase because the two are complementary, thereby preventing this mRNA from being translated. The antisense RNA thus inhibits the expression of the polygalacturonase gene and inhibits the overripening process. In this way, the FlavrSavr tomato has the trait of not overripening as quickly as traditional tomatoes.

FIGURE **21-10**

Transgenic plants that are resistant to glyphosate.

GENES → TRAITS: This field of soybean plants has been treated with glyphosate. The plants on the left have been engineered genetically to contain a herbicide resistance gene. They are resistant to killing by glyphosate. By comparison, the dead or stunted plants in the row with the orange stick do not contain this gene.

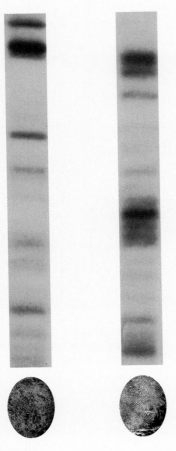

FIGURE 21-12

A comparison of two DNA fingerprints and two human fingerprints. The chromosomal DNA from two different individuals was subjected to DNA fingerprinting. Following the hybridization of a radiolabeled probe, their DNA appears as a series of bands on a gel. It is the dissimilarity in the pattern of these bands that distinguishes different individuals, much as the differences in physical fingerprint patterns can be used for identification.

DNA FINGERPRINTING

DNA fingerprinting is a technology that identifies particular individuals using properties of their DNA. Like the human fingerprint, the DNA of each individual is a distinctive characteristic that provides a means of identification. The application of DNA fingerprinting to forensics has captured the most public attention. Samples of blood and semen can be subjected to DNA fingerprinting to determine that a particular individual has been at a crime scene. In addition, DNA fingerprinting can determine if two individuals are related genetically. For example, DNA fingerprinting is used commonly in paternity testing. In this section, we will examine how DNA fingerprinting is conducted and consider several of its applications.

DNA fingerprinting provides a unique banding pattern for every individual

When subjected to DNA fingerprinting, the chromosomal DNA gives rise to a series of bands on a gel (Figure 21-12). The order of the bands on a gel is an individual's DNA fingerprint. It is the unique pattern of these bands that makes it possible to distinguish individuals.

The development of DNA fingerprinting relied on the identification of DNA fragments that are quite variable among members of the human population. This variability is needed so that each individual will have a unique DNA fingerprint. In the 1980s, Alec Jeffries and his colleagues at the University of Leicester in England found that certain locations within human chromosomes are particularly variable in their lengths. These loci contain tandemly repeated sequences called **minisatellites**. Within the human population, the number of tandem repeats at each minisatellite varies substantially. Therefore, another name for minisatellites is loci with a **variable number of tandem repeats (VNTRs)**.

The occurrence of VNTRs is shown schematically in Figure 21-13. In this figure, the chromosomal DNA from two individuals is compared. The diagram emphasizes two features of the chromosomal DNA. First, it shows the sites recognized by a particular restriction enzyme (designated R). As a matter of chance, these sites are interspersed throughout the genome of both individuals. Second, VNTRs (designated by sequences of Vs) are also found. At some sites, the two individuals have VNTRs of similar length. At other sites, the VNTRs of the two individuals differ substantially. Therefore, when the chromosomal DNAs are digested with the restriction enzyme, these differences will yield a distinct pattern of DNA fragments when analyzed via gel electrophoresis.

If all the chromosomal DNA from a sample of cells were digested with a restriction enzyme, this would yield too many DNA fragments to analyze. In DNA fingerprinting, the DNA probes used will hybridize specifically to the repeat sequence located within selected VNTRs. Investigators have found several different VNTRs in the human genome. DNA fingerprinting probes recognize a selected VNTR sequence found at a relatively small number of sites (say, 20 to 40 sites) in the human genome. A DNA fingerprinting analysis examines the length of the DNA fragments at the sites of the chosen VNTR sequence. This, therefore, involves analysis of the 20 to 40 bands that correspond to the VNTR sites.

Figure 21-14 outlines the steps in a DNA fingerprinting experiment. This procedure is a Southern blot using a radiolabeled probe complementary to a selected VNTR sequence. The chromosomal DNA is isolated from a sample and digested with a restriction enzyme. The resulting DNA fragments are then separated by gel electrophoresis. The fragments in the gel are blotted onto a nitrocellulose filter, the DNA is denatured, and the filter is exposed to the radiolabeled probe. Since the probe is complementary to the VNTR sequence, it hybridizes to approximately 20 to 40 fragments of DNA that contain this sequence and thereby labels 20 to 40 bands. This is called a *multilocus probe (MLP)*. In Figure 21-14, the results of a DNA

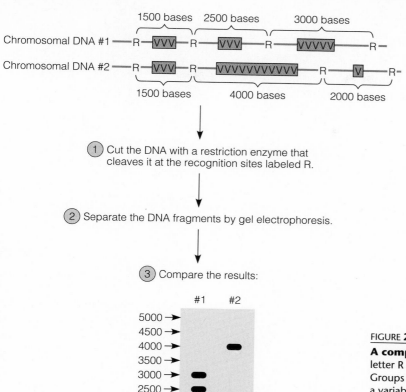

① Cut the DNA with a restriction enzyme that cleaves it at the recognition sites labeled R.

② Separate the DNA fragments by gel electrophoresis.

③ Compare the results:

FIGURE **21-13**

A comparison of VNTRs between two individuals. The letter R represents restriction sites found in both individuals. Groups of the letter V represent a sequence that is repeated a variable number of times between the two individuals. The variation in the number of repeats affects the sizes of the molecular fragments produced when the DNA is digested with the restriction enzyme that cleaves it at the designated sites.

fingerprinting experiment on two samples are compared. Since the pattern of band sizes does not match between the two samples, it is concluded that the samples came from different individuals.

In certain situations, the amount of DNA within a sample may be too small to obtain an accurate DNA fingerprint. For example, at a crime scene, only a small stain of blood may be available for DNA analysis. In such a case, the amount of DNA can be increased via polymerase chain reaction (PCR). This technique was described in Chapter 19 (pp. 522–524).

DNA fingerprinting is used for identification and relationship testing

Within the past decade or so, the use of DNA fingerprinting has expanded in many ways. DNA fingerprinting is gaining acceptance as a valid method of identification. In many criminal cases, it has provided critical evidence that an individual was at a crime scene.

When a sample taken from a crime scene matches the DNA fingerprint of an individual, the probability that this match could occur simply by chance can be calculated. Each VNTR size is given a probability score based on its observed frequency within a reference human population (e.g., Caucasian). A DNA fingerprint contains many bands, and the probability scores for each band can be multiplied together to arrive at the likelihood that a particular pattern of bands would be observed. For example, if a DNA fingerprint contains 20 bands, and the probability of each band is 1/4, the likelihood of having that pattern would be $(1/4)^{20}$, or roughly one in one trillion. Therefore, a match between two samples is rarely a matter of random chance. However, when evaluating DNA fingerprinting evidence, other factors

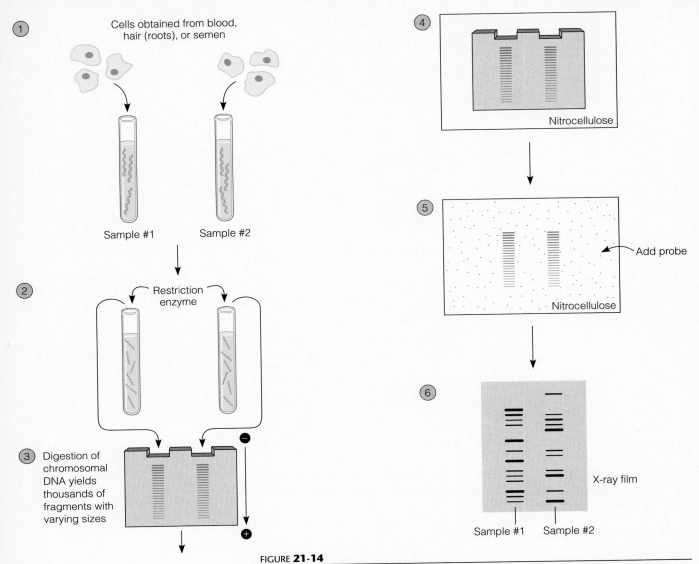

FIGURE **21-14**

Protocol for DNA fingerprinting. (1) Isolate chromosomal DNA from a sample of cells. This is often a blood sample, hair roots, or semen. In this experiment, we are comparing two different samples. **(2)** Cut the DNA with a restriction enzyme. **(3)** Separate the DNA fragments by gel electrophoresis. **(4)** Blot the gel to a nitrocellulose (or nylon) filter. **(5)** Denature the DNA, and add a radiolabeled probe that is complementary to a selected VNTR sequence. Allow the probe to hybridize, and then wash away the excess probe. **(6)** Expose the filter to X-ray film, and compare the results. The outcome of this experiment indicates that the two samples came from different individuals.

such as the potential mishandling of samples and the misinterpretation of banding patterns must also be considered.

Another important application of DNA fingerprinting is relationship testing. Persons who are related genetically will have some bands in common. The number of bands they share depends on the closeness of their genetic relationship. For example, an offspring is expected to receive half of their VNTRs from one parent and the rest from the other. Figure 21-15 shows the DNA fingerprint of an offspring, mother, and two potential fathers. In paternity testing, the offspring's DNA fingerprint is first compared with that of the mother. The bands that the offspring has in common with the mother are depicted in blue. The bands that are not similar between the offspring and the mother must have been inherited from the father.

These bands are depicted in red. In Figure 21-15, male #2 does not have many of the paternal bands. Therefore, he can be excluded as being the father of this child. However, male #1 has all of the paternal bands. Using a calculation similar to that just described, one can compute the likelihood that male #1 could have all of these bands by chance alone. This yields a rather small probability, and it is thus fairly certain that he is the biological father.

HUMAN GENE THERAPY

Throughout this text, we have considered examples in which mutant genes cause human diseases. As will be discussed in Chapter 23, these include rare inherited disorders and common diseases such as cancer. Since mutant genes cause disease, geneticists are actively pursuing the goal of using normal, cloned genes to compensate for defects in mutant genes. **Gene therapy** is the introduction of cloned genes into living cells in an attempt to cure disease. It represents a new, potentially powerful means of treating a wide variety of illnesses.

Many current research efforts in gene therapy are aimed at alleviating inherited human diseases. There are more than 4000 human genetic diseases that involve a single gene abnormality. Common examples include cystic fibrosis, sickle cell disease, and hemophilia. In addition, gene therapies have also been aimed at treating diseases, such as cancer and cardiovascular disease, that may occur later in life. Some researchers are even pursuing research that will use gene therapy to combat infectious diseases such as AIDS.

Human gene therapy is still at an early stage of development. Relatively few patients have been treated with gene therapy. Nevertheless, some of the initial results are promising. Table 21-4 describes several types of diseases that are being investigated as potential targets for gene therapy. In this section, we will examine the approaches to gene therapy and how it is being used to treat human disease.

Adenosine deaminase deficiency was the first inherited disease treated with gene therapy

Adenosine deaminase (ADA) is an enzyme involved in purine metabolism. If both copies of the ADA gene are defective, deoxyadenosine will accumulate within the cells of the individual. At high concentration, deoxyadenosine is particularly toxic to cells in the immune system, namely, T-cells and B-cells. In affected individuals, the destruction

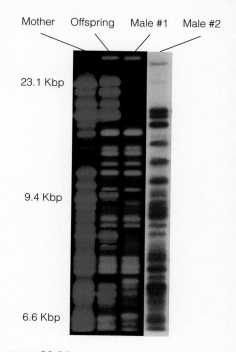

FIGURE **21-15**

The use of DNA fingerprinting to establish paternity. The bands are schematically colored blue for maternal and red for paternal. Because of matching bands with the offspring, male #1 is the likely father.

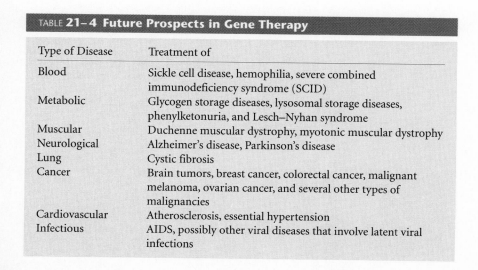

TABLE 21-4 Future Prospects in Gene Therapy	
Type of Disease	Treatment of
Blood	Sickle cell disease, hemophilia, severe combined immunodeficiency syndrome (SCID)
Metabolic	Glycogen storage diseases, lysosomal storage diseases, phenylketonuria, and Lesch–Nyhan syndrome
Muscular	Duchenne muscular dystrophy, myotonic muscular dystrophy
Neurological	Alzheimer's disease, Parkinson's disease
Lung	Cystic fibrosis
Cancer	Brain tumors, breast cancer, colorectal cancer, malignant melanoma, ovarian cancer, and several other types of malignancies
Cardiovascular	Atherosclerosis, essential hypertension
Infectious	AIDS, possibly other viral diseases that involve latent viral infections

of T- and B-cells leads to a severe combined immunodeficiency disease (SCID). If left untreated, SCID is typically fatal at an early age (generally, 1 to 2 years old), because the compromised immune system of these individuals cannot fight infections.

Three approaches can be used to treat ADA. In some cases, it is possible to receive a bone marrow transplant from a compatible donor. A second method is to treat SCID patients with purified ADA that is coupled to polyethylene glycol (PEG). This PEG–ADA is taken up by lymphocytes and can correct the ADA deficiency. Unfortunately, these two approaches are not always available and/or successful. A third, more recent approach is to treat ADA patients with gene therapy.

On September 14, 1990, the first human gene therapy was approved for a young girl suffering from ADA deficiency. This work was carried out by a large team of researchers composed of R. Michael Blaese, Kenneth Culter, W. French Anderson, and several collaborators at the National Institutes of Health (NIH). Prior to this clinical trial, the normal gene for ADA had been cloned into a viral vector that can infect lymphocytes. The general aim of this therapy was to remove lymphocytes from the blood of the person with SCID, introduce the normal ADA gene into the cells, and then return them to her bloodstream.

Figure 21-16 outlines the protocol for her experimental treatment. Lymphocytes (i.e., T-cells) were removed and cultured in a laboratory. The lymphocytes were then transfected with a nonpathogenic retrovirus that had been engineered genetically to contain the normal ADA gene. During the life cycle of a retrovirus, the retroviral genetic

FIGURE **21-16**

The first human gene therapy, for adenosine deaminase (ADA) deficiency.

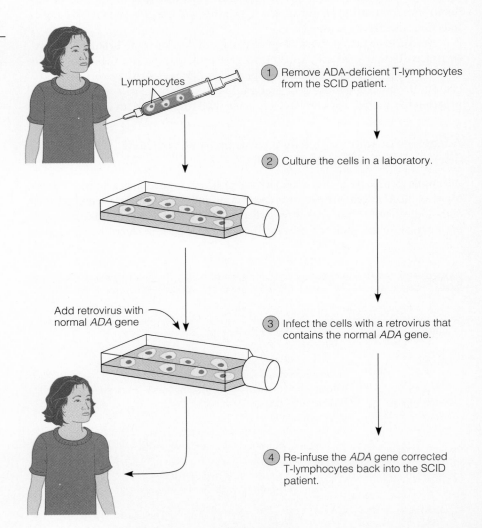

Lymphocytes

1 Remove ADA-deficient T-lymphocytes from the SCID patient.

2 Culture the cells in a laboratory.

Add retrovirus with normal *ADA* gene

3 Infect the cells with a retrovirus that contains the normal *ADA* gene.

4 Re-infuse the *ADA* gene corrected T-lymphocytes back into the SCID patient.

material is inserted into the host cell's DNA. Therefore, since this retrovirus contains the normal ADA gene, this gene also was inserted into the chromosomal DNA of the girl's lymphocytes. After this has occurred in the laboratory, the cells were reintroduced back into the patient. This is called an *ex vivo* approach because the genetic manipulations occur outside of the body, yet the products are re-introduced into the body.

Within 5 to 6 months of beginning gene therapy, the T-cell counts in this patient increased and stabilized in the normal range. Furthermore, they remained relatively stable even after the two year period when gene therapy was ended. In addition, ADA enzyme activity, which was nearly undetectable prior to gene therapy, increased progressively to a level about half that found in a heterozygous carrier. ADA enzyme activity has also remained stable even after gene therapy was ended. This patient, who had been kept in relative isolation in her home for her first 4 years, was enrolled in public kindergarten after 1 year on the protocol and has missed no more school than her classmates or siblings. She is considered to be normal by her parents. Overall, these first studies offer promise that gene therapy may be applied to treat certain diseases.

Aerosol sprays may be used to treat cystic fibrosis

Cystic fibrosis (CF) is one of the most common recessive inherited disorders with debilitating consequences. It is caused by a defect in a gene which encodes a protein that functions in the transport of ions across the plasma membrane of epithelial cells (e.g., cells lining the respiratory tract and the intestinal tract). A defect in membrane transport leads to an abnormality in salt and water balance that causes a variety of symptoms, particularly overaccumulation of mucus in the lungs. Even though great strides have been made in the treatment of the symptoms of CF, this disease remains associated with repeated lung infections and a shortened life. In most cases, mortality results from chronic lung infections.

CF has been the subject of much gene therapy research. As mentioned earlier in this chapter, CF gene knockouts in mice have generated a CF animal model. These animal models have been used to test the efficacy of gene therapy approaches. More recently, clinical trials have tested the ability of gene therapy to improve the condition of patients suffering from CF. To implement CF-gene therapy, it is necessary to deliver the normal CF gene to the lung cells. Unlike ADA-gene therapy in which the lymphocytes can be treated *ex vivo*, it is not possible to remove lung epithelial cells and then put them back into the individual. Instead, it is necessary to design innovative approaches that can target the CF gene directly to the lung cells.

To achieve this goal, two exciting CF-gene therapy methods involve the use of an aerosol spray. In one protocol, the normal CF gene is cloned into an adenovirus. Adenoviruses normally infect lung epithelial cells and cause a lung infection. This adenovirus, however, has been engineered so that it will gain entry into the epithelial cells but not cause a lung infection. In addition, the adenovirus has been engineered to contain the normal CF gene. A second approach is liposome delivery of the CF-gene. In this therapy, the normal CF gene is contained within liposomes (i.e., lipid-membrane vesicles). When inhaled by the patient via an aerosal spray, liposomes and adenoviruses are taken up by the lung epithelial cells.

Like ADA-gene therapy, CF-gene therapy is at an early stage of development. It is hoped that gene therapy eventually will become an effective method to alleviate the symptoms associated with this disease.

The connection between *biotechnology* and molecular genetics has become an important interface. Molecular geneticists are analyzing the microorganisms that we use for food fermentation, *biological control*, *bioremediation*, and the production of medicines, enzymes, and fuels. This research yields a better understanding of the

CONCEPTUAL
AND EXPERIMENTAL
SUMMARY

molecular mechanisms that underlie the beneficial features of these microorganisms. In many cases, this knowledge can be used to produce microbial strains better suited to serve humans. Some of these are recombinant strains that can make products not normally made by bacteria.

In addition to microorganisms, molecular tools are also being directed toward agriculturally important plants and animals. Several examples of *transgenic* species have been made and tested. A few are commercially available. In the future, it is expected that the production and commercial success of transgenic species will become more prevalent. This will provide exciting career opportunities for well-trained molecular geneticists.

Molecular genetic tools are also directed toward humans. Molecular genetics has produced a technology, *DNA fingerprinting*, that can be used as a tool for human identification. As a forensic tool, this technology can determine whether an individual may have been at a crime scene based on samples of blood, hair (roots), or semen. In addition, it can be used to evaluate possible genetic relationships among individuals.

Very recently, recombinant DNA technology is being used to combat human diseases via *gene therapy*. In this approach, a cloned gene is introduced into somatic cells as a way to compensate for a defective gene. In the future, this may become an important way to alleviate many human disorders.

PROBLEM SETS

SOLVED PROBLEMS

1. Describe the strategy for producing human proteins in the milk of livestock.

Answer: Milk proteins are encoded by genes with promoters and regulatory sequences that direct the expression of these genes within the cells of the mammary gland. To get other proteins expressed in the mammary gland, the strategy is to link the promoter and regulatory sequences from a milk-specific gene to the coding sequence of the gene that encodes the human protein of interest. In some cases, it is also necessary to add a signal sequence to the amino terminal end of the target protein. A signal sequence is a short polypeptide that directs the secretion of a protein from a cell. If the target protein does not already have a signal sequence, it is possible to use a signal sequence from a milk-specific gene to promote the secretion of the target from the mammary cells and into the milk. During this process, the signal sequence is cleaved from the secreted protein.

CONCEPTUAL AND EXPERIMENTAL QUESTIONS

1. What is a recombinant microorganism? Discuss some current commercial successes of recombinant microbes.

2. Recombinant bacteria can produce hormones that are normally produced in humans. Briefly describe how this is accomplished.

3. In Experiment 21A, why did the plasmid with β-galactosidase in the wrong orientation fail to produce somatostatin?

4. What is a biological control agent? Briefly describe three examples.

5. *Bacillus thuringiensis* can make toxins that kill insects. This toxin must be applied several times during the growth season to prevent insect damage. As an alternative to repeated applications, one strategy is to apply bacteria directly to leaves. However, *B. thuringiensis* does not survive very long in the field. Other bacteria, such as *Pseudomonas syringae*, do. Propose a way to alter *P. syringae* so that it could be used in the battle against insects. Discuss advantages and disadvantages of this approach compared with the repeated applications of the insecticide from *B. thuringiensis*.

6. A conjugation-deficient strain of *Agrobacterium radiobacter* is used to combat crown gall disease. Explain how this bacterium prevents the disease and the advantage of a conjugation-deficient strain.

7. What is bioremediation? What is the difference between biotransformation and biodegradation?

8. Explain how it is possible to select for homologous recombination in mice.

9. What phenotypic marker is used to readily identify chimeric mice?

10. What is a mouse model for human disease?

11. Here are the DNA fingerprints of five people: a child, mother, and three potential fathers:

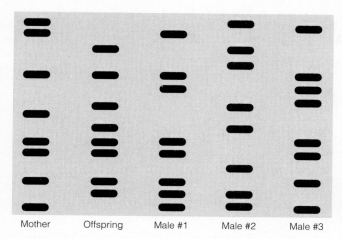

Mother   Offspring   Male #1   Male #2   Male #3

Which males can be ruled out as being the father? Explain your answer. If one of the males could be the father, explain the general strategy for calculating the likelihood that he could match the offspring's DNA fingerprint by chance alone.

12. What is a transgenic organism? Describe three examples.

13. To produce transgenic plants, plant tissue is exposed to *A. tumefaciens* and then grown in media containing kanamycin, carbenicillin, and plant growth hormones. Explain the purpose behind each of these three agents. What would happen if you left out the kanamycin?

14. What part of the *A. tumefaciens* DNA gets transferred to the genome of a plant cell during infection?

15. List and briefly describe five methods for the introduction of cloned genes into plants.

16. Discuss some of the worthwhile traits that can be modified in transgenic plants.

17. What is molecular pharming? Compared with the production of proteins by bacteria, why might it be advantageous?

18. What is DNA fingerprinting? How can it be used in human identification and relationship testing?

19. What is the difference between gene replacement and gene addition?

20. What roles do PCR and Southern blotting play in the analysis of DNA fingerprints?

21. What is a VNTR? Discuss the relationship between VNTRs and DNA fingerprinting.

22. Treatment of adenine deaminase deficiency (ADA) is an example of *ex vivo* gene therapy. Why is this therapy called *ex vivo*? Can *ex vivo* gene therapy be used to treat all human diseases? Explain.

23. Describe the targeting methods used in cystic fibrosis gene therapy. Provided the CF gene gets to the patient's lung cells, would you expect this to be a permanent cure for the patient, or would it be necessary to perform this gene therapy on a regular basis (say, monthly)?

QUESTIONS FOR STUDENT DISCUSSION/COLLABORATION

1. Discuss the advantages and disadvantages of gene therapy. Since a limited amount of funding is available for gene therapy research, make a priority list of the three top diseases that you would fund. Discuss your choices.

2. A commercially available strain of *P. syringae* marketed as Frostban B is used to combat frost damage. This is a naturally occurring *Ice⁻* strain. Discuss the advantages and disadvantages of using this strain compared with a recombinant version.

3. Make a list of the types of traits you would like to see altered in transgenic plants and animals. Suggest ways (i.e., what genes would you use?) to accomplish these alterations.

CHAPTER 22

COMPUTER ANALYSIS
OF GENETIC SEQUENCES

**GENERAL CONCEPTS IN
SEQUENCE ANALYSIS**

**IDENTIFICATION OF
FUNCTIONAL SEQUENCES**

HOMOLOGY

STRUCTURE PREDICTION

Geneticists use computers to collect, store, manipulate, and analyze data. Molecular genetic data, which comes in the form of a DNA, RNA, or amino acid sequence, is particularly amenable to computer analysis. In recent years, the marriage between genetics and biocomputing has yielded an important branch of science. Several scientific journals are devoted largely to this topic.

As the amount of genetic information has expanded rapidly, the scientific community has created large **databases** that serve as common storage facilities for genetic sequences. Most importantly, the ability of computers to analyze data at a rate of millions or even billions of operations per second has made it possible to solve problems concerning genetic information that were thought intractable a few decades ago. In this chapter, we will explore how geneticists utilize computer technology to analyze genetic data.

GENERAL CONCEPTS IN SEQUENCE ANALYSIS

Computer analysis of genetic sequences usually relies on three basic components: a computer, a computer program, and some type of data. In genetic research, the data are typically a particular genetic sequence (or several sequences) that a researcher or clinician wants to study. For example, this could be a DNA sequence derived from a cloned DNA fragment. A sequence of interest may be relatively short or thousands to millions of nucleotides in length. Experimentally, DNA sequences and related data are obtained using the techniques described in Chapter 19.

A **computer program** is a defined series of operations that can analyze data in a desired way. For example, a computer program might be designed to take a DNA sequence and translate it into an amino acid sequence. In this section, we will learn how computer programs are made, and how they are used to analyze genetic data. In later sections, we will consider how computer programs can identify functional genetic sequences, discover homology between evolutionarily related sequences, and help us to predict the structure of DNA, RNA, and proteins.

Genetic sequences are entered into a computer file

A first step in the computer analysis of genetic data is the creation of a **computer data file** to store the data. This file is simply a collection of information in a form suitable for storage and manipulation on a computer. In genetic studies, a computer data file typically might contain an experimentally obtained DNA, RNA, or amino acid sequence. For example, a file could contain the DNA sequence of the *lacY* gene from *Escherichia coli*, shown here:

1	ATGTACTATT	TAAAAAACAC	AAACTTTTGG	ATGTTCGGTT	TATTCTTTTT
51	CTTTTACTTT	TTTATCATGG	GAGCCTACTT	CCCGTTTTTC	CCGATTTGGC
101	TACATGACAT	CAACCATATC	AGCAAAGTG	ATACGGGTAT	TATTTTGCC
151	GCTATTTCTC	TGTTCTCGCT	ATTATTCCAA	CCGCTGTTTG	GTCTGCTTC
201	TGACAAACTC	GGGCTGCGCA	AATACCTGCT	GTGGATTATT	ACGGCATGT
251	TAGTCATGTT	TGCGCCGTTC	TTTATTTTA	TCTTCGGGCC	ACTGTTACAA
301	TACAACATTT	TAGTAGGATC	GATTGTTGGT	GGTATTTATC	TAGGCTTTTG
351	TTTTAACGCC	GGTGCGCCAG	CAGTAGAGGC	ATTTATTGAG	AAAGTCAGCC
401	GTCGCAGTAA	TTTCGAATTT	GGTCGCGCGC	GGATGTTTGG	CTGTGTTGGC
451	TGGGCGCTGT	GTGCCTGAT	TGTCGGCATC	ATGTTCACCA	TCAATAATCA
501	GTTTGTTTTC	TGGCTGGGCT	CTGGCTGTGC	ACTCATCCTC	GCCGTTTTAC
551	TCTTTTTCGC	CAAAACGGAT	GCGCCCTCTT	CTGCCACGGT	TGCCAATGCG
601	GTAGGTGCCA	ACCATTCGGC	ATTTAGCCTT	AAGTGGCAC	TGGAACTGTT
651	CAGACAGCCA	AAACTGTGGT	TTTGTCTACT	GTATGTTATT	GGCGTTTCCT
701	GCACCTACGA	TGTTTTTGAC	CAACAGTTTG	CTAATTTCTT	TACTTCGTTC
751	TTTGCTACCG	GTGAACAGGG	TACGCGGGTA	TTTGGCTACG	TAACGACAAT
801	GGGCGCAATTA	CTTAACGCCT	CGATTATCTT	CTTGCGCCA	CTGATCATTA
851	ATCGCATCGG	TGGGAAAAAC	GCCCTGCTGC	TGGCTGGCAC	TATTATCTCT
901	CTACGTATTA	TTGGCTCATC	GTTCGCCACC	TCAGCGCTGG	AAGTGGTTAT
951	TCTGAAAACG	CTGCATATGT	TTGAAGTACC	GTTCCTGCTG	GTGGGCTGCT
1001	TTAAATATAT	TACCAGCCAG	TTTGAAGTGC	GTTTTTCAGC	GACGATTTAT
1051	CTGGTCTGTT	TCTGCTTCTT	TAAGCAACTG	GCATGATTT	TTATGTCTGT
1101	ACTGGCGGGC	AATATGTATG	AAAGCATCGG	TTTCCAGGGC	GCTTATCTGG
1151	TGCTGGGTCT	GGTGGCGCTG	GGCTTCACCT	TAATTTCCGT	GTTCACGCTT
1201	AGCGGCCCCG	GCCCGCTTTC	CCTGCTGCGT	CGTCAGGTGA	ATGAAGTCGC
1251	TTAA				

To store data, such as this sequence, in a computer data file, a scientist creates the file (if it does not already exist) and enters the data. This may be done simply by using a keyboard to enter (i.e., type) the data into the file. For example, a scientist could obtain a DNA sequence by reading a sequencing ladder like the one shown in Figure 19-11 and then keyboarding the data into a computer file.

Entry via a keyboard is a common way to create a small computer data file, say for short genetic sequences (e.g., less than a few thousand nucleotides in length). For very long genetic sequences, like those obtained in genome sequencing projects, this would be a tedious and error-prone process. As an alternative, many laboratory instruments (e.g., densitometers and fluorometers) are designed with the capability to read data such as a sequencing ladder and enter the information directly into a computer file.

Sequence files are analyzed by computer programs

The purpose of making a computer file that contains a genetic sequence is to take advantage of the swift speed with which computers can analyze this information. Genetic sequence data in a computer file can be investigated in many different ways corresponding to the myriad questions a researcher might ask about the sequence and its functional significance. These include:

1. Does a sequence contain a gene?
2. Where are functional sequences such as promoters, regulatory sites, and splice sites located within a particular gene?
3. Does a sequence encode a polypeptide? If so, what is the amino acid sequence of the polypeptide?
4. Does a sequence predict certain structural features for DNA, RNA, or proteins? For example, is a DNA sequence likely to be in a B-DNA conformation? What is the secondary structure of an RNA sequence or polypeptide sequence?
5. Is a sequence homologous to any other known sequences?
6. What is the evolutionary relationship between two or more genetic sequences?

To answer these and many other questions, computer programs have been written to analyze genetic sequences in particular ways. These programs have been devised by theoreticians who understand basic genetic principles and can design computational strategies for analyzing genetic sequences. The process of developing a computer program is outlined here:

Flowchart for the Development of a Computer Program

1. A theoretician or group of theoreticians decide on the goal that a program is meant to fulfill.
↓

2. The theoretician(s) develop computational strategies that serve to analyze genetic sequences. This consists of a "plan of operations."
↓

3. The plan of operations is encoded in a computer language. In some cases, the encoding is done by computer programmers.
↓

4. The computer program is tested to see if it can properly analyze the data in a sequence file(s).
↓

5. The program is debugged until it meets the desired goal.
↓

6. The program is tried by users to evaluate how easy it is to operate. In some cases, the program may be modified to make it more user-friendly.

When constructing a computer program, a theoretician has a goal that the program is meant to fulfill. For example, the theoretician may wish to write a program to translate a DNA sequence into an amino acid sequence. Based on knowledge of the genetic code, a theoretician can devise computational procedures, or **algorithms**, that relate a DNA sequence to an amino acid sequence.

In a computer program, these procedures are executed as a stepwise "plan of operations" that manipulates the data in a sequence file. For a computer to perform the plan of operation, the program instructions must be written in a programming language the computer can decipher. After this has been accomplished, the program is tested to see if it works. Usually, errors or bugs are found in the program that prevent it from working. After these bugs are overcome, in a process called *debugging*, the program is ready for use. An ideal computer program is one that accurately manipulates data and at the same time is easy to use (user-friendly).

Let's consider how a computer program to translate a DNA sequence into an amino acid sequence might work in practice. The operation of the program as it would appear on a computer screen is shown in Figure 22-1. The geneticist (i.e., the user) has a DNA sequence file that he/she may want to have translated into an amino acid sequence. The user is sitting at a computer; this computer contains the program that can translate a DNA sequence into an amino acid sequence. In this hypothetical example, the program is named TRANSLATION.

On the computer screen, the user sees the C prompt (i.e., C:\>). (Note: This prompt may be different for different computers.) To begin running the program, the user types the name of the program, TRANSLATION. As it runs, the program will ask the user a series of questions. These are depicted on the screen. The first question asks which sequence file the user wants translated. In this case, the user wants the *lacY* sequence (refer back to p. 601) translated into an amino acid sequence. The name of this file is LacY.SEQ. The program retrieves this file and then asks the user what part of this file the user wants translated. The user decides to begin the translation at the first nucleotide in the sequence file and end the translation at nucleotide number 1251.

The program is now ready to translate the *lacY* DNA sequence into an amino acid sequence. It will place the translated amino acid sequence into another data file. The program asks the user to provide a name for the file that will contain the translated sequence. The user decides to name it LacY.PEP. (Note: The symbols PEP provide a reminder that it is a poly*pep*tide sequence). The translation program proceeds

FIGURE **22-1**

The operation of a computer program as it would appear on a computer screen.

to translate the *lacY* sequence and stores the amino acid sequence in a file called LacY.PEP. The contents of the LacY.PEP file are shown here:

1	MYYLKNTNFW	MFGLFFFFYF	FIMGAYFPFF	PIWLHDINHI	SKSDTGIIFA
51	AISLFSLLFQ	PLFGLLSDKL	GLRKYLLWII	TGMLVMFAPF	FIFIFGPLLQ
101	YNILVGSIVG	GIYLGFCFNA	GAPAVEAFIE	KVSRRSNFEF	GRARMFGCVG
151	WALCASIVGI	MFTINNQFVF	WLGSGCALIL	AVLLFFAKTD	APSSATVANA
201	VGANHSAFSL	KLALELFRQP	KLWFLSLYVI	GVSCTYDVFD	QQFANFFTSF
251	FATGEQGTRV	FGYVTTMGEL	LNASIMFFAP	LIINRIGGKN	ALLLAGTIMS
301	VRIIGSSFAT	SALEVVIKLT	LHMFEVPFLL	VGCFKYITSQ	FEVRFSATIY
351	LVCFCFFKQL	AMIFMSVLAG	NMYESIGFQG	AYLVLGLVAL	GFTLISVFTL
401	SGPGPLSLLR	RQVNEVA			

In this file, which was created by the computer program TRANSLATION, each of the 20 amino acids is given a single letter abbreviation (see Figure 12-11, p. 333). If you like, you can compare the nucleotide sequence on page 601 and the translated sequence shown here to see if the computer has made any mistakes! The advantages of running this program are speed and accuracy. It can translate a relatively long genetic sequence within seconds. By comparison, it would probably take you several minutes or hours to look each codon up in the genetic code table and write the sequence out in the correct order.

In genetic research, computer programs (i.e., *software*) are frequently sold as parts of large software packages. A genetics software package contains many computer programs that can analyze genetic sequences in different ways. For example, one program can translate a DNA sequence into an amino acid sequence, another program can locate introns within genes, and there are many other capabilities. These software packages commonly are sold to universities, hospitals, and industry. At these locations, a central computer with substantial memory and high-speed computational abilities is responsible for running the software. Individuals with personal computers can connect to this central computer. In this way, many individuals can use the programs within a software package. In addition, many programs are available on the Internet.

Although some programs are relatively simple, many programs are more complex and difficult to run. It is usually beneficial for students and researchers to enroll in courses that familiarize them with the available programs and teach them the basic computer skills needed to run the programs. Nevertheless, proficiency at running many programs often comes with struggle and experience.

The scientific community has collected sequence files and stored them in large computer databases

In Chapter 19, we considered how researchers can clone and sequence genes. Likewise, in Chapter 20, we learned how scientists are investigating the genetic sequences of entire genomes from several species, including humans. The amount of genetic information that is generated by researchers has become enormous. The Human Genome Project, for example, will yield more data than any other undertaking in the history of biology. With these advances, scientists realize that another critical use of computers is to store the staggering amount of data produced from genetic research.

When a large number of computer data files are collected and stored in a single location, this is called a computer database. The scientific community has collected the genetic information from thousands of research labs and created several large databases. Table 22-1 describes some of the major genetic databases in use worldwide. These databases enable researchers to access and compare genetic sequences that are obtained by many laboratories. Later in this chapter, we will learn how researchers can use databases to analyze genetic sequences.

TABLE 22–1 Examples of Major Computer Databases	
Name[a]	Contents
GenBank	Nucleic acid sequences from published articles and sequences submitted directly
EMBL	Nucleic acid sequences from published articles and sequences submitted directly
SWISS-PROT	Amino acid sequences from published articles and translated sequences from EMBL
NBRF/PIR	Amino acid sequences from published articles
PROSITE	Sequences of protein motifs
PDB	Protein and nucleic acid three-dimensional structures from X-ray crystallographic, NMR, and molecular modeling data

[a] EMBL = European Molecular Biology Laboratory Data Library; SWISS-PROT = Swiss Protein Database; NBRF/PIR = National Biomedical Research Foundation/Protein Information Resource; PDB = Protein Data Bank.

From: Boguski, M. S. (1992) Computational sequence analysis revisited: New databases, software tools, and the research opportunities they engender. *J. Lipid Res. 33*, 957–974.

Some databases contain a enormous collection of sequence files. For example, at the beginning of 1998, GenBank contained more than 700 million base pairs of DNA sequence. A significant proportion of the sequences have come from human sequencing data, although sequence data from more than 10,000 different species are contained within GenBank. In addition to a nucleotide sequence, each sequence file contains a concise description of the sequence and the name of the organism from which the sequence was obtained. The file may also describe other features of significance and provide a published reference that contains the sequence.

The databases described in Table 22-1 collect genetic information from many different species. Scientists have also created more specialized databases, called **genome databases**, that focus on the genetic characteristics of a single species. Genome databases have been created for species of bacteria (e.g., *E. coli*), yeast (e.g., *Saccharomyces cerevisiae*), worms (e.g., *Caenorhabditis elegans*), fruit flies (e.g., *Drosophila melanogaster*), plants (e.g., *Arabidopsis thaliana*), and mammals (e.g., mice and humans). The primary aim of genome databases is to organize the information from sequencing and mapping projects for a single species. Genome databases identify the known genes within an organism and describe their map locations in the genome. In addition, a genome database may provide information concerning gene alleles, bibliographic information, a directory of researchers who study the species, and other pertinent information.

IDENTIFICATION OF FUNCTIONAL SEQUENCES

At the molecular level, the function of the genetic material is based largely on specific genetic sequences that play distinct roles. For example, codons are three-base sequences that specify particular amino acids, and promoters are sequences that provide a binding site for RNA polymerase to initiate transcription. Computer programs can be designed to scan very long sequences and locate meaningful features within the sequence. To illustrate this concept, let's first consider the following sequence file, which contains an alphabetic sequence of 54 letters:

Sequence file:

GJTYLLAMAQLHEOGYLTOBWENTMNMTORXXXTGOODNTHEQALL
YTLSTORE

We will now compare how three different computer programs can analyze this sequence to identify meaningful features. The goal of our first program is to locate all of the English words within this sequence. If we ran this program, we would obtain the following result:

GJTY<u>LLAMA</u>Q<u>L</u><u>HE</u>OGYL<u>TO</u>B<u>WENT</u>MNM<u>TOR</u>XXXT<u>GOOD</u>N<u>THE</u>Q<u>ALL</u>
YTL<u>STORE</u>

In this case, a computer program has identified locations where the sequence of letters forms a word. Several words (which are underlined) have been located within this sequence.

A second computer program could be aimed at locating a series of words that are organized in the correct order to form a grammatically logical English sentence. If we used our sequence file and ran this program, we would obtain the following result:

GJTYLLAMAQL<u>HE</u>OGYLTOB<u>WENT</u>MNM<u>TO</u>RXXXTGOODN<u>THE</u>QALL
YTL<u>STORE</u>

The second program has identified five words that form a logical sentence.

Finally, a computer program might be used to identify patterns of letters, rather than words or sentences. For example, a computer program could locate a pattern of five letters that occurs in both the forward and reverse directions. If we applied this program to our sequence file, we would obtain the following:

GJ<u>TYLLA</u>MAQLHEOGYLTOBWENTMNMTORXXXTGOODNTHEQ<u>ALL</u>
<u>YTL</u>STORE

In this case, the program has identified a pattern where five letters are found in both the forward and reverse directions.

In the three previous examples, we can distinguish between **sequence recognition** (as in our first example) and **pattern recognition** (as in our third example). In sequence recognition, the program has the information that a specific sequence of symbols has a specialized meaning. This information must be supplied to the computer program. For example, the first program would have access to the information from a dictionary with all known English words. With this information, the first program can identify sequences of letters that make words. By comparison, the third program does not rely on specialized sequence information. Rather, it is looking for a pattern of symbols that can occur within any group of symbol arrangements.

Overall, the simple programs we have considered illustrate three general types of identification strategies:

1. Locate specialized sequences within a very long sequence. A specialized sequence is called a **sequence element**. Sequence elements are predefined and are contained explicitly within the computer program.
2. Locate an organization of sequences. As shown in the second program, this could be an organization of sequence elements. Alternatively, it could be an organization of a pattern of sequences.
3. Locate a pattern of sequences. The third program is an example.

The great power of computer analysis is that these types of operations can be performed with great speed and accuracy on sequences of symbols that may be enormously long. In this section, we will examine how sequence and pattern recognition can be applied to genetic sequences.

Computer programs can locate short specialized sequences within much longer genetic sequences

As we have discussed throughout this text, there are many short nucleotide sequences that play specialized roles in the structure or function of genetic material. In general, a genetic sequence with a particular function is called a sequence element or **sequence motif**. Table 22-2 lists some examples. A geneticist may want to locate a short sequence element within a longer nucleotide sequence in a data file. For example, a sequence of chromosomal DNA might be tens of thousands of nucleotides in length, and a geneticist may want to know whether a sequence element, such as a TATA box, is found at one or more sites within the chromosomal DNA. To do so, a researcher could visually examine the long chromosomal DNA sequence in search of a TATA sequence. Of course, this would be tedious and prone to error. By comparison, the appropriate computer program can locate a sequence element within seconds. Therefore, computers are very useful for this type of application.

By comparing the amino acid sequences and known functions of proteins in thousands of cases, researchers have also found amino acid motifs that carry out specialized functions within proteins. For example, researchers have determined that the amino acid motif asparagine–X–serine (where X is any amino acid except proline) within eukaryotic proteins is a glycosylation site (i.e., it has a carbohydrate attached to it). The PROSITE database (refer back to Table 22-1) contains a collection of all amino acid sequence motifs known to be functionally important. Researchers can use computer programs to determine whether an amino acid sequence contains any of the motifs found in the PROSITE database. This may help them to elucidate the role of a newly found protein of unknown function.

Several computer-based approaches can be used to identify structural genes within a nucleotide sequence

As was described in Chapter 12, a structural gene is composed of nucleotide sequences organized in a particular way. A typical gene contains a promoter, followed by a start codon, a coding sequence, a stop codon, and a transcriptional termination site. In addition, many genes contain regulatory sequences (e.g., eukaryotic response elements or prokaryotic operator sites), and eukaryotic genes are likely to contain introns. After researchers have sequenced a long segment of chromosomal DNA, they frequently want to know if the sequence contains any

---

TABLE 22–2 Short Sequence Elements that Can Be Identified by Computer Analysis

Type of Sequence	Examples[a]
Promoter	Many *E. coli* promoters contain TTGACA (−35 site) and TATAAT (−10 site) Eukaryotic core promoters may contain CAAT boxes, GC boxes, TATA boxes, etc.
Response elements	Glucocorticoid response element (AGR ACA), cAMP response element (GTGACGTRA)
Start codons	AUG
Stop codons	UAA, UAG, UGA
Splice site	GURAGU————YNYURAC(Y)$_n$AG
Polyadenylation signal	AAUAAA
Highly repetitive sequences	Relatively short sequences that are repeated many times throughout a genome
Transposable elements	Usually characterized by a pattern in which direct repeats flank inverted repeats

[a] R = purine (A or G); Y = pyrimidine (T or C); N = A, T, G, or C; U in RNA = T in DNA.

genes. In an attempt to answer this question, geneticists can use computer programs that are aimed at identifying genes in large genomic DNA sequences.

Computer programs can employ different strategies to locate genes. A **search by signal** approach relies on known sequences such as promoters, start and stop codons, and splice sites to help predict whether or not a DNA sequence contains a structural gene. This type of method is analogous to the second program described previously (p. 606). The program tries to locate an organization of known sequence elements that normally are found within a gene. It would try to locate a region that contains a promoter sequence, followed by a start codon, coding sequence, a stop codon, and a transcriptional terminator.

A second strategy is a **search by content** approach. The goal here is to identify sequences with a nucleotide content that differs significantly from a random distribution. Within structural genes, this occurs primarily due to codon usage. Although there are 64 codons, most organisms display a **codon bias** within structural genes. This means that certain codons are used much more frequently than others. For example, UUA, UUG, CUU, CUC, CUA, and CUG all specify leucine. In yeast, however, the UUG codon is used for that purpose 80% of the time. Codon bias allows organisms to more efficiently rely on a smaller population of tRNA molecules. A search by content strategy, therefore, attempts to locate coding regions by identifying regions where the nucleotide content displays certain types of bias.

Another way to locate coding regions within a DNA sequence is to examine translational reading frames. As discussed in Chapter 14, each codon contains three nucleotides. In a new DNA sequence, researchers must consider that the reading of codons (in groups of three nucleotides) could begin with the first nucleotide (Reading frame #1), the second nucleotide (Reading frame #2), or the third nucleotide (Reading frame #3). An **open reading frame (ORF)** is a nucleotide sequence that does not contain any stop codons. Since most proteins are several hundred amino acids in length, a relatively long reading frame is required to encode them. In prokaryotic species, long ORFs are contained within the chromosomal gene sequences. In eukaryotic genes, however, the chromosomal coding sequence may be interrupted by introns. One way to determine eukaryotic ORFs is to clone and sequence cDNA, which is complementary to mRNA (see Chapter 19). A computer program can then translate a cDNA sequence in all three reading frames, seeking to identify a long ORF. In Figure 22-2, a DNA sequence has been translated in all three reading frames. Only one of the three reading frames (#3) contains a very long, open reading frame without any stop codons, suggesting that this DNA sequence encodes a protein.

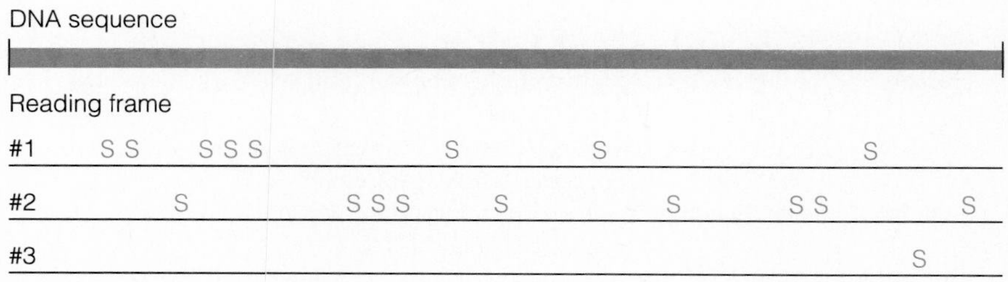

DNA sequence

Reading frame

#1   S S   S S S   S   S   S

#2   S   S S S   S   S   S S   S

#3   S

FIGURE **22-2**

Translation of a DNA sequence in all three reading frames. The top line indicates a DNA segment, and below it are lines representing the translation of the sequence in each of three reading frames; the reading frames proceed from left to right. The letter S indicates the location of a stop codon. Reading frame #3 has a very long open reading frame, suggesting that it may be the reading frame for a structural gene. Reading frames #1 and #2 are not likely to be the reading frames for a structural gene, because they contain many stop codons. During the cloning of DNA, the orientation of a gene may become flipped so that the coding sequence is inverted. Therefore, when analyzing many cloned DNA fragments, six reading frames (i.e., three forward and three reverse) are evaluated. Only the three forward frames are shown here.

HOMOLOGY

The ability to sequence DNA allows geneticists to examine evolutionary relationships at the molecular level. When comparing genetic sequences, researchers sometimes find two or more sequences that are similar. For example, the sequence of the *lacY* gene that encodes the lactose permease in *E. coli* is similar to that of the gene that encodes the lactose permease in a closely related bacterium, *Klebsiella pneumoniae*. As shown here, when segments of the two *lacY* genes are lined up, approximately 60% of their bases are a perfect match:

	151				200
E. coli	TCTTTTTCTT	TTACTTTTTT	ATCATGGGAG	CCTACTTCCC	GTTTTTCCCG
K.pneumoniae	TCTTTTTCTT	TTACTATTTC	ATTATGTCAG	CCTACTTTCC	TTTTTTTCCG

	201				250
	ATTTGGCTAC	ATGACATCAA	CCATATCAGC	AAAAGTGATA	CGGGTATTAT
	GTGTGGCTGG	CGGAAGTTAA	CCATTTAACC	AAAACCGAGA	CGGGTATTAT

In this case, the two sequences are similar because the genes are **homologous** to each other. This means they have been derived from the same ancestral gene. This idea is shown schematically in Figure 22-3. An ancestral *lacY* gene was located in a bacterium that preceded the evolutionary divergence between *E. coli* and *K. pneumoniae*. After these two bacteria had diverged from each other, their *lacY* genes accumulated distinct mutations that produced somewhat different base sequences for this gene. Therefore, in these two species of bacteria, the *lacY* genes are similar but not identical.

Two or more homologous genes can also be found within a single organism. As was discussed in Chapter 8, this can occur because abnormal gene duplication events can produce multiple copies of a gene and ultimately lead to the formation of a gene family. A gene family consists of two or more copies of homologous genes within the genome of a single organism.

It is important not to confuse the terms homology and similarity. **Homology** implies a common ancestry. **Similarity** means that two sequences have similar

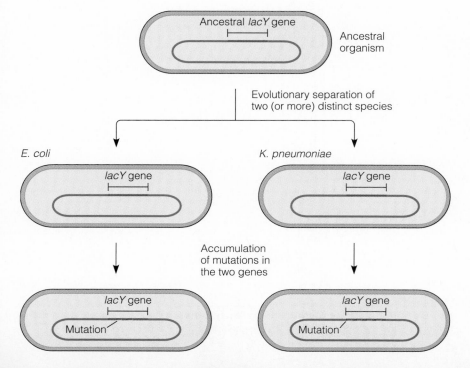

FIGURE **22-3**

The origin of homologous *lacY* genes in *Escherichia coli* and *Klebsiella pneumoniae*. This figure emphasizes a single gene within an ancestral organism. During evolution, the ancestral organism split into two different species, *E. coli* and *K. pneumoniae*. After this divergence, the *lacY* gene in the two separate species accumulated mutations, yielding *lacY* genes with different sequences.

GENES → TRAITS: After two organisms diverge evolutionarily, their genes will accumulate different random mutations. This example concerns the *lacY* gene, which encodes lactose permease. In both species, the function of lactose permease is to transport lactose into the cell. The *lacY* gene in these two species has accumulated different mutations that slightly alter the amino acid sequence of the protein. Researchers have determined that these two species transport lactose at significantly different rates. Therefore, the changes in gene sequences have affected the ability of these two species to utilize lactose, an important trait for survival.

sequences. In many cases, such as the *lacY* example, similarity is due to homology. However, this is not always the case. Short genetic sequences may be similar to each other even though two genes are not related evolutionarily. For example, many non-homologous bacterial genes contain similar promoter sequences at the −35 and −10 regions. In this section, we will examine how homology is detected, and how it can provide important clues regarding the function of genetic sequences.

A simple dot matrix can compare the degree of similarity between two sequences

To evaluate the similarity between two sequences, a matrix can be constructed. Figure 22-4 illustrates the use of a simple dot matrix. In Figure 22-4a, the sequence

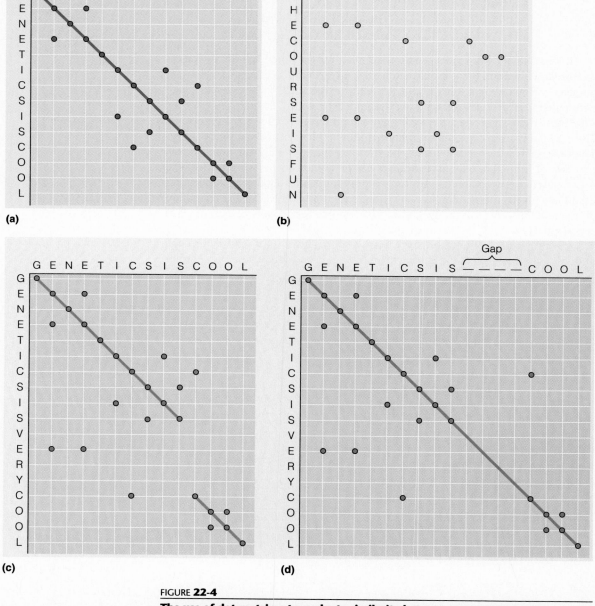

FIGURE **22-4**

The use of dot matrices to evaluate similarity between two sequences. When a dot matrix contains many points that fall on a diagonal line, this indicates regions of similarity.

GENETICSISCOOL is compared with itself. Each point in the grid corresponds to one position of each sequence. The matrix allows all such pairs to be compared simultaneously. Dots are placed where the same letter occurs at the two corresponding positions. The key observation is that regions of similarity are distinguished by the occurrence of many dots along a diagonal line within the matrix. In contrast, Figure 22-4b compares two unrelated sequences: GENETICS-ISCOOL and THECOURSEISFUN. In this comparison, no diagonal lines are seen.

In some cases, two sequences may be related to each other but differ in length. Figure 22-4c compares the sequences GENETICSISCOOL and GENET-ICSISVERYCOOL. In this example, the second sequence is four letters longer than the first. In the dot matrix, two diagonal lines occur. To align these two lines, a **gap** must be created in the first sequence. Figure 22-4d shows the insertion of a gap that aligns the two sequences. Now the two diagonal lines fall along the same line.

Overall, Figure 22-4 illustrates two important features of matrix methods. First, regions of homology are recognized by a series of dots that lie along a diagonal line. Second, gaps can be inserted to align sequences that are of unequal length but still have some similarity.

Homologous sequences can be compared in a multiple sequence alignment

EXPERIMENT 22A

Dot matrices, such as the ones described in Figure 22-4, may contain patterns that indicate sequence similarity. In the cases of Figure 22-4, direct human observation could detect such patterns. Furthermore, in these simple cases, it was obvious where gaps were needed to align two sequences.

Unfortunately, for long genetic sequences, a simple dot matrix approach is not adequate. Instead, dynamic programming methods are used to identify similarities between genetic sequences. This approach was proposed originally by Saul Needleman and Christian Wunsch at Northwestern University and the V.A. Hospital in Chicago. Dynamic programming methods are theoretically similar to a dot matrix, but they involve mathematical operations that are beyond the scope of this text.

In their original work of 1970, Needleman and Wunsch demonstrated that whale myoglobin and human β-hemoglobin have similar sequences. Since then, this approach has been extended to compare more than two genetic sequences. Newer computer programs can align several genetic sequences and sensibly put in gaps. This produces a **multiple sequence alignment**.

To illustrate the usefulness of multiple sequence alignment, let's use the general methods of Needleman and Wunsch and apply them to the globin gene family. As you may recall from Chapter 8, hemoglobin is a protein found in red blood cells; it is responsible for carrying oxygen through the bloodstream. In humans, nine homologous globin genes are functionally expressed. (There are also several pseudogenes that are not expressed and one myoglobin gene.) The nine globin genes fall into two categories: the α-chains and the β-chains. The α-chain genes are α_1, α_2, θ, and ξ; the β-chain genes are β, δ, γ_A, γ_G, and ε. Each hemoglobin protein is composed of two α-chains and two β-chains.

Since the globin genes are expressed at different stages of human development, the composition of hemoglobin changes during the course of growth. For example, the ξ and ε genes are expressed during early embryonic development, whereas the α and β genes are expressed in the adult. Insights into the structure and function of the hemoglobin polypeptide chains can be gained by comparing their sequences. In the computer-based experiment described here, the sequences of the globin polypeptides are compared in a multiple sequence alignment.

THE HYPOTHESIS

Dynamic programming methods can align several genetic sequences. An inspection of a multiple sequence alignment may reveal important features concerning the similarities and differences within a multigene family. In this example, this approach is applied to the globin gene family in humans.

TESTING THE HYPOTHESIS

Starting material: The amino acid sequences of the nine globin genes are contained within sequence files. The user has access to a computer program that can produce a multiple sequence alignment. We will call this program ALIGN.

> C:\>
>
> C:\> ALIGN
>
> C:\> *What sequences do you want aligned?*
>
> C:\> globin-α_1. PEP
> globin-α_2. PEP
> globin-θ. PEP
> globin-ξ. PEP
> globin-β. PEP
> globin-δ. PEP
> globin-γ_A. PEP
> globin-γ_G. PEP
> globin-ϵ. PEP
>
> C:\> *What is the name of the file that you want the multiple sequence alignment placed in?*
>
> C:\> globin.align

THE DATA

	1				50
β beta	VHLTPEEEKSA	VTALWGKV . .	NVDEVGGEAL	GRLLVVYPWT	QRFFESFGDL
δ delta	VHLTPEEEKSA	VNALWGKV . .	NVDAVGGEAL	GRLLVVYPWT	QRFFESFGDL
γ_A gamma-A	GHFTEEDKAT	ITSLWGKV . .	NVEDAGGEAL	GRLLVVYPWT	QRFFESFGDL
γ_G gamma-G	GHFTEEDKAT	ITSLWGKV . .	NVEDAGGEAL	GRLLVVYPWT	QRFFESFGDL
ε epsilon	VHFTAEEKAA	VTSLWSKM . .	NVEEAGGEAL	GRLLVVYPWT	QRFFESFGDL
α_1 alpha-1	VLSPADKTN	VKAAWGKVGA	HAGEGAEAL	ERMFLSFPTT	KTYFPHF . DL
α_2 alpha-2	VLSPADKTN	VKAAWGKVGA	HAGEGAEAL	ERMFLSFPTT	KTYFPHF . DL
θ theta	ALSAEDRAL	VRALWKKLGS	NVGVYTTEAL	ERTFLAFPAT	KTYFSHL . DL
ξ zeta	SLTKTERTI	IVSMWAKIST	QADTIGTETL	ERLFLSHPQT	KTYFPHF . DL
	51				100
β beta	STPDAVMGNP	KVKAHGKKVL	GAFSDGLAHL	DNLKGTFATL	SELHCDKLHV
δ delta	SSPDAVMGNP	KVKAHGKKVL	GAFSDGLAHL	DNLKGTFSQL	SELHCDKLHV
γ_A gamma-A	SSASAIMGNP	KVKAHGKKVL	TSLGDAIKHL	DDLKGTFAQL	SELHCDKLHV
γ_G gamma-G	SSASAIMGNP	KVKAHGKKVL	TSLGDAIKHL	DDLKGTFAQL	SELHCDKLHV
ε epsilon	SSPSAILGNP	KVKAHGKKVL	TSFGDAIKNM	DNLKPAFAKL	SELHCDKLHV
α_1 alpha-1	SHGSA.	QVKGHGKKVA	DALTNAVAHV	DDMPNALSAL	SDLHAHKLRV
α_2 alpha-2	SHGSA	QVKGHGKKVA	DALTNAVAHV	DDMPNALSAL	SDLHAHKLRV
θ theta	SPGSS	QVKAHGQKVA	DALSLAVERL	DDLPHALSAL	SHLHACQLRV
ξ zeta	HPGSA	QLRAHGSKVV	AAVGDAVKSI	DDIGGALSKL	SELHAYILR

↑   ↑

	101				148
β beta	DPENFRLLGN	VLVCVLAHHF	GKEFTPPVQA	AYQKVVAGVA	NALAHKYH
δ delta	DPENFRLLGN	VLVCVLARNF	GKEFTPQMQA	AYQKVVAGVA	NALAHKYH
γ_A gamma-A	DPENFRLLGN	VLVCVLAIHF	GKEFTPEVQA	SWQKMVTAVA	SALSSRYH
γ_G gamma-G	DPENFRLLGN	VLVCVLAIHF	GKEFTPEVQA	SWQKMVTAVA	SALSSRYH
ε epsilon	DPENFRLLGN	VMVIILATHF	GKEFTPEVQA	AWQKLVSAVA	IALAHKYH
α_1 alpha-1	DPVNFKLLSH	CLLVTLAAHL	PAEFTPAVHA	SLDKFLASVS	TVLTSKYR
α_2 alpha-2	DPVNFKLLSH	CLLVTLAAHL	PAEFTPAVHA	SLDKFLASVS	TVLTSKYR
θ theta	DPASFQLLGH	CLLVTLARHL	PGDFSPALQA	SLDKFLSHSVI	SALVSEYR
ξ zeta	DPVNFKLLSH	CLLVTLAARF	PADFTAEAHA	AWDKFLSVVS	SVLTEKYR

INTERPRETING THE DATA

A careful examination of a multiple sequence alignment may reveal many important trends within a gene family. In this alignment, dots are shown where it is necessary to create gaps to keep the sequences aligned. As we can see, the sequence similarity is very high between α_1, α_2, θ, and ξ. In fact, the α_1 and α_2 amino acid sequences are identical. This suggests that the four types of α-chains likely carry out very similar functions. Likewise, the β-chains encoded by the β, δ, γ_A, γ_G, and ε genes are very similar to each other. In the globin gene family, the α-chains are much more similar to each other than they are to the β-chains, and vice versa.

In general, amino acids that are highly conserved within a gene family are more likely to be important functionally. The arrows in the multiple sequence alignment point to histidine amino acids that are conserved in all nine members of the hemoglobin gene family. These histidine residues are involved in the necessary function of binding the heme molecule to the globin polypeptides.

Overall, the experiment described here illustrates the type of information that can be derived from a multiple sequence alignment. In this case, multiple sequence alignment has shown that a group of nine genes falls into two closely related groups. The alignment has also identified particular amino acids within the proteins' sequences that are highly conserved. This conservation is consistent with an important role in protein function.

A database can be searched to identify homologous sequences

Homologous genes usually carry out similar or identical functions. For example, the members of the globin gene family are all involved with carrying and transporting oxygen. Likewise, the *lacY* genes in *E. coli* and *K. pneumoniae* both encode lactose permeases that transport lactose across the bacterial cell membrane.

In general, there is a strong correlation between homology and function. In many cases, the first indication of the function of a newly determined sequence is through homology to known sequences in a database. An example is the gene that is altered in cystic fibrosis patients. After this gene was identified in humans, a database search revealed that it is homologous to several genes found in other species. Moreover, the homologous genes were known to encode proteins that function in the transport of ions and small molecules across the plasma membrane. This observation provided an important clue that cystic fibrosis involves a defect in membrane transport.

The ability of computer programs to identify homology between genetic sequences provides a powerful tool for predicting the function of genetic sequences. Computer programs can start with a particular genetic sequence and then locate homologous sequences within a database. Since there are only four bases but 20 amino acids, homology among protein sequences is easier to identify than is DNA sequence homology. Among proteins, sequences that diverged more than 2.5 billion years ago can still be correlated. By comparison, usually only homologous DNA sequences that diverged less than 100 million years ago can be identified.

TABLE **22–3 Example of a Database Search**

Sequence in the Database	Species	Percent Identity with Human ς-Globin[a]
δ-Globin	Human	100.0
δ-Globin	Chimpanzee	99.3
ß-Globin	Hanuman langur	95.2
δ-Globin	Colobus polykomos	95.9
ß-Globin	Common gibbon	94.5
ß-Globin	Green monkey	94.5
ß-Globin	Colobus polykomos	93.8
ß-Globin	Mandrill	93.2
δ-Globin	Black-handed spider monkey	93.8
ß-Globin	Japanese macaque	93.2

[a] The rightmost column gives the percentages of amino acids that are identical with those of human δ-globin. The entries in this table are listed in the order of their degree of similarity to human δ-globin. In some cases (e.g., compare hanuman langur and colobus polykomos), a sequence with a lower percentage identity may be judged as more similar to human δ-globin because the alignment may require fewer gaps in the sequence.

Table 22-3 describes the results of a database search that started with the amino acid sequence for human δ (delta) globin. This sequence was used as a "query sequence" by a computer program to search the SWISS-PROT database, which contains hundreds of thousands of different protein sequences. With a high-speed computer and the appropriate computer program, the human δ-globin gene sequence can be compared with hundreds of thousands of polypeptide sequences in the SWISS-PROT database; the program can determine which sequences are the closest matches to human δ-globin (Table 22-3). Since the human δ-globin sequence is already in the SWISS-PROT database, its closest match is to itself. This search identified many other sequences that are very similar to human δ-globin. These include β-globin and δ-globin sequences from several primate species.

The results shown in Table 22-3 illustrate the remarkable computational abilities of current computer technology. In minutes, the human δ-globin sequence can be compared with hundreds of thousands of different sequences.

STRUCTURE PREDICTION

The function of macromolecules such as DNA, RNA, and proteins relies on their structure. The three-dimensional structure of these macromolecules, in turn, depends on the linear sequences of their building blocks. In the case of DNA and RNA, this means a linear sequence of nucleotides; proteins are composed of a linear sequence of amino acids. Currently, the three-dimensional structure of macromolecules is determined primarily through the use of biophysical techniques such as X-ray crystallography and nuclear magnetic resonance (NMR). These methods are technically difficult and very time consuming.

DNA sequencing, by comparison, requires much less effort. Therefore, since the three-dimensional structure of macromolecules depends ultimately on the linear sequence of their building blocks, it would be far easier if we could predict the structure (and function) of DNAs, RNAs, and proteins from their sequence of building blocks. The last few decades have seen great advances in our abilities to predict macromolecular structure using sequence data. The future holds great promise. In this section, we will examine the types of structural predictions that can be made using the information contained within genetic sequences.

Alterations in DNA conformation may be predicted from its primary sequence

As we discussed in Chapter 9, the predominant form of the DNA double helix is B-DNA. This conformation is a right-handed helix with a wide major groove and a narrow minor groove. By comparison, A-DNA, which is also a right-handed helix, has a broader, shallower major groove and a deeper, narrower minor groove. Z-DNA is a left-handed helix with a broad, flat major groove and an extremely narrow minor groove. Since DNA is a dynamic structure, short segments of DNA may change from one conformation to another.

Though many determinants influence DNA conformation, one important factor is the sequence of nucleotides. B-DNA is the favored form of the double helix, and repeats of AA or TT tend to hold DNA in a B-DNA conformation. However, repeated CG sequences can (under certain conditions) cause B-DNA to change its conformation to Z-DNA. Alternatively, repeats of GG or CC tend to change B-DNA to A-DNA. Many studies have suggested that B to Z transitions and possibly B to A transitions may be important biologically, since they affect the ability of proteins to bind within the major or minor groove.

Based on the primary sequence, computer programs can predict how likely such transitions are to occur. However, these predictions are not completely reliable. For this reason, structural predictions must be combined with other experimental evidence. For example, a computer program may predict that a segment of DNA is likely to be found in a Z-DNA conformation. To confirm this prediction, antibodies that specifically recognize Z-DNA (and not B-DNA or A-DNA) can be used. If the DNA is really in the Z-DNA conformation, the antibodies will bind this molecule. This can be measured experimentally using radiolabeled antibodies.

Computer programs can predict RNA secondary structure

As discussed in Chapter 9, RNA molecules typically are folded into a secondary structure, which commonly contains double-stranded regions. These secondary structure regions are further folded and twisted to adopt a tertiary conformation. As mentioned in many places throughout the text, such structural features of RNA molecules are functionally important. For example, the folding of RNA into secondary structures such as stem-loops affects transcriptional termination and other regulatory events. Therefore, geneticists are interested in the secondary and tertiary structures that RNA molecules can adopt.

Many approaches are available for investigating RNA structure. In addition to biophysical and biochemical techniques, computer modeling of RNA structure has become an important tool. Modeling programs can consider different types of information. For example, the known characteristics of RNA secondary structure, such as the ability to form double-stranded regions, can provide parameters for use in a modeling program.

A comparative approach can also be used in RNA structure prediction. This method assumes that RNAs of similar function and sequence have a similar structure. For example, the genes that encode certain types of RNAs, such as the 16S rRNAs, have been sequenced from many different species. Among different species, the 16S rRNAs have similar but not identical sequences. Computer programs can compare many different 16S rRNA sequences to aid in the prediction of secondary structure. Figure 22-5 illustrates a secondary structural model for 16S rRNA based on a comparative sequence analysis. This large RNA contains 45 stem-loop regions. As you can imagine, it would be rather difficult to deduce such a model without the aid of a computer!

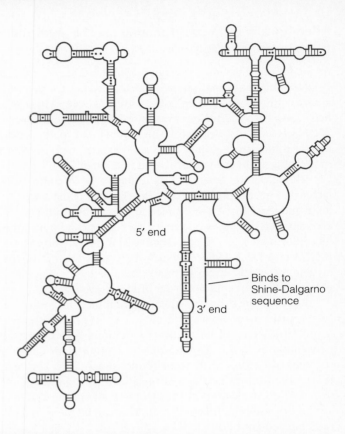

5′ end

Binds to
Shine-Dalgarno
sequence

3′ end

FIGURE **22-5**

A secondary structural model for *E. coli* 16S rRNA.

Computer programs can predict protein secondary structure

As was described in Chapter 12, proteins contain repeating secondary structural patterns known as α-helices and β-sheets. Several computer-based approaches attempt to predict secondary structure from the primary amino acid sequence. These programs base their predictions on different types of parameters. Some programs rely on the physical and energetic properties of the amino acids and the polypeptide backbone. More commonly, however, secondary structure predictions are based on the statistical frequency of amino acids within secondary structures that have been crystallized.

For example, Peter Chou and Gerald Fasman at Brandeis University have compiled X-ray crystallographic data to calculate the likelihood that an amino acid will be found in an α-helix or a β-sheet. Certain amino acids, such as glutamate and alanine, are likely to be found in an α-helix; others, such as valine and isoleucine, are more likely to be found in a β-sheet structure. Such information can be used to predict whether a sequence of amino acids within a protein is likely to be folded in an α-helix or β-sheet conformation.

Secondary structure prediction is correct for approximately 60 to 70% of all sequences. While this degree of accuracy is promising, it is generally not sufficient to predict protein secondary structure reliably. Therefore, one must be cautious in interpreting the results of a secondary structure prediction program.

In recent years, an exciting computer methodology known as *neural nets* has been applied to protein secondary structure prediction. A computer neural network is a large number of calculation units organized into interconnected layers; this structure is reminiscent of the organization of neurons in the brain. The input layer receives data and may (or may not) transmit that information to the next layer.

Neural networks can adjust the parameters that define the interconnections among its units in response to data; they can thus be trained to identify very complex patterns coming from the input data. For example, the amino acid sequences of proteins with known crystal structures can be used to train a network (i.e., adjust its parameters) to predict secondary structures for new amino acid sequences. Thus far, neural networks have yielded small improvements in secondary structure predictions. In the future, a combination of innovative predictive approaches and increased information concerning the biophysical properties of amino acids may make secondary structure prediction a reliable strategy.

Protein tertiary structure may be predicted using a comparative approach

The three-dimensional structure of a protein is extremely difficult to predict solely from its amino acid sequence. However, researchers have had some success in predicting tertiary structure using a comparative approach. This strategy requires that the protein of interest be homologous to another protein, the tertiary structure of which already has been solved by X-ray crystallography. In this situation, the crystal structure of the known protein can be used as a starting point to model the three-dimensional structure of the protein of interest. This approach is known as **homology-based modeling**, **knowledge-based modeling**, or **comparative homology**.

In 1987 and 1989, two research groups predicted the structure of a protein encoded by HIV. The protein, known as HIV protease, is homologous to other proteases with structures that have been solved by X-ray crystallography. Two similar models of the HIV protease were predicted before its actual crystal structure was determined by X-ray crystallography. Both models turned out to be fairly accurate representations of the actual structure.

The computer has become an important tool in genetic studies. Genetic sequences contained within *computer files* can be analyzed by many different *programs*. For example, short sequence elements can be located within very long genetic sequences. Likewise, various strategies can be used to find structural genes within DNA sequences. In addition to analyzing genetic sequences, computers are used as storage facilities for genetic information. The scientific community has created large *databases* that contain genetic data from thousands of laboratories.

Homology has become an important concept in sequence analysis. Two genes are *homologous* if they have been derived from a common ancestral gene. Homologous genes have similar sequences and are very likely to carry out similar or identical functions. Computer programs can produce a *multiple sequence alignment* in which several homologous gene sequences are compared. In addition, geneticists may take newly identified sequences and search a database to identify homologous sequences. In many cases, the identification of homologous genes within a database can provide important clues concerning the function of a recently determined sequence.

The three-dimensional structure of DNA, RNA, and proteins ultimately depends on the sequence of their building blocks. Therefore, much research is being directed toward predicting the structure and function of these macromolecules from their genetic sequences. Computer programs have been developed to predict the secondary and/or tertiary structure of DNA, RNA, and proteins. These programs rely on known biochemical and biophysical properties of the macromolecules and may also incorporate data concerning homologous sequences of known structure. Some predictive programs are more reliable than others. Nevertheless, future research is expected to improve the predictive capabilities of such computer programs.

PROBLEM SETS

SOLVED PROBLEMS

1. Using a comparative sequence analysis, the secondary structures of rRNAs have been predicted. Among many homologous rRNAs, one stem-loop usually has the following structure:

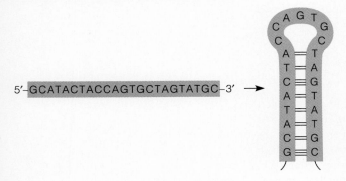

5′–GCATACTACCAGTGCTAGTATGC–3′ →

You are sequencing a homologous rRNA from a new species and have obtained most of its sequence, but you cannot read the last five bases on your sequencing gel:

5′–GCATTCTACCAGTGCTAG?????–3′

Of course, you will eventually repeat this experiment to determine the last five bases. However, before you get around to doing this, what do you expect will be the sequence of the last five bases?

Answer: AATGC–3′

This will also form a similar stem-loop structure.

2. How can codon bias be used to search for structural genes within uncharacterized genetic sequences?

Answer: Most species exhibit a bias in the codons they use within the coding sequence of structural genes. This causes the base content within coding sequences to differ significantly from that of noncoding DNA regions. By knowing the codon bias for a particular species, researchers can use a computer to locate regions that display this bias and thereby identify what are likely to be the coding regions of structural genes.

CONCEPTUAL AND EXPERIMENTAL QUESTIONS

1. What are the contents of a computer file?

2. In molecular genetics, what is the general goal of a computer program? Give three examples.

3. What is a user-friendly program?

4. What is a database? What types of information are stored within a database? Where does the information come from?

5. Discuss the objectives of a genome database.

6. Identify all the start codons in the following DNA sequence:

5′–ATGGGTATACCCGATCACTTTATTTATGGGCACGATAG
ATATATGCCATGCTATGATCATACATACATGCACATACA
TAAACGGATAATCCATAATAATGACAATGTTTTACGTA
ATGCATACATACATACATGGCCATACATGCACTATACA
ATGA–3′

Time how long it takes for you to find them. (A computer would take less than a millisecond.)

7. Discuss why computer technology is useful in identifying functional sequences.

8. Discuss the distinction between sequence recognition and pattern recognition (i.e., the first and third programs described on p. 606).

9. Besides the examples listed in Table 22-2, list five short sequences that a geneticist might want to locate within a DNA sequence.

10. Fill in the following dot matrix:

```
      THISQUESTIONISVERYANNOYING
T
H
I
S
Q
U
E
S
T
I
S
A
N
N
O
Y
I
N
G
```

How many regions of similarity can you identify? Make a second dot matrix that aligns the two sequences.

11. Discuss the strategies that can be used to identify a structural gene.

12. A multiple sequence alignment of five homologous proteins is shown here:

```
     1                                                                            50
#1   MLAFLNQVRK    PTLDLPLEVR    RKMWFKPFM.    QSYLVVFIGY    LTMYLIRKNF
#2   MLAFLNQVRK    PTLDLALDVR    RKMWFKPFM.    QSYLVVFIGY    LTMYLIRKNF
#3   MLPFLKAPAD    APL.MTDKYE    IDARYRYWRR    HILLTIWLGY    ALFYFTRKSF
#4   MLSFLKAPAN    APL.ITDKHE    VDARYRYWRR    HILITIWLGY    ALFYFTRKSF
#5   MLSIFKPAPH    KAR.LPAA.E    IDPTYRRLRW    QIFLGIFFGY    AAYYLVRKNF


     51                                                                          100
#1   NIAQNDMIST    YGLSMTQLGM    IGLGFSITYG    VGKTLVSYYA    DGKNTKQFLP
#2   NIAQNDMIST    YGLSMTELGM    IGLGFSITYG    VGKTLVSYYA    DGKNTKQFLP
#3   NAAVPEILAN    GVLSRSDIGL    LATLFYITYG    VSKFVSGIVS    DRSNARYFMG
#4   NAAAPEILAS    GILTRSDIGL    LATLFYITYG    VSKFVSGIVS    DRSNARYFMG
#5   ALAMPYLVEQ    .GFSRGDLGF    ALSGISIAYG    FSKFIMGSVS    DRSNPRVFLP
```

Discuss some of the interesting features that this alignment reveals.

13. What is an open reading frame (ORF)? In the chromosomal DNA of prokaryotes and eukaryotes, how long would you expect ORFs to be? How long would they be in cDNA?

14. From random chance alone, how often would you expect to find a stop codon in a DNA sequence? (Hint: There are 3 stop codons and a total of 64 codons.)

15. What is a motif? Why is it useful for computer programs to identify functional motifs within amino acid sequences?

16. What is the difference between similarity and homology?

17. Discuss why it is useful to search a database to identify sequences that are homologous to a newly determined sequence.

18. The secondary structure of 16S rRNA has been predicted using a computer-based sequence analysis. In general terms, discuss what type of information is used in a comparative sequence analysis, and explain what assumptions are made concerning the structure of homologous RNAs.

19. Discuss the basis for secondary structure prediction in proteins. How reliable is it?

20. To reliably predict the tertiary structure of a protein based on its amino acid sequence, what type of information must be available?

21. In sequence alignment, what is meant by the term gap? Why is it necessary to have gaps within a multiple sequence alignment?

QUESTIONS FOR STUDENT DISCUSSION/COLLABORATION

1. Let's suppose that you are in charge of organizing and publicizing a genomic database for the mouse genome. Make a list of innovative strategies that you would initiate to make the mouse genome database useful and effective.

2. Let's suppose that a 5-year-old told you that she was interested in pursuing a career studying the three-dimensional structure of proteins. (Okay, so she's a bit precocious.) Would you advise her to become a geneticist, a mathematical theoretician, or a biophysicist?

3. If you have access to the necessary computer software, make a sequence file and analyze it in the following ways: What is the translated sequence in all three reading frames? What is the longest open reading frame? Is the sequence homologous to any known sequences? If so, does this provide any clues about the function of the sequence? Note: You can make a sequence file by getting a sequence from someone you know who works in a research laboratory, or you can use the sequence on the following page.

Sequence for Question 3

```
   1  GGCATAATAA  TATGGAATAT  TCACAATGAA  ACTCTCTGAA  CTCGCGCCAC
  51  GCGAACGGCA  TAACTTTATT  TATTTCATGC  TGTTCTTTTT  CTTTTACTAT
 101  TTCATTATGT  CAGCCTACTT  TCCTTTTTTT  CCGGTGTGGC  TGGCGGAAGT
 151  TAACCATTTA  ACCAAAACCG  AGACAGGGAT  CGTATTCTCC  TGCATTTCGC
 201  TATTCGCCAT  CATTTTCCAG  CCGGTATTTG  GCCTGATTTC  CGATAAGCTC
 251  GGCCTGCGCA  AGCATCTGCT  GTGGACGATT  ACGATATTAT  TAATCCTGTT
 301  TGCCCCCTTC  TTTATTTTTG  TTTTCTCGCC  ATTGCTGCAG  ATGAATATCA
 351  TGGCGGGCGC  GCTGGTGGGC  GGTGTATATC  TGGGGATCGT  TTTCTCCAGC
 401  CGCTCCGGGG  CGGTAGAAGC  CTATATTGAA  CGCGTCAGCC  GCGCCAACCG
 451  TTTTGAATAC  GGTAAAGTGC  GCGTCTCAGG  CTGCGTCGGC  TGGGCGCTGT
 501  GCGCCTCCAT  CACCGGTATT  TTGTTTAGTA  TCGACCCCAA  TATTACCTTC
 551  TGGATCGCCT  CCGGTTTCGC  GCTGATCCTC  GGCGTGCTGC  TGTGGGTCTC
 601  AAAACCGGAG  AGCAGCAATA  GCGCTGAGGT  TATTGACGCC  CTGGGCGCCA
 651  ACCGTCAGGC  CTTCTCAATG  CGTACCGCCG  CCGAGCTTTT  CCGGATGCCG
 701  CGCTTCTGGG  GCTTTATTAT  ATACGTGGTT  GGCGTCGCCA  GCGTCTATGA
 751  CGTTTTCGAC  CAGCAGTTCG  CCAACTTTTT  TAAAGGCTTC  TTCTCCAGCC
 801  CACAGCGCGG  CACCGAAGTC  TTTGGCTTCG  TGACCACCGG  TGGGGAATTA
 851  CTCAATGCGC  TGATCATGTT  CTGCGCGCCG  GCGATTATTA  ACCGAATTGG
 901  CGCCAAGAAT  GCCCTGTTAA  TTGCCGGGTT  GATTATGTCA  GTGCCAATTT
 951  TAGGCTCGTC  TTTCGCCACC  TCGGCGGTGG  AAGTCATTAT  ATTAAAAATG
1001  CTGCATATGT  TTGAGATCCC  GTTCCTGCTG  GTCGGCACCT  TTAAATATAT
1051  CTCCTCGGCA  TTTAAGGGAA  AACTCTCGGC  GACGCTGTTC  CTGATCGGCT
1101  TTAATTTATC  GAAGCAGCTT  TCAAGCGTGG  TGCTCTCGGC  GTGGGTAGGA
1151  CGGATGTATG  ACACCGTCGG  CTTCCATCAG  GCTTATCTGA  TCCTGGGCTG
1201  TATCACCCTG  AGCTTTACCG  TTATTTCGCT  GTTTACCCTG  AAAGGCAGCA
1251  AAACGCTGCT  GCCGGCCACG  GCATAAACAC  AAGGCCGCCG  GAAAGCCTCC
1301  CTC
```

Part VI
Genetic Analysis of Individuals and Populations

CHAPTER 23

GENETICS OF CANCER AND OTHER HUMAN DISEASES

In this chapter, we will focus our attention on certain genes that when mutated, contribute to human disease. In the first part of the chapter, we will explore the molecular basis of several genetic disorders and possible patterns of inheritance. Many examples will be described in which defective genes cause phenotypic abnormalities. We will also examine how genetic screening can determine if an individual carries a defective allele. The second part of the chapter will concern cancer, a disease that involves the uncontrolled growth of somatic cells. We will examine the underlying genetic basis for this disease, and discuss the roles that many different genes may play in the development of cancer.

GENETIC ANALYSIS
OF HUMAN DISEASES

GENETIC BASIS
OF CANCER

GENETIC ANALYSIS OF HUMAN DISEASES

The genetic analysis of humans is a topic that is hard to resist. Almost everyone who looks at a newborn is tempted to speculate whether the baby resembles the mother, father, or perhaps a distant relative. In this section, we will focus primarily on human genetic diseases rather than the normal traits found in the general population. Even so, we can consider a human genetic disease to be a trait that deviates from the prevailing (wild-type) trait found among most people.

Hemophilia, for example, can be viewed as an abnormal trait in which the individual's blood will not clot properly. The wild-type trait, which is found in most people, is normal blood clotting. By analyzing people with hemophilia, researchers have identified genes, and the proteins they encode, that participate in the normal process of blood clotting. Therefore, when we study the inheritance of genetic diseases, we often learn a great deal about the genetic basis for normal traits as well.

Thousands of human diseases have an underlying genetic basis. Therefore, human genetic analysis is of great medical importance. In this section, we will primarily examine the causes and inheritance patterns of human genetic diseases that result from defects in single genes. As we will learn, the abnormal genes that cause these diseases often follow simple Mendelian inheritance patterns.

A genetic basis for a human trait may be suggested from a variety of observations

When we view the characteristics of human beings, we usually think that some traits are inherited while others are caused by environmental factors. For example, when the facial features of two related individuals look strikingly similar, we think that this similarity has a genetic basis. The profound resemblance between identical twins is a perfect example. By comparison, other traits are governed by the environment. If we see a person with purple hair, we likely suspect that he or she has used hair dye as opposed to carrying a genetic abnormality.

For normal human traits, as well as for diseases, geneticists would like to know the relative contributions from genetics and environment. For example, is a disease caused by a pathogenic microorganism or is it caused by a faulty gene? Unlike the case with experimental organisms, we cannot conduct human crosses to elucidate the genetic basis for traits and diseases. Instead, we must rely on analyses of families that already exist. As described in the following list, several observations are consistent with the idea that a trait is caused, at least in part, by the inheritance of genes. When the occurrence of a trait or disease correlates with several of these observations, a geneticist will become increasingly confident that the disease has a genetic basis.

1. *When an individual exhibits a trait, this trait is more likely to occur in genetic relatives than in the general population.* For example, if someone has cystic fibrosis, it is more likely that his or her relatives will also have this disease than will randomly chosen members of the general population.
2. *Identical twins share the trait more often than nonidentical twins.* Identical twins, also called **monozygotic twins**, are genetically identical to each other, because they were formed from the same sperm and egg. By comparison, nonidentical twins, also called fraternal or **dizygotic twins**, are formed from separate pairs of sperm and egg cells. Dizygotic twins share 50% of their genetic material. When a trait has a genetic component, both identical twins are more likely to exhibit the trait than are nonidentical twins.
3. *The trait does not spread to individuals sharing similar environmental situations.* Genetic traits cannot spread from person to person. The only way genetic traits can be transmitted is from parent to offspring during sexual reproduction.
4. *Different populations tend to have different frequencies of the trait.* Due to evolutionary forces that we will discuss in Chapter 26, the frequencies of traits usually vary among different populations of humans. For example, the frequency of the disease sickle cell anemia is highest among African populations and relatively low in other parts of the world (see Figure 26-11).
5. *The trait tends to develop at a characteristic age.* Many genetic traits exhibit a characteristic **age of onset** at which the trait appears. Many mutant genes exert their effects during embryonic and fetal development, so that their effects are apparent at birth. Other genetic traits, such as male pattern baldness, tend to develop much later in life.
6. *The human trait may resemble a trait in an animal that is already known to have a genetic basis.* In animals, where we can conduct experimentation, various traits are known to be governed by genes. For example, the albino phenotype is found in humans as well as in many animals (Figure 23-1).

(a) A human family that includes some albino members

(b) An albino wildebeest

FIGURE **23-1**

The albino phenotype in humans and other mammals.

GENES ⟶ TRAITS: Certain enzymes (encoded by genes) are necessary for the production of pigment. A homozygote with two defective alleles in these pigmentation genes exhibits an albino phenotype. This phenotype can occur in humans and other animals.

7. *A correlation is observed between a trait and a mutant human gene or a chromosomal alteration.* A particularly convincing piece of evidence that a trait has a genetic basis is the identification of altered genes or chromosomes that occur only in people exhibiting the trait. When comparing two individuals, one with a trait and one without, we expect them to have differences in their genetic material, if the trait has a genetic component. We can discover alterations in gene sequences by gene cloning and DNA sequencing techniques (as was described in Chapter 19); we can detect alterations in chromosome structure and number by the microscopic examination of chromosomes.

(c) An albino squirrel

Inheritance patterns of human diseases may be determined via pedigree analysis

When a human trait is governed by a single gene, the pattern of inheritance can be deduced by analyzing human pedigrees. To use this method, a geneticist must obtain data from many large pedigrees containing several individuals who exhibit the trait. To appreciate the basic features of pedigree analysis, we will examine a few large pedigrees that involve diseases inherited in different ways. You may wish to review Figure 2-7 for the organization and symbols of pedigrees.

The pedigree shown in Figure 23-2 concerns a genetic disorder called **Tay–Sachs disease (TSD)**. TSD was first described by Warran Tay, a British ophthalmologist, and Bernard Sachs, an American neurologist, in the 1880s. Affected individuals appear normal at birth but then develop neurodegenerative symptoms at 4–6 months of age. The primary characteristics are cerebral degeneration, blindness, and loss of motor function. Individuals with TSD typically die in the third or fourth year of life. This disease is particularly prevalent in Ashkenazi (eastern European) Jewish populations, in which it has a frequency of about 1 in 3600 births. It is over 100 times less frequent in most other human populations.

At the molecular level, the mutation that causes TSD is in a gene that encodes the enzyme hexosaminidase A (hexA). HexA is responsible for the breakdown of a category of lipids called GM_2-gangliosides. This lipid is particularly prevalent in the cells of the central nervous system. A defect in the ability to break down this lipid, as in TSD, leads to an excessive accumulation of this lipid in nerve cells and eventually causes neurodegeneration and the symptoms just described.

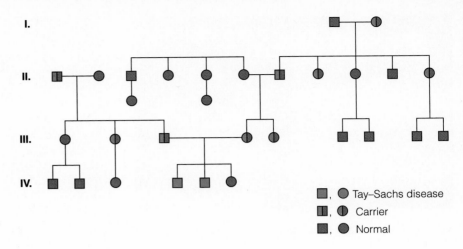

FIGURE **23-2**

A family pedigree of Tay–Sachs disease.

As illustrated in Figure 23-2, Tay–Sachs disease is inherited in an autosomal, recessive manner. The four common features of autosomal recessive inheritance are as follows:

1. *Frequently, an affected offspring will have two unaffected parents.* For rare recessive traits, the parents are usually unaffected. For deleterious alleles that cause early death or infertility, the two parents must be unaffected. This is always the case in TSD.
2. *When two unaffected heterozygotes have children, the percentage of affected children is (on average) 25%.*
3. *Two affected individuals will have 100% affected children.* This observation can only be made when a recessive trait produces fertile, viable individuals. In the case of TSD, the affected individual dies in early childhood, and so it is not possible to observe crosses between two affected people.
4. *The trait occurs with the same frequency in both genders.*

Autosomal recessive inheritance is a common mode of transmission for genetic disorders, particularly those that involve defective enzymes. The heterozygous carrier has 50% of the normal enzyme, which is sufficient for a normal phenotype. There are hundreds of human genetic diseases that are inherited in this way. In many cases, mutant genes that cause disease have been cloned and/or mapped. Several of these are described in Table 23-1.

TABLE **23–1 Examples of Human Diseases Inherited in an Autosomal, Recessive Manner**

Trait/Disease	Chromosome Location	Gene Product	Description
Adenosine deaminase deficiency	20q	Adenosine deaminase	Defective immune system, and skeletal and neurological abnormalities
Albinism (Type I)	11q	Tyrosinase	Inability to synthesize melanin, white skin, hair, etc.
Cystic fibrosis	7q	CF transmembrane regulator	Water imbalance in tissues of the pancreas, intestine, sweat glands, and lungs; due to impaired ion transport
Phenylketonuria (PKU)	12q	Phenylalanine hydroxylase	Foul-smelling urine and neurological abnormalities; may be remedied by diet modification at birth
Sickle cell anemia	11p	ß-Globin	Anemia, blockages in blood circulation
Tay–Sachs disease	15q	Hexosaminidase A	Similar to Sandhoff disease, progressive neurodegeneration

Now let's examine a human pedigree involving an autosomal dominant trait (Figure 23-3). In this example, the affected individuals have a disorder called **Huntington disease** (also called Huntington chorea). The major symptom of the disease is a degeneration of certain types of neurons in the brain, leading to personality changes, dementia, and early death, which usually occur during middle age. In 1993, the gene for Huntington disease was identified and sequenced. It encodes a protein, called huntingtin, that is expressed in neurons but is also found in some cells not affected in Huntington disease. Future research will be needed to elucidate the molecular relationship between a defect in the huntingtin protein and the disease symptoms. Four common features of autosomal dominant inheritance are as follows:

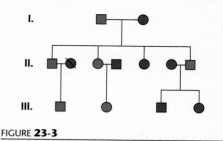

FIGURE **23-3**

A family pedigree of Huntington disease.

1. *An affected offspring has one or both affected parents.* However, this is not always the case. Some other dominant traits show incomplete penetrance (see Chapter 4, p. 87), so that a heterozygote may not exhibit the trait even though it may be passed along to offspring who do exhibit the trait. Also, a dominant mutation may occur during gametogenesis, so that two unaffected parents may produce an affected offspring.
2. *An affected individual, with only one affected parent, is expected to produce 50% affected offspring (on average).*
3. *It is possible for two affected individuals to have unaffected offspring.*
4. *The trait occurs with the same frequency in both genders.*

Many dominant diseases have been identified in humans (Table 23-2). In these disorders, 50% of the normal protein is not sufficient to produce a normal phenotype. Dominant diseases may involve defects in enzymes, but it is also common for them to involve alterations in structural proteins or membrane receptors. For example, one inherited form of osteoporosis is a missense mutation in a gene encoding a subunit of collagen, a constituent of bone and cartilage. In this genetic disorder, a mutation in the Type1 A2 collagen gene substitutes a serine residue for a glycine, producing a kink in collagen, which disrupts the linear triple α-helix. This alteration in collagen conformation leads to a weakened bone structure.

A third pattern of inheritance common in humans is X-linked inheritance (Table 23-3). With regard to recessive genetic diseases, X-linked inheritance poses a special problem for males. Most X-linked genes lack a counterpart on the Y-chromosome; males thus have a single copy of (i.e., are hemizygous for) these genes. Therefore, a female heterozygous for an X-linked recessive gene will pass this trait to half of her sons.

TABLE **23–2 Examples of Human Diseases Inherited in an Autosomal, Dominant Manner**

Trait/Disease	Chromosome Location	Gene Product	Description
Achondroplasia	4p	Fibroblast Growth Factor Receptor	A common form of dwarfism associated with a defect in the growth of long bones
Osteoporosis	7q	Collagen (Type 1 A2)	Brittle, weakened bones
Familial hypercholesterolemia	19p	LDL Receptor	Characterized by very high serum levels of low-density lipoprotein; a predisposing factor in heart disease
Huntington disease	4p	Huntingtin	Neurodegeneration that occurs relatively late in life, usually in middle age
Neurofibromatosis I	17q	Tumor suppressor gene	Individuals may exhibit spots of abnormal pigmentation (café-au-lait spots) and growth of noncancerous tumors in the nervous system

Trait/Disease	Gene Product	Description
Duchenne muscular dystrophy	Dystrophin	A progressive degeneration of muscles that begins in early childhood
Hemophilia A	Clotting Factor VIII	Defect in blood clotting
Hemophilia B	Clotting Factor IX	Defect in blood clotting
Testicular feminization	Androgen receptor	Missing the male steroid hormone receptor; XY individuals have external features that are feminine but internally have undescended testes and no uterus

TABLE 23–3 **Examples of Human Traits and Diseases Inherited in an X-linked Manner**

We show this idea in a Punnett square for the trait called hemophilia (X^{h-A} is the chromosome that carries the mutant allele causing hemophilia):

	X^H	Y
X^H	$X^H X^H$ Normal female	$X^H Y$ Normal male
X^{h-A}	$X^H X^{h-A}$ Carrier female	$X^{h-A} Y$ Male with hemophilia

Hemophilia is a disorder in which the blood cannot clot properly when a wound occurs. For individuals with this trait, a minor cut may bleed for a very long time, and small bumps can lead to large bruises because internal broken capillaries may leak blood profusely before they are repaired. For hemophiliacs, common accidental injuries pose a threat of severe internal or external bleeding. Hemophilia A (also called classical hemophilia) is caused by a defect in an X-linked gene that encodes the clotting protein Factor VIII. This disease has also been called the "Royal disease," because it has affected many members of European royal families. The pedigree shown in Figure 23-4 illustrates the prevalence of hemophilia A among the descendants of Queen Victoria of England. The pattern of X-linked recessive inheritance is revealed by the following observations:

1. *Males are much more likely to exhibit the trait.*
2. *The mothers of affected males often have brothers or fathers who are affected with the same trait.*
3. *The daughters of affected males will produce, on average, 50% affected sons.*

Many genetic disorders are heterogeneous

Hemophilia can be used to illustrate another concept in genetics, called **heterogeneity**. This term refers to the phenomenon that a particular type of disease may be caused by mutations in different genes. For example, blood clotting involves the participation of several different proteins. These proteins take part in a cellular cascade that leads to the formation of a clot. Therefore, a defect in any of the proteins needed for clot formation can cause hemophilia. Hemophilia B is caused by a defect in the blood clotting protein Factor IX; it is also X-linked. In addition, several other forms of hemophilia are caused by defects in autosomal genes that also encode clotting factor proteins.

In hemophilia, genetic heterogeneity arises from the participation of several proteins in a common cellular process. Another mechanism that may lead to genetic

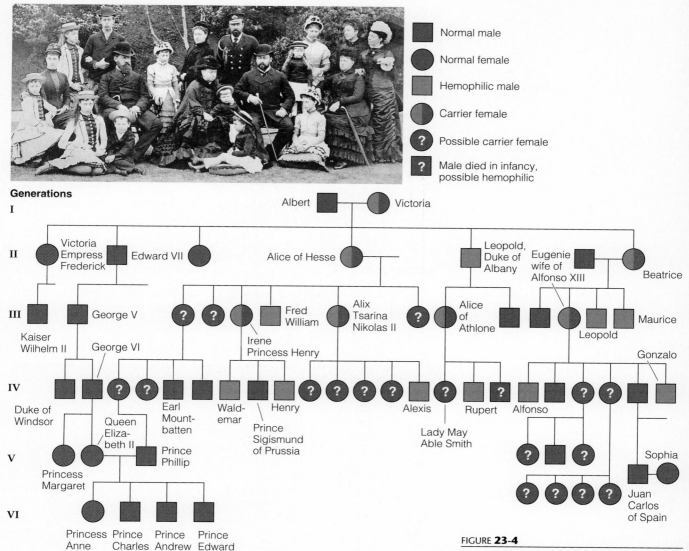

Generations

I — Albert ☐ ⬤ Victoria

II — Victoria Empress Frederick — Edward VII | Alice of Hesse | Leopold, Duke of Albany | Eugenie wife of Alfonso XIII | Beatrice

III — Kaiser Wilhelm II | George V | ? | ? | Irene Princess Henry | Fred William | Alix Tsarina Nikolas II | ? | Alice of Athlone | Leopold | Maurice

IV — Duke of Windsor | Queen Elizabeth II | George VI | ? | ? | Earl Mountbatten | Waldemar | Henry Prince Sigismund of Prussia | ? | ? | ? | ? | Alexis | Lady May Able Smith | Rupert | Alfonso | Gonzalo | ? | ?

V — Princess Margaret | Prince Phillip | ? | ? | ? | Sophia

VI — Princess Anne | Prince Charles | Prince Andrew | Prince Edward | ? | ? | ? | ? | Juan Carlos of Spain

Present generation and their children are free from hemophilia.

Legend:
- Normal male
- Normal female
- Hemophilic male
- Carrier female
- ? Possible carrier female
- ? Male died in infancy, possible hemophilic

FIGURE 23-4

The inheritance pattern of hemophilia A in the royal families of Europe. Pictured are Queen Victoria and Prince Albert of Great Britain, with some of their descendants.

heterogeneity occurs when proteins are composed of two or more different subunits, with each subunit being encoded by a different gene. The disease **thalassemia** presents an example of genetic heterogeneity caused by a mutation in a protein composed of multiple subunits. This potentially life-threatening disease involves defects in the ability of the red blood cells to transport oxygen. The underlying cause is an alteration in hemoglobin. In adults, hemoglobin is a tetrameric protein composed of two α-chains and two ß-chains; α-globin and ß-globin are encoded by separate genes (namely, the α- and ß-globin genes). Two main types of thalassemia have been discovered in human populations: α-thalassemia, in which the α-globin subunit is defective; and β-thalassemia, in which the ß-globin subunit is defective.

Unfortunately, genetic heterogeneity may greatly confound pedigree analysis. For example, a human pedigree might contain individuals with X-linked hemophilia and other individuals with autosomally linked hemophilia; a geneticist who assumed that all affected individuals had defects in the same gene would be unable to explain the pattern of inheritance. For disorders such as hemophilia and thalassemia, pedigree analysis is not a major problem, since the biochemical basis for these diseases is well understood. However, for rare diseases that are poorly understood at the molecular level, genetic heterogeneity may profoundly obscure the pattern of inheritance.

Genetic testing can screen for many inherited human diseases

Because genetic abnormalities occur in the human population at a significant rate, people have sought ways to determine whether individuals possess disease-causing alleles. The term **genetic screening** refers to the use of testing methods to discover if an individual is a heterozygous carrier for or has a genetic disease. Testing methods are available for the detection of many genetic abnormalities, but certainly not all. Table 23-4 describes some strategies for detecting genetic abnormalities.

In many cases, single gene mutations that affect the function of cellular proteins can be screened at the protein level. If a gene encodes an enzyme, biochemical assays to measure that enzyme's activity may be available. As mentioned earlier, Tay–Sachs disease involves a defect in the enzyme hexA. Enzymatic assays for this enzyme involve the use of an artificial substrate in which 4-methylumbelliferone (MU) is covalently linked to N-acetylglucosamine (GlcNAc). HexA cleaves this covalent bond and releases MU, which is fluorescent:

$$\text{MU–GlcNAc} \xrightarrow{\text{HexA}} \text{MU} + \text{GlcNAc}$$
$$\text{(nonfluorescent)} \qquad\qquad \text{(fluorescent)}$$

To perform this assay, a blood sample is collected and incubated with MU–GlcNAc, and the fluorescence is measured with a device called a *fluorometer*. Individuals affected with Tay–Sachs will produce little or no fluorescence, whereas nor-

TABLE 23–4 Screening Methods for Genetic Abnormalities

Method	Description
Protein Level	
Biochemical	As described for HexA, the enzymatic activity of a protein can be assayed *in vitro*.
Immunological	The presence of a protein can be detected by using antibodies that specifically recognize the protein. An example of this type of technique, Western blotting, is described in Chapter 19.
DNA or Chromosomal Level	
RFLP linkage	As described in Experiment 20B, certain inherited alleles are linked to RFLPs.
Altered restriction site	A gene mutation may alter the presence of a particular restriction enzyme site. A sample of DNA from an individual can be subjected to restriction enzyme digestion and gel electrophoresis (see Chapter 19) to determine whether or not the restriction enzyme site is present.
DNA sequencing	If the normal gene has already been cloned, it is possible to design PCR primers that can amplify the gene from a sample of cells. The amplified DNA segment can then be subjected to DNA sequencing as described in Chapter 19.
In situ hybridization	A DNA probe that hybridizes to a particular gene can be used to determine if the gene is present in an individual. This approach can be used to determine if a gene or segment of a gene has been deleted or altered. The technique of fluorescence *in situ* hybridization (FISH) is described in Chapter 20.
Karyotyping	The chromosomes from a sample of cells can be stained and then analyzed microscopically for abnormalities in chromosome structure and number.

mal individuals will produce a high level of fluorescence. Heterozygotes, who have 50% hexA activity, are expected to produce intermediate levels of fluorescence.

An alternative approach is to detect single gene mutations at the DNA level. To apply this screening strategy, researchers must have previously identified the mutant gene using molecular techniques. While this is often a difficult task, progress in the cloning and sequencing of human genes is advancing rapidly. The cloning of several human genes, such as those involved in Duchenne muscular dystrophy, cystic fibrosis, and Huntington disease, has made it possible to screen for affected individuals or those who may be carriers of these diseases. Table 23-4 describes several ways to screen for gene mutations by analyzing DNA sequences. The laboratory techniques involved were described in Chapters 19 and 20.

As was discussed in Chapter 8, many human genetic abnormalities involve changes in chromosome structure and/or number. In fact, changes in chromosome number are the most common class of human genetic abnormality. Most of these result in spontaneous abortions. However, approximately 1 in 200 live births are aneuploid or have unbalanced chromosomal alterations. About 5% of infant and childhood deaths are related to such genetic abnormalities. With regard to testing, changes in chromosome number and many changes in chromosome structure can be detected by karyotyping in which the chromosomes are examined with a light microscope.

In the United States, screening for some specific genetic abnormalities has become common medical practice. For example, pregnant women over 35 years old often have tests conducted to see if their fetuses are carrying chromosomal abnormalities. These tests are indicated because the rate of such defects increases with the age of the mother (see Figure 8-16). Another example is the widespread screening for phenylketonuria (PKU). As was described in Chapter 2, an inexpensive test can determine if newborns have this disease. Those who test positive can then be given a phenylalanine-free diet, thereby avoiding PKU's devastating effects.

Genetic screening also is conducted on specific populations in which a genetic disease is prevalent. For example, in 1971, community-based screening for heterozygous carriers of Tay–Sachs disease was begun among local Ashkenazi Jewish populations. Over the course of one generation, the incidence of TSD births has been reduced by 90%.

For most rare genetic abnormalities, however, genetic screening is not routine practice. Rather, genetic screening is performed only when family history reveals a strong likelihood that a couple may produce an affected child. This typically involves a couple who already has an affected child, or has other relatives afflicted by a genetic disease.

Genetic screening can be performed prior to birth. There are two common ways of obtaining cellular material from a fetus for the purpose of genetic testing: **amniocentesis** and **chorionic villus sampling**. In amniocentesis, a doctor removes amniotic fluid containing fetal cells using a needle that is passed through the abdominal wall (see Figure 23-5). The cells are cultured for several weeks and then karyotyped to determine the number of chromosomes per cell and whether changes in chromosome structure have occurred. In chorionic villus sampling, a small piece of the chorion (the fetal part of the placenta) is removed, and a karyotype is prepared directly from the collected cells. Chorionic villus sampling can be performed earlier during pregnancy than amniocentesis (around the 8th to 10th week, compared with the 14th to 16th week for amniocentesis). Weighed against this advantage, however, is that chorionic villus sampling may pose a slightly greater risk of miscarriage.

Genetic screening is a medical practice with many social and ethical dimensions. For example, people who learn they are carriers of genetic diseases, such as Huntington disease or familial breast cancer (described later in this chapter), are affected profoundly by the news. Some argue that people have a right to know about their genetic makeup; others assert that it does more harm than good. Another issue is privacy. Will an understanding of our genetic makeup subject us to discrimination by

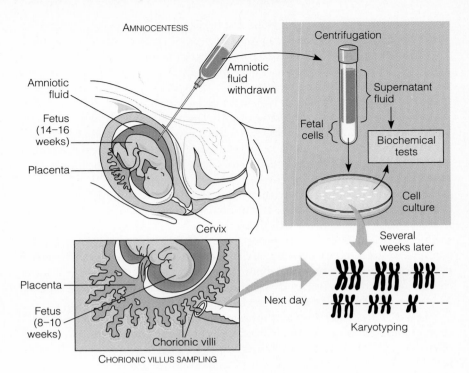

FIGURE **23-5**

Techniques to determine genetic abnormalities before birth. In amniocentesis, amniotic fluid is withdrawn and fetal cells are collected by centrifugation. The cells are then allowed to grow in laboratory culture media for several weeks prior to karyotyping. In chorionic villus sampling, a small piece of the chorion is removed. These cells can be prepared directly for karyotyping.

employers or medical insurance companies? During the 21st century, it is likely that we will gain an ever-increasing awareness of our genetic makeup and the underlying causes of genetic diseases. As a society, it will be necessary, yet very difficult, to establish guidelines for the uses of genetic testing.

Prions are infectious particles that alter protein function posttranslationally

Before we end this section on inherited diseases, let's consider an unusual mechanism in which agents known as **prions** cause disease. As shown in Table 23-5, prions have been hypothesized to cause several types of neurodegenerative diseases affecting humans and livestock. Recent evidence also suggests that prions may exist in yeast. Stanley Prusiner at the University of California, San Francisco, proposed that prions can act as infectious agents composed entirely of protein. (The term emphasizes the prion's unusual character as a *pro*teinaceous *in*fectious agent.) Before the discovery of prions, all known infectious agents such as viruses and bacteria contained their own genetic material (either DNA or RNA). The idea of prion-based disease is still somewhat controversial, though gaining much experimental support.

According to the prion hypothesis, such diseases arise from the ability of the prion protein to exist in two conformational states: a normal form PrP^C, which does not cause disease, and an abnormal form PrP^{Sc}, which does. The normal prion protein is encoded by normal individuals' genomes, and the protein is expressed at low levels. The abnormal conformation of the prion protein can come from two sources. An individual can be infected with the abnormal protein by direct contact with an affected individual or by eating raw meat from an organism with the disease. Alternatively, some people carry alleles of the *PrP* gene that cause the normal prion protein to convert spontaneously to the abnormal conformation at a very low rate. These individuals have an inherited predisposition to developing a prion-related disease. An example of an inherited prion disease is fatal familial insomnia (Table 23-5).

Let's now consider the molecular mechanism by which prions cause disease. The prion protein is predicted to exist in two conformations, PrP^{Sc} and PrP^C. As shown in Figure 23-6, the abnormal conformation, PrP^{Sc}, acts as a catalyst to

TABLE 23-5 Neurodegenerative Diseases Caused by Prions

Disease	Description
Infectious Diseases	
Kuru	A human disease that was once common in New Guinea. It begins with a loss of coordination, usually followed by dementia. Infection was spread by cannibalism, a practice that ended in 1958.
Scrapie	A disease of sheep and pigs characterized by intense itching (in which the animals tend to scrape themselves against trees, etc.), followed by neurodegeneration.
Mad cow disease	Begins with changes in posture and temperament, followed by loss of coordination and neurodegeneration.
Human Inherited Diseases	
Creutzfeldt–Jakob disease	Characterized by loss of coordination and dementia. On rare occasions, this disease may be spread by direct exchange of bodily fluid between an infected and uninfected individual.
Gerstmann–Straussler–Scheinker disease	Characterized by loss of coordination and dementia.
Familial fatal insomnia	Begins with sleeping and autonomic nervous system disturbances, followed by insomnia and dementia.

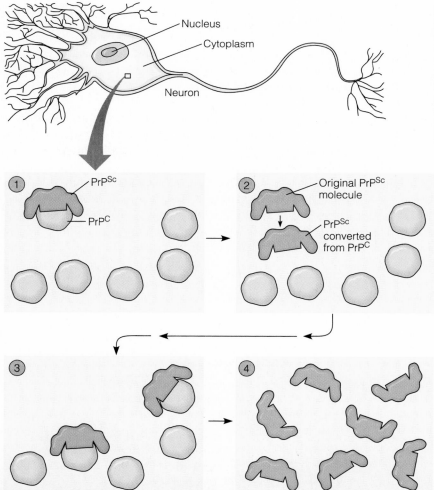

FIGURE **23-6**

A proposed molecular mechanism of prion diseases. A healthy neuron normally contains only the PrP^C conformation of the prion protein. **(1)** When a PrP^{Sc} conformation is found within the cell, it will **(2)** catalyze the conversion of PrP^C into PrP^{Sc}. **(3)** Then, both of the proteins in the PrP^{Sc} conformation will bind to PrP^C proteins and convert them to the PrP^{Sc} conformation. **(4)** Over time, the PrP^{Sc} conformation will accumulate to high levels, leading to symptoms of the prion diseases.

convert normal prion proteins within the cell into the abnormal conformation. Therefore, once an abnormal prion protein is found with a cell, the normal prion proteins already present will be converted to the abnormal conformation. As a prion disease progresses, the PrP^{Sc} protein is deposited as dense aggregates in the cells of the brain and peripheral nervous tissues. This deposition is thought to be related to the disease symptoms affecting the nervous system. Some of the abnormal prion protein is also excreted from infected cells, where it can travel through the bloodstream and be taken up by uninfected cells. In this way, a prion disease can spread through the body just like many viral diseases.

GENETIC BASIS OF CANCER

Cancer is a disease characterized by uncontrolled cell growth. It is a genetic disease at the cellular level. There are more than one hundred kinds of human cancers. These are classified according to the type of cell that has become cancerous. Though cancer is a diverse collection of many diseases, some characteristics are common to all cancers:

1. Most cancers originate in a single cell. This single cell, and its line of daughter cells, undergo a series of genetic changes that accumulate during cell division.In this regard, a cancerous growth can be considered to be clonal in origin. A hallmark of a cancer cell is that it divides to produce two daughter cancer cells.
2. At the cellular and genetic level, cancer usually is a multistep process that begins with a precancerous genetic change (i.e., a **benign** growth), and following additional genetic changes, progresses to cancerous cell growth (Figure 23-7).
3. Once a cellular growth has become cancerous or **malignant**, the cancer cells are invasive (i.e., they can invade normal tissues) and **metastatic** (i.e., they can migrate to other parts of the body).

Approximately one million Americans are diagnosed with cancer each year; about half that number will die from the disease. In 5 to 10% of cancers, a higher predisposition to develop the cancer is an inherited trait. We will examine some inherited forms of cancer later in this chapter.

Most cancers, though, perhaps 90 to 95%, are not passed from parent to offspring. Rather, cancer is usually an acquired condition that typically occurs later in life. While some cancers are caused by spontaneous mutations and viruses, at least 80% of all human cancers are related to exposure to agents that promote genetic changes in somatic cells. These environmental agents, such as UV light and chemical carcinogens, are mutagens that alter the DNA. These DNA alterations can lead to changes in gene expression that ultimately affect cell division. Since the DNA is modified, these somatic cell alterations are transmitted during cell division.

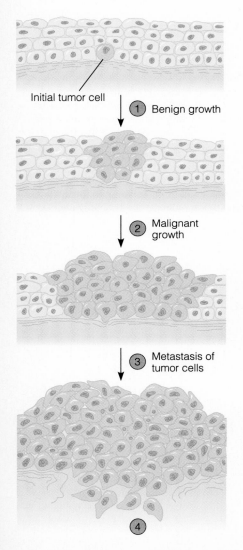

Initial tumor cell
① Benign growth
② Malignant growth
③ Metastasis of tumor cells
④

FIGURE **23-7**

Progression of cellular growth leading to cancer.

GENES → TRAITS:(1) In a healthy individual, an initial gene mutation converts a healthy cell into a tumor cell. (2) This tumor cell divides to produce a benign tumor. (3) Additional genetic changes in the tumor cells may occur, leading to malignant growth. (4) At a later stage in malignancy, the tumor cells will invade surrounding tissues, and some malignant cells may metastasize by traveling through the bloodstream to other parts of the body. As a trait, cancer can be viewed as a series of genetic changes that eventually lead to uncontrolled cell growth.

In this section, we will begin by discussing some early experimental observations that suggested genes play a role in cancer. We will then explore how genetic abnormalities, which affect the functions of particular cellular proteins, can lead to cancer.

Certain viruses can cause cancer by carrying viral oncogenes into the cell

As mentioned, most types of cancers are caused by mutagens that alter the structure and expression of genes. A few viruses, however, are known to cause cancer in plants, animals, and humans. We begin our discussion here, because early studies of cancer-causing viruses identified the first genes known to play a role in cancer. Many of these viruses can also infect laboratory-grown normal cells and convert them into malignant cells. The process of converting a normal cell into a malignant cell is called **transformation**.

Most cancer-causing viruses are not very potent at inducing cancer. An organism must be infected for a long period of time before a tumor actually develops. Furthermore, most viruses are inefficient at or unable to transform normal cells grown in the laboratory. By comparison, some viruses rapidly induce tumors in animals and efficiently transform cells in culture. These are called **acutely transforming viruses (ACVs)**. About 40 ACVs have been isolated from chickens, turkeys, mice, rats, cats, and monkeys. The first virus of this type, the *Rous sarcoma virus (RSV)*, was isolated from chicken sarcomas by Peyton Rous at Rockefeller University in 1911.

During the 1970s, RSV research led to the first identification of a gene that promotes cancer—an **oncogene**. Researchers investigated RSV by infecting chicken fibroblast cells. This causes the chicken fibroblasts to grow like cancer cells. Researchers identified mutant RSV strains that infected and proliferated within chicken cells without transforming them into malignant cells. These RSV strains were determined to contain a defective viral gene, designated *src* (for *sarcoma*, the type of cancer it causes). The *src* gene (also known as a v-*src*, because it is found within a viral genome) was the first example of a **viral oncogene**, a gene in a viral genome that promotes cancer.

Since the viral oncogene of RSV is not necessary for viral replication, researchers were curious why the virus has this v-*src* oncogene. Harold Varmus and Michael Bishop, in collaboration with Peter Vogt, soon discovered that normal host cells contain a copy of the *src* gene in their chromosomes. This normal copy of the *src* gene in the host is termed a **proto-oncogene**. This proto-oncogene is also termed c-*src*, because it is found in the chromosome.

The *c-src* gene does not cause cancer. However, once incorporated into a viral genome, this gene can become a viral oncogene that promotes cancer. There are two possible explanations of this phenomenon. First, the many copies of the virus made during viral replication may lead to an overexpression of the *src* gene. Alternatively, the v-*src* gene may accumulate additional mutations that convert it to an oncogene.

RSV has acquired the *src* gene by capturing it from a host cell's chromosome. This can occur during the RSV life cycle. RSV is a retrovirus, with an RNA genome. During its life cycle, a retrovirus uses reverse transcriptase to make a DNA copy of its genome, which becomes integrated as a **provirus** into the host cell genome. This integration may occur next to a proto-oncogene. During transcription of the proviral DNA, the neighboring proto-oncogene may be included in the RNA transcript. This RNA transcript can then recombine with an RNA retroviral genome within the cell to yield a retrovirus that contains an oncogene.

Since these early studies of RSV, many other retroviruses carrying oncogenes have been investigated. The characterization of their oncogenes has led to the identification of several genes with oncogenic potential. Besides retroviruses, several viruses with DNA genomes cause tumors (see Table 23-6). Some of these are known to cause cancer in humans.

TABLE 23–6 Examples of Viruses that Cause Cancer

Virus	Description
Retroviruses	
Rous sarcoma virus	Causes sarcomas in chickens.
Simian sarcoma virus	Causes sarcomas in monkeys.
Abelson leukemia virus	Causes leukemia in mice.
Hardy–Zuckerman-4 feline sarcoma virus	Causes sarcomas in cats.
DNA Tumor Viruses	
Hepatitis B	Causes liver cancer in several species, including humans.
SV40, Polyomavirus	Does not cause cancer in their natural hosts, but can transform cells in culture.
Papillomavirus	Causes benign tumors and malignant carcinomas in several species, including humans. Causes cervical cancer in humans.
Adenovirus	Does not cause cancer in their natural hosts, but can transform cells in culture.
Herpesvirus	Causes carcinoma in frogs and T-cell lymphoma in chickens. A human herpesvirus, Epstein–Barr virus, is a causative agent in Burkitt's lymphoma, which occurs primarily in immunosuppressed individuals such as AIDS patients.

EXPERIMENT 23A

In 1979, Weinberg found that DNA isolated from malignant mouse cells can transform normal mouse cells into malignant cells

The study of retroviruses and other tumor-producing viruses led to the identification of a few dozen viral oncogenes. These were the first genes implicated in causing cancer. However, most cancers are caused not by viruses, but by environmental mutagens that alter the expression of normal cellular genes. Therefore, researchers also wanted to identify cellular genes that have been altered in a way that leads to malignancy. Methods developed in the study of viral oncogenes were to prove valuable in this search. In particular, Miroslav Hill and Jana Hillova in Villejuif, France, showed in 1971 that the purified DNA of RSV could be taken up by chicken fibroblasts and would transform them into malignant cells.

In 1979, Robert Weinberg and his colleagues at the Massachusetts Institute of Technology wanted to determine if chromosomal DNA purified from cells that have become malignant due to mutagens can transform normal cells into malignant cells. If so, this would be the first step in the identification of chromosomally located genes that had been converted to oncogenes.

Before we discuss this experiment, let's consider how a researcher can identify malignant cells in a laboratory. A widely used assay relies on the ability to recognize a clump of transformed cells as a distinct **focus** that grows over a monolayer of cells on a culture dish (Figure 23-8). Unlike normal cells, which grow as a simple monolayer on culture dishes, a malignant focus piles up to form a mass of cells. The focus of cells is derived from a single cell that has become malignant and then proliferated. Also, the malignant cells of the focus or clump have altered morphologies.

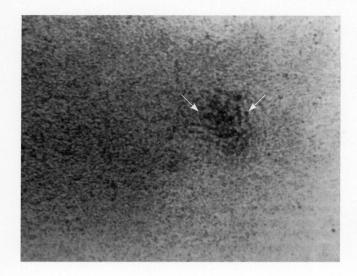

FIGURE **23-8**

A photograph showing the malignant growth of cells. The growth of normal cells leads to the formation of a monolayer, while certain transformed cells grow as a focus [see arrows] or a raised pile of cells.

THE HYPOTHESIS

Cellular DNA isolated from malignant cells will be taken up by normal cells and transform them into malignant cells.

TESTING THE HYPOTHESIS

Starting material: Several mouse cell lines. Some of the cell lines were normal, whereas others were malignant due to exposure to chemical or physical mutagens. It was known that none of the cell lines in this experiment were infected with oncogenic viruses.

Experimental Level

Conceptual Level

1. Extract the chromosomal DNA from normal or malignant cell lines.

    Homogenize the cells.

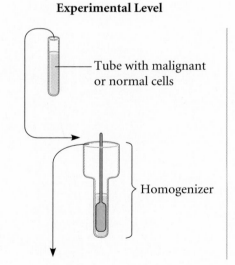

Tube with malignant or normal cells

Homogenizer

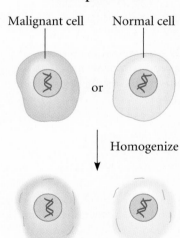

Malignant cell   Normal cell

or

Homogenize

(continued)

TESTING THE HYPOTHESIS

(continued)

Isolate the nuclei by differential centrifugation. See the appendix.

Dissolve the nuclear membrane with detergent, and isolate the chromosomal DNA.

2. Mix the DNA with normal mouse fibroblast cells that are growing on a tissue culture plate.

3. Add a buffer containing calcium ions and phosphate ions. This buffer makes the cells permeable to DNA.

4. Incubate for 14–20 days.

5. Examine the plates under a light microscope for cells growing as transformed foci.

   Note: Normal cells in culture grow as a monolayer and very rarely form foci.

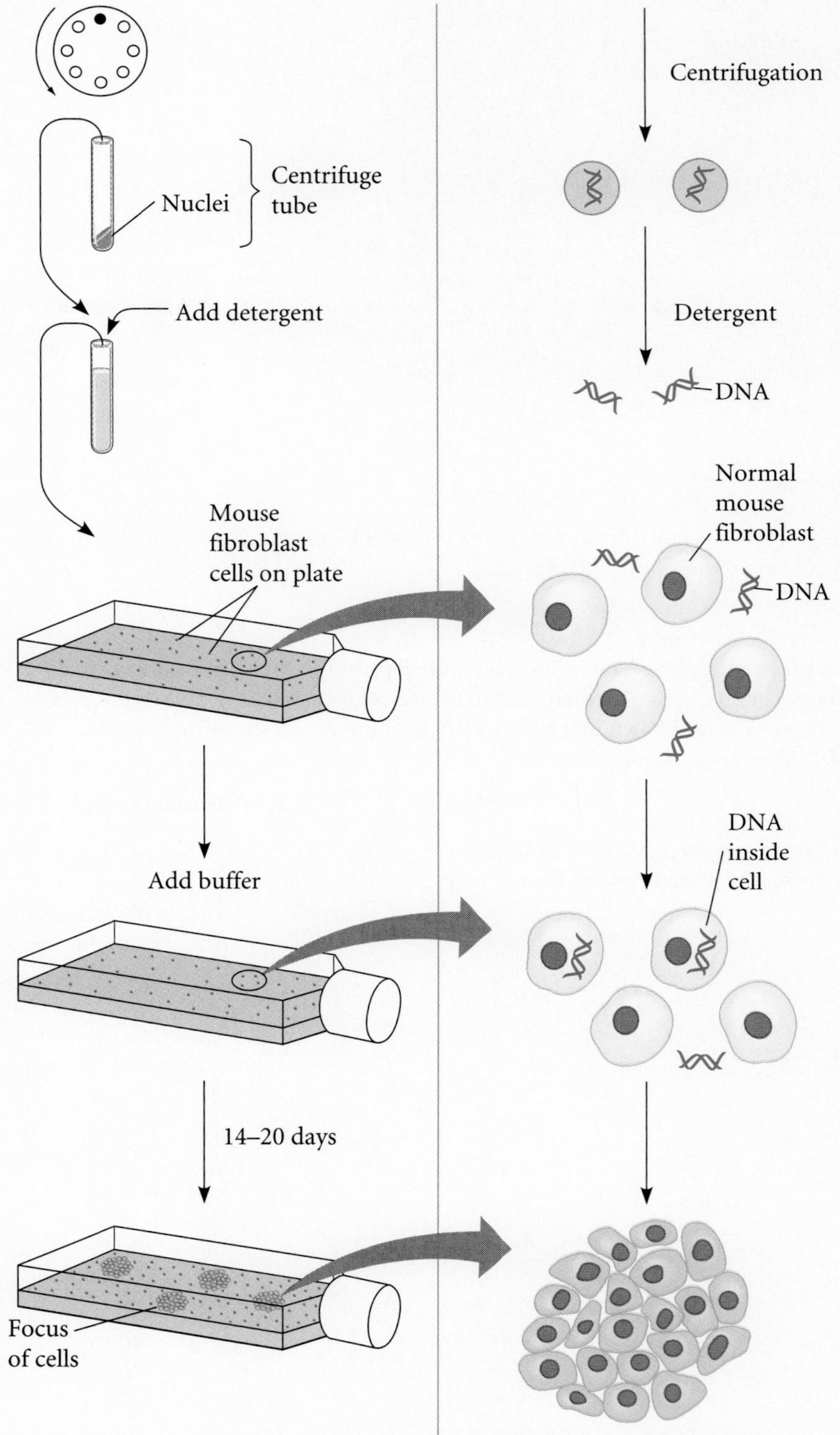
Centrifugation
Nuclei
Centrifuge tube
Add detergent
Detergent
DNA
Mouse fibroblast cells on plate
Normal mouse fibroblast
DNA
Add buffer
DNA inside cell
14–20 days
Focus of cells

Malignant cells derived from a normal cell that had taken up DNA containing an oncogenic mutation.

THE DATA

Source of DNA	Recipient Cells	Number of Malignant Foci Found on 12 Plates
Malignant Cell Lines		
MC5-5-0	NIH3T3 (normal fibroblasts)	48*
MCA16	"	5
MB66 MCA ad 36	"	8
MB66 MCA ACL6	"	0
MB66 MCA ACL13	"	0
Normal Cell Lines		
NIH3T3	"	<1
C3H10T1/2	"	0

*In this experiment, 2 of the plates were contaminated, so this is 48 foci on 10 plates.

INTERPRETING THE DATA

As shown in the data, the DNA isolated from some (but not all) malignant cell lines could transform normal mouse cells which proliferated and produced malignant foci. These results are consistent with the hypothesis that oncogenes had been taken up by and expressed in the normal mouse cells, converting them into malignant cells. By comparison, the DNA isolated from normal cells did not cause a significant amount of transformation. From these experiments, it is not clear why the DNA from two of the malignant cell lines (namely, MB66 MCA ACL6 and MB66 MCA ACL13) could not transform the normal mouse cells. One possibility is that some malignancies are caused by dominant oncogenes, while others involve genes that act recessively. Recessive genes would be unable to transform normal mouse cells that already contain the normal (nonmalignant) dominant allele. Later in this chapter, we will learn how another category of genes involved in cancer, called tumor-suppressor genes, act recessively.

At about the same time as Weinberg's work, Geoffrey Cooper and his colleagues at the Dana–Farber Cancer Institute in Boston showed that the DNA from normal cells could be isolated *in vitro* and activated to become oncogenic. Taken together, these early studies provided the first demonstration that organisms contain **cellular oncogenes** that cause malignancy.

Just two years later, in 1981, the laboratories of Weinberg and Cooper identified the first cellular oncogene in humans. They extracted DNA from human bladder carcinoma cells and showed that it could transform mouse cells *in vitro*. This observation paved the way for the isolation of many human cellular oncogenes.

Many oncogenes have abnormalities that affect proteins involved in cell division pathways

As researchers began to identify oncogenes, they wanted to understand how these abnormal genes cause cancer. In parallel with cancer research, cell biologists have studied the roles that normal cellular proteins play in cell division. In the somatic cells of the body, the cell cycle is regulated in part by polypeptide hormones known as *growth factors*. These growth factors bind to cell surface receptors and initiate a cascade of cellular events that lead eventually to cell division. Many different cellular proteins participate in this cascade. Most, but not all, oncogenes encode proteins that function in cell growth signaling pathways (see Table 23-7). These include growth factors, growth factor receptors, proteins involved in the intracellular signaling pathways, and transcription factors.

TABLE **23–7 Examples of Oncogenes**

Gene	Cellular Function
Growth factors[a]	
sis	Platelet-derived growth factor
int-2	Fibroblast growth factor
Receptors	
erbB	Growth factor receptor for EGF (epidermal growth factor)
trk	Growth factor receptor for NGF (nerve growth factor)
fms	Growth factor receptor for CSF-1 (cytostatic factor)
K-sam	Growth factor receptor for FGF (fibroblast growth factor)
erbA	Steroid receptor (which functions as a transcription factor)
Intracellular Signaling Proteins	
ras	GTP/GDP-binding protein
raf	Serine kinase
src	Tyrosine kinase
abl	Tyrosine kinase
gsp	G-protein α subunit
Transcription Factors	
jun	Transcription factor
fos	Transcription factor
myc	Transcription factor
gli	Transcription factor

[a]The genes described in this table are found in humans as well as in other eukaryotic species. Many of these genes were initially identified in retroviruses. Most of the genes have been given three-letter names that are abbreviations for the type of cancer the oncogene causes or the type of virus in which the gene was first identified.

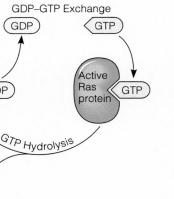

GDP–GTP Exchange

GDP   GTP

Inactive Ras protein — GDP

Active Ras protein — GTP

GTP Hydrolysis

P$_i$

FIGURE **23-9**

Functional cycle of the Ras protein.
When GTP is bound to Ras, the latter is in the active form that promotes cell division. The hydrolysis of GTP to GDP and P$_i$ converts the active form to an inactive form.

An oncogene may promote cancer by keeping the cell growth signaling pathway in a permanent "on" position. This can occur in two ways. In some cancers, the oncogene is overexpressed, yielding too much of the encoded protein. In 1982, research groups headed by Robert Gallo at the NIH and Mark Groudine at the University of Washington Hospital in Seattle showed that a *myc* gene, designated c-*myc*, was amplified about tenfold in a human promyelocytic leukemia cell line. Since that time, it has been found that *myc* genes are overexpressed in many forms of cancer, including breast cancer, lung cancer, and colon carcinoma. Presumably, the overexpression of this transcription factor leads to the transcriptional activation of genes that promote cell division.

Rather than being overexpressed in the amount of protein, certain oncogenes produce a functionally aberrant protein. For example, mutations that alter the amino acid sequence of the Ras protein have been shown to cause functional abnormalities. The Ras protein is a GTPase that hydrolyzes GTP to GDP + P$_i$ (Figure 23-9). Therefore, after it has been activated, the Ras protein returns to its inactive state by hydrolyzing GTP. Mutations that convert normal *ras* into an oncogenic *ras* either decrease the GTPase activity of the Ras protein or they increase the rate of exchange of bound GDP for GTP. Both of these functional changes result in a greater amount of the active GTP-bound form of the Ras protein. In this way, these mutations keep the signaling pathway turned on.

Genetic changes in cellular proto-oncogenes can convert them to cellular oncogenes

A fundamental problem in cancer biology is to identify the specific genetic alteration that converts a normal proto-oncogene into an abnormal oncogene. By isolating and studying oncogenes at the molecular level, researchers have discovered four main ways that this occurs (Table 23-8). These changes can be categorized as missense mutations, gene amplifications, chromosomal translocations, and retroviral integrations.

As mentioned previously, changes in the structure of the Ras protein can cause it to become permanently activated. This is caused by a missense mutation in the *ras* gene. The human genome contains four different, but evolutionarily related *ras* genes: *ras*H, *ras*N, *ras*K-4a, and *ras*K-4b. All four genes encode proteins with very similar amino acid sequences containing a total of 188 or 189 amino acids. Missense mutants in these normal *ras* genes are associated with particular forms of cancer. For example, a missense mutation in *ras*H that changes a glycine to a valine is responsible for the conversion of *ras*H into an oncogene:

	1	2	3	4	5	6	7	8	9	10	11	12	13		188	189
Normal	Met	Thr	Glu	Tyr	Lys	Leu	Val	Val	Val	Gly	Ala	Gly	Gly		Leu	Ser
Human *ras*H	ATG	ACG	GAA	TAT	AAG	CTG	GTG	GTG	GTG	GGC	GCC	GGC	GGT.....		CTC	TCC

$$\downarrow$$

	1	2	3	4	5	6	7	8	9	10	11	GTC	13		188	189
Oncogenic *ras*H	Met	Thr	Glu	Tyr	Lys	Leu	Val	Val	Val	Gly	Ala	Val	Gly.....		Leu	Ser

Experimentally, chemical carcinogens have been shown to cause these missense mutations and thereby lead to cancer.

Another genetic event that occurs in cancer cells is an abnormal increase in the copy number of a proto-oncogene. Gene amplification does not take place normally in mammalian cells, but it is a common occurrence in cancer cells. As mentioned previously, Gallo and Groudine discovered that c-*myc* was amplified in a human leukemia cell line. Many human cancers are associated with the amplification of particular oncogenes. In some cases, the extent of oncogene amplification is correlated closely with the progression of tumors to increasing malignancy. These include the amplification of N-*myc* in neuroblastomas and *erbB-2* in breast carcinomas. In other types of

TABLE 23–8 Genetic Changes that Convert Proto-oncogenes into Oncogenes

Type of Change	Description and Examples
Missense mutation	A change in the amino acid sequence of a proto-oncogene protein may cause it to function in an abnormal way. Missense mutations can convert *ras* genes into oncogenes.
Gene amplification	The copy number of a proto-oncogene may be increased by gene duplication. *Myc* genes have been amplified in human leukemias, breast, stomach, lung, and colon carcinomas, and neuroblastomas and glioblastomas. *ErbB* genes have been amplified in glioblastomas, squamous cell carcinomas, and breast, salivary gland, and ovarian carcinomas.
Chromosomal translocations	A piece of chromosome may be translocated to another chromosome and affect the expression of genes at the breakpoint site. In Burkitt's lymphoma, a region of chromosome 8 is translocated to either chromosome 2, 14, or 22. The breakpoint in chromosome 8 causes the overexpression of the c-*myc* gene.
Retroviral integration	When a virus integrates into the chromosome, it may enhance the expression of nearby proto-oncogenes. In avian lymphomas, the integration of the avian leukosis virus can enhance the transcription of the c-*myc* gene.

Adapted from: Cooper, G. M. (1995) *Oncogenes,* 2nd ed. Jones and Bartlett Publishers, Boston.

malignancies, gene amplification is more random and may be a secondary event that increases the expression of oncogenes previously activated by other genetic changes.

A third type of genetic alteration that can lead to cancer is a chromosomal translocation. While structural abnormalities are common in cancer cells, very specific types of chromosomal translocations have been identified in certain types of tumors. In 1960, Peter Nowell at the University of Pennsylvania Medical School discovered that chronic myelogenous leukemia was correlated with the presence of a shortened version of chromosome 22, which he called the **Philadelphia chromosome** after the city where it was discovered. Rather than a deletion, this shortened chromosome is the result of a reciprocal translocation between chromosomes 9 and 22. Later studies revealed that this translocation activates a proto-oncogene, *abl*, in an unusual way (Figure 23-10). The reciprocal translocation involves breakpoints within the *bcr* and *abl* genes. Following the reciprocal translocation, the first part of the *bcr* gene fuses with the *abl* gene. This yields an oncogene that encodes an abnormal fusion protein, which contains the polypeptide sequences encoded from both genes.

In other forms of cancer, chromosomal translocations cause an overexpression of an oncogene. In Burkitt's lymphoma, for example, a region of chromosome 8 is translocated to either chromosome 2, 14, or 22. The breakpoint in chromosome 8 is near the c-*myc* gene, and the sites on chromosomes 2, 14, and 22 correspond to locations of different immunoglobulin genes that are normally expressed in lymphocytes. The translocation of the c-*myc* gene near the immunoglobulin genes leads to the overexpression of the c-*myc* gene and thereby promotes malignancy in lymphocytes.

Tumor-suppressor genes play a role in preventing the proliferation of cancer cells

Thus far, we have considered how oncogenes promote cancer. An oncogene is an abnormally activated gene that leads to uncontrolled cell growth. We will now turn our attention to a second category of genes, called **tumor-suppressor genes** or **anti-oncogenes**. As the name suggests, the role of a tumor-suppressor gene is to prevent cancerous growth. Therefore, when a tumor-suppressor gene becomes inactivated by mutation, it becomes more likely that cancer will occur.

The first identification of a human tumor-suppressor gene involved studies of *retinoblastoma*, a tumor that occurs in the retina of the eye. Some people have an inherited predisposition to develop this disease within the first few years of life. By

FIGURE **23-10**

The reciprocal translocation commonly found in people with chronic myelogenous leukemia.

GENES ──▶ TRAITS: In healthy individuals, the *bcr* gene is located on chromosome 22, the *abl* gene on chromosome 9. In certain forms of myelogenous leukemia, a reciprocal translocation causes the first part of *bcr* to fuse with *abl*. This combined gene encodes an abnormal fusion protein, which leads to leukemia.

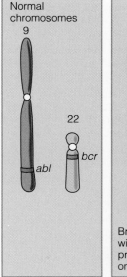
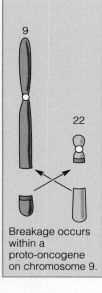
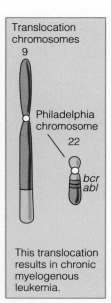

comparison, the noninherited form of retinoblastoma, which is caused by environmental agents, is more likely to occur later in life.

Based on these differences, in 1971, Alfred Knudson at the University of Texas in Houston proposed a "two-hit" model for retinoblastoma. According to this idea, retinoblastoma requires two mutations to occur. People with the inherited form already have received one mutation from one of their parents. They only need one additional mutation to develop the disease. Since there are more than one million cells in the retina, it is not unlikely that a mutation may occur in one of these cells at an early age, leading to the disease. However, people with the noninherited form of the disease must have two mutations in the same retinal cell to cause the disease. Since two rare events are much less likely to occur than a single such event, the noninherited form of this disease is expected to occur much later in life, and only rarely. Therefore, this hypothesis explains the different populations typically affected by the inherited and noninherited forms of retinoblastoma.

Since 1971, molecular studies have confirmed the two-hit hypothesis. The gene that causes this disease is designated *rb* (for *retinoblastoma*). *Rb* is found on the long arm of chromosome 13. Most people have two normal copies of this gene. Persons with hereditary retinoblastoma have inherited one functionally defective copy. Therefore, in nontumorous cells throughout the body, they have one normal copy and one defective copy of *rb*. However, in retinal tumor cells, the normal *rb* gene has also suffered the second hit (i.e., a mutation) that renders it defective. These observations verify the two-hit theory of Knudson.

More recent studies have revealed how the Rb protein prevents the proliferation of cancer cells (Figure 23-11). The Rb protein regulates a transcription factor called E2F. This factor activates genes required for cell cycle progression. The binding of the Rb protein to E2F inhibits its activity. This prevents cell division. When a normal cell is supposed to divide, cyclins bind to cyclin-dependent protein kinases, leading to the phosphorylation of the Rb protein. The phosphorylated form of the Rb protein is released from E2F. This allows E2F to activate genes needed to progress through the cell cycle. By comparison, it is easy to see how the cell cycle becomes unregulated without functional Rb protein. When both copies of Rb are defective, the E2F protein is always active. This explains why uncontrolled cell division can occur.

The vertebrate *p53* gene is a master tumor-suppressor gene that senses DNA damage

After the *rb* gene, the second tumor-suppressor gene discovered was the *p53* gene. *p53* is the most commonly altered gene in human cancers. About 50% of all human

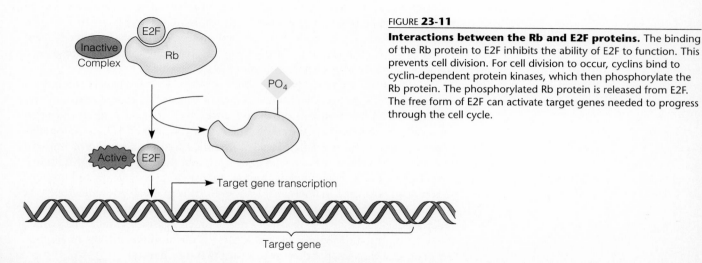

FIGURE **23-11**

Interactions between the Rb and E2F proteins. The binding of the Rb protein to E2F inhibits the ability of E2F to function. This prevents cell division. For cell division to occur, cyclins bind to cyclin-dependent protein kinases, which then phosphorylate the Rb protein. The phosphorylated Rb protein is released from E2F. The free form of E2F can activate target genes needed to progress through the cell cycle.

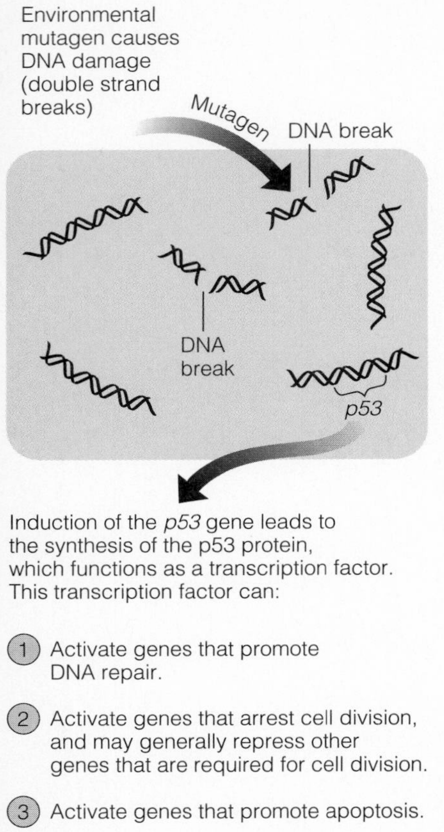

Environmental mutagen causes DNA damage (double strand breaks)

Induction of the *p53* gene leads to the synthesis of the p53 protein, which functions as a transcription factor. This transcription factor can:

(1) Activate genes that promote DNA repair.

(2) Activate genes that arrest cell division, and may generally repress other genes that are required for cell division.

(3) Activate genes that promote apoptosis.

FIGURE **23-12**

Central role of *p53* in preventing the proliferation of cancer cells. Expression of the *p53* gene, which encodes a transcription factor, is induced by agents that cause double-stranded DNA breaks. The p53 transcription factor may **(1)** activate genes that promote DNA repair, **(2)** activate genes that arrest cell division and repress genes required for cell division, and **(3)** activate genes that promote apoptosis.

cancers are associated with defects in this gene. This includes malignant tumors of the lung, breast, esophagus, liver, bladder, and brain as well as sarcomas, lymphomas, and leukemias. For this reason, an enormous amount of research has been aimed at elucidating the function of the p53 protein.

As we have discussed, environmental mutagens play a causative role in the development of most cancers. In addition to the removal of malignant cells via the immune system, animal cells have three different defense mechanisms to thwart the proliferation of cancer cells. First, when confronted with a DNA-damaging agent, the cell can try to repair its DNA. This may prevent the accumulation of mutations that activate oncogenes or inactivate tumor-suppressor genes. Second, if the cell is in the process of dividing, it can arrest itself in the cell cycle. By stopping its cell cycle, the cell has more time to repair its DNA and avoid producing two mutant daughter cells.

The third, and most drastic thing, is that a cell can initiate a series of events called **apoptosis** or programmed cell death. In response to DNA-damaging agents, a cell may program its own death. Apoptosis is an active process that involves cell shrinkage, chromatin condensation, and eventually DNA degradation. Apoptosis occurs in a small number of cells as a normal process of embryonic development. In addition, it is beneficial for an adult organism to kill an occasional cell with cancer-causing potential!

Figure 23-12 illustrates how *p53* plays a central role in all three processes. As shown here, the expression of the *p53* gene is induced by agents that damage the DNA. The inducing signal appears to be double-stranded DNA breaks. The p53 protein functions as a transcription factor. It contains a DNA-binding domain and a transcriptional activation domain. Experimental studies have shown that p53 can activate the transcription of several specific target genes. As shown in Figure 23-12, it can activate genes that promote DNA repair, arrest the cell cycle, and promote apoptosis. In addition, p53 appears to act as a negative regulator by interacting with general transcription factors. This decreases the general expression of many other structural genes. This inhibition may also help prevent the cell from dividing.

Overall, p53 activates the expression of a few specific cellular genes and, at the same time, inhibits the expression of many other genes in the cell. Through the regulation of gene transcription, the p53 protein can prevent the proliferation of cells that have incurred DNA damage.

Many types of cancer involve multiple genetic changes leading to malignancy

The discovery of oncogenes and tumor-suppressor genes, along with molecular techniques that can detect genetic alterations, has enabled researchers to study the progression of certain forms of cancer at the molecular level. Many cancers begin with a benign genetic alteration that, over time and with additional mutations, leads to malignancy. Furthermore, a malignancy can continue to accumulate genetic changes that make it even more difficult to treat.

In 1990, Eric Fearon and Bert Vogelstein at Johns Hopkins University proposed a series of genetic changes that leads to colorectal cancer, the second most common cancer in the United States. As shown in Figure 23-13, colorectal cancer is derived from cells in the mucosa of the colon. The loss of function of *APC*, a tumor-suppressor gene on chromosome 5, leads to an increased proliferation of mucosal cells and the development of a benign polyp, a noncancerous growth. Additional genetic changes involving the loss of other tumor-suppressor genes and the activation of an oncogene (namely, *ras*) leads eventually to the development of a carcinoma. In Figure 23-13, the genetic changes that lead to colon cancer occur in an orderly sequence. While the growth of a tumor often begins with mutations in *APC*, the order of mutations shown in Figure 23-13 is not absolute. It is the total number of genetic changes, not their exact order, that is important.

Inherited forms of cancers usually are caused by defects in tumor-suppressor or DNA repair genes

As we have just seen, cancer cells usually are generated by multiple genetic alterations that cause the activation of oncogenes and the inactivation of tumor-suppressor genes. As mentioned earlier in this chapter, about 5 to 10% of all cancers involve inherited (germ-line) mutations. People who have inherited such mutations have a predisposition to developing cancer. This does not mean they will definitely get cancer, but they are more likely to develop the disease than are individuals in the general population.

Most inherited forms of cancer involve a defect in a tumor-suppressor gene (Table 23-9). Therefore, the individual is heterozygous, with one normal and one inactive allele. As was described earlier for retinoblastoma, a mutation in the normal allele may occur in a somatic cell, thereby producing a cell with two inactive copies of the tumor-suppressor gene. This cell may then progress to a malignant state as it accumulates additional genetic changes. At the phenotypic level, a predisposition for developing cancer is inherited in a dominant fashion, since a heterozygote exhibits this tendency. Examined at the cellular level, the mutation is recessive, because a second mutation that results in two abnormal alleles is necessary to promote malignancy.

As noted in Table 23-9, not all hereditary forms of cancer are due to defective tumor-suppressor genes. Two different genes (*MSH*2 and *MLH*1) identified in nonpolyposis (i.e., non-polyp-forming) colorectal cancer are defects in enzymes required for mismatch DNA repair. A defect in DNA repair may lead to a higher rate of mutation and thereby increase the likelihood of cancer-causing mutations. Likewise, as was discussed in Chapter 17, some human diseases, such as xeroderma pigmentosum, involve inherited defects in genes that are involved with nucleotide excision repair. These individuals have many premalignant lesions and a predisposition to developing skin cancer.

TABLE 23–9 Examples of Inherited Mutant Genes that Confer a High Predisposition to Develop Cancer

Gene	Type of Cancer	Protein Function
Tumor Suppressor Genes		
APC	Familial colon cancer	The exact function is not known, but its inheritance suggests that it is a tumor suppressor gene.
rb	Retinoblastoma	Rb protein binds to E2F and prevents it from activating genes that promote cell division.
p53	Li–Fraumeni syndrome	p53 is a transcription factor that positively regulates a few specific target genes and negatively regulates others in a general manner.
BRCA-1	Familial breast cancer	The BRCA-1 protein contains a zinc finger domain, which suggests that it may function as a transcription factor.
*WT*1	Wilm's tumor	The WT-1 protein functions as a transcriptional repressor.
*NF*1	Neurofibromatosis	The NF1 protein stimulates *Ras* to hydrolyze its GTP to GDP.
*MTS*1	Melanoma	The MTS1 protein acts as an inhibitor of cyclin-dependent protein kinases.
DNA Repair Genes		
*MSH*2, *MLH*1	Nonpolyposis colorectal cancer	Enzymes that function in DNA mismatch repair.
Oncogenes		
ret	Multiple endocrine neoplasia (Type 2)	An abnormally activated receptor protein that functions as a tyrosine kinase.

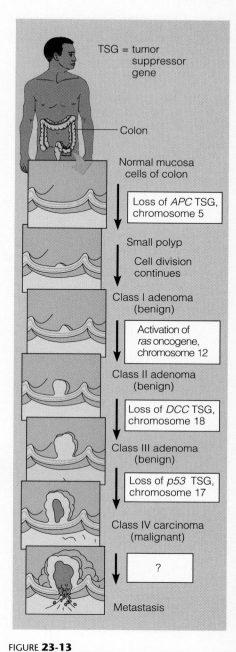

FIGURE **23-13**

Multiple genetic changes leading to colorectal cancer.

CONCEPTUAL SUMMARY

Throughout this chapter, we have discussed many human diseases that have a genetic basis, such as *Tay–Sachs disease*, *Huntington disease*, *hemophilia*, and *thalassemia*. In some cases, mutant genes are passed from parents to offspring, causing inherited abnormalities. The pattern of inheritance of these mutant genes can often be deduced from a pedigree analysis, although this may be difficult if the disease has a *heterogeneous* genetic origin. In the case of single gene defects, some genetic diseases are recessive, some are dominant, and some are X-linked. In addition, irregularities in chromosome structure and/or number can cause phenotypic abnormalities. These types of genetic defects are more common in humans than defects in single genes.

Cancer is also a human disease with a genetic basis. Cancer can be caused by oncogenic viruses, such as RSV and other *ACVs*, or by the inheritance of mutant genes. However, most forms of cancer are due to environmentally induced mutations in somatic cells that lead to uncontrolled cell growth. These mutations may involve two types of genes: *oncogenes* and *tumor-suppressor genes*. An oncogene, which is derived from a normal *proto-oncogene*, is an abnormally activated gene that stimulates cell growth. At the molecular level, a proto-oncogene may be converted to an oncogene by a missense mutation, a chromosome translocation, gene amplification, or retroviral integration. By comparison, a tumor-suppressor gene normally inhibits cell growth, but if it is rendered inactive by mutation, unconstrained cell growth may ensue. A master tumor-suppressor gene, called *p53*, plays a critical role in monitoring DNA damage and preventing the accumulation of cells that have been damaged.

EXPERIMENTAL SUMMARY

Researchers cannot experiment with humans, as they can with model organisms such as *Drosophila* and mice. Nevertheless, researchers can gain knowledge of human genetic disorders by analyzing the occurrence of diseases within pedigrees and conducting *in vitro* tests on human cell samples.

Often, when a human disease is newly identified, researchers would like to know if it is caused, wholly or in part, by a mutation in a gene. Various observations point to a genetic cause. These include a higher frequency of affected individuals among family members and in identical versus fraternal twins, the inability of the disease to be spread by contact, different rates of the disease among different populations, a characteristic age of onset, a similar disease of known genetic origin in animals, and a correlation between a disease and a mutant gene.

Once genetics has become established as a cause of disease, researchers attempt to determine the pattern of transmission from parents to offspring. Many, but not all, genetic diseases follow a simple Mendelian pattern of inheritance. A pedigree analysis may indicate whether a disease is transmitted recessively or dominantly, and if it is X-linked or autosomal.

To understand the relationship between genetics and disease, researchers investigate how mutant genes lead to the disease symptoms at the cellular and organismal levels. There are many experimental approaches to investigating this question; most of these have not been described here. At the molecular level, the ultimate goal is to identify the function of the protein encoded by the normal gene, and the defect resulting from the disease-causing mutation. In this chapter, we have considered many examples where a disease-causing allele has been identified, and the function of the normally encoded protein is known. With this information, molecular and cell biologists can begin to unravel how a gene mutation affects protein function, cellular processes, and ultimately the phenotype of the individual.

Much research will continue to be aimed at understanding how mutant genes lead to abnormalities in people. In addition, our identification of an increasing number of mutant genes has provided an impetus to determine via genetic screening tests if individuals are carriers of a genetic lesion, or have the potential of transmitting a genetic lesion to their children.

Cancer is a collection of human diseases that receives a large amount of attention. Experimentally, our understanding of cancer has been aided greatly by the characterization of viruses that can cause cancer in animals or transform laboratory-grown cells. The study of these tumor-causing viruses has led to the identification of many oncogenes that play a central role in promoting cancerous cell growth. In addition, experiments in which DNA is used to transform cells *in vitro* have also enabled researchers to identify oncogenes.

Researchers continue to investigate the specific ways that mutations alter cancer-causing genes and the combinations of mutations that lead to malignant growth. By studying the DNA at the cellular and molecular level, scientists have discovered that an oncogene can be created by a missense mutation, a chromosome translocation, gene amplification, or retroviral integration, which lead to an abnormally high quantity and/or activity of a growth-related protein. By comparison, mutations that inactivate tumor-suppressor genes may render an individual more vulnerable to the proliferation of cancer.

In the future, scientists will continue to explore how oncogenes and tumor-suppressor genes influence the cell division cycle. They would ultimately like to understand this relationship in hopes of discovering how we can prevent cancer after these mutations have occurred.

PROBLEM SETS

SOLVED PROBLEMS

1. The pedigree shown here concerns a human disease known as familial hypercholesterolemia.

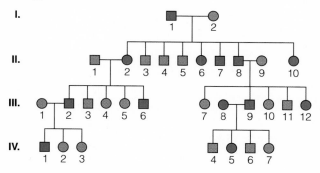

This disorder is characterized by an elevation of serum cholesterol in the blood. This genetic disorder can be a contributing factor to heart attacks. However, this disorder is relatively rare. At the molecular level, this disease is caused by a defective gene that encodes a protein called the low-density lipoprotein receptor (LDLR). In the bloodstream, serum cholesterol is bound to a carrier protein known as low-density lipoprotein (LDL). LDL binds to LDLR so that the cells of your body can absorb cholesterol. When LDLR is defective, it becomes more difficult for the cells to absorb cholesterol. This explains why the levels of blood cholesterol remain high. Explain the pattern of inheritance of this disorder.

Answer: The pedigree is consistent with a dominant pattern of inheritance. An affected individual always has an affected parent. Also, individuals III-8 and III-9, who are both affected, produced unaffected offspring. If this trait was recessive, two affected parents should always produce affected offspring. However, since the trait is dominant, two heterozygous parents can produce homozygous unaffected offspring. The ability of two affected parents to have unaffected offspring is a striking characteristic of dominant inheritance. On average, we would expect that two heterozygous parents should produce 25% unaffected offspring. In the family containing IV-4, IV-5, IV-6, and IV-7, there are actually 3 out of 4 unaffected offspring. This higher than expected proportion of affected offspring is not too surprising, since the family is a very small population and may deviate substantially from the expected value due to random sampling error.

2. One way to identify a human cellular oncogene is to use human DNA from a malignant cell to transform a mouse cell. A mouse cell that has been transformed by human DNA will have a human oncogene incorporated into its genome; it also may have Alu sequences that are closely linked to this human oncogene. (Note: As was discussed in Chapter 10, the human genome contains Alu sequences interspersed every 5000 to 6000 bp. Alu sequences are not found in the mouse genome.) Discuss how the Alu sequence can provide a way to clone human oncogenes.

Answer: One approach is analogous to transposon tagging, described in Chapter 18. When a mouse cell is transformed with a human DNA fragment containing an oncogene, that fragment is likely to contain an Alu sequence as well. To clone the human oncogene, the chromosomal DNA can be isolated from the transformed mouse cells, digested with a restriction enzyme, and cloned into vectors to create a library of DNA fragments. The members of the library that carries the human oncogene can be identified by using a probe complementary to the Alu sequence, since this sequence is not found in the mouse genome. Using this strategy, researchers have identified several human cellular oncogenes. In human bladder carcinoma, for example, a human cellular oncogene called *ras* has been identified this way.

CONCEPTUAL QUESTIONS

1. In regard to pedigree analysis, make a list of the patterns that distinguish recessive, dominant, and X-linked genetic diseases from each other.

2. Explain, at the molecular level, why human genetic diseases often follow a simple Mendelian pattern of inheritance, whereas most normal traits (e.g., the shape of your nose or the size of your liver) are governed by multiple gene interactions.

3. Many genetic disorders are heterogeneous. What does this mean? Give two examples of genetic heterogeneity. How does genetic heterogeneity confound a pedigree analysis?

4. In general, why do changes in chromosome structure and/or number affect an individual's phenotype? Explain why some changes in chromosome structure, such as reciprocal translocations, do not.

5. What is the difference between an oncogene and a tumor-suppressor gene? Give two examples.

6. What is a proto-oncogene? What are the typical functions of proteins encoded by proto-oncogenes? At the level of protein function, what are the general ways that proto-oncogenes can be converted into oncogenes?

7. What is a retroviral oncogene? Is it necessary for viral infection and proliferation? How have retroviruses acquired oncogenes?

8. A genetic predisposition to develop cancer is usually inherited as a dominant trait. At the level of cellular function, are the genes involved actually dominant? Explain why some individuals who have inherited these dominant alleles do not develop cancer during their lifetimes.

9. Describe the types of genetic changes that commonly convert a proto-oncogene into an oncogene. Give three examples. Explain how the genetic changes are expected to alter the activity of the gene product.

10. Relatively few inherited forms of cancer involve the inheritance of mutant oncogenes. Instead, most inherited forms of cancer are defects in tumor-suppressor genes or DNA repair genes. Give two or more reasons that we seldom see inherited forms of cancer involving activated oncogenes.

11. The *rb* gene encodes a protein that inhibits E2F, a transcription factor that activates several genes involved in cell division. Mutations in *rb* are associated with certain forms of cancer, such as retinoblastoma. Under each of the following conditions, would you expect cancer to occur?

    A. One copy of *rb* is defective; both copies of *E2F* are normal.
    B. Both copies of *rb* are defective; both copies of *E2F* are normal.
    C. Both copies of *rb* are defective; one copy of *E2F* is defective.
    D. Both copies of *rb* and *E2F* are defective.

12. A *p53* knockout mouse (i.e., both copies of *p53* are defective) has been produced by researchers. This type of mouse appears normal at birth. However, it is highly sensitive to UV light. Based on your knowledge of *p53*, explain the normal appearance at birth and the high sensitivity to UV light.

EXPERIMENTAL QUESTIONS

1. At the beginning of this chapter, we discussed the types of experimental observations that suggest a disease is inherited. Which of these observations do you find the least convincing? The most convincing? Explain your answer.

2. What is meant by the term genetic screening? What is different between screening at the protein level versus screening at the DNA level? Describe five different techniques used in genetic screening.

3. A particular disease is found in a group of South American Indians. During the 1920s, many of these people migrated to Central America. In the Central American group, the disease is never found. Discuss whether or not you think the disease has a genetic component. What types of further observations would you make?

4. What is a transformed cell? Describe three different methods to transform cells in a laboratory.

5. In Experiment 23A, what would be the results if the DNA sample had been treated with either DNase, RNase, or protease prior to the treatment with calcium and phosphate ions?

6. Explain how the experimental study of oncogenic viruses has increased our understanding of cancer.

7. Discuss ways to distinguish whether a particular form of cancer involves an inherited predisposition or is due strictly to (postzygotic) somatic mutations. In your answer, you should consider that only one mutation may be inherited, but the cancer might develop only after several somatic mutations.

QUESTIONS FOR STUDENT DISCUSSION/COLLABORATION

1. Make a list of the benefits that may arise from genetic screening as well as its possible negative consequences. Discuss the items on your list.

2. Our government has finite funds to devote to cancer research. Discuss which aspects of cancer biology you would spend the most money pursuing:

    A. Identifying and characterizing oncogenes and tumor-suppressor genes

    B. Identifying agents in our environment that cause cancer
    C. Identifying viruses that cause cancer
    D. Devising methods aimed at killing cancer cells in the body
    E. Informing the public of the risks involved in exposure to carcinogens

    In the long run, which of these areas would you expect to be the most effective in decreasing mortality due to human cancer?

CHAPTER 24

DEVELOPMENTAL GENETICS

M ulticellular organisms begin their lives with a fairly simple organization (namely, a fertilized egg) and then proceed step by step to a much more complex arrangement. As this occurs, cells divide, migrate, and change their characteristics as they become highly specialized units within a multicellular individual. In an adult, each cell plays its own particular role for the good of the entire individual. In animals, for example, muscle cells allow an organism to move, while intestinal cells facilitate the absorption of nutrients. This division of labor among the various cells and organs of the individual works collectively to promote its survival.

Developmental genetics, currently one of the hottest fields in molecular biology, is concerned with the roles genes play in orchestrating the changes that occur during development. In this chapter, we will examine how the sequential actions of genes provide a program for the development of an organism from a fertilized egg to an adult.

The last couple of decades has seen staggering advances in our understanding of developmental genetics at the molecular level. Scientists have chosen a few experimental organisms, such as the fruit fly, nematode, frog, mouse, and *Arabidopsis*, and worked toward the identification and characterization of the genes required for running their developmental programs. In certain organisms, notably the fruit fly, most of the genes that play a critical role in the early stages of development have been identified. Researchers are now exploring how the proteins encoded by these genes control the course of development. In this chapter, we will consider several examples in which geneticists understand how the actions of genes govern the developmental process.

INVERTEBRATE DEVELOPMENT

VERTEBRATE DEVELOPMENT

PLANT DEVELOPMENT

INVERTEBRATE DEVELOPMENT

We will begin our discussion of multicellular development by considering two model organisms, *Drosophila melanogaster* and the nematode *Caenorhabditis elegans*, that have been pivotal in our understanding of developmental genetics. As we have seen throughout this text, *Drosophila* has been a favorite subject of geneticists since 1910, when Thomas Hunt Morgan isolated his first white-eyed mutant (see Experiment 3A). It has been used to determine many of the fundamental principles of genetics, including the chromosome theory of inheritance, the random mutation theory, and linkage mapping, to mention a few.

In developmental biology, *Drosophila* is also useful, for a variety of reasons. First, researchers have identified many mutant strains with altered developmental pathways. The techniques for generating and analyzing mutants in this organism are more advanced than in any other animal. Second, at the larval stage, *Drosophila* is large enough to conduct transplantation experiments, yet small enough to determine where particular genes are expressed at critical stages of development.

By comparison, *C. elegans* is used by developmental geneticists because of its simplicity. The adult organism is a small transparent worm composed of only about 1000 somatic cells. Starting with the fertilized egg, the pattern of cell division and the fate of each cell within the embryo are completely known.

In this section, we will begin by describing the general features of *Drosophila* development. We will then focus our attention on embryonic development (embryogenesis), because it is during this stage of development that the overall body plan is determined. We will see how the expression of particular genes and the localization of gene products within the embryo influences the developmental process. We will then briefly consider development in *C. elegans*. In this organism, we will examine how the timing of gene expression plays a key role in determining the developmental fate of particular cells.

The early stages of embryonic development determine the pattern of structures in the adult organism

Multicellular development follows a body plan or pattern. The term pattern refers to the spatial arrangement of different regions of the body. At the cellular level, the body pattern is due to the arrangement of cells and their specialization.

The progressive growth of a fertilized egg into an adult organism involves four types of cellular events: cell division, cell movement, cell differentiation, and cell death. It is the coordination of these four events that leads to the formation of a body with a particular pattern. As we will see, the temporal expression of genes and the localization of gene products at precise regions in the fertilized egg and early embryo are the critical phenomena that underlie this coordination.

Figure 24-1 illustrates the general sequence of events in *Drosophila* development. The oocyte is the most critical cell in determining the pattern of development in the adult organism. It is an elongated cell with preestablished axes and a well-defined cytoplasmic organization (Figure 24-1a). After fertilization takes place, the zygote goes through a series of nuclear divisions that are not accompanied by cytoplasmic division. Initially, the resulting nuclei are scattered throughout the yolk, but eventually they migrate to the periphery. This is the syncytial blastoderm stage (Figure 24-1b).

After the nuclei have lined up along the cell membrane, individual cells are formed as portions of the cell membrane envelop each nucleus; this creates a structure called a cellular blastoderm (Figure 24-1c). This structure is composed of a sheet of cells on the outside with yolk in the center. In this arrangement, the cells are distributed asymmetrically. At the posterior end are a group of cells called the pole cells. These are the primordial germ cells that eventually will give rise to gametes in the adult organism.

**Progression of
Developmental Events**

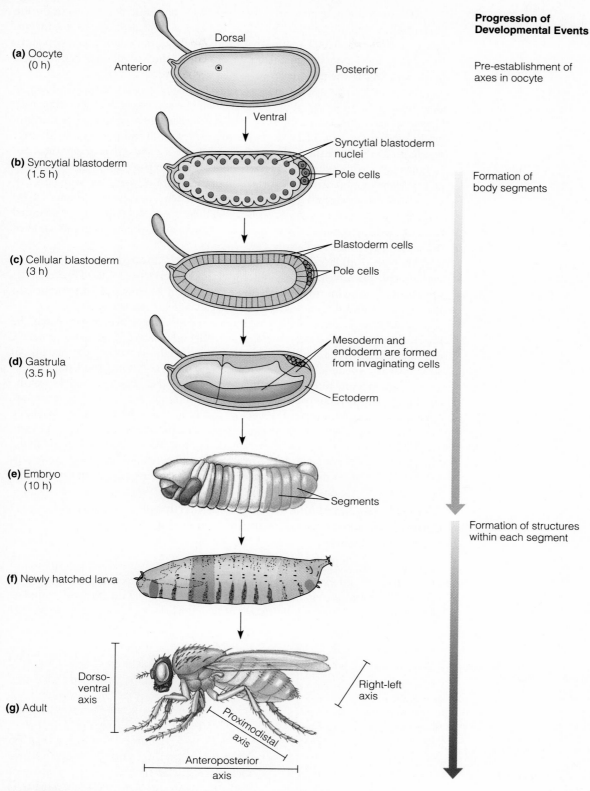

(a) Oocyte
(0 h)

Dorsal

Anterior   Posterior

Ventral

Pre-establishment of
axes in oocyte

(b) Syncytial blastoderm
(1.5 h)

Syncytial blastoderm
nuclei

Pole cells

Formation of
body segments

(c) Cellular blastoderm
(3 h)

Blastoderm cells

Pole cells

(d) Gastrula
(3.5 h)

Mesoderm and
endoderm are formed
from invaginating cells

Ectoderm

(e) Embryo
(10 h)

Segments

Formation of structures
within each segment

(f) Newly hatched larva

(g) Adult

Dorso-
ventral
axis

Right-left
axis

Proximodistal
axis

Anteroposterior
axis

FIGURE **24-1**

Developmental stages of the fruit fly *Drosophila*. **(a)** Oocyte. **(b)** Syncytial blastoderm. **(c)** Cellular blastoderm. **(d)** Gastrula. **(e)** Embryo at 10 hours. **(f)** Newly hatched
larva. **(g)** Adult.

After blastoderm formation is complete, some dramatic changes occur during **gastrulation** (Figure 24-1d). This stage involves a great deal of **cell migration**, which produces three cell layers known as the ectoderm, mesoderm, and endoderm. In general, the ectoderm remains on the outside of the gastrula, the endoderm is on the inside, and the mesoderm is wedged in the middle.

As this process occurs, the embryo begins to be subdivided into morphologically detectable units. Initially, shallow grooves divide the embryo into 14 **parasegments**. However, this is a transient condition. A short time later, these grooves disappear, and new boundaries are formed that divide the embryo into morphologically discrete **segments**. Figure 24-1e shows the segmented pattern of a *Drosophila* embryo at about 10 hours postfertilization. Later in this section, we will explore how the coordination of gene expression underlies the formation of these parasegments and segments.

At the end of **embryogenesis**, a larva will hatch from the egg (Figure 24-1f) and begin feeding on its own. In *Drosophila*, there are three larval stages, separated by molts. During molting, the larva sheds its cuticle, a hardened extracellular shell that is secreted by the epidermis.

After the third larval stage, *Drosophila* proceeds through a process known as **metamorphosis**. Groups of cells called **imaginal disks** were produced earlier in development. During metamorphosis, these imaginal disks grow and differentiate into the structures found in the adult fly (e.g., head, wings, legs, abdomen).

In metazoa (i.e., animals that are more complex than unicellular protozoa), the final result of development commonly is an adult body organized along three axes: the **dorsoventral axis**, the **anteroposterior axis**, and the **right–left axis** (Figure 24-1g). An additional axis, used mostly for designating limb parts, is the **proximodistal axis**.

Although many interesting events occur during the three larval stages and the adult stage, we will focus most of our attention on the genetic events that occur during embryonic development. Even before hatching, the embryo develops the basic body plan that will be found in the adult organism. In other words, during the early stages of development, the embryo is divided into segments that correspond to the segments of the larva and adult. Therefore, an understanding of how these segments form in the embryo is critical to our understanding of pattern formation.

The study of *Drosophila* mutants with disrupted development patterns has identified genes that control development

Mutations that alter the course of *Drosophila* development have contributed greatly to our understanding of the normal process. For example, Figure 24-2 shows an example of mutations in a complex of genes called the **bithorax** complex. This mutant fly has four wings instead of two; the halteres (a balancing organ that

FIGURE **24-2**

The bithorax mutation in *Drosophila*.

GENES ⟶ TRAITS: A normal fly contains two wings on the second thoracic segment, and two halteres on the third thoracic segment. However, this mutant fly contains mutations in a complex of genes called the bithorax complex. (The function of the bithorax complex is discussed later in this chapter.) In this fly, the third thoracic segment has the same characteristics as the second thoracic segment, thereby producing a fly with four wings instead of the normal number of two.

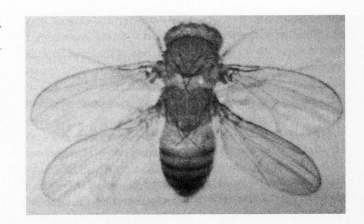

resembles a miniature wing), which are found on the third thoracic segment, are changed into wings, normally found on the second thoracic segment. The term bithorax refers to the observation that the characteristics of the second thoracic segment are duplicated.

Edward Lewis at the California Institute of Technology, a pioneer in the genetic study of development, became interested in the bithorax phenotype and began investigating it in 1946. He discovered that the mutant chromosomal region actually contained a complex of three genes involved in specifying developmental pathways in the fly. A gene that plays a central role in specifying the final outcome of a body region is called a **homeotic gene**. We will discuss particular examples of homeotic genes later in this chapter.

During the 1960s and 1970s, interest in the relationship between genetics and embryology blossomed as biologists began to appreciate the role of genetics at the molecular and cellular levels. It soon became clear that the genome of multicellular organisms contains groups of genes that initiate a program of development involving networks of gene regulation. By identifying mutant alleles that disrupt development, geneticists have begun to unravel the pattern of gene expression that underlies the normal pattern of multicellular development.

Early in development, a category of genes known as **segmentation genes** plays a role in the formation of body segments. The expression of segmentation genes in specific regions of the embryo causes it to become segmented into the pattern shown in Figure 24-1e. In the 1970s, while working at the European Molecular Biology Laboratory in Germany, Christiane Nusslein-Volhard and Eric Wieschaus undertook a systematic search for *Drosophila* mutants with disrupted segmentation patterns. It was their pioneering effort that led to the identification of most of the genes required for the embryo to develop into a segmented pattern.

Figure 24-3 illustrates a few of the interesting phenotypic effects that Nusslein-Volhard and Wieschaus observed when a particular segmentation gene is defective. The gray boxes indicate the regions that are missing in the resulting larvae. The segments adjacent to the deleted regions exhibit a mirror-image duplication. They identified three classes of segmentation genes: **gap genes**, **pair-rule genes**, and **segment-polarity genes**. When a mutation inactivates a gap gene, a contiguous section of the embryo is missing (Figure 24-3a). In other words, there is a gap of several segments. For example, when the *Krüppel* gene is defective, about eight segments are missing from the embryo. By comparison, a defect in a pair-rule gene causes alternating parasegments to be deleted (Figure 24-3b). For example, when the *even-skipped* gene is defective, the even-numbered parasegments are missing. Finally, segment-polarity mutations cause portions of segments to be missing either an anterior or a posterior region. Figure 24-3c shows a mutation in a segmentation gene known as *gooseberry*. When this gene is defective, the anterior portion of each segment is missing from the larva.

Overall, the phenotypic effects of mutant segmentation genes provide geneticists with important clues regarding the roles of these genes in the developmental process of segmentation. Later in this section, we will examine when and where the segmentation genes are expressed, and how their pattern of expression leads to the segmentation of the embryo.

The generation of a body pattern depends on the positional information that each cell receives during development

The identification of mutant alleles that disrupt the developmental process has permitted great insight into the genes controlling pattern formation. Before we discuss how these genes function, let's consider a central concept in developmental biology known as **positional information**. For an organism to develop into a segmented pattern with unique morphological and cellular features, each cell of the body must somehow know its position relative to the other cells. A cell may respond to

(a) Gap

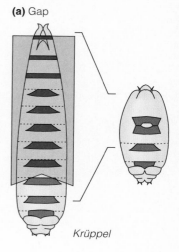

Krüppel

(b) Pair-rule

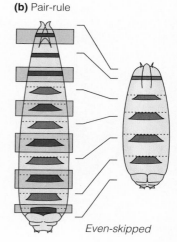

Even-skipped

(c) Segment-polarity

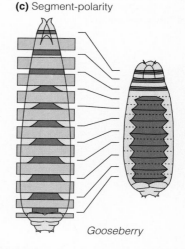

Gooseberry

FIGURE **24-3**

Phenotypic effects in *Drosophila* larvae that have mutations in segmentation genes. Effects shown are caused by defects in **(a)** gap genes, **(b)** pair-rule genes, and **(c)** segment-polarity genes.

(a) Asymmetric distribution of morphogens in the oocyte

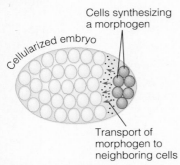

(b) Asymmetric synthesis and extracellular distribution of a morphogen

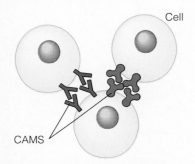

(c) Cell-to-cell contact conveys positional information

FIGURE **24-4**

Three molecular mechanisms of positional information.

positional information in various ways. For example, positional information may stimulate a cell to divide into two daughter cells. Or it may cause a cell to differentiate into a particular cell type. Positional information may cause a cell or group of cells to migrate in a particular direction from one region of the embryo to another. Finally, positional information may tell a cell that it is supposed to die. This process, known as **apoptosis**, is a necessary event during normal development. For example, in the later stages of development of the retina in *Drosophila*, excess cells are eliminated by apoptosis.

Molecules that convey positional information and promote developmental changes are known as **morphogens**. A morphogen influences the developmental fate of a cell. It provides the positional information that stimulates a cell to divide, differentiate, migrate, or die. For example, as we will learn later, when *Drosophila* cells are exposed to a high concentration of the morphogen known as Bicoid, they differentiate into structures characteristic of the anterior region of the body.

Within an oocyte and during embryonic development, morphogens typically are distributed along a concentration gradient (Figure 24-4). In other words, the concentration of a morphogen varies from low to high in different regions of the organism. A key feature of morphogens is that they act in a concentration-dependent manner. At a high concentration, a morphogen will restrict a cell into a particular developmental pathway; at a lower concentration, it will not. There is often a critical **threshold concentration** above which the morphogen will exert its effects but below which it is ineffective.

During the earliest stages of development, several morphogenic gradients are preestablished within the oocyte (Figure 24-4a). At the cellular blastoderm stage, the zygote subdivides into many smaller cells. Due to the preestablished gradient of morphogens within the oocyte, these smaller cells have higher or lower concentrations of morphogens, depending on their location. In this way, the morphogen gradients in the oocyte can provide positional information that is important in establishing the general polarity of an embryo along two main axes: the anteroposterior axis and the dorsoventral axis. This topic will be described in greater detail later in this chapter.

A morphogen gradient can also be established by cell secretion and transport. A certain cell or group of cells may synthesize and secrete a morphogen at a specific stage of development. After secretion, the morphogen may be transported to neighboring cells. The concentration of the morphogen is usually highest near the cells that secrete it (Figure 24-4b). The morphogen may then influence the developmental fate of the cells that are exposed to it. When a cell or group of cells governs the developmental fate of neighboring cells by producing a morphogen, this process is known as **induction**.

In addition to morphogens, positional information is conveyed by **cell adhesion** (Figure 24-4c). Each cell makes its own collection of surface receptors; these receptors cause it to adhere to other cells and/or to the extracellular matrix (ECM), which consists primarily of carbohydrates and fibrous proteins. Such receptors are known as **cell adhesion molecules (CAMs)**. A cell may gain positional information via the combination of contacts it makes with other cells or with the ECM.

The phenomenon of cell adhesion, and its role in multicellular development, was first recognized by H. V. Wilson in 1907 while working at the University of North Carolina in Chapel Hill. He took multicellular sponges and disaggregated them into individual cells. Remarkably, the cells actively migrated until they adhered to one another to form a new sponge, complete with the chambers and canals that characterize a sponge's internal structure! When sponge cells from different species were mixed, they sorted themselves properly, adhering only to cells of the same species. Overall, these results indicate that cell adhesion plays an important role in governing the position that a cell will adopt during development.

Figure 24-4 provides an overview of the general phenomena that underlie positional information during embryonic development. With these ideas in mind, we can begin to examine how specific genes in *Drosophila* encode morphogens that convey positional information. In this organism, the establishment of the body axes and division of the fly into segments involves the participation of a few dozen genes. Table 24-1 lists many of the important genes governing pattern formation during embryonic development. These genes are often given interesting names based on the phenotypic effects when they are mutant. It is beyond the scope of this text to examine how all of these genes exert their effects during embryonic development. Instead, we will consider a few examples that illustrate how the expression of a particular gene and the localization of its gene product have a defined effect on the pattern of development.

The gene products of maternal effect genes are deposited asymmetrically into the oocyte and establish the anteroposterior and dorsoventral axes at a very early stage of development

The first stage in *Drosophila* embryonic pattern development is establishment of the body axes. This occurs before the embryo becomes segmented. In fact, the morphogens necessary to establish these axes are distributed prior to fertilization. During oogenesis, certain gene products, which are important in early developmental stages, are deposited asymmetrically within the egg. Later, after the egg has been fertilized and development begins, these gene products will establish independent developmental programs that govern the formation of the four major body regions of the embryo. These are the anterior, posterior, terminal, and dorsoventral regions.

TABLE 24–1 Examples of *Drosophila* Genes that Play a Role in Pattern Development

Description	Examples
Some genes play a role in determining the axes of development. Also, certain genes govern the formation of the extreme terminal (anterior and posterior) regions.	Anterior: *bicoid, exuperantia, hunchback, swallow* Posterior: *nanos, cappuccino, oskar, pumilio, spire, staufen, tudor, vasa* Terminal: *torso, torsolike* Dorsoventral: *toll, cactus, dorsal easter, gurken, nudel, pelle, pipe, snake, spatzle*
Some genes play a role in promoting the subdivision of the embryo into segments. These are called segmentation genes. As described later, there are three types of segmentation genes, known as gap genes, pair-rule genes, and segment-polarity genes.	Gap genes: *empty spiracles, huckebein hunchback, knirps, Kruppel, tailless* Pair-rule genes: *even-skipped, hairy, runt, fushi tarazu, odd-paired, odd-skipped, paired, sloppy paired* Segment-polarity genes: *engrailed, hedgehog, wingless, gooseberry*
Some genes play a role in determining the fate of particular segments. These are known as homeotic genes. *Drosophila* has two clusters of homeotic genes, known as the *antennapedia* complex and the *bithorax* complex.	Antennapedia complex: *labial, proboscipedia, deformed, sex combs reduced, antennapedia* Bithorax complex: *ultrabithorax, abdominal A, abdominal B*

Adapted from: Kalthoff, K. (1996) *Analysis of Biological Development.* McGraw-Hill, New York.

In each of these regions, the products of several different genes ensure that proper development will occur.

As shown in Figure 24-5, a few gene products act as key morphogens, or receptors for morphogens, that initiate changes in embryonic development. As shown here, these gene products are deposited asymmetrically in the egg. For example, the product of the **bicoid** gene is necessary to initiate development of the anterior structures of the organism. During oogenesis, the mRNA for *bicoid* accumulates in the anterior region of the oocyte. In contrast, the mRNA from the **nanos** gene accumulates in the posterior end. Later in development, the *nanos* mRNA will be translated into protein, which functions to influence posterior development. *Nanos* is required for the formation of the abdomen.

In addition to *bicoid* and *nanos*, the development of the structures at the extreme anterior and posterior ends of the embryo are regulated in part by a receptor protein called **Torso**. This receptor, which is activated only at the anterior and posterior ends of the egg, is necessary for the formation of the head and abdomen. The fourth major system is the dorsoventral system. A receptor protein known as **Toll** is activated along the ventral midline of the egg and initiates the establishment of the dorsoventral axis.

Let's now take a closer look at the molecular mechanism of one of these modulators, namely *bicoid*. The *bicoid* gene got its name because a larva defective in this gene develops with two posterior ends (Figure 24-6). This allele exhibits a maternal effect pattern of inheritance (see Chapter 7). A female fly that is phenotypically normal (because its mother was heterozygous for the normal *bicoid* allele), but genotypically homozygous for an inactive *bicoid* allele (because it inherited the in-

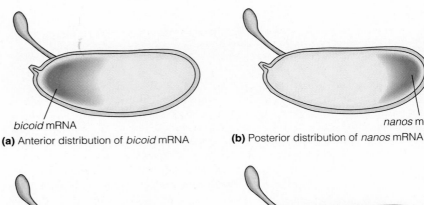

bicoid mRNA
(a) Anterior distribution of *bicoid* mRNA

nanos mRNA
(b) Posterior distribution of *nanos* mRNA

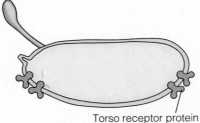

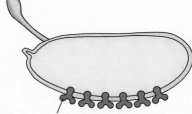

Torso receptor protein
(c) Terminal distribution of Torso receptor protein

Toll receptor protein
(d) Ventral distribution of Toll receptor protein

FIGURE **24-5**

The establishment of the axes of polarity in the *Drosophila* embryo. This figure shows some of the gene products that are critical in the establishment of the anteroposterior, terminal, and dorsoventral axes. **(a)** *Bicoid* mRNA is distributed in the anterior end of the oocyte and promotes the formation of anterior structures. **(b)** *Nanos* mRNA is localized to the posterior end and promotes the formation of posterior structures. **(c)** The Torso receptor protein is found in the membrane at either end of the oocyte and causes the formation of structures that are found only at the ends of the organism. **(d)** The Toll receptor protein is localized in the ventral side of the oocyte and establishes the dorsoventral axis.

active allele from its mother and father), will produce 100% affected offspring even when mated to a male that is homozygous for the normal *bicoid* allele. In other words, the genotype of the mother determines the phenotype of the offspring. This occurs because the *bicoid* gene product is provided to the oocyte via the nurse cells.

In the ovaries of female flies, the nurse cells are localized asymmetrically toward the anterior end of the oocyte. During oogenesis, gene products are transferred into the oocyte via cell-to-cell connections called cytoplasmic bridges. Maternally encoded gene products enter one side of the oocyte (Figure 24-7a). This side will eventually become the anterior end of the embryo. The *bicoid* gene is actively transcribed in the nurse cells, and *bicoid* mRNA is transported into the anterior end of the oocyte. The 3′ end of *bicoid* mRNA contains a signal that is recognized by binding proteins thought necessary for the transport of this mRNA into the oocyte. After it enters the oocyte, the *bicoid* mRNA is trapped near the anterior side.

Figure 24-7b shows an *in situ* hybridization experiment in which a *Drosophila* egg was examined via a probe complementary to the *bicoid* mRNA. (The technique of *in situ* hybridization was described in Chapter 20, pp. 542–544.) As seen here, the *bicoid* mRNA is highly concentrated near the anterior pole of the egg cell. When the *bicoid* mRNA subsequently is translated, a gradient of Bicoid protein is established as shown in Figure 24-7c.

After fertilization occurs, the Bicoid protein functions as a transcription factor. A remarkable feature of this protein is that its ability to influence gene expression is tuned exquisitely to its concentration. Depending on the distribution of the Bicoid protein, this transcription factor will only activate genes in certain regions of the embryo. For example, Bicoid stimulates a gene called *hunchback* in the anterior half of the embryo, but it does not activate the *hunchback* gene in the posterior half.

Gap, pair-rule, and segment-polarity genes act sequentially to divide the *Drosophila* embryo into segments

After the anteroposterior, dorsoventral, and terminal regions of the embryo have been established by maternal effect genes, the next developmental process is to organize the embryo transiently into parasegments and then permanently into segments. The segmentation pattern of the embryo is shown in Figure 24-8. This pattern of positional

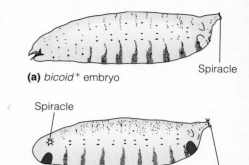

(a) *bicoid*⁺ embryo

(b) *bicoid*⁻ embryo

FIGURE **24-6**

The *bicoid* mutation in *Drosophila*. (a) A normal, *bicoid*⁺ embryo. **(b)** A *bicoid*⁻ embryo, in which both ends of the larva develop posterior structures. For example, both ends develop a spiracle, which normally is found only at the posterior end.

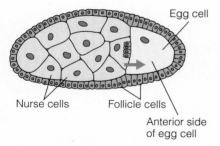

(a) Transport of maternal effect gene products into the oocyte

(b) *In situ* hybridization of *bicoid* mRNA

FIGURE **24-7**

Asymmetrical localization of gene products during oogenesis in *Drosophila*. **(a)** The nurse cells transport gene products into the anterior end of the developing oocyte. **(b)** An *in situ* hybridization experiment showing that the *bicoid* mRNA is trapped near the anterior end. **(c)** The *bicoid* mRNA is translated into protein soon after fertilization. The location of the Bicoid protein is revealed by immunostaining using an antibody that specifically recognizes this protein.

(c) Immunostaining of Bicoid protein

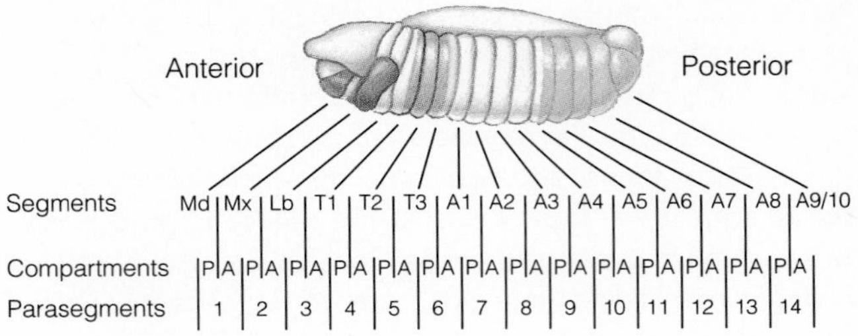

FIGURE **24-8**

**A comparison of segments and parasegments in the
Drosophila embryo.** Note that the parasegments and segments
are out of register. The posterior (P) and anterior (A) regions are
shown for each segment.

information will be maintained, or "remembered," throughout the rest of development. In other words, each of the segments in the embryo will give rise to unique morphological features in the adult. For example, T2 will become a thoracic segment with a pair of legs and a pair of wings, and A8 will become a segment of the abdomen.

As shown in Figure 24-8, the boundaries of the segments are out of register with the boundaries of the parasegments. An appreciation of this feature is critical to our understanding of segmentation. From the viewpoint of genes, we will see that the parasegments are the locations where gene expression is controlled spatially. Figure 24-8 shows the overlapping relationship between parasegments and segments. The anterior compartment of each segment overlaps with the posterior region of a parasegment; the posterior compartment of a segment overlaps with an anterior region of the next parasegment. The pattern of gene expression that occurs in the anterior region of one parasegment and the posterior region of an adjacent parasegment results in the formation of a segment.

Now that we have a general understanding of the way the *Drosophila* embryo is subdivided, we can examine how particular genes cause it to become segmented into this pattern. As mentioned, the genes that play a role in the formation of body segments are called segmentation genes; there are three classes of segmentation genes: gap genes, pair-rule genes, and segment-polarity genes. The expression and activation patterns of these genes in specific regions of the embryo cause it to become segmented.

A partial, simplified scheme of the genetic hierarchy that leads to a segmented pattern in the *Drosophila* embryo is shown in Figure 24-9. This figure presents the general sequence of events that occurs during the early stages of embryonic development. (As indicated in Table 24-1, many more genes are actually involved in this process.) As described in this figure, the following steps occur:

1. Maternal effect gene products, such as *bicoid* mRNA, are deposited asymmetrically into the oocyte. These gene products form a gradient that will later influence the formation of axes, such as the anteroposterior axis.

2. After fertilization, maternal effect gene products activate **zygotic genes**. In contrast to maternal effect genes, which are expressed during oogenesis, zygotic genes are expressed after fertilization. The first set of zygotic genes to be activated are the gap genes. As shown in step 2 of Figure 24-9, gap genes are activated as broad bands within particular regions of the embryo. These bands do not correspond to parasegments or segments within the embryo.

3. The gap genes and maternal effect genes then activate the pair-rule genes. The photograph shown in step 3 of Figure 24-9 illustrates the alternating pattern of *ftz* and *even-skipped* gene expression. Note that these pair-rule genes are expressed in particular parasegments. *Ftz* is expressed in the odd-numbered parasegments, and *even-skipped* is expressed in the even-numbered segments.

4. Once the pair-rule genes are activated in an alternating banding arrangement, their gene products then regulate the segment-polarity genes. As shown in step 4 of Figure 24-9, the segment-polarity gene *engrailed* is expressed in the anterior region of each parasegment. Another segment-polarity gene, *wingless*, is expressed in the posterior region. Later in devel-

① Maternal effect genes establish the anteroposterior and dorsoventral axes. They activate gap genes.

① Asymmetric localization of Bicoid

Examples

Asymmetric localization of Bicoid protein. Other maternal effect gene products (not shown) are also asymmetrically localized.

② Gap gene products act as genetic regulators of pair-rule genes. They bind to stripe-specific enhancers that are located adjacent to pair-rule genes.

② Gap gene expression

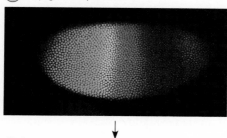

Gap gene expression occurs as broad bands in the embryo. In this photo, Krüppel protein is shown in red and Hunchback in green. Their region of overlap is yellow. Other gap genes (not shown) are also expressed.

③ The expression of pair-rule genes in a stripe defines the boundary of a parasegment. Pair-rule gene products regulate the expression of segment-polarity genes.

③ Pair-rule gene expression

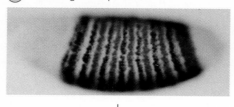

Pair-rule genes are expressed in alternating stripes. Each stripe corresponds to a parasegment. In this photo, the Even-skipped gene product is shown in gray and occupies the even parasegments. The Ftz gene product is labeled in brown and occupies odd-numbered parasegments.

④ Segment-polarity genes define the anterior or posterior compartment of each parasegment.

④ Segment-polarity gene expression

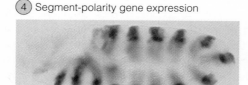

Segment-polarity genes are expressed in either an anterior or posterior compartment. This photo shows the product of the *engrailed* gene in the anterior compartment of each parasegment.

FIGURE **24-9**

Overview of pattern formation in *Drosophila*.

opment, the anterior end of one parasegment and the posterior end of another parasegment will develop into a segment with particular morphological characteristics.

The expression of homeotic genes controls the phenotypic characteristics of segments

We have considered how the *Drosophila* embryo becomes organized along axes and then into a segmented body pattern. Now we will examine how each segment develops its unique morphological features. Geneticists often use the term **cell fate** to describe the ultimate morphological features that a cell or group of cells will adopt. For example, the fate of the cells in segment T2 in the *Drosophila* embryo is to develop into a thoracic segment containing two legs and two wings. In *Drosophila*, the cells in each segment of the body have their fate determined at a very early stage of embryonic development, long before the morphological features become apparent.

FIGURE 24-10

Expression pattern of homeotic genes in *Drosophila*. The order of homeotic genes, *labial* (*lab*), *proboscipedia* (*pb*), *deformed* (*dfd*), *sex combs reduced* (*scr*), *antennapedia* (*antp*), *ultrabithorax* (*ubx*), *abdominal A* (*abdA*), and *abdominal B* (*abdB*), correlates with the order of expression in the embryo. The expression pattern of four of these genes is shown. *Lab* (purple) is expressed in the region that will eventually give rise to mouth structures. *Dfd* (green) is expressed in the region that will form much of the head. *Antp* is expressed in embryonic segments that give rise to thoracic segments, and *abdB* (orange) is expressed in posterior segments that will form the abdomen.

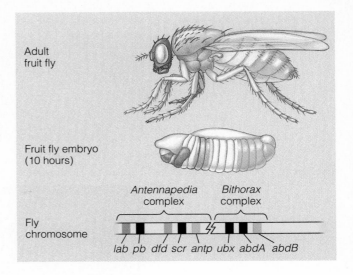

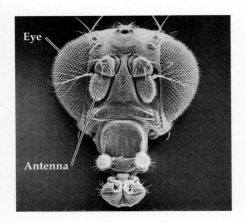

(a) Normal fly

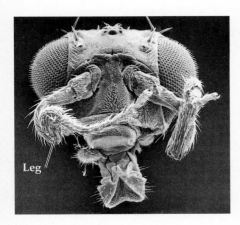

(b) *Antennapedia* mutation

FIGURE 24-11

The *Antennapedia* mutation in *Drosophila*.

GENES ⟶ TRAITS: (a) A normal fly with antennae. (b) This mutant fly has a gain of function mutation in which the antp gene is expressed in the embryonic segment that normally gives rise to antennae. The expression of antp causes this region to have legs rather than antennae.

Our understanding of developmental fate has been aided greatly by the identification of mutant genes that alter cell fates. In animals, the first mutant of this type was described by the German entomologist G. Kraatz in 1876. He observed a sawfly (*Climbex axillaris*) in which part of an antenna was replaced with a foot. During the late 19th century, the English zoologist William Bateson collected many of these types of observations and published them in 1894 in a book entitled *Materials for the Study of Variation Treated with Especial Regard to Discontinuity in the Origin of Species*. In this book, Bateson coined the term **homeotic** to describe mutant alleles in which one body part is replaced by another.

As mentioned earlier in this chapter, Edward Lewis at Caltech began to study strains of *Drosophila* having homeotic mutations. This work, which began in 1946, was the first systematic study of homeotic genes. Each homeotic gene controls the fate of a particular region of the body. *Drosophila* contains two clusters of homeotic genes called the *bithorax* complex and the *antennapedia* complex. Figure 24-10 shows the organization of genes within these complexes. The antennapedia complex contains five genes, designated *lab*, *pb*, *dfd*, *scr*, and *antp*. The bithorax complex has three genes, *ubx*, *abdA*, and *abdB*. Both of these complexes are located on chromosome 3, but a large segment of DNA separates them.

As noted in Figure 24-10, the order of these genes along chromosome 3 correlates with the anteroposterior axis of the body. For example, *lab* is expressed in the anterior segment and governs the formation of mouth structures. The *antp* gene is expressed strongly in the thoracic region during embryonic development and controls the formation of thoracic structures. Transcription of the *abdB* gene occurs in the posterior region of the embryo; this gene controls the formation of the posteriormost abdominal segments.

The role of homeotic genes in determining the identity of particular segments has been revealed by mutations that alter their function. As shown in Figure 24-11, the antennapedia mutation is a **gain of function mutation** in the *antp* gene that causes it to be expressed in an additional place in the embryo. In this case, the *antp* gene is expressed abnormally in the anterior segment that normally gives rise to the antennae. In other words, there has been a gain of *antp* function in this segment. The abnormal expression of *antp* in this region causes the antennae to be converted into legs!

Investigators have also studied many loss of function alleles in homeotic genes. When a particular homeotic gene is defective, the region that it normally governs will be controlled by the homeotic gene that acts in the adjacent anterior region. For example, the *ubx* gene normally functions within parasegments 5 and 6. If this gene is missing, this section of the fly becomes converted to the structures that are found normally in parasegment 4.

The homeotic genes are part of the genetic hierarchy that produces the morphological characteristics of the fly. They are regulated in a very complex way. Their expression is controlled by gap genes and pair-rule genes, and they are also regulated by interactions among themselves. In addition, a group of genes known as the *polycomb* genes represses the expression of homeotic genes in regions where they should not act. Overall, the concerted actions of many gene products cause the homeotic genes to be expressed only in the appropriate region of the embryo as shown in Figure 24-10.

Since they are part of a genetic hierarchy, it is not too surprising that homeotic genes encode transcription factors. The coding sequence of homeotic genes contains a 180 base pair consensus sequence, known as a **homeobox** (Figure 24-12). This sequence was first discovered in the *antp* and *ubx* genes, and it has since been found in all *Drosophila* homeotic genes and in some other genes affecting pattern development, such as *bicoid*. The protein domain encoded by the homeobox is called a **homeodomain**. The arrangement of α-helices within the homeodomain promotes the binding of the protein to the major groove of DNA. In this way, homeotic proteins can bind to DNA in a sequence-specific manner. In addition to DNA-binding ability, homeotic proteins also contain a transcriptional activation domain that functions to activate the genes to which the homeodomain can bind.

The transcription factors encoded by homeotic genes activate the next category of genes, collectively known as **realizator genes**. These genes produce the morphological characteristics of each segment. Much current research attempts to identify realizator genes and determine how their expression in particular regions of the embryo leads to morphological changes in the embryo, larva, and adult.

In some cases, realizator genes also encode transcription factors. Presumably, such realizator genes control the expression of other sets of genes that will alter the morphological characteristics of cells. In other cases, realizator genes encode proteins involved in cell-to-cell signaling pathways. The activation of these signaling pathways is thought to play a key role in the ability of cells and groups of cells to adopt their correct morphologies. It is expected that research during the next few decades will shed considerable light on the pathways by which realizator genes control morphological changes in the fruit fly.

The developmental fate of each cell in the nematode *Caenorhabditis elegans* is known

We now turn our attention to another invertebrate, *C. elegans*, that has been the subject of numerous studies in developmental genetics. As does *Drosophila*, this

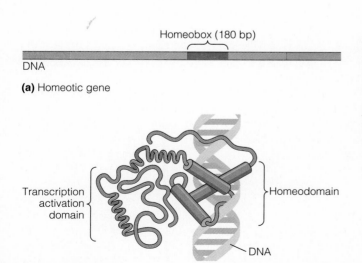

Homeobox (180 bp)

DNA

(a) Homeotic gene

Transcription activation domain

Homeodomain

DNA

(b) Homeotic protein bound to DNA

FIGURE 24-12

Molecular features of homeotic proteins. (a) A homeotic gene (shown in blue) contains a 180 bp sequence called the homeobox (shown in red). **(b)** When a homeotic gene is expressed, it produces a protein that functions as a transcription factor. The homeobox encodes a region of the protein called a homeodomain, which functions in binding to the major groove of DNA. These DNA-binding sites are found within genetic regulatory elements (i.e., enhancers), which are found next to genes. The homeotic protein also contains a transcription activation domain, which will activate the transcription of a gene after the homeodomain has bound to the DNA.

worm begins its development as a fertilized egg. The embryo develops within the egg shell and hatches when it reaches a size of 558 cells. After hatching, it continues to grow and mature as it passes through four successive molts. It takes about three days for a fertilized egg to develop into an adult worm.

With regard to gender, *C. elegans* can be a male (and only produce sperm) or a hermaphrodite (capable of producing sperm and egg cells). An adult male is composed of 1031 somatic cells and produces about 1000 sperm. A hermaphrodite consists of 959 somatic cells and produces about 2000 gametes (both sperm and eggs).

A remarkable feature of this organism is that the pattern of cellular development is extremely invariant from worm to worm. In the early 1960s, Sydney Brenner pioneered the effort to study the pattern of cell division in *C. elegans*. To do so, a researcher can identify a particular cell at an embryonic stage, follow that cell as it divides, and observe where its descendant cells will be located in the adult.

Because *C. elegans* is transparent and composed of relatively few cells, researchers can follow cell division step by step, beginning with a fertilized egg and ending with an adult worm. An illustration that depicts how cell division proceeds is called a **fate map**. It describes the fate of any cell's descendants. Figure 24-13 shows a partial fate map for a *C. elegans* hermaphrodite. At the first cell division, the egg divides to produce two cells, called AB and P_1. AB then divides into two cells, ABa and ABp; and P_1 divides into two cells, EMS and P_2. As noted in Figure 24-13, the EMS cell then divides into two cells, called MS and E. The cellular descendants of the E cell give rise to the worm's intestine. In other words, the fate of the E cell's descendants is to develop into intestinal cells. This diagram also illustrates the concept of a **cell lineage**. This term refers to a series of cells that are derived from each other by cell division. For example, the EMS cell, E cell, and the intestinal cells are all part of the same cell lineage.

Having a cellular fate map for an organism is an important experimental advantage. It allows researchers to investigate how gene expression in any cell, at any stage of development, may affect the outcome of a cell's fate. In the experiment described next, we will see how the timing of gene expression is an important parameter in the fate of a cell's descendants.

FIGURE **24-13**

A cell fate map of the nematode *Caenorhabditis elegans*. This partial fate map illustrates how the cells divide to produce different regions of the adult. The fate of the intestinal cell lineage is shown in greater detail than that of other cell lineages. A complete fate map is known for this organism, although its level of detail is beyond the scope of this text.

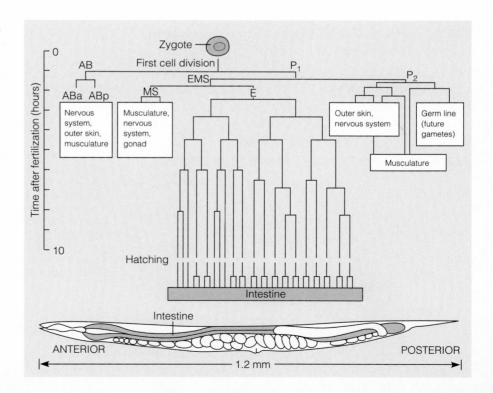

Horvitz and Ambros found that heterochronic mutations disrupt the timing of developmental changes in *C. elegans*

Our discussion of *Drosophila* development focused on how the spatial expression and localization of gene products can lead to a particular pattern of embryonic development. Another important issue in development is timing. The cells of a multicellular organism must know when to divide and when to differentiate into a particular cell type. If the timing of these processes is not coordinated, certain tissues will develop too early or too late, disrupting the developmental process.

In *C. elegans*, the timing of developmental events can be examined carefully at the cellular level. As mentioned, the fate of each cell has been determined. Using a microscope (with Nomarski differential interference contrast optics), a researcher can focus on a particular cell within this transparent worm and watch it divide into two cells, then four cells, and so forth. Therefore, a scientist can judge whether a cell is behaving as it should during the developmental process.

To identify genes that play a role in the timing of cell fates, researchers have searched for mutant alleles that disrupt the normal timing process. In a collaboration in the late 1970s, H. Robert Horvitz at the Massachusetts Institute of Technology and John Sulston at the MRC Laboratory of Molecular Biology in England set out to identify mutant alleles in *C. elegans* that disrupt cell fates or the timing of cell fates.

Prior to this work, they did not know what phenotypic effects to expect from a mutation that altered the fate of cells within a cell lineage. Using a microscope, they screened thousands of worms for altered morphologies that might indicate an abnormality in development. During this screening process, one of the phenotypic abnormalities they found was an *egg-laying defective* phenotype. They reasoned that since the egg-laying system depended on a large number of cell types (vulval cells, muscle cells, and nerve cells) an abnormality in any of the cell lineages leading to these cell types might cause an inability to lay eggs.

In *C. elegans*, an egg-laying defective phenotype is easy to identify, because the hermaphrodite will be able to fertilize its own eggs but will be unable to lay them. When this occurs, the eggs actually hatch within the hermaphrodite's body. This leads to the death of the hermaphrodite as it becomes filled with hatching worms. This egg-laying defective phenotype, in which the hermaphrodite becomes filled with its own offspring, is called a "bag of worms":

Normal *C. elegans*

C. elegans with egg-laying defective phenotype, the "bag of worms"

Eventually, the newly hatched larva eat their way out and can be saved for further study.

In their initial study, published in 1980, the egg-laying defective phenotype yielded several mutant strains that were defective in particular cell lineages. In 1984, in the experiment described here, Victor Ambros and Horvitz took this same approach and were able to identify genes that play a key role in the timing of cell fate.

THE HYPOTHESIS

Mutations that cause an egg-laying defective phenotype may affect the timing of cell lineages.

TESTING THE HYPOTHESIS

Starting material: Prior to this work, many laboratories had screened thousands of *C. elegans* worms and identified many different mutant strains that were egg-laying defective. (Note: There are many different genes that when mutated may cause an egg-laying defective phenotype. Only some of them are expected to be genes that alter the timing of cell fate within a particular cell lineage.)

Experimental Level	**Conceptual Level**

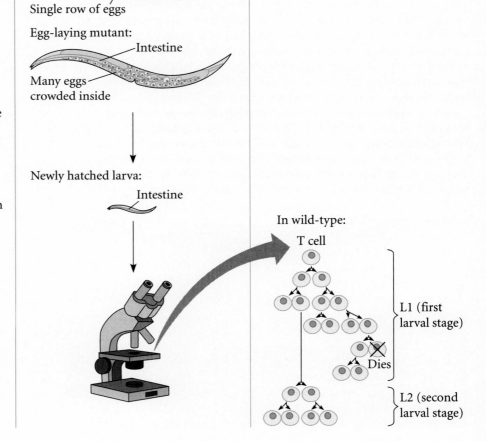

1. Obtain a large number of *C. elegans* strains that have an egg-laying defective phenotype. The wild-type strain was also studied as a control.

Normal adult worm:
Intestine
Single row of eggs

Egg-laying mutant:
Intestine
Many eggs crowded inside

2. Right after hatching, observe the fate of particular cells via microscopy. This involves long hours of viewing specific cells within a worm and watching to see if they divide at the appropriate time. Typically, the viewer looks at the cell nuclei (which are relatively easy to see in this transparent worm) and keeps track of when they divide. A researcher can watch and time the division of the cell nuclei as a way to monitor cell division. In this example, a researcher began watching a cell called the T cell, and monitored its division pattern, and the pattern of subsequent daughter cells, during the first and second larval stages. These patterns were examined in both wild-type and egg-laying defective worms.

Newly hatched larva:
Intestine

In wild-type:
T cell

L1 (first larval stage)

Dies

L2 (second larval stage)

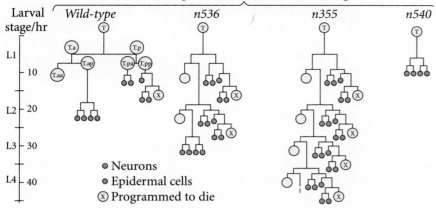

T cell lineages in different strains of *C. elegans*

These data show the division pattern of one particular cell lineage. The top of this lineage begins with a cell called the T cell. It is part of the cell lineage that includes the ABp cell (see Figure 24-13). The cells at the left of each lineage are located anteriorly in the worm; those on the right are located posteriorly. Neurons are labeled in blue, epidermal cells in red.

As shown in the data, the wild-type strain shows a particular pattern of cell division for the T cell lineage, which occurs at specific times during the L1 and L2 larval stages. In the normal strain, the T cell divides during the L1 larval stage to produce a T.a and T.p cell. The T.a cell also divides during L1 to produce a T.aa and T.ap cell. The T.p cell divides during L1 to produce T.pa and T.pp. These cells also divide during L1, eventually producing five neurons (labeled in blue) and one cell that is programmed to die (designated with an X). During the L2 larval stage, the T.ap cell resumes division to produce four cells: three epidermal cells (labeled in red) and one neuron.

The other T cell lineages are from worms that carry mutations in a gene called *lin-14*. The allele designated *n536* has caused the reiteration of the normal events of L1 during the L2 larval stage. In L2, the only cell of this lineage that is supposed to divide is T.ap. In worms carrying the *n536* allele, however, this cell behaves as if it were a T cell, rather than a T.ap cell. It produces a group of cells that are identical to what a T cell produces during the L1 stage. In the L3 stage, the cells in the *n536* strain behave as if they were in L2. Besides the egg-laying defect, the phenotypic outcome of this irregularity in the timing of cell fates is a worm that has a few more cells and goes through five or six larval stages instead of the normal four.

A more severe allele that causes multiple reiterations is the *n355* allele. This strain continues to reiterate the normal events of L1 during the L2, L3, and L4 stages. In contrast, the *n540* allele has an opposite effect on the T cell lineage. During the L1 larval stage, the T cell behaves as if it were a T.ap cell in the L2 stage. In this case, it skips the divisions and cell fates of the L1 and proceeds directly to cell fates that occur during the L2 stage.

The types of mutations described here are called **heterochronic mutations**. This term refers to the fact that the timing of fates for particular cell lineages is not synchronized with the development of the rest of the organism. More recent molecular data have shown that this is due to an irregular pattern of gene expression. In wild-type worms, the Lin-14 protein accumulates during the L1 stage and promotes the T cell division pattern shown for the wild-type. During L2, the Lin-14 protein diminishes to negligible levels. The *n536* and *n355* alleles are examples of gain of function mutations. In these alleles, the Lin-14 protein persists during later larval stages. By comparison, the *n540* allele is a loss of function mutation. This allele causes Lin-14 to be inactive during L1, so that it cannot promote the normal L1 pattern of cell division and cell fate.

Overall, the results described in this experiment are consistent with the idea that the precise timing of *lin-14* expression during development is necessary to correctly control the fates of particular cells in *C. elegans*. Mutations that alter the expression of *lin-14* lead to phenotypic abnormality, namely the inability to lay eggs. These results illustrate the importance of the correct timing of developmental events.

VERTEBRATE DEVELOPMENT

Embryologists have studied the morphological features of development in many vertebrate species. Historically, amphibians and birds have been studied extensively, because their eggs are rather large and easy to manipulate. For example, the early developmental stages of the chicken and frog (*Xenopus laevis*) have been described in great detail. In more recent times, the successes obtained in *Drosophila* have shown the great power of genetic analyses in elucidating the underlying molecular mechanisms that govern biological development. With this knowledge, many researchers are attempting to understand the genetic pathways that govern the development of the more complex body structure found in vertebrate organisms.

Several vertebrate species have been the subject of genetic studies of development. These include the mouse, the chicken, *Xenopus*, and the small aquarium zebrafish (*Brachydanio rerio*). In this section, we will discuss primarily the genes that are important in mammalian development, particularly those that have been characterized in the mouse. Among mammals, the most extensive genetic analyses have been performed on the mouse. As we will see, several genes affecting its developmental pathways have been cloned and characterized. In this section, we will examine how these genes affect the course of mouse development.

Researchers have identified homeotic genes in vertebrates

In most vertebrates, which have long generation times and produce relatively few offspring, it is not practical to screen large numbers of embryos or offspring in search of mutant phenotypes with developmental defects. As an alternative, the most successful way of identifying genes that affect vertebrate development has been the use of molecular techniques to identify vertebrate genes similar to those that control development in simpler organisms such as *Drosophila*.

As discussed in Chapters 22 and 27, species that are related evolutionarily to each other often contain genes with similar DNA sequences. When two or more genes have similar sequences because they are derived from the same ancestral gene, they are called **homologous genes**. Since they have similar sequences, a DNA strand from one gene will hybridize to a complementary strand of a homologue.

The general approach of using cloned *Drosophila* genes as probes to identify homologous vertebrate genes has been quite successful. Using this method, researchers have found complexes of homeotic genes in many vertebrate species that bear striking similarities to those in the fruit fly. In the mouse, these groups of adjacent homeotic genes are called **Hox complexes**. As shown in Figure 24-14, the mouse has four *Hox* complexes, designated *HoxA* (on chromosome 6), *HoxB* (on chromosome 11), *HoxC* (on chromosome 15), and *HoxD* (on chromosome 2). There are a total of 38 genes in the four complexes. There are 13 different gene types within the four *Hox* complexes, although each complex contains less than all 13 types of genes. Among the first 6 types of genes, 5 of them are homologous to genes found in the *antennapedia* complex of *Drosophila*. Among the last 7, 3 are homologous to the genes of the *bithorax* complex.

Like the *bithorax* and *antennapedia* complexes in *Drosophila*, the arrangement of *Hox* genes along the mouse chromosome reflects their pattern of expression from

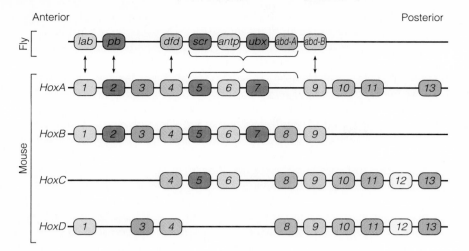

Anterior | Posterior

FIGURE **24-14**

A comparison of homeotic genes in *Drosophila* and the mouse. The mouse contains four gene clusters, *Hox A–D*, that correspond to the homeotic genes found in *Drosophila*. Thirteen different types of homeotic genes are found in the mouse, although each *Hox* gene cluster does not contain all 13 genes. In this drawing, homologous genes are aligned in columns. For example, *lab* is homologous to *Hox A-1*, *Hox B-1*, and *Hox D-1*; *ubx* is homologous to *Hox A-7* and *Hox B-7*.

the anterior to the posterior end (Figure 24-15a). This phenomenon is seen in more detail in Figure 24-15b, which shows the results of the expression pattern for a group of *HoxB* genes in a mouse embryo. Overall, these results are consistent with the idea that the *Hox* genes play a role in determining the fates of segments along the anteroposterior axis.

Currently, researchers are trying to understand the functional roles of the genes within the *Hox* complexes in vertebrate development. In the fly, great advances in developmental genetics have been made by studying mutant alleles in genes that control development. In mice, however, there are few natural mutations

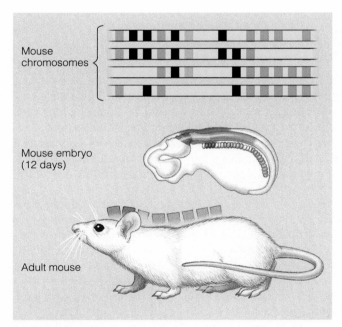

(a) Correlation between *Hox* gene arrangement and expression

(b) Anterior expression boundaries for a series of *HoxB* genes

FIGURE **24-15**

Expression pattern of *Hox* genes in the mouse. (a) A schematic illustration of the *Hox* gene expression in the embryo and the corresponding regions in the adult. **(b)** A more detailed description of *Hox B* expression in a mouse embryo. The arrows indicate the anteriormost boundaries for the expression of *Hox B-3–Hox B-9*. From K. Kalthoff, *Analysis of Biological Development*.©1996 McGraw-Hill, Inc. Reproduced with permission of the McGraw-Hill Companies.

affecting development. This has made it difficult to understand the role that genetics plays in the development of the mouse and other vertebrate organisms.

To circumvent this problem, geneticists are taking an approach known as **reverse genetics**. In this strategy, researchers first identify the wild-type gene using cloning methods. In this case, the *Hox* genes in vertebrates have been cloned using *Drosophila* genes as probes. The next step is to create a mutant version of a *Hox* gene *in vitro*. This mutant allele is then reintroduced into a mouse using techniques described in Chapter 21 (see pp. 583–585). When the function of the wild-type gene is thereby eliminated, this is called a **gene knockout**. In this way, researchers can determine how the mutant allele affects the phenotype of the mouse.

The term reverse genetics reflects that the experimental steps occur in an order opposite to that in the conventional approach used in *Drosophila*. In the fly, the mutant alleles were identified by their phenotype first, and then they were cloned. In the mouse, the genes were cloned first, the mutations were made *in vitro*, and then these were introduced into the mouse to observe their phenotypic effects.

During the 1990s, several laboratories have used a reverse genetics approach to understand how the *Hox* genes affect vertebrate development. In *Drosophila*, loss of function alleles for homeotic genes usually show an anterior transformation. This means that the segment where the defective homeotic gene is expressed now exhibits characteristics that resemble the adjacent anterior segment. Similarly, certain gene knockouts (e.g., *Hox A-2*, *B-4*, and *C-8*) also show anterior transformations within particular regions of the mouse. However, knockouts of other *Hox* genes (e.g., *A-5* and *A-11*) have posterior transformations, and knockouts of *A-3* and *A-1* exhibit abnormalities in morphology but no clear homeotic transformations. Overall, the current picture indicates that the *Hox* genes in vertebrates play an important role in homeotic transformations. Nevertheless, additional research will be necessary to understand the individual roles that each of the 38 *Hox* genes play during embryonic development.

Genes that encode transcription factors also play a key role in cell differentiation

Throughout most of this chapter, we have focused our attention on patterns of gene expression that occur during the very early stages of development. These genes control the basic body plan of the organism. At later stages of development, cells reach their predetermined destinations and eventually become **differentiated**. This means that a cell's morphology and function have changed, usually permanently, into a highly specialized cell type. For example, an undifferentiated mesodermal cell may differentiate into a specialized muscle cell, or an ectodermal cell may differentiate into a nerve cell.

At the molecular level, the profound morphological differences between muscle cells and nerve cells arise from gene regulation. Though nerve and muscle cells contain the same genetic material (i.e., the same set of genes), they regulate the expression of their genes in very different ways. Certain genes that are transcriptionally active in muscle cells are completely inactive in nerve cells, and vice versa. Therefore, nerve and muscle cells express different proteins, which affect the morphological and physiological characteristics of the respective cells in distinct ways. In this manner, differential gene regulation underlies cell differentiation.

We learned earlier that in *Drosophila* a hierarchy of gene regulation is responsible for establishing the body pattern. Maternal effect genes control the expression of gap genes, which control the expression of pair-rule genes, and so forth. A similar type of hierarchy is thought to underlie cell differentiation. Researchers have identified specific genes, realizator genes, the expression of which causes cells to differentiate into particular cell types. These genes cause undifferentiated cells to differentiate into their proper cell fates.

In 1987, Harold Weintraub and his colleagues at the Hutchinson Cancer Research Center in Seattle identified a gene, which they called *myoD*. This gene plays a key role in skeletal muscle cell differentiation. Experimentally, when the cloned *myoD* gene was expressed in fibroblast cells in a laboratory, the fibroblasts differentiated into skeletal muscle cells. This result was particularly remarkable, since fibroblasts normally differentiate into osteoblasts (bone cells), chrondrocytes (cartilage cells), adipocytes (fat cells), and smooth muscle cells, but they never differentiate, *in vivo*, into skeletal or cardiac muscle cells.

Since this initial discovery, researchers have found that *myoD* belongs to a small group of genes that play a role in initiating muscle development. Besides *myoD*, these include *myogenin*[+], *myf-5*[+], and *MRF-4*[+]. All four of these genes encode transcription factors that contain a **basic** domain and a **helix–loop–helix** domain (bHLH). The basic domain is responsible for DNA binding and the activation of skeletal muscle cell–specific genes. The HLH domain is necessary for dimer formation between transcription factor proteins. Because of their common structural features and their role in muscle differentiation, myoD, myogenin, myf-5, and MRF-4 are called **myogenic bHLH proteins**. They are found in all vertebrates, and have been identified in several invertebrates, such as *Drosophila* and *C. elegans*. In all cases, the myogenic bHLH genes are activated during skeletal muscle cell development.

In addition to myogenic bHLH genes, there are other bHLH genes that are not muscle cell specific. These other bHLH genes encode transcription factors known as **E proteins**. E proteins are synthesized in a wide variety of tissues. They are also important in cell differentiation.

At the molecular level, two key features enable myogenic bHLH proteins to promote muscle cell differentiation. First, the basic domain binds specifically to a muscle cell–specific enhancer sequence; this sequence is adjacent to genes that are expressed only in muscle cells (Figure 24-16). Therefore, when bHLH proteins are activated, they can bind to these enhancers and activate the expression of many different muscle cell–specific genes. In this way, myogenic bHLH proteins act as master switches that activate the expression of many muscle-specific genes. When the encoded proteins are synthesized, they change the characteristics of an undifferentiated cell into those of a highly specialized skeletal muscle cell.

Another important aspect of myogenic bHLH proteins is that their activity is regulated by dimerization. As shown in Figure 24-16, heterodimers may be activating or inhibitory. When a heterodimer forms between a myogenic bHLH protein and an E protein, which also contains a bHLH domain, the heterodimer binds to the DNA and activates gene expression (Figure 24-16a). However, when a heterodimer forms between a myogenic bHLH protein and a protein called Id (for *in*hibitor of *d*ifferentiation), the heterodimer cannot bind to DNA, because the Id protein lacks a basic domain (Figure 24-16b). The Id protein is produced during early stages of development and prevents myogenic bHLH proteins from promoting muscle differentiation too soon. At later stages of development, the amount of Id protein falls, and myogenic bHLH proteins can then combine with E proteins to induce muscle differentiation.

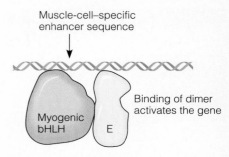

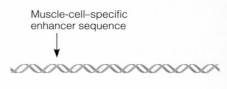

(a) Action of bHLH–E dimer

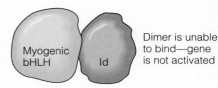

(b) Action of bHLH–Id dimer

FIGURE **24-16**

Regulation of muscle cell–specific genes by myogenic bHLH proteins. (a) A heterodimer formed from a myogenic bHLH protein and an E protein can bind to a muscle cell–specific enhancer sequence and activate gene expression. **(b)** When a myogenic bHLH protein forms a heterodimer with an Id protein, it cannot bind to the DNA and therefore does not activate gene transcription.

PLANT DEVELOPMENT

In developmental plant biology, the model organism for genetic analysis is *Arabidopsis thaliana*. Unlike most plants, which have long generation times and large genomes, *Arabidopsis* has a generation time of about two months and a genome size of 7×10^7 bp, which is similar to *Drosophila* and *C. elegans*. A flowering *Arabidopsis* plant is small enough to be grown in the laboratory, and it produces a large number of seeds. Like *Drosophila*, *Arabidopsis* can be subjected to mutagens such as X-rays to generate mutations that alter developmental processes. The small

FIGURE **24-17**

Photograph of *Arabidopsis*. The plant is relatively small, making it easy to grow many of them in the laboratory.

genome size of this organism makes it relatively easy to map these mutant alleles and eventually clone the relevant genes (as described in Chapters 20 and 21).

The morphological patterns of growth are markedly different between plants and animals. As described previously, animal embryos become organized along anteroposterior, dorsoventral, and lateral axes, and then they subdivide into segments. By comparison, the form of higher plants has two key features. The first is the root–shoot axis. Most plant growth occurs via cell division near the tips of the shoots and the bottoms of the roots.

Second, this growth occurs in a well-defined radial pattern. For example, early in *Arabidopsis* growth, a rosette of leaves is produced from leaf buds that emanate in a spiral pattern directly from the main shoot (Figure 24-17). Later, the shoot generates branches that will also produce leaf buds as they grow. Overall, the radial pattern in which a plant shoot gives off the buds that give rise to branches, leaves, and flowers is an important mechanism that determines much of the general morphology of the plant.

At the cellular level too, plant development differs markedly from animal development. For example, cell migration does not occur during plant development. In addition, the development of a plant does not rely on morphogens that are deposited asymmetrically in the oocyte. In plants, an entirely new individual can be regenerated from most types of somatic cells. In other words, most plant cells are **totipotent**, meaning that they have the ability to produce an entire individual. By comparison, animal development invariably relies on the organization within an oocyte as a starting point for development.

In spite of these apparent differences, the underlying molecular mechanisms of pattern development in plants still share some similarities with those in animals. In this section, we will consider a few examples in which genes encoding transcription factors play a key role in plant development.

Plant growth occurs from meristems that are formed during embryonic development

In animal development, the earliest developmental stages serve to subdivide the embryo into segments that eventually will give rise to adult structures. The growth of the embryo into the adult is simply an expansion of the embryo body pattern. However, plants do not grow like this. Instead, the parts of adult plants are formed from groups of dividing cells known as **apical meristems**. A meristem is an organized group of actively dividing cells. The *shoot apical meristem* grows upward and gives rise to the shoot structures, while *root apical meristems* grow downward to produce the roots. As they grow, a meristem produces offshoots of proliferating cells. On the shoot, for example, these offshoots or buds give rise to structures such as leaves and flowers.

The organization of a plant into separate root and shoot meristems occurs during the early embryonic stages. Figure 24-18 illustrates a common sequence of events that occurs in the development of seed plants such as *Arabidopsis*. After fertilization, the first cellular division is asymmetrical and produces a smaller cell, called the terminal cell, and a larger basal cell. The terminal cell will give rise to most of the embryo, and it will later develop into the shoot of the plant. The basal cell will give rise to the root, along with extraembryonic tissue that is required for seed formation. At the heart stage, which is composed of only about 100 cells, the basic organization of the plant has been established. As shown in Figure 24-18, the shoot apical meristem will arise from a group of cells located between the cotyledons. The root apical meristem is located at the opposite side of the embryo.

In *Arabidopsis*, geneticists are also identifying genes that play a role during development. G. Jürgens and his colleagues at the University of München in Germany

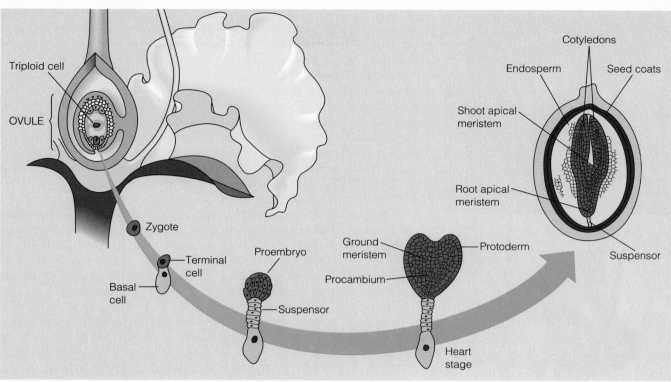

FIGURE 24-18

Developmental steps in the formation of a plant embryo.

have identified a category of genes, known as the **apical-basal-patterning genes**, that are important in early stages of development. As shown in Figure 24-19, defects in the apical-basal-patterning genes cause the plant embryo to develop abnormally. For example, the *gurke* gene is necessary for apical development. When it is defective, the embryo lacks apical structures (Figure 24-19a). The *fackel* gene influences the formation of the central region of the embryo. When it is defective, this region of the embryo is missing (Figure 24-19b). The *monopterous* gene is necessary for the formation of basal structures. As shown in Figure 24-19c, the basal region of the plant embryo is missing when this gene is defective. Finally, certain genes, like *gnom*, are necessary for the formation of the terminal regions of the embryo. When this gene is defective, these regions of the embryo do not form (Figure 24-19d).

Plant homeotic genes control flower development

Although the term homeotic was coined by William Bateson to describe homeotic mutations in animals, the first known homeotic genes were described in plants. In ancient Greece and Rome, for example, double flowers in which stamens were

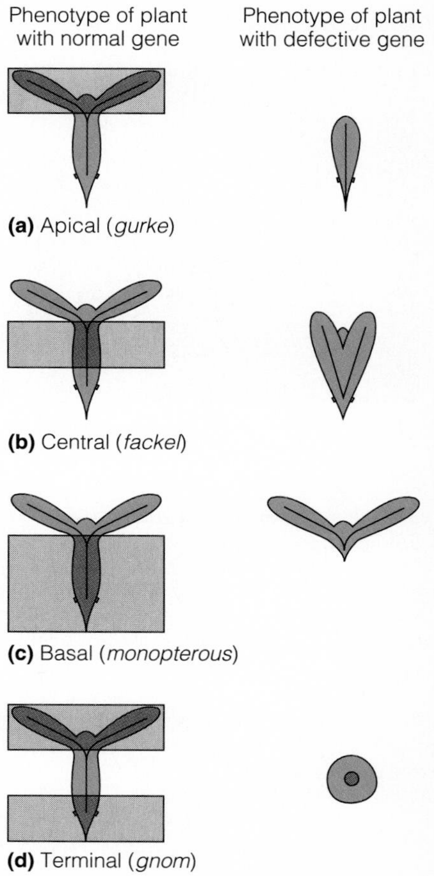

FIGURE 24-19

The effects of mutations in the apical-basal-patterning genes of *Arabidopsis*.
(a) A defect in the *gurke* gene causes a loss of apical structure. **(b)** Mutations in the *fackel* gene cause the central region of the plant embryo to be missing. **(c)** When the *monopterous* gene is defective, the basal structure that gives rise to the root is missing. **(d)** A mutation in the *gnom* gene causes defects in both terminal regions of the plant embryo. From K. Kalthoff, *Analysis of Biological Development*. ©1996 McGraw-Hill, Inc. Reproduced with permission of the McGraw-Hill Companies.

Phenotype of plant with normal gene

Phenotype of plant with defective gene

(a) Apical (*gurke*)

(b) Central (*fackel*)

(c) Basal (*monopterous*)

(d) Terminal (*gnom*)

(a) Normal flower

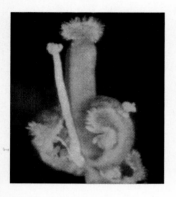

(b) Single homeotic mutant

(c) Triple mutant

FIGURE **24-20**

Examples of homeotic mutations in
Arabidopsis.

GENES ⟶ TRAITS: **(a)** A normal flower.
It is composed of four concentric whorls of
structures: sepals (Se), petals (Pe), stamens
(St), and carpel (Ca). **(b)** A homeotic mutant in which the sepals have been transformed into carpels, and the petals have
been transformed into stamens. **(c)** A triple
mutant in which all of the whorls have
been changed into leaves.

replaced by petals were noted. In current research, geneticists have been studying
these types of mutations to better understand developmental pathways in plants.
Many homeotic mutations affecting flower development have been identified in
Arabidopsis and also in the snapdragon (*Antirrhinum majus*).

Examples in *Arabidopsis* are shown in Figure 24-20. Part (a) shows a normal
Arabidopsis flower. It is composed of four concentric whorls of structures. The
outer whorl contains four sepals, which protect the flower bud before it opens. The
second whorl is composed of four petals, and the third whorl contains six stamens.
The stamens are the structures that make the male gametophyte, pollen. Finally, the
innermost whorl contains two carpels, which are fused together. The carpel produces the female gametophyte. The homeotic mutants shown in Figure 24-20 have
undergone transformations of particular whorls. For example, in Figure 24-20b, the
sepals have been transformed into carpels, the petals into stamens.

By analyzing the effects of many different homeotic mutations in *Arabidopsis*,
Elliot Meyerowitz at the California Institute of Technology and his colleagues have
proposed the *ABC* model for flower development. In this model, three classes of
genes, called *A*, *B*, and *C*, govern the formation of sepals, petals, stamens, and carpels.
Figure 24-21 illustrates how these genes affect normal flower development and what
happens when the gene products are defective, as in homeotic mutants. In the outermost whorl (whorl 1), the gene *A* products are made. This promotes sepal formation. In whorl 2, both gene *A* and gene *B* products are made, which promotes
petal formation. In whorl 3, the expression of genes *B* and *C* causes stamens to be
made, and in whorl 4, only gene *C* is expressed, which promotes carpel formation.

Now let's take a look at what happens in certain homeotic mutants. In this
model, genes *A* and *C* repress each other's expression, and gene *B* functions independently. In a mutant defective in gene *A* expression, gene *C* will also be expressed
in whorls 1 and 2. This produces a carpel–stamen–stamen–carpel arrangement.
When gene *B* is defective, a flower cannot make petals or stamens. Therefore, a gene
B defect yields a flower with a sepal–sepal–carpel–carpel arrangement. Finally,
when gene *C* is defective, gene *A* is expressed in all four whorls. This results in a
sepal–petal–petal–sepal pattern.

Overall, it appears that the types of genes described in Figure 24-21 promote
cell differentiation that leads to either sepal, petal, stamen, or carpel structures. But
what happens if all three types of genes are defective? As shown in Figure 24-20c,
this produces a "flower" that is composed of leaves! These results indicate that the
leaf structure is the default pathway and that the *A*, *B*, and *C* genes cause development to deviate from a leaf structure in order to make something else. In this regard, the sepals, petals, stamens, and carpels can be viewed as modified leaves.

In *Arabidopsis*, there are two types of gene *A* (*apetala1* and *apetala2*), two types
of gene *B* (*apetala3* and *pistillata*), and one type of gene *C* (*agamous*). All of these
plant homeotic genes encode transcription factor proteins. The proteins contain a
DNA-binding domain and a dimerization domain. However, the *Arabidopsis*
homeotic genes do not contain a sequence similar to the homeobox found in animal homeotic genes. Instead, most of them (except for *apetala2*) contain a **MADS
box** (an acronym of the first four plant genes of this type that were identified:
MCM1, *AG*, *DEF*, and *SRF*). In the transcription factor proteins, the MADS domain
promotes the binding of the transcription factor to specific DNA sequences.

Like the *Drosophila* homeotic genes, plant homeotic genes are thought to be
part of a hierarchy of gene regulation. Genes that are expressed within the flower
bud primordium produce proteins that activate the expression of these homeotic
genes. Once they are transcriptionally activated, the homeotic genes then regulate
the expression of realizator genes, the products of which promote the formation of
sepals, petals, stamens, or carpels.

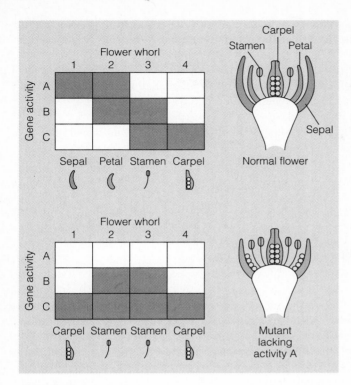

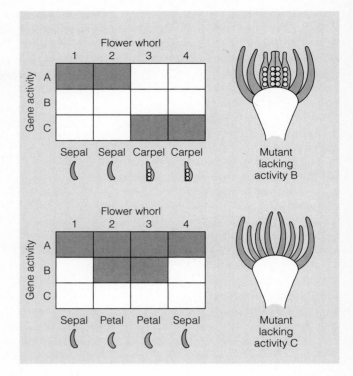

FIGURE **24-21**

The *ABC* model of homeotic gene action in *Arabidopsis*.

In this chapter, we have examined the role genetics plays in the *development* of multicellular organisms. Each organism has its own developmental program, driven by a hierarchy of genes that encode transcription factors and other types of regulatory proteins. In *Drosophila*, the developmental program begins in the oocyte. It is here that a few key gene products act as *morphogens* that are asymmetrically located or activated. This provides *positional information* that initiates development of the body plan or pattern in this organism. The maternal effect gene products serve to organize the *anteroposterior* and *dorsoventral axes* soon after fertilization. Once these axes have been established, the next step is to divide the embryo into *segments*.

Three categories of *segmentation genes*, known as *gap genes*, *pair-rule genes*, and *segment-polarity genes*, act sequentially to promote segmentation. The pair-rule genes control the identity of each *parasegment*, and the segment-polarity genes divide each parasegment into anterior and posterior regions. A segment, which is a morphological feature, is composed of the anterior region of one parasegment and the posterior region of an adjacent parasegment. The role of *homeotic genes* is to dictate the morphological characteristics of each segment. To do so, they activate *realizator genes*, which express proteins that determine the morphological characteristics of cells.

In *C. elegans*, the availability of a *fate map* has allowed geneticists to identify *heterochronic mutations* that alter the timing of *cell fate*. These mutations lead to abnormalities in morphology. This phenomenon illustrates the importance of coordinated cell division in determining cell fate during development.

In vertebrates, much less is known about the programs that promote development. Nevertheless, they appear to bear many similarities to the programs in simpler invertebrates. For example, the *Hox complex* in mice contains groups of genes that are *homologous* to homeotic genes in *Drosophila*. Furthermore, experiments with *gene knockouts* of particular *Hox* genes suggest that they act as homeotic genes in vertebrates.

A well-studied feature of vertebrate development is cell *differentiation*. Researchers have identified genes involved in the differentiation of undifferentiated cells into highly specialized cell types. For example, the *myogenic bHLH proteins* play a critical role in the differentiation of skeletal muscle cells.

Morphologically, plant development differs markedly from animal development. Plants are organized along a root–shoot axis and growth occurs along *apical meristems*. Plant geneticists have identified some of the genes that play a role in the general organization of the plant embryo. Geneticists have also identified many homeotic genes that dictate the morphology of particular structures later in development. As was described in this chapter, three classes of homeotic genes, genes *A*, *B*, and *C*, are involved in the formation of the four whorls that make up a flower. These results show that plant development is dictated by a genetic program that bears some similarities to those found in animals.

EXPERIMENTAL SUMMARY

In this chapter, we have seen that a genetic approach (namely, the identification of mutant alleles) has been essential in our understanding of development. Researchers, particularly those studying invertebrates, have identified many different mutant alleles that alter key steps in the developmental process. In *Drosophila*, for example, these include mutations in maternal effect genes, segmentation genes, and homeotic genes. The molecular characterization of these genes has shown that many of them encode transcription factors that influence the expression of other genes. The study of developmental mutations has enabled scientists to piece together a genetic hierarchy that underlies many developmental programs.

The characterization of developmental mutants has also been correlated with the spatial localization of gene products using techniques such as *in situ* hybridization. This enables researchers to determine when and where a gene product is made during oogenesis or embryogenesis. This approach has shown that some mutations that alter development are gain of function alleles, which cause a gene to be expressed in the wrong place or at the wrong time. Other mutations are loss of function alleles, which cause a defect in gene expression. In *C. elegans*, the availability of a fate map has also allowed researchers to examine in detail how mutations (e.g., gain of function and loss of function alleles) can affect the timing of developmental steps.

The key experimental advantage of studying invertebrates is the ability to identify mutations that alter developmental steps. This approach is not as easy in vertebrates, but a reverse genetic approach is proving to be successful. Using invertebrate genes as probes, researchers have identified vertebrate genes, such as the *Hox* genes in mice, that are homologous to the homeotic genes of invertebrates. To understand their role, the wild-type *Hox* genes have been mutated *in vitro* and reintroduced into mice. The results of this method are consistent with the idea that the *Hox* genes function as the mouse homologues of the *Drosophila* homeotic genes. However, further research will be needed to understand the individual roles of these 38 genes.

In plants, a genetic approach is also proving effective in elucidating the development process. Early in development, a category of genes, known as the apical-basal-patterning genes, appear necessary for the formation of the apical and basal regions of the embryo. These genes were identified by investigating the effects of mutations on the developing plant embryo. Similarly, mutations in plant homeotic genes have been discovered by their ability to abnormally transform one plant structure into another (e.g., sepals into petals). A comparison of single, double, and triple homeotic mutations has provided the framework for the *ABC* model of flower development.

PROBLEM SETS

SOLVED PROBLEMS

1. Discuss and distinguish the functional roles of the maternal effect genes, gap genes, pair-rule genes, and segment-polarity genes in *Drosophila*.

Answer: All of these types of genes are involved in promoting the segmentation of the *Drosophila* embryo. The asymmetric distribution of maternal effect gene products in the oocyte establishes the anteroposterior and dorsoventral axes. These gene products also control the expression of the gap genes, which are expressed as broad bands in certain regions of the embryo. The overlapping expression of maternal effect genes and gap genes controls the pair-rule genes, which are expressed in alternating stripes. A stripe corresponds to a parasegment. Within each parasegment, the expression of segment-polarity genes defines an anterior and posterior compartment. With regard to morphology, an anterior compartment of one parasegment and the posterior compartment of an adjacent parasegment will form a segment of the fly.

2. With regard to genes affecting development, what are the phenotypic effects of gain of function mutations versus loss of function mutations?

Answer: Gain of function mutations cause a gene to be expressed in the wrong place or at the wrong time. When they are expressed in the wrong place, that region may develop into an inappropriate structure. For example, when *antp* is abnormally expressed in an anterior segment, this segment develops legs in place of antennae. When gain of function mutations cause a gene to be expressed at the wrong time, this can also disrupt the development process. Gain of function mutations in heterochronic genes cause cell lineages to be reiterated and thereby alter the course of development. By comparison, loss of function mutations result in a defect in the expression of a gene. This usually will disrupt the developmental process, because the cells in the region where the gene is supposed to be expressed will not be directed to develop along the correct pathway.

CONCEPTUAL QUESTIONS

1. Discuss the morphological differences between the parasegments and segments of *Drosophila*. Discuss the evidence, providing specific examples, that suggests the parasegments of the embryo are the subdivisions for the organization of gene expression.

2. Here are some schematic diagrams of mutant larvae.

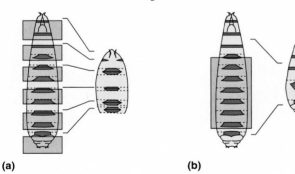

(a) **(b)**

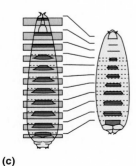

(c)

The left side of each pair shows a wild-type larva, with gray boxes showing the sections that are missing in the mutant larva. Which type of gene is defective in each larva: a gap gene, a pair-rule gene, or a segment-polarity gene?

3. Describe what a morphogen is and how it exerts its effects. What do you expect will happen when a morphogen is expressed in the wrong place in an embryo? List five examples of morphogens that function in *Drosophila*.

4. What is the meaning of the term positional information? Discuss three different ways that cells can obtain positional information. Which of these three ways do you think is the most important for the formation of a segmented body pattern in *Drosophila*?

5. Gradients of morphogens can be preestablished in the oocyte. Also, later in development, morphogens can be secreted from cells. How are these two processes similar and different?

6. Discuss how the anterior portion of the anteroposterior axis is established. What aspects of oogenesis are critical in establishing this axis? What do you think would happen if the *bicoid* mRNA was not trapped at the anterior end, but instead diffused freely throughout the oocyte?

7. Describe the function of the Bicoid protein. Explain how its ability to exert its effects in a concentration-dependent manner is a critical feature of its function.

8. With regard to development, what are the roles of the maternal effect genes versus the zygotic genes? Which types of genes are needed earlier in the development process?

9. What is meant by a genetic hierarchy? The following is an incomplete hierarchy beginning with *bicoid* and ending with *labial*.

bicoid (oocyte/anterior end) → ????????????? → *labial* (blastoderm/first parasegment)

Fill in the missing steps. More than one answer is possible. Estimate where (which regions of the embryo) and when (e.g., oocyte, preblastoderm, blastoderm) these gene products are needed.

10. Discuss the role of homeotic genes in development. Explain what happens to the phenotype of a fly when a gain of function homeotic gene mutation occurs in an abnormal region of the embryo. What are the consequences of a loss of function mutation in such a gene?

11. Describe the molecular features of the homeobox and homeodomain. Explain how these features are important in the function of homeotic genes.

12. Although we do not know a lot about realizator genes, developmental geneticists propose that they are the targets of homeotic gene products. Explain the role of realizator genes during development and the types of gene products they are likely to encode.

13. Predict the phenotypic consequences of each of the following mutations:

    A. *apetala1* defective
    B. *pistillata* defective
    C. *apetala1* and *pistillata* defective

14. Discuss the similarities and differences between the *bithorax* and *antennapedia* complexes in *Drosophila* and the *Hox* gene complexes in mice.

15. What is cell differentiation? Discuss the role of bHLH proteins in the differentiation of muscle cells. Explain how they work at the molecular level. In your answer, you should indicate how protein dimerization is a key feature in gene regulation.

16. Discuss the morphological differences between plants and animals. How are they different at the cellular level? How are they similar at the genetic level?

17. What is a heterochronic mutation? How does it affect the phenotypic outcome of an organism? What phenotypic effects would you expect if a heterochronic mutation affected the cell lineage that determines the fates of the intestinal cells?

EXPERIMENTAL QUESTIONS

1. Researchers have used the cloning methods described in Chapter 19 to clone the *bicoid* gene and express large amounts of the Bicoid protein. The Bicoid protein was then injected into the posterior end of a zygote immediately after fertilization. What phenotypic results would you expect to occur? What do you think would happen if the Bicoid protein was injected into a segment of a larva?

2. Compare and contrast the experimental advantages of *Drosophila* and *C. elegans* in the study of developmental genetics.

3. What is meant by the term cell fate? What is a fate map? Discuss the experimental advantage of having a fate map. What is a cell lineage?

4. Explain why a fate map is necessary in order to determine if a mutation is heterochronic.

5. Experimentally, describe two strategies you might follow to identify a vertebrate gene involved in pattern formation. How would you create mutant genes that alter the phenotype of the organism? Can you think of ways to make gain of function mutations?

6. Why have geneticists been forced to use reverse genetics to study the genes involved in vertebrate development? Explain how this strategy differs from traditional genetic analyses like those done by Mendel.

7. Explain the rationale behind the use of the "bag of worms" phenotype as a way to identify heterochronic mutations.

8. Here is the result of cell lineage analysis of hypodermal cells in wild-type and mutant strains of *C. elegans*:

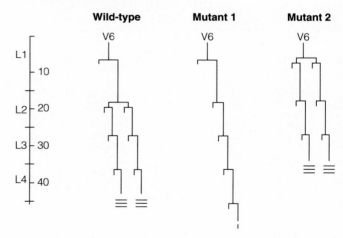

Explain the nature of the mutations in the altered strains.

QUESTIONS FOR STUDENT DISCUSSION/COLLABORATION

1. Compare and contrast the experimental advantages and disadvantages of *Drosophila*, *C. elegans*, mammals, and *Arabidopsis*.

2. It seems that developmental genetics boils down to a complex network of gene regulation. Try to draw how this network is structured for *Drosophila*. How many genes do you think are necessary to describe a complete developmental network for the fruit fly? How many genes do you think are needed for a network to specify one segment? Do you think it is more difficult to identify genes that are involved in the beginning, middle, or end of this network? Knowing what you

know about *Drosophila* development, suppose you were trying to identify all of the genes needed for development in a chicken. Would you first try to identify genes that are necessary for early development, or would you begin by identifying genes involved in cell differentiation?

3. At the molecular level, how do you think a gain of function mutation in a developmental gene might cause it to be expressed in the wrong place or at the wrong time? Explain what type of DNA sequence would be altered.

CHAPTER 25

QUANTITATIVE GENETICS

Quantitative genetics is the study of traits that can be described in a quantitative way. In humans, quantitative traits include height, the shape of our noses, the rate at which we metabolize food, and our I.Q, to name a few examples. The features of these traits can be characterized and analyzed with numbers.

Quantitative genetics is an important branch of genetics for several reasons. In agriculture, most of the key characteristics considered by plant and animal breeders are quantitative traits. These include traits such as weight, resistance to disease, and the ability to withstand harsh environmental conditions. As we will see later in this chapter, quantitative genetic techniques have improved our ability to develop strains of agriculturally important species with desirable quantitative traits.

Another important reason to study quantitative genetics is its relationship to evolution. Many of the traits that allow a species to adapt to its environment are quantitative. Examples include the long neck of a giraffe, the swift speed of the cheetah, and the sturdy branches of trees in windy climates. The importance of quantitative traits in the evolution of species will be discussed in Chapter 27. In this chapter, we will examine how genes and the environment contribute to the phenotypic expression of quantitative traits.

QUANTITATIVE TRAITS

When we compare characteristics among members of the same species, the differences are often quantitative rather than qualitative. Humans, for example, all have the same basic anatomical features (two eyes, two ears, etc.), but they differ in

TABLE 25-1 Types of Quantitative Traits

Trait	Examples
Anatomical traits	Height, weight, number of bristles in *Drosophila,* ear length in corn, the degree of pigmentation in flowers and skin
Physiological traits	Metabolic traits, speed of running and flight, tolerance of harsh temperatures, milk production in mammals
Behavioral traits	I.Q., mating calls, courtship rituals, ability to learn a maze, the ability to move toward light
Complex diseases	Diabetes, hypertension, arthritis, obesity

quantitative ways. People vary with regard to height, weight, the shape of facial features, pigmentation, and many other characteristics. As shown in Table 25-1, quantitative traits can be categorized as anatomical, physiological, and behavioral. In addition, a few human diseases exhibit characteristics and inheritance patterns analogous to those of quantitative traits.

In many cases, quantitative traits are easily measured and described numerically. For example, height and weight can be measured in centimeters and kilograms. Speed can be measured in kilometers per hour, and metabolic rate can be assessed as the grams of glucose burned per minute. Behavioral traits can also be quantified. For example, a mating call can be evaluated with regard to its duration, sound level, and pattern. The ability to learn a maze can be described as the time and/or repetitions it takes to master the skill. Finally, complex diseases such as diabetes can also be studied and described via numerical parameters. For example, the severity of the disease can be assessed by the age of onset or by the amount of insulin needed to prevent adverse symptoms.

From a scientific viewpoint, the measurement of quantitative traits is essential when comparing individuals or evaluating groups of individuals. It is not very informative to say that two people are very tall. Instead, we are better informed if we know that one person is 6′4″ and the other is 6′7″. In this branch of genetics, the measurement of a quantitative trait is how we describe the phenotype.

In the early 1900s, Francis Galton in England and his student Karl Pearson showed clearly that many traits in humans and domesticated animals are quantitative in nature. To understand the underlying genetic basis of these traits, they founded what became known as the **biometric field** of genetics. During this period, Galton and Pearson developed various statistical tools for studying the variation of quantitative traits within groups of individuals; many of these tools are still in use today. In this section, we will examine how quantitative traits are measured, and how statistical tools are used to analyze their variation within groups.

Quantitative traits exhibit a continuum of phenotypic variation that may follow a normal distribution

In Part II of this text, we considered many traits that fell into discrete categories. For example, fruit flies might have white eyes or red eyes, and pea plants might have wrinkled or smooth seeds. The alleles that govern these traits affect the phenotype in a qualitative way. In analyzing crosses involving these types of traits, each offspring can be put into a particular phenotypic category. Such attributes are called **discontinuous traits**.

In contrast, **quantitative traits** show a continuum of phenotypic variation within a group of individuals. For such traits, it is often impossible to place organisms into a discrete phenotypic class. For example, Figure 25-1a is a classic photograph showing the range of heights of 175 students at the Connecticut Agricultural

Number of individuals	1	0	0	1	5	7	7	22	25	26	27	17	11	17	4	4	1
Height in inches	58	59	60	61	62	63	64	65	66	67	68	69	70	71	72	73	74

(a)

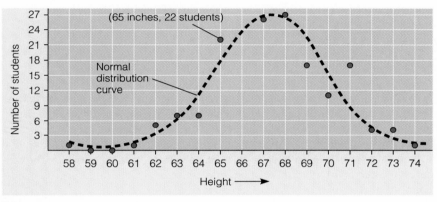

(b)

FIGURE **25-1**

Normal distribution of a quantitative trait. (a) A classic photograph showing the distribution of heights in 175 students at the Connecticut Agricultural College. **(b)** A frequency distribution for the heights of students shown in part **(a)**.

College. Though there is a minimum and maximum height, the range of heights between these minimum and maximum values is fairly continuous.

Since quantitative traits do not naturally fall into a small number of discrete categories, an alternative way to describe them is a **frequency distribution**. To construct a frequency distribution, the trait is divided arbitrarily into a number of convenient discrete phenotypic categories. For example, in Figure 25-1, the range of heights is partitioned into 1-inch intervals. Then a graph is made that shows the numbers of individuals found in each of several phenotypic categories.

Figure 25-1b shows a frequency distribution for the heights of students pictured in part (a). The measurement of the height is plotted along the x-axis, and the number of individuals who exhibit that phenotype is plotted on the y-axis. The values along the x-axis are divided into the discrete 1-inch intervals that define the phenotypic categories, even though height is essentially continuous within a group of individuals. For example, in Figure 25-1a, 22 students were between 64.5 and 65.5 inches in height, which is plotted as the point (65 inches, 22 students) on the graph in Figure 25-1b. This type of analysis can be conducted on any group of individuals that vary with regard to a quantitative trait.

The dotted line in the frequency distribution depicts a **normal distribution curve**, a distribution for an infinite sample in which the trait of interest varies in a symmetric way around an average value. The distribution of measurements of many biological characteristics is approximated by a symmetrical bell curve like the dotted line in Figure 25-1b. We will consider the significance of this type of distribution next.

Statistical methods are used to evaluate a frequency distribution quantitatively

Statistical tools can be used to analyze a normal distribution in a number of ways. One measure that you are probably familiar with is a parameter called the **mean**. The mean is the sum of all the values in the group divided by the number of individuals in the group. It is computed using the following formula:

$$\overline{X} = \frac{\Sigma x}{N}$$

where

$\overline{X}$ is the mean

Σx is the sum of all the values in the group

N is the number of individuals in the group

For example, suppose a bushel of corn cobs had the following lengths (rounded to the nearest centimeter): 15, 14, 13, 14, 15, 16, 16, 17, 15, and 15. Then

$$\overline{X} = \frac{15 + 14 + 13 + 14 + 15 + 16 + 16 + 17 + 15 + 15}{10}$$

$$= 15 \text{ cm}$$

In genetics, we are often interested in the amount of phenotypic variation that exists in a group. As we will see later in this chapter, and in Chapters 26 and 27, variation lies at the heart of breeding experiments and evolution. Without variation, selective breeding is not possible, and evolution cannot favor one type over another. A common way to evaluate variation within a population is a statistic called the **variance**. The variance is the sum of the squared deviations from the mean divided by the degrees of freedom (*df* equals $N - 1$; see Chapter 2, p. 40 for a review of *df*):

$$V_X = \frac{\Sigma(X - \overline{X})^2}{N - 1}$$

where

V_X is the variance

$X - \overline{X}$ is the difference between each value and the mean

N equals the number of observations

For example, if we use the values given previously for corn cob lengths, the variance in this group is calculated as follows:

$$\Sigma(X - \overline{X})^2 = (15 - 15)^2 + (14 - 15)^2 + (13 - 15)^2 + (14 - 15)^2 + (15 - 15)^2 +$$

$$(16 - 15)^2 + (16 - 15)^2 + (17 - 15)^2 + (15 - 15)^2 + (15 - 15)^2$$

$$= 0 + 1 + 4 + 1 + 0 + 1 + 1 + 4 + 0 + 0$$

$$= 12 \text{ cm}^2$$

$$V_X = \frac{\Sigma(X - \overline{X})^2}{N - 1}$$

$$= \frac{12 \text{ cm}^2}{9}$$

$$= 1.33 \text{ cm}^2$$

Since the variance is computed from squared deviations, it is a statistic that may be difficult to understand intuitively. For example, weight can be measured in grams; the corresponding variance would be measured in square grams. Even so, variances are centrally important in the analysis of quantitative traits, because they are additive under certain conditions. Later in this chapter, we will examine how this property is useful in predicting the outcome of genetic crosses.

To gain an intuitive grasp for variation, we can take the square root of the variance. This statistic is called the **standard deviation (S.D.)**. Again, using the same values for corn cob length, the standard deviation is:

$$\text{S.D.} = \sqrt{V_X} = \sqrt{1.33 \text{ cm}^2}$$
$$= 1.15 \text{ cm}$$

If the values in a population follow a normal distribution, then it is easy to appreciate the amount of variation by considering the standard deviation. Figure 25-2 illustrates the relationship between the standard deviation and the percentages of individuals who deviate from the mean. Approximately 68% of all individuals have values within one standard deviation from the mean, either in the positive or negative direction. About 95% are within two standard deviations, and 99.7% are within three standard deviations. When a quantitative characteristic follows a normal distribution, less than 0.3% of the individuals will have values that are more or less than three standard deviations away from the mean of the population. In our corn cob example, three standard deviations equals 3.45 cm. Therefore, we would expect that fewer than 0.3% of the corn cobs would be less than 11.55 cm or greater than 18.45 cm, assuming corn cob length follows a normal distribution.

Some statistical methods compare two variables with each other

In many biological problems, it is useful to compare two different variables. For example, we may wish to compare the occurrence of two different phenotypic traits. Do obese animals have larger hearts? Are brown eyes more likely to occur in people with dark skin pigmentation? A second type of comparison is between traits and environmental factors. Does insecticide resistance occur more frequently in areas that have been exposed to insecticides? Is heavy body weight more prevalent in colder climates? Finally, a third type of comparison is between traits and genetic relationships. Do tall parents tend to produce tall offspring? Do smart women have smart brothers?

To gain insight into such questions, a statistic known as the **correlation** is often applied. To calculate this statistic, we first need to determine the **covariance**, which

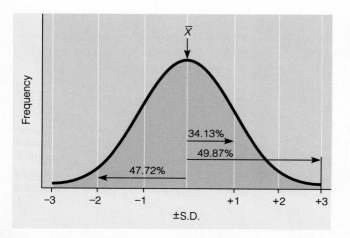

FIGURE **25-2**

The relationship between the standard deviation and the proportions of individuals in a normal distribution. For example, 34.13% of the individuals in a population are between the mean and one standard deviation above the mean; 47.72% are between the mean and two standard deviations below the mean.

describes the degree of variation between two variables within a group. The co-variance is similar to the variance, except that we multiply together the deviations of two different variables rather than squaring the deviations from a single factor:

$$\text{CoV}_{(X,Y)} = \frac{\Sigma[(X - \bar{X})(Y - \bar{Y})]}{N - 1}$$

where,

X are the values for one variable and $\bar{X}$ is the mean value in the group

Y are values for another variable and $\bar{Y}$ is the mean value in that group

N is the total number of pairs of observations

As an example, let's consider the weight of cattle at five years of age between parents and offspring. A farmer might be interested in this relationship to determine if genetic factors play a role in the weight of cattle. The data here describe the five-year weights for ten different pairs of cows and their female offspring:

Mother's Weight (kg)	Offspring's Weight (kg)	$X - \bar{X}$	$Y - \bar{Y}$	$(X - \bar{X})(Y - \bar{Y})$
570	568	−26	−30	780
572	560	−24	−38	912
599	642	3	44	132
602	580	6	−18	−108
631	586	35	−12	−420
603	642	7	44	308
599	632	3	34	102
625	580	29	−18	−522
584	605	−12	7	−84
575	585	−21	−13	273
$\bar{X}$ = 596	$\bar{Y}$ = 598			Σ = 1373
SD_X = 21.1	SD_Y = 30.5			

$$\text{CoV}_{(X,Y)} = \frac{\Sigma\left[(X - \bar{X})(Y - \bar{Y})\right]}{N - 1}$$

$$= \frac{1373}{10 - 1}$$

$$= 152.6$$

After we have calculated the covariance, we can evaluate the strength of the association between the two variables by calculating a **correlation coefficient (r)**:

$$r_{(X,Y)} = \frac{\text{CoV}_{(X,Y)}}{\text{SD}_X \text{SD}_Y}$$

This value, which ranges between +1 and −1, indicates how two factors vary in relation to each other. A value of +1 is a perfect correlation. It means that two factors vary in a completely predictable way relative to each other; as one factor increases, the other will increase with it. A value of zero indicates there is no detectable way in which the two factors vary relative to each other; the values of the two factors are not related. Finally, an inverse correlation, in which the correlation coefficient is negative, indicates that the two factors tend to vary in opposite ways to each other; as one factor increases, the other will decrease.

Let's use the data of five-year weights for mother and offspring to calculate a correlation coefficient:

$$r_{(X,Y)} = \frac{152.6}{(21.1)(30.5)}$$
$$= 0.237$$

The result is a positive correlation between the five-year weights of mother and offspring. In other words, the positive correlation value suggests that heavy mothers tend to have heavy offspring and lighter mothers, lighter offspring.

However, after a correlation coefficient has been calculated, one must evaluate whether the r value represents a true association between the two variables, or whether it could be simply due to chance. To accomplish this, we can test the hypothesis that there is no real correlation (i.e., the null hypothesis): The r value differs from zero only as a matter of random sampling error. This is the same approach followed in the chi square analysis described in Chapter 2 (see pp. 37–40). Like the chi square value, the significance of the correlation coefficient is related directly to sample size and the degrees of freedom (df). In testing the significance of correlation coefficients, df equals $N - 2$, which is one less than the degrees of freedom of variance (i.e., df for variance equals $N - 1$). Table 25-2 shows the relationship between the r values and degrees of freedom at the 5% and 1% significance levels.

This approach, though, is only valid if several assumptions are met. First, the values of X and Y in the study must have been obtained by an unbiased sampling of the entire population. In addition, this approach assumes that the scores of X and Y follow a normal distribution (like Figure 25-1) and that the relationship between X and Y is linear.

To illustrate the use of Table 25-2, let's consider the correlation we have just calculated for five-year weights of mother and offspring. In this case, we obtained a value of 0.237 for r, and the value of N was 10. Under these conditions, df equals 8. According to Table 25-2, it is fairly likely that this value could have occurred as a matter of random sampling error. Therefore, we cannot conclude that the positive correlation is due to a true association between the weights of mother and offspring (i.e., we cannot reject the null hypothesis).

In an actual experiment, however, a researcher would examine many more pairs of mothers and offspring, perhaps 500 to 1000. If a correlation of 0.237 was observed for $N = 1000$, the value would be significant at the 1% level. We would therefore reject the null hypothesis that weights are not associated with each other. Instead, we would conclude that there is a real association between the weights of mothers and their offspring. In fact, these kinds of experiments have been done for cattle weights, and the correlations between parents and offspring have been found to be significant.

Whenever a statistically significant correlation is obtained, one must be very cautious in interpreting its meaning. An r value that is statistically significant need not imply a cause and effect relationship. When parents and offspring display a significant correlation for a trait, one should not jump to the conclusion that genetics is the underlying cause of the positive association. In many cases, parents and offspring share similar environments, so that the positive association might be rooted in environmental factors. In general, correlations are quite useful in identifying positive or negative associations between two variables. Even so, this statistic by itself cannot prove that the association is due to a cause and effect.

TABLE 25–2 Values of *r* at the 5% and 1% Significance Levels

Degrees of Freedom (*df*)	5%	1%
1	.997	1.000
2	.950	.990
3	.878	959
4	.811	.917
5	.754	.874
6	.707	.834
7	.666	.798
8	.632	.765
9	.602	.735
10	.576	.708
11	.553	.684
12	.532	.661
13	.514	.641
14	.497	.623
15	.482	.606
16	.468	.590
17	.456	.575
18	.444	.561
19	.433	.549
20	.423	.537
21	.413	.526
22	.404	.515
23	.396	.505
24	.388	.496
25	.381	.487
26	.374	.478
27	.367	.470
28	.361	.463
29	.355	.456
30	.349	.449
35	.325	.418
40	.304	.393
45	.288	.372
50	.273	.354
60	.250	.325
70	.232	.302
80	.217	.283
90	.205	.267
100	.195	.254
125	.174	.228
150	.159	.208
200	.138	.181
300	.113	.148
400	.098	.128
500	.088	.115
1000	.062	.081

Note: *df* equals N - 2.

Taken from: Spence, J.T. et al (1976) *Elementary Statistics*, Prentice-Hall, Inc., Englewood Cliffs, New Jersey.

POLYGENIC INHERITANCE

In the preceding section, we saw that quantitative traits tend to show a continuum of variation, which can be analyzed with various statistical tools. At the beginning of the 1900s, there was great debate concerning the inheritance of quantitative traits. The biometric school, founded by Galton and Pearson, argued that these types of traits are not controlled by single discrete genes that affect phenotypes in a predictable way. To some extent, the biometric school favored a blending theory of inheritance, which had been proposed earlier (see Chapter 2, p. 18). Alternatively, the followers of Mendel, led by William Bateson in England and William Castle in the United States, held firmly to the idea that traits are governed by genes, which are inherited as discrete units. As we know now, Bateson and Castle were correct. However, as we will see in this section, the difficulty of studying quantitative traits lies in the fact that these traits are influenced by multiple genes and substantial environmental factors.

Most quantitative traits are polygenic and exhibit a continuum of phenotypic variation. The term **polygenic inheritance** refers to the transmission of traits that are governed by two or more genes. The locations on chromosomes where these genes reside are called **quantitative trait loci (QTLs)**.

Just a few years ago, it was extremely difficult for geneticists to determine the inheritance patterns for genes underlying polygenic traits, particularly those determined by three or more genes having multiple alleles for each gene. Recently, however, molecular genetic tools (described in Chapters 19 and 20) have enhanced greatly our ability to find regions in the genome where QTLs are likely to reside. This has been a particularly exciting advance in the field of quantitative genetics. In some cases, the identification of QTLs may allow the improvement of quantitative traits in agriculturally important species.

Polygenic inheritance creates overlaps between genotypes and phenotypes

The first experiment demonstrating that continuous variation is related to polygenic inheritance was conducted by the Swedish geneticist Herman Nilsson-Ehle in 1909. He studied the inheritance of red pigment in the hull of wheat, *Triticum aestivum* (Figure 25-3a). When true-breeding plants with white hulls were crossed to a variety with red hulls, the F_1 generation had an intermediate color. When the F_1 generation was allowed to self-fertilize, there was a great variation in redness in the F_2 generation, ranging from white, light pink, and pink to intermediate red, medium red, basic red, and dark red. An unsuspecting observer might conclude that this F_2 generation displayed a continuous variation in hull color. However, as shown in Figure 25-3b, Nilsson-Ehle carefully categorized the colors of the hulls and discovered that they fell into a $1:6:15:20:15:6:1$ ratio. He concluded that this species is diploid for three different genes that control hull color, each gene existing in a red or white allelic form. He hypothesized that these three loci must contribute additively to the color of the hull. The results of Nilsson-Ehle make perfect sense, because we now know that strains of *T. aestivum* are actually hexaploid and therefore have six copies of each gene.

Nilsson-Ehle categorized wheat hull colors into several discrete genotypic categories. However, for many polygenically inherited quantitative traits, this is difficult or impossible. In general, as the number of genes controlling a trait increases, and the influence of the environment increases, the categorization of phenotypes into discrete genotypic classes becomes increasingly difficult if not impossible.

Figure 25-4 illustrates how genotypes and phenotypes may overlap for polygenic traits. In this example, the environment (sunlight, soil conditions, and so forth) may affect the phenotypic outcome of a trait in plants (namely, seed weight). Part (a) considers a situation where seed weight is controlled by one gene existing in light (*w*) and heavy (*W*) alleles. A heterozygous plant (*Ww*) is allowed

(a) Red and white hulls of wheat

	R1 R2 R3	R1 R2 r3	R1 r2 r3	R1 r2 R3	r1 R2 R3	r1 r2 R3	r1 R2 r3	r1 r2 r3
R1 R2 R3	R1R1 R2R2 R3R3 — Dark red	R1R1 R2R2 R3r3 — Basic red	R1R1 r2R2 r3R3 — Medium red	R1R1 r2R2 R3R3 — Basic red	r1R1 R2R2 R3R3 — Basic red	r1R1 r2R2 R3R3 — Medium red	r1R1 R2R2 r3R3 — Medium red	r1R1 r2R2 r3R3 — Intermediate red
R1 R2 r3	R1R1 R2R2 R3r3 — Basic red	R1R1 R2R2 r3r3 — Medium red	R1R1 r2R2 r3r3 — Intermediate red	R1R1 r2R2 R3r3 — Medium red	r1R1 R2R2 R3r3 — Medium red	r1R1 r2R2 R3r3 — Intermediate red	r1R1 R2R2 r3r3 — Intermediate red	r1R1 r2R2 r3r3 — Pink
R1 r2 r3	R1R1 R2r2 R3r3 — Medium red	R1R1 R2r2 r3r3 — Intermediate red	R1R1 r2r2 r3r3 — Pink	R1R1 r2r2 R3r3 — Intermediate red	r1R1 R2r2 R3r3 — Intermediate red	r1R1 r2r2 R3r3 — Pink	r1R1 R2r2 r3r3 — Pink	r1R1 r2r2 r3r3 — Light pink
R1 r2 R3	R1R1 R2r2 R3R3 — Basic red	R1R1 R2r2 r3R3 — Medium red	R1R1 r2r2 r3R3 — Intermediate red	R1R1 r2r2 R3R3 — Medium red	r1R1 R2r2 R3R3 — Medium red	r1R1 r2r2 R3R3 — Intermediate red	r1R1 R2r2 r3R3 — Intermediate red	r1R1 r2r2 r3R3 — Pink
r1 R2 R3	R1r1 R2R2 R3R3 — Basic red	R1r1 R2R2 r3R3 — Medium red	R1r1 r2R2 r3R3 — Intermediate red	R1r1 r2R2 R3R3 — Medium red	r1r1 R2R2 R3R3 — Medium red	r1r1 r2R2 R3R3 — Intermediate red	r1r1 R2R2 r3R3 — Intermediate red	r1r1 r2R2 r3R3 — Pink
r1 r2 R3	R1r1 R2r2 R3R3 — Medium red	R1r1 R2r2 r3R3 — Intermediate red	R1r1 r2r2 r3R3 — Pink	R1r1 r2r2 R3R3 — Intermediate red	r1r1 R2r2 R3R3 — Intermediate red	r1r1 r2r2 R3R3 — Pink	r1r1 R2r2 r3R3 — Pink	r1r1 r2r2 r3R3 — Light pink
r1 R2 r3	R1r1 R2R2 R3r3 — Medium red	R1r1 R2R2 r3r3 — Intermediate red	R1r1 r2R2 r3r3 — Pink	R1r1 r2R2 R3r3 — Intermediate red	r1r1 R2R2 R3r3 — Intermediate red	r1r1 r2R2 R3r3 — Pink	r1r1 R2R2 r3r3 — Pink	r1r1 r2R2 r3r3 — Light pink
r1 r2 r3	R1r1 R2r2 R3r3 — Intermediate red	R1r1 R2r2 r3r3 — Pink	R1r1 r2r2 r3r3 — Light pink	R1r1 r2r2 R3r3 — Pink	r1r1 R2r2 R3r3 — Pink	r1r1 r2r2 R3r3 — Light pink	r1r1 R2r2 r3r3 — Light pink	r1r1 r2r2 r3r3 — White

(b) R1r1 R2r2 R3r3 × R1r1 R2r2 R3r3

FIGURE **25-3**

The Nilsson-Ehle experiment studying how continuous variation is related to polygenic inheritance in wheat. (a) Red and white pigments in the hulls of wheat, *Triticum aestivum*. **(b)** Nilsson-Ehle carefully categorized the colors of the hulls in the F$_2$ generation and discovered that they fell into a 1 : 6 : 15 : 20 : 15 : 6 : 1 ratio.

GENES ⟶ TRAITS: In this example, there are three genes, with two alleles each (red and white), that govern hull color. Offspring can display a range of colors, depending on how many copies of the red allele they inherit. If an offspring is homozygous for the red allele of all three genes, it will have very dark red hulls. By comparison, if it carries five red alleles and one white allele, it will be basic red (which is not quite as deep in color). In this way, this polygenic trait can exhibit a range of phenotypes from very deep red to completely white.

to self-hybridize. When the weight is only slightly influenced by variation in the environment, as seen on the left, the heavy, light, and intermediate seeds fall into separate, well-defined categories. When the environmental variation has a greater impact on seed weight, as shown on the right, there is more phenotypic variation in seed weight within each genotypic class. Even so, it is still possible to classify most individuals into the three main categories.

Low environmental effect

High environmental effect

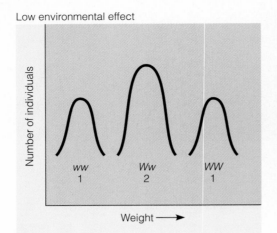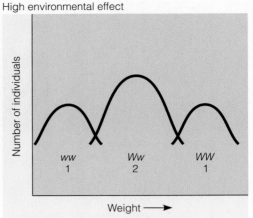

(a) *Ww* × *Ww*

FIGURE **25-4**

How genotypes and phenotypes may overlap for polygenic traits. (a) Situations in which seed weight is controlled by one gene, existing in light (*w*) and heavy (*W*) alleles. **(b)** Situations in which seed weight is governed by three genes instead of one, each existing in light and heavy alleles.

GENES ⟶ TRAITS: The ability of geneticists to correlate genotype and phenotype depends on how many genes are involved and how much the environment causes the phenotype to vary. In part (a), a single gene influences weight. To the left, the environment does not cause much variation in weight. This makes it easy to distinguish the three genotypes. There is no overlap in the weights of *ww*, *Ww*, and *WW* individuals. To the right of part (a), the environment causes more variation in weight. In this case, a few individuals with *ww* genotypes will have the same weight as a few individuals with *Ww* genotypes; and a few *Ww* genotypes will have the same weight as *WW* genotypes. As shown in part (b), it becomes even more difficult to distinguish genotype based on phenotype when three genes are involved. The overlaps are minor when the environment does not cause much weight variation. However, when the environment causes substantial phenotypic variation, the overlaps between genotypes and phenotypes are very pronounced and greatly confound genetic analysis.

By comparison, Figure 25-4b illustrates a situation where seed weight is governed by three genes instead of one, each existing in light and heavy alleles. When the environmental variation is low and/or plays a minor role in the outcome of this trait, the expected 1 : 6 : 15 : 20 : 15 : 6 : 1 ratio is observed. As shown in the upper illustration in part (b), nearly all of the individuals fall within a phenotypic category that corresponds to their genotypes. When the environment has a more significant effect on phenotype, as shown in the lower illustration, the situation becomes more ambiguous. For example, individuals with five *W* alleles and one *w* allele have phenotypes that overlap with individuals having six *W* alleles or four *W* alleles and two *w* alleles. Therefore, it becomes difficult to categorize each phenotype into a unique genotypic class. Instead, the trait displays a continuum ranging from light to heavy seed weight.

When the overlap between phenotypic classes is so great that separating them is not possible, another way to identify the number of genes affecting the polygenic inheritance of a quantitative trait is to look for linkage between these genes and genes affecting discontinuous traits. While this type of approach can be conducted on any organism, it was studied earlier in organisms, such as *Drosophila melanogaster*, in which many alleles affecting discontinuous traits had been identified and mapped to particular chromosomes.

Low environmental effect

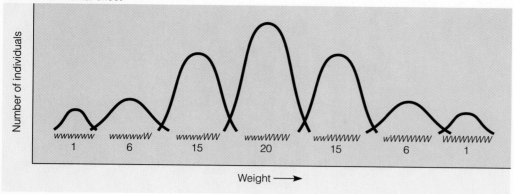

High environmental effect

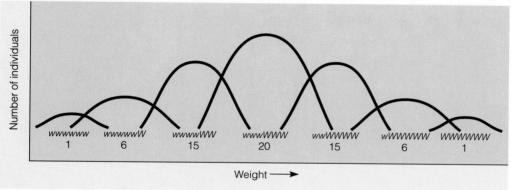

(b) *WWWwww × WWWwww*

Figure 25-5 shows the results of a 1957 study, by James Crow and his colleagues at the University of Wisconsin, Madison, that applied this general strategy to identify the loci of polygenic inheritance in the fruit fly. Crow's goal in this experiment was to determine the genetic basis for insecticide resistance in *Drosophila melanogaster* Many mutations were already known in this species, and these were used as markers for each of its chromosomes. Their strategy in identifying QTLs was to cross two different strains that had distinct genetic markers on chromosomes 2, 3, and the X-chromosome, and also differed in the quantitative trait of DDT resistance. This cross produced an F_1 generation that was heterozygous for the genetic markers on each chromosome and had an intermediate level of resistance. The next step was to backcross the F_1 offspring to the parental strains. This backcross produced a population of offspring that differed in their combinations of parental chromosomes. By analyzing the phenotypes of the offspring with regard to known chromosomal markers, the researchers could determine whether the chromosomes were inherited from the DDT-resistant or DDT-sensitive strain. These same offspring were analyzed with regard to their ability to survive in the presence of DDT. The results of this analysis are shown in Figure 25-5. As a matter of chance, some offspring contained all of the chromosomes from one original parental strain or the other, but most offspring contained a few chromosomes from one parental strain and the rest from the other.

The data in Figure 25-5 show that each copy of the X-chromosome, and chromosomes 2 and 3, confer a significant amount of insecticide resistance. When a fly contains all of its X-chromosomes, and chromosomes 2 and 3 from the insecticide-resistant parental strain, it has maximal insecticide resistance. Even when only a single chromosome is derived from the insecticide-sensitive strain, the resistance to

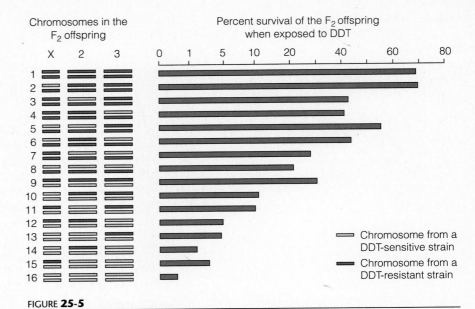

FIGURE **25-5**

The location of DDT resistance genes on the X-chromosome and chromosomes 2 and 3 in *Drosophila melanogaster*. A parental strain that had a high level of DDT resistance was crossed to a sensitive strain, and then the F_1 offspring were backcrossed to the parental strains. The population of F_2 offspring inherited differing combinations of chromosomes from the sensitive and resistant strains. The F_2 offspring were exposed to DDT, and the percentage of survivors was recorded.

GENES ⟶ TRAITS: When F_2 offspring carried one or more chromosomes from the DDT-resistant strain, they were more resistant to DDT than the original sensitive strain. Furthermore, the level of resistance generally correlated with the number of chromosomes from the DDT-resistant strain. Offspring that inherited all of their chromosomes from the sensitive strain were most sensitive to DDT, while those receiving all of their chromosomes from the DDT-resistant strain were the most resistant. Offspring carrying one chromosome from the DDT-resistant strain were slightly more resistant than the sensitive strain. And so on. These results indicate that DDT resistance is due to the cumulative effects of several genes located on the X-chromosome and on chromosomes 2 and 3.

DDT is less than maximal. These results are consistent with the hypothesis that insecticide resistance is a polygenic trait involving genes that reside on the X-chromosome and on chromosomes 2 and 3.

Quantitative trait loci (QTLs) can now be mapped by RFLP analysis

In the past few years, restriction fragment length polymorphisms (RFLPs; see Chapter 20) and other molecular markers have been identified in the genomes of many different organisms. These markers have been used to construct genetic maps of several species' genomes. Once a genome map is obtained, it becomes much easier to analyze the genetic contributions of complex traits. In addition to model organisms such as *Drosophila*, *Arabidopsis*, *Caenorhabditis elegans*, and mice, detailed molecular maps have been obtained for species of agricultural importance. These include crops such as corn, rice, and tomatoes, as well as livestock such as cattle, pigs, and sheep.

When a genetic map is available, geneticists can determine the location of a gene within a genome using RFLP mapping. The characteristics of RFLPs and the general methods of RFLP mapping were described in Chapter 20 (see pp. 549–554). In 1989, Eric Lander and David Botstein extended this technique to identify QTLs that govern a quantitative trait. The basis of QTL detection is the association between genetically determined phenotypes (e.g., quantitative traits) and molecular markers such as RFLPs. For example, in a species of grain, high yield might be associated with five RFLPs that map to different regions of the genome. These results would indicate that phenotypic variation in yield is influenced by five different loci.

The general strategy for RFLP mapping begins by mating organisms that are very dissimilar in their RFLP markers and also different for quantitative traits. For example, in 1988, Andrew Paterson at Cornell University and his colleagues examined quantitative trait inheritance in the tomato. They studied a domestic strain of tomato and a South American green-fruited variety. These two strains differed in their RFLPs, and they also exhibited dramatic differences in three agriculturally important characteristics: fruit mass, soluble solids content (in the fruit), and fruit pH. The researchers crossed the two strains together, and then they backcrossed the offspring to the domestic tomato. A total of 237 plants were then examined with regard to 70 known RFLP markers. In addition, between 5 and 20 tomatoes from each plant were analyzed with regard to fruit mass, soluble solids content, and fruit pH. Using this approach, they were able to map the genes contributing variation in these traits to particular intervals along the tomato chromosomes. They identified six loci causing variation in fruit mass, four affecting soluble solids content, and five with effects on fruit pH.

This method of analyzing quantitative traits via RFLP mapping has provided an important tool in the analysis and manipulation of quantitative traits. In addition, other types of genetic markers such as sequence-tagged sites (STSs; see Chapter 20, pp. 565–566), which are identified by PCR methods, can be used in mapping studies of QTLs. These types of studies are advancing our understanding of how quantitative traits are influenced by genes. In some cases, a quantitative trait may be influenced by many genes, with only a few having major importance in determining the phenotypic outcome. Once the principal genes have been identified, researchers or breeders can determine if two or more different genes act additively to affect a quantitative trait. Also, the mapping of QTLs can allow geneticists to identify alleles within the same gene and determine how they affect the trait in a homozygous or heterozygous state. In the future, it is clear that RFLP mapping will enhance greatly our basic understanding of quantitative traits and may provide exciting applications in the field of agriculture.

HERITABILITY

As we have just seen, recent approaches in molecular mapping have enabled researchers to identify the genes that contribute to a quantitative trait. The other key factor that affects the phenotypic outcomes of quantitative traits is the environment. All traits of biological organisms are influenced by genetics and the environment, and this is particularly pertinent in the study of quantitative traits.

A geneticist, however, can never actually determine the relative amount of a quantitative trait that is controlled by genetics, or the amount that is governed by the environment. When you think about it, a researcher cannot raise an organism without an environment and discover the amount of a trait that is governed by genetics. Likewise, it is impossible to rear an organism without its genes and conclude how much the environment contributes to the outcome of a trait. Instead, the focus of our analysis of quantitative traits is to explain how variation, both genetic and environmental, will affect the phenotypic results.

The term **heritability** refers to the amount of phenotypic variation within a group of individuals that is due to genetic factors. If all of the phenotypic variation in a group were due to genetic variation, the heritability would have a value of 1. If all the variation were due to environmental effects, the heritability would equal 0. For most groups of organisms, the heritability for a given trait lies between these two extremes. For example, both genes and diet affect the size that an individual will attain. Some individuals will inherit genes that tend to make them large, and a proper diet will also promote larger size. Other individuals will inherit genes that make them small, and an inadequate diet may contribute further to small size. Taken together, both genetics and the environment influence the phenotypic results.

In the study of quantitative traits, a primary goal is to determine how much of the phenotypic variation arises from genetic factors, and how much comes from environmental factors. In this section, we will examine how geneticists analyze the genetic and environmental components that affect quantitative traits. As we will see, this approach has been applied successfully in agricultural breeding strategies to produce domesticated species with desirable characteristics.

Phenotypic variance is due to the additive effects of genetic variance and environmental variance

Earlier in this chapter, we examined the amount of phenotypic variation within a group by calculating the variance. In studying quantitative trait variation, the first step is to partition this variation into components that are attributable to different causes. If we assume that genetic and environmental factors are the only two components that determine a trait, and if genetic and environmental factors are independent of each other, then the total variance for a trait in a group of individuals is

$$V_T = V_G + V_E$$

where V_T is the total variance. It reflects the amount of variation that is measured at the phenotypic level.

V_G is the relative amount of variance due to genetic factors.

V_E is the relative amount of variance due to environmental factors.

The partitioning of variance into genetic and environmental components allows us to estimate their relative importance in influencing the variation within a group. If V_G is very high and V_E is very low, genetics plays a greater role in promoting variation within a group. Alternatively, if V_G is low and V_E is high, environmental causes underlie much of the phenotypic variation. As will be described later in this chapter, a livestock breeder might want to apply selective breeding if V_G for an important (quantitative) trait is high. In this way, the characteristics of the herd may be improved. Alternatively, if V_G is negligible, it would make more sense to investigate (and manipulate) the environmental causes of phenotypic variation.

With experimental animals, one possible way to determine V_G and V_E is by comparing the variation in traits between genetically identical and genetically disparate groups. For example, researchers have developed genetically homogeneous strains of mice. After many generations of brother–sister matings, these strains have become monomorphic for all of their genes. Within such a strain of mice, V_G equals zero. Therefore, all phenotypic variation is due to V_E. When studying quantitative traits such as weight, an experimenter might want to know the genetic and environmental variance for a different, genetically heterogeneous group of mice. To do so, the genetically homogeneous and heterogeneous mice could be raised under the same environmental conditions, and their weights measured. The phenotypic variance for weight could then be calculated as described earlier. Let's suppose we obtained the following results:

$V_T = 0.30$ sq oz   for the group of genetically homogeneous mice

$V_T = 0.52$ sq oz   for the group of genetically heterogeneous mice

In the case of the homogeneous mice, $V_T = V_E$, because V_G equals zero. Therefore, V_E equals 0.30 sq oz. To estimate V_G for the heterogeneous group of mice, we assume that V_E (i.e., the environmentally produced variance) is the same for them as it is for the homogeneous mice, since the two groups were raised in identical environments. This assumption allows us to calculate the genetic variance for the heterogeneous mice:

$$V_T = V_G + V_E$$
$$0.52 = V_G + 0.30$$
$$V_G = 0.22 \ \text{sq oz}$$

This result tells us that some of the phenotypic variance in the genetically heterogeneous group is due to the environment (namely, 0.30 sq oz) and some (0.22 sq oz) is due to genetic variation in alleles that affect the weight.

Heritability is the relative amount of phenotypic variation that is due to genetic factors

Another way to view variance is to focus our attention on the genetic contribution to phenotypic variation. Heritability is the proportion of the phenotypic variance that is attributable to genetic factors. If we assume again that environment and genetics are the only two components, then

$$H_B^2 = V_G / V_T$$

where H_B^2 is the heritability in the broad sense

$\quad V_G$ is the variance due to genetics

$\quad V_T$ is the total phenotypic variance

The heritability defined here, H_B^2, is called the **true heritability in the broad sense**. It takes into account all genetic factors that may affect the phenotype.

As we have seen throughout this text, genes can affect phenotypes in various ways. As described earlier in this chapter, the Nilsson-Ehle experiment showed that the alleles determining hull color in wheat affect the phenotype in an additive way. Alternatively, alleles affecting other traits may show a dominant–recessive relationship. In this case, the alleles are not strictly additive, because the heterozygote has a phenotype closer to, or perhaps the same as, the homozygote containing two copies of the dominant allele. In addition, another complicating factor is epistasis (which was described in Chapter 4). The alleles for one gene may influence the phenotypic expression of the alleles of another gene. To account for these differences, geneticists usually subdivide V_G into these three different genetic factors:

$$V_G = V_A + V_D + V_I$$

where V_A is the variance due to additive alleles

$\quad V_D$ is the variance due to alleles that follow a dominant/recessive pattern of inheritance

$\quad V_I$ is the variance due to genes that interact in an epistatic manner

In analyzing quantitative traits, geneticists often focus on V_A and neglect the contributions of V_D and V_I. This is done for scientific as well as practical reasons. For many quantitative traits, the additive effects of alleles often dominate the phenotypic outcome in genetic crosses. In addition, when the alleles behave additively, we can predict the outcomes of crosses based on the quantitative characteristics of the parents. The heritability of a trait due to the additive effects of alleles is called the **narrow sense heritability**:

$$h_N^2 = V_A / V_T$$

The narrow sense heritability (h_N^2) may be an inaccurate measure of the true heritability if V_D and V_I are not small values. However, for many quantitative traits, geneticists have found that the value of V_A is very large compared with V_D and V_I. In such cases, the true heritability in the broad sense and the narrow sense heritability are similar to each other.

There are several ways to estimate narrow sense heritability. A common strategy is to measure a quantitative trait among groups of genetically related individuals. For example, agriculturally important traits, such as egg weight in poultry, can be analyzed in this way. To calculate the heritability, one would determine the observed egg weights between individuals whose genetic relationships are known, such as a mother and her female offspring. These data could then be used to compute a correlation between the parent and offspring using the methods described earlier in this chapter. The narrow sense heritability is then calculated as

$$h_N^2 = r_{obs}/r_{exp}$$

where r_{obs} is the observed phenotypic correlation between related individuals

r_{exp} is the expected correlation based on the known genetic relationship

In our example, r_{obs} is the observed phenotypic correlation between parent and offspring. In research studies, the observed phenotypic correlation for egg weights between mothers and daughters has been found to be about 0.25 (although this will vary among strains). The expected correlation, r_{exp}, is based on the known genetic relationship. A parent and child share 50% of their genetic material, so that r_{exp} equals 0.50. So,

$$h_N^2 = r_{obs}/r_{exp}$$
$$= 0.25/0.50 = 0.50$$

Note: For siblings, $r_{exp} = 0.50$; for identical twins, $r_{exp} = 1.0$; and for an uncle–niece relationship, $r_{exp} = 0.25$.

According to this calculation, about 50% of the phenotypic variation in egg weight is due to genetic factors; the other half is due to the environment.

When calculating heritabilities from correlation coefficients, keep in mind that this computation assumes that genetics and the environment are independent variables. This is not always the case. The environments of parents and offspring are often more similar to each other than they are to those of unrelated individuals. There are several ways to avoid this confounding factor. First, in human studies, one may analyze the heritabilities from correlations between adopted children and their biological parents. Alternatively, one can examine a variety of relationships (uncle–niece, identical twins versus fraternal twins, etc.) and see if the heritability values are roughly the same in all cases. This approach was applied in the study to be described next.

EXPERIMENT 25A

Holt found that the heritability of dermal ridge count in human fingerprints is very high

Fingerprints are inherited as a quantitative trait. It has been long known that identical twins have fingerprints that are very similar, whereas fraternal twins show considerably less agreement. Galton was the first researcher to study fingerprint patterns, but this trait became more amenable to genetic studies when Kristine Bonnevie at the University of Kristiania in Norway developed a method in 1924 for counting the number of ridges within a human fingerprint.

As shown in Figure 25-6, human fingerprints can be categorized as having an arch, a loop, or a whorl (or a combination of these patterns). The primary difference among these patterns is the number of triple junctions, each known as a *triradius* (Figure 25-6b and c). At a triradius, a ridge emanates in three directions. An arch has zero triradii, a loop has one, and a whorl has two. In Bonnevie's method of counting, a line is drawn from a triradius to the center of the fingerprint. The ridges that touch this line are then counted. (Note: The triradius ridge itself is not counted, and the last ridge is not counted if it forms the center of the fingerprint.) With this

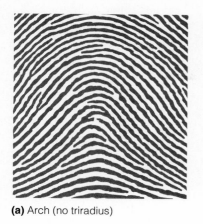

(a) Arch (no triradius)

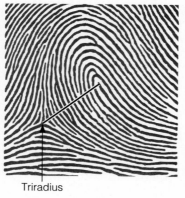

Triradius

(b) Loop (one triradius)

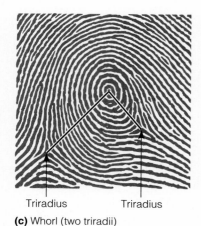

Triradius   Triradius

(c) Whorl (two triradii)

FIGURE **25-6**

Human fingerprints and the ridge count method of Bonnevie. (a) This print has an arch rather than a triradius. The ridge count is zero. **(b)** This print has one triradius. A straight line is drawn from the triradius to the center of the print. The number of ridges dissecting this straight line is 13. **(c)** This print has two triradii. Straight lines are drawn from both triradii to the center. There are 15 ridges touching the left line and 6 touching the right line, giving a total ridge count of 21.

method, one can obtain a ridge count for all ten fingers. Bonnevie conducted a study on a small population and found that ridge count correlations were relatively high in genetically related individuals.

In the experiment described here, published in 1961, Sarah Holt at the University College of London, who was also interested in the inheritance of this quantitative trait, carried out a more exhaustive study of ridge counts in a British population. In groups of 825 males or 825 females, the ridge count varied from 0 to 300, with mean values of approximately 145 for males (standard deviation equaled 51.1) and 127 for females (standard deviation 52.5):

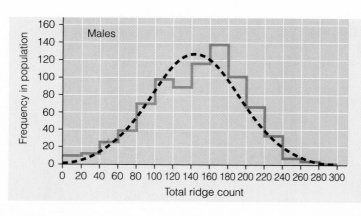

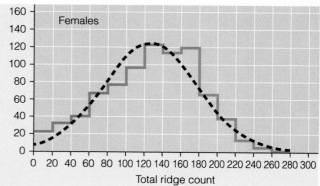

Dermal ridge count has a genetic component. The goal of this experiment is to determine the contribution of genetics in the variation of dermal ridge counts.

Starting material: A group of human subjects from Great Britain.

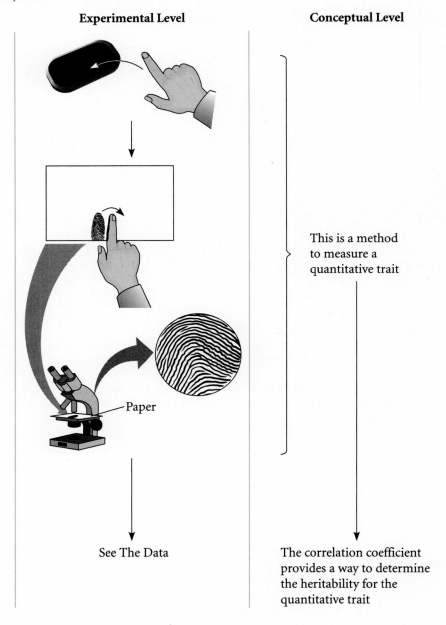

Experimental Level   **Conceptual Level**

1. Take a person's finger and blot it onto an ink pad.

2. Roll the person's finger onto a recording surface to obtain a print.

This is a method to measure a quantitative trait

3. With a low-power binocular microscope, count the number of ridges using the Bonnevie method described previously.

Paper

4. Calculate the correlation coefficients between different pairs of individuals as described earlier in this chapter.

See The Data

The correlation coefficient provides a way to determine the heritability for the quantitative trait

Type of Relationship	Number of Pairs Examined	Correlation Coefficient	Heritability r_{obs}/r_{exp}
Parent–child	810	$0.48 \pm 0.04^{*}$	0.96
Parent–parent	200	0.05 ± 0.07	—[†]
Sibling–sibling	642	0.50 ± 0.04	1.00
Identical twins	80	0.95 ± 0.01	0.95
Fraternal twins	92	0.49 ± 0.08	0.98
			0.97 average heritability

[*] $\pm$ Standard error of the mean.

[†]Note: We cannot calculate a heritability value, because the value for r_{exp} is not known. Nevertheless, the value for r_{obs} is very low, suggesting that there is a negligible correlation between unrelated individuals.

Adapted from: Holt, S. B. (1961) Quantitative Genetics of Finger-Print Patterns. *British Med. Bull. 17*, 247–250.

When we look at the data shown here, it becomes apparent that genetics plays the major role in explaining the variation in this trait. Genetically unrelated individuals (namely, parent-parent relationships) have a negligible correlation for this trait. By comparison, individuals who are related genetically have a substantially higher correlation. When we divide the observed correlation coefficient by the expected correlation coefficient based on the known genetic relationships, the average heritability value is 0.97, which is very close to 1.0. These values indicate that nearly all of the variation in fingerprint pattern is genetic variation. Significantly, fraternal and identical twins have substantially different correlation coefficients, even though we expect that they have been raised in very similar environments.

These results support the idea that genetics is playing the major role in promoting variation and that the results are not biased heavily by environmental similarities that may be associated with genetically related individuals. From an experimental viewpoint, the results show us how the determination of correlation coefficients between related and unrelated individuals can provide insight regarding the relative contributions of genetics and environment to the variation of a quantitative trait.

Heritability values are only relevant to particular groups raised in a particular environment

Table 25-3 describes some heritability values that have been calculated for particular populations. Unfortunately, heritability is a widely misunderstood concept. Heritability describes the amount of phenotypic variation due to genetic factors *for a particular population raised in a particular environment.* The words "variation," "particular population," and "particular environment" cannot be overemphasized. For example, in one population of cattle the heritability for milk production may be 0.35, while in another group (with less genetic variation) the heritability may be 0.1. Second, if a group displays a heritability of 1.0 for a particular trait, this does not mean that the environment is unimportant in affecting the outcome of the trait. A heritability value of 1.0 only means that the amount of variation within this group is due to genetics. Perhaps, the group has been raised in a relatively homogeneous environment, so that the environment has not caused a significant amount of variation. Nevertheless, the environment may be quite important. It just is not causing much variation within this particular group.

TABLE 25–3 Examples of Heritabilities for Quantitative Traits

Trait	Heritability Value[a]
Humans	
Stature	0.65
Cattle	
Body weight	0.65
Butterfat, %	0.40
Milk yield	0.35
Mice	
Tail length	0.40
Body weight	0.35
Litter size	0.20
Poultry	
Body weight	0.55
Egg weight	0.50
Egg production	0.10

[a]As emphasized in the text, these values apply to particular groups raised in particular environments.

Taken from: Falconer, D. S. (1989) *Introduction to Quantitative Genetics*, 3rd edition. Longman, Essex, England.

Saint Bernard

German shepard

Bulldog

Chihuahua

As a hypothetical example, let's suppose that we take a species of rodent and raise a group on a poor diet; we find their weights range from 1.5 to 2.5 pounds, with a mean weight of 2 pounds. We allow them to mate, and then raise their off-spring on a healthy diet of rodent chow. The weights of the offspring range from 2.5 to 3.5 pounds, with a mean weight of 3 pounds.

In this hypothetical experiment, we might find a positive correlation in which the small parents tended to produce small offspring, and the large parents, large off-spring. The correlation of weights between parent and offspring might be, say, 0.5. In this case, the heritability for weight would be calculated as r_{obs}/r_{exp}, which equals 0.5/0.5, or 1.0. The value of 1.0 means that all of the variation within the groups is due to genetics. The offspring vary from 2.5 to 3.5 pounds because of genetic vari-ation, and also the parents range from 1.5 to 2.5 because of genetics. However, as we see here, environment has played an important role. Presumably, the mean weight of the offspring is higher because of their better diet.

This example is meant to emphasize the point that heritability only tells us the rel-ative contributions of genetics and environment in influencing phenotypic *variation* in a *specific group* under a *specific set of conditions*. Heritability may not indicate the relative importance of these two factors in determining the outcomes of traits.

Selective breeding of species can alter quantitative traits dramatically

The term **selective breeding** refers to programs and procedures designed to mod-ify phenotypes in species of economically important plants and animals. This phe-nomenon, also called **artificial selection**, is related to natural selection, which will be discussed in Chapter 26. In fact, in forming his theory of natural selection, Charles Darwin was influenced by his observations of selective breeding by pigeon fanciers and other breeders. The primary difference between artificial and natural selection is how the parents are chosen. Natural selection is due to natural varia-tion in reproductive success; in artificial selection, the breeder chooses individuals who possess traits that are desirable from a human perspective.

For centuries, humans have been practicing selective breeding to obtain domes-tic species with interesting or agriculturally useful characteristics. The common breeds of dogs and cats have been obtained by selective breeding strategies (Figure 25-7). As shown here, it is very striking how selective breeding can modify the quantitative traits in a species. When comparing a Chihuahua with a Saint Bernard, the magnitude of the differences is fairly amazing. They hardly look like members of the same species.

Likewise, most of the food we eat is obtained from species that have been mod-ified profoundly by selective breeding strategies. This includes products such as grains, fruit, vegetables, meat, milk, and juices. Figure 25-8 illustrates how certain characteristics in the wild mustard plant have been modified by selective breeding to create several species of domesticated crops. As seen here, certain quantitative traits in the domestic strains differ considerably from those of the original wild species.

FIGURE **25-7**

Some common breeds of dogs that have been obtained by selective breeding.

GENES ⟶ TRAITS: By selecting parents carrying the alleles that influence cer-tain quantitative traits in a desired way, dog breeders have produced breeds with distinctive sets of traits. For example, the bulldog has alleles that give it short legs and a flat face. By comparison, the corresponding genes in a German shepherd are found in alleles that produce longer legs and a more pointy snout. All of the dogs shown in this figure carry the same kinds of genes (e.g., many genes that affect their sizes, shapes, and fur color). However, the alleles for many of these genes are different among these dogs, thereby producing breeds with strikingly different phenotypes.

FIGURE **25-8**

Crop plants developed by selective breeding of the wild mustard plant.

GENES ⟶ TRAITS: The wild mustard plant carries a large amount of genetic (i.e., allelic) variation, which was used by plant breeders to produce modern strains that are agriculturally desirable. For example, by selecting for alleles that promote the formation of large lateral buds, the strain of Brussels sprouts was created. By selecting for alleles that alter the leaf morphology, kale was developed. Although these six agricultural plants look quite different from each other, they carry many of the same alleles as the wild mustard. However, they differ in alleles that affect the formation of flowers, buds, stems, and leaves.

The concept that underlies selective breeding is variation. Within a group of individuals, there may be allelic variation that affects the outcomes of quantitative traits. The fundamental strategy of the selective breeder is to choose parents who will pass on advantageous alleles to their offspring. In general, this means that the breeder will choose parents with desirable phenotypic characteristics. For example, if a breeder wants large cattle, he/she will choose the largest members of the herd as parents for the next generation. Presumably, these large cattle will transmit alleles to their offspring that confer large size. The breeder will often choose genetically related individuals (e.g., brothers and sisters) as the parental stock. The practice of mating between genetically related individuals is known as **inbreeding**. Some of the consequences of inbreeding will be described in Chapter 26.

Figure 25-9 shows a common outcome when selective breeding is conducted for a quantitative trait. Figure 25-9a shows the results of a program begun at the Illinois Experiment Station in 1896, even before the rediscovery of Mendel's laws. This experiment began with 163 ears of corn with an oil content ranging from 4 to 6%. In each of 80 succeeding generations, corn plants were divided into two separate groups. In one group, members with the highest oil content were chosen as parents

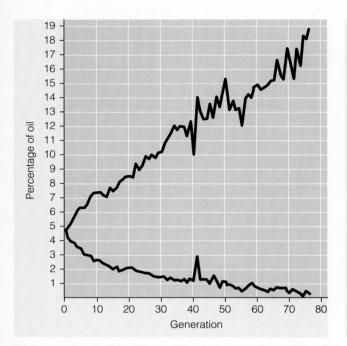

(a) Results of selective breeding for and against oil content in corn

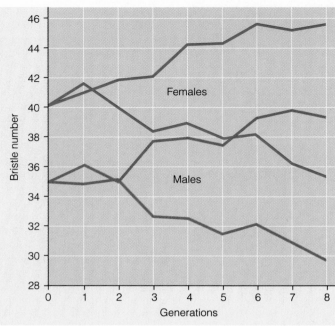

(b) Results of selective breeding for low and high bristle number in flies

FIGURE **25-9**

Common results of selective breeding for a quantitative trait. (a) Selection for high and low oil content in corn. **(b)** Selection for high and low bristle number in flies.

of the next generation. In the other group, members with the lowest oil content were chosen. After many generations, the oil content in the first group rose to over 18%; in the other group, it dropped to less than 1%. These results show that selective breeding can modify quantitative traits in a very directed manner.

Similar results have been obtained for many other quantitative traits. Figure 25-9b shows an experiment in 1941, by K. Mather of the John Innes Horticultural Institution, in which flies were selected on the basis of their bristle number. The starting group had an average of 40 bristles for females and 35 bristles for males. After eight generations, the group selected for high bristle number had an average of 46 bristles for females and 40 for males, while the group selected for low bristle number had an average of 36 bristles for females and 30 for males.

When comparing the curves in Figure 25-9, some general observations can be made. Quantitative traits are often at an intermediate value in unselected populations. Therefore, artificial selection can be used to increase or decrease the magnitude of the trait. Oil content can go up or down, and bristle number can increase or decrease. Nevertheless, there is a limit to how far this can continue. Presumably, the starting population possesses a large of amount of genetic variation, which contributes to the diversity in phenotypes. In other words, at the beginning of these experiments, the heritability for the trait is fairly high. By carefully choosing the parents, each succeeding generation has a higher proportion of the desirable alleles. However, after many generations, the population will eventually become monomorphic for all or most of the desirable alleles that affect the trait of interest. At this point, additional selective breeding will have no effect. When this occurs, the heritability for the trait is near zero, because nearly all genetic variation has been eliminated from the population.

In artificial selection experiments, the response to selection is the most common way to estimate the narrow sense heritability in a starting population.

The heritability measured in this way is also called the **realized heritability**. It is calculated as

$$h_{\text{N}}^2 = \frac{R}{S}$$

where,

R is the response in the offspring

S is the selection differential in the parents

Here,

$$R = \overline{X_{\text{O}}} - \overline{X}$$
$$S = \overline{X_{\text{P}}} - \overline{X}$$

where,

$\overline{X}$ is the mean of the starting population

$\overline{X_{\text{O}}}$ is the mean of the offspring

$\overline{X_{\text{P}}}$ is the mean of the parents

So,

$$h_{\text{N}}^2 = \frac{\overline{X_{\text{O}}} - \overline{X}}{\overline{X_{\text{P}}} - \overline{X}}$$

As an example, let's suppose we began with a population of fruit flies in which the average bristle number for both genders was 37.5. The parents chosen from this population had an average bristle number of 40. The offspring of the next generation had a average bristle number of 38.7. With these values, the realized heritability is

$$h_{\text{N}}^2 = \frac{38.7 - 37.5}{40 - 37.5}$$
$$= \frac{1.2}{2.5} = 0.48$$

This result tells us that about 48% of the phenotypic variation is due to the additive effects of alleles.

An important aspect of narrow sense heritabilities is their ability to predict the outcome of selective breeding. Solved Problem 1 at the end of this chapter illustrates this idea.

Heterosis may be explained by dominance or overdominance

As we have just seen, selective breeding can be used to alter the phenotypes of domestic species in a highly directed way. An unfortunate consequence of inbreeding, however, is that it tends to decrease the overall genetic variation within a group and may unknowingly promote homozygosity for deleterious alleles. In agriculture, it is widely observed that when two different inbred strains are crossed to each other, the resulting offspring are more vigorous (e.g., larger, longer-lived) than either of the inbred parental strains. This phenomenon is called **heterosis** or **hybrid vigor**. In modern agricultural breeding practices, many strains of plants and animals are hybrids produced by crossing two different inbred lines. In fact, much of the success in agricultural breeding programs is founded in heterosis. In rice, for example, hybrid strains have a 15-20% yield advantage over the best conventional inbred varieties under similar cultivation conditions.

The genetic basis for heterosis has been debated for more than 80 years, and the controversy is still not resolved. There are two major hypotheses to explain heterosis:

A. The Dominance Hypothesis to Explain Heterosis

Inbred		*Inbred*
Strain 1		*Strain 2*
AAbb	×	*aaBB*

The recessive alleles (*a* and *b*) are slightly harmful in the homozygous condition.

↓

AaBb   The hybrid offspring is more vigorous, because the harmful effects of the recessive alleles are masked by the dominant alleles.

B. The Overdominance Hypothesis to Explain Heterosis

Inbred		*Inbred*
Strain 1		*Strain 2*
A−1A−1	×	*A−2A−2*

Neither the A−1 nor A−2 allele is recessive.

↓

A−1A−2   The hybrid offspring is more vigorous, because the heterozygous combination of alleles exhibits overdominance. This means that the *A−1A−2* heterozygote is more vigorous than either the *A−1A−1* or *A−2A−2* homozygote.

In 1908, the **Dominance Hypothesis** was proposed by Charles Davenport at Cold Spring Harbor. He suggested that highly inbred strains have become homozygous for one or more recessive genes that are somewhat deleterious (but not lethal). This undesirable homozygosity can happen randomly as result of inbreeding or genetic drift (see Chapter 26). Because the homozygosity occurs by chance, two different inbred strains are likely to be homozygously recessive for different genes. Therefore, when they are crossed to each other, the resulting hybrids are heterozygous and do not suffer the consequences of homozygosity for deleterious recessive alleles. In other words, the dominance of the beneficial alleles explains the observed heterosis.

In 1908, a second hypothesis, known as the **Overdominance Hypothesis**, was proposed by George Shull at Cold Spring Harbor and Edward East at the Connecticut Agricultural Experimental Station. As was described in Chapter 4, overdominance occurs when the heterozygote is more vigorous than either corresponding homozygote. According to this idea, heterosis occurs because the resulting hybrids are heterozygous for one or more genes that display overdominance. A classic case of overdominance involves the β-globin allele for sickle cell anemia. In areas where malaria is prevalent, the heterozygote has a higher survival rate than both the wild-type homozygote and the homozygote affected with sickle cell anemia (see Chapter 4).

As mentioned earlier in this chapter, the recent advent of RFLP linkage mapping has made it possible to identify the QTLs that underlie heterosis. In corn, Charles Stuber at the U.S. Department of Agriculture in Raleigh, North Carolina, and his colleagues have found that heterozygotes of most QTLs for grain yield had higher production than the corresponding homozygotes. These results support the Overdominance Hypothesis. The researchers noted, however, that overdominance is very difficult to distinguish from **pseudo-overdominance**, a phenomenon initially suggested by James Crow in 1952. Pseudo-overdominance is really the same as dominance, except that the two (or more) genes involved are very closely linked. In our last example, *a* and *B* may be closely linked in one strain while *A* and *b* are closely linked in another strain. The hybrid is really heterozygous (*AaBb*) for two different genes, but this may be difficult to discern in mapping experiments because the genes are so close together. Therefore, without very fine mapping, which is currently difficult to do for QTLs, it is hard to distinguish between overdominance and pseudo-overdominance. By comparison, Steven Tanksley at Cornell University, working with colleagues in China, found that heterosis in rice is due to dominance rather than overdominance. Over the next few years, it will be interesting to see if heterosis among agriculturally important species is usually explained by dominance, overdominance, or a combination of the two.

Quantitative genetics is the study of traits that vary in a continuous, quantitative way in populations. Such *quantitative traits* can be anatomical, physiological, or behavioral, and some diseases exhibit characteristics and inheritance patterns analogous to those of quantitative traits.

Within populations, quantitative traits commonly exhibit a continuum of phenotypic variation that may follow a *normal distribution*. Statistical methods can be used to analyze such a distribution. The *mean* describes the average value among the population; the *standard deviation (S.D.)* provides a measure of the amount of variation in the group. The *variance* is a measure of the squared deviations from the mean. Variances are useful statistics, because they are additive. Also, some statistical methods compare two variables with each other. The *correlation coefficient* evaluates the strength of association between two variables. In genetics, it is often used to see if there are phenotypic correlations between genetically related individuals.

Most quantitative traits are *polygenic*. The chromosomal locations of genes that influence quantitative traits are called *quantitative trait loci (QTLs)*. The reasons most quantitative traits show a continuum are that each trait is governed by several genes existing in two or more alleles, and the environment contributes a substantial amount of variation to the phenotypic outcome.

All traits of biological organisms are influenced by genetics and the environment, particularly quantitative traits. Phenotypic variance is due to the additive effects of genetic variance and environmental variance. *Heritability* is the fraction of the phenotypic variance that is attributable to genetic factors; it describes the amount of phenotypic variation that is due to genetic factors for a particular population raised in a particular environment. *Heritability in the broad sense* takes into account all of the genetic factors that could affect the phenotype. By comparison, the heritability of a trait due to the additive effects of alleles is called the *narrow sense heritability*. For many quantitative traits, the broad sense and narrow sense heritabilities are similar.

Geneticists measure the quantitative traits of an individual and describe them numerically. In a population, experimentally obtained numerical values from each individual are used to calculate statistics such as the mean, standard deviation, variance, and correlation.

More than one method can be used to determine the number of genes or QTLs that influence a quantitative trait. For example, QTLs can be identified via their linkage to genes on chromosomes. More recently, RFLPs are being used as molecular markers to determine the number of QTLs for a given quantitative trait. In some cases, this approach enables plant and animal breeders to understand the genetic nature of quantitative traits and to develop useful breeding strategies.

A central issue for quantitative geneticists is the relative contributions of genetics and the environment that underlie the phenotypic variation seen in quantitative traits. In this chapter, we have considered two experimental methods to measure the heritability of quantitative traits. One method, which was illustrated in Experiment 25A, is to compare the correlation coefficients among related and unrelated individuals.

A second method is selective breeding, which is frequently used to improve quantitative traits in agriculturally important species. Quantitative traits are often at an intermediate value in unselected populations. Therefore, artificial selection can be used to increase or decrease the magnitude of the trait. The response to selection can be used as a way to estimate the heritability in the starting population. The heritability measured in this way is called the realized heritability.

Selective breeding sometimes involves the repeated breeding of related individuals to produce strains that are highly inbred. An unfortunate consequence of inbreeding is that it decreases the overall genetic variation within a group and may inadvertently promote homozygosity for deleterious alleles. When two different inbred strains are crossed to each other, the resulting offspring are often more vigorous than either of the inbred parental strains. This phenomenon, called heterosis or hybrid vigor, may be due to the dominant effects of masking harmful recessive alleles or to overdominance of alleles in the heterozygous state.

PROBLEM SETS

SOLVED PROBLEMS

1. The narrow sense heritability for potato weight in a starting population of potatoes is 0.42, and the mean weight is 1.4 lb. If a breeder crosses two individuals with average potato weights of 1.9 and 2.1 lb, respectively, what is the predicted average weight of potatoes in the offspring?

Answer: The mean weight of the parents is 2.0 lb. To solve for the mean weight of the offspring:

$$h_N^2 = \frac{\overline{X}_O - \overline{X}}{\overline{X}_P - \overline{X}}$$

$$0.42 = \frac{\overline{X}_O - 1.4}{2.0 - 1.4}$$

$$\overline{X}_O = 1.65 \text{ lb}$$

2. A farmer wants to increase the average body weight in a herd of cattle. She begins with a herd having a mean weight of 595 kg and chooses individuals to breed that have a mean weight of 625 kg. Twenty offspring were obtained, having the following weights in kg: 612, 587, 604, 589, 615, 641, 575, 611, 610, 598, 589, 620, 617, 577, 609, 633, 588, 599, 601, and 611. Calculate the realized heritability in this herd with regard to body weight.

Answer:

$$h_N^2 = \frac{R}{S}$$

$$= \frac{\overline{X}_O - \overline{X}}{\overline{X}_P - \overline{X}}$$

We already know the mean weight of the starting herd (595 kg) and the mean weights of the parents (625 kg). We need to calculate the mean weights of the offspring and then substitute this value in the formula:

$$\overline{X}_O = \frac{\text{sum of the offspring' s weights}}{\text{number of offspring}}$$

$$= 604 \text{ kg}$$

$$h_N^2 = \frac{604 - 595}{625 - 595}$$

$$= 0.3$$

3. The following are data that describe the six-week weights of mice and their offspring of the same gender:

Parent (g)	Offspring (g)
24	26
21	24
24	22
27	25
23	21
25	26
22	24
25	24
22	24
27	24

Calculate the correlation coefficient.

Answer: To calculate the correlation coefficient, we first need to calculate the means and standard deviation for each group:

$$\overline{X}_{\text{parents}} =$$

$$\frac{24 + 21 + 24 + 27 + 23 + 25 + 22 + 25 + 22 + 27}{10} = 24$$

$$\overline{X}_{\text{offspring}} =$$

$$\frac{26 + 24 + 22 + 25 + 21 + 26 + 24 + 24 + 24 + 24}{10} = 24$$

$$\text{S.D.}_{\text{parents}} =$$

$$\sqrt{\frac{0 + 9 + 0 + 9 + 1 + 1 + 4 + 1 + 4 + 9}{9}} = 2.1$$

$$\text{S.D.}_{\text{offspring}} =$$

$$\sqrt{\frac{4 + 0 + 4 + 1 + 9 + 4 + 0 + 0 + 0 + 0}{9}} = 1.6$$

Next, we need to calculate the covariance:

$$\text{CoV}_{(\text{parents,offspring})} = \frac{\Sigma[(X_p - \overline{X}_p)(X_o - \overline{X}_o)]}{N - 1}$$

$$= \frac{0 + 0 + 0 + 3 + 3 + 2 + 0 + 0 + 0 + 0}{9}$$

$$= 0.9$$

Finally, we calculate the correlation coefficient:

$$r_{(\text{parent,offspring})} = \frac{\text{CoV}_{(p,o)}}{\text{SD}_p \, \text{SD}_o}$$

$$= \frac{0.9}{(2.1)(1.6)}$$

$$= 0.27$$

CONCEPTUAL QUESTIONS

1. Give several examples of quantitative traits. How are these quantitative traits described within groups of individuals?

2. At the molecular level, explain why quantitative traits often exhibit a continuum of phenotypes within a population. How does the environment help produce this continuum?

3. What is a normal distribution? Discuss this curve with regard to quantitative traits within a population. What is the relationship between the standard deviation and the normal distribution?

4. The variance for weight in a particular herd of cattle is 424 lb². How heavy would an animal have to be if it was in the top 2.5% of the herd? The bottom 0.15%?

5. Two different strains of potatoes both have the same mean weight of 1.5 pounds. One group has a very low variance, and the other has a much higher variance.

  A. Discuss the possible reasons for the differences in variance.
  B. If you were a potato farmer, would you rather raise a strain with a low or high variance? Explain your answer from a practical point of view.
  C. If you were a potato breeder, and you wanted to develop potatoes with a heavier weight, would you choose the strain with a low or high variance? Explain your answer.

6. Discuss the meaning of the correlation coefficient. If an *r* value equals 0.5 and *N* = 4, would you conclude that there is a positive correlation between the two variables? Explain your answer. What if *N* = 500?

7. In a particular strain of pigs, the correlation among littermates for weight is 0.15. What is the narrow sense heritability for this trait?

8. Here are data for height and weight among ten male college students:

Height (cm)	Weight (kg)
159	48
162	50
161	52
175	60
174	64
198	81
172	58
180	74
161	50
173	54

  A. Calculate the correlation coefficients for this group.
  B. Is the correlation coefficient statistically significant? Explain.

9. What does it mean when a correlation coefficient is negative? Can you think of examples?

10. When a correlation coefficient is statistically significant, what do you conclude about the two variables? What do the results mean with regard to cause and effect?

11. What is polygenic inheritance? Discuss the issues that make polygenic inheritance difficult to study.

12. The broad sense heritability for a trait equals 1.0. In your own words, explain what this value means. Would you conclude that the environment is unimportant in the outcome of this trait? Explain your answer.

13. In a newly developed hybrid strain of tomato plants, the phenotypic variance for tomato weight is 3.2 g². In another strain of highly inbred tomatoes raised under the same environmental conditions, the phenotypic variance is 2.2 g². With regard to the hybrid strain:

A. Estimate V_G.
B. What is H_B^2?
C. Assuming all the genetic variance is additive, what is h_N^2?

14. Compare and contrast the Dominance and Overdominance Hypotheses. Based on your knowledge of mutations and genetics, which do you think is more likely?

15. Selective breeding is a common strategy for altering quantitative traits. Let's suppose that a plant breeder began with a wild population of blueberry plants and subjected them to many generations of selective breeding with the goal of obtaining plants that bear larger blueberries. How do you expect the heritability of this trait to change from the original wild population to the population of plants obtained after many generations of selective breeding? Explain.

16. The average thorax length in a *Drosophila* population is 1.01 mm. You want to practice selective breeding to make larger *Drosophila*. To do so, you choose 10 parents (five males and five females) of the following sizes: 0.97, 0.99, 1.05, 1.06, 1.03, 1.21, 1.22, 1.17, 1.19, and 1.20. You mate them and then analyze the thorax sizes of 30 offspring (half male and half female):

0.99, 1.15, 1.20, 1.33, 1.07, 1.11, 1.21, 0.94, 1.07, 1.11, 1.20, 1.01, 1.02, 1.05, 1.21, 1.22, 1.03, 0.99, 1.20, 1.10, 0.91, 0.94, 1.13, 1.14, 1.20, 0.89, 1.10, 1.04, 1.01, 1.26

Calculate the realized heritability in this group of flies.

17. In a strain of mice, the average six-week body weight is 25 g and the narrow sense heritability for this trait is 0.21.

A. What would be the average weight of the offspring if parents with a mean weight of 27 g were chosen?
B. What weight of parents would you have to choose in order to obtain offspring with an average weight of 26.5 g?

18. What is hybrid vigor (also known as heterosis)? Give examples that you might find in a vegetable garden.

19. A herd of cattle has a high phenotypic variance for meat production. Even so, the broad sense heritability is estimated to be very low. Describe the strategy you would follow to improve the meat yield of the herd.

20. Two tomato strains, A and B, both produce fruit that weighs, on average, one pound each. All of the variance is due to V_G. When these two strains are crossed to each other, the F₁ offspring display heterosis with regard to fruit weight, with an average weight of two pounds. You take these F₁ offspring and backcross them to strain A. You then grow several plants from this cross and measure the weights of their fruit. What would be the expected results for each of the following scenarios?

A. Heterosis is due to a single overdominant gene.
B. Heterosis is due to two dominant genes, one in each strain.
C. Heterosis is due to two overdominant genes.
D. Heterosis is due to dominance of several genes each from strains A and B.

EXPERIMENTAL QUESTIONS

1. Using the same strategy as Figure 25-5, the following data are the survival of F₂ offspring obtained from backcrosses to an insecticide resistent and control strains:

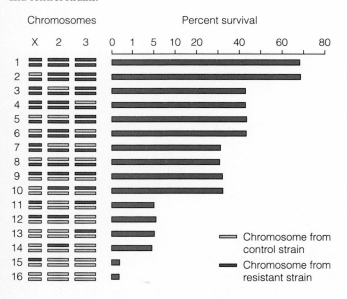

Interpret these results with regard to the locations of QTLs.

2. In one strain of cabbage, you conduct an RFLP analysis with regard to head weight; you determine that seven QTLs affect this trait.

In another strain of cabbage, you find that only four QTLs affect this trait. Note that both strains of cabbage are from the same species, although they may have been subjected to different degrees of inbreeding. Explain how one strain can have seven QTLs and another strain four QTLs for exactly the same trait. Is the second strain missing three genes?

3. From an experimental viewpoint, what does it mean to say that an RFLP is associated with a trait? Let's suppose that two strains of pea plants differ in two RFLPs that are linked to two genes governing pea size. RFLP-1 is found in 2000 bp and 2700 bp bands, and RFLP-2 is found in 3000 bp and 4000 bp bands. The plants producing large peas have RFLP-1 (2000 bp) and RFLP-2 (3000 bp); those producing small peas have RFLP-1 (2700 bp) and RFLP-2 (4000 bp). A cross is made between these two strains, and the F₁ offspring are allowed to self-fertilize. Five phenotypic classes are observed: small peas, small-medium peas, medium peas, medium-large peas, and large peas. We assume that each of the two genes makes an equal contribution to pea size, and that the genetic variance is additive. Draw a gel, and explain what RFLP banding patterns you would expect to observe for these five phenotypic categories. Note: Certain phenotypic categories may have more than one possible banding pattern.

4. Let's suppose that two strains of pigs differ in 500 RFLPs. One strain is much larger than the other. The pigs are crossed to each other, and the F₁ generation is also crossed among themselves to produce an F₂ generation. Three distinct RFLPs are associated with F₂ pigs that are larger, but the other 497 RFLPs are distributed randomly among small and large pigs. How would you interpret these results?

5. Outline the steps you would follow to determine the number of genes that influence the yield of rice. Describe the results you might get if rice yield is governed by six different genes.

6. A danger in computing heritability values from studies involving genetically related individuals is the possibility that these individuals share more similar environments than do unrelated individuals. In Experiment 25A, which data are the most compelling evidence that ridge count is not caused by genetically related individuals sharing common environments? Explain.

QUESTIONS FOR STUDENT DISCUSSION/COLLABORATION

1. Discuss why heritability is an important phenomenon in agriculture. Discuss how it is misunderstood.

2. From a biological viewpoint, speculate as to why many traits seem to fit a normal distribution. Students with a strong background in math and statistics may want to explain how a normal distribution is generated, and what it means. Can you think of biological examples that do not fit a normal distribution?

3. What is heterosis? Discuss whether it is caused by a single gene or several genes. Discuss the two major hypotheses proposed to explain heterosis. Which do you think is more likely to be correct?

CHAPTER 26

POPULATION GENETICS

GENES IN POPULATIONS

THE HARDY–WEINBERG EQUILIBRIUM

FACTORS THAT CHANGE ALLELE FREQUENCIES IN POPULATIONS

The central issue in **population genetics** is genetic variation. Population geneticists want to know the extent of genetic variation within populations, why it exists, and how it changes over the course of many generations. Population genetics emerged as a branch of genetics in the 1920s and 1930s. Its mathematical foundations were developed by theoreticians who extended the principles of Mendel and Darwin by deriving formulae to explain the occurrence of genotypes within populations. These foundations can be attributed largely to three individuals: Sir Ronald Fisher, at the Rothamsted Experimental Station at Harpenden, England; Sewall Wright at Cold Spring Harbor; and J. B. S. Haldane at Cambridge University. As we will see, support for their mathematical theories was provided by several researchers who analyzed the genetic composition of natural and experimental populations. More recently, population geneticists have used laboratory techniques to probe genetic variation at the molecular level. In addition, the staggering improvement in computer technology has aided population geneticists in the analysis of their genetic theories and data. In this chapter, we will explore the genetic variation that occurs in populations.

GENES IN POPULATIONS

Population genetics may seem like a significant departure from other topics in this text, but it is a direct extension of our understanding of Mendel's laws of inheritance, molecular genetics, and the ideas of Darwin. The focus is shifted away from the individual and toward the population of which the individual is a member.

Conceptually, all the genes in a population make up the **gene pool**, defined as the totality of all genes within a particular population. In this regard, each member of the population is viewed as receiving its genes from the gene pool. Furthermore, if an individual reproduces, it contributes to the gene pool of the next generation. Population geneticists study the genetic variation within the gene pool, and how this variation changes from one generation to the next. In this introductory section, we will examine some of the general features of populations and gene pools.

A population is a group of interbreeding individuals who share a gene pool

In genetics, the term population has a very specific meaning. A **population** is a group of individuals of the same species that can interbreed with one another. Many species occupy a wide geographic range and are divided into discrete populations. For example, distinct populations of a given species may be located on different continents.

A large population usually is composed of smaller groups called **subpopulations**, **local populations**, or **demes**. The members of a subpopulation are far likelier to breed among themselves than with other members of the general population. Subpopulations are often separated from each other by moderate geographic barriers. Figure 26-1 presents an example. As shown here, two populations of Douglas fir (*Pseudotsuga menziesii*) are separated by a wide river bottom, where the fir trees are unlikely to grow. The groups of trees on opposite sides of this river bottom constitute local populations. Interbreeding is much more apt to occur among members of each local population than between members of neighboring populations. On relatively rare occasions, however, pollen can be blown across the river bottom, which allows interbreeding between these two local populations.

Populations typically are dynamic units that change from one generation to the next. A population may change its size, geographic location, and genetic composition. With regard to size, natural populations commonly go through cycles of "feast or famine," during which the population swells or shrinks. In addition, natural predators or disease may periodically decrease the size of a population to significantly lower levels; the population later may rebound to its original size. Populations or individuals within populations may migrate to a new site and establish a distinct population in this location. This new geographic location may differ in environment from the original site. As population sizes and locations change, their genetic composition generally changes as well. As will be described later in this

FIGURE **26-1**

Two local populations of Douglas fir. A wide river bottom separates the two populations. There is a much higher frequency of mating among members of the same population than of mating between the two populations.

chapter, population geneticists have developed mathematical theories that predict how the gene pool will change in response to fluctuations in size, migration, and new environments.

Some genes are monomorphic, and others are polymorphic

In population genetics, the term **polymorphism** (meaning many forms) refers to the observation that many traits display variation within a population. Historically, polymorphism first referred to the variation in traits that are observable with the naked eye. Polymorphisms in color and pattern have long attracted the attention of population geneticists. These include studies of melanism in the peppered moth and of variation in snail color, which are discussed later in this chapter. Figure 26-2 illustrates a striking example of polymorphism in the Hawaiian happy-face spider. All of the individuals shown in this figure are from the same species, but they differ in alleles that affect color and pattern.

At the DNA level, polymorphism is due to two or more alleles that influence the phenotype of the individual who inherits them. In other words, it is due to genetic variation. Geneticists also use the term polymorphic to describe the variation in genes that govern a polymorphic trait. A gene that commonly exists as two or more alleles in a population is described as polymorphic. By comparison, a **monomorphic** gene exists predominantly as a single allele in a population. By convention, when a single allele is found in at least 99% of all instances of a gene, the gene is considered monomorphic. To be judged polymorphic, a gene must have one or more additional alleles that make up at least 1% of the alleles in the population.

During the 1960s, molecular techniques became available that could assess variation in alleles that influence enzyme structure. In 1966, John Hubby and Richard Lewontin, at the University of Chicago, studied allelic variation in populations of the fruit fly *Drosophila pseudoobscura*. This species of fruit fly was chosen because (unlike *D. melanogaster*) *D. pseudoobscura* does not share its habitat with humans, and so it is considered truly wild. In addition, it has a wide geographic distribution in North and Central America. Hubby and Lewontin studied many structural genes that were known to encode enzymes. At the molecular level, genetic variation in a structural gene may result in the production of an enzyme with slight differences in its amino acid sequence compared with the wild-type enzyme. Based on previous work, Hubby and Lewontin knew that these differences in amino acid sequence can be detected as changes in mobility of the enzymes during gel electrophoresis. Two enzymes with alterations in their gel mobilities due to small differences in their amino acid sequences are called **allozymes**. In other words, they are *alle*les *o*f the same en*zyme*.

In their study, Hubby and Lewontin looked for genetic variation in 18 different enzymes in five different fruit fly populations. On average, 30% of the enzymes were found as two or more allozymes. This means that the genes encoding these enzymes had slightly different DNA sequences; these differences resulted in alleles that encoded slightly different amino acid sequences. This study tends to underestimate genetic variability, since some amino acid substitutions do not alter protein mobility during gel electrophoresis. Nevertheless, the important conclusion from this work was that the genetic variability within these populations is quite high. Around the time of this work, Harry Harris of University College in London also found that approximately 30% of human genes encoding enzymes are polymorphic. More recently, many population geneticists have investigated genetic variation using DNA sequencing methods (described in Chapter 19).

In most natural populations, a substantial percentage of genes are polymorphic. Some examples are shown in Figure 26-3. However, in small populations that are near extinction, genetic variation is expected to be low, since the gene pool

FIGURE **26-2**

Polymorphism in the Hawaiian happy-face spider.

GENES → TRAITS: These three spiders are members of the same species and carry the same genes. However, several genes that affect pigmentation patterns are polymorphic, meaning that there is more than one allele for each gene within the population. This polymorphism within the Hawaiian happy-face spider population produces members that look quite different from each other.

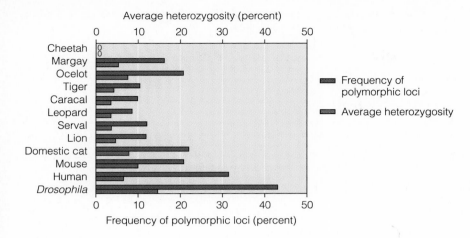

FIGURE **26-3**

A comparison of the level of polymorphisms in different species. The upper purple bars indicate the percentage of genes that are polymorphic within an entire species; the lower blue bars indicate the average percentage of genes that are heterozygous within any given individual (average heterozygosity). This information was derived from the analysis of allozymes by gel electrophoresis.

is derived from a small number of individuals. An extreme example is the African cheetah population, which has a genetic variation near zero. Later in this chapter, we will examine how small population size contributed to this problem. By comparison, approximately 30% of human genes are polymorphic. Since humans have approximately 100,000 different genes, this means that roughly 30,000 genes can be found in two or more different alleles.

Keep in mind that polymorphism refers to the diversity of genes in a population. Within a single individual, genetic variation will be less, because an individual may be homozygous for polymorphic genes. For example, the ABO blood type is determined by three alleles, designated I^A, I^B, and i. A person with type O blood is homozygous, ii, and therefore has less genetic variation than the population as a whole. In human populations, about 30% of all genes are polymorphic, but within any individual less than 10% of all genes are heterozygous.

Population genetics is concerned with allele and genotype frequencies

As we have seen, population geneticists want to understand the prevalence of polymorphic genes within populations. Much of their work involves evaluating the prevalence of genes in a quantitative way. Two fundamental calculations are central to population genetics: **allele frequencies** and **genotype frequencies**. The allele and genotype frequencies are defined as

$$\frac{\text{allele}}{\text{frequency}} = \frac{\text{number of copies of an allele in a population}}{\text{total number of all alleles for that gene in a population}}$$

$$\frac{\text{genotype}}{\text{frequency}} = \frac{\text{number of individuals with a particular genotype in a population}}{\text{total number of individuals in a population}}$$

Though these two frequencies are related, a clear distinction between them must be kept in mind. As an example, let's consider a population of 100 pea plants with the following genotypes:

64 tall plants with the genotype *TT*

32 tall plants with the genotype *Tt*

4 dwarf plants with the genotype *tt*

When calculating an allele frequency, homozygous individuals have two copies of an allele, whereas heterozygotes only have one. For example, in tallying the *t* allele, each of the 32 heterozygotes has one copy of the *t* allele, and each dwarf plant has two copies. The allele frequency for *t* equals

$$t = \frac{32 + (2)(4)}{(2)(64) + (2)(32) + (2)(4)}$$

$$= \frac{40}{200} = 0.2, \text{ or } 20\%$$

This result tells us that the allele frequency of *t* is 20%. In other words, 20% of the alleles for this gene in the population are the *t* allele.

Let's now calculate the genotype frequency of *tt* (dwarf) plants:

$$tt = \frac{4}{64 + 32 + 4}$$

$$= \frac{4}{100} = 0.04, \text{ or } 4\%$$

We see that 4% of the individuals in this population are dwarf plants.

Allele and genotype frequencies are always less than or equal to 1 (i.e., less than or equal to 100%). If a gene is monomorphic, the allele frequency for the single allele will equal (or be very close to) a value of 1.0. For polymorphic genes, if we add up the frequencies for all of the alleles in the population, we should obtain a value of 1.0. In our pea plant example, the allele frequency of *t* equals 0.2. The frequency of the other allele, *T*, equals 0.8. If we add the two together, we obtain a value of 0.2 + 0.8 = 1.0. In the next section, we will learn how allele and genotype frequencies within populations are related.

THE HARDY–WEINBERG EQUILIBRIUM

Now that we have a general understanding of genes in populations, we can begin to relate these concepts to mathematical expressions in order to examine whether allele and genotype frequencies will change over the course of many generations. In 1908, Godfrey Harold Hardy, professor of mathematics at Cambridge University, and Wilhelm Weinberg, a physician in Stuttgart, Germany, independently derived a simple mathematical expression that predicted stability of allele and genotype frequencies from one generation to the next. This expression, known as the **Hardy–Weinberg equation**, relates allele and genotype frequencies within a population. It is also called an equilibrium, because (under a given set of conditions, as described later) the allele and genotype frequencies do not change over the course of many generations. This relationship established a framework on which to understand genetic stability.

In subsequent decades, as the field of population genetics developed, it became apparent that genetic stability is not always present in natural populations. Furthermore, genetic change underlies the theory of evolution. Therefore, in the 1920s and 1930s, population geneticists turned their efforts largely toward understanding how genetic stability is altered. In this section, we will examine the Hardy–Weinberg equilibrium and the conditions that must be met for it to be valid. At the end of this chapter, we will learn how natural populations usually violate the Hardy–Weinberg equilibrium and thereby cause allele frequencies to change.

The Hardy–Weinberg equation can be used to calculate genotype frequencies based on allele frequencies

The Hardy–Weinberg equation is a simple mathematical expression that relates genotype and allele frequencies. Let's first examine the mathematical components of the Hardy–Weinberg equation, and then we will look at the conditions necessary for an equilibrium to be achieved.

Let's begin by considering a situation in which a gene is polymorphic and exists as two different alleles, A and a. If the allele frequency of A is denoted by the variable p, and the allele frequency of a by q, then:

$$p + q = 1$$

For example, if $p = 0.8$, then q must be 0.2. In other words, if the allele frequency of A equals 80%, the remaining 20% of alleles must be a, because together they equal 100%.

The Hardy–Weinberg equation states that:

$$p^2 + 2pq + q^2 = 1 \qquad \text{(Hardy–Weinberg equation)}$$

If this equation is applied to a gene that exists in alleles designated A and a, then

p^2 equals the genotype frequency of AA

$2pq$ equals the genotype frequency of Aa

q^2 equals the genotype frequency of aa

If $p = 0.8$ and $q = 0.2$, then

$$AA = p^2 = (0.8)^2 = 0.64$$

$$Aa = 2pq = 2(0.8)(0.2) = 0.32$$

$$aa = q^2 = (0.2)^2 = 0.04$$

In other words, if the allele frequency of A is 80% and the allele frequency of a is 20%, the genotype frequency of AA is 64%, Aa 32%, and aa 4%.

To see the relationship between allele frequencies and genotypes, Figure 26-4 compares the Hardy–Weinberg equation with the Punnett square approach (see Chapter 2, pp. 29–30 for a description of Punnett squares). As seen here, the Hardy–Weinberg equation results from the way gametes combine randomly with each other to produce offspring. In a population, the frequency of a gamete carrying a particular allele is equal to the allele frequency in that population. For example, the frequency of a gamete carrying the A allele equals 0.8.

We can use the product rule to determine the frequency of genotypes. For example, the frequency of producing an AA homozygote is $0.8 \times 0.8 = 0.64$, or 64%.

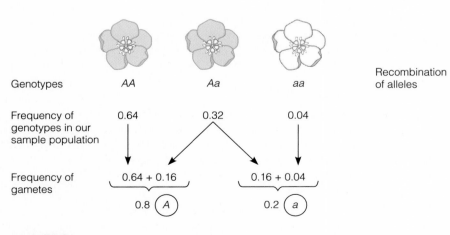

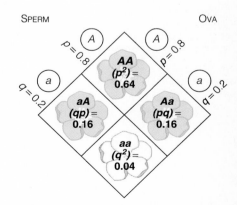

FIGURE **26-4**

A comparison between the Hardy–Weinberg equation and the Punnett square approach.

Likewise, the probability of inheriting the *a* allele is $0.2 \times 0.2 = 0.04$, or 4%. In our Punnett square, there are two different ways to produce heterozygotes (see Figure 26-4). An offspring could inherit the *A* allele from its father and *a* from its mother, or *A* from its mother and *a* from its father. Therefore, the frequency of heterozygotes is $pq + pq$, which equals $2pq$; in our example, this is $2(0.8)(0.2) = 0.32$, or 32%.

The Hardy–Weinberg equation predicts an equilibrium (i.e., unchanging allele and genotype frequencies) if a certain set of conditions are met in a population. These are as follows:

1. The population is so large that allele frequencies do not change due to random sampling effects. (See Chapter 2, pp. 33–34 for a description of random sample errors.)
2. The members of the population mate with each other without regard to their phenotypes and genotypes.
3. There is no migration between different populations.
4. There is no survival or reproductive advantage for any of the genotypes. In other words, no natural selection occurs.
5. No new mutations occur within the population.

According to the equilibrium, the Hardy–Weinberg equation provides a quantitative relationship between allele and genotype frequencies in a population. Figure 26-5 describes this relationship for different allele frequencies of *A* and *a*. As expected, when the allele frequency of *A* is very low, the *aa* genotype predominates; when the *A* allele frequency is high, the *AA* homozygote is most frequent in the population. When the allele frequencies of *A* and *a* are intermediate in value, the heterozygote predominates.

In reality, no population satisfies the Hardy–Weinberg equilibrium completely. Nevertheless, in large natural populations with little migration and negligible natural selection, the Hardy–Weinberg equilibrium may be nearly approximated. In addition, as discussed later in this chapter, deviations from the Hardy–Weinberg equilibrium help us understand how populations are changing from one generation to the next.

Nonrandom mating may occur in natural and human populations

As mentioned earlier, one of the conditions required to establish the Hardy–Weinberg equilibrium is random mating. This means that individuals choose their mates

FIGURE **26-5**

The relationship between allele frequencies and genotype frequencies according to Hardy–Weinberg equilibrium. This graph assumes that *A* and *a* are the only two alleles for this gene. From E.B. Speiss, *Genes in Populations*. ©1977 John Wiley & Sons, Inc. Reprinted by permission of John Wiley & Sons, Inc.

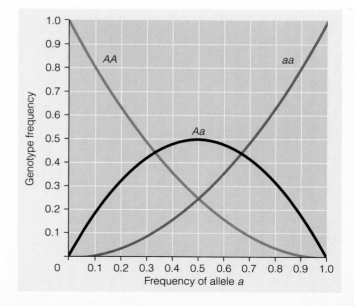

irrespective of their genotypes and phenotypes. In many cases, particularly human populations, this condition is violated frequently.

When two individuals are more likely to mate due to similar phenotypic characteristics, this is known as **assortative mating**. The opposite situation, where dissimilar phenotypes mate preferentially, is called **disassortative mating**. In addition, individuals may choose a mate who is part of the same genetic lineage. The mating of two genetically related individuals (e.g., cousins) with each other is called **inbreeding**. This sometimes occurs in human societies and is more likely to take place in nature when population size becomes very limited. In Chapter 25, we also learned that inbreeding is a useful strategy for developing agricultural breeds or strains with desirable characteristics. Conversely, **outbreeding**, which involves mating between unrelated individuals, can be used to create hybrids that are heterozygous for many genes (see Chapter 25, pp. 697–699).

In the absence of other evolutionary processes, inbreeding and outbreeding do not affect allele frequencies in a population. However, these patterns of mating do disrupt the balance of genotypes that is predicted by the Hardy–Weinberg equilibrium. Let's consider inbreeding in a family pedigree. Figure 26-6 illustrates a human pedigree involving a mating between cousins. Individuals III-2 and III-3 are cousins and have produced the son labeled IV-1. This offspring is said to be **inbred**, because his parents are related genetically to each other.

During inbreeding, the gene pool is smaller, because the parents are related genetically. In the 1910s to 1930s, Wright and Fisher developed methods to quantify the degree of inbreeding. A **coefficient of inbreeding** (F) can be computed by analyzing the degree of relatedness within a pedigree. As an example, let's find the coefficient of inbreeding for individual IV-1. To begin this problem, we must first identify all of the common ancestors that this individual has. A **common ancestor** is anyone who is an ancestor to *both* of an individual's parents. In Figure 26-6, IV-1 has one common ancestor, I-2, his great-grandfather. (I-2 is the grandfather of III-2 and III-3.)

Our next step is to determine the inbreeding paths. An **inbreeding path** for an individual is the shortest path through the pedigree that includes both parents and the common ancestor. In a pedigree, there is an inbreeding path for each common ancestor. The length of each inbreeding path then is calculated by adding together all of the individuals in the path except the individual of interest. In this case, there is only one path, since there is only one common ancestor. To add the members of the path, we begin with individual IV-1, but we do not count him. We then move to his father (III-2), to his grandfather (II-2), to I-2, his great-grandfather (the common ancestor), and back down to his other grandmother (II-3), and finally to his mother (III-3). This path has five members in it. Finally, to calculate the inbreeding coefficient, we use the following formula:

$$F = \Sigma (1/2)^n (1 + F_A)$$

where,

F is the inbreeding coefficient of the individual of interest
n is the number of individuals in the inbreeding path, excluding the inbred offspring
F_A is the inbreeding coefficient of the common ancestor
Σ indicates that we add together $(1/2)^n (1 + F_A)$ for each inbreeding path

In this case, there is only one common ancestor. Also, we do not know anything about the heritage of the common ancestor, and so we assume that F_A is zero. Thus, in our example of Figure 26-6,

$$F = \Sigma (1/2)^n (1 + 0)$$

$$= (1/2)^5 = 1/32 = 3.125\%$$

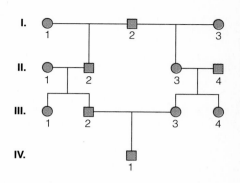

FIGURE **26-6**

A human pedigree containing inbreeding. Individual IV-1 is the result of a consanguineous mating.

Our inbreeding coefficient, 3.125%, tells us the probability that a gene in the inbred individual (IV-1) is homozygous due to its inheritance from a common ancestor (I-2). In this case, therefore, each gene in individual IV-1 has a 3.125% chance of being homozygous because he has inherited the same allele twice from his great-grandfather (I-2), once through each parent. For example, let's suppose that the common ancestor (I-2) is heterozygous for the allele that causes cystic fibrosis. There is a 3.125% probability that the inbred individual (IV-1) is homozygous for this gene because he has inherited both copies from his great-grandfather. He has a 1.56% probability of inheriting the normal allele twice and a 1.56% probability of inheriting the mutant allele twice.

The coefficient of inbreeding is also called the *fixation coefficient*. That is why the inbreeding coefficient is denoted by the letter F (for *Fixation*). It is the probability that an allele will be fixed in the homozygous condition. The term fixation signifies that the homozygous individuals can pass only one type of allele to their offspring.

In other pedigrees, an individual may have two or more common ancestors. In this case, the inbreeding coefficient F is calculated as the sum of the inbreeding paths. Such an example is described in Solved Problem 2 at the end of this chapter.

Inbreeding can have both positive and negative consequences in a population. From an agricultural viewpoint, it results in a higher proportion of homozygotes, which may exhibit a desirable trait. For example, an animal breeder may use inbreeding to produce animals that are larger because they have become homozygous for alleles promoting larger size. On the negative side, many genetic diseases are inherited in a recessive manner (see Chapter 23). For these rare recessive disorders, inbreeding increases the likelihood that an individual will be homozygous and therefore afflicted with the disease.

FACTORS THAT CHANGE ALLELE FREQUENCIES IN POPULATIONS

The Hardy–Weinberg equilibrium and the high prevalence of polymorphisms within natural populations seem, at first glance, to contradict each other. On the one hand, the Hardy–Weinberg equilibrium predicts that allele frequencies will not change from one generation to the next. On the other hand, research has shown that many genes are polymorphic, indicating that two or more alleles can attain a significant percentage within a population. This leads to a central question in population genetics: How do genetic polymorphisms originate, and why are they maintained in natural populations?

For very rare, detrimental alleles, the explanation is straightforward. Random mutations are far likelier to produce deleterious alleles than beneficial ones, because the intricate functioning of proteins will almost certainly be disrupted rather than improved by a haphazard change. Therefore, mutation tends to favor the accumulation of deleterious alleles in a population. However, as we will learn in this section, natural selection promotes the elimination of harmful alleles in a population. So, in a population, the low rate of mutation produces detrimental alleles, and the process of natural selection works to eliminate them. The opposing actions of these two processes maintain detrimental alleles at very low frequencies.

However, many genetic polymorphisms exist in which two or more alleles are found at substantial frequencies in a population. In these cases, the allele percentages commonly observed in genetic polymorphisms cannot be explained by new mutations; the allele frequencies are too high, the rate of new mutations too low. For example, let's use our previous example of a polymorphic gene existing in alleles A and a at respective frequencies of 0.8 and 0.2. One possibility is that this current population is derived from an ancestral population that was originally monomorphic, containing only the A allele in its gene pool. Over time, new mutations could have converted the A allele into the a allele in members of the population. However, the present-day frequency of 0.2 for the *a* allele is unlikely to be merely the result of the gradual accumulation of new mutations of *A* to *a*, because the rate of such mutations is exceedingly low. Therefore, though new mutations provide a source of allelic variation, they do not occur fast enough to generate the allele frequencies observed in many polymorphisms. After that new mutation occurred, other evolutionary forces must have increased the frequency of the *a* allele from near zero to a frequency of 0.2. But what are these other evolutionary forces, and how do they work?

We can divide evolutionary processes into two categories: neutral and adaptive (Table 26-1). A major debate in population genetics centers on which of these two types of processes accounts for most of the genetic variation in natural populations. **Neutral processes** (or neutral forces) alter allele frequencies in a random manner. In other words, neutral processes change allele frequencies without any regard to the survival of the individual. They can occur as a matter of chance. The two key neutral processes are migration and genetic drift.

By comparison, **adaptive forces** are processes in which alleles that confer greater survival or reproductive success increase in frequency. Natural selection produces individuals that are well adapted to their environment. Those individuals who happen to possess these beneficial alleles are more likely to survive, reproduce, and contribute to the gene pool of the next generation. In this section, we will examine the contributions of neutral and adaptive processes in creating and maintaining genetic variation.

Mutations provide the source of genetic variation

As was discussed in Chapters 8 and 17, mutations can involve changes in gene sequences, chromosome structure, and/or chromosome number. They are random events that can occur spontaneously at a low rate, or can be caused by mutagens at a higher rate. In this chapter, we will be concerned primarily with the effects of gene mutations on the frequency of new alleles within a population. Even so, as we will learn in Chapter 27, changes in chromosome structure and number are important events in the evolutionary process.

In 1926, the Russian geneticist Sergei Tshetverikov was the first to suggest that mutational variability provides the raw material for evolution, but does not constitute evolution itself. In other words, mutation can provide new alleles to a population

TABLE 26–1 Processes that Alter Allele Frequencies

Mutation	Mutation introduces new alleles into populations, but at a very low rate. New mutations may be beneficial, neutral, or deleterious. For alleles to rise to a significant percentage in a population, it is necessary for other evolutionary processes (namely, genetic drift, migration, and/or natural selection) to operate.
Neutral Forces	
Random genetic drift	A change in allele frequencies due to random survival and reproduction. In other words, allele frequencies may change as a matter of chance from one generation to the next. This occurs to a greater degree in smaller populations.
Migration	Migration can occur between two populations having different allele frequencies. The introduction of migrants into the recipient population may change the latter's allele frequencies.
Adaptive Forces	
Natural selection	Natural selection is the enhanced survival or reproduction of individuals having genotypes that confer beneficial phenotypes. Likewise, natural selection is also the diminished survival and reproduction of members with unfavorable phenotypes.

but does not act as a major force in dictating the final balance of allele frequencies. Tshetverikov concluded that populations in nature absorb mutations "like a sponge" and retain them in a heterozygous condition, thereby providing a source of variability for future change.

In population genetics, it is most useful to consider new mutations as they affect the survival and reproductive potential of the individual who inherits them. In this regard, a new mutation may be beneficial, neutral, or deleterious. For structural genes that encode proteins, the effects of new mutations will depend on their impact on protein function. Neutral and deleterious mutations are far likelier to occur than beneficial mutations. For example, alleles can be altered in many different ways that render an encoded protein defective. Deletions, frameshift mutations, missense mutations, and nonsense mutations all may cause a gene to express a protein that is nonfunctional or less functional than the wild-type protein (see Chapter 17, pp. 456–457). Neutral mutations can also occur in several different ways. For example, a neutral mutation can change the wobble base without affecting the amino acid sequence of the encoded protein, or it can be a missense mutation that has no effect on protein function. Such mutations are point mutations at specific sites within the coding sequence. Neutral mutations can also occur within noncoding sequences of genes, such as introns. By comparison, beneficial mutations are relatively uncommon. To be advantageous, a new mutation could alter the amino acid sequence of a protein to yield a better-functioning product. While such mutations do occur, they are expected to be very rare for a population in a stable environment.

The **mutation rate** is the probability that a gene will be altered by a new mutation. It is expressed as the number of new mutations in a given gene per generation. It is commonly thought to be in the range of 1 in 100,000 to 1 in 1,000,000, or 10^{-5} to 10^{-6} per generation. However, studies of mutation rates typically follow the change of a normal (functional) gene to a deleterious (nonfunctional) allele. The mutation rate producing beneficial alleles is expected to be substantially less.

It is clear that new mutations provide genetic variability. Population geneticists also want to know how the mutation rate affects the allele frequencies in a population over time. To appreciate this idea, let's take the simple case where a gene exists in a functional allele, A; the allele frequency of A is denoted by the variable p. A deleterious mutation can convert the A allele into a nonfunctional allele, a. The allele frequency of a is designated by q. The conversion of the A allele into the a allele by mutation will occur at a rate that we call u. If we assume that the rate of the reverse mutation (a to A) is negligible, the increase in the frequency of the a allele after one generation will be

$$\Delta q = up$$

For example, let's consider the following conditions:

$p = 0.8$ (i.e., frequency of A is 80%)

$q = 0.2$ (i.e., frequency of a is 20%)

$u = 10^{-5}$ (i.e., the mutation rate of converting A to a)

$\Delta q = (10^{-5})(0.8) = (0.00001)(0.8) = 0.000008$

Therefore, in the next generation,

$$q_{n+1} = 0.2 + 0.000008 = 0.200008$$
$$p_{n+1} = 0.8 - 0.000008 = 0.799992$$

As we can see from this calculation, new mutations do not significantly alter the allele frequencies in a single generation.

We can use the following equation to calculate the change in allele frequency after any number of generations:

$$(1 - u)^t = \frac{P_t}{P_0}$$

where,

    u is the mutation rate of the conversion of A to a

    t is the number of generations

    p_0 is the allele frequency of A in the starting generation

    p_t is the allele frequency of A after t generations

As an example, let's suppose that the allele frequency of A is 0.8, $u = 10^{-5}$, and we want to know what the allele frequency will be after 1000 generations ($t = 1000$). Plugging these values into the preceding equation and solving for p_t,

$$(1 - 0.00001)^{1000} = \frac{p_t}{0.8}$$

$$p_t = 0.792$$

Therefore, after 1000 generations the frequency of A has dropped only from 0.8 to 0.792. Again, these results point to how slowly the mutation rate effects changes in allele frequencies. In natural populations, the rate of new mutation is rarely a significant catalyst in shaping allele frequencies. Instead, other processes such as migration, genetic drift, and natural selection have far greater effects on allele frequencies. We will examine how these other factors work in the remainder of this chapter.

In small populations, allele frequencies can be altered by random genetic drift

During the 1930s, Sewall Wright played a large role in developing the concept of **random genetic drift**, which refers to random changes in allele frequencies due to sampling error. In other words, allele frequencies may drift from generation to generation as a matter of chance. This situation is similar to flipping a coin. If we flipped a coin 10 times, we might get 4 heads and 6 tails; the frequency of heads is 0.4. If we do it again, we might get 5 heads and 5 tails for a frequency of 0.5. The frequency of heads is drifting due to random sampling error.

In populations, genetic drift favors either the loss or the fixation of an allele. The rate at which an allele is lost or becomes fixed at 100% depends on the population size. Figure 26-7 illustrates the potential consequences of genetic drift in one large

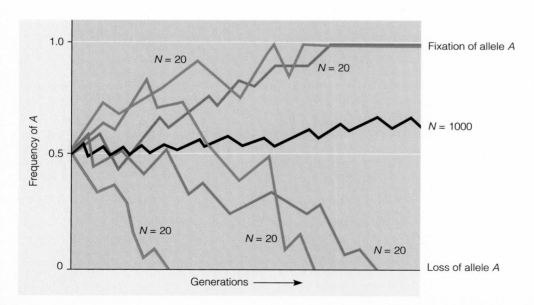

FIGURE **26-7**

A hypothetical simulation of random genetic drift. In all cases, the starting allele frequencies are $A = 0.5$ and $a = 0.5$. The colored lines illustrate five populations in which $N = 20$; the black line shows a population in which $N = 1000$.

($N = 1000$) and five small ($N = 20$) populations. At the beginning of this hypothetical simulation, all of these populations have identical allele frequencies: $A = 0.5$ and $a = 0.5$. In the five small populations, this allele frequency fluctuates substantially from generation to generation. Eventually, in these small populations, one of the alleles is eliminated and the other is fixed at 100%. At this point, an allele has become monomorphic and cannot fluctuate any further. By comparison, the allele frequencies in the large population fluctuate much less, since random sampling error is expected to have a smaller impact on allele frequencies. Nevertheless, genetic drift will lead to homozygosity even in large populations, but this will take many more generations to occur.

An important controversy in population genetics centers on whether most genetic variation is due to the accumulation of neutral mutations by genetic drift, or to mutations with a survival advantage that are promoted by natural selection. As we will see in Chapter 27, the DNA sequencing of genes has revealed that both processes have been important in the evolution of species.

Now let's ask two questions:

1. How many new mutations do we expect in a natural population?
2. How likely is it that a new mutation will be either fixed in or eliminated from a population due to random genetic drift?

With regard to the first question, the average number of new mutations depends on the mutation rate (u) and the number of individuals in a population (N). If each individual has two copies of the gene of interest, then the expected number of new mutations in this gene is:

$$\text{expected number of new mutations} = 2Nu$$

From this, we can see that a new mutation is more likely to occur in a large population than in a small one. This makes sense, since the larger population has more copies of the gene to be mutated. If a new mutation does occur, it may be eliminated or fixed in a population due to random genetic drift. If a mutation is neutral, the probability of fixation of a newly arising allele is:

$$\text{probability of fixation} = 1/2N \quad \text{(assuming equal numbers of males and females contribute to the next generation)}$$

In other words, the probability of fixation is the same as the allele frequency in the population. For example, if $N = 20$, the probability of fixation equals $1/(2 \times 20)$, or 2.5%. Conversely, a new allele may be lost from the population:

$$\text{probability of elimination} = 1 - \text{probability of fixation}$$
$$= 1 - 1/2N$$

As you may have noticed, the value of N has opposing effects with regard to mutations and their eventual fixation in a population. When N is very large, new mutations are much more likely to occur. Each new mutation, however, has a greater chance of being eliminated from the population due to random genetic drift. On the other hand, when N is small, the probability of new mutations is also small, but if they occur, the likelihood of fixation is relatively large.

Now that we have an appreciation for the phenomenon of genetic drift, we can ask a third question:

3. If fixation does occur, how many generations is it likely to take?

Again, this will depend on the number of individuals in the population:

$$\bar{t} \approx 4N$$

where,

$\bar{t}$ equals the average number of generations to achieve fixation

N equals the number of individuals in the population, assuming that males and females contribute equally to each succeeding generation

As you may have expected, allele fixation will take much longer in large populations. If there are 1 million breeding members in a population, it will take 4 million generations, perhaps an insurmountable period of time, to reach fixation. In a small group of 100 individuals, however, fixation will only take 400 generations. As we will see in Chapter 27, the drifting of neutral alleles among different populations and species provides a way to measure the rate of evolution and can be used to determine evolutionary relationships.

Our preceding discussion of random genetic drift has emphasized two important points. First, genetic drift ultimately operates in a directional manner with regard to allele frequency. It leads to either allele fixation or elimination. Second, its impact is more significant in smaller populations. In nature, there are several interesting ways that small population size and genetic drift affect the genetic composition of a species. For example, some species occupy wide ranges in which small populations become geographically isolated from the rest of the species. The allele frequencies within these small populations are more susceptible to genetic drift. Since this is a random process, small isolated populations tend to become more genetically disparate in relation to other populations.

A second example of genetic drift is called the **bottleneck effect**. In nature, a population can be reduced dramatically in size by events such as earthquakes, floods, drought, and the human destruction of habitat. These types of events may randomly eliminate most of the members of the population without regard to genetic composition. The period of the bottleneck, when the population size is very small, may be influenced by genetic drift. First, the original surviving members may have allele frequencies that differ from those of the original population. In addition, allele frequencies are expected to drift substantially during the generations when the population size is small. In extreme cases, alleles may even be eliminated. Eventually, the bottlenecked population may regain its original size (Figure 26-8a). However, the new population will have less genetic variation than the original one. As mentioned earlier in this chapter, the African cheetah population has lost nearly all of its genetic variation (Figure 26-8b). This is due to a bottleneck effect. An analysis by population geneticists has suggested that a severe bottleneck occurred approximately 10,000–12,000 years ago, reducing the population size to near extinction. The population eventually rebounded, but the bottleneck reduced the genetic variation to very low levels.

A third interesting case of genetic drift is the **founder effect**. This refers to the phenomenon in which a small group of individuals separates from a larger population and establishes a colony in a new location. For example, a few individuals may migrate away from a large continental population and become the founders of an island population. The founder effect has two important consequences. First, the founding population is expected to have less genetic variation than the original population from which it was derived. Second, as a matter of chance, the allele frequencies in the founding population may differ markedly from those of the original population.

Population geneticists have studied many examples where isolated populations have been derived via colonization of members of another population. In the 1960s, Victor McKusick of Johns Hopkins University studied allele frequencies in the Old Order Amish of Lancaster County, Pennsylvania. This is a group of about 8000 people, descended from just three couples who immigrated to the United States in 1770. Among this population of 8000, a genetic disease known as the

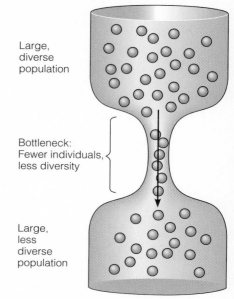

(a) Bottleneck effect

(b) An African cheetah

FIGURE **26-8**

The bottleneck effect, an example of genetic drift. (a) A representation of the bottleneck effect. **(b)** The African cheetah. The modern species is monomorphic for nearly all of its genes. This low genetic variation is due to a genetic bottleneck that is thought to have occurred about 10,000 to 12,000 years ago.

Ellis–vanCreveld syndrome (a recessive form of dwarfism) was found at a frequency of 0.07 or 7%. By comparison, this disorder is extremely rare in other human populations, even the population from which the founding members had originated. The high frequency in the Lancaster County population is a chance occurrence due to the founder effect.

Migrations between two populations can alter allele frequencies

We have just seen how migration to a new location by a relatively small group can result in a founding population with an altered genetic composition due to genetic drift. In addition, migration between two different established populations can alter allele frequencies. For example, a species of bird may occupy two geographic regions that are separated by a large body of water. On rare occasions, the prevailing winds may allow birds from the western population to fly over this body of water and become members of the eastern population. If the two populations have different allele frequencies, and if migration occurs in sufficient numbers, this may alter the allele frequencies in the eastern population.

After migration has occurred, the new (eastern) population is called a **conglomerate**. To calculate the allele frequencies in the conglomerate, we need two kinds of information. First, we must know the original allele frequencies in the donor and recipient populations. Second, we must know the proportion of the conglomerate population that is due to migrants. With these data, we can calculate the change in allele frequency in the conglomerate population using the following equation:

$$\Delta p_C = m(p_D - p_R)$$

where,

Δp_C is the change in allele frequency in the conglomerate population

p_D is the allele frequency in the donor population

p_R is the allele frequency in the original recipient population

m is the proportion of migrants that make up the conglomerate population:

$$m = \frac{\text{number of donor individuals in the conglomerate population}}{\text{total number of individuals in the conglomerate population}}$$

As an example, let's suppose the allele frequency of A is 0.7 in the donor population and 0.3 in the recipient population. A group of 20 individuals migrate and join the recipient population, which originally had 80 members. Thus,

$$m = \frac{20}{20 + 80}$$

$$= 0.2$$

$$\Delta p_c = m(P_D - P_R)$$

$$= 0.2(0.7 - 0.3)$$

$$= 0.08$$

We can now calculate the allele frequency in the conglomerate:

$$p_C = p_R + \Delta p_C$$
$$= 0.3 + 0.08 = 0.38$$

Therefore, in the conglomerate population, the allele frequency of A has changed from 0.3 (its value before migration) to 0.38. This increase in allele frequency arises from the higher allele frequency for A in the donor population. Since population geneticists are concerned with allele frequencies rather than the migration of individuals, this phenomenon is called **gene flow**. It occurs whenever individuals migrate between populations having different allele frequencies.

In our previous example, we considered the consequences of a unidirectional migration from a donor to a recipient population. In nature, it is common for individuals to migrate in both directions. This bidirectional migration has two important consequences. Depending on its rate, migration tends to reduce differences in allele frequencies between neighboring populations. In fact, population geneticists can analyze allele frequencies in two different populations to evaluate the rate of migration between them. Populations that frequently mix their gene pools via migration tend to have similar allele frequencies, whereas isolated populations are expected to be more disparate. In addition, migration can enhance genetic diversity within a population. As discussed earlier in this chapter, new mutations are relatively rare events. Therefore, a particular mutation may only arise in one population. Migration may then introduce this new allele into neighboring populations.

Natural selection favors the survival of the fittest

In the 1850s, Charles Darwin and Alfred Russell Wallace independently proposed the theory of **natural selection**. We will discuss the phenotypic consequences of natural selection in greater detail in Chapter 27. According to this idea, there is a "struggle for existence." The conditions found in nature result in the selective survival and reproduction of individuals whose characteristics make them well adapted to their environment. These surviving individuals are more likely to reproduce and contribute offspring to the next generation. More recently, population geneticists have realized that natural selection can be related not only to differential survival, but also to mating efficiency and fertility.

A modern description of natural selection can relate our knowledge of molecular genetics to the phenotypes of individuals:

1. Within a population, there is genetic variation arising from differences in DNA sequences. Distinct alleles may encode proteins of differing function.
2. Some alleles may encode proteins that enhance an individual's survival or reproductive capability as compared with other members of the population. For example, an allele may produce a protein that is more efficient at a higher temperature, conferring on the individual a greater probability of survival in a hot climate.
3. Individuals with beneficial alleles are more likely to survive and contribute to the gene pool of the next generation.
4. Over the course of many generations, allele frequencies of many different genes may change through natural selection, thereby significantly altering the characteristics of a species. The net result of natural selection is to make a population better adapted to its environment and/or more successful at reproduction.

Natural selection acts on phenotypes (which are derived from an individual's genotype). With regard to quantitative traits, there are three ways that natural selection may operate (Figure 26-9):

1. **Directional selection** favors the survival of one extreme of a phenotypic distribution that is better adapted to an environmental condition. For example, in a dimly lit forest, directional selection would favor snails with darker shells, because they would be better camouflaged and less susceptible to predation (Figure 26-9a).

FIGURE **26-9**

The three patterns whereby natural selection acts on phenotypes.

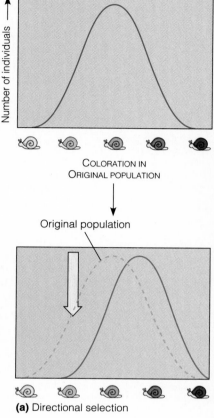

(a) Directional selection

OR

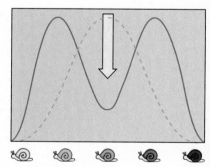

(b) Disruptive selection

OR

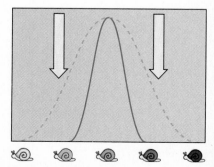

(c) Stabilizing selection

2. **Disruptive (or diversifying) selection** favors the survival of two (or more) different phenotypic classes of individuals. For example, disruptive selection favors the survival of dark-shelled and light-shelled snails that occupy a heterogeneous environment (Figure 26-9b). Disruptive selection is likely to occur in species that occupy diverse environments so that some members of the species will survive in each type of environment.

3. **Stabilizing selection** favors the survival of individuals with intermediate phenotypes. For example, moderately colored snails may survive better in an environment that has an intermediate level of light, such as a tall grassland (Figure 26-9c).

During the 1920s and 1930s, Haldane, Fisher, and Wright developed mathematical relationships to explain the theory of natural selection. As our knowledge of natural selection increased, it became apparent that there are many different ways for natural selection to operate. In this text, we will consider a few examples of natural selection involving a single gene that exists in two alleles. In reality, however, natural selection acts on populations of individuals, in which many genes are polymorphic and each individual contains thousands or tens of thousands of different genes.

To begin our quantitative discussion of natural selection, we must examine the concept of **Darwinian fitness**. Fitness is the relative likelihood that a phenotype will survive and contribute to the gene pool of the next generation as compared with another phenotype. Although this property often correlates with physical fitness, the two ideas should not be confused. Darwinian fitness is a measure of reproductive superiority. An extremely fertile phenotype may have a higher Darwinian fitness than a less fertile phenotype that appears more physically fit.

To consider Darwinian fitness in a quantitative way, let's use our example of two genes existing in the A and a alleles. We can assign fitness values to each of the three genotypic classes according to their relative reproductive potential. For example, let's suppose that in a population the average reproductive success of the three genotypes are as follows:

AA produce 5 offspring

Aa produce 4 offspring

aa produce 1 offspring

By convention, the genotype with the highest reproductive ability is given a fitness value of 1.0. Fitness values are denoted by the variable W. The fitness values of the other genotypes are assigned values relative to this 1.0 value:

Fitness of AA: $W_{AA} = 1.0$

Fitness of Aa: $W_{Aa} = 4/5 = 0.8$

Fitness of aa: $W_{aa} = 1/5 = 0.2$

Keep in mind that the differences in reproductive achievement among these three genotypes could stem from different reasons. A very common reason is that the most fit phenotype is more likely to survive. Of course, an individual must survive into adulthood to reproduce. A second possibility is that the most fit phenotype is more likely to mate. For example, a bird with brightly colored feathers may have an easier time attracting a mate than a bird with duller plumage. Finally, a third possibility is that the fittest phenotype may be more fertile. It may produce a higher number of gametes, or gametes that are more successful at fertilization.

An opposite parameter that population geneticists also use in their calculations is the *selection coefficient (s)*. This measures the degree to which a genotype is selected against:

$$s = 1 - W$$

By convention, of the existing genotypes, the one with the highest fitness has an s value of zero. Genotypes at a selective disadvantage have s values that are greater than zero but less than or equal to 1.0. An extreme case is a recessive lethal allele. It would have an s value of 1.0 in the homozygote, while the s value in the heterozygote could be zero.

   Natural selection alters allele frequencies in a step-by-step, generation-per-generation way. To appreciate how this occurs, let's take a look at how fitness affects the Hardy–Weinberg equilibrium and allele frequencies. Again, let's suppose that a gene exists in two alleles, A and a. The three fitness values are

$$W_{AA} = 1.0$$
$$W_{Aa} = 0.8$$
$$W_{aa} = 0.2$$

In the next generation, we expect that the Hardy–Weinberg equilibrium will be modified in the following way:

Frequency of AA: p^2W_{AA}
Frequency of Aa: $2pqW_{Aa}$
Frequency of aa: q^2W_{aa}

   In a population that is changing due to natural selection, these three terms will not add up to 1.0, as they would in the Hardy–Weinberg equilibrium. Instead, the three terms sum to a value known as the **mean fitness of the population:**

$$p^2W_{AA} + 2pqW_{Aa} + q^2W_{aa} = \overline{W}$$

Dividing both sides of the equation by the mean fitness of the population,

$$\frac{p^2W_{AA}}{\overline{W}} + \frac{2pqW_{Aa}}{\overline{W}} + \frac{q^2W_{aa}}{\overline{W}} = 1$$

Using this equation, we can calculate the expected genotype and allele frequencies after one generation of natural selection:

Frequency of AA genotype: $\dfrac{p^2W_{AA}}{\overline{W}}$

Frequency of Aa genotype: $\dfrac{2pqW_{Aa}}{\overline{W}}$

Frequency of aa genotype: $\dfrac{q^2W_{aa}}{\overline{W}}$

Allele frequency of A: $p_A = \dfrac{p^2W_{AA}}{\overline{W}} + \dfrac{pqW_{Aa}}{\overline{W}}$

Allele frequency of a: $q_a = \dfrac{q^2W_{aa}}{\overline{W}} + \dfrac{pqW_{Aa}}{\overline{W}}$

   As an example, let's suppose that the starting allele frequencies are $A = 0.5$ and $a = 0.5$, and use fitness values of 1.0, 0.8, and 0.2 for the three genotypes, AA, Aa, and aa, respectively. We begin by calculating the mean fitness of the population:

$$p^2W_{AA} + 2pqW_{Aa} + q^2W_{aa} = \overline{W}$$
$$\overline{W} = (0.5)^2(1) + 2(0.5)(0.5)(0.8) + (0.5)^2(0.2)$$
$$= 0.25 + 0.4 + 0.05 = 0.7$$

After one generation of selection, we get,

Frequency of AA genotype: $\dfrac{p^2 W_{AA}}{\overline{W}} = \dfrac{(0.5)^2(1)}{0.7} = 0.36$

Frequency of Aa genotype: $\dfrac{2pq W_{Aa}}{\overline{W}} = \dfrac{2(0.5)(0.5)(0.8)}{0.7} = 0.57$

Frequency of aa genotype: $\dfrac{q^2 W_{aa}}{\overline{W}} = \dfrac{(0.5)^2(0.2)}{0.7} = 0.07$

Allele frequency of A: $\quad p_A = \dfrac{p^2 W_{AA}}{\overline{W}} + \dfrac{pq W_{Aa}}{\overline{W}}$

$$= \dfrac{(0.5)^2(1)}{0.7} + \dfrac{(0.5)(0.5)(0.8)}{0.7} = 0.64$$

Allele frequency of a: $\quad q_a = \dfrac{q^2 W_{aa}}{\overline{W}} + \dfrac{pq W_{Aa}}{\overline{W}}$

$$= \dfrac{(0.5)^2(0.2)}{0.7} + \dfrac{(0.5)(0.5)(0.8)}{0.7} = 0.36$$

After one generation, the allele frequency of A has increased from 0.5 to 0.64, while the frequency of a has decreased from 0.5 to 0.36. This is because the AA genotype has the highest fitness, while the Aa and aa genotypes have progressively lower fitness values. Another interesting feature of natural selection is that it raises the mean fitness of the population. If we assume that the individual fitness values are constant, the mean fitness of this next generation is

$$\overline{W} = p^2 W_{AA} + 2pq W_{Aa} + q^2 W_{aa}$$

$$= (0.64)^2(1) + 2(0.64)(0.36)(0.8) + (0.36)^2(0.2)$$

$$= 0.80$$

The mean fitness of the population has increased from 0.7 to 0.8. This population is more well adapted to its environment than the previous one. Another way of viewing this calculation is that the subsequent population has a greater reproductive potential than the previous one. We could perform the same types of calculations to find the allele frequencies and mean fitness value in the next generation. If we assume that the individual fitness values remain constant, the frequencies of A and a in the next generation are 0.85 and 0.15, respectively, and the mean fitness increases to 0.931. As we can see, the general trend is to increase A, decrease a, and increase the mean fitness of the population.

In the previous example, we considered the effects of natural selection by beginning with allele frequencies at intermediate levels (namely, $A = 0.5$ and $a = 0.5$). Figure 26-10 illustrates what would happen if a new mutation introduced the A allele into a population that was originally monomorphic for the a allele. As before, the AA homozygote has a fitness of 1.0, the Aa heterozygote 0.8, and the recessive aa homozygote 0.2. Initially, the A allele is at a very low frequency in the population. If it is not lost initially due to genetic drift, its frequency slowly begins to rise and then, at intermediate values, rises much more rapidly.

Eventually, this type of natural selection may lead to the fixation of the beneficial allele. However, this new beneficial allele is in a precarious situation when its frequency is very low. As mentioned earlier in this chapter, random genetic drift is likely to eliminate new neutral mutations due to sampling error. Genetic drift can also eradicate beneficial alleles, although this is not as likely, because natural selection favors them.

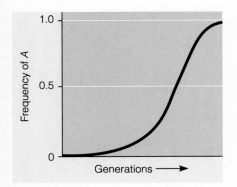

FIGURE **26-10**

The fate of a beneficial dominant allele that is introduced as a new mutation into a population. This allele is beneficial in the homozygous condition: $W_{AA} = 1.0$. The corresponding heterozygote, Aa ($W_{Aa} = 0.8$), and recessive homozygote, aa ($W_{aa} = 0.2$), have lower fitness values.

Balanced polymorphisms may exist due to heterozygote superiority or heterogeneous environments

As we have just seen, selection for an allele that is beneficial in a homozygote may lead to its eventual fixation in a population. Likewise, genetic drift also eliminates or fixes neutral alleles. So how do we explain the high levels of polymorphisms observed for genes in natural populations?

One possibility is that many of these polymorphisms involve neutral alleles not subject to natural selection. The genes become polymorphic due to genetic drift, and there has not been enough time for fixation to occur. (As we have seen, fixation can require vast amounts of time.) In other cases, a polymorphism may have reached an equilibrium where opposing selective forces have reached a balance; the population is not evolving toward fixation or allele elimination. This situation, known as a **balanced polymorphism**, can occur for several different reasons.

In some cases, a balanced polymorphism occurs when the heterozygote is at a selective advantage. The higher fitness of the heterozygote is balanced by the lower fitnesses of both corresponding homozygotes. Let's consider the following case of relative fitness:

$$W_{AA} = 0.7$$
$$W_{Aa} = 1.0$$
$$W_{aa} = 0.4$$

This is an example of **heterozygote advantage**, also called **overdominance**. The selection coefficients are:

$$s_{AA} = 1 - 0.7 = 0.3$$
$$s_{Aa} = 1 - 1.0 = 0$$
$$s_{aa} = 1 - 0.4 = 0.6$$

Under these conditions, the population will reach an equilibrium in which

$$\text{Allele frequency of } A = \frac{s_{aa}}{s_{AA} + s_{aa}}$$

$$= \frac{0.6}{0.3 + 0.6} = 0.67$$

$$\text{Allele frequency of } a = \frac{s_{AA}}{s_{AA} + s_{aa}}$$

$$= \frac{0.3}{0.3 + 0.6} = 0.33$$

Balanced polymorphisms can sometimes explain the high frequency of alleles that are deleterious in a homozygous condition. A classic example is the H^S allele of the human β-globin gene. A homozygous $H^S H^S$ individual displays sickle cell anemia, a disease that leads to the sickling of the red blood cells. The $H^S H^S$ homozygote has a lower fitness than a homozygote with two normal copies of the β-globin gene, $H^A H^A$. However, the heterozygote, $H^A H^S$, has the highest level of fitness in areas where malaria is endemic (see Figure 26-11). Compared with normal, $H^A H^A$, homozygotes, the heterozygotes have a 10–15% better chance of survival if infected by the malarial parasite, *Plasmodium falciparum*. Therefore, the H^S allele is maintained in populations where malaria is prevalent, even though the allele is detrimental in the homozygous state.

Besides sickle cell anemia, several other gene mutations that cause human disease in the homozygous state are thought to be prevalent because of heterozygote advantage. These include cystic fibrosis, in which the heterozygote is resistant to diarrheal disease; PKU, in which the heterozygous fetus is resistant to abortion caused by a fungal toxin; and Tay–Sachs disease, in which the heterozygote is resistant to tuberculosis.

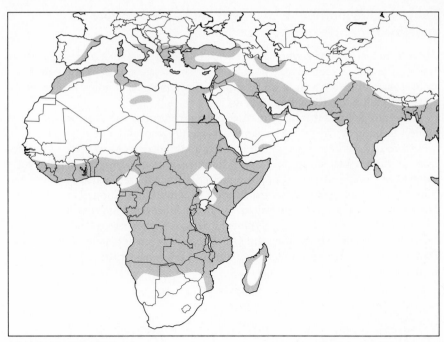

(a) Malaria prevalence

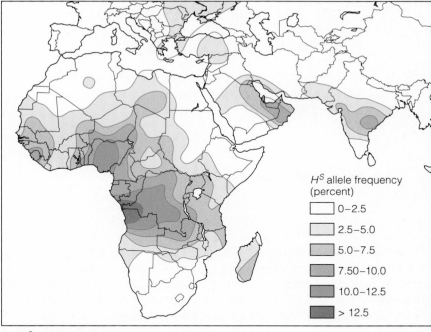

FIGURE **26-11**

The geographical relationship between malaria and the frequency of the sickle cell allele in human populations. (a) The geographical prevalence of malaria in Africa and surrounding areas. **(b)** The frequency of the H^S allele in the same areas.

GENES ──→ TRAITS: The sickle cell allele of the β-globin gene is maintained in human populations as a balanced polymorphism. In areas where malaria is prevalent, the heterozygote carrying one copy of the H^S allele has a greater fitness than either of the corresponding homozygotes ($H^A H^A$ and $H^S H^S$). Therefore, even though the $H^S H^S$ homozygotes suffer the detrimental consequences of sickle cell anemia, this negative aspect is balanced by the beneficial effects of malarial resistance in the heterozygotes.

H^S allele frequency (percent)

☐ 0–2.5
☐ 2.5–5.0
☐ 5.0–7.5
☐ 7.50–10.0
☐ 10.0–12.5
☐ > 12.5

(b) H^S allele frequency

Balanced polymorphism also may occur when a species occupies a region that contains heterogeneous environments. One environment may favor a particular allele, while a neighboring environment may favor a different allele. Figure 26-12a shows an example. This is a photograph of land snails, *Cepaea nemoralis*, that live in woods and open fields. This snail is polymorphic in color and banding patterns. In 1954, A. J. Cain and P. M. Sheppard of the University of Oxford found that snail color was correlated with the environment. As shown in Figure 26-12b, the high-

(a) Land snails

FIGURE **26-12**

Polymorphism in the land snail *Cepaea nemoralis*.
(a) This species of snail can exist in several different colors and banding patterns. **(b)** Coloration of the snails is correlated with the specific environments where they are located. Taken from: Cain, A. J., and Sheppard, P. M. (1954) Natural selection in *Cepaea. Genetics 39,* 89–116.

GENES $\longrightarrow$ TRAITS: Snail coloration is an example of genetic polymorphism due to heterogeneous environments; the genes governing shell coloration are polymorphic. The predation of snails is correlated with their ability to be camouflaged in their natural environment. Snails in the beechwoods, where the soil is dark, have the highest frequency of brown shell color. Pink snails are usually found in the leaf litter of forest floors in deciduous woods. And yellow snails are prevalent in the sunny, unwooded areas of hedgerows and rough herbage.

Habitat	Brown	Pink	Yellow
Beechwoods	0.23	0.61	0.16
Deciduous woods	0.05	0.68	0.27
Hedgerows	0.05	0.31	0.64
Rough herbage	0.004	0.22	0.78

(b) Frequency of snail color

est frequency of brown shell color was found in snails in the beechwoods, where there are wide expanses of dark soil. Pink snails are most common in the leaf litter of forest floors in deciduous woods. The yellow snails are most abundant in the sunny, unwooded areas of hedgerows and rough herbage. It has been proposed that disruptive selection has favored the survival of these different phenotypes. Migration between the snail populations keeps the polymorphism in balance among these different environments.

Kettlewell studied industrial melanism in the moth *Biston betularia* as a modern example of natural selection

The concept of natural selection provides a framework for explaining how allele frequencies are maintained in populations; it also explains how species become adapted to the environments in which they live. As scientists, we tend to examine the ultimate result of natural selection. In other words, we can observe species in their native environments, and in some cases, it seems obvious how certain characteristics provide an organism with a survival advantage. The smell of a skunk and the ability of a cactus to retain water are classic examples. To support the theory of natural selection, population geneticists would like to witness the process in a species over time.

The first observed example of modern natural selection occurred in the decades following the Industrial Revolution (which began in the latter half of the 19th century). As a result of industrialization, large areas of the earth's surface are blanketed by the fallout of smoke particles. In and around industrial areas, this fallout is dramatically more evident than in undeveloped areas. The smoke particles tend to kill vegetative lichens on the trunks and boughs of trees. Rain washes the pollutants down the tree trunks until they are bare and black. By comparison, tree trunks in unpolluted areas are much lighter and are speckled with growing lichens. During the early 1900s, scientists in Europe and North America began to notice the proliferation of melanic moth varieties that are darker in color than the previously abundant varieties. Examples are shown here:

Light variety (left) and dark variety (right) of *Biston betularia.*

This moth, *Biston betularia*, is polymorphic. One variety, *carbonaria*, is very dark, whereas the *typical* variety is much lighter in color. As seen here, the *carbonaria* morph is easily seen on a lichen-covered tree trunk, and the *typical* morph is readily visible on dark trunks. These observations led geneticists to propose that the Industrial Revolution was selecting for the survival of *carbonaria*. It was hypothesized that the darkened color of the *carbonaria* morph made it less susceptible to predation by birds. Supporting this idea, a field study in England showed that the *carbonaria* morph is more prevalent in polluted areas of Birmingham, while the *typical* morph is predominant in unpolluted areas (Figure 26-13).

In the experiment described here, conducted in 1956, H. Bernard Kettlewell, a British biologist, set out to test the hypothesis that the coloration polymorphism in *B. betularia* is due to natural selection involving predation by birds.

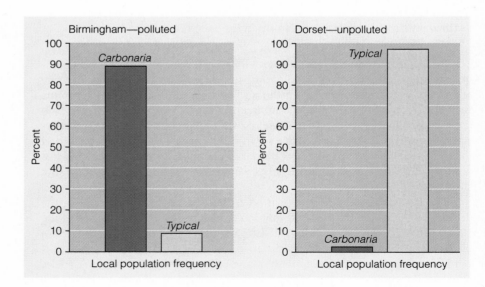

FIGURE **26-13**

A comparison of the prevalence of the *carbonaria* morph and the *typical* morph of the moth *Biston betularia* in polluted and unpolluted areas.

THE HYPOTHESIS

The coloration of *B. betularia* affects the probability that birds will see and eat them. This natural selection leads to the survival of those moths with coloration more closely matching that on which they rest, whether that be in polluted or unpolluted woods.

TESTING THE HYPOTHESIS

Starting material: Samples of *B. betularia*, both *carbonaria* and *typical*, were collected from the wild. In some cases, they were bred in the laboratory, to be released later into the wild.

	Experimental Level	**Conceptual Level**
1. Mark the underside of *carbonaria* and *typical* moths with cellulose paint.	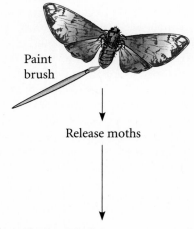 Paint brush	Marking provides identification of released moths versus those already present.
2. Release equal numbers of marked *carbonaria* and *typical* moths in either polluted woods near an industrial area of Birmingham, or in an unpolluted woods in Dorset, England.	Release moths	
3. On several consecutive days, recapture moths that have been marked. This can be done by attracting them to a mercury vapor light. Record the number of recaptured moths.	Recapture moths: 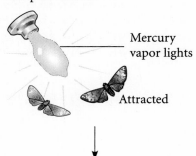Mercury vapor lights Attracted Record number of *carbonaria* and *typical* moths	Recapture: Only surviving moths will be recaptured Compare the numbers of *carbonaria* and *typical* moths: Natural selection may be occurring due to differences in predation by birds
4. Also, sit in a blind and observe the moths on a tree trunk. Record the numbers and types of sitting moths that are eaten by birds.		Bird predation may be higher for moths that are not well camouflaged.

THE DATA

Location	Number Eaten by Birds		Percentage of Marked Moths that Were Recaptured	
	Carbonaria	Typical	Carbonaria	Typical
Dorset: unpolluted woods	43	15	7.0	12.5
Birmingham: polluted wood	26	164	27.5	13.1

Data taken from: Kettlewell, H. B. D. (1956) Further Selection Experiments on Industrial Melanism in the *Lepidoptera. Heredity 10,* 287–301.

INTERPRETING THE DATA

As shown in the data, the *carbonaria* moths were more likely to be eaten by birds in an unpolluted woods. This observation was corroborated by the lower percentage of *carbonaria* moths that were recaptured in this area. Conversely, the *typical* moths were much more likely to be eaten in a polluted woods. Again, this observation was supported by the lower percentage of recapture for *typical* moths in Birmingham. Taken together, these results supported the hypothesis that *carbonaria* have a greater chance of surviving in polluted woods, *typicals* in unpolluted woods. In other words, the consequences of the Industrial Revolution have resulted in natural selection for the *carbonaria* variety. As stated by Kettlewell, "These experiments showed that birds act as selective agents and that the melanic forms of *betularia* are at a cryptic advantage in an industrial area such as Birmingham."

Genetic load is the negative consequence of genetic variation

As we have just seen, genetic variation in a population can lead to adaptation via natural selection. In general, we often think of genetic variation in a positive light. It allows species to adapt to their environment and provides the raw material for evolution. Furthermore, genetic diversity adds interest and beauty to our perception of biology. On the negative side, however, certain kinds of genetic variation are detrimental to the survival of a population. The term **genetic load (L)** refers to genetic variation that decreases the average fitness of a population as compared with a (theoretical) maximum or optimal value:

$$L = \frac{W_{max} - W}{W_{max}}$$

where,

L equals the genetic load

W equals the average fitness of a population

W_{max} equals the maximal fitness of a population

There are many factors that contribute to the genetic load in populations. Several of these are listed here:

1. *Mutation*: Most new mutations are deleterious. Therefore, the accumulation of new mutations in a population contributes to the genetic load.
2. *Segregation*: As mentioned earlier in this chapter, certain mutations are beneficial in the heterozygous condition but detrimental to homozygotes. Therefore, the segregation of alleles to create homozygotes causes a genetic load in the population.
3. *Recombination*: Sometimes, linked genes occur in optimal combinations. For example, two genes may be found in *A* and *a* alleles and *B* and *b* alleles. Let's suppose that the *AB* and *ab* combinations are beneficial, but the *Ab* and *aB* combinations are less fit. Although natural selection may select for

chromosomes containing the *AB* and *ab* pairs of genes, recombination (i.e., crossing over) may create *Ab* or *aB* combinations and thereby decrease the fitness of the population.

4. *Heterogeneous environments*: As mentioned earlier in this chapter (namely, for snail coloration), some species occupy two or more types of environments. This creates polymorphisms that are suited to different environments. Some members of the population will have genotypes that are not well suited to their environment. For example, a darkly colored snail may be located in a sunny environment.

5. *Meiotic drive/gamete selection*: Certain types of alleles and chromosomal rearrangements are transmitted preferentially to gametes even though they may be detrimental to the survival of the organism. For example, some alleles favor gamete viability even though they may not be beneficial in the adult.

6. *Maternal–fetal incompatibility*: In a few cases, the genotype of the mother may interact in an unfavorable way with that of a fetus. A classic example is the Rh factor, a type of red blood cell antigen. If a mother is Rh⁻ and a fetus is Rh⁺, the mother's body will occasionally mount an immunological reaction against the fetus, thereby diminishing its chances for survival.

7. *Finite population*: In a small population, genetic drift causes allele frequencies to deviate from their optimal fitness values.

8. *Migration*: When individuals from one environment migrate to a new environment, they usually do not possess the optimal combination of alleles for survival. Therefore, migration tends to decrease the fitness of a population.

While it is beyond the scope of this text, population geneticists have derived mathematical formulae that describe the amount of genetic load that these various factors may contribute to a population.

In this chapter, we have surveyed some of the critical issues in *population genetics*. The primary focus of this field is understanding genetic variation within *populations*. A substantial proportion (e.g., about 30% in humans) of genes in natural populations are *polymorphic*, existing in two or more alleles. Population geneticists want to understand why this variation in the *gene pool* exists and how it may change.

According to the *Hardy–Weinberg equilibrium*, *allele frequencies* remain stable as long as several conditions are met. These include no new mutations, random mating, large population size, no migration, and no natural selection. While these conditions are never truly met, they can be approximated in large, well-adapted populations. Therefore, the Hardy–Weinberg equilibrium predicts *genotype frequencies*, and genetic stability when conditions are appropriate.

However, we know that many processes in nature promote genetic change. As we will see in the following chapter, these forces lead to the evolution of species. The initial source of genetic variation is mutation. However, the *mutation rate* is too low to account for the allele frequencies observed for most genes in natural populations. Instead, mutation provides the source of variation, and other evolutionary forces act to increase the frequency of some new alleles. In small populations, *neutral forces* such as *random genetic drift* can alter allele frequencies dramatically. Simply due to random sampling error, a new neutral mutation may be eliminated from a population or it may become fixed in it. This can occur via a *founder effect*, when a small population moves to a new site, or a *bottleneck effect*, when a population is reduced in size due to negative environmental events. Allele frequencies can also be altered by migration. When two populations differ in their allele frequencies, migration can alter the existing allele frequencies. This is known as *gene flow*.

Finally, *adaptive forces*, such as *natural selection*, make a species better adapted to its environment. Natural selection favors individuals with the greatest reproductive potential (i.e., the greatest *Darwinian fitness*). These may be individuals who are more likely to survive, more likely to mate, or more fertile. When natural selection favors one homozygote, it selects for the fixation of the favorable allele. However, when the heterozygote has the highest fitness, or when two or more phenotypes in an environmentally diverse region have the highest fitnesses, a *balanced polymorphism* will result. In Chapter 27, we will learn how the combined evolutionary processes of genetic drift, migration, and natural selection lead to the formation of new species. We will also examine how these processes change the sequences of genes among evolutionarily related species.

EXPERIMENTAL SUMMARY

A central approach in population genetics is to measure variation experimentally, and then examine the mathematical relationships between genetic variation and the occurrence of genotypes and phenotypes within populations. In this chapter, we have focused on how population geneticists explain phenotypic variation via mathematical theories. Experimentally, phenotypic variation can be measured by tallying the organisms within a population that display particular phenotypes. For example, we can count the number of dark snails and light snails that live in a particular location. This provides a measure of the phenotypic variation within this population. We also learned that variation can be studied at the molecular level by identifying allozymes. In Chapter 27, we will see that many other molecular approaches, particularly DNA sequencing, allow population geneticists to measure the genetic variation within populations.

After variation is measured, population geneticists attempt to develop and apply mathematical relationships that provide insight into the causes and dynamics of the variation. Although geneticists understand that mutation is the source of new variation, their mathematical formulae indicate that the rate of new mutations is too low to explain the experimentally observed variation in natural populations. Instead, other processes, such as genetic drift, migration, and natural selection must alter the frequency of new mutations after mutations have occurred. The mathematical theories of population geneticists tell us that these forces may be directional, leading to the eventual fixation or elimination of an allele, or they may promote a balanced polymorphism within a population. These theories also show us that variation may have negative consequences, known as the genetic load.

PROBLEM SETS

SOLVED PROBLEMS

1. The phenotypic frequency of individuals who cannot taste phenylthiocarbamide (PTC) is approximately 0.3. The inability to taste this bitter substance is due to a recessive allele. If we assume that there are only two alleles in the population (namely, tasters, *T*, and nontasters, *t*) and that the population is in Hardy–Weinberg equilibrium, calculate the frequencies of these two alleles.

Answer: Let's take p = allele frequency of the taster allele and q = the allele frequency of the nontaster allele. The frequency of nontasters is 0.3. This is the frequency of the genotype *tt*, which in this case is equal to q^2:

$$q^2 = 0.3$$

To determine the frequency q of the nontaster allele, we take the square root of both sides of this equation:

$$\sqrt{q^2} = \sqrt{0.3}$$
$$q = 0.55$$

With this value, we can calculate the frequency p of the taster allele:

$$p = 1 - q$$
$$= 1 - 0.55 = 0.45$$

2. In the pedigree shown here, answer the following questions with regard to individual VII-1:

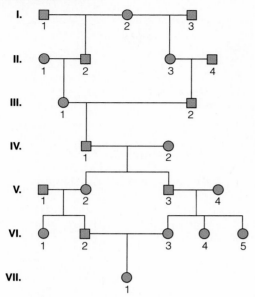

A. Who are the common ancestors?
B. What is the inbreeding coefficient?

Answer:

A. The common ancestors are IV-1 and IV-2. They are the grandparents of VI-2 and VI-3, who are the parents of VII-1.
B. The inbreeding coefficient is calculated using the formula

$$F = \Sigma(1/2)^n (1 + F_A)$$

In this case there are two common ancestors, IV-1 and IV-2. Also, IV-1 is inbred, because I-2 is a common ancestor to both of IV-1's parents. So, the first step is to calculate F_A, the inbreeding coefficient for this common ancestor. The inbreeding loop for IV-1 contains five people, III-1, II-2, I-2, II-3, and III-2. Therefore,

$$n = 5$$
$$F_A = (1/2)^5 = 0.03$$

Now we can calculate the inbreeding coefficient for VII-1. Each inbreeding loop contains five people: VI-2, V-2, IV-1, V-3, and VI-3; and VI-2, V-2, IV-2, V-3, and VI-3.Thus,

$$F = (1/2)^5 (1 + 0.03) + (1/2)^5 (1 + 0)$$
$$= 0.032 + 0.031 = 0.063$$

3. Let's suppose that pigmentation in a species of insect is controlled by a single gene existing in two alleles, *D* for dark and *d* for light. The heterozygote *Dd* is intermediate in color. In a heterogeneous environment, the allele frequencies are *D* = 0.7 and *d* = 0.3. This polymorphism is maintained because the environment contains some dimly lit forested areas and some sunny fields. During a hurricane, a group of 1000 insects is blown to a completely sunny area. In this environment, the fitness values are *DD* = 0.3, *Dd* = 0.7, and *dd* = 1.0. Calculate the allele frequencies in the next generation.

Answer: The first step is to calculate the mean fitness of the population:

$$p^2 W_{DD} + 2pq W_{Dd} + q^2 W_{dd} = \overline{W}$$
$$\overline{W} = (0.7)^2(0.3) + 2(0.7)(0.3)(0.7) + (0.3)^2(1.0)$$
$$= 0.15 + 0.29 + 0.09 = 0.53$$

After one generation of selection, we get

Allele frequency of *D*:

$$p_D = \frac{p^2 W_{DD}}{\overline{W}} + \frac{pq W_{Dd}}{\overline{W}}$$
$$= \frac{(0.7)^2(0.3)}{0.53} + \frac{(0.7)(0.3)(0.7)}{0.53}$$
$$= 0.55$$

Allele frequency of *d*:

$$q_d = \frac{q^2 W_{dd}}{\overline{W}} + \frac{pq W_{Dd}}{\overline{W}}$$
$$= \frac{(0.3)^2(1.0)}{0.53} + \frac{(0.7)(0.3)(0.7)}{0.53}$$
$$= 0.45$$

After one generation, the allele frequency of *D* has decreased from 0.7 to 0.55, while the frequency of *d* has increased from 0.3 to 0.45.

CONCEPTUAL QUESTIONS

1. What is the gene pool? How is a gene pool described in a quantitative way?

2. In genetics, what does the term population mean? Pick any species you like and describe how its population might change over the course of many generations.

3. What is a genetic polymorphism? What is the source of genetic variation?

4. What evolutionary forces cause genetic polymorphisms to attain their levels of allele frequencies in natural populations? Discuss the relative importance of each type of process.

5. In the term genetic drift, what is drifting? Why is this an appropriate term to describe this phenomenon?

6. Cystic fibrosis is a recessive, autosomal trait. In certain Caucasian populations, the number of people born with this disorder is about 1 in 2500. Assuming a Hardy–Weinberg equilibrium for this trait:

A. What are the frequencies for the normal and CF alleles?
B. What are the genotypic frequencies of homozygous normal, heterozygous, and homozygous affected individuals?
C. Assuming random mating, what is the probability that two phenotypically normal heterozygous carriers will choose each other as mates?

7. Does inbreeding affect allele frequencies? Why or why not? How does it affect genotype frequencies? With regard to rare recessive diseases, what are the consequences of inbreeding in human populations?

8. In the pedigree here, answer the following questions for individual VI-1:

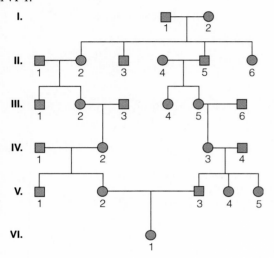

A. Is this individual inbred?
B. If so, who are her common ancestor(s)?
C. Calculate the inbreeding coefficient for VI-1.
D. Are the parents of VI-1 inbred?

9. What is the difference between a neutral and an adaptive evolutionary process? Describe two or more examples. At the molecular level, explain how mutations can be neutral or adaptive.

10. Let's suppose the mutation rate for converting a *B* allele into a *b* allele is 10^{-4}. The current allele frequencies are *B* = 0.6 and *b* = 0.4. How long will it take for the allele frequencies to equal each other, assuming that no genetic drift is taking place?

11. For a gene existing in two alleles, what are the allele frequencies when the heterozygote frequency is at its maximum value? What if there are three alleles?

12. For each of the following, state whether it is an example of an allele, genotype, and/or phenotype frequency:

A. Approximately 1 in 2500 Caucasians are born with cystic fibrosis.
B. The percentage of carriers of the sickle cell allele in West Africa is approximately 13%.

C. The number of new mutations for achondroplasia is approximately 5×10^{-5}.

13. Why is genetic drift more significant in small populations? Why does it take longer for genetic drift to cause allele fixation in large populations than in small ones?

14. A group of four birds fly to a new location and initiate the formation of a new colony. Three of the birds are homozygous *AA*, and one bird is heterozygous *Aa*.

A. What is the probability that the *a* allele will become fixed in the population?
B. If fixation occurs, how long will it take?
C. How will the growth of the population, from generation to generation, affect the answers to parts A and B? Explain.

15. Describe what happens to allele frequencies during the bottleneck effect. Discuss the relevance of this effect with regard to species that are approaching extinction.

16. When two populations frequently intermix due to migration, what are the long-term consequences with regard to allele frequencies and genetic variation?

17. In a donor population, the allele frequencies for the normal (H^A) and sickle cell alleles (H^S) are 0.9 and 0.1, respectively. A group of 550 individuals migrates to a new population containing 10,000 individuals; in the recipient population, the allele frequencies are $H^A = 0.99$ and $H^S = 0.01$.

A. Calculate the allele frequencies in the conglomerate population.
B. Assuming that the donor and recipient populations are each in Hardy–Weinberg equilibrium, calculate the genotype frequencies in the conglomerate population prior to further mating between the donor and recipient populations.
C. What will be the genotype frequencies of the conglomerate population in the next generation, assuming that it achieves Hardy–Weinberg equilibrium in one generation?

18. Discuss the similarities and differences among directional, disruptive, and stabilizing selection.

19. What is Darwinian fitness? What types of characteristics can promote high fitness values? Give several examples.

20. A recessive lethal allele has achieved a frequency of 0.22 due to genetic drift in a very small population. Based on natural selection, how would you expect the allele frequencies to change in the next three generations? Note: Your calculation can assume that genetic drift is not altering allele frequencies in either direction.

21. What is the intuitive meaning of the mean fitness of a population? How does its value change in response to natural selection?

22. What is the genetic load of a population? How does genetic variation contribute to the genetic load?

EXPERIMENTAL QUESTIONS

1. Experimentally, how are allozymes typically detected in population genetics studies? Does this technique underestimate or overestimate the frequency of allozymes in populations? Explain. What technique should be able to detect all allozymes within a population?

2. Let's suppose that a species of sparrow is found in two different geographic locations. By gel electrophoresis, you analyze 24 different enzymes to see if they are polymorphic. In one location, 2 out of 24 exist as two or more allozymes. In the other location, 11 out of 24 exist as two or more allozymes. With regard to the historical events that produced these two populations, propose at least two different hypotheses that would be consistent with these data.

3. The human *M–N* blood group is determined by two codominant alleles, *M* and *N*. The following data were obtained from various human populations:

| Population | Place | *Percentages* | | |
		M	*MN*	*N*
Eskimo	East Greenland	83.5	15.6	0.9
Navaho Indians	New Mexico	84.5	14.4	1.1
Finns	Karajala	45.7	43.1	11.2
Russians	Moscow	39.9	44.0	16.1
Aborigines	Queensland	2.4	30.4	67.2

Data from: Speiss, E.B. (1990) *Genes in Populations,* 2nd edition, Wiley-Liss, New York.

A. Calculate the allele frequencies in these five populations.
B. Which populations appear to be in Hardy–Weinberg equilibrium?
C. Which populations do you think have had significant intermixing due to migration?

4. Resistance to the poison warfarin is a genetically determined trait in rats. Homozygotes carrying the resistance allele (W^rW^r) have a lower fitness because they suffer from vitamin K deficiency, but heterozygotes (W^rw) do not. However, the heterozygotes are still resistant to warfarin. In an area where warfarin is applied, the heterozygote has a survival advantage. Due to warfarin resistance, the heterozygote is more fit than the normal homozygote (ww). The heterozygote is also more fit than the warfarin-resistant homozygote (W^rW^r), because the homozygote suffers from vitamin K deficiency. If the relative fitness values for W^rw, W^rW^r, and ww individuals are 1.0, 0.37, and 0.19 in areas where warfarin is applied, calculate the allele frequencies at equilibrium. How would this equilibrium be affected if the rats were no longer exposed to warfarin?

5. Why was it necessary (in Experiment 26A) for Kettlewell to mark the undersides of the moths he released?

6. Carry out a chi square analysis to determine if Kettlewell's results are significant. Carefully state the hypothesis you are testing, and then determine whether the chi square value allows you to accept or reject it.

7. Describe, in as much experimental detail as possible, how you would test the hypothesis that snail color distribution is due to predation.

QUESTIONS FOR STUDENT DISCUSSION/COLLABORATION

1. Discuss examples of assortative and disassortative mating in natural populations, human populations, and agriculturally important species.

2. Discuss the role of mutation in the origin of genetic polymorphisms. Suppose that a genetic polymorphism has two alleles at frequencies of 0.45 and 0.55. Describe three different scenarios to explain these observed allele frequencies. You can propose that the alleles are neutral, beneficial, or deleterious.

3. Most new mutations are detrimental, yet rare beneficial mutations can be adaptive. Discuss whether you think it is more important for natural selection to select against detrimental alleles or for beneficial alleles. Which do you think is more significant in human populations?

CHAPTER 27

EVOLUTIONARY GENETICS

ORIGIN OF SPECIES

MOLECULAR EVOLUTION

Biological evolution is a genetically based, heritable change in one or more characteristics of a population or species over the course of many generations. Evolution can be viewed on a small scale as it relates to a single gene, or it can be viewed on a larger scale as it relates to the formation of new species. At the end of Chapter 26, we examined several factors that cause allele frequencies to change in populations. This process, also known as **microevolution**, concerns the changing composition of gene pools with regard to particular alleles. As we have seen, several evolutionary forces, such as mutation, inbreeding, genetic drift, migration, and natural selection, affect the allele and genotypic frequencies within natural populations.

In the first part of this chapter, we will be concerned with evolution on a larger scale, which leads to the origin of new species. This phenomenon was central to the development of evolutionary theory. In more advanced courses, you may also learn about evolution on an even grander scale. The term **macroevolution** refers to evolutionary changes above the species level. It concerns the diversity of organisms established over long periods of time through the accumulated evolution and extinction of many species. Macroevolution involves relatively large changes in form and function that are sufficient to produce new species and higher taxa.

At the end of this chapter, we will tie together molecular genetics and the evolution of species. The development of techniques for analyzing chromosomes and DNA sequences has enhanced greatly our understanding of evolutionary processes. The term **molecular evolution** refers to the molecular changes in genetic material that underlie the phenotypic changes associated with evolution. The topic of molecular evolution is a fitting way to end our discussion of genetics, since it integrates the ongoing theme of this text—the relationship between molecular genetics and traits—in the grandest and most profound ways. Theodosius Dobzhansky, an influential evolutionary scientist, once said, "Nothing in biology makes sense except in the light of evolution." The extraordinarily diverse and seemingly bizarre array of species that exist on our planet are explained naturally within the context of evolution. An examination of molecular evolution allows us to make sense of the existence of these species at both the populational and the molecular levels.

ORIGIN OF SPECIES

The theory of evolution is attributed largely to Charles Darwin, a British naturalist, who was born in 1809. Like many great scientists, Darwin had a very broad background in science, which enabled him to see connections among different disciplines. His thinking was influenced greatly by theories of geology indicating that the Earth is very old, and that slow geological processes can lead eventually to substantial changes in the Earth's characteristics.

A second important influence on Darwin was his own experimental observations. His famous voyage on the *Beagle*, which lasted from 1832 to 1836, involved a careful examination of many different species. He observed the similarities among many discrete species, yet noted the differences that enabled them to be adapted to their environmental conditions. He was particularly struck by the distinctive adaptations of island species. For example, the finches found on the Galapagos Islands had unique phenotypic characteristics as compared with similar finches found on the mainland.

Finally, a third important impact on Darwin was a paper published in 1798, *Essay on the Principle of Population*, by Thomas Malthus, an English economist. Malthus asserted that the population size of humans can, at best, increase arithmetically due to increased land usage and improvements in agriculture, while the reproductive potential of humans can increase geometrically. He argued that famine, war, and disease will limit population growth, especially among the poor.

With these three ideas in mind, Darwin had largely formulated his theory of evolution by the mid-1840s. He then spent several years studying barnacles without having published his ideas. The geologist Charles Lyell, who had greatly influenced Darwin's thinking, strongly encouraged Darwin to publish his theory of evolution. In 1856, Darwin began to write a long book to explain his ideas. In 1858, however, Alfred Wallace, a naturalist working in the East Indies, sent Darwin an unpublished manuscript to read prior to its publication. In it, Wallace proposed the same ideas concerning evolution. Darwin therefore quickly excerpted some of his writings on this subject, and two papers, one by Darwin and one by Wallace, were published in the Linnaean Society of London. These papers were not widely recognized. A short time later, however, Darwin finished his book, *The Origin of Species*, which expounded his ideas in greater detail, and with experimental support. This book, which received high praise from many scientists and scorn from others, started a great debate concerning evolution. Although there were some flaws in his ideas, Darwin's work represents one of the most important contributions to our understanding of biology.

The theory of evolution was called "the theory of descent with modification through variation and natural selection" by Charles Darwin. As its name suggests, it is based on two fundamental principles: genetic variation and natural selection. A modern interpretation of evolution can view these principles with regard to speciation and microevolution:

1. *Genetic variation at the species level:* As we have seen in Chapter 26, genetic variation is a consistent feature of most natural populations. Darwin observed that many species exhibit a great amount of phenotypic variation. Although the theory of evolution preceded Mendel's pioneering work in

genetics, Darwin observed that offspring resemble their parents more than they do unrelated individuals. Therefore, he assumed that some phenotypic variation is passed from parent to offspring.

At the microevolutionary level: Genetic variation results from differences in DNA sequences that create alleles, changes in chromosome structure, and alterations in chromosome number. These differences are caused by random mutations. Alternative alleles may affect the functions of the proteins they encode and thereby affect the phenotype of the organism. Likewise, changes in chromosome structure and number may affect gene expression and thereby influence the phenotype of the individual.

2. *Natural selection at the species level:* Darwin agreed with Malthus that most species produce many more offspring than survive and reproduce. This creates a "*struggle for existence*" that results in a "*survival of the fittest*." Over the course of many generations, those individuals who happen to possess the most favorable traits will dominate the composition of the population. The ultimate result of natural selection is to make a species better adapted to its environment and/or more efficient at reproduction.

At the microevolutionary level: Some alleles may encode proteins that provide the individual with a selective advantage. Over time, natural selection may change the allele frequencies of genes and thereby lead to the fixation of beneficial alleles and the elimination of detrimental alleles.

In this section, we will examine the features of evolution as it occurs in natural populations over time.

A biological species is a group of reproductively isolated individuals

Perhaps it is now obvious that the study of evolution requires a dialogue among naturalists, taxonomists, and geneticists. However, after Darwin's death, and during the first three decades of the 20th century, naturalists and taxonomists hardly interacted with geneticists. This situation changed dramatically due to the efforts of Theodosius Dobzhansky.

Dobzhansky was a naturalist and a taxonomist (with an interest in insects) who starting working with Thomas Hunt Morgan in 1927 (a pioneer in studying genetic mutations in *Drosophila*; see Chapters 3 and 4). Dobzhansky's profound perception was that an understanding of evolution must be rooted in the discreteness and discontinuity among different species. A key contribution of Dobzhansky was the idea that discrete species exist due to *isolating mechanisms*. He proposed that species are reproductively isolated from other species. In 1937, he published a book, *Genetics and the Origin of Species*, which laid the foundation for our modern understanding of evolution. Unlike Darwin (who thought that environmental variation was somehow responsible for genetic variation), he argued that random mutations provide the basis of variation, but he agreed with Darwin that natural selection is responsible for creating adaptive changes in species. In addition, Dobzhansky acknowledged that forces such as genetic drift and migration could also play important roles.

Ernst Mayr at Harvard University expanded on the ideas of Dobzhansky to provide a biological definition of a species. According to Mayr's **biological species concept**, a **species** is defined as a group of individuals whose members have the potential to interbreed with one another in nature to produce viable, fertile offspring, but who cannot successfully interbreed with members of other species. There are several different ways to achieve reproductive isolation (Table 27-1). These can be classified as prezygotic mechanisms that prevent the formation of a zygote, and postzygotic mechanisms that prevent the development of a viable individual after fertilization has taken place. Reproductive isolation in nature may be circumvented when species are kept in captivity. For example, different species of the genus

TABLE 27–1 Types of Reproductive Isolation between Different Species

Mechanism	Description
Prezygotic Mechanisms	
Habitat isolation	Species may occupy different habitats, so that they never come in contact with each other.
Temporal isolation	Species have different mating or flowering seasons or times of day, or become sexually mature at different times of the year.
Sexual isolation	Sexual attraction between males and females of different animal species is limited due to differences in behavior, physiology, or morphology.
Mechanical isolation	The anatomical structures of genitalia prevents mating between different species.
Gametic isolation	Gametic transfer takes place, but the gametes fail to unite with each other. This can occur because the male and female gamete fail to attract, because they are unable to fuse, or because the male gametes are inviable in the female reproductive tract of another species.
Postzygotic Mechanisms	
Hybrid inviability	The egg of one species is fertilized by the sperm from another species, but the fertilized egg fails to develop past early embryonic stages.
Hybrid sterility	The interspecies hybrid survives, but it is sterile. For example, the mule is a cross between a female horse *(Equus caballus)* and a male donkey *(Equus asinus)*.
Hybrid breakdown	The F_1 interspecies hybrid is viable and fertile, but succeeding generations (i.e., F_2, etc.) become increasing inviable. This is usually due to formation of less-fit genotypes by genetic recombination.

Drosophila rarely mate with each other in nature. In the laboratory, however, it is fairly easy to produce interspecies hybrids.

Hugh Paterson at the University of Witwatersrand in South Africa proposed an alternative way to view a species known as the **species recognition concept**. This idea agrees with the biological species concept in that it recognizes that species are reproductive communities that are isolated from other species. The difference in the species recognition concept concerns the mechanisms by which new species arise. According to Paterson, the sexual recognition of members of the same species acts as a positive force in evolution. There is variation in the ability of mates to locate each other and participate in a productive fertilization. Therefore, natural selection favors the survival of members who can recognize each other and mate successfully. In this regard, recognition can have a very broad meaning. It could involve factors such as visual recognition, behavioral recognition, or even cellular recognition (e.g., sperm and egg cells).

The species recognition concept is consistent with many examples of secondary sexual traits that lead to reproductive isolation. The term **sexual selection**, originally described by Darwin, refers to "the advantage which certain individuals have over others of the same sex and species solely in respect of reproduction." In many species of animals, sexual selection affects male characteristics more intensely than it does female. Unlike females, who tend to be fairly uniform in their reproductive success, male success tends to be more variable, with some males mating with many females and others not mating at all. There are many examples of male traits that appear to be solely for the purpose of recognition by the female of the species. Figure 27-1 illustrates the morphological differences among several species of male hummingbirds. These differences allow the females to recognize and choose an appropriate mate. By comparison, the females of these species look rather similar to each other.

Speciation usually occurs via a branching process called cladogenesis

By examining the fossil record, evolutionary biologists have found two different patterns of speciation (Figure 27-2). During **anagenesis** (Figure 27-2a; from the Greek, *ana*, meaning "up," and *genesis*, meaning "origin"), a single species is transformed into a different species over the course of many generations. During this process, the characteristics of the species change due to both neutral evolutionary forces and adaptive forces promoted by natural selection. As a result of natural selection, the new species may be better adapted to survive in its original environment, or the environment may have changed so that the new species is better adapted to the new surroundings.

By comparison, **cladogenesis** (Figure 27-2b; from the Greek *clados*, meaning "branch") involves the division of a species into two or more species. This is the most common form of speciation. Although cladogenesis is usually thought of as a branching process, it commonly occurs as a budding process in which a single species divides into the original species plus a new species with different characteristics. If we view evolution as a tree, the new species buds from the original species and develops characteristics that prevent it from interbreeding with the original one.

Divergent evolution can be allopatric, parapatric, or sympatric

The divergence of one species into two or more discrete species is the most common form of speciation. Depending on the geographic locations of the evolving population(s), speciation is categorized as allopatric, parapatric, or sympatric. Table 27-2 presents a overview of these types of speciation patterns. The following discussion will describe some examples.

Allopatric speciation (from the Greek *allos*, "other," and Latin *patria*, "homeland") is thought to be the most prevalent way for a species to diverge. It occurs when members of a species become geographically separated from the other members. This form of speciation can occur by the geographic subdivision of large populations by geological processes. For example, a mountain range may emerge and split a species that occupies the lowland regions. Or a creeping glacier may divide a population. Figure 27-3 shows an interesting example in which geological separation promoted speciation. Two species of antelope squirrels occupy opposite rims of the Grand Canyon. On the south rim is Harris's antelope squirrel (*Ammospermophilus harrisi*), while a closely related white-tailed antelope squirrel (*Ammospermophilus leucurus*) is found on the north rim. Presumably, these two species evolved from a common species that existed before the canyon was formed. Over time, the accumulation of genetic changes in the two populations led to the formation of two

Blue throated *Lampornis clemenciae*

Broad billed *Cynanthus latirostris*

Rufous hummingbird *Selasphorus rufus*

FIGURE **27-1**

Species of male hummingbirds exhibit great differences in their feather plumage.

GENES ⟶ TRAITS: One way that natural selection works is by promoting the sexual recognition of members of the same species. Natural selection favors the survival of members with phenotypes that make it easier for them to find each other and mate successfully. The male hummingbirds shown in this figure carry alleles that affect their plumage size and coloration. These distinctive features make it easier for males and females of the same species to recognize and mate with each other.

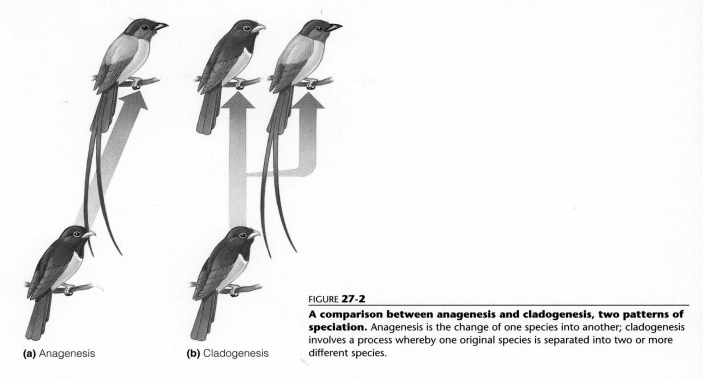

(a) Anagenesis   (b) Cladogenesis

FIGURE **27-2**

A comparison between anagenesis and cladogenesis, two patterns of speciation. Anagenesis is the change of one species into another; cladogenesis involves a process whereby one original species is separated into two or more different species.

morphologically distinct species. Interestingly, birds that can easily fly across the canyon have not diverged into different species on the opposite rims.

Allopatric speciation can also occur via a second mechanism, known as the *founder effect*, which is thought to be more rapid and frequent than allopatric speciation caused by geological events. The founder effect, which was discussed in Chapter 26, occurs when a small group migrates to a new location that is geographically separated from the main population. For example, a storm may force

TABLE **27–2 Common Genetic Mechanisms that Underlie Allopatric, Parapatric, and Sympatric Speciation**

Type of Speciation	Common Genetic Mechanisms Responsible for Speciation
Allopatric: two large populations are separated by geographic barriers.	Many small genetic differences may accumulate over a long period of time, leading to reproductive isolation. Some of these genetic differences may be adaptive; others will be neutral.
Allopatric: a small founding population separates from the main population.	Genetic drift and shifts in adaptive peaks may lead to the rapid formation of a new species. If the group has moved to an environment that differs from its previous environment, natural selection is expected to favor beneficial alleles and eliminate harmful alleles.
Parapatric: two populations occupy overlapping ranges so that a limited amount of interbreeding occurs.	A new combination of alleles or chromosomal rearrangement may limit the amount of gene flow between neighboring populations due to the very low fitness of hybrid offspring.
Sympatric: within a population occupying a single habitat in a continuous range, a small group evolves into a reproductively isolated species.	An abrupt genetic change leads to reproductive isolation. For example, a mutation may affect gamete recognition. In plants, the formation of a polyploid often leads to the formation of new species, because the interspecies hybrid (e.g., diploid × tetraploid) is triploid and sterile.

A. leucurus

A. harrisi

FIGURE 27-3

An example of allopatric speciation: two closely related species of antelope squirrels that occupy opposite rims of the Grand Canyon.

GENES $\longrightarrow$ TRAITS: Harris's antelope squirrel (*Ammospermophilus harrisi*) is found on the south rim of the Grand Canyon, while the white-tailed antelope squirrel (*Ammospermophilus leucurus*) is found on the north rim. These two species evolved from a common species that existed before the canyon was formed. After it was formed, the two separated populations accumulated genetic changes due to mutation, natural selection, and genetic drift that led eventually to the formation of two distinct species.

a small group of birds from the mainland to a distant island. In this case, migration between the island and the mainland populations is a very infrequent event. In a relatively short period of time, the founding population on the island may evolve into a new species. Several evolutionary forces may contribute to this rapid evolution. First, as discussed in Chapter 26, genetic drift may quickly lead to the random fixation of certain alleles and the elimination of the other alleles from the population. Another factor is natural selection. The environment on an island may differ significantly from the mainland environment.

To explain how natural selection may act on the phenotypic effects of many alleles, Sewall Wright developed the concept of **adaptive peaks** to describe combinations of many alleles that provide an optimal fitness for individuals in a stable environment. As shown in Figure 27-4, there are several different combinations of alleles that may lead to highly fit individuals. If a small change occurs in the frequency of alleles at one or a few loci, it will drive the population off an adaptive peak and into a (less fit) valley; natural selection, though, will tend to push the population back to the original peak.

When, however, a small group migrates to a new environment, the adaptive landscape is likely to be changed. Allele combinations that were highly adaptive in the original environment may be much less so in the new environment. If this occurs, natural selection may lead to a shift toward a new adaptive peak in which the frequencies of many alleles have been changed. In other words, natural selection will act to change the morphological features of the organism to make it better adapted to its new environment. Similarly, in a species occupying a large geographic range,

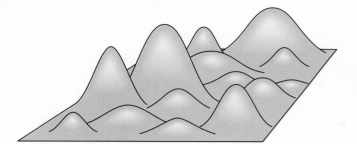

FIGURE **27-4**

The concept of adaptive peaks. This illustration uses a geographic metaphor to symbolically represent genotypes. It suggests that certain genotypic combinations are favorable and form peaks of high fitness values, while other genotypic combinations are unfavorable and form valleys of low fitness values.

the adaptive peaks of local populations may differ due to variation in the environmental conditions within the range.

Parapatric speciation (from the Greek *para*, "beside") occurs when members of a species are separated only partially or when a species is very sedentary. In these cases, the geographic separation is not complete. For example, a mountain range may divide a species into two populations, but with breaks in the range where the two groups are connected physically. In these zones of contact, the members of two populations can interbreed, although this tends to occur infrequently. Likewise, parapatric speciation may occur among very sedentary species even though no geographic isolation exists. Certain organisms are so sedentary that 100 to 1000 meters may be sufficient to limit the interbreeding between neighboring groups. Plants, terrestrial snails, rodents, grasshoppers, lizards, and many flightless insects may speciate in a parapatric manner.

Prior to parapatric speciation, the zones where two populations can interbreed are known as **hybrid zones**. For speciation to occur, the amount of gene flow within the hybrid zones must become very limited. In other words, there must be selection against the offspring produced in the hybrid zone. One way that this can happen is that each of the two parapatric populations may accumulate different neutral chromosomal rearrangements, such as inversions and balanced translocations. As was discussed in Chapter 8, if an individual has one chromosome with a large inversion and one that does not carry the inversion, crossing over during meiosis can lead to the production of grossly abnormal chromosomes. Therefore, such an individual is substantially less fertile. By comparison, an individual who is homozygous for two normal chromosomes or for two chromosomes carrying the same inversion will be fertile, because crossing over can proceed normally.

Finally, **sympatric speciation** (from the Greek *sym*, "together") occurs when members of a species initially occupy the same habitat within the same range. In plants, a common way for sympatry to occur is by the formation of polyploids. As was discussed in Chapter 8 (see pp. 220–223), a complete nondisjunction of chromosomes during gamete formation can increase the number of chromosome sets within the same species (autopolyploidy) or between different species (allopolyploidy). Polyploidy is so frequent in plants that it is a major form of speciation. In ferns and flowering plants, about 30 to 35% of the species are polyploid. By comparison, polyploidy is much less common in animals, but it can occur. For example, roughly 30 species of reptiles and amphibians have been identified that are polyploids derived from diploid relatives.

The formation of a polyploid can lead abruptly to reproductive isolation. As an example, let's consider the probable formation of a natural species of common hemp nettle known as *Galeopsis tetrahit*. As described in Figure 27-5, this species is thought to be an allopolyploid between two diploid species, *Galeopsis pubescens* and *Galeopsis speciosa*. These two diploid species contain 16 chromosomes each ($2n = 16$), while *G. tetrahit* contains 32. Figure 27-5 illustrates what would happen in crosses between the allopolyploid and the diploid species. The allopolyploid crossed to another allopolyploid produces an allopolyploid. The allopolyploid is

Galeopsis tetrahit

G. pubescens

G. speciosa

Haploid gametes:

Chromosomal composition of F₁ offspring:

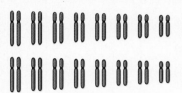

Fertile

G. tetrahit × *G. tetrahit*

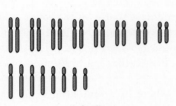

Infertile

G. tetrahit × *G. pubescens*

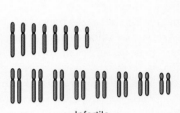

Infertile

G. tetrahit × *G. speciosa*

FIGURE **27-5**

A comparison of crosses between three natural species of hemp with different ploidy levels. *Galeopsis tetrahit* is an allopolyploid that is thought to be derived from *Galeopsis pubescens* and *Galeopsis speciosa*. If *G. tetrahit* is mated with the other two species, the F₁ hybrid offspring will be monoploid for one chromosome set and diploid for the other set. The F₁ offspring are likely to be sterile, because they will produce highly aneuploid gametes.

fertile, because all of its chromosomes occur in homologous pairs that can segregate evenly during meiosis. However, a cross between an allopolyploid and a diploid produces an offspring that is monoploid for one chromosome set and diploid for the other set. These offspring are expected to be sterile, because they will produce highly aneuploid gametes that have incomplete sets of chromosomes. This hybrid sterility causes the allopolyploid to be reproductively isolated from the diploid species.

EXPERIMENT 27A

Müntzing was able to re-create an allopolyploid species by selective breeding

In studying biological evolution, we cannot look into the past and know with complete certainty how events occurred. An alternative way to explore past events is called **retrospective testing**. In this procedure, a researcher formulates a hypothesis and then collects observations as a way to confirm or refute the hypothesis. Therefore, this approach relies on the observation and testing of already existing materials, which themselves are the result of past events. The careful study or manipulation of these materials may then be used to support or refute ideas concerning how evolution may have occurred in the past.

An interesting example of this approach is provided by a study in the 1930s by Arne Müntzing, a Swedish botanist. It was Müntzing who first proposed that *Galeopsis tetrahit* is an allopolyploid that arose from an interspecies cross between *G. pubescens* and *G. speciosa* (see Figure 27-5). To confirm this hypothesis, he attempted to "reenact" evolution by re-creating an "artificial" *G. tetrahit* by crossing *G. pubescens* and *G. speciosa* in a way that produced an allopolyploid.

THE HYPOTHESIS

G. tetrahit is an allopolyploid that contains a diploid chromosomal composition from *G. pubescens* and *G. speciosa*.

TESTING THE HYPOTHESIS

Starting material: Specimens of *G. pubescens*, *G. speciosa*, and *G. tetrahit*.

	Experimental Level	**Conceptual Level**

1. Cross *G. pubescens* to *G. speciosa*.

   Note: In most cases, this will produce alloploids that have 16 chromosomes.

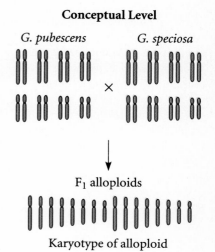

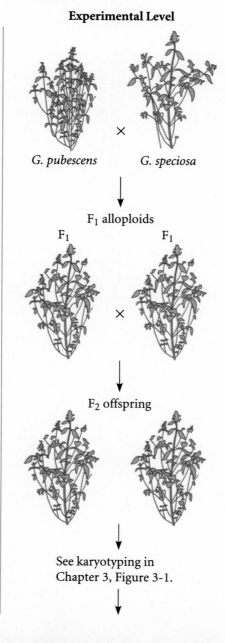

2. Make an $F_1 \times F_1$ cross. The fertility of the alloploids is very low, because the *G. pubescens* chromosomes usually do not pair properly with the *G. speciosa* chromosomes during meiosis. For this reason, the F_2 offspring obtained are expected to have abnormal numbers of chromosomes.

3. Examine the chromosome compositions of 200 F_2 offspring.

G. pubescens × *G. speciosa*

F_1 alloploids

F_1 × F_1

F_2 offspring

See karyotyping in Chapter 3, Figure 3-1.

G. pubescens   *G. speciosa*

F_1 alloploids

Karyotype of alloploid

(continued)

TESTING THE HYPOTHESIS

(continued)

4. Among 200 offspring examined, one of them had 16 chromosomes from *G. speciosa* and 8 chromosomes from *G. pubescens*. Presumably, this occurred because of nondisjunction of the *G. speciosa* chromosomes. Let's call this strain the "intermediate F₂ strain."

5. Mate the intermediate F₂ strain to *G. pubescens*.

6. Only one seed was obtained! It was diploid for the chromosomes of both species. Presumably, it arose from a normal *G. pubescens* gamete fusing with a gamete from the intermediate F₂ strain that had resulted from complete nondisjunction. This allopolyploid strain was called "artificial" *G. tetrahit*.

7. Examine the morphological features of the "artificial" *G. tetrahit* and compare them with the natural species.

8. Explore the ability of the "artificial" *G. tetrahit* to be crossed with itself and with the natural *G. tetrahit* species.

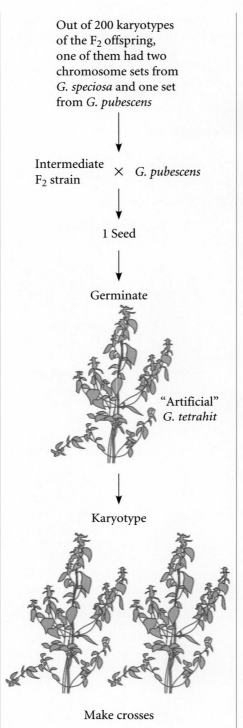

Out of 200 karyotypes of the F₂ offspring, one of them had two chromosome sets from *G. speciosa* and one set from *G. pubescens*

Intermediate F₂ strain   ×   *G. pubescens*

1 Seed

Germinate

"Artificial" *G. tetrahit*

Karyotype

Make crosses

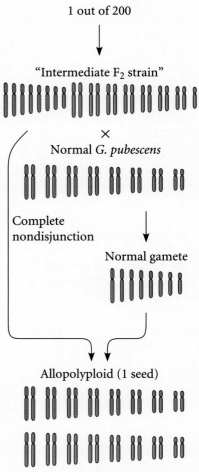

1 out of 200

"Intermediate F₂ strain"

×

Normal *G. pubescens*

Complete nondisjunction

Normal gamete

Allopolyploid (1 seed)

Karyotype of "artificial" *G. tetrahit*

To see if the natural and "artificial" *G. tetrahit* look the same

To see if the natural and "artificial" *G. tetrahit* produce fertile offspring

	Artificial G. tetrahit	Natural G. tetrahit	G. pubescens	G. speciosa
Chromosomal composition	32	32	16	16
Mean flower size (mm)	19	19	29	27
Calyx length (mm)	15	15	12	18
Degree of stem hairiness	Moderate	Moderate	Low	High
Fertility:	The artificial *G. tetrahit* could produce fertile offspring when it was crossed to another artificial *G. tetrahit* and it could produce fertile offspring when it was crossed to a natural strain of *G. tetrahit*. Microscopically, normal chromosome pairing was observed during meiosis in the artificial and natural *G. tetrahit* hybrid.			

Adapted from: Müntzing, A. (1932) Cyto-genetic Investigations on Synthetic *Galeopsis tetrahit*. *Hereditas 16*, 105–154.

Before we interpret the data, let's examine how the "artificial" *G. tetrahit* was produced. Müntzing knew that each of these two species occasionally produces diploid gametes (by abnormal nondisjunction), which could fuse together and produce an allopolyploid in a single step. However, he expected that this would be an extremely rare event. Therefore, rather than hoping for such a rare event to occur, he decided to make the allopolyploid in a two-step procedure.

Müntzing first produced a strain that was diploid for *G. speciosa* chromosomes and monoploid for the *G. pubescens* chromosomes. To develop such a strain, he first crossed *G. pubescens* to *G. speciosa*. Müntzing assumed that the F_1 alloploids would have a higher frequency of nondisjunction because they would contain one set of chromosomes from two different species. He hoped that the $F_1 \times F_1$ cross would tend to produce gametes that exhibited nondisjunction. An occasional offspring might be perfectly diploid for the chromosomes of one species (e.g., *G. speciosa*) and perfectly monoploid for the other species (e.g., *G. pubescens*). Out of 200 F_2 offspring he examined, he found one F_2 strain that contained 16 chromosomes from *G. speciosa* and 8 chromosomes from *G. pubescens* (which we have named the intermediate F_2 strain).

Due to its unusual chromosomal composition, Müntzing knew that the intermediate F_2 strain would usually make highly aneuploid gametes that could not produce viable seeds. He reasoned that the intermediate F_2 strain might occasionally produce a gamete that would be the result of complete nondisjunction. Such a gamete would be diploid for the *G. speciosa* chromosomes and monoploid for the *G. pubescens* chromosomes. If this gamete fused with a normal haploid gamete from *G. pubescens*, this would result in a viable polyploid offspring containing two sets of chromosomes from each species—the composition of *G. tetrahit*. Therefore, to obtain an offspring with the chromosomal composition of *G. tetrahit*, he crossed the intermediate F_2 strain to a normal *G. pubescens* strain. Since the intermediate F_2 strain usually produced aneuploid gametes, it was almost unable to make any viable seeds. However, Müntzing did obtain one seed that had the chromosomal composition he wanted: it had a diploid number of chromosomes from both species.

This seed was the artificial *G. tetrahit* Müntzing had hoped for. Based on morphological characteristics, the artificial *G. tetrahit* species confirmed his hypothesis. It had traits that were similar to the natural *G. tetrahit* species but different from *G. pubescens* and *G. speciosa*. Furthermore, the artificial and natural *G. tetrahit* strains could be mated to each other to produce fertile offspring. Overall, Müntzing came to the conclusion that:

> "not only the artificial tetraploid but probably also natural *G. tetrahit* represents a synthesis of *pubescens*- and *speciosa*-genomes. Consequently, this case might be regarded as the first instance where a species already existing in nature [as a result of evolution] has been synthesized from the genomes of two other species."

Evolution can proceed gradually or be punctuated by periods of rapid change

As we have seen, there are many different genetic mechanisms that give rise to new species. For this reason, the rates of evolutionary change are not constant, although the degree of inconstancy has been debated since the time of Darwin. Even Darwin himself suggested that evolution can occur at fast and slow paces. Figure 27-6 compares the two opposing views concerning the rates of evolutionary change. The **gradualism** hypothesis suggests that each new species evolves continuously over long spans of time (Figure 27-6a). The principal idea is that large phenotypic differences that produce the divergence of species are due to the accumulation of many small genetic changes.

In 1972, Niles Eldrege at the American Museum of Natural History in New York and Stephen Jay Gould at Harvard University proposed an alternative hypothesis, called **punctuated equilibrium** (Figure 27-6b). According to this proposal, species exist relatively unchanged for many generations. During this period, the species is in equilibrium with its environment. These long periods of equilibrium are punctuated by relatively short periods (i.e., on a geological time scale) during which evolution occurs at a far more rapid rate.

The phenomenon of punctuated equilibrium is largely supported by the fossil record. Paleontologists rarely find a gradual transition of fossil forms. Instead, it is much more common to observe species appearing as new forms rather suddenly in a layer of rocks, persisting relatively unchanged for a very long period of time, and then suddenly becoming extinct. It is presumed that the transition period during which a previous species evolved into a new, morphologically distinct species was so short that few, if any, of the transitional members were preserved as fossils. Therefore, the fossil record primarily contains representatives from the long equilibrium periods.

As was discussed earlier, rapid evolutionary change can be explained by genetic phenomena. As we have seen throughout this text, single gene mutations can have dramatic effects on phenotypic characteristics. Therefore, only a small number of new mutations may be required to alter phenotypic characteristics, eventually producing a group of individuals that make up a new species. Likewise, genetic events such as changes in chromosome structure (e.g., inversions and translocations) or chromosome number may abruptly create individuals with new phenotypic traits. On an evolutionary time scale, these types of events can be rather rapid, since one or only a few genetic changes can have a major impact on the phenotype of the organism.

In conjunction with genetic changes, species may also be subjected to sudden environmental shifts that quickly drive the gene pool in a particular direction via natural selection. For example, a small group may migrate to a new environment

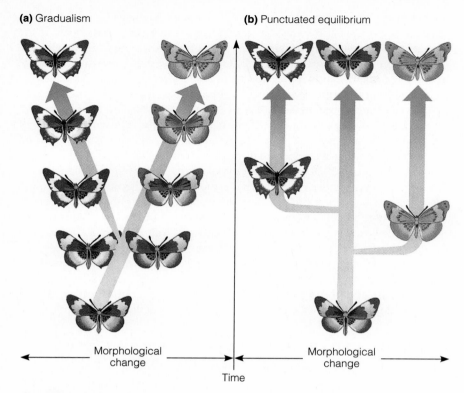

(a) Gradualism

(b) Punctuated equilibrium

Morphological change ←——→ | ←——→ Morphological change

Time

FIGURE **27-6**

A comparison of gradualism and punctuated equilibrium. (a) During gradualism, the morphology of a species gradually changes due to the accumulation of small genetic changes. **(b)** During punctuated equilibrium, species exist essentially unchanged for long periods of time during which they are in equilibrium with their environment. These equilibrium periods are punctuated by short periods of evolutionary change during which morphological features may change rapidly.

in which different alleles provide better adaptation to the surroundings. Alternatively, a species may be subjected to a relatively sudden environmental event that has a major impact on survival. There may be a change in climate, or a new predator may have infiltrated the geographic range of the species. To ensure survival, these kinds of events may necessitate the rapid evolution of the gene pool by natural selection so that the new members of the population can survive the climatic change or have phenotypic characteristics that allow them to avoid the predator.

Overall, the fossil record and known genetic phenomena tend to support the punctuated equilibrium model of evolution, although the debate is by no means resolved. Rapid evolutionary changes are quite common in many phylogenetic lines, leading to the rapid branching of evolutionary trees. However, gradual changes are sometimes observed in certain species over long periods of time. Furthermore, the gradual accumulation of mutations has been revealed by molecular analyses of DNA. We will explore this topic in the next section.

MOLECULAR EVOLUTION

Historically, taxonomists have categorized different species primarily on the basis of phenotypes. Initially, morphological differences were used to construct evolutionary trees in which species that are more similar in appearance tended to be placed closer together on the tree. In addition, species have been categorized based on differences in physiology, biochemistry, and even behavior. However, since evolution is based on genetic changes, it makes sense to categorize species based on the properties of their genetic material (i.e., their genotypes). Those species that are closely related evolutionarily are expected to have greater similarities in their genetic material than are distantly related species.

The advent of molecular approaches for analyzing genetic material and gene products has revolutionized the field of evolution. In the past few decades, molecular genetics has taken on the dominant role in facilitating our understanding of speciation and evolution. Differences in nucleotide sequences are quantitative and can be analyzed using mathematical principles in conjunction with computer programs. Evolutionary changes at the DNA level can be objectively compared among different species to establish evolutionary relationships. Furthermore, this approach can be used to compare any two existing organisms, no matter how greatly they differ in their morphological traits. For example, we can compare DNA sequences between humans and bacteria, or between plants and fruit flies. Such comparisons would be very difficult at a morphological level.

There are many ways to compare the genetic material and gene products among different species. Table 27-3 lists the most common molecular approaches that geneticists use to evaluate evolutionary relationships. In this section, we will focus most of our attention on the evolution of gene sequences.

Homologous genes are derived from a common ancestral gene

Two genes are said to be **homologous** if they are derived from the same ancestral gene (see Chapter 22, pp. 609–614). Genes can exhibit **interspecies homology** and

TABLE **27–3 Experimental Approaches Used to Compare Species at the Molecular Level**	
Technique	Explanation
DNA Level	
DNA sequencing	The degree of similarities and differences between DNA sequences are used to establish evolutionary relationships. This approach, along with the analysis of the deduced amino acid sequences of structural genes, is the most common and reliable method in use.
Hybridization	Genetic material from two different species is allowed to hybridize (see Chapter 10, p. 271). The rate of hybridization will be faster for closely related species.
Restriction mapping	Segments of DNA are isolated from different species and subjected to restriction mapping (see Figure 19-3). Closely related species will have more similar restriction maps.
Cytogenetic analysis	The chromosomes of different species are analyzed microscopically. Closely related species will have identical or similar chromosome numbers and banding patterns.
Protein Level	
Deduced amino acid sequence	DNA sequencing is technically much easier to perform than amino acid sequencing. Therefore, to determine an amino acid sequence of a protein, the coding region of a structural gene is determined, and the amino acid sequence is then deduced using the genetic code. The amino acid sequences of a given protein will be more similar between closely related species.
Immunological methods	Antibodies that recognize specific macromolecules, usually on the cell surface, are tested on different species. Antibodies that recognize macromolecules from one species will often recognize closely related species, but not from distantly related species.

intraspecies homology. Figure 27-7 shows examples involving the globin gene family. Hemoglobin is an oxygen-carrying protein found in all vertebrate species. Adult hemoglobin is composed of two different types of subunits, encoded by the α-globin and β-globin genes. Figure 27-7 shows the deduced amino acid sequences of these genes. The sequences are homologous between humans and horses because of their evolutionary relationship. Therefore, the homology between human and horse globin genes is an example of interspecies homology.

Figure 27-7 also illustrates intraspecies homology. Within a particular vertebrate species, several globin genes are homologous to each other. For example, the human α-globin gene is homologous to the human β-globin gene. As discussed in Chapter 8 (see pp. 205–206), this has occurred due to a gene duplication event. On rare occasions, an unequal crossover event will add an extra copy of a gene to a chromosome. Over time, the two or more copies may accumulate distinct mutations that cause their sequences to diverge.

Variation in gene sequences can be used to construct phylogenetic trees

A **phylogenetic tree** is a diagram that describes the evolutionary relationships among different species. In 1963, Linus Pauling and Emile Zukerkandl were the first to suggest that molecular data should be used to establish evolutionary relationships. With advances in DNA sequencing technology, the amount of phylogenetic information in

Hemoglobin

β	1	2	3	4	5	6	7	8	9	10	11	12	13	14	15	16	17	18	—	—	19	20	21	22	23	24	25	26	27	28	29
Human	Val	His	Leu	Thr	Pro	Glu	Glu	Lys	Ser	Ala	Val	Thr	Ala	Leu	Trp	Gly	Lys	Val	—	—	Asn	Val	Asp	Glu	Val	Gly	Gly	Glu	Ala	Leu	Gly
Horse	Val	Gln	Leu	Ser	Gly	Glu	Glu	Lys	Ala	Ala	Val	Leu	Ala	Leu	Trp	Asp	Lys	Val	—	—	Asn	Glu	Glu	Glu	Val	Gly	Gly	Glu	Ala	Leu	Gly
α	1	—	2	3	4	5	6	7	8	9	10	11	12	13	14	15	16	17	18	19	20	21	22	23	24	25	26	27	28	29	30
Human	Val	—	Leu	Ser	Pro	Ala	Asp	Lys	Thr	Asn	Val	Lys	Ala	Ala	Trp	Gly	Lys	Val	Gly	Ala	His	Ala	Gly	Glu	Tyr	Gly	Ala	Glu	Ala	Leu	Glu
Horse	Val	—	Leu	Ser	Ala	Ala	Asp	Lys	Thr	Asn	Val	Lys	Ala	Ala	Trp	Ser	Lys	Val	Gly	Gly	His	Ala	Gly	Glu	Val	Gly	Ala	Glu	Ala	Leu	Glu

β	30	31	32	33	34	35	36	37	38	39	40	41	42	43	44	45	46	47	48	49	50	51	52	53	54	55	56	57	58	59	60
Human	Arg	Leu	Leu	Val	Val	Tyr	Pro	Trp	Thr	Gln	Arg	Phe	Phe	Glu	Ser	Phe	Gly	Asp	Leu	Ser	Thr	Pro	Asp	Ala	Val	Met	Gly	Asn	Pro	Lys	Val
Horse	Arg	Leu	Leu	Val	Val	Tyr	Pro	Trp	Thr	Gln	Arg	Phe	Phe	Asp	Ser	Phe	Gly	Asp	Leu	Ser	Asn	Pro	Gly	Ala	Val	Met	Gly	Asn	Pro	Lys	Val
α	31	32	33	34	35	36	37	38	39	40	41	42	43	44	45	46	—	47	48	49	50	—	—	—	—	—	51	52	53	54	55
Human	Arg	Met	Phe	Leu	Ser	Phe	Pro	Thr	Thr	Lys	Thr	Tyr	Phe	Pro	His	Phe	—	Asp	Leu	Ser	His	—	—	—	—	—	Gly	Ser	Ala	Gln	Val
Horse	Arg	Met	Phe	Leu	Gly	Phe	Pro	Thr	Thr	Lys	Thr	Tyr	Phe	Pro	His	Phe	—	Asp	Leu	Ser	His	—	—	—	—	—	Gly	Ser	Ala	Gln	Val

β	61	62	63	64	65	66	67	68	69	70	71	72	73	74	75	76	77	78	79	80	81	82	83	84	85	86	87	88	89	90	91
Human	Lys	Ala	His	Gly	Lys	Lys	Val	Leu	Gly	Ala	Phe	Ser	Asp	Gly	Leu	Ala	His	Leu	Asp	Asn	Leu	Lys	Gly	Thr	Phe	Ala	Thr	Leu	Ser	Glu	Leu
Horse	Lys	Ala	His	Gly	Lys	Lys	Val	Leu	His	Ser	Phe	Gly	Glu	Gly	Val	His	His	Leu	Asp	Asn	Leu	Lys	Gly	Thr	Phe	Ala	Ala	Leu	Ser	Glu	Leu
α	56	57	58	59	60	61	62	63	64	65	66	67	68	69	70	71	72	73	74	75	76	77	78	79	80	81	82	83	84	85	86
Human	Lys	Gly	His	Gly	Lys	Lys	Val	Ala	Asp	Ala	Leu	Thr	Asn	Ala	Val	Ala	His	Val	Asp	Asp	Met	Pro	Asn	Ala	Leu	Ser	Ala	Leu	Ser	Asp	Leu
Horse	Lys	Ala	His	Gly	Lys	Lys	Val	Gly	Asp	Ala	Leu	Thr	Leu	Ala	Val	Gly	His	Leu	Asp	Asp	Leu	Pro	Gly	Ala	Leu	Ser	Asp	Leu	Ser	Asn	Leu

β	92	93	94	95	96	97	98	99	100	101	102	103	104	105	106	107	108	109	110	111	112	113	114	115	116	117	118	119	120	121	122
Human	His	Cys	Asp	Lys	Leu	His	Val	Asp	Pro	Glu	Asn	Phe	Arg	Leu	Leu	Gly	Asn	Val	Leu	Val	Cys	Val	Leu	Ala	His	His	Phe	Gly	Lys	Glu	Phe
Horse	His	Cys	Asp	Lys	Leu	His	Val	Asp	Pro	Glu	Asn	Phe	Arg	Leu	Leu	Gly	Asn	Val	Leu	Ala	Val	Val	Leu	Ala	Arg	His	Phe	Gly	Lys	Asp	Phe
α	87	88	89	90	91	92	93	94	95	96	97	98	99	100	101	102	103	104	105	106	107	108	109	110	111	112	113	114	115	116	117
Human	His	Ala	His	Lys	Leu	Arg	Val	Asp	Pro	Val	Asn	Phe	Lys	Leu	Leu	Ser	His	Cys	Leu	Leu	Val	Thr	Leu	Ala	Ala	His	Leu	Pro	Ala	Glu	Phe
Horse	His	Ala	His	Lys	Leu	Arg	Val	Asp	Pro	Val	Asn	Phe	Lys	Leu	Leu	Ser	His	Cys	Leu	Leu	Ser	Thr	Leu	Ala	Val	His	Leu	Pro	Asn	Asp	Phe

β	123	124	125	126	127	128	129	130	131	132	133	134	135	136	137	138	139	140	141	142	143	144	145	146
Human	Thr	Pro	Pro	Val	Gln	Ala	Ala	Tyr	Gln	Lys	Val	Val	Ala	Gly	Val	Ala	Asn	Ala	Leu	Ala	His	Lys	Tyr	His
Horse	Thr	Pro	Glu	Leu	Gln	Ala	Ser	Tyr	Gln	Lys	Val	Val	Ala	Gly	Val	Ala	Asn	Ala	Leu	Ala	His	Lys	Tyr	His
α	118	119	120	121	122	123	124	125	126	127	128	129	130	131	132	133	134	135	136	137	138	139	140	141
Human	Thr	Pro	Ala	Val	His	Ala	Ser	Leu	Asp	Lys	Phe	Leu	Ala	Ser	Val	Ser	Thr	Val	Leu	Thr	Ser	Lys	Tyr	Arg
Horse	Thr	Pro	Ala	Val	His	Ala	Ser	Leu	Asp	Lys	Phe	Leu	Ser	Ser	Val	Ser	Thr	Val	Leu	Thr	Ser	Lys	Tyr	Arg

FIGURE **27-7**

A comparison of the α- and β-globin genes from humans and horses. This figure shows the deduced amino acid sequences that were obtained by sequencing the exon portions of the corresponding genes. The gaps indicate where additional amino acids are found in the sequence of myoglobin, another member of this gene family.

genetic sequences is immense. Molecular evolutionists want to analyze the information within genetic sequences to obtain a better understanding of evolutionary links.

Nucleotide and amino acid sequence data are particularly well suited to the construction of evolutionary trees, because genetic sequences change over the course of many generations due to the accumulation of mutations. Therefore, when comparing homologous genes in different organisms, the DNA sequences from closely related organisms are more similar to each other than are the sequences from distantly related species. In a sense, the relatively constant rate of neutral mutation acts as a "molecular clock" on which to measure evolutionary time. According to this idea, neutral mutations will become substituted at an average rate equal to the rate of mutation per generation. On this basis, the genetic divergence between species that is due to neutral mutations reflects the time elapsed since their latest common ancestor.

For molecular evolutionary studies, the DNA sequences of many genes have been obtained from a wide range of sources. Several different types of gene sequences have been used to construct evolutionary trees. One very commonly analyzed gene is that encoding 16S rRNA, a small subunit rRNA. This gene has been sequenced from thousands of different species. It is as reliable a molecular measure of phylogenetic relationships among organisms as is now available. Because rRNA is universal in all living organisms, its function was established at an early stage in the evolution of life on this planet, and its sequence has changed fairly slowly. Presumably, most mutations in this gene are deleterious, so that few neutral or beneficial alleles can occur. This limitation causes this gene sequence to change very slowly during evolution. Furthermore, 16S rRNA is a rather large molecule, and so it contains a large amount of sequence information.

Figure 27-8 illustrates a phylogenetic tree of all life on Earth based on a sequence analysis of the gene encoding the 16S rRNA. This *three-domain* phylogenetic tree has been championed by Carl Woese at the University of Illinois. The current model links all life on our planet. It proposes three main evolutionary branches: the eubacteria, the archaebacteria, and the eukaryotes. From these types of genetic analyses, it has become apparent that all living organisms are connected through a complex evolutionary tree.

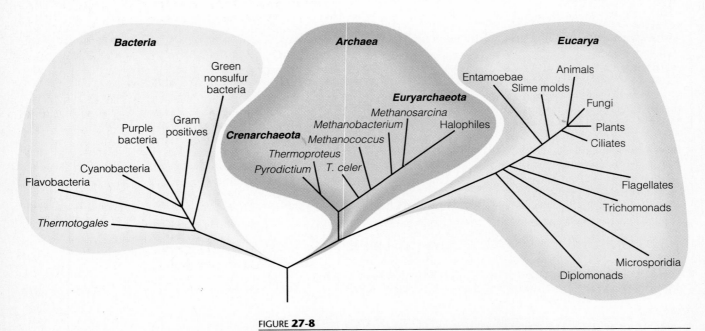

FIGURE **27-8**

A phylogenetic tree of all life on Earth based on 16S rRNA data.

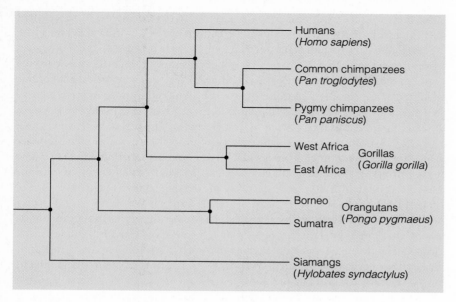

FIGURE **27-9**

A phylogenetic tree of closely related species of hominoids, including humans.
This tree is based on a comparison of mitochondrial gene sequences of cytochrome oxidase subunit II.

By comparing sequence similarities among many genes in myriad species, researchers have found that the rates of sequence change differ dramatically for various genes. Slowly changing genes such as the gene that encodes rRNA are useful for evaluating distant evolutionary relationships and also in assessing evolutionary relationships among bacterial species that evolve fairly rapidly due to their short generation times.

By comparison, other genes have changed more rapidly because of a greater tolerance of neutral mutations. In addition, the mitochondrial genome and DNA sequences within introns tend to have less negative selective pressure, and so their sequences change fairly rapidly during evolution. More rapidly changing DNA sequences have been used to elucidate more recent evolutionary relationships, particularly among eukaryotic species that have long generation times and tend to evolve more slowly. In these cases, slowly evolving genes may not be very useful for establishing evolutionary relationships, because two closely related species likely have identical or nearly identical DNA sequences for such genes. Instead, it is easier to find sequence differences among closely related species when the DNA sequences are more rapidly changing. For example, Figure 27-9 illustrates possible evolutionary relationships among hominoid species, including humans. This tree was derived by comparing DNA sequences in a mitochondrial gene that encodes a protein called cytochrome oxidase subunit II.

Scientists can examine the relationships between living and extinct flightless birds by analyzing ancient DNA and then comparing DNA sequences

EXPERIMENT 27B

The vast majority of our knowledge concerning molecular evolution has come from the analysis of DNA samples collected from living species. Using this approach, we can infer the prehistoric changes that gave rise to present-day DNA sequences. As an alternative, scientists have discovered that it is occasionally possible to obtain DNA sequence information from species that have lived in the past. In

1984, the first successful attempt at determining DNA sequences from extinct species was accomplished by groups at the University of California at Berkeley and the San Diego Zoo, including Russell Higuchi, Barbara Bowman, Mary Freiberger, Oliver Rynder, and Allan Wilson. They obtained a sample of dried muscle from a museum specimen of the quagga (*Equus quagga*), a zebra-like species that became extinct in 1883. This piece of muscle tissue was obtained from an animal that had died 140 years ago. A sample of its skin and muscle had been preserved in salt in the Museum of Natural History at Mainz, Germany. The researchers were able to extract DNA from the sample, clone pieces of it into vectors, and then sequence hybrid vectors containing the quagga DNA. At about the same time, Svante Paabo at the University of Munich obtained DNA sequences from an Egyptian mummy that was more than 2000 years old. These pioneering studies opened the field of **ancient DNA analysis**.

Since the mid-1980s, many researchers have become excited about the information that might be derived from sequencing DNA obtained from older specimens. Currently there is debate concerning how long DNA can remain significantly intact after an organism has died. Over time, the structure of DNA is degraded by hydrolysis and the loss of purines. Nevertheless, under certain conditions (e.g., cold temperature, low oxygen, and so forth), DNA samples may be stable as long as 50,000 to 100,000 years.

In most studies involving prehistoric specimens (in particular, much older than the salt-preserved quagga sample), the ancient DNA is extracted from bone, dried muscle, or preserved skin. These samples are often obtained from museum specimens that have been gathered by archaeologists. However, it is unlikely that enough DNA will be extracted to enable a researcher to directly clone the DNA into a vector. Since 1985, however, the advent of PCR technology (see Figure 19-4) has made it possible to amplify the very small amounts of DNA using PCR primers that flank a region within the 12S rRNA gene, a slowly changing gene. In recent years, this approach has been used to elucidate the phylogenetic relationships between modern and extinct species.

In the experiment described here, published in 1992, Alan Cooper, Cecile Mourer-Chauvire, Geoffrey Chambers, Arndt von Haeseler, Allan Wilson, and Svante Paabo investigated the evolutionary relationships between some extinct and modern species of flightless birds. Two groups of flightless birds, the kiwis and the moas, existed in New Zealand during the Pleistocene. The moas are now extinct, although 11 species were formerly present. In this study, the researchers investigated the phylogenetic relationships between the kiwis and moas of New Zealand, and several other (nonextinct) species of flightless birds. These included the emu and the cassowary (found in Australia and New Guinea), the ostrich (found in Africa and formerly Asia), and two rheas (found in South America).

THE HYPOTHESIS

Because DNA is a relatively stable molecule, it can be PCR-amplified from a preserved sample of a deceased organism and subjected to DNA sequencing. A comparison of DNA sequences with modern species may help elucidate the phylogenetic relationships between extinct and modern species.

TESTING THE HYPOTHESIS

Starting material: Tissue samples from four extinct species of moas were obtained from museum specimens. Tissue samples were also obtained from three species of kiwis, one ostrich, one cassowary, one emu, and two species of rhea.

Experimental Level

Conceptual Level

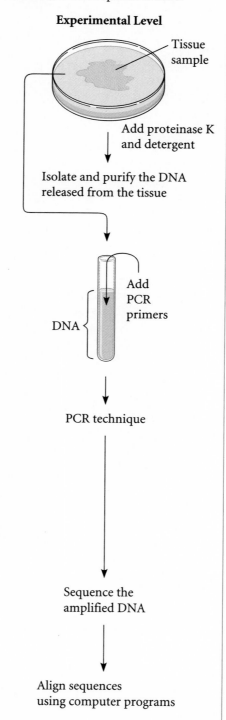
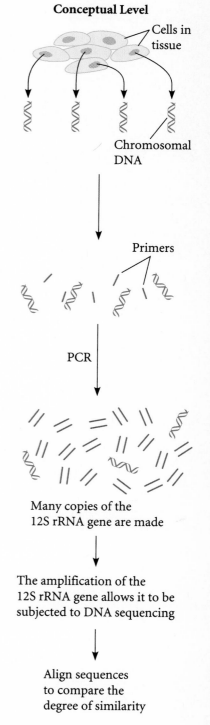

1. For soft tissue samples, treat with proteinase K (which digests protein) and a detergent that dissolves cell membranes. This releases the DNA from the cells.

Tissue sample

Add proteinase K and detergent

Isolate and purify the DNA released from the tissue

Cells in tissue

Chromosomal DNA

2. Individually, mix the DNA samples with a pair of PCR primers that are complementary to the 12S rRNA gene.

   Note: Primers recognize the 12S rRNA gene.

DNA

Add PCR primers

Primers

PCR

3. Subject the samples to PCR. See Chapter 19 (Figure 19-4) for a description of PCR.

PCR technique

Many copies of the 12S rRNA gene are made

4. Subject the amplified DNA fragments to DNA sequencing. See Chapter 19 for a description of DNA sequencing.

Sequence the amplified DNA

The amplification of the 12S rRNA gene allows it to be subjected to DNA sequencing

5. Align the DNA sequences to each other. Methods of DNA sequence alignment are described in Chapter 22, Experiment 22A.

Align sequences using computer programs

Align sequences to compare the degree of similarity

THE DATA

```
MOA 1      GCTTAGCCCTAAATCCAGATACTTACCCTACACAAGTATCCGCCCGAGAACTACGAGCACAAACGCTTAAAACTCTAAGGACTTGGCGGTGCCCCAAACCCACCTAGAGGAGCCTGTTCTATAATCGATAATCCACGATA
MOA 2      ...............................................................................................................................................
MOA 3      ...G.........T...............................T.................................................................................................
MOA 4      .....................C...............................................................................................................C...T.....
KIWI 1     .............T.G......GT...CT...C............................................................T.........................................C.......
KIWI 2     .............T.G.G....AT...CT...C............................................................T.........................................C.......
KIWI 3     .............T.G.G.AT...C...C................................................................T.........................................C.......
EMU        ............TT...C..T...CAG..C...T...........................................................T.........................................C.......
CASSOWARY  ............TT...CG.TA...CTG.................................................................T.........................................C.......
OSTRICH    ......T....AT.....C..CT......................................................................T........................................T
RHEA 1     ..............T...C..CT......................................................................T.........................................C.......
RHEA 2     ..............C...C.C........................................................................T.........................................C.......

MOA 1      CACCCGACCATCCCTCGCCCGT-GCAGCCTACATACCGGCCGTCCCCAGCCCGCCT--AATGAAAG-AACAATAGCGAGCACAACAGCCCTCCCCCGCTAACAAGACAGGTCAAGGTATAGCATATGAGATGGAAGAAATG
MOA 2      ................A...........................................TCA-...............................................................................
MOA 3      ......T.T..A-----...............TA---T.........................................................................................................
MOA 4      ......T.T..A-----...............T..AC--........................................................................................................
KIWI 1     ...A....T.T..AAC-A.......T......G...T...AA...G.-----..C...A....TA..-..A................................C........................................
KIWI 2     ...A....T.T..AAC-A.......T......G...T...AA...G.-----..C...A....TA..-..A................................C........................................
KIWI 3     ...A....T.T..AAC-A.......G......AA...-----..GC.......TACA--A.......................................CC.C.....G..................................
EMU        ...AG...T.T..AA--A.......G.-----........T..AC--TT.............G....................................
CASSOWARY  ...A....T.T..AA.TA.......G..-----.G..G........T...AC--T.........................................................................................
OSTRICH    ...A...C...T..A-T.......G....C----G.........T...A...............................GAG.............................................................
RHEA 1     ......T.T..A-----..........TA.G.-----..C..AG..T.T..TA.........................G..............................................
RHEA 2     ...........TA...G.-----..C..A..T.T..TA-----...G................................................................................................

MOA 1      GGCTACATTTTCTAACATAGAACACCC-------------ACGAAAGAGAAGGTGAAACCCTCCTCAAAAGGCGGATTTAGCAGTAAAATAGAACAAGAATGCCTATTTTAAGCCCGGCCCTGGGGC
MOA 2      ...............................................A....T....G.......................................................T.....................
MOA 3      .............T-------------......G.............C...C...C......T.........
MOA 4      .............................A..........G.............G...C..C.......T....
KIWI 1     .....A.....T.T-------------A.GGT...T.-C..T.G........C...T...GA.T...............T...A...
KIWI 2     .....A.....T.T-------------A.GGT...T.-C..T.G........C...T...GA.T..........T...A...
KIWI 3     .....A...T.T-------------A.GGTA..T.-C...T.G..A.......C...T...A.T...........A...
EMU        ...........T.T-------------AG.T...T.AC.T..G........C...T...GA.T.........A.......A...
CASSOWARY  ...............T-------------A..G.T...T.A...T.G........C...GA.T.......A.........A...
OSTRICH    ...........T.A-------------G.TA..T.A..G.........T...GA.T........T---T...T.A...
RHEA 1     ...........TC....A-------------G....GGCA......AC...CG........G...G.TC...A..C.C........A...
RHEA 2     ...........GTC...G-------------GGCA......AC...CG........G.G.G.TC...A..C.C........A...
```

INTERPRETING THE DATA

The data illustrate a multiple sequence alignment of the amplified DNA sequences. The first line shows the DNA sequence of one of the extinct moa species. Underneath it are the sequences of the other species. When the other sequences are identical to the first sequence, a dot is placed in the corresponding position. When the sequences are different, the nucleotide base (A, T, G, or C) is placed there. In a few regions, the genes are different lengths. In these cases, a dash is placed at the corresponding position.

As you can see from the large number of dots, the sequences among all of these flightless birds are very similar. To establish evolutionary relationships, we need to focus on the few differences that occur. Some surprising results were obtained. The sequences from the kiwis (a New Zealand species) are actually more similar to the sequence from the ostrich (an African species) than they are to those of the moas, which were once found in New Zealand. Likewise, the kiwis are more similar to the emu and cassowary (found in Australia and New Guinea) than they are to the moas. Contrary to their original expectations, the authors concluded that the kiwis are more closely related to Australian and African flightless birds than they are to the moas. They proposed that New Zealand was colonized twice by ancestors of flightless birds.

As shown here, the researchers constructed a new evolutionary tree that illustrates the relationships among these modern and extinct species:

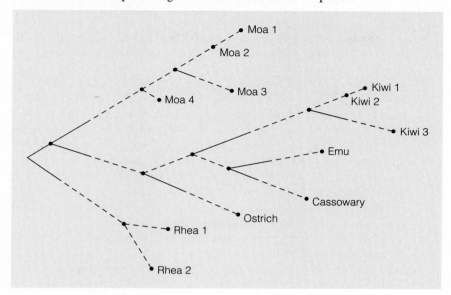

This tree was made by analyzing the sequence differences among the various species, using computer programs that are beyond the scope of this text. If you would like a general description of how computer programs analyze genetic sequences, see Chapter 22. In any case, the experiment described here illustrates the exciting possibilities that can be realized when ancient DNA sequences are obtained.

Genetic variation at the molecular level commonly is associated with neutral changes in gene sequences; nonneutral changes are acted on by natural selection

In Chapter 26, we learned that genetic variation is prevalent among natural populations. During the last few decades, a great debate has occurred among population and evolutionary geneticists. The debate centers on the reason for genetic variation in natural populations. Is it due primarily to mutations that are favored by natural selection, or to random genetic events?

A **nonneutral mutation** is one that affects the phenotype of the organism and can be acted on by natural selection. A nonneutral mutation may only subtly alter the phenotype of an organism, or it may have a major impact. According to Darwin, natural selection is the agent that leads to evolutionary change in populations. It selects for the survival of the fittest and thereby promotes the establishment of beneficial alleles and the elimination of deleterious ones. Therefore, many geneticists have assumed that natural selection is the dominant force in changing the genetic composition of natural populations, thereby leading to variation.

In 1968, Motoo Kimura, at the National Institute of Genetics in Mishima, Japan, proposed the **neutral theory of evolution**. According to this idea, most genetic variation observed in natural populations is due to the accumulation of neutral mutations. **Neutral mutations** do not affect the phenotype of the organism; neutral alleles are not acted on by natural selection. For example, a mutation within a structural gene that changes a glycine codon from GGG to GGC would not affect the amino acid sequence of the encoded protein. Since neutral mutations do not affect phenotype, they spread throughout a population according to their frequency of appearance and to genetic drift. This theory has been called the "survival of the luckiest" and also **non-Darwinian evolution** to contrast it with Darwin's "survival of the fittest." Kimura agreed with Darwin that natural selection is responsible for adaptive changes in a species during evolution. His main argument is that most modern variation in gene sequences is explained by neutral variation rather than adaptive variation. In 1974, Kimura, along with his colleague Tomoko Ohta, suggested five principles that govern the evolution of genes at the molecular level:

1. For each protein, the rate of evolution, in terms of amino acid substitutions, is approximately constant with regard to neutral substitutions that do not affect protein structure or function.

   Evidence: As an example, the amount of genetic variation between the coding sequence of the human α-globin and β-globin genes is approximately the same as the difference between the α-globin and β-globin genes in carp. This type of comparison holds true among many different genes compared among many different species.

2. Proteins that are functionally less important for the survival of an organism, or parts of a protein that are less important for its function, tend to evolve faster than more important proteins or regions of a protein. In other words, during evolution, less important proteins will accumulate amino acid substitutions more rapidly than important proteins.

   Evidence: Certain proteins are critical for survival, and their structure is exquisitely tuned to their function. An example are the histone proteins necessary for nucleosome formation in eukaryotes. Histone genes tolerate very few mutations and have evolved extremely slowly. By comparison, fibrinopeptides, which bind to fibrinogen to form a blood clot, evolve very rapidly. Presumably, the sequence of amino acids in this polypeptide is not very important in allowing it to aggregate and form a clot. Another example concerns the amino acid sequences of enzymes. It is known that amino acid substitutions are very rare within the active site (which is critical for function), but are more frequent in other parts of the protein.

3. Those mutant amino acid substitutions that do not disrupt the existing structure and function of a protein (conservative substitutions) occur more frequently in evolution than disruptive amino acid changes.

   Evidence: When examining the rate of change of the coding sequence within structural genes, nucleotide substitutions are more likely to occur in the wobble base than in the first or second base within a codon. Mutations in the wobble base are often silent (i.e., do not change the amino acid sequence of the protein). In addition, conservative substitutions (i.e., a similar amino acid substitution, such as a nonpolar amino acid for another nonpolar amino acid) are fairly common. By comparison, nonconservative substitutions are less frequent, although they do occur. Nonsense and frameshift mutations are very rare within the coding sequences of genes. Also, intron sequences evolve more rapidly than exon sequences.

4. Gene duplication must always precede the emergence of a gene having a new function.

   Evidence: When a single copy of a gene exists in a species, it usually plays a functional role similar to that of the homologous gene found in another species. Gene duplications have created gene families in which each family member can evolve somewhat different functional roles. An example is the hemoglobin family described in Chapter 8 (see pp. 205–206).

5. Selective elimination of definitely deleterious mutations and random fixation of selectively neutral or very slightly deleterious alleles occur far more frequently in evolution than Darwinian selection of definitely advantageous mutants.

   Evidence: As mentioned in #3, silent and conservative mutations are much more common than nonconservative substitutions. Presumably these nonconservative mutations usually have a negative effect on the phenotype of the organism, so that they are effectively eliminated from the population by natural selection. On rare occasions, however, an amino acid substitution due to a mutation may have a beneficial effect on the phenotype. For example, a nonconservative mutation in the β-globin gene produces H^S, which gives an individual resistance to malaria in the heterozygous condition.

In general, the DNA sequencing of hundreds of thousands of different genes from thousands of species has provided compelling support for these five principles of gene evolution at the molecular level. However, the argument is by no means resolved. There are some geneticists, called **selectionists**, who oppose the neutralist theory. They often can offer persuasive theoretical arguments in favor of natural selection as the primary factor promoting genetic variation. In any case, the argument is largely a quantitative rather than a qualitative one. Each school of thought accepts that genetic drift and natural selection both play key roles in evolution. The neutralists argue that most genetic variation arises from neutral genetic mutations and genetic drift, whereas the selectionists argue that beneficial mutations and natural selection are primarily responsible.

Evolution is associated with changes in chromosome structure and number

In this section, we have emphasized mutations that alter the DNA sequences within genes. In addition to gene mutations, however, other types of changes, such as gene duplications, inversions, translocations, and changes in chromosome number, are important features of evolution.

As was discussed earlier in this chapter, changes in chromosome structure and/or number may not always be adaptive, but they can lead to reproductive isolation and the origin of new species. As an example of variation of chromosome structure among closely related species, Figure 27-10 compares the banding pattern

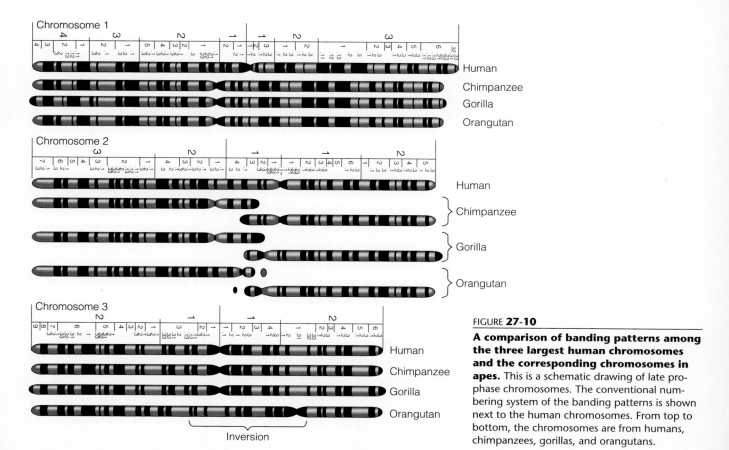

FIGURE **27-10**

A comparison of banding patterns among the three largest human chromosomes and the corresponding chromosomes in apes. This is a schematic drawing of late prophase chromosomes. The conventional numbering system of the banding patterns is shown next to the human chromosomes. From top to bottom, the chromosomes are from humans, chimpanzees, gorillas, and orangutans.

of the three largest chromosomes in humans, and the corresponding chromosomes in chimpanzees, gorillas, and orangutans. The banding patterns are strikingly similar, because these species are closely related evolutionarily. However, there are some interesting differences. Humans have one large chromosome 2, but this chromosome is divided into two separate chromosomes in the other three species. This explains why humans have 23 types of chromosomes while the apes have 24. This may have occurred by a fusion of the two smaller chromosomes during the development of the human lineage. Another interesting change in chromosome structure is seen in chromosome 3. The banding patterns among humans, chimpanzees, and gorillas are very similar, but the orangutan has a large inversion that flips the arrangement of bands in the centromeric region.

CONCEPTUAL SUMMARY

Biological evolution involves heritable changes in one or more characteristics in a population or species over the course of many generations. In the first part of this chapter, we were concerned primarily with how evolution results in the formation of new *species*. To become a separate species, a population must be reproductively isolated from all other species. This enables their gene pool to evolve as a single unit. There are several ways that reproductive isolation can occur. The *biological species concept* emphasizes reproductive isolation as an important event that leads to speciation, whereas the *species recognition concept* emphasizes *sexual selection* as a positive force during speciation.

Speciation is usually a branching process (*cladogenesis*), although *anagenesis* (transformation of a single species) occasionally occurs. Depending on geographic barriers, divergent speciation may be *allopatric*, *parapatric*, or *sympatric*. Allopatric speciation, which involves geographic barriers, is thought to be the most widespread form of speciation. It can occur slowly, due to the gradual formation of a geographic barrier, or rapidly, due to the founder effect. Parapatric speciation is similar to allopatric, except that the geographic barriers are less complete, with ever-decreasing levels of interbreeding taking place in *hybrid zones*. Sympatric speciation does not require geographic barriers. Instead, an abrupt genetic event may lead to reproductive isolation. In plants, the formation of polyploids is a common form of sympatric speciation. Polyploids are reproductively isolated, because they produce sterile hybrids when crossed to nonpolyploid species. In general, the fossil record suggests that speciation is often a rapid process that punctuates long periods during which the species is in equilibrium with its environment. This *punctuated equilibrium* hypothesis is contrasted with *gradualism*, which proposes a slower but steady rate of evolution due to the accumulation of many small genetic changes.

Molecular evolution is the study of the molecular changes in the genetic material during evolution. There are many ways to investigate the genetic material, but the analysis of DNA sequences and (the deduced) amino acid sequences are the most commonly used methods and perhaps the most informative. *Homologous* genes are derived from a common ancestral gene. *Interspecies homology* can be used to reconstruct *phylogenetic trees*. When two species are closely related evolutionarily, they will tend to have gene sequences that are more similar to each other.

Kimura and Ohta proposed five principles of molecular evolution that are consistent with our present knowledge of gene sequences. The *neutral theory of evolution* argues that most variation in gene sequences is neutral. In the case of structural genes, *neutral mutations* are not expected to significantly alter the structure and function of the encoded protein; many *nonneutral mutations* are highly deleterious and are thus eliminated quickly from the gene pool; other nonneutral mutations are adaptive and are acted on by natural selection to change the characteristics of a species. According to the neutral theory of evolution, though, these adaptive changes represent a small proportion of the total number of genetic changes that occur during evolution. In addition to changes at the gene level, it is also common for changes in chromosome structure and number to occur. These events are often instrumental in leading to reproductive isolation.

In a broad sense, evolutionary biologists would like to know how genetic changes have led to the phenotypic characteristics of present-day species. At the heart of this question is speciation: How do we explain the existence of so many different species? To answer this question, biologists have tried to understand how two different, but closely related species, have become reproductively isolated. By observing species in nature, they have identified several prezygotic and postzygotic mechanisms that prevent the production of viable, interspecies offspring. Researchers have also correlated reproductive isolation with experimentally observable genetic changes. For example, parapatric speciation may be associated with the accumulation of chromosomal inversions that prevent the production of fertile offspring within hybrid zones of contact. Sympatric speciation can occur via the formation of allopolyploids. As was discussed in Experiment 27A, this speciation mechanism can be confirmed by producing "artificial" polyploids.

Evolutionary biologists are also interested in the rates of evolutionary change. Experimentally, the fossil record suggests that it is more common for evolution to follow a pattern of punctuated equilibrium, in which a species is well adapted to its environment for a long period of time (the equilibrium), which is punctuated by short periods of rapid evolutionary change. These short periods may occur due to abrupt genetic changes (e.g., allopolyploidism or new alleles that have a dramatic effect on phenotype), or there may be a short period of strong selective pressure (e.g., the founder effect, a new predator in the region, or a significant environmental change).

At the molecular level, evolution is due to alterations in DNA sequences, and changes in chromosome structure and number. Geneticists have found a great amount of variation within most species. As was discussed in Table 27-3, there are several experimental methods for detecting genetic variation at the DNA and protein levels. The earliest methods involved the study of allozymes; DNA sequencing is now the most common way to study molecular evolution. By comparing homologous gene sequences within a species and among different species, evolutionary biologists have been able to probe the relationship between organismal evolution and changes in DNA sequences. They have also constructed evolutionary trees that describe the phylogenetic relationships among many species. More recently, it has even been possible to analyze DNA sequences from some extinct species.

With all of this sequence information available, researchers have debated the origin of genetic variation in modern species. The data suggest that most genetic variation occurs through neutral changes that accumulate due to genetic drift. Assuming that the rate of new mutation is essentially constant, the accumulation of neutral changes provides a biological clock to measure the time scale of evolution. However, not all variation is neutral. Adaptive changes, which are acted on by natural selection, must also occur and alter the phenotypic characteristics of species over time, thereby leading to the evolution of new, better adapted species.

PROBLEM SETS

SOLVED PROBLEMS

1. A codon for leucine is UUA. A mutation causing a single base substitution in a gene can change this codon in the transcribed mRNA into GUA (valine), AUA (isoleucine), CUA (leucine), UGA (stop), UAA (stop), UCA (serine), UUG (leucine), UUC (phenylalanine), or UUU (phenylalanine). According to the neutral theory, which of these mutations would you expect to see within the genetic variation of a natural population? Explain.

Answer: The neutral theory proposes that neutral mutations will accumulate to the greatest extent in a population. Leucine is a nonpolar amino acid. For a UUA codon, single base changes of CUA and UUG are silent, and so they would be the most likely to occur in a natural population. Likewise, conservative substitutions to other nonpolar amino acids such as isoleucine (AUA), and valine (GUA), and phenylalanine (UUC and UUU) may not affect protein structure and function, and so they may also occur and not be eliminated rapidly by natural selection. The polar amino acid serine (UCA) is a nonconservative substitution; one would predict that it is more

likely to disrupt protein function. Therefore, it may be less likely to be found. Finally, the stop codons, UGA and UAA, would be expected to diminish or eliminate protein function, particularly if they occur early in the coding sequence. These types of mutations are selected against and therefore are not usually found in natural populations.

2. Explain why homologous genes have sequences that are similar but not identical.

Answer: Homologous genes are derived from the same ancestral gene. Therefore, as a starting point, they had identical sequences. Over time, however, each gene accumulated random mutations that the other homologous genes did not acquire. These random mutations changed the gene from its original sequence. Therefore, much of the sequences between homologous genes remains identical, but some of the sequence will be altered due to the accumulation of independent random mutations.

CONCEPTUAL QUESTIONS

1. Discuss the two principles on which evolution is based.

2. What is meant by the term reproductive isolation? Give several examples. Compare and contrast the biological species concept and the species recognition concept with regard to reproductive isolation.

3. What is sexual selection? For most animals, which gender does the selecting, the male or female? Give examples. Describe and give examples of how sexual selection can alter the traits of animals.

4. Distinguish between anagenesis and cladogenesis. Which type of speciation is more prevalent. Why?

5. Describe three or more genetic mechanisms that may lead to the rapid evolution of a new species. Which of these genetic mechanisms are influenced by natural selection, and which are not?

6. Explain the type of speciation (allopatric, sympatric, or parapatric) that is most likely to occur under each of the following conditions:

  A. A pregnant female rat is transported by an ocean liner to a new continent.
  B. A meadow containing several species of grasses is exposed to a pesticide that promotes nondisjunction.
  C. In a very large lake containing several species of fish, the water level gradually falls over the course of several years. Eventually, the large lake becomes subdivided into smaller lakes, some of which are connected by narrow streams.

7. Alloploids are created by crosses involving two different species. Explain why alloploids are reproductively isolated from the two original species from which they were derived. Explain why alloploids are usually sterile, whereas allopolyploids (containing a diploid set from each species) are commonly fertile.

8. Discuss the major goals in the field of molecular evolution.

9. Discuss the evidence in favor of the punctuated equilibrium theory of evolution. What mechanisms could account for this pattern of evolution? In contrast, what type of genetic changes are consistent with gradualism?

10. The following are two DNA sequences from homologous genes:

TTGCATAGGCATACCGTATGATATCGAAAACTAGAAAAATA-
GGGCGATAGCTA

GTATGTTATCGAAAAGTAGCAAAATAGGGCGATAGCTACCCA-
GACTACCGGAT

  The two sequences, however, do not begin and end at the same location. Try to line them up according to the their homologous regions.

11. Why do some genes evolve at a fast rate and others at a slow rate? Explain how differences in the rate of molecular evolution can be useful with regard to establishing distant evolutionary relationships versus close evolutionary relationships.

12. Explain why the neutral theory of evolution is sometimes called non-Darwinian evolution.

13. As was discussed in Chapter 26, genetic variation is prevalent in natural populations. This variation is revealed in the electrophoresis of allozymes and in the DNA sequencing of genes. According to the neutral theory, discuss the relative importance of natural selection against detrimental mutations, natural selection in favor of beneficial mutations, and neutral mutations, in accounting for the genetic variation we see in natural populations.

14. If you were comparing the karyotypes of species that are closely related evolutionarily, what types of similarities and differences would you expect to find?

15. Would the rate of deleterious or beneficial mutations be a good molecular clock? Why or why not?

EXPERIMENTAL QUESTIONS

1. In Chapter 8, we discussed the use of colchicine to promote nondisjunction. If this drug had been available to Arne Müntzing in 1932, how might he have conducted the experiment described in Experiment 27A differently?

2. In Experiment 27A, what do the results mean considering that the artificial *G. tetrahit* could produce fertile offspring with the natural *G. tetrahit*? What would the results have meant if these two strains had not produced fertile offspring?

3. Sympatric speciation by allopolyploidy has been proposed as a common mechanism for speciation. Let's suppose you were interested in the origin of grass species in southern California. Experimentally, how would you go about determining if some of the grass species are the results of allopolyploidy?

4. Two diploid species of closely related frogs, which will we call species A and species B, were analyzed with regard to genes that encode an enzyme called hexokinase. Species A has two distinct copies of this gene: *A1* and *A2*. In other words, this diploid species is *A1A1 A2A2*. The other species has three copies of the hexokinase gene, which we will call *B1*, *B2*, and *B3*. A diploid individual of species B would be *B1B1 B2B2 B3B3*. These hexokinase genes from the two species were subjected to DNA sequencing, and the percentage of sequence identity was compared among these genes. The results are shown here:

Percentage of DNA Sequence Identity

	A1	A2	B1	B2	B3
A1	100	62	54	94	53
A2	62	100	91	49	92
B1	54	91	100	67	90
B2	94	49	67	100	64
B3	53	92	90	64	100

If we assume that hexokinase genes were never lost in the evolution of these frog species, how many distinct hexokinase genes do you think there were in the most recent ancestor that preceded the divergence of these two species? Explain your answer. Also explain why species B has three distinct copies of this gene while species A only has two.

5. Ancient samples often contain minute amounts of DNA. What technique can be used to increase the amount of DNA in an ancient sample? Explain how this technique is performed, and how it increases the amount of a specific region of DNA.

6. In Experiment 27B, explain how we know that the kiwis are more closely related to the emu and cassowary than to the moas. Cite particular regions in the sequences that support your answer.

7. In Chapter 20, we learned about a technique called *in situ* hybridization, during which a cloned piece of DNA is hybridized to a set of chromosomes (see pp. 542–544). Let's suppose that we cloned a piece of DNA from *G. pubescens* and used it as a probe for *in situ* hybridization. What would you expect to happen if we hybridized it to the *G. speciosa*, the natural *G. tetrahit*, or the artificial *G. tetrahit* strains? Draw your expected results.

8. A team of researchers has obtained a dinosaur bone (*Tyrannosauris rex*) and has attempted to extract ancient DNA from it. Using primers to the 12S rRNA gene, they have used PCR and obtained a DNA segment that yields a sequence homologous to crocodile DNA. Other scientists are skeptical that this sequence is really from the dinosaur; they believe instead that it may be due to contamination from more recent DNA, such as the remains of a reptile that lived much more recently. What criteria might you use to establish the credibility of the dinosaur sequence?

QUESTIONS FOR STUDENT DISCUSSION/COLLABORATION

1. The raw material for evolution is random mutation. Discuss whether or not you view evolution as a random process.

2. Compare the forms of speciation that are slow with those that occur more rapidly. Make a list of the slow and fast forms. With regard to mechanisms of genetic change, what features do slow and rapid speciation have in common? What features are different?

3. Do you think that Darwin would object to the neutral theory of evolution?

EXPERIMENTAL TECHNIQUES

METHODS OF CELL GROWTH

MICROSCOPY

SEPARATION METHODS

METHODS TO MEASURE CONCENTRATIONS OF SMALL MOLECULES, AND DETECT RADIOISOTOPES AND ANTIGENS

METHODS OF CELL GROWTH

Researchers often grow cells in a laboratory as a way to study them. This is known as **cell culture**. Cell culturing offers several technical advantages. The primary advantage is that the growth medium is defined and can be controlled. Minimal growth medium contains the bare essentials for cell growth: salts, a carbon source, an energy source, essential vitamins, amino acids, and trace elements. In their experiments, geneticists often compare strains that can grow in *minimal media* with mutant strains that cannot grow unless the medium is supplemented with additional components. A *rich growth medium* contains many more components than are required for growth.

Researchers also add substances to the culture medium for other experimental reasons. For example, radioactive isotopes can be added to the culture medium to radiolabel cellular macromolecules. Or an experimenter could add a hormone to the growth medium and then monitor the cells' response to the hormone. In all of these cases, cell culturing is advantageous because the experimenter can control and vary the composition of the growth medium.

The first step in creating a cell culture is the isolation of a cell population that the researchers wish to study. For bacteria, such as *Escherichia coli*, and eukaryotic microorganisms such as yeast and *Neurospora*, the researchers simply obtain a sample of cells from a colleague or a stock center. For animal or plant tissues, the procedure is a bit more complicated. When cells are contained within a complex tissue, they must first be dispersed by treating the tissue with agents that separate it into individual cells to create a *cell suspension*.

Once a desired population of cells has been obtained, researchers can grow them in a laboratory (i.e., *in vitro*) either *suspended* in a liquid growth medium or attached to a solid surface such as agar. Both methods have been used commonly in the ex-

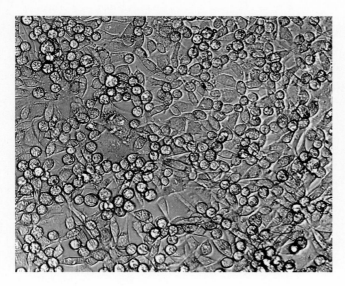

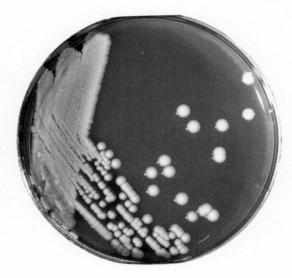

(a) Fibroblast (animal cell) culture   **(b)** Bacterial colonies

FIGURE **A-1**

Growth of cells on solid growth media. (a) Normal fibroblasts grow as a monolayer on the growth medium. **(b)** Bacterial cells form colonies that are a clonal population of cells derived from a single cell.

periments considered throughout this text. Liquid culture is often used when re-searchers want to obtain a large quantity of cells and isolate individual cellular components, such as nuclei or DNA. By comparison, Figure A-1 shows micrographs of animal cells and bacteria cells that are grown on solid growth media. As discussed in Experiment 23A, solid media are used to study cancer cells, since such cells can be distinguished by the formation of foci in which malignant cells pile up on top of each other. In gene cloning experiments with bacteria and yeast, solid media are also used. Each colony of cells is a clone of cells that is derived from a single cell that di-vided many times (Figure A-1b). As discussed in Chapter 19, a solid medium is used in the isolation of individual clones that contain a desired gene.

MICROSCOPY

As you probably already know, **microscopy** is a technique to observe things that are not visible (or are hardly visible) with the naked eye. A key concept in microscopy is **resolution**, which is defined as the minimum distance between two objects that can be seen as separate from each other. The ability to resolve two points as being separate depends on several factors, including the wavelength of the illumination source (light or electron beam), the medium in which the sample is immersed, and structural features of the microscope (which are beyond the scope of this text).

As shown in Figure A-2, there are two widely used kinds of microscopes, the optical (light) microscope and the transmission electron microscope (TEM). The light microscope is used to resolve cellular structure to a limit of approximately 0.3 μm. (For comparison, a typical bacterium is about 1 μm long.) At this resolution, the individual cell organelles in eukaryotic cells can be discerned easily, and chromosomes are also visible. Karyotyping is accomplished via light microscopy after the chromosomes have been treated with stains. A variation of light microscopy known as fluorescence microscopy is often used to highlight a particular feature of a chromosome or cellular structure. The technique of florescence *in situ* hybridization (FISH; see Chapter 20) makes use of this type of

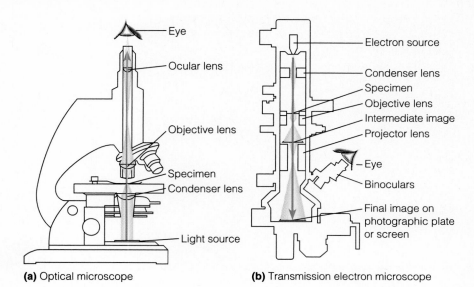

FIGURE **A-2**

Design of (a) optical and (b) transmission electron microscopes.

(a) Optical microscope

(b) Transmission electron microscope

microscope. Also, optical modifications in certain light microscopes (e.g., phase contrast and differential interference) can be used to exaggerate the differences in densities between neighboring cells or cell structures. These kinds of light microscopes are useful in monitoring cell division in living (unstained) cells or in transparent worms (as in Experiment 24A).

The structural details of large macromolecules such as DNA and ribosomes are not observable by light microscopy. The coarse topology of these macromolecules can be determined by electron microscopy. Electron microscopes have a limit of resolution of about 2 nm, which is about 100 times finer than the best light microscopes. The primary advantage of electron microscopy over light microscopy is its better resolution. Disadvantages include a much higher expense and more extensive sample preparation. In transmission electron microscopy, the sample is bombarded with an electron beam. This requires that the sample be dried, fixed, and usually coated with a heavy metal that absorbs electrons.

SEPARATION METHODS

Biologists often wish to take complex systems and separate them into less complex components. For example, the cells within a complex tissue can be separated into individual cells, or the macromolecules within cells can be separated from the other cellular components. In this section, we will focus primarily on methods aimed at separating and ultimately purifying macromolecules.

Disruption of cellular components

In many experiments described in this text, researchers have obtained a sample of cells and then wish to isolate particular components within the cell. For example, a researcher may want to purify a protein that functions as a transcription factor. To do so, the researcher would begin with a sample of cells that synthesize this protein and then break open the cells using one of the methods described in Table A-1. In eukaryotes, the breakage of cells releases the soluble proteins from the cell; it also dissociates the cell organelles that are bounded by membranes. This mixture of proteins and cell organelles can then be isolated and purified by centrifugation and chromatographic methods that are described next.

TABLE **A–1 Common Methods of Cell Disrption**

Method	Description
Sonication	The exposure of cells to intense sound waves, which breaks the cell membranes.
French press	The passage of cells through a small orifice under high pressure, which breaks the cell membranes and cell wall.
Homogenization	Cells are placed in a tube that contains a pestle. When the pestle is spun, the cells are squeezed through the small space between the pestle and the glass wall of the tube, thereby breaking them.
Osmotic shock	The transfer of cells into a hypoosmotic medium. The cells take up water and eventually burst.

Centrifugation

Centrifugation is a method that is commonly used to separate cell organelles and macromolecules. A **centrifuge** contains a motor, which causes a rotor holding centrifuge tubes to spin very rapidly. As the rotor spins, particles will move toward the bottom of the centrifuge tube; the rate at which they move depends on several factors, including their densities, sizes, shapes, and the viscosity of the medium. The rate at which a macromolecule or cell organelle sediments to the bottom of a centrifuge tube is called its **sedimentation coefficient**, which is normally expressed in Svedberg units (S). A sedimentation coefficient has the units of seconds: $1 S = 1 \times 10^{-13}$ seconds.

When a sample contains a mixture of macromolecules or cell organelles, it is likely that different components will sediment at different rates. This phenomenon, known as **differential centrifugation**, is shown in Figure A-3. As seen here, particles with large sedimentation coefficients reach the bottom of the tube more quickly than those with smaller coefficients. There are two ways that researchers can use differential centrifugation as a separation technique. One way is to separate the *supernatant* from the *pellet* following centrifugation. The pellet is a collection of particles found at the bottom of the tube, and the supernatant is the liquid found above the pellet. In Figure A-3, if the experimenter had separated the supernatant and pellet at stage 3, most of the particles with large sedimentation coefficients would be found in the pellet while most of the particles with small coefficients would be found in the supernatant. Therefore, a separation of the supernatant and the pellet at this point would provide a way to segregate these three types of particles.

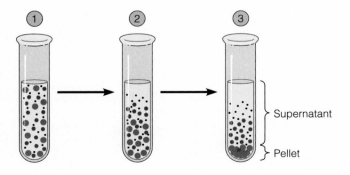

- ● Particles with large sedimentation coefficients
- • Particles with intermediate sedimentation coefficients
- · Particles with small sedimentation coefficients

FIGURE **A-3**

The method of differential centrifugation. (1) A sample containing a mixture of particles with different sedimentation coefficients is placed in a centrifuge tube. The tube is then spun for a short **(2)** and a longer **(3)** period of time. After the longer time period, the particles with a large sedimentation coefficient are found in the pellet, while those with a small sedimentation coefficient are in the liquid supernatant.

An alternative way to separate particles using centrifugation is to collect **fractions**. A fraction is a portion of the liquid contained within a centrifuge tube. The collection of fractions is done when the solution within the centrifuge tube contains a gradient. For example, as shown in Figure A-4, the solution at the top of the tube may have a lower concentration of sucrose than that at the bottom. In this experiment, a sample containing a mixture of cell organelles is layered on the top of the gradient and then centrifuged. In this example, the lysosomes and mitochondria separate from each other, because they have different sedimentation coefficients. The experimenter then punctures the bottom of the tube and collects fractions. The mitochondria, which are heavier, come out of the tube in the earlier fractions; the lysosomes will be collected in later fractions.

A type of gradient centrifugation that may also be used to separate macromolecules and organelles is *equilibrium density centrifugation*. In this method, the particles will sediment through the gradient, reaching a position where the density of the particle matches the density of the solution. At this point, the particle is at equilibrium and so does not move any farther toward the bottom of the tube.

Gel electrophoresis and chromatography

Chromatography is a method to separate different macromolecules and small molecules based on their chemical and physical properties. In this method, a sam-

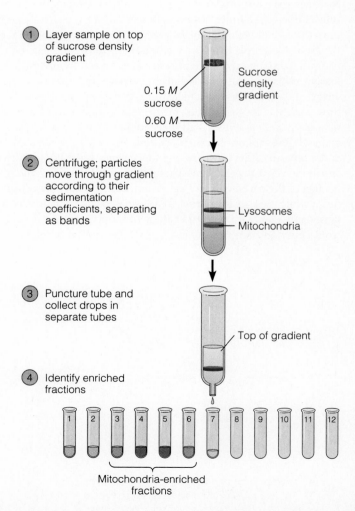

FIGURE **A-4**

Gradient centrifugation and the collection of fractions.

1. Layer sample on top of sucrose density gradient

    Sucrose density gradient

    0.15 *M* sucrose

    0.60 *M* sucrose

2. Centrifuge; particles move through gradient according to their sedimentation coefficients, separating as bands

    Lysosomes
    Mitochondria

3. Puncture tube and collect drops in separate tubes

    Top of gradient

4. Identify enriched fractions

    1 2 3 4 5 6 7 8 9 10 11 12

    Mitochondria-enriched fractions

ple is dissolved in a liquid solvent and exposed to some type of matrix, such as a column containing beads or a thin strip of paper. The degree to which the molecules interact with the matrix will depend on their chemical and physical characteristics. For example, a positively charged molecule will bind tightly to a negatively charged matrix, while a neutral molecule will not.

Figure A-5 illustrates how *column chromatography* can be used to separate molecules that differ in their charges. Prior to this experiment, a column is packed with beads that are negatively charged. There is plenty of space between the beads for small molecules to flow from the top of the column to the bottom. However, if the molecules are positively charged, they will spend some of their time binding to the negative charges on the surface of the beads. In the example shown in Figure A-5, the purple particles are neutral and, therefore, flow rapidly from the top of the column to the bottom. They emerge in the fractions that are collected early in this experiment. The green particles, however, are positively charged and tend to bind to the beads. The binding of the green particles to the beads can be disrupted by increasing the ionic strength or lowering the pH of the solution that is added to the column. Eventually, the green particles will be eluted (i.e., leave the column) in later fractions.

There are many variations of chromatography used by researchers to separate molecules and macromolecules. The type shown in Figure A-5 is called *ion-exchange chromatography*, because its basis for separation depends on the charge of the molecules. In another type of column chromatography, known as *gel filtration* chromatography, the beads are porous. Small molecules are temporarily trapped within the beads, while large molecules flow between the beads. In this way, gel filtration separates molecules on the basis of size. To separate different types of macromolecules such as proteins, researchers may use another type of bead; this bead has a preattached molecule that binds specifically to the protein they want to purify. For example, if a

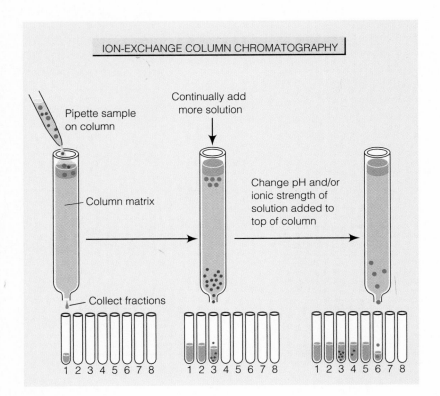

FIGURE **A-5**

Ion-exchange chromatography.

transcription factor binds a particular DNA sequence as part of its function, the beads within a column may have this DNA sequence preattached to them. Therefore, the transcription factor will bind tightly to DNA attached to these beads, while all other proteins will be eluted rapidly from the column. This form of chromatography is called *affinity chromatography*, because the beads have a special affinity for the macromolecule of interest.

Besides column chromatography, in which beads are packed into a column, there are other ways to make a matrix. In *paper chromatography*, molecules pass through a matrix composed of paper. The rate of movement of molecules through the paper will depend on their degree of interaction with the solvent and paper. In *thin-layer chromatography*, a matrix is spread out as a very thin layer on a rigid support such as a glass plate. In general, paper and thin-layer chromatography are effective at separating small molecules, whereas column chromatography is used to separate macromolecules such as DNA fragments or proteins.

Gel electrophoresis combines chromatography and electrophoresis to separate molecules and macromolecules. As its name suggests, the matrix used in gel electrophoresis is composed of a gel. As shown in Figure A-6, samples are loaded in wells at one end of the gel, and an electric field is applied across the gel. This electric field causes charged molecules to migrate from one side of the gel to the other. The migration of molecules in response to an electric field is called **electrophoresis**. In the examples of gel electrophoresis found in this text, the macromolecules within the sample migrate toward the positive end of the gel. In most forms of gel electrophoresis, a mixture of macromolecules is separated according to their molecular weights. Small proteins or DNA fragments move to the bottom of the gel more quickly than larger ones. Since the samples are loaded as bands within a well at the top of the gel, the molecules within the sample are separated into bands within the gel. These bands of separated macromolecules can be visualized with stains. For example, ethidium bromide is a stain that binds to DNA and RNA and can be seen under ultraviolet light.

The two most commonly used gels are polymers made from agarose or acrylamide. Proteins typically are separated on polyacrylamide gels, whereas DNA fragments are separated on agarose gels. Occasionally, researchers will use polyacrylamide gels to separate DNA fragments that are relatively small (namely, less than 1000 bp in length).

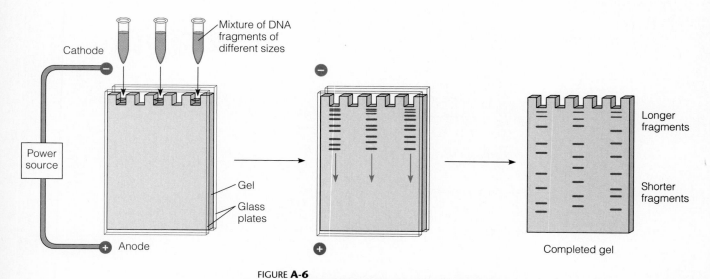

FIGURE **A-6**

Acrylamide gel electrophoresis of DNA fragments.

METHODS TO MEASURE CONCENTRATIONS OF SMALL MOLECULES, AND DETECT RADIOISOTOPES AND ANTIGENS

To understand the structure and function of cells, researchers often need to detect the presence of molecules and macromolecules within living cells, and to measure their concentrations. In this section, we will consider a variety of methods to detect and measure the concentrations of biological molecules and macromolecules.

Spectroscopy

Macromolecules found in living cells, such as proteins, DNA, and RNA, are fairly complex molecules that can absorb radiation (i.e., light). Likewise, small molecules such as amino acids and nucleotides can also absorb light. A device known as a **spectrophotometer** is used by researchers to determine how much radiation at various wavelengths a sample can absorb. The amount of absorption can be used to determine the concentration of particular molecules within a sample, because each type of molecule or macromolecule has its own characteristic wavelength(s) of absorption, called its absorption spectrum.

A spectrophotometer typically has two light sources, which can emit ultraviolet or visible light. As shown in Figure A-7, the light source is passed through a monochromator, which emits the light at a desired wavelength. This incident light then strikes a sample contained within a cuvette. Some of the incident light will be absorbed, and some will not. The unabsorbed light passes through the sample and is detected by the spectrophotometer. The amount of light which strikes the detector is subtracted from the amount of incident light, yielding the measure of absorption. In this way, the spectrophotometer provides an absorption reading for the sample. This reading can be used to calculate the concentration of particular molecules or macromolecules in a sample.

Detection of radioisotopes

A **radioisotope** is an unstable form of an atom that decays to a more stable form by emitting α-, β-, or γ-rays, which are types of ionizing radiation. In research, radioisotopes that are β and/or γ emitters are commonly used. A β-ray is an emitted electron, and a γ-ray is a an emitted photon. Some radioisotopes commonly used in biological experiments are shown in Table A-2.

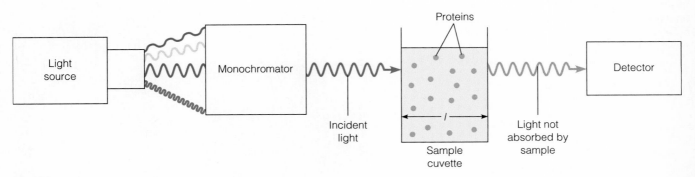

FIGURE **A-7**

Design of a spectrophotometer.

TABLE A–2 Some Useful Isotopes in Genetics

Isotope	Stable or Radioactive	Emission	Half-life
^2H	Stable		
^3H	Radioactive	β	12.1 years
^{13}C	Stable		
^{14}C	Radioactive	β	5568 years
^{15}N	Stable		
^{18}O	Stable		
^{24}Na	Radioactive	β (and γ)	15 hours
^{32}P	Radioactive	β	14.2 days
^{35}S	Radioactive	β	87 days
^{45}Ca	Radioactive	β	164 days
^{59}Fe	Radioactive	β (and γ)	45 days
^{131}I	Radioactive	β (and γ)	8.1 days

Experimentally, radioisotopes are used frequently, because they are easy to detect. Therefore, if a particular compound is radiolabeled, its presence can be detected specifically throughout the course of the experiment. For example, if a nucleotide is radiolabeled with ^{32}P, a researcher can determine whether the isotope becomes incorporated into newly made DNA or whether it remains as the free nucleotide. Researchers commonly utilize two different methods to detect radioisotopes: scintillation counting and autoradiography.

The technique of **scintillation counting** permits a researcher to count the number of radioactive emissions from a sample containing a population of radioisotopes. In this approach, the sample is dissolved in a solution (called the *scintillant*) that contains organic solvents and one or more compounds known as *fluors*. When radioisotopes emit ionizing radiation, the energy will be absorbed by the solvent. This excites solvent molecules, causing their electrons to be boosted to higher energy levels. The excited electrons will return to lower, more stable energy levels by releasing photons of light. When a fluor is struck by ionizing radiation, it also absorbs the energy and then releases a photon of light within a particular wavelength range. The role of a device known as a scintillation counter is to count the photons of light that are emitted by the fluor. Figure A-8 shows a scintillation counter. To use this device, a researcher dissolves his/her sample in a scintillant and then places the sample in a scintillation vial. The vial is then placed in the scintillation counter, which detects the amount of radioactivity. The scintillation counter has a digital meter that displays the amount of radioactivity in the sample, and it also provides a printout of the amount

FIGURE **A-8**

Photograph of a scintillation counter. The researcher is holding a blue rack that contains rows of scintillation vials in which the radioactivity will be counted.

of radioactivity in counts per minute. A scintillation counter contains several rows for the loading and analysis of many scintillation vials. After they have been loaded, the scintillation counter will count the amount of radioactivity in each vial, and provide the researcher with a printout of the amount of radioactivity in each vial.

A second way to detect radioisotopes is via **autoradiography**. This technique is not as quantitative as scintillation counting, because it does not provide the experimenter with a precise measure of the amount of radioactivity in counts per minute. However, autoradiography has the great advantage that it can detect the location of radioisotopes as they are found in macromolecules or cells. For example, autoradiography is used to detect a particular band on a gel, or to map the location of a gene within an intact chromosome.

To conduct autoradiography, a sample containing a radioisotope is fixed and usually dried. If it is a cellular sample, it also may be thin sectioned. The sample is then pressed next to X-ray film (in the dark), and placed in a lightproof cassette. When a radioisotope decays, it will emit a β- or γ-ray, which may strike a thin layer of photoemulsion next to the film. The photoemulsion contains silver salts such as AgBr. When a radioactive particle is emitted and strikes the photoemulsion, a silver grain is deposited on the film. This produces a dark spot on the film, which correlates with the original location of the radioisotope in the sample. In this way, the dark image on the film reveals the location(s) of the radioisotopes in the sample. Figure 11-3 shows how autoradiography can be used to visualize the process of bacterial chromosome replication. In this case, radiolabeled nucleotides were incorporated into the DNA, making it possible to picture the topology of the chromosome as two replication forks pass around the circular chromosome.

Detection of antigens by radioimmunoassay

Antibodies, also known as **immunoglobulins**, are proteins that are used to ward off infection by foreign substances; they are produced by cells of the immune system. Antibodies bind to structures on the surface of foreign substances known as **epitopes**; the foreign substance is called an **antigen**. A particular antibody binds to a particular antigen with a very high degree of specificity. For this reason, antibodies have been used extensively by researchers to detect particular antigens. For example, a human protein such as hemoglobin can be injected into a rabbit. Human hemoglobin is a foreign substance in the rabbit's bloodstream. Therefore, the rabbit will make antibodies that specifically recognize human hemoglobin and are designed to destroy it. Researchers can isolate and purify these antibodies from a sample of the rabbit's blood, and then use them to detect human hemoglobin in their experiments.

A **radioimmunoassay** is a method to measure the amount of an antigen in a biological sample. The steps in this method are shown in Figure A-9a. The researcher begins with two tubes that have a known amount of radiolabeled antigen (shown in purple). An unknown amount of the same antigen, which is not radiolabeled (shown in orange), is added to the tube on the right. The nonradiolabeled antigen comes from a biological sample; the goal of this experiment is to determine how much of this antigen is contained within the sample. In step 2, a known amount of antibody is added to each of the two tubes. The amount of the antibody is less than the amount of the antigen, and so the nonlabeled and radiolabeled antigens compete with each other for binding to the antibody. After binding, a precipitating agent such as an anti-immunoglobulin antibody is added, and the precipitate is centrifuged to the bottom of the tube. The radioactivity in the precipitate is then determined by scintillation counting.

To calculate the amount of antigen in the sample being assayed, the researcher must determine the percentage of antibody that has bound to nonlabeled antigen. To do so, a second component of the experiment is to develop a standard curve in which

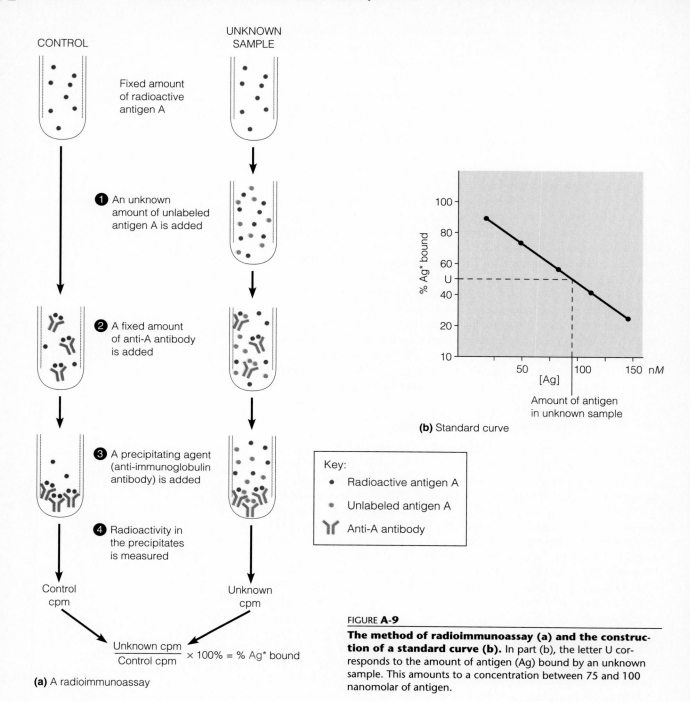

CONTROL

UNKNOWN
SAMPLE

Fixed amount
of radioactive
antigen A

❶ An unknown
amount of unlabeled
antigen A is added

❷ A fixed amount
of anti-A antibody
is added

❸ A precipitating agent
(anti-immunoglobulin
antibody) is added

❹ Radioactivity in
the precipitates
is measured

Control
cpm

Unknown
cpm

$$\frac{\text{Unknown cpm}}{\text{Control cpm}} \times 100\% = \% \text{ Ag* bound}$$

(a) A radioimmunoassay

Key:
- Radioactive antigen A
- Unlabeled antigen A
- Anti-A antibody

(b) Standard curve

FIGURE **A-9**

The method of radioimmunoassay (a) and the construction of a standard curve (b). In part (b), the letter U corresponds to the amount of antigen (Ag) bound by an unknown sample. This amounts to a concentration between 75 and 100 nanomolar of antigen.

a fixed amount of radiolabeled antigen is mixed with varying amounts of unlabeled antigen (Figure A-9b). Using this calibration curve, a researcher can determine how much antigen is found in the unknown sample. For example, as shown in the dashed line, if the unknown sample had about 50% of the antibody bound, then the concentration of antigen in the sample would be between 75 and 100 nanomolar.

Radioimmunoassays are used to determine the concentrations of many different kinds of antigens. This includes small molecules such as hormones (as in Experiment 21A) or macromolecules such as proteins.

GLOSSARY

A: an abbreviation for adenine.

acentric: describes a chromosome without a centromere.

acrocentric: describes a chromosome with the centromere significantly off center, but not at the very end.

activator: a transcriptional regulatory protein that increases the rate of transcription.

adaptive peak: a combination of many alleles that provides an optimal fitness for individuals in a stable environment.

adaptor hypothesis: a hypothesis that proposes a tRNA has two functions: recognizing a three-base codon sequence in mRNA and carrying an amino acid that is specific for that codon.

adenine: a purine base found in DNA and RNA. It base pairs with thymine in DNA.

age of onset: for alleles that cause genetic diseases, the time of life at which disease symptoms appear.

alkaptonuria: a human genetic disorder involving the accumulation of homogentisic acid due to a defect in homogentisic acid oxidase.

allele: an alternative form of a specific gene.

allele frequency: the number of copies of a particular allele in a population divided by the total number of all alleles for that gene in the population.

allopatric speciation (Greek, *allos*, "other"; Latin, *patria*, "homeland"): an evolutionary phenomenon in which speciation occurs when members of a species become geographically separated from the other members.

alloploid: an organism that contains chromosomes from two (or more) different species.

allopolyploid: an organism that contains two (or more) sets of chromosomes from two (or more) species.

allosteric enzyme: an enzyme that contains two binding sites— a catalytic site and a regulatory site.

allozymes: two or more enzymes (encoded by the same type of gene) with alterations in their amino acid sequences, which may affect their gel mobilities.

α-helix: a type of secondary structure found in proteins.

alternative splicing: refers to the phenomenon in which a pre-mRNA can be spliced in more than one way.

amber: a stop codon with the sequence UAG.

Ames test: a test using strains of a bacterium, *Salmonella typhimurium*, to determine if a substance is a mutagen.

amino acid: a building block of polypeptides and proteins. It contains an amino group, a carboxyl group, and a side chain.

aminoacyl-tRNA synthetase: an enzyme that catalyzes the attachment of a specific amino acid to the correct tRNA.

amino terminus: the location of the first amino acid in a polypeptide chain. The amino acid at the amino terminus still retains a free amino group that is not covalently attached to the second amino acid.

amniocentesis: a method of obtaining cellular material from a fetus for the purpose of genetic testing.

anagenesis (Greek, *ana*, "up," and *genesis*, "origin"): the evolutionary phenomenon in which a single species is transformed into a different species over the course of many generations.

anaphase: the third stage of M phase. As anaphase proceeds, half of the chromosomes move to one pole and the other half move to the other pole.

ancient DNA analysis: analysis of DNA that is extracted from the remains of extinct species.

aneuploid: not euploid. Refers to a variation in chromosome number such that the total number of chromosomes is not an exact multiple of a set or *n* number.

antibodies: proteins produced by the B-cells of the immune system that recognize foreign substances (namely, viruses, bacteria, and so forth) and target them for destruction.

anticipation: the phenomenon in which the severity of an inherited disease tends to get worse in future generations.

anticodon: a three-nucleotide sequence in tRNA that is complementary to a codon in mRNA.

antigens: foreign substances that are recognized by antibodies.

anti-oncogene: see *tumor-suppressor gene*.

antiparallel: an arrangement in a double helix in which one strand is running in the 5′ to 3′ direction while the other strand is 3′ to 5′.

antisense RNA: an RNA strand that is complementary to a strand of mRNA.

apical–basal-patterning gene: one of several plant genes that play a role in embryonic development.

apoptosis: programmed cell death.

artificial selection: see *selective breeding*.

ascus (pl. **asci**): a sac that contains haploid spores of fungi (i.e., yeast or molds).

asexual reproduction: the way that some unicellular organisms produce new individuals. In this process, a preexisting cell divides to produce two new cells.

assortative mating: breeding in which individuals with similar phenotypes preferentially mate with each other.

A–T/G–C rule: in DNA, the phenomenon in which an adenine base in one strand always hydrogen bonds with a thymine base in the opposite strand, and a guanine base always hydrogen bonds with a cytosine.

attenuation: a mechanism of genetic regulation, seen in the *trp* operon, in which a short RNA is made but its synthesis is terminated before RNA polymerase can transcribe the rest of the operon.

AU-rich element (ARE): a sequence found in many short-lived mRNAs that contains the consensus sequence AUUUA.

autopolyploid: a polyploid produced within a single species due to nondisjunction.

autosomes: chromosomes that are not sex chromosomes.

backbone: the portion of a DNA or RNA strand that is composed of the repeated covalent linkage of the phosphates and sugar molecules.

bacteriophages (or **phages**): viruses that infect bacteria.

Barr body: a structure in the interphase nuclei of somatic cells of female mammals that is a highly condensed X-chromosome.

basal transcription apparatus: the minimum number of proteins that is needed to transcribe a gene.

base substitution: a point mutation in which one base is substituted for another base.

benign: refers to a tumor that is not malignant.

β-sheet: a type of secondary structure found in proteins.

bidirectional replication: the phenomenon in which two DNA replication forks emanate in both directions from an origin of replication.

binary fission: the physical process whereby a bacterial cell divides into two daughter cells. During this event, the two daughter cells become divided by the formation of a septum.

biological control: the use of microorganisms or products from microorganisms to alleviate plant diseases or damage from undesirable environmental conditions (e.g., frost damage).

biological evolution: the accumulation of genetic changes in a species or population over the course of many generations.

biological species concept: definition of a species as a group of individuals whose members have the potential to interbreed with one another in nature to produce viable, fertile offspring, but who cannot interbreed successfully with members of other species.

bioremediation: the use of microorganisms to decrease pollutants in the environment.

biotechnology: technologies that involve the use of living organisms, or products from living organisms, as a way to benefit humans.

blending theory of inheritance: an early, incorrect theory of heredity. According to this view, the seeds that dictate hereditary traits are able to blend together from generation to generation. The blended traits would then be passed on to the next generation.

bottleneck effect: a type of genetic drift that occurs when most of the members of population are eliminated without any regard to their genetic composition.

box: in genetics, a term used to describe a sequence with a specialized function.

C: an abbreviation for cytosine.

cancer cell: a cell that has lost its normal growth control. Cancer cells are invasive (i.e., they can invade normal tissues) and metastatic (i.e., they can migrate to other parts of the body).

CAP: an abbreviation for the catabolite activator protein, a genetic regulatory protein found in bacteria.

capping: the covalent attachment of a 7-methylguanosine nucleotide to the 5′ end of mRNA in eukaryotes.

carboxyl terminus: the location of the last amino acid in a polypeptide chain. The amino acid at the carboxyl terminus still retains a free carboxyl group that is not covalently attached to another amino acid.

catabolite repression: the phenomenon in which a catabolite (such as glucose) represses the expression of other genes (such as the *lac* operon).

catenane: interlocked circular molecules.

cDNA: see *complementary DNA*.

cDNA library: a DNA library made from a collection of cDNAs.

cell cycle: in eukaryotic cells, a series of stages through which a cell progresses in order to divide. The phases are G for growth, S for synthesis (of the genetic material), and M for mitosis. There are two G phases, G_1 and G_2.

cell fate: the final morphological features that a cell or group of cells will adopt.

cell fusion: describes the process in which individual cells are mixed together and made to fuse.

cell lineage: a series of cells that are descended from a cell or group of cells by cell division.

centimorgans (cM): (same as a *map unit*) a unit of map distance obtained from genetic crosses. Named in honor of Thomas Hunt Morgan.

centromere: a segment of eukaryotic chromosomal DNA that provides an attachment site for the kinetochore.

character: in genetics, this word has the same meaning as trait.

Chargaff's rule: see *A–T/G–C rule*.

charged tRNA: a tRNA that has an amino acid attached to its 3′ end by an ester bond.

chiasma (pl. chiasmata): the site where crossing over occurs between two chromosomes. It resembles the Greek letter chi, χ.

chimera: an organism composed of cells that are embryonically derived from two different individuals.

chi square (χ²) test: a commonly used statistical method to determine the goodness of fit. This method can be used to analyze population data in which the members of the population fall into different categories. It is particularly useful for evaluating the outcome of genetic crosses, because these usually produce a population of offspring that differ with regard to phenotypes.

chorionic villus sampling: a method to obtain cellular material from a fetus for the purpose of genetic testing.

chromatin: the association between DNA and proteins that is found within chromosomes.

chromocenter: the central point where polytene chromosomes aggregate.

chromomere: a dark band within a polytene chromosome.

chromosome: the structures within living cells that contain the genetic material. Genes are physically located within the structure of chromosomes. Biochemically, chromosomes contain a very long segment of DNA, which is the genetic material, and proteins, which are bound to the DNA and provide it with an organized structure.

chromosome mutation: a substantial change in chromosome structure that may affect more than a single gene.

chromosome sorting: a technique in which chromosomes are dispersed into individual droplets, and the droplets carrying the desired chromosome are sorted into a separate tube.

chromosome theory of inheritance: a theory of Sutton and Boveri, which indicated that the inheritance patterns of traits can be explained by the transmission patterns of chromosomes during gametogenesis and fertilization.

chromosome walking: a common method used in positional cloning in which a mapped gene or RFLP marker provides a starting point to molecularly walk toward a gene of interest via overlapping clones.

cis effect: an effect on gene expression due to genetic sequences immediately adjacent to the gene.

cistron: refers to the smallest genetic unit that produces a positive result in a complementation experiment. A cistron is equivalent to a gene.

cladogenesis (Greek, *clados*, "branch"): during evolution, a form of speciation that involves the division of a species into two or more species.

clone: the general meaning of this term is to make many copies of something. In genetics, this term has several meanings: (1) a single cell that has divided to produce a colony of genetically identical cells; (2) an individual who has been produced from a somatic cell of another individual, such as the sheep Dolly; (3) many copies of a DNA fragment that are propagated within a vector or produced by PCR.

closed complex: the complex between transcription factors, RNA polymerase, and a bacterial promoter before the DNA has denatured to form an open complex.

closed conformation: a tightly packed conformation of chromatin that cannot be transcribed.

cM: an abbreviation for centimorgans; also see *map unit*.

coding strand: the strand in DNA that is not used as a template for mRNA synthesis.

codominance: a pattern of inheritance in which two alleles are both expressed in the heterozygous condition. For example, a person with the genotype $I^A I^B$ will have the blood type AB, and will express both surface antigens A and B.

codon: a sequence of three nucleotides in mRNA that functions in translation. A start codon, which usually specifies methionine, initiates translation, and a stop codon terminates translation. The other codons specify the amino acids within a polypeptide sequence according to the genetic code.

codon bias: the phenomenon that, in a given species, certain codons are used more frequently than other codons.

colinearity: the correspondence between the sequence of codons in the DNA coding strand and the amino acid sequence of a polypeptide.

colony hybridization: a technique in which a probe is used to identify bacterial colonies that contain a hybrid vector with a gene of interest.

common ancestor: someone who is an ancestor to both of an individual's parents.

competent cells: cells that can be transformed by extracellular DNA.

complementary: describes sequences in two DNA strands that match each other according to the A–T/G–C rule. For example, if one strand has the sequence of ATGGCGGATTT, then the complementary strand must be TACCGCCTAAA.

complementary DNA (cDNA): DNA that is made from an RNA template by the action of reverse transcriptase.

complementation: a phenomenon in which the presence of two different mutant alleles in the same organism produces a wild-type phenotype. It usually happens because the two mutations are in different genes, so that the organism carries one copy of each mutant allele and one copy of each wild-type allele.

computer data file: a file (a collection of information) stored by a computer.

computer program: a series of operations that can analyze data in a defined way.

conditional alleles: alleles in which the phenotypic expression depends on the environmental conditions. An example are temperature-sensitive alleles, which only affect the phenotype at a particular temperature.

conjugation: a form of genetic transfer between bacteria that involves direct physical interaction between two bacterial cells. One bacterium acts as donor and transfers genetic material to a recipient cell.

consensus sequence: the most commonly occurring bases within a sequence element.

constitutive gene: a gene that is not regulated and has essentially constant levels of expression over time.

constitutive heterochromatin: regions of chromosomes that are always heterochromatic and are permanently transcriptionally inactive.

contig: a series of clones that contain overlapping pieces of chromosomal DNA.

corepressor: a small effector molecule that binds to a repressor protein, thereby causing the repressor protein to bind to DNA and inhibiting transcription.

core promoter: a DNA sequence that is absolutely necessary for transcription to take place. It provides the binding site for general transcription factors and RNA polymerase.

correlation coefficient (*r*): a statistic with a value that ranges between -1 and 1. It describes how two factors vary with regard to each other.

$$r_{(X,Y)} = \frac{CoV_{(X,Y)}}{SD_X \, SD_Y}$$

cosmid: a vector that is a hybrid between a plasmid vector and phage λ. Cosmid DNA can replicate in a cell like a plasmid or be packaged into a protein coat like a phage. Cosmid vectors can accept fragments of DNA that are typically tens of thousands of base pairs in length.

cotransduction: the phenomenon in which bacterial transduction that transfers a piece of DNA carrying a certain gene also transfers a closely linked gene.

cotranslational: events that occur during translation.

covariance: a statistic that describes the degree of variation between two variables within a group:

$$CoV_{(X,Y)} = \frac{\Sigma[(X - \bar{X})(Y - \bar{Y})]}{N - 1}$$

cpDNA: an abbreviation for chloroplast DNA.

CREB protein (cAMP response element-binding protein): a transcription factor that becomes activated in response to specific cell-signaling molecules.

cross: a mating between two distinct individuals. An analysis of their offspring may be conducted to understand how traits are passed from parent to offspring.

cross-fertilization: same meaning as *cross*. It requires that the male and female gametes come from separate individuals.

crossing over: a physical exchange of chromosome pieces that most commonly occurs during prophase of meiosis I.

C-terminus: see *carboxyl terminus.*

cyclin: a type of protein that plays a role in the regulation of the eukaryotic cell cycle.

cyclin-dependent protein kinases (CDKs): enzymes that are regulated by cyclins and can phosphorylate other cellular proteins by covalently attaching a phosphate group.

cytogenetics: the field of genetics that involves the microscopic examination of chromosomes.

cytokinesis: the division of a single cell into two cells. The two nuclei produced in M phase are segregated into separate daughter cells during cytokinesis.

cytoplasmic inheritance (also known as *extranuclear inheritance*): refers to the inheritance of genetic material that is not found within the cell nucleus.

cytosine: a pyrimidine base found in DNA and RNA. It base pairs with guanine in DNA.

Darwinian fitness: the relative likelihood that a phenotype will survive and contribute to the gene pool of the next generation as compared with other phenotypes.

database: a computer storage facility that stores many data files such as those containing genetic sequences.

deficiency: condition in which a segment of chromosomal material is missing.

degeneracy: in genetics, this term means that more that one codon specifies the same amino acid. For example, the codons GGU, GGC, GGA, and GGG all specify the amino acid glycine.

degrees of freedom: in a statistical analysis, the number of categories that are independent of each other.

deletion: condition in which a segment of DNA is missing.

deletion mapping: the use of strains carrying deletions within a defined region to map a mutation of unknown location.

deoxyribonucleic acid (DNA): the genetic material. It is a double-stranded structure, with each strand composed of repeating units of deoxynucleotides.

developmental genetics: the area of genetics concerned with the roles of genes in orchestrating the changes that occur during development.

diauxic growth: the sequential use of two sugars by a bacterium.

dicentric: describes a chromosome with two centromeres.

dideoxynucleotides: a nucleotide used in DNA sequencing that is missing the 3′–OH group. If a dideoxynucleotide is incorporated into a DNA strand, it stops any further growth of the strand.

dihybrid cross: a cross in which an experimenter follows the outcome of two different traits.

diploid: a organism or cell that contains two copies of each type of chromosome.

disassortative mating: breeding in which individuals with unlike phenotypes preferentially mate with each other.

discontinuous trait: a trait in which each offspring can be put into a particular phenotypic category.

dizygotic twins: also known as fraternal twins; twins formed from separate pairs of sperm and egg cells.

DNA: the abbreviation for *deoxyribonucleic acid.*

dnaA box sequence: serves as a recognition site for the binding of the dnaA protein, which is involved in the initiation of bacterial DNA replication.

DNA fingerprinting: a technology to identify a particular individual based on the properties of their DNA.

DNA footprinting: a method to study protein–DNA interactions in which the binding of a protein to DNA protects the DNA from digestion by DNaseI.

DNA helicase: an enzyme that separates the two strands of DNA.

DNA library: a collection of many hybrid vectors, each vector carrying a particular fragment of DNA from a larger source. For example, each hybrid vector in a DNA library might carry a small segment of chromosomal DNA from a particular species.

DNA ligase: an enzyme that catalyzes a covalent bond between two DNA fragments.

DNA methylation: the phenomenon in which an enzyme covalently attaches a methyl group (-CH$_3$) to a base (usually adenine or cytosine) in DNA.

DNA-N-glycosylase: an enzyme that can recognize an abnormal base and cleave the bond between it and the sugar in the DNA backbone.

DNA polymerase: an enzyme that catalyzes the covalent attachment of nucleotides together to form a strand of DNA.

DNA replication: the process in which original DNA strands are used as templates for the synthesis of new DNA strands.

DNaseI: an endonuclease that cleaves DNA.

DNA sequencing: a method to determine the base sequence in a segment of DNA.

DNA supercoiling: the formation of additional coils in DNA due to twisting forces.

domain: a segment of a protein that has a specific function.

dominant: describes an allele that determines the phenotype in the heterozygous condition. For example, if a plant is *Tt* and has a tall phenotype, the *T* (tall) allele is dominant over the *t* (dwarf) allele.

dosage compensation: refers to the phenomenon that in species with sex chromosomes, one of the sex chromosomes is altered so that males and females will have similar levels of gene expression, even though they do not contain the same complement of sex chromosomes.

double helix: the arrangement in which two strands of DNA (and sometimes RNA) interact with each other to form a double-stranded helical structure.

down-regulation: genetic regulation that leads to a decrease in gene expression.

duplication: the copying of a segment of DNA.

editosome: a complex that catalyzes RNA editing.

egg cell: also known as an ovum; it is a female gamete that is usually very large and nonmotile.

electroporation: the use of electric current that creates transient pores in the plasma membrane of a cell to allow entry of DNA.

empirical approach: a strategy in which experiments are designed to determine quantitative relationships as a way to derive laws that govern biological, chemical, or physical phenomena.

empirical laws: laws that are discovered using an empirical (observational) approach.

endonuclease: an enzyme that can cut in the middle of a DNA strand.

endosymbiosis: a symbiotic relationship in which the symbiont actually lives inside ("*endo*") the larger of the two species.

endosymbiosis theory: the theory that the ancient origin of plastids and mitochondria was the result of certain species of bacteria taking up residence within a primordial eukaryotic cell.

enhancer: a DNA sequence that functions as a regulatory element. The binding of a regulatory transcription factor to the enhancer increases the level of transcription.

enzyme: a protein that functions to accelerate chemical reactions within the cell.

enzyme adaptation: the phenomenon in which a particular enzyme only appears within a living cell after the cell has been exposed to the substrate for that enzyme.

epigenetic inheritance: an inheritance pattern in which a modification to a nuclear gene or chromosome alters gene expression in an organism, but the expression is not changed permanently over the course of many generations.

episome: a segment of bacterial DNA that can exist as an F factor and also integrate into the chromosome.

epistasis: an inheritance pattern where one gene can mask the phenotypic effects of a different gene.

epitope: the structure on the surface of an antigen that is recognized by an antibody.

essential gene: a gene that is essential for survival.

euchromatin: DNA that is not highly compacted and may be transcriptionally active.

eukaryotes (Greek, "true nucleus"): organisms whose cells contain nuclei bounded by cell membranes. Some simple eukaryotic species are single-celled protists and yeast; more complex multicellular species include fungi, plants, and animals.

euploid: describes an organism in which the chromosome number is an exact multiple of a chromosome set.

evolution: see *biological evolution*.

exon: a segment of RNA that is contained within the RNA after splicing has occurred. In mRNA, the coding sequence of a polypeptide is contained within the exons.

exon shuffling: the phenomenon that exons have been transferred between different genes during evolution. Some researchers believe that transposable elements have played a role in this phenomenon.

exonuclease: an enzyme that digests an RNA or DNA strand from the end.

expressivity: the degree to which a trait is expressed. For example, flowers with deep red color would have a high expressivity of the red allele.

extranuclear inheritance (also known as *cytoplasmic inheritance*): refers to the inheritance of genetic material that is not found within the nucleus.

facultative heterochromatin: heterochromatin that is derived from the conversion of euchromatin to heterochromatin.

fate map: a diagram that depicts how cell division proceeds in an organism.

feedback inhibition: the phenomenon in which the final product of a metabolic pathway inhibits an enzyme that acts early in the pathway.

fertilization: the union of sperm and egg to begin the life of a new organism.

F factor: a fertility factor found in certain strains of bacteria in addition to their circular chromosome. Strains of bacteria that contain an F factor are designated F⁺; strains without F factors are F⁻. F factors carrying pieces of chromosomal DNA are known as F′ factors.

F_1 generation: the offspring produced from a cross of the parental generation.

F_2 generation: the offspring produced from a cross of the F_1 generation.

fidelity: a term used to describe the accuracy of a process. If there are few mistakes, a process has a high fidelity.

fine structure mapping: also known as *intragenic mapping*; the aim of fine structure mapping is to ascertain the distances between two (or more) different mutations within the same gene.

fitness: see *Darwinian fitness*.

fluorescence *in situ* hybridization (FISH): *in situ* hybridization in which the probe is fluorescent.

footprinting: see *DNA footprinting*.

fork: see *replication fork*.

forward mutation: a mutation that changes the wild-type genotype into some new variation.

founder effect: changes in allele frequencies that occur when a small group of individuals separates from a larger population and establishes a colony in a new location.

frameshift mutation: a mutation that involves the addition or deletion of nucleotides not in a multiple of three and thereby shifts the reading frame of the amino acid sequence downstream from the mutation.

frequency distribution: a graph that describes the numbers of individuals that are found in each of several phenotypic categories.

G: an abbreviation for guanine.

gain of function mutation: a mutation that causes a gene to be expressed in an additional place where it is not normally expressed, or during a stage of development when it is not normally expressed.

gametogenesis: the production of gametes (i.e., sperm or egg cells).

gametophyte: the haploid generation of plants.

G bands: the chromosomal banding pattern that is observed when the chromosomes have been treated with the chemical dye Giemsa.

gel retardation assay: a technique to study protein–DNA interactions in which the binding of protein to a DNA fragment retards it mobility during gel electrophoresis.

gene: a unit of heredity that may influence the outcome of an organism's traits.

gene amplification: an increase in the copy number of a gene.

gene conversion: the phenomenon in which one allele is converted to another allele due to genetic recombination and DNA repair.

gene dosage effect: when the number of copies of a gene affects the phenotypic expression of a trait.

gene expression: the process in which the information within a gene is accessed, first to synthesize RNA (and proteins), and eventually to affect the phenotype of the organism.

gene family: two or more different genes that are homologous to each other because they were derived from the same ancestral gene.

gene flow: changes in allele frequencies due to migration.

gene interaction: when two or more different genes influence the outcome of a single trait.

gene knockout: when both copies of a normal gene have been replaced by an inactive mutant gene.

gene mutation: a relatively small mutation that only affects a single gene.

gene pool: the totality of all genes within a particular population.

gene rearrangement: a rearrangement in segments of a gene, as occurs in antibody precursor genes.

gene redundancy: the phenomenon in which an inactive gene is compensated for by another gene with a similar function.

gene regulation: the phenomenon in which the level of gene expression can vary under different conditions.

gene therapy: the introduction of cloned genes into living cells in an attempt to cure or alleviate disease.

genetic code: the correspondence between a codon (i.e., a sequence of three bases in an mRNA molecule) and the functional role that the codon plays during translation. Each codon specifies a particular amino acid or the end of translation.

genetic cross: a mating between two individuals and the analysis of their offspring in an attempt to understand how traits are passed from parent to offspring.

genetic drift: random changes in allele frequencies due to sampling error.

genetic load (*L*): genetic variation that decreases the average fitness of a population as compared with a (theoretical) maximum or optimal value.

genetic mapping: any method used to determine the linear order of genes as they are linked to each other along the same chromosome. This term is also used to describe the use of genetic crosses to determine the linear order of genes.

genetic mosaic: see *mosaicism*.

genetic recombination: the process in which chromosomes are broken and then rejoined to form a novel genetic combination.

genetics: the study of heredity.

genetic screening: the use of testing methods to determine if an individual is a heterozygous carrier for or has a genetic disease.

genetic transfer: describes the physical transfer of genetic material from one bacterial cell to another.

genetic variation: genetic differences among members of the same species or among different species.

genome: all of the chromosomes and DNA sequences that an organism can possess.

genome database: a database that focuses on the genetic sequences and characteristics of a single species.

genome mutation: a change in chromosome number.

genomic clone: a clone made from the digestion and cloning of chromosomal DNA.

genomic imprinting: a pattern of inheritance that involves a change in a single gene or chromosome during gamete formation. Depending on whether the modification occurs during spermatogenesis or oogenesis, imprinting governs whether an offspring will express a gene that has been inherited from its mother or father.

genomic library: a DNA library made from chromosomal DNA fragments of a single species.

genomics: the molecular analysis of the entire genome of a species.

genotype: the genetic composition of an individual, especially in terms of the alleles for particular genes.

genotype frequency: the number of individuals with a particular genotype in a population divided by the total number of individuals in the population.

germ cells: the gametes (i.e., sperm and egg cells).

goodness of fit: the degree to which the observed data and expected data are similar to each other. If the observed and predicted data are very similar, the goodness of fit is high.

gradualism: an evolutionary hypothesis suggesting that each new species evolves continuously over long spans of time. The principal idea is that large phenotypic differences that cause the divergence of species are due to the accumulation of many small genetic changes.

grande: normal (large-sized) yeast colonies.

grooves: in DNA, the indentations where the atoms of the bases are in contact with the surrounding water. In B-DNA there is a smaller minor groove and a larger major groove.

guanine: a purine base found in DNA and RNA. It bases pairs with cytosine in DNA.

guide RNA: in trypanosome RNA editing, an RNA molecule that directs the addition of uracil residues into the mRNA.

gyrase: also known as topoisomerase II; an enzyme that introduces negative supercoils into DNA using energy from ATP. Gyrase can also relax positive supercoils when they occur.

haploid: describes the phenomenon that gametes contain half the genetic material found in somatic cells. For a species that is diploid, a haploid gamete contains a single set of chromosomes.

Hardy–Weinberg equilibrium: the phenomenon that under certain conditions allele frequencies will be maintained in a stable condition and genotypes can be predicted according to

$$p^2 + 2pq + q^2 = 1 \qquad \text{(Hardy–Weinberg equation)}$$

helicase: see *DNA helicase*.

hemizygous: describes the single copy of an X-linked gene in the male. A male mammal is said to be hemizygous for X-linked genes.

heritability: the amount of phenotypic variation within a particular group of individuals that is due to genetic factors.

heterochromatin: highly compacted DNA. It is usually transcriptionally inactive.

heterochronic mutation: a mutation that alters the timing of expression of a gene, and thereby alters the timing of cell fates.

heteroduplex: a double-stranded region of DNA that contains one or more base mismatches.

heterogametic sex: in species with two types of sex chromosomes, the heterogametic sex is the gender that produces two types of gametes. For example, in mammals, the male is the heterogametic sex, because a sperm can contain either an X- or a Y-chromosome.

heterogamous: describes a species that produces two morphologically different types of gametes (i.e., sperm and eggs).

heterogeneity: refers to the phenomenon that a particular type of disease may be caused by mutations in two or more different genes.

heterogeneous nuclear RNA (hnRNA): same as pre-mRNA.

heterokaryon: a cell produced from cell fusion that contains two separate nuclei.

heterosis: the phenomenon in which hybrids display traits superior to either corresponding parental strain. Heterosis is usually different from overdominance, because the hybrid may be heterozygous for many genes, not just a single gene, and because the superior phenotype may be due to the masking of deleterious recessive alleles.

heterozygote: an individual who is heterozygous.

heterozygote advantage: a pattern of inheritance in which a heterozygote is more vigorous than either of the corresponding homozygotes.

heterozygous: describes a diploid individual who has different copies (i.e., two different alleles) of the same gene.

Hfr strain (for High frequency of recombination): a bacterial strain in which an F factor has become integrated into the bacterial chromosome. During conjugation, an Hfr strain can transfer segments of the bacterial chromosome.

histones: a group of proteins involved in forming the nucleosome structure of eukaryotic chromatin.

hnRNA: an abbreviation for heterogeneous nuclear RNA.

homeobox: a 180 base pair consensus sequence found in homeotic genes.

homeodomain: the protein domain encoded by the homeobox. The homeodomain promotes the binding of the protein to the DNA.

homeologous: describes the analogous chromosomes from evolutionarily related species.

homeotic gene: a gene that functions in governing the developmental fate of a particular region of the body.

homoallelic: describes two or more alleles in different organisms that are due to mutations at exactly the same base within a gene.

homogametic sex: in species with two types of sex chromosomes, the homogametic sex is the gender that produces only one type of gamete. For example, in mammals, the female is the homogametic sex, because an egg can only contain an X-chromosome.

homologous: in the case of genes, this term describes two genes that are derived from the same ancestral gene. Homologous genes have similar DNA sequences. In the case of chromosomes, the two homologues of a chromosome pair are said to be homologous to each other.

homologous recombination: recombination between DNA segments that are homologous to each other.

homologue: in a diploid species, each member of a pair of chromosomes is referred to as a homologue of the other.

homozygous: describes a diploid individual who has two identical copies of the same allele.

hot spots: sites within a gene that are more likely to be mutated than other locations.

housekeeping gene: a gene that encodes a protein required in most cells of a multicellular organism.

***Hox* genes:** mammalian genes that play a role in development. They are homologous to homeotic genes found in *Drosophila*.

Human Genome Project: a worldwide collaborative project that aims to provide a detailed map of the human genome, and eventually obtain a complete DNA sequence of the human genome.

hybrid: (1) an offspring obtained from a hybridization experiment; (2) a cell produced from a cell fusion experiment in which the two separate nuclei have fused to make a single nucleus.

hybrid dysgenesis: a syndrome involving defective *Drosophila* offspring, due to the phenomenon that P elements can transpose freely.

hybridization: (1) the mating of two organisms of the same species with different characteristics; (2) the phenomenon in which two single-stranded molecules renature together to form a hybrid molecule.

hybrid vigor: see *heterosis*.

hypothesis testing: using statistical tests to determine if the data from genetic crosses are consistent with a hypothesis regarding a particular pattern of inheritance.

immunoglobulin (IgG): see *antibody*.

imprinting: see *genomic imprinting*.

inborn error of metabolism: a genetic disease that involves a defect in a metabolic enzyme.

inbreeding: the practice of mating between genetically related individuals.

incomplete dominance: a pattern of inheritance in which a heterozygote that carries two different alleles exhibits a phenotype that is intermediate to the corresponding homozygous individuals. For example, an *Rr* heterozygote may be pink, while the *RR* and *rr* homozygotes are red and white, respectively.

incomplete penetrance: a pattern of inheritance in which a dominant allele does not always control the phenotype of the individual.

induced mutation: a mutation caused by environmental agents.

inducer: a small effector molecule that binds to a genetic regulatory protein and thereby increases the rate of transcription.

infective particle: genetic material found within the cytoplasm of eukaryotic cells that differs from the genetic material normally found in cell organelles.

inhibitor: a small effector molecule that binds to an activator protein, causing the protein to be released from the DNA and thereby inhibiting transcription.

insertion sequences: the simplest transposable elements. They are commonly found in bacteria.

***in situ* hybridization:** a technique used to cytologically map the locations of genes or other DNA sequences within large eukaryotic chromosomes. In this method, a complementary probe is used to detect the location of a gene within a set of chromosomes.

integrase: an enzyme that functions in the integration of viral DNA or retroelements into the host chromosome.

interference: see *positive interference*.

interphase: the series of phases G_1, S, and G_2, during which a cell spends most of its life.

interrupted mating: a method used in conjugation experiments in which the length of time that the bacteria spend conjugating is stopped by a blender treatment or other type of harsh agitation.

intragenic mapping: see *fine structure mapping*.

intron: intervening sequences that are found between exons. Introns are spliced out of the RNA prior to translation.

inversion: a change in the orientation of genetic material along a chromosome.

inverted repeats: DNA sequences found in transposable elements that are identical (or very similar) but run in the opposite directions.

isogamous: describes a species that makes morphologically similar gametes.

karyotype: a photographic representation of all the chromosomes within a cell. It reveals how many chromosomes are found within an actively dividing somatic cell.

kinetochore: a group of cellular proteins that attach to the centromere during meiosis and mitosis.

knockout: see *gene knockout*.

lagging strand: a strand during DNA replication that is synthesized as short Okazaki fragments in the direction away from the replication fork.

lariat: a circular intron structure.

leading strand: a strand during DNA replication that is synthesized continuously toward the replication fork.

lethal allele: an allele that causes the death of an organism.

library: see *DNA library*.

ligase: see *DNA ligase*.

LINEs: in mammals, long interspersed elements that are usually 1 to 5 kbp in length and found in 20,000 to 100,000 copies per genome.

linkage: refers to the occurrence of two or more genes along the same chromosome.

linkage groups: a group of genes that are linked together because they are found on the same chromosome.

locus (pl. **loci**): the physical location of a gene within a chromosome.

locus control region (LCR): a segment of DNA that is involved in the regulation of chromatin opening and closing.

lod score method: a method that analyzes pooled data from a large number of pedigrees or crosses to determine the probability that two genetic markers exhibit a certain degree of linkage. A lod score value of >3 or higher is usually accepted as strong evidence that two markers are linked.

long terminal repeat (LTR) sequences: sequences containing many short segments that are tandemly repeated. They are found in retroviruses and viral-like retroelements.

loss of function allele: an allele of a gene that encodes an RNA or protein that is nonfunctional or compromised in function.

LTRs: see *long terminal repeat sequences*.

Lyon hypothesis: a hypothesis to explain the pattern of X-inactivation seen in mammals. Initially, both X-chromosomes are active. However, at an early stage of embryonic development, one of the two X-chromosomes is randomly inactivated in each somatic cell.

lysogenic cycle: a type of growth cycle for a phage in which the phage integrates its genetic material into the chromosome of the bacterium. This integrated phage DNA can exist in a dormant state for a long time, during which no new bacteriophages are made.

lytic cycle: a type of growth cycle for a phage in which the phage directs the synthesis of many copies of the phage genetic material and coat proteins. These components then assemble to make new phages. When synthesis and assembly is completed, the bacterial host cell is lysed and the newly made phages are released into the environment.

macroevolution: evolutionary changes above the species level involving relatively large changes in form and function that are sufficient to produce new species and higher taxa.

malignant: describes a tumor composed of cancerous cells.

map unit (mu): a unit of map distance obtained from genetic crosses. One map unit is equivalent to 1% recombinant offspring in a test cross.

maternal effect: an inheritance pattern for certain nuclear genes in which the genotype of the mother directly determines the phenotypic traits of her offspring.

maternal inheritance: inheritance of DNA that occurs through the cytoplasm of the egg.

maturase: an enzyme that enhances the rate of splicing of Group I and II introns.

mean: the sum of all the values in a group divided by the number of individuals in the group.

meiosis: a form of nuclear division in which the sorting process results in the production of haploid gametes from a diploid germ cell.

meiotic nondisjunction: condition in which chromosomes do not segregate equally into the gametes during meiosis.

Mendelian inheritance: the common pattern of inheritance observed by Mendel, which involves the transmission of eukaryotic genes that are located on the chromosomes found within the cell nucleus.

Mendel's law of independent assortment: two different genes will randomly assort their alleles during gamete formation (if they are not linked).

Mendel's law of segregation: the two copies of a gene segregate from each other during transmission from parent to offspring.

meristem: in plants, an organized group of actively dividing cells.

merozygote: a partial diploid strain of bacteria containing F′ factor genes.

messenger RNA (mRNA): a type of RNA that contains the information for the synthesis of a polypeptide.

metacentric: describes a chromosome with the centromere in the middle.

metaphase: the second phase of M phase. The chromosomes align along the center of the spindle apparatus and the formation of the spindle apparatus is complete.

metastatic: describes cancer cells that migrate to other parts of the body.

methylation: see *DNA methylation*.

methyl-directed mismatch repair: a DNA repair system in *E. coli* that detects a mismatch and specifically removes the segment from the newly made daughter strand.

microevolution: changes in the gene pool with regard to particular alleles that occur over the course of many generations.

microsatellite: short simple sequences that are interspersed throughout a genome and are quite variable in length among different individuals.

minichromosomes: a structure formed from many copies of circular DNA molecules.

minute: a unit of measure in bacterial conjugation experiments. This unit refers to the relative time it takes for genes to first enter an F⁻ recipient strain during conjugation.

missense mutation: a base substitution that leads to a change in the amino acid sequence of the encoded polypeptide.

mitochondrial DNA: the DNA found within mitochondria.

mitosis: a type of nuclear division into two nuclei, such that each daughter cell will receive the same complement of chromosomes.

mitotic nondisjunction: an event in which chromosomes do not segregate equally during mitosis.

mitotic recombination: recombination that occurs during mitosis.

molecular evolution: the molecular changes in the genetic material that underlie the phenotypic changes associated with evolution.

molecular genetics: an examination of DNA structure and function at the molecular level.

molecular marker: a segment of DNA that is found at a specific site in the genome and has properties that enable it to be uniquely recognized using molecular tools such as gel electrophoresis.

molecular pharming: a recombinant technology that involves the production of medically important proteins in the mammary glands of livestock.

monohybrid cross: a cross in which an experimenter is only following the outcome of a single trait.

monomorphic: a term used to describe a gene that is essentially only found as one allele in a population.

monoploid: an organism with a single set of chromosomes within its somatic cells.

monozygotic twins: twins that are genetically identical to each other because they were formed from the same sperm and egg.

morph: a form or phenotype in a population. For example, red eyes and white eyes are different eye color morphs.

morphogen: a molecule that conveys positional information and promotes developmental changes.

mosaicism: when the cells of part of an organism differ genetically from the rest of the organism.

motif: the name given to a domain or amino acid sequence that functions in a similar manner in many different proteins.

M phase: a general name given to nuclear division that can apply to mitosis or meiosis. It is divided into prophase, metaphase, anaphase, and telophase.

mRNA: see *messenger RNA*.

mtDNA: an abbreviation for *mitochondrial DNA*.

multiple alleles: when the same gene exists in two or more alleles within a population.

multiple sequence alignment: an alignment of two or more genetic sequences based on their homology to each other.

mutagen: an agent that causes alterations in the structure of DNA.

mutant alleles: alleles that have been altered by mutation.

mutation: a permanent change in the genetic material that can be passed from cell to cell or from parent to offspring.

mutation frequency: the number of mutant genes divided by the total number of genes within the population.

mutation rate: the likelihood that a gene will be altered by a new mutation.

***n*:** an abbreviation that designates the number of chromosomes in a set. In humans, $n = 23$ and an individual has $2n = 46$ chromosomes.

natural selection: refers to the process whereby differential fitness acts on the gene pool: when a mutation creates a new allele that is beneficial, the allele may become prevalent within future generations because the individuals possessing this allele are more likely to survive and/or reproduce and pass the beneficial allele to their offspring.

negative control: transcriptional regulation by repressor proteins.

neutral mutation: a mutation that has no detectable effect on protein function and/or no detectable effect on the survival of the organism.

neutral theory of evolution: the theory that most genetic variation observed in natural populations is due to the accumulation of neutral mutations.

nick translation: the phenomenon in which DNA polymerase uses its 5′ to 3′ exonuclease activity to remove a region of DNA and, at the same time, replaces it with new DNA.

noncomplementation: the phenomenon in which two mutant alleles in the same organism do not produce a wild-type phenotype.

non-Darwinian evolution: see *neutral theory of evolution*.

nondisjunction: event in which chromosomes do not segregate properly during mitosis or meiosis.

nonessential genes: genes that are not absolutely required for survival, although they are likely to be beneficial to the organism.

nonneutral mutation: a mutation that affects the phenotype of the organism and can be acted on by natural selection.

nonparental: refers to combinations of alleles or traits that are not found in the parental generation.

nonsense mutation: a mutation that involves a change from a normal codon to a stop codon.

normal distribution: a distribution for an infinite sample in which the trait of interest varies in a symmetrical way around an average value.

Northern blotting: a technique used to detect a specific RNA within a mixture of many RNA molecules.

N-terminus: see *amino terminus*.

nuclear genes: genes that are located on chromosomes found in the cell nucleus of eukaryotic cells.

nuclear matrix (or **nuclear scaffold**): a group of proteins that anchor the loops found in eukaryotic chromosomes.

nucleic acid: RNA or DNA. A macromolecule that is composed of repeating nucleotide units.

nucleoid: a darkly staining region that contains the genetic material of mitochondria, chloroplasts, or bacteria.

nucleolus: a region within the nucleus of eukaryotic cells where the assembly of ribosomal subunits occurs.

nucleoside: structure in which a base is attached to a sugar, but no phosphate is attached to the sugar.

nucleosome: the repeating structural unit within eukaryotic chromatin. It is composed of double-stranded DNA wrapped around an octamer of histone proteins.

nucleotide: the repeating structural unit of nucleic acids, composed of a sugar, phosphate, and base.

nucleotide excision repair (**NER**): a DNA repair system in which several nucleotides in the damaged strand are removed from the DNA and the undamaged strand is used as a template to resynthesize a normal strand.

nucleus: a membrane-bound organelle in eukaryotic cells where the linear sets of chromosomes are found.

ocher: a stop codon with the sequence UGA.

octad: a group of eight fungal spores contained within an ascus.

Okazaki fragments: short segments of DNA that are synthesized in the lagging strand during DNA replication.

oncogene: a gene that promotes cancer.

oogenesis: the production of egg cells.

opal: a stop codon with the sequence UAA.

open complex: the region of separation of two DNA strands produced by RNA polymerase during transcription.

open conformation: a loosely packed chromatin structure that is capable of transcription.

open reading frame (ORF): a reading frame of a DNA sequence that does not contain any stop codons.

operator (or **operator site**): a sequence of nucleotides in a bacterial DNA that provides a binding site for a genetic regulatory protein.

operon: an arrangement in DNA where two or more structural genes are found within a regulatory unit that is under the transcriptional control of a single promoter.

ORF: see *open reading frame.*

origin of replication: a nucleotide sequence that functions as an initiation site for the assembly of several proteins required for DNA replication.

origin of transfer: the location on an F factor or within the chromosome of an Hfr strain that is the initiation site for transfer of the DNA from one bacterium to another during conjugation.

outbreeding: mating between genetically unrelated individuals.

overdominance: an inheritance pattern in which a heterozygote is more vigorous than either of the corresponding homozygotes.

p: an abbreviation for the short arm of a chromosome.

pangenesis: an incorrect theory of heredity. It suggested that hereditary traits could be modified depending on the lifestyle of the individual. For example, it was believed that a person who practiced a particular skill would produce offspring that would be better at that skill.

paracentric inversion: an inversion in which the centromere is found outside of the inverted region.

parapatric speciation (Greek, *para*, "beside"; Latin, *patria*, "homeland"): a form of speciation that occurs when members of a species are only partially separated or when a species is very sedentary.

parasegments: transient subdivisions that occur in the *Drosophila* embryo prior to the formation of segments.

particulate theory of inheritance: a theory proposed by Mendel. It states that traits are inherited as discrete units that remain unchanged as they are passed from parent to offspring.

paternal leakage: the phenomenon that in species where maternal inheritance is generally observed, the male parent may, on rare occasions, provide mitochondria or chloroplasts to the zygote.

PCR: see *polymerase chain reaction.*

pedigree analysis: a genetic analysis using information contained within family trees. In this approach, the aim is to determine the type of inheritance pattern that a gene follows.

peptide bond: a covalent bond formed between the carboxyl group in the last amino acid in the polypeptide chain and the amino group in the next amino acid to be added to the chain.

peptidyltransferase: a complex that functions during translation to catalyze the formation of a peptide bond between the amino acid in the A site of the ribosome and the growing polypeptide chain.

pericentric inversion: an inversion in which the centromere is located within the inverted region of the chromosome.

petites: mutant strains of yeast that form small colonies due to defects in mitochondrial function.

PFGE: see *pulsed-field gel electrophoresis.*

P generation: the parental generation in a genetic cross.

pharming: see *molecular pharming.*

phenotype: the outward appearance of an organism.

phenylketonuria (PKU): a human genetic disorder arising from a defect in phenylalanine hydroxylase.

phosphodiester linkage: in a DNA or RNA strand, a linkage in which a phosphate group connects two sugar molecules together.

phylogenetic tree: a diagram that describes the evolutionary relationships among different species.

PKU: see *phenylketonuria.*

plaque: a clear zone within a bacterial lawn on a petri plate. It is due to repeated cycles of viral infection and bacterial lysis.

plasmid: a general name used to describe circular pieces of DNA that exist independently of the chromosomal DNA. Some plasmids are used as vectors in cloning experiments.

point mutation: a change in a single base pair within DNA.

polyadenylation: the process of attachment of a string of adenine nucleotides to the 3′ end of eukaryotic mRNAs.

polyA tail: the string of adenine nucleotides at the 3′ end of eukaryotic mRNAs.

polygenic inheritance: refers to the transmission of traits that are governed by two or more different genes.

polymerase chain reaction (PCR): the method to amplify a DNA region involving the sequential use of oligonucleotide primers and *taq* polymerase.

polymorphic: a term used to describe a trait or gene that is found in two or more forms in a population.

polymorphism: (1) the prevalence of two or more phenotypic forms in a population; (2) the phenomenon in which a gene exists in two or more alleles within a population.

polypeptide: a sequence of amino acids that is the product of a mRNA translation. One or more polypeptides will fold and associate with each other to form a functional protein.

polyploid: an organism or cell with three or more sets of chromosomes.

polyribosome: an mRNA transcript that has many bound ribosomes in the act of translation.

polytene chromosome: chromosomes that are found in certain cells, such as *Drosophila* salivary cells, in which the chromosomes have replicated many times and the copies lie side by side.

population: a group of individuals of the same species that are capable of interbreeding with one another.

population genetics: the field of genetics that is primarily concerned with the prevalence of genetic variation within populations.

positional cloning: a cloning strategy in which a gene is cloned based on its mapped position along a chromosome.

positional information: chemical substances and other environmental cues that enable a cell to deduce its position relative to other cells.

position effect: a change in phenotype that occurs when the position of a gene is changed from one chromosomal site to a different location.

positive control: genetic regulation by activator proteins.

positive interference: the phenomenon in which a crossover occurs in one region of a chromosome and decreases the probability that another crossover will occur nearby.

posttranslational: describes events that occur after translation is completed.

pre-mRNA: in eukaryotes, the transcription of structural genes produces a long transcript known as pre-mRNA, which is located within the nucleus. This pre-mRNA is usually altered by splicing and other modifications before it exits the nucleus.

primase: an enzyme that synthesizes a short RNA primer during DNA replication.

primosome: a multiprotein complex composed of DNA helicase, primase, and several accessory proteins.

prion: an infectious particle that causes several types of neurodegenerative diseases affecting humans and livestock. It is thought to be composed entirely of protein.

probability: the chance that an event will occur in the future.

processive enzyme: RNA and DNA polymerase are processive enzymes: they do not dissociate from the template strand as they catalyze the covalent attachment of nucleotides.

product rule: the probability that two or more independent events will occur in a particular order is equal to the products of their individual probabilities.

prokaryotes (Greek, "prenucleus"): another name for bacteria. The term refers to the fact that their chromosomes are not contained within a separate nucleus of the cell.

promoter: a sequence within a gene that initiates (i.e., promotes) transcription.

proofreading: the ability of DNA polymerase to remove mismatched bases from a newly made strand.

prophage: phage DNA that has been integrated into the bacterial chromosome.

prophase: the first phase of M phase. The chromosomes have already replicated and begin to condense.

protein: a functional unit composed of one or more polypeptides.

proto-oncogene: a normal cellular gene that does not cause cancer, but which may incur a mutation or become incorporated into a viral genome and thereby lead to cancer.

protoplast: a plant cell without a cell wall.

pseudoautosomal inheritance: the inheritance pattern of genes that are found on both the X- and Y-chromosomes. Even though such genes are located physically on the sex chromosomes, their pattern of inheritance is identical to that of autosomal genes.

pseudodominance: a pattern of inheritance that occurs when a single copy of a recessive allele is phenotypically expressed because the second copy of the gene has been deleted from the homologous chromosome.

pulsed-field gel electrophoresis (PFGE): a method of gel electrophoresis used to separate small chromosomes or very large pieces of chromosomes.

punctuated equilibrium: an evolutionary theory proposing that species exist relatively unchanged for many generations. These long periods of equilibrium are punctuated by relatively short periods during which evolution occurs at a relatively rapid rate.

Punnett square: a diagrammatic method in which the gametes that two parents can produce are aligned next to a square grid as a way to predict the types of offspring the parents will produce and in what proportions.

purine: a type of nitrogenous base that has a double-ring structure. Examples are adenine and guanine.

_p_ value: (e.g., in a chi square table) the probability that the deviations between observed and expected values are a matter of random chance alone.

pyrimidine: a type of nitrogenous base that has a single-ring structure. Examples are cytosine, thymine, and uracil.

q: an abbreviation for the long arm of a chromosome.

quantitative genetics: the area of genetics concerned with traits that can be described in a quantitative way.

quantitative trait loci (QTLs): the locations on chromosomes where the genes that influence quantitative traits reside.

random sampling error: the deviation between the observed and expected outcomes due to chance.

realizator gene: a developmental gene that plays a role in promoting the morphological characteristics of a body segment.

recessive: describes a trait or gene that is masked by the presence of a dominant trait or gene.

reciprocal crosses: a pair of crosses in which the traits of the two parents differ with regard to gender. For example, one cross could be a red-eyed female fly and a white-eyed male fly and the reciprocal cross would be a red-eyed male fly and a white-eyed female fly.

recombinant: (1) refers to combinations of alleles or traits that are not found in the parental generation; (2) describes DNA molecules that are produced by molecular techniques in which segments of DNA are joined to each other in ways that differ from their original arrangement in their native chromosomal sites. The cloning of DNA into vectors is an example.

recombinant DNA technology: the use of _in vitro_ molecular techniques to isolate and manipulate different pieces of DNA.

recombination: see _genetic recombination_.

redundancy: see _gene redundancy_.

repetitive sequences: DNA sequences that are present in many copies in the genome.

replication: see _DNA replication_.

replication fork: the region in which two DNA strands have separated and new strands are being synthesized.

replisome: a putative complex that contains a primosome and DNA polymerase.

repressor: a regulatory protein that binds to DNA and inhibits transcription.

resolution: the last stage of homologous recombination, in which the entangled DNA strands become resolved into two separate structures.

resolvase: an endonuclease that makes the final cuts in DNA during the resolution phase of recombination.

restriction endonuclease (or **restriction enzyme**): an endonuclease that cleaves DNA. The restriction enzymes used in cloning experiments bind to specific base sequences and then cleave the DNA backbone at two defined locations, one in each strand.

restriction fragment length polymorphism (RFLP): genetic variation within a population in the lengths of DNA fragments that are produced when chromosomes are digested with particular restriction enzymes.

restriction mapping: a technique to determine the locations of restriction endonuclease sites within a segment of DNA.

retroelement: a type of transposable element that moves via an RNA intermediate.

retroposon: see _retroelement_.

retrospective testing: a procedure in which a researcher formulates a hypothesis and then collects observations as a way to confirm or refute the hypothesis. This approach relies on the observation and testing of already existing materials, which themselves are the result of past events.

retrotransposon: see _retroelement_.

reverse genetics: an experimental strategy in which researchers first identify the wild-type gene using cloning methods. The next step is to make a mutant version of the wild-type gene, introduce it into an organism, and see how the mutant gene affects the phenotype of the organism.

reverse transcriptase: an enzyme that uses an RNA template to make a complementary strand of DNA.

reversion: a mutation that returns a mutant allele back to the wild-type allele.

RFLP: see _restriction fragment length polymorphism_.

rho (ρ): a protein that is involved in transcriptional termination for certain bacterial genes.

ribonucleic acid (RNA): a nucleic acid that is composed of ribonucleotides. In living cells, RNA is synthesized via the transcription of DNA.

ribosome: a large macromolecule structure that acts as the catalytic site for polypeptide synthesis. The ribosome allows the mRNA and tRNAs to be positioned correctly as thepolypeptide is made.

ribozyme: a RNA molecule with enzymatic activity.

R loop: experimentally, a DNA loop that is formed because RNA is displacing it from its complementary DNA strand.

RNA: see *ribonucleic acid*.

RNA editing: the process in which a change occurs in the nucleotide sequence of an RNA molecule that involves additions or deletions of particular bases, or a conversion of one type of base to a different type.

RNA polymerase: an enzyme that synthesizes a strand of RNA using a DNA strand as a template.

RNase P: a bacterial enzyme that is an endonuclease and cuts precursor tRNA molecules. RNase P is a ribozyme, which means that its catalytic ability is due to the action of RNA.

RNA splicing: the process in which pieces of RNA are removed and the remaining pieces are covalently attached to each other.

satellite DNA: in a density centrifugation, a peak of DNA that is separated from the majority of the chromosomal DNA. It is usually composed of highly repetitive sequences.

science: a way of knowing about our natural world. The science of genetics allows us to understand how the expression of genes produces the traits of an organism.

scientific method: a basis for conducting science. It is a process that scientists typically follow so that they may reach verifiable conclusions about the world in which they live.

segments: anatomical subdivisions that occur during the development of species such as *Drosophila*

selectable marker: a gene that provides a selectable phenotype in a cloning experiment. Many selectable markers are genes that confer antibiotic resistance.

selectionists: scientists who oppose the neutral theory of evolution.

selective breeding: refers to programs and procedures designed to modify the phenotypes in economically important species of plants and animals.

self-fertilization: fertilization that involves the union of male and female gametes derived from the same parent.

selfish DNA theory: the idea that transposable elements exist because they possess characteristics that allow them to multiply within the host cell DNA and inhabit the host without offering any selective advantage.

semiconservative replication: refers to the net result of DNA replication: the DNA contains one original strand and one newly made strand.

semilethal alleles: lethal alleles that kill only some individuals but not all.

sequence element: in genetics, a sequence with a specialized function.

sequence-tagged site (STS): a short segment of DNA, usually between 100 and 400 base pairs long, the base sequence of which is found to be unique within the entire genome. Sequence-tagged sites are identified by PCR.

sequencing: see *DNA sequencing*.

sex chromosomes: a pair of chromosomes (e.g., X and Y in mammals) that determines gender in a species.

sex-influenced inheritance: an inheritance pattern in which an allele is dominant in one gender but recessive in the opposite gender. In humans, pattern baldness is an example of a sex-influenced trait.

sex-limited traits: traits that occur in only one of the two genders.

sex linkage: the phenomenon that certain genes are found on one of the two types of sex chromosomes but not both.

sex pilus (pl. pili): a structure on the surface of bacterial cells that acts as an attachment site to promote the binding of bacteria to each other. The sex pilus provides a passageway for the movement of DNA during conjugation.

sexual reproduction: the process whereby parents make gametes (i.e., sperm and egg) that fuse with each other in the process of fertilization to begin the life of a new organism.

sexual selection: natural selection that acts to promote characteristics that give individuals a greater chance of reproducing.

Shine–Dalgarno sequence: a sequence in bacterial mRNAs that functions as a ribosomal binding site.

sigma factor: a transcription factor that recognizes bacterial promoter sequences and facilitates the binding of RNA polymerase to the promoter.

silencer: a DNA sequence that functions as a regulatory element. The binding of a regulatory transcription factor to the silencer decreases the level of transcription.

silent mutation: a mutation that does not alter the amino acid sequence of the encoded polypeptide even though the nucleotide sequence has changed.

SINEs: in mammals, short interspersed elements that are less than 500 base pairs in length.

single-strand binding protein: a protein that binds to both of the single strands of DNA during DNA replication and prevents them from re-forming a double helix.

sister chromatid exchange (SCE): the phenomenon in which crossing over occurs between sister chromatids, thereby exchanging identical genetic material.

sister chromatids: pairs of replicated chromosomes that are attached to each other at the centromere. Sister chromatids are genetically identical to each other.

site-directed mutagenesis: a technique that enables scientists to change the sequence of cloned DNA segments.

somatic cell: refers to any cell of the body that is not a gamete.

sorting signal: an amino acid sequence or posttranslational modification that directs a protein to the correct region of the cell.

Southern blotting: a technique used to detect the presence of a particular genetic sequence within a mixture of many chromosomal DNA fragments.

species: see *biological species concept* and *species recognition concept*.

species recognition concept: similar to the biological species concept, but emphasizes that the sexual recognition of members of the same species acts as a positive force in evolution.

spermatids: immature sperm cells produced from spermatogenesis.

spermatogenesis: the production of sperm cells.

sperm cell: a male gamete. Sperm are small and usually travel relatively far distances to reach the female gamete.

spliceosome: a multisubunit complex that functions in the splicing of eukaryotic pre-mRNA.

splicing: see *RNA splicing*.

spontaneous mutation: a change in DNA structure that results from random abnormalities in biological processes.

spores: haploid cells that are produced by certain species such as fungi (i.e., yeast and molds).

sporophyte: the diploid generation of plants.

standard deviation: a statistic that is computed as the square root of the variance.

steroid receptor: a category of transcription factors that respond to steroid hormones. An example is the glucocorticoid receptor.

strand: in DNA or RNA, nucleotides covalently linked together to form a long, linear polymer.

structural gene: a gene that encodes the amino acid sequence within a particular polypeptide or protein.

STS: see *sequence-tagged site.*

subcloning: the procedure of making smaller DNA clones from a larger one.

submetacentric: describes a chromosome in which the centromere is slightly off center.

sum rule: the probability that one of two or more mutually exclusive events will occur is equal to the sum of their individual probabilities.

supercoiling: see *DNA supercoiling.*

suppressor (or **suppressor mutation**): a mutation at a second site that suppresses the phenotypic effects of another mutation.

sympatric speciation (Greek, *sym*, "together"; Latin, *patria*, "homeland"): a form of speciation that occurs when members of a species diverge while occupying the same habitat within the same range.

synapsis: the event in which homologous chromosomes recognize each other and then align themselves along their entire lengths.

T: an abbreviation for thymine.

tandem array (or **tandem repeat**): a short nucleotide sequence that is repeated many times in a row.

taq polymerase: a thermostable form of DNA polymerase used in PCR experiments.

T-DNA: a segment of DNA found within a Ti plasmid that is transferred from a bacterium to infected plant cells. The T-DNA from the Ti plasmid becomes integrated into the chromosomal DNA of the plant cell by recombination.

TE: see *transposable element.*

telocentric: describes a chromosome with its centromere at one end.

telomerase: the enzyme that recognizes telomeric sequences at the ends of eukaryotic chromosomes and synthesizes additional numbers of telomeric repeat sequences.

telomeres: specialized DNA sequences found at the ends of eukaryotic, linear chromosomes.

telophase: the fourth phase of M phase. The chromosomes have reached their respective poles and decondense.

temperate phage: a bacteriophage that usually exists in the lysogenic life cycle.

template strand: a DNA strand that is used for the synthesis of a new DNA or RNA strand.

terminator: a sequence within a gene that signals the end of transcription.

testcross: an experimental cross between a recessive individual and an individual whose genotype the experimenter wishes to determine.

tetrad: (1) the association among four sister chromatids during meiosis; (2) a group of four fungal spores contained within an ascus.

thermocycler: a device that automates the timing of temperature changes in each cycle of a PCR experiment.

thymine: a pyrimidine base found in DNA. It bases pairs with adenine in DNA.

thymine dimer: a mutation involving a covalent linkage between two adjacent thymine bases in a DNA strand.

Ti plasmid: a tumor-inducing plasmid found in *Agrobacterium tumefaciens*. It is responsible for inducing tumor formation after a plant has been infected.

tissue-specific gene: a gene that is highly regulated and is expressed in a particular cell type.

topoisomerase: an enzyme that alters the degree of supercoiling in DNA.

topoisomers: DNA conformations that differ with regard to supercoiling.

totipotent: a cell that possesses the genetic potential to produce an entire individual. A somatic plant cell or a fertilized egg is totipotent.

traffic signal: see *sorting signal.*

trait: any characteristic that an organism displays. Morphological traits affect the appearance of an organism. Physiological traits affect the ability of an organism to function. A third category of traits are those that affect an organism's behavior (behavioral traits).

transcription: the process of synthesizing RNA from a DNA template.

transcription factors: a broad category of proteins that influence the ability of RNA polymerase to transcribe DNA into RNA.

transduction: a form of genetic transfer between bacterial cells in which a bacteriophage transfers bacterial DNA from one bacterium to another.

trans effect: an effect on gene expression that occurs even though two DNA segments are not physically adjacent to each other. Trans effects are mediated through diffusible genetic regulatory proteins.

transfection: when a viral vector is introduced into a host cell.

transfer RNA (tRNA): a type of RNA used in translation that carries an amino acid. The anticodon in tRNA is complementary to a codon in the mRNA.

transformation: (1) when a plasmid vector or segment of chromosomal DNA is introduced into a host cell; (2) when a normal cell is converted into a malignant cell.

transgenic: an organism that has DNA from another organism incorporated into its genome via recombinant DNA techniques.

transition: a point mutation involving a change of a pyrimidine to another pyrimidine (e.g., C → T), or a purine to another purine (e.g., A → G).

translation: the synthesis of a polypeptide using the codon information within mRNA.

translocation: when one segment of a chromosome breaks off and becomes attached to a different chromosome.

transposable element (TE): a small genetic element that can move to multiple locations within the host's chromosomal DNA.

transposase: the enzyme that catalyzes the transposition of transposable elements.

transposition: the phenomenon of transposon movement.

transposon: see *transposable element.*

transposon tagging: a technique for cloning genes in which a transposon inserts into a gene and inactivates it. The transposon-tagged gene is then cloned using a complementary transposon as a probe to identify the gene.

transversion: a point mutation in which a purine is interchanged with a pyrimidine.

trihybrid cross: a cross in which an experimenter follows the outcome of three different traits.

trinucleotide repeat expansion (TNRE): a type of mutation that involves an increase in the number of tandemly repeated trinucleotide sequences.

triplex DNA: a double-stranded DNA that has a third strand wound around it to form a triple-stranded structure.

triploid: a organism or cell that contains three sets of chromosomes.

tRNA: see *transfer RNA*.

true-breeding line: a strain of a particular species that continues to produce the same trait after several generations of self-fertilization (in plants) or inbreeding.

tumor-suppressor gene: a gene that functions to inhibit cancerous growth.

U: an abbreviation for uracil.

universal: in genetics, this terms refers to the phenomenon that nearly all organisms use the same genetic code with just a few exceptions.

up-regulation: genetic regulation that leads to an increase in gene expression.

uracil: a pyrimidine base found in RNA.

UTR: an abbreviation for the untranslated region of mRNA.

U tube: a U-shaped tube that has a filter at bottom of the U. The pore size of the filter allows the passage of small molecules (e.g., DNA molecules) from one side of the tube to the other, but restricts the passage of bacterial cells.

variance: the sum of the squared deviations from the mean divided by the degrees of freedom.

variants: individuals of the same species who exhibit different traits. An example is tall and dwarf pea plants.

vector: a small segment of DNA that is used as a carrier of another segment of DNA. Vectors are used in DNA cloning experiments.

virus: a small infectious particle that contains nucleic acid as its genetic material, surrounded by a capsid of proteins. Some viruses also have an envelope consisting of a membrane embedded with spike proteins.

VNTRs: segments of DNA that are located in several places in a genome and have a variable number of tandem repeats. The pattern of VNTRs is often used in DNA fingerprinting.

Western blotting: a technique used to detect a specific protein among a mixture of proteins.

wild-type allele: the allele that is most prevalent in a natural population. For polymorphic genes, there may be more than one wild-type allele.

wobble base: the third base in an anticodon. This term suggests that the third base in the anticodon can wobble a bit to recognize more than one type of base in the mRNA.

X-inactivation: a process in which mammals essentially equalize the expression of X-linked genes by randomly turning off one X-chromosome in the somatic cells of females.

X-inactivation center (Xic): a site on the X-chromosome that appears to play a critical role in X-inactivation.

X-linked genes (alleles): genes (or alleles of genes) that are physically located within the X-chromosome.

yeast artificial chromosome (YAC): a cloning vector propagated in yeast that can reliably contain very large insert fragments of DNA.

Y-linked genes (alleles): genes (or alleles of genes) that are only located on the Y-chromosome.

zygote: a cell formed from the union of a sperm and egg.

zygotic gene: a gene that is expressed after fertilization.

SOLUTIONS TO EVEN-NUMBERED PROBLEMS

CHAPTER 2

Conceptual Questions

2. In the case of plants, cross-fertilization is when the sperm cells (within the pollen) and eggs come from different plants; in self-fertilization, they come from the same plant.

4. A homozygote.

6. Diploid organisms contain two copies of each type of gene. When they make gametes, only one copy of each gene is found in a gamete. Two alleles cannot stay together within the same gamete.

8. Genotypes:1:1, *Tt* and *tt*. Phenotypes: 1:1, Tall and dwarf.

10. *c* is the recessive allele for constricted pods; *Y* is the dominant allele for yellow color. The cross is *ccYy* × *CcYy*.

	cY	*cY*	*cy*	*cy*
CY	*CcYY*	*CcYY*	*CcYy*	*CcYy*
Cy	*CcYy*	*CcYy*	*Ccyy*	*Ccyy*
cY	*ccYY*	*ccYY*	*ccYy*	*ccYy*
cy	*ccYy*	*ccYy*	*ccyy*	*ccyy*

The genotype ratio is 2 *CcYY* : 4 *CcYy* : 2 *Ccyy* : 2 *ccYY* : 4 *ccYy* : 2 *ccyy*. This 2:4:2:2:4:2 ratio could be reduced to a 1:2:1:1:2:1 ratio.
The phenotype ratio is 6 inflated pods/yellow seeds : 2 inflated pods/green seeds : 6 constricted pods/yellow seeds : 2 constricted pods/green seeds.

This 6:2:2:6 ratio could be reduced to a 3:1:1:3 ratio.

12. The dwarf parent with terminal flowers must be homozygous for both genes, since it is expressing these two recessive traits:*ttaa*, where *t* is the recessive dwarf allele, and *a* is the recessive allele for terminal flowers. The phenotype of the other parent is dominant for both traits. However, since this parent was able to produce dwarf offspring with terminal flowers, it must have been heterozygous for both genes: *TtAa*.

14. **(A)** It behaves like a recessive trait, because unaffected parents sometimes produce affected offspring. In other words, we think that the unaffected parents are heterozygous carriers.

(B) It behaves like a dominant trait. An affected offspring always has an affected parent.

16. The parents must be heterozygotes, and so the probability is 1/4.

18. We are assuming that round and yellow are dominant traits.

(A) 3/16.

(B) (9/16)(9/16)(9/16) = 0.18.

(C) We are looking for a specific order of plants in a row. We can use the product rule to determine this order: (9/16)(9/16)(3/16)(1/16)(1/16) = 243/1,048,576 = 0.00023, or 0.023%.

(D) Another way of looking at this is that the probability it will have round, yellow seeds is 9/16. The probability it will not is 1 − 9/16 = 7/16.

20. This problem is a bit unwieldy, but we can solve it using the multiplication rule:

For height, the ratio is 3 tall : 1 dwarf.
For seed texture, the ratio is 1 round : 1 wrinkled.
For seed color, they are all yellow.
For flower location, the ratio is 3 axial : 1 terminal.

Thus, the product is

(3 tall + 1 dwarf)(1 round + 1 wrinkled)(1 yellow)(3 axial + 1 terminal)

Multiplying this out, the answer is

9 tall, round, yellow, axial
9 tall, wrinkled, yellow, axial
3 tall, round, yellow, terminal
3 tall, wrinkled, yellow, terminal
3 dwarf, round, yellow, axial
3 dwarf, wrinkled, yellow, axial
1 dwarf, round, yellow, terminal
1 dwarf, wrinkled, yellow, terminal

22. Our hypothesis is that blue flowers and purple seeds are dominant traits and are governed by two genes that assort independently. According to this hypothesis, the F_2 generation should yield a ratio of 9 blue flowers, purple seeds : 3 blue flowers, green seeds : 3 white flowers, purple seeds : 1 white flower, green seeds. Since there are a total of 300 offspring produced, the expected numbers are

$9/16 \times 300 = 169$ blue flowers, purple seeds
$3/16 \times 300 = 56$ blue flowers, green seeds
$3/16 \times 300 = 56$ white flowers, purple seeds
$1/16 \times 300 = 19$ white flowers, green seeds.

These are the expected values. We then plug the observed and expected values into our chi square equation:

$$X^2 = \Sigma \frac{(O - E)^2}{E}$$

$$= \frac{(103 - 169)^2}{169} + \frac{(49 - 56)^2}{56} + \frac{(44 - 56)^2}{56} + \frac{(104 - 19)^2}{19}$$

$$= 409.5$$

If we look this value up in the chi square table under three degrees of freedom, the value is much higher than would be expected 1% of the time by chance alone. Therefore, we reject our hypothesis. The idea that the two genes are assorting independently seems to be incorrect. The F_1 generation supports the idea that blue flowers and purple seeds are dominant traits.

Experimental Questions

2. The experimental difference depends on where the pollen comes from. In self-fertilization, the pollen and eggs come from the same plant. In cross-fertilization, they come from different plants.

4. According to Mendel's law of segregation, the genotypic ratio should be 1 homozygote dominant : 2 heterozygotes : 1 homozygote recessive. This data table only considers the plants with a dominant phenotype. The genotypic ratio should be 1 homozygote dominant : 2 heterozygotes. The homozygote dominant would be true breeding, while the heterozygotes would not be true breeding. This 1:2 ratio is very close to what Mendel observed.

Questions for Student Discussion/Collaboration

2. It is a tricky combination of operations. We first use the product rule to calculate the probability that the first three offspring will be tall with axial flowers, (3/8)(3/8)(3/8), and the probability that the first three will be dwarf with terminal flowers, (1/8)(1/8)(1/8). We then use the sum rule to calculate the probability the either of these events could occur:

(3/8)(3/8)(3/8) + (1/8)(1/8)(1/8) = 27/512 + 1/512 = 0.055, or 5.5%

The probability that the fourth offspring will be tall with axial flowers is 3/8.

So, gathering our information, 0.055 is the probability that the first three offspring will be tall with axial flowers or dwarf with terminal flowers, and 3/8 is the probability that the fourth offspring will be tall with axial flowers. To calculate the probability that these two events will occur in this order, we use the product rule:

(0.055)(3/8) = 0.021, or 2.1% likelihood that this will occur

CHAPTER 3

Conceptual Questions

2. The term homologue refers to the members of a chromosome pair. Homologues are usually the same size and carry the same types of genes. They may differ in that the genes they carry may be different alleles.

4. Metaphase is the organization phase, and anaphase is the separation phase.

6. It would be probably be lost and degraded because it would not migrate to a pole. Therefore, it would not become enclosed in a nuclear membrane after telophase. If left out in the cytoplasm, it eventually would be degraded.

8. The reduction occurs because there is a single DNA replication event, but two cell divisions. Because of the nature of separation during anaphase I, each cell receives one copy of each type of chromosome.

10. It means that the arrangement of the maternally derived and paternally derived chromosomes is random along the metaphase plate during metaphase I.

12. There are three pairs of chromosomes. So the possible number of arrangements equals 2^3, which is 8. This is assuming no crossing over, which would greatly increase the number of possibilities.

14. It would be much lower, because pieces of maternal chromosomes would be likely to be mixed with the paternal chromosomes. Therefore, it would be more difficult to inherit a chromosome that was completely paternally derived.

16. (A) Dark males and light females; reciprocal:all dark offspring.

   (B) All dark offspring; reciprocal: dark females and light males.

   (C) All dark offspring; reciprocal: dark females and light males.

   (D) All dark offspring; reciprocal: dark females and light males.

18. To produce sperm, a spermatogonial cell first goes through mitosis to produce two cells. One of these will remain a spermatogonial cell, and the other will progress through meiosis. In this way, the testis continues to maintain a population of spermatogonial cells.

20. (A) 1 ABC : 1 ABc : 1 AbC : 1 Abc : 1 aBC : 1 abC : 1 aBc : 1 abc

   (B) 1 ABC : 1 AbC

   (C) 1 ABC : 1 ABc : 1 aBC : 1 aBc

   (D) 1 Abc : 1 abc

22. (A) X-linked recessive (unaffected mothers transmit the trait to sons).

   (B) Autosomal recessive (affected daughters and sons are produced from unaffected parents).

24. The student should first set up a Punnett square, where X^h represents the hemophilia allele:

	X^H	Y
X^H	$X^H X^H$	$X^H Y$
X^h	$X^H X^h$	$X^h Y$

   (A) 1/4

   (B) (3/4)(3/4)(3/4)(3/4) = 81/256

   (C) Use the sum rule:1/4 + 1/4 + 1/4 = 3/4.

   (D) Use the binomial expansion: the answer is 0.263, which equals 26.3%.

Experimental Questions

2. I would say that the observation was that all of the white-eyed flies of the F_2 generation were males; not a single female was observed. This

suggests a link between gender determination and the inheritance of this trait. Since gender determination in fruit flies is determined by the number of X-chromosomes, this suggests a relationship between the inheritance of the X-chromosome and the inheritance of this trait.

4. Basically, the same way, except that only the male can transmit the trait. If it is Y-linked, it will only be passed from father to son. Beginning with an affected male and crossing it to a wild-type female, you would observe affected sons in the F_1 generation. If it is X-linked, you would observed affected sons in the F_2 generation, but not in F_1.

6. The female could be $X^{w+}X^{w+}X^{w+}$, $X^{w+}X^{w+}X^{w}$, $X^{w+}X^{w}X^{w}$, $X^{w+}X^{w+}X^{w-e}$, $X^{w+}X^{w-e}X^{w-e}$, or $X^{w+}X^{w}X^{w-e}$. The 3:1 sex ratio occurs because the female produces 50% XX gametes and 50% X gametes. The XX gametes will produce XXX and XXY females; the X gametes will produce XX females and XY males. The original female was $X^{w+}X^{w-e}X^{w-e}$:

	X^w	Y
$X^{w+}X^{w-e}$	$X^{w+}X^{w-e}X^{w}$ Red female	$X^{w+}X^{w-e}Y$ Red female
$X^{w+}X^{w-e}$	$X^{w+}X^{w-e}X^{w}$ Red female	$X^{w+}X^{w-e}Y$ Red female
$X^{w-e}X^{w-e}$	$X^{w-e}X^{w-e}X^{w}$ Eosin female	$X^{w-e}X^{w-e}Y$ Eosin female
X^{w+}	$X^{w+}X^{w}$ Red female	$X^{w+}Y$ Red male
X^{w-e}	$X^{w-e}X^{w}$ Light-eosin female	$X^{w-e}Y$ Light-eosin male
X^{w-e}	$X^{w-e}X^{w}$ Light-eosin female	$X^{w-e}Y$ Light-eosin male

Questions for Student Discussion/Collaboration

2. I cannot really give a written answer to this question, but the point is for students to be able to draw chromosomes in different configurations. The chromosomes may or may not be (1) in homologous pairs; (2) connected as sister chromatids; (3) lined up in metaphase; (4) moving toward the poles.

CHAPTER 4
Conceptual Questions

2. A gene dosage effect means that the number of gene copies influences the phenotype of a trait. Gene dosage effects occur when a single copy of allele is not sufficient to produce a certain phenotypic trait. For example, one copy of a pigmentation allele may not be sufficient to produce a maximum amount of pigment. Therefore, two copies of the allele will alter the phenotype by producing a darker pigmentation.

4. Both genetics and the environment can influence expressivity. Modifier genes can alter the level of expression of another gene. This was observed in Experiment 4B. In addition, environmental influences can have a great impact. For example, sunlight and nutrients in the soil can greatly affect plant traits, such as height and flower color.

6. If the normal allele is dominant, it tells us that one copy of the gene produces a sufficient amount of the encoded protein to produce its phenotypic effects. Having twice as much of this protein, as in the normal homozygote, does not alter the phenotype. If the allele is incompletely dominant, this means that one copy of the normal allele is not sufficient to produce its phenotypic effects.

8. The red and white seed packs should be from true-breeding (homozygous) strains. The pink pack should be seeds from a cross between white- and red-flowered plants.

10. Types O and AB.

12. **(A)** 1/4

   (B) 0

   (C) (1/4)(1/4)(1/4) = 1/64

   (D) Use the binomial expansion:

$$P = \frac{n!}{x!\,(n-x)!}\,p^x q^{(n-k)}$$

$$n = 3,\ p = 1/4,\ q = 1/4,\ x = 2$$

$$P = 0.047 = 4.7\%$$

14. Perhaps it should be called codominant at the "hair level," because one or the other allele is dominant with regard to a single hair. However, this is not the same as codominance in blood types, in which every cell can express both alleles.

16. All of the F_1 generation will be white, because they have inherited the dominant white allele from their leghorn parent.

   For the F_2 generation, construct a Punnett square. Let W and w represent alleles of one gene, where W is dominant and causes a white phenotype. Let A and a represent alleles of the second gene, where the recessive allele causes a white phenotype in the homozygous condition. The genotype of the F_1 birds is $WwAa$. The phenotypic ratio will be 13 white : 3 brown. The only brown birds will be 2 $wwAa$ and 1 $wwAA$.

18. There may be two redundant genes that are involved in feathering. The unfeathered Buff Rocks are homozygous recessive for the two genes. The Black Langhans are homozygous dominant for both genes. In the F_2 generation (which is a double heterozygote crossed to another double heterozygote), 1 out of 16 offspring will be doubly homozygous for both recessive genes. All of the others will have at least one dominant allele for one of the two (redundant) genes.

20. On a normal diet, it would be 50% white and 50% yellow. On a xanthophyll-free diet, it would be 100% white.

22. The sandy variation may be due to a homozygous recessive allele at two different genes in these two varieties of sandy pigs. Let's call them genes A and B. One variety of sandy pig could be $aaBB$, and the other $AAbb$. The F_1 generation in this cross will be heterozygotes for both genes and all will be red. This tells us that the A and B alleles are dominant. In the F_2 generation, 6 out of 16 will be homozygous for either the aa or bb alleles and become sandy. One out of 16 will be doubly homozygous and be white. The remaining 9 will contain at least one dominant allele for both genes.

24. Since the parents have walnut combs, we know that they must have one R allele and one P allele. If both of the parents were $RrPp$, we would expect all four phenotypic categories in the offspring. However, we never get offspring with single or pea combs. This tells us that we cannot produce offspring with an rr genotype. Therefore, one of the parents must be RR. Since we do get offspring with rose combs, both parents must be Pp. Therefore, in this cross, one parent must be $RRPp$ and the other parent could be $RRPp$ or $RrPp$.

26. The parents were $WwGG$ and $wwgg$. They would produce 50% $WwGg$ (white) and 50% $wwGg$ (green) offspring.

28. **(A)** Could be.

   (B) No, because an unaffected father has an affected daughter.

   (C) No, because unaffected parents have affected children.

   (D) No, because an unaffected father has an affected daughter.

   (E) No, because both genders exhibit the trait.

   (F) Could be.

30. I would look at the pattern within families and over the course of many generations. For a recessive trait, we expect 50% of the offspring to be affected (if one of them is affected). I would look at many families and see if this 50% value is approximately true. Incomplete penetrance would not necessarily predict 50% affected offspring in a single family. Also, for alleles that are found at a low frequency in the population, an inheritance pattern due to incomplete penetrance would probably have a much higher frequency of affected parents producing affected offspring. Finally, the most informative pedigrees would be situations where two affected parents produce children. If they can produce an unaffected offspring, this would indicate incomplete penetrance. If all of their offspring were affected, this would be consistent with recessive inheritance.

Experimental Questions

2. I might expect the alleles with intermediate pigmentation to exhibit a gene dosage effect, because these are the ones that are not making a maximum amount of pigment. Alleles such as ivory, pearl, and apricot would be the best candidates. The alleles producing a darker reddish phenotype may not show a gene dosage effect, because a large amount of pigment is already being made.

4. The interpretation of the data proposes that the F_1 cross is $Cc^aX^{W+}Y \times Cc^aX^{W+}X^{w-e}$. It also proposes that c^a is recessive and that it only modifies X^{w-e}. Based on these assumptions, the expected ratio is 8 females/red eyes : 4 males/red eyes : 3 males/eosin eyes : 1 male/cream eyes.

A total of 209 flies were produced. The expected numbers are

(8/16)(209) = 105 females/red eyes
(4/16)(209) = 53 males/red eyes
(3/16)(209) = 39 males with eosin eyes
(1/16)(209) = 13 males with cream eyes.

Plugging these values into our chi square formula,

$$X^2 = \Sigma \frac{(O-E)^2}{E}$$

$$= \frac{(104-105)^2}{105} + \frac{(47-53)^2}{53} + \frac{(44-39)^2}{39} + \frac{(14-13)^2}{13}$$

$$= 0.01 + 0.7 + 0.6 + 0.08$$

$$= 1.39$$

If we look this value up in the chi square table under three degrees of freedom, the value is well within our expected range. The chi square would have to be above 7.815 to reject our hypothesis. Therefore, we accept that our hypothesis is correct.

Questions for Student Discussion/Collaboration

2. The easiest way to solve this problem is to take one trait at a time. With regard to combs, all of the F_1 generation would be $RrPp$ or walnut comb. With regard to shanks, they would all be feathered, because they would inherit one dominant copy of a feathered allele. With regard to hen- or cock-feathering, this cross would produce a ratio of 1 male/cock-feathered : 1 male/hen-feathered : 2 female/hen-feathered. Overall then, we would have a 1 : 1 : 2 ratio of

walnut comb/feathered shanks/cock-feathered males
walnut comb/feathered shanks/hen-feathered males
walnut comb/feathered shanks/hen-feathered females

CHAPTER 5

Conceptual Questions

2. There are four phenotypic categories for the F_2 offspring:brown fur/short tails, brown fur/long tails, white fur/short tails, and white fur/long tails. The recombinants are brown fur/long tails and white fur/short tails. The F_2 offspring will occur in a 1:1:1:1 ratio if the two genes are not linked. In other words, there will be 25% of each of the four phenotypic categories. If the genes are linked, there will be a lower percentage of the recombinant offspring.

4. An independent assortment hypothesis is used because it enables us to calculate the expected values based on Mendel's ratios. Using the observed and expected values, we can calculate whether or not the deviations between the observed and expected values are too large to be expected to occur as a matter of chance. If the deviations are very large, we reject the hypothesis of independent assortment.

6. (A) Since they are 12 map units apart, we expect 12% (or 120) recombinant offspring. This would be approximately 60 *Aabb* and 60 *aaBb*, plus 440 *AaBb* and 440 *aabb*.

(B) 60 *AaBb*, 60 *aabb*, 440 *Aabb*, and 440 *aaBb*.

8. Map distance:

$$= \frac{11+9}{144+150+11+9} \times 100$$

$$= 6.4 \text{ mu}$$

10.

If the chromosomes labeled 2 and 4 move into one daughter cell, that will lead to a patch that is albino and has long hair. The other cell will receive chromosomes 1 and 3, which will produce a patch that has dark, short hair.

12. (A) If we hypothesize two genes independently assorting, then the predicted ratio is 1:1:1:1. There are a total of 402 offspring. The expected number of offspring in each category is about 101. Plugging the values into our chi square formula,

$$X^2 = \frac{(104-101)^2}{101} + \frac{(102-101)^2}{101} + \frac{(97-101)^2}{101} + \frac{(99-101)^2}{101}$$

$$= 0.09 + 0.01 + 0.16 + 0.04$$

$$= 0.3$$

Looking this value up in the chi square table under three degrees of freedom, we find that we cannot reject our hypothesis. The chi square value would have to be above 7.815 to reject the hypothesis at the 5% confidence level. Instead, we accept the hypothesis that the genes are assorting independently.

(B) It is not valid to calculate the map distance between these genes, because they appear to be assorting independently.

14. Since the two genes are 18 mu apart, 18% of the offspring will be recombinant. In this case, 9% will have normal wings and normal legs, and 9% will have dumpy wings and short legs.

16. A single crossover produces ABC, Abc, aBC, and abc.

(A) Between 2 and 3, between genes *B* and *C*.

(B) Between 1 and 4, between genes *A* and *B*.

(C) Between 1 and 4, between genes *B* and *C*.

(D) Between 2 and 3, between genes *A* and *B*.

18. *Ass-1* ←43→ *Sdh-1* ←5→ *Hdc* ←9→ *Hao-1* ←6→ *Odc-2* ←8→ *Ada*.

20. The basic strategy is first to determine the percentage of recombination between the different gene pairs. The percentage of recombinants between the green/yellow and wide/narrow is 7% or 0.07; there will be 3.5% of the green/narrow and 3.5% of the yellow/wide. The remaining 93% parentals will be 46.5% green/wide and 46.5% yellow/narrow. The third gene assorts independently. There will be 50% long and 50% short with respect to each of the other two genes. To calculate the number of offspring out of a total of 800, we multiply 800 by the percentages of each category:

(0.465 green/wide)(0.5 long)(800) = 186 green/wide/long
(0.465 yellow/narrow)(0.5 long)(800) = 186 yellow/narrow/long
(0.465 green/wide)(0.5 short)(800) = 186 green/wide/short

(0.465 yellow/narrow)(0.5 short)(800) = 186 yellow/narrow/short
(0.035 green/narrow)(0.5 long)(800) = 14 green/narrow/long
(0.035 yellow/wide)(0.5 long)(800) = 14 yellow/wide/long
(0.035 green/narrow)(0.5 short)(800) = 14 green/narrow/short
(0.035 yellow/wide)(0.5 short)(800) = 14 yellow/wide/short

22. A tetrad contains four spores; an octad contains eight. In a tetrad, meiosis produces four spores. In an octad, meiosis produces four cells, and then they all go through mitosis to double the number to eight cells.

24. (A) The first thing to do is to determine which asci are parental ditypes, nonparental ditypes, and tetratypes. A parental ditype will contain a 2:2 combination of spores with the same genotype as the original haploid parents. The combination of 502 asci are the parental ditypes. The nonparental ditypes are those containing a 2:2 combination of genotypes that are unlike the parentals. The combination of 4 asci fits this description. Finally, the tetratypes contain a 1:1:1:1 arrangement of genotypes, half of which have a parental genotype and half of which do not. There are 312 tetratypes in this case. Computing the map distance:

$$\text{Map distance} = \frac{\text{Nonparental ditypes} + (1/2)(\text{Tetratypes})}{\text{Total number of asci}} \times 100$$

$$= \frac{4 + (1/2)(312)}{818}$$

$$= 19.6 \text{ mu}$$

If we use the more accurate equation:

$$\text{Map distance} = \frac{T + 6NPD}{\text{Total number of asci}} \times 0.5 \times 100$$

$$= \frac{312 + 6(4)}{818}$$

$$= 20.5 \text{ mu}$$

(B) The frequency of single crossovers is 0.205 if we use the more accurate equation.

(C) Nonparental ditypes are produced from a double crossover. To compute the expected number, we multiply 0.205 × 0.205 = 0.042, or 4.2%. Since we had a total of 818 asci, we would expect 34.3 asci to be the product of a double crossover. However, as described in Figure 5-16, only 1/4 of them would be a nonparental ditype. Therefore, we multiply 34.3 by 1/4, obtaining a value of 8.6 nonparental ditypes due to a double crossover. Since we only observed 4, this calculation tells us that positive interference is occurring.

26. A.

								B. *Number*
pro-1	pro-1	pro-1	pro-1	pro⁺	pro⁺	pro⁺	pro⁺	402
pro⁺	pro⁺	pro⁺	pro⁺	pro-1	pro-1	pro-1	pro-1	402
pro⁺	pro⁺	pro-1	pro-1	pro⁺	pro⁺	pro-1	pro-1	49
pro-1	pro-1	pro⁺	pro⁺	pro-1	pro-1	pro⁺	pro⁺	49
pro⁺	pro⁺	pro-1	pro-1	pro-1	pro-1	pro⁺	pro⁺	49
pro-1	pro-1	pro⁺	pro⁺	pro⁺	pro⁺	pro-1	pro-1	49

Experimental Questions

2. He determined this by analyzing the data in gene pairs. This analysis revealed that there were fewer recombinants between certain gene pairs (e.g., body color and eye color) than between other gene pairs (e.g., eye color and wing shape). From this comparison, he hypothesized that genes that are close together on the same chromosome will produce fewer recombinants than genes that are farther apart.

4. A double crossover.

6. As described in Figure 5-10, the relationship between map distance and the percentage of recombinant offspring becomes very inaccurate at map distances approaching or exceeding 50 map units. This may explain the underestimate of the map distance, determined from other genes that are closer together, by the percentages of recombinant offspring. There is also some inaccuracy due to random sampling error.

Questions for Student Discussion/Collaboration

2. The X- and Y-chromosomes are not completely distinct linkage groups. One might describe them as overlapping linkage groups having some genes in common, but most genes of which are not common to both.

CHAPTER 6

Conceptual Questions

2. It is not a form of sexual reproduction whereby two distinct parents produce gametes that unite to form a new individual. However, conjugation is similar to sexual reproduction in the sense that the genetic material from two cells are somewhat mixed. In conjugation, there is not the mixing of two genomes, one from each gamete. Instead, there is a transfer of genetic material from one cell to another. This transfer can alter the combination of genetic traits in the recipient cell.

4. An *F⁺* strain contains a separate, circular piece of DNA that has its own origin of transfer. An Hfr strain has an F factor integrated into the bacterial chromosome. An *F⁺* strain can only transfer the DNA contained on the F factor. If given enough time, an Hfr strain can actually transfer the entire bacterial chromosome to the recipient cell.

6. (A) If we extrapolate this line back to the x-axis, *hisE* intersects at about 3 minutes and *pheA* intersects at about 24 minutes. These are the values for the times of entry. Therefore, the distance between these two genes is 21 minutes (i.e., 24 minus 3).

(B)

↑	4	↑	17	↑
hisE		*pabB*		*pheA*

8. Briefly, the lytic cycle involves the production of new viruses. The phage takes over the synthetic machinery of the cells to make many more additional virus particles. Eventually, the cell lyses, releasing these new particles. The lysogenic cycle involves the integration of the viral DNA into the host cell's chromosome. This integrated DNA is called a prophage. It will be replicated whenever the host chromosome is replicated. At a later time, a prophage can be induced to progress through the lytic cycle.

10.
$$\text{Cotransduction frequency} = (1 - d/L)^3$$
$$= (1 - 0.6 \text{ minutes}/2 \text{ minutes})^3$$
$$= 0.34, \text{ or } 34\%$$

12. I would conclude that the two genes are further apart than the length of 2% of the bacterial chromosome.

14. The transfer of F factor DNA.

16. The term refers to the outcome in which two mutations are combined in a single individual or cell, with the individual or cell showing no phenotypic effect from either mutation. For example, if two defective mutations combined in a single individual result in a nondefective phenotype, complementation has occurred. This is because the two mutations are in different genes, and the corresponding wild-type versions are dominant. If complementation does not occur, the defective mutations are in the same gene; when they occur together in the same individual, complementation cannot occur, because the individual does not have a normal copy of the gene.

18. He first determined the individual nature of each gene by showing that mutations within the same gene did not complement each other. He then could map the distance between two mutations within the same gene. The map distances defined each gene as a linear, divisible unit. In this regard, the gene is divisible due to crossing over.

Experimental Questions

2. Mix the two strains together, and then put some of them on minimal plates containing streptomycin and some of them on plates without streptomycin. If mated colonies are present on both types of plates, then the P^+ and T^+ genes were transferred to the $B^+M^+P^-T^-$ strain. If colonies are only found on the plates that lack streptomycin, then the B^+ and M^+ genes are being transferred to the $B^-M^-P^+T^+$ strain. This answer assumes that the str^r gene is not readily transferred between the two strains.

4. Mutations in *rIIA*: r47, r101, r103, r104, r105, r106. Mutations in *rIIB*: r51, r102.

6. The time of entry is the time it takes for a gene to be initially transferred from one bacterium to another. To determine this time, we make many measurements at various lengths of time and then extrapolate these data back to the *x*-axis.

8. Benzer could use this observation as a way to evaluate if intragenic recombination had occurred. If two rII mutations recombined to make a wild-type gene, the phage would produce plaques in this *E. coli* strain. This makes it possible to identify recombinants that occur very rarely within the population.

Questions for Student Discussion/Collaboration

2. Epistasis is another example. When two different genes affect flower color, the white allele of one gene may be epistatic to the purple allele of another gene. A heterozygote for both genes would have purple flowers. In other words, the two purple alleles are complementing the two white alleles. Other similar examples of complementation are discussed in Chapter 4, in the section on Gene Interactions.

CHAPTER 7

Conceptual Questions

2. A maternal effect gene is one in which the genotype of the mother determines the phenotype of the offspring. At the cellular level, this occurs because maternal effect genes are expressed in the nurse cells and the gene products then are transported into the oocyte. These gene products play key roles in the early steps of embryonic development.

4. The genotype of the mother must be *bicoid⁻bicoid⁻*. That is why she produces abnormal offspring. Since the mother is alive and able to produce offspring, her mother (the maternal grandmother) must have been *bicoid⁺bicoid⁻* and passed the *bicoid⁻* allele to her daughter (the mother in this problem). The maternal grandfather also must have passed the *bicoid⁻* allele to his daughter. He could be either *bicoid⁺bicoid⁻* or *bicoid⁻bicoid⁻*.

6. A Barr body is a mammalian X-chromosome that is highly condensed. It is found in somatic cells that have two or more X-chromosomes. Nearly all of the genes on the Barr body are transcriptionally inactive.

8. X-inactivation occurs during embryonic stages. In heterozygous females, this produces a mosaic pattern of gene expression. Since X-inactivation occurs randomly during embryonic development, certain patches of tissue have one X-chromosome inactivated and other patches have the other X-chromosome inactivated. In the case of a female that is heterozygous for a gene that affects pigmentation of the fur, this produces a variegated pattern of coat color. Two different heterozygous females will have different patterns, because the process of X-inactivation is random, so that the choice of the X-chromosome to be inactivated in different regions of the embryo will differ among individuals. A variegated coat pattern is not expected in marsupials, because the paternal X-chromosome is always inactivated.

10. The male is XXY. This person is male because of the presence of the Y-chromosome. Due to the counting of Xics, one of the X-chromosomes is inactivated to produce a Barr body.

12. The *de novo* methylase would be inactive in somatic cells and active in germ cells. Somehow, the methylase must be regulated so that it is targeted to methylate particular genes, because we know that some genes are imprinted in males but not in females, whereas other genes are imprinted in females but not males.

14. Extranuclear inheritance is the transmission of genetic material (in eukaryotes) that is not located in the cell nucleus. The two most important examples involve genetic material in mitochondria and plastids. Less common examples are infectious particles that produce traits such as killer paramecia, and the sex ratio trait in *Drosophila*.

16. Extranuclear inheritance does not always occur via the female gamete. Sometimes it occurs via the male gamete. Even in species where maternal inheritance is prevalent, paternal leakage can also occur. With regard to cytoplasmic inheritance, maternal inheritance is probably the most common, because the female gamete is relatively large and more likely to contain cell organelles.

18. The phenotype of a petite mutant is that it forms small colonies on growth media containing an energy source that does not require mitochondrial function. These mutants cannot grow on an energy source that does require mitochondrial function. Since nuclear and mitochondrial genes are necessary for mitochondrial function, it is possible for a petite mutation to involve a gene in the nucleus or in the mitochondrial genome. Neutral petites lack most of their mitochondrial DNA; suppressive petites usually lack small segments of the mitochondrial genetic material.

20. The mitochondrial and chloroplast genomes are composed of a circular chromosome found in one or more copies. These copies are located in a region of the organelle known as the nucleoid. The number of genes per chromosome varies from species to species. Mitochondria tend to have fewer genes than do chloroplasts. See Table 7-3 for examples of the variation among mitochondrial and chloroplast genomes.

Experimental Questions

2. The first type of observation was cytological. The presence of the Barr body in female cells was consistent with the idea that one of the X-chromosomes was highly condensed. The second type of observation was genetic. A variegated phenotype that is only found in females is consistent with the idea that certain patches express one allele and other patches express the other allele. This variegated phenotype would only occur if the inactivation happened at an early stage of embryonic development and was inherited permanently thereafter.

4. The biopsy must have been large enough so that not all of the tissue was derived from the same embryonic cell. If a small biopsy was taken, and all of the tissue was derived from the same embryonic cell in which X-inactivation had already occurred, then the cells from that biopsy would only produce one of the two G-6-PD alleles.

6. In the absence of UV, we would expect all *sm^r* offspring. With UV light, we would expect a greater percentage of *sm^s* offspring.

Questions for Student Discussion/Collaboration

2. An infective particle is something in the cytoplasm that contains its own genetic material and is not an organelle. Some symbiotic infective particles, such as those found in killer paramecia, are similar to mitochondria and chloroplasts in that they contain they own genomes and are known to be bacterial in origin. The observation that these endosymbiotic relationships can be initiated in modern species tells us that endosymbiosis can occur spontaneously. Therefore, it is reasonable that it happened a long time ago and led to the evolution of mitochondria and plastids.

CHAPTER 8

Conceptual Questions

2. Small deletions and duplications are less likely to affect phenotype, simply because they usually involve fewer genes. If a small deletion did have a phenotypic effect, I would conclude that the gene or genes in this region are required in two functional copies in order to have a normal phenotype.

4. A gene family is a group of genes derived from the process of gene duplication. They have similar sequences, but with some differences due to the accumulation of mutations over many generations. The members of a gene family usually encode proteins with similar but specialized functions. The specialization may occur in different cells, or at different stages of development.

6. It is a pericentric inversion.

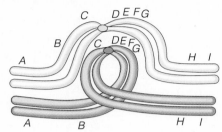

8. There are four products from meiosis. One would be a normal chromosome, and one would contain the inversion. The other two chromosomes would be

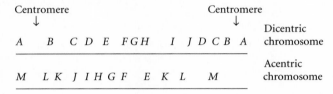

10. In the absence of crossing over, alternate segregation would yield half the gametes with two normal chromosomes and half the gametes with a balanced translocation. For adjacent-1 segregation, two gametes would be

A B   C D E   +   A I J K L M
_____   _____

And the other two gametes would be

H B   C D E   +   H I J K L M
_____   _____

12. One of the parents probably carries a balanced translocation between chromosomes 5 and 7. The phenotypically abnormal offspring has inherited an imbalanced translocation due to the segregation of translocated chromosomes during meiosis.

14. (A) 16
    (B) 9
    (C) 7
    (D) 12
    (E) 17

16. One parent is probably normal, while the other parent has one normal copy of chromosomes 14 and 21 and one chromosome 14 that has most of chromosome 21 attached to it.

18. Imbalances in aneuploidy, deletions, and duplications are related to the copy number of genes. For many genes, the level of gene expression is directly related to the number of genes per cell. If there are too many copies, as in trisomy, or too few, as in monosomy, the level of gene expression will be too high or too low, respectively. It is difficult to say why deletions and monosomies are more detrimental, although one could speculate that having too little of a gene product causes more cellular problems that having too much of a gene product. Also, deleterious recessive alleles are more likely to exert their effects when chromosomes are monosomic.

20. Most human aneuploidies are lethal during embryonic development and lead to early spontaneous abortions.

22. Trisomy-13, -18, and -21 survive because the chromosomes are small and probably contain fewer genes than do the larger chromosomes. Individuals with abnormal numbers of X-chromosomes can survive because the extra copies are converted to transcriptionally inactive Barr bodies. The other aneuploidies are lethal because they cause a great amount of imbalance between the level of gene expression on the normal diploid chromosomes and the level of gene expression from the chromosomes that are trisomic or monosomic.

24. Endopolyploidy means that a particular somatic tissue is polyploid even though the rest of the organism is not. The biological significance is not entirely understood, although it has been speculated that an increase in euploidy may enable the cell to make more gene products that the cell needs.

26. In certain types of cells, such as salivary cells, the homologous chromosomes pair with each other and then replicate about nine times to produce a polytene chromosome. The centromeres from each type of chromosome associate with each other at the chromocenter. This structure has six arms, which are due to one arm of two telomeric chromosomes (the X and 4) and two arms each from chromosomes 2 and 3.

28. Nondisjunction is a mechanism whereby the chromosomes do not segregate equally into the two daughter cells. This can occur during meiosis to produce gametes with altered numbers of chromosomes, or it can occur during mitosis to produce an individual that is a genetic mosaic. A third way to alter chromosome number is by interspecies matings to produce an alloploid.

30. (1) A mitotic nondisjunction in which the two chromosomes carrying the *b* allele went to one cell and the two chromosomes carrying the *B* allele went to the other daughter cell. (2) The chromosome carrying the *B* allele could be lost. (3) The *B* allele could have mutated.

32. Homeologous chromosomes are chromosomes from evolutionarily related species that are analogous to each other. For example, chromosome 1 in chimpanzees and gorillas are homeologous; they carry the same types of genes.

Experimental Questions

2. They can be viewed in much greater detail under the microscope. This makes it much easier to detect very small changes in chromosome structure.

4. You could begin with a normal diploid strain and first use anther culture. This would create a haploid strain. This haploid strain could then be treated on two successive occasions with colchicine to first produce a diploid strain and then a tetraploid strain. Since this tetraploid strain would be derived from a haploid strain, it would be homozygous for all of its genes.

6. Cell fusion techniques can be used to create hybrids between strains or different species that cannot readily interbreed. Any two types of cells can be made to fuse in the laboratory, even interspecies hybrids that could never interbreed and produce viable offspring.

Questions for Student Discussion/Collaboration

2. There are lots of possibilities. The students could look in agriculture and botany books to find many examples. In the insect world, there are interesting examples of euploidy affecting gender determination. Among amphibians and reptiles, there are also several examples of closely related species having euploid variation.

4. (1) Polyploids are often more robust and disease resistant. (2) Allopolyploids may have useful combinations of traits. (3) Hybrids are often more vigorous; they can be generated from monoploids. (4) Strains with an odd number of chromosome sets (e.g., triploids) are usually seedless.

CHAPTER 9

Conceptual Questions

2. (1) It binds to the cell surface. (2) It penetrates the cell wall/cell membrane. (3) It enters the cytoplasm. (4) It recombines with the chromosome (although students should not be expected to know this). (5) The genes within the DNA are expressed (i.e., transcription and translation). (6) The gene products create a capsule. That is, they are enzymes that synthesize a capsule using cellular molecules as building blocks.

4. Guanine is the free base. The structure is given in the chapter. Guanosine is the base attached to the number 1 carbon on ribose. Deoxyguanosine triphosphate is a guanosine with three phosphates attached to the number 5 carbon on ribose.

6. The bases conform to the A–T/G–C rule of complementarity. There are two hydrogen bonds between A and T, and three hydrogen bonds between G and C. The planar rings of the bases stack on top of each other within the helical structure to provide even more stability.

8. The sequence of nucleotide bases.

10. A- and B-DNA are right-handed helices and the backbones are relatively helical, whereas Z-DNA is left-handed and the backbone is rather zig-zagged. A-DNA and Z-DNA have the bases tilted relative to the central axis, whereas they are perpendicular in B-DNA. There are also minor differences in the number of bases per turn.

12. DNA has deoxyribose as its sugar, while RNA has ribose. DNA has the base thymine, while RNA has uracil. DNA is a double-helical structure. RNA is single-stranded, although parts of it may form double-stranded regions.

14. 5′–<u>GAUCCCU</u>AAAC<u>GG</u>AUCCCAG GACUCCCA<u>CG</u>UUUA<u>GGGAUC</u>–3′

The complementary stem regions are underlined.

16. The helix in part A is more difficult to separate, because it has a higher percentage of GC base pairs than the one in part B.

18. Complementarity is important in several ways. First, it is needed to copy genetic information. This occurs during replication, when new DNA strands are made, and also during transcription, when RNA strands are made. Complementarity is also important during translation for codon–anticodon recognition. It also allows RNA molecules to form secondary structure and to recognize each other.

20. I would conclude it is single-stranded, since A does not equal U.

22. One possibility is a sequential mechanism. First, the double helix could unwind and copy itself via a semiconservative mechanism. Next, the third strand (bound in the major groove) could copy itself via a semiconservative mechanism. This double-stranded molecule could unwind yielding the third strand, along with its complementary strand. The complementary strand of the third strand could be used to make a another copy of the third strand. At this point, you would have two double helices and two copies of the third strand. These could assemble to make two triple helices.

24. Lysines and arginines, and also polar amino acids.

Experimental Questions

2. (1) Isolate DNA from resistant bacteria. (2) In three separate tubes, add DNase, RNase, or protease. (3) Use these three separate tubes to try to transform sensitive bacteria. (4) Plate on petri plates containing tetracycline. *Expected results:* Tetracycline-resistant colonies should only grow when the DNA has been exposed to RNase and protease, but not to DNase.

4.

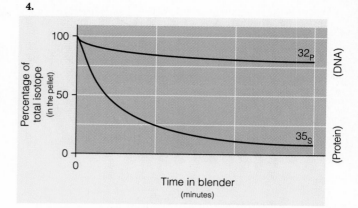

6. (1) Perhaps the shearing just is not strong enough to remove all of the phages. (2) Perhaps the tail fibers remain embedded in the bacterium and only the head region is sheared off.

8. (1) You can make lots of different shapes. (2) You can use mathematical formulae to fit things together in a systematic way. (3) Computers are very fast. (4) You can store the information you have obtained from model building in a computer file.

10. Probably not. The strength of his data was that all species appeared to conform to the A–T/G–C rule, suggesting that this is an inherent feature of DNA structure. In a single species, the observation that A = T and G = C could occur as a matter of chance.

Questions for Student Discussion/Collaboration

2. Again, there are lots of possibilities. You could use a DNA-specific chemical and show that it causes heritable mutations. Perhaps you could inject an oocyte with a piece of DNA and produce a mouse with a new trait.

CHAPTER 10

Conceptual Questions

2. The bacterial nucleoid is a region in a bacterial cell that contains a compacted circular chromosome. Unlike eukaryotic nuclei, a nucleoid is not surrounded by a membrane.

4.

6. Centromeres are found in eukaryotic chromosomes. They are necessary so that the chromosomes are sorted (i.e., segregated) during mitosis and meiosis. They are most important during M phase.

8. It would depend on several factors. First, the base content of the Alu sequence would have to be different from that of the chromosomal DNA. Also, the chromosomal DNA would have to be broken into fairly small pieces to separate the Alu sequences from their adjacent chromosomal pieces. If these two conditions were met, a satellite peak would be observed.

10. This type of experiment gives the relative proportions of highly repetitive, moderately repetitive, and unique DNA sequences within the genome. The highly repetitive sequences renature at a fast rate, the moderately repetitive sequences renature at an intermediate rate, and the unique sequences renature at a slow rate.

12. In the G_1 phase, and during the rest of interphase (S and G_2), the chromatin is less compact than in M phase. In interphase, there are euchromatic regions that are less compact than heterochromatic regions. During M phase, all of the chromatin becomes very compact in preparation for chromosome sorting and nuclear division.

14.

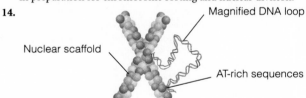

16. During interphase, the chromosomes are found within the cell nucleus. They are less tightly packed and are transcriptionally active. Segments of chromosomes are anchored to the nuclear membrane. During M phase, the chromosomes become highly condensed and the nuclear membrane is fragmented into vesicles. The chromosomes become attached to the spindle apparatus via microtubules that attach to the kinetochore, which is bound to the centromere.

Experimental Questions

2. Supercoiled DNA would look all curled up into a relatively compact structure. You could add different purified topoisomerases and see how they affect the structure. For example, gyrase relaxes positive supercoils, whereas topoisomerase I relaxes negative supercoils. If we added topoisomerase I to a DNA preparation and it became less compacted, then the DNA was negatively supercoiled.

4. Yes, if its base composition is similar to the main chromosomal DNA.

6.

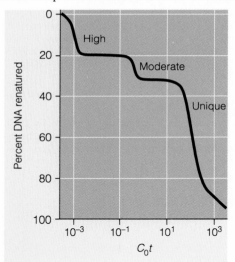

8. Lots of possibilities. You could digest it with DNase and see if it gives multiples of 200 base pairs or so. You could try to purify proteins from the sample and see if eukaryotic proteins or prokaryotic proteins are present.

Questions for Student Discussion/Collaboration

2. You need to have the DNA compact to fit in the cell. Compaction allows organisms to store more information. Without this ability, organisms would not have been able to evolve to the high levels they have achieved. The downside is that it is necessary to copy and access this information, which is difficult in a compact structure. Therefore, you need to loosen the structure. Overall, the structure of chromosomes is extremely dynamic.

4. The DNA structure is the same, except the degree of supercoiling. The protein compositions are very different. Bacteria compact their chro-

mosomes by supercoiling and looping. Eukaryotes wrap their DNA around histones to form nucleosomes. The nucleosomes associate with one another to form fibers, and these fibers are supported on a scaffold.

CHAPTER 11

Conceptual Questions

2. Bidirectionality refers to the idea that two replication forks emanate from an origin of replication. There are four DNA strands being made in a direction radiating outward from the origin.

4. If we assume that the *polIII* synthesizes at a rate of 850 nucleotides per second, it would take 92.2 minutes to replicate the entire chromosome. Since it is bidirectional, however, we can divide by two, for a value of 46.1 minutes. This is in fact a bit longer than it really takes.

For each 1000 *E. coli*, there would be 1000×4.7 million nucleotides synthesized. If the error rate is 1 in 100,000,000, there would be 47 mistakes made, on average.

6. DNA polymerase would slide from right to left. The new strand would be 3′–CTAGGGCTAGGCGTATGTAAATGGTCTAGTGGTGG–5′

8. DNA methylation is the covalent attachment of methyl groups to bases in the DNA. Immediately after replication, there has not been sufficient time to attach methyl groups to the bases in the newly made daughter strand. The time delay in DNA methylation helps to prevent premature DNA replication prior to cell division.

10. It would be difficult to delay DNA replication after cell division, because the dilution of the DnaA protein is one mechanism that regulates replication. One might expect that such a strain would have more copies of the chromosome per cell than a normal strain.

12. An Okazaki fragment is a short segment of newly made DNA in the lagging strand. It is necessary to make short fragments, because the fork is exposing the lagging strand in the 3′ to 5′ direction but DNA polymerase can only work in the 5′ to 3′ direction. Therefore, the lagging strand is made in short pieces, in the direction away from the replication fork.

14. The active site of DNA polymerase has the ability to recognize a distortion in the newly made strand and remove it. This occurs by a 3′ to 5′ exonuclease activity. After the mistake is removed, DNA polymerase resumes DNA synthesis.

16. Topoisomerase, helicase, and primase.

18. It is necessary to have the correct number of chromosomes per cell. If DNA replication is too slow, daughter cells may not receive any chromosomes. If it is too fast, they may receive too many chromosomes.

20. The opposite strand is made in the conventional way by DNA polymerase using the telomerase-added strand as a template.

Experimental Questions

2. All of the DNA double helices would be 1/8 heavy.

4. You would need to add a primer, dNTPs, and DNA polymerase. If the DNA was double-stranded, you would also need helicase, single-stranded binding protein, and perhaps topoisomerase.

6. Adding double-stranded DNA would be appropriate, since telomerase adds segments of DNA to the ends of double-stranded chromosomes. You would also have to add telomerase and dNTPs.

Questions for Student Discussion/Collaboration

2. Basically, the idea is to only add certain combinations of enzymes and see what happens. For example, you could add helicase to double-stranded DNA and then view it by microscopy. Perhaps you could see regions where the DNA has become single-stranded. You could add helicase and primase to double-stranded DNA along with radiolabeled ribonucleotides. Under these conditions, you would make short (radiolabeled) primers, and if you added DNA polymerase, you would make DNA strands. Alternatively, if you added DNA polymerase plus an RNA primer, you would not need to add primase.

CHAPTER 12
Conceptual Questions

2. I would basically agree. The primary role of the DNA is to provide a blueprint for the activities of the cells. In general, the DNA is "acted on" by proteins to make proteins. Likewise, proteins are needed to replicate DNA so that it can be passed from parent to offspring and from cell to cell. Proteins carry out the biosynthetic, structural, and metabolic activities of the cell.

4. The word transcription means the act of making a copy. During DNA transcription, a nucleotide language in DNA is copied into a nucleotide language of RNA. The word translation means the act of converting one series of words or symbols into another. For example, one could translate English into French. During RNA translation, the nucleotide language in mRNA is converted into an amino acid language in a polypeptide.

6. (A) You would not get any translation, because there would not be any transcript.

   (B) It may have no effect.

   (C) Translation would be inhibited.

   (D) Translation would be altered. In some cases, the next start codon downstream from the original one would be used.

   (E) The polypeptide would be longer than normal.

   (F) This would alter when and/or how much transcription and translation takes place.

8. To specify the first amino acid in a polypeptide. This amino acid is typically formylmethionine in prokaryotes and methionine in eukaryotes (see Chapter 14).

10. It has not been altered. This is why DNA can store information indefinitely.

12. If one of the introns was not spliced out, it would remain in the mRNA, and the codon sequence within the intron would be translated. The codon sequence within the intron would be inappropriate and might even contain a stop codon. It is likely that the translated protein would not function correctly.

14. An anticodon is found in the second stem-loop of tRNA molecules. Its sequence is complementary to a codon sequence in mRNA. At its 3′ end, the tRNA has an amino acid attached that corresponds to the codon in the mRNA that will be recognized. For example, if the codon is 5′–UAC–3′, the anticodon in the tRNA will be 3′–AUG–5′ and will have a tyrosine attached to it. The function of the tRNA is to carry the correct amino acid to the ribosome, based on the codon sequence in the mRNA.

16. It means that more than one codon can specify the same amino acid. For example, CUU and CUC both specify the amino acid leucine.

18. The three-dimensional structure of a polypeptide is determined by its amino acid sequence. The amino acid sequence is ultimately determined by the gene sequence. As a polypeptide is being synthesized, certain regions of the amino acids initially fold into secondary structures such as an α-helix or a β-sheet. These secondary structures, which are connected by random coiled regions, then fold together with each other to form the tertiary structure of the polypeptide. Disulfide bond formation can also affect structure.

Experimental Questions

2. The purpose of using X-rays or UV light is to increase the frequency of mutations in particular genes. These agents will alter the sequence of the DNA and thereby alter the properties of the encoded proteins. In some cases, these alterations may cause the protein to function improperly or to be made in negligible amounts.

4. When they obtained mutant strains that could not grow on minimal media, Beadle and Tatum then checked to determine the nature of the defect. In all cases, the strains were defective only in a requirement for a single vitamin, not many different vitamins. They reasoned that each mutation was probably disrupting a gene encoding an enzyme that was necessary for the synthesis of the required vitamin.

6. RNA has a fairly short half-life. If you treat the cell extract with DNase, the DNA is digested, so that new RNA cannot be made by transcription. After a short time, all the preexisting RNA will be degraded. At that point, when new RNA is made synthetically using polynucleotide phosphorylase, the added synthetic RNA will be the only RNA in the extract.

Questions for Student Discussion/Collaboration

2. Students, particularly those with a strong cell biology background, should be able to think of examples where cell structure/function is altered to affect phenotype. They need to be aware that visible characteristics are due to the traits of cells, even though we cannot see individual cells with our naked eyes.

CHAPTER 13
Conceptual Questions

2. A consensus sequence is the most common version of a nucleotide sequence with a particular functional role. An example is the −35 and −10 consensus sequences that are found in bacterial promoters. At −35, it is TTGACA, but it can differ in one or two nucleotides and still function efficiently as a promoter. The −35 site is primarily for the recognition by the sigma factor. The −10 site, also known as the Pribnow box, is the site where the DNA will begin to unwind to allow transcription to occur.

4. It will not affect transcription. However, it will shorten the length of the encoded polypeptide during translation.

6. Sigma factor can slide along the major groove of the DNA. In this way, it can recognize base sequences that are exposed in the groove. Hydrogen bonding between the bases and the sigma factor protein can promote a tight and specific interaction.

8. G in DNA to C in RNA
C in DNA to G in RNA
A in DNA to U in RNA
T in DNA to A in RNA

The template strand is 3′–CCGTACGTAATGCCGTAGTGTGATCCCTAG–5′, and the coding strand is 5′–GGCATGCATTACGGCATCACACTAGGGATC–3′.

10. Transcriptional termination occurs when the hydrogen bonding is broken between the DNA and the part of the newly made RNA transcript in the open complex.

12. (A) Ribosomal RNA (5.8S, 18S, and 28S).

    (B) All of the mRNA and a few other genes.

    (C) All of the tRNAs and the 5S rRNA.

14. (A) RNA polymerase would not be bound to the TATA box.

    (B) TFIID is the TATA-binding protein. If it were missing, RNA polymerase would not bind to the TATA box either.

    (C) The formation of the open complex would not take place.

16. If there are no other splice donor sites within the second intron, the mature mRNA would only have intron 1 spliced out.

18. As described in the chapter, the spliceosome is composed of multiple protein subunits and some RNAs. The function of the spliceosome is to cut the RNA in two places, hold the ends together, and then catalyze cova-

lent bond formation between the two ends. The role of snRNAs during this process could be the following:(1) It could be involved in binding the pre-mRNA, much like the role of rRNA in the binding of mRNA to the ribosome. (2) It could be catalytic, like the RNA in RNase P.

20. A gene is colinear when the sequence of bases in the sense strand of the DNA corresponds to the sequence of bases in the mRNA. Most bacterial genes and many eukaryotic genes are colinear. Therefore, you can look at the gene sequence in the DNA and predict the amino acid sequence in the polypeptide. Many eukaryotic genes, however, are not colinear. They contain introns that are spliced out of the pre-mRNA. An example given is this chapter is the gene that encodes α-tropomyosin.

22. In eukaryotes, pre-mRNA may be trimmed (at the 5′ and 3′ ends), capped, tailed, spliced, and then exported out of the nucleus.

24. Alternative splicing occurs when exons are spliced out or alternative splice sites are used at intron–exon boundaries. The biological significance is that two or more polypeptide sequences can be derived from a single gene. This is a more efficient use of the genetic material. In multicellular organisms, alternative splicing is often used in a tissue-specific or developmentally specific manner.

Experimental Questions

2. It is a loop of DNA that is formed by its interaction with a segment of RNA. While the RNA is hydrogen bonding to one of the DNA strands, the other DNA strand does not have a partner to hydrogen bond with, and so it bubbles out as a loop.

4.

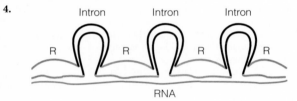

Questions for Student Discussion/Collaboration

2. RNA transcripts come in two basic types:those that function as RNA (e.g., tRNA, rRNA), and those that are translated (namely, mRNA). As described in this chapter, they play a variety of functional roles. RNAs that form complexes with proteins carry out some interesting roles. In some cases, the role is to bind other types of RNA molecules. For example, rRNA in bacteria plays a role in binding mRNA. In other cases, the RNA plays a catalytic role. An example is RNase P.

CHAPTER 14
Conceptual Questions

2. The assembly process is very complex at the molecular level. In eukaryotes, 33 proteins and 1 rRNA assemble to form a 40S subunit, and 49 proteins and 3 rRNAs assemble to form a 60S subunit. This assembly occurs within the nucleolus.

4. 3′–CUU–5′ or 3′–CUC–5′.

6. It can recognize 5′–GGU–3′, 5′–GGC–3′, and 5′–GGA–3′. All of these specify glycine.

8. All tRNA molecules have some basic features in common. They all have a cloverleaf structure with three stem-loop structures. The second stem-loop contains the anticodon sequence that recognizes the codon sequence in mRNA. At the 3′ end, there is an acceptor site, with the sequence CCA, that serves as an attachment site for an amino acid. Most tRNAs also have base modifications that occur within their nucleotide sequences.

10. The role of aminoacyl-tRNA synthetase enzymes is to specifically recognize tRNA molecules and attach the correct amino acid to them. They are sometimes described as the second genetic code, because

the specificity of their attachment is a critical step in deciphering the genetic code. For example, if a tRNA has a 3′–GGG–5′ anticodon, it will recognize a 5′–CCC–3′ codon, which should specify proline. It is essential that the prolyl-tRNA-synthetase recognizes this tRNA and attaches proline to the 3′ end. The other aminoacyl-tRNA synthetases should not recognize this tRNA.

12. Bases that have been chemically modified can occur at various locations throughout the tRNA molecule. The significance of all of these modifications is not entirely known. However, within the anticodon region, base modification alters base pairing to allow the anticodon to recognize two or more different bases within the codon.

14. No, it is not. Due to the wobble rules, the 5′ base in a tRNA can sometimes recognize two or more bases in the third (3′) position of the mRNA. Therefore, any given cell type synthesizes far fewer than 61 types of tRNAs.

16. Most bacterial mRNAs contain a Shine–Dalgarno sequence, which is necessary for the binding of the mRNA to the small ribosomal subunit. This sequence, AGGAGGU, is complementary to a sequence in the 16S rRNA. Due to this complementarity, these sequences will hydrogen bond to each other during the initiation stage of translation.

18. The ribosome binds at the 5′ end of the mRNA and then scans in the 3′ direction in search of an AUG start codon. If it finds one that reasonably obeys Kozak's rules, it will begin translation at that site. Aside from an AUG start codon, the two other features are a purine at the −3 position and a guanosine at the +4 position.

20. The A site is the acceptor site. It is the location where a tRNA initially "floats in" and recognizes a codon in the mRNA. The only exception is the initiator tRNA, which binds to the P site. The P site is the next location where the tRNA moves. When it first moves to the P site, it carries with it the polypeptide chain. In each round of elongation, the polypeptide chain is transferred from the tRNA in the P site to the amino acid attached to the tRNA in the A site. The third site is the E site. During translocation, the empty tRNA in the P site is transferred to the E site. It exits or is released from this site.

22. Sorting signals provide an address that sends the protein to the correct location (i.e., compartment) within the cell. Proteins destined for the ER, Golgi, lysosomes, plasma membrane, or secretion have an SRP sorting signal at their amino terminal end. Nuclear proteins have an NLS sequence, etc. These sorting signals are recognized by cellular proteins/complexes that then traffic the proteins to their correct destination.

24. Cotranslational sorting occurs via the SRP signal. This form of sorting is necessary for proteins that are destined for the ER, Golgi, lysosomes, plasma membrane, or secretion. The SRP recognizes the signal as the protein is being translated. It then directs the ribosome to proteins within the ER membrane, so that the protein is synthesized into the ER. By comparison, posttranslational sorting occurs after the protein has been completely made. In this case, the protein contains an amino acid sequence that traffics it to a particular cellular organelle. Examples of these types of sorting signals include mitochondrial, chloroplast, and nuclear sorting signals.

Experimental Questions

2. There could have been other choices, but this template would be predicted to contain a cysteine codon, UGU, but would not contain any alanine codons.

4. To solve this problem, the first thing you need to do is to determine which codons would be present in the mRNA, and in what amounts. If there is 50% C and 50% A, a random polymer would contain 12.5% each of the following codons:CCC (proline), CCA (proline), CAC (histidine), CAA (glutamine), AAC (asparagine), AAA (lysine), ACC (threonine), and ACA (threonine). This yields 25% proline, 12.5% histidine, 12.5% glutamine, 12.5% asparagine, 12.5% lysine, and 25% threonine. In the resulting polypeptides, these percentages match, except that we have 25% serine instead of 25% threonine. Therefore, serine has been

substituted for threonine. Based on their structures, a demethylation of threonine has occurred. In other words, the methyl group on threonine has been replaced with a hydrogen, yielding a serine residue.

6. The antibodies could be used to determine where specific proteins are located. For example, one could add a radiolabeled antibody to a purified ribosome preparation and then make an autoradiograph that reveals where the antibody is binding. When viewed under the EM, one would observe dark spots that correspond to where the antibody is bound. Since the antibody is binding specifically to a particular protein, the dark spots would also reveal where that protein is located within the ribosomal complex.

8. The idea here is that the antibody looks like a round bump where it binds to a particular protein. The EM pictures would look the same as described in the chapter, except a round bump would be found (close to the DNA), corresponding to where RNA polymerase is transcribing the template strand.

Questions for Student Discussion/Collaboration

2. This could be a very long list. There are similarities along several lines: (1) There is a lot of molecular recognition going on, either between two nucleic acid molecules or between proteins and nucleic acid molecules. Students may see these as similarities or differences, depending on their point of view. (2) There is biosynthesis taking place in both processes. Small building blocks are being connected together. This requires an input of energy. (3) There are genetic signals that determine the beginning and ending of these processes.

There are also many differences: (1) Transcription produces a molecule with a similar structure to the DNA, whereas translation produces a polypeptide with a very different structure. (2) Depending on your point of view, it seems that translation is more biochemically complex, requiring more proteins and RNA molecules to accomplish the task.

CHAPTER 15

Conceptual Questions

2. In bacteria, gene regulation greatly enhances the efficiency of cell growth. It takes a lot of energy to transcribe and translate genes. Therefore, a cell is much more efficient and better at competing in its environment if it only expresses genes when the gene product is needed. For example, a bacterium will only express the genes needed for lactose metabolism when a bacterium is exposed to lactose. When the environment is missing lactose, these genes are transcriptionally inactive. Similarly, when tryptophan levels are high within the cytoplasm, the genes required for tryptophan biosynthesis are repressed.

4. (A) No transcription would take place. The *lac* operon could not be expressed.

   (B) No regulation would take place. The operon would be continuously turned on.

   (C) The operon would function normally, except that none of the transacetylase would be made.

6. In this case, an activator protein and inhibitor molecule are involved. The binding of the inhibitor molecule to the activator protein would prevent it from binding to the DNA and thereby inhibit its ability to activate transcription.

8. The term enzyme adaptation means that particular enzymes are only made when a cell is exposed to the substrate for that enzyme. This occurs because the genes that encode the enzymes involved in the metabolism of the substrate are only expressed when the cells have been exposed to the substrate.

10. It would be impossible to turn the *lac* operon on even in the presence of lactose, because the repressor protein would remain bound to the operator site.

12. AraC protein binds to the *araI*, *araO$_1$*, and *araO$_2$* operator sites. The binding of the AraC protein to the *araO$_1$* and *araO$_2$* sites activates the transcription of the *araC* gene. In contrast, the AraC proteins bound at *araO$_2$* and *araI* repress the arabinose operon. The AraC proteins at *araO$_2$* and *araI* can bind to each other by causing a loop in the DNA. This DNA loop prevents RNA polymerase from transcribing the *ara* (arabinose) operon. In absence of arabinose, the arabinose operon is turned off. When cells are exposed to arabinose, however, arabinose binds to the AraC protein and breaks the interaction between the AraC proteins at the *araO$_2$* and *araI* sites, thereby breaking the DNA loop. In addition, a second AraC protein binds at the *araI* site. This AraC dimer at the *araI* operator site activates transcription.

14. (A) Attenuation will occur because a 3–4 loop will form.

    (B) Attenuation will occur because 2–3 cannot form, so that 3–4 will form.

    (C) Attenuation will not occur, because 3–4 cannot form.

    (D) Attenuation will not occur because 3–4 cannot form.

16. A translation repressor protein is one that inhibits translation, usually by binding directly to mRNA and preventing it from being translated. It may prevent translation by causing the mRNA to form a secondary structure (e.g., a stem-loop) that cannot be recognized by the ribosome, or it may sterically block the ribosomal initiation site (i.e., the Shine–Dalgarno sequence).

18. The term feedback inhibition refers to the phenomenon whereby the product of a enzymatic pathway binds to an enzyme (usually at the beginning of the pathway) and inhibits its activity. This acts as a feedback control that indicates the cell has made enough of the product. The product of an enzymatic pathway may also bind to genetic regulatory proteins and thereby inhibit the expression of genes that encode the relevant enzymes.

20. (1) Phosphorylation:Attachment of a phosphate group. (2) Methylation: Attachment of a methyl group. (3) Glycosylation: Attachment of a carbohydrate.

    A reversible modification can modulate protein function if the activity of the protein is affected by the modification. For example, phosphorylation of a protein may activate it, and the subsequent dephosphorylation will inactivate it.

22. In the lytic cycle, the virus directs the bacterial cell to make more virus particles, until eventually the cell lyses and releases them. In the lysogenic cycle, the viral genome is incorporated into the host cell's genome as a prophage. It remains there in a dormant state, until some stimulus causes it to switch to the lytic cycle.

24. The λ repressor acts as a repressor by binding to P_R and P_L. Binding to P_R inhibits transcription of *cro* and thereby favors the lysogenic cycle. The cro protein binds to P_L and P_{RM}. Binding to P_{RM} inhibits transcription of the λ repressor and thereby favors the lytic cycle.

26. It would act to increase the amount of cro protein, so that the lytic cycle would be favored.

Experimental Questions

2. You could mate a strain that has a normal *lac* operon on an F′ factor to this mutant strain. Since the mutation is in the operator site, you will continue to get expression of β-galactosidase, even in the absence of lactose.

4. In this case, things are more complex, because araC acts as a repressor and an activator protein. If araC was missing due to mutation, there would not be repression or activation of the *ara* operon in the presence or absence of arabinose. It would be expressed constitutively at low levels. The introduction of a normal *araC* gene into the bacterium on an F′ factor would restore normal regulation (i.e., a trans effect).

6. The results suggest that there is a mutation in the araC protein, so that it cannot bind arabinose although it still binds to the operator sites correctly.

Questions for Student Discussion/Collaboration

2. As mentioned in the answer to Experimental Question 5, the bend may expose the base sequence that the sigma factor of RNA polymerase is recognizing. The bend may open up the major groove, so that this base sequence is more accessible to the binding by sigma factor and RNA polymerase.

CHAPTER 16
Conceptual Questions

2. DNA amplification is the phenomenon in which a particular segment of DNA, such as a particular gene, is copied to a much higher number than the rest of the genome. Examples include rRNA genes during oogenesis, and genes, such as the gene encoding dihydrofolate reductase, that provide resistance to drugs such as methotrexate.

4. Perhaps this methylase is supposed to methylate certain genes in muscle cells so that they will be inactive. The mutation may have caused a defect in the methylase causing the muscle genes to remain active.

6. The 30 nm fiber is the predominant form of chromatin during interphase. The chromatin must be converted from a closed conformation to an open conformation for transcription to take place. This opening process involves less packing of the chromatin and may involve changes in the locations of histone proteins.

8. The position of nucleosomes can affect transcription, because they may affect the ability of transcription factors and RNA polymerase to bind to response elements or promoter sequences, and/or to initiate transcription. In some cases, the nucleosome structure will inhibit these processes because the DNA is tightly wound around histone proteins, making it more difficult for other proteins to bind to the DNA and unwind it for transcription.

10. The translocation breakpoint occurred between the β-globin gene and the locus control region. Therefore, the β-globin gene is not expressed, because it needs the locus control region to allow the chromatin to be in an accessible conformation.

12. A regulatory transcription factor can bind specifically to base sequences that are exposed within the major and minor grooves of the DNA. Amino acid side chains within the transcription factor proteins hydrogen bond and form a perfect fit with bases within the grooves of DNA. This specificity is necessary so that regulatory transcription factors will only regulate the particular genes that contain the specific DNA sequences recognized by the transcription factor.

14. Transcription factor modulation refers to the different ways that the function of transcription factors can be regulated. The three general ways are the binding of an effector molecule, protein–protein interactions, and posttranslational modifications.

16. For the glucocorticoid receptor to bind to a GRE, the cell must be exposed to the hormone and it must enter the cell. It binds to the receptor, which releases hsp98. After hsp98 is released, the receptor dimerizes and enters the nucleus. Once inside the nucleus, it will recognize a GRE and bind to it, thereby leading to the activation of specific genes that have GREs next to them.

18. Phosphorylation of the CREB protein causes it to act as a transactivator. The unphosphorylated CREB protein can still bind to CREs, but it does not stimulate transcription.

20. The function of SR proteins is to promote the selection of splice sites in RNA. In certain tissues, the concentration of particular SR proteins is higher than in other tissues. The high concentration of an SR protein promotes the selection of particular splice sites and thereby leads to tissue-specific splicing.

22. If mRNA stability is low, this means that it is degraded more rapidly. Therefore, low stability results in a low mRNA concentration. The length of the polyA tail is one factor that affects stability: a longer tail makes mRNA more stable. Certain mRNAs have particular sequences that affect their half-lives. For example, AREs are found in many short-lived mRNAs. The AREs are recognized by cellular proteins that cause the mRNAs to be degraded rapidly.

Experimental Questions

2. A DNaseI hypersensitive site is one that is very easily cleaved by DNaseI as compared with most other sites in the chromosomal DNA. In other words, its conformation is very accessible to cleavage.

4. The genes encoding the following proteins may have been rendered defective by mutation: the adrenalin receptor, G-protein, adenylate cyclase, protein kinase A, or CREB protein. We could do a Western blot using antibodies against these proteins to determine if they are being made. We could use ^{32}P-labeled ATP and see if radiolabeled cAMP is being made, or if CREB is being phosphorylated by protein kinase A. To determine if the adrenalin receptor is defective, we could try to activate adenylate cyclase with a different hormone (that binds to a different receptor). If adenylate cyclase can be activated by the other hormone but is not activated by adrenalin, we would suspect a defect in the adrenalin receptor.

6. Certain individuals had defects in globin gene expression even though the structural gene itself was intact. This led researchers to suspect that other regions nearby were important in controlling the chromosomal conformation so as to make the globin genes accessible to RNA polymerase and transcription factors.

Questions for Student Discussion/Collaboration

2. Probably the most efficient method would be to systematically make deletions of progressively smaller sizes. For example, we could begin by deleting 20,000 base pairs on either side of the gene and see if that affects transcription. If we found that only the deletion on the 5′ end of the gene had an effect, we could then start making deletions in from the 5′ end, perhaps in 10,000 or 5,000 base pair increments, until we localized response elements. We would then make smaller deletions in the putative region, until it was down to a hundred or a few dozen nucleotides. At this point, we might conduct site-directed mutagenesis, as described in Chapter 19, to specifically identify the response element sequence. Also, you may wish to consider the possibility that there may be more than one enhancer, each of which promotes an incremental increase in the transcription rate.

CHAPTER 17
Conceptual Questions

2. A gene mutation is a relatively small mutation that is localized to a particular gene. A chromosome mutation is a large enough change in the genetic material so that it can be seen with a light microscope. Genome mutations are changes in chromosome number.

4. It is a gene mutation, a point mutation, a base substitution, a transition mutation, a forward mutation, a deleterious mutation, a mutant allele, a nonsense mutation, a conditional mutation, and a temperature-sensitive (lethal) mutation.

6. **(A)** It would probably inhibit protein function, particularly if it was not near the end of the coding sequence. Shorter proteins are usually degraded rapidly.

(B) It may or may not affect protein function, depending on the nature of the amino acid substitution and whether the substitution is in a critical region of the protein.

(C) It would increase the amount of functional protein.

(D) It may affect protein function if the alteration in splicing adds an exon to the mRNA that results in a protein with a perturbed structure.

8. **(A)** Causes G → A and T → C mutations, which are transition mutations.

(B) 5-Bromouracil causes G → A mutations, which are transitions.

(C) Proflavin causes small additions or deletions, which may result in frameshift mutations.

10. If a nonsense suppressor (e.g., amber suppressor) was very efficient, many of the proteins in the cell (those with the type of stop codon that the suppressor recognized, such as an amber stop codon) would be too long, because there would be readthrough of their normal stop codons until another stop codon (ocher or opal) was reached. The abnormally long proteins may not function properly.

12. (A) Silent, because the same amino acid (glycine) is encoded by GGA and GGT.

(B) Missense, because a different amino acid is encoded by CGA than by GGA.

(C) Missense, because a different amino acid is encoded by GTT than by GAT.

(D) Frameshift, because an extra base is inserted into the sequence.

14. A spontaneous mutation originates from within a living cell. It may be due to errors in DNA replication or products of normal metabolism that may alter the structure of DNA. The causes of induced mutations originate from outside of the cell. They may be physical agents such as UV light or X-rays, or chemicals that acts as mutagens. Both spontaneous and induced mutations may cause a harmful phenotype such as a cancer. In many cases, induced mutations are avoidable if the individual can prevent their exposure to the environmental agent that acts as a mutagen.

16. Nitrous acid can change a cytosine to uracil. An alkyltransferase could directly repair this defect. Excision repair systems could also remove the defect and replace it with the correct base.

18. One possibility is that a translocation may move a gene next to a heterochromatic region of another chromosome and thereby diminish its expression; or it could be moved next to a euchromatic region and increase its expression. Another possibility is that the translocation breakpoint may move the gene next to a new promoter or regulatory sequences that may then influence the gene's expression.

20. Recombinational repair occurs when the DNA is being replicated. If a mutation occurred in exactly the same location in both of the original DNA strands, recombinational repair could not work, because it relies on the use of one original DNA strand as a template to repair the damaged DNA strand.

22. In E. coli, the TRCF recognizes when RNA polymerase is stalled on the DNA. This stalling may be due to a mutation such as a thymine dimer. The TRCF removes RNA polymerase and recruits the excision DNA repair system to the region in a way that causes it to repair the template strand of DNA.

Experimental Questions

2. Absence of mutagen: $17/10,000,000 = 17 \times 10^{-7} = 1.7 \times 10^{-6}$.
   Presence of mutagen: $2017/10,000,000 = 2.0 \times 10^{-4}$.

   The mutagen increases the frequency of mutation more than 100-fold; actually; it is $2017/17 = 118$-fold higher.

4. Müller's experiment is not measuring the mutation rate within a single gene. There are many genes along the X-chromosome that may mutate to produce a lethal phenotype. Müller's experiment is a measure of the rate at which X-linked mutations in many different genes can occur to produce a lethal phenotype.

6. If a crossover occurred during gametogenesis in the female offspring, it could happen that two different X-chromosomes, each one carrying a different lethal mutation, could recombine to produce an X-chromosome without any lethal mutation and an X-chromosome with two mutations. The chromosome without any lethal mutations could produce viable male offspring. Therefore, crossing over would mask

the effects of the mutagen, so that you would obtain fewer crosses where a female fly was unable to produce male offspring.

8. The results suggest that the strain is defective in excision repair. If we compare the normal and mutant strains that have been incubated for two hours at 37°C, much of the radioactivity in the normal strain has been transferred to the supernatant because it has been excised. In the mutant strain, however, no significant radioactivity has been transferred to the supernatant (i.e., compare 940 cpm for the wild-type with less than 40 cpm for the mutant), suggesting that it cannot remove thymine dimers. That would explain why it is sensitive to UV, which causes thymine dimers.

Questions for Student Discussion/Collaboration

2. The worst time to be exposed to mutagens would be during very early stages of embryonic development. An early embryo is most sensitive to mutation, because a mutation will affect a large region of the body. Adults must also worry about mutagens for several reasons. Mutations in somatic cells can cause cancer, a topic to be discussed in Chapter 23. Also, adults should be careful to avoid mutagens that may affect the ovaries or testes, since the resulting mutations could be passed along to offspring.

CHAPTER 18

Conceptual Questions

2. Usually, the overall net effect is not to create any new mutations in particular genes. However, homologous recombination does rearrange the combinations of alleles along particular chromosomes. This can be viewed as a mutation, since the sequence of a chromosome has been altered.

4. It depends on which way the breaks occur in the DNA strands. If the two breaks occur in the crossed DNA strands (see steps 6a to 7a, Figure 18-3), nonrecombinant chromosomes result. If the two breaks occur in the uncrossed strands (steps 6b to 7b), the result is a recombinant chromosome.

6. Holliday: proposes two breaks, one in each chromatid, and then both strands exchange a single strand of DNA. Radding: proposes a single break; one strand in the broken chromosome initially migrates and replaces a strand in the homologue. Double-stranded break model: proposes two breaks, both in the same chromatid; as in the Holliday model, single-strand migration occurs between both homologues.

8. The RecA protein binds to single-stranded DNA and forms a filament. The formation of the filament promotes the sliding of the filament along another DNA region until it recognizes homologous sequences. This recognition process may involve the formation of triplex DNA. After recognition has occurred, RecA protein mediates the movement of the invading DNA strand and the displacement of the original strand.

10. RecG, RuvAB, and RuvC bind to Holliday junctions. Based on this observation, they do not necessarily recognize a DNA sequence, but instead could merely recognize a region of crossing over at the molecular level. Although not discussed in this chapter, RuvC actually does recognize a specific base sequence for cleavage.

12. The function of the recombinase is to splice together the V and J sections of the light-chain genes and the V, J, and D sections of the heavy-chain genes. This is DNA splicing. Segments of the genes are cut out and then ligated together. This creates different combinations of the V, J, (D), and constant regions, thereby creating a large amount of diversity in the encoded antibodies.

14. The ends of the element would be flanked by short (a few nucleotides) direct repeats. This is a universal characteristic of all transposable elements. In addition, many elements contain inverted repeats or LTRs that are involved in the transposition process. In transposons, one

might also look for the presence of a transposase gene, although this is not an absolute requirement since the transposase gene is missing in incomplete elements. In retroviral elements, one might also locate a reverse transcriptase gene, but this also may not be present.

16. The drawing should parallel Figure 18-13, but with added detail regarding a composite transposon. Transposase should cut out the transposon by cutting adjacent to the inverted repeats (i.e., between the inverted and direct repeats). It makes staggered cuts in the host DNA. After insertion and gap repair synthesis, the new site will have new direct repeats.

18. Retroelements have the greatest potential for proliferation, because the element is transcribed into RNA as an intermediate. Many copies of this RNA could be transcribed and then copied into DNA by reverse transcriptase. Theoretically, many copies of the element could be inserted into the genome in this way.

Experimental Questions

2. I would conclude that the substance is a mutagen. Substances that damage DNA tend to increase the level of genetic exchange such as sister chromatid exchange.

4. When she started with a colorless strain containing *Ds*, she identified 20 cases where *Ds* had moved to a new location to produce red kernels. This identification was possible because the 20 strains had a higher frequency of chromosomal breaks at a specific site, and because of the mutability of particular genes. She also had found a strain where *Ds* had inserted into the red-color-producing gene, so that its transposition out of the gene would produce a red phenotype. Overall, her analysis of the data showed that the sectoring (i.e., mutability) phenotype was consistent with the transposition of *Ds*.

6. Transposon tagging is an experimental method that is aimed at cloning genes. In this approach, a transposon is introduced into a strain, and the experimenter tries to identify individuals in which the gene of interest has lost its function. In many cases, the loss of function has occurred because the transposon has been inserted into the gene. If so, one can make a library and then use a labeled transposon to identify clones in which the transposon has been inserted. Not all clones will contain the gene of interest, because the transposon may be inserted at multiple sites within the genome. Nevertheless, some clones may involve an insertion of the transposon into the gene of interest and thereby provide a way to clone this particular gene. Later, one would use the transposon-inserted clone to screen a library from an individual with an active version of the gene. In this way, one could then identify a normal copy (non-transposon-inserted) of the gene.

Questions for Student Discussion/Collaboration

2. *Beneficial consequences:* You would not get (as many) translocations, inversions, and the accumulation of selfish DNA.

   Harmful consequences: The level of genetic diversity would be decreased, because linked combinations of alleles would not be able to recombine. You would not be able to produce antibody diversity in the same way. Gene duplication could not occur, and so the evolution of new genes would be greatly inhibited.

CHAPTER 19

Conceptual Questions

2. A restriction enzyme recognizes a DNA sequence and then cleaves the (covalent) phosphoester bond in each of two DNA strands.

4. GGGCCCATATATATGGGCCC
   CCCGGGTATATATACCCGGG

6. A dideoxynucleotide is missing the 3′–OH group. When the 5′ end of a dideoxynucleotide is added to a growing strand of DNA, another phosphoester bond cannot be formed at the 3′ position. Therefore, the

dideoxynucleotide terminates any further addition of nucleotides to the (previously growing) strand of DNA.

Experimental Questions

2. All vectors must have the ability to replicate when introduced into a living cell. This ability is due to a DNA sequence known as an origin of replication. Modern vectors also contain convenient restriction sites for the insertion of DNA fragments. These vectors also contain selectable markers, which are genes that confer some selectable advantage for the host cell that carries them. The most common selectable markers are antibiotic resistance genes, which confer resistance to antibiotics that normally would inhibit the growth of the host cell.

4. First, the chromosomal DNA that contains the source of the gene that you want to clone must be isolated from a cell (tissue) sample. A vector must also be obtained. The vector and chromosomal DNA are digested with the same restriction enzyme. They are mixed together to allow the sticky ends of the DNA fragments to bind to each other, hopefully to create a hybrid vector. DNA ligase is then added to promote covalent bonds. The DNA is then transformed or transfected into a living cell. Only host cells that have taken up the vector will survive, because the vector has a selectable marker such as an antibiotic resistance gene. For a plasmid vector, it will replicate and the cells will divide to produce a colony of cells that contain "cloned" DNA pieces. To identify colonies that contain the gene you wish to clone, you must use a probe that will identify specifically a colony containing the correct hybrid vector. A probe may be a DNA probe that is complementary to the gene you want to clone; or it could be an antibody that recognizes the protein that is encoded by the gene.

6. A DNA library is a collection of hybrid vectors that contain different pieces of DNA from a source of chromosomal DNA. Since it is a diverse collection of many different DNA pieces, the name library seems appropriate.

8. Conditions of high stringency require a close match in the complementarity between the probe and unlabeled DNA fragment. Conditions of low stringency will allow a lesser degree of complementarity and thereby may identify homologous genes within a gene family.

10. The purpose of gel electrophoresis is to separate the many DNA fragments, RNA molecules, or proteins that were obtained from the sample you want to probe. This separation is based on molecular weight and allows you to identify the molecular weight of the DNA fragment, RNA molecule, or protein that is being recognized by the probe.

14. You would first make a DNA library using chromosomal DNA from pig cells. You would then radiolabel the human β-globin gene and use it as a probe in a colony hybridization experiment; each colony would contain a different cloned piece of the pig genome. You would identify "hot" colonies that hybridize to the human β-globin probe. You would then go back to the master plate and pick these hot colonies, and grow them in a test tube. You would then isolate the plasmid DNA from the hot colonies and subject the DNA to sequencing. By comparing the DNA sequences of the human β-globin gene and the putative clones, you could determine if the putative clones were homologous to the human clone and likely to be the pig homologue of the β-globin gene.

16. In this case, the transcription factor binds to the response element when the hormone is present. Therefore, the hormone promotes binding of the transcription factor to the DNA and thereby promotes transactivation.

18. The rationale behind a gel retardation assay is that a segment of DNA with a protein bound to it will migrate more slowly through a gel than will the same DNA without any bound protein. A shift in a DNA band to a higher molecular weight provides a way to identify DNA-binding proteins.

20. The rationale behind a footprinting experiment has to do with accessibility. If a protein is bound to the DNA, it will cover up the part of

the DNA where it is bound. This region of the DNA will be inaccessible to the actions of chemicals or enzymes that cleave the DNA, such as DNaseI.

22. Histones, scaffolding proteins, transcription factors, and so forth.

24. The two approaches are similar in that they used DNaseI as a probe to detect the presence of proteins that interact with DNA. In a footprinting experiment, the DNA is labeled at one end, so that you figure out the region of DNA that is protected from that end. This can be done by comparing the digestion patterns between protected and unprotected DNA samples. In Experiment 10B, the goal is to look for a pattern within the DNA structure that is due to the binding of histone proteins at regular intervals. In this case, the sizes of the DNA fragments reveal information about the repeating nucleosome structure.

26. To determine the sequence in the 5′ to 3′ direction, you start at the bottom and read the bands in each lane: CTAGCAAGGGAATTG-TACGATAGATTT.

28. The muscle cells are making a large amount of the myosin protein, while only a small amount is made in red blood cells and spleen cells. The different levels of expression could be due to a variety of factors, including the regulation of transcription, translation, or protein degradation.

30. There are lots of different strategies one could follow. For example, you could mutate every other base and see what happens. It would be best to make very nonconservative mutations, such as a purine for a pyrimidine or a pyrimidine for a purine. If the mutation prevents protein binding in a gel retardation assay, then the mutation is probably within the response element. If the mutation has no effect on protein binding, it probably is outside the response element.

32. Change the GGC codon to GAA or GAG.

Primers could be   3′–TACGGGCTACTTTAGCTAAA–5′
        or   3′–TACGGGCTACTCTAGCTAAA–5′

Questions for Student Discussion/Collaboration

2. (1) Does a particular amino acid within a protein sequence play a critical role in the protein's structure or function? (2) Does a DNA sequence function as a promoter? (3) Does a DNA sequence function as a regulatory site? (4) Does a DNA sequence function as a splicing junction? (5) Is a sequence important for correct translation? (6) Is a sequence important for RNA stability? *And many others.*

CHAPTER 20

Conceptual and Experimental Questions

2. One would conclude that she has a deletion of the gene that the probe recognizes. To clone this gene, one could begin with a marker that is known to be near band p11 and walk in either direction. This walking experiment would be done on the DNA from a normal person and compared with the DNA from the person described in the problem. At some point, the walk would yield a clone that contained a deletion in the abnormal person, but the DNA would be present in a normal person. This DNA fragment in the normal person should also hybridize to the probe.

4. **(A)** Chromosome 7.

   (B) If the cell line was from a homozygote that did not produce any of the protein, there would be no protein product produced by any of the hybrid. If the cell line was from a heterozygote, all hybrids that produced the protein would contain chromosome 7, but some hybrids would contain the chromosome 7 that carried the faulty CF gene and would therefore not produce the protein.

6. A contig is a collection of clones that contain overlapping segments of DNA that span a particular region of a chromosome. To determine if two clones are overlapping, one could conduct a Southern blotting experiment. In this approach, one of the clones is used as a probe. If

it is overlapping with the second clone, it will bind to it in a Southern blot. Therefore, the second clone is run on a gel and the first clone is used as a probe. If the band corresponding to the second clone is labeled, this means that the two clones are overlapping.

8. A polymorphism refers to genetic variation at a particular locus within a population. If the polymorphism occurs within gene sequences, this is allelic variation. Genetic markers that do not encode genes can also be polymorphic. An example of a polymorphic marker such as this is provided by RFLPs, which are DNA fragments that vary in their lengths when chromosomal DNA is subjected to restriction enzyme digestion. The molecular basis for an RFLP is that two distinct individuals will have variation in their DNA sequences and some of the variation may affect the locations of restriction enzyme sites. Since this occurs relatively frequently between unrelated individuals, many RFLPs can be identified. They can be detected by restriction digestion and agarose gel electrophoresis and then Southern blotting. They are useful in mapping studies, because it is relatively easy to find many of them along a chromosome. They can be used in gene cloning as a starting point for a chromosomal walk.

10. If the genes were unlinked, we would expect a 1:1:1:1 ratio among the four combinations of offspring. Since there are a total of 272 offspring, there are expected to be 68 in each category according to independent assortment. Plugging the data into our chi square formula,

$$X^2 = \Sigma \frac{(O - E)^2}{E}$$

$$= \frac{(40 - 68)^2}{68} + \frac{(98 - 68)^2}{68} + \frac{(97 - 68)^2}{68} + \frac{(37 - 68)^2}{68}$$

$$= 51.2$$

With three degrees of freedom, this high chi square value would be expected to occur by chance less than 1% of the time. Therefore, we reject the hypothesis that the RFLPs are independently assorting.

Now, in this example, the recombinant offspring contain the 5200, 4500, and 2100, and 4500, 2100, and 3200 RFLPs. Therefore,

$$\text{Map distance} = \frac{40 + 37}{40 + 98 + 97 + 37} \times 100$$

$$= 28.3 \text{ mu}$$

12. A microsatellite is a repetitive sequence that is found in a few locations in the genome. The lengths of the tandem repeats are often variable. This variation can be used as a molecular marker in mapping studies. Microsatellites are usually made by PCR amplification. Microsatellites and RFLPs are similar in that they are identified as DNA fragments that differ in length between two individuals. In the case of RFLPs, the different lengths are due to the presence or absence of particular restriction enzyme sites. In the case of microsatellites, it is due to variation in the number of repeats.

14. Children 1 and 3 belong to father 2.
    Children 2 and 4 belong to father 1.
    Child 5 could belong to either father.

16.   Deduced Outcome

                          STSs

              4   6   7   9   3   2   8   5   1
    YAC
      3   ⟵――――――⟶
      1   ⟵――――――――――⟶
      5   ⟵―――――――――⟶
      4   ⟵―――――――⟶
      2   ⟵―――――――⟶

18. *In situ* hybridization is a cytological method of mapping. A probe that is complementary to a gene sequence is used to locate the gene

microscopically within a mixture of many different chromosomes. Therefore, it can be used to map cytologically the location of a gene sequence. When more than one probe is used, the order of genes along a particular chromosome can be determined.

20. They are complementary to each other.

22. The first piece of information you would start with is the location of a gene or marker that is known to be close to the gene of interest by previous mapping studies. You would begin with a clone containing this marker (or gene) and follow the procedure of chromosome walking to eventually reach the gene of interest. A contig would make this much easier, because you would not have to conduct a series of subcloning experiments to reach your gene. Instead, you could simply analyze the members of the contig.

Questions for Student Discussion/Collaboration

2. A molecular marker is a segment of DNA, not usually within a structural gene, that has a known location within a particular chromosome. It marks the location of a site along a chromosome. RFLPs and STSs are examples. It is easier to use these types of markers, because they can be identified readily by molecular techniques such as restriction digestion analysis and PCR. The locations of functional genes are usually more difficult to map, because doing this relies on conventional genetic mapping approaches whereby allelic differences in the gene are mapped by making crosses or following a pedigree. For monomorphic genes, this approach does not work.

CHAPTER 21

Conceptual and Experimental Questions

2. The human gene that encodes the hormone is manipulated *in vitro* and then transformed into a bacterium. In many cases, the human hormone gene is fused with a bacterial gene to prevent the rapid degradation of the human hormone. The bacteria then express the fusion protein, and the hormone is separated by cyanogen bromide cleavage.

4. A biological control agent is an organism that prevents the harmful effects of some other agent in the environment. Examples include *Bacillus thuringiensis*, a bacterium that synthesizes compounds that act as toxins to kill insects, *Ice⁻* bacteria that inhibit the proliferation of *Ice⁺* bacteria, and the use of *Agrobacterium radiobacter* to prevent crown gall disease (caused by *Agrobacterium tumefaciens*).

6. *A. radiobacter* synthesizes an antibiotic that kills *A. tumefaciens*. The genes necessary for antibiotic biosynthesis and resistance are plasmid encoded and can be transferred during interspecies matings. If *A. tumefaciens* received this plasmid during conjugation, it would be resistant to killing. Therefore, the conjugation-deficient strain prevents the occurrence of *A. tumefaciens*–resistant strains.

8. Basically, one can follow the strategy described in Figure 21-5. If homologous recombination occurs, only the *NeoR* gene is incorporated into the genome. The cells will be neomycin resistant and also resistant to gancyclovir. If gene addition occurs, the cells will be sensitive to gancyclovir. By growing the cells in the presence of neomycin and gancyclovir, one can select for homologous recombinants.

10. A mouse model is a strain of mice that carry a mutation in a mouse gene that is analogous to a mutation in a human gene. For example, after the mutation causing cystic fibrosis was identified, the analogous gene was mutated in the mouse. Mice with mutations in this gene have symptoms similar to the human symptoms. These mice can be used in experiments to study the disease and to test therapeutic agents.

12. A transgenic organism is one that has recombinant DNA incorporated into its genome. The FlavrSavr tomato is an example of a transgenic plant. Plants resistant to glyphosate are another example. The large mouse shown in Figure 21-3 contains a transgene from a rat. Sheep that express human hormones in their milk are also transgenics.

14. The T-DNA gets transferred to the plant cell; it then is incorporated into the plant cell's genome. In plant gene transfer, a gene is cloned into the T-DNA region (i.e., between the T-DNA borders).

16. There are a variety of traits that can be modified in transgenic plants, yielding strains, for example, that are resistant to insects, disease, and herbicides or that have higher yield or better food quality.

18. DNA fingerprinting is a method of identification based on the properties of DNA. VNTR sequences are variable with regard to size in natural populations. This variation can seen when DNA fragments are subjected to gel electrophoresis. Within a population, any two individuals (who are not genetically identical) will display a different pattern of DNA fragments, which is called their DNA fingerprint.

20. PCR is used to amplify DNA if there is only a small amount of it (e.g., a small sample at a crime scene). Southern blotting, using a probe that is complementary to a VNTR, is needed to specifically identify a limited number of bands (20 or so) that are variable within the population.

22. *Ex vivo* therapy involves the removal of living cells from the body and their modification after they have been removed. The modified cells are then reintroduced into a person's body. This approach works well for cells, such as blood cells, that are easily removed and replaced. By comparison, this approach would not work very well for many cell types. For example, lung cells cannot be removed and put back again. In this case, *in vivo* approaches must be sought.

Questions for Student Discussion/Collaboration

2. From a genetic viewpoint, the recombinant and nonrecombinant strains are very similar. The main difference is their history. The recombinant strain has been subjected to molecular techniques to eliminate a particular gene. The nonrecombinant strain has the advantage of good public relations. People are less worried about releasing nonrecombinant strains into the environment.

CHAPTER 22

Conceptual and Experimental Questions

2. The general goal of a computer program is to analyze the data in a computer file. For example, a computer program may translate a DNA sequence into an amino acid sequence, it may identify functional motifs, or it may locate genes. Computer programs can also be used to predict the secondary and tertiary structure of DNA, RNA, and proteins, and to establish evolutionary relationships.

4. A database is a collection of many computer files in a single location. In genetics research, these data are usually DNA, RNA, or protein sequences. The data come from the contributions of many research labs.

6.

5′–<u>ATG</u>GGTATACCCGATCACTTTATTT<u>ATG</u>GGCACGATAGAT
AT<u>ATG</u>CC<u>ATG</u>CT<u>ATG</u>ATCATACATAC<u>ATG</u>CACATACATAAACG
GATAATCCATAATA<u>ATG</u>ACA<u>ATG</u>TTTTACGTA<u>ATG</u>CATACATA
CATAC<u>ATG</u>GCCATAC<u>ATG</u>CACTATACA<u>ATGA</u>–3′

8. Sequence recognition usually involves the recognition of a particular sequence of an already-known function. For example, a program could locate start codons (ATG) within a DNA sequence. By comparison, pattern recognition relies on a pattern of arrangement of symbols but is not restricted to particular sequences.

10.

```
THISQUESTIONISVERYANNOYING
T
H
I
S
Q
U
E
```

S
T
I
S

← **You need to insert a four-space gap here to align the sequences**

A
N
N
O
Y
I
N
G

12. There are a few interesting trends. Sequences 1 and 2 are similar to each other, as are sequences 3 and 4. There are a few places where amino acid residues are conserved among all five sequences. These amino acids may be particularly important with regard to function.

14. The chance of a stop codon is 3/64. Therefore, by random chance alone, about 1 in 21.3 codons will be a stop codon. Since a codon is 3 nucleotides in length, this means that a stop codon will appear about every 64 nucleotides. If one observes an open reading frame that is much longer than 64 nucleotides, one begins to suspect that the region encodes a protein.

16. In the analysis of genetic sequences, the term similarity means that two sequences are similar to each other. Homology means that two genetic sequences have evolved from the same ancestral sequence. Homologous sequences are similar to each other, but not all (short) similar sequences are due to homology.

18. In a comparative approach, one uses the sequences of many homologous genes. This method assumes that RNAs of similar function and sequence have a similar structure. Computer programs can compare many different 16S rRNA sequences to aid in the prediction of secondary structure.

20. The three-dimensional structure of a homologous protein must already be solved before one can attempt to predict the three-dimensional structure of a protein based on its amino acid sequence.

Questions for Student Discussion/Collaboration

2. This is a very difficult question. In 20 years, we may have enough predictive information so that the structure of macromolecules can be predicted from their genetic sequences. If so, it would be better to be a mathematical theoretician with some genetics background. If not, it is probably better to be a biophysicist with some genetics background.

CHAPTER 23
Conceptual Questions

2. When a disease-causing allele affects a trait, it is causing a deviation from normality, but the gene involved is not usually the only gene that governs the trait. For example, an allele causing hemophilia prevents the normal blood clotting pathway from operating correctly. However, normal blood clotting is due to the actions of many genes.

4. Changes in chromosome number and unbalanced changes in chromosome structure tend to affect phenotype, because they create an imbalance of gene expression. For example, in Down syndrome, there are three copies of chromosome 21 and, therefore, three copies of all of the genes on chromosome 21. This leads to a relative overexpression of genes that are located on chromosome 21 compared with those on other chromosomes. Balanced translocations and inversions often are without phenotypic consequences, because the total amount of the genetic material is not altered, and the level of gene expression is not significantly changed.

6. A proto-oncogene is a normal cellular gene that typically plays a role in cell division. It can be altered by mutation to become an oncogene

and thereby lead to cancer. At the level of protein function, a proto-oncogene can become an oncogene by synthesizing too much of a protein or by synthesizing a protein that is abnormally active.

8. The predisposition to develop cancer is inherited in a dominant fashion, because the heterozygote has the higher predisposition. It is not a dominant gene, however. The mutant allele is usually recessive at the cellular level. But since we have so many cells in our bodies, it becomes relatively likely that a defective mutation will occur in the normal gene and lead to a cancerous cell. Some heterozygous individuals may not develop the disease as a matter of chance. They may be lucky and never get a defective mutation in the normal gene. Or, perhaps their immune system is better at destroying cancerous cells once they arise.

10. If an oncogene was inherited, it may cause uncontrolled cell growth at an early stage of development and thereby cause an embryo to develop improperly. This could lead to an early spontaneous abortion and thereby explain why we do not observe individuals with inherited oncogenes. Another possibility is that inherited oncogenes may adversely affect gamete survival, which would make it difficult for them to be passed from parent to offspring. A third possibility would be that oncogenes could affect the fertilized zygote in a way that would prevent the correct implantation in the uterus.

12. The role of *p53* is to sense DNA damage and prevent damaged cells from proliferating. Perhaps, prior to birth, the fetus is in a protected environment so that DNA damage may be minimal. In other words, the fetus may not really need *p53*. After birth, agents such as UV light may cause DNA damage. At this point, *p53* is important. A *p53* knockout is more sensitive to UV light, because it cannot repair its DNA properly in response to this DNA-damaging agent, and it cannot kill cells that have become irreversibly damaged.

Experimental Questions

2. The term genetic screening refers to the use of laboratory tests to determine if an individual is a carrier of or affected by a genetic disease. Screening at the protein level means that the amount or activity of the protein is assayed. Screening at the DNA level means that the researcher tries to detect the mutant allele at the molecular or chromosomal level. Examples of approaches are described in Table 23-4.

4. A transformed cell is one that has become malignant. In a laboratory, this can be done in three ways. First, the cells could be treated with a mutagen that can convert a proto-oncogene into an oncogene. Second, cells could be exposed to the DNA from a malignant cell line. Under the appropriate conditions, this DNA can be taken up by the cells, and integrated into their genome so that they become malignant. A third way to transform cells is by exposure to an oncogenic virus.

6. By comparing oncogenic viruses with strains that have lost their oncogenicity, researchers have been able to identify particular genes that cause cancer. This has led to the identification of many oncogenes. From this work, researchers have also learned that normal cells contain proto-oncogenes, which usually play a role in a cell division. This suggests that oncogenes carry out that role by affecting the cell division process. In particular, it appears that oncogenes are abnormally active and keep the cell division cycle in a permanent "on" position.

Questions for Student Discussion/Collaboration

2. I do not imagine there is a clearly correct answer to this question, but it should stimulate a large amount of discussion.

CHAPTER 24
Conceptual Questions

2. **(A)** This is a mutation in *runt*, which is a pair-rule gene.

(B) This is a mutation in *knirps*, which is a gap gene.

(C) This is a mutation in *patch*, which is a segment-polarity gene.

4. Positional information refers to the phenomenon whereby the spatial locations of morphogens and cell adhesion molecules (CAMs) provide a cell with information regarding its position relative to other cells. In *Drosophila*, the formation of a segmented body pattern relies initially on the spatial location of maternal gene products. These gene products lead to the sequential activation of the segmentation genes.

6. The anterior portion of the anteroposterior axis is established by the action of *bicoid*. During oogenesis, the *bicoid* mRNA enters the anterior end of the oocyte and is sequestered there to establish an anterior (high) to posterior (low) gradient. Later, when the mRNA is translated, the Bicoid protein in the anterior region establishes a genetic hierarchy that leads to the formation of anterior structures. If the Bicoid protein was able to freely diffuse, this would prevent *bicoid* mRNA from attaining a concentration gradient, and this would block the ability of the Bicoid protein to activate only genes in the anterior portion of the embryo.

8. Maternal gene products influence the formation of the main body axes, including the anteroposterior, dorsoventral, and terminal regions. They are needed very early in development. Zygotic genes, particularly the three classes of the segmentation genes, are necessary after the axes have been established. The segmentation genes are expressed after fertilization during embryogenesis.

10. A homeotic gene governs the final fate of particular segments in the adult animal. A gain of function mutation is due to an aberrant expression of a homeotic gene in the wrong place or at the wrong time. This will cause the region to develop inappropriate characteristics. A loss of function allele will usually cause a segment to develop characteristics that are normally found in the anterior adjacent segment.

12. Realizator genes are located near the end of the developmental genetic hierarchy. The expression of realizator genes leads to the expression of genes that encode cell-specific proteins. For example, the *myoD* gene causes cells to express muscle-specific proteins and thereby realize the muscle-specific phenotype.

14. *Drosophila* has eight homeotic genes located in two clusters (*antennapedia* and *bithorax*) on chromosome 3. The mouse has four *Hox* complexes, designated *HoxA* (on chromosome 6), *HoxB* (on chromosome 11), *HoxC* (on chromosome 15), and *HoxD* (on chromosome 2). There are a total of 38 genes in the four complexes. There are 13 different gene types within the four *Hox* complexes, although each complex contains less than all 13 types of genes. Among the first 6 types of genes, 5 of them are homologous to genes found in the *antennapedia* complex of *Drosophila*. Among the last 7, 3 of them are homologous to the genes of the *bithorax* complex. Like the *bithorax* and *antennapedia* complexes in *Drosophila*, the arrangement of *Hox* genes along the mouse chromosome reflects their pattern of expression from the anterior to the posterior end. With regard to differences, the mouse has a larger number of homeotic genes, and gene knockouts do not always lead to transformations that resemble the anterior adjacent segment.

16. Animals begin their development from an egg, and then they usually become segmented along anteroposterior and dorsoventral axes. The formation of an adult organism is an expansion of the embryonic body plan. Plants grow primarily from two meristems, a shoot and root meristem. At the cellular level, plant development is different in that it does not involve cell migration, and most plant cells are totipotent. Animals require the organization within an oocyte to begin development. At the genetic level, however, animal and plant development are similar in that they involve a genetic hierarchy of transcription factors that govern pattern formation and cell specialization.

Experimental Questions

2. *Drosophila* is more advanced from the perspective that many more mutant alleles have been identified that alter development in specific ways. The hierarchy of gene regulation is particularly well understood in the fruit fly. *C. elegans* has the advantage of simplicity and a complete knowledge of cell fates. This enables researchers to explore how the timing of gene expression is critical to the developmental process.

4. To determine that a mutation is affecting the timing of developmental decisions, a researcher needs to know the normal time or stage of development when cells are supposed to divide, and what type of cells will be produced. With this information (i.e., a fate map), one can then determine if particular mutations alter the timing of cell division.

6. Geneticists who are interested in mammalian development have used reverse genetics, because it has been difficult for them to identify mutations in developmental genes based on phenotypic effects in the embryo. This is because it's difficult to screen a large number of mammalian embryos in search of abnormal ones that carry mutant genes. Instead, it is easier to clone the normal gene based on its homology to invertebrate genes and then produce mutations *in vitro*. These mutations can be introduced into a mouse to create a gene knockout. This strategy is opposite to that of Mendel, who characterized genes by first identifying phenotypic variants (e.g., tall vs. dwarf, green seeds vs. yellow seeds).

8. Mutant 1 is a gain of function allele; it keeps reiterating the L1 pattern of division. Mutant 2 is a loss of function allele; it skips the L1 pattern and immediately follows the L2 pattern of division.

Questions for Student Discussion/Collaboration

2. In this problem, the students should try to make a flow diagram that begins with maternal effect genes, then gap genes, pair-rule genes, and segment-polarity genes. These genes then lead to homeotic genes and finally realizator genes. It is almost impossible to make an accurate flow diagram, because there are so many gene interactions, but it is instructive to think about developmental genetics in this way. It is probably easier to identify mutant phenotypes that affect later stages of development, because they are less likely to be lethal. However, modern methods can screen for conditional mutants as described in Chapter 12. To identify all of the genes necessary for chicken development, I would prefer to begin with early genes, but this assumes that I have some way to identify them. If they had been identified, I would then try to identify the genes that they stimulate or repress. This could be done using molecular methods described in Chapters 15, 16, and 19.

CHAPTER 25

Conceptual Questions

2. At the molecular level, quantitative traits often exhibit a continuum of phenotypic variation, because they are usually influenced by multiple genes that exist as multiple alleles. A large amount of environmental variation will also increase the overlaps among different genotypic categories.

4. To be in the top 2.5% is above 2 standard deviation units. If we take the square root of the variance, the standard deviation would be 20.6 lb. To be in the top 2.5%, an animal would have to weigh 41.2 lb heavier than the mean. To be in the bottom 0.15%, an animal would have to be 3 standard deviations lighter, which would be 61.8 lb lighter than the mean.

6. A statistically significant correlation suggests that there is a positive or negative association between two variables. In this example, there is a positive correlation, but it could have occurred as a matter of chance alone. You would need to conduct more experimentation, such as examining a greater number of pairs of individuals, to determine if there is a significant correlation. If $N = 500$, the results would be statistically significant.

8. **(A)** We first need to calculate the standard deviations for height and weight, and then the covariance for both traits:

Height:variance = 140.0; standard deviation = 11.8

Weight:variance = 121.43; standard deviation = 11.02
Covariance = 123.3

$$r_{(X,Y)} = \frac{\text{CoV}_{(X,Y)}}{\text{SD}_X \text{SD}_Y}$$

$$= 123.3/(11.8)(11.02)$$

$$= 0.948$$

(B) With eight degrees of freedom, this value is statistically significant. This means the association between these two variables occurs more frequently than would be expected by random sampling error. This does not necessarily imply cause and effect.

10. When a correlation coefficient is statistically significant, it means that the association is likely to have occurred for reasons other than random sampling error. It may indicate cause and effect, but not necessarily. For example, large parents may have large offspring due to genetics (cause and effect). However, the correlation may be related to the sharing of similar environments rather than genetic cause and effect.

12. If the broad sense heritability equals 1.0, all of the variation in the population is due to genetic factors rather than environmental ones. It does not mean that the environment is unimportant in the outcome of the trait. Under another set of environmental conditions, the environment could contribute greatly to the phenotypic variation of the trait.

14. According to the Dominance Hypothesis, heterosis occurs because each of two strains are homozygous recessive for one or more genes that have a negative impact on their phenotypes. When mated together, the offspring are heterozygous and, therefore, do not display these negative effects. The Overdominance Hypothesis suggests that the offspring are more vigorous because certain genes display overdominance when they are found in the heterozygous condition. Thus far, data has supported both hypotheses to some extent. In general, it is known that inbred strains tend to become homozygous for alleles, which would support the Dominance Hypothesis. Also, relatively few genes are known in which two alleles cause overdominance in the heterozygous condition, again supporting the Dominance Hypothesis. However, the situation is by no means resolved.

16.

$$H_R = \frac{R}{S}$$

$$R = \overline{X}_O - \overline{X}$$

$$S = \overline{X}_P - \overline{X}$$

In this problem,
$\overline{X}$ equals 1.01 (as given in the problem)
$\overline{X}_P$ equals 1.11 (by calculating the mean for the parents)
$\overline{X}_O$ equals 1.09 (by calculating the mean for the offspring)

$$R = 1.09 - 1.01 = 0.08$$

$$S = 1.11 - 1.01 = 0.10$$

$H_R = 0.08/0.10 = 0.8$ (which is a pretty high heritability value)

18. Hybrid vigor is the phenomenon in which an offspring produced from two inbred strains is more vigorous than the corresponding parents. Tomatoes and corn are often the products of hybrids.

20. In this problem, you need to set up Punnett squares based on genotypes.

(A) Let's call the alleles *A1* and *A2*. The F$_1$ would be genotypically *A1A2*. The backcross to strain A1 would produce a 1:1 ratio of *A1A1* to *A1A2*. Since *A1A1* plants produce 1 pound fruit and *A1A2* produce 2 pound fruit, this would yield a phenotypic ratio of 50% 1 pounders to 50% 2 pounders.

(B) Let's call the alleles *A* and *a*, and *B* and *b* and assume they contribute additively. The F$_1$ offspring will be *AaBb*. If we assume strain A is *AAbb*, the backcross ratio will be 1 *AaBb* (2 pounds) : 1 *aaBb* (1 pound) : 1 *Aabb* (1 pound) : 1 *aabb* (1 pound). However, this last

genotypic category (*aabb*) might be less than 1 pound, since it does not have either dominant allele.

The phenotypic ratio will be 1:3, or possibly 1:2:1 (if the *aabb* strain is less than one pound).

(C) Let's call the alleles *A1* and *A2*, and *B1* and *B2* and assume they contribute additively. The F$_1$ offspring will be *A1A2B1B2*. If we assume strain A is *A1A1B1B1*, the backcross will yield

1 *A1A1B1B1* (1 pound) : 1 *A1A1B1B2* (1.5 pounds) : 1 *A1A2B1B1* (1.5 pounds) : 1 *A1A2B1B2* (2 pounds).

The phenotypic ratio would be 1:2:1.

(D) You cannot calculate a precise ratio. As the number of genes increases, it becomes more unlikely to be heterozygous at all of the loci, so that it becomes less likely to produce 2 pounders. With a very large number of genes, most of the offspring would be in the intermediate (1.5 pound) range.

Experimental Questions

2. The heritability reflects the amount of genetic variation between two strains, usually the strain of interest and some other tester strain. In the strain with seven QTLs, this means there are seven loci that display genetic variation relative to the tester strain. In the other strain of interest, there are the same types of genes, but three of them probably have the same alleles as the tester strain. Therefore, relative to the tester strain, this other strain only exhibits four QTLs because the other three genes do not contribute to the variation in the outcome of the trait.

4. These results suggest that there are (at least) three different genes the variation of which influences the size of pigs. This is a minimum estimate, because some QTLs may actually involve two or more closely linked genes. Also, it is possible that the large and small strains have the same RFLP band associated with one or more of the genes that affect size.

6. The identical and fraternal twins, which probably share a very similar environment, but which differ in the amount of genetic material they share, are a strong argument against an environmental bias.

Questions for Student Discussion/Collaboration

2. Most traits depend on the influence of many genes. Also, genetic variation is a common phenomenon in most populations. Therefore, most individuals have a variety of alleles that contribute to a given trait. For quantitative traits, some alleles may make the trait bigger and other alleles may make the trait turn out smaller. If a population contains many different genes and alleles that govern a quantitative trait, most individuals will have an intermediate phenotype, because they will have inherited some large and some small alleles. Fewer individuals will inherit a predominance of large alleles or a predominance of small alleles. An example of a quantitative trait that does not fit a normal distribution is snail pigmentation. The dark snail and light snails are favored rather than the intermediate colors, because such snails are less susceptible to predation (see Chapter 26).

CHAPTER 26

Conceptual Questions

2. A population is a group of interbreeding individuals. Let's consider a squirrel population in a forested area. Over the course of many generations, several things could happen to this population that may change its gene pool. A forest fire, for example, could dramatically decrease the number of individuals and thereby cause a bottleneck. This would decrease the genetic diversity of the population. A new predator may enter the region, and natural selection may select for the survival of squirrels that are best able to evade the predator. Another possibility is that a group of squirrels within the population may migrate to a new region and found a new squirrel population.

4. Migration, genetic drift, and natural selection are the driving forces that alter allele frequencies within a population. Natural selection acts to eliminate harmful alleles and promote beneficial alleles. Genetic drift involves random changes in allele frequencies that may eventually lead to elimination or fixation of alleles; it is thought to be important in the establishment of neutral alleles in a population. Migration is important because it introduces new alleles into neighboring populations.

6. (A) The genotype frequency for the CF homozygote is 1/2500, or 0.004. This is equal to q^2. The allele frequency is the square root of this value, which equals 0.02. The frequency of the corresponding normal allele is $1 - 0.02 = 0.98$.

   (B) The frequency of the CF homozygote is 0.004. The frequency of the normal homozygote is $(0.98)^2 = 0.96$. The frequency of the heterozygote is $2(0.98)(0.02) = 0.039$.

   (C) If a person is already identified as a heterozygous carrier, the chance that the person will choose another person as a mate who is a heterozygous carrier is equal to the frequency of heterozygous carriers in the population, which is 0.039 or 3.9%. In the population as a whole, to calculate the chances that two heterozygous carriers will choose each other as mates, we use the product rule: $0.039 \times 0.039 = 0.0015 = 0.15\%$.

8. (A) Yes.

   (B) The common ancestors are I-1 and I-2.

   (C) $$F = \Sigma(1/2)^n(1 + F^A)$$
   $$= (1/2)^9 + (1/2)^9$$
   $$= 1/512 + 1/512 = 2/512 = 0.0039$$

   (D) Not that we can tell.

10. $$(1 - u)^t = \frac{p_t}{p_0}$$
    $$(1 - 10^{-4})^t = 0.5 / 0.6 = 0.833$$
    $$(0.9999)^t = 0.833$$
    $$t = 1827 \text{ generations}$$

12. (A) Phenotype frequency and genotype frequency.

    (B) Genotype frequency.

    (C) Allele frequency.

14. (A) Probability of fixation $= 1/2N = 1/(2)(4) = 1/8$, or 0.125.

    (B) $t \cong 4N = 4(4) = 16$ generations.

    (C) The preceding calculations assume a constant population size. If the population grows after it has been founded by these four individuals, the probability of fixation will be lower and the time it takes to reach fixation will be longer.

16. When two populations intermix, both populations tend to have more genetic variation, because each population introduces new alleles into the other population. In addition, the two populations tend to have similar allele frequencies, particularly when there is a large proportion of individuals who migrate.

18. Directional selection favors the phenotype at one phenotypic extreme. Over time, natural selection is expected to favor the fixation of the alleles that cause these phenotypic characteristics. Disruptive selection favors two or more phenotypic categories. It will lead to a population with a balanced polymorphism for the trait. Stabilizing selection favors individuals with intermediate phenotypes. It will also tend to promote polymorphism, since the favored individuals may be heterozygous for particular alleles.

20. Let's assume that the relative fitness values are 1.0 for the dominant homozygote and the heterozygote, and 0 for the recessive homozygote. The first thing we need to do is to calculate the mean fitness for the population:

$$\overline{W} = p^2W_{AA} + 2pqW_{Aa} + q^2W_{aa}$$
$$= (0.78)^2 + 2(0.78)(0.22)$$
$$= 0.95$$

The allele frequency in the next generation for A is:

$$\frac{p^2W_{AA}}{\overline{W}} + \frac{pqW_{Aa}}{\overline{W}} = 0.82$$

So, $A = 0.82$ and $a = 0.18$

For the second generation, we first need to calculate the mean fitness of the population, which now equals 0.97. The genotype frequency of AA in the second generation equals 0.69 and the allele frequency equals 0.845. The frequency of the recessive allele in the second generation would equal 0.155 and the mean fitness now equals 0.976. The genotype frequency of AA in the third generation is 0.73 and the allele frequency is 0.866. The frequency of the recessive allele is 0.134.

22. The genetic load refers to all the genetic mutations within a population that are deleterious. It is a common consequence of genetic variation, since sometimes there are genetic balances that occur. For example, heterozygote advantage can lead to the prevalence of alleles that are deleterious in the homozygous condition. In general, it happens that alleles, or combinations of alleles, that are advantageous in one circumstance are not advantageous in another. Since populations shuffle their genes from one generation to the next, and may move to new environments, genetic load occurs.

Experimental Questions

2. (1) One hypothesis is that the population having only two allozymes was founded by a small group that left the other population. When the small founding group left, it had less genetic diversity than in the original population. (2) The population with more genetic diversity may be in a more diverse environment, so that it may select for a greater variety of phenotypes. (3) It may just be a matter of chance that one population has accumulated more neutral alleles than the other.

4.
$$\text{Warfarin } (W^r) \text{ allele frequency} = \frac{S_{ww}}{S_{w^rw^r} + S_{w^rw}}$$
$$= \frac{0.19}{0.19 + 0.37} = 0.34$$

$$\text{Normal } (w) \text{ allele frequency} = \frac{S_{w^rw^r}}{S_{w^rw^r} + S_{w^rw}}$$
$$= \frac{0.37}{0.19 + 0.37} = 0.66$$

If the rats are not exposed to warfarin, the equilibrium will no longer exist, and natural selection will tend to eliminate the warfarin resistance allele because the homozygotes are vitamin K deficient and have lower fitness.

6. Let's use the data for bird predation, but we could also carry out a chi square analysis for the percentage of recapture.

Hypothesis: Color has nothing to do with predation by birds. (Note: We need to propose this hypothesis to obtain expected values.) According to this hypothesis, we would expect an equal number of dark and light moths to be eaten by birds. We need to calculate the sum

$$X^2 = \Sigma \frac{(O - E)^2}{E}$$

In the Dorset woods, there were $43 + 15 = 58$ moths that were eaten. We would expect 29 to be *carbonaria* and 29 to be *typical* according to our hypothesis. In the Birmingham woods, there were 26 + 164 = 190 moths eaten, and so we would expect 95 to be *carbonaria* and 95 to be *typical*. Plugging the values into our chi square formula,

$$X^2 = \frac{(43 - 29)^2}{29} + \frac{(15 - 29)^2}{29} + \frac{(26 - 95)^2}{95} + \frac{(164 - 95)^2}{95}$$
$$= 113.7$$

If we look in the chi square table in Chapter 2 under three degrees of freedom, this rather high chi square value is very unlikely to occur as a matter of chance (less than 1% of the time). Therefore, we reject our hypothesis that color does not affect predation. As an alternative, we would propose that the color of the moths does have a significant effect on their likelihood of predation.

Questions for Student Discussion/Collaboration

2. Mutation is responsible for creating new alleles, but the rate of new mutation is so low that it cannot explain allele frequencies in this range. Let's call the two alleles *B* and *b* and assume that *B* was the original allele and *b* is a more recent allele that arose from mutation. Three scenarios to explain the allele frequencies are: (1) The *b* allele is neutral and reached its present frequency by genetic drift. It has not reached extinction or fixation yet. (2) The *b* allele is beneficial, and its frequency is increasing due to natural selection. However, there has not been enough time to reach fixation. (3) The *Bb* heterozygote is at a selective advantage, leading to a balanced polymorphism.

CHAPTER 27

Conceptual Questions

2. Reproductive isolation occurs when two species do not mate and produce viable offspring. As discussed in Table 27-1, there are several prezygotic and postzygotic mechanisms that can prevent interspecies matings. According to the biological species concept, reproductive isolation is the underlying cause of speciation. Speciation occurs when a group becomes reproductively isolated from another group. The recognition species concept is similar, except that it places a greater emphasis on the forces of natural selection in promoting reproductive isolation. According to the recognition species concept, natural selection promotes reproductive isolation by selecting for groups of individuals who are reproductively compatible.

4. Anagenesis is the evolution of one species into another; cladogenesis is the divergence of one species into two or more species. Cladogenesis is more prevalent. There may be many reasons for this. It is common for an abrupt genetic change such as alloploidy to produce a new species from a preexisting one. Also, migrations of a few members of species into a new region may lead to the formation of a new species in the new region (i.e., allopatric speciation).

6. **(A)** Allopatric.

   (B) Sympatric.

   (C) At first, it may involve parapatric speciation with a low level of intermixing. Eventually, when smaller lakes are formed, allopatric speciation will occur.

8. One of the major goals in the field of molecular evolution is to understand, at the molecular level, how changes in the genetic material have led to the formation of present-day species. Along these same lines, molecular biologists would like to understand why genetic variation is prevalent within a single species, and also to examine the degree of differences in genetic variation among different species. Comparisons of the genetic material at the molecular level can help to elucidate evo-

lutionary relationships. The field of molecular evolution also is aimed at understanding how genes change at the molecular level. Geneticists would like to know the rate of genetic change and whether changes are neutral or adaptive.

10. Line up the sequences where the two Gs are underlined.

    TTGCATAGGCATACC<u>G</u>TATGATATCGAAAACTAGAAAAAT
    AGGGCGATAGCTA
    <u>G</u>TATGTTATCGAAAAGTAGCAAAATAGGGCGATAGCTACC
    CAGACTACCGGAT

12. The neutral theory of evolution is sometimes called non-Darwinian evolution, because it is not based on natural selection, which was a central tenet in Darwin's theory. It is difficult to say if Darwin would object, but I doubt it. During Darwin's time, the random nature of mutations was not understood. Therefore, neutral change was not something that he would have considered. The neutral theory does not refute Darwin's major idea, which was that natural selection leads to adaptation. The neutral theory simply suggests that the percentage of adaptive changes in DNA sequences are relatively small compared with the percentage of neutral ones.

14. Generally, one would expect a similar number of chromosomes with very similar banding patterns. However, there may be a few notable differences. An occasional translocation could change the size or chromosomal number between two different species. Also, an occasional inversion may alter the banding pattern between two species.

Experimental Questions

2. Since the artificial and natural *G. tetrahit* strains can produce fertile offspring, it means that the chromosomes in these strains form homologous sets. When the offspring make gametes, each chromosome has a homologous partner to pair with. Therefore, the offspring can make gametes that have a complete set of chromosomes rather than making highly aneuploid gametes, which would be inviable.

4. There were probably two hexokinase genes in the previous common ancestor, because *A1* is very similar to *B2*, and *A2* is very similar to *B1* and *B3*. In species B, there was probably a gene duplication that created *B1* and *B3*; this occurred after the formation of species A and B. This gene duplication did not occur in species A (after the divergence).

6. As shown in the sequence alignment on the next page, there are regions, which are boxed, where the kiwis and cassowary are more similar to each other than to the moas.

8. One way to establish credibility would be to analyze *Tyrannosauris rex* DNA from samples that have been obtained from different locations. Samples from different locations would be less likely to be contaminated with the same kind of DNA.

Questions for Student Discussion/Collaboration

2. The founder effect and allopolyploidy are examples of rapid forms of evolution. In addition, some single gene mutations may have a great impact on phenotype and lead to rapid evolution of new species by cladogenesis. Geological processes may promote the slower accumulation of alleles and alter the characteristics of a species more gradually. In this case, it is the accumulation of many phenotypically minor genetic changes that ultimately leads to reproductive isolation. Slow and fast mechanisms of evolution have the common theme that they result in reproductive isolation. This is a prerequisite for the evolution of new species. Fast mechanisms tend to involve small populations and a low number of genetic changes. Slower mechanisms may involve larger populations and the accumulation of many genetic changes that each contribute in a small way.

MOA 1   GCTTAGCCCTAAATCCAGATACTTACCCTACACAAGTATCCGCCCGAGAACTACGAGCACAAACGCTTAAAACTCTAAGGACTTGGCGGTGCCCCAAACCCACCTAGAGGAGCCTGTTCTATAATCGATAATCCACGATA

MOA 1   CACCCGACCATCCCTCGCCCGT-GCAGCCTACATACCGCCGTCCCCAGCCCGCCT--AATGAAAG-AACAATAGCGAGCACAACAGCCCTCCCCCGCTAACAAGACAGGTCAAGGTATAGCATATGAGATGGAAGAAATG

MOA 1   GGCTACATTTTCTAACATAGAACACCC--------------ACGAAAGAGAAGGTGAAACCCTCCTCAAAAGGCGGATTTAGCAGTAAAATAGAACAAGAATGCCTATTTTAAGCCCGGCCCTGGGGC

CREDITS

Chapter 25 25-01: Courtesy of the Library of Congress 25-03a: PhotoDisc 25-03b: PhotoDisc 25-07a: Gerard Lacz/Peter Arnold, Inc. 25-07b: © Ron Kimball 25-07c: Ron Kimball 25-07d: Gerard Lacz/Peter Arnold, Inc. 25-08a: Michael Gadomski/Photo Researchers, Inc. 25-08b: © Dwight R. Kuhn 25-08c: © Dwight R. Kuhn 25-08d: Animals Animals/© Jack Wilburn 25-08e: © Dwight R. Kuhn 25-08f: © David Cavagnaro/Peter Arnold, Inc. 25-08g: © Clyde H. Smith/Peter Arnold, Inc.
Chapter 26 26-01: PhotoDisc. 26-02 Courtesy of Oxford and Gilespie, *J. of Heredity* Vol 76. Reprinted with permission 26-08b: S.R. Maglione/Photo Researchers, Inc. 26-12a: Animals Animals/© O. S. F. Experiment 26a1: © M. W. Tweedie/Photo Researchers, Inc. Experiment 26a2: © Irene Vandermolen/Photo Researchers, Inc.
Chapter 27 27-01a: © Archive Photos 27-01b: Lambert/© Archive Photos 27-01c: © Archive Photos 27-03a: Michael Fogden/Bruce Coleman Inc. 27-03b: © Larry Ulrich/DRK PHOTO 27-03c: John Shaw/Bruce Coleman Inc.
Appendix APP-01a: K. K. Sanford APP-01b: © Biophoto Associates/Photo Researchers, Inc. APP-08: Lara Hartley/Terraphotographics.

ILLUSTRATIONS

The following figures are adapted from Neil A. Campbell, *Biology,* 4th ed. (Menlo Park, CA: Benjamin/Cummings, 1996). © 1996 The Benjamin/Cummings Publishing Company:
Figures 1-09, 2-02c, 2-03, 2UN.03, 3-01a, 3-07, 3-13, 3-16, 4.EXPA.F01, 4.EXPB.F01, 4-02, 4-03, 4-16, 6-04, 6-08, 6-09, 8-21, 9-10, 9-11a, 9-15, 9-20, 10-18, 11-02, 11-06, 12.UN.01, 12-06, 12-07, 12-12, 13-01, 19-02, 21-07, 23-05, 24-01 e & g, 24-07a, 24-10, 24-13, 24-15a, 24-21, 26-04, 27-02, 27-06, APP-06.
The following figures are adapted from Wayne M. Becker, Jane B. Reece, and Martin F. Poenie, *The World of the Cell,* 3rd ed. (Menlo Park, CA: Benjamin/Cummings, 1996). © 1996 The Benjamin/Cummings Publishing Company:
Figures 1-01b, 3-02, 3-09, 3-10, 3-12, 3-14, 3-15, 5-01, 9.ExpA.F2, 9-07, 9-09, 10-02, 10-13a, 11-20, 12-01, 13-02, 13-06, 13-11, 13-16, 14-03, 14-07b, 14-49, 14-17, 16-01, 16.12.01, 16-13, 16-20, 17-11, 18-01, 19-04, 20-15, 23-13, 24-12, APP-03, APP-04.
The following figures are adapted from Neil A. Campbell, Lawrence G. Mitchell, and Jane B. Reece, *Biology: Concepts and Connections,* 2nd ed. (Menlo Park, CA: Benjamin/Cummings, 1997). © 1997 The Benjamin/Cummings Publishing Company:
Figures 2-04, 2-05, 2EXPA.01, 3-06a, 24.18, 26-09.
The following figures are adapted from Robert A. Wallace, Gerald P. Sanders, Robert J. Ferl, *Biology: The Science of Life,* 4th ed. (Menlo Park, CA: Addison Wesley Educational Publishers, 1996). © 1996 HarperCollins Publishers Inc.
Figures: 2EXPB.01, 4-11, 9-12, 17-05, 23-04b, 23-10.
The following figures are adapted from Gerard J. Tortora, Berdell R. Funke, and Christine L. Case, *Microbiology: An Introduction,* 5th ed. (Redwood City, CA: Benjamin/Cummings, 1995). © 1995 The Benjamin/Cummings Publishing Company.
Figures: 3-03, 6-05, 9-02, 9-04a, 9-05, 10-01, 11-03b, 17-07, 19-01, 19-07, 20-12,
The following figures are adapted from C.K. Mathews and K.E. van Holde, *Biochemistry,* 2nd ed. (Menlo Park, CA: Benjamin/Cummings, 1996). © 1996 The Benjamin/Cummings Publishing Company.
Figures: 8-07, 9-16a, 10-09, 10-10a, 10-13b, 10-20, 11-05, 11-07, 11-12, 11-13a, Table 12-2, 12-08, 13-03, 13-04, 13-12, 13-17, 13-18, 14-02, 14-08, 14-10, 14-12, 14-13, 14-14, 15-03, 15-09, 15-12, 15-13, 17-13, 17-16, 17-17, 18-08, 18-09, 18-14, 19-05, 19-11, 19-12, APP-02, APP-07, APP-09.
The following figures are adapted from Lewis J. Kleinsmith and Valerie M. Kish, *Principles of Cell and Molecular Biology,* 2nd ed. (Menlo Park, CA: Addison Wesley Educational Publishers, 1995). ©1995 HarperCollins College Publishers Inc.
Figures: 3-08, 3-11, 7-03b, 9-17b, 10-15b, 13-08, 13-15, 14-11, 15-04,15-06, 22-05, APP-05.
4-15 Adapted from *An Introduction to Genetics* by David J. Merrell. Copyright ©1975 by W.W. Norton & Company, Inc. Reprinted by permission of W.W. Norton & Company, Inc.
6-01 Adapted from Tatum and Lederberg (1947) *J. of Bacteriology* 53: 673-684. Reprinted with permission.
6-12 Adapted from Birge, *Bacterial and Bacteriophage Genetics,* 3e. ©1994 Springer/Verlag. Reprinted with permission.
6-19 Adapted from Benzer (1961) *Proc. Natl. Acad. Sci. USA* 47: 403-415. Reprinted with permission of the author.
7-11 With permission, Adapted from the *Annual Review of Biochemistry,* Volume 61 ©1992 by Annual Reviews.
7-12 Adapted from N.W. Gilham, *Organelle Genes and Genomes.* ©1994 Oxford University Press/Academic Press. Reprinted with permission.
9-13 Adapted from J.D. Watson, *The Double Helix,* 1968, New York: Atheneum Press. Reprinted with permission of Cold Spring Harbor Laboratory Archives.
9-14a Adapted from J.D. Watson, *The Double Helix,* 1968, New York: Atheneum Press. Reprinted with permission of Cold Spring Harbor Laboratory Archives.
9-17a Adapted from R.E. Dickerson (1983) *Scientific American,* Dec.: 100-104. Illustration ©Irving Geis. Reprinted with permission.
9-17c Courtesy of Dr. A.H.-J. Wang, *Cold Spring Harbor Symp. Quant. Biol.* (1982) 47:41.
9-18 Adapted from A.S. Moffat (1991) *Nature* 252: 1374-1375. Copyright ©1991 Macmillan Magazines Limited. Reprinted with permission.
9-21 With permission, adapted from the *Annual Review of Biochemistry,* Volume 62 © 1993 by Annual Reviews.
9-22a Adapted from Holbrook et al (1978) *J. Mol. Biol.* 123: 931-660. Reprinted by permission of the publisher, Academic Press Limited, London.
10-15c Adapted from *J. of Cell Biol.*131: 365-76. © 1996 Rockefeller University Press. Reprinted with permission.
11.01a Adapted from Elaine N. Marieb, *Human Anatomy and Physiology,* 3rd ed. (Redwood City, CA: Benjamin/Cummings, 1995). © 1995 The Benjamin/Cummings Publishing Company.
11-13b Adapted from L. Beese et al (1993) *Science* 260: 352. Reprinted with permission.

13-13 Adapted from P.A. Sharp (1994) *Cell* 77: 805-815. Copyright © 1993 Nobel Foundation. Reprinted with permission.
13-19 Adapted from B.K. Adler and S.L. Hajduk (1994) *Current Opinions in Gen. and Dev.* 4: 316-322. Reprinted with permission.
14-04 Adapted from L. Beese et al (1993) *Science* 260:352. Reprinted with permission.
15-16 Adapted from Voet and Voet, *Biochemistry,* © 1990 John Wiley & Sons. Reprinted with permission of John Wiley & Sons.
15-18 Adapted from Ptashne (1992) *A Genetic Switch: Phase Lambda in Higher Organisms.* Cambridge, MA: Blackwell Scientific Publishing. Reprinted by permission of Blackwell Science, Inc.
16-05 Adapted from A. Razin and A. Riggs (1980) *Science* 210: 604-609. Reprinted with permission.
16-08 Adapted from Townes and Behringer *Trends in Genetics* 6: 219-223. Reprinted with permission Adapted from Elsevier Science.
16-15 Adapted from E. Wingender (1993) *Gene Regulation in Eukaryotes.* Wiley-VCH: Weinheim (FRG). Reprinted with permission.
17-18 Adapted from A. Sancar and A. Selby (1993) *Science* 260: 53-58. Reprinted with permission.
18-04 Adapted from Kowalczykowski et al. (1994) *Microbiol. Rev.* 58: 401-465. Reprinted with permission.
20-01 Adapted from R.D. Fleishmann et al (1995) *Science* 269: 28. Reprinted with permission.
20-02 Adapted from (1935) *J. of Heredity* 26: 60-64. © The American Genetic Association. Reprinted with permission of Oxford U. Press.
20-05 Illustration by Bunji Tagawa.
20-10 Adapted from C. Chang et al (1988) *Proc. Natl. Acad. Sci. USA* 85: 6856-6860. Reprinted with permission.
20-18 Adapted from Cooper, *Human Genome Project* (Sausalito, CA: University Science Books,1994). Reprinted with permission.
20-20 Adapted from Cooper, *Human Genome Project* (Sausalito, CA: University Science Books,1994). Reprinted with permission.
21-01 Adapted from RECOMBINANT DNA 2/E by Watson, Gilman, Witkowski, Zoller © 1992 by James D. Watson, Michael Gilman, Jan Witkowski, and Mark Zoller. Used with permission of W.H. Freeman and Company.
23-02 Adapted from Volk, *Tay-Sachs Disease* (New York: Grune & Stratton, 1969). Reprinted with permission of W.B. Saunders Company.
23-03 Adapted from S.E. Folstein, *Huntington's Disease, a Disorder of Families* (Baltimore: The Johns Hopkins University Press,1989). © 1989 The Johns Hopkins University Press. Reprinted with permission.
23-06 Dimitry Schidlovsky
23-07 Adapted from Cooper, *Oncogenes* 2nd ed. (Sudbury, MA: Jones and Bartlett Publishers www.jbpub.com, 1995). Reprinted with permission.
23-11 Adapted from Cooper, *Oncogenes* 2nd ed. (Sudbury, MA: Jones and Bartlett Publishers www.jbpub.com, 1995). Reprinted with permission.
24.ExpAF.2 Adapted from Ambros and Horvitz (1984) *Science* 226: 409-416. Reprinted with permission.
24-UNa Adapted from Nusslein-Volhard et al (1980) *Nature* 287: 795-801. Reprinted with permission.
24-UNc Adapted from A. Ambros and H.R. Horvitz (1984) *Science* 226: 409-416. Reprinted with permission.
24-01a-d Adapted from D. St Johnston and C. Nusslein-Volhard (1992) *Cell* 68, 201-219. Reprinted with permission.
24-03 Adapted from Nusslein-Volhard et al (1980) *Nature* 287: 795-801. Reprinted with permission.
24-06 Adapted from Lawrence, *The Making of a Fly* (Oxford: Blackwell Scientific Pub, 1992). Reprinted with permission.
24-08 Adapted from Martinez-Arias and P. Lawrence (1985) *Nature* 313: 639-642
24-14 Adapted from M.P. Scott (1992) *Cell* 71: 551-553.
24-20a Adapted from Elliott Meyerowitz and John Bowman (1991) *Development* 112: 1-20. Reprinted with permission of Company of Biologists, Ltd.
25.UN.02 Adapted from Crow (1957) *Ann. Review of Entomology* 2: 227-246.
25-05 Adapted from Crow (1957) *Ann. Review of Entomology* 2:227-246.
25-06 Adapted from S.B. Holt (1961) *Brit Med Bulletin* 17: 247-250.
25-09 Adapted from A. Falconer (1989) *Introduction to Quantitative Genetics* 3rd ed (Longman Publishing Company. © 1989 Longman Publishing, Inc. Reprinted with permission.)
26-03 Patricia J. Wynne
26-11 Adapted from Cavalli-Sfarza (1974) *Scientific American* 231: 80-89. Illustration by Eric Mose.
26-12 Adapted from A.J. Cain and P.M. Sheppard (1954) *Genetics* 39: 89-116. © Genetics Society of America. Reprinted with permission.
26-13 Adapted from Kettlewell (1956), *J. of Heredity* 10: 287-301. Reprinted with permission.
27-ExpBF2 Adapted from Cooper et al (1992) *Proc. Natl. Acad. Sci. USA* 89: 8741-8744. Reprinted with permission of the author.
27-UNa Adapted from Cooper et al (1992) *Proc. Natl. Acad. Sci. USA* 89: 8741-8744. Reprinted with permission of the author.
27-08 Adapted from Olsen & Woesl (1993) *FASEB Journal* 7: 113-123. Reprinted with permission.
27-09 Adapted from Ruovolo (1994) *Proc. Natl. Acad. Sci. USA* 91: 8900-8904. Copyright 1994 National Academy of Sciences, U.S.A. Reprinted with permission.
27-10 Adapted from Yunis and Prakash (1982) *Science* 215:1525-1530. Reprinted with permission.

INDEX

Note: Italicized page numbers refer to information in figures and tables.